INSECT PATHOLOGY

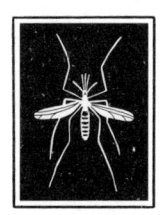

INSECT PATHOLOGY

Yoshinori Tanada
Department of Entomological Sciences
University of California, Berkeley
Berkeley, California

Harry K. Kaya
Department of Nematology
University of California, Davis
Davis, California

ACADEMIC PRESS, INC.
Harcourt Brace Jovanovich, Publishers
San Diego   New York   Boston
London   Sydney   Tokyo   Toronto

This book is printed on acid-free paper. ∞

Copyright © 1993 by ACADEMIC PRESS, INC.
All Rights Reserved.
No part of this publication may be reproduced or transmitted in any form or by any means, electronic or mechanical, including photocopy, recording, or any information storage and retrieval system, without permission in writing from the publisher.

Academic Press, Inc.
1250 Sixth Avenue, San Diego, California 92101-4311

United Kingdom Edition published by
Academic Press Limited
24–28 Oval Road, London NW1 7DX

Library of Congress Cataloging-in-Publication Data

Tanada, Yoshinori.
    Insect pathology / Yoshinori Tanada, Harry K. Kaya.
      p. cm.
    Includes index.
    ISBN 0-12-683255-2
    1. Insects–Diseases.   I. Kaya, Harry K.   II. Title.
SB942.T35   1992
632'.7–dc20   92-14551
                                                                                                    CIP

PRINTED IN THE UNITED STATES OF AMERICA
92 93 94 95 96 97   QW   9 8 7 6 5 4 3 2 1

DEDICATION

We dedicate this text to two men who played prominent roles in the evolution of insect pathology. Professor Harry S. Smith, renowned worldwide as a leader in biological control, recognized insect pathology as an integral part of biological control and strongly supported this discipline. As chairman of the statewide Division of Beneficial Insect Investigations, Professor Smith established the Laboratory of Insect Pathology at the University of California, Berkeley, with Dr. E. A. Steinhaus in charge. Professor Steinhaus developed the laboratory into a department acknowledged internationally in teaching and research. For his accomplishments, his peers have called Professor Steinhaus the founder of modern insect pathology and father of invertebrate pathology.

Professor Harry S. Smith
(1883–1957)

Professor Edward A. Steinhaus
(1914–1969)

CONTENTS

Preface		xi
1	INTRODUCTION	1
	I. Scope of Insect Pathology	1
	II. Application	1
	III. A Brief History	2
	References	9
2	ASSOCIATIONS BETWEEN INSECTS AND NONPATHOGENIC MICROORGANISMS	12
	I. Symbiosis	13
	II. External Microbiota	14
	III. Insect-Cultivated Fungi	17
	IV. Internal Microbiota	21
	V. Internal Microbiota of Termites	27
	VI. Intrahemocoelic and Intracellular Microbiota	30
	VII. Role of Symbionts	37
	VIII. Reproductive Incompatibility	37
	IX. Methods of Obtaining Aposymbiotic Hosts	38
	X. Transmission of Symbionts through Host Generations	39
	References	40
3	AMICROBIAL AND MICROBIAL AGENTS	52
	I. Amicrobial Diseases	53
	II. Microbial Diseases	65
	References	76
4	BACTERIAL INFECTIONS: BACILLACEAE	83
	I. Symptoms and Pathology	84
	II. Portals of Entry	84
	III. Types of Entomopathogenic Bacteria	85
	IV. Family Bacillaceae	86
	References	122

5 OTHER BACTERIAL INFECTIONS 147

- I. Pseudomonadaceae Infections — 147
- II. Enterobacteriaceae Infections — 148
- III. Vibrionaceae Infections — 151
- IV. Streptococcaceae Infections — 152
- V. Actinomycetes Infections — 153
- VI. Rickettsial Infections — 153
- VII. Class Mollicutes — 158
- References — 162

6 DNA-VIRAL INFECTIONS: BACULOVIRIDAE 171

- I. Viral Structure and Replication — 172
- II. Insect Viral Classification — 173
- III. Hosts and Tissue Culture — 174
- IV. Types of Insect Viruses — 174
- V. Family Baculoviridae — 176
- VI. Nuclear Polyhedrosis Viruses — 182
- VII. Granulosis Viruses — 207
- VIII. Nonoccluded Viruses — 216
- References — 220

7 OTHER DNA-VIRAL INFECTIONS 245

- I. Polydnaviridae—dsDNA, Enveloped — 245
- II. Poxviridae—dsDNA, Enveloped — 249
- III. Ascoviridae—dsDNA, Enveloped — 253
- IV. Iridoviridae—dsDNA, Nonenveloped — 255
- V. Parvoviridae—ssDNA, Nonenveloped — 261
- References — 264

8 RNA-VIRAL INFECTIONS: REOVIRIDAE 275

- I. Cytoplasmic Polyhedrosis Viruses — 276
- II. *Muscareovirus*—House Fly Virus — 287
- References — 288

9 OTHER RNA-VIRAL INFECTIONS 296

- I. Birnaviridae—Bisegmented, dsRNA — 296
- II. Rhabdoviridae—ssRNA, Enveloped — 297
- III. Picornaviridae—ssRNA, Nonenveloped — 300
- IV. Caliciviridae—ssRNA, Nonenveloped — 307
- V. Tetraviridae (*Nudaurelia* β Virus Group)—ssRNA, Nonenveloped — 308
- VI. Nodaviridae—ssRNA, Nonenveloped — 309
- References — 310

10	FUNGAL INFECTIONS		318
	I.	Classification	319
	II.	Structure and Reproduction	319
	III.	Hosts	321
	IV.	Pathogenicity	327
	V.	Signs and Symptoms	327
	VI.	Effect of Environmental Conditions	328
	VII.	Subdivision Mastigomycotina	329
	VIII.	Subdivision Zygomycotina, Class Zygomycetes, Order Mucorales	336
	IX.	Subdivision Zygomycotina, Class Zygomycetes, Order Entomophthorales	336
	X.	Subdivision Ascomycotina	347
	XI.	Subdivision Basiodiomycotina	356
	XII.	Subdivision Deuteromycotina	356
		References	366

11	PROTOZOAN INFECTIONS: ZOOMASTIGINA, RHIZOPODA, AND CILIOPHORA		388
	I.	Historical Aspects	389
	II.	Relation of Protozoa to Insects	391
	III.	Portals of Entry	391
	IV.	Pathogenesis, Signs, and Symptoms	392
	V.	Zoomastigina (Flagellate) Infections	392
	VI.	Rhizopoda (Amoebic) Infections	399
	VII.	Ciliophora (Ciliate) Infections	402
		References	407

12	PROTOZOAN INFECTIONS: APICOMPLEXA, MICROSPORA		414
	I.	Apicomplexa, Gregarinia	415
	II.	Apicomplexa, Coccidia	423
	III.	Microspora, Microsporea	425
		References	447

13	NEMATODES, NEMATOMORPHS, AND PLATYHELMINTHES		459
	I.	Historical Review	460
	II.	Taxonomy	461
	III.	General Life Cycle of Nematodes	461
	IV.	Associations with Insect Hosts	462
	V.	Types of Insect–Nematode Associations	463
	VI.	Host Specificity	464
	VII.	Mode of Infection	465
	VIII.	Host Resistance	466
	IX.	Pathology	467
	X.	Biology and Life Cycle of Selected Nematodes	471

	XI.	Nematomorpha	482
	XII.	Platyhelminthes	483
		References	483

14 HOST RESISTANCE 492

	I.	Resistance at the Species Level	493
	II.	Insect Age and Stage	495
	III.	Nutritional Factors	497
	IV.	Physical Factors	499
	V.	Morphological and Physiological Defenses	500
	VI.	Hematology	507
	VII.	Blood Clotting	511
	VIII.	Cellular Immunity	512
	IX.	Insect Plasma and Humoral Immunity	521
	X.	Innate Humoral Immunity	522
	XI.	Acquired Immunity	524
	XII.	Passively Acquired Immunity	533
		References	533

15 MICROBIAL CONTROL 554

	I.	Approaches to Microbial Control	555
	II.	Economic Threshold	556
	III.	Factors Affecting Efficacy	557
	IV.	Safety	561
	V.	Molecular Genetics	563
	VI.	Microbial Pesticides	565
	VII.	Mass Production	566
	VIII.	Formulations	567
	IX.	Standardization	568
	X.	Storage	569
	XI.	Application Technology	569
	XII.	Advantages and Disadvantages of Microbial Control	570
	XIII.	Examples of Long-Term Control	571
	XIV.	Examples of Short-Term Control	577
	XV.	Examples of the IPM Approach	582
	XVI.	Destruction of Symbionts	584
		References	584

16 EPIZOOTIOLOGY 595

	I.	Definition of Some Epizootiological Terms	596
	II.	Modeling	597
	III.	Key Factors in Epizootiology	600
	IV.	The Environment	612
	V.	Present Status	622
		References	622

Index 633

P R E F A C E

Insect pathology dates back to 2700 B.C., when the Chinese recorded diseases of the silkworm. Through the following centuries, insect pathology developed slowly as diseases, mainly of the silkworm and honey bee, were described by early scientests. The nineteenth century was the renaissance for insect pathology. In the mid-1830s, Agostino Bassi, the father of insect pathology, demonstrated that a microorganism (fungus) caused disease in an animal (silkworm). In the 1860s, silkworms throughout the world were ravaged by diseases, in particular a protozoan disease (pébrine). Louis Pasteur, the renowned chemist and microbiologist, developed a practical method of rearing healthy silkworms and thereby saved the silk industry. Bassi and Pasteur recognized the potential for using microorganisms to control pestiferous insects, and by the 1870s this notion of microbial control was expressed by several prominent entomologists. It was Elie Metchnikoff who demonstrated that an entomopathogenic fungus could infect an insect pest in the soil and applied the fungus for insect control. This beginning led to an era of using fungi for insect control; but the successes were few, underscoring the need to obtain basic information on the pathology and pathogenesis of these fungi and on the ecological conditions for their success.

Insect pathology emerged and matured as a science in the twentieth century with many benchmarks. Except for the identity of humoral factors (e.g., attacins and cecropins), advances in the general principles of pathology have not been as noteworthy as those made on the biology of the pathogens. Some of the major biological accomplishments with pathogens are the discovery of heteroecism in the fungus *Coelomomyces,* the clarification of sexuality in the microsporida, and the application of molecular biology to *Bacillus thuringiensis* and especially to baculoviruses, which are used as vectors for the expression of eucaryotic genes to produce pharmaceutical products. Moreover, technological advances made with, among others, electron and scanning microscopes, computers, electrophoresis, and DNA sequencing have revolutionized the science. There are many other significant accomplishments, too numerous to document here, which are discussed in the following chapters.

As the twenty-first century approaches, the growth of insect pathology has accelerated and diversified into many highly specialized disciplines. No longer are students trained as general insect pathologists. Historically, students and teachers were cognizant of most, if not all, pathogen groups. Today, there are only a few scientists who fit the general insect pathologists' category, and most have specialized into subdisciplines within insect pathology such as molecular biology, virology, bacteriology, protozoology, mycology, or nematology. The advantage of specialization is the ability to focus and to advance the knowledge of a particular pathogen system. The price of specialization is a loss of both the understanding of the roots of insect pathology and the awareness of the commonality that exists among pathogens. To minimize the loss, we have attempted to focus on basic principles of insect pathology and at the same time cover recent advances in the subdisciplines.

This book is designed for a broad spectrum of readers. It should be useful to students, lecturers, and researchers requiring information about the principles in insect pathology and the biology of pathogens. It should serve as a resource for specialists to learn about other insect pathogen systems, for generalists to become aware of advances in insect pathology, and for scientists and students, beginning or otherwise, interested in learning about insect pathology.

This book was originally intended to update the 1949 text by E. A. Steinhaus entitled "Principles of Insect Pathology." Our purpose for this book was twofold: To serve (1) as a text for an insect pathology and/or biological control class and (2) as a comprehensive reference source. Because this book summarizes much of the available information, its usefulness as a textbook for an insect pathology class is apparent. Although the literature citations are extensive, they are far from complete. The literature in insect pathology is voluminous and for the past decade has been expanding at an almost exponential rate. A complete review of the literature is beyond the scope of this book, and an omission of a reference does not preclude its importance. Our citations, however, should serve as a good starting point for those who wish to obtain further information. We have attempted to cover equally all subdisciplines, but shortcomings are unavoidable. For these, we take full responsibility.

The continual changes in scientific names of the pathogens and arthropods may cause problems to the reader. Taxonomic and nomenclature changes are inevitable. We have used the names as cited in the references and, where possible, have incorporated the current names.

The inception of this book goes back many years, and we are indebted to the students who assisted with the references, the individuals and the institutions that graciously provided photographs, and the administrative staff who typed the manuscripts. We thank J. J. Becnel (protozoa), W. M. Brooks (protozoa), J. R. Fuxa (epizootiology), S. Maeda (virology), V. Sprague (protozoa), G. S. Thurston (insect resistance), and T. Yamamoto (*Bacillus thuringiensis*) for assistance in their specialties. We are especially grateful to Academic Press for its patience and to our families for their understanding.

Y. Tanada
H. K. Kaya

CHAPTER 1

INTRODUCTION

I. SCOPE OF INSECT PATHOLOGY
II. APPLICATION
III. A BRIEF HISTORY

I. SCOPE OF INSECT PATHOLOGY

Insects, through their diversity in types, numbers, life cycles, and habitats, expose themselves to a wide range of pathologies. Whatever goes wrong with an insect may be considered a pathology or disease. This aspect will be expanded on in the following chapters, where we shall discuss the different types of diseases in insects.

Insect pathology is a branch of invertebrate pathology. However, its scope and development are much more extensive than other areas of invertebrate pathology because two-thirds of the invertebrates are insects and insects are economically and medically more important to humans than other invertebrates. Consequently, a larger number of pathologists work in insect pathology than in any other area of invertebrate pathology.

As in the case of plant and vertebrate pathologies, the general principles of pathology apply to diseases of insects. They involve the etiology (cause), symptomatology, and epizootiology, and the structural, chemical, and functional alterations that result from disease. In addition to entomology, other sciences, such as virology, bacteriology, mycology, protozoology, nematology, pathology, immunology, are closely interrelated in the study of insect diseases. This implies that an insect pathologist should have a very broad background, but one which a single individual could not acquire in depth. In practice, an insect pathologist specializes in one or few areas, but he or she should have a general understanding of the different types of diseases in insects.

II. APPLICATION

The importance of insect pathology to human welfare has been demonstrated in the fields of general biology, medicine, and agriculture. In general biology and medicine, the study of diseases in insects has made significant contributions in many areas, such as vector transmission, immunity, infectious processes, and mutualism. In agriculture, populations of insect pests can be devastated by natural outbreaks of pathogens. Entomologists, as early as the mid-nineteenth century, were aware of outbreaks of diseases caused by pathogens, and they attempted to control the insect pests with the use of pathogens. There are many recent examples of the effectiveness of pathogens when used against insect pests, and the United States Environmental Protection Agency and the Department of Agriculture have already approved at least one pathogen each from viruses, bacteria, fungi, and protozoa for commercial production. Nematodes are exempt from the registration process and are available commercially. In sericulture and apiculture, the control of diseases of the silkworm and the honey bee is a requirement for the successful production of their products.

The principal application of insect pathology in entomology is in economic and medical entomology. This area of microbial control is considered a branch of biological control. Pathogens have been introduced into insect pest populations, similar to the colonizations with insect parasitoids or predators. Pathogens have been applied directly to insect-infested plants with spray equipment, just as in the case of chemical insecticides. Insect pathology has played important roles in the control of diseases in beneficial insects, such as the silkworm, honey bee, parasitoids, and predators, and in insects that are being cultured or mass produced in the laboratory for test purposes.

Insect pathology is also involved in other fields of entomology. The epizootiology of insect diseases is an important aspect in ecology. Insects in nature may be susceptible to one or more pathogens that may play a role in the regulation of their populations. The study of nutritional and metabolic diseases and the role of symbionts touch upon insect physiology. Abnormalities that result from diseases to cells, tissues, and morphology are part of insect cytology and morphology. In taxonomy, the study of insect diseases has revealed strains and races of insects that differ in their susceptibility to pathogens. In the case of termites and lac insects, internal symbionts have been used to separate species or subspecies. In addition, the presence of similar mutualistic protozoa in the termite and wood roach is one of the bases supporting the common origin of these insects. The study of the toxic effects of insecticides and other chemicals that concern toxicologists is of interest to insect pathologists. This also applies to the hereditary abnormalities that arise in insects. In medical entomology and parasitology, the association of the insect vector and the pathogen may be considered as a section of insect pathology. Insect pathology, therefore, is involved in all major disciplines within entomology. Traditionally, however, an insect pathologist's research has focused on pathogens.

A recent applied use of insect pathogens is the development of baculoviruses as vectors for the expression of foreign genes. The expression products are useful to humans and vertebrates [e.g., antibodies, immune factors (interleukins and interferons), vaccines, and pharmaceutical products]. These products are almost identical to those produced in the source animals and biologically as effective. An increasing number of institutional and commercial organizations are involved.

III. A BRIEF HISTORY

Insect pathology has a very long and detailed development up to its present status. The treatment here is rather concise and is concerned primarily with the period up to

1950 with emphasis on the use of pathogens for insect control (microbial control), which developed very rapidly subsequent to 1950. Areas such as the isolation and systematics of various pathogens, and their biology and pathogenesis, will not be considered or alluded to only briefly. Students who are interested in the historical aspects are encouraged to read the detailed and fascinating account presented by Steinhaus (1956a, 1975).

When humans first began to raise the silkworm and honey bee, they detected abnormalities in these insects. In the Far East, the Chinese recorded diseases in the silkworm as early as the seventh century B.C. (Wang 1965). The ancient Chinese identified fungal species of *Cordyceps* and *Isaria* from the cadavers of the silkworm and cicada, respectively (McCoy et al. 1988). These cadavers bore conspicuous fungal growths designated as "vegetable growths." Some of these growths of *Cordyceps*, the so-called Chinese plantworms, were used in folk pharmacopoeia to treat a large number of ailments, and they are still being used today.

In the western world, diseases of the honey bee were also recorded prior to the Christian era. Aristotle (384–322 B.C.) in his Historia Animalium observed that the honey bee developed abnormalities or diseases (Steinhaus 1956a). Such diseases occurred in seasons of drought or when flowers were covered with mildew. In his description of an affliction by "rusts," he might have been referring to a disease known today as foulbrood.

There are more early reports on the diseases of the silkworm than those of the honey bee because of economic and cultural considerations. Numerous reports appeared during the eighteenth and nineteenth centuries in the Orient and in European countries where silkworm rearing had been introduced. One of the earliest scientific treatises on silkworm diseases was by P. H. Nysten in 1808. Gradually reports appeared on diseases of insects other than the silkworm and the honey bee.

The study of insect disease was formally introduced in a chapter entitled "Diseases of Insects" written by W. Kirby in the famous text, *An Introduction to Entomology: Or Elements of the Natural History of Insects,* by Kirby and Spence (1826). Kirby divided the diseases of insects into two classes, those caused by accidental external injury or internal derangement, and those produced by vegetable and animal parasitic organisms. In the second class, the vegetable parasites included the fungi, and the animal parasites were nematodes, mites, and parasitic insects. Kirby and other workers suspected that fungi might be causing diseases of insects. Although the parasitization of insects by other insects and mites is generally not considered to be a part of insect pathology, Kirby was right that, in a broad sense, such parasitization is a manifestation of a disease in an insect.

The first experimental demonstration that an insect pathogen, a fungus, caused an infectious disease in insects was made by Agostino Bassi (also called Bassi de Lodi), whose great work, *Del Mal del Segno, Calcinaccio o Moscardino,* was published in 1835 and 1836 (Steinhaus 1956a, 1975) (Fig. 1–1). Through ingenious experimentation, Bassi showed that the fungus grew and developed in the living silkworm, that it caused the insect's death, and that it could be transmitted by inoculation, contact, and contaminated food. He noted that the white covering that developed around the dead silkworm was the fungus mass containing the "seeds" that were disseminated and caused the disease in other individuals. This demonstration was not only the first for the diseases of insects, but also the first for those of an animal. The fungus was later named in his honor as *Beauveria bassiana*, commonly known as the white-muscardine fungus. Bassi is recognized as the father of insect

4   1   INTRODUCTION

FIGURE 1-1

Two pioneer insect pathologists. (A) Agostino Bassi (1773–1856), who first demonstrated the parasitic nature of muscardine (fungal) disease of the silkworm. [From Steinhaus (1956) and courtesy of the University of California Agricultural Experiment Station.] (B) Louis Pasteur (1822–1895), who worked with the diseases of the silkworm and developed a practical method for the control of pébrine (protozoan infection). (Provided by the Museum of the Pasteur Institute, Paris.)

pathology. His work is included with others that eventually overthrew the doctrine of spontaneous generation.

During the mid-nineteenth to early twentieth centuries, numerous taxonomic studies appeared on fungi pathogenic to insects. Some studies also considered the biology and pathology of the fungi. Among the outstanding studies were the monumental treatises on Entomophthorales and Laboulbeniales by Roland Thaxter (1888, 1896, 1908, 1924, 1926, 1931).

Louis Pasteur (Steinhaus 1956a, 1975) played a dramatic role in rescuing the silkworm from certain diseases, particularly the protozoan disease called pébrine, which was devastating sericulture in nearly every major silkworm rearing nation (Fig. 1–1). During Pasteur's time, sericulture was a major industry in France and other European countries. Pasteur was approached by an influential politician, Senator Jean-Baptiste Dumas, who was a great chemist and his former teacher, and asked to find a method of controlling the diseases of the silkworm. Pasteur at that time was already an international celebrity in chemistry through his works on racemic mixture of tartaric acid and fermentation. He at first refused to take on the assignment, because as he frankly admitted he had never so much as seen or touched a silkworm cocoon. Dumas, who had become minister of agriculture, replied, "All the better, you will have no preconceived ideas" (Hemphill 1977). Pasteur was persuaded and commissioned in 1865 by the French government. He applied sound experimental methods in his study and established that a pathogen caused pébrine (subsequently, others identified the pathogen as a protozoan), and that the pathogen was transmitted by way of the egg (transovarial transmission), by the ingestion of contaminated food, and by contact with diseased silkworms. He developed a practical method, which is being used even to this day, of screening out infected eggs and infected adults, thereby providing to the sericulturists a supply of healthy insects. By using Pasteur's method or a modification of it, the silkworm industry was saved not only in France but throughout the world.

Pasteur studied another important disease, the flacherie of the silkworm. He published his findings in 1870 in his famous two-volume memoirs, *Etudes sur la Maladie des Vers à Soie*. Steinhaus (1956a) considers that Pasteur, through his studies with silkworm diseases, may have developed the interest, expertise, and incentive to study human and animal diseases. This resulted in his outstanding discoveries on anthrax, rabies, cholera, septicemia, and other infectious diseases of vertebrates. Marie-Louise Hemphill (1977), daughter of the nephew and close associate of Pasteur, concurs with Steinhaus on the significance of Pasteur's study on the diseases of the silkworm. She states, "According to the method he had used with crystals or detected corpuscles in the silkworm moths, he now examined the rabid brains sent on from Paris by Viala at a time when corpus negri had not yet been discovered." Hemphill points out further that Pasteur by the age of 12 was a good artist and that his paintings are displayed at the Louvre. She believes that this talent to look for details was most important among his other great qualities in achieving his remarkable discoveries. Together with Bassi, Pasteur shares the credit for initiating the scientific development of insect pathology.

By the mid-nineteenth century, there was a general understanding that microorganisms cause diseases in insects and that disease outbreaks occur in insect colonies in the laboratory and as epizootics in nature. Both Bassi and Pasteur (1874) had suggested the use of pathogens for insect pest control but no attempts had been made up to this time. Workers began to demonstrate the infectiousness of the white-muscardine fungus, not only to the silkworm, but also to other insects, some of them being important pests of agricultural crops. For example, Audoin (1837) infected several species of beetles with the white-muscardine fungus obtained from the silkworm and was able to reinfect the silkworm with the spores from the diseased beetles. He also tested the infectiousness of the fungal growth extracted from the hemocoel of an infected insect. He found that such growths when inoculated into the hemocoel of a healthy silkworm larva resulted in a shorter period for the infection to develop.

LeConte (1874) made a distinct recommendation in his paper "Hints for the Promotion of Economic Entomology" to study the epizootics of insects, especially those caused by fungi, and to investigate the most effective means of introducing and transmitting such pathogens. Another American entomologist, H. A. Hagen (1879), strongly advocated the use of fungi for insect control and even outlined a procedure to conduct field experiments with a yeast fungus. He remarked with perspicacity, "Nature uses always to attain its purposes the most simple and most effectual ways; therefore it is always the safest way to follow nature." Within this period, several workers made some attempts with commercially available yeasts and generally obtained negative results.

A Russian associate of Pasteur, Elie Metchnikoff, studied the diseases of the wheat cockchafer, *Anisoplia austriaca*, in the laboratory and in nature (Metchnikoff 1879). Among the three distinct diseases that he found was one caused by a fungus, which he named *Entomophthora anisopliae* (presently known as *Metarhizium anisopliae*), and he referred to it as the "green muscardine" from the dark green color of its sporulating colony. Metchnikoff studied the fungus along mycological and pathological lines. He observed that healthy larvae became infected when placed together in the soil with infected larvae or with fungal spores (Metchnikoff 1879). For the control of the wheat cockchafer, he suggested applications of (1) fungus-killed larvae, (2) free spores, or (3) the soil in which the diseased larvae had been found. Metchnikoff propagated the fungus on sterilized beer mash and recognized the importance of such "artificial" production for insect control. Furthermore, he noted that the sugar-beet weevil, *Cleonus punctiventris*, was naturally infected by the green-muscardine fungus, which destroyed 40% of the weevil population. Like Pasteur, Metchnikoff is known for his outstanding contributions in vertebrate pathology, and was awarded the Nobel prize in 1908 for his discovery of phagocytosis and his work on immunity.

Following the recommendation of Metchnikoff, Krassilstchik (1888) mass-produced spores of the green-muscardine fungus in a small production plant and applied the fungus in the field for the control of the sugar-beet weevil. He reported a mortality of 55 to 80% of the larvae.

During the late nineteenth and early twentieth centuries, there was an increased interest in the use of fungi against insect pests. In Europe, attempts were made against such insects as flies, the nun moth (*Lymantria monacha*), grasshoppers, cockchafers (*Melolontha* spp.), and others. In the United States, there were two major efforts to use fungi for insect control. They were applied in the midwestern states to control the chinch bug, *Blissus leucopterus*, on cereal crops (Lugger 1888; Snow 1890; Forbes 1895; Billings and Glenn 1911) and in Florida to control citrus pests, such as scale insects and whiteflies. Several fungi were used on the chinch bug, but the most widely used fungus was *Beauveria bassiana* (syn. *Sporotrichum globuliferum*) (Forbes 1895; Kelly and Parks 1911). These attempts mostly failed.

In Florida, the control of the scale insects and whitefly on citrus was conducted with the so-called "friendly fungi" (*Aschersonia, Aegerita, Verticillium, Podonectria, Nectria, Myriangium*, etc.). The control of the whitefly was successful (McCoy et al. 1988), but the attempts against scale insects produced incomplete results primarily because of the lack of basic information on the pathology and pathogenesis of these fungi and the failure to recognize the importance of epizootiological or ecological conditions, both biotic and physical, of the environment (Steinhaus 1949). These discouraging results on the chinch bug and scale insects greatly reduced interest in the use of fungi, even though spectacular natural outbreaks of fungi often decimated insect pest populations. It was not until the mid-twentieth century that fungi again were used in serious attempts to control insect pests.

Among the first attempts to use bacteria for insect control occurred in the early twentieth century. Felix d'Herelle (1911) observed in Yucatan, Mexico, epizootics in populations of the locust *Schistocerca americana*. He isolated from infected locusts a bacterium that he named *Coccobacillus acridiorum* (this genus is invalid and the status of the bacterium is still unknown). He considered this bacterium to be the causal agent of the epizootic (d'Herelle 1914). In 1912, he applied the bacterium successfully in Argentina and Colombia, but less effectively in other countries. His reports stimulated others in the use of this bacterium. Some investigators confirmed his results, but most failed in their attempts, which led to the abandonment of this bacterium as a control agent. The cause of the failure, which has not been clarified even to this day, has been associated with host resistance, loss in virulence and pathogenicity of the bacterium, and even to the improper isolation of the true pathogen (Glaser 1918). Bucher (1959) obtained a culture of *C. acridiorum* from the Pasteur Institute and identified it as Cloaca Type A, a common saprophyte in the insect's gut. Nonetheless, d'Herelle is recognized as being the codiscoverer, along with F. W. Twort, of the bacteriophage that he had observed in *C. acridiorum* (Duckworth 1976).

Up to this period, most of the attempts to use insect pathogens for the control of insect pests failed because of a lack of understanding of the fundamental principles in pathology (Paillot 1933; Steinhaus 1949). Errors and misjudgments were made in the isolation, characterization, virulence, and pathogenicity of the pathogen, and the interaction of the pathogen with the insect host and the environment. Predictable successes in microbial control became possible only after these factors were studied carefully in the laboratory and field.

In 1927, the International Livestock Exposition organized the International Corn Borer Investigations to develop methods for the control of the European corn borer, *Pyrausta nubilalis*, and the results of four years of studies were reported in four volumes (Ellinger 1928, 1929, 1930, 1931). The studies were conducted mainly in Europe, and the collaborating countries were France, Germany, Denmark, Sweden, Hungary, Romania, Russia, Yugoslavia, the United States, and Canada. Investigations were made on broad fundamental bases. They involved the biology and ecology of the corn borer and its host plants; its parasitoids, predators, and pathogens; and control measures, both biotic and chemical. The study should be regarded as one of the early attempts at developing pest management programs because of the wide scope of the investigation, even though it did not completely fulfill its goals. Pathogens, such as fungi, nematodes, and bacteria, were studied and tested for the control of the corn borer. Vouk (1931) reported that the microbiological methods gave the best and most decisive results. The pathogens *Bacillus thuringiensis* and the green muscardine fungus, *Metarhizium anisopliae*, were effective against the corn borer, but for some unknown reasons the study was not continued to develop practical methods that could be recommended to growers.

In addition to the corn borer, *B. thuringiensis* was found to be highly pathogenic for the gypsy moth (*Lymantria dispar*), *Aporia crataegi*, and *Vanessa urticae* but was not effective against grasshoppers, mosquitoes, and beetles (Metalnikov and Chorine 1929). The effectiveness of several bacteria against the corn borer led to tests against the pests of cotton and cabbage (Metalnikov 1930; Metalnikov and Metalnikov 1933, 1935). In spite of the promising results, especially with *B. thuringiensis*, the use of bacteria was discontinued. Thus, the full potential of *B. thuringiensis* was not developed until Steinhaus (1951) in 1942 obtained a culture of the bacterium from Nathan R. Smith, and through his research brought this organism to the attention of insect pathologists.

The accidental introduction of the Japanese beetle, *Popillia japonica*, in 1916 in a nursery in southern New Jersey led to the isolation of a number of pathogens of the beetle (St. Julian and Bulla 1973; Fleming 1976). The beetle became a serious pest and fed on almost 300 species of plants. The United States Bureau of Entomology established the Japanese Beetle Laboratory in 1917 at Riverton, New Jersey, to study its biology, ecology, and control measures. The laboratory moved to Moorestown, New Jersey, in 1927.

The larva of the Japanese beetle was known to be susceptible to certain diseases as early as 1921, but serious studies on these diseases apparently began in 1926

(Smith and Hadley 1926). Bacteria, fungi, and protozoa were found infecting the larvae. Major emphasis was placed on bacterial diseases, which included the "white group" or "milky diseases" caused by bacteria. Dutky (1940) described the organisms of the Types A and B milky diseases as *Bacillus popilliae* and *B. lentimorbus*, respectively. Emphasis was placed on *B. popilliae* because it offered the most promise for the control of the beetle. Excellent research was conducted by the staff on the causative agent, mode of infection by the bacterium, time of development of disease, effect of disease on molting and metamorphosis, development of the disease in the field, and mass production and application of the bacterium. *B. popilliae* was propagated in live beetle grubs and a bacterial powder was formulated by mixing ground-up, infected larvae and talc. Federal and state government agencies developed liaison programs to distribute the bacterium. This has resulted in one of the outstanding successes in the microbial control of insect pests. *Bacillus popilliae* is the first insect pathogen to be approved by the United States federal government for use against an insect pest, and has been produced commercially.

A nematode, *Steinernema* (syn. *Neoaplectana*) *glaseri*, was isolated by Glaser and Fox (1930) from the Japanese beetle. The nematode was used with some local successes (Glaser and Farrell 1935; Glaser et al. 1940), but it was not widely used apparently because of the remarkable achievement of the milky disease organisms in controlling the Japanese beetle and problems associated with the mass production of the nematode.

In 1945, Professor Harry S. Smith, chairman of the statewide Division of Beneficial Insect Investigations (later the Division of Biological Control), established the Laboratory of Insect Pathology in Berkeley, California, with Edward A. Steinhaus in charge (Steinhaus 1975). Steinhaus made important contributions to the fundamental and applied aspects of insect pathology. Among his major contributions in applied aspects was the first practical application of an insect virus for insect control (Steinhaus and Thompson 1949; Thompson and Steinhaus 1950). He resurrected the highly potent insect pathogen *Bacillus thuringiensis* and demonstrated its effectiveness against many insect pests. He was mainly instrumental in arousing the interest of commercial firms in the mass production of this bacterium (Steinhaus 1956b). *Bacillus thuringiensis* is at present the most important insect pathogen from a commercial standpoint because of its ease in production and effectiveness against many different insect pests.

In addition to his research contributions in insect pathology, Steinhaus was the first to offer a formal course in 1947 in Berkeley. A second course at the University of Hawaii was taught by Y. Tanada in 1957, and this was followed by a third course taught by I. M. Hall in 1962 at the University of California, Riverside. Subsequently, the number of courses concerned with insect and invertebrate pathologies proliferated throughout the United States. Similar courses are also offered in a number of major countries throughout the world.

Steinhaus has been acclaimed the founder of modern insect pathology (Weiser 1958) and the father of invertebrate pathology (Knipling 1971) by his peers because of his significant contributions in research and teaching and his leadership in establishing insect pathology as a viable, distinct discipline.

During the late 1930s and into the 1940s, serious outbreaks of the spruce budworm, *Choristoneura fumiferana*, developed in Canada, and J. J. de Gryse, chief of the Forest Biology Division, Entomology Branch, Canada Department of Agriculture, conceived the idea of a forest entomology research center to be established in Sault Ste. Marie (Cameron 1973). This suggestion was strongly supported when the viral disease of the European spruce sawfly, *Diprion hercyniae*, began to destroy spectacularly the sawfly populations under natural conditions during the early 1940s (extracted from a letter by J. J. de Gryse to E. A. Steinhaus dated February 20, 1949). Insect pathology was initiated in Canada in 1946 when K. Graham was appointed the first insect pathologist (Cameron 1973). Graham resigned after a year to accept an appointment at the University of British Columbia. In 1947, F. T. Bird and G. H. Bergold, both of whom worked with insect viruses, arrived in Sault Ste. Marie. In 1950, the Insect Pathology Research Institute was established as an autonomous national institute under the direction of J. W. M. Cameron (Anon. 1975). Field tests were conducted by the staff of the institute with the nuclear polyhedrosis viruses against the European spruce (*Diprion hercyniae*) and pine (*Neodiprion pini*) sawflies with marked successes.

Up to 1950, four of the five major groups of pathogens, viruses, bacteria, fungi, and nematodes, were used in microbial control attempts. The fifth, the protozoa, was not used until after 1950. Applications of insect pathogens have accelerated since 1950 and resulted in the successful control of many insect pests. The readers should be reminded again that these successes were based on sound preparatory studies of the biology, ecology, and pathogenesis of the pathogens, and on a clear understanding of the biology and ecology of insect pests.

REFERENCES

Anonymous. 1975. In memoriam James William MacBain Cameron. *J. Invertebr. Pathol.* 26, 273–274.

Audoin, V. 1837. Nouvelles expériences sur la nature de la maladie contagieuse qui attaque les vers à soie, et qu'on désigne sous le nom de *Muscardine*. *C.R. Acad. Sci. Paris 5*, 712–717.

Billings, F. H., and Glenn, P. A. 1911. Results of the artificial use of the white-fungus disease in Kansas: With notes on approved methods of fighting chinch bugs. *U.S. Dep. Agric. Bur. Entomol. Bull.* 107.

Bucher, G. E. 1959. The bacterium *Coccobacillus acridiorum* d'Herelle: Its taxonomic position and status as a pathogen of locusts and grasshoppers. *J. Insect Pathol. 1*, 331–346.

Cameron, J. W. M. 1973. Insect pathology. *Annu. Rev. Entomol. 18*, 285–306.

d'Herelle, F. 1911. Sur une épizootie de nature bactérienne sévissant sur les sauterelles au Mexique. *C.R. Acad. Sci. Paris Ser. D 152*, 1413–1415.

d'Herelle, F. 1912. Sur la propagation, dans la République Argentine, de l'épizootie des sauterelles du Mexique. *C.R. Acad. Sci. Paris Ser. D 154*, 623–625.

d'Herelle, F. 1914. Le coccobacille des sauterelles. *Ann. Inst. Pasteur Paris 28*, 280–328, 387–407.

Duckworth, D. H. 1976. "Who discovered bacteriophage?" *Bacteriol. Rev. 40*, 793–802.

Dutky, S. R. 1940. Two new spore-forming bacteria causing milky diseases of Japanese beetle larvae. *J. Agric. Res. 61*, 57–68.

Ellinger, T. 1928. International corn borer investigations. Vol. 1. Int. Livestock Exposition, Union Stock Yards, Chicago.

Ellinger, T. 1929. International corn borer investigations. Vol. 2. Int. Livestock Exposition, Union Stock Yards, Chicago.

Ellinger, T. 1930. International corn borer investigations. Vol. 3. Int. Livestock Exposition, Union Stock Yards, Chicago.

Ellinger, T. 1931. International corn borer investigations. Vol. 4. Int. Livestock Exposition, Union Stock Yards, Chicago.

Fleming, W. E. 1976. Integrating control of the Japanese beetle—a historical review. *U.S. Dep. Agric. Tech. Bull.* 1545.

Forbes, S. A. 1895. Experiments with the muscardine disease of the chinch-bug, and with the trap and barrier method for the destruction of that insect. *Univ. Ill. Agric. Exp. Stn. Bull. 38*, 25–86.

Glaser, R. W. 1918. A systematic study of the organisms distributed under the name of *Coccobacillus acridiorum* d'Herelle. *Ann. Entomol. Soc. Am. 11*, 19–42.

Glaser, R. W., and Farrell, C. C. 1935. Field experiments with the Japanese beetle and its nematode parasite. *J. N.Y. Entomol. Soc. 43*, 345–371.

Glaser, R. W., and Fox, H. 1930. A nematode parasite of the Japanese beetle (*Popillia japonica* Newm.). *Science 71*, 16–17.

Glaser, R. W., McCoy, E. E., and Girth, H. B. 1940. The biology and economic importance of a nematode parasitic in insects. *J. Parasitol. 26*, 479–495.

Hagen, H. 1879. Obnoxious pests—suggestions relative to their destruction. *Can Entomol. 11*, 110–114.

Hemphill, M.-L. 1977. Pasteur at Arbois. *Am. Soc. Microbiol. News. 43*, 298–299.

Kelly, E. O. G., and Parks, T. H. 1911. Chinch-bug investigations west of the Mississippi River. *U.S. Dept. Agric. Div. Entomol. Bull.* 95(3), 23–52.

Kirby, W., and Spence, W. 1826. Diseases of insects. *In* "An Introduction to Entomology: Or Elements of the Natural History of Insects." Vol. 4, pp. 197–232. Longman, Hurst, Rees, Orme, and Brown, London.

Knipling, E. F. 1971. The role of microbial agents in the development of alternative means of insect control. *Proc. Fourth Int. Coll. Insect Pathol.* 2–8. College Park, Maryland. 25–28 August 1970.

Krassilstchik, J. 1888. La production industrielle des parasites végétaux pour la destruction des insectes nuisibles. *Bull Sci. Fr. Belg. 19*, 461–472.

LeConte, J. L. 1874. Hints for the promotion of economic entomology. *Proc. Am. Assoc. Adv. Sci. 22*, 10–22.

Lugger, O. 1888. Fungi which kill insects. *Univ. Minn. Coll. Agric. Bull. 4*, 26–41.

McCoy, C. W., Samson, R. A., and Boucias, D. G. 1988. Entomogenous fungi. *In* "CRC Handbook of Natural Pesticides. Microbial Insecticides, Part A. Entomogenous Protozoa and Fungi." (C. M. Ignoffo, ed.), Vol. 5, pp. 151–236. CRC Press, Boca Raton, Florida.

Metalnikov, S. 1930. Utilisation des microbes dans la lutte contre *Lymantria* et autres insectes nuisibles. *C.R. Soc. Biol. 105*, 535–537.

Metalnikov, S., and Chorine, V. 1929. On the infection of the gypsy moth and certain other insects with *Bacterium thuringiensis*. A preliminary report. *In* "International Corn Borer Investigation." (T. Ellinger, ed.) Vol. 2, pp. 60–61. Int. Livestock Exposition, Union Stock Yards, Chicago.

Metalnikov, S., and Metalnikov, S. S. 1933. Utilisation des bactéries dans la lutte contre les insectes nuisibles aux Cotonniers. *C.R. Soc. Biol. Paris 113*, 169–172.

Metalnikov, S., and Metalnikov, S. S. 1935. Utilisation des microbes dans la lutte contre les insectes nuisibles. *Ann. Inst. Pasteur. 55*, 709–760.

Metchnikoff, E. 1879. Diseases of the larvae of the grain weevil. Insects harmful to agriculture [series]. Issue III. "The grain weevil." Published by the Commission attached to the Odessa Zemstvo Office. (From Steinhaus 1949).

Nysten, P. H. 1808. "Recherches sur les maladies des vers à soie et les moyens de les prévenir." De imprimerie Impéarile, Paris.

Paillot, A. 1933. "L'Infection chez les insectes." G. Patissier, Trévoux.

Pasteur, L. 1870. "Études sur la Maladie des Vers a Soie." Tome I and II. Gauthier-Villars, Paris.

Pasteur, L. 1874. (On the use of fungi against Phylloxera. Discussion). *C.R. Acad. Sci. Paris.* 79, 1233–1234.

Smith, L. B., and Hadley, C. H. 1926. The Japanese Beetle. *U.S. Dep. Agric. Circ.* 363.

Snow, F. H. 1890. Experiments for the destruction of chinch bugs. *Twenty First Annu. Rep. Entomol. Soc. Ontario.* 93–97.

St. Julian, G., and Bulla, L. A., Jr. 1973. Milky disease. *In* "Current Topics in Comparative Pathobiology." (T. C. Cheng, ed.), Vol. 2, pp. 57–87. Academic Press, New York.

Steinhaus, E. A. 1949. "Principles of Insect Pathology." McGraw-Hill, New York.

Steinhaus, E. A. 1951. Possible use of *Bacillus thuringiensis* Berliner as an aid in the biological control of the alfalfa caterpillar. *Hilgardia* 20, 359–381.

Steinhaus, E. A. 1956a. Microbial control—the emergence of an idea. *Hilgardia* 26, 107–160.

Steinhaus, E. A. 1956b. Potentialities for microbial control of insects. *Agric. Food Chem.* 4, 676–680.

Steinhaus, E. A. 1975. "Disease in a Minor Chord." Ohio State University Press, Columbus, Ohio.

Steinhaus, E. A., and Thompson, C. G. 1949. Preliminary field tests using a polyhedrosis virus to control the alfalfa caterpillar. *J. Econ. Entomol.* 42, 301–305.

Thaxter, R. 1888. The Entomophthoreae of the United States. *Mem. Boston Soc. Nat. Hist.* 4, 133–201.

Thaxter, R. 1896. Contribution towards a monograph of the Laboulbeniaceae, Part I. *Mem. Am. Acad. Arts Sci.* 12, 187–249.

Thaxter, R. 1908. Contribution toward a monograph of the Laboulbeniaceae, Part II. *Mem. Am. Acad. Arts Sci.* 13, 217–469

Thaxter, R. 1924. Contribution towards a monograph of the Laboulbeniaceae, Part III. *Mem. Am. Acad. Arts Sci.* 14, 309–426.

Thaxter, R. 1926. Contribution towards a monograph of the Laboulbeniaceae, Part IV. *Mem. Am. Acad. Arts Sci.* 15, 427–580.

Thaxter, R. 1931. Contribution towards a monograph of the Laboulbeniaceae, Part V. *Mem. Am. Acad. Arts Sci.* 16, 1–435.

Thompson, C. G., and Steinhaus, E. A. 1950. Further tests using a polyhedrosis virus to control the alfalfa caterpillar. *Hilgardia* 19, 411–445.

Vouk, V. 1931. Four years of international corn borer investigations. *In* "International Corn Borer Investigations." (T. Ellinger, ed.), Vol. 4, pp. 92–96. Int. Livestock Exposition, Union Stock Yards, Chicago.

Wang, Z. 1965. Knowledge on the control of silkworm disease in ancient China. *Symp. Sci. Hist.* (China) 8, 15–21.

Weiser, J. 1958. Address by the secretary of the conference, J. Weiser. *Trans. First Int. Conf. Insect Pathol. Biol. Control.* 27–30. Praha, Czechoslovakia.

CHAPTER 2
ASSOCIATIONS BETWEEN INSECTS AND NONPATHOGENIC MICROORGANISMS

I. SYMBIOSIS
II. EXTERNAL MICROBIOTA
III. INSECT-CULTIVATED FUNGI
    A. Ambrosia Beetles
    B. Wood Wasps
    C. Ants
    D. Termites
IV. INTERNAL MICROBIOTA
    A. Bacteria
    B. Fungi
    C. Viruses
    D. Protozoa
    E. Effect of the Host on Internal Microbiota
V. INTERNAL MICROBIOTA OF TERMITES
    A. Flagellates
    B. Bacteria
    C. Termite Nutrition
VI. INTRAHEMOCOELIC AND INTRACELLULAR MICROBIOTA
    A. Intracellular Mutualism
    B. Insect Hosts
    C. Loss of Endosymbionts
VII. ROLE OF SYMBIONTS
VIII. REPRODUCTIVE INCOMPATIBILITY
IX. METHODS OF OBTAINING APOSYMBIOTIC HOSTS
X. TRANSMISSION OF SYMBIONTS THROUGH HOST GENERATIONS

The diversity and ubiquitousness of insects provide ample opportunities for them to come into contact with microorganisms. Without a knowledge of the types of microorganisms associated with normal insects, the insect pathologist cannot hope to gain an accurate understanding of pathogenic microbiota of insects. A microorganism may be nonpathogenic in an insect under one situation and may be pathogenic in others, or both pathogenic and nonpathogenic forms may share a common relationship and, at times, cannot be easily and distinctly separated. In spite of the importance of the normal microbiota of healthy insects, there are relatively few studies on them (aside from mutualistic forms) as compared to those on pathogenic microorganisms.

    The association of insects with nonpathogenic microorganisms is a fascinating study in itself, and because we are able to treat it only briefly, the readers are encouraged to read some of the numerous publications that have appeared, especially in the area of mutualism. For the general aspects of microbe–insect associations, the readers are referred to Steinhaus (1942, 1946), Henry (1967); and for mutualism to Buchner (1965), Henry (1967), Graham (1967), Batra (1979), Houk and Griffiths (1980), Dasch et al. (1984), Wilding et al. (1989), and Schwemmler and Gassner (1989).

I. SYMBIOSIS

The term *symbiosis* was first used by De Bary (1879) for the living together of dissimilar organisms regardless of the result of such an association. The term *symbiont* (or *symbiote*) may apply to both organisms, but it is customary to refer to the larger, phylogenetically younger organism, in our case the insect, as the host, and the smaller, older organism, the symbiont (Hertig et al. 1937).

On the basis of De Bary's definition, symbiosis may be divided into inquilinism, commensalism, mutualism, and parasitism. This is a rather simplistic classification, and the reader should refer to Starr (1975) regarding the complicated variations that may occur in the interactions of organisms. In inquilinism, the microorganism is accidentally associated with the insect and causes no harm to or obtains no benefit from the insect; in commensalism, the microorganism benefits from the association but does no harm to the host; in mutualism, the microorganism and insect mutually benefit from the assocation, even though each may sacrifice a portion of its energy, space, etc., to the other organism; in parasitism, the microorganism lives at the expense and detriment of the host.

Mutualism occurs when (1) there is a close and intimate association between the host and symbiont, and usually neither can survive without the other; (2) the symbiont causes no apparent injury or harm to the host, and the changes or effects that it causes are not pathological; (3) the symbiont is transmitted by an elaborate and complicated manner from generation to generation of the host; (4) the host rigidly controls the numbers of the symbionts so that it is not overcome by their numbers; and (5) the symbiont provides nutrition or other beneficial products to the host and, in return, it obtains protection, food, lodging, and a means of distribution. These aspects will be touched upon in our discussion on mutualism.

The classification of insects that are mutualistically associated with microorganisms is difficult because these insects are scattered throughout the different insect orders. Nearly all of them have one thing in common: they have a diet that is incomplete in one or more nutritional factors (Dadd 1985). In addition, some microorganisms produce pheromones which affect insect behavior.

The location of mutualistic symbionts associated with insects may vary from external to the host, to internal in the digestive tracts and other chambers, to within the hemocoel and within the cells. In a close association between the insect and microorganism, the anatomy of the insect is usually modified, and the microorganism may be structurally atypical.

The location of the microorganisms in insects may be divided into two large groups: (1) the extracellular microbiota (exosymbiont or ectosymbiont), and (2) the intrahemocoelic and intracellular microbiota (endosymbiont). The extracellular forms may be further divided into (1) the external microbiota found on the exterior of the insect's body, and (2) the internal microbiota present in the interior of the insect's body, such as the lumens of the digestive tract and trachea and in special cavities and ducts. Some authors refer to the fungi that are cultured and eaten, especially by termites, ants, and beetles, as external symbionts (Wilding et al. 1989).

In the intrahemocoelic habitat, the symbionts occur free in the hemolymph or attached to structures in the hemocoel. Some symbionts exist extracellularly, whereas others spend part of their time within the insect cells, and still others live entirely intracellularly (endocytobiosis). They may be found in single individual cells or in a complex arrangement of cells to form specialized structures or organs. Šulc (1910) proposed the term *mycetome* for the specialized structures because he believed that the microorganisms in the cells were fungi (yeasts). The individual cells that contain the symbionts are called *mycetocytes*. Some workers differentiate the cells with bacterial symbionts as bacteriocytes, which when grouped together form bacteriotomes. At present, the terms *mycetocyte* and *mycetome* are often used regardless of the type of microorganisms. Mycetomes have also been called "pseudovitelli," "green bodies," and "symbiotic organs."

Mycetomes vary in sizes, shapes, and colors. Some are complex and contain different types of symbionts, which may be segregated and confined to distinct zones within the mycetome. The mycetocytes and mycetomes occur in specific regions of the insect's body. Some insects have several mycetomes that may have different symbionts. The mycetomes are hereditary structures and develop even in the absence of symbionts. They provide the symbionts with nutrients and protection. According to the endocytobiotic concept, there is a mutual gene exchange between the host and symbiont (Schwemmler 1983a, 1983b, 1989). An intracellular symbiont may act as an organelle within the mycetocyte.

II. EXTERNAL MICROBIOTA

The numbers and types of microorganisms associated with insects on the external and internal surfaces are dependent mainly on the habitat of the insect. Soil insects will have a large number of soil-inhabiting microorganisms, those in sewage or decomposing media will have microorganisms commonly associated with the breakdown of organic substances, and those in an aquatic habitat will have aquatic microorganisms. The external features of an insect, such as the plumosity, hairiness, dryness of the cuticle, modifications of appendages, etc., may play a role in determining or selecting the associated microorganisms. Most of the external microbiota occur on insects by accident or by chance, and their numbers vary. There are some, however, that are intimately associated in a commensalistic or mutualistic manner.

An interesting example of external microbiota is the occurrence of microflora and microfauna on the backs of large flightless weevils in wet forests of New Guinea (Gressitt et al. 1965). These weevils have structural modifications on their dorsal surfaces where a veritable microecosystem of fungi, lichens, algae, mites, nematodes, rotifers, psocids, and diatoms are found. This interaction may be mutualistic as in the case of marine arthropods and algae.

Some bacteria and fungi found on the external surface of an insect may interact with certain pathogens. In the beetle *Hylobius pales* the presence of fungal and bacterial contaminants on the surface of the beetle inhibits the conidial germination of the fungus *Metarhizium anisopliae*, a case of antibiosis (Schabel 1978).

The bacteria reported on the external surfaces of insects are predominantly gram-positive spore formers (Steinhaus 1949). In laboratory-reared insects, many are cocci and short rods. The number, however, is relatively few when compared with those found in the internal surfaces. This is possibly related to the unfavorable nature of the cuticle of many insects, such as the dryness and oiliness, and to the more favorable and protected environment found in internal cavities and lumens.

In mosquitoes, the bacteria serve as food or as a stimulant to egg hatching by providing an environment low in dissolved oxygen around the egg (Gillett et al. 1977). In the onion maggot, *Hylemya antiqua*, the oviposition of the adults is enhanced by microorganisms, present on onion seedlings, which convert chemical precursors to volatile stimulatory compounds (Hough et al. 1981, 1982). The fly may help in the dispersal of the bacteria. Certain bacteria (e.g., *Streptococcus* spp. and *Micrococcus* spp.) found in the epicuticular wax layer of the blowfly *Calliphora erythrocephala* are able to metabolize the chlorinated hydrocarbon insecticide, dieldrin (Singh 1981).

All major classes of fungi have been reported from the insect's external surface, but aside from the pathogenic fungi, very few can complete their development solely in this habitat. The classic example of a direct and intimate external association is that between Laboulbeniales (class Ascomycotina) and insects. The Laboulbeniales have been considered to be commensals or mutualists, although an increasing number have been found to be pathogenic and they will be treated in Chapter 10.

The fungus *Laboulbeniopsis termitarius*, at one time included in the Laboulbeniales, has been removed from this taxon (Blackwell and Kimbrough 1976a). This fungus has a foot cell with no haustorium and does not penetrate the integument of the termite. However, its occurrence in large numbers on legs, mouthparts, and antennae may hinder the movement of the termite. *Coreomycetopsis oedipus* occurs on the termite exoskeleton and also does not penetrate the integument (Blackwell and Kimbrough 1976b, 1978). Blackwell and Kimbrough (1978) have prepared a list of fungi found on the external surfaces of termites.

Aquatic insects are commonly associated with fungi, especially with the Trichomycetes (Steinhaus and Tanada 1971). Although most of them are commensals, *Amoebidium parasiticum* causes increased mortality when the larvae of *Aedes aegypti* are stressed by starvation (Kuno 1973).

The genus *Septobasidium* (class Basidiomycotina, family Auriculariaceae), which includes about 300 species, is the best known genus of basidiomycetes associated with insects (Fig. 2–1). Their unique association with scale insects has been thoroughly studied by Couch (1931, 1938). About 300 species are known. They form patches that range in size from 3 mm to over 20 cm, and are usually brownish but may be brilliantly colored in shades of gold, yellow, red, or purple. A patch consists of top and bottom layers between which are numerous tunnels and chambers, many of which are connected to the outside. The scale insects are found within the chambers, usually one per chamber. The fungus exists superficially and does not penetrate into plant tissues.

The association between *Septobasidium* and scale insects should be considered on a colonial basis rather than on the individual scale insects. Some young scale insects are infected by the fungus and others are left alone to reproduce crawlers, which may settle down beneath the same fungus or emerge from the protective fungal covering and spread to other parts of the plant (Fig. 2–1B). The infected scale insect is dwarfed, does not reproduce, and usually does not form a covering over its body. This is a case of parasitism, but Couch (1931) maintains that a state of mutualism occurs between the two organisms. The fungus cannot survive in nature

FIGURE 2–1

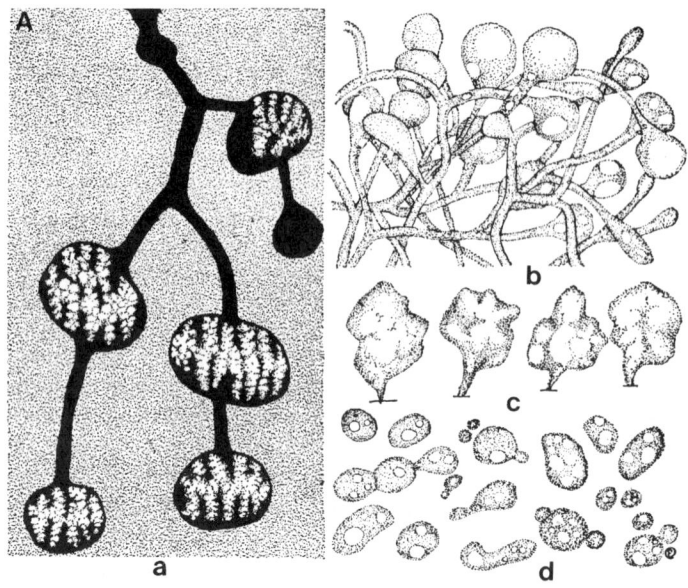

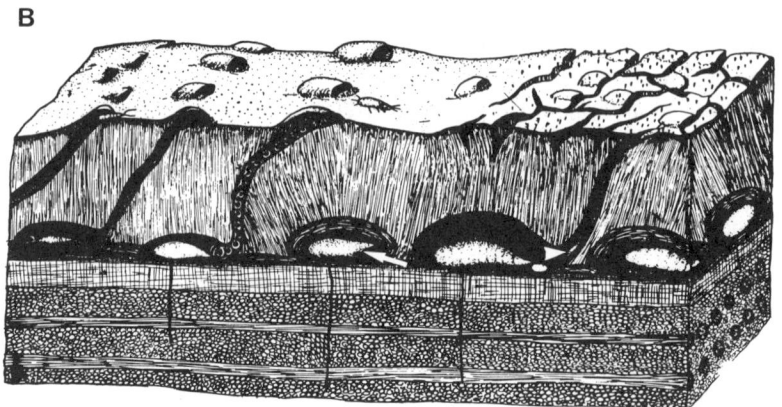

Relationships between fungi and insects. (A) Fungi cultivated by fungus-growing ants. (*a*) Diagram of a large nest of an ant (*Trachmyrmex*), showing in five chambers with pendent fungal gardens and a newly excavated chamber in which the garden has not yet been started. (*b,c*) Modified mycelia (bromatia) of two different fungi cultivated by different ant species. (*d*) Yeastlike cells that make up the bromatia shown in (*c*). (B) Diagram of the relationship of a *Septobasidium* sp. and the associated scale insect. Note two young scale insects entering crescent-shaped entrances to insect houses; older insects with stylets penetrating into plant tissues; insect in center covered by fungus and parasitized (*arrow*); adult insect on the right giving birth to young (*arrowhead*). [From Steinhaus (1949) and courtesy of McGraw-Hill Book Company.]

without the insect. It provides a protective home for the scale insect against predators and in return, some of the insects serve as its food, and the scale crawlers distribute the fungus. Subsequent workers have confirmed the mutualism between the fungi and scales (Watson et al. 1960).

Septobasidium spp. are associated with a large number of genera of scale insects [e.g., *Aspidiotus*, *Cerococcus*, *Kermes* (syn. *Chermes*), *Chionaspis*, *Chrysomphalus*, and *Lepidosaphes*] (Steinhaus 1949). Most of these fungi occur with several species of scale insects, but some are found only with a single-scale species.

Protozoans, especially sessile ciliates and suctorians, are commonly attached to the external surfaces of aquatic insects. Some of these forms are highly specialized and associated with a few or even a single-insect species. Among the ciliates, the highly specialized species of *Opercularia* occur only on Heteroptera and those of *Orbopercularia* only on Coleoptera (Steffan 1967). The trend in specialization in the sessile ciliates is in the colonization of agile, fast-moving host species, such as the beetle families Haliplidae, Dytiscidae, and Gyrinidae; the trend in the suctorians is directed to slowly moving hosts, such as in the families Hydrophilidae, Hydraenidae, and Elmidae.

The nonpathogenic nematodes associated externally with insects are generally phoretic in nature, with most of them being accidental but some are obligatory associations. The insects serve as passive carriers. Many nematode juveniles excrete a drop of adhesive fluid from their mouths and attach themselves securely to the hosts. They generally leave the hosts when they arrive in a suitable environment.

III. INSECT-CULTIVATED FUNGI

Members of nearly all insect orders are associated with fungi (Hammond and Lawrence 1989). Certain insects, such as beetles, ants, termites, and gall-inhabiting cecidomyiid flies, cultivate fungi for food (mycophagy) and other purposes, such as maintaining a moist, suitable environment (Wilding et al. 1989). In some cases, specific fungi are cultivated under complex conditions. The insects control the type, growth, and form of the fungi cultured in their gardens. In the absence of insects, the fungal garden is soon overrun by bacteria and other microorganisms, and the cultured fungus is replaced by other fungi. The insects appear to maintain and cultivate their fungal gardens through oral, anal, and epidermal excretions and through the continuous harvest of the fungal crop. The insect's anatomy is modified for carrying and dispersing the fungus in order to ensure its transmission from one insect generation to the next.

A. AMBROSIA BEETLES

Insect–fungus relationships have been extensively studied in the forest ecosystem (Dowding 1984). This section will deal primarily with the ambrosia beetles, but the widest range and distribution of symbiotic associations among insects are found in Coleoptera (Dasch et al. 1984; Lawrence 1989). Beetles have developed feeding strategies for handling the various types of fungal substrates and specific morphological adaptations (mainly mouth parts) to carry out these strategies (Lawrence 1989).

The ambrosia beetles are found mainly in the two families Scolytidae and Platypodidae and also in Lymexylidae. They differ from bark beetles in their feeding habit; they feed on fungi while the latter feed primarily on wood and are serious forest pests.

In 1836, Schmidberger observed a beetle larva feeding on a white, glistening substance, which he called "ambrosia" (Steinhaus 1946). Initially the ambrosia was considered to be a mixture of insect spittle and crusts of coagulated sap, but subsequently it was found to be fungal in nature. Ambrosia fungi are not able to exist without the beetles. They are transported and cultured for food by the beetles (Beaver 1989). The fungi invade the phloem and wood and may spread widely but generally do not kill the tree. They form "ambrosial" layers covering the walls of the beetle galleries. These layers are browsed upon by the larval and adult beetles.

True ambrosia fungi are highly specialized forms and are not easily isolated and cultured. Most of them are known only in the imperfect state (Deuteromycotina); and where the perfect states are known, they belong mainly to Ascomycotina and a few to Basidiomycotina (Francke-Grosmann 1967; Beaver 1989; Lawrence 1989). The commom genera of imperfect fungi are *Fusarium*, *Cephalosporium*, and *Ambrosiella*; those of ascomycetes are *Endomycopsis*, *Ascoidea*, and *Ceratocystis*. The ambrosia fungi are pleomorphic and may form hyphae and an extended mycelium, or produce ambrosia, or become yeastlike in form.

How the beetles prevent contamination by other fungi and microorganisms has not been resolved, but it may be associated with secretions from the mycangial and oral glands (Beaver 1989). Contrary to early belief, the ambrosia fungi are not species-specific to ambrosia beetles, and other beetles may feed on them. Moreover, the ambrosia beetles and their larvae probably feed on a complex of fungi and bacteria (Haanstad and Norris 1985).

The ambrosia fungi are carried in specialized integumental organs (mycangia or mycetangia) of the beetles (Francke-Grosmann 1956; Abrahamson and Norris 1966, 1969; Baker and Norris 1968). The location of the organs is mainly in the head and thorax and rarely on the elytra and legs (Beaver 1989). The organs are found generally in only the female beetle. Some beetles, however, lack these organs and carry the fungi in their digestive tracts or carry the spores in association with fatty or oily secretions on their body surfaces.

The fungus provides complete nutrition for growth and reproduction of the beetle (Norris and Baker 1968; Norris 1972; Beaver 1989). The fungal diet may also affect the behavioral patterns and initial reproduction of the beetle (Kingsolver and Norris 1977). Moreover, the fungus produces sulcatol, a major aggregation pheromone, in the ambrosia beetle genus *Gnathotrichus* (Berryman 1989).

The bark beetles, which feed along the cambial layers of trees, do not cultivate fungi, but the adults possess mycangia in which fungi occur (Whitney and Farris 1970; Happ et al. 1976; Paine and Birch 1983). The mycangium of the southern pine beetle, *Dendroctonus frontalis*, contains an unidentified basidiomycete, which favors the increase and survival of the beetle brood (Bridges 1983).

The ambrosia and bark beetles react to pheromones, which are comparable to hormones except that they are concerned with the coordination of individuals in a population rather than with an individual organism. These pheromones can act as sex and/or aggregation attractants but sometimes can serve as repellents (Berryman 1989; Hunt and Borden 1990). Both the beetles and microorganisms produce these pheromones. The terpene alcohol verbenol is an attractant produced mainly by the beetle (Conn et al. 1984; Hunt and Borden 1989) and also by bacteria (Brand et al. 1975). Microorganisms, such as mycangial yeasts, convert the verbenol to the repellent verbenone (Brand et al. 1976; Brand and Barras 1977; Hunt and Borden 1990). The pheromones play a role in determining whether a tree is attacked and the severity of the attack by the beetles.

B. WOOD WASPS

Wood wasps are in the families Siricidae and Xiphydriidae. They resemble ambrosia beetles in attacking mainly weakened trees or freshly cut logs, but their mutualism with fungi differs from that of ambrosia beetles. Both fungi and wasp may cause serious damage to trees and timber.

The female wood wasp has intersegmental pouches at the base of the ovipositor (Francke-Grosmann 1967). These pouches are filled with mucus and fungus oidia or arthrospores that are discharged when an egg is laid (Fig. 2–2). The fungus develops in the wood surrounding the oviposition holes and larval tunnels.

As in the case of other mutualistic fungi, those associated with wood wasps are difficult to isolate and culture. They belong to wood-destroying fungi, and the species assigned to the genus *Amylostereum* (syn. *Stereum*) have been isolated (Talbot 1977). The fungi may serve as food for the wood wasp, but unlike ambrosia fungi, their growths are sparse and form only a small part of the insect's diet. Some workers believe that the fungi may soften the cell walls and make it easier for the larva to gnaw the wood. Kukor and Martin (1983) report that the larva of the wood wasp, *Sirex cyaneus*, obtains C_x-cellulases and xylanases through the ingestion of the fungal symbiont *Amylostereum chailletii* present in the wood on which the larva feeds.

C. ANTS

Certain ants of the tribe Attini subfamily Myrmicinae cultivate fungi in a manner similar to those of the fungus-growing termites of the old world (Cherrett et al. 1989). There are about 190 ant species found in the Nearctic and predominantly neotropical biogeographic regions. Each species or genus of ants maintains different species of fungi. The substrates, which are used to grow the fungi, vary with the ant species. The primitive ant genera use insect feces and dead vegetable matter, the intermediate genera use plant debris, leaf fragments, and flowers, and the complex genera (*Acromyrmex* and *Atta*) use exclusively fresh leaves and flowers cut from live plants.

Ants in the genus *Atta* build large subterranean cavities in which fungi are cultivated on pieces of foliage and flowers that the ants have harvested from live plants. This trait has given them the name of "leaf-cutting" ants, and since they carry the pieces of leaves over their heads, they are also called "parasol" ants. In some areas, these ants are polyphagous and are serious pests (Cherrett et al. 1989).

The fungus cultivated in the garden produces bromatia (staphylae) consisting of sterile hyphae with swollen terminal cells called "kohlrabi globules" or "gongylidia" that are eaten by the ants or fed to the larvae (Cherrett et al. 1989) (Fig. 2–1). The fungus, eaten by an ant (*Atta* spp.), provides enzymes that are found in the fecal fluid of the ant (Martin et al. 1975; Boyd and Martin 1975). The fecal fluid, when applied to plant fragments, catalyzes the degradation of proteins and polysaccharides, providing a medium that enhances fungal growth on plant tissues. Thus, the ant serves as a carrier of enzymes from the fungus to plant substrates. In return, the fungus provides the ant with low molecular weight nutrients that can be absorbed readily without further digestion. However, the ants have enzymes of their own or enzymes obtained from the plant sap that may help to digest the plant substrates (Cherrett et al. 1989).

FIGURE 2-2

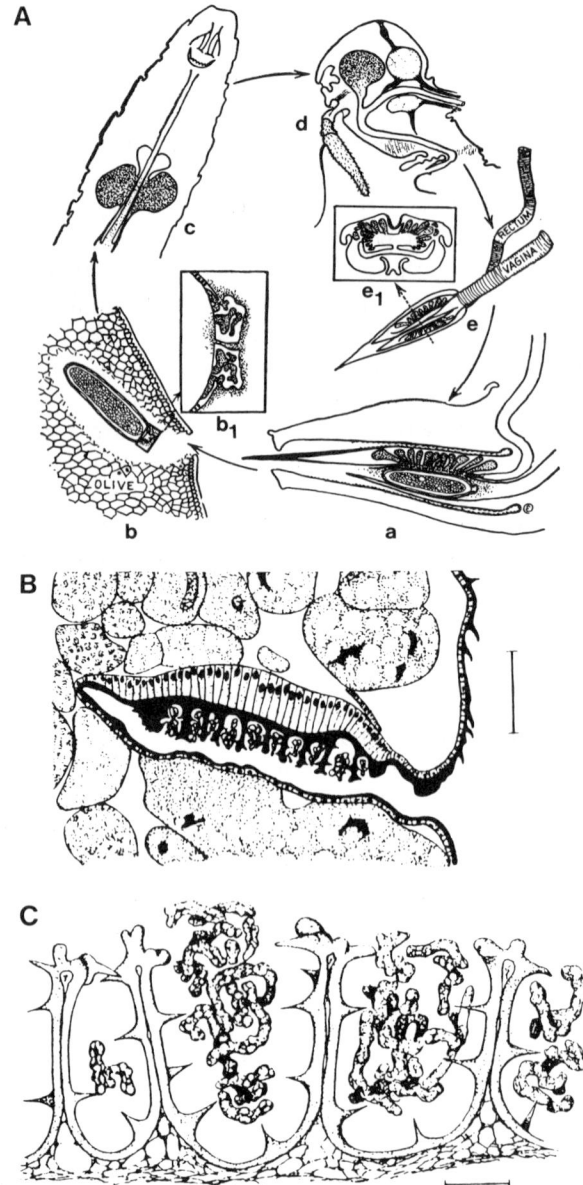

Extracellular mutualistic bacteria and fungi associated with insects. (A) Transmission of bacteria in the olive fruit fly, *Dacus oleae*, from one generation to the next. (*a*) Longitudinal section through the ovipositor of adult fly showing the bacteria-filled pouches connected to the oviduct by longitudinal slit (dotted line above egg). On passing out, the egg is pressed against the pouches and smeared with bacteria. (*b*) Egg laid in an olive fruit. (b_1) Section through the micropyle of an egg through which the bacteria enter and "infect" the embryo. (*c*) Anterior end of a mature larva showing a longitudinal section of two of the four spherical caeca (filled with bacteria) near the fore part of the midgut. (*d*) Longitudinal section through the head of an adult fly showing the bulbous diverticulum of the esophagus filled with bacteria retained from the pupal stage. Later, the entire digestive tract of an adult is contaminated from this organ. (*e*) Female sex apparatus showing the location of the pouches that have acquired the bacteria from the digestive tract. (e_1) Cross section of the pouch area. [From Steinhaus (1949) and courtesy of McGraw-Hill Book Company.] (B) Longitudinal section through a mycangium of *Sirex cyaneus*. (C) Section through a mycangium of *Sirex gigas* showing fungus-filled crypts of a 10-mm larva. [From Buchner (1965) and courtesy of E. A. Parkin and John Wiley and Sons.]

The anatomy of the ants is modified for carrying and spreading the fungus. The virgin queen ant ingests a mass of hyphae and leaf tissues, which is held as a pellet in her infrabuccal pouch. After mating, she uses the pellet to start a new fungal garden, which is fertilized with her feces or an occasional egg. The initial workers gather more foliage and tend to the fungal garden. The queen henceforth becomes an egg-producing "factory." The fungal garden increases in size and numbers, and at this stage, the ant colony may become an economically devastating pest when it is near a cultivated area (Wheeler 1910; Weber 1937).

Very little is known about the fungi cultivated by ants. The fungi cultured by *Atta* and *Acromyrmex* belong to the class Basidiomycotina (Weber 1966; Martin 1970; Hervey et al. 1977; Weber 1979). The identified fungal species is *Attamyces bromatificus* and other proposed species in *Lepiota*, *Leucocoprinus*, *Leucoagaricus*, or *Agaricus* are either synonyms or contaminants (Cherrett et al. 1989). *Attamyces bromatificus* has not been found outside of leaf-cutting ant nests and the ants are completely dependent on it. Stradling and Powell (1986) have suggested that the highly productive fungal strains are a factor in the speciation of the ants.

D. TERMITES

Termites form the third largest group of insects that cultivate fungi for food, but not all termites cultivate fungi. Those that do belong to the family Termitidae ("higher termites") (Wood and Thomas 1989). They usually live in large nests or "termitaria," which are built above or below ground. The termitarium contains special chambers in which the fungus is cultivated on "fungus combs" that are derived from termite fecal material. The surfaces of the combs are covered sparsely by fungal mycelium, spherical or nodulelike conidia, and conidiophores. The mycelia permeate the entire comb and may help in maintaining the required humidity within the nest (Zoberi 1979). The fungal combs have high nutritional value (Rohrmann 1978), and the termite cannot survive without the combs (Sands 1969).

Some termite species seem to have their own species of fungus. The fungal genus commonly associated with termites is *Termitomyces* (Basidiomycotina) (Wood and Thomas 1989). The fungus is carried by the female or male alate which ingests the asexual spores and carries them as a bolus in its digestive tract. The spores pass through the digestive tract and inoculate the fecal deposits of foraging workers.

Higher termites, unlike lower forms, do not rely on mutualistic protozoans for the production of digestive enzymes. The adult workers (subfamily Macrotermitinae) produce, in their midgut epithelium and salivary glands, digestive enzymes to digest cellulose (Wood and Thomas 1989). Bacteria in the digestive tract of the termite may produce digestive enzymes. The cellulases are produced also by the fungus and are acquired by the termites feeding on the fungal combs.

A mutualistic association occurs between the fungus *Termitomyces* and the termite by providing the termite with digestive enzymes and nourishment and the termite maintains the fungal garden free of invaders (Wood and Thomas 1989). *Termitomyces* has no active existence outside of the fungal combs.

IV. INTERNAL MICROBIOTA

Internal microbiota of insects are found in the lumens of the trachea and the digestive tract; in cavities, pouches, ducts, etc.; in the hemocoel (intrahemocoelic); and within cells (intracellular). They are generally greater in numbers and more varied in types than the external microbiota. Some microorganisms may persist in both the external and internal surfaces of insects (e.g., the fungi found in the mycangia of ambrosia beetles).

As in the case of the external microbiota, the environment of the insect plays an important role in determining the quality and quantity of the internal microbiota. The microorganisms in the digestive tract are generally bacteria or bacteria-like organisms and occasionally yeasts. The anatomy of an insect may also govern the types and quantities of microorganisms. Insects with a simple, straight digestive tract usually have adventitious and saprophytic forms. Those with complex tracts consisting of pouches, diverticula, caeca, etc., have a wide assortment of microorganisms because the environment of these specialized structures varies from region to region, especially in pH, enzymes, food, etc. In certain insects, a great variety of microorganisms occur with some confined only to specific areas of the gut. Diversity in the structure of the digestive tracts may occur even within a single family of insects in which a series can be traced from little apparent modification, to simple pouches, to complex diverticula, and to organs (mycetomes) attached to the gut. This suggests an evolutionary trend for the accommodation of symbionts.

Observations with the scanning and transmission electron microscopes have confirmed previous reports that certain microbiota (e.g., bacteria and protozoa) are directly attached to or very intimately associated with the gut epithelium (Fig. 2–3) (Foglesong et al. 1975; Breznak and Pankratz 1977; Bracke et al. 1979). Bignell (1984a) has described the suitability of the arthropod digestive tract as an environment for microorganisms on the basis of the physical, biochemical, and physiological properties of the gut and its lumen.

Certain internal microbiota in the insect's digestive tract may become pathogenic and cause infections when the condition of the insect is weakened by stress, such as starvation, high humidity, temperature, etc. This sometimes occurs when insects are reared in the laboratory.

Another important factor that affects the type of internal microbiota is the insect's food. Stored food products, dead animals and their products, plant sap, and animal blood greatly determine the selection of microorganisms. Some insects (e.g., termites and cockroaches) depend on the enzymes produced by the internal microbiota for the digestion of food, and others obtain such enzymes from the ingested external microbiota as in the case of fungus-growing ants (Martin 1984). Certain plants contain antibacterial substances that prevent some types of bacteria from growing in the insect's gut (Kushner and Harvey 1962). Some plant varieties resist the attack of insect pests (antibiosis) and such food plants may affect the internal microbiota. Some insects may have sterile guts (e.g., insects that suck sap or blood). However, only certain parts of the digestive tract may be sterile.

The composition of the internal microbiota is governed to some extent by the presence of certain principal microorganisms. *Streptococcus faecalis* is the predominant bacteria in the guts of the greater wax moth, *Galleria mellonella*, and of the cutworm *Scotia* (syn. *Agrotis*) *segetum*. The bacteriolytic properties of *S. faecalis* restrict the microbial flora of the gut and may even confer a natural defense to the insect (Jarosz 1975; Charpentier et al. 1978). In *Drosophila melanogaster*, the normal bacteria in two wild stocks are gram-positive forms (Bakula 1969). When the gram-negative *Escherichia coli* is introduced into the digestive tracts of these flies, it persists only in the absence of the native normal bacteria.

FIGURE 2–3

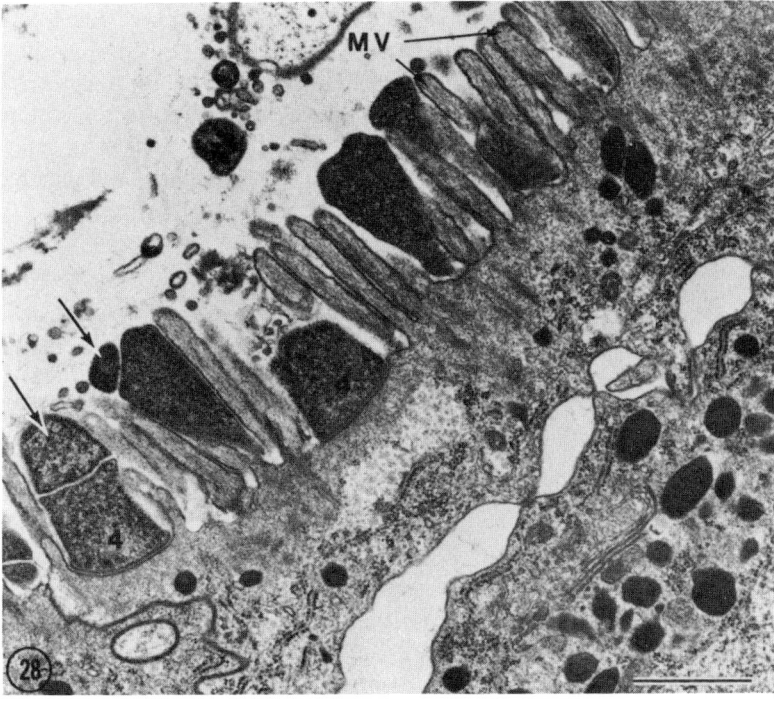

Thin section of bacteria present among the microvilli (*MV*) of the midgut of the worker termite, *Reticulitermes flavipes*. Some of the bacteria are present as "doublets" (*unlabeled arrows*), which remain together after cell division. (Bar = 1 μm.) [Courtesy of Breznak and Pankratz (1977) and American Society for Microbiology.]

A. BACTERIA

The most common bacterial flora found in digestive tracts of healthy insects are gram-negative small rods and, in this respect, resemble the flora of higher animals. Coliform bacteria are frequently found followed by micrococci and spore-forming bacilli. Spirilli are rarely found in the normal insect's gut with the exception of certain termites. Many of these bacteria are saprophytes and occur ubiquitously in nature; others are obligate anaerobes that are specific to the insect. Specialized appendages, such as "gastric caeca," attached to the gut may be filled with a large number of bacteria that are characteristic for the insect species harboring them. These cases are found especially in the higher Hemiptera. The bacteria, in different insect species, may range from very small rods to large spirochete-like forms, and they are difficult to culture on artificial media. Some of these bacteria are intimately attached to the midgut epithelium and to the chitinous wall of the hindgut (Cruden and Markovetz 1984) (Fig. 2–3). Adult insects pass some bacteria from generation to generation in association with the egg. The bacteria appear to play a mutualistic role with the insect host by providing the host with nutrients and other substances and obtaining in return a home and means of transmission.

In general, the internal microbiota that are acquired during the immature stages are not retained by the adult insect, and they may even differ between sexes (Burges et al. 1979). The loss of such microbiota occurs in some muscoid flies that live in media containing an abundance of bacteria. In the house fly, *Musca domestica*, and the blow fly, *Phenicia* (syn. *Lucilia*) *sericata*, some of the newly emerged adults have sterile digestive tracts, while other species of newly emerged blow flies have greatly reduced bacterial counts as compared with those in the mature maggots (Greenberg 1960). There are, however, exceptions where some bacteria are passed from the adult through the egg to the next generation. The plant pathogenic bacterium *Erwinia carotovora*, which causes potato black leg, survives in the digestive tract of the puparium, and is transmitted to the adult of the seedcorn maggot, *Hylemya platura*. The adult smears the bacteria on the egg surface and the emerging larva ingests the bacteria, which may be transmitted to the potato (Dasch et al. 1984).

Bacteria of the internal microbiota are transmitted from generation to generation of the insect by several methods: (1) within the egg, (2) on the surface of the egg, (3) by the egg being laid in an environment containing the bacteria, and (4) by invading the ovary. In *Drosophila melanogaster*, two wild-type stocks have different gram-positive bacteria in the digestive tracts and these bacteria are transmitted through all stages of the fly development (Bakula 1969). The egg becomes contaminated by feces and the maggot acquires the bacteria by feeding on the chorion. Bakula (1969) considers the relationship between the bacteria and *Drosophila* to be mutualistic and that it may even be genetically determined. A mutualistic association occurs between bacteria and the mosquito *Culex pipiens* (Awahmukalah and Brooks 1983). The presence of symbiotic rickettsia and fortuitous microorganisms in the larval diet is needed for the oviposition of viable eggs by the adult mosquito.

A classic example of the complex transmission of mutualistic bacteria in insects is that of the olive fruit fly, *Dacus oleae* (Petri 1909) (Fig. 2–2). The female adult fly has a small series of bacteria-filled little pouches in the anal tract. When eggs are being laid, they are pressed against the pouches and their surfaces are smeared with bacteria. The bacteria enter an egg through the micropyle. The larva acquires the bacteria on hatching, and the bacteria enter the gut lumen and into four spherical caeca near the anterior midgut where they multiply. The pupa has a bulblike diverticulum that branches off the esophagus anterior to the brain. The bacteria accumulate in the diverticulum and during the adult stage, they spread throughout the gut and enter the anal pouches. The transmission is repeated when the eggs are laid.

The bacterium *Pseudomonas savastanoi* plays an important role in the nutrition of the olive fruit fly (Hagen 1966). The bacterium hydrolyzes the protein in the olive and synthesizes methionine and threonine. The mutualistic bacterium has been described as *Pseudomonas syringae* pathovar *savastanoi*, but this has been questioned by Yamvrias et al. (1970), who believe that other gram-negative bacteria may be the true symbiont.

Other fruit flies have similar mutualistic associations with bacteria as the olive fruit fly. Hellmuth (1956) examined 43 species belonging to three subfamilies of Tephritidae (syn. Trypetidae) and reported that all were associated with bacteria. There were various levels of coevolution between the hosts and bacteria with the most advanced association developed in the genus *Dacus*.

The tsetse fly, *Glossina morsitans*, has bacteria in the fore part of the gut. By eliminating these, and feeding the flies on an artificial diet, Nogge (1975) demonstrated an increased need for B vitamins. The bacteria are essential for the reproduction but not the survival of the flies (Nogge 1978).

Bacteria of the internal microbiota may favor the insect host by serving as food, assisting in the digestion of food, producing enzymes, synthesizing vitamins, fixing nitrogen, producing pheromones, and preventing contamination of the insect's midgut by other microorganisms. We have already touched upon these areas except in the case of nitrogen fixation. We shall also expand on pheromone production.

Certain bacteria in the guts of termites are capable of fixing atmospheric nitrogen (Benemann 1973; Breznak et al. 1973; French 1975). The following nitrogen-fixing bacteria have been isolated: *Citrobacter freundii* from *Coptotermes lacteus*, *Mastotermes darwiniensis*, and *Nasutitermes exitiosus* (French et al. 1976), and *Enterobacter agglomerans* from *Coptotermes formosanus* (Potrikus and Breznak 1977). These bacteria may be important in the nitrogen economy of the termite. Prior to the discovery of the nitrogen-fixing bacteria, the general belief was that nitrogen is conserved in termites by recycling the nitrogen in cadavers of dead protozoa and other dead microbiota, and by the termites feeding on dead termites. Uricolytic bacteria, *Streptococcus* sp., *Bacteroides termitidis*, and *Citrobacter* sp., are found in the termite *Reticulitermes flavipes* (Potrikus and Breznak 1980a,b). These bacteria can convert uric acid under the anaerobic condition of the digestive tract into products usable by the termite for carbon, nitrogen, energy, or all three. However, the importance of the nitrogen-fixing bacteria in termite nutrition is still to be established. For example, Nazarczuk et al. (1981) have observed that the gut flora and fauna of *Nasutitermes exitiosus* and *Coptotermes lacteus* may not fix nitrogen under short-term conditions.

Pheromones, which function as sexual attractants, in communication, maintenance of colony structure, etc., are generally produced by various ectodermal glands of insects. However, internal bacterial symbionts have been found producing pheromones. A phenol-producing bacterium in the colleterial glands that are located near the vagina of the female grass grub beetle, *Costelytra zealandica*, is associated with the production of the insect's sex pheromone, possibly a phenol (Hoyt et al. 1971). In bark beetles (e.g., *Ips paraconfusus*, *I. grandicollis*, and *Dendroctonus* spp.) species of *Bacillus* (in particular, *B. cereus*) have been reported to convert certain precursors, such as α-pinene, to pheromones (verbenol), which are found in the beetle's frass (Brand et al. 1975, 1976). Byers and Wood (1981), however, have not been able to confirm the role of *B. cereus* but believe that the unidentified symbiotic bacteria that are sensitive to streptomycin are involved in the conversion of host plant terpene, such as myrcene, to male specific pheromones, ipsenol and ipsdienol, in the bark beetle *I. paraconfusus*. As pointed out previously, certain symbiotic fungi of bark beetles are also capable of producing pheromones. The quantity of pheromones found in the feces of the adult boll weevil, *Anthonomus grandis*, is greatly reduced in laboratory-reared insects contaminated by many species of bacteria (Gueldner et al. 1977). In the migratory locust, *Locusta migratoria migratorioides*, bacteria in the digestive tract appear to convert lignin to produce locustol a pheromone that regulates aggregation (Nolte 1977).

Insects are sometimes associated with bacteria pathogenic to humans, such as those causing typhoid, tuberculosis, dysentery, and anthrax. In some cases, such bacteria are found on the external surfaces, but in others, they are members of the internal microbiota. Fecal coliform bacteria may occur in some aquatic and nonaquatic insects and may introduce these bacteria into the water (Sarai 1976). Some insects serve as vectors of such diseases as plague and tularemia. The readers should be aware of this type of insect association with pathogenic microorganisms of animals and plants and refer to texts in medical entomology and transmission of plant diseases by insects.

B. FUNGI

Relatively few fungi are found in the internal microbiota of a normal insect as compared with those found on the external surfaces or those pathogenic to insects. Occasionally saprophytic fungi, which are mostly soil forms, may be detected in the gut. Those that occur commonly in insects are yeasts or yeastlike organisms, although as mentioned in the earlier portion of this chapter, the members of the order Trichomycetes are usually found in the hindguts of aquatic insects. Yeasts or yeastlike forms are often found in insects reared on artificial diets and in some cases their excessive numbers may cause pathologies (Martignoni and Iwai 1969). Under natural conditions, insects (e.g., *Drosophila*) that frequent fermenting fruits and vegetables feed on yeasts and other microorganisms in addition to plant products.

Some yeasts are important in the diets of certain species of lace wings (*Chrysopa* spp.) whose adults feed on honeydew under field conditions (Hagen et al. 1970; Hagen and Tassan 1972). The yeast *Torulopsis* sp. resides in the esophageal diverticulum (crop) of the adult lace wing, *Chrysopa carnea*, and provides the essential amino acids that may be missing in the honeydews and pollens eaten by the adult. The newly emerged adult does not possess the *Torulopsis* yeast, and acquires it through its food or by trophallaxis when regurgitated liquid is exchanged from an infected to an uninfected adult prior to mating. In the honey bee, the yeast flora develops under conditions of stress, such as in caged colonies, deficient diets, and herbicide treatment of the field (Gilliam et al. 1977).

Yeastlike organisms occur in the hemolymph, digestive tract, and poison gland of the encyrtid parasitoid, *Comperia merceti* (Lebeck 1989). They are concentrated in the poison gland reservoir of the female parasitoid and are injected into the ootheca of the brown-banded cockroach to ensure transmission to the next generation of the parasitoid. The yeast is not pathogenic to the parasitoid.

C. VIRUSES

Nonpathogenic viruses, which occur as internal microbiota of normal insects, are usually bacteriophages and pathogens of plants and higher animals. Insects are well-known vectors of plant viruses (Maramorosch and Jensen 1963). In some cases, these viruses enter the insect's hemocoel before being transmitted to their hosts, and they may or may not cause harm to their insect vectors.

House flies in nature or bred on nonsterile media invariably contain bacteriophages active against *Salmonella typhosa*, *S. paratyphi*, *Escherichia coli*, and *Micrococcus muscae* (Shope 1927; Glaser 1938). Other insects also have bacteriophages, but the role of these phages within the gut or its effect on the enteric bacteria is unknown.

D. PROTOZOA

Many protozoa are members of the internal microbiota of insects. They include all classes of Protozoa. In most cases, their biological relationships with insects are not well known. The largely pathogenic protozoa, the members of the former class

Sporozoa, will be considered in Chapter 12, including the gregarines because they do damage and destroy the midgut cells to which they are attached. Flagellates are common in termites and may be in mutualistic association with them.

E. EFFECT OF THE HOST ON INTERNAL MICROBIOTA

There is very little information on the effect of the host on the physiology of the symbionts. A very interesting effect occurs in the wood cockroach *Cryptocercus*, which can govern the type of reproduction of the mutualistic flagellates in its digestive tract. When the cockroach molts, the flagellates become sexually reproductive and undergo gametogenesis (Cleveland 1957). This is brought about by the molting and developmental hormone, ecdysone (Cleveland et al. 1960). When an adult cockroach or an intermolt nymph, which has no ecdysone, is inoculated with ecdysone, the protozoa in such a host undergo gametogenesis. Gametogenesis is followed by cyst formation, enabling the flagellates to survive in the fecal matter. Since, unlike termites, the wood cockroaches molt in synchrony and lose their symbionts at the same time, proctodaeal feeding cannot reestablish the infection, and feeding on fecal material containing cysts is the behavior that ensures the relationship.

V. INTERNAL MICROBIOTA OF TERMITES

The rich protozoan faunule of termites may have been first noticed by Lespés in 1856 when he observed that the intestines were ordinarily filled with a living agglomeration of infusorians. The protozoan biomass in the termite during the intermolt period may consist of one-seventh to one-third of the body weight of the termite. When students in insect pathology examine a fecal droplet of a worker termite in a wet mount, they are fascinated by the vast milieu of animalculi in constant motion. Most of the microorganisms are flagellates and bacteria.

All families of termites have members containing protozoa, and over 300 species of protozoa have been described. Steinhaus (1946) prepared a list of various species of protozoa and their termite hosts. Mutualistic bacteria and unique protozoa are found mainly in the five families of "lower" termites: Mastotermitidae, Hodotermitidae, Kalotermitidae, Rhinotermitidae, and Serritermitidae (Breznak 1982, 1984a). They are absent or rare in the "higher" family Termitidae; those that are found appear not to be associated mutualistically with these termites. The wood-eating cockroach *Cryptocercus punctulatus* has a protozoan faunule similar to that of the lower termites.

A. FLAGELLATES

The mutualistic protozoa in the lower termites and the wood-eating cockroach are unique flagellates that are restricted to these hosts. They are oxymonad, trichomonad, and hypermastigote flagellates.

Protozoa have not evolved as rapidly as the speciation that has occurred in termites, and it seems probable that many flagellates known today may have existed in ancestors of termites (Kirby 1937, 1949). The close similarity between the protozoan faunule in the lower termites and in the wood-eating cockroach is one of the bases for the proposal that the termite and the cockroach evolved from a preblattid ancestor.

Certain flagellates are restricted to a single species or species complex of termites, and workers have suggested their use in diagnosing or identifying species, especially those of lower termites, and in some cases of higher termites (Kirby 1937, 1949; Honigberg 1970). The presence of internal flagellates appears to have induced a rather high rate of race-forming process in the termites. Mannesmann (1974) proposed establishing subspecies of termites based on their flagellate symbionts because he detected differences in symbionts in colonies of *Reticulitermes* spp., which cannot be otherwise differentiated.

1. Location in Termite

In the termite, the bulk of the microbiota including the protozoans is confined to the thin-walled enlarged hindgut, which unlike that of most insects, plays the major role in the absorption of nutrients (Breznak 1984a). Absorption also occurs in the midgut, but this is limited to soluble nutrients present in food or those liberated from food by termite-secreted enzymes or by bacteria that may colonize the midgut.

Termite castes may harbor the same species of flagellates, although the workers, nymphs, and alates may have larger numbers of flagellates than the soldiers (Lai et al. 1984). In some termites, each flagellate species occupies a more or less distinct location in the hindgut of the worker. This suggests that different flagellate species have specific roles in termite nutrition and some are more directly involved than others in the breakdown of cellulose.

2. Reinfestation

The protozoa are absent in newly hatched termite nymphs. Most or all of the protozoa are lost at times of molting, except in the last molt to adults, and are not found in certain of the reproductive castes (Cleveland 1925c). Termite nymphs have to be reinfested with protozoa in order to survive. The reinfestation is accomplished by feeding on the exudates from the anus of infested individuals (proctodaeal feeding) and by accepting food from nursing workers (stomodaeal feeding).

3. Mutualism

The biological association between termites and flagellates is a classic example of mutualism (Cleveland 1924). The flagellates provide enzymes for the digestion of cellulose (Cleveland 1925a; Trager 1932). Even though axenically cultured flagellates, in the absence of bacteria and fungi, can produce the enzymes (Mauldin et al. 1972; Yamin 1980, 1981; Yamin and Trager 1979; Odelson and Breznak 1985), the flagellates cannot survive for long periods in the absence of bacteria that help to maintain conditions favorable for the flagellates (Yamaoka 1989). Thus, the mutualism is a complex interaction of termite, flagellate, and bacteria.

Cellulose digesting enzymes are found in both the lower and higher termites. In the lower termites the cellulolytic activity occurs in the enlarged hindgut from enzymes secreted by the termite and the symbiotic protozoa (O'Brien et al. 1979; Breznak 1982; O'Brien and Slaytor 1982; Yamaoka 1989), by bacteria (French

1975), and by the fungus *Termitomyces* (O'Brien and Slaytor 1982). In the higher termites of the family Termitidae, there are no mutualistic protozoa, and the cellulose digesting enzymes are secreted in their midguts. In addition, the enzymes may be obtained from cultivated fungi (e.g., *Termitomyces*) (O'Brien and Slator 1982) and from microbial faunule in the termite's digestive tract, such as actinomycetes (Pasti and Belli 1985) and spirochaetes (Eutick et al. 1978).

The main difficulty in establishing the complex nature of mutualism in the termite is the removal and isolation of each symbiont without affecting others. The physiological interdependence between termites and their flagellates has been demonstrated by eliminating the flagellates from the termites. In his classical studies, Cleveland (1924, 1925b) eliminated the flagellates from the termites by rearing them at high temperatures (36°C for 24 h), by starving the termites, and by exposing them to high oxygen pressures. Only the anaerobic flagellates are destroyed under these conditions. These studies demonstrated that the flagellates are essential for the survival of the termites. The role of the flagellate has been confirmed by the breakthrough in the axenic culture of the protozoa (Yamin 1978).

Not all of the intestinal protozoa can metabolize cellulose. Some flagellates may have different nutritional roles (Smythe and Mauldin 1972). Others appear to be commensals. The large amoebas in the higher termites are capable of digesting wood, but their role in the physiological interdependence with termites has not been established.

The lower termites, which are mutualistically associated with flagellates, feed on sound or nearly sound wood, grass, herbs, and straw and can survive on paper, cotton cellulose, and a lignin-cellulose complex. The higher termites of the family Termitidae, with no internal flagellates or only a few of them, live on a varied diet of wood, dried grass, leaves, or most other vegetable material, and they may ingest soil to extract from it useful nutrients (Honigberg 1970). Although many feed on wood, this wood is much more decayed than that eaten by the lower termites.

B. BACTERIA

Bacteria are associated intimately with the termite flagellates, and are located on their surfaces or within them. Those on the surface (epibiotic) are attached at specialized regions, and those within (endobiotic) the protozoa occur in the cytoplasm and possibly in the nucleus (Smith and Arnott 1974; Breznak 1975). In some flagellate species, the location of the epibiotic bacteria is specific and their arrangement forms characteristic patterns on the flagellates (Kirby 1941). The surface bacteria may have specialized structures (e.g., spirochaetes and rod-shaped forms with membrane specialization) for attachment to flagellates of *Reticulitermes* and the wood cockroach *Cryptocercus* (Bloodgood and Fitzharris 1976).

The internal bacteria in the hindgut of the lower family Rhinotermitidae (*Reticulitermes flavipes* and *Coptotermes formosanus*) are intimately associated with each other and with the gut epithelium of the termite by means of holdfast elements (Breznak and Pankratz 1977) (Fig. 2–3). The paunch epithelium of the termite is also modified with cuplike indentations where the bacteria aggregate. The bacteria, therefore, appear to be an integral part of the digestive tracts of termites. Spirochaetes and protozoa play an important role in maintaining the anaerobic condition and in regulating the bacterial flora of the digestive tracts of both the lower and higher termites (Veivers et al. 1982). Breznak (1984b) has treated in detail the spirochaetes of termites and of the wood-eating cockroach *Cryptocercus punctulatus*.

In the higher termite *Nasutitermes exitiosus*, the bacterial flora includes spirochaetes that differ from those found in the lower termites (Czolij et al. 1985). The spirochaetes are essential to the termite (Eutick et al. 1978). The soil-feeding higher termites *Procubitermes aburiensis* and *Cubitermes severus* have filamentous actinomycetelike bacteria (Bignell et al. 1979, 1980a,b). Nonfilamentous bacteria, chiefly rods, colonize the walls of the proctodaeal segments and colon. These bacteria may act on soil organic matter and provide nourishment to the termites.

C. TERMITE NUTRITION

The type of food consumed by termites may be governed by the reducing capacity (redox potential) of the hindgut. In *Zootermopsis nevadensis*, the hindgut has a strong reducing condition for the anaerobic fermentation of wood (Bignell 1984b). A more mildly reducing condition occurs in the hindgut of *Cubitermes severus*, which feeds on humus in the soil.

The quantitative analysis of cellulose digestion is complicated because of the metabolic activity of the intestinal bacteria, and because the flagellates, when undergoing cytolysis, release unidentified products into the substrate. At present, the evidence indicates that the end product of cellulose digestion is acetate, the major usable end product of cellulose metabolism available for the nutrition of the termite (Hungate 1939; Yamin 1980). The pathway of acetate metabolism has been traced in the termite to fatty acids and lipids (Mauldin 1977, 1982; Odelson and Breznak 1983) and the acetate may also be a precursor for amino acids, cuticular hydrocarbons, and terpenes (Breznak 1982).

Since nitrogen and vitamins are critically deficient in wood, the termite must obtain these nutrients from other sources. Fungi provide vitamins and also nitrogen that has been concentrated by utilizing the uric acid wastes of termites. In addition, the nitrogen in termites is recycled through the digestion of the dead internal microbes and their nitrogenous waste products and by the termite feeding on termite exoskeleton, dead termites, and feces (coprophagy) (Breznak 1984a). As mentioned earlier, certain bacteria isolated from the guts of termites are capable of fixing atmospheric nitrogen, and others convert uric acid to products usable to the termites. The nitrogenous products of such bacteria may be a major source of nitrogen for some termites.

VI. INTRAHEMOCOELIC AND INTRACELLULAR MICROBIOTA

The association between insects and microorganisms increases in complexity as the habitat or environment of the microorganisms changes from the external to the internal and to the intrahemocoelic and intracellular states. The intrahemocoelic and intracellular associations are intimate and the host and symbiont are nearly always vitally dependent on each other (Steinhaus 1946; Buchner 1965). The intracellular associations of symbionts in insects are extremely diversified at the morphological and physiological levels (Nardon 1988). The associations may occur in single cells or in highly complex mycetomes (Fig. 2–4). The simple to complex mycetomes in Coleoptera suggest an evolutionary development (Nardon and Grenier 1989).

In 1665, Robert Hooke with his crude and primitive optics saw the symbiont-

FIGURE 2–4

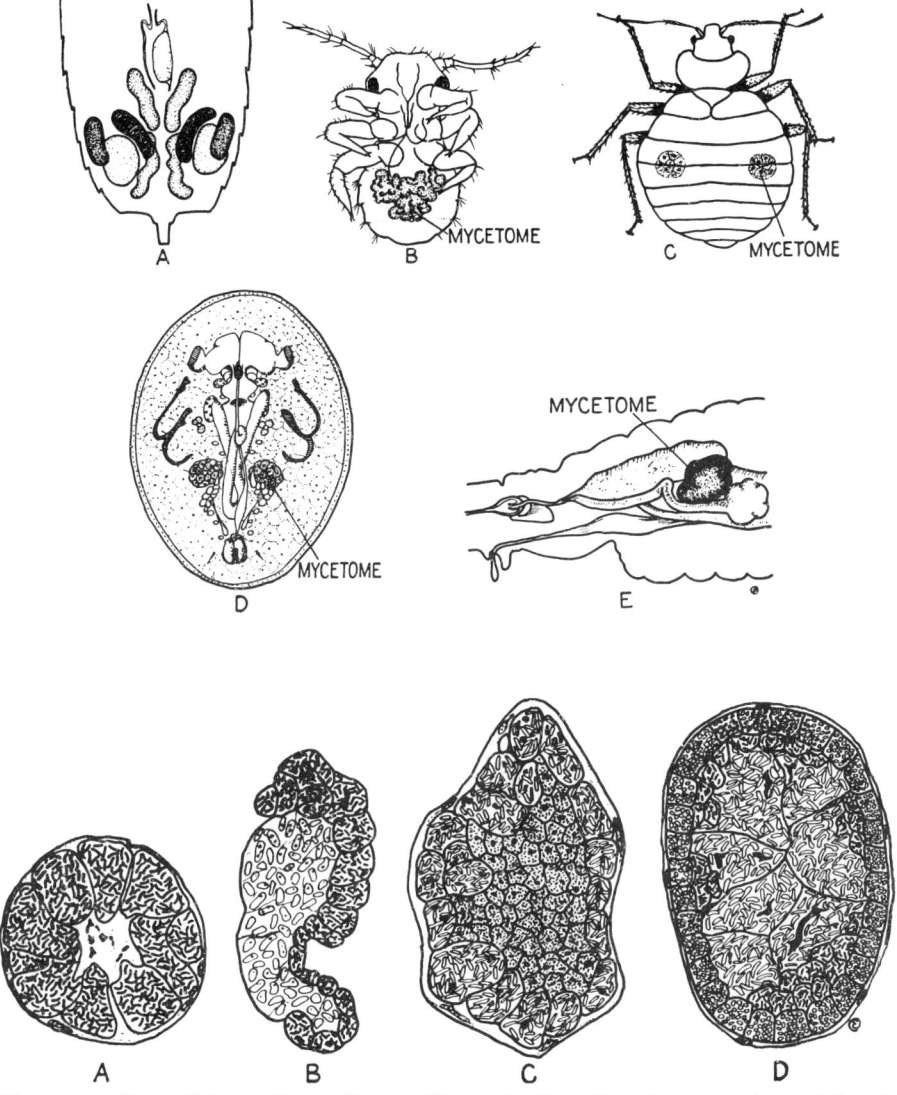

Mycetomes of intracellular symbionts of insects. (*Upper plate*) Locations of mycetome in certain insects. (A) Four different mycetomes in the female Cixiinae. (B) Nymph of *Psylla buxi* showing the mycetome in the abdomen. (C) Bedbug, *Cimex lectularis*, showing the paired mycetomes. (D) Greenhouse whitefly, *Trialeurodes vaporariorum*, showing the paired mycetomes. (E) Longitudinal section of the greenhouse whitefly showing one of the mycetomes. (*Lower plate*) Diagrams of principal types of mycetome–symbiont arrangement in Homoptera. (A) Singly formed mycetome harboring one kind of symbiont. (B) Mycetome consisting of two more or less loosely joined tissues, not enclosed in a common epithelium, and harboring two different kinds of symbionts. (C) Mycetome consisting of two types of tissues surrounded by a common epithelium and harboring two different symbionts. (D) Mycetome with three different types of cells (mycetocytes) enclosed by a common epithelium and harboring three different types of symbionts. [From Steinhaus (1949) and courtesy of McGraw-Hill Book Company.]

bearing organ of the human louse and suggested that it may be the "liver" (Koch 1967). Others also noted the presence of peculiar bodies within insect cells as early as the mid-nineteenth century, and they speculated for many years on the nature of these bodies. The intracellular bodies were believed to be sporozoa, products of metabolism, or yolk spheres. Blochmann (1888) observed rod-shaped particles in fat cells and eggs of cockroaches and was fully convinced that they were microorganisms, if not bacteria, and coined the term "bacteroids." It has been confirmed by later workers that the bacteroids are microorganisms, but there are still some who question whether all similar intracellular bodies are microorganisms.

Intracellular bodies, which are considered to be microorganisms, have been called protoplastoids, bacteria, rickettsia, mycoplasma, chlamydia, actinomycetes, and yeasts, but usually they cannot be identified. Some forms are pleomorphic and change in size and shape depending on the stage of the insect (Müller 1942; Buchner 1965; Korner 1976). Some plant viruses are involved mutualistically with their insect vectors (Maramorosch and Jensen 1963). Such viruses multiply in their insect vectors but cause no apparent harm to their hosts. The viruses, in turn, may render the infected plants more susceptible or nutritionally more favorable to the insect.

A. INTRACELLULAR MUTUALISM

Intracellular mutualism is common in blood-sucking and plant-sucking insects and in specialized feeders but are rare or absent in insects with adequate diets and detritus scavengers, such as in the primitive orders Protura (telsontails), Thysanura (silverfish), and Collembola (springtails); and in the more advanced orders Dermaptera (earwigs), Embioptera (webspinners), Psocoptera (booklice and barklice), Zoraptera (termitelike zorapterans), and Mecoptera (scorpionflies) (Dasch et al. 1984). In addition, insects with aquatic nymphs that have either phytophagous, scavenging, or predaceous habits do not harbor symbionts [e.g., Plecoptera (stoneflies), Neuroptera (dobsonflies, alderflies), and Trichoptera (caddisflies)]. Intracellular associations of symbionts are also uncommon among the endoparasitic Strepsiptera (stylopids) and the predaceous members of Thysanoptera (thrips), and Hymenoptera (vespids, spider wasps).

Intracellular symbiosis has been referred to as endosymbiosis or endocytobiosis (Schwemmler 1983a; Nardon 1988; Schwemmler and Gassner 1989). Taylor and Harrison (1983) have reviewed the intracellular and extracellular ecosystems in regards to symbiosis. The intracellular symbiont (endocytobiont or endosymbiont) exists in a minute ecosystem, the cell (cytocosm), that is the smallest, clearly definable ecosystem. The obligatory endosymbiont must be "inherited" generally through the maternal line (Buchner 1965; Schwemmler 1983a). How the symbionts enter the ovary and ovum vary in different insects (Tremblay 1989; Nardon and Grenier 1989). In some cases, the symbionts may occur within a cell, the mycetocyte, at one stage in their development and occur outside of the cell in the hemolymph or even in the external and internal environments at other stages in their development. The mycetocytes may form a complex structure, the mycetome (Fig. 2–4). Some insects have only one type of intracellular symbiont (monosymbiosis), but many have more than one (plurisymbiosis) in a single insect (Fig. 2–4).

In spite of the complexity of the mutualism between intracellular symbionts and their hosts, the association probably dates back to the Carboniferous period about 300 million years ago (Buchner 1965). The evolution of symbiosis is still continuing (e.g., in certain mosquitoes). According to some authorities, intracellular symbiosis has resulted in the incorporation of symbionts in eukaryotic cells to form organelles, such as mitochondrion, chloroplast, microtubule, etc. This is referred to as the serial endosymbiosis theory (Schwemmler 1974; Taylor 1979, 1983). On the other hand, the intracellular symbiosis may have caused the evolution of species and higher taxa of insects (Margulis and Bermudes 1985).

1. Location

As in the case of modifications of the insect's gut to accommodate the internal symbionts, a series can also be constructed for the accommodation of the symbionts in the hemocoels of related insects. The symbionts, especially yeasts, may lie free and extracellular in the hemolymph, or in fat cells, or in a syncytium of cells or an organ, the mycetome, which may be simple to highly complex. The locations and numbers of symbionts may be genetically controlled by the insect [e.g., in the rice weevil, *Sitophilus oryzae* (Nardon and Wicker 1983)].

2. Isolation of Symbionts

It is difficult to isolate and culture the symbionts from mycetocytes in the fat body of cockroaches and early reports of such isolation appear to be erroneous (Brooks and Richards 1966). The isolated symbionts and the isolated bacteriocyte have not been cultured separately at the present time (Philippe and Landureau 1989). The symbionts have muramic acid, a distinctive component of bacterial cell wall, and possess cytological and ultrastructural features of gram-positive and -negative bacteria as well as those of rickettsia (Daniel and Brooks 1967). Brooks (1970, 1974) has concluded that only *Blattabacterium cuenoti* has been identified as a true symbiont of Blattidae. Absence of the symbionts results in deranged protein metabolism, modified behavior, shorter survival rate, and failure to reproduce in the cockroaches (Brooks and Kringen 1972; Philippe and Landureau 1989).

3. Regulation of Symbiont Population

An interesting question in mutualism is how the host regulates the numbers of symbionts, especially intracellular forms. Unlike the uncontrolled multiplication of pathogenic microorganisms, the reproduction of symbionts is not detrimental to an insect. The host has control of the mutualistic association (Buchner 1965). It regulates the symbionts in regard to their rate of propagation, their form of development, the suppression of their virulence, their strict spatial limitations, and their transmission to the next host generation. The endosymbionts may be regulated by phagocytes, lysozymes in the hemolymph, hormones, and antiproteins (Hinde 1971; Houk and Griffiths 1980; Dasch et al. 1984; Philippe and Landureau 1989; Tiivel 1989). Within the mycetocytes and mycetomes, workers have observed the destruction of symbionts by lysosomal activity (cytolysomes) (Musgrave and Grinyer 1968; Müller 1972; Hinde 1971; Griffiths and Beck 1973; Chang 1975). Schwemmler 1983b), in his study with the cicada *Euscelus incisus* (syn. *E. plebejus*), suggests that the host controls the cellular specificity of the endosymbionts by α- and/or t-specific systems of membrane receptor proteins and by the amount of antibacterial lysozyme that is produced in the cell. Furthermore, the host seems to regulate the endosymbiotic infectious forms by direct coupling to its sexual hormone system. In cockroaches, the development of the endosymbionts, as well as the mitochondria, complies with the common hormonal signals of regulation generated by the host cell (Philippe and Landureau 1989).

4. Protection of Endosymbionts in Host

Intracellular symbionts appear to be protected from the immune mechanisms of the host because they remain for the most part inside special cells (mycetocytes) that have lost the ability to recognize the symbionts as "non-self" (Ermin 1939; Malke and Schwartz 1966a; Nardon 1988; Gupta 1989). When they occur outside of such cells, they may be attacked by lytic agents or phagocytized by insect hemocytes. In some leafhoppers and scale insects, however, the intracellular symbionts spend a stage free in the hemolymph before they are incorporated into the mycetocytes (Nasu 1965; Korner 1976; Tremblay, 1989).

B. INSECT HOSTS

Within insect groups as large as tribes, subfamilies, and even families, there may be uniformity in the types of mycetomes and in the kinds of symbionts. However, it cannot be assumed that symbiosis in the three major groups (leafhoppers, aphids, and coccids) is similar or that it is essentially the same within each group (Dasch et al. 1984). Some of the most advanced stages in symbiosis occur in cicadas. Certain species of lac and wax insects possess characteristic yeast symbionts and they can be differentiated on the basis of their symbionts (Mahdihassan 1928, 1929).

1. Homopteran and Hemipteran Hosts

Most studies with intracellular symbiosis have been conducted with homopteran and hemipteran insects, especially aphids and leafhoppers (Buchner 1965; Houk and Griffiths 1980; Dasch et al. 1984; Tiivel 1989). These insects generally have two different forms of symbionts: (1) an infectious form that is smaller and optically denser, and transmitted extracellularly to the next generation; (2) a vegetative form that is larger, optically more transparent, and reproduces intracellularly (Schwemmler 1983a; Nardon 1988). The homopteran symbionts have morphogical features, characteristic of prokaryotes usually Eubacteriales, that vary from typical bacterial rods to bizarre polymorphic forms (Schwemmler 1983a,b; Tiivel 1989) considered to be rickettsia (Unterman and Baumann 1990) (Fig. 2–5). Nearly all of them have two endogenous cell membranes and a third membrane of host origin.

a. Aphids

Several species of aphids have two distinct types of endosymbionts, a rod-shaped secondary symbiont that is morphologically bacterialike found outside of a mycetocyte and an oval primary symbiont found within a mycetocyte (McLean and Houk 1973; Houk and Griffiths 1980). The endosymbionts of aphids have been isolated and cultured (Houk and McLean 1974; Ishikawa 1982). Recently, the genes coding for the 16S ribosomal RNAs of the two symbionts have been cloned and sequenced (Unterman et al. 1989; Unterman and Baumann 1990). Sequence comparisons indicate that the rod-shaped symbiont is a member of family Enterobac-

FIGURE 2-5

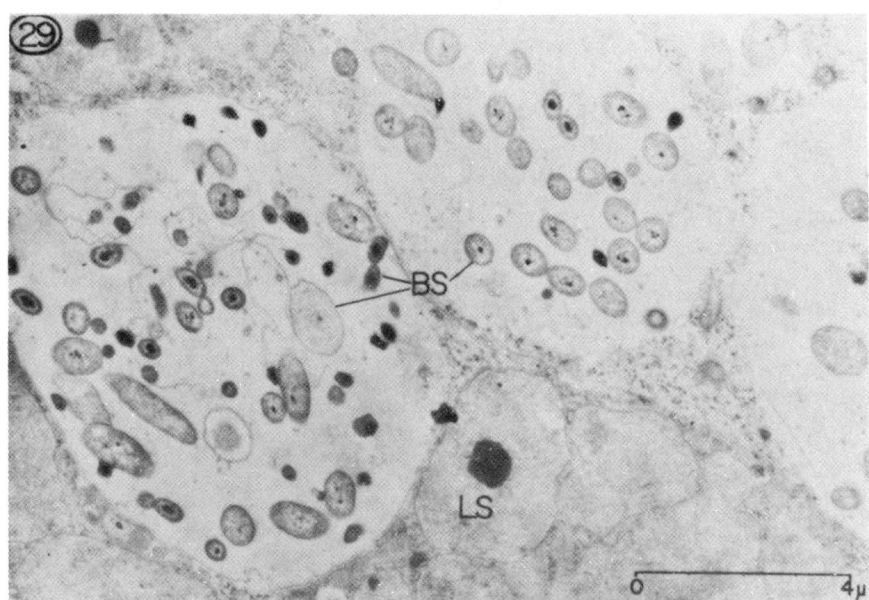

Electron micrograph of central layer of mycetome of an adult leafhopper, *Nephotettix cincticeps*, showing different morphotypes of the bacterial symbiont (*BS*) and L-symbiont (*LS*). [Courtesy of Nasu (1965) and the Japanese Society of Applied Entomology and Zoology.]

teriaceae, in class Proteobacteria (proposed by Stackebrandt et al. 1988), and the primary symbiont is also in the same class but of a distinct lineage. The primary symbiont has been placed in a new genus *Buchnera* (Munson et al. 1991b). The primary symbionts from four aphid families appear to be monophyletic and the symbiotic association was established in a common ancestor (Munson et al. 1991a).

The role of the aphid endosymbionts is still not completely known (Unterman and Baumann 1990). From the endosymbionts, the aphids may obtain nutritious substances required for growth, fecundity, and longevity (Dadd 1985) and possibly plant cell wall degrading enzymes (Campbell and Dreyer 1985; Eisenbach and Mittler 1987). A protein, symbionin (MW 63,000), is produced by endosymbionts and is essential for the growth of the aphids *Acyrthosiphon pisum* and *A. kondoi* (Ishikawa 1984a; Ishikawa and Yamaji 1985a,b; Ohtaka and Ishikawa 1991). The endosymbionts synthesize symbionin in the aphid embryo and the protein is metabolized during postembryonic development (Ishikawa and Yamaji 1985c). The endosymbionts are capable of producing a number of proteins in aged aphids; this indicates that the aphid restricts the endosymbionts to produce only symbionin during the postembryonic up to adult maturity (Ishikawa 1983, 1984b, 1989). The restriction may result from protein-inhibiting factors in the aphid (Ishikawa and Yamaji 1985c). Ishikawa (1989) speculates that symbionin is encoded by the insect genome and the mRNA is transferred to the endosymbiont for translation.

b. Scale Insects

The endosymbionts of scale insects are commonly described as bacteria- or yeastlike microorganisms, but several have been isolated and classified as *Coryne-*

bacterium, Flavibacterium, or *Achromobacter* (Tremblay 1989). The yeastlike forms have not been identified and some are considered to be conidial forms of ascomycetes.

2. Other Insects

A yeastlike endosymbiont found in the mycetocytes of the planthopper *Laodelphax striatellus* is transferred from the fat body to the ovary and penetrates into the eggs as a symbiotic ball via the ovarial pedicel. Mitsuhashi (1975) has succeeded in isolating the symbiotic ball and culturing the endosymbiont *in vitro*. Antigenically similar yeastlike forms occur in the egg and adult of the brown planthopper, *Nilaparvata lugens*, suggesting a close relationship between these endosymbionts (Nasu et al. 1981).

Middledorf and Ruthmann (1984) have reported the first case of a yeastlike endosymbiont in Hymenoptera. The endosymbiont occurs in various tissues of both sexes of the ichneumonid wasp, *Pimpla turionellae*. It is transmitted to the next wasp generation through the oocytes.

In addition to bacteria and yeasts, rickettsialike and viruslike particles have been detected in insects. Endosymbionts of the order Rickettsiales are in three genera, *Wolbachia*, *Symbiontes*, and *Blattabacterium* (Tiivel 1984). The green rice leafhopper, *Nephotettix cincticeps*, harbors three types, two of which (*a*- and *t*-symbionts) occur in mycetomes and a third, which is rickettsialike, is found in almost all tissues of the leafhopper (Nasu 1965; Mitsuhashi and Kono 1975). The widely distributed, nonspecific, rickettsialike endosymbiont is found most frequently in the nuclei but also occurs in the cytoplasm.

Schwemmler and Vago (1970) have prepared organ cultures from the embryo and larva of the leafhopper *Euscelus incisus* and have infected the cultures with the endosymbionts *a* and *t*. The unidentified endosymbionts have been called protoplastoids (Schwemmler et al. 1973). The *a* and *t* symbionts differ in the ultrastructures of their surface membranes (Louis et al. 1976). The membrane of the *t* endosymbiont suggests that this endosymbiont may play a role of a "kidney"; may use the host catabolic products, such as uric acid, xanthene, etc.; and may regulate the osmotic pressure in the cells.

In the *Culex pipiens* complex of mosquitoes, *Wolbachia* endosymbionts occur in the male and female gonads (Gassner 1989). In *Culex australicus* and *C. globocoxitus*, there are viruslike particles in the cytoplasm of ovaries and associated tissues (Irving-Bell 1974).

Viral particles (polydnavirus) have been reported from the reproductive tracts of braconid and ichneumonid parasitoids of lepidopteran larvae (Stoltz et al. 1976; Stoltz and Vinson 1977, 1979). They are transmitted to the host larva during oviposition by the parasitoid and are taken up by the host cells. There is no evidence of viral replication in the lepidopteran larva. The successful development of the parasitoid egg within the host depends on the virus, which acts to suppress the host's immune response (encapsulation) toward the egg (Edson et al. 1981). This appears to be the first example of obligatory mutualism between a virus and an insect.

C. LOSS OF ENDOSYMBIONTS

Some insects may lose their endosymbionts at certain stages in their life cycle. In some species of Cerambycidae and Curculionidae, the endosymbionts of the larval gut are lost in the adult male beetle and the female has endosymbionts only in her germ cells (Dasch et al. 1984; Nardon and Grenier 1988). In the human body louse, *Pediculus humanus*, the endosymbionts in the larva are found in the mycetome (stomach disk) and are destroyed in the male adult (Dasch et al. 1984). In the female adult, the endosymbionts migrate to the ovary and into the developing eggs. The endosymbiont undergoes, in its life cycle, three changes in tissue location, three cycles of proliferation, and regulated changes in size or degeneration.

VII. ROLE OF SYMBIONTS

The complete role of symbionts is generally unestablished because these microorganisms are difficult to isolate and culture. In many cases, their roles are speculations. Their absence is manifested in many different ways, especially in the marked reduction in insect growth, difficulty in molting, deranged metamorphosis, reduction or loss in reproductive capability, and death in severe cases. The mutualism is based primarily on the exchange of metabolites between the insect and the symbiont. In general, the major role of the symbiont is to provide nutrition, growth factors, source of energy, and possibly hormones to their hosts (Buchner 1965; Houk and Griffiths 1980; Ishikawa 1989). Some symbionts appear to function in the catabolism of the host, such as the breakdown of urates, in the regulation of pH, osmotic pressure, and certain endogenous rhythms of the host, and in the fixation of atmospheric nitrogen. In addition, Schwemmler (1980) suggests the possibility of genetic interchange between the host and symbiont. This suggestion has been expanded by Rautian and Ossipov (1985) in treating the relationships between the genetic systems of the symbiotic partners from the viewpoint of population genetics and ontogenetics.

VIII. REPRODUCTIVE INCOMPATIBILITY

Symbionts may cause failure in insect reproduction. In certain mosquitoes, (e.g., *Culex* and *Aedes*), strains are often incompatible and this was explained in genetic terms as cytoplasmic incompatibility. The cause of the incompatibility is rickettsiae. In *Culex pipiens* the rickettsia is *Wolbachia pipientis* (Hertig and Wolbach 1924; Yen and Barr 1973), and in *Aedes* spp. possibly several species of *Wolbachia* (Beckett et al. 1978; Wright and Wang 1980). The rickettsiae are transmitted maternally (i.e., the pole cells are infected at the time of formation and the rickettsiae are transmitted to the germ cells, which develop later) (Wright and Barr 1981). Yen and Barr (1973) have removed the symbionts by treating the mosquito with tetracycline antibiotics and have cultured the mosquito free from symbionts (aposymbiotic state) for many generations. In crosses of symbiotic and aposymbiotic mosquitoes, an aposymbiotic male is compatible with both the normal symbiotic and the aposymbiotic females, resulting in the production of fertile eggs. The aposymbiotic female, on the other hand, is compatible with the aposymbiotic male but not with the normal symbiotic male. The eggs laid by the aposymbiotic female mated to a symbiotic male show no development, even though the female is inseminated by the male. The endosymbionts appear to play a role in the fertilization of the egg.

Awahmukalah and Brooks (1983) challenged the concept that cytoplasmic incompatibility in *Culex pipiens* was caused by the symbiont neutralizing the sperm. Since they found that the sperm was not neutralized in an egg containing rickettsiae, they suggested that the major significance of the rickettsial agent was in the embryonic development of the mosquito.

Similar cases of *Wolbachia*-induced reproductive incompatibility occur in the almond moth, *Ephestia cautella* (Brower 1976; Kellen et al. 1981), and in the alfalfa weevil, *Hypera postica* (Hsiao and Hsiao 1985). In both insect species, the rickettsiae are maternally transmitted. The rickettsia is unidentified in the almond moth. It causes incompatibility between crosses of aposymbiotic female and symbiotic male moths. Even though it occurs in abundance in hypertrophied spermatids, it appears to cause no deleterious effects on the production and activity of the sperms. Thus, the mechanism for the failure of egg fertilization is unknown. In the alfalfa weevil, the endosymbiont has been identified as *Wolbachia postica* (Hsiao and Hsiao 1985). Reproductive isolation is also induced in the flour beetle, *Tribolium confusum*, but the microorganism has not been identified (Wade and Stevens 1985).

A reproductive situation that is unlike those of the mosquito occurs in the ambrosia beetle. A gram-positive bacterial endosymbiont, *Staphylococcus* sp., in the ambrosia beetle, *Xyleborus ferrugineus*, enables the beetle to reproduce parthenogenetically (Peleg and Norris 1972). The bacteria invade the oocytes and initiate nuclear division in the oocytes even though they are unfertilized by the sperm (arrhenotokous parthenogenesis). The oocyte does not develop beyond the primary oocyte stage in the absence of the bacteria endosymbiont (Norris 1972). Thus, the reproduction of the beetle is primarily dependent on the bacterium.

When the protoplastoids of *Euscelis incisus* are removed from eggs, the aposymbiotic eggs give rise to embryos without abdomen, "head embryos" (Schwemmler et al. 1973). The endosymbionts are necessary for normal embryogenesis. Moreover, the partial reduction in endosymbiont numbers reduces the fecundity and growth of the insect; complete elimination causes death. The endosymbionts play essential roles in the anabolism and catabolism of the nymph and adult insects.

IX. METHODS OF OBTAINING APOSYMBIOTIC HOSTS

In order to study the role of microorganisms in mutualistic associations with insects, they are usually removed partially or completely from the host and the reaction of the aposymbiotic host observed. Such removal has been accomplished by various methods (Houk and Griffiths 1980; Dasch et al. 1984). Symbionts may be removed directly by dissection, which is not difficult when they are segregated in mycetomes. In the human body louse, the mycetome is an organ called *Magenscheibe* (stomach disk) by German authors and it can be excised, or the eggs can be centrifuged to displace the stomach disks of the embryos, which results in aposymbiotic insects (Aschner 1934). Another method is to destroy the symbionts by placing the insects under an environment that is more unfavorable to the symbiont than to the host, such as high temperatures or high oxygen pressures (Cleveland 1924, 1925b). When the symbionts are transmitted on the egg surface, they can be removed mechanically or by surface sterilization of the egg with chemical disinfectants. Antibiotics, sulfa drugs, and lysozyme are also used in removing symbionts. Methyl bromide fumigation destroys the symbionts in the granary weevil, *Sitophilus granarius* (Musgrave et al. 1965). The absence of certain minerals, such as Mn and Zn, and of unsaturated fatty acids in the diets of the German cockroach produces aposymbiotic offspring (Brooks 1960). The inoculation of 1% solution of lysozyme into abdomens of the cockroaches causes lysis of symbionts (Malke and Schwartz 1966b). The above methods, however, may not completely remove the symbionts, and this results in erroneous conclusions, especially in attempts to reintroduce the symbionts into presumably aposymbiotic hosts.

X. TRANSMISSION OF SYMBIONTS THROUGH HOST GENERATIONS

Insects have developed diverse means, some very complex, to ensure the transmission of symbionts from generation to generation of the host (Koch 1960). These involve elaborate insect structures and behavior. In some cases, especially when the symbionts occur on the external surfaces or in the lumens of the fore- and hindguts and the trachea, they are lost to the insect upon ecdysis. The newly molted individuals have to be reinfected as in the case of termites.

Many endosymbionts penetrate the ovary and egg by way of the follicular lining or nurse cells and are transmitted within the egg. The endosymbionts enter the egg by diverse routes in different insect orders, such as by the anterior or posterior poles of the eggs, or may involve the polar bodies as in the scale insects (Tremblay 1989). In the German cockroach, the ovaries of the embryo are symbiont-free until the embryo hatches from the ootheca and the bacteriocytes migrate from the fat body to the ovaries (Sacchi et al. 1985). The bacteriocytes penetrate the ovarioles, adhere, and infect the oocytes. The yeastlike symbionts of the smaller brown planthopper, *Laodelphax striatellus*, move into the primary oocyte through the epithelial plug from the posterior pole of the adult female ovary (Noda 1977). The bacterial symbiont of the human body lice moves from the mycetome to the lateral oviducts of the nymph undergoing adult molt (Eberle and McLean 1983).

Some symbionts are incorporated into the chorion or smeared on the surface of the egg during oviposition. In some insects, the digestive tracts and reproductive organs are greatly modified to ensure the proper placement of the symbionts with the egg (e.g., the olive fly). The symbionts may enter the egg through the micropyle and become ingested by the hatching insect, or they may remain on or in the egg shell and be consumed when the emerging larva feeds on the chorion.

Symbionts may be transmitted through anal secretions. In the case of termites, the newly molted individual acquires the flagellates by feeding on anal secretions of normal individuals (proctodaeal feeding). The transmission through anal secretion is more complex in cases where there are adaptations of maternal and filial behaviors. In some plant-feeding bugs (e.g., *Brachypelta aterrima*) the eggs are laid in loose sand, and the female remains in the vicinity of the eggs until they hatch (Schorr 1957). The nymphs crawl on their mother's abdomen and consume the droplets of pure bacterial suspension eliminated from the mother's rectum. In the bug *Coptosoma scutellatum*, the female encloses packets of symbiotic bacteria in a secretion from the gut, which hardens to form little bacterial cocoons or "sausages" (Schneider 1940). One of these packets is deposited between each pair of eggs as they are laid. The emerging nymphs pierce the packets and suck up a dose of bacteria.

Although the transmission of symbionts is carried out usually by the female, the male is also involved in at least one case. In bostrychid beetles, the symbionts are transmitted through the seminal fluid of the male beetle and deposited at copulation into the female, and the symbionts enter the egg through the micropyle (Mansour 1934).

REFERENCES

Abrahamson, L. P., and Norris, D. M. 1966. Symbiontic interrelationships between microbes and ambrosia beetles. I. The organs of microbial transport and perpetuation of *Xyloterinus politus*. *Ann. Entomol. Soc. Am.* 59, 877–880.

Abrahamson, L. P., and Norris, D. M. 1969. Symbiontic interrelationships between microbes and ambrosia beetles. IV. Ambrosial fungi associated with *Xyloterinus politus*. *J. Invertebr. Pathol.* 14, 381–385.

Aschner, M. 1934. Studies on the symbiosis of the body louse. I. Elimination of the symbionts by centrifugalisation of the eggs. *Parasitology* 26, 309–314.

Awahmukalah, D. S. T., and Brooks, M. A. 1983. Reproduction of an inbred strain of *Culex pipiens* prevented by loss of *Wolbachia pipientis*. *J. Invertebr. Pathol.* 41, 184–190.

Baker, J. M., and Norris, D. M. 1968. A complex of fungi mutualistically involved in the nutrition of the ambrosia beetle *Xyleborus ferrugineus*. *J. Invertebr. Pathol.* 11, 246–250.

Bakula, M. 1969. The persistence of a microbial flora during postembryogenesis of *Drosophila melanogaster*. *J. Invertebr. Pathol.* 14, 365–374.

Batra, L. R. 1979. "Insect-Fungus Symbiosis. Nutrition, Mutualism, and Commensalism." Allanheld, Osmun, Montclair, New Jersey.

Batra, S. W. T., and Batra, L. R. 1967. The fungus gardens of insects. *Sci. Am.* 217(5), 112–120.

Beaver, R. A. 1989. Insect-fungus relationships in the bark and ambrosia beetles. In "Insect-Fungus Interactions." (N. Wilding, N. M. Collins, P. M. Hammond, and J. F. Webber, eds.), pp. 121–143. Academic Press, London.

Beckett, E. B., Boothroyd, B., and MacDonald, W. W. 1978. A light and electron microscope study of rickettsia-like organisms in the ovaries of mosquitoes of the *Aedes scutellaris* group. *Ann. Trop. Med. Parasitol.* 72, 277–283.

Benemann, J. R. 1973. Nitrogen fixation in termites. *Science 181*, 164–165.

Berryman, A. A. 1989. Adaptive pathways in scolytid-fungus associations. In "Insect-Fungus Interactions." (N. Wilding, N. M. Collins, P. M. Hammond, and J. F. Webber, eds.), pp. 145–159. Academic Press, London.

Bignell, D. E. 1984a. The arthropod gut as an environment for microorganisms. In "Invertebrate-Microbial Interactions." (J. M. Anderson, A. D. M. Rayner, and D. W. H. Walton, eds.), pp. 205–227. Cambridge University Press, Cambridge, England.

Bignell, D. E. 1984b. Direct potentiometric determination of redox potentials of the gut contents in the termites *Zootermopsis nevadensis* and *Cubitermes severus* and in three other arthropods. *J. Insect Physiol.* 30, 169–174.

Bignell, D. E., Oskarsson, H., and Anderson, J. M. 1979. Association of actinomycete-like bacteria with soil-feeding termites (Termitidae, Termitinae). *Appl. Environ. Microbiol.* 37, 339–342.

Bignell, D. E., Oskarsson, H., and Anderson, J. M. 1980a. Colonization of the epithelial face of the peritrophic membrane and the ectoperitrophic space by actinomycetes in a soil-feeding termite. *J. Invertebr. Pathol.* 36, 426–428.

Bignell, D. E., Oskarsson, H., and Anderson, J. M. 1980b. Distribution and abundance of bacteria in the gut of a soil-feeding termite *Procubitermes aburiensis* (Termitidae, Termitinae). *J. Gen. Microbiol.* 117, 393–403.

Blackwell, M., and Kimbrough, J. W. 1976a. Ultrastructure of the termite-associated fungus *Laboulbeniopsis termitarius*. *Mycologia* 68, 541–550.

Blackwell, M., and Kimbrough, J. W. 1976b. A developmental study of the termite-associated fungus *Coreomycetopsis oedipus*. *Mycologia* 68, 551–558.

Blackwell, M., and Kimbrough, J. W. 1978. *Hormiscioideus filamentosus* gen. et sp. nov., a termite-infesting fungus from Brazil. *Mycologia 70*, 1274–1280.

Blochmann, F. 1888. Ueber das reqelmässige Vorkommen von bakterienähnlichen Gebilden in den Geweben und Eiern veschiedener Insecten. *Z. Biol. 24*, 1–15.

Bloodgood, R. A., and Fitzharris, T. P. 1976. Specific associations of prokaryotes with symbiotic flagellate Protozoa from the hindgut of the termite *Reticulitermes* and the wood-eating roach *Cryptocercus*. *Cytobios 17*, 103–122.

Boyd, N. D., and Martin, M. M. 1975. Faecal proteinases of the fungus-growing ant, *Atta texana*: Their fungal origin and ecological significance. *J. Insect Physiol. 21*, 1815–1820.

Bracke, J. W., Cruden, D. L., and Markovetz, A. J. 1979. Intestinal microbial flora of the American cockroach, *Periplaneta americana L*. *Appl. Environ. Microbiol. 38*, 945–955.

Brand, J. M., and Barras, S. J. 1977. The major volatile constituents of a basidiomycete associated with the southern pine beetle. *Lloydia 40*, 398–400.

Brand, J. M., Bracke, J. W., Markovetz, A. J., Wood, D. L., and Browne, L. E. 1975. Production of verbenol pheromone by a bacterium isolated from bark beetles. *Nature 254*, 136–137.

Brand, J. M., Bracke, J. W., Britton, L. N., Markovetz, A. J., and Barras, S. J. 1976. Bark beetle pheromones: Production of verbenone by a mycangial fungus of *Dendroctonus frontalis J. Chem. Ecol. 2*, 195–199.

Breznak, J. A. 1975. Symbiotic relationships between termites and their intestinal microbiota. *In* "Symbiosis." (D. H. Jennings and D. L. Lee, eds.), (Society for Experimental Biology Symposium Ser. no. 29). pp. 559–580. Cambridge University Press, Cambridge, England.

Breznak, J. A. 1982. Intestinal microbiota of termites and other xylophagous insects. *Annu. Rev. Microbiol. 36*, 323–343.

Breznak, J. A. 1984a. Biochemical aspects of symbiosis between termites and their intestinal microbiota. *In* "Invertebrate-Microbial Interactions." (J. M. Anderson, A. D. M. Rayner, and D. W. H. Walton, eds.), pp. 173–203. Cambridge University Press, Cambridge, England.

Breznak, J. A. 1984b. Hindgut spirochaetes of termites and *Cryptocercus punctulatus*. *In* "Bergey's Manual of Systematic Bacteriology." (N. R. Krieg and J. G. Holt, eds.), Vol. 1, pp. 67–70. Williams and Wilkins, Baltimore.

Breznak, J. A., and Pankratz, H. S. 1977. In situ morphology of the gut microbiota of wood-eating termites [*Reticulitermes flavipes* (Kollar) and *Coptotermes formosanus* Shiraki]. *Appl. Environ. Microbiol. 33*, 406–426.

Breznak, J. A., Brill, W. J., Mertins, J. W., and Coppel, H. C. 1973. Nitrogen fixation in termites. *Nature 244*, 577–580.

Bridges, J. R. 1983. Mycangial fungi of *Dendroctonus frontalis* (Coleoptera: Scolytidae) and their relationship to beetle population trends. *Environ. Entomol. 12*, 858–861.

Brooks, M. A. 1960. Some dietary factors that affect ovarial transmission of symbiotes. *Proc. Helminthol. Soc. Wash. 27*, 212–220.

Brooks, M. A. 1970. Comments on the classification of intracellular symbiotes of cockroaches and a description of the species. *J. Invertebr. Pathol. 16*, 249–258.

Brooks, M. A. 1974. Genus IX. *Blattabacterium* Hollande and Favre 1931, 754. *In* "Bergey's Manual of Determinative Bacteriology.) (R. E. Buchanan and N. E. Gibbons, eds.), Eighth Ed., p. 901. Williams and Wilkins, Baltimore.

Brooks, M. A., and Kringen, W. B. 1972. Polypeptides and proteins as growth factors for aposymbiotic *Blatella germanica* (L.). *In* "Insect and Mite Nutrition." (J. G. Rodriguez, ed.), pp. 353–364. North-Holland, Amsterdam.

Brooks, M. A., and Richards, K. 1966. On the in vitro culture of intracellular symbiotes of cockroaches. *J. Invertebr. Pathol. 8*, 150–157.

Brower, J. H. 1976. Cytoplasmic incompatibility: Occurrence in a stored-product pest *Ephestia cautella*. *Ann. Entomol. Soc. Am. 69*, 1011–1015.

Buchner, P. 1965. "Endosymbiosis of Animals with Plant Microorganisms. Interscience Publishers, John Wiley and Sons, New York.

Burges, H. D., Grove, J. F., and Pople, M. 1979. The internal microbial flora of the elm bark beetle, *Scolytus scolytus*, at all stages of its development. *J. Invertebr. Pathol. 34*, 21–25.

Byers, J. A., and Wood, D. L. 1981. Antibiotic-induced inhibition of pheromone synthesis in a bark beetle. *Science 213*, 763–764.

Campbell, B. C., and Dreyer, D. L. 1985. Host-plant resistance of sorghum: Differential hydrolysis of sorghum pectic substances by polysaccharases of greenbug biotypes (*Schizaphis graminum*, Homoptera: Aphididae). *Arch. Insect Biochem. Physiol. 2*, 203–215.

Chang, K.-P. 1975. Haematophagous insect and haemoflagellate as hosts for prokaryotic endosymbionts. *Symp. Soc. Exp. Biol. 29*, 407–428.

Charpentier, R., Charpentier, B., and Zethner, O. 1978. The bacterial flora of the midgut of two Danish populations of healthy fifth instar larvae of the turnip moth, *Scotia segetum*. *J. Invertebr. Pathol. 32*, 59–63.

Cherrett, J. M., Powell, R. J., and Stradling, D. J. 1989. The mutualism between leaf-cutting ants and their fungus. *In* "Insect-Fungus Interactions." (N. Wilding, N. M. Collins, P. M. Hammond, and J. F. Webber, eds.), pp. 93–120. Academic Press, London.

Cleveland, L. R. 1924. The physiological and symbiotic relationships between the intestinal protozoa of termites and their host, with special reference to *Reticulitermes flavipes* Kollar. *Biol. Bull. Marine Biol. Lab. Woods Hole, Massachusetts 46*, 178–227.

Cleveland, L. R. 1925a. The ability of termites to live perhaps indefinitely on a diet of pure cellulose. *Biol. Bull. Marine Biol. Lab. Woods Hole, Massachusetts 48*, 289–293.

Cleveland, L. R. 1925b. The effects of oxygenation and starvation on the symbiosis between the termite, *Termopsis*, and its intestinal flagellates. *Biol. Bull Marine Biol. Lab. Woods Hole, Massachusetts 48*, 309–326.

Cleveland, L. R. 1925c. The feeding habit of termite castes and its relation to their intestinal flagellates. *Biol. Bull. Marine Biol. Lab. Woods Hole, Massachusetts 48*, 295–308.

Cleveland, L. R. 1957. Correlation between the molting period of *Cryptocercus* and sexuality in its protozoa. *J. Protozool. 4*, 168–175.

Cleveland, L. R., Burke, A. W., Jr., and Karlson, P. 1960. Ecdysone induced modifications in the sexual cycles of the protozoa of *Cryptocercus*. *J. Protozool. 7*, 229–239.

Conn, J. E., Borden, J. H., Hunt, D. W. A., Holman, J., Whitney, H. S., Spanier, O. J., Pierce, H. D., Jr., and Oehlschlager, A. C. 1984. Pheromone production by axenically reared *Dendroctonus ponderosae* and *Ips paraconfusus* (Coleoptera: Scolytidae). *J. Chem. Ecol. 10*, 281–290.

Couch, J. N. 1931. The biological relationship between *Septobasidium retiforme* (B. & C.) Pat. and *Aspidiotus Osborni* New. and Ckll. *Q. J. Microsc. Sci. 74*, 383–437.

Couch, J. N. 1938. "The Genus *Septobasidium*." Univ. North Carolina Press, Chapel Hill, North Carolina.

Cruden, D. L., and Markovetz, A. J. 1984. Microbial aspects of the cockroach hindgut. *Arch. Microbiol. 138*, 131–139.

Czolij, R., Slaytor, M., and O'Brien, R. W. 1985. Bacterial flora of the mixed segment and the hindgut of the higher termite *Nasutitermes exitiosus* Hill (Termitidae, Nasutitermitinae). *Appl. Environ. Microbiol. 49*, 1226–1236.

Dadd, R. H. 1985. Nutrition: Organisms. *In* "Comprehensive Insect Physiology Biochemistry and Pharmacology." (G. A. Kerkut and L. I. Gilbert, eds.), Vol. 4, pp. 313–390. Pergamon Press, Oxford, England.

Daniel, R. S., and Brooks, M. A. 1967. Chromatographic evidence for murein from the bacteroid symbiotes of *Periplaneta americana* (L.). *Experientia 23*, 499–502.

Dasch, G. A., Weiss, E., and Chang, K.-P. 1984. Endosymbionts of insects. *In* "Bergey's Manual of Systematic Bacteriology." (N. R. Krieg and J. G. Holt, eds.), Vol. 1, pp. 811–833. Williams and Wilkins, Baltimore, Maryland.

De Bary, A. 1879. "Die Erscheinung der Symbiose." Karl J. Trubner Publisher, Strassburg.

Dowding, P. 1984. The evolution of insect-fungus relationships in the primary invasion of forest timber. *In* "Invertebrate-Microbial Interactions." (J. M. Anderson, A. D. M. Rayner and D. W. H. Walton, eds.), pp. 133–153. Cambridge University Press, Cambridge, England.

Eberle, M. W., and McLean, D. L. 1983. Observation of symbiote migration in human body lice with scanning and transmission electron microscopy. *Can. J. Microbiol.* 29, 755–762.

Edson, K. M., Vinson, S. B., Stoltz, D. B., and Summers, M. D. 1981. Virus in a parasitoid wasp: Suppression of the cellular immune response in the parasitoid's host. *Science 211*, 582–583.

Eisenbach, J., and Mittler, T. E. 1987. Extra-nuclear inheritance in a sexually produced aphid: The ability to overcome host plant resistance by biotype hybrids of the greenbug, *Schizaphis graminum. Experientia 43*, 332–334.

Ermin, R. 1939. Über Bau und Funktion der Lymphocyten bei Insekten (*Periplaneta americana* L.). *Z. Zellforsch. Mikrosk. Anat.* 29, 613–669.

Eutick, M. L., Veivers, P., O'Brien, R. W., and Slaytor, M. 1978. Dependence of the higher termite, *Nasutitermes exitiosus* and the lower termite, *Coptotermes lacteus* on their gut flora. *J. Insect Physiol.* 24, 363–368.

Foglesong, M. A., Walker, D. H., Jr., Puffer, J. S., and Markovetz, A. J. 1975. Ultrastructural morphology of some prokaryotic microorganisms associated with the hindgut of cockroaches. *J. Bacteriol.* 123, 336–345.

Francke-Grosmann, H. 1956. Hautdrüsen als Träger der Pilzsymbiose bei Ambrosiakäfern. *Z. Morphol. Okol. Tiere 45*, 275–308.

Francke-Grosmann, H. 1967. Ectosymbiosis in wood-inhabiting insects. *In* "Symbiosis." (S. M. Henry, ed.), Vol. 2, pp. 141–205. Academic Press, New York.

French, J. R. J. 1975. The role of termite hindgut bacteria in wood decomposition. *Mater. Org.* 10, 1–13.

French, J. R. J., Turner, G. L., and Bradbury, J. F. 1976. Nitrogen fixation by bacteria from the hindgut of termites. *J. Gen. Microbiol.* 95, 202–206.

Gassner, G. 1989. Dipteran endocytobionts. *In* "Insect Endocytobiosis: Morphology, Physiology, Genetics, Evolution." (W. Schwemmler and G. Gassner, eds.), pp. 217–232. CRC Press, Boca Raton, Florida.

Gillett, J. D., Roman, E. A., and Phillips, V. 1977. Erratic hatching in *Aedes* eggs: A new interpretation. *Proc. R. Soc. London B 196*, 223–232.

Gilliam, M., Morton, H. L., Prest, D. B., Martin, R. D., and Wickerham, L. J. 1977. The mycoflora of adult worker honeybees, *Apis mellifera*: Effects of 2,4,5-T and caging of bee colonies. *J. Invertebr. Pathol.* 30, 50–54.

Glaser, R. W. 1938. Test of a theory on the origin of bacteriophage. *Am. J. Hyg.* 27, 311–315.

Graham, K. 1967. Fungal-insect mutualism in trees and timber. *Annu. Rev. Entomol.* 12, 105–126.

Greenberg, B. 1960. Host-contaminated biology of muscoid flies: I. Bacterial survival in the pre-adult stages and adults of four species of blow flies. *J. Insect Pathol.* 2, 44–54.

Gressitt, J. L., Sedlacek, J., and Szent-Ivany, J. J. H. 1965. Flora and fauna on backs of large Papuan moss-forest weevils. *Science 150*, 1833–1835.

Griffiths, G. W., and Beck, S. D. 1973. Intracellular symbiotes of the pea aphid, *Acyrthosiphon pisum. J. Insect Physiol.* 19, 75–84.

Gueldner, R. C., Sikorowski, P. P., and Wyatt, J. M. 1977. Bacterial load and pheromone production in the boll weevil, *Anthonomus grandis. J. Invertebr. Pathol.* 29, 397–398.

Gupta, A. P. 1989. Insect host immune system and endocytobionts: Their avoidance strategies. *In* "Insect Endocytobiosis: Morphology, Physiology, Genetics, Evolution." (W. Schwemmler and G. Gassner, eds.), pp. 11–20. CRC Press, Boca Raton, Florida.

Haanstad, J. O., and Norris, D. M. 1985. Microbial symbiotes of the ambrosia beetle, *Xyloterinus politus*. *Microb. Ecol. 11*, 267–276.

Hagen, K. S. 1966. Dependence of the olive fly, *Dacus oleae*, larvae on symbiosis with *Pseudomonas savastanoi* for the utilization of olive. *Nature 209*, 423–424.

Hagen, K. S., and Tassan, R. L. 1972. Exploring nutritional roles of extracellular symbiotes on the reproduction of honeydew feeding adult chrysopids and tephritids. *In* "Insect and Mite Nutrition." (J. G. Rodriguez, ed.), pp. 323–351. North Holland Publishing Co., Amsterdam.

Hagen, K. S., Tassan, R. L., and Sawall, E. F., Jr. 1970. Some ecophysiological relationships between certain *Chrysopa*, honeydews and yeasts. *Boll. Lab. Entomol. Agraria Portici 28*, 113–134.

Hammond, P. M., and Lawrence, J. F. 1989. Appendix: Mycophagy in insects: A summary. *In* "Insect-Fungus Interactions." (N. Wilding, N. M. Collins, P. M. Hammond, and J. F. Webber, eds.), pp. 275–324. Academic Press, London.

Happ, G. M., Happ, C. M., and Barras, S. J. 1976. Bark beetle-fungal symbiosis. II. Fine structure of a basidiomycetous ectosymbiont of the southern pine beetle. *Can. J. Bot. 54*, 1049–1062.

Hellmuth, H. 1956. Untersuchungen zur Bakteriensymbiose der Trypetiden (Diptera). *Z. Morphol. Okol. Tierre 44*, 483–517.

Henry, S. M. 1967. "Symbiosis." Vol. 2. Academic Press, New York.

Hertig, M., and Wolbach, S. B. 1924. Studies on rickettsia-like micro-organisms in insects. *J. Med. Res. 44*, 329–374.

Hertig, M., Taliaferro, W. H., and Schwartz, B. 1937. The term *symbiosis*, *symbiont* and *symbiote*. *J. Parasitol. 23*, 326–329.

Hervey, A., Rogerson, C. T., and Leong, I. 1977. Studies on fungi cultivated by ants. *Brittonia 29*, 226–236.

Hinde, R. 1971. The control of the mycetome symbiotes of the aphids *Brevicoryne brassicae*, *Myzus persicae*, and *Macrosiphum rosae*. *J. Insect Physiol. 17*, 1791–1800.

Honigberg, B. M. 1970. Protozoa associated with termites and their role in digestion. *In* "Biology of Termites." (K. Krishna and F. M. Weesner, eds.), Vol. 2, pp. 1–36. Academic Press, New York.

Hough, J. A., Harman, G. E., and Eckenrode, C. J. 1981. Microbial stimulation of onion maggot oviposition. *Environ. Entomol. 10*, 206–210.

Hough, J. A., Eckenrode, C. J., and Harman, G. E. 1982. Nonpathogenic bacteria affecting oviposition behavior in the onion fly. *Environ. Entomol. 11*, 585–589.

Houk, E. J., and Griffiths, G. W. 1980. Intracellular symbiotes of the Homoptera. *Annu. Rev. Entomol. 25*, 161–187.

Houk, E. J., and McLean, D. L. 1974. Isolation of the primary intracellular symbiote of the pea aphid, *Acyrthosiphon pisum*. *J. Invertebr. Pathol. 23*, 237–241.

Hoyt, C. P., Osborne, G. O., and Mulcock, A. P. 1971. Production of an insect sex attractant by symbiotic bacteria. *Nature 230*, 472–473.

Hsiao, C., and Hsiao, T. H. 1985. Rickettsia as the cause of cytoplasmic incompatibility in the alfalfa weevil, *Hypera postica*. *J. Invertebr. Pathol. 45*, 244–246.

Hungate, R. E. 1939. Experiments on the nutrition of *Zootermopsis*. III. The anaerobic carbohydrate dissimilation by the intestinal protozoa. *Ecology 20*, 230–245.

Hunt, D. W. A., and Borden, J. H. 1989. Terpene alcohol pheromone production by *Dendroctonus ponderosae* and *Ips paraconfusus* (Coleoptera: Scolytidae) in the absence of readily culturable microorganisms. *J. Chem. Ecol. 15*, 1433–1463.

Hunt, D. W. A., and Borden, J. H. 1990. Conversion of verbenols to verbenone by yeasts isolated from *Dendroctonus ponderosae* (Coleoptera: Scolytidae). *J. Chem. Ecol. 16*, 1385–1397.

Irving-Bell, R. J. 1974. Cytoplasmic factors in the gonads of *Culex pipiens* complex mosquitoes. *Life Sci. 14*, 1149–1151.
Ishikawa, H. 1982. Isolation of the intracellular symbionts and partial characterizations of their RNA species of the elder aphid, *Acyrthosiphon magnoliae*. *Comp. Biochem. Physiol. 72B*, 239–247.
Ishikawa, H. 1983. Biochemistry of aphid symbionts. *Endocytobiology 2*, 759–769.
Ishikawa, H. 1984a. Characterization of the protein species synthesized *in vivo* and *in vitro* by an aphid endosymbiont. *Insect Biochem. 14*, 417–425.
Ishikawa, H. 1984b. Age-dependent regulation of protein synthesis in an aphid endosymbiont by the host insect. *Insect Biochem. 14*, 427–433.
Ishikawa, H. 1989. Biochemical and molecular aspects of the aphid endocytobiosis. *In* "Insect Endocytobiosis: Morphology, Physiology, Genetics, Evolution." (W. Schwemmler and G. Gassner, eds.), pp. 123–143. CRC Press, Boca Raton, Florida.
Ishikawa, H., and Yamaji, M. 1985a. Symbionin, an aphid endosymbiont-specific protein—I. Production of insects deficient in symbiont. *Insect Biochem. 15*, 155–163.
Ishikawa, H., and Yamaji, M. 1985b. Protein synthesis by intracellular symbionts in two closely interrelated aphid species. *BioSystems 17*, 327–335.
Ishikawa, H., and Yamaji, M. 1985c. Protein synthesis by an endosymbiont in the aphid embryo. *Endocytobiosis Cell Res. 2*, 119–126.
Ishikawa, H., Yamaji, M., and Hashimoto, H. 1985. Symbionin, an aphid endosymbiont-specific protein—II. Diminution of symbionin during post-embryonic development of aposymbiotic insects. *Insect Biochem. 15*, 165–174.
Jarosz, J. 1975. Lysozymelike lytic enzyme of *Streptococcus faecalis* and its role in the larval development of wax moth, *Galleria mellonella*. *J. Invertebr. Pathol. 26*, 275–281.
Kellen, W. R., Hoffmann, D. F., and Kwock, R. A. 1981. *Wolbachia* sp. (Rickettsiales: Rickettsiaceae) a symbiont of the almond moth, *Ephestia cautella*: Ultrastructure and influence on host fertility. *J. Invertebr. Pathol. 37*, 273–283.
Kingsolver, J. G., and Norris, D. M. 1977. The interaction of *Xyleborus ferrugineus* (Coleoptera: Scolytidae) behavior and initial reproduction in relation to its symbiotic fungi. *Ann. Entomol. Soc. Am. 70*, 1–4.
Kirby, H. 1937. Host–parasite relations in the distribution of protozoa in termites. *Univ. Calif. Berkeley Publ. Zool. 41*, 189–211.
Kirby, H. 1941. Devescovinid flagellates of termites. I. The genus *Devescovina*. *Univ. Calif. Berkeley Publ. Zool. 45*, 1–91.
Kirby, H. 1949. Systematic differentiation and evolution of flagellates in termites. *Rev. Soc. Mex. Hist. Nat. 10*, 57–79.
Koch, A. 1960. Intracellular symbiosis in insects. *Annu. Rev. Microbiol. 14*, 121–140.
Koch, A. 1967. Insects and their endosymbionts. *In* "Symbiosis." (S. M. Henry, ed.), Vol. 2, pp. 1–106. Academic Press, New York
Körner, H. K. 1976. On the host–symbiont–cycle of a leafhopper (*Euscelis plebejus*) endosymbiosis. *Experientia 32*, 463–464.
Kukor, J. J., and Martin, M. M. 1983. Acquisition of digestive enzymes by siricid woodwasps from their fungal symbiont. *Science 220*, 1161–1163.
Kuno, G. 1973. Biological notes of *Amoebidium parasiticum* found in Puerto Rico. *J. Invertebr. Pathol. 21*, 1–8.
Kushner, D. J., and Harvey, G. T. 1962. Antibacterial substances in leaves: Their possible role in insect resistance to disease. *J. Insect Pathol. 4*, 155–184.
Lai, P. Y., Tamashiro, M., and Fujii, J. K. 1983. Abundance and distribution of the three species of symbiotic protozoa in the hindgut of *Coptotermes formosanus* (Isoptera: Rhinotermitidae). *Proc. Hawaii. Entomol. Soc. 24*, 271–276.
Lawrence, J. F. 1989. Mycophagy in the Coleoptera: Feeding strategies and morphological adaptations. *In* "Insect–Fungus Interactions." (N. Wilding, N. M. Collins, P. M. Hammond, and P. F. Webber, eds.), pp. 1–23. Academic Press, London.

Lebeck, L. M. 1989. Extracellular symbiosis of a yeast-like microorganism within *Comperia merceti* (Hymenoptera: Encyrtidae). *Symbiosis* 7, 51–66.

Lespés, C. 1856. Recherches sur l'organisation et les moeurs du termite lucifuge. *An. Sci. Nat. Zool.* 5, 227–282.

Louis, C., Nicolas, G., and Pouphile, M. 1976. Ultrastructure of the endocellular procaryotes of arthropods as revealed by freeze-etching. II.—"t" type endosymbionts of the leafhopper *Euscelis plebejus* Fall. (Homoptera, Jassidae). *J. Microsc. Biol. Cell.* 27, 53–58.

Mahdihassan, S. 1928. Symbionts specific of wax and pseudo lac insects. *Arch. Protistenkd.* 63, 18–22.

Mahdihassan, S. 1929. The microorganisms of red and yellow lac insects. *Arch. Protistenkd.* 68, 613–624.

Malke, H., and Schwartz, W. 1966a. Untersuchungen über die Symbiose von Tieren mit Pilzen und Bakterien. XI. Die Rolle des Wirtslysozyms in der Blattidensymbiose. *Arch. Mikrobiol.* 53, 17–32.

Malke, H., and Schwartz, W. 1966b. Untersuchungen über die Symbiose von Tieren mit Pilzen und Bakterien. XII. Die Bedeutung der Blattiden-Symbiose. *Z. Allg. Mikrobiol.* 6, 34–68.

Mannesmann, R. 1974. Qualitative und quantitative Untersuchung der Darmfaunen mehrerer Populationen von *Reticulitermes* (Isopt., Rhinotermitidae). *Z. Ang. Entomol.* 76, 86–97.

Mansour, K. 1934. On the intracellular micro-organisms of some bostrychid beetles. *Q. J. Microsc. Sci. N.S.* 77, 243–253.

Maramorosch, K., and Jensen, D. D. 1963. Harmful and beneficial effects of plant viruses in insects. *Annu. Rev. Microbiol.* 17, 495–530.

Margulis, L., and Bermudes, D. 1985. Symbiosis as a mechanism of evolution: Status of cell symbiosis theory. *Symbiosis* 1, 101–123.

Martignoni, M. E., and Iwai, P. J. 1969. A candidiasis in larvae of the douglas-fir tussock moth, *Hemerocampa pseudotsugata*. *J. Invertebr. Pathol.* 14, 108–110.

Martin, M. M. 1970. The biochemical basis of the fungus-attine ant symbiosis. *Science* 169, 16–20.

Martin, M. M. 1984. The role of ingested enzymes in the digestive processes of insects. *In* "Invertebrate-Microbial Interactions." (J. M. Anderson, A. D. M. Rayner, and D. W. H. Walton, eds.), pp. 155–172. Cambridge University Press, Cambridge, England.

Martin, M. M., Boyd, N. D., Gieselmann, M. J., and Silver, R. G. 1975. Activity of faecal fluid of a leaf-cutting ant toward plant cell wall polysaccharides. *J. Insect Physiol.* 21, 1887–1892.

Mauldin, J. K. 1977. Cellulose catabolism and lipid synthesis by normally and abnormally faunated termites, *Reticulitermes flavipes*. *Insect Biochem.* 7, 27–31.

Mauldin, J. K. 1982. Lipid synthesis from [^{14}C]-acetate by two subterranean termites, *Reticulitermes flavipes* and *Coptotermes formosanus*. *Insect Biochem.* 12, 193–199.

Mauldin, J. K., Smythe, R. V., and Baxter, C. C. 1972. Cellulose catabolism and lipid synthesis by the subterranean termite, *Coptotermes formosanus*. *Insect Biochem.* 2, 209–217.

McLean, D. L., and Houk, E. J. 1973. Phase contrast and electron microscopy of the mycetocytes and symbiotes of the pea aphid, *Acyrthosiphon pisum*. *J. Insect Physiol.* 19, 625–633.

Middledorf, J., and Ruthmann, A. 1984. Yeast-like endosymbionts in an ichneumonid wasp. *Z. Naturforsh.* 39, 322–326.

Mitsuhashi, J. 1975. Cultivation of intracellular yeast-like organisms in the smaller brown planthopper, *Laodelphax striatellus* Fallén (Hemiptera, Delphacidae). *Appl. Entomol. Zool.* 10, 243–245.

Mitsuhashi, J., and Kono, Y. 1975. Intracellular microorganisms in the green rice leafhopper,

Nephotettix cincticeps Uhler (Hemiptera: Deltocephalidae). *Appl. Entomol. Zool. 10*, 1–9.

Müller, H. J. 1942. Formende Einflüsse des tierischen Wirtskörpers auf symbiontische Bakterien. *Forsch. Fortschr. 18*, 193–197.

Müller, J. 1972. Die intrazellulare Symbiose der Zikaden mit Mikroorganismen. *Biol. Rundsch. 10*, 46–57.

Munson, M. A., Baumann, P., Clark, M. A., Baumann, L., Moran, N. A., Voegtlin, D. J., and Campbell, B. C. 1991a. Evidence for the establishment of aphid–eubacterium endosymbiosis in an ancestor of four aphid families. *J. Bacteriol. 173*, 6321–6324.

Munson, M. A., Baumann, P., and Kinsey, M. G. 1991b. *Buchnera* gen. nov. and *Buchnera aphidicola* sp. nov., a taxon consisting of the mycetocyte-associated, primary endosymbionts of aphids. *Int. J. Syst. Bacteriol. 41*, 566–568.

Musgrave, A. J., and Grinyer, I. 1968. Membranes associated with the disintegration of mycetomal micro-organisms in *Sitophilus zea-mais* (Mots.) (Coleoptera). *J. Cell Sci. 3*, 65–70.

Musgrave, A. J., Monro, H. A. U., and Upitis, E. 1965. Apparent elimination of symbiotes in successive generations of *Sitophilus* (Coleoptera) fumigated with methyl bromide. *J. Invertebr. Pathol. 7*, 506–511.

Nardon, P. 1988. Cell to cell interactions in insect endocytobiosis. In "Cell to Cell Signals in Plant, Animal and Microbial Symbiosis." (S. Scannerini, D. Smith, P. Bonfante-Fasolo, and V. Gianinazzi-Pearson, eds.), pp. 85–100. Springer-Verlag, Berlin.

Nardon, P., and Grenier, A. M. 1988. Genetical and biochemical interactions between the host and its endocytobiotes in the weevils *Sitophilus* (Coleoptera, Curculionidae) and other related species. In "Cell to Cell Signals in Plant, Animal and Microbial Symbiosis." (S. Scannerini, D. Smith, P. Bonfante-Fasolo, and V. Gianinazzi-Pearson, eds.), pp. 255–270. Springer-Verlag, Berlin.

Nardon, P., and Grenier, A. M. 1989. Endocytobiosis in Coleoptera: Biological, biochemical, and genetic aspects. In "Insect Endocytobiosis: Morphology, Physiology, Genetics, Evolution." (W. Schwemmler and G. Gassner, eds.), pp. 175–216. CRC Press, Boca Raton, Florida.

Nardon, P., and Wicker, C. 1983. Genetic control of symbiotes by the host in the insect *Sitophilus oryzae* L. (Coleoptera, Curculionidae). *Endocytobiology 2*, 727–731.

Nasu, S. 1965. Electron microscopic studies on transovarial passage of rice dwarf virus. *Jpn. J. Appl. Entomol. Zool. 9*, 225–237.

Nasu, S., Kusumi, T., Suwa, Y., and Kita, H. 1981. Symbiotes of planthoppers: II. isolation of intracellular symbiotic microorganisms from the brown planthopper, *Nilaparvata lugens* Stål, and immunological comparison of the symbiotes associated with rice planthoppers (Hemiptera: Delphacidae). *Appl. Entomol. Zool. 16*, 88–93.

Nazarczuk, R. A., O'Brien, R. W., and Slaytor, M. 1981. Alteration of the gut microbiota and its effect on nitrogen metabolism in termites. *Insect Biochem. 11*, 267–275.

Noda, H. 1977. Histological and histochemical observation of intracellular yeastlike symbiotes in the fat body of the smaller brown planthopper, *Laodelphax striatellus* (Homoptera: Delphacidae). *Appl. Entomol. Zool. 12*, 134–141.

Nogge, G. 1975. Zur Bedeutung der Endosymbionten für Tsetsefliegen Glossina morsitans Westwood. *Verh. Dtsch. Zool. Ges. 68*, 159.

Nogge, G. 1978. Aposymbiotic tsetse flies, *Glossina morsitans morsitans* obtained by feeding on rabbits immunized specifically with symbionts. *J. Insect Physiol. 24*, 299–304.

Nolte, D. J. 1977. The action of locustol. *J. Insect Physiol. 23*, 899–903.

Norris, D. M. 1972. Dependence of fertility and progeny development of *Xyleborus ferrugineus* upon chemicals from its symbiotes. In "Insect and Mite Nutrition." (J. G. Rodriguez, ed.), pp. 299–310. North-Holland, Amsterdam.

O'Brien, R. W., and Slaytor, M. 1982. Role of microorganisms in the metabolism of termites. *Aust. J. Biol. Sci. 35*, 239–262.

O'Brien, G. W., Veivers, P. C., McEwen, S. E., Slaytor, M., and O'Brien, R. W. 1979. The origin and distribution of cellulase in the termites, *Nasutitermes exitiosus* and *Coptotermes lacteus*. *Insect Biochem. 9*, 619–625.

Odelson, D. A., and Breznak, J. A. 1983. Volatile fatty acid production by the hindgut microbiota of zylophagous termites. *Appl. Environ. Microbiol. 45*, 1602–1613.

Odelson, D. A., and Breznak, J. A. 1985. Cellulase and other polymer-hydrolyzing activities of *Trichomitopsis termopsidis*, a symbiotic protozoan from termites. *Appl. Environ. Microbiol. 49*, 622–626.

Ohtaka, C., and Ishikawa, H. 1991. Effects of heat treatment on the symbiotic system of an aphid mycetocyte. *Symbiosis 11*, 19–30.

Paine, T. D., and Birch, M. C. 1983. Acquisition and maintenance of mycangial fungi by *Dendroctonus brevicomis* LeConte (Coleoptera: Scolytidae). *Environ. Entomol. 12*, 1384–1386.

Pasti, M. B., and Belli, M. L. 1985. Cellulolytic activity of actinomycetes isolated from termites (Termitidae) gut. *FEMS Microbiol. Lett. 26*, 107–112.

Peleg, B., and Norris, D. M. 1972. Symbiotic interrelationships between microbes and ambrosia beetles. VII. Bacterial symbionts associated with *Xyleborus ferrugineus*. *J. Invertebr. Pathol. 20*, 59–65.

Petri, L. 1909. Ricerche sopra i batteri intestinali della mosca olearia. *Mem. Real. Staz. Patol. Veg.*, Tipografia Nazionale de Giovanni Bertero E.C., Rome.

Philippe, C., and Landureau, J.-C. 1989. Blattid endocytobiosis: *In vitro* tissue culture model. *In* "Insect Endocytobiosis: Morphology, Physiology, Genetics, Evolution." (W. Schwemmler and G. Gassner, eds.), pp. 89–109. CRC Press, Boca Raton, Florida.

Potrikus, C. J., and Breznak, J. A. 1977. Nitrogen-fixing *Enterobacter agglomerans* isolated from guts of wood-eating termites. *Appl. Environ. Microbiol. 33*, 392–399.

Potrikus, C. J., and Breznak, J. A. 1980a. Uric acid-degrading bacteria in guts of termites [*Reticulitermes flavipes* (Kollar)]. *Appl. Environ. Microbiol. 40*, 117–124.

Potrikus, C. J., and Breznak, J. A. 1980b. Anaerobic degradation of uric acid by gut bacteria of termites. *Appl. Environ. Microbiol. 40*, 125–132.

Rautian, M. S., and Ossipov, D. V. 1985. Genetic effects of the intracellular symbiosis. *Tsitologiya 27*, 124–135.

Rohrmann, G. F. 1978. The origin, structure, and nutritional importance of the comb in two species of Macrotermitinae (Insecta, Isoptera). *Pedobiologia 18*, 89–98.

Sacchi, L., Grigolo, A., Laudani, U., Ricevuti, G., and Dealessi, F. 1985. Behavior of symbionts during oogenesis and early stages of development in the German cockroach, *Blattella germanica* (Blattodea). *J. Invertebr. Pathol. 46*, 139–152.

Sands, W. A. 1969. The association of termites and fungi. *In* "Biology of Termites." (K. Krishna and F. M. Weesner, eds.), Vol. 1, pp. 495–524. Academic Press, New York.

Sarai, D. S. 1976. Total and fecal coliform bacteria in some aquatic and other insects. *Environ. Entomol. 5*, 365–367.

Schabel, H. G. 1978. Percutaneous infection of *Hylobius pales* by *Metarrhizium anisopliae*. *J. Invertebr. Pathol. 31*, 180–187.

Schneider, G. 1940. Beiträge zur Kenntnis der symbiontischen Einrichtungen der Heteropteren. *Z. Morphol. Okol. Tiere 36*, 595–644.

Schorr, H. 1957. Zur Verhaltensbiologie und Symbiose von *Brachypelta aterrima* Först. (Cydnidae, Heteroptera). *Z. Morphol. Okol. Tiere 45*, 561–602.

Schwemmler, W. 1974. Zikadenendosymbiose: Ein Model für die Evolution höherer Zellen? *Acta Biotheor. 23*, 132–169.

Schwemmler, W. 1980. Endocytobiosis: General principles. *BioSystems 12*, 111–122.

Schwemmler, W. 1983a. Endocytobiosis as an intracellular ecosystem. *Endocytobiology 2*, 363–411.

Schwemmler, W. 1983b. Analysis of possible gene transfer between an insect host and its bacteria-like endocytobionts. *Int. Rev. Cytol. Suppl. 14*, 247–266.

Schwemmler, W. 1989. Introduction. *In* "Insect Endocytobiosis: Morphology, Physiology, Genetics, Evolution." (W. Schwemmler and G. Gassner, eds.), pp. 3–8. CRC Press, Boca Raton, Florida.

Schwemmler, W., and Gassner, G. 1989. "Insect Endocytobiosis: Morphology, Physiology, Genetics, Evolution." CRC Press, Boca Raton, Florida.

Schwemmler, W., and Vago, C. 1970. Infection expérimentale avec des bactéries symbiotiques, des cultures de cellules et d'organes de l'Homoptère *Euscelis plebejus* Fall. (Cicadina). *C.R. Acad. Sci. Paris Ser. D* **270**, 1644–1647.

Schwemmler, W., Duthoit, J.-L., Kuhl, G., and Vago, C. 1973. Sprengung der Endosymbiose von *Euscelis plebejus* F. und Ernährung aposymbiontischer Tiere mit synthetischer Diät (Hemiptera, Cicadidae). *Z. Morph. Tiere* **74**, 297–322.

Shope, R. E. 1927. Bacteriophage isolated from the common house fly (*Musca domestica*). *J. Exp. Med.* **45**, 1037–1044.

Singh, G. J. P. 1981. Studies on the role of microorganisms in the metabolism of dieldrin in the epicuticular wax layer of blowflies of *Calliphora erythrocephala*. *Pest. Biochem. Physiol.* **16**, 256–266.

Smith, H. E., and Arnott, H. J. 1974. Epi- and endobiotic bacteria associated with *Pyrsonympha vertens*, a symbiotic protozoon of the termite *Reticulitermes flavipes*. *Trans. Am. Microsc. Soc.* **93**, 180–194.

Smythe, R. V., and Mauldin, J. K. 1972. Soldier differentiation, survival, and wood consumption by normally and abnormally faunated workers of the Formosan termite, *Coptotermes formosanus*. *Ann. Entomol. Soc. Am.* **65**, 1001–1004.

Stackebrandt, E., Murray, R. G. E., and Trüper, H. G. 1988. *Proteobacteria* classis nov., a name for the phylogenetic taxon that includes the "purple bacteria and their relatives." *Int. J. Syst. Bacteriol.* **38**, 321–325.

Starr, M. P. 1975. A generalized scheme for classifying organismic associations. *In* "Symbiosis." (D. H. Jennings and D. L. Lee, eds.). pp. 1–20. Cambridge University Press, Cambridge, England.

Steffan, A. W. 1967. Ectosymbiosis in aquatic insects. *In* "Symbiosis." (S. M. Henry, ed.), Vol. 2, pp. 207–289. Academic Press, New York.

Steinhaus, E. A. 1942. "Catalogue of bacteria associated extracellularly with insects and ticks." Burgess, Minneapolis, Minnesota.

Steinhaus, E. A. 1946. "Insect Microbiology." Comstock Publishing Company, Ithaca, New York.

Steinhaus, E. A. 1949. "Principles of Insect Pathology." McGraw-Hill, New York.

Steinhaus, E. A., and Tanada, Y. 1971. Diseases of the insect integument. *In* "Current Topics in Comparative Pathobiology." (T. C. Cheng, ed.), Vol. 1, pp. 1–86. Academic Press, New York.

Stoltz, D. B., and Vinson, S. B. 1977. Baculovirus-like particles in the reproductive tracts of female parasitoid wasps. II: The genus *Apanteles*. *Can J. Microbiol.* **23**, 28–37.

Stoltz, D. B., and Vinson, S. B. 1979. Viruses and parasitism in insects. *Adv. Virus Res.* **24**, 125–171.

Stoltz, D. B., Vinson, S. B., and MacKinnon, E. A. 1976. Baculovirus-like particles in the reproductive tracts of female parasitoid wasps. *Can. J. Microbiol.* **22**, 1013–1023.

Stradling, D. J., and Powell, R. J. 1986. The cloning of more highly productive fungal strains: A factor in the speciation of fungus-growing ants. *Experientia* **42**, 962–964.

Šulc, K. 1910. "Pseudovitellus" und ähnliche Gewebe der Homopteren sind Wohnstätten symbiotischer Saccharomyceten. *Česká Spolecnost. Nauk. Trida Math.-Prirod. Věstnik.* **3**, 1–39.

Talbot, P. H. B. 1977. The *Sirex-Amylostereum-Pinus* association. *Annu. Rev. Phytopathol.* **15**, 41–54.

Taylor, F. J. R. 1979. Symbionticism revisited: A discussion of the evolutionary impact of intracellular symbioses. *Proc. R. Soc. London Ser. B* **204**, 267–286.

Taylor, F. J. R. 1983. Some eco-evolutionary aspects of intracellular symbiosis. *Int. Rev. Cytol. Suppl. 14*, 1–28.

Taylor, F. J. R., and Harrison, P. J. 1983. Ecological aspects of intracellular symbiosis. *Endocytobiology 2*, 827–842.

Tiivel, T. 1984. Ultrastructural aspects of endocytobiosis in leafhopper (*Insecta: Cicadinea*) cells. *Proc. Acad. Sci. Estonian S.S.R. Biol. 33*, 244–255.

Tiivel, T. 1989. Endocytobiosis of leafhoppers with prokaryotic microorganisms. In "Insect Endocytobiosis: Morphology, Physiology, Genetics, Evolution." (W. Schwemmler and G. Gassner, eds.), pp. 111–122. CRC Press, Boca Raton, Florida.

Trager, W. 1932. A cellulase from the symbiotic intestinal flagellates of termites and of the roach, *Cryptocercus punctulatus*. *Biochem. J. 26*, 1762–1771.

Tremblay, E. 1989. Coccoidea endocytobiosis. In "Insect Endocytobiosis: Morphology, Physiology, Genetics, Evolution." (W. Schwemmler and G. Gassner, eds.), pp. 145–173. CRC Press, Boca Raton, Florida.

Unterman, B. M., and Baumann, P. 1990. Partial characterization of the ribosomal RNA operons of the pea–aphid endosymbionts: Evolutionary and physiological implications. In "Aphid–Plant Genotype Interactions." (R. K. Campbell and R. D. Eikenbary, eds.), pp. 329–350. Elsevier, Amsterdam.

Unterman, B. M., Baumann, P., and McLean, D. L. 1989. Pea–aphid–symbiont relationships established by analysis of 16S rRNAs. *J. Bacteriol. 171*, 2970–2974.

Veivers, P. C., O'Brien, R. W., and Slaytor, M. 1982. Role of bacteria in maintaining the redox potential in the hindgut of termites and preventing entry of foreign bacteria. *J. Insect Physiol. 28*, 947–951.

Wade, M. J., and Stevens, L. 1985. Microorganism mediated reproductive isolation in flour beetles (genus *Tribolium*). *Science 227*, 527–528.

Watson, W. Y., Underwood, G. R., and Reid, J. 1960. Notes on *Matsucoccus macrocicatrices* Richards (Homoptera: Margarodidae) and its association with *Septobasidium pinicola* Snell in eastern Canada. *Can. Entomol. 92*, 662–667.

Weber, N. A. 1937. The biology of the fungus-growing ants. Part II. Nesting habits of the bachac (*Atta cephalotes* L.). *Trop. Agric. 14*, 223–226.

Weber, N. A. 1966. Fungus-growing ants. *Science 153*, 587–604.

Weber, N. A. 1979. Historical note on culturing attine-ant fungi. *Mycologia 71*, 633–634.

Wheeler, W. M. 1910. "Ants, Their Structure, Development and Behavior." Columbia University Press, New York.

Whisler, H. C. 1968. Experimental studies with a new species of Stigmatomyces (Laboulbeniales). *Mycologia 60*, 65–75.

Whitney, H. S., and Farris, S. H. 1970. Maxillary mycangium in the mountain pine beetle. *Science 167*, 54–55.

Wilding, N., Collins, N. M., Hammond, P. M., and Webber, J. F. 1989. "Insect–Fungus Interactions." Academic Press, London.

Wood, T. G., and Thomas, R. J. 1989. The mutualistic association between Macrotermitinae and *Termitomyces*. In "Insect–Fungus Interactions." (N. Wilding, N. M. Collins, P. M. Hammond, and J. F. Webber, eds.), pp. 69–92. Academic Press, London.

Wright, J. D., and Barr, A. R. 1981. *Wolbachia* and the normal and incompatible eggs of *Aedes polynesiensis* (Diptera: Culicidae). *J. Invertebr. Pathol. 38*, 409–418.

Wright, J. D., and Wang, B.-T. 1980. Observations on Wolbachiae in mosquitoes. *J. Invertebr. Pathol. 35*, 200–208.

Yamaoka, I. 1989. Termite endosymbiosis. In "Insect Endocytobiosis: Morphology, Physiology, Genetics, Evolution." (W. Schwemmler and G. Gassner, eds.), pp. 77–87. CRC Press, Boca Raton, Florida.

Yamin, M. A. 1978. Axenic cultivation of the cellulolytic flagellate *Trichomitopsis termopsidis* (Cleveland) from the termite *Zootermopsis*. *J. Protozool. 25*, 535–538.

Yamin, M. A. 1980. Cellulose metabolism by the termite flagellate *Trichomitopsis termopsidis*. *Appl. Environ. Microbiol. 39*, 859–863.

Yamin, M. A. 1981. Cellulose metabolism by the flagellate *Trichonympha* from a termite is independent of endosymbiotic bacteria. *Science 211*, 58–59.

Yamin, M. A., and Trager, W. 1979. Cellulolytic activity of an axenically-cultivated termite flagellate, *Trichomitopsis termopsidis*. *J. Gen. Microbiol. 113*, 417–420.

Yamvrias, C., Panagopoulos, C. G., and Psallidas, P. G. 1970. Preliminary study of the internal bacterial flora of the olive fruit fly (*Dacus oleae* Gmelin). *Ann. Inst. Phytopathol. Benaki, N.S. 9*, 201–206.

Yen, J. H., and Barr, A. R. 1973. The etiological agent of cytoplasmic incompatibility in *Culex pipiens*. *J. Invertebr. Pathol. 22*, 242–250.

Zoberi, M. H. 1979. The ecology of some fungi in a termite hill. *Mycologia 71*, 537–545.

CHAPTER 3

AMICROBIAL AND MICROBIAL AGENTS

I. AMICROBIAL DISEASES
   A. Injuries by Mechanical Agents
   B. Injuries by Physical Agents
   C. Injuries by Chemical Agents
   D. Injuries by Biological Agents
   E. Injuries by Genetic Factors
   F. Nutritional Diseases
   G. Hormonal Disruption
II. MICROBIAL DISEASES
   A. Potential, Facultative, and Obligate Pathogens
   B. Microbial Toxins
   C. Portals of Entry
   D. Infectivity
   E. Virulence and Pathogenicity
   F. Dosage of the Pathogen
   G. Signs, Symptoms, and Syndromes
   H. Course of Infection
   I. Latent, Chronic, and Acute Infections
   J. Interactions among Microorganisms
   K. Koch's Postulates
   L. Diagnosis

An insect is in a diseased condition when abnormality occurs because of physical or physiological derangements. Disease literally means "lack of ease." It is the result of an injury or insult or whatever that goes wrong with an insect (Steinhaus 1949). The disease may be caused by living and nonliving agents and is the result of the interaction between the insect and the causal agent. A common error is to say that a disease infects an insect. Correctly, it is the pathogen that infects the insect and this infection is expressed as a disease. In the case of nonliving agents, their absence or noxious presence produces abnormalities in insects.

Diseases may be classified on the following basis (Steinhaus 1949):

1. The presence or absence of an infectious microorganism (e.g., diseases caused by infectious or noninfectious agents).
2. The extent of the disease in an insect (e.g., local, focal, or systemic diseases).
3. The location or site of the disease (e.g., intestinal, fat body, hypodermal, blood, etc.).

4. The course of the disease (e.g., chronic, subacute, or acute).
5. The source of the infectious agent (e.g., exogenous, endogenous, and idiopathic or hidden).
6. The type of etiological or causal agent (e.g., bacterium, virus, fungus, protozoa, or nematode).
7. The distribution or prevalence of the disease in an insect population (e.g., sporadic, enzootic, or epizootic).
8. The method of transmission (e.g., direct contact, vector, per os, transovum, or transovarial).
9. The basis of sequence (e.g., primary, secondary, attenuated, progressive, mixed, or multiple).

We shall classify diseases in two broad categories, diseases caused by amicrobial (noninfectious) agents and those caused by microbial (infectious) agents. Microbial diseases will be further subdivided on the basis of the causal or etiologic agents.

I. AMICROBIAL DISEASES

The amicrobial diseases are those in which no living microorganism is involved. Such diseases are often considered as injuries. The various types are caused by (1) mechanical agents, (2) physical agents, (3) chemical agents, (4) biological agents, such as insect parasitoids and predators, (5) genetic factors, (6) nutritional deficiencies, and (7) deranged metabolism. We shall consider these various types very briefly, but this does not mean that they are of less importance or of lesser scope than the microbial diseases. Traditionally, however, insect pathologists have focused their research on microbial agents and not on the amicrobial ones.

A. INJURIES BY MECHANICAL AGENTS

Mechanical agents cause two types of diseases: (1) distension and (2) trauma. Distension results when a duct or hollow viscera, such as the digestive tract or Malpighian tubules, is obstructed to prevent the flow of its contents. Trauma results from the cutting, crushing, and tearing of tissues and organs. The wounds or injuries usually affect the insect by (1) damage to the tissues and organs, (2) loss of hemolymph, and (3) exposure to pathogens at the site of injury. There are various kinds of trauma, such as "bruises" from injurious contact with a blunt object, "concussions" through jarring, "crushing" due to excessive pressure, "cutting" by sharp instruments, "tearing" by the forceful pulling apart of structures, and "puncturing" from sharp or pointed objects.

When an insect is wounded, it responds to the injury. For example, when the pupa of the *Cecropia* silkworm is incised or punctured, there is a rapid increase in oxygen consumption, a release of an "injury factor" into the hemolymph, and DNA and RNA syntheses leading to the production of proteins (Harvey and Williams 1961; Yeaton 1983). The injury factor is released until the wound is sealed off by a deposit of hemocytes. The factor causes a speeding up of the heart beat and a mobilization of hemocytes into the circulating hemolymph and provokes chemical and cytological changes in most other cells. It also stimulates RNA synthesis in all tissues except the gonads (Berry et al. 1967). The wound factor, named haemokinin (hemokinin) by Cherbas (1973), may occur in epidermal, blood, and possibly other cells of an insect and cause the hemocytes to increase in motility and adhesiveness.

B. INJURIES BY PHYSICAL AGENTS

The common physical agents that cause injuries are high and low temperatures, abundance or lack of moisture, radiation, and possibly the excess or absence of gases, such as oxygen and carbon dioxide (Day and Oster 1963). Insects vary in their reactions to unfavorable physical agents as well as in their tolerance to these agents. The disease, resulting from a physical agent, is generally manifested in the coagulation of proteins in the cell, in the malfunction of enzymes and hormones, and in the mechanical alteration of the noncellular parts of insects.

1. High Temperature

Insects have a range of optimum temperatures and beyond the upper and lower temperature limits, injuries and even death may result. The upper lethal limits are usually between 40 and 50°C (Bursell 1974), but some insects, such as stored-product and desert insects, can withstand temperatures in the neighborhood of 60°C. An insect can usually withstand high temperatures better in a dry atmosphere than in a moist atmosphere.

The effect of high temperature results in burns, irritation or discomfort, paralysis, heat stupor or rigor, and death. At the cellular levels, there may be denaturation of proteins and the melting of lipids and phosphatides.

As the temperature rises, the permeability of the waxy epicuticular layer of the integument changes abruptly at the transition temperature, or at the point where the cuticular wax becomes disoriented (Beament 1959). This temperature ranges from 33 to 64°C in different insect species and also varies with the age and instar of the insect. With the disruption of the epicuticular layer, the insect is exposed to rapid desiccation.

The extremes of temperatures more readily affect the reproductive capacities of insects than most other physiological functions (Bursell 1974). In *Dahlbominus fuscipennis*, an apparently harmless amount of heat can have a significant effect on reproduction (Riordan 1957). The males are sterilized permanently by heat because of the sensitivity of their spermatozoa to high temperature. Females are more resistant than males to both the lethal and sterilizing effects of such treatments at 43°C.

2. Low Temperature

Some insects can withstand extremely low temperatures, even below the freezing point of water, to as low as −70°C in the laboratory (Danks 1978; Sømme 1982; Duman and Horwath 1983). Insects can be separated into three categories depending on their cold-hardiness (Bursell 1974): (1) Those living in warm environments and dying at temperatures above freezing, (2) those surviving until their body fluids freeze, and (3) those surviving despite the freezing of body fluids. In the case of insects that require high temperatures or occur in warm tropical habitats, the cause of death is not clear but may involve a critical upset of normal metabolism. Insects that do not survive low temperatures above freezing include the honey bee, tsetse fly, and the mosquito *Aedes aegypti*.

Some insects prepare themselves for low temperatures, such as the onset of winter, by adjusting their hemolymphs and cellular contents, so that the tissues can reach supercooling temperatures. Insects that are sensitive to freezing cannot survive the formation of internal ice, and they prevent this occurrence by several mechanisms: (1) The depression of the supercooling points of body fluids to $-20°C$ or more, (2) the elimination or replacement of ice nucleation sites in the body, and (3) the accumulation of cryoprotectants ("antifreeze" compounds) (Sømme 1982; Storey and Storey 1983a,b).

How insects prepare themselves to meet the onset of freezing temperatures in winter has not been resolved. Both photoperiod (circadian system) and temperature and their interaction are important in controlling the seasonal pattern of hemolymph antifreeze protein levels in larvae of the beetle *Dendroides canadensis* (Horwath and Duman 1983, 1984a,b). The overwintering stage of this beetle has the ability to survive freezing (freezing tolerance) and to avoid freezing by supercooling (freezing susceptibility) (Horwath and Duman 1984b).

Active insects have ice nucleating components present within their gut contents and/or in intracellular compartments, where they appear to be associated with the cellular matrix (Zachariassen 1982). Such nucleators will cause injurious internal freezing. In freeze-sensitive insects that hibernate in a supercooled state, these nucleating components are either removed (e.g., emptying of gut contents) or are masked by sequestering within the cellular matrix. Some insects prevent intracellular freezing by producing ice nucleators in the hemolymph, where ice formation may take place without injuring the cells (Duman 1984a; Duman et al. 1985). The nucleators in the hemolymph and intracellular components appear to be proteins or peptides.

Many overwintering insects produce natural "antifreeze" compounds (cryoprotectants) that are small hydrophilic compounds (e.g., glycerol, sorbitol, and polyhydric alcohols) (Storey and Storey 1983a,b), amino acids (Male and Storey 1982), and proteins (Duman and Horwath 1983). The cryoprotectants are believed to protect insects from extremely low temperatures by preventing or restricting the growth of ice crystals within the insects' tissues (Duman 1977; Husby and Zachariassen 1980; Zachariassen and Husby 1982). The overwintering ability of an insect, however, is complex and may occur with or without the cryoprotectants. For successful overwintering, it is the behavioral, ecological, and physiological interrelationships that enable the overwintering stages to maintain activities, normally restricted to "above-freezing" situations (Danks 1978; Ring and Tesar 1981; Baust 1981).

When tissues freeze, most insects die because of tissue dehydration or mechanical injury by ice crystals. If the ice crystals are confined to the extracellular fluids, many insects can survive the low temperatures. The growth of ice crystals results in a gradual withdrawal of water from cells, and prolonged freezing may cause cellular dehydration and death. Generally, extracellular freezing precedes intracellular ice formation. Some insects can withstand ice crystal formation within their cells. On the other hand, insects that are susceptible to freezing are killed rapidly by intracellular freezing. Shortly thereafter, the insects' cells are covered with extracellular ice and the insects darken almost instantaneously. In quick freezing, rapid formation of numerous minute ice crystals occurs and is called "flashing" (Asahina 1969).

The degree of resistance to freezing varies with the tissues even within the same insect. The probability of intracellular ice formation depends on factors that control the rate of removal of water from the cell during the early stage of extracellular freezing, such as the cooling rate, surface-volume ratio, membrane permeability to water, and the temperature coefficient of the permeability constant. Thus, large spherical cells (e.g., fat cells) are more susceptible to intracellular freezing than small slender-form cells (e.g., muscle).

When cells with intracellular ice crystals are thawed, there is a destruction of cell structure, such as vacuolation and swelling of the protoplasm for light freezing to a network of coagulated cytoplasm for longer and more severe freezing. In both cases, there are changes in the protoplasmic patterns of the cell.

Frozen insects may not die immediately after thawing. They may live for a brief period with their hearts beating, but their hemolymph has a brown coloration. Apparently freezing the cells results in melanin production. Those that survive freezing often have difficulty in molting or completing their metamorphosis, which are the most common sublethal injuries from freezing.

3. Moisture

The requirement of insects for moisture is closely correlated to that of temperature. Moisture alone, nonetheless, is important. Most terrestrial arthropods are at least 70% water by weight (Wharton and Arlian 1972). Insects vary in their moisture requirements for normal growth, development, and reproduction. Excess moisture or humidity may cause (1) waterlogging of tissues, (2) drowning or suffocation, (3) intestinal upset caused by feeding on wet food, and (4) increased susceptibility to pathogens. Most adverse effects of moisture are due to its scarcity or absence, and many insects are irreversibly injured when their moisture content drops 20%. One insect, the larva of an African chironomid midge, *Polypedilum vanderplanki*, can tolerate dehydration and suspension of metabolism (cryptobiosis) for several years (Hinton 1977).

Insects lose most of their moisture through their tracheal and excretory systems. The loss of water through the tracheal system is controlled by the types of spiracles and by their opening and closing. The amount of moisture lost through the integument is generally small because of the waxy epicuticular covering. Removal of the waxy layer results in rapid desiccation and death of the insect. This is the basis for the control of some household insects by the use of abrasive and sorptive dusts (Ebeling 1961).

Desiccation at the time of molting and pupation may cause deformities in the insect following ecdysis. Remnants of the previous larval cuticle may remain on the emerging larva, and as the larva develops, such cuticle ligatures the larva. Adults emerging from desiccated pupae may fail to free themselves from the pupal shell or may have deformed wings. Desiccation may also cause the hardening of the chorion of some insect eggs and such physically hardened eggs prevent the emergence of the larvae.

4. Oxygen and Carbon Dioxide

Insects can tolerate a wide range of oxygen and carbon dioxide. Some insects can survive several days in the absence of oxygen by reducing their metabolic rates and utilizing the oxygen stored in their tissues. Other insects, when reared under low-oxygen conditions, alter their tracheal system; growth of the trachea may vary inversely with the oxygen tension.

Excess carbon dioxide causes varied reactions in different insects. It is commonly used to anesthetize insects in the laboratory. Some insects can live in an atmosphere with high carbon dioxide for several days. However, an excess of this gas in the atmosphere causes growth retardation in insects, such as silkworm and cockroaches. If the environment is high in carbon dioxide, the spiracles of insects tend to remain open, which may lead to excessive water loss. In *Drosophila melanogaster* infected with the sigma virus, the flies do not recover when anesthetized with carbon dioxide.

5. Radiation

Light affects insects in correlation with temperature and humidity. Direct sunshine may injure or kill an exposed insect largely because of heat and desiccation. Ultraviolet and infrared radiations, electric shocks, and supersonic waves result in injuries and can kill insects. Generally, the longer wavelength radiations produce heating effects and the shorter wavelength radiations cause chemical effects, including the ionization of the atoms (Nelson 1967). These agents have been used for the control of certain pests in stored food products and in wood. Irradiation with X-rays, gamma rays, and ultraviolet rays causes deformities, tumors, mutations, and sterility. Mutations are mostly detrimental to insects and produce abnormalities that may cause their death.

Little is known about the histopathology of somatic tissues that have been injured by radiation. Typically the effect is on the abdomen, especially in the midgut (Grosch 1974). The regenerative and secretory cells are killed, and often the entire midgut epithelium is destroyed. The cause of death is one or more of the following: starvation, fluid leakage, toxin infiltration, microbial invasion, etc.

Much more information is available on the effect of radiation on the germinal tissues, the ovary and testis. The susceptibility of the different tissues in oogenesis and spermatogenesis varies with the insect species. Extensive studies on the effect of radiation on the male insect and the use of sterile males for the eradication of insect pests have been made (Knipling 1955, 1967; LaChance et al. 1967).

C. INJURIES BY CHEMICAL AGENTS

Chemical agents that cause injuries to insects are chemical insecticides and toxins or poisons produced by plants, by microorganisms, by disintegration of necrotic tissues, possibly by the suppressed function of certain tissues, and by the perverted metabolism of body cells. The study of the effect of chemical agents falls in the realm of insect toxicology, but the insect pathologist should be aware of the pathological conditions and diseases that develop in the poisoned insects. Insect pathologists will frequently receive insects for diagnosis that have been killed by chemical insecticides. Such specimens may cause considerable confusion among unwary insect pathologists.

At one time, the chemical insecticides were classified on the basis of their modes of entry as follows: stomach poisons, contact poisons through the integument, and fumigants or respiratory poisons. Although these terms are still in use, the present-day insect toxicologists prefer to group insecticides according to their molecular structures and their biological origin because many insecticides involve more than one mode of entry (Matsumura 1985) (Table 3–1). For example, the chlorinated hydrocarbon, DDT, acts through contact and ingestion and the organophosphate, parathion, acts by ingestion, contact, and fumigation.

The chemical insecticides may also be classified on their modes of action on insects as (1) physical poisons, (2) protoplasmic poisons, (3) metabolic inhibitors, (4) neuroactive agents, (5) hormone mimics, and (6) stomach poisons (Matsumura 1985). The poisons may also be grouped broadly on their effect on cells and tissues as nerve and tissue poisons (Brown 1963).

When insects are poisoned by chemical insecticides, certain cells often display an alteration of staining properties; some cells may separate from each other in the tissue and may break down (cytolysis). Within the nucleus the chromatin granules clump together (pycnosis), and in the nerve cells the Nissl bodies dissolve. Vacuoles appear in the cytoplasm. Certain inorganic insecticides that contain heavy metals, such as lead and mercury, cause necrosis in the midgut epithelium and the breakdown of cells.

The common insecticides that act on respiration are the arsenicals (sodium arsenite, Paris green, lead arsenate, etc.), rotenone, and cyanide (Corbett 1975). The arsenite ions kill the insects by inhibiting the pyruvate or the α-ketoglutarate dehydrogenase system, or both systems during respiration. Rotenone, a natural plant product extracted from various legumes, inhibits the mitochondrial electron transport during respiration. It is also considered to act on the respiratory chain of tissues including the nerve and muscle (Yamamoto 1970). Cyanide is not a specific poison and acts on a large number of enzymes, but its major effect is believed to be on cytochrome oxidase, an enzyme active in respiration.

The primary target of most synthetic insecticides is the insect's nervous system. Nerve poisons characteristically induce the appearance of symptoms in four stages: excitation, convulsion, paralysis, and death. A typical neurotoxic fumigant results in only three steps: excitation, paralysis, and death. Disturbance in the nervous systems often affects other metabolic systems, such as the respiratory, muscular, and circulatory. Ultimately, the insect dies from an alteration or malfunction of some physiological and metabolic processes. For example, an insect larva immobilized by a nerve poison may die from starvation because it is incapable of ingesting its food, or a mosquito larva paralyzed by a nerve poison may die from suffocation because of its inability to rise to the water surface to obtain oxygen.

Chlorinated hydrocarbons, such as DDT and its analogues, gamma benzene hexachloride (lindane), pyrethroids, and cyclodienes, are neurotoxins, but their exact modes of action are still not established. They are believed to act by the release of acetylcholine in the nervous system. Acetylcholine is the chemical mediator that conducts the message between the gaps (synapsis) of two nerve endings or between the nerve endings and an organ or a muscle. Disturbance in the acetylcholine concentrations results in erratic muscular contractions. The insect responds to DDT poisoning by displaying the familiar "DDT jitters." DDT analogues and pyrethroids act at similar or identical target sites on the membranes of axon nerves (Yamamoto 1970; Beeman 1982).

The organophosphorus and carbamate insecticides act as anticholinesterases, but the two groups have certain differences, especially since the mode of cholinesterase inhibition by carbamates is apparently reversible (Matsumura 1985).

TABLE 3–1 Some General Groupings of Insecticides[a]

Class	Origin	Chemical or biological group	Examples
Insecticides and acaricides	Synthetic—organic	Chlorinated hydrocarbons	DDT derivatives
			BHC
			Cyclodienes
		Organophosphorus compounds	Aliphatics
			Aryl compounds
		Carbamates	Naphthyl compounds
			Phenyl compounds
		Organofluorine compounds	Fluoroacetate
		Fumigants	Methyl bromide
	Synthetic—inorganic	Arsenicals	
		Fluorides	
		Mercurials	
	Natural products—organic	Botanicals	Nicotinoids
			Pyrethroids
			Rotenoids
		Microbials	Toxins
			Bacteria, viruses, fungi, protozoa, nematodes
Activators or synergists	Mostly synthetic—organic	Synergists	Methylenedioxyphenyl compounds
Carrier or bulk material	Natural—organic	Petroleum products	
	Natural—inorganic	Dusts	
Growth regulators	Synthetic—organic	Hormone mimics	Methoprene
Chitin synthesis inhibitors	Synthetic—organic		Diflubenzuron

[a]Modified after Matsumura (1985).

These insecticides are neurotoxins, and they affect the concentration of acetylcholine in the nervous tissue through the enzyme cholinesterase. Insects poisoned by them show initial hyperactivity, followed by convulsive and uncoordinated movements, and ending in paralysis and death. They may also act on neurohormones that are responsible for the control of urinary output and the plasticity of the cuticle (Corbett 1975). In addition, the organophosphorus compounds have been used as ovicides to control insects in the egg stage where there is no acetylcholinesterase activity apparent in the embryo; how these compounds kill the eggs is unknown.

Chemicals inhibiting chitin formation or causing a breakdown of chitin are effective insecticides (Matsumura 1985). Because chitin is found in insects and other arthropods and in fungi, and does not occur in vertebrates and higher plants, these pesticides are considered to have a selective toxicity (Corbett 1975). Examples of chitin-inhibiting substances in insects are the benzoylphenylureas (Hajjar and Casida 1978). These insecticides may prevent the synthesis of insect chitin by interfering with the proteolytic activation of the chitin synthetase zymogen (i.e., block the conversion of chitin synthetic zymogen into active enzyme) (Leighton et al. 1981).

Insects are exposed to natural poisons produced by microorganisms and plants. We shall discuss the poisons or toxins produced by microorganisms in the section on microbial diseases. Higher plants synthesize substantial quantities of "protective" substances that are repellent, poisonous, or inhibitory to other organisms, including insects (see Arnason et al. 1989; Hedin 1991). These substances are usually secondary plant products, such as phenolic, terpinoid or alkaloid compounds, or of smaller groups including organic cyanides. Extensive research is being conducted to employ these naturally occurring plant compounds for insect pest suppression.

D. INJURIES BY BIOLOGICAL AGENTS

The attack or invasion of an insect through parasitization and predation by other insects and arthropods is a disease. Insect parasitoids, which occur on the external surface (ectoparasitoids) or within the insects (endoparasitoids), are comparable to microorganisms in causing pathologies to their hosts. Parasitoids and predators cause mechanical injuries, such as destruction and irritation, and physiological injuries that result in disruption and malfunction from the loss of nutrients and the introduction of toxic substances (Doutt 1963; Coudron 1991). Mechanical irritation is caused by ectoparasitoid larva that obtains nutrients through the insect's integument. The piercing of the integument by the parasitoid's ovipositor causes minor wounds. Some ectoparasitic adults puncture the integument to imbibe the body fluids of the host. Internal parasitoids may feed directly on the tissues or organs of their hosts, but most of them live on nonvital structures.

In some cases, the presence of internal parasitoids suppresses completely or partially the reproductive capacity of the insect ("parasite castration") and may also cause a change in the secondary sexual characters. The insect's reaction to parasitoids varies: by encapsulating the immature parasitoids or resisting the attack of the adults, by developing ingrowths of the body wall as in the case of certain tachinid-fly attacks, by a reduction in size and structure, or by an acceleration of its development. As a rule, when the parasitoid completes its development and emerges from the host, the insect dies.

In some solitary Hymenoptera, the adult wasp immobilizes the insect host with

E. INJURIES BY GENETIC FACTORS

Hereditary or genetic diseases in insects are well known, especially those occurring in *Drosophila*, silkworm, honey bee, and mosquitoes. These diseases are genetically determined and may involve one or more genes, and may be sex linked. The diseases are expressed as biochemical, physiological, and morphological characters that are harmful for the insect. Some hereditary factors cause sterility and others cause death. Most manifestations of the inherited diseases are dramatic alterations of the insect's body structures, especially the appendages. The malformations may occur in almost every organ (e.g., deformed body, no bristles or scales, deformed or too small eyes, or no eyes, vestigial or crumpled wings, deformed mouthparts or no mouthparts, crippled legs, and supernumerary legs or legs in place of antennae). Some of these mutants are fully viable in the laboratory but would not survive in nature. Certain malformations, such as wingless adults, may be advantageous and therefore favored on an island where winged insects are blown out to sea. Some workers consider this to be the case in the evolution of endemic wingless insects in Hawaii.

In *Drosophila*, there are tumors, both malignant and nonmalignant, that are hereditary (Harker 1963; Gateff 1978a,b; Sparrow 1978). Most tumors in insects, however, lack malignant characteristics and seem to result from defense reactions to injury, pathogens, and imbalance in the endocrine system (see Chapter 14). Malignant neoplasms in *Drosophila* and vertebrates show striking similarities in exhibiting fast, autonomous growth, loss of the capacity for differentiation, increased mobility and invasiveness, lethality *in situ* and after transplantation, and fine structural and nuclear abnormalities (Gateff 1978a,b). Some neoplasms that occur in larval mutants of *Drosophila* affect the adult optic neuroblasts and ganglion-mother cells in the larval brain, the imaginal discs, and the hematopoietic organs.

Mutations may be produced by irradiations (X-rays or other ionizing rays) and by mutagenic chemicals. In nature, mutations occur in all living organisms, and the total rate of mutation per generation is at least 5% (Benz 1963). Most mutations or genetic aberrations are more or less harmful to insects, but this may depend on the characteristics of the insect and the environment.

Attempts at controlling insect pests have been made by means of engineering the hereditary mechanisms. There is an arsenal of techniques in genetic control, and we shall discuss some of them (Pal and LaChance 1974; Whitten and Foster 1975; Fitz-Earle 1976). Genetic control gained impetus since the initial eradication of the screwworm fly, *Cochliomyia hominivorax*, on the island of Curacao in the Caribbean Sea by the use of the sterile-male technique (Lindquist 1955; Knipling 1955, 1967). In this technique, the males are sterilized by irradiation or by chemical mutagens and then released into the wild populations.

The genetic manipulation in insect control also utilizes the principles of genetic incompatibility and hybrid sterility (Fitz-Earle 1976). In incompatibility, the insect strains from different geographical areas when mated result in complete or partial sterility. This is associated with some component in the egg that prevents the fusion of its nucleus with an alien sperm. Sterility in F_1 hybrids between sibling species occurs in certain insects, such as those in the mosquito and the Hessian fly, *Mayetiola destructor*.

Inherited sterility can be produced in some insects by manipulating the chromosomes with radiation so that one piece of chromosome is exchanged with that of another to form a rearranged chromosome. One type of chromosomal rearrangement is known as translocation. In certain translocations, one half of the offspring do not survive. Since one half of the survivors contain individuals with translocated chromosomes, this "fatal" character is maintained in the population.

A method in genetic control is to replace an entire pest population containing standard chromosomes with another possessing rearranged chromosomes that act as carriers for factors desirable from the human's point of view (Fitz-Earle 1976). Thus, mosquito strains that cannot transmit vertebrate pathogens may replace the vector strains; chemical insecticide resistant strains may be replaced by susceptible ones; temperature-sensitive lethal mutants may replace those tolerant to wide temperature fluctuations; nondiapausing strains may replace those that diapause, etc. The major advantage of these methods is that the noxious population is not eliminated but is replaced by an innocuous one, or the pest population is reduced to economically acceptable levels.

F. NUTRITIONAL DISEASES

The relationship between food and feeding habits of insects is complex and depends on two basic conditions: (1) The food possesses characteristic properties that peculiarly attract and induce a particular insect to feed and (2) the food contains certain substances that fulfill the nutritional requirements of this insect (House 1974). The first pertains to feeding requirements, such as odor, texture, and other qualities of food; the second involves the nutritionally essential substances that are required to nourish the insect (i.e., to build body tissues and to serve as a source of energy). An "essential nutrient" is one that is needed for normal growth, development, and reproduction of the insect. Deviations from these basic conditions may result in nutritional disease.

Diseases caused by faulty nutrition and by deranged physiology or metabolism often are difficult to separate because nutritional problems usually result in abnormal metabolism and vice versa. Nutritional diseases arise from the general deficiency or the improper proportion of nutritional substances, such as proteins, amino acids, fats, carbohydrates, vitamins, inorganic ions, etc. The disease is manifested through abnormalities in growth, behavior, development, and reproduction of the insect. However, there is difficulty in producing characteristic effects from a deficiency of an essential nutrient in insects because usually such a deficiency causes merely cessation of growth and prolonged survival (Gordon 1959).

1. Starvation

Insects vary in their ability to withstand starvation, which is probably the most common dystrophy in nature (House 1963). When insects are starved, they utilize mainly their stored carbohydrates and fats, and the tissue proteins may or may not be affected. Limited feeding reduces growth, development, and reproduction.

2. Dietary Components

For normal growth and development, most insects require in their diet the common 10 essential amino acids, six or more B vitamins, a sterol such as cholesterol, inorganic salts, carbohydrates, and certain fatty acids. Some require components of nucleic acids, and a few need fat-soluble vitamins, vitamin C, and miscellaneous, unidentified substances. The importance of these foods and accessory food substances varies between and within species. The requirements of the larvae often differ greatly from those of the adult. The basic qualitative differences, however, are few, and the peculiar or restricted natural dietetics of many insects are caused by phagostimulatory behavior, dependence on symbionts, compartmentalization of nutrients between different developmental stages, or critical proportionalities between common basic nutrients (Dadd 1973).

Proteins (or amino acids) are needed principally to build tissues and enzymes. In addition to the 10 essential amino acids (arginine, lysine, leucine, isoleucine, tryptophan, histidine, phenylalanine, methionine, valine, and threonine), supplementary amino acids, such as glutamic and aspartic acids, are needed in some cases (Rock 1972). The protein requirements of insects vary with the qualitative and quantitative compositions of the amino acids and with the stage of the insect.

Deficiency in proteins affects the growth, development, and reproduction of the insect. It may also affect the characteristic and development of the skeletal structure. In the honey bee, *Apis mellifera*, that is fed a low-protein diet, the nitrogenous reserves are depleted mostly from the integument, resulting in a brittle integument and in general paralysis (Butler 1943). The lack of protein in the European corn borer, *Ostrinia* (syn. *Pyrausta*) *nubilalis*, delays molting or causes supernumerary molts (Beck 1950). Yolk synthesis is prevented in adults of *D. melanogaster* (Sang and King 1961) and *Protophormia terrae-novae* (Harlow 1956) by the lack of various amino acids. Excessive amounts of various amino acids, including tryptophan, in the diet of *Drosophila melanogaster* produce melanotic tumors. There is apparently a close relationship between tryptophan metabolism and tumors, eye color, and other abnormalities (House 1974).

Insects utilize carbohydrates mainly for energy, but the carbohydrates also serve other metabolic functions. The carbohydrates are the major dietary constituents for most insects. In addition to their calorific value, the deficiency in carbohydrates may affect molting in the European corn borer (Beck 1950) and retard the ovarian development in *Phormia regina* (Rasso and Fraenkel 1954).

Lipids or fats (including fatty acids), as in the case of carbohydrates, are mainly used for energy, but they also have specific metabolic roles other than calorific. Fats are important sources of metabolic water, especially during periods of desiccation. They also serve as components of phospholipids, which form essential structural units of cellular membranes. Most fatty acids can be synthesized from proteins and carbohydrates by all insects, but others, such as linoleic and linolenic acids, are required by some insects. Diets deficient in these polyunsaturated fatty acids may lead to decreased growth, deformed wings, abnormal molting, and other abnormalities in some moths and locusts (Fraenkel and Blewett 1946; Vanderzant et al. 1957; Dadd 1960). In the German cockroach, linoleic acid deficiency causes first-generation females to abort their egg cases. Any nymphs produced from these females walk erratically, fall, and lie on their backs with weak agitation of legs and antennae, and die in a few days (Gordon 1959).

Sterols are essential to insects and no insect, except those with symbionts, has been found to be independent of an exogenous sterol source (Clayton 1970). The requirement of insects for dietary sterols is one of the unique differences between insect and mammalian nutrient requirements (Vanderzant 1974). Cholesterol, with a few exceptions, fulfills the sterol requirement in insects. The structural role of sterols in their universal association with cytoplasmic membranes is quantitatively their most important function, but the sterols also provide the starting material for the synthesis of ecdysone, the molting hormone in insects (Clayton 1970).

Vitamins are essential in metabolism and, in general, act as constituents of the enzyme systems that are required in various metabolic activities of insects. The B vitamins act as enzyme cofactors in basic metabolic transformation. Thiamine and riboflavin are involved in carbohydrate metabolism. Pyridoxine acts as a coenzyme in tryptophan metabolism. Some vitamins are essential for the fecundity of adult insects, especially in oogenesis. Most plant-feeding insects require vitamin C (ascorbic acid). At least some insects need the fat-soluble vitamins A and E. Vitamin E (α-tocopherol), which plays a role in the reproduction of mammals, also is essential in the reproduction of the insect predator *Cryptolaemus montrouzieri* and the insect parasitoid *Agria affinis* (House 1974). Vitamin A accelerates the growth and development of *A. affinis*, especially the male.

Some of the nutritional problems associated with dietary studies stem from the imbalance of nutrients. Many nutrient requirements are dependent on the requirements of other nutrients. Such imbalances may affect the growth rate or produce covert stress and cause decreased viability, increased tumor prevalence, or abnormal lethargy (Dadd 1973). Furthermore, an unbalanced set of nutrients may be spurned by the insect in favor of a better balance of the same nutrients.

A nutritionally well-fed insect is less likely to be afflicted with diseases caused by noninfectious and infectious agents. A healthy insect is generally more resistant to pathogens than a nutritionally deficient one. In addition to the nutritional requirement, certain plant foods have antimicrobial factors that may reduce the likelihood of the insect becoming infected by pathogens. This aspect of the effect of nutrition on the susceptibility to infectious agents will be discussed in Chapter 14.

G. HORMONAL DISRUPTION

A malfunction of the endocrine hormonal metabolism brings about disease in an insect. The neuroendocrine system regulates a number of metabolic processes, all of which contribute to the basal metabolism of the insect. These include O_2 uptake, the syntheses of trehalose, lipids, and proteins, and the reproduction-related metabolism in adult female insects (Keeley 1972). The effective endocrine system is the brain–corpora cardiaca complex. The insect can adjust its endocrine output for optimal survival to variations in the environment, such as nutrient quantity and quality, humidity, photoperiod, and temperature (Keeley 1972).

Hormones essential for growth and metamorphosis are the molting (ecdysone) and the juvenile hormones. Not only the quantities of these hormones, but also a delicate balance between the two hormones are required for a normal insect life. In the absence of ecdysone, the insect does not undergo metamorphosis to pupa and adult, and in the absence of juvenile hormone, the larva reacts to ecdysone by undergoing precocious metamorphosis to form a miniature pupa. An excess of juvenile hormone causes derangement of metamorphosis and prevents the maturation of the larva to an adult, or inhibits embryonic development beyond a certain egg stage. Since these adverse effects can develop with topical application, the juvenile hormone has been proposed as a chemical insecticide against insects (Williams 1967). In certain insects, the males transmit the hormone to females and cause them to become sterile. The use of such venereal transmission of the hormone to produce sterility can be considered as comparable to the sterile male technique (Williams 1970). There are reports, however, that certain insects, such as mosquitoes, beetles, and plant bugs, are capable of developing resistance to the insect growth regulators (Brown et al. 1978).

II. MICROBIAL DISEASES

In microbial diseases, pathogenic microorganisms generally invade and multiply in an insect and spread to infect other insects. Pathogens are transmitted to insects by contact, ingestion, and vectors, and from the parents to their offspring. The pathogens are noncellular, parasitic infectious agents (viruses), prokaryotic forms (organisms without a true nucleus and nuclear membrane, such as bacteria), and eukaryotic forms (organisms with a true nucleus enclosed by a nuclear membrane, such as fungi and protozoa). In addition, we have included the nematodes even though they are often larger and more complex than microorganisms. Nematodes have characteristics of both parasitoids–predators and microbial pathogens, but they are placed with pathogens because they have no functional response and often produce pathologies similar to pathogens. A brief description of the major pathogen groups is provided in Chapters 4 and 5 (Bacteria), Chapters 6 through 9 (Viruses), Chapter 10 (Fungi), Chapters 11 and 12 (Protozoa), and Chapter 13 (Nematodes).

The term parasite has been used, at times, interchangeably or synonymously with pathogen. We restrict the use of the terms parasite and parasitoid to the larger and more complex organism that generally does not multiply on or in the insect. There are exceptions. Polyembryonic parasitoids lay a single egg in an insect host and the egg subsequently divides to produce from a few to several hundred embryos. The mite *Acarapis woodi*, causal agent of the Isle of Wight disease in the honey bee, enters the adult bee through the first thoracic spiracles and into the trachea. It may multiply to such great numbers and plug the trachea almost completely. Its mouthparts penetrate through the trachea and injure the surrounding muscle and nerve tissues (Anderson 1928). On the other hand, some gregarines (protozoa) do not multiply when they enter their insect host. Their multiplication occurs in a stage outside of the host. Thus, there is some overlapping in the definitions of parasite and pathogen, but in general, the pathogens are minute and multiply on or in the host. For nematodes, some (e.g., mermithids) are referred to as parasites; whereas others (e.g., steinernematids) are referred to as pathogens.

A successful infection of an insect by a microorganism depends on the properties and characteristics possessed by both organisms. Since the pathogenic properties of the microorganism are pitted against the resistive properties of the host, the basis for infection, at times, is difficult to determine whether it is due to the properties of the pathogen or those of the host. Additionally, environmental factors may enhance or reduce the chances for infection (see Chapter 16).

A. POTENTIAL, FACULTATIVE, AND OBLIGATE PATHOGENS

Not all microorganisms cause infection even after they gain entrance into the insect's hemocoel. This lack of infection may be due to the resistant characteristics of the host or to the inability of the microorganisms to survive and replicate in the host's environment. Infectious microorganisms can be separated broadly into potential, facultative, and obligate pathogens. Potential pathogens are those microorganisms that are incapable of invading the host, either through the body wall or through the digestive tract, without the assistance of external factors that lower the insect's resistance or enhance the ability of the microorganism to invade the insect. Facultative pathogens are those microorganisms that do not require an insect weakened by external factors to cause infection. Moreover, the survival of both types of pathogens is not dependent entirely on the insect, and they can live and multiply independently from the insect. These pathogens are readily reared on nonliving media in the laboratory. Most of them are bacteria and fungi. Even though many pathogens can be easily grown in the laboratory, if they do not reproduce independently in nature, they should be considered as obligate pathogens. Obligate pathogens require live insect hosts for survival and replication. They may occur outside of the insect in a dormant stage, such as spore, cyst, viral occlusion body, etc. The obligate pathogens are viruses, microsporidia (protozoa), most nematodes, and certain fungi and bacteria.

B. MICROBIAL TOXINS

A disease can be brought about in a susceptible host by the pathogen through the effects of chemical or toxic substances, the mechanical destruction of cells and tissues, and the combination of these two actions. There are two general types of toxins produced by entomopathogenic organisms, catabolic and anabolic substances. The catabolic toxins result from decomposition brought about by the activity of the pathogen. They may arise from the substrate or from the decomposition of the pathogen itself. For example, the breakdown of proteins, carbohydrates, and lipids by the pathogen may produce toxic alcohols, acids, mercaptans, alkaloids, etc.

Anabolic toxins are substances synthesized by the pathogen. These may be classified as exotoxins and endotoxins. The exotoxins (ectotoxins) are true toxins or soluble toxins. They are excreted or passed out of the cell of the pathogen. Exotoxins have been detected from entomogenous pathogens, especially among bacteria and fungi.

The endotoxins, produced by the pathogen, are not excreted but are confined to the cell. These toxins are liberated when the pathogen dies and degenerates, or as in the case of *Bacillus thuringiensis*, when it sporulates and the sporangial wall disintegrates. The most widely investigated endotoxin of an entomopathogen is the δ-endotoxin (crystalline endotoxin) of *Bacillus thuringiensis*. The toxin is contained within a proteinaceous crystal that is formed during sporulation and lies adjacent to the spore in the bacterial sporangium.

The difference between exotoxin and endotoxin may not be distinct at times and causes confusion because exotoxins are formed inside the cell and some endotoxins are found on the cell surface. Some toxins remain attached to the cell membrane, yet they can be extracted by certain solvents or solutions and have been called exotoxins. The major toxins produced by entomopathogenic microorganisms are discussed within the pathogen chapters.

C. PORTALS OF ENTRY

The portals of entry are the points or sites through which a pathogen invades or gains entrance into an insect. Invasion may take place through the integument,

FIGURE 3-1

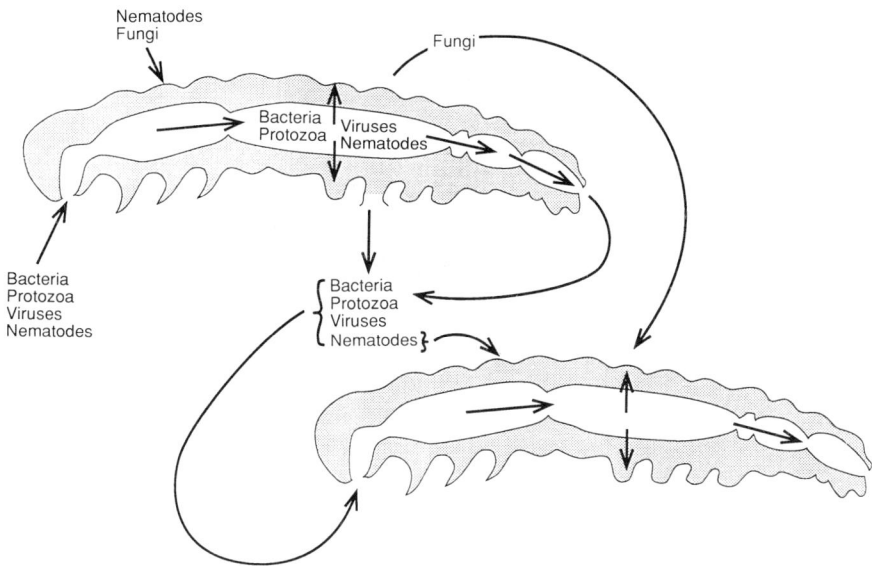

Diagram of a lepidopteran larva showing the portals of entry for infectious microorganisms and their dissemination to another host. (Modified after Steinhaus 1949.)

which may be intact or broken by injury, by way of the mouth (per os) and digestive tract, by congenital passage through the ova or spermatozoa, through the anus, and through the spiracles and other body openings. Parasitic insects and other vectors may also transmit the pathogen into a susceptible host insect. Passage through the integument and mouth are the most common portals of entry (Fig. 3-1). Unlike humans and other vertebrates where respiratory tract infections are common occurrences, the spiracles and tracheae of insects are rarely invaded by pathogens.

The method of entry varies with the type of pathogen. Fungi and nematodes generally invade through the integument, whereas viruses, bacteria, and protozoa enter their hosts primarily through the mouth. There are more cases of congenital transmission with viruses and protozoa than with other pathogens. Such congenital transmission most frequently occurs by way of the egg, either through the ovary (transovarial transmission) or on the surface of the egg (transovum transmission). The term *transovum transmission* is used to refer to the passage of a pathogen on or within the egg. A special case of transovum transmission is transovarial transmission in which the pathogen enters the egg while it is still in the ovary. There are only a few reported cases of transmission by way of the male (i.e., venereal transmission in insects). The sigma virus of *Drosophila melanogaster* is usually transmitted through the egg, but it can also be passed to the progeny via the sperm (Seecof 1968). The microsporidia *Nosema fumiferanae* and *N. plodiae* are transmitted to their offspring by adult males of the spruce budworm, *Choristoneura fumiferana*, and of the Indianmeal moth, *Plodia interpunctella*, respectively (Thomson 1958; Kellen and Lindegren 1971).

D. INFECTIVITY

Infectivity is the ability of a microorganism to produce infection (i.e., to enter the body of a susceptible insect). Generally, an infection results in detectable pathologic effects, such as injuries or dysfunctions. An infection, however, may not be harmful; such as when a mutualistic symbiont infects and multiplies in a host, there is no pathology involved. Thus, infection may result in a nondiseased or a diseased condition. In an infection that results in disease, there are two main factors: invasiveness and pathologies of abnormalities or dysfunctions. In some cases a pathogen does not need to invade an insect's body cavity or hemocoel to cause diseases. This is the case with the European foulbrood of the honey bee, in which the causal bacterium *Melissococcus pluton* is confined to the lumen of the digestive tract of the larva (Bailey 1963). In the tent caterpillars *Malacosoma* spp., the bacteria *Clostridium* spp., which are also confined to the midgut lumen, produce dysentery and cause the larva to shrink in size (brachytosis), followed by death (Bucher 1961). In all of these cases, the insects appear to be affected by toxins produced by the bacteria.

When an infectious agent is naturally transmitted by direct contact, the resultant disease is designated as "contagious" or "communicable." Contagious diseases are common in insects and occur frequently in laboratory insect cultures. Infection is to be differentiated from "contamination." A susceptible insect may be contaminated or be harboring a pathogen without being infected. A nonsusceptible insect or other organism or objects may be in contact or contaminated with the pathogen. An "intoxication" is brought about by the activity of the pathogen in the form of toxins, but the pathogen does not need to be present to cause the disease. For example, the δ-endotoxin of *B. thuringiensis* alone can cause disease independent of the bacterial spore or rod.

E. VIRULENCE AND PATHOGENICITY

"Virulence" is the disease producing power of a microorganism. This is the ability of a microorganism to invade and cause injury to the host. It is the relative capacity of a microorganism to overcome the host defense mechanism and is often defined in relation to the resistance of the host. A pathogen may be highly virulent because of the low resistance or high susceptibility of the host, and conversely, a pathogen may have low virulence because of the high resistance or low susceptibility of the host. "Pathogenicity" is nearly synonymous to virulence in that it refers to the disease-producing ability of the microorganism. The difference is that pathogenicity is applied to groups or species of microorganisms; whereas, virulence is used in the sense of degree of pathogenicity within the group or species. Pathogenicity is sometimes regarded as a genetically determined ability to produce disease, and virulence as not being genetically produced. Thus, we may say that the pathogenicity of *Bacillus thuringiensis* is high for lepidopteran species, but its virulence may differ depending on conditions, such as methods of cultivation, storage, formulation, and environmental factors. In this regard, pathogenicity is primarily the basis used to separate *B. thuringiensis*, an insect pathogen, and *B. anthracis*, a vertebrate pathogen, from *B. cereus*.

Virulence is a variable property and can be intensified or reduced. The process of decreasing the virulence or the disease-producing power of a pathogen is called "attenuation"; that of increasing the resistance of the host is "immunization." There are several ways by which the virulence of an insect pathogen can be increased: (1) Passing the pathogen through a series of susceptible insects or possibly other organisms, (2) dissociate and isolate the pathogen into more virulent strains, (3) culture the pathogen with nutrients that favor its virulence, (4) combine the pathogen with substances (mucin, boric acid, etc.) that increase its invasive power, and (5) associate the pathogen with other pathogens to render it more capable of invading the insect. In general, procedures similar to the above are used to decrease the virulence of a pathogen as follows: (1) Passing the pathogen through a series of insects unfavorable to its growth and development, (2) dissociate and isolate the pathogen into strains of low virulence, (3) culture the pathogen under abnormal conditions, such as high temperatures and abnormal nutrients, and (4) associate the pathogen with other pathogens to reduce or inhibit its invasive power. All of these methods have been used with entomogenous pathogens (entomopathogens). We shall briefly discuss and cite a few examples of manipulating the virulence of insect pathogens.

The intensification of virulence has been observed with nearly all insect pathogens. This is especially the case with bacteria and fungi, which, under repeated culture in artificial media, tend to lose their virulence; but their virulence can be restored by passage through one or more series of susceptible hosts. Increase in virulence also occurs with obligate pathogens, such as viruses (Aizawa 1971, 1975). The virulence of a pathogen is generally enhanced when passed through habitual or normal hosts as compared with passage through alternate or factitious hosts. However, with the microsporidia *Nosema acridophagus* and *N. cuneatum*, propagation in the factitious host, the corn earworm, *Heliothis zea*, increases the virulence to their habitual hosts, the grasshoppers (Henry et al. 1979). Smirnoff (1963) passed the nuclear polyhedrosis virus (NPV) of the poplar sawfly, *Trichiocampus viminalis*, to another sawfly, *T. irregularis*. The virulence of the virus for the second host increased after each passage up to the third passage. When the NPV of *Autographa californica* was propagated several times in *Spodoptera exigua* larvae and then passed through *Trichoplusia ni* larvae, the virus became more virulent for the larvae of *T. ni* (Tompkins et al. 1981). When the NPV of *T. ni* was propagated several times in *T. ni* larvae and then passaged through *S. exigua* larvae, the virus was more virulent for *T. ni* larvae than the virus produced in *T. ni* larvae. In these cases, passage through an alternate host, *S. exigua*, enhanced the virulence of the viruses.

As mentioned previously, a decrease in virulence generally results when pathogens are passed through continuously in alternate hosts or *in vitro* culture systems. The *in vivo* transmission of baculoviruses to alternate hosts may result in alterations of the occlusion bodies (e.g., icosahedral to cuboidal polyhedra in the NPVs of the silkworm) (Aizawa 1962; Watanabe et al. 1975) and of the almond moth, *Cadra cautella* (Hunter et al. 1973). NPV also becomes less infective after prolonged passage through insect-cell cultures (MacKinnon et al. 1974; Hirumi et al. 1975; Yamada et al. 1982). The morphology and infectivity of the virus are greatly altered.

Under certain nutrient conditions, the virulence of a pathogen is higher than with other nutrients. The same strain of *Bacillus thuringiensis* cultured with different nutrients produces organisms that differ greatly in their virulence (Dulmage and Rhodes 1971). The virulence of the fungus *Verticillium lecanii* varies with the method of conidial production (Galani 1979; Hall 1980). Less virulent conidia are obtained with fermented media as compared with shaken liquid or solid media.

All entomopathogenic microorganisms have strains of greater or lesser virulence for insects. In many instances, such strains can be isolated under laboratory conditions on nonliving media. Some of these microorganisms are distinct and can be characterized by their physical and physiological properties. For example, the white-muscardine fungus, *Beauveria bassiana*, has many strains, which at one time were considered to be distinct species because of different morphological characteristics. Such species are presently reduced to varieties or strains of *B. bassiana* (MacLeod 1954). The strains, however, vary in their virulence depending on the susceptible insect species. The above examples of increased or decreased virulence of viruses after passage through a series of susceptible hosts may be a selection of strains rather than increasing the virulence of a single strain. On the other hand, mutants of baculoviruses have been developed with the use of mutagens and they are much more virulent than the wild isolates (Reichelderfer and Benton 1973; Wood et al. 1981).

Substances, such as mucin, are known to increase the virulence of bacteria pathogenic for vertebrates. Mucin has been used with the nonspore-forming bacterium *Pseudomonas aeruginosa* to enhance its infection to grasshoppers (Stephens 1959). Boric acid at 1% concentration when combined with *Bacillus thuringiensis* and fed to the gypsy moth larva, *Lymantria* (syn. *Porthetria*) *dispar*, increases larval mortalities (Doane and Wallis 1964). Greater larval mortality also occurs in *Spodoptera litura* larvae when boric acid is added to *B. thuringiensis* (Govindarajan et al. 1976). Boric and sorbic acids also enhance the infectivity of the NPV in the gypsy moth (Shapiro and Bell 1982).

F. DOSAGE OF THE PATHOGEN

A single infective cell or stage may theoretically cause an infection, but generally, its invasion is inhibited or it does not cause a disease. In nearly all cases, there is a minimal number of infective stages of a pathogen passing through the portal of entry in order to cause infection. This infective dosage is expressed quantitatively in terms of host susceptibility, as lethal dose (LD), effective dose (ED), infectious dose (ID), lethal time (LT), etc. The 50% level of response is usually assigned as LD_{50}, ED_{50}, ID_{50}, LT_{50}, etc. The range of doses obtained from bioassay when graphed against host response (mortality, infection, etc.) gives an S-shaped curve, which is transformed to a straight line by converting host response to a probit scale and dose to logs. Students interested in the bioassay of insect pathogens should refer to the excellent treatment by Burges and Thomson (1971).

In the case of pathogens that invade through the mouth, the number of infective stages, usually in the form of spores, occlusion bodies, resting stages, vegetative forms, etc., is quantitatively measured and the prevalence of infection observed among the insects. Pathogens penetrating through the integument can also be applied at various dosages to the integument. In some cases, there are problems and difficulties in determining the dosages quantitatively, as with insects living in an aqueous habitat or the use of certain pathogens whose resting stages cannot be easily applied to an insect.

G. SIGNS, SYMPTOMS, AND SYNDROMES

During an infection, the insect exhibits characteristic aberrations or dysfunctions, which are designated as signs or symptoms. When it is a functional or behavioral disturbance, the term symptom is used; when it is a physical or structural abnormality, the term sign is used. A symptom may be expressed by abnormal movement, abnormal response to stimuli, digestive disturbance, inability to mate, etc. A sign is indicated by abnormalities in the morphology or structure, changes in color, malformation of appendages, integument, etc. Some workers do not differentiate between these two terms and use symptom to refer to the two types of abnormal expressions.

A particular disease has a group of characteristic signs and symptoms, which is known as the syndrome of the disease. The syndrome refers to a system complex or a particular combination, set, or sequence of signs and symptoms. At times, the syndrome is very characteristic and specific for a disease caused by a pathogen. More often, the same syndrome may arise from many different causes. For example, the flacherie of silkworm may be caused by environmental conditions, bacteria, or viruses. Diarrhea in insects may develop from poisons, toxins, bacteria, viruses, and protozoa. When a silkworm larva ingests the spore and endotoxin of *Bacillus thuringiensis* subspecies *sotto*, it stops feeding in a few minutes, becomes sluggish in about 10 min, the pH of the blood becomes higher, and within an hour the larva collapses and dies. The syndrome is distinct and brief; whereas, in the cabbage looper, *Trichoplusia ni*, the larva feeding on *B. thuringiensis* subspecies *thuringiensis* stops feeding momentarily, continues to feed, and develops diarrhea; its midgut epithelium is destroyed; the bacteria invade the hemocoel, multiply, and produce a septicemia leading to death. The syndrome in the cabbage looper is more complex and prolonged than in the silkworm. In the silkworm, death is caused by the endotoxin, but in the cabbage looper, the bacteria and/or the toxin participate in causing a lethal infection.

H. COURSE OF INFECTION

After a pathogen invades an insect, it multiples and the course of infection can be separated into the following periods or stages: the incubation period, the beginning of disease, the height of disease, and the termination of disease (Fig. 3–2). The incubation period is the period from the entrance or introduction of the pathogen into an insect's body until the development of signs and symptoms. The beginning of the disease is the first appearance of signs or symptoms until the disease is fully developed. The height of the disease is the period when the signs and symptoms attain its maximum and may begin to fall or reach a steady state. At the height of disease, the microorganism is most active, the lesions caused by it are greatest in extent, the signs and symptoms are most severe, and if toxins are produced, they are present in the greatest amount. However, many pathogens, in particular bacteria and fungi, continue to grow and reproduce after the death of an insect. The termination of the disease may involve the recovery of an insect due to immune responses or the death of an insect in the absence of such responses. In cases where the pathogen has low pathogenicity, the insect may have a chronic infection and survive to reproduce.

A common error in the literature is to refer to the incubation period as the period from the entrance of the pathogen to the time of insect death. Such usage does not meet the definition of incubation period, which refers to the brief period from invasion to the development of signs or symptoms, and does not involve the period up to insect death. The proper term to apply to the period up to insect death is the period of lethal infection or, if the insect recovers, the period of infection. A short period of lethal infection indicates a pathogen of high pathogenicity, and a long period indicates one of low pathogenicity. The end result of infection is either

FIGURE 3–2

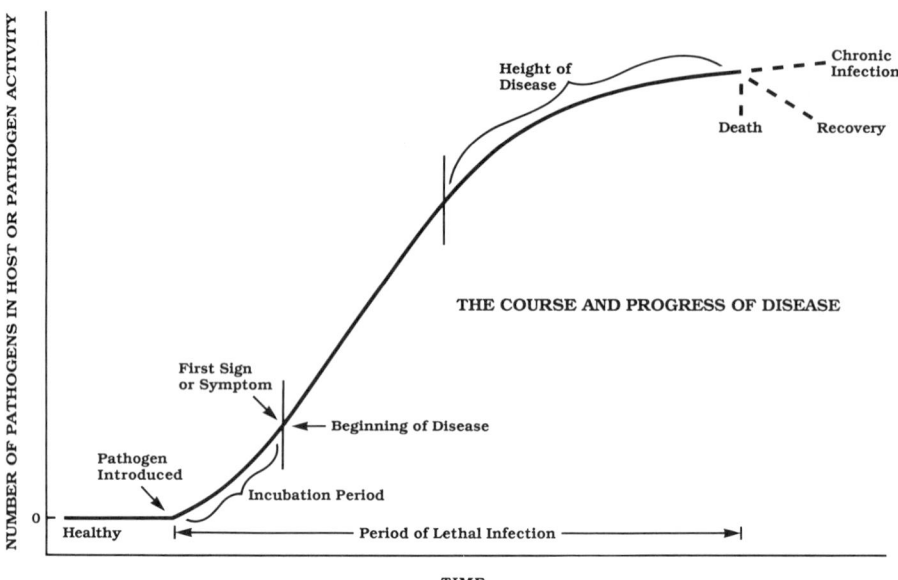

The course of infection by microorganisms in an insect host.

recovery (partial or full) or death. Some workers use the adjectives "frank" and "patent" to emphasize a successful infection.

I. LATENT, CHRONIC, AND ACUTE INFECTIONS

The infection in insects may vary from latent, chronic, and acute. Acute infections are most obvious because of the distinct and characteristic response of an insect in developing a disease. Chronic infections are often overlooked because their manifestations are not striking and apparent to the observer. Latent infections may occur with various types of pathogens that are present in the insect in a "dormant" or nonactive stage and produce no signs or symptoms. Latent infections in insects have been observed primarily with virus. In such cases, the term latent or occult viral infection is used, and the virus is referred to as an occult virus and not as a latent virus. Most of the difficulty in studies with occult virus is the differentiation of the state of infection, whether it is a latent (occult) or a chronic infection.

Chronic and latent infections can become acute infections by the action of stressors. The activation or incitation by stressors (force, condition, or circumstance) in or on an insect or on one of its physiological or anatomical systems results in a syndrome or bodily changes (Steinhaus 1960). A "stressor" acts on the insect host, and an "incitant" acts on the pathogen. Sometimes it is difficult to differentiate between these two, particularly when the force acts on both the host and pathogen. There are many factors, both physical and biological, that act as stressors and incitants. We have already touched upon stressors and incitants in Section II.E., Virulence and Pathogenicity.

J. INTERACTIONS AMONG MICROORGANISMS

Different pathogenic and nonpathogenic microorganisms are often found within the same insect and they may interact with each other (Vago 1963). The interaction may result in independent coexistence, complementation (synergism), and interference. Unfortunately, most of the cases of interactions have neither been studied in detail nor supported by adequate quantitative data. The interactions are affected by such factors as larval age, dosages of the pathogens, sequence and time of feeding of each pathogen, and environmental conditions, especially temperature. We shall briefly consider a few cases of complementation and interference.

1. Complementation

In complementation or synergism, the pathogens are associated with each other in such a way as to render one or both of them more pathogenic for the host. The enhanced pathogenicity by the combined pathogens is greater than the sum of the pathogenicity of each pathogen.

When the armyworm, *Pseudaletia unipuncta*, is grown under axenic conditions, the addition of another bacterium to the bacterium *Serratia marcescens* increases the infection of the latter (Goodwin 1968). The microsporidium *Nosema lymantriae* infects the nun moth, *Lymantria monacha*, and another microsporidium, *Thelohania similis*, infects not only the nun moth, but also *Euproctis chrysorrhoea* (Weiser 1961). *Nosema lymantriae* is able to infect *Euproctis chrysorrhoea* only when it is combined with *T. similis*. The basis for this enhancement has not been established. A synergistic interaction occurs between a granulosis virus (GV) and a nuclear polyhedrosis virus (NPV) in the larva of the armyworm, *Pseudaletia unipuncta* (Tanada 1959). In this association, the GV is the synergist and enhances the infection of the NPV. The GV contains an enhancing factor, a lipoprotein, within the capsule (viral occlusion body) (Hara et al. 1976; Yamamoto and Tanada 1978). The virulence of the NPV is enhanced as much as 56 times by the GV (Tanada and Hukuhara 1971). The nematode *Steinernema* (syn. *Neoaplectana*) *carpocapsae* is associated in mutualism with the bacterium *Xenorhabdus* (syn. *Achromobacter*) *nematophilus* (Dutky 1959; Poinar and Thomas 1966). The bacterium, by itself, cannot enter the insect. It is carried in the intestinal lumen of the infective stage nematode and is released through the anus when the nematode invades the insect's hemocoel. The nematode serves as a means of dispersal and transmission of the bacterium into host insects, and in turn, the bacterium provides the necessary environment for the development and reproduction of the nematode within the host.

2. Interference

Interference or antagonism may occur at the cellular and organismal levels. Most observations have been made at the organismal level, but in some cases where no interference is observed at the organismal level, it occurs at the cellular level where both microorganisms do not occupy the same cell. Interference occurs commonly, especially among microorganisms present in the midgut. Since bacteria are among the most prevalent microorganisms in this habitat, they are generally involved in the interactions that may be antagonistic or complementary (Tanada and Fuxa 1987).

Interference between the bacterium *Pseudomonas aeruginosa* and other bacteria has been demonstrated in armyworm larvae maintained under axenic conditions where the infectivity of the *Pseudomonas* is inhibited (Goodwin 1968). The enteric bacteria also commonly interfere with the infections of microsporidia and viruses fed to the larvae.

Interference among viruses has been studied in much more detail than among other types of pathogens. In the cytoplasmic polyhedrosis viruses, interference may occur among various strains that are distinguished by the shapes of their occlusion bodies (Tanada 1971). Such interference develops even when the antagonistic virus is inactivated by heat (Aruga et al. 1963) and ultraviolet radiation (Aruga and Hashimoto 1965). In the case of baculoviruses, interference occurs between strains of NPVs of the fall webworm, *Hyphantria cunea* (Aruga et al. 1961), of the armyworm, *Pseudaletia unipuncta* (Tanada et al. 1969; Ritter and Tanada 1978), and between NPV and GV of the spruce budworm, *Choristoneura fumiferana* (Bird 1959). In the case of the armyworm and the spruce budworm, the less pathogenic virus must be given an advantage in time or dosage before a multiple or mixed infection can occur in a larva.

K. KOCH'S POSTULATES

One of the basic requirements for establishing the etiologic or causal agent of a disease is the application of Koch's postulates. Robert Koch (1843–1910), a German physician, is considered as one of the founders of bacteriology. He received the Nobel Prize in 1905 in physiology and medicine for his work on tuberculosis. He made his brilliant discoveries on the causal agents of tuberculosis, anthrax, cholera, etc., through the application of the postulates.

The postulates are

1. A specific pathogenic organism must be seen in all cases of a disease.
2. This organism must be attained in pure culture.
3. The organism from the pure culture must reproduce the disease in experimental animals.
4. The same organism must be recoverable from the experimental animal.

There is, at times, difficulty in fulfilling the postulates. One or more steps cannot be followed or can be performed only under very difficult conditions. Some organisms are difficult to detect, and even to this day the cause of certain plant and animal diseases has not been established. The electron microscope has made it possible to detect submicroscopic organisms, such as viruses, mycoplasma, spiroplasma, etc., but the mere detection of these organisms does not prove them to be causal agents without further experimentation following Koch's postulates. Step two cannot be met with certain pathogens whose cultures are not possible with the presently available laboratory methods. They are mainly obligate pathogens requiring living cells or tissues. Insect tissue cultures can be used to fulfill step two provided that they are not contaminated with other organisms. The experimenter should be aware that different pathogens may produce similar signs, symptoms, and syndromes and may mistakenly designate a wrong organism as the etiologic agent. Moreover, the causal agent may produce symptoms in the laboratory that differ from those in nature.

L. DIAGNOSIS

Diagnosis is a fundamental branch of insect pathology and should be allotted a similar position of importance as in the case for human, vertebrate, and plant pathologies (Steinhaus 1963, 1964). It is defined as the process by which one disease is distinguished from another. The mere identification of the etiologic or causal agent alone is not diagnosis, but only one of a series of operations. In order to accomplish a proper diagnosis, a study has to be made of the etiology, symptomatology, pathogenesis, pathologies, and epizootiology of the disease. Thus, diagnosis is one of the most complex branches of insect pathology. It is fundamental not only to the basic aspects but also to the applications of insect pathology: microbial control and the control of diseases of beneficial insects. Steinhaus (1963) stressed that "The importance of diagnosis in insect pathology lies in the fact that one must know the nature of the disease and what ails or has killed an insect before the disease can be properly studied, controlled or suppressed, used as a microbial control measure, its potentialities for natural spread determined, or its role in the ecological life of an insect species ascertained."

There are various types of diagnostic operations, such as symptomatic diagnosis, laboratory diagnosis, postmortem diagnosis, etc. "Differential diagnosis" compares and contrasts the signs and symptoms and postmortem changes of different diseases in a systematic manner to distinguish one disease from another. This is one of the more common type of diagnosis. "Preliminary diagnosis" is the first cursory examination of a diseased insect. It is generally an educated guess of the type and cause of disease. "Tentative diagnosis" is made after general macroscopic and microscopic examinations have been conducted on a diseased specimen. It often involves routine laboratory tests. "Definitive diagnosis" completes the diagnosis, and a final conclusion is reached after all pertinent data and facts have been gathered, tabulated, analyzed, and adjusted. The conclusion is based on information derived from the laboratory, field, and literature.

In diagnosis, there are two procedures that should be performed properly: (1) collection of facts and data and (2) analysis of these facts and data. The facts and data that should be collected are outlined as follows (Steinhaus 1963):

1. History of disease
2. Physical examination
    (a) General inspection
    (b) Macroscopic examination
    (c) Microscopic examination
3. Laboratory and ancillary examinations
    (a) Isolation and possible study of microbial pathogens
    (b) Assay for nonmicrobial factors (e.g., poisons, metabolic diseases, etc.)
    (c) Detailed study of hosts; biochemical and biophysical tests of tissues and fluids
    (d) Observation of course of disease

The history of the disease is comparable to the "medical" or "clinical" history of human diseases. The diagnostician must prepare the history through his or her observation and on field data provided by the collector. The physical examination entails general examination of the insect's body and examinations made with instruments, such as the optical and electron microscopes. Laboratory tests are conducted to isolate, culture and study the pathogen and its infectivity or to assay for nonmicrobial agents, such as poisons, metabolic diseases, etc. The tests also include a detailed study of the host based on biochemical and biophysical tests of fluids and tissues. When live specimens are available, careful observation is made of the course followed by the disease from its onset to termination in death or recovery. Whenever possible, Koch's postulates should be applied to establish the causal agent.

A common question is how to submit an insect specimen for diagnosis. This is an important matter because a successful diagnosis is not possible unless the specimen is submitted in a satisfactory condition. It is preferable to submit fresh specimens. The specimens should not be killed and fixed in preservatives, such as alcohol or formaldehyde, unless they are to be sectioned for histological examination. If available, the following types of specimens should be submitted: uninfected individuals, individuals at an early stage of infection, at a late stage of infection, and moribund specimens. At times, a diagnosis can be rendered with dead, dried specimens, but usually this is not possible. The specimens should be placed individually in clean vials or containers and stoppered with cotton or nonairtight cover. If sterilizing equipment, such as autoclave or oven, is not available, the vials or containers should be new, unused ones or they should be cleaned with disinfectants and detergents. Unless the specimens are live, aquatic forms, they should be placed in the vial alone or with small portions of their food. During shipment, any hard material in the vial or container may injure the specimens.

The specimens should be shipped by the most direct and fastest available means of transportation, usually by air mail. In most countries, there are quarantine regulations that must be met and the submitter should first check with the diagnostician or the receiving laboratory concerning such regulations. Many specimens have been destroyed by quarantine officials whose responsibility is to restrict or inhibit the movement of animal and plant pests into their country.

The specimens should be submitted with pertinent data that will be needed by the diagnostician and also help in his diagnosis. These data are (1) common and scientific names of the insect, (2) name of the collector and his accession number or identifying number, if any, (3) name of the submitter if different from that of the collector, (4) date and location of collection, (5) name of the host plant or animal, (6) nature of the environment in which the diseased insect was found, (7) extent of the disease outbreak or epizootic, (8) conditions under which the disease outbreak occurred, (9) abundance of insect, (10) abnormal behavior or appearance, if any, of insects when found, and (11) whether or not the insect was exposed to chemical insecticides.

An increasing number of laboratories in insect pathology are being formed in universities and other institutions and at the state, federal, and private levels to which specimens may be sent. The insect pathologist should be contacted before specimens are sent for diagnosis because the protocol may be slightly different and individuals and addresses are subject to change.

REFERENCES

Aizawa, K. 1962. Infection of the greater wax moth, *Galleria mellonella* (Linnaeus), with the nuclear polyhedrosis virus of the silkworm, *Bombyx mori* (Linnaeus). *J. Insect Pathol.* 4, 122–127.

Aizawa, K. 1971. Strain improvement and preservation of virulence of pathogens. *In* "Microbial Control of Insects and Mites." (H. D. Burges and N. W. Hussey, eds.), pp. 655–672. Academic Press, New York.

Aizawa, K. 1975. Selection and strain improvement of insect pathogenic micro-organisms for microbial control. *In* "Approaches to Biological Control." (K. Yasumatsu and H. Mori, eds.), *JIBP Synthesis 7*, 99–105.

Anderson, E. J. 1928. The pathological changes in honeybees infested with the Isle of Wight disease. *J. Econ. Entomol. 21*, 404–407.

Arnason, J. T., Philogene, B. J. R., and Morand, P. 1989. "Insecticides of Plant Origin." Am. Chem. Soc., Washington, D.C.

Aruga, H., and Hashimoto, Y. 1965. Interference between the UV-inactivated and active cytoplasmic-polyhedrosis viruses in the silkworm, *Bombyx mori* (Linnaeus). *J. Seric. Sci. Jpn. 34*, 351–354.

Aruga, H., Yoshitake, N., Watanabe, H., Hukuhara, T., Nagashima, E., and Kawai, T. 1961. Further studies on polyhedroses of some Lepidoptera. *Jpn. J. Appl. Entomol. Zool. 5*, 141–144.

Aruga, H., Watanabe, H., and Nagano, H. 1963. Interference by the heat-inactivated virus on the active virus of the cytoplasmic polyhedrosis in the silkworm, *Bombyx mori* L. *J. Seric. Sci. Jpn., 32*, 51–57.

Asahina, E. 1969. Frost resistance in insects. *In* "Advances in Insect Physiology." (J. W. L. Beament, J. E. Treherne, and V. B. Wigglesworth, eds.), Vol. 6, pp. 1–49. Academic Press, New York.

Bailey, L. 1963. "Infectious Diseases of the Honey-bee." Land Books, London.

Baust, J. G. 1981. Biochemical correlates to cold hardening in insects. *Cryobiology 18*, 186–198.

Beament, J. W. L. 1959. The waterproofing mechanism of arthropods. I. The effect of temperature on cuticle permeability in terrestrial insects and ticks. *J. Exp. Biol. 36*, 391–422.

Beck, S. D. 1950. Nutrition of the European corn borer, *Pyrausta nubilalis* (Hbn.). II. Some effects of diet on larval growth characteristics. *Physiol. Zool. 23*, 353–361.

Beeman, R. W. 1982. Recent advances in mode of action of insecticides. *Annu. Rev. Entomol. 27*, 253–281.

Benz, G. 1963. Genetic diseases and aberrations. *In* "Insect Pathology: An Advanced Treatise." (E. A. Steinhaus, ed.), Vol. 1, pp. 161–189. Academic Press, New York.

Berry, S. J., Krishnakumaran, A., Oberlander, H., and Schneiderman, H. A. 1967. Effects of hormones and injury on RNA synthesis in saturniid moths. *J. Insect Physiol. 13*, 1511–1537.

Bird, F. T. 1959. Polyhedrosis and granulosis viruses causing single and double infections in the spruce budworm, *Choristoneura fumiferana* Clemens. *J. Insect Pathol. 1*, 406–430.

Brown, A.W.A. 1963. Chemical injuries. *In* "Insect Pathology: An Advanced Treatise." (E. A. Steinhaus, ed.), Vol. 1, pp. 65–131. Academic Press, New York.

Brown, T. M., DeVries, D. H., and Brown, A. W. A. 1978. Induction of resistance to insect growth regulators. *J. Econ. Entomol. 71*, 223–229.

Bucher, G. E. 1961. Artificial culture of Clostridium brevifaciens n.sp. and C. malacosomae n.sp., the causes of brachytosis of tent caterpillars. *Can. J. Microbiol. 7*, 641–655.

Burges, H. D., and Thomson, E. M. 1971. Standardization and assay of microbial insecticides. *In* "Microbial Control of Insects and Mites." (H. D. Burges and N. W. Hussey, eds.), pp. 591–622. Academic Press, New York.

Bursell, E. 1974. Environmental aspects - humidity. *In* "The Physiology of Insecta." (M. Rockstein, ed.), Second Ed., Vol. 2, pp. 43–84. Academic Press, New York.

Butler, C. G. 1943. Bee paralysis, May-sickness, etc. *Bee World 24*, 3–7.

Cherbas, L. 1973. The induction of an injury reaction in cultured haemocytes from saturniid pupae. *J. Insect Physiol. 19*, 2011–2023.

Clayton, R. B. 1970. The chemistry of nonhormonal interactions: terpenoid compounds in ecology. *In* "Chemical Ecology." (E. Sondheimer and J. B. Simeone, eds.), pp. 235–280. Academic Press, New York.

Corbett, J. R. 1975. "The biochemical mode of action of pesticides." Academic Press, New York. Second printing.

Coudron, T. A. 1991. Host-regulating factors associated with parasitic Hymenoptera. *In* "Naturally Occurring Pest Bioregulators." (P. A. Hedin, ed.), pp. 41–65. Am. Chem. Soc., Washington, D.C.

Dadd, R. H. 1960. The nutritional requirements of locusts—I. Development of synthetic diets and lipid requirements. *J. Insect Physiol. 4*, 319–347.

Dadd, R. H. 1973. Insect nutrition: Current developments and metabolic implications. *Annu. Rev. Entomol. 18*, 381–420.

Danks, H. V. 1978. Modes of seasonal adaptation in the insects. I. Winter survival. *Can. Entomol. 110*, 1167–1205.

Day, M. F., and Oster, I. I. 1963. Physical injuries. *In* "Insect Pathology: An Advanced Treatise." (E. A. Steinhaus, ed.), Vol. 1, pp. 29–63. Academic Press, New York.

Doane, C. C., and Wallis, R. C. 1964. Enhancement of the action of *Bacillus thuringiensis* var. *thuringiensis* Berliner on *Porthetria dispar* (Linnaeus) in laboratory tests. *J. Insect Pathol. 6*, 423–429.

Doutt, R. L. 1963. Pathologies caused by insect parasites. *In* "Insect Pathology: An Advanced Treatise." (E. A. Steinhaus, ed.), Vol. 2, pp. 393–422. Academic Press, New York.

Dulmage, H. T., and Rhodes, R. A. 1971. Production of pathogens in artificial media. *In* "Microbial Control of Insects and Mites." (H. D. Burges and N. W. Hussey, eds.), pp. 507–540. Academic Press, New York.

Duman, J. G. 1977. The role of macromolecular antifreeze in the darkling beetle, *Meracantha contracta*. *J. Comp. Physiol. B115*, 279–286.

Duman, J. G. 1984a. Change in overwintering mechanism of the cucujid beetle, *Cucujus clavipes*. *J. Insect Physiol. 30*, 235–239.

Duman, J. G. 1984b. Thermal hysteresis antifreeze proteins in the midgut fluid of overwintering larvae of the beetle *Dendroides canadensis*. *J. Exp. Zool. 230*, 355–361.

Duman, J., and Horwath, K. 1983. The role of hemolymph proteins in the cold tolerance of insects. *Annu. Rev. Physiol. 45*, 261–270.

Duman, J. G., Neven, L. G., Beals, J. M., Olson, K. R., and Castellino, F. J. 1985. Freeze-tolerance adaptations, including haemolymph protein and lipoprotein nucleators, in the larvae of the cranefly *Tipula trivittata*. *J. Insect Physiol. 31*, 1–8.

Dutky, S. R. 1959. Insect microbiology. *Adv. Appl. Microbiol. 1*, 175–200.

Ebeling, W. 1961. Physicochemical mechanisms for the removal of insect wax by means of finely divided powders. *Hilgardia 30*, 531–564.

Fitz-Earle, M. 1976. Insect population control using genetic engineering. *Bull. Entomol. Soc. Am. 22*, 11–14.

Fraenkel, G., and Blewett, M. 1946. Linoleic acid, vitamin E and other fat-soluble substances in the nutrition of certain insects, *Ephestia kuehniella*, *E. elutella*, *E. cautella*, and *Plodia interpunctella* (Lep.). *J. Exp. Biol. 22*, 172–190.

Galani, G. 1979. Studies on the variation of the pathogenicity of *Verticillium lecanii* (Zimm.) Viégas to the larvae of *Trialeurodes vaporariorum* Westw. *An. ICPP, Bucarest 15*, 243–248.

Gateff, E. 1978a. Malignant neoplasms of genetic origin in *Drosophila melanogaster*. *Science 200*, 1448–1459.

Gateff, E. 1978b. Malignant and benign neoplasms of *Drosophila melanogaster*. *In* "The Genetics and Biology of *Drosophila*." (M. Ashburner and T. R. F. Wright, eds.), Vol. 2b, pp. 181–275. Academic Press, New York.

Goodwin, R. H. 1968. Nonsporeforming bacteria in the armyworm, *Pseudaletia unipuncta*, under gnotobiotic conditions. *J. Invertebr. Pathol. 11*, 358–370.

Gordon, H. T. 1959. Minimal nutritional requirements of the German roach, *Blattella germanica* L. *Ann. N.Y. Acad. Sci. 77*, 290–351.

Govindarajan, R., Jayaraj, S., and Narayanan, K. 1976. Mortality of the tobacco caterpillar, *Spodoptera litura* (F.), when treated with *Bacillus thuringiensis* combinations with boric acid and insecticides. *Phytoparasitica 4*, 193–196.

Grosch, D. S. 1974. Environmental aspects: Radiation. *In* "The Physiology of Insecta." (M. Rockstein, ed.), Second Ed., Vol. 2, pp. 85–126. Academic Press, New York.

Hajjar, N. P., and Casida, J. E. 1978. Insecticidal benzoylphenyl ureas: Structure–activity relationships as chitin synthesis inhibitors. *Science 200*, 1499–1500.

Hall, R. A. 1980. Effect of repeated subculturing on agar and passaging through an insect host on pathogenicity, morphology, and growth rate of *Verticillium lecanii*. *J. Invertebr. Pathol. 36*, 216–222.

Hara, S., Tanada, Y., and Omi, E. M. 1976. Isolation and characterization of a synergistic enzyme from the capsule of a granulosis virus of the armyworm, *Pseudaletia unipuncta*. *J. Invertebr. Pathol. 27*, 115–124.

Harker, J. E. 1963. Tumors. *In* "Insect Pathology: An Advanced Treatise." (E. A. Steinhaus, ed.), Vol. 1, pp. 191–213. Academic Press, New York.

Harlow, P. M. 1956. A study of ovarial development and its relation to adult nutrition in the blowfly *Protophormia terrae-novae* (R.D.). *J. Exp. Biol. 33*, 777–797.

Harvey, W. R., and Williams, C. M. 1961. The injury metabolism of the Cecropia silkworm—I. Biological amplification of the effects of localized injury. *J. Insect Physiol. 7*, 81–99.

Hedin, P. A. 1991. "Naturally Occurring Pest Bioregulators." Am. Chem. Soc., Washington, D.C.

Henry, J. E., Oma, E. A., and Onsager, J. A. 1979. Infection of the corn earworm, *Heliothis zea*, with *Nosema acridophagus* and *Nosema cuneatum* from grasshoppers: Relative virulence and production of spores. *J. Invertebr. Pathol. 34*, 125–132.

Hinton, H. E. 1977. Enabling mechanisms. Proc. Fifteenth Int. Cong. Entomol. Aug. 19–27, 1976. pp. 71–83, Washington, D.C.

Hirumi, H., Hirumi, K., and McIntosh, A. H. 1975. Morphogenesis of a nuclear polyhedrosis virus of the alfalfa looper in a continuous cabbage looper cell line. *Ann. N.Y. Acad. Sci. 266*, 302–326.

Horwath, K. L., and Duman, J. G. 1983. Photoperiodic and thermal regulation of antifreeze protein levels in the beetle *Dendroides canadensis*. *J. Insect Physiol. 29*, 907–917.

Horwath, K. L., and Duman, J. G. 1984a. Further studies on the involvement of the circadian system in photoperiodic control of antifreeze protein production in the beetle *Dendroides canadensis*. *J. Insect Physiol. 30*, 947–955.

Horwath, K. L., and Duman, J. G. 1984b. Yearly variations in the overwintering mechanisms of the cold-hardy beetle Dendroides canadensis. *Physiol. Zool. 57*, 40–45.

House, H. L. 1963. Nutritional diseases. *In* "Insect Pathology: An Advanced Treatise.) (E. A. Steinhaus, ed.), Vol. 1, pp. 133–160. Academic Press, New York.

House, H.L. 1974. Nutrition. *In* "The Physiology of Insecta." (M. Rockstein, ed.), Second Ed., Vol. 5, pp. 1–62. Academic Press, New York.

Hunter, D. K., Hoffmann, D. F., and Collier, S. J. 1973. Cross-infection of a nuclear polyhedrosis virus of the almond moth to the Indian meal moth. *J. Invertebr. Pathol. 22*, 186–192.

Husby, J. A., and Zachariassen, K. E. 1980. Antifreeze agents in the body fluid of winter active insects and spiders. *Experientia 36*, 963–964.

Keeley, L. L. 1972. Neuroendocrine regulation of insect metabolism and the influence of nutrition. *In* "Insect and Mite Nutrition." (J. G. Rodriguez, ed.), pp. 541–554. North Holland, Amsterdam.

Kellen, W. R., and Lindegren, J. E. 1971. Modes of transmission of *Nosema plodiae* Kellen and Lindegren, a pathogen of *Plodia interpunctella* (Hübner). *J. Stored Prod. Res. 7*, 31–34.

Knipling, E. F. 1955. Possibilities of insect control or eradication through the use of sexually sterile males. *J. Econ. Entomol. 48*, 459–462.

Knipling, E. F. 1967. Sterile technique—principles involved, current application, limitations, and future applications. *In* "Genetics of Insect Vectors of Disease." (J. W. Wright and R. Pal, eds.), pp. 587–616. Elsevier, Amsterdam.

LaChance, L. E., Schmidt, C. H., and Bushland, R. C. 1967. Radiation-induced sterilization. *In* "Pest Control. Biological, Physical, and Selected Chemical Methods." (W. W. Kilgore and R. L. Doutt, eds.), pp. 147–196. Academic Press, New York.

Leighton, T., Marks, E., and Leighton, F. 1981. Pesticides: Insecticides and fungicides are chitin synthesis inhibitors. *Science 213*, 905–907.

Lindquist, A. W. 1955. The use of gamma radiation for control or eradication of the screwworm. *J. Econ. Entomol. 48*, 467–469.

MacKinnon, E. A., Henderson, J. F., Stoltz, D. B., and Faulkner, P. 1974. Morphogenesis of nuclear polyhedrosis virus under conditions of prolonged passage *in vitro*. *J. Ultrastruct. Res. 49*, 419–435.

MacLeod, D. M. 1954. Investigations on the genera *Beauveria* Vuill. and *Tritirachium* Limber. *Can. J. Bot. 32*, 818–890.

Male, K. B., and Storey, K. B. 1982. Purification and properties of glutamate dehydrogenase from the cold–hardy gall fly larva, *Eurosta solidaginis*. *Insect Biochem. 12*, 507–514.

Matsumura, F. 1985. "Toxicology of Insecticides." Second Ed. Plenum Press, New York.

Nelson, S. O. 1967. Electromagnetic energy. *In* "Pest Control. Biological, Physical, and Selected Chemical Methods." (W. W. Kilgore and R. L. Doutt, eds.), pp. 89–145. Academic Press, New York.

Pal, R., and LaChance, L. E. 1974. The operational feasibility of genetic methods for control of insects of medical and veterinary importance. *Annu. Rev. Entomol. 19*, 269–291.

Poinar, G. O., Jr., and Thomas, G. M. 1966. Significance of *Achromobacter nematophilus* Poinar and Thomas (Achromobacteraceae: Eubacteriales) in the development of the nematode, DD-136 (*Neoaplectana* sp. Steinernematidae). *Parasitology 56*, 385–390.

Rasso, S. C., and Fraenkel, G. 1954. The food requirements of the adult female blow-fly, *Phormia regina* (Meigen), in relation to ovarian development. *Ann. Entomol. Soc. Am. 47*, 636–645.

Reichelderfer, C. F., and Benton, C. V. 1973. The effect of 3-methylcholanthrene treatment on the virulence of a nuclear polyhedrosis virus of *Spodoptera frugiperda*. *J. Invertebr. Pathol. 22*, 38–41.

Ring, R. A., and Tesar, D. 1981. Adaptations to cold in Canadian artic insects. *Cryobiology 18*, 199–211.

Riordan, D. F. 1957. Effects of a high temperature on the fertility of *Dahlbominus fuscipennis* (Zett.) (Hymenoptera: Chalcidoidea). *Can. J. Zool. 35*, 603–608.

Ritter, K. S., and Tanada, Y. 1978. Interference between two nuclear polyhedrosis viruses of the armyworm, *Pseudaletia unipuncta* [Lep.: Noctuidae]. *Entomophaga 23*, 349–359.

Rock, G. C. 1972. Optimal proportions of dietary amino acids. *In* "Insect and Mite Nutrition." (J. G. Rodriguez, ed.), pp. 183–197. North Holland, Amsterdam.

Sang, J. H., and King, R. C. 1961. Nutritional requirements of axenically cultured *Drosophila melanogaster* adults. *J. Exp. Biol. 38*, 793–809.

Seecof, R. 1968. The sigma virus infection of Drosophila melanogaster. *Curr. Top. Microbiol. Immunol. 42*, 59–93.

Shapiro, M., and Bell, R. A. 1982. Enhanced effectiveness of *Lymantria dispar* (Lepidoptera: Lymantriidae) nucleopolyhedrosis virus formulated with boric acid. *An. Entomol. Soc. Am. 75*, 346–349.

Smirnoff, W. A. 1963. Adaptation of a nuclear-polyhedrosis virus of *Trichiocampus viminalis* (Fallén) to larvae of *Trichiocampus irregularis* (Dyar). *J. Insect Pathol. 5*, 104–110.

Sømme, L. 1982. Supercooling and winter survival in terrestrial arthropods. *Comp. Biochem. Physiol. 73A*, 519–543.
Sparrow, J. C. 1978. Melanotic "tumours". *In* "The Genetics and Biology of *Drosophila*." (M. Ashburner and T. R. F. Wright, eds.), Vol. 2b, pp. 277–313. Academic Press, New York.
Steinhaus, E. A. 1949. "Principles of Insect Pathology." McGraw-Hill, New York.
Steinhaus, E. A. 1959. *Serratia marcescens* Bizio as an insect pathogen. *Hilgardia 28*, 351–380.
Steinhaus, E. A. 1960. The importance of environmental factors in the insect microbe ecosystem. *Bacteriol. Rev. 24*, 365–373.
Steinhaus, E. A. 1963. Background for the diagnosis of insect diseases. *In* "Insect Pathology: An Advanced Treatise." (E. A. Steinhaus, ed.), Vol. 2, pp. 549–589. Academic Press, New York.
Steinhaus, E. A. 1964. Diagnosis: A central pillar of insect pathology. Entomophaga; Colloq. Int. Pathol. Insectes, Memoire No. 2. pp. 7–21. Paris, 1962.
Stephens, J. M. 1959. Mucin as an agent promoting infection by Pseudomonas aeruginosa (Schroeter) Migula in grasshoppers. *Can. J. Microbiol. 5*, 73–77.
Storey, K. B., and Storey, J. M. 1983a. Biochemistry of freeze tolerance in terrestrial insects. *Trends Biochem. Sci. 8*, 242–245.
Storey, J. M., and Storey, K. B. 1983b. Regulation of cryoprotectant metabolism in the overwintering gall fly larva, *Eurosta solidaginis*: Temperature control of glycerol and sorbitol levels. *J. Comp. Physiol. B149*, 495–502.
Tanada, Y. 1959. Synergism between two viruses of the armyworm, *Pseudaletia unipuncta* (Haworth) (Lepidoptera, Noctuidae). *J. Insect Pathol. 1*, 215–231.
Tanada, Y. 1971. Interactions of insect viruses, with special emphasis on interference. *In* "The Cytoplasmic-polyhedrosis Virus of the Silkworm." (H. Aruga and Y. Tanada, eds.), pp. 185–200. University of Tokyo Press, Tokyo.
Tanada, Y., and Fuxa, J. R. 1987. The pathogen population. *In* "Epizootiology of Insect Diseases." (J. R. Fuxa and Y. Tanada, eds.), pp. 113–157. John Wiley and Sons, New York.
Tanada, Y., and Hukuhara, T. 1971. Enhanced infection of a nuclear-polyhedrosis virus in larvae of the armyworm, *Pseudaletia unipuncta*, by a factor in the capsule of a granulosis virus. *J. Invertebr. Pathol. 17*, 116–126.
Tanada, Y., Hukuhara, T., and Chang, G. Y. 1969. A strain of nuclear-polyhedrosis virus causing extensive cellular hypertrophy. *J. Invertebr. Pathol. 13*, 394–409.
Thomson, H. M. 1958. Some aspects of the epidemiology of a microsporidian parasite of the spruce budworm, *Choristoneura fumiferana* (Clem.). *Can. J. Zool. 36*, 309–316.
Tompkins, G. J., Vaughn, J. L., Adams, J. R., and Reichelderfer, C. F. 1981. Effects of propagating *Autographa californica* nuclear polyhedrosis virus and its *Trichoplusia ni* variant in different hosts. *Environ. Entomol. 10*, 801–806.
Vago, C. 1963. Predispositions and interrelations in insect diseases. *In* "Insect Pathology: An Advanced Treatise." (E. A. Steinhaus, ed.), Vol. 1, pp. 339–379. Academic Press, New York.
Vanderzant, E. S. 1974. Development, significance, and application of artificial diets for insects. *Annu. Rev. Entomol. 19*, 139–160.
Vanderzant, E. S., Kerur, D., and Reiser, R. 1957. The role of dietary fatty acids in the development of the pink bollworm. *J. Econ. Entomol. 50,* 606–608.
Watanabe, H., Aratake, Y., and Kayamura, T. 1975. Serial passage of a nuclear polyhedrosis virus of the silkworm, *Bombyx mori*, in larvae of rice stem borer, *Chilo suppressalis*. *J. Invertebr. Pathol. 25*, 11–17.
Weiser, J. 1961. Die Mikrosporidien als Parasiten der Insekten. *Monogr. Angew. Entomol.* Vol. 17. Paul Parey, Hamburg.
Wharton, G. W., and Arlian, L. G. 1972. Utilization of water by terrestrial mites and insects.

In "Insect and Mite Nutrition." (J. G. Rodriguez, ed.), pp. 153–165. North-Holland, Amsterdam.

Whitten, M. J., and Foster, G. G. 1975. Genetical methods of pest control. *Annu. Rev. Entomol. 20*, 461–476.

Williams, C. M. 1967. Third-generation pesticides. *Sci. Am. 217*, 13–17.

Williams, C. M. 1970. Hormonal interactions between plants and insects. *In* "Chemical Ecology." (E. Sondheimer and J. B. Simeone, eds.), pp. 103–132. Academic Press, New York.

Wood, H. A., Hughes, P. R., Johnston, L. B., and Langridge, W. H. R. 1981. Increased virulence of *Autographa californica* nuclear polyhedrosis virus by mutagenesis. *J. Invertebr. Pathol. 38*, 236–241.

Yamada, K., Sherman, K. E., and Maramorosch, K. 1982. Serial passage of *Heliothis zea* singly embedded nuclear polyhedrosis virus in a homologous cell line. *J. Invertebr. Pathol. 39*, 185–191.

Yamamoto, I. 1970. Mode of action of pyrethroids, nicotinoids, and rotenoids. *Annu. Rev. Entomol. 15*, 257–272.

Yamamoto, T., and Tanada, Y. 1978. Phospholipid, an enhancing component in the synergistic factor of a granulosis virus of the armyworm, *Pseudaletia unipuncta*. *J. Invertebr. Pathol. 31*, 48–56.

Yeaton, R. W. 1983. Wound responses in insects. *Am. Zool. 23*, 195–203.

Zachariassen, K. E. 1982. Nucleating agents in cold-hardy insects. *Comp. Biochem. Physiol. 73A*, 557–562.

Zachariassen, K. E., and Husby, J. A. 1982. Antifreeze effect of thermal hysteresis agents protects highly supercooled insects. *Nature 298*, 865–867.

CHAPTER 4

BACTERIAL INFECTIONS: BACILLACEAE

I. SYMPTOMS AND PATHOLOGY
II. PORTALS OF ENTRY
III. TYPES OF ENTOMOPATHOGENIC BACTERIA
IV. FAMILY BACILLACEAE
    A. *Bacillus cereus*
    B. *Bacillus thuringiensis*
    C. *Bacillus sphaericus*
    D. Milky Diseases of Scarabaeidae
    E. *Bacillus moritai*
    F. *Bacillus larvae*
    G. Other Entomopathogenic Bacilli
    H. Clostridial Pathogens

Bacteria are unicellular (also designated as acellular) organisms, small in size (less than 1 μm to several μm in length), and lack a defined nucleus (procaryotes). Those with rigid cell walls may be spherical (cocci), rod-shaped (bacilli), or spiral (spirilla, spirochaetes), whereas those without cell walls (mollicutes) are pleomorphic. Bacteria may occur in regular or irregular aggregations, may develop chains or packets of individual cells, and may be motile. They reproduce by binary fission and, in some situations, may reproduce through sexual reproduction (i.e., conjugation). Some species develop in the presence of oxygen (aerobes) and others in its absence (anaerobes). Some species produce resistant endospores. Many have plasmids that introduce genes into the bacteria. Bacteria are the most common microorganisms associated with insects. Many cause no pathologies to insects, but a significant number are important pathogens.

The taxonomy of bacteria has undergone considerable revision during the past decade. The number of microorganisms that are considered to be related to bacteria has increased greatly. There are proposals to place in higher taxa such forms as mycoplasma, spiroplasma, acholeplasma, etc.

At the species level, the problem in bacterial taxonomy lies in the definition of a species (Gordon 1978). We have selected the eighth edition of *Bergey's Manual of Determinative Bacteriology* (Buchanan and Gibbons 1974; Holt 1977) and the *Bergey's Manual of Systematic Bacteriology* (Holt 1984, 1986, 1989a, 1989b) as the basis for the systematics and nomenclature of bacteria used in the text.

I. SYMPTOMS AND PATHOLOGY

Bacterial infections in insects can be broadly classified as bacteremia, septicemia, and toxemia. Bacteremia occurs when the bacteria multiply in the insect's hemolymph without the production of toxins. This situation occurs in the case of bacterial symbionts and rarely occurs with bacterial pathogens. Septicemia occurs most frequently with pathogenic bacteria, which invade the hemocoel, multiply, produce toxins, and kill the insect. Toxemia occurs when the bacteria produce toxins and the bacteria are usually confined to the gut lumen, as in the case of brachytosis of the tent caterpillar.

Since the syndromes produced by different bacteria vary, we shall describe first a generalized type of infection. Pathogenic bacteria, upon ingestion by a susceptible insect, multiply and produce toxins in the midgut lumen. The insect loses its appetite, becomes diarrheic, discharges watery feces, and may vomit. The invasion of the bacteria into the hemocoel results in septicemia and death of the insect. The bacteria, in general, are extracellular pathogens except for the pathogenic rickettsia and mollicutes.

Insects killed by bacteria, especially in the larval stages, rapidly darken in color and are often very soft. The internal tissues and organs are rapidly broken down to a viscid consistency, accompanied sometimes by a putrid odor. The integument remains intact. There is an abundance of bacteria in an insect shortly after death. The cadaver shrivels, dries, and hardens.

II. PORTALS OF ENTRY

Bacteria infect insects mostly through the mouth and digestive tract, and less commonly through the egg, integument, and trachea. They may also enter an insect by means of parasitoids and predators. Within the digestive tract, the bacteria produce enzymes (e.g., lecithinase, proteinase, and chitinase) that act on the midgut cell and enable the bacteria to enter the hemocoel. The mechanism by which this takes place, however, has not been explained in most cases. Lecithinase is produced by many bacteria including some varieties of *Bacillus cereus* and *Bacillus thuringiensis*. There are subspecies of *B. thuringiensis* capable of producing chitinase, but there is little or no information about the role of this enzyme on the pathogenicity of these forms (de Barjac and Cosmao Dumanoir 1975; Chigaleichik 1976; Smirnoff and Valero 1977). Daoust and Gunner (1979) have isolated a chitinolytic bacterium from healthy gypsy moth larvae. The chitinase does not act in the highly alkaline digestive tract of the larva since it has an optimum pH activity of 5.5. When the chitinolytic bacterium, however, is combined with another nonpathogenic bacterium that is an acid producer in healthy larvae, there is an enhanced increase in mortality.

Chitinase and phospholipase A in the bacterium *Aeromonas punctata* may facilitate in producing wounds that form black lesions in the cuticle and in the digestive tract of a larva of the mosquito *Anopheles annulipes* (Kalucy and Daniel 1972). The bacterium invades into the hemocoel through these wounds.

Bacterial exotoxins and endotoxins play a role in the invasion of bacteria through the digestive tract. This has been studied most thoroughly with *Bacillus thuringiensis* and will be discussed later. In general, toxins are produced in the initial stage of an infection. They may damage the gut wall and enable the bacteria to enter the hemocoel and may act on tissues in the hemocoel.

In some cases, the bacteria multiply in the gut to large numbers before invading into the hemocoel. This occurs when a larva is placed under stress conditions, such as abnormal nutrition, unfavorable temperature and humidity, or other microbial infections.

Only a few cases are known of direct bacterial invasion through the trachea and integument. Northrup (1914) has reported that *Micrococcus nigrofaciens* invades the June beetles (*Lachnosterna* spp.) through the joints, spiracles, and white soft portions of the integument, which turn black and oily. Most bacterial invasions, however, occur at sites of wounds and injuries. In the bark beetle *Scolytis multistriatus*, several species of bacteria are transmitted through a larva biting other larvae under crowded conditions (Doane 1960). Most of these bacteria are poor invaders through the digestive tract.

Transmission through the egg occurs when the bacteria are found on the surface or within an egg. The common non-spore-forming bacterium *Serratia marcescens* is transmitted within the egg of the brown locust *Locustana pardalina* (Prinsloo 1960).

Some bacteria are transmitted to insects by parasitoids and predators. Sometimes, the adult parasitoids puncture the digestive tract and thereby enable the bacteria in the lumen to enter the hemocoel. An example of parasitoid transmission is the ichneumon *Itoplectis conquisitor*, which can inoculate with its ovipositor *Serratia marcescens* and *Proteus mirabilis* from infected to healthy pupae of *Galleria mellonella* (Bucher 1963). Moreover, a wasp infected by the bacteria may transmit the bacteria through its ovipositor, which is contaminated by its contagious feces. The wasp may retain the infection for life, and when infected, both its host and its own progeny are doomed.

III. TYPES OF ENTOMOPATHOGENIC BACTERIA

To classify bacteria on the basis of pathogenicity is difficult. Some bacteria are obligate pathogens, but the majority are facultative pathogens and a few are potential pathogens which may show a certain degree of pathogenicity. Moreover, some very virulent bacterial species (e.g., *B. thuringiensis*) may have strains of little or no pathogenicity and others of low virulence may have strains of high pathogenicity. Lysenko (1983) examined 300 samples of 46 species of diseased insects and detected over 200 strains of bacteria. Most of the bacteria were facultative pathogens whose pathogenicity depended on stress conditions affecting the host insect. These were mainly the bacterial genera *Enterobacter*, *Serratia*, *Pseudomonas*, and *Proteus*. Under stress conditions, normally nonpathogenic bacteria in the digestive tract (e.g., *Streptococcus faecalis* and other enterococci) may exhibit pathogenicity.

Most of the insect pathogenic bacteria occur in the families Bacillaceae, Pseudomonadaceae, Enterobacteriaceae, Streptococcaceae, and Micrococcaceae; in the order Rickettsiales and class Mollicutes. Members of Bacillaceae, particularly *B. thuringiensis* and *B. popilliae*, have received considerable attention as microbial control agents. Recently, the Mollicutes are under wide investigation because some of them (e.g., the mycoplasmas and spiroplasmas) have members that infect insects, plants, and vertebrates, including humans.

IV. FAMILY BACILLACEAE

Members of the family Bacillaceae produce endospores and are gram-positive, motile or nonmotile rods. The ancestral form is "clostrida" (Fox et al. 1980). The insect pathogens occur in the genera *Bacillus* and *Clostridium*. *Bacillus* is aerobic or a facultative aerobe and usually produces catalase (Claus and Berkeley 1986), whereas *Clostridium* is anaerobic and does not reduce sulfate to sulfite (Cato et al. 1986). Both genera form rod-shaped cells, which sometimes occur in chains. The two genera are separated mainly on oxygen requirements; but there are some members of *Bacillus* with limited oxygen requirements (e.g., *B. popilliae* has been considered by some bacteriologists as being more of a *Clostridium* than a *Bacillus*).

The vegetative cell, when it begins to produce a spore, is called a sporangium. During sporulation, some members of *Bacillus* produce one or more inclusions or parasporal bodies in a sporangium. The genus *Bacillus* is the most promising of the bacteria for insect control.

A. *BACILLUS CEREUS*

Bacillus cereus is widely distributed and commonly found in soil. It is considered to be closely related to *B. thuringiensis*, an insect pathogen, and to *B. anthracis*, a vertebrate pathogen. The relationship is considered to be sufficiently close by some authorities as to place these pathogens as varieties of *B. cereus* (Gordon et al. 1973; Logan and Berkeley 1984).

Inasmuch as the question of speciation is still unsettled, we shall consider the above three bacteria as distinct species as given in Volume 2 of Bergey's Manual (Claus and Berkeley 1986). *Bacillus cereus* differs from *B. thuringiensis* mainly in the absence of a protoxin parasporal body and is therefore acrystalliferous. However, there are acrystalliferous mutants of *B. thuringiensis*. Most varieties or subspecies of *B. thuringiensis* are specific insect pathogens, whereas many of those of *B. cereus* are not insect pathogens and some produce pathologies in humans and domesticated animals (Turnbull 1981). Furthermore, the numerical classification reveals *B. thuringiensis* strains in separate clusters from those of *B. cereus* (Priest et al. 1988).

Bacillus cereus has been isolated from over 30 insect species, mainly in the orders Coleoptera, Hymenoptera, and Lepidoptera. Heimpel (1955) has obtained significant correlation between the pathogenicity of various strains of *B. cereus* for the larch sawfly, *Pristiphora erichsonii*, and their ability to produce the toxin lecithinase (phospholipase C). In addition, the susceptibility of various insect species is related to their midgut pH. Since the optimum pH range for the phospholipase of *B. cereus* is from 6.6 to 7.4, the midguts of susceptible insects have a low pH. High midgut pH also inhibits spore germination and vegetative reproduction during which the lecithinase is produced. *Bacillus cereus* produces the antibiotic cerecine (Goze 1972) and a number of exotoxins known to affect vertebrates (Turnbull 1981; Kramer 1984).

B. *BACILLUS THURINGIENSIS*

At present, the most widely used bacterium in microbial control is *Bacillus thuringiensis* (Fig. 4–1). It is being applied in many countries throughout the world.

FIGURE 4–1

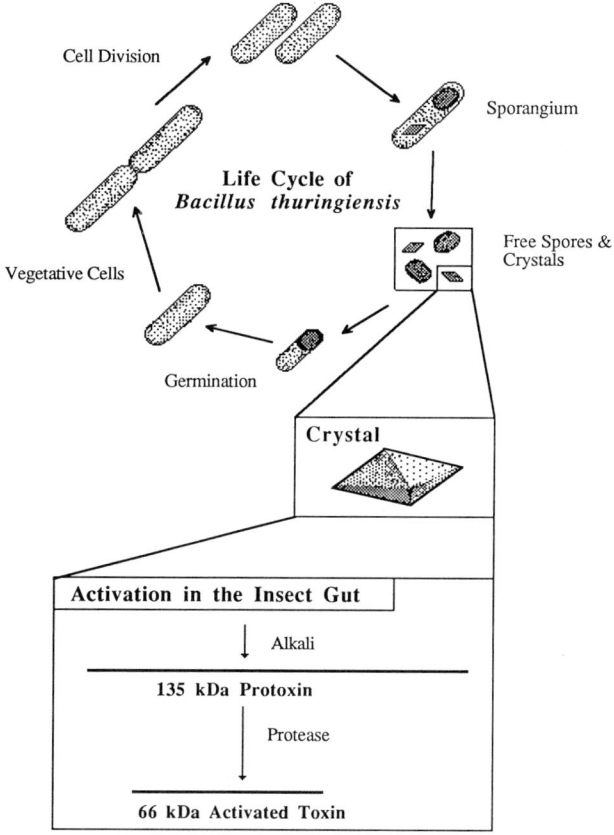

Life cycle of *Bacillus thuringiensis*, showing the production of bipyramidal crystal and the endotoxin. (Prepared by T. Yamamoto.)

The bacillus is not as common in the soil as *B. cereus*, and may represent about 35% of the *B. cereus–B. thuringiensis* group (Ohba and Aizawa 1986a). Nonetheless, Martin and Travers (1989) have obtained over 1000 isolates of *B. thuringiensis* from soils of five continents: Africa, Asia, Europe, and North and South America. Over 60% of the isolates are toxic for lepidopterons or dipterons.

According to Steinhaus (1960), Louis Pasteur between 1865 and 1870 in his study on the silkworm flacherie may have dealt with two species of *Bacillus*, one of which was *B. thuringiensis*. Pasteur found *B. thuringiensis* in dusts from silkworm-rearing menageries. Ohba et al. (1984) also have isolated numerous subspecies of *B. thuringiensis* from the litter of sericultural farms in Japan.

In 1901, Ishiwata isolated a bacillus from diseased silkworm larvae. He found the bacillus so highly pathogenic to larval silkworms that he named it "sotto disease bacillus" (sudden-collapse bacillus). He attributed the cause of the disease to a toxin that was located either within the spore or in close contact with the spore (Ishiwata 1905, 1906). In 1908, Iwabuchi called the bacterium *Bacillus sotto* Ishiwata. This name, however, cannot be used since Ishiwata did not apply a binomial name (Steinhaus 1961). Aoki and Chigasaki (1915) confirmed Ishiwata's conjecture that the pathogenicity of the "sotto" strain was due to a preformed toxin present in sporulating cultures. Mitani and Watarai (1916) obtained an active toxic filtrate from the sotto strain.

In 1911, Berliner isolated from the Mediterranean flour moth, *Anagasta kuehniella*, a bacillus, which he named in 1915 as *Bacillus thuringiensis* after the province Thuringia in Germany from which the infested flour was obtained. The bacterium was reisolated by Mattes in 1927. Both Berliner and Mattes reported the high pathogenicity of the bacillus for the flour moth larvae. They both noted the parasporal bodies in the sporangia but did not consider them as being a source of toxins.

None of the above early workers attempted field tests with *B. thuringiensis*. The first attempt was made by Husz (1929) under an international program developed to control the European corn borer, *Ostrinia nubilalis*. He obtained promising results with the culture that he had received from Mattes. Others followed with field tests and also reported favorable results against the corn borer and other insect pests. For some unknown reason, interest in *B. thuringiensis* diminished for over a decade.

In 1942, Steinhaus (1975) obtained a culture of the bacillus from Nathan Smith, who had received it from J. R. Porter, who in turn had gotten it from O. Mattes in 1936. Steinhaus attracted the attention of insect pathologists and industrial organizations to *B. thuringiensis* through his basic and applied studies that demonstrated the potential of this bacterium in insect pest control.

Hannay (1953) described the properties of the parasporal body of *B. thuringiensis*, such as its structure, staining properties, and solubility in dilute alkali but not in organic solvents. In 1955, Hannay and Fitz-James reported on the proteinaceous nature of the crystal that had at least 17 amino acids. Angus (1954, 1956a,b) demonstrated that the parasporal body was the source of the toxin for the silkworm and his results confirmed the observations of the Japanese workers.

The readers are referred to the comprehensive treatments of *B. thuringiensis*, in particular the δ-endotoxin, in the reviews of Carlton and González (1985), Aronson et al. (1986), Whiteley and Schnepf (1986), Andrews et al. (1987), and Höfte and Whiteley (1989).

1. Subspecies

During the past decade, a large number of subspecies of *Bacillus thuringiensis* have been isolated from insects. Workers have designated the subspecies also as strains, varieties, serotypes, serovars, biovars, pathovars, or crystovars. The subspecies have been differentiated by various methods, such as biochemical tests (Heimpel and Angus 1958), H serotypes (flagella antigen) (de Barjac and Bonnefoi 1962; de Barjac and Frachon 1990), parasporal body (crystal) antigen (Pendleton and Morrison 1967; Krywienczyk et al. 1978; Lynch and Baumann 1985), esterase production (Norris and Burges 1963), antibiotic production (Toumanoff and Lapied 1954; Pendleton 1969), enzymes (de Barjac and Bonnefoi 1973), phages (Rautenstein et al. 1975; Vasantharajan and Munirathnamma 1980; Jones et al. 1983), and lectin grouping (DeLucca 1984).

There is considerable agreement in classifying the subspecies by biochemical tests, esterase production, and flagella antigen (H serotype) (de Barjac and Bonnefoi 1962; de Barjac and Frachon 1990). At present, the use of H serotypes, supplemented in some cases by specific biochemical tests, is universally accepted for designating subspecies (Table 4–1).

Another serological method uses the heat-stable somatic (O) antigens of the vegetative cells of *B. thuringiensis* to differentiate 17 subspecies into 13 groups (Sekijima and Ono 1982). Padua et al. (1980) have applied the O-antigenic test to three isolates of *B. thuringiensis darmstadiensis* (serotype 10). The isolates differ not only in O antigenicity but also in their pathogenicity to insects (e.g., lepidopterons and mosquitoes). Their results suggest that O antigen may be useful in supplementing the H antigen. Crystal antigens also have been used to differentiate subspecies of *B. thuringiensis* (Smith 1987). It is useful in separating the isolates on the basis of toxicity towards insects in different orders.

a. Pathotypes

Early in the discovery of the various subspecies of *B. thuringiensis*, it became apparent that the subspecies differed in their insect host spectrum (Table 4–2). Krieg et al. (1983) have proposed separating the subspecies as pathotypes. Pathotype A includes subspecies that are pathogenic for Lepidoptera; pathotype B for those pathogenic for Diptera; and pathotype C for Coleoptera. The problems associated with the use of pathotypes have been discussed by de Barjac and Frachon (1990). The pathotype fails to separate subspecies that are pathogenic to two or more orders of insects [e.g., *kurstaki* produces two types of toxins, one for Lepidoptera and another for Diptera (Yamamoto and McLaughlin 1981; Yamamoto and Iizuka 1983)]. Even isolates in the same serotype (e.g., as noted above in *darmstadiensis* serotype 10) differ in their pathognicity for insects. The subspecies *fukuokaensis* (serotype 3a3d3e) has two pathotypes, one pathogenic for mosquitoes and the other for lepidopterons (Ohba and Aizawa 1990). The two subspecies *israelensis* (serotype 14) (Goldberg and Margalit 1977) and *kyushuensis* (serotype 11a11c) (Ohba and Aizawa 1979) do not affect lepidopterons but are pathogenic for a wide range of aquatic dipterons (Guillet and de Barjac 1979; Garcia et al. 1980).

The serotype 8a8b has three pathotypes: pathotype A has subspecies *morrisoni* and strain G2, which are pathogenic only for lepidopterons; pathotype B has strain PG-14, which is pathogenic for aquatic dipterons; and pathotype C, with strains of subspecies *tenebrionis*, is pathogenic for coleopterons (Krieg et al. 1983, 1987; Ferro and Gelernter 1989). The subspecies *tenebrionis* (Krieg et al. 1983) and *san diego* (Herrnstadt et al. 1986) are considered to be the same and as a pathotype of *morrisoni* (de Barjac and Frachon 1990). The subspecies *tenebrionis* is toxic also to the predatory phytoseid mite, *Metaseiulus occidentalis* (Chapman and Hoy 1991).

b. Subspecies *Israelensis*

The subspecies *israelensis* has aroused considerable interest because of its promising potential for the control of a broad spectrum of aquatic Diptera that are vectors of vertebrate pathogens and for its pathogenicity to vertebrates (de Barjac and Sutherland 1990) (Fig. 4–2). The first isolate was collected from the Negrev Desert of Israel (Goldberg and Margalit 1977). This subspecies is pathogenic for members of Culicidae (mosquitoes), Simuliidae (black flies), Dixidae, Ceratopogonidae, Chironomidae (midges) (Garcia et al. 1980), Sciaridae (Cantwell and Cantelo 1984), *Haematobia* (horn flies) (Temeyer 1984), and Psychodidae (*Phlebotomus* flies) (de Barjac et al. 1981). Isolates of subspecies *israelensis* differ in insecticidal

TABLE 4–1 Classification of *Bacillus thuringiensis* Strains According to the H Serotype[a]

H Serotype	Serovar	First mention[b] and first valid description
1	thuringiensis	Berliner 1915; Heimpel, Angus 1958
2	finitimus	Heimpel, Angus 1958
3a3c	alesti	Toumanoff, Vago 1951; Heimpel, Angus 1958
3a3b3c	kurstaki	de Barjac, Lemille 1970
3a3d	sumiyoshiensis	Ohba, Aizawa 1989[c]
3a3d3e	fukuokaensis	Ohba, Aizawa 1989[c]
4a4b	sotto	Ishiwata 1905; Heimpel, Angus 1958
4a4c	kenyae	Bonnefoi, de Barjac 1963
5a5b	galleriae	Shvetsova 1959; de Barjac, Bonnefoi 1962
5a5c	canadensis	de Barjac, Bonnefoi 1972
6	entomocidus	Heimpel, Angus 1958
7	aizawai	Bonnefoi, de Barjac 1963
8a8b	morrisoni	Bonnefoi, de Barjac 1963
8a8c	ostriniae	Gaixin, Ketian, Minghua, Xingmin 1975
8b8d	nigeriensis	Rajagopalan et al. (not published)
9	tolworthi	Norris 1964; de Barjac, Bonnefoi 1968
10	darmstadiensis	Krieg, de Barjac, Bonnefoi 1968
11a11b	toumanoffi	Krieg 1969
11a11c	kyushuensis	Ohba, Aizawa 1979
12	thompsoni	de Barjac, Thompson 1970
13	pakistani	de Barjac, Cosmao Dumanoir, Shaik, Viviani 1977
14	israelensis	de Barjac 1978
15	dakota	DeLucca, Simonson, Larson 1979
16	indiana	DeLucca, Simonson, Larson 1979
17	tohokuensis	Ohba, Aizawa, Shimizu 1981
18	kumamotoensis	Ohba, Ono, Aizawa, Iwanami 1981
19	tochigiensis	Ohba, Ono, Aizawa, Iwanami 1981
20a20b	yunnanensis	Wan-yu, Qi-fang, Xue-ping, You-wei 1979
20a20c	pondicheriensis	Rajagopalan et al. (not published)
21	colmeri	DeLucca, Palmgren, de Barjac 1984
22	shandongiensis	Ying, Jie, Xichang 1986
23	japonensis	Ohba, Aizawa 1986
24	neoleonensis	Rodriguez-Padilla, Galan-Wong, de Barjac, Dulmage, Tamez-Guerra, Roman Calderon 1988

(continued)

TABLE 4-1 (*Continued*)

H Serotype	Serovar	First mention[b] and first valid description
25	*coreanensis*	Lee et al. (not published)
26	*silo*	de Barjac, Lecadet (not published)
27	*mexicanensis*	Rodriguez-Padilla, Galan-Wong 1988
28	*monterrey*	(not published)
29	*amagiensis*	(not published)
30	*medellin*	(not published)
31	*toguchini*	(not published)
32	*cameroun*	(not published)
33	*leesis*	Lee et al. (not published)
34	*konkukian*	Lee et al. (not published)

[a]Courtesy of H. de Barjac.
[b]See de Barjac 1990 for cited references.
[c]Ohba, M., and Aizawa, K. 1989. *J. Invertebr. Pathol. 54,* 208–212.
N.B. Other described biovars or pathovars : *dendrolimus* H4a4b 1956–1963, *subtoxicus* H6 1958–1963, *tenebrionis* H8a8b 1983, *wuhanensis* nonmotile 1976.

activity. Some are much more active than the original isolate (Brownbridge and Margalit 1986) and others, lacking the 130-kDa protoxin, are noninsecticidal (Ohba et al. 1988).

c. Reference Standards

In order to obtain a uniform basis for comparing the pathogenicity of the various subspecies of *B. thuringiensis* and the susceptibility of various insect species, several subspecies have been selected as reference standards (e.g., *kurstaki* for lepidopterons, *israelensis* for aquatic dipterons, and *tenebrionis* for coleopterons) (see Chapter 15). The subspecies that have been used in commercial formulations throughout the world are *thuringiensis, kurstaki, dendrolimus, galleriae, israelensis,* and *aizawai*.

2. Toxins

Heimpel (1967), recognizing that an increasing number of toxins might be discovered from *B. thuringiensis*, proposed designating the toxins by the Greek alphabet. He listed four toxins: α-exotoxin, β-exotoxin, γ-exotoxin, and δ-endotoxin. Krieg and Lysenko (1979) modified the classification and added additional toxic candidates. Heimpel (1967) named the γ-exotoxin after an egg-yolk clearing factor detected in the bacillus. This factor appears not to be toxic to insects and γ-exotoxin should not be used for this factor (Krieg and Lysenko 1979).

a. α-Exotoxin and Phospholipase C

The α-exotoxin is a proteinaceous, thermolabile exotoxin produced by some varieties of *B. thuringiensis* and *B. cereus* (Krieg 1971a). It is highly toxic to certain

TABLE 4-2 Comparison of Spectra of Activities of Six Subspecies of *Bacillus thuringiensis*[a]

		Typical potency versus species				
Subspecies	Crystal type	*Trichoplusia ni*	*Heliothis virescens*	*Hyphantria cunea*	*Bombyx mori*	Mosquito
thuringiensis	thu	4,000	1,000	30,000	1,200	na[b]
alesti	ale	na	na	51,000	30,000	na
kurstaki	k-1	18,000	7,600	6,000–100,000	16,000–40,000	+
galleriae	g-9	5,000	600	48,000	7,000	±
aizawai	aiz	11,000	900	80,000	46,000	na
israelensis	—	na	na	na	na	+++

[a]Courtesy of Dulmage et al. 1990, © Rutgers University Press.
[b]Not active.

FIGURE 4-2

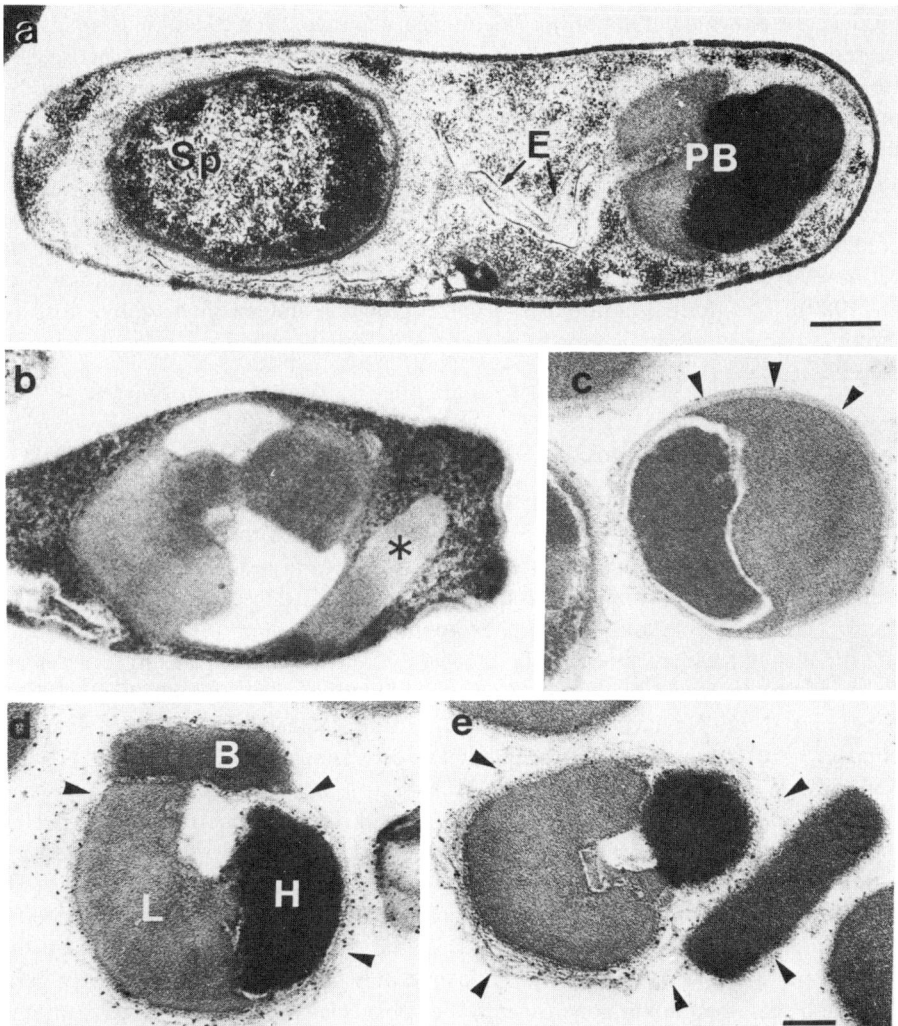

Spore and parasporal bodies of *Bacillus thuringiensis israelensis*. (a) Developing spore (*Sp*) and parasporal body (*PB*). (b) Completely formed parasporal body just prior to the lysis of the sporangium. Note the inclusions of different densities and the bar-shaped inclusion (∗) adjacent to the main body. (c) Parasporal body recently released from a sporangium, showing the multilayered envelope (*arrow heads*) tightly bound around the inclusions. (d) Parasporal body illustrating the three different inclusion types: (*L*) large inclusion of low electron density; (*B*) bar-shaped body; (*H*) inclusion of high electron density. (e) Parasporal body in which the envelope is partially disrupted. Arrowheads indicate envelope. (Bar in (a) = 250 nm; (b)–(e) are approximately the same magnification, with the bar in (e) = 250 nm.) [Courtesy of Federici et al. (1990), © Rutgers University Press.]

insects by oral and intrahemocoelic inoculations. In the hemocoel, it causes the degeneration and lysis of hemocytes (Krieg 1971b). It is toxic to mice and other vertebrates (Krieg 1971a; Fluer et al. 1981). It is destroyed by trypsin and urea and at pHs exceeding 10 and less than 3.5. These properties differ from those of phospholipase C.

Phospholipase C or lecithinase is a thermolabile enzyme toxic to insects (Heimpel 1967). Some subspecies of *B. thuringiensis* and varieties of *B. cereus* produce the enzyme but not others (lecithinase negative). Since lecithinase-negative subspecies produce α-exotoxin, lecithinase cannot be α-exotoxin as proposed by Heimpel (Krieg 1971a; Krieg and Lysenko 1979). Lecithinase is produced also by other bacteria, such as *Pseudomonas chlororaphis* and *Clostridium perfringes*, but the molecular weights of the enzyme vary with different bacterial species (Lysenko 1974).

The phospholipase C is believed to act by penetrating between the cell membranes and attacking the cell substrate (Heimpel 1967). *In vitro*, the enzyme inhibits cell growth and causes cellular deformation and internal deterioration (Ikezawa et al. 1989). The normal fibroblasts become globular and swollen with a loss of protoplasmic extensions. Lysosomes increase in numbers and vacuoles develop in the cells.

b. β-Exotoxin—Thuringiensin

McConnell and Richards (1959) first detected in cultures of *B. thuringiensis* a heat-stable (autoclavable for at least 15 min), dialyzable toxin that killed insects when inoculated into the hemocoel. Subsequently, Burgerjon and de Barjac (1960) detected in the bacillus a soluble heat-stable toxin that killed insects when it was fed to them. The heat-stable toxin is the β-exotoxin.

Various names have been applied to the β-exotoxin (thuringiensin) (e.g., heat-stable toxin, thermostable toxin, fly factor, fly toxin, etc.) (Bond et al. 1971). Šebesta et al. (1981) have adopted the name "thuringiensin" proposed by Kim and Huang (1970). This toxin has been reviewed comprehensively by Faust (1973), Vaňková (1978), and Šebesta et al. (1981).

The general consensus is that the β-exotoxin (thuringiensin) is formed during the vegetative growth phase of the bacteria and is secreted into the medium. Very little or none is produced during sporulation (Šebesta et al. 1973; Holmberg et al. 1980). The exotoxin production is not connected with that of the parasporal body. Some strains that do not produce a parasporal body are known to produce the β-exotoxin. A 62-MDa (megadalton) plasmid may be carrying the β-exotoxin gene (*exo*) and the δ-endotoxin gene (*cry*) in the subspecies *darmstadiensis* (Ozawa and Iwahana 1986).

Not all subspecies of *B. thuringiensis* produce thuringiensin. Its production is a strain-specific property rather than a serotype-specific property (Ohba et al. 1981). It is formed by isolates of serotypes 1 (*thuringiensis*), 4a4c (*kenyae*), 4a4b (*sotto*), 5a5b (*galleriae*), 7 (*aizawai*), 8a8b (*morrisoni*), 9 (*tolworthi*), 10 (*darmstadiensis*), 11a11b (*toumanoffi*), and 12 (*thompsoni*) (Bond et al. 1971; Šebesta et al. 1981). The possibility exists, however, that isolates of other serotypes may produce thuringiensin but at a low level (Vǎnková 1978).

Thuringiensin affects a broad host spectrum both by per os and parenteral inoculations of insects as well as other invertebrates and vertebrates. The susceptible insect species are found in the orders Diptera, Lepidoptera, Hymenoptera, Coleoptera, Isoptera, Orthoptera, Hemiptera, and Neuroptera. In Hemiptera, Herbert and Harper (1986) have found that the nymphs, but not the adults, of the predaceous lygaeid bug, *Geocoris punctipes*, are affected by the topical applications of thuringiensin. Bugs feeding on *Heliothis zea* larvae intoxicated with thuringiensin have been killed; surviving nymphs have produced adults with reduced longevity. Larvae of the lace wing, *Chrysopa carnea*, also are killed after consuming the eggs of *Sitotroga cerealella* treated with the exotoxin (Kiselev 1975).

The toxicity of thuringiensin is greater through hemocoelic inoculation than through the mouth (Schmid and Benz 1969). The exotoxin is deactivated in the digestive tract very likely through enzymatic activity (Schmid and Benz 1969; Vănková et al. 1974; Vănková and Horská 1975). Most insects, however, are susceptible if given adequately high doses of thuringiensin (Bond et al. 1971).

Thuringiensin acts as a feeding deterrent with the black cutworm (*Agrotis ipsilon*), fall armyworm (*Spodoptera frugiperda*), and the European corn borer (*Ostrinia nubilalis*) (Mohd-Salleh and Lewis 1982). This feeding inhibition is evident with high concentrations of the toxin, and even though only a small quantity is ingested, the larva dies after a period of 7 to 14 days. In mosquitoes, the β-exotoxin acts as a larvicide and adulticide when ingested (Larget-Thiery et al. 1984). Sublethal concentration of the exotoxin causes a delay in larval molting and induces teratological effects in larvae and pupae. After a sublethal application of the exotoxin at the larval stage, the next generation of mosquito is much more susceptible to this toxin. When applied to piggeries, compost toilets, and hen houses, the exotoxin has effectively controlled *Musca domestica* (house fly) and less effectively controlled *Fannia canicularis* and *Stomoxys calcitrans* (stable fly) (Carlberg 1986).

Besides teratological effects, when fed daily to *Drosophila*, thuringiensin reduces the longevity and fecundity of the flies and decreases egg size (David and Vago 1967). The fecundity of noctuid adults developing from larvae fed the exotoxin is also reduced (Hornby and Gardner 1987). In the citrus red mite (*Panonychus citri*), the tetranychid mite (*Tetranychus pacificus*), and the phytoseid mite (*Metaseiulus occidentalis*), the exotoxin kills larvae and adults and to a lesser extent the eggs (Hall et al. 1971; Hoy and Ouyang 1987). The exotoxin acts on the mites through ingestion and by contact.

The insecticidal activity of thuringiensin is high [e.g., the LD_{50} by intrahemocoelic inoculations in larvae of *Galleria mellonella* is from 0.055 to 0.5 μg/larva, about 1000-fold more toxic than DDT (Bond et al. 1971)]. It is less toxic than the δ-endotoxin derived from the crystalline parasporal body [e.g., it is about one-fifth as toxic to neonate larvae of *Heliothis virescens* as the δ-endotoxin with a potency of 16,000 international units per milligram (IU/mg) (Wolfenbarger et al. 1972)]. When thuringiensin is combined with the subspecies *kurstaki*, a potentiation or enhancement develops against larvae of the beet armyworm, *Spodoptera exigua* (Moar et al. 1986).

Workers have speculated that more than one toxin are involved in the thermostable component. However, Bond et al. (1971) have concluded that all evidence point to a single heat-stable exotoxin, thuringiensin. It is a toxic metabolite, a yellow solid, acidic in nature, and with a molecular weight (MW) 701 (Kim and Huang 1970; Farkaš et al. 1977). Its molecular structure is related to adenosine triphosphate and is composed of adenine, ribose, glucose, and allaric acid with a phosphate group (Farkaš et al. 1977; Šebesta et al. 1981). Thuringiensin is a specific inhibitor of DNA-dependent RNA polymerases. At high dosages, it may affect the biosynthesis of proteins and DNA.

Thuringiensin, through its interference with RNA transcription, affects cell mitosis, particularly during molting and metamorphosis (Šebesta et al. 1981). If larvae are not killed by the toxin, they transform to deformed pupae and adults with teratologies, such as abnormal antennae, wings, legs, or mouthparts. Affected adults are infertile or have reduced fecundity and longevity.

The presence of thuringiensin in *B. thuringiensis* is generally determined by a bioassay of susceptible insects. Microbiological and biochemical methods have also been developed. The microbiological method is based on the growth inhibition of *Sarcina flava* (Rosenberg et al. 1971). The biochemical determination, based on the inhibition of bacterial RNA polymerase by the exotoxin, is much more precise than insect feeding tests (de Barjac and Lecadet 1976).

As in the case with insects, thuringiensin is less toxic to vertebrates through the mouth than by inoculation into the body (de Barjac and Riou 1969). It affects mammals by producing lesions mainly in the liver, kidney, and adrenal gland (Šebesta et al. 1981). Chickens, when fed the toxin, exhibit a loss of vigor, reduced feeding, and undersized eggs (Faust 1973; Barker and Anderson 1975). Because of the vertebrate toxicity, most commercial preparations of *B. thuringiensis* are composed of subspecies that do not produce thuringiensin (β-exotoxin).

c. δ-Endotoxin

Bacillus thuringiensis produces in its sporangial cell one or more parasporal bodies or crystals that are proteinaceous (Fig. 4–1). When Heimpel (1967) designated the parasporal body as the δ-endotoxin, there was no knowledge that (1) the protein of the parasporal body was a protoxin and (2) some strains produced parasporal bodies with no known biological activity. In most cases, the crystal protein is a protoxin and to designate the crystal as the δ-endotoxin is etymologically incorrect. Since the term δ-endotoxin is widely used in the literature, we propose to continue its usage but to define δ-endotoxin as a class of toxic proteins produced from protoxins (crystal proteins) that are degraded by proteolytic enzymes to form smaller toxic peptides, the endotoxins. The δ-endotoxin, the first toxin detected in *B. thuringiensis*, is at present the most widely studied of all entomocidal toxins.

3. Parasporal Body—Crystal

The parasporal body or crystal is often referred to as the crystal inclusion, crystal endotoxin, or crystalloid. It varies in shapes and sizes, and in the mode of its formation depending on the subspecies of the bacillus (Norris 1971; Mikkola et al. 1982). It is generally bipyramidal in shape (diamond-shaped) in subspecies that are mainly active against lepidopterons, and may be triangular, cuboidal, flat, bar-shaped, or amorphous in other subspecies (Figs. 4–1, 4–2).

The toxic property of the crystal is denatured by heat (Cooksey 1971). The crystal is insoluble in water and organic solvents but is soluble in a highly alkaline solution, above pH 12. At a lower pH, a reducing agent is required to dissolve the crystal. The crystal dissolves in an insect's midgut at a lower pH (from 10 to 11) through the activity of proteolytic enzymes and, at the same time, is digested to form the endotoxin.

a. Formation

The parasporal body is produced during sporulation, and the general belief is that the crystal protein is not formed in vegetative cells. However, Lüthy et al. (1970)

have detected, with bioassay on *Bombyx mori* larvae, small quantities of the endotoxin bound to cell particles, probably membranes, during the vegetative growth phase of subspecies *sotto*. In certain strains of subspecies *yunnanensis*, the crystals are formed only in asporogenous cells, but other strains form crystals together with spores (Ohba and Aizawa 1986c). Moreover, crystal production has been obtained in vegetative cells of *B. subtilis* and *B. megaterium* by inserting the crystal protein genes into these species (Shivakumar et al. 1986; Mettus and Macaluso 1990). In non-spore-forming bacteria, such as *E. coli*, with recombinant plasmids carrying the crystal protein gene, it appears that the crystal protein is produced at all stages of the bacterial growth (Wong et al. 1983; Oeda et al. 1989).

The crystal formation is initiated during the early stage of sporulation and coincides with the formation of the forespore septum (Sommerville 1971; Bulla et al. 1980; Lüthy 1980). The crystal is nearly complete by the time the exosporium is formed. The crystal is formed external to the exosporium and, when the cell wall is lysed, the mature spore and crystal are released separately. The subspecies *finitimus* is an exception. It produces two types of crystals, one formed within the exosporium so that spore and crystal remain together when the cell wall lyses (Huber and Lüthy 1981), and a second crystal formed external to the exosporium as in a typical *B. thuringiensis* (Debro et al. 1986).

During sporulation, the expression of the crystal protein gene for crystal formation depends on the transcription by one or more sporulation-specific RNA polymerases (Whiteley and Schnepf 1986; Höfte and Whiteley 1989). The expression may be influenced by other unidentified sporulation-regulated mechanism and by factors encoded on plasmids (Aronson and Dunn 1985). The accumulation of large amounts of crystal protein may be a result of the action of a strong terminator, the presence of an efficient ribosome-binding site, and high gene copy number.

Spore formation is not essential for crystal production since non-spore-forming mutants also produce crystals as the normal wild strains of *B. thuringiensis* (Nishiitsutsuji-Uwo et al. 1975; Wakisaka et al. 1982; Yousten 1986). The spore coat, however, contains proteins that are related to, if not identical with, the crystal protein, very likely from the over production of this protein. This similarity of the spore-coat protein and the crystal protein has been suggested by serological and biochemical homologies (Delafield et al. 1968; Somerville et al. 1968; Tyrell et al. 1981), by ferritin-labeled and peroxidase-labeled antibodies and freeze etching (Short et al. 1974), and by the presence of toxic crystal protein (δ-endotoxin) in membrane fractions of the outer layers of the spores (Scherrer and Somerville 1977).

There are mutants of *B. thuringiensis* that form spores but do not produce crystals. Clones that have lost their capability for crystal formation cannot be distinguished from *B. cereus* (Huber and Lüthy 1981). Acrystalliferous mutants, however, have been converted to forms producing crystals (Perlak et al. 1979).

b. Number and Type

There is usually one crystal per cell, with each serological group of *B. thuringiensis* generally having one crystal type. When several crystals occur in a sporangium, they may differ in sizes and shapes (Sharpe and Baker 1979; Aronson et al. 1986) (Fig. 4–2), and the crystal proteins may differ in biochemical and toxic properties. Some strains of subspecies *kurstaki* (serotype 3a3b3c) have two types of crystals, bipyramidal and ovoidal, which differ in toxicity to insects and in antigenic properties (Krywienczyk et al. 1978; Sharpe and Baker 1979). The bipyramidal crystal, present in many subspecies, is active against lepidopterons. The ovoidal or cuboidal crystal is active against lepidopterons and dipterons (Yamamoto and McLaughlin 1981; Tojo et al. 1986). In the subspecies *israelensis* (serotype 14), several inclusions are bound together by a laminated netlike envelope of unknown composition to form the basically spherical parasporal body of about 1 μm (Federici et al. 1990) (Fig. 4–2). The inclusions are relatively small (from 0.1 to 0.5 μm) and may be cuboidal, bipyramidal, ovoidal to amorphous in shape (Charles and de Barjac 1982; Mikkola et al. 1982; Yamamoto et al. 1983). Besides sizes and shapes, the inclusions can be differentiated by electron density and lattice spacing. The largest inclusion is round to polyhedral, a second type is bar shaped, and a third inclusion is highly electron dense and is hemispherical to spherical (Federici et. al. 1990). The solubilization of each crystal varies with the pH and reducing agents used.

The subspecies *tenebrionis* (syn. *san diego*), a coleopteran pathogen, has crystals that are square to rectangular flat plates or chunky blocks (Krieg et al. 1983; Herrnstadt et al. 1986; Garfield and Stout 1988). The amino acid composition of the crystal is similar to that of crystals of other subspecies.

Not all types of crystals have proteins that are protoxins. For example, an isolate of subspecies *darmstadiensis* produces large irregular-shaped crystals that are nontoxic to Lepidoptera and Diptera, although the crystals of other isolates of this subspecies are toxic to insects of these two orders (Padua et al. 1980; Iizuka et al. 1983). Isolates with crystals, nontoxic to the common test insects, appear to predominate in the natural environment (Ohba and Aizawa 1986b). However, these isolates may be active against other untested insects.

4. Crystal Protein

In the bipyramidal crystal, the subunit is an ellipsoidal or dumbbell-shaped protein molecule about 5×15.5 nm (Norris 1971; Cooksey 1971). The protein molecules are assembled in the crystal through a -S-S linkage (Dastidar and Nickerson 1979; Huber et al. 1981; Lecadet and Lereclus 1984). The subunits are considered to be glycoproteins (Aronson and Arvidson 1987; Pfannenstiel et al. 1987, 1990), although the presence of carbohydrate is not consistently reported by various investigators.

Unlike the crystals of other subspecies, those of *tenebrionis* have subunits that appear not to be linked by disulfide linkages because the crystals can be dissolved in the absence of reducing agents (Bernhard 1986). The crystals are soluble in aqueous solutions of NaBr at neutral pH and in water with pH above 10.

The estimated molecular weights of the crystal subunit vary greatly depending on the method of crystal dissolution, the removal or inactivation of the protease in the crystal, the method of analysis, and with different strains of the bacillus. The estimates for the proteins of bipyramidal crystals range from 120,000 to 230,000 daltons. The crystal subunit of approximately 230 kDa (kilodaltons) is presently considered to be a dimer of two approximately 130-kDa proteins (Holmes and Munro 1965; Nagamatsu et al. 1978). When the crystal is dissolved in alkali *in vitro* through the reduction of the disulfide linkages, the dimer (230 kDa) is converted to the lepidopteran protoxin (from 130 to 140 kDa). The protein is proteolytically degraded into a toxin of 64 to 71 kDa (Fig. 4–1). In the digestive tracts of lepidopteran insects, the dissolution and digestion of the crystal take place simultaneously.

The proteolytic enzymes acting on the dissolved crystal proteins may be of endogenous or exogenous origins. An endogenous protease is associated with the crystal and it may play a role in releasing or activating the toxin (Chilcott et al. 1981; Thurley et al. 1985), but this enzyme is not essential (Pfannenstiel et al. 1984). The enzymes in the insect's midgut are mainly responsible for the degradation of the crystal proteins to form the endotoxins.

Even with the same crystal protein, the proteolytic enzymes from different insects may produce endotoxins with different host specificities. For example, a crystal protein (130-kDa polypeptide) of the subspecies *aizawai* can be active against either lepidopteran or dipteran cell lines. When this protein is exposed to tryptic digestion, it is reduced to a protease-resistant 55-kDa protein, which is toxic for lepidopteran but not dipteran cells (Haider and Ellar 1987). However, when the protein is treated sequentially, first with trypsin and then with gut proteases of a mosquito, the resulting smaller 53-kDa polypeptide is only toxic to dipteran and not to lepidopteran cells. Thus, the dual host specificity of *aizawai* is due to differential proteolytic processing of a single protein.

The proteolytic enzymes activate the protoxin of a lepidopteron-specific crystal by removing the first 28 amino acids from the amino-terminus (N-terminus) and the entire hydrophilic carboxy-terminal (C-terminal) half to form a toxin of about 66,000 in molecular weight (Nagamatsu et al. 1984; Bietlot et al. 1989; Arvidson et al. 1989). The overall secondary structure of the activated toxin (66 kDa) remains similar to that of the protoxin (Choma et al. 1990). It is a highly folded protein with approximately 70% of the amino acids involved in forming α-helices and β-sheet structures. The activated toxin is believed to consist of two structural units: an N-terminal half forms the toxic or lytic region relatively rich in α-helices and a C-terminal half, with alternating β-strands and coil structures, forms the binding domain for the attachment to cellular receptor sites (Covents et al. 1990). The sequence characteristic of the binding domain is widely believed to affect the host specificity of the endotoxin (Höfte and Whiteley 1989).

5. Genetics of Crystal Proteins

The genetics of crystal proteins are complex because an increasing number of genes are being recorded as producing the various crystal proteins. Moreover, many subspecies possess multiple protoxin genes. Until 1989, there was no uniform nomenclature for the genes and their products. Höfte and Whiteley (1989) have proposed a nomenclature and classification scheme based on the insect specificity (toxicity) and the primary structure of the proteins (Table 4–3). They have established four major classes of related crystal protein genes (*cry*): (I) Lepidoptera-specific, (II) Lepidoptera- and Diptera-specific, (III) Coleoptera-specific, and (IV) Diptera-specific genes. A fifth category is the totally unrelated cytolytic gene (*cyt* A) from the subspecies *israelensis*.

a. Gene Location

The *cry* genes occur mostly on plasmids (extrachromosomal DNAs) and in some subspecies also on bacterial chromosomes (Klier et al. 1982). A large number of

TABLE 4-3 Overview of Crystal Protein Gene Sequences Reported in Subspecies of *Bacillus thuringiensis*[a]

Crystal protein gene	B. thuringiensis subsp. and/or strain	Number of amino acid differences from holotype sequence[b]	
		Protoxin	Toxin[c]
*cry*IA(a)	kurstaki HD-1	H	H
	aizawai	3	2
	kurstaki HD-1	1	0
	sotto	24	3
	entomocidus	1	0
*cry*IA(b)	berliner 1715	H	H
	berliner 1715	2	0
	kurstaki HD-1	2	2
	kurstaki HD-1	5	4
	aizawai IPL-7	4	2
	kurstaki HD-1	6	2
	kurstaki NRD-12	10	6
	aizawai IC-1	4	4
*cry*IA(c)	kurstaki HD-73	H	H
*cry*IB	thuringiensis HD-2	H	H
	entomocidus HD-110	1	1
*cry*IC	entomocidus 601	H	H
	aizawai HD-137	7	7
	entomocidus HD-110	2	2
*cry*ID	aizawai HD-68	H	H
*cry*IIA	kurstaki HD-263	H	
	kurstaki HD-1	0	0
*cry*IIb	kurstaki HD-1	H	
*cry*IIIa	san diego	H	
	tenebrionis	0	0
	EG2158	0	0
*cry*IVA	israelensis	H	H
	israelensis	4	1
*cry*IVB	israelensis	H	H
	israelensis	1	1
	israelensis	3	3
	israelensis	97	78
*cry*IVC	israelensis	H	H
*cry*IVD	israelensis	H	
*cyt*A	israelensis	H	
	morrisoni PG-14	1	1
	israelensis	0	0
	morrisoni PG-14	1	1

[a]Specific references are listed in the original article by Höfte and Whiteley (1989). [Slightly modified, courtesy of Höfte and Whiteley (1989) and Microbiological Reviews.]
[b]The first reported sequence of a gene type is considered the holotype (H) sequence. Subsequently reported amino acid sequences are defined by the number of amino acid changes with respect to the holotype sequence.
[c]Toxin, the N-terminal half of the crystal protein, delineated by the C-terminal amino acid of the conserved amino acid sequence block 5.

plasmids are associated with the subspecies of *B. thuringiensis*. Among the 21 subspecies and strains surveyed by Carlton and González (1985), the number of plasmids varied from 2 to 12 per strain with sizes ranging from approximately 1.4 to 150 MDa. All of the crystalliferous strains contained at least one large plasmid of 30 MDa or more. Most strains harbored complex plasmid arrays of three or more size classes. Strains of the same subspecies usually had similar plasmid arrays and strains of different subspecies had distinct plasmid arrays. The plasmids with crystal protein genes are from 30 to over 225 MDa in size and in most subspecies that are toxic to lepidopterons, the genes are on the large plasmids (Lecadet and Lereclus 1984; Aronson et al. 1986). The association of different types and numbers of plasmids among strains of a subspecies may be complex (Iizuka et al. 1983).

Initial studies, demonstrating the presence of crystal protein genes on plasmids, involved the presence of certain plasmids in crystalliferous and their absence in acrystalliferous strains (Stahly et al. 1978; González and Carlton 1980). The plasmids could vector, during cell mating (transconjugation), the crystal protein gene to acrystalliferous and other strains or subspecies of *B. thuringiensis* and to *B. subtilis*, *B. cereus*, and *B. anthracis* (González et al. 1982; Klier et al. 1983a; Battisti et al. 1985; Honigman et al. 1986). The crystals produced in these recipient bacteria resembled those produced in *B. thuringiensis*. However, the direct evidence that crystal protein genes were located on plasmids came from gene cloning and hybridization (Southern blotting) (Kronstad et al. 1983).

Both plasmid and chromosomal crystal protein genes may occur in the same subspecies (Klier et al. 1982; Kronstad et al. 1983; Sekar 1987). Some subspecies, however, may have only chromosomal crystal protein genes [e.g., *dendrolimus* (Klier et al. 1982; Klier and Rapoport 1984) and *wuhanensis* (Kronstad et al. 1983)]. The plasmid and chromosomal genes are not identical and may have different modes of regulation (Aronson et al. 1986; Whiteley and Schnepf 1986). In some cases, the chromosomal gene is cryptic and is expressed only in the presence of the plasmid gene (i.e., a plasmic-encoded gene regulates the expression of a chromosomal crystal protein gene). Certain crystal protein genes are mobile and are on a transposable element (Kronstad and Whiteley 1984; Lereclus et al. 1985, 1986; Aronson et al. 1986). Mahillon et al. (1985) have reported an insertion element IS231 that codes for a transposase.

b. Biochemical Properties of Genes

Höfte and Whiteley (1989) have summarized the 42 nucleotide sequences of the crystal protein genes that have been cloned from various isolates and subspecies of *B. thuringiensis*. Some sequences are nearly identical and may represent a group of variants of a single gene. Significant differences in the sequences are thought to cause variations in the toxicity of the crystal proteins. A comparison of the amino acid sequences of the crystal protein genes of several subspecies suggests that homologous recombinations between different genes have occurred during the evolution of *B. thuringiensis* (Thorne et al. 1986; Höfte et al. 1986).

Physical maps have been prepared with restriction endonucleases for the crystal protein genes from plasmids and bacilli (Wong et al. 1983; Kronstad and Whiteley 1984; Adang et al. 1985; Schnepf et al. 1985; Lereclus et al. 1985). The gene maps differ indicating that portions of the crystal protein can vary and yet the protoxin retains its potential activity. The promoter regions of cloned protoxin genes, especially the type IA genes, from several subspecies have been sequenced and are virtually identical (Wong et al. 1983; Kronstad and Whiteley 1984; Adang et al. 1985; Schnepf et al. 1985; Aronson et al. 1986). The *in vitro* transcription of the cloned chromosomal crystal gene has provided evidence for a specific recognition between a sporulation-specific RNA polymerase (form II enzyme) and the promoter site of the crystal gene (Klier et al. 1978, 1983b). The transcriptional translational start sites of the crystal protein gene have been established (Wong et al. 1983; Klier et al. 1983b; Schnepf et al. 1985; Waalwijk et al. 1985; Aronson et al. 1986). There are different transcription start sites recognized by *E. coli* and *B. thuringiensis*. The transcription start sites vary with the stage of sporulation.

c. Recombinant Studies

An increasing number of recombinant strains have been formed. The transfer of the genetic material has been achieved by transduction (Thorne 1978; Lecadet et al. 1980; Barsomian et al. 1984), transformation of protoplasts and by conjugation (Aronson et al. 1986), plasmid transformation of vegetative cells (Heierson et al. 1987), and electroporation to introduce DNA (plasmid) into the cells (Bone and Ellar 1989; Mahillon et al. 1989). Shuttle plasmid vectors have been constructed with *B. thuringiensis* plasmids to transfer a crystal protein gene from one *B. thuringiensis* strain to another (Baum et al. 1990).

The first cloned crystal protein gene was from subspecies *kurstaki* (Schnepf and Whiteley 1981). The gene was introduced into *E. coli*, which produced a protein that reacted with the antibodies to the crystal. The protein was toxic when fed to neonate larvae of the tobacco hornworm, *Manduca sexta* (Schnepf and Whiteley 1981; McLinden et al. 1985). Others have transformed *B. megaterium*, *B. cereus*, and *B. subtilis* with the recombinant plasmids of several subspecies of *B. thuringiensis* (Sakanyan et al. 1982; González et al. 1982; Sekar and Carlton 1985; Calogero et al. 1989).

Recombinant DNA technology offers promise to develop super strains of *B. thuringiensis* for insect control. Bassand et al. (1989) have hybridized the subspecies *kurstaki*, *aizawai*, *entomocidus*, *israelensis*, and *tenebrionis*. The hybrid from *kurstaki* and *tenebrionis* has crystal protein genes of both parents and their endotoxins are active against lepidopterons and dipterons. Furthermore, the two truncated crystal protein genes, belonging to classes *cry*IA(b) and *cry*IC, have been translationally fused into a recombinant plasmid (Honée et al. 1990). The fused gene in *E. coli* expressed a biologically active protein with a toxicity spectrum that overlapped those of both contributing crystal protein genes. These examples show that, with genetic manipulations, *B. thuringiensis* can be modeled for specific or broad-ranged insect control, for more efficient production of crystals and spores, or to form a "super" organism in regard to virulence, pathogenicity, and persistence.

Through genetic engineering, the crystal protein genes have been introduced into root-colonizing strains of pseudomonads and *Agrobacterium* and into crop plants (Obukowicz et al. 1986; Meeusen and Warren 1989; Delannay et al. 1989). The expressed toxic proteins in the plants have been effective as an insecticidal agent, especially on insects that are highly susceptible to the endotoxin.

d. Genes of Different Crystals

The genes designated *cry*I and *cry*II for the proteins of the bipyramidal and cuboidal types of crystals have been cloned and sequenced (Höfte and Whiteley 1989). The two proteins designated P1 and P2 by Yamamoto and McLaughlin (1981) display 87% identity in amino acid sequence of the corresponding domains, but they differ in toxin specificities. The *cry*II gene product (P2 protein of cuboidal crystal), a 65-kDa protein, is toxic to both lepidopterons and dipterons; whereas the *cry*I gene product (P1 protein of bipyramidal crystal), a 135-kDa protein, acts only against lepidopterons (Yamamoto and McLaughlin 1981). Further investigations in gene cloning have revealed three *cry*I genes [*cry*IA(a), *cry*IA(b), and *cry*IA(c)] and two *cry*II genes (*cry*IIA and *cry*IIB). In strains of subspecies *kurstaki*, the bipyramidal crystal may contain three proteins and the cuboidal crystal may have two proteins (Höfte and Whiteley 1989). Isolates from other subspecies (e.g., *thuringiensis*, *tolworthi*, *galleriae*, and *kenyae*) also produce cuboidal crystals.

An isolate, similar to the strain of subspecies *tenebrionis* reported by Krieg et al. (1983), has two crystal types: the main type, rhomboid-shaped, contains a 73-kDa protein toxic to beetles, and a second flat, diamond-shaped crystal with a 30-kDa protein (Donovan et al. 1988; Rhim et al. 1990).

The vegetative cells of subspecies *israelensis* contain toxins (Walther et al. 1986), but the parasporal bodies provide the major larvicidal activity. The genes encoding the toxic proteins reside on a 75-MDa plasmid (González and Carlton 1984; Bourgouin et al. 1986). The expressed proteins produce several inclusions that are enclosed together in an envelope to form a spherical parasporal body. The crystal proteins range in size from approximately 20 to 130 kDa (Federici et al. 1990) (Table 4–4). The presently detected four major proteins are 135, 128, 65, and 27 kDa in size (Bourgouin et al. 1986; Federici et al. 1990). The 27-kDa protein is from the largest, round to polyhedral inclusion, the 65-kDa protein is from the bar-shaped inclusion, and the 128- and 135-kDa proteins are from the highly electron

TABLE 4–4 Major Mosquiticidal Proteins in the Parasporal Body of *Bacillus thuringiensis* Subsp. *israelensis* and Nomenclature for the Encoding Genes[a]

Mass (kDa) of encoded protein (protoxin)	Proteolytic cleavage product (toxin)	Gene nomenclature	Other nomenclature for the gene	Reference[b]
134.4	53–67	*cry*IVA	130-kDa endotoxin gene	Ward, Ellar 1988
			125-kDa protein gene	Bourgouin et al. 1988
			ISRH4	Sen et al. 1988
			pCH 130	Ward, Ellar 1988
127.8	53–67	*cry*IVB	130-kDa protein gene	Sekar 1986
			135-kDa protein gene	Bourgouin et al. 1988
			Bt8	Chunjatupornchai et al. 1988
			135-kDa protein gene	Delecluse et al. 1988
			ISRH3	Sen et al. 1988
			pPC 130	Ward, Ellar 1988
			130-kDa endotoxin gene	Yamamoto et al. 1988
72.4	30–38	*cry*IVD	Cry D gene	Donovan et al. 1988
27.4	25	*cyt*A	27-kDa toxin gene	Waalwijk et al. 1985

[a]Courtesy of Federici et al. (1990), © Rutgers University Press.
[b]See original article for cited references.

dense and hemispherical to spherical inclusion. The genes encoding these proteins are given in Table 4–4.

In addition to its insecticidal activity, the alkaline-dissociated crystal proteins of subspecies *israelensis* are cytolytic to cultured cells, lethal when inoculated parenterally to suckling mice (Thomas and Ellar 1983a,b; Mayes et al. 1989), neurotoxic, and myotoxic (Singh and Gill 1985; Singh et al. 1986; Cheung et al. 1987b). Most workers agree that the 25-kDa protein derived from the 27-kDa protein accounts for most of the cytolytic activity. The mosquitocidal activity has been ascribed to the 65-kDa protein (Hurley et al. 1985; Lee et al. 1985), or the 130- to 135-kDa proteins (Bourgouin et al. 1986; Sekar 1986; Visser et al. 1986; Yoshida et al. 1989). Others report that the 25-kDa fragment derived from the 27-kDa peptide is the active toxin responsible for insecticidal, cytolytic, hemolytic, and mouse-lethal activities (Davidson and Yamamoto 1984; Armstrong et al. 1985; Gill and Hornung 1987; Cheung et al. 1987a), and neurotoxic property (Cheung et al. 1987b). The current belief is that all four types of proteins are mosquitocidal but not at the same level. The 27-kDa protein is much less toxic than the other proteins (Wu and Chang 1985; Chilcott and Ellar 1988). When the 27-kDa protein is combined with the 65- and/or 130-kDa proteins, the proteins react synergistically and develop high activity.

Aside from the 27-kDa protein gene, the mosquitocidal genes of the subspecies *israelensis* are homologous to the lepidopteron-specific *cry*I-type genes (Höfte and Whiteley 1989). The similarities suggest an evolutionary relationship between one of the dipteran toxins and the lepidopteran protoxin.

Other lepidopteron active subspecies have strains with mosquitocidal activity (e.g., *aizawai*, *darmstadiensis*, *entomocidus*, *galleriae*, *kurstaki*, *kyushuensis*, *morrisoni*, and *tolworthi*) (Larget and de Barjac 1981; Kim et al. 1984; Padua et al. 1984; Haider et al. 1987). Some of these strains have crystals and crystal proteins that are similar to those of subspecies *israelensis*. Two strains of *darmstadiensis*, 73-E-10–2 and 73-E-10–16, affecting mosquitoes and not lepidopterons, contain parasporal bodies similar to those of *israelensis* (Padua et al. 1980). The crystal of strain 73-E-10–2 has a 23-kDa protein, derived from the 28-kDa protoxin, that is hemolytic and cytotoxic to mosquito and lepidopteran cell lines (Drobniewski and Ellar 1989). Strain PG-14 of *morrisoni*, which contains the typical bipyramidal parasporal body and a body similar to that of *israelensis*, is as toxic to mosquito larvae as *israelensis* (Padua et al. 1984). The crystal proteins of the PG-14 strain consist of a 144-kDa peptide (presumably derived from the bipyramidal crystal) that is not toxic to mosquitoes but substantially toxic for the lepidopteron *Trichoplusia ni*, and a mixture of polypeptides (27, 65, 128, and 135 kDa) typical of those of *israelensis* crystals (Ibarra and Federici 1986; Padua and Federici 1990). The 25-kDa protein derived from the 27-kDa protoxin, as in the case of *israelensis*, is highly hemolytic (Yu et al. 1989).

6. Pathological and Cytological Effects of δ-Endotoxin

The activity of the δ-endotoxin is confined to the insects' digestive tracts. Susceptible insects have high midgut pH that is required for the crystal dissolution and the subsequent action of proteolytic enzymes to produce the endotoxin. Reducing substances also play a role in crystal dissolution. *Papilio demoleus*, which is highly susceptible to *B. thuringiensis*, has not only a high pH (from 9.7 to 11.0) but also has large quantities of reducing substances and high proteolytic activities in the midgut (Narayanan et al. 1976). This is in contrast to the resistant *Spodoptera litura* with a low midgut pH (from 8.2 to 8.5), small amounts of reducing substances, and low proteolytic activity. In the Japanese beetle, *Popillia japonica*, the ingested food regulates the pH of the digestive tract (Sharpe and Detroy 1979a). In the presence of food, the midgut pH ranges from 9.2 to 10.2, sufficiently high to dissolve the crystals, and in the absence of food the pH is from 6.8 to 8.3.

The first pathological reaction is paralysis of the gut and mouthparts, leading to a cessation of feeding (Heimpel and Angus 1959; Cooksey 1971). The midgut epithelium of *Manduca sexta* shows some signs, when exposed to low doses of *B. thuringiensis* spores, that are similar to those of fasting (deLello et al. 1984). The midgut columnar cells respond cytologically as in fasting, but the goblet cells become swollen in response to the bacillus and they are unaffected by fasting. The site of action of the endotoxin is considered to be specific for the midgut because the fore- and hindguts continue to contract long after all locomotor activity and heartbeat have stopped (Nishiitsutsuji-Uwo and Endo 1980a).

The δ-endotoxin produces the following aberrant conditions to the midgut epithelium and the hemolymph. There is a ballooning, exfoliation, and breakdown of the insect's midgut epithelium (Cooksey 1971). The ultrastructual changes in the affected midgut cells are an enlargement of the nuclei, alteration of the endoplasmic reticula to vacuole-like configurations, the disintegration of microvilli, and the deformation of the basal infoldings of the epithelium (Endo and Nishiitsutsuji-Uwo 1981; Oron et al. 1985). In *Pieris* and *Bombyx*, the mitochondria are condensed while in *Lymantria*, *Ephestia*, and *Galleria*, they are intact.

Physiological and biochemical analyses of intoxicated larvae indicate marked alteration in midgut permeability that is, the movement of carbonate is accelerated, but that of glucose is first enhanced and then inhibited (Fast and Angus 1965; Fast and Donaghue 1971; Faust et al. 1974); a dramatic increase in K^+ and Na^+ in the hemolymph (Ramakrishnan 1968; Pendleton 1970; Nishiitsutsuji-Uwo and Endo 1980a); an increase in hemolymph pH and a decrease in midgut pH (Heimpel and Angus 1959; Nishiitsutsuji-Uwo and Endo 1980a); a modification of the ion transport properties of the midgut (Angus 1968; Nickerson and Schnell 1983); an effect on the ion regulation of the cell membrane (Nishiitsutsuji-Uwo et al. 1979; Himeno et al. 1985); inhibition of the K^+ pump, resulting in changes in the intracellular pH, Ca^{2+}, etc. (Gupta et al. 1985); and the enhancement of oxygen consumption and the uncoupling of phosphorylation, which may account for the change in the uptake of glucose by the midgut (Faust et al. 1974; Travers et al. 1976).

The endotoxin affects the fine structure of a cell within one or a few minutes after the ingestion of the crystal by a susceptible insect, such as the silkworm larva. The action takes place on the surface proteins of the cell membrane (Ebersold et al. 1977; Fast et al. 1978; Percy and Fast 1983). The columnar cells show the most pronounced reactions, and the goblet cells display few, if any, changes (Lüthy 1980; Percy and Fast 1983). Within 5 to 10 min after the ingestion of the crystals, the cisternae of rough endoplasmic reticulum enlarge, become denuded of ribosomes, and undergo vacuolization. The mitochondria and Golgi complexes become swollen. In the mitochondria, the cristae disintegrate and the contents dissolve leaving only membranous fragments. The swollen mitochondria become vacuolated but do not condense as reported by some workers (Nishiitsutsuji-Uwo et al. 1979). The increase in the number of lytic vacuoles suggests that part of the cytopathology is caused by autolysis (Lüthy and Ebersold 1981). Concomitant with the intracellular damage, the microvilli become less consistently uniform in diameter and their organized microfilaments are disrupted and disappear (Percy and Fast 1983). The nuclei of midgut cells display little pathological signs (Lüthy and Ebersold 1981).

The effect of the crystal protein toxin to hemocoelic-inoculated larvae and to cultured cell lines is greatly less than that of feeding the toxin to the larvae (Faust and Dougherty 1969; Lüthy and Ebersold 1981). The endotoxin does not affect all types of cells and some resistant cells occur in an insect cell line (Murphy et al. 1976) and in certain vertebrate cell lines (Geiser 1979; Nishiitsutsuji-Uwo et al. 1980).

As in the case with insect larvae, the activity of crystal proteins from subspecies *kurstaki* is restricted to lepidopteran cell lines and that from subspecies *israelensis* to dipteran cell lines (Johnson and Davidson 1984; Laurent and Charles 1984). An exception is the cell line of the Indianmeal moth, *Plodia interpunctella*, which does not respond to either of the crystal proteins. Indianmeal moth larva, however, responds to preparations containing *B. thuringiensis* spores (Johnson and McGaughey 1984).

7. Mode of Action of δ-Endotoxin

Workers have proposed a number of hypotheses on the mode of action of the δ-entotoxins (Chilcott et al. 1990) (Table 4–5). These hypotheses are based largely on observations of the cytopathology of insect cells *in vitro*. As in the case of the larval midgut cells, the cultured cells swell, form vesicles, and undergo lysis within

TABLE 4–5 Theories for *Bacillus thuringiensis* Toxin Mechanism of Action[a]

Target of action	Reference[b]
Plasma membrane (Ionophore)	Angus 1968 Nickerson, Schnell 1983
Mitochondria	Travers, Faust, Reichelderfer 1976
Plasma membrane (General breakdown)	Lüthy, Ebersold 1981 Nishiitsutsuji-Uwo, Endo, Himeno 1979
Goblet cell K^+ pump	Griego, Moffett, Spence 1979 Harvey, Wolfersberger 1979
Plasma membrane (*B.t.i.*) (detergent-like action)	Thomas, Ellar 1983b
Neuromuscular system	Chilcott, Kalmakoff, Pillai 1984 Cheung et al. 1985, 1987 Singh, Schouest, Gill 1986
Na^+ and/or K^+ transport	Himeno et al. 1985 Gupta et al. 1985 Sacchi et al. 1986
Plasma membrane (colloid–osmotic lysis)	Knowles, Ellar 1987 Haider, Ellar 1987 Drobniewski, Ellar 1988a

[a]Courtesy of Chilcott et al. (1990), © Rutgers University Press.
[b]See original article for cited references.

a minute (Murphy et al. 1976; Nishiitusutsuji-Uwo et al. 1979; Tojo 1986). The rapid cytopathology suggests that the toxin does not need to be internalized and that it causes a cytolytic mode of action (Lüthy and Ebersold 1981). Accordingly, the toxin appears to affect the structure and permeability of the cell membrane and it does not directly affect an internal cellular structure.

The hypotheses that closely apply to the above cellular reactions to the toxin are (1) inhibition of the ion (K^+) pump in the cell (Harvey et al. 1983) and (2) formation of small holes or pores in the plasma membrane that results in colloid–osmotic lysis (Knowles and Ellar 1987; Chilcott et al. 1990). In the first hypothesis, the K^+ pump, located at the terminal ends of midgut epithelial cells, is irreversibly inhibited by the δ-endotoxin (Harvey and Wolfersberger 1979; Harvey et al. 1983). This inhibition prevents the movement of ions in the cell, leading to cytolysis.

The second hypothesis is based on (1) the attachment of the endotoxin to receptor sites on the midgut epithelial cells, (2) the creation of pores in the cell membrane, and (3) the free movement of ions and molecules through the permeable membrane. This activity disturbs the colloid–osmotic equilibrium and results in cell lysis. Specific plasma membrane receptors for the δ-endotoxin have been detected (Knowles et al. 1984; Knowles and Ellar 1986; Haider and Ellar 1987). The attached toxin generates small pores by inserting the N-terminus into the plasma membrane (McPherson et al. 1987; Knowles et al. 1989; Ahmad and Ellar 1990). The formation of pores causes the loss of semipermeability of the plasma membrane, resulting in a net inflow of ions, accompanied by the influx of water, cell swelling, and cell lysis. Most of the evidence for this hypothesis are based on the 25-kDa toxin of subspecies *israelensis*, but the same mechanism of action may account for the effects of other δ-endotoxins, since receptor sites have also been reported for the δ-endotoxins of other subspecies (Van Rie et al. 1989; Wolfersberger 1990; Chilcott et al. 1990). Moreover, the receptor sites differ with the various endotoxins and this may explain the variations in host specificity of the subspecies of *B. thuringiensis*.

8. Host Range and Syndrome

Insects susceptible to *B. thuringiensis* have increased in numbers with the isolation of a large number of bacterial subspecies. Susceptible insects are mainly lepidopterons, dipterons, and coleopterons. Different bacterial isolates, even in the same subspecies, may differ in their pathogenicity to a specific insect, primarily because of differences in toxins produced by the strains (Dulmage 1970; Dulmage et al. 1990). The host range phenomenon with regard to the crystal protein may be explained by (1) differences in the crystal proteins of the subspecies, (2) differences in the midguts of insects that affect the processing or binding of the toxin, and/or (3) the production of additional toxins (e.g., exotoxins) by the bacteria.

The structures of endotoxins that are activated by proteolytic digestion affect host specificity. Variations in the N-terminal halves, which are involved in pore formation in the plasma membrane, result in differences in host activity. Highly active endotoxins are more apt to form pores (Arvidson et al. 1989). The C-terminal domain, involved in toxin binding, also determines host specificity (Haider and Ellar 1989; Chow et al. 1989). The host, in turn, has specific receptor sites located on the plasma membranes of midgut epithelial cells (Hoffmann et al. 1988). A heterogeneity of binding sites may occur, with one population of sites shared by several endotoxins and others are limited to a specific toxin (Van Rie et al. 1989).

Multiple crystal protein genes are found in certain subspecies of *B. thuringiensis* and they encode for protoxins that are specific for certain groups of insects, lepidopterons, dipterons, and coleopterons. The genes may also prescribe specific proteins that are toxic to specific insect species. For example, in the crystal of subspecies *aizawai*, there is a protein encoded by the *cry*IA(b) gene that is specifically active against *Pieris brassicae* and *Choristoneura fumiferana* and another protein of the *cry*IC gene is effective against *Spodoptera littoralis* and *S. frugiperda* (Lecadet et al. 1988; Knowles and Ellar 1988).

The syndrome in insects infected with *B. thuringiensis* may vary with the different insect species, the diverse toxins in the bacillus subspecies, and the different susceptible sites in the hosts. In addition, the spores in some cases play a distinct or a collaborative role.

The syndromes in lepidopteran larvae display two major signs: general paralysis, which usually results in death, and gut paralysis, from which the insect may recover or die. In some insects, both types of paralysis occur. General paralysis has a striking effect and may cause death in a highly susceptible insect, such as the silkworm, within 80 min. The paralysis is associated with a progressive increase in blood pH (Angus and Heimpel 1959; Heimpel and Angus 1959).

Gut paralysis occurs in most insects susceptible to the δ-endotoxin. A larva upon ingesting the crystal stops feeding. This has been cleverly demonstrated by Heimpel and Angus (1959) with the use of $BaSO_4$ in the diet and recording with X-rays its movement in the digestive tract. In a larva fed the barium salt, the movement of the salt stopped upon the ingestion of the crystal. The gut paralysis is associated with the ballooning and exfoliation of the midgut epithelium and the relaxation and fenestration of the midgut muscles caused by injury.

The first recognizable effect of the δ-endotoxin on the movement of the silkworm larval midgut is a reduction in the frequency and strength of contraction in 30 min at the anterior one-fourth of the midgut (second and third segments) (Hukuhara et al. 1984). However, when the crystal protein has been activated by treating it with the larval gut juice and then is fed to a larva, the contraction is inhibited earlier and over a wider portion of the midgut. The time of the development of midgut paralysis decreases with an increase in toxin dosage. At very low dosages, the larva recovers from midgut paralysis. This has also been shown by Dulmage et al. (1978) in the larva of *Heliothis virescens* through X-ray examination with barium salts. The various subspecies of *B. thuringiensis* differ in the optimum conditions for activating the toxin and in their pathogenicity to the silkworm (e.g., the δ-endotoxins of subspecies *dendrolimus* are the most toxic followed by those of *sotto*, whereas those of subspecies *darmstadiensis* are the least toxic) (Hukuhara et al. 1984).

There are few reports on the germination of the *B. thuringiensis* spores in the midgut lumen. High vegetative replication in the lumen before extensive midgut damage has been observed in the Mediterranean flour moth, *Anagasta kuehniella* (Mattes 1927), in the European corn borer, *Ostrinia nubilalis* (Sutter and Raun 1967), and in coleopteran larvae infected with the subspecies *tenebrionis* (Huger et al. 1986). On the other hand, Berliner (1915) reported extensive bacterial multiplication only after midgut destruction. In the imported cabbageworm, *Artogeia rapae*, occasional vegetative rods, without much replication in the midgut lumen, penetrate through the damaged basement membrane into the hemocoel and cause septicemia (Tanada 1953). Berliner (1915), Mattes (1927), and Huger et al. (1986) also reported bacterial replication in the hemolymph, but Sutter and Raun (1967) did not observe vegetative rods in the hemocoel of live larvae.

Heimpel and Angus (1959) classified the lepidopterons susceptible to *B. thuringiensis* into three types. Type I species are lethally stricken by general paralysis caused by the δ-endotoxin alone. There is an increase in blood pH of 1.0 to 1.5 units from the leakage of alkaline midgut contents into the poorly buffered hemolymph. Type I insects are the silkworm, *Bombyx mori*; the Chinese oak silkworm, *Antheraea pernyi*; the tobacco hornworm, *Manduca* (syn. *Protoparce*) *sexta*; the tomato hornworm, *M. quinquefasciatus* (Heimpel and Angus 1959), *Philosamia ricini* (Nishiitsutsuji-Uwo and Endo 1980a); the black fly, *Simulium vittatum* (Lacey and Federici 1979); and mosquito larvae (de Barjac 1978).

Both types I and II insects suffer from midgut paralysis a few minutes after ingesting the crystal, but type II insects do not develop general paralysis and die without either blood pH change or paralysis. Most susceptible lepidopterons belong to type II.

Type III insects do not develop general paralysis. They are not killed by the ingestion of the crystal alone and require the presence of spores. Representatives of type III insects are *Anagasta kuehniella* (Heimpel and Angus 1959) and *Galleria mellonella* (Burges et al. 1976). In these insects, the gut pH is just above neutral and the crystals do not dissolve, but the spores are able to germinate and produce conditions favorable for the δ-endotoxin and other enzymes (Lüthy and Ebersold 1981). For example, the *A. kuehniella* larva is susceptible to the δ-endotoxin if the crystals are first dissolved prior to ingestion (Yamvrias 1962).

The classification of insect types, however, varies with the dosage and variety of the bacillus. *Nymphalis antiopa*, a type II insect, will exhibit general paralysis of a type I insect when fed massive doses of the endotoxin (Heimpel and Angus 1959). When tested with the bacteria of serotype 5, *Galleria mellonella* reacts as a type III (Burges et al. 1976); but with the bacteria of serotype 7, it acts as a type I host insect (Nishiitsutsuji-Uwo and Endo 1981). Moreover, a sequential series of insects displaying the syndromes of types I to III can be assembled and the three types cannot be differentiated distinctly.

The silkworm, a type I insect, after ingesting the endotoxin crystal, displays the following syndrome: (1) cessation of feeding, (2) abrupt increase in blood pH and K^+, (3) slight sluggishness without midgut peristalsis, (4) vomiting, (5) diarrhea, (6) extreme sluggishness, (7) display of reflex movements only, (8) complete paralysis and loss of reflex movements, (9) cessation of heart beat, (10) death, (11) cessation of foregut contractions, (12) dark spots on the integument and midgut, and (13) cessation of hindgut contractions (Nishiitsutsuji-Uwo and Endo 1980a).

The syndromes produced by the δ-endotoxins in coleopteran (Sharpe 1976; Krieg et al. 1983; Ferro and Gelernter 1989) and in aquatic dipteran larvae (Lacey et al. 1978; Guillet and de Barjac 1979) are similar to those in the lepidopteran larvae.

The major cause of pathology by *B. thuringiensis* in most susceptible insects is the δ-endotoxin, and other toxins produce other forms of syndromes. The β-exotoxin causes abnormal molt or metamorphosis and abnormal appendages.

In some cases, the spores, especially in the presence of the δ-endotoxin, are responsible for the death of the insect through septicemia (Somerville et al. 1970; Burges et al. 1976; Fast 1977; McGaughey 1978). Together with the δ-endotoxin, the spores may greatly enhance the lethal effect of the bacillus (Nishiitsutsuji-Uwo and Endo 1980b). Spores also contain on their spore coats small amounts of the δ-endotoxin or a related substance. In the larva of *Spodoptera littoralis*, which has a midgut with high alkaline pH, the spores of *B. thuringiensis* do not germinate until the pH is reduced after larval death (Sneh and Schuster 1981). In the hornworm

Manduca sexta, both spores and crystals produce similar mortality curves (Schesser and Bulla 1978).

At times, the infection with *B. thuringiensis* does not kill the larvae but affects the surviving adults [e.g., adults of *Heliothis virescens*, developing from larvae fed spores and crystals in their diets, exhibit reduced fertility and fecundity (Dulmage and Martinez 1973; Salama et al. 1981)]. The direct feeding of *B. thuringiensis* subspecies *kurstaki* to the adults of the tobacco budworm, *Heliothis virescens*, also causes reduced longevity and fecundity (Ali and Watson 1982). In an unusual case, Morris (1969) has reported the transstadial passage of the bacillus from larvae to pupae and even to adults of *Vanessa cardui*.

In general, the cadavers of insects that die from *B. thuringiensis* infections in the laboratory show no obvious presence of spore development. There are a few reported cases of sporulation and crystal formation in cadavers (Berliner 1915; Mattes 1927; Prasertphon et al. 1973; Hamed 1978). Lüthy and Ebersold (1981) suggest that the absence of sporulation is due to a lack of oxygen. In infected larvae of *Dendrolimus spectabilis*, sporulation occurs more frequently in larvae held at 34°C than at room temperatures (Katagiri and Shimazu 1974). In subspecies *israelensis*, the spores produced in infected *Aedes* larvae can germinate in live and moribund larvae and in cadavers (Aly 1985).

The failure of the subspecies of *B. thuringiensis* to sporulate in infected insects appears to be responsible for the low prevalence of epizootics in nature. Persistence by vegetative rods appears unlikely because their numbers, when inoculated into natural soils, are reduced rapidly in 1 to 2 days (West et al. 1984; Akiba 1986). The few reported cases of epizootics are in stored-product insects (Burges and Hurst 1977), in the field in the Siberian silkworm, *Dendrolimus sibiricus* (Talalaev 1958; Gukasyan 1961), in a lepidopteran forest pest, *Selenephera lunigera* (Talalayeva 1967), and in the horn fly, *Haematobia irritans* (Gingrich 1984).

9. Other Exotoxins and Factors with Insecticidal Activity

Krieg and Lysenko (1979) have listed hemolysins, proteases, lecithinase, and chitinase as toxins produced by *B. thuringiensis*. They act primarily during bacterial proliferation in the hemocoel, but chitinase appears to act on chitin in the peritrophic membrane. Very little is known about the role and effectiveness of these enzymes.

Sidén et al. (1979) have isolated from the bacillus an immune inhibitor (InA), a polypeptide of 78 kDa, that when inoculated into *Drosophila* causes muscular contractions and paralysis before killing the flies. Even though InA neutralizes the initial host defense system and enables the bacteria to grow, it appears late in the course of infection that it may not be a factor in the virulence of *B. thuringiensis* (Steiner 1985). The bacillus produces bacteriocin (antibiotic) and thuricin (Krieg 1970), which appears to be identical to cerecine of *B. cereus* (Goze 1972). A number of intracellular enzymes have been detected in the prospore, sporangia, and vegetative cells of crystalliferous and acrystalliferous strains of *B. thuringiensis* (Krainova et al. 1983). The role of these enzymes in the pathogenicity of the bacillus has not been clarified.

Thuringiolysin is a sulfhydryl-dependent cytolytic protein excreted by *B. thuringiensis* at the end of the exponential growth phase (Pendleton et al. 1973;

Fatou-Rakotobe and Alouf 1984). The toxic protein has MW 55,000. It is lethal to mice at 4 to 5 μg when injected intravenously.

The biting lice (e.g., *Bovicola bovis*, *B. crassipes*, *B. limbatus*, and *B. ovis*) are susceptible to the β-exotoxin. Gingrich et al. (1974) have detected another toxin, louse factor, in the spore–δ-endotoxin complex.

A polypeptide with MW 78,000 is an immune inhibitor that inhibits the killing of *E. coli* by the immune hemolymph. The inhibitor is produced by *B. thuringiensis* and is a toxin that causes muscular contractions and paralysis when injected into the abdomens of *Drosophila* flies (Sidén et al. 1979).

10. Insect Resistance

Laboratory studies have shown that insects can develop resistance to the major toxins, the β-exotoxin and δ-endotoxin, of *B. thuringiensis* (see Chapters 14 and 15). House flies and *Drosophila melanogaster* have developed resistance to the β-exotoxin (thuringiensin) (Harvey and Howell 1965; Wilson and Burns 1968; Carlberg and Lindström 1987). This resistance is not related to the resistance of the flies to chemical insecticides (Galichet 1967; Carlberg and Lindström 1987). Resistance against the δ-endotoxin has developed in the stored-grain insect *Plodia interpunctella* (McGaughey 1985). This trait is inherited in *Plodia* as a recessive factor. The resistant *Plodia* strain, however, can be controlled with five other serotypes (McGaughey and Johnson 1987).

The tobacco budworm, *Heliothis virescens*, has developed resistance to the δ-endotoxin produced in a recombinant *Pseudomonas fluorescens* (Stone et al. 1989). Under continuous exposure to *P. fluorescens*, the resistance increased 24-fold by generation seven and fluctuated by 13- and 20-fold thereafter. The resistance is a stable trait and has been maintained for two generations.

11. Pathogenicity to Vertebrates

Thus far, there is no distinct evidence that subspecies of *Bacillus thuringiensis* are pathogenic for vertebrates through oral ingestion or topical applications, but there are cases of pathology when certain strains are injected into mammals. As described previously, the β-exotoxin is toxic to unicellular and multicellular organisms. The δ-endotoxin that affects lepidopterons appears to be innocuous to vertebrates and to normal vertebrate tissue cultures (Fisher and Rosner 1959; Lüthy 1980), but it has been reported to be cytotoxic to certain vertebrate tumor cells (Prasad and Shetna 1976; Rao et al. 1979). Nishiitsutsuji-Uwo et al. (1980), however, have not observed such cytotoxic effects on normal and malignant mammalian cells.

The δ-endotoxin of subspecies *israelensis* acts through oral ingestion primarily in aquatic dipterons, but it is toxic when inoculated into mice and insects of Lepidoptera, Orthoptera, Coleoptera, Hemiptera, and Diptera (Thomas and Ellar 1983b; Cheung et al. 1985). It also causes toxic effects on various mammalian cells. Inoculation into mice produces paralysis and death. Chilcott et al. (1984) have isolated from the crystals of this subspecies two proteins, A and B. Protein A is hemolytic for mammalian erythrocytes and is neurotoxic for cockroach ganglion.

According to Warren et al. (1984), the subspecies *israelensis* and *kurstaki* are fatal to rats when inoculated intracerebrally. They have reported also that a mixture of subspecies *israelensis* and *Acinetobacter calcoaceticus* var. *anitratus*, which is common on human skin, has caused discoloration, swelling, lymphagitis, and other reactions when accidentally inoculated into a human finger. Neither bacteria alone is lethal to mice in intravenous inoculation, but a mixture of the two bacteria is lethal.

C. *BACILLUS SPHAERICUS*

Bacillus sphaericus is a strict aerobe that produces a round spore located in a terminal or subterminal position in a swollen sporangium. It is unable to use sugars as a carbon source for growth (Claus and Berkeley 1986; Russell et al. 1989). The species is a genetically heterogeneous group composed of at least five DNA homology groups that may deserve species status (Krych et al. 1980; Alexander and Priest 1990). The strains of *B. sphaericus* are mostly saprophytes, but several are vertebrate pathogens (Siegel and Shadduck 1990), and over 50 isolates are mosquito pathogens. Like *B. thuringiensis*, it occurs world-wide in soil and aquatic habitats.

Bacillus sphaericus was not considered to be an insect pathogen until 1965, when Kellen et al. isolated this species from the mosquito *Culiseta incidens*. They reported that it caused septicemia in the mosquito. Singer (1973) was the first to suggest that a toxin was involved and others verified this suggestion (Davidson et al. 1975; Myers and Yousten 1978). With increased knowledge, *B. sphaericus* has come to resemble *B. thuringiensis* in certain properties and pathogenicity. Both species are promising candidates for the control of mosquitoes, particularly *B. sphaericus* because of its persistence in the field.

1. Strains

The first isolate of *B. sphaericus* had a low larvicidal activity, but interest in this bacillus was aroused when highly pathogenic strains were isolated from mosquitoes (Singer 1990). The strains that are principal candidates for microbial control are 1593, 2297, and 2362. Most studies on *B. sphaericus* have been conducted with these strains. However, strains with greater activity are being isolated (Brownbridge and Margalit 1986, 1987; de Barjac et al. 1988). The highly mosquitocidal strains contain parasporal inclusions or crystals. Not all insect pathogenic forms have been isolated from mosquitoes and some were obtained from lepidopterons, grasshoppers, and black flies (Weiser 1984; Lysenko et al. 1985). Isolates from the nonmosquito sources are not pathogenic for the insects from which they have been isolated.

Attempts have been made to classify the pathogenic and nonpathogenic strains of *B. sphaericus* by bacteriophage typing (Yousten et al. 1980; Yousten 1984a,b), flagella antigen serotyping (de Barjac 1990; de Barjac et al. 1980, 1985), surface proteins of the bacterial cells (Lewis et al. 1987), and numerical taxonomy based on DNA homology (Alexander and Priest 1990). There is a close correlation among these methods in separating the clusters of strains, some of which are pathogenic for mosquitoes. The most active strains belong to phage-type group 3 (Yousten et al. 1980; Yousten 1984a) and in serotype H5a5b, with the exception of 2297 in serotype H25 (phage group 4) (de Barjac et al. 1985). At present, the serotypes have not been assigned the status of variety or subspecies. Furthermore, the clusters of *B. sphaericus* strains are not readily distinguishable by universally positive and negative traits and cannot be given species designations (Baumann et al. 1991).

2. Toxins

Two types of toxins have been proposed for the pathogenesis of *B. sphaericus*: toxins from vegetative cells and toxins from parasporal crystals. Initial studies with strain SSII-1 indicated that a toxin is released when the vegetative cells are digested by a mosquito larva (Davidson et al. 1975; Myers and Yousten 1978). A second strain 1593, which is more toxic than SSII-1, and unlike SSII-1, develops an increase in activity as the bacterium begins to sporulate (Myers et al. 1979). The toxin is localized in the cytoplasm, cell wall, and spore (Myers and Yousten 1980; Davidson 1982). Realizing that the parasporal crystals in *B. thuringiensis* are protoxins, these workers examined the strains SSII-1 and 1593 for crystals but without success (Myers et al. 1979). To further confuse the issue, the vegetative cells of acrystalliferous strains are also toxic to mosquitoes (Davidson and Myers 1981).

The toxin from vegetative cells is considered to be a protein of 100 to 125 kDa isolated from cell surfaces (Davidson 1982; Baumann et al. 1985; Bourgouin et al. 1984a; Broadwell and Baumann 1986). It is produced during vegetative growth. A 122-kDa protein is the precursor of a toxic 110-kDa protein (Bowditch et al. 1989). The gene for the precursor protein has been cloned and sequenced. Both the precursor and toxic proteins are absent from the parasporal crystal. The chemical nature and exact location of the toxin have not been determined (Davidson and Yousten 1990).

In spite of the early failure to detect parasporal crystals in *B. sphaericus*, subsequent extensive examinations revealed that they are present in some strains and that they act, as in *B. thuringiensis*, in producing toxins (Wickremesinghe and Mendis 1980; Davidson and Myers 1981). More research has been conducted with the crystal toxin than the vegetative toxin because the former is much more toxic to mosquitoes. The crystal has the shape of a parallelepiped (de Barjac and Charles 1983). It appears during early sporulation (Yousten and Davidson 1982; Kalfon et al. 1984). Unlike most subspecies of *B. thuringiensis*, the crystal of *B. sphaericus* is partially enclosed in an elongated exosporium (Yousten 1984b). This keeps the crystal and spore together after cell lysis.

The activation of the crystal, through dissolution and degradation at a high pH, produces toxic proteins of 35 to 63 kDa (Tinelli and Bourgouin 1982; Davidson 1983; Bourgouin et al. 1984b; Baumann et al. 1985; Narasu and Gopinathan 1986). Higher molecular weight precursors (110- and 125-kDa proteins) have been proposed for the toxic proteins. This proposal, however, is incorrect since cloning and sequencing of the toxin genes show that the genes express 42-(41.9-) and 51-kDa proteins, which form the major components of the crystal (Berry and Hindley 1987; Hindley and Berry 1988; Arapinis et al. 1988; Baumann and Baumann 1989; Bowditch et al. 1989). The amino acid sequences of the two proteins are distinctly different from those of the toxic proteins of *B. thuringiensis*, thereby indicating that the toxic proteins of the two bacteria belong to separate families of insecticidal toxins (Baumann et al. 1988; Baumann et al. 1991).

Both 42- and 51-kDa proteins are degraded in the larval midgut. The 42-kDa protein is a protoxin (Broadwell et al. 1990a,b; Clark and Baumann 1990). It is degraded by proteolytic enzymes to a toxic 39-kDa polypeptide. The 51-kDa protein is degraded to a 43- or 44-kDa polypeptide that is nontoxic. Both the 42- and 51-kDa proteins are required to produce toxicity, a case of a binary toxin (Arapinis et al. 1988; Broadwell et al. 1990c; Clark and Baumann 1990; Davidson et al. 1990). The role of the 43-kDa protein derived from the 51-kDa protein has not been established. It may be involved in the attachment or penetration of the toxic 39-kDa protein in midgut target cells or it may serve to modify some barrier (e.g., peritrophic membrane) to enable the 39-kDa protein to have access to target cells (Baumann et al. 1991). Baumann et al. (1991) have reviewed the evidence for the unique binary toxin.

The crystal protein genes have been cloned and sequenced (Louis et al. 1984; Baumann et al. 1987, 1988) and they differ from those of *B. thuringiensis*. Moreover, their products also differ. The binary toxin of *B. sphaericus* (Broadwell et al. 1990a,b) has the toxic portion of the protein in the carboxyl-terminal half rather than in the amino-terminal half as in *B. thuringiensis* (Baumann et al. 1987).

The location of the crystal genes of *Bacillus sphaericus*, either on chromosomes or on plasmids, has not been definitively established. Baumann et al. (1991) question the conclusion of Ganesan et al. (1983) and Louis et al. (1984) that the genes are on the chromosomes. Plasmids are found in this bacillus, but their role in pathogenicity, as in the case of *B. thuringiensis*, has not been established. A number of plasmid vectors have been developed for cloning the crystal protein genes in several bacterial species (Burke and Orzech 1990; Taylor and Burke 1990).

The toxin of *B. sphaericus* has been introduced into a cyanobacterium, *Anacystis nidulans* (Tandeau de Marsac et al. 1987). It is active against *Culex* when the larvae feed on the cyanobacterium.

3. Host and Pathology

The pathogenicity of *B. sphaericus* is restricted to mosquitoes, which vary in their susceptibility to this bacillus (Wraight et al. 1987). The decreasing order of susceptibility is *Culex*, *Anopheles*, *Mansonia*, and *Aedes* (Yap 1990). In contrast, the order for *B. thuringiensis* subspecies *israelensis* is *Aedes*, *Culex*, *Anopheles*, and *Mansonia*. The major difference is in *Culex* and *Aedes*. *Culex* is much more susceptible to the toxin of *B. sphaericus* than *Aedes*. This difference in susceptibility is not due to differences in the rates of ingestion, dissolution, or the specificity of the proteases (Aly et al. 1989). The difference is due to the effective binding of the *B. sphaericus* toxin to the glycoprotein receptors on the larval midgut cells of the highly susceptible *Culex* and not to those of *Aedes* (Davidson 1988, 1989).

The initial pathology of *B. sphaericus* in mosquitoes, as with *B. thuringiensis* subspecies *israelensis*, occurs in the larval midgut, but there are specific differences between the two bacteria. In general, the pathology is not as fast and there does not appear to be a general dissolution of midgut cells as in the case of subspecies *israelensis*. The primary changes occur in the cells of the gastric cecum and the posterior midgut with some cells swelling in 30 min and showing vacuolation, and the enlargement of mitochondria and endoplasmic reticula (Davidson 1979; Singh and Gill 1988; Charles 1987). The midgut cells separate from each other. Peristalsis of the digestive tract ceases and the midgut expands and comes to lie on the body wall (Yousten 1984b). Necrosis develops in the muscle and nerve tissues and causes paralysis (Singh and Gill 1988). Larvae react to the toxin as early as 2–12 h and die within 2 days (Singer 1973, 1981; Davidson et al. 1975). Sublethal dosages of the toxin cause long-term effects in the mosquito population (e.g., continued mortality among the larvae, delayed pupation, and reduced adult longevity) (Lacey et al. 1987).

The multiplication of *B. sphaericus* is not essential to cause larval death. The bacteria are contained within the peritrophic membrane and some of them are digested in the midgut (Davidson 1979). After a larva dies, the bacteria invade the cadaver, recycle, and the spore numbers are amplified (Davidson et al. 1975; Charles and Nicolas 1986). Recycling may occur in nature (Hertlein et al. 1979). Numerous studies have confirmed the persistence of *B. sphaericus* in field applications and the recycling of the bacteria may be a factor (Davidson et al. 1984; Lacey 1990; Davidson 1990).

The mode of action of *B. sphaericus* toxin has not been established. Similar or identical receptor sites for the toxic protein are found on the susceptible and resistant tissue culture cells of *Anopheles gambiae*, *Aedes dorsalis*, and *Culex quinquefasciatus* (Davidson et al. 1987; Broadwell and Baumann 1987). The midgut proteolytic enzymes from these mosquitoes similarly convert the protoxin (42-kDa protein) to a toxic protein. The toxin is internalized in both susceptible and resistant mosquito cells but is not effective in the resistant cells (Schroeder et al. 1989). Thus, the binding and internalization are not the basis for the activity of the *B. sphaericus* toxin.

D. MILKY DISEASES OF SCARABAEIDAE

A number of bacilli multiply and sporulate in the hemolymph of scarabaeid larvae. The hemolymph becomes turbid and the posterior portion of the larval body turns opaque and milky white; thus the name, milky disease (Fig. 4–3). The syndrome was first described in grubs of the Japanese beetle, *Popillia japonica*.

The Japanese beetle was first reported in New Jersey (U.S.A.) in 1916. It became such a serious pest that the Bureau of Entomology of the United States Department of Agriculture established a Japanese beetle laboratory in New Jersey. In 1926, G. E. Spencer was the first to study the diseases of the beetle (Hawley and White 1935), and the milky disease was first noted in 1933 (Klein 1981). In 1935, Hawley and White reported high mortality of beetle larvae caused by three groups of pathogens: (1) the black group, in which the larva turned black and the isolated bacteria were easily cultured; (2) the white group, in which the larva turned white and the bacteria were difficult to culture; and (3) the fungal group. The white group, which was most common in the field (Hadley 1938), was further separated into type A and type B milky diseases (White and Dutky 1940). Dutky (1940) described the bacteria causing the type A disease as *Bacillus popilliae* and the type B disease as *B. lentimorbus*. *Bacillus popilliae* is the most important pathogen isolated for the control of the beetle.

1. *Bacillus popilliae*

Bacillus popilliae is the most thoroughly studied of the various bacilli causing milky disease in scarabaeids. Its host range within the scarabaeids is fairly broad infecting about 70 insect species. Some of them resist infection through the mouth, but most are susceptible through intrahemocoelic inoculation.

The ingested spores germinate in the larval digestive tract and the rods penetrate by phagocytosis into midgut cells (Splittstoesser and Kawanishi 1981). The

FIGURE 4-3

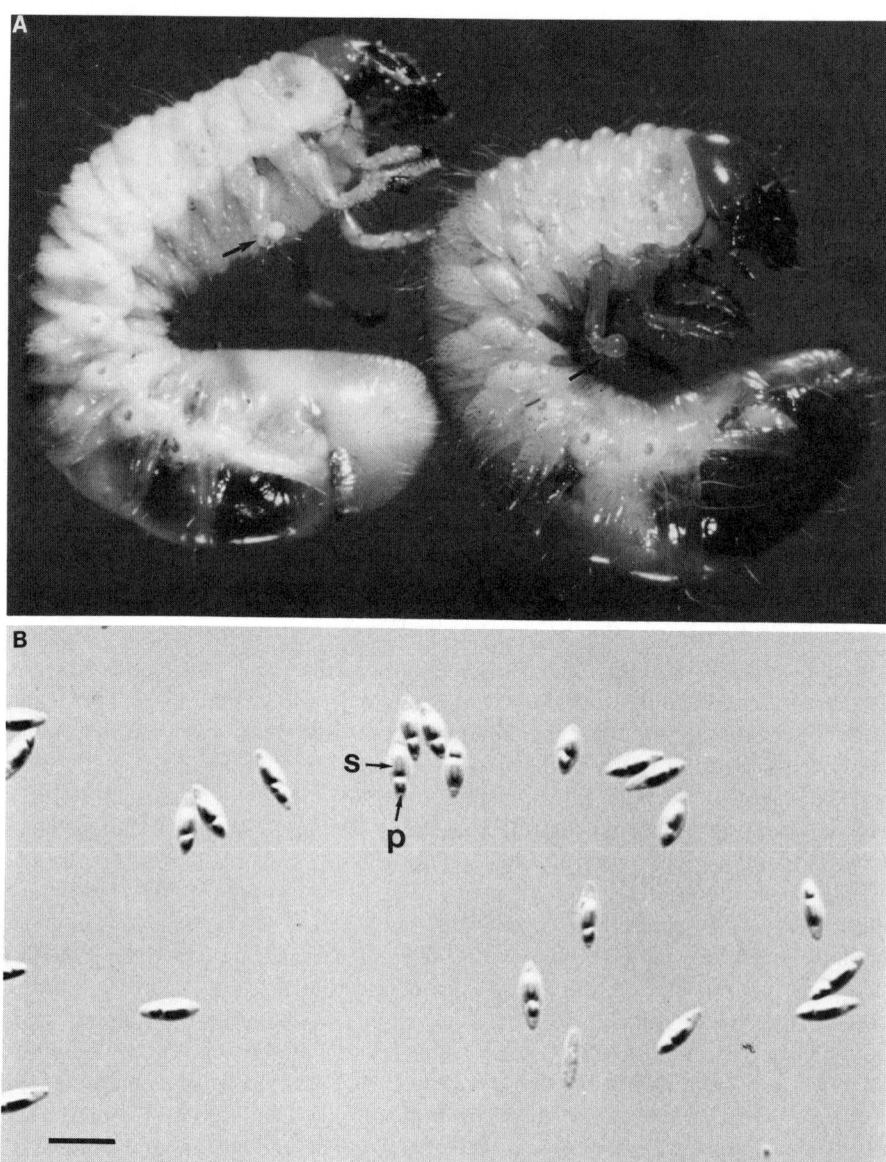

Milky disease of scarabaeids. (A) *Cyclocephala* sp. larva (*left*) infected with *Bacillus popilliae* and larva (*right*) healthy. Arrows point to milky hemolymph (*left*) compared with the normal hemolymph (*right*). (Provided by M. Klein.) (B) *Bacillus popilliae* spores from *Cyclocephala hirta*. (s) Spore. (p) Parasporal body. (Bar = 5 μm.) (Provided by B. A. Jaffee.)

first bacterial cell produced by a germinating spore has an arrowhead-like structure at the emerging end (Steinkraus and Tashiro 1967). This structure may function in the penetration of the bacterial cell into the hemocoel. The bacteria multiply initially on the luminal side of the basement membrane in the regenerative nidi areas of the midgut and, after repeated multiplication, invade into the hemocoel where multiplication continues (Kawanishi et al. 1978).

When spores are introduced into a larval hemocoel, they germinate in centers of hemocytic capsules. During sporulation, the bacilli are concentrated in connective tissue sheaths or in close association with hemocytes. Septicemia develops in the insect, but because of limited toxemia, the infection is almost a bacteremia. The rods multiply and sporulate in waves.

In the laboratory, milky disease undergoes four developmental phases (St. Julian et al. 1972; Klein 1981). Larvae die during all phases but mostly at phases 2 and 3. In phase 1, all bacterial cells are in the vegetative stage; in phase 2, 90% of the cells are vegetative; in phase 3, 65–76% are vegetative and 17–28% are spore-bearing cells; in phase 4, 95% of the cells contain spores. The insect turns milky white in phase 4, and less than 30% of the infected larvae reach this phase. In the type A milky disease, about 60 million spores are needed to produce turbidity in the blood; in type B disease, about 100 million spores are needed (Dutky 1963). Young infected larvae, however, may die without developing milky disease symptoms (Milner 1976).

Infected larvae remain active until just prior to death when they become sluggish and moribund. The infection is generally long drawn, except when the infection occurs in the first instar larva or after a larva receives a massive bacterial dose. When infected in a given instar, the larva generally does not molt and metamorphose (Beard 1945; Dutky 1963). The milky-white hemolymph, when exposed to air, does not darken as in the case of a normal, uninfected hemolymph (Beard 1945). The cause of death is unknown and is considered to be caused by a depletion of essential nutrients and the presence of heat-labile toxins (Dutky 1963) or by a disturbance of oxidative enzymes (Beard 1945). Sharpe and Detroy (1979b) have suggested that the depletion of fat-body reserves is the cause of failure to pupate. The adult beetle is also susceptible and, when infected, its life span is reduced (Langford et al. 1942).

The sporangium of *B. popilliae* is characteristic in that the spore and parasporal (refractile) body in the sporangium form a footprint appearance (Fig. 4–3B). The function of the parasporal body is unknown (Klein 1981). It has been considered to be nontoxic, but Weiner (1978) has shown that the solubilized parasporal bodies are twice as toxic as intact bodies when inoculated into an insect's hemocoel.

Spore production of *B. popilliae* is associated with host nutrition since larvae receiving poor nutrition contain only a small amount of spores (Dutky 1963). The average number of spores produced per larva is about 2 billion (Langford et al. 1942; Beard 1945). The bacillus is fastidious and does not multiply or sporulate in dead insects. In this regard, a virulent strain that kills the host at an early stage would have difficulty in persisting in the environment due to the lack of spores (Dutky 1963).

Although sporulation in *B. popilliae* has been studied extensively, no satisfactory method has been developed, thus far, to mass produce spores under *in vitro* conditions. The bacillus grows poorly in the usual bacteriological media. In special media, abundant vegetative growth is obtained, but sporulation is limited. The bacillus grows best under a nearly anaerobic condition, which is comparable to that for members of *Clostridium* (Dutky 1963; Klein et al. 1976). It has been considered to be a facultative anaerobe and catalase negative (Aronson et al. 1986) or a strict aerobe (Rhodes 1965).

Limited sporulation can be obtained when the bacillus is transferred from a complete to a starvation medium and cultured at a high temperature (37°C)

(Steinkraus and Tashiro 1955). Other systems for inducing sporulation involve the addition of activated carbon to a liquid medium (Haynes and Rhodes 1966), a shaken and still culture procedure (Sharpe et al. 1970), and the addition of specific lots of yeast extract (Sharpe and Rhodes 1973). In spite of these attempts, none of the *in vitro* methods are as productive or efficient in producing spores of *B. popillae* and *B. lentimorbus* as *in vivo*. The spores produced *in vitro* are also less virulent, especially by per os inoculation, than those produced in the larvae (Steinkraus and Tashiro 1967).

Spores of *B. popilliae* persist for long periods in the soil and this is mainly responsible for the effectiveness of the bacillus to control the Japanese beetle (White and McCabe 1950; Dutky 1963). Once established, spores accumulate in the soil in enormous numbers, especially in the upper levels, which may contain the equivalent of nearly 100 billion spores per kilogram of soil (Dutky 1963). The use of *B. popilliae* to control the Japanese beetle is an outstanding example of microbial control (see Chapter 15).

2. *Bacillus lentimorbus*

The type B milky-disease organism, *Bacillus lentimorbus*, does not produce parasporal body and the sporangium is more spindle-shaped than that of *B. popilliae*. Unlike the type A disease, when the larva is infected by *B. lentimorbus*, it continues to grow and molt (Dutky 1963). Type B syndrome resembles that of type A except for the overwintering larvae that turn milky-brown and remain active for some time before dying. The darkening is due to extensive blood clots, which are absent in the type A disease. Blood clots accumulate in the legs and block circulation causing a gangrenous condition.

3. Other Milky Disease Bacilli

A milky disease syndrome develops in the European cockchafer, *Melolontha melolontha* (Hurpin 1955). It is caused by *Bacillus popilliae* var. *melolontha* (Hurpin 1967), which is serologically related to *Bacillus popilliae* and *B. lentimorbus* (Krywienczyk and Lüthy 1974).

In 1956, Wille isolated from milky-diseased larvae of *M. melolontha*, a bacillus that he named *Bacillus fribourgensis*. The sporangium contains a spore and parasporal body (Lüthy 1968). The spores germinate readily in larval digestive juice, but the vegetative rods develop only in the hindgut juice. In the hemolymph, the rods are protected by a capsule against phagocytosis. The capsules remain attached to freshly harvested spores, and germination occurs only after the capsules are removed. *Bacillus fribourgensis* is considered to be identical to *B. popilliae* var. *melolontha* (Lüthy and Krywienczyk 1972). Wyss (1971) places *B. popilliae*, *B. lentimorbus*, and *B. fribourgensis* as varieties of one species [i.e., *B. popilliae* var. *popilliae*, *B. popilliae* var. *lentimorbus*, and *B. popilliae* var. *melolontha* (syn. *melolontha*)].

An unusual bacillus causes milky-disease syndrome in larvae of *Aphodius tasmaniae* and other scarabaeids (Milner 1981). It has an elongated sporangium with a small central spore. It is closely related to a bacillus reported from the crane fly *Tipula paludosa* and may belong to a different species and genus from *B. popilliae* (Milner and Beaton 1981).

E. BACILLUS MORITAI

An acrystalliferous bacterium, *Bacillus moritai*, resembling *B. cereus*, has been isolated from the soil (Aizawa and Fujiyoshi 1968; Aizawa et al. 1974). The bacillus infects the house fly and other muscoid flies, such as *Lucilia* sp., *Ophyra leucostoma*, *Stomoxys* sp., and *Hylemya* sp. It is not pathogenic for *Fannia* sp. and *Sarcophaga* sp.

At high dosages of *B. moritai*, the house fly larvae are killed; at low dosages, the larvae pupate but usually die. If adults develop, they have malformed wings or their fecundity and longevity are reduced. Death is caused by the live bacteria. No toxin has been detected in the liquid culture medium.

Since *B. moritai* is much more effective against the house fly than *B. thuringiensis*, it is a promising candidate for the control of muscoid flies. Moreover, it is nonpathogenic to pig, cattle, birds, fish, silkworm, honey bee, and humans (Aizawa et al. 1974).

F. BACILLUS LARVAE

The term foulbrood is applied to several diseases affecting the honey bee brood (Shimanuki 1977) (Fig. 4–4). The American foulbrood, first detected in bees about 1885 (White 1920), is caused by *Bacillus larvae* (Fig. 4–4A). In spite of its common name, the disease occurs world-wide. Other common names applied to the disease are "black brood," "ropy brood," "diseased brood," "foulbrood," or simply as "foul." Europeans have used such names as "Brutpest," "Faulbrut," "Bipest," "loque americaine." American foulbrood, a persistent disease, is the most destructive of brood diseases (Burnside and Sturtevant 1949). Its prevalence is not seasonal (Bailey 1963). The bacillus weakens the colony and may destroy it. The colony, however, may suppress infection to a low level, especially during periods of heavy nectar flow.

Bacillus larvae is a facultative anaerobic (for growth at least), gram-positive, spore former, and is catalase negative. It is pleomorphic. The vegetative rods are motile and occur singly or in chains. The bacterium is fastidious and requires thiamine and other growth factors. Spores remain viable in food, soil, and larval cadavers for many years (White 1920).

Bakhiet and Stahley (1985) studied the transfection and transformation of protoplasts of *B. larvae*, *B. subtilis*, and *B. popilliae*. They succeeded in transfecting the protoplasts of mutants of *B. larvae* and *B. subtilis* with the DNA from *B. larvae* bacteriophage but not those of *B. popilliae*.

Bacillus larvae infects only the larvae of all castes and not the adult bees. Spores of the bacillus germinate in the gut lumen within 24 h (Woodrow and Holst 1942; Bamrick 1964). The vegetative rods enter the midgut cell by phagocytosis (Davidson 1973). Some bacteria are destroyed in phagocytic vacuoles, but others survive. The invaded cells lyse and the bacteria enter the hemocoel, multiply, and cause systemic bacteremia. An infected larva often matures and the brood cell is capped (White 1920; Burnside and Sturtevant 1949). The adult removes the caps from a large but varying proportion of cells containing dead brood. Period of lethal infection is about 7 days. A larva dies, usually after it has been sealed in its cell, as a mature larva and less often as a pupa. Sporulation occurs in the decomposing larva or pupa. Aside from the lysis of an invaded midgut cell, there seems to be no toxin

FIGURE 4-4

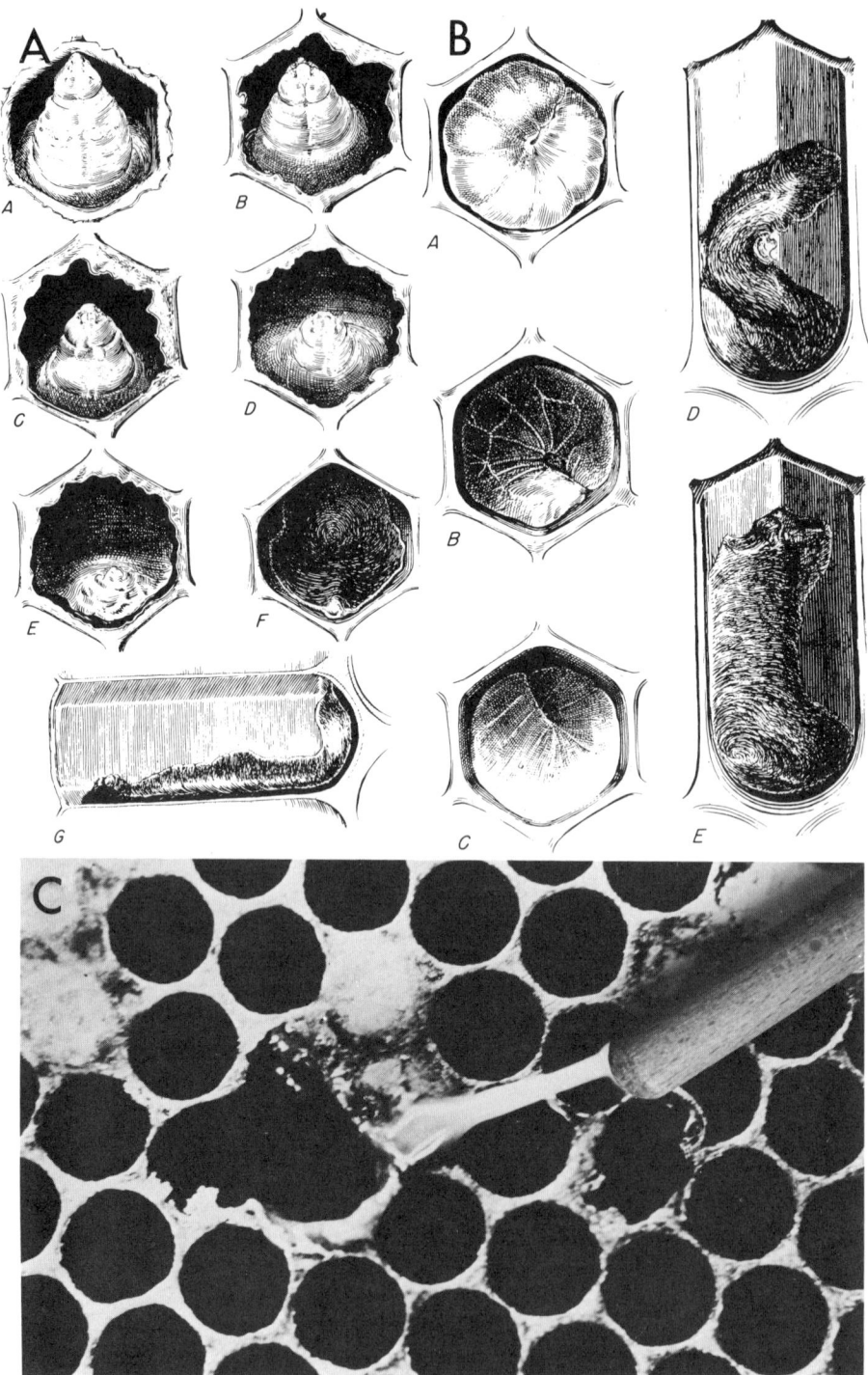

Bacterial infections of the honey bee. Syndromes of (A) the American foulbrood and (B) the European foulbrood. (C) Gluelike thread ("ropy stage"), a characteristic of American foulbrood. [Courtesy of Shimanuki (1977) and U.S.D.A., A.R.S., Beltsville.]

involved and the disease is a systemic bacteremia (Davidson 1973; Splittstoesser and Kawanishi 1981).

After death, the normally white larva becomes dark brown and settles to the bottom of the cell. Its body wall is easily ruptured. The body contents increase in viscosity and adhere to an object, such as a toothpick, and string out in a thread of gummy substance (ropiness) for a considerable distance (Fig. 4–4C).

A dead larva or pupa dries to a "scale" that adheres tenaciously to the brood cell wall. The mouthparts of a dead pupa may protrude as a fine thread, commonly referred to as the "stuck up tongue" of the pupa. This is a symptom characteristic of the American foulbrood. The caps over the cells, containing infected brood, are usually sunken, discolored, and perforated by worker bees to give the comb a "pepper box" appearance. Dead bees give off a characteristic odor of gluepot or burnt glue.

The Holst test for the detection of *B. larvae* spores in dried scales (Holst 1946) and immunodiffusion and immunofluorescence techniques for American foulbrood diagnosis have been developed (Peng and Peng 1979), but they have not been widely accepted in apiculture.

Larvae of all three castes are most susceptible to *B. larvae* during the first 24 h after hatching (Woodrow 1942; Bamrick 1964). The basis for the maturation immunity is not completely known. It may be caused by the thickening of the peritrophic membrane as the larva matures (Davidson 1973) or the unfavorable environment in the digestive tracts of older larvae may prevent spore germination (Bamrick 1967). Types of food may affect the susceptibility of the larvae. Vegetative rods of *B. larvae* are not infective when fed to larvae possibly because of bactericidal substances present in the larval food (McCleskey and Melampy 1939). When fed to larvae prior to the exposure of *B. larvae* spores, pollen reduces the rate of infection (Rinderer et al. 1974). The resistance in the various castes is associated with the types of food (pollen and honey) [i.e., drone larvae fed mostly pollen are more resistant, worker larvae fed moderate amounts are intermediate, and queen bee larvae fed the least amount of pollen are most susceptible (Rinderer and Rothenbuhler 1969)].

The natural spread of *B. larvae* is low, contrary to popular belief mainly because most spores are removed from circulation by the adult bee behavior, and because only the youngest larvae are susceptible. Resistant colonies tend to remove infected larvae much faster than susceptible colonies (Woodrow and Holst 1942). In addition, adult bees have a structure called the honey stopper in the proventriculus, which restricts spore contamination in the liquid food fed to the larvae (Sturtevant and Revell 1953). Resistant colonies are efficient in removing the spores.

Adult bees are resistant when fed spores of *B. larvae* (Wilson 1971). They eliminate most spores, but some remain in the digestive tracts for over 2 months; during this period the spores are gradually eliminated with the feces (Wilson 1972). Such contaminated bees serve as carriers and disseminators of the bacillus.

G. OTHER ENTOMOPATHOGENIC BACILLI

At one time, *Bacillus alvei* was believed to cause European foulbrood in the honey bee (Steinhaus 1949), but at present *Melissococcus pluton* (syn. *Streptococcus pluton*) is considered to be the primary agent (Bailey 1983) (Fig. 4–4B). Balaraman et al. (1979) found *B. alvei* and *B. brevis* to be highly infectious for larvae of the mosquitoes *Culex fatigans*, *Anopheles stephensi*, and *Aedes aegypti*. *Bacillus alvei* was more effective than *B. brevis* in controlling the mosquitoes. A toxin is involved in the pathogenicity of *B. alvei* (Singer 1973, 1981).

Gilliam and Dunham (1978) have isolated *Bacillus pulvifaciens* from the powdery scales of honey bee larvae and they suspect that it may be responsible for larval death. *Bacillus megaterium* is pathogenic for eggs of the European corn borer, *Ostrinia nubilalis*. According to Lynch et al. (1976), the bacillus has limited ability to invade the egg chorion, but when present on the egg surface, it produces enzymes and toxins that affect the embryo.

Bacillus laterosporus has strains that are pathogenic to mosquitoes (*Culex* and *Aedes*) and to the black fly (*Simulium bivittatum*) but not to the lepidopteron *Trichoplusia ni* (Favret and Yousten 1985). The bacillus is crystalliferous and forms a canoe-shaped parasporal body attached to the side of the spore (Claus and Berkeley 1986). The parasporal body is not toxic. The bacillus acts through a toxin associated with the cell particulate fraction rather than with the soluble fraction of triturated bacteria, suggesting that the toxin occurs in the cell wall or cytoplasmic membrane.

Bacillus firmus is pathogenic for the lepidopteron *Ailanthus triphysa* (Varma and Ali 1986). It is morphologically similar to *B. subtilis*. An infected larva dies from septicemia and shows a wilt syndrome.

H. CLOSTRIDIAL PATHOGENS

Members of the genus *Clostridium* are usually gram-positive, motile, or nonmotile (Cato et al. 1986). Most species are anaerobes and are usually catalase negative. The bulging spore causes the sporangium to have a swollen appearance. Bucher (1957, 1961) was the first to isolate an insect pathogenic *Clostridium*. He described two species, *Clostridium brevifaciens* and *C. malacosomae*, from the larval digestive tracts of the tent caterpillar *Malacosoma pluviale*. *Clostridium brevifaciens* is usually motile but not *C. malacosomae*.

Bucher (1957, 1961) called the disease in the tent caterpillar "brachytosis," since the infected larvae greatly shrink in size at death. *Clostridium brevifaciens* is the primary cause of the disease in nature. The bacteria are restricted to the larval digestive tract and do not invade the hemocoel. Ingested spores germinate and the rods multiply in 16 to 24 h in all areas of the digestive tract except in the hindgut and rectum. From 36 to 48 h, the infection is at its height.

The tent caterpillar larva becomes extremely irritable and regurgitates readily on the second day after spore ingestion. Feeding is reduced during the third day, and the feces is wet, stringy, and reddish brown. The larva shrinks in size, is sluggish, and less responsive to stimuli. On the fifth or sixth day, the larva empties its digestive tract, becomes moribund, and dies in 1 to 4 h. The dry cadaver is short, mummified, and resistant to putrefaction. Bucher (1957) speculates that toxins are produced at sporulation because of the onset of symptoms at this time. In addition to *M. pluviale*, the clostridia infect *M. disstria* and *M. americanum*.

REFERENCES

Adang, M. J., Staver, M. J., Rocheleau, T. A., Leighton, J., Barker, R. F., and Thompson, D. V. 1985. Characterized full-length and truncated plasmid clones of the crystal protein of *Bacillus thuringiensis* subsp. *kurstaki* HD-73 and their toxicity to *Manduca sexta*. Gene 36, 289–300.

Ahmad, W., and Ellar, D. J. 1990. Directed mutagenesis of selected regions of a *Bacillus thuringiensis* entomocidal protein. *FEMS Microbiol. Lett. 68*, 97–104.

Aizawa, K., and Fujiyoshi, N. 1968. Selection and breeding of bacteria for control of insect pests in the sericultural countries. Proc. Joint U.S.–Jpn. Sem. Microb. Control Insect Pests. pp. 79–83. Fukuoka, April 21–23, 1967.

Aizawa, K., Kawamura, A., Fujiyoshi, N., and Maehashi, H. 1974. Human feeding test using the bacterial insecticide—*Bacillus moritai* (*Bacillus moritai* Series No. 3). *Jpn. J. Hyg. 29*, 275–280.

Akiba, Y. 1986. Microbial ecology of *Bacillus thuringiensis* VI. Germination of *Bacillus thuringiensis* spores in the soil. *Appl. Entomol. Zool. 21*, 76–80.

Alexander, B., and Priest, F. G. 1990. Numerical classification and identification of *Bacillus sphaericus* including some strains pathogenic for mosquito larvae. *J. Gen. Microbiol. 136*, 367–376.

Ali, A-S. A., and Watson, T. F. 1982. Effects of *Bacillus thuringiensis* var. *kurstaki* on tobacco budworm (Lepidoptera: Noctuidae) adult and egg stages. *J. Econ. Entomol. 75*, 596–598.

Aly, C. 1985. Germination of *Bacillus thuringiensis* var. *israelensis* spores in the gut of *Aedes* larvae (Diptera: Culicidae). *J. Invertebr. Pathol. 45*, 1–8.

Aly, C., Mulla, M. S., and Federici, B. A. 1989. Ingestion, dissolution, and proteolysis of the *Bacillus sphaericus* toxin by mosquito larvae. *J. Invertebr. Pathol. 53*, 12–20.

Andrews, R. E., Jr., Faust, R. M., Wabiko, H., Raymond, K. C., and Bulla, L. A., Jr. 1987. The biotechnology of *Bacillus thuringiensis*. *CRC Crit. Rev. Biotechnol. 6*, 163–232.

Angus, T. A. 1954. A bacterial toxin paralysing silkworm larvae. *Nature 73*, 545–546.

Angus, T. A. 1956a. Association of toxicity with protein-crystalline inclusions of Bacillus sotto Ishiwata. *Can. J. Microbiol. 2*, 122–131.

Angus, T. A. 1956b. Extraction, purification, and properties of Bacillus sotto toxin. *Can. J. Microbiol. 2*, 416–426.

Angus, T. A. 1968. Similarity of effect of valinomycin and *Bacillus thuringiensis* parasporal protein in larvae of *Bombyx mori*. *J. Invertebr. Pathol. 11*, 145–146.

Angus, T. A., and Heimpel, A. M. 1959. Inhibition of feeding, and blood pH changes, in lepidopterous larvae infected with crystal-forming bacteria. *Can. Entomol. 91*, 352–358.

Aoki, K., and Chigasaki, Y. 1915. Uber die Pathogenitat der sog. Sottobacillen (Ishiwata) bei Seidenraupen. *Bull. Imp. Seric. Exp. Stn. Jpn. 1*, 97–139.

Arapinis, C., de la Torre, F., and Szulmajster, J. 1988. Nucleotide and deduced amino acid sequence of the *Bacillus sphaericus* 1593M gene encoding a 51.4-kDa polypeptide which acts synergistically with the 42-kDa protein for expression of the larvicidal toxin. *Nucleic Acids Res. 16*, 7731.

Armstrong, J. L., Rohrmann, G. F., and Beaudreau, 1985. Delta endotoxin of *Bacillus thuringiensis* subsp. *israelensis*. *J. Bacteriol. 161*, 39–46.

Aronson, J. N., and Arvidson, H. C. 1987. Toxic trypsin digest fragment from the *Bacillus thuringiensis* parasporal protein. *Appl. Environ. Microbiol. 53*, 416–421.

Aronson, A. I., and Dunn, P. E. 1985. Regulation of protoxin synthesis in *Bacillus thuringiensis*: Conditional synthesis in a variant is suppressed by D-cycloserine. *FEMS Microbiol. Lett. 27*, 237–241.

Aronson, A. I., Beckman, W., and Dunn, P. 1986. *Bacillus thuringiensis* and related insect pathogens. *Microbiol. Rev. 50*, 1–24.

Arvidson, H., Dunn, P. E., Strnad, S., and Aronson, A. I. 1989. Specificity of *Bacillus thuringiensis* for lepidopteran larvae: Factors involved *in vivo* and in the structure of a purified protoxin. *Mol. Microbiol. 3*, 1533–1543.

Bailey, L. 1963. "Infectious Diseases of the Honey-bee." Land Books, London.

Bailey, L. 1983. *Melissococcus pluton*, the cause of European foulbrood of honey bees (*Apis* spp.). *J. Appl. Bacteriol. 55*, 65–69.

Bakhiet, N., and Stahly, D. P. 1985. Studies on transfection and transformation of protoplasts of *Bacillus larvae*, *Bacillus subtilis*, and *Bacillus popilliae*. *Appl. Environ. Microbiol. 49*, 577–581.

Balaraman, K., Rao, U. S. B., and Rajagopalan, P. K. 1979. Bacterial pathogens of mosquito larvae—Bacillus alvei (Cheshire and Cheyene) and Bacillus brevis (Migula)—isolated in Pondicherry. *Indian J. Med. Res. 70*, 615–619.

Bamrick, J. F. 1964. Resistance to American foulbrood in honey bees. V. Comparative pathogenesis in resistant and susceptible larvae. *J. Insect Pathol. 6*, 284–304.

Bamrick, J. F. 1967. Resistance to American foulbrood in honey bees. VI. Spore germination in larvae of different ages. *J. Invertebr. Pathol. 9*, 30–34.

Barker, R. J., and Anderson, W. F. 1975. Evaluation of β-exotoxin of *Bacillus thuringiensis* Berliner for control of flies in chicken manure. *J. Med. Entomol. 12*, 103–110.

Barsomian, G. D., Robillard, N. J., and Thorne, C. B. 1984. Chromosomal mapping of *Bacillus thuringiensis* by transduction. *J. Bacteriol. 157*, 746–750.

Bassand, D., Jellis, C. L., and Piot, J.-C. 1989. Application des techniques d'échange et de génie génétiques à l'amélioration des propriétés insecticides de *Bacillus thuringiensis*. *C.R. Acad. Agric. Fr. 75*, 127–134.

Battisti, L., Green, B. D., and Thorne, C. B. 1985. Mating system for transfer of plasmids among *Bacillus anthracis*, *Bacillus cereus*, and *Bacillus thuringiensis*. *J. Bacteriol. 162*, 543–550.

Baum, J. A., Coyle, D. M., Gilbert, M. P., Jany, C. S., and Gawron-Burke, C. 1990. Novel cloning vectors for *Bacillus thuringiensis*. *Appl. Environ. Microbiol. 56*, 3420–3428.

Baumann, L., and Baumann, P. 1989. Expression in *Bacillus subtilis* of the 51- and 42-kilodalton mosquitocidal toxin genes of *Bacillus sphaericus*. *Appl. Environ. Microbiol. 55*, 252–253.

Baumann, P., Unterman, B. M., Baumann, L., Broadwell, A. H., Abbene, S. J., and Bowditch, R. D. 1985. Purification of the larvicidal toxin of *Bacillus sphaericus* and evidence for high-molecular-weight precursors. *J. Bacteriol. 163*, 738–747.

Baumann, P., Baumann, L., Bowditch, R. D., and Broadwell, A. H. 1987. Cloning of the gene for the larvicidal toxin of *Bacillus sphaericus* 2362: Evidence for a family of related sequences. *J. Bacteriol. 169*, 4061–4067.

Baumann, L., Broadwell, A. H., and Baumann, P. 1988. Sequence analysis of the mosquitocidal toxin genes encoding 51.4- and 41.9-kilodalton proteins from *Bacillus sphaericus* 2362 and 2297. *J. Bacteriol. 170*, 2045–2050.

Baumann, P., Clark, M. A., Baumann, L., and Broadwell, A. H. 1991. *Bacillus sphaericus* as a mosquito pathogen: Properties of the organism and its toxins. *Microbiol. Rev. 55*, 425–436.

Beard, R. L. 1945. Studies on the milky disease of Japanese beetle larvae. *Conn. Agric. Exp. Stn. New Haven Bull. 491*.

Berliner, E. 1911. Die "Schlaffsucht" der Mehlmottenraupe. *Z. Gesamte Getreidewes. 3*, 63–70.

Berliner, E. 1915. Über die Schlaffsucht der Mehlmottenraupe (*Ephestia kühniella* Zell.) und ihren Erreger *Bacillus thuringiensis* n. sp. *Z. Angew. Entomol. 2*, 29–56.

Bernhard, K. 1986. Studies on the delta-endotoxin of *Bacillus thuringiensis* var. *tenebrionis*. *FEMS Microbiol. Lett. 33*, 261–265.

Berry, C., and Hindley, J. 1987. *Bacillus sphaericus* strain 2362: Identification and nucleotide sequence of the 41.9kDa toxin gene. *Nucleic Acids Res. 15*, 5891.

Bietlot, H., Carey, P. R., Choma, C., Kaplan, H., Lessard, T., and Pozsgay, M. 1989. Facile preparation and characterization of the toxin from *Bacillus thuringiensis* var. *kurstaki*. *Biochem. J. 260*, 87–91.

Bond, R. P. M., Boyce, C. B. C., Rogoff, M. H., and Shieh, T. R. 1971. The thermostable exotoxin of *Bacillus thuringiensis*. *In* "Microbial Control of Insects and Mites." (H. D. Burges and N. W. Hussey, eds.), pp. 275–303. Academic Press, London.

Bone, E. J., and Ellar, D. J. 1989. Transformation of *Bacillus thuringiensis* by electroporation. *FEMS Microbiol. Lett. 58*, 171–177.

Bourgouin, C., Tinelli, R., Bouvet, J.-P., and Pires, R. 1984a. *Bacillus sphaericus*: 1593-4 purification of fractions toxic for mosquito larvae. *In* "Bacterial Protein Toxins.", (J. E. Alouf, F. J. Fehrenbach, J. H. Freer, and J. Jeljaszewicz, eds.), pp. 387–388. Academic Press, London.

Bourgouin, C., Larget-Thiery, I., and de Barjac, H. 1984b. Efficacy of dry powders from *Bacillus sphaericus*: RB80, a potent reference preparation for biological titration. *J. Invertebr. Pathol. 44*, 146–150.

Bourgouin, C., Klier, A., and Rapoport, G. 1986. Characterization of the genes encoding the haemolytic toxin and the mosquitocidal delta-endotoxin of *Bacillus thuringiensis israelensis*. *Mol. Gen. Genet. 205*, 390–397.

Bowditch, R. D., Baumann, P., and Yousten, A. A. 1989. Cloning and sequencing of the gene encoding a 125-kilodalton surface-layer protein from *Bacillus sphaericus* 2362 and of a related cryptic gene. *J. Bacteriol. 171*, 4178–4188.

Broadwell, A. H., and Baumann, P. 1986. Sporulation-associated activation of *Bacillus sphaericus* larvicide. *Appl. Environ. Microbiol. 52*, 758–764.

Broadwell, A. H., and Baumann, P. 1987. Proteolysis in the gut of mosquito larvae results in further activation of the *Bacillus sphaericus* toxin. *Appl. Environ. Microbiol. 53*, 1333–1337.

Broadwell, A. H., Baumann, L., and Baumann, P. 1990a. The 42- and 51-kilodalton mosquitocidal proteins of *Bacillus sphaericus* 2362: Construction of recombinants with enhanced expression and in vivo studies of processing and toxicity. *J. Bacteriol. 172*, 2217–2223.

Broadwell, A. H., Clark, M. A., Baumann, L., and Baumann, P. 1990b. Construction by site-directed mutagenesis of a 39-kilodalton mosquitocidal protein similar to the larva-processed toxin of *Bacillus sphaericus* 2362. *J. Bacteriol. 172*, 4032–4036.

Broadwell, A. H., Baumann, L., and Baumann, P. 1990c. Larvicidal properties of the 42 and 51 kilodalton *Bacillus sphaericus* proteins expressed in different bacterial hosts: Evidence for a binary toxin. *Curr. Microbiol. 21*, 361–366.

Brownbridge, M., and Margalit, J. 1986. New *Bacillus thuringiensis* strains isolated in Israel are highly toxic to mosquito larvae. *J. Invertebr. Pathol. 48*, 216–222.

Brownbridge, M., and Margalit, J. 1987. Mosquito active strains of *Bacillus sphaericus* isolated from soil and mud samples collected in Israel. *J. Invertebr. Pathol. 50*, 106–112.

Buchanan, R. E., and Gibbons, N. E. 1974. "Bergey's Manual of Determinative Bacteriology." Eighth Ed. Williams and Wilkins, Baltimore.

Bucher, G. E. 1957. Disease of the larvae of tent caterpillars caused by a sporeforming bacterium. *Can. J. Microbiol. 3*, 695–709.

Bucher, G. E. 1961. Artificial culture of *Clostridium brevifaciens* n. sp. and *C. malacosomae* n. sp., the causes of brachytosis of tent caterpillars. *Can. J. Microbiol. 7*, 641–655.

Bucher, G. E. 1963. Transmission of bacterial pathogens by the ovipositor of a hymenopterous parasite. *J. Insect Pathol. 5*, 277–283.

Bulla, L. A., Jr., Bechtel, D. B., Kramer, K. J., Shethna, Y. I., Aronson, A. I., and Fitz-James, P. C. 1980. Ultrastructure, physiology, and biochemistry of *Bacillus thuringiensis*. *CRC Crit. Rev. Microbiol. 8*, 147–204.

Burgerjon, A., and de Barjac, H. 1960. Nouvelles données sur le rôle de la toxine soluble thermostable produite par *Bacillus thuringiensis* Berliner. *C.R. Acad. Sci. Paris Ser. D 251*, 911–912.

Burges, H. D., and Hurst, J. A. 1977. Ecology of *Bacillus thuringiensis* in storage moths. *J. Invertebr. Pathol. 30*, 131–139.

Burges, H. D., Thomson, E. M., and Latchford, R. A. 1976. Importance of spores and δ-endotoxin protein crystals of *Bacillus thuringiensis* in *Galleria mellonella*. *J. Invertebr. Pathol. 27*, 87–94.

Burke, W. F., Jr., and Orzech, K. A. 1990. Genetics of *Bacillus sphaericus*. *In* "Bacterial Control of Mosquitoes & Black Flies." (H. de Barjac and D. J. Sutherland, eds.), pp. 256–271. Rutgers University Press, New Brunswick.

Burnside, C. E., and Sturtevant, A. P. 1949. Diagnosing bee diseases in the apiary. *U.S. Dep. Agric. Circ.* 392.

Calogero, S., Albertini, A. M., Fogher, C., Marzari, R., and Galizzi, A. 1989. Expression of a cloned *Bacillus thuringiensis* delta-endotoxin gene in *Bacillus subtilis*. *Appl. Environ. Microbiol.* 55, 446–453.

Cantwell, G. E., and Cantelo W. W. 1984. Effectiveness of *Bacillus thuringiensis* var. *israelensis* in controlling a sciarid fly, *Lycoriella mali*, in mushroom compost. *J. Econ. Entomol.* 77, 473–475.

Carlberg, G. 1986. *Bacillus thuringiensis* and microbial control of flies. *Mircen J.* 2, 267–274.

Carlberg, G., and Lindström, R. 1987. Testing fly resistance to thuringiensin produced by *Bacillus thuringiensis* serotype H-1. *J. Invertebr. Pathol.* 49, 194–197.

Carlton, B. C., and González, J. M., Jr. 1985. Plasmids and delta-endotoxin production in different subspecies of *Bacillus thuringiensis*. *In* "Molecular Biology of Microbial Differentiation." (J. A. Hoch and P. Setlow, eds.), pp. 246–252. Am. Soc. Microbiol., Washington, D.C.

Cato, E. P., George, W. L., and Finegold, S. M. 1986. Genus *Clostridium* Prazmowski 1880, 23. *In* "Bergey's Manual of Systematic Bacteriology." (P. H. A. Sneath, ed.), Vol. 2, pp. 1141–1200. Williams and Wilkins, Baltimore.

Chapman, M. H., and Hoy, M. A. 1991. Relative toxicity of *Bacillus thuringiensis* var. *tenebrionis* to the two-spotted spider mite (*Tetranychus urticae* Koch) and its predator *Metaseiulus occidentalis* (Nesbitt) (Acari, Tetranychidae and Phytoseiidae). *J. Appl. Entomol.* 111, 147–154.

Charles, J.-F. 1987. Ultrastructural midgut events in Culicidae larvae fed with *Bacillus sphaericus* 2297 spore/crystal complex. *Ann. Microbiol.* (Inst. Pasteur) 138A, 471–484.

Charles, J.-F., and de Barjac, H. 1982. Sporulation et cristallogenèse de *Bacillus thuringiensis* var. *israelensis* en microscopie électronique. *Ann. Microbiol.* (Inst. Pasteur) 133A, 425–441.

Charles, J.-F., and Nicolas, L. 1986. Recycling of *Bacillus sphaericus* 2362 in mosquito larvae: A laboratory study. *Ann. Microbiol.* (Inst. Pasteur) 137B, 101–111.

Cheung, P. Y. K., Roe, R. M., Hammock, B. D., Judson, C. L., and Montague, M. A. 1985. The apparent *in vivo* neuromuscular effects of the δ-endotoxin of *Bacillus thuringiensis* var. *israelensis* in mice and insects of four orders. *Pesti. Biochem. Physiol.* 23, 85–94.

Cheung, P. Y. K., Buster, D., and Hammock, B. D. 1987a. Lack of mosquitocidal activity by the cytolytic protein of the *Bacillus thuringiensis* subsp. *israelensis* parasporal crystal. *Curr. Microbiol.* 15, 21–23.

Cheung, P. Y. K., Buster, D., Hammock, B. D., Roe, R. M., and Alford, A. R. 1987b. *Bacillus thuringiensis* var. *israelensis* δ-endotoxin: Evidence of neurotoxic action. *Pesti. Biochem. Physiol.* 27, 42–49.

Chigaleichik, A. G. 1976. Chitinase of *Bacillus thuringiensis*. *Microbiology* (USSR) 45, 966–972.

Chilcott, C. N., and Ellar, D. J. 1988. Comparative toxicity of *Bacillus thuringiensis* var. *israelensis* crystal proteins *in vivo* and *in vitro*. *J. Gen. Microbiol.* 134, 2551–2558.

Chilcott, C. N., Kalmakoff, J., and Pillai, J. S. 1981. The biological significance of proteases in *Bacillus thuringiensis* var. *israelensis* crystals. *Proc. Univ. Otago Med. Sch.* 59, 40–41.

Chilcott, C. N., Kalmakoff, J., and Pillai, J. S. 1984. Neurotoxic and haemolytic activity of a protein isolated from *Bacillus thuringiensis* var. *israelensis* crystals. *FEMS Microbiol. Lett.* 25, 259–263.

Chilcott, C. N., Knowles, B. H., Ellar, D. J., and Drobniewski, F. A. 1990. Mechanism of action of *Bacillus thuringiensis israelensis* parasporal body. *In* "Bacterial Control of Mosquitoes & Black Flies." (H. de Barjac and D. J. Sutherland, eds.), pp. 45–65. Rutgers University Press, New Brunswick.

Choma, C. T., Surewicz, W. K., Carey, P. R., Pozsgay, M., and Kaplan, H. 1990. Secondary structure of the entomocidal toxin from *Bacillus thuringiensis* subsp. *kurstaki* HD-73. *J. Protein Chem.* 9, 87–94.

Chow, E., Singh, G. J. P., and Gill, S. S. 1989. Binding and aggregation of the 25-kilodalton toxin of *Bacillus thuringiensis* subsp. *israelensis* to cell membranes and alteration by monoclonal antibodies and amino acid modifiers. *Appl. Environ. Microbiol.* 55, 2779–2788.

Clark, M. A., and Baumann, P. 1990. Deletion analysis of the 51-kilodalton protein of the *Bacillus sphaericus* 2362 binary mosquitocidal toxin: Construction of derivatives equivalent to the larva-processed toxin. *J. Bacteriol.* 172, 6759–6763.

Claus, D., and Berkeley, R. C. 1986. Genus *Bacillus* Cohn 1872, 174. *In* "Bergey's Manual of Systematic Bacteriology." (P. H. A. Sneath, ed.), Vol. 2, pp. 1105–1141. Williams and Wilkins, Baltimore.

Convents, D., Houssier, C., Lasters, I., and Lauwereys, M. 1990. The *Bacillus thuringiensis* δ-endotoxin. Evidence for a two domain structure of the minimal toxic fragment. *J. Biol. Chem.* 265, 1369–1375.

Cooksey, K. E. 1971. The protein crystal toxin of *Bacillus thuringiensis*: Biochemistry and mode of action. *In* "Microbial Control of Insects and Mites." (H. D. Burges and N. W. Hussey, eds.), pp. 247–274. Academic Press, London.

Daoust, R. A., and Gunner, H. B. 1979. Microbial synergists pathogenic to *Lymantria dispar*: Chitinolytic and fermentative bacterial interactions. *J. Invertebr. Pathol.* 33, 368–377.

Dastidar, P. G., and Nickerson, K. W. 1979. Interchain crosslinks in the entomocidal *Bacillus thuringiensis* protein crystal. *FEBS Lett.* 108, 411–414.

David, J., and Vago, C. 1967. Influence des toxines de *Bacillus thuringiensis* sur divers caractères physiologiques de Drosophiles adultes. *Entomophaga* 12, 153–159.

Davidson, E. W. 1973. Ultrastructure of American foulbrood disease pathogenesis in larvae of the worker honey bee, *Apis mellifera*. *J. Invertebr. Pathol.* 21, 53–61.

Davidson, E. W. 1979. Ultrastructure of midgut events in the pathogenesis of *Bacillus sphaericus* strain SSII-1 infections of *Culex pipiens quinquefasciatus* larvae. *Can. J. Microbiol.* 25, 178–184.

Davidson, E. W. 1982. Purification and properties of soluble cytoplasmic toxin from the mosquito pathogen *Bacillus sphaericus* strain 1593. *J. Invertebr. Pathol.* 39, 6–9.

Davidson, E. W. 1983. Alkaline extraction of toxin from spores of the mosquito pathogen, *Bacillus sphaericus* strain 1593. *Can. J. Microbiol.* 29, 271–275.

Davidson, E. W. 1988. Binding of the *Bacillus sphaericus* (Eubacteriales: Bacillaceae) toxin to midgut cells of mosquito (Diptera: Culicidae) larvae: Relationship to host range. *J. Med. Entomol.* 25, 151–157.

Davidson, E. W. 1989. Variation in binding of *Bacillus sphaericus* toxin and wheat germ agglutinin to larval midgut cells of six species of mosquitoes. *J. Invertebr. Pathol.* 53, 251–259.

Davidson, E. W. 1990. Microbial control of vector insects. *In* "New Directions in Biological Control: Alternatives for Suppressing Agricultural Pests and Diseases." (R. R. Baker and P. E. Dunn, eds.), pp. 199–212. Alan R. Liss, New York.

Davidson, E. W., and Myers, P. 1981. Parasporal inclusions in *Bacillus shaericus*. *FEMS Microbiol. Lett.* 10, 261–265.

Davidson, E. W., and Yamamoto, T. 1984. Isolation and assay of the toxic component from the crystals of *Bacillus thuringiensis* var. *israelensis*. *Curr. Microbiol.* *11*, 171–174.

Davidson, E. W., and Yousten, A. A. 1990. The mosquito larval toxin of *Bacillus sphaericus*. *In* "Bacterial Control of Mosquitoes & Black Flies." (H. de Barjac and D. J. Sutherland, eds.), pp. 237–255. Rutgers University Press, New Brunswick.

Davidson, E. W., Singer, S., and Briggs, J. D. 1975. Pathogenesis of *Bacillus sphaericus* strain SSII-1 infections in *Culex pipiens quinquefasciatus* (=*C. pipiens fatigans*) larvae. *J. Invertebr. Pathol.* *25*, 179–184.

Davidson, E. W., Urbina, M., Payne, J., Mulla, M. S., Darwazeh, H., Dulmage, H. T., and Correa, J. A. 1984. Fate of *Bacillus sphaericus* 1593 and 2362 spores used as larvicides in the aquatic environment. *Appl. Environ. Microbiol.* *47*, 125–129.

Davidson, E. W., Shellabarger, C., Meyer, M., and Bieber, A. L. 1987. Binding of the *Bacillus sphaericus* mosquito larvicidal toxin to cultured insect cells. *Can. J. Microbiol.* *33*, 982–989.

Davidson, E. W., Oei, C., Meyer, M., Bieber, A. L., Hindley, J., and Berry, C. 1990. Interaction of the *Bacillus sphaericus* mosquito larvicidal proteins. *Can J. Microbiol.* *36*, 870–878.

de Barjac, H. 1978. Une nouvelle variété de *Bacillus thuringiensis* très toxique pour les Moustiques: *B. thuringiensis* var. *israelensis* sérotype 14. *C.R. Acad. Sci. Paris Ser. D* *286*, 797–800.

de Barjac, H. 1990. Classification of *Bacillus sphaericus* strains and comparative toxicity to mosquito larvae. *In* "Bacterial Control of Mosquitoes & Black Flies.) (H. de Barjac and D. J. Sutherland, eds.), pp. 228–236. Rutgers University Press, New Brunswick.

de Barjac, H., and Bonnefoi, A. 1962. Essai de classification biochimique et sérologique de 24 souches de *Bacillus* du type *B. thuringiensis*. *Entomophaga* *7*, 5–31.

de Barjac, H., and Bonnefoi, A. 1973. Mise au point sur la classification des *Bacillus thuringiensis*. *Entomophaga* *18*, 5–17.

de Barjac, H., and Charles, J.-F. 1983. Une nouvelle toxine active sur les Moustiques, présente dans des inclusions cristallines produites par *Bacillus sphaericus*. *C.R. Acad. Sci. Paris Ser.* III *296*, 905–910.

de Barjac, H., and Cosmao Dumanoir, V. 1975. Répartition des pouvoirs chitinolytiques et lipolytiques chez les divers sérotypes de *Bacillus thuringiensis*. *Entomophaga* *20*, 43–48.

de Barjac, H., and Frachon, E. 1990. Classification of *Bacillus thuringiensis* strains. *Entomophaga* *35*, 233–240.

de Barjac, H., and Lecadet, M.-M. 1976. Dosage biochimique de l'exotoxine thermostable de *B. thuringiensis* d'après l'inhibition d'*ARN*-polymérases bactériennes. *C.R. Acad. Sci. Paris Ser. D* *282*, 2119–2122.

de Barjac, H., and Riou, J.-Y. 1969. Action de la toxine thermostable de B. thuringiensis var. thuringiensis administrée à des souris. *Rev. Pathol. Comp. Méd. Exp.* *69*, 367–374.

de Barjac, H., and Sutherland, D. J. 1990. "Bacterial Control of Mosquitoes & Black Flies." Rutgers University Press, New Brunswick.

de Barjac, H., Véron, M., and Cosmao Dumanoir, V. 1980. Caractérisation biochimique et sérologique de souches de *Bacillus sphaericus* pathogènes ou non pour les moustiques. *Ann. Microbiol.* (Inst. Pasteur) *131B*, 191–201.

de Barjac, H., Larget, I., and Killick-Kendrick, R. 1981. Toxicité de *Bacillus thuringiensis* var. *israelensis* serotype H_{14}, pour les larves de phlébotomes, vecteurs de leishmanioses. *Bull. Soc. Pathol Exot.* *74*, 485–489.

de Barjac, H., Larget-Thíery, I., Cosmao Dumanoir, V., and Ripouteau, H. 1985. Serological classification of *Bacillus sphaericus* strains on the basis of toxicity to mosquito larvae. *Appl. Microbiol. Biotechnol.* *21*, 85–90.

de Barjac, H., Thiery, I., Cosmao-Dumanoir, V., Frachon, E., Laurent, P., Charles, J.-F., Hamon, S., and Ofori, J. 1988. Another *Bacillus sphaericus* serotype harbouring strains very toxic to mosquito larvae: Serotype H6. *Ann. Microbiol.* (Inst. Pasteur) *139*, 363–377.

Debro, L., Fitz-James, P. C., and Aronson, A. 1986. Two different parasporal inclusions are produced by *Bacillus thuringiensis* subsp. *finitimus*. *J. Bacteriol.* 165, 258–268.

Delafield, F. P., Somerville, H. J., and Rittenberg, S. C. 1968. Immunological homology between crystal and spore protein of *Bacillus thuringiensis*. *J. Bacteriol.* 96, 713–720.

Delannay, X., LaVallee, B. J., Proksch, R. K., Fuchs, R. L., Sims, S. R., Greenplate, J. T., Marrone, P. G., Dodson, R. B., Augustine, J. J., Layton, J. G., and Fischhoff, D. A. 1989. Field performance of transgenic tomato plants expressing the *Bacillus thuringiensis* var. *kurstaki* insect control protein. *Biotechnology 7*, 1265–1269.

deLello, E., Hanton, W. K., Bishoff, S. T., and Misch, D. W. 1984. Histopathological effects of *Bacillus thuringiensis* on the midgut of tobacco hornworm larvae (*Manduca sexta*): Low doses compared with fasting. *J. Invertebr. Pathol.* 43, 169–181.

DeLucca, A. J., II. 1984. Lectin grouping of *Bacillus thuringiensis* serovars. *Can. J. Microbiol.* 30, 1100–1104.

Doane, C. 1960. Bacterial pathogens of *Scolytus multistriatus* Marsham as related to crowding. *J. Insect Pathol.* 2, 24–29.

Donovan, W. P., Gonzalez, J. M., Jr., Gilbert, M. P., and Dankocsik, C. 1988. Isolation and characterization of EG2158, a new strain of *Bacillus thuringiensis* toxic to coleopteran larvae, and nucleotide sequence of the toxin gene. *Mol. Gen. Genet.* 214, 365–372.

Drobniewski, F. A., and Ellar, D. J. 1989. Purification and properties of a 28-kilodalton hemolytic and mosquitocidal protein toxin of *Bacillus thuringiensis* subsp. *darmstadiensis* 73-E10-2. *J. Bacteriol.* 171, 3060–3067.

Dulmage, H. T. 1970. Production of the spore-δ-endotoxin complex by variants of *Bacillus thuringiensis* in two fermentation media. *J. Invertebr. Pathol.* 16, 385–389.

Dulmage, H. T., and Martinez, E. 1973. The effects of continuous exposure to low concentrations of the δ-endotoxin of *Bacillus thuringiensis* on the development of the tobacco budworm, *Heliothis virescens*. *J. Invertebr. Pathol.* 22, 14–22.

Dulmage, H. T., Graham, H. M., and Martinez, E. 1978. Interactions between the tobacco budworm, *Heliothis virescens*, and the δ-endotoxin produced by the HD-1 isolate of *Bacillus thuringiensis* var. *kurstaki*: Relationship between length of exposure to the toxin and survival. *J. Invertebr. Pathol.* 32, 40–50.

Dulmage, H. T., Correa, J. A., and Gallegos-Morales, G. 1990. Potential for improved formulations of *Bacillus thuringiensis israelensis* through standardization and fermentation development. *In* "Bacterial Control of Mosquitoes & Black Flies." (H. de Barjac and D. J. Sutherland, eds.), pp. 110–133. Rutgers University Press, New Brunswick.

Dutky, S. R. 1940. Two new spore-forming bacteria causing milky diseases of Japanese beetle larvae. *J. Agric. Res.* 61, 57–68.

Dutky, S. R. 1963. The milky diseases. *In* "Insect Pathology: An Advanced Treatise." (E. A. Steinhaus, ed.), Vol. 2, pp. 75–115. Academic Press, New York.

Ebersold, H. R., Luethy, P., and Mueller, M. 1977. Changes in the fine structure of the gut epithelium of Pieris brassicae induced by the δ-endotoxin of Bacillus thuringiensis. *Mitt. Schweiz. Entomol. Ges. Bull. Soc. Entomol. Suisse 50*, 269–276.

Endo Y., and Nishiitsutsuji-Uwo, J. 1981. Mode of action of *Bacillus thuringiensis* δ-endotoxin: Ultrastructural changes of midgut epithelium of *Pieris*, *Lymantria* and *Ephestia* larvae. *Appl. Entomol. Zool.* 16, 231–241.

Farkaš, J., Šebesta, K., Horská, K., Samek, Z., Dolejš, L., and Šorm, F. 1977. Structure of thuringiensin, the thermostable exotoxin from *Bacillus thuringiensis*. *Collect. Czeck. Chem. Commun.* 42, 909–929.

Fast, P. G. 1977. *Bacillus thuringiensis* δ-endotoxin: On the relative roles of spores and crystals in toxicity to spruce budworm (Lepidoptera: Tortricidae). *Can. Entomol.* 109, 1515–1518.

Fast, P. G., and Angus, T. A. 1965. Effects of parasporal inclusions of *Bacillus thuringiensis* var. *sotto* Ishiwata on the permeability of the gut wall of *Bombyx mori* (Linnaeus) larvae. *J. Invertebr. Pathol. 7*, 29–32.

Fast, P. G., and Donaghue, T. P. 1971. The δ-endotoxin of *Bacillus thuringiensis*. II. On the mode of action. *J. Invertebr. Pathol. 18*, 135–138.

Fast, P. G., Murphy, D. W., and Sohi, S. S. 1978. *Bacillus thuringiensis* δ-endotoxin: Evidence that toxin acts at the surface of susceptible cells. *Experientia 34*, 762–763.

Fatou-Rakotobe, E., and Alouf, J. E. 1984. Purification and properties of thuringiolysin, a sulfhydryl-activated cytolytic toxin of *Bacillus thuringiensis*. *In* "Bacterial Protein Toxins." (J. E. Alouf, F. J. Fehrenbach, J. H. Freer, and J. Jeljaszewicz, eds.), p. 265. Academic Press, London.

Faust, R. M. 1973. The *Bacillus thuringiensis* β-exotoxin: Current status. *Bull. Entomol. Soc. Am. 19*, 153–156.

Faust, R. M., and Dougherty, E. M. 1969. Effects of the B. t. δ-endotoxin produced by *Bacillus thuringiensis* var. *dendrolimus* on the hemolymph of the silkworm, *Bombyx mori*. *J. Invertebr. Pathol. 13*, 155–157.

Faust, R. M., Travers, R. S., and Hallam, G. M. 1974. Preliminary investigations on the molecular mode of action of the δ-endotoxin produced by *Bacillus thuringiensis* var. *alesti*. *J. Invertebr. Pathol. 23*, 259–261.

Favret, M. E., and Yousten, A. A. 1985. Insecticidal activity of *Bacillus laterosporus*. *J. Invertebr. Pathol. 45*, 195–203.

Federici, B. A., Lüthy, P., and Ibarra, J. E. 1990. Parasporal body of *Bacillus thuringiensis israelensis*. *In* "Bacterial Control of Mosquitoes & Black Flies." (H. de Barjac and D. J. Sutherland, eds.), pp. 16–44. Rutgers University Press, New Brunswick.

Ferro, D. N., and Gelernter, W. D. 1989. Toxicity of a new strain of *Bacillus thuringiensis* to Colorado potato beetle (Coleoptera: Chrysomelidae). *J. Econ. Entomol. 82*, 750–755.

Fisher, R., and Rosner, L. 1959. Insecticide safety. Toxicology of the microbial insecticide, Thuricide. *Agric. Food Chem. 7*, 686–688.

Fluer, F. S., Ivinskene, V. L., and Zajačkauskas, P. A. 1981. Detection of thermolabile *Bacillus thuringiensis* exotoxin and its separation from phospholipase C. *Zh. Mikrobiol. Epidemiol. Immunobiol. 8*, 81–85.

Fox, G. E., Stackebrandt, E., Hespell, R. B., Gibson, J., Maniloff, J., Dyer, T. A., Wolfe, R. S., Balch, W. E., Tanner, R. S., Magrum, L. J., Zablen, L. B., Blakemore, R., Gupta, R., Bonen, L., Lewis, B. J., Stahl, D. A., Luehrsen, K. R., Chen, K. N., and Woese, C. R. 1980. The phylogeny of prokaryotes. *Science 209*, 457–463.

Galichet, P. F. 1967. Sensitivity to the soluble heat-stable toxin of *Bacillus thuringiensis* of strains of *Musca domestica* tolerant to chemical insecticides. *J. Invertebr. Pathol. 9*, 261–262.

Ganesan, S., Kamdar, H., Jayaraman, K., and Szulmajster, J. 1983. Cloning and expression in *Escherichia coli* of a DNA fragment from *Bacillus sphaericus* coding for biocidal activity against mosquito larvae. *Mol. Gen. Genet. 189*, 181–183.

Garcia, R., Des Rochers, B., and Tozer, W. 1980. Studies on the toxicity of *Bacillus thuringiensis* var. *israelensis* against organisms found in association with mosquito larvae. *Proc. Calif. Mosq. Vector Control Assoc. 48*, 33–36.

Garfield, J. L., and Stout, C. D. 1988. Crystallization and preliminary X-ray diffraction studies of a toxic crystal protein from a subspecies of *Bacillus thuringiensis*. *J. Biol. Chem. 263*, 11800–11801.

Geiser, P. 1979. Versuche in vitro zum Nachweis einer Wirkung des δ-Endotoxin von *Bacillus thuringiensis*. Doctoral dissertation, Eidgenössische Technischen Hochschule Zürich. Juris, Zürich.

Gill, S. S., and Hornung, J. M. 1987. Cytolytic activity of *Bacillus thuringiensis* proteins to insect and mammalian cell lines. *J. Invertebr. Pathol. 50*, 16–25.

Gilliam, M., and Dunham, D. R. 1978. Recent isolations of *Bacillus pulvifaciens* from powdery scales of honey bee, *Apis mellifera*, larvae. *J. Invertebr. Pathol. 32*, 222–223.

Gingrich, R. E. 1984. Control of the horn fly, *Haematobia irritans*, with *Bacillus thuringiensis*. *In* "Comparative Pathobiology," (T. C. Cheng, ed.), Vol. 7, pp. 47–57. Plenum Press, New York.

Gingrich, R. E., Allan, N., and Hopkins, D. E. 1974. *Bacillus thuringiensis*: Laboratory tests against four species of biting lice (Mallophaga: Trichodectidae). *J. Invertebr. Pathol. 23*, 232–236.

Goldberg, L. J., and Margalit, J. 1977. A bacterial spore demonstrating rapid larvicidal activity against *Anopheles sergentii*, *Uranotaenia unguiculata*, *Culex univitattus*, *Aedes aegypti* and *Culex pipiens*. *Mosq. News 37*, 355–358.

González, J. M., Jr., and Carlton, B. C. 1980. Patterns of plasmid DNA in crystalliferous and acrystalliferous strains of *Bacillus thuringiensis*. *Plasmid 3*, 92–98.

González, J. M., Jr., and Carlton, B. C. 1984. A large transmissible plasmid is required for crystal toxin production in *Bacillus thuringiensis* variety *israelensis*. *Plasmid 11*, 28–38.

González, J. M., Jr., Brown, B. J., and Carlton, B. C. 1982. Transfer of *Bacillus thuringiensis* plasmids coding for δ-endotoxin among strains of *B. thuringiensis* and *B. cereus*. *Proc. Natl. Acad. Sci. U.S.A. 79*, 6951–6955.

Gordon, R. E. 1978. A species definition. *Int. J. Syst. Bacteriol. 28*, 605–607.

Gordon, R. E., Haynes, W. C., and Pang, C. H.-N. 1973. The genus *Bacillus*. *U.S. Dep. Agric. Hand. 427*.

Goze, A. 1972. Thuricines et cérécines moléculaires. *C.R. Soc. Biol. Paris 166*, 200–204.

Guillet, P., and de Barjac, H. 1979. Toxicité de *Bacillus thuringiensis* var. *israelensis* pour les larves de Simulies vectrices de l'Onchocercose. *C.R. Acad. Sci. Paris Ser. D 289*, 549–552.

Gukasyan, A. B. 1961. Prospective control method for the Siberian silkworm. *Akad. Nauk S.S.S.R. Vestn. 31*, 58–59.

Gupta, B. L., Dow, J. A. T., Hall, T. A., and Harvey, W. R. 1985. Electron probe X-ray microanalysis of the effects of *Bacillus thuringiensis* var. *kurstaki* crystal protein insecticide on ions in an electrogenic K^+-transporting epithelium of the larval midgut in the lepidopteran, *Manduca sexta*, in vitro. *J. Cell. Sci. 74*, 137–152.

Hadley, C. H. 1938. Progress of Japanese beetle investigations. *J. N.Y. Entomol. Soc. 46*, 203–216.

Haider, M. Z., and Ellar, D. J. 1987. Characterization of the toxicity and cytopathic specificity of a cloned *Bacillus thuringiensis* crystal protein using insect cell culture. *Mol. Microbiol. 1*, 59–66.

Haider, M. Z., and Ellar, D. J. 1989. Mechanism of action of *Bacillus thuringiensis* insecticidal δ-endotoxin: Interaction with phospholipid vesicles. *Biochim. Biophys. Acta 978*, 216–222.

Haider, M. Z., Ward, E. S., and Ellar, D. J. 1987. Cloning and heterologous expression of an insecticidal delta-endotoxin gene from *Bacillus thuringiensis* var. *aizawai* IC1 toxic to both lepidoptera and diptera. *Gene 52*, 285–290.

Hall, I. M., Hunter, D. K., and Arakawa, K. Y. 1971. The effect of the β-exotoxin fraction of *Bacillus thuringiensis* on the citrus red mite. *J. Invertebr. Pathol. 18*, 359–362.

Hamed, A. R. 1978. Zur Wirkung von Bacillus thuringiensis auf Yponomeuta evonymellus (L.) and Y. padellus (L.) (Lep., Yponomeutidae). *Z. Angew. Entomol. 85*, 392–412.

Hannay, C. L. 1953. Crystalline inclusions in aerobic spore-forming bacteria. *Nature 172*, 1004–1006.

Hannay, C. L., and Fitz-James, P. 1955. The protein crystals of *Bacillus thuringiensis* Berliner. *Can. J. Microbiol. 1*, 694–710.

Harvey, T. L., and Howell, D. E. 1965. Resistance of the house fly to *Bacillus thuringiensis* Berliner. *J. Invertebr. Pathol.* 7, 92–100.

Harvey, W. R., and Wolfersberger, M. G. 1979. Mechanism of inhibition of active potassium transport in isolated midgut of *Manduca sexta* by *Bacillus thuringiensis* endotoxin. *J. Exp. Biol.* 83, 293–304.

Harvey, W. R., Cioffi, M., Dow, J. A. T., and Wolfersberger, M. G. 1983. Potassium ion transport ATPase in insect epithelia. *J. Exp. Biol.* 106, 91–117.

Hawley, I. M., and White, G. F. 1935. Preliminary studies on the diseases of larvae of the Japanese beetle (*Popillia japonica* Newm.). *J. N.Y. Entomol. Soc.* 43, 405–412.

Haynes, W. C., and Rhodes, L. J. 1966. Spore formation by *Bacillus popilliae* in liquid medium containing activated carbon. *J. Bacteriol.* 91, 2270–2274.

Heierson, A., Landén, R., Lövgren, A., Dalhammar, G., and Boman, H.G. 1987. Transformation of vegetative cells of *Bacillus thuringiensis* by plasmid DNA. *J. Bacteriol.* 169, 1147–1152.

Heimpel, A. M. 1955. Investigations of the mode of action of strains of *Bacillus cereus* Fr. and Fr. pathogenic for the larch sawfly, *Pristiphora erichsonii* (Htg.). *Can. J. Zool.* 33, 311–326.

Heimpel, A. M. 1967. A critical review of *Bacillus thuringiensis* var. *thuringiensis* Berliner and other crystalliferous bacteria. *Annu. Rev. Entomol.* 12, 287–322.

Heimpel, A. M., and Angus, T. A. 1958. The taxonomy of insect pathogens related to Bacillus cereus Frankland and Frankland. *Can. J. Microbiol.* 4, 531–541.

Heimpel, A. M., and Angus, T. A. 1959. The site of action of crystalliferous bacteria in Lepidoptera larvae. *J. Insect Pathol.* 1, 152–170.

Herbert, D. A., and Harper, J. D. 1986. Bioassay of a beta-exotoxin of *Bacillus thuringiensis* against *Geocoris punctipes* (Hemiptera: Lygaeidae). *J. Econ. Entomol.* 79, 592–595.

Herrnstadt, C., Soares, G. G., Wilcox, E. R., and Edwards, D. L. 1986. A new strain of *Bacillus thuringiensis* with activity against coleopteran insects. *Biotechnology* 4, 305–308.

Hertlein, B. C., Levy, R., and Miller, T. W., Jr. 1979. Recycling potential and selective retrieval of *Bacillus sphaericus* from soil in a mosquito habitat. *J. Invertebr. Pathol.* 33, 217–221.

Himeno, M., Koyama, N., Funato, T., and Komano, T. 1985. Mechanism of action of *Bacillus thuringiensis* insecticidal delta-endotoxin on insect cells *in vitro*. *Agric. Biol. Chem.* 49, 1461–1468.

Hindley, J., and Berry, C. 1988. *Bacillus sphaericus* strain 2297: Nucleotide sequence of 41.9kDa toxin gene. *Nucleic Acids Res.* 16, 4168.

Hofmann, C., Vanderbruggen, H., Höfte, H., Van Rie, J., Jasens, S., and Van Mellaert, H. 1988. Specificity of *Bacillus thuringiensis* δ-endotoxin is correlated with the presence of high-affinity binding sites in the brush border membrane of target insect midguts. *Proc. Natl. Acad. Sci. U.S.A.* 85, 7844–7848.

Höfte, H., and Whiteley, H. R. 1989. Insecticidal crystal proteins of *Bacillus thuringiensis*. *Microbiol. Rev.* 53, 242–255.

Höfte, H., de Greve, H., Seurinck, J., Jansen, S., Mahillon, J., Ampe, C., Vandekerckhove, J., Vanderbruggen, H., van Montagu, M., Zabeau, M., and Vaeck, M. 1986. Structural and functional analysis of a cloned delta endotoxin of *Bacillus thuringiensis berliner* 1715. *Eur. J. Biochem.* 161, 273–280.

Holmberg, A., Sievänen, R., and Carlberg, G. 1980. Fermentation of *Bacillus thuringiensis* for exotoxin production: Process analysis study. *Biotechnol. Bioeng.* 22, 1707–1724.

Holmes, K. C., and Monro, R. E. 1965. Studies on the structure of parasporal inclusions from *Bacillus thuringiensis J. Mol. Biol.* 14, 572–581.

Holst, E. C. 1946. A simple field test for American foulbrood. *Bee World* 27, 13–14.

Holt, J. G. 1977. "The Shorter Bergey's Manual of Determinative Bacteriology." Eighth Ed., Williams and Wilkins, Baltimore, Maryland.

Holt, J. G. 1984, 1986, 1989a, 1989b. "Bergey's Manual of Systematic Bacteriology." Vol. 1 to 4. Williams and Wilkins, Baltimore, Maryland.

Honée, G., Vriezen, W., and Visser, B. 1990. A translation fusion product of two different insecticidal crystal protein genes of *Bacillus thuringiensis* exhibits an enlarged insecticidal spectrum. *Appl. Environ. Microbiol.* 56, 823–825.

Honigman, A., Nedjar-Pazerini, G., Yawetz, A., Oron, U., Schuster, S., Broza, M., and Sneh, B. 1986. Cloning and expression of the lepidopteran toxin produced by *Bacillus thuringiensis* var. *thuringiensis* in *Escherichia coli*. *Gene* 42, 69–77.

Hornby, J. A., and Gardner, W. A. 1987. Dosage/mortality response of *Spodoptera frugiperda* (Lepidoptera: Noctuidae) and other noctuid larvae to *beta*-exotoxin of *Bacillus thuringiensis*. *J. Econ. Entomol.* 80, 925–929.

Hoy, M. A., and Ouyang, Y.-L. 1987. Toxicity of the β-exotoxin of *Bacillus thuringiensis* to *Tetranychus pacificus* and *Metaseiulus occidentalis* (Acari: Tetranychidae and Phytoseiidae). *J. Econ. Entomol.* 80, 507–511.

Huber, H. E., and Lüthy, P. 1981. *Bacillus thuringiensis* delta-endotoxin: Composition and activation. *In* "Parthogenesis of Invertebrate Microbial Diseases." (E. W. Davidson, ed.), pp. 209–234. Allanheld, Osmun; Totowa, New Jersey.

Huber, H. E., Lüthy, P., Ebersold, H.-R., and Cordier, J.-L. 1981. The subunits of the parasporal crystal of *Bacillus thuringiensis*: Size, linkage and toxicity. *Arch. Microbiol.* 129, 14–18.

Huger, A. M., Krieg, A., Langenbruch, G. A., and Schnetter, W. 1986. Discovery of a new strain of *Bacillus thuringiensis* effective against Coleoptera. *In* "Symposium in Memorian Dr. Ernst Berliner on the Occasion of the 75th Anniversary of Primary Description of *Bacillus thuringiensis*." Darmstadt, 25 August 1986. (A. Krieg and A. M. Huger, eds.), Mitt. Biol. Bundesanst. Land- und Forstwirtsch. Berlin-Dahlem No. 233, pp. 83–96. Paul Parey, Berlin.

Hukuhara, T., Midorikawa, M., and Iwahana, H. 1984. The effect of δ-endotoxin of *Bacillus thuringiensis* on the gut movements of the silkworm, *Bombyx mori*. *Appl. Entomol. Zool.* 19, 221–226.

Hurley, J. M., Lee, S. G., Andrews, R. E., Jr., Klowden, M. J., and Bulla, L. A., Jr. 1985. Separation of the cytolytic and mosquitocidal proteins of *Bacillus thuringiensis* subsp. *israelensis*. *Biochem. Biophys. Res. Commun.* 126, 961–965.

Hurpin, B. 1955. Sur une "maladie laiteuse" des larves de *Melolontha melolontha* L. (Coléopt. Scarabeidae). *C.R. Soc. Biol.* 149, 1966–1967.

Hurpin, B. 1967. Recherches épizootiologiques sur la maladie laiteuse à *Bacillus popilliae* "Melolontha". *Ann. Epiphyt.* 18, 127–173.

Husz, B. 1929. On the use of *Bacillus thuringiensis* in the fight against the corn borer. *Int. Corn Borer Invest. Sci. Rep.* 2, 99–110.

Ibarra, J. E., and Federici, B. A. 1986. Parasporal bodies of *Bacillus thuringiensis* subsp. *morrisoni* (PG-14) and *Bacillus thuringiensis* subsp. *israelensis* are similar in protein composition and toxicity. *FEBS Microbiol. Lett.* 34, 79–84.

Iizuka, T., Faust, R. M., and Ohba, M. 1983. Comparative profiles of plasmid DNA and morphology of parasporal crystals in four strains of *Bacillus thuringiensis* subsp. *darmstadiensis*. *Appl. Entomol. Zool.* 18, 486–494.

Ikezawa, H., Hashimoto, A., Taguchi, R., Nakabayashi, T., and Himeno, M. 1989. Release of PI-anchoring enzymes and other effects of phosphatidylinositol-specific phospholipase C from *Bacillus thuringiensis* on TN-368 cells from a moth ovary. *Toxicon* 27, 637–645.

Ishiwata, S. 1901. On a severe flacherie (sotto disease). *Dainihon Sanshi Kaiho 114*, 1–5.

Ishiwata, S. 1905. On Sotto Bacillus. *Dainihon Sanshi Kaiho 160*, 24–28; *161*, 24–26.

Ishiwata, S. 1906. Investigations on Sotto Bacillus. *Kyoto Sangyo Koshujo Sanji Hokoku 10*, 1–20.

Iwabuchi, H. 1908. "Popular Silkworm Pathology." First Ed. Meibundo, Tokyo.

Johnson, D. E., and Davidson, L. I. 1984. Specificity of cultured insect tissue cells for bioassay of entomocidal protein from *Bacillus thuringiensis*. *In Vitro 20*, 66–70.

Johnson, D. E., and McGaughey, W. H. 1984. Insecticidal activity of spore-free mutants of *Bacillus thuringiensis* against the Indian meal moth and almond moth. *J. Invertebr. Pathol. 43*, 156–159.

Jones, D. R., Karunakaran, V., and Burges, H. D. 1983. Phages naturally associated with the *aizawai* variety of insect pathogen *Bacillus thuringiensis* and their relevance to strain identification. *J. Appl. Bacteriol. 54*, 373–377.

Kalfon, A., Charles, J.-F., Bourgouin, C., and de Barjac, H. 1984. Sporulation of *Bacillus sphaericus* 2297: An electron microscope study of crystal-like inclusion biogenesis and toxicity to mosquito larvae. *J. Gen. Microbiol. 130*, 893–900.

Kalucy, E. C., and Daniel, A. 1972. The reaction of *Anopheles annulipes* larvae to infection by *Aeromonas punctata*. *J. Invertebr. Pathol. 19*, 189–197.

Katagiri, K., and Shimazu, M. 1974. Sporulation of *Bacillus thuringiensis* in the cadavers of insect hosts. *J. Jpn. For. Soc. 56*, 325–331.

Kawanishi, C. Y., Splittstoesser, C. M., and Tashiro, H. 1978. Infection of the European chafer, *Amphimallon majalis*, by *Bacillus popilliae*: Ultrastructure. *J. Invertebr. Pathol. 31*, 91–102.

Kellen, W. R., Clark, T. B., Lindegren, J. E., Ho, B. C., Rogoff, M. H., and Singer, S. 1965. *Bacillus sphaericus* Neide as a pathogen of mosquitoes. *J. Invertebr. Pathol. 7*, 442–448.

Kim, Y. T., and Huang, H. T. 1970. The β-exotoxins of *Bacillus thuringiensis*. I. Isolation and characterization. *J. Invertebr. Pathol. 15*, 100–108.

Kim, K.-H., Ohba, M., and Aizawa, K. 1984. Purification of the toxic protein from *Bacillus thuringiensis* serotype 10 isolate demonstrating a preferential larvicial activity to the mosquito. *J. Invertebr. Pathol. 44*, 214–219.

Kiselev, E. V. 1975. The effect of biopreparations on insect parasites and predators. *Zash. Rast. 12*, 23.

Klein, M. G. 1981. Advances in the use of *Bacillus popilliae* for pest control. *In* "Microbial Control of Pests and Plant Diseases 1970–1980." (H. D. Burges, ed.), pp. 183–192. Academic Press, London.

Klein, M. G., Johnson, C. H., and Ladd, T. L., Jr. 1976. A bibliography of the milky disease bacteria (*Bacillus* spp.) associated with the Japanese beetle, *Popilla japonica* and closely related Scarabaeidae. *Bull. Entomol. Soc. Am. 22*, 305–310.

Klier, A., and Rapoport, G. 1984. Cloning and heterospecific expression of the crystal protein genes from *Bacillus thuringiensis*. *In* "Bacterial Protein Toxins." (J. E. Alouf, F. J. Fehrenbach, J. H. Freer, and J. Jeljaszewicz, eds.), pp. 65–72. Academic Press, London.

Klier, A., Lecadet, M.-M., and Rapoport, G. 1978. Transcription in vitro of sporulation-specific mRNA's by RNA polymerase from *Bacillus thuringiensis*. *In* "Spores VII." (G. Chambers and J. C. Vary, eds.), pp. 205–212. American Society of Microbiology, Washington, D.C.

Klier, A., Fargette, F., Ribier, J., and Rapoport, G. 1982. Cloning and expression of the crystal protein genes from *Bacillus thuringiensis* strain *berliner* 1715. *EMBO J. 1*, 791–799.

Klier, A., Bourgouin, C., and Rapoport, G. 1983a. Mating between *Bacillus subtilis* and *Bacillus thuringiensis* and transfer of cloned crystal genes. *Mol. Gen. Genet. 191*, 257–262.

Klier, A., Parsot, C., and Rapoport, G. 1983b. *In vitro* transcription of the cloned chromosomal crystal gene from *Bacillus thuringiensis*. *Nucleic Acids Res. 11*, 3973–3987.

Knowles, B. H., and Ellar, D. J. 1986. Characterization and partial purification of a plasma membrane receptor for *Bacillus thuringiensis* var. *kurstaki* lepidopteran-specific δ-endotoxin. *J. Cell Sci. 83*, 89–101.

Knowles, B. H., and Ellar, D. J. 1987. Colloid-osmotic lysis is a general feature of the mechanism of action of *Bacillus thuringiensis* δ-endotoxins with different insect specificity. *Biochem. Biophys. Acta 924*, 509–518.

Knowles, B. H., and Ellar, D. J. 1988. Differential specificity of two insecticidal toxins from *Bacillus thuringiensis* var. *aizawai*. *Mol. Microbiol. 2*, 153–157.

Knowles, B. H., Thomas, W. E., and Ellar, D. J. 1984. Lectin-like binding of *Bacillus thuringiensis* var. *kurstaki* lepidopteran-specific toxin is an initial step in insecticidal action. *FEBS Lett. 168*, 197–202.

Knowles, B. H., Blatt, M. R., Tester, M., Horsnell, J. M., Carroll, J., Menestrina, G., and Ellar, D. J. 1989. A cytolytic δ-endotoxin from *Bacillus thuringiensis* var. *israelensis* forms cation-selective channels in planar lipid bilayers. *FEBS Lett. 244*, 259–262.

Krainova, O. A., Kosareva, N. I., and Shevtsov, V. V. 1983. Activity of intracellular enzymes in prospores and sporangia of *Bacillus thuringiensis*. *Microbiology 52*, 37–39.

Kramer, J. M. 1984. *Bacillus cereus* exotoxins: Production, isolation, detection and properties. *In* "Bacterial Protein Toxins." (J. E. Alouf, F. J. Fehrenbach, J. H. Freer, and J. Jeljaszewicz, eds.), pp. 385–386. Academic Press, London.

Krieg, A. 1970. Thuricin, a bacteriocin produced by *Bacillus thuringiensis*. *J. Invertebr. Pathol. 15*, 291.

Krieg, A. 1971a. Concerning α-exotoxin produced by vegetative cells of *Bacillus thuringiensis* and *Bacillus cereus*. *J. Invertebr. Pathol. 17*, 134–135.

Krieg, A. 1971b. Is the potential pathogenicity of bacilli for insects related to production of α-exotoxin? *J. Invertebr. Pathol. 18*, 425–426.

Krieg, A., and Lysenko, O. 1979. Toxine und Enzyme bei einigen *Bacillus*-Arten unter besonderer Berücksichtigung der *B. cereus-thuringiensis*-Gruppe. *Zentralbl. Bakteriol. Parasitenkd. Infektionskr. Hyg. Abt. 2. 134*, 70–88.

Krieg, A., Huger, A. M., Langenbruch, G. A., and Schnetter, W. 1983. *Bacillus thuringiensis* var. *tenebrionis*: Ein neuer gegenüber Larven von Coleopteren wirksamer Pathotyp. *Z. Angew. Entomol. 96*, 500–508.

Krieg, A., Schnetter, W., Huger, A. M., and Langenbruch, G. A. 1987. *Bacillus thuringiensis* subsp. *tenebrionis*, strain BI 256–82: A third pathotype within the H-serotype 8a8b. *System. Appl. Microbiol. 9*, 138–141.

Kronstad, J. W., and Whiteley, H. R. 1984. Inverted repeat sequences flank a *Bacillus thuringiensis* crystal protein gene. *J. Bacteriol. 160*, 95–102.

Kronstad, J. W., and Whiteley, H. R. 1986. Three classes of homologous *Bacillus thuringiensis* crystal-protein genes. *Gene 43*, 29–40.

Kronstad, J. W., Schnepf, H. E., and Whiteley, H. R. 1983. Diversity of locations for *Bacillus thuringiensis* crystal protein genes. *J. Bacteriol. 154*, 419–428.

Krych, V. K., Johnson, J. L., and Yousten, A. A. 1980. Deoxyribonucleic acid homologies among strains of *Bacillus sphaericus*. *Int. J. System. Bacteriol. 30*, 476–484.

Krywienczyk, J., and Lüthy, P. 1974. Serological relationship between three varieties of *Bacillus popilliae*. *J. Invertebr. Pathol. 23*, 275–279.

Krywienczyk, J., Dulmage, H. T., and Fast, P. G. 1978. Occurrence of two serologically distinct groups within *Bacillus thuringiensis* serotype 3 ab var. *kurstaki*. *J. Invertebr. Pathol. 31*, 372–375.

Lacey, L. A. 1990. Persistence and formulation of *Bacillus sphaericus*. *In* "Bacterial Control of Mosquitoes & Black Flies." (H. de Barjac and D. J. Sutherland, eds.), pp. 284–294. Rutgers University Press, New Brunswick.

Lacey, L. A., and Federici, B. A. 1979. Pathogenesis and midgut histopathology of *Bacillus thuringiensis* in *Simulium vittatum* (Diptera: Simuliidae). *J. Invertebr. Pathol. 33*, 171–182.

Lacey, L. A., Mulla, M. S., and Dulmage, H. T. 1978. Some factors affecting the pathogenicity of *Bacillus thuringiensis* Berliner against blackflies. *Environ. Entomol. 7*, 583–588.

Lacey, L. A., Day, J., and Heitzman, C. M. 1987. Long-term effects of *Bacillus sphaericus* on *Culex quinquefasciatus. J. Invertebr. Pathol. 49*, 116–123.

Langford, G. S., Vincent, R. H., and Cory, E. N. 1942. The adult Japanese beetle as host and disseminator of Type A milky disease. *J. Econ. Entomol. 35*, 165–169.

Larget, I., and de Barjac, H. 1981. Activité comparée de 22 variétés de *Bacillus thuringiensis* sur 3 espèces de *Culicidae. Entomophaga 26*, 143–148.

Larget-Thiery, I., Hamon, S., and de Barjac, H. 1984. Sensibilité des *Culicidae* à la β-exotoxine de *Bacillus thuringiensis. Entomophaga 29*, 95–108.

Laurent, P., and Charles, J.-F. 1984. Action comparée des cristaux solubilisés des sérotypes H-14 et H-1 de *Bacillus thuringiensis* sur des cultures de cellules de *Aedes aegypti. Ann. Microbiol.* (Inst. Pasteur) *135A*, 473–484.

Lecadet, M.-M., and Lereclus, D. 1984. Structure and activity of the *B. thuringiensis* δ-endotoxin recent development in genetics. *In* "Bacterial Protein Toxins." (J. E. Alouf, F. J. Fehrenbach, J. H. Freer, and J. Jeljaszewicz, eds.), pp. 147–154. Academic Press, London.

Lecadet, M.-M., Blondel, M.-O., and Ribier, J. 1980. Generalized transduction in *Bacillus thuringiensis* var. *berliner* 1715 using bacteriophage CP-54Ber. *J. Gen. Microbiol. 121*, 203–212.

Lecadet, M.-M., Sanchis, V., Menou, G., Rabot, P., Lereclus, D., Chaufaux, J., and Martouret, D. 1988. Identification of a δ-endotoxin gene product specifically active against *Spodoptera littoralis* Bdv. among proteolysed fractions of the insecticidal crystals of *Bacillus thuringiensis* subsp. *aizawai* 7.29. *Appl. Environ. Microbiol. 54*, 2689–2698.

Lee, S. G., Eckblad, W., and Bulla, L. E., Jr. 1985. Diversity of protein inclusion bodies and identification of mosquitocidal protein in *Bacillus thuringiensis* subsp. *israelensis. Biochem. Biophys. Res. Commun. 126*, 953–960.

Lereclus, D., Lecadet, M. M., Klier, A., Ribier, J., Rapoport, G., and Dedonder, R. 1985. Recent aspects of genetic manipulation in *Bacillus thuringiensis. Biochimie 67*, 91–99.

Lereclus, D., Mahillon, J., Menou, G., and Lecadet, M.-M. 1986. Identification of Tn*4430*, a transposon of *Bacillus thuringiensis* functional in *Escherichia coli. Mol. Gen. Genet. 204*, 52–57.

Lewis, L. O., Yousten, A. A., and Murray, R. G. E. 1987. Characterization of the surface protein layers of the mosquito-pathogenic strains of *Bacillus sphaericus. J. Bacteriol. 169*, 72–79.

Logan, N. A., and Berkeley, R. C. W. 1984. Identification of *Bacillus* strains using the API system. *J. Gen. Microbiol. 130*, 1871–1882.

Louis, J., Jayaraman, K., and Szulmajster, J. 1984. Biocide gene(s) and biocidal activity in different strains of *Bacillus sphaericus*. Expression of the gene(s) in *E. coli* maxicells. *Mol. Gen. Genet. 195*, 23–28.

Lüthy, P. 1968. Untersuchungen an *Bacillus fribourgensis* Wille. *Zentralbl. Bakteriol. Parasitenkd. Infektionskr. Hyg. 122*, 671–711.

Lüthy, P. 1980. Insecticidal toxins of *Bacillus thuringiensis. FEMS Microbiol. Lett. 8*, 1–7.

Lüthy, P., and Ebersold, H.R. 1981. *Bacillus thuringiensis* delta-endotoxin: Histopathology and molecular mode of action. *In* "Pathogenesis of Invertebrate Microbial Diseases." (E. W. Davidson, ed.), pp. 235–267. Allanheld, Osmun; Totowa, New Jersey.

Lüthy, P., and Krywienczyk, J. 1972. Serological comparison of three milky disease isolates. *J. Invertebr. Pathol. 19*, 163–165.

Lüthy, P., Hayashi, Y., and Angus, T. A. 1970. Presence of endotoxin in vegetative cells of *Bacillus thuringiensis* var. *sotto. Can. J. Microbiol. 16*, 905–906.

Lynch, M. J., and Baumann, P. 1985. Immunological comparisons of the crystal protein from strains of *Bacillus thuringiensis. J. Invertebr. Pathol. 46*, 47–57.

Lynch, R. E., Lewis, L. C., and Brindley, T. A. 1976. Bacteria associated with eggs and first-instar larvae of the European corn borer: Isolation techniques and pathogenicity. *J. Invertebr. Pathol. 27*, 325–331.

Lysenko, O. 1974. Bacterial exoenzymes toxic for insects; proteinase and lecithinase. *J. Hyg. Epidemiol. Microbiol. Immunol. 18*, 347–352.

Lysenko, O. 1983. Report on diagnosis of bacteria from insects (1963–1981). *Acta Entomol. Bohemoslov. 80*, 473–478.

Lysenko, O., Davidson, E. W., Lacey, L. A., and Yousten, A. A. 1985. Five new mosquito larvicidal strains of *Bacillus sphaericus* from non-mosquito origins. *J. Am. Mosq. Control Assoc. 1*, 369–371.

Mahillon, J., Seurinck, J., Van Rompuy, L., Delcour, J., and Zabeau, M. 1985. Nucleotide sequence and structural organization of an insertion sequence element (IS*231*) from *Bacillus thuringiensis* strain berliner 1715. *Embo J. 4*, 3895–3899.

Mahillon, J., Chungjatupornchai, W., Decock, J., Dierickx, S., Michiels, F., Peferoen, M., and Joos, H. 1989. Transformation of *Bacillus thuringiensis* by electroporation. *FEMS Microbiol. Lett. 60*, 205–210.

Martin, P. A. W., and Travers, R. S. 1989. Worldwide abundance and distribution of *Bacillus thuringiensis* isolates. *Appl. Environ. Microbiol. 55*, 2437–2442.

Mattes, O. 1927. Parasitäre Krankheiten der Mehlmottenlarven und Versuche über ihre Verwendbarkeit als biologisches Bekämpfungsmittel. (Zugleich ein Beitrag zur Zytologie der Bakterien.). *Ges. Beford. Gesamte Naturwiss. Marburg 62*, 381–417.

Mayes, M. E., Held, G. A., Lau, C., Seely, J. C., Roe, R. M., Dauterman, W. C., and Kawanishi, C. Y. 1989. Characterization of the mammalian toxicity of the crystal polypeptides of *Bacillus thuringiensis* subsp. *israelensis*. *Fundam. Appl. Toxicol. 13*, 310–322.

McCleskey, C. S., and Melampy, R. M. 1939. Bactericidal properties of royal jelly of the honeybee. *J. Econ. Entomol. 32*, 581–587.

McConnell, E., and Richards, A. G. 1959. The production by Bacillus thuringiensis Berliner of a heat-stable substance toxic for insects. *Can. J. Microbiol. 5*, 161–168.

McGaughey, W. H. 1978. Response of *Plodia interpunctella* and *Ephestia cautella* larvae to spores and parasporal crystals of *Bacillus thuringiensis*. *J. Econ. Entomol. 71*, 687–688.

McGaughey, W. H. 1985. Insect resistance to the biological insecticide *Bacillus thuringiensis*. *Science 229*, 193–194.

McGaughey, W. H., and Johnson, D. E. 1987. Toxicity of different serotypes and toxins of *Bacillus thuringiensis* to resistant and susceptible Indianmeal moths (Lepidoptera: Pyralidae). *J. Econ. Entomol. 80*, 1122–1126.

McLinden, J. H., Sabourin, J. R., Clark, B. D., Gensler, D. R., Workman, W. E., and Dean, D. H. 1985. Cloning and expression of an insecticidal k-73 type crystal protein gene from *Bacillus thuringiensis* var. *kurstaki* into *Escherichia coli*. *Appl. Environ. Microbiol. 50*, 623–628.

McPherson, A., Jurnak, F., Singh, G. J. P., and Gill, S. S. 1987. Preliminary X-ray diffraction analysis of crystals of *Bacillus thuringiensis* toxin, a cell membrane disrupting protein. *J. Mol. Biol. 195*, 755–757.

Meeusen, R. L., and Warren, G. 1989. Insect control with genetically engineered crops. *Annu. Rev. Entomol. 34*, 373–381.

Mettus, A.-M., and Macaluso, A. 1990. Expression of *Bacillus thuringiensis* δ-endotoxin genes during vegetative growth. *Appl. Environ. Microbiol. 56*, 1128–1134.

Mikkola, A. R., Carlberg, G. A., Vaara, T., and Gyllenberg, H. G. 1982. Comparison of inclusions in different *Bacillus thuringiensis* strains. An electron microscope study. *FEMS Microbiol. Lett. 13*, 401–408.

Milner, R. J. 1976. A laboratory evaluation of the pathogenicity of *Bacillus popilliae* var. *rhopaea*, the agent of milky disease in *Rhopaea verreauxi* (Coleoptera: Scarabaeidae). *J. Invertebr. Pathol. 28*, 185–190.

Milner, R. J. 1981. A novel milky disease organism from Australian scarabaeids: Field occurrence, isolation, and infectivity. *J. Invertebr. Pathol. 37*, 304–309.

Milner, R. J., and Beaton, C. D. 1981. A novel milky disease organism from Australian scarabaeids: Ultrastructure. *J. Invertebr. Pathol. 37*, 310–318.

Mitani, K., and Watarai, J. 1916. A new method to isolate the toxin of *Bacillus sotto* Ishiwata by passing through a bacterial filter and a preliminary report on the toxic action of this toxin to the silkworm larva. *Aichi Gensanshu Seizojo Hokoku 3*, 33–42.

Moar, W. J., Osbrink, W. L. A., and Trumble, J. T. 1986. Potentiation of *Bacillus thuringiensis* var. *kurstaki* with thuringiensin on beet armyworm (Lepidoptera: Noctuidae). *J. Econ. Entomol. 79*, 1443–1446.

Mohd-Salleh, M. B., and Lewis, L. C. 1982. Feeding deterrent response of corn insects to β-exotoxin of *Bacillus thuringiensis*. *J. Invertebr. Pathol. 39*, 323–328.

Morris, O. N. 1969. Susceptibility of several forest insects of British Columbia to commercially produced *Bacillus thuringiensis*. II. Laboratory and field pathogenicity tests. *J. Invertebr. Pathol. 13*, 285–295.

Murphy, D. W., Sohi, S. S., and Fast, P. G. 1976. *Bacillus thuringiensis* enzyme-digested delta endotoxin: Effect on cultured insect cells. *Science 194*, 954–956.

Myers, P., and Yousten, A. A. 1978. Toxic activity of *Bacillus sphaericus* SSII-1 for mosquito larvae. *Infect. Immun. 19*, 1047–1053.

Myers, P. S., and Yousten, A. A. 1980. Localization of a mosquito-larval toxin of *Bacillus sphaericus* 1593. *Appl. Environ. Microbiol. 39*, 1205–1211.

Myers, P., Yousten, A. A., and Davidson, E. W. 1979. Comparative studies of the mosquito-larval toxin of *Bacillus sphaericus* SSII-1 and 1593. *Can. J. Microbiol. 25*, 1227–1231.

Nagamatsu, Y., Tsutsui, R., Ichimaru, T., Nagamatsu, M., Koga, K., and Hayashi, K. 1978. Subunit structure and toxic component of δ-endotoxin from *Bacillus thuringiensis*. *J. Invertebr. Pathol. 32*, 103–109.

Nagamatsu, Y., Itai, Y., Hatanaka, C., Funatsu, G., and Hayashi, K. 1984. A toxic fragment from the entomocidal crystal protein of *Bacillus thuringiensis*. *Agric. Biol. Chem. 48*, 611–619.

Narasu, M. L., and Gopinathan, K. P. 1986. Purification of larvicidal protein from *Bacillus sphaericus* 1593. *Biochem. Biophys. Res. Commun. 141*, 756–761.

Narayanan, K., Jayaraj, S., and Govindarajan, R. 1976. Further observations on the mode of action of *Bacillus thuringiensis* on *Papilio demoleus* and *Spodoptera litura*. *J. Invertebr. Pathol. 28*, 269–270.

Nickerson, K. W., and Schnell, D. J. 1983. Toxicity of cyclic peptide antibiotics to larvae of *Aedes aegypti*. *J. Invertebr. Pathol. 42*, 407–409.

Nishiitsutsuji-Uwo, J., and Endo, Y. 1980a. Mode of action of *Bacillus thuringiensis* δ-endotoxin: General characteristics of intoxicated *Bombyx* larvae. *J. Invertebr. Pathol. 35*, 219–228.

Nishiitsutsuji-Uwo, J., and Endo, Y. 1980b. Mode of action of *Bacillus thuringiensis* δ-endotoxin: Relative roles of spores and crystals in toxicity to *Pieris, Lymantria* and *Ephestia* larvae. *Appl. Entomol. Zool. 15*, 416–424.

Nishiitsutsuji-Uwo, J., and Endo, Y. 1981. Mode of action of *Bacillus thuringiensis* δ-endotoxin: Changes in hemolymph pH and ions of *Pieris, Lymantria* and *Ephestia* larvae. *Appl. Entomol. Zool. 16*, 225–230.

Nishiitsutsuji-Uwo, J., Wakisaka, Y., and Eda, M. 1975. Sporeless mutants of *Bacillus thuringiensis*. *J. Invertebr. Pathol. 25*, 355–361.

Nishiitsutsuji-Uwo, J., Endo, Y., and Himeno, M. 1979. Mode of action of *Bacillus thuringiensis* δ-endotoxin: Effect on TN-368 cells. *J. Invertebr. Pathol. 34*, 267–275.

Nishiitsutsuji-Uwo, J., Endo, Y., and Himeno, M. 1980. Effects of *Bacillus thuringiensis* δ-endotoxin on insect and mammalian cells *in vitro*. *Appl. Entomol. Zool. 15*, 133–139.

Norris, J. R. 1971. The protein crystal toxin of *Bacillus thuringiensis*: Biosynthesis and physical structure. *In* "Microbial Control of Insects and Mites." (H. D. Burges and N. W. Hussey, eds.), pp. 229–246. Academic Press, London.

Norris, J. R., and Burges, H. D. 1963. Esterases of crystalliferous bacteria pathogenic for insects: Epizootiological applications. *J. Insect Pathol. 5*, 460–472.

Northrup, Z. 1914. A bacterial disease of the larvae of the June beetle, Lachnosterna spp. *Mich. Agric. Exp. Stn. Tech. Bull.* 18.

Obukowicz, M. G., Perlak, F. J., Kusano-Kretzmer, K., Mayer, E. J., and Watrud, L. S. 1986. Integration of the delta-endotoxin gene of *Bacillus thuringiensis* into the chromosome of root-colonizing strains of pseudomonads using Tn5. *Gene 45*, 327–331.

Oeda, K., Inouye, K., Ibuchi, Y., Oshie, K., Shimizu, M., Nakamura, K., Nishioka, R., Takada, Y., and Ohkawa, H. 1989. Formation of crystals of the insecticidal proteins of *Bacillus thuringiensis* subsp. *aizawai* IPL7 in *Escherichia coli*. *J. Bacteriol. 171*, 3568–3571.

Ohba, M., and Aizawa, K. 1979. A new subspecies of *Bacillus thuringiensis* possessing 11a:11c flagellar antigenic structure: *Bacillus thuringiensis* subsp. *kyushuensis*. *J. Invertebr. Pathol. 33*, 387–388.

Ohba, M., and Aizawa, K. 1986a. Distribution of *Bacillus thuringiensis* in soils of Japan. *J. Invertebr. Pathol. 47*, 277–282.

Ohba, M., and Aizawa, K. 1986b. Insect toxicity of *Bacillus thuringiensis* isolated from soils of Japan. *J. Invertebr. Pathol. 47*, 12–20.

Ohba, M., and Aizawa, K. 1986c. Crystals of *Bacillus thuringiensis* subsp. *yunnanensis* are produced only in asporogenous cells. *J. Invertebr. Pathol. 48*, 254–256.

Ohba, M., and Aizawa, K. 1990. Occurrence of two pathotypes in *Bacillus thuringiensis* subsp. *fukuokaensis* (flagellar serotype 3a:3d:3e). *J. Invertebr. Pathol. 55*, 293–294.

Ohba, M., Tantichodok, A., and Aizawa, K. 1981. Production of heat-stable exotoxin by *Bacillus thuringiensis* and related bacteria. *J. Invertebr. Pathol. 38*, 26–32.

Ohba, M., Aizawa, K., and Sudo, S. 1984. Distribution of *Bacillus thuringiensis* in sericultural farms of Fukuoka Prefecture, Japan. *Proc. Assoc. Plant Prot. Kyushu 30*, 152–155.

Ohba, M., Yu, Y. M., and Aizawa, K. 1988. Occurrence of non-insecticidal *Bacillus thuringiensis* flagellar serotype 14 in the soil in Japan. *System. Appl. Microbiol. 11*, 85–89.

Oron, U., Sokolover, M., Yawetz, A., Broza, M., Sneh, B., and Honigman, A. 1985. Ultrastructural changes in the larval midgut epithelium of *Spodoptera littoralis* following ingestion of δ-endotoxin of *Bacillus thuringiensis* var. *entomocidus*. *J. Invertebr. Pathol. 45*, 353–355.

Ozawa, K., and Iwahana, H. 1986. Involvement of a transmissible plasmid in heat-stable exotoxin and delta-endotoxin production in *Bacillus thuringiensis* subspecies *darmstadiensis*. *Curr. Microbiol. 13*, 337–340.

Padua, L. E., and Federici, B. A. 1990. Development of mutants of the mosquitocidal bacterium *Bacillus thuringiensis* subspecies *morrisoni* (PG-14) toxic to lepidopterous or dipterous insects. *FEMS Microbiol. Lett. 66*, 257–262.

Padua, L. E., Ohba, M., and Aizawa, K. 1980. The isolates of *Bacillus thuringiensis* serotype 10 with a highly preferential toxicity to mosquito larvae. *J. Invertbr. Pathol. 36*, 180–186.

Padua, L. E., Ohba, M., and Aizawa, K. 1984. Isolation of a *Bacillus thuringiensis* strain (serotype 8a:8b) highly and selectively toxic against mosquito larvae. *J. Invertebr. Pathol. 44*, 12–17.

Pendleton, I. R. 1969. Ecological significance of antibiotics of some varieties of *Bacillus thuringiensis*. *J. Invertebr. Pathol. 13*, 235–240.

Pendleton, I. R. 1970. Sodium and potassium fluxes in *Philosamia ricini* during *Bacillus thuringiensis* protein crystal intoxication. *J. Invertebr. Pathol. 16*, 313–314.

Pendleton, I. R., and Morrison, R. B. 1967. Antigenic analysis of the digests of the crystal toxins of *Bacillus thuringiensis*. *J. Appl. Bacteriol. 30*, 402–405.

Pendleton, I. R., Bernheimer, A. W., and Grushoff, P. 1973. Purification and partial characterization of hemolysins from *Bacillus thuringiensis*. *J. Invertebr. Pathol. 21*, 131–135.

Peng, Y., and Peng, K. 1979. A study on the possible utilization of immunodiffusion and immunofluorescence techniques as the diagnostic methods for American foulbrood of honeybees (*Apis mellifera*). *J. Invertebr. Pathol. 33*, 284–289.

Percy, J., and Fast, P. G. 1983. *Bacillus thuringiensis* crystal toxin: Ultrastructural studies of its effect on silkworm midgut cells. *J. Invertebr. Pathol. 41*, 86–98.

Perlak, F. J., Mendelsohn, C. L., and Thorne, C. B. 1979. Converting bacteriophage for sporulation and crystal formation in *Bacillus thuringiensis*. *J. Bacteriol. 140*, 699–706.

Pfannenstiel, M. A., Ross, E. J., Kramer, V. C., and Nickerson, K. W. 1984. Toxicity and composition of protease-inhibited *Bacillus thuringiensis* var. *israelensis* crystals. *FEMS Microbiol. Lett. 21*, 39–42.

Pfannenstiel, M. A., Muthukumar, G., Couche, G. A., and Nickerson, K. W. 1987. Amino sugars in the glycoprotein toxin from *Bacillus thuringiensis* subsp. *israelensis*. *J. Bacteriol. 169*, 796–801.

Pfannenstiel, M. A., Cray, W. C., Jr., Couche, G. A., and Nickerson, K. W. 1990. Toxicity of protease-resistant domains from the delta-endotoxin of *Bacillus thuringiensis* subsp. *israelensis* in *Culex quinquefasciatus* and *Aedes aegypti* bioassays. *Appl. Environ. Microbiol. 56*, 162–166.

Prasad, S. S. S. V., and Shethna, Y. I. 1976. Antitumor immunity against Yoshida ascites sarcoma after treatment with the proteinaceous crystal of *Bacillus thuringiensis* var. *thuringiensis*. *Indian J. Exp. Biol. 14*, 285–288.

Prasertphon, S., Areekul, P., and Tanada, Y. 1973. Sporulation of *Bacillus thuringiensis* in host cadavers. *J. Invertebr. Pathol. 21*, 205–207.

Priest, F. G., Goodfellow, M., and Todd, C. 1988. A numerical classification of the genus *Bacillus*. *J. Gen. Microbiol. 134*, 1847–1882.

Prinsloo, H. E. 1960. Parasitiese mikro-organismes by die bruinsprinkaan *Locustana pardalina* (Walk.). *S. Afr. Tydskr. Landbouwet. 3*, 551–560.

Ramakrishnan, N. 1968. Observations on the toxicity of *Bacillus thuringiensis* for the silkworm, *Bombyx mori*. *J. Invertebr. Pathol. 10*, 449–450.

Rao, A. S., Amonkar, S. V., and Phondke, G. P. 1979. Cytotoxic activity of the δ-endotoxin of *Bacillus thuringiensis* var. *thuringiensis* (Berliner) on fibrosarcoma in Swiss mice. *Indian J. Exp. Biol. 17*, 1208–1212.

Rautenstein, Ya. I., Talalaeva, G. B., Blokhina, T. P., Krukovskaya, G. E., Pokrovskaya, L. A., and Ilyina, T. V. 1975. Specific phages used for differentiation of entomopathogenic bacterium Bacillus thuringiensis. *Microbiologia 44*, 1081–1085.

Rhim, S.-L., Jahn, N., Schnetter, W., and Geider, K. 1990. Heterologous expression of a mutated toxin gene from *Bacillus thuringiensis* subsp. *tenebrionis*. *FEMS Microbiol. Lett. 66*, 95–99.

Rhodes, R. A. 1965. Symposium on microbial insecticides. II. Milky disease of the Japanese beetle. *Bacteriol. Rev. 29*, 373–381.

Rinderer, T. E., and Rothenbuhler, W. C. 1969. Resistance to American foulbrood in honey bees. X. Comparative mortality of queen, worker, and drone larvae. *J. Invertebr. Pathol. 13*, 81–86.

Rinderer, T. E., Rothenbuhler, W. C., and Gochnauer, T. A. 1974. The influence of pollen on the susceptibility of honey-bee larvae to *Bacillus larvae*. *J. Invertebr. Pathol. 23*, 347–350.

Rosenberg, G., Carlberg, G., and Gyllenberg, H. G. 1971. Microbiological assay of the β-exotoxin of *Bacillus thuringiensis*. *J. Appl. Bacteriol. 34*, 417–423.

Russell, B. L., Jelley, S. A., and Yousten, A. A. 1989. Carbohydrate metabolism in the mosquito pathogen *Bacillus sphaericus* 2362. *Appl. Environ. Microbiol. 55*, 294–297.

Sakanyan, V. A., Tsoi, T. V., Sezonov, G. V., and Alikhanian, S. I. 1982. Expression of enterobacterial gene for antibiotic resistance under control of regulatory signals of Bacillus thuringiensis in gram-negative and gram-positive bacteria. *Genetika 18*, 1825–1834.

Salama, H. S., Foda, M. S., El-Sharaby, A., Matter, M., and Khalafallah, M. 1981. Development of some lepidopterous cotton pests as affected by exposure to sublethal levels of endotoxins of *Bacillus thuringiensis* for different periods. *J. Invertebr. Pathol.* 38, 220–229.

Scherrer, P. S., and Somerville, H. J. 1977. Membrane fractions from the outer layers of spores of *Bacillus thuringiensis* with toxicity to lepidopterous larvae. *Eur. J. Biochem.* 72, 479–490.

Schesser, J. H., and Bulla, L. A., Jr. 1978. Toxicity of *Bacillus thuringiensis* spores to the tobacco hornworm, *Manduca sexta*. *Appl. Environ. Microbiol.* 35, 121–123.

Schmid, E., and Benz, G. 1969. Oral and parenteral toxicity of *Bacillus thuringiensis* "exotoxin", and its inactivation in larvae of *Galleria mellonella*. *Experientia* 25, 96–98.

Schnepf, H. E., and Whiteley, H. R. 1981. Cloning and expression of the *Bacillus thuringiensis* crystal protein gene in *Escherichia coli*. *Proc. Natl. Acad. Sci. U.S.A.* 78, 2893–2897.

Schnepf, H. E., Wong, H. C., and Whiteley, H. R. 1985. The amino acid sequence of a crystal protein from *Bacillus thuringiensis* deduced from the DNA base sequence. *J. Biol. Chem.* 260, 6264–6272.

Schroeder, J. M., Chamberlain, C., and Davidson, E. W. 1989. Resistance to the *Bacillus sphaericus* toxin in cultured mosquito cells. *In Vitro Cell. Dev. Biol.* 25, 887–891.

Šebesta, K., Horská, K., and Vaňková, J. 1973. Estimation of exotoxin production by different strains of *Bacillus thuringiensis* using ^{32}P-labelled exotoxin. *Coll. Czech. Chem. Commun.* 38, 298–303.

Šebesta, K., Farkaš, J., Horská, K., and Vaňková, J. 1981. Thuringiensin, the beta-exotoxin of *Bacillus thuringiensis*. In "Microbial Control of Pests and Plant Diseases 1970–1980." (H. D. Burges, ed.), pp. 249–281. Academic Press, London.

Sekar, V. 1986. Biochemical and immunological characterization of the cloned crystal toxin of *Bacillus thuringiensis* var. *israelensis*. *Biochem. Biophys. Res. Commun.* 137, 748–751.

Sekar, V. 1987. Location of the crystal toxin gene of *Bacillus thuringiensis* var. *aizawai*. *Curr. Microbiol.* 14, 301–304.

Sekar, V., and Carlton, B. C. 1985. Molecular cloning of the delta-endotoxin gene of *Bacillus thuringiensis* var. *israelensis*. *Gene* 33, 151–158.

Sekijima, Y., and Ono, K. 1982. Grouping of *Bacillus thuringiensis* by heat-stable somatic antigens. *Appl. Entomol. Zool.* 17, 393–397.

Sharpe, E. S. 1976. Toxicity of the parasporal crystal of *Bacillus thuringiensis* to Japanese beetle larvae. *J. Invertebr. Pathol.* 27, 421–422.

Sharpe, E. S., and Baker, F. L. 1979. Ultrastructure of the unusual crystal of the HD-1 isolate of *Bacillus thuringiensis* var. *kurstaki*. *J. Invertebr. Pathol.* 34, 320–322.

Sharpe, E. S., and Detroy, R. W. 1979a. Susceptibility of Japanese beetle larvae to *Bacillus thuringiensis*: Associated effects of diapause, midgut pH, and milky disease. *J. Invertebr. Pathol.* 34, 90–91.

Sharpe, E. S., and Detroy, R. W. 1979b. Fat body depletion, a debilitating result of milky disease on Japanese beetle larvae. *J. Invertebr. Pathol.* 34, 92–94.

Sharpe, E. S., and Rhodes, R. A. 1973. The pattern of sporulation of *Bacillus popilliae* in colonies. *J. Invertebr. Pathol.* 21, 9–15.

Sharpe, E. S., St. Julian, G., and Crowell, C. 1970. Characteristics of a new strain of *Bacillus popilliae* sporogenic in vitro. *Appl. Microbiol.* 19, 681–688.

Shimanuki, H. 1977. Identification and control of honey bee diseases. *U.S. Dep. Agric. Farmers' Bull.* 2255.

Shivakumar, A. G., Gundling, G. J., Benson, T. A., Casuto, D., Miller, M. F., and Spear, B. B. 1986. Vegetative expression of the δ-endotoxin genes of *Bacillus thuringiensis* subsp. *kurstaki* in *Bacillus subtilis*. *J. Bacteriol.* 166, 194–204.

Short, J. A., Walker, P. D., Thomson, R. O., and Somerville, H. J. 1974. The fine structure of *Bacillus finitimus* and *Bacillus thuringiensis* spores with special reference to the location of crystal antigen. *J. Gen. Microbiol. 84*, 261–276.

Sidén, I., Dalhammar, G., Telander, B., Boman, H. G., and Somerville, H. 1979. Virulence factors in *Bacillus thuringiensis*: Purification and properties of a protein inhibitor of immunity in insects. *J. Gen. Microbiol. 114*, 45–52.

Siegel, J. P., and Shadduck, J. A. 1990. Mammalian safety of *Bacillus sphaericus*. *In* "Bacterial Control of Mosquitoes & Black Flies." (H. de Barjac and D. J. Sutherland, eds.), pp. 321–331. Rutgers University Press, New Brunswick.

Singer, S. 1973. Insecticidal activity of recent bacterial isolates and their toxins against mosquito larvae. *Nature 244*, 110–111.

Singer, S. 1981. Potential of *Bacillus sphaericus* and related spore-forming bacteria for pest control. *In* "Microbial Control of Pests and Plant Diseases 1970–1980." (H. D. Burges, ed.), pp. 283–298. Academic Press, London.

Singer, S. 1990. Introduction to the study of *Bacillus sphaericus* as a mosquito control agent. *In* "Bacterial Control of Mosquitoes & Black Flies." (H. de Barjac and D. J. Sutherland, eds.), pp. 221–227. Rutgers University Press, New Brunswick.

Singh, G. J. P., and Gill, S. S. 1985. Myotoxic and neurotoxic activity of *Bacillus thuringiensis* var. *israelensis* crystal toxin. *Pesti. Biochem. Physiol. 24*, 406–414.

Singh, G. J. P., and Gill, S. S. 1988. An electron microscope study of the toxic action of *Bacillus sphaericus* on *Culex quinquefasciatus* larvae. *J. Invertebr. Pathol. 52*, 237–247.

Singh, G. J. P., Schouest, L. P., Jr., and Gill, S. S. 1986. Action of *Bacillus thuringiensis* subsp. *israelensis* δ-endotoxin on the ultrastructure of the house fly larva neuromuscular system in vitro. *J. Invertebr. Pathol. 47*, 155–166.

Smirnoff, W. A., and Valero, J. 1977. Determination of the chitinolytic activity of nine subspecies of *Bacillus thuringiensis*. *J. Invertebr. Pathol. 30*, 265–266.

Smith, R. A. 1987. Use of crystal serology to differentiate among varieties of *Bacillus thuringiensis*. *J. Invertebr. Pathol. 50*, 1–8.

Sneh, B., and Schuster, S. 1981. Recovery of *Bacillus thuringiensis* and other bacteria from larvae of *Spodoptera littoralis* previously fed *B. thuringiensis*-treated leaves. *J. Invertebr. Pathol. 37*, 295–303.

Somerville, H. J. 1971. Formation of the parasporal inclusion of *Bacillus thuringiensis*. *Eur. J. Biochem. 18*, 226–237.

Somerville, H. J., Delafield, F. P., and Rittenberg, S. C. 1968. Biochemical homology between crystal and spore protein of *Bacillus thuringiensis*. *J. Bacteriol. 96*, 721–726.

Somerville, H. J., Tanada, Y., and Omi, E. M. 1970. Lethal effect of purified spore and crystalline endotoxin preparations of *Bacillus thuringiensis* on several lepidopterous insects. *J. Invertebr. Pathol. 16*, 241–248.

Splittstoesser, C. M., and Kawanishi, C. Y. 1981. Insect diseases caused by bacilli without toxic mediated pathologies. *In* "Pathogenesis of Invertebrate Microbial Diseases." (E. W. Davidson, ed.), pp. 189–208. Allanheld, Osmun; Totowa, New Jersey.

St. Julian, G., Bulla, L. A., Jr., and Adams, G. L. 1972. Milky disease development in field-infected Japanese beetle larvae. *J. Invertebr. Pathol. 20*, 109–113.

Stahly, D. P., Dingman, D. W., Bulla, L. A., Jr., and Aronson, A. I. 1978. Possible origin and function of the parasporal crystals in *Bacillus thuringiensis*. *Biochem. Biophys. Res. Commun. 84*, 581–588.

Steiner, H. 1985. Role of the exoprotease InA in the pathogenicity of *Bacillus thuringiensis* in pupae of *Hyalophora cecropia*. *J. Invertebr. Pathol. 46*, 346–347.

Steinhaus, E. A. 1949. "Principles of Insect Pathology." McGraw-Hill, New York.

Steinhaus, E. A. 1960. Insect pathology: Challenge, achievement, and promise. *Bull. Entomol. Soc. Am. 6*, 9–16.

Steinhaus, E. A. 1961. On the correct author of *Bacillus sotto*. *J. Insect Pathol. 3*, 97–100.

Steinhaus, E. A. 1975. "Disease in a Minor Chord." Ohio State University Press, Columbus.

Steinkraus, K. H., and Tashiro, H. 1955. Production of milky-disease spores (*Bacillus popilliae* Dutky and *Bacillus lentimorbus* Dutky) on artificial media. *Science 121*, 873–874.

Steinkraus, K. H., and Tashiro, H. 1967. Milky disease bacteria. *Appl. Microbiol. 15*, 325–333.

Stone, T. B., Sims, S. R., and Marrone, P. G. 1989. Selection of tobacco budworm for resistance to a genetically engineered *Pseudomonas fluorescens* containing the δ-endotoxin of *Bacillus thuringiensis* subsp. *kurstaki*. *J. Invertebr. Pathol. 53*, 228–234.

Sturtevant, A. P., and Revell, I. L. 1953. Reduction of Bacillus larvae spores in liquid food of honey bees by action of the honey stopper, and its relation to the development of American foulbrood. *J. Econ. Entomol. 46*, 855–860.

Sutter, G. R., and Raun, E. S. 1967. Histopahology of European-corn-borer larvae treated with *Bacillus thuringiensis*. *J. Invertebr. Pathol. 9*, 90–103.

Talalaev, E. V. 1958. Induction of epizootic septicemia in the caterpillars of Siberian silkworm moth, Dendrolimus sibiricus Tschtv. (Lepidoptera, Lasiocampidae). *Entomol. Rev.* (USSR) *37*, 557–567.

Talalayeva, G. B. 1967. A case of bacterial epizooty in a larval population of Selenephera lunigera Esp. (Lepidoptera, Lasiocampidae). *Entomol. Rev.* (USSR) *46*, 191–192.

Tanada, Y. 1953. Susceptibility of the imported cabbageworm to *Bacillus thuringiensis* Berliner. *Proc. Hawaii. Entomol. Soc. 15*, 159–166.

Tandeau de Marsac, N., de la Torre, F., and Szulmajster, J. 1987. Expression of the larvicidal gene of *Bacillus sphaericus* 1593 M in the cyanobacterium *Anacystis nidulans* R2. *Mol. Gen. Genet. 209*, 396–398.

Taylor, L. D., and Burke, W. F., Jr. 1990. Transformation of an entomopathic strain of *Bacillus sphaericus* by high voltage electroporation. *FEMS Microbiol. Lett. 66*, 125–127.

Temeyer, K. B. 1984. Larvicidal activity of *Bacillus thuringiensis* subsp. *israelensis* in the dipteran *Haematobia irritans*. *Appl. Environ. Microbiol. 47*, 952–955.

Thomas, W. E., and Ellar, D. J. 1983a. Mechanism of action of *Bacillus thuringiensis* var. *israelensis* insecticidal δ-endotoxin. *FEBS Lett. 154*, 362–368.

Thomas, W. E., and Ellar, D. J. 1983b. *Bacillus thuringiensis* var. *israelensis* crystal δ-endotoxin: Effects on insect and mammalian cells *in vitro* and *in vivo*. *J. Cell Sci. 60*, 181–197.

Thorne, C. B. 1978. Transduction in *Bacillus thuringiensis*. *Appl. Environ. Microbiol. 35*, 1109–1115.

Thorne, L., Garduno, F., Thompson, T., Decker, D., Zounes, M., Wild, M., Walfield, A. M., and Pollock, T. J. 1986. Structural similarity between the Lepidoptera- and Diptera-specific insecticidal endotoxin genes of *Bacillus thuringiensis* subsp. "*kurstaki*" and "*israelensis*". *J. Bacteriol. 166*, 801–811.

Thurley, P., Chilcott, C. N., Kalmakoff, J., and Pillai, J. S. 1985. Characterization of proteolytic activity associated with *Bacillus thuringiensis* var. *darmstadiensis* crystals. *FEMS Microbiol. Lett. 27*, 221–225.

Tinelli, R., and Bourgouin, C. 1982. Larvicidal toxin from *Bacillus sphaericus* spores. Isolation of toxic components. *FEBS Lett. 142*, 155–158.

Tojo, A. 1986. Mode of action of bipyramidal δ-endotoxin of *Bacillus thuringiensis* subsp. *kurstaki* HD-1. *Appl. Environ. Microbiol. 51*, 630–633.

Tojo, A., Samasanti, W., Yoshida, N., and Aizawa, K. 1986. Effects of the three proteases from gut juice of the silkworm, *Bombyx mori*, on the two morphologically different inclusions of δ-endotoxin produced by *Bacillus thuringiensis kurstaki* HD-1 strain. *Agric. Biol. Chem. 50*, 575–580.

Toumanoff, C., and Lapied, M. 1954. L'effet des antibiotiques sur les souches entomophytes ou non de *Bacillus cereus* Frank et Frank. *Ann. Inst. Pasteur 87*, 370–374.

Travers, R. S., Faust, R. M., and Reichelderfer, C. F. 1976. Effects of *Bacillus thuringiensis* var. *kurstaki* δ-endotoxin on isolated lepidopteran mitochondria. *J. Invertebr. Pathol.* 28, 351–356.

Turnbull, P. C. B. 1981. *Bacillus cereus* toxins. *Pharmacol. Ther.* 13, 453–505.

Tyrell, D. J., Bulla, L. A., Jr., Andrews, R. E., Jr., Kramer, K. J., Davidson, L. I., and Nordin, P. 1981. Comparative biochemistry of entomocidal parasporal crystals of selected *Bacillus thuringiensis* strains. *J. Bacteriol.* 145, 1052–1062.

Vaňková, J. 1978. The heat-stable exotoxin of *Bacillus thuringiensis*. *Folia Microbiol.* 23, 162–174.

Vaňková, J., and Horská, K. 1975. The activity of gut phosphatases in some insect species and the effects of gut homogenates on the exotoxin of *Bacillus thuringiensis* in vitro. *Acta Entomol. Bohemoslov.* 72, 7–12.

Vaňková, J., Horská, K., and Šebesta, K. 1974. The fate of exotoxin of *Bacillus thuringiensis* in *Galleria mellonella* caterpillars. *J. Invertebr. Pathol.* 23, 209–212.

Van Rie, J., Jansens, S., Höfte, H., Degheele, D., and Van Mellaert, H. 1989. Specificity of *Bacillus thuringiensis* δ-endotoxins. Importance of specific receptors on the brush border membrane of the mid-gut of target insects. *Eur. J. Biochem.* 186, 239–247.

Varma, R. V., and Ali, M. I. M. 1986. *Bacillus firmus* as a new insect pathogen on a lepidopteran pest of *Ailanthus triphysa*. *J. Invertebr. Pathol.* 47, 379–380.

Vasantharajan, V. N., and Munirathnamma, N. 1980. Studies on silkworm diseases. Phage and serotyping of *Bacillus thuringiensis* strains occurring in the sericultural tracts of Karnataka. *Curr. Sci.* 49, 248–249.

Visser, B., van Workum, M., Dullemans, A., and Waalwijk, C. 1986. The mosquitocidal activity of *Bacillus thuringiensis* var. *israelensis* is associated with M_r 230,000 and 130,000 crystal proteins. *FEMS Microbiol. Lett.* 30, 211–214.

Waalwijk, C., Dullemans, A. M., van Workum, M. E. S., and Visser, B. 1985. Molecular cloning and the nucleotide sequence of the M_r 28 000 crystal protein gene of *Bacillus thuringiensis* subsp. *israelensis*. *Nucleic Acid Res.* 13, 8207–8217.

Wakisaka, Y., Masaki, E., Koizumi, K., Nishimoto, Y., Endo, Y., Nishimura, M. S., and Nishiitsutsuji-Uwo, J. 1982. Asporogenous *Bacillus thuringiensis* mutant producing yields of δ-endotoxin. *Appl. Environ. Microbiol.* 43, 1498–1500.

Walther, C. J., Couche, G. A., Pfannenstiel, M. A., Egan, S. E., Bivin, L. A., and Nickerson, K. W. 1986. Analysis of mosquito larvicidal potential exhibited by vegetative cells of *Bacillus thuringiensis* subsp. *israelensis*. *Appl. Environ. Microbiol.* 52, 650–653.

Warren, R. E., Rubenstein, D., Ellar, D. J., Kramer, J. M., and Gilbert, R. J. 1984. Bacillus thuringiensis var israelensis: Protoxin activation and safety. *Lancet 1*, 678–679.

Weiner, B. A. 1978. Isolation and partial characterization of the parasporal body of *Bacillus popilliae*. *Can. J. Microbiol.* 24, 1557–1561.

Weiser, J. 1984. A mosquito-virulent *Bacillus sphaericus* in adult *Simulium damnosum* from Northern Nigeria. *Zentralbl. Mikrobiol.* 139, 57–60.

West, A. W., Crook, N. E., and Burges, H. D. 1984. Detection of *Bacillus thuringiensis* in soil by immunofluorescence. *J. Invertebr. Pathol.* 43, 150–155.

White, G. F. 1920. American foulbrood. *U.S. Dep. Agric. Bull. 809.*

White, R. T., and Dutky, S. R. 1940. Effect of the introduction of milky diseases on populations of Japanese beetle larvae. *J. Econ. Entomol.* 33, 306–309.

White, R. T., and McCabe, P. J. 1950. The effect of milky disease on Japanese beetle populations over a ten-year period. *U.S. Bur. Entomol. Plant Quarantine Publ. E-801.*

Whiteley, H. R., and Schnepf, H. E. 1986. The molecular biology of parasporal crystal body formation in *Bacillus thuringiensis*. *Annu. Rev. Microbiol.* 40, 549–576.

Wickremesinghe, R. S. B., and Mendis, C. L. 1980. *Bacillus sphaericus* spore from Sri Lanka demonstrating rapid larvicidal activity on *Culex quinquefasciatus*. *Mosq. News* 40, 387–389.

Wille, H. 1956. Bacillus fribourgensis, *n.sp.*, Erreger einer "milky disease" im Engerling von Melolontha melolontha L. *Mitt. Schweiz. Entomol. Ges. 29*, 271–282.

Wilson, B. H., and Burns, E. C. 1968. Induction of resistance to *Bacillus thuringiensis* in a laboratory strain of house flies. *J. Econ. Entomol. 61*, 1747–1748.

Wilson, W. T. 1971. Resistance to American foulbrood in honey bees. XI. Fate of *Bacillus larvae* spores ingested by adults. *J. Invertebr. Pathol. 17*, 247–255.

Wilson, W. T. 1972. Resistance to American foulbrood in honey bees. XII. Persistence of viable *Bacillus larvae* spores in the feces of adults permitted flight. *J. Invertebr. Pathol. 20*, 165–169.

Wolfenbarger, D. A., Guerra, A. A., Dulmage, H. T., and Garcia, R. D. 1972. Properties of the β-exotoxin of *Bacillus thuringiensis* IMC 10,001 against the tobacco budworm. *J. Econ. Entomol. 65*, 1245–1248.

Wolfersberger, M. G. 1990. The toxicity of two *Bacillus thuringiensis* δ-endotoxins to gypsy moth larvae is inversely related to the affinity of binding sites on midgut brush border membranes for the toxins. *Experientia 46*, 475–477.

Wong, H. C., Schnepf, H. E., and Whiteley, H. R. 1983. Transcriptional and translational start sites for the *Bacillus thuringiensis* crystal protein gene. *J. Biol. Chem. 258*, 1960–1967.

Woodrow, A. W. 1942. Susceptibility of honeybee larvae to individual inoculations with spores of *Bacillus larvae*. *J. Econ. Entomol. 35*, 892–895.

Woodrow, A. W., and Holst, E. C. 1942. The mechanism of colony resistance to American foulbrood. *J. Econ. Entomol. 35*, 327–330.

Wraight, S. P., Molloy, D. P., and Singer, S. 1987. Studies on the culicine mosquito host range of *Bacillus sphaericus* and *Bacillus thuringiensis* var. *israelensis* with notes on the effects of temperature and instar on bacterial efficacy. *J. Invertebr. Pathol. 49*, 291–302.

Wu, D., and Chang, F. N. 1985. Synergism in mosquitocidal activity of 26 and 65 kDa proteins from *Bacillus thuringiensis* subsp. *israelensis* crystal. *FEBS Lett. 190*, 232–236.

Wyss, C. 1971. Sporulationsversuche mit drei Varietäten von *Bacillus popilliae* Dutky. *Zentralbl. Bakteriol. Parasitenkd. Infektionskr. Hyg. II. 126*, 461–492.

Yamamoto, T., and Iizuka, T. 1983. Two types of entomocidal toxins in the parasporal crystals of *Bacillus thuringiensis kurstaki*. *Arch. Biochem. Biophys. 227*, 233–241.

Yamamoto, T., and McLaughlin, R. E. 1981. Isolation of a protein from the parasporal crystal of Bacillus thuringiensis var. kurstaki toxic to the mosquito larva Aedes taeniorhynchus. *Biochem. Biophys. Res. Comm. 103*, 414–421.

Yamamoto, T., Iizuka, T., and Aronson, J. N. 1983. Mosquitocidal protein of *Bacillus thuringiensis* subsp. *israelensis*: Identification and partial isolation of the protein. *Curr. Microbiol. 9*, 279–284.

Yamvrias, C. 1962. Contribution à l'étude du mode d'action de *Bacillus thuringiensis* Berliner vis-à-vis de la teigne de la farine *Anagasta* (*Ephestia*) *kühniella* Zeller (Lépidoptère). *Entomophaga 7*, 101–159.

Yap, H.-H. 1990. Field trials of *Bacillus sphaericus* for mosquito control. *In* "Bacterial Control of Mosquitoes & Black Flies." (H. de Barjac and D. J. Sutherland, eds.), pp. 307–320. Rutgers University Press, New Brunswick.

Yoshida, K., Matsushima, Y., Sen, K., Sakai, H., and Komano, T. 1989. Insecticidal activity of a peptide containing the 30th to 695th amino acid residues of the 130-kDA protein of *Bacillus thuringiensis* var. *israelensis*. *Agric. Biol. Chem. 53*, 2121–2127.

Yousten, A. A. 1984a. Bacteriophage typing of mosquito pathogenic strains of *Bacillus sphaericus*. *J Invertebr. Pathol. 43*, 124–125.

Yousten, A. A. 1984b. *Bacillus sphaericus*: Microbiological factors related to its potential as a mosquito larvicide. *In* "Advances in Biotechnological Process." (A. Mizrahi and A. L. van Wezel, eds.), Vol. 3, pp. 315–343. Alan R. Liss, New York.

Yousten, A. A. 1986. Isolation and characterization of oligosporogenic, paraspore-forming mutants of *Bacillus thuringiensis* serovar *israelensis*. *FEMS Microbiol. Lett. 33*, 59–63.

Yousten, A. A., and Davidson, E. W. 1982. Ultrastructural analysis of spores and parasporal crystals formed by *Bacillus sphaericus* 2297. *Appl. Environ. Microbiol. 44*, 1449–1455.

Yousten, A. A., de Barjac, H., Hedrick, J., Cosmao Dumanoir, V., and Myers, P. 1980. Comparison between bacteriophage typing and serotyping for the differentiation of *Bacillus sphaericus* strains. *Ann. Microbiol.* (Inst. Pasteur) *131B*, 297–308.

Yu, Y. M., Ohba, M., and Aizawa, K. 1989. The 25-kilodalton hemolytic protein affinity-purified from parasporal inclusions of *Bacillus thuringiensis* strain PG-14 (serotype 8a:8b). *Curr. Microbiol. 18*, 243–246.

CHAPTER 5

OTHER BACTERIAL INFECTIONS

I. PSEUDOMONADACEAE INFECTIONS
II. ENTEROBACTERIACEAE INFECTIONS
    A. *Serratia* Infections
    B. Infections by Other Enterobacteriaceae
III. VIBRIONACEAE INFECTIONS
IV. STREPTOCOCCACEAE INFECTIONS
    A. European Foulbrood of Honey Bee
    B. *Streptococcus faecalis*
V. ACTINOMYCETES INFECTIONS
VI. RICKETTSIAL INFECTIONS
VII. CLASS MOLLICUTES
    A. Mycoplasma
    B. Spiroplasma

Most insect pathogenic bacteria are the non-spore-formers. They have low pathogenicity, but once they enter the insect's hemocoel, they multiply rapidly and cause death within a brief period. Such forms with low invasive property require stress conditions applied to the insect host or other external factors (e.g., parasitoids or predators) to invade the hemocoel. Some non-spore-formers, however, can cause severe pathologies without invading the hemocoel, such as brachytosis caused by *Streptococcus* spp.

I. PSEUDOMONADACEAE INFECTIONS

Members of the family Pseudomonadaceae are strictly aerobic, gram-negative, straight or curved rods with polar flagella. A number of species are pathogens and others are found commonly in the digestive tracts of insects as commensals (Bucher 1963b). Within this family, *Pseudomonas aeruginosa* is most often isolated from insects. It produces water-soluble, blue or green pigments, although some strains are colorless. It occasionally is found in wounds and sores in vertebrates. It has attracted wide attention in hospitals because strains have developed resistance to the commonly used antibiotics.

Pseudomonas aeruginosa has low invasive power through the insect's midgut, but once within the hemocoel, it is highly pathogenic. In the grasshoppers *Melanoplus bivittatus* and *Camnula pellucida*, the LD_{50} through the mouth is from 8000 to 29,000 rods per insect, while that of intrahemocoelic inoculation is from 10 to 20 rods per insect (Bucher and Stephens 1957). The grasshoppers remain alive and active while supporting as many as 10^9 bacteria per individual but succumb shortly after this concentration is reached. The bacteria do not multiply to large numbers in the digestive tract and are rapidly eliminated, but a few survive in some unknown areas, probably in pockets or crypts. Since it has been found in the foam of eggs and in the soil surrounding the eggs, it may be ingested by newly emerged grasshopper nymphs. This means of transmission probably occurs in grasshoppers under natural conditions.

Bucher (1963b) commented that *P. aeruginosa* has never been recorded to cause epizootics in field insect populations and is only destructive under laboratory conditions, especially at high humidities and temperatures. Dorn (1976) reported that the bacterium spread rapidly in the laboratory-reared milkweed bug, *Oncopeltus fasciatus*, and killed nearly all of the insects in 1 to 3 days. The bug acquired the bacterium probably through its mouth and died from septicemia.

Pseudomonas aeruginosa produces a number of toxic exoenzymes for insects (Lysenko and Kučera 1968). The pathogenicity of the bacterium is correlated with the production of proteolytic enzymes (Bucher 1960) and to the toxicity and the clotting of the insect's hemolymph by proteases (Kučera and Lysenko 1977). The proteases cause degenerative changes in hemocytes (Madziara-Borusiewicz and Lysenko 1971), and digest certain specific insect hemolymph proteins (Kučera and Lysenko 1971; Lysenko 1974). Their effect may vary with different insect species. For example, the larvae of *Mamestra brassicae* are 100 times more susceptible to the protease than those of *Galleria mellonella* (Lysenko and Kučera 1971).

Strains of *P. aeruginosa* differ in their virulence to insects. Jarrell and Kropinski (1982) have observed that a rough strain deficient in lipopolysaccharide is less virulent than the wild strain, a difference of 10,000-fold. This suggests that the lipopolysaccharide plays an important role in the infection of this bacterium. On the other hand, Dunphy et al. (1986a,b) have found no correlation between pathogenicity and lipopolysaccharide production in mutants (smooth and rough strains) of *P. aeruginosa* for *G. mellonella* larvae. Øgaard et al. (1985) have correlated the adhesion of *P. aeruginosa* to host cell surfaces with bacterial virulence. The adhesion factors are located extracellularly on the bacterial surface and the loss of adhesiveness upon subculturing is restored by adding the extracellular factors from recent isolates.

Pseudomonas septica is pathogenic to a number of insects, including scarabaeids (Hurpin and Vago 1958). In the house fly, the pathogenicity is due mainly to an exotoxin (Amonkar et al. 1967). Other entomopathogenic pseudomonads are *P. chlororaphis*, *P. fluorescens*, *P. putida*, and *P. aureofaciens* (Krieg 1987).

II. ENTEROBACTERIACEAE INFECTIONS

The family Enterobacteriaceae contains many species that are found associated with plants and animals. Some are saprophytes and others are pathogens. Many live in the digestive tracts of animals, including insects. They are non-spore-forming, gram-negative, facultative anaerobic, rod-shaped bacteria with peritrichous flagella. The entomopathogenic forms generally are highly virulent when they enter the insect's hemocoel, but they usually lack invasive ability to penetrate the midgut wall.

Grasshoppers are known to develop bacterial dysentery and septicemia. In 1910, d'Herelle (1911) isolated and named a small, motile, gram-negative bacterium, *Coccobacillus acridiorum*, that he believed to be highly virulent and to cause epizootics in outbreaks of a Mexican locust, *Schistocerca pallens*. The bacterium infected the digestive tract. It rapidly lost its virulence in successive cultivation *in vitro* and had to be passed through the grasshopper to regain its virulence. d'Herelle (1914) claimed successful control of diverse species of grasshoppers in Argentina and Colombia with the use of the bacterium. Some workers confirmed his results, but most failed to do so, and interest in this bacterium dissipated.

Bucher (1959a) obtained from the Lister Institute the culture 2163 *Coccobacillus acridiorum* d'Herelle, which he diagnosed as Cloaca Type A, family Enterobacteriaceae, a common inhabitant of the digestive tracts of many insects. He pointed out that Cloaca Type A is frequently isolated from dead insects but is not pathogenic when fed to insects even at high dosages, and its virulence is not enhanced by passage through grasshoppers. He questioned the effectiveness of this bacterium on grounds of faulty techniques and premises.

Since the genus *Coccobacillus* is not valid, Steinhaus (1949) proposed in its place *Aerobacter aerogenes* var. *acridiorum*, and Lysenko (1958) suggested *Cloaca cloacae* var. *acridiorum*, which is now designated as *Enterobacter cloacae* (Buchanan and Gibbons 1974).

A. *SERRATIA* INFECTIONS

Members of the genus *Serratia* are ubiquitous in nature and are commonly found as saprophytes in water, soil, milk, and other foods (Grimont and Grimont 1978). The colonies are generally characteristically pink or red from a water-insoluble pigment, prodigiosin, but some strains produce white colonies that have been mistaken for species in other genera. The white strains, in particular, cause infections in vertebrates, including humans.

Grimont et al. (1979) presented a detailed report on the species and biotype identification of *Serratia* strains associated with insects. They studied 48 strains and reported no evidence that a particular *Serratia* species or biotype was associated with a specific insect. The major group associated with insects was *S. marcescens* biotype A2a of the red-pigmented strains and *S. liquefaciens* biotype Cla of the nonpigmented strains.

Serratia species are more commonly detected in laboratory-reared rather than in field-collected insects (Steinhaus 1959; Bucher 1963b; Krieg 1987). They infect over 70 insect species in Orthoptera, Coleoptera, Hymenoptera, Lepidoptera, and Diptera. Many of the achromogenic strains have been misidentified as other bacterial species. At one time, the septicemia in the hornworms *Protoparce sexta* and *P. quinquemaculata* was attributed to *Bacillus sphingidis* and that in the honey bee to *B. apisepticus*, but they are presently considered to be caused by *Serratia marcescens* (Grimont et al. 1979). Strains of *Serratia liquefaciens* have been identified as *Bacillus melolonthae liquefaciens* infecting *Melolontha melolontha*, as *Para-*

colobactrum rhyncoli from *Rhyncolus porcatus* and *Scolytus* spp., as *Bacillus noctuarum* and *Pseudomonas noctuarum* from cutworms.

Serratia marcescens is considered to be a potential or a facultative pathogen (Bucher 1960, 1963b). It is the most widely studied species of *Serratia*. It produces two proteases, a metaloprotease and a serine protease (Kaška 1976; Kaška et al 1976), and a chitinase (Lysenko 1976) that are toxic to *Galleria* larvae when inoculated into the hemocoel.

Serratia marcescens and other *Serratia* species generally are not pathogenic to insects when present in the digestive tract, but once they enter the hemocoel, they multiply rapidly and cause death in 1 to 3 days. In the adult boll weevil, *Anthonomus grandis*, the LD_{50} of *S. marcescens* is 5.1 ± 1 rods per weevil when inoculated into the hemocoel, whereas the adult is not susceptible to infection through the mouth except when fed grossly contaminated food (Slatten and Larson 1967). Similarly in the gypsy moth, the ED_{50} by intrahemocoelic inoculations into third and fourth instar larvae is 7.5 and 14.5 rods per larva, respectively; but the per os inoculation varies greatly and requires high dosages of the bacterium (Podgwaite and Cosenza 1976). On the other hand, in the tsetse flies *Glossina* spp., *S. marcescens* is capable of invading into the hemocoel to cause fatal septicemia (Poinar et al. 1979).

A *Serratia* that offers promise for insect control has been isolated from the New Zealand grass grub, *Costelytra zealandica* (Trought et al. 1982). An infected grub develops amber disease (previously called honey disease). The etiological agents are *S. entomophila*, the main causative pathogen, and *S. proteamaculans* (syn. *S. liquefaciens*) (Stucki et al. 1984; Grimont et al. 1988). *Serratia entomophila* has several serotypes (Allardyce et al. 1991). An infected larva stops feeding within a few days and empties its digestive tract with the midgut developing an amber discoloration. The larva gradually loses weight and dies in 4 to 6 weeks. *Serratia entomophila* has not been isolated from animals (other than insects) or plants. It has been applied for the control of the New Zealand grass grub (Jackson et al. 1986).

A strain of Cloaca Type B infects locusts in western Canada (Bucher 1959b). This strain has been reclassified as *Serratia liquefaciens* (Grimont et al. 1979). It has low invasive power, but it causes fatal septicemia once it enters a hemocoel through the physical rupture of the digestive tract or damage by intestinal gregarines and at molting. This or similar strains infect *Scolytus scolytus*, *Malacosoma* sp., and the silkworm (Bucher 1963a).

In an unusual case, a hymenopteran (ichneumonid) parasitoid, *Exeristes comstockii*, acquires an infection of *S. marcescens* through its larva (Bracken and Bucher 1967). The parasitoid's host, *Galleria mellonella*, carries a small number of the bacteria in the digestive tract. The bacteria multiply in a heat-killed larva that is used to rear the parasitoid larva. A parasitoid larva ingests the bacteria without ill effects. When the larva matures, the bacteria, which are retained in the adult's digestive tract, are transmitted to other adults and cause fatal septicemia.

The growth of certain bacteria in the digestive tracts of insects creates an environment that is favorable for species of *Serratia*. The invasion of *Serratia piscatorum* into the hemocoel of a larval silkworm is enhanced by the presence of *Streptococcus faecalis–S. faecium* in the larval midgut (Kodama and Nakasuji 1971). The rapid growth of *S. faecalis–S. faecium* in the midgut lowers the pH because of increases in lactates and acetates. This favors *Serratia piscatorum* to invade the hemocoel and produce a fatal septicemia. The presence of a *Staphylococ-*

cus species in the digestive tract of a silkworm larva also enhances the infection of *S. marcescens* (Vasantharajan and Munirathnamma 1978).

Certain strains of *S. marcescens* have a repellent effect on insects. The strain isolated from the eggs of the cabbage looper, *Trichoplusia ni*, repels the larvae of this species when the bacteria are added to a meridic diet (Bell 1969). The repellency, however, is overcome when the bacterium is applied to collard leaves.

B. INFECTIONS BY OTHER ENTEROBACTERIACEAE

Proteus species are found in the digestive tracts of many animals, including insects. Three species, *P. vulgaris*, *P. mirabilis*, and *P. rettgeri*, are potential pathogens when present in insects (Bucher 1963b). As in the case of *Serratia* and *Pseudomonas*, these bacteria are highly virulent when they gain entrance into the hemocoel. By intrahemocoelic inoculation in grasshoppers, the LD_{50}s are from 50 to 100 bacteria per insect for *P. vulgaris*, from 25 to 500 for *P. mirabilis*, and from 300 to 1000 for *P. rettgeri* (Bucher 1959b). The pathogenicity of these bacteria in grasshoppers is correlated with their proteolytic activity (Bucher 1960).

Insect pathogens occur among species of *Enterobacter*, such as *Enterobacter aerogenes* (syn. *Aerobacter aerogenes*), which in association with *Proteus mirabilis* results in enhanced infection in larvae of the geometrid *Boarmia selenaria* (Wysoki and Raccah 1980).

The genus *Xenorhabdus* (syn. *Achromobacter*) contains entomopathogenic bacteria that are closely associated with entomogenous nematodes (Thomas and Poinar 1983; Akhurst 1986a,b). They are large rods, gram-negative, facultative anaerobes. They are negative for catalase and nitrate reductase. Their normal habitat is the intestinal lumen of a nematode, and they are introduced by the nematode into the hemocoel of a host insect. There are several species associated with different nematode species. For example, the type species *X. nematophilus* is associated with *Steinernema* (syn. *Neoaplectana*) *carpocapsae* and another species, *X. luminescens*, is associated with *Heterorhabditis* spp. (Akhurst 1986b). The *Xenorhabdus* species differ in biochemical properties and in pathogenicity. Paracrystalline inclusions occur in several species of *Xenorhabdus* (Boemare et al. 1983). The inclusions vary in shape and structure depending on the species and are formed at the exponential growth phase. They resemble the parasporal crystals of *B. thuringiensis*, but their role is unknown.

The genus *Xenorhabdus* is mutualistically associated with insect pathogenic nematodes of families Steinernematidae and Heterorhabditidae (see Chapter 13). The nematode assists the bacteria in penetrating into the insect's hemocoel, and the bacteria provide nutrients essential to the nematode.

III. VIBRIONACEAE INFECTIONS

The Vibrionaceae consists of gram-negative, straight or curved rods with polar flagella. Members are primarily aquatic inhabitants of sea- and freshwater and in association with aquatic animals.

Aeromonas punctata causes black lesions in larvae of *Anopheles annulipes* (Kalucy and Daniel 1972). The bacteria invade the hemocoel through the integument and the intestinal epithelium possibly through the activity of chitinase and phospholipase A. Black lesions are found at the sites of invasion. In a silkworm larva with signs of flacherie, an *Aeromonas* sp. causes the dissolution of the peritrophic membrane (Ono and Kato 1968).

IV. STREPTOCOCCACEAE INFECTIONS

Members of Streptococcaceae are gram-positive, aerobic to anaerobic, coccoid bacteria in pairs or in chains, and are nonmotile or rarely motile.

A. EUROPEAN FOULBROOD OF HONEY BEE

The etiology of the European foulbrood of the honey bee was confused for many years because of the large number of different bacteria isolated from bees presumably displaying this disease (Steinhaus 1949). Among the bacteria considered as the causal agent are the following: *Bacillus alvei*, *Streptococcus apis*, *Achromobacter* (*Bacterium*) *eurydice*, *Bacillus laterosporus*, and *Bacillus pluton*. Bailey (1963) conducted a thorough study and confirmed the conclusion by White (1912, 1920) that the causal agent was *Streptococcus pluton* (syn. *Bacillus pluton*). This species has been reclassified into a new genus and the proposed name is *Melissococcus pluton* (Bailey and Collins 1982). Strains of *M. pluton* are known and are mostly difficult to culture, but the Brazilian strain has multiplied well under anaerobic condition on chemically defined media buffered with potassium phosphate (Bailey 1984).

The infection in the honey bee larva takes place by way of the alimentary tract, where the bacterium multiplies and is confined in the peritrophic membrane (White 1920). Death apparently results from the toxic products of the bacterium, which diffuse through the intestinal wall to the vital tissues. Adult bees are not affected.

European foulbrood occurs in honey bees throughout the world, including the United States. It is not as devastating a disease as the American foulbrood, although at times it can result in a serious loss of brood and even destroy a colony. It occurs most commonly in the spring at the height of brood rearing and subsides usually by midsummer (Burnside et al. 1949). The disease has also been referred to as "melting brood" because of the condition in which the dead larva "melts" away from its tracheal system.

An infected larva initially loses the plumpness and glistening white color of a healthy larva and the intersegmental lines of the body become less prominent. The larva turns white and, subsequently, to a faint yellow color, which is an important symptom (White 1920; Burnside et al. 1949). As the infection progresses, the larva shows abnormal movements and occupies unnatural positions in a cell. At death, most larvae are coiled at the bottoms of open cells, and few are fully extended (Fig. 4–4B). The faint yellow color of the larva later turns brownish and then a dark brown. The cells are not capped when the larvae die before maturity.

In dead larvae, the tracheae are shown more clearly as radiating white lines than in uninfected larvae. The prominence of the tracheae is a valuable but not an absolutely dependable sign (Burnside et al. 1949). A dead larva dries down to a scale, the color of which depends on whether the larva dies before the cell is capped. In an uncapped cell, the scale is light colored, but in a capped cell in which drying takes place slowly, it is dark brown or nearly black. The scale does not cling closely to the cell wall and is easily removed unlike that of the American foulbrood. The body contents of larvae that die in sealed cells are sometimes ropy and resemble those of larvae with the American foulbrood. A sour odor or an odor of spoiled meat is characteristic of the disease. Diagnosis of European foulbrood is more difficult after the larva is decayed and dried.

B. *STREPTOCOCCUS FAECALIS*

Certain strains of *Streptococcus faecalis* infect a number of insect species, and others are nonpathogenic. Distinct serotypes of *S. faecalis* occur and they are serologically different from noninsect strains (Doane and Redys 1970). The antigenic differences may serve to separate insect pathogenic strains. In the gypsy moth, *S. faecalis* multiplies in the larval digestive tract and causes diarrhea. The digestive tract, especially the foregut, is distended with liquid from the hemocoel. An infected larva does not feed, and as the fluid continues to be lost from diarrhea, its body shrinks in size. Death occurs in 3 to 15 days. The shrunken cadaver becomes mummified and desiccated. The syndrome resembles the brachytosis of the tent caterpillar caused by *Clostridium*. Doane (1971) induced brachytosis by applying the streptococcus to a gypsy moth population on apple trees and prevented defoliation in an area of high insect population.

Strain AD-4 of *S. faecalis*, unlike the above streptococcus in the gypsy moth, causes septicemia in silkworm larvae (Iizuka 1972). From 2 to 3 days after ingestion, the bacterium multipies and forms large colonies attached to the larval peritrophic membrane. After 6 days, the peritrophic membrane dissolves, and the goblet and cylindrical midgut cells are vacuolated. The larva dies from the digestive tract fluids passing through the vacuolated midgut cells into the hemolymph and from septicemia.

V. ACTINOMYCETES INFECTIONS

A bacterial septicemia causes a chalky disease to develop in the cicada *Okanagana rimosa* (Lüthy and Soper 1969). The etiologic agent is called *Corynebacterium okanaganae*, but the species may not belong to *Corynebacterium* (Collins and Jones 1981). Whether the bacterium gains entrance through the piercing-sucking mouth parts of the nymph is not known. A chalky-white color develops at the extremities and on the thorax. In nature, the prevalence of the disease is low, about 0.1%. The disease is characteristically slow in development and results in death mainly from the general exhaustion and depletion of nutritive substances rather than from tissue destruction.

VI. RICKETTSIAL INFECTIONS

The pathogenic rickettsiae differ mainly from bacteria in that they are gram-negative, obligate intracellular pathogens with typical bacterial cell walls and no flagella (Weiss and Moulder 1984). Many are nonpathogenic to insects and occur as commensals or mutualists within the tissues. The entomogenous rickettsiae are in the family Rickettsiaceae, tribe Wolbachieae. The following two genera are recognized: *Wolbachia*, which are seldom pathogenic, and *Rickettsiella*, which are commonly pathogenic. We shall be concerned largely with the genus *Rickettsiella*. A number of *Rickettsiella* species have been placed in synonymy with *R. popilliae*, the type species, in Bergey's Manual (Weiss 1974). These species are *R. melolonthae*,

R. tipulae, *R. grylli*, *R. chironomi*, *R. blattae*, *R. tenebrionis*, *R. schistocercae*, *R. cetonidarum*, and *R. armadillidii*. Since there is some doubt concerning these synonymies, we shall consider some of the species separately.

Rickettsiella species have not been cultivated in host cell-free media. Some of them produce characteristic crystals. They infect mainly the fat and blood cells and produce, in general, a prolonged, chronic infection in an insect. In addition, *Rickettsiella blattae* attacks other tissues and causes a systemic infection in the cockroach through its invasion of the hypodermis, trachea, Malpighian tubule, muscle, nerve, blood cell, ovariole, intestinal epithelium, etc. (Huger 1964).

Certain *Rickettsiella*, which are pathogens of insects and other invertebrates and are not transmitted to vertebrates, are separable into three distinct serological groups (Croizier and Meynadier 1971; Croizier et al. 1975). The insect pathogenic forms that produce crystals (e.g., *R. melolonthae*, *R. tipulae*, and *R. cetonidarum*), are in one group; *R. grylli* on crickets in a second, and the rickettsiae of scorpion in a third group. These arthropod rickettsiae are not related antigenically to rickettsiae and chlamydia that infect vertebrates. On the other hand, *Rickettsiella chironomi* appears morphologically and developmentally more closely related to chlamydia than to rickettsiae (Federici 1980).

The entomopathogenic *Rickettsiella* possess a more complex developmental cycle than most bacteria, and their forms change from the typical elementary form (rod, coccoid, or kidney-shaped) to spherical and giant forms (Fig. 5–1). In addition, *R. chironomi* produces disc-shaped bodies (Götz 1972; Federici 1980). The life cycles are not well established and may vary with the rickettsial species. Moreover, the different terminology applied to rickettsial developmental stages has caused difficulty in determining the relationship among various isolates (Henry et al. 1986).

Huger and Krieg (1967) described the life cycles of *Rickettsiella melolonthae*, *R. tipulae*, *R. tenebrionis*, and *R. blattae* (Fig. 5–2). When ingested, the infectious elementary forms penetrate the insect's midgut wall and replicate in tissues, especially the fat body, of the hemocoel, including the gonads (Devauchelle et al. 1972; Krieg 1987). The small rickettsiae within cytoplasmic vacuoles undergo transformation to bacteria-like forms that multiply by binary fission. Some cells are pleomorphic and form the characteristic protein crystals. The bacteria-like forms may enlarge to giant cells or to a rickettsiogenic stroma that reverts to small rickettsiae at the end of reproduction. Eventually, all infected cells undergo lysis, and masses of rickettsiae, crystals, and rickettsia-filled vacuoles are released in the hemolymph. Larvae at an advanced stage of infection may be blue-grayish in color. Those with sublethal infections develop into infected pupae and adults. Infected adults transmit the rickettsia to their offspring through the egg.

Rickettsiella popilliae infects the Japanese beetle, *Popillia japonica*, and other beetles (Dutky and Gooden 1952; Dutky 1959). An infected larva turns bluish and hence the name "blue disease." The infected fat body contains crystals. When ingested by a larva, the rickettsiae cause death in about 45 days, and when inoculated into the hemocoel, death occurs in 25 days. Large doses of *R. popilliae* are required to cause infection when fed to a larva, but less than six rickettsiae per larva is the LD_{50} by intrahemocoelic inoculation. The longevity of an infected larva is prolonged when the amount of available food is increased, but this does not prevent its death.

Rickettsiella melolonthae infects *Melolontha melolontha* and other scarabaeid beetles (Krieg 1955). The disease in *Melolontha vulgaris* was discovered by Wille

FIGURE 5–1

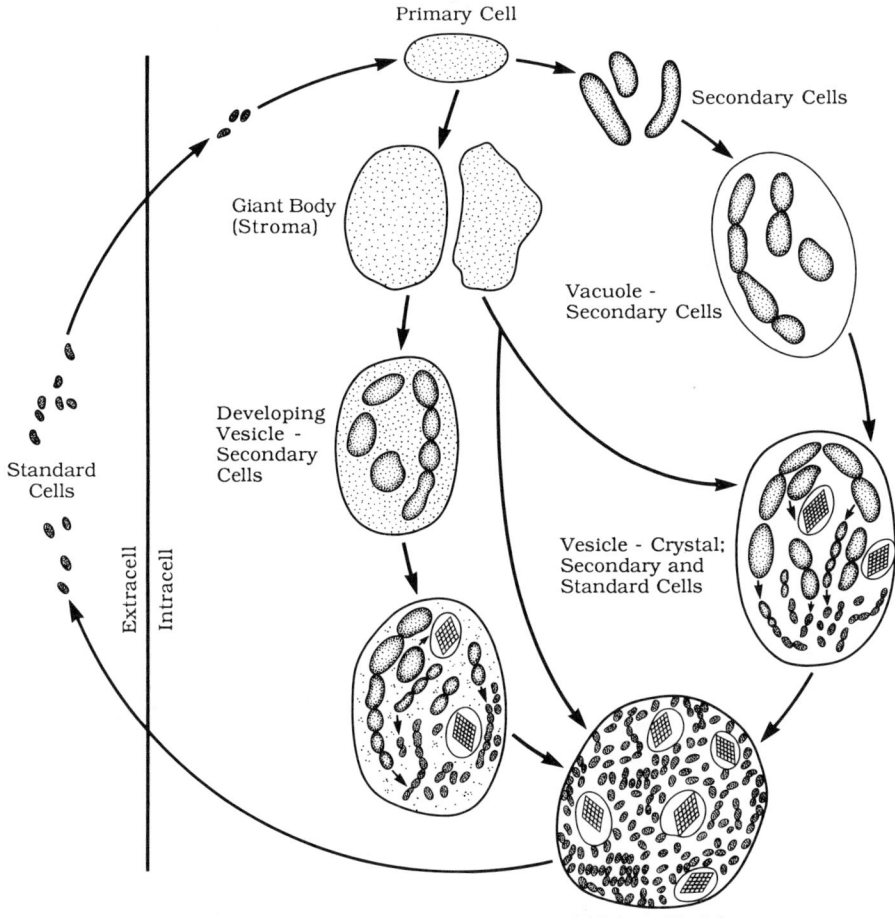

Life cycle of an insect pathogenic rickettsia as proposed by A. M. Huger and A. Krieg. (Drawn by Christina L. Jordan.)

and Martignoni (1952), and it is called "Lorscher Krankheit." *Rickettsiella melolonthae* causes a chronic infection with apparent symptoms appearing after 2 to 3 months. Death occurs in 4 to 6 months. An infected larva develops white-bluish discoloration from the mass of rickettsiae in the hemolymph. It is reduced in turgor and its reflexes diminish. An abnormal nervous reaction develops, and an infected larva tends to rise to the surface of the soil when the temperature drops, whereas an uninfected larva tends to burrow deeper in the soil (Wille and Martignoni 1952; Niklas 1957). Such movement and death of infected larvae near the soil surface may play an important role in epizootics through the transmission of the rickettsiae to young larvae that emerge from eggs laid in the upper region of the soil. Moreover, the wide host range, including many coleopteran families and Tipulidae (Diptera), may be important in the persistence of the rickettsiae in the host habitat (Krieg 1987).

Rickettsial infection in chironomid was first reported in 1949 from larvae of *Camptochironomus tentans* (Weiser 1949). Weiser (1966) described the pathogen as

FIGURE 5-2

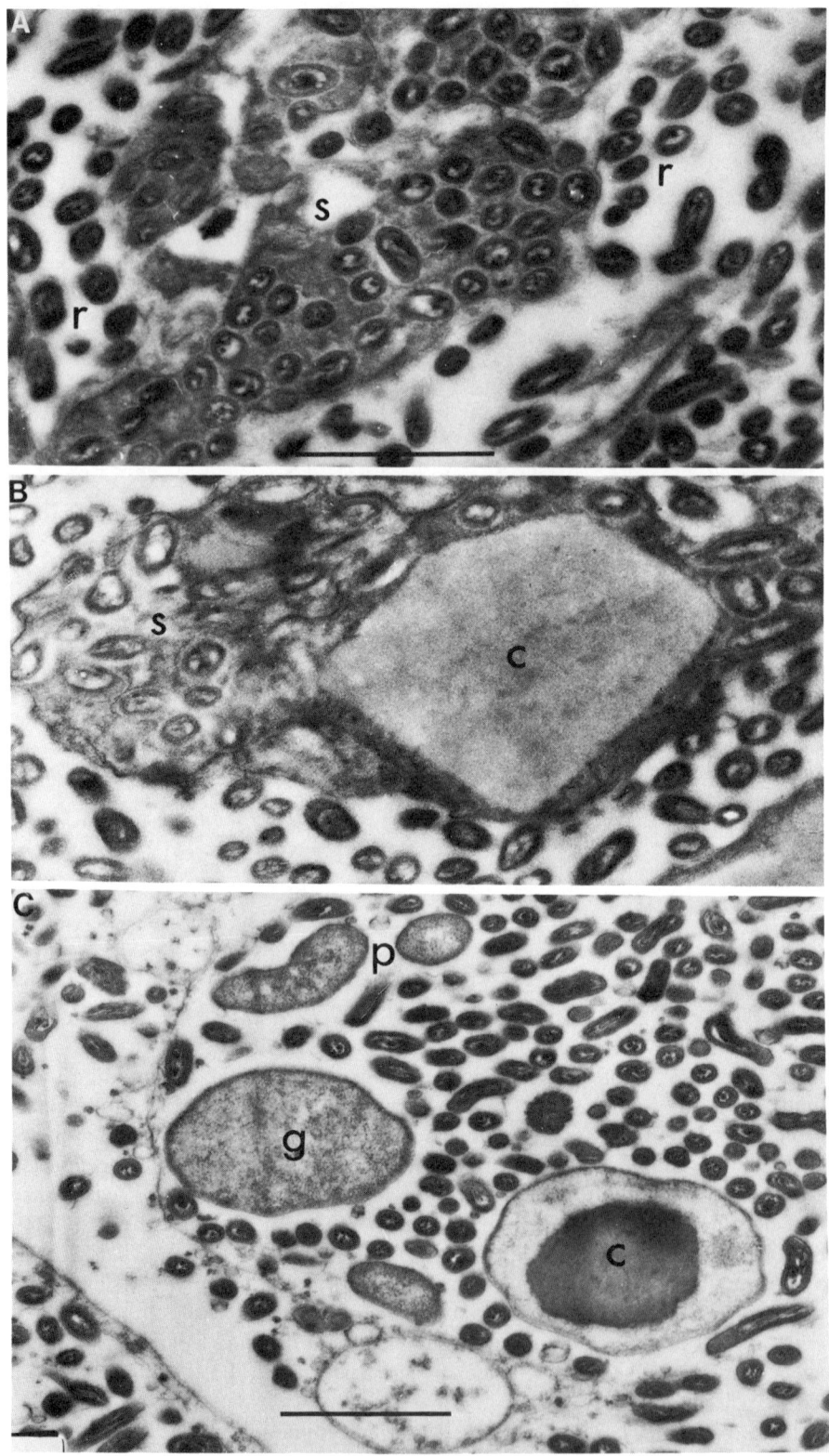

Rickettsia infection in (A, B) *Tipula paludosa* and in (C) *Melolontha melolontha*. (A) Stroma material (*s*) and standard cells (*r*). (B) Stroma material (*s*) and crystal (*c*). (C) Primary cells (*p*) and giant body (*g*) transforming into crystal (*c*). (Bars = 1 μm.) [Courtesy of Huger and Krieg (1967) and Academic Press.]

Rickettsiella chironomi, and this identification as a rickettsia was later confirmed by electron microscopical studies (Weiser and Žižka 1968). The species was again isolated from *Chironomus thummi* (Götz 1972), and from *C. decorus* and *C. frommeri* (Federici 1980). Federici (1980) pointed out that, because of differences in the ultrastructure and morphogenesis, in particular the elementary bodies being flat and disclike rather than bacilliform, *R. chironomi* was an atypical rickettsia and was more closely allied to chlamydia.

In the life cycle of *R. chironomi*, Götz (1972) reported three forms: (1) small spherical cells (0.6 μm in diameter), (2) large spherical cells (from 1 to 3 μm in diameter), and (3) disclike cells (0.6 μm in diameter, 0.06 μm in thickness); whereas, Federici (1980) described four cell types: (1) medium-sized spherical initial bodies (approximately 1 μm in diameter), (2) large spherical initial bodies (from 1.5 to 2 μm), (3) spherical intermediate bodies (600 to 700 nm), and (4) disc-shaped elementary bodies (60 × 600 nm). Development of these stages occurs in vesicles or vacuoles in the cytoplasm of fat cells. Replication is by binary fission of the small spherical cells (medium-sized initial bodies) to form the large spherical cells or other initial or intermediate bodies.

According to Götz (1972), the disclike elementary cells are formed in a large spherical cell in association with a "matrix," which is not comparable to the virogenic stroma. Within the large spherical cells, up to 30 disclike cells are formed simultaneously in a *de novo* synthesis; the replication represents multiple cell division. On the other hand, Federici (1980) maintains that the disclike elementary bodies are produced by binary fission of the medium-sized initial bodies, with each body condensing to form a single disclike body. He believes that multiple division in large spherical bodies to form the elementary bodies, if it occurs at all, is not a major mode of reproduction.

The disclike elementary bodies of *R. chironomi* infect a chironomid larva through the mouth and attack the fat body. The infection resembles that of *R. melolonthae* in being restricted to fat cells and in being chronic with a long period of lethal infection. The development of *R. melolonthae*, however, appears to be much more complex, with several pathways to form elementary cells. In *R. chironomi*, replication is mainly by binary fission, and multiple fission occurs infrequently or not at all. The major differences lie in the shape and ultrastructure of the elementary, standard bodies, which are rod-shaped in *R. melolonthae* and disclike (similar in shape to a human erythrocyte) in *R. chironomi*. Whether these differences justify placing *R. chironomi* with the chlamydia awaits further study.

Rickettsiella stethorae was isolated from larvae of five species of the coccinellid *Stethorus* (Hall and Badgley 1957). The elementary bodies (0.4 by 1.0 μm) are ovoid to elliptical in shape with tapered and rounded ends, in contrast to the kidney-shaped bodies with rounded ends of *R. popilliae*. It grows preferentially in the cytoplasm of larval midgut cells. It is highly virulent and can endanger the mass production of coccinellids.

Entomopathogenic rickettsiae or rickettsia-like forms have been reported from several Lepidoptera, such as the saturniid moths (e.g., *Samia cynthia* x *ricini*, *Hyalophora cecropia*) (Entwistle and Robertson 1968a,b); the navel orangeworm, *Paramyelois transitella* (Kellen et al. 1972); the cabbage looper, *Trichoplusia ni* (Browning et al. 1982); and the clover cutworm, *Scotogramma trifolii* (Federici 1982). In the saturniid moths, the rickettsiae multiply by binary fission in the cytoplasm of muscle cells adjacent to the midgut epithelium and to a lesser extent in the cytoplasm of fat and hypodermal cells. This rickettsia is serologically related to *R. melolonthae* (Entwistle and Robertson 1968a). Crystals are produced in the navel orangeworm, but none has been reported from the saturniid moths.

Rickettsiella pathogens of insects can infect mammals, mainly in their lungs (Croizier and Meynadier 1972). Workers should exercise care in working with these bacteria.

Members of *Wolbachia* are a heterogeneous group of small-rod or coccoid, gram-negative forms that stain well with Giemsa stain but poorly with most bacterial stains (Weiss et al. 1984). They occur extracellularly or intracellularly but seldom develop inside mycetomes. They are rarely pathogenic to insects except that some induce reproductive incompatibility in their hosts.

In insects, *Wolbachia* species occur in the reproductive organs of Diptera, Lepidoptera, and Coleoptera and may cause cytoplasmic incompatibility. Later, we shall discuss other cases of reproductive incompatibility by bacteria (e.g., in *Drosophila* and a wasp).

VII. CLASS MOLLICUTES

Wall-less prokaryotic microorganisms surrounded with a unit membrane and related to bacteria are included in the class Mollicutes (Razin and Freundt 1984). These microorganisms have received wide attention because many infect vertebrates, plants, and invertebrates, including insects, but some are commensals or saprophytes. They resemble L-phase bacteria, which are also wall-less, but differ in being incapable of producing the cell-wall components, peptidoglycan and its precursors. They are gram-negative. Cells of mollicutes are small, highly pleomorphic, varying from coccoid to helical to filamentous, and ranging in sizes from 50 to 150 nm. No resting stages are known. They can be grown on cell-free media. The colony on solid agar medium has a typical "fried-egg" appearance. Most require sterol and are resistant to penicillin and its analogues but are susceptible to tetracycline antibiotics (e.g., aureomycin and streptomycin) and to tylosin tartrate.

The class Mollicutes in the division Tenericutes has four families, Mycoplasmataceae, Acholeplasmataceae, Spiroplasmataceae, and Anaeroplasmataceae (Razin 1985). Mycoplasmataceae requires sterol for growth but not Acholeplasmataceae. Spiroplasmataceae also requires sterols but differs from Mycoplasmataceae in exhibiting helical symmetry and motility. Anaeroplasmataceae has forms that may or may not require sterols, but all are strict anaerobes and are distinct according to DNA hybridization. The following genera, *Mycoplasma*, *Spiroplasma*, and *Acholeplasma*, multiply in insects (Whitcomb et al. 1973, 1974). However, the pathogenicity of *Acholeplasma* in insects has not been established (Tully 1984).

The mycoplasmas are the smallest and simplest self-replicating procaryotes (Razin 1985). The first mycoplasma-like organism found to produce lethal infections in insects appears to be the infectious agent of peach-yellow leaf roll strains of Western X-disease, which causes premature death to the leafhopper vector *Colladonus montanus* (Jensen 1959; Whitcomb et al. 1968). In *C. montanus*, in addition to this plant pathogenic form, another mycoplasma occurs in healthy leafhoppers (Nasu et al. 1970). Other insect vectors are also susceptible to the mycoplasma-like organisms that cause plant diseases (Whitcomb and Davis 1970).

A. MYCOPLASMA

A mycoplasma pathogenic only for insects was detected in the fat and cardiac cells of *Melolontha melolontha* larvae (Devauchelle et al. 1969). It causes "léthargie" in this beetle. The mycoplasma has been propagated in liquid and solid cell-free media (Giannotti et al. 1978) and in an insect cell line of *Lymantria dispar* (Louis et al. 1978). It multiplies intracellularly in vacuoles. It produces cytopathic effects in cultured cells that subsequently rupture. In the beetle, the mycoplasma appears to undergo a developmental cycle from a replicating globular stage to a stable, resting rod-shaped stage (Louis 1979). *In vitro*, it is helical in shape, and *in vivo*, it assumes various forms (e.g., giant, globular, bacillar, and vermicular) depending on its developmental site in an insect (Giannotti et al. 1981). Serological tests and electrophoretically obtained protein spectra of these forms indicate that all of them belong to the same mollicute. It can be transmitted to *Drosophila melanogaster* and causes reduced longevity and fecundity (Louis and Plus 1979). Infection of the nervous system results in death.

A mycoplasma-like symbiote causes infertility in hybrids of the semispecies of *Drosophila paulistorum* complex (Kernaghan and Ehrman 1970a,b; Ehrman and Kernaghan 1971). In hybrids, the females are fertile, but the males are sterile. The hybrid sterility is believed to result from the discordance between a cytoplasmic mycoplasma-like symbiont and the host genotype. The discordant factor is inherited maternally. Sterility is partially alleviated by treating females with antimycoplasmal antibiotics, such as tetracycline HCl and tylosin tartrate. In treated females, the symbiont is missing. In male flies, the symbiont occurs in the testis and vas deferens and causes the degeneration of spermatid bundles, resulting in sterility. This symbiont, when introduced into larvae of *Ephestia kuehniella*, is pathogenic and may cause death (Gottlieb et al. 1977, 1981). Extracts from infected *E. kuehniella*, when inoculated into *D. paulistorum*, produce the syndrome of sterility expressed by the mycoplasma-like organism.

B. SPIROPLASMA

Spiroplasmas were first recognized in 1972 (Whitcomb 1980). They are wall-less forms with helical configurations and are similar to but distinctly different from spirochetes in lacking cell components (muramic acid and peptidoglycan) and periplasmic fibrils (axial filaments) (Davis 1979; Whitcomb 1981; Whitcomb and Tully 1984). They are motile with flectional and twitching movements, and often show an apparent rotatory motility. One strain of *Spiroplasma citri* is nonhelical and nonmotile (Townsend et al. 1980). Spiroplasmas have been placed in the family Spiroplasmataceae (Whitcomb 1980). A serogroup classification scheme is used for the species in *Spiroplasma* (Whitcomb et al. 1987).

At present, only a few *Spiroplasma* species are recognized, but the various isolates from plants and insects may belong to different species (Junca et al. 1980; Coomaraswamy and Gumpf 1984). The spiroplasmas in insects may be pathogens, commensals, and possibly mutualists (Clark and Whitcomb 1984). Some spiroplasmas have been found infecting hosts in four orders: Hymenoptera, Hemiptera, Diptera, and Coleoptera (Clark 1982; Clark and Whitcomb 1984). They are found predominantly in the gut lumen and/or hemolymph of insects. Some spiroplasma diseases of plants cause pathologies to insect vectors (Granados and

Meehan 1975; Clark and Whitcomb 1984). A spiroplasma that causes corn stunt also infects the leafhopper vector (Davis et al. 1981). It causes subcellular disorganization and degeneration of axons. It reduces the longevity and fecundity of the vector.

Spiroplasmas may occur in the vascular tissues of plants and on the flower surfaces. Some plant spiroplasmas (e.g., *S. citri*) are capable of replicating and causing lethal infections when inoculated into insects (Giannotti et al. 1980; Eskafi et al. 1987; Klein and Purcell 1987). *Spiroplasma citri* has been detected with DNA probes both in infected periwinkle plants and infected leafhoppers (Nur et al. 1986).

A spiroplasma heavily infects workers and drones of the honey bee (Clark 1977). This form is serologically related to *S. citri* but is distinct (Tully et al. 1980; Archer and Best 1980; Davis 1981). Infection in the honey bee takes place through the mouth and by inoculation into the hemocoel. Infected bees are sluggish and die usually within a week, or as early as 3 days and as long as 20 days. The disease seems to be of seasonal occurrence with the highest prevalence occurring in May and June and declining to zero by mid-July. The spiroplasma or an antigenically closely related form occurs on flower surfaces from which the honey bee apparently acquires the microorganism (Clark 1977; Davis et al. 1979; Raju et al. 1981). This organism is called *Spiroplasma melliferum* (Clark et al. 1985). Another distinct spiroplasma that causes lethal infection in the honey bee ("May disease") has been named *Spiroplasma apis* (Mouches et al. 1983, 1984). The strain from the honey bee, however, is less pathogenic when inoculated into the hemocoel of a *Galleria mellonella* larva than the strains that are epiphytic on flowers (Dowell et al. 1981). Other mycoplasmas isolated from the honey bee are not pathogenic to bees (Junca et al. 1980).

A spiroplasma, which infects the digestive tract of the Colorado potato beetle, *Leptinotarsa decemlineata*, may be transmitted to uninfected insects that feed on contaminated plants (Clark 1982). This fastidious spiroplasma has been cultured in tissue culture (Hackett and Lynn 1985).

A mollicute affects the sex ratio of the *Drosophila* offspring by eliminating the male progeny (Williamson 1965; Williamson and Poulson 1979). This disease differs from the hybrid sterility in *Drosophila* caused by a mycoplasma-like symbiont described by Kernaghan and Ehrman (1970b). The sex-ratio disease was known for many years and was previously called the "sex-ratio" condition (Malogolowkin and Poulson 1957; Malogolowkin 1958). The etiological agent was at one time believed to be a spirochete but is presently referred to as a *Spiroplasma* or the sex-ratio organism (SRO) (Fig. 5–3).

The hemolymph of flies infected with SRO is highly infectious, and when inoculated into a number of species and strains of *Drosophila*, produces infection (Sakaguchi and Poulson 1961a,b; 1963). The SRO selectively kills the males primarily as embryos, and to a lesser extent other developmental stages up to the adult (Counce and Poulson 1962; Williamson and Poulson 1979). The embryos, however, with two or more X chromosomes survive, whereas those with a single X chromosome are killed regardless of whether they were phenotypically males, intersexes, or females (Miyamoto and Oishi 1975). The primary target tissue of the spiroplasma is the nervous system (Koana and Miyake 1983). Occasionally, the infected individuals survive as gynandromorphs (Tsuchiyama et al. 1978).

The SRO had not been cultured for more than 25 years until Hackett et al.

FIGURE 5–3

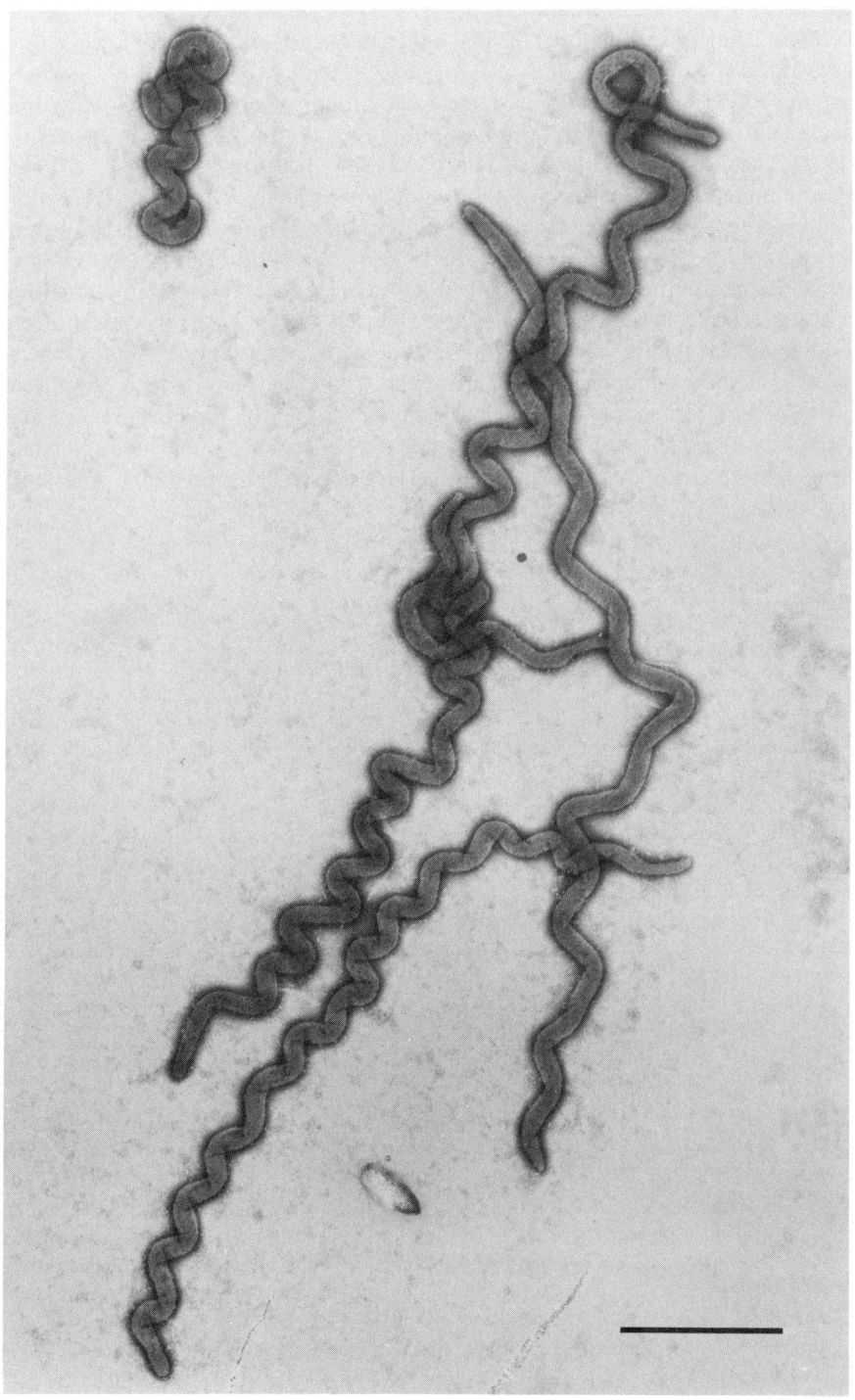

Sex-ratio spiroplasmas from the hemolymph of female *Drosophila pseudoobscura*. (Bar = 1 μm.) (Provided by D. L. Williamson.)

(1986) succeeded in growing the *Spiroplasma* in a tissue-culture medium containing an embryo-derived lepidopteran cell line. Subsequently, the spiroplasma has been adapted to a cell-free medium. However, only transient proliferation of the SRO occurs in primary cell cultures of SRO-infected *Drosophila* embryonic cells (Ueda et al. 1989).

Yamada et al. (1982) have isolated a sex-ratio mutant that has lost its male-specific killing activity in *Drosophila*. Spiroplasma not associated with the sex-ratio trait also occurs in *Drosophila* (Whitcomb and Williamson 1979).

A similar case of sex-ratio distortion, but not by a mollicute, occurs in *Nasonia vitripennis*, a gregarious wasp parasitizing fly pupae (Huger et al. 1985; Werren et al. 1986). Three extrachromosomal factors cause sex-ratio distortion in this wasp; one of them being a pleomorphic bacterium that has characteristics of Enterobacteriaceae. The bacterium kills male eggs, but not female eggs, by preventing the development of unfertilized eggs (in Hymenoptera, unfertilized eggs develop into males). The infected female wasp transmits the son-killer trait during oviposition by way of mechanical and/or transovum transmission. The chronic infection originates in the larval midgut and spreads to other tissues including the brain, fat body, muscles, eyes, and hemocytes. The bacterium is passed transtadially. The female reproductive tract is often infected but not that of the male.

REFERENCES

Akhurst, R. J. 1986a. *Xenorhabdus nematophilus* subsp. *poinarii*: Its interaction with insect pathogenic nematodes. *Syst. Appl. Microbiol. 8*, 142–147.

Akhurst, R. J. 1986b. *Xenorhabdus nematophilus* subsp. *beddingii* (*Enterobacteriaceae*): A new subspecies of bacteria mutualistically associated with entomopathogenic nematodes. *Int. J. Syst. Bacteriol. 36*, 454–457.

Allardyce, R. A., Keenan, J. I., O'Callaghan, M., and Jackson, T. A. 1991. Serological identification of *Serratia entomophila*, a bacterial pathogen of the New Zealand grass grub (*Costelytra zealandica*). *J. Invertebr. Pathol. 57*, 250–254.

Amonkar, S. V., Kalle, G. P., and Nair, K. K. 1967. Mechanism of pathogenicity of *Pseudomonas* in the house fly. *J. Invertebr. Pathol. 9*, 235–240.

Archer, D. B., and Best, J. 1980. Serological relatedness of spiroplasmas estimated by enzyme-linked immunosorbent assay and crossed immunoelectrophoresis. *J. Gen. Microbiol. 119*, 413–422.

Bailey, L. 1963. "Infectious diseases of the honey-bee." Land Books, London.

Bailey, L. 1984. A strain of *Melissococcus pluton* cultivable on chemically defined media. *FEMS Microbiol. Lett. 25*, 139–141.

Bailey, L., and Collins, M. D. 1982. Reclassification of '*Streptococcus pluton*' (White) in a new genus *Melissococcus*, as *Melissococcus pluton* nom. rev.; comb. nov. *J. Appl. Bacteriol. 53*, 215–217.

Bell, J. V. 1969. *Serratia marcescens* found in eggs of *Heliothis zea*: Tests against *Trichoplusia ni*. *J. Invertebr. Pathol. 13*, 151–152.

Boemare, N., Louis, C., and Kuhl, G. 1983. Etude ultrastructurale des cristaux chez *Xenorhabdus* spp., bactéries inféodées aux nématodes entomophages *Steinernematidae* et *Heterorhabditidae*. *C.R. Soc. Biol. 177*, 107–115.

Bracken, G. K., and Bucher, G. E. 1967. Mortality of hymenopterous parasite caused by *Serratia marcescens*. *J. Invertebr. Pathol. 9*, 130–132.

Browning, H. W., Federici, B. A., and Oatman, E. R. 1982. Occurrence of a disease caused by a rickettsia-like organism in a larval population of the cabbage looper, *Trichoplusia ni*, in southern California. *Environ. Entomol. 11*, 550–554.

Buchanan, R. E., and Gibbons, N. E. 1974. "Bergey's Manual for Determinative Bacteriology." Eighth Ed. Williams & Wilkins, Baltimore.

Bucher, G. E. 1959a. The bacterium *Coccobacillus acridiorum* d'Herelle: Its taxonomic position and status as a pathogen of locusts and grasshoppers. *J. Insect Pathol. 1*, 331–346.

Bucher, G. E. 1959b. Bacteria of grasshoppers of western Canada: III. Frequency of occurrence, pathogenicity. *J. Insect Pathol. 1*, 391–405.

Bucher, G. E. 1960. Potential bacterial pathogens of insects and their characteristics. *J. Insect Pathol. 2*, 172–195.

Bucher, G. E. 1963a. Transmission of bacterial pathogens by the ovipositor of a hymenopterous parasite. *J. Insect Pathol. 5*, 277–283.

Bucher, G. E. 1963b. Nonsporulating bacterial pathogens. *In* "Insect Pathology: An Advanced Treatise." (E. A. Steinhaus, ed.), Vol. 2., pp. 117–147. Academic Press, New York.

Bucher, G. E., and Stephens, J. M. 1957. A disease of grasshoppers caused by the bacterium Pseudomonas aeruginosa (Schroeter) Migula. *Can. J. Microbiol. 3*, 611–625.

Burnside, C. E., Sturtevant, A. P., and Holst, E. C. 1949. Diagnosing bee diseases in the apiary. *U.S.D.A. Circ. No. 392*.

Clark, T. B. 1977. *Spiroplasma* sp., a new pathogen in honey bees. *J. Invertebr. Pathol. 29*, 112–113.

Clark, T. B. 1982. Spiroplasmas: Diversity of arthropod reservoirs and host–parasite relationships. *Science 217*, 57–59.

Clark, T. B., and Whitcomb, R. F. 1984. Pathogenicity of mollicutes for insects: Possible use in biological control. *Ann. Microbiol.* (Inst. Pasteur) *135A*, 141–150.

Clark, T. B., Whitcomb, R. F., Tully, J. G., Mouches, C., Saillard, C., Bové, J. M., Wróblewski, H., Carle, P., Rose, D. L., Henegar, R. B., and Williamson, D. L. 1985. *Spiroplasma melliferum*, a new species from the honeybee (*Apis mellifera*). *Int. J. Syst. Bacteriol. 35*, 296–308.

Collins, M. D., and Jones, D. 1981. Lipid composition of the entomopathogen *Corynebacterium okanaganae* (Lüthy). *FEMS Microbiol. Lett. 10*, 157–159.

Coomaraswamy, U., and Gumpf, D. J. 1984. Immuno-double-diffusion serological relationships among spiropasma isolates. *Int. J. Syst. Bacteriol. 34*, 173–176.

Counce, S. J., and Poulson, D. F. 1962. Developmental effects of the sex-ratio agent in embryos of *Drosophila willistoni*. *J. Exp. Zool. 151*, 17–31.

Croizier, G., and Meynadier, G. 1971. Recherche d'antigènes de groupe chez des rickettsies de la tribu des *Wolbachiae* par la technique d'agglutination des corps élémentaires. *Entomophaga 16*, 11–17.

Croizier, G., and Meynadier, G. 1972. Étude en immunofluorescence de l'infection expérimentale de la souris par *Rickettsiella grylli*. *Ann. Rech. Veter. 3*, 373–380.

Croizier, G., Meynadier, G., Morel, G., and Capponi, M. 1975. Comparaison immunologique de quelques *Wolbachiae* et recherche de communauté antigénique avec d'autres membres de l'ordre des Rickettsiales. *Bull. Soc. Pathol. Exot. 68*, 133–141.

Davis, R. E. 1979. Spiroplasmas: Newly recognized arthropod-borne pathogens. *In* "Leafhopper Vectors and Plant Disease Agents." (K. Maramorosch and K. F. Harris, eds.), pp. 451–484. Academic Press, New York.

Davis, R. E. 1981. The enigma of the flower spiroplasmas. *In* "Mycoplasma Diseases of Trees and Shrubs." (K. Maramorosch and S.P. Raychaudhuri, eds.), pp. 259–279. Academic Press, New York.

Davis, R. E., Lee, I.-M., and Basciano, L. K. 1979. Spiroplasmas: Seriological grouping of strains associated with plants and insects. *Can. J. Microbiol. 25*, 861–866.

Davis, R. E., Chen, T.-A., and Worley, J. F. 1981. Corn stunt spiroplasma. *In* "Virus and Viruslike Diseases of Maize in the United States." (D. T. Gordon, J. K. Knoke, and G. E. Scott, eds.). *South. Coop. Ser. Bull. 247*. Ohio Agricultural Research Developmental Center, Wooster, Ohio.

Devauchelle, G., Vago, C., Giannotti, J., and Quiot, J. M. 1969. Micro-organismes de type mycoplasme chez les coléoptères. *Entomophaga 14*, 457–461.

Devauchelle, G., Meynadier, G., and Vago, C. 1972. Étude ultrastructurale du cycle de multiplication de *Rickettsiella melolonthae* (Krieg), Philip, dans les hémocytes de son hôte. *J. Ultrastruct. Res. 38*, 134–148.

d'Herelle, F. 1911. Sur une épizootie de nature bactérienne sévissant sur les sauterelles au Mexique. *C.R. Acad. Sci. Paris 152*, 1413–1415.

d'Herelle, F. 1914. Le coccobacille des sauterelles. *Ann. Inst. Pasteur Paris 28*, 280–328, 387–407.

Doane, C. C. 1971. Field application of a *Streptococcus* causing brachyosis in larvae of *Porthetria dispar*. *J. Invertebr. Pathol. 17*, 303–307.

Doane, C. C., and Redys, J. J. 1970. Characteristics of motile strains of *Streptococcus faecalis* pathogenic to larvae of the gypsy moth. *J. Invertebr. Pathol. 15*, 420–430.

Dorn, A. 1976. *Pseudomonas aeruginosa* causes epidemic disease in the milkweed bug, *Oncopeltus fasciatus* Dallas (Insecta, Heteroptera). *Experientia 32*, 599–600.

Dowell, R. V., Basham, H. G., and McCoy, R. E. 1981. Influence of five spiroplasma strains on growth rate and survival of *Galleria mellonella* (Lepidoptera: Pyralidae) larvae. *J. Invertebr. Pathol. 37*, 231–235.

Dunphy, G. B., Morton, D. B., and Chadwick, J. M. 1986a. Pathogenicity of lipopolysaccharide mutants of *Pseudomonas aeruginosa* for larvae of *Galleria mellonella*: Hemocyte interaction with the bacteria. *J. Invertebr. Pathol. 47*, 56–64.

Dunphy, G. B., Morton, D. B., Kropinski, A., and Chadwick, J. M. 1986b. Pathogenicity of lipopolysaccharide mutants of *Pseudomonas aeruginosa* for larvae of *Galleria mellonella*: Bacterial properties associated with virulence. *J. Invertebr. Pathol. 47*, 48–55.

Dutky, S. R. 1959. Insect microbiology. *Adv. Appl. Microbiol. 1*, 175–200.

Dutky, S. R., and Gooden, E. L. 1952. *Coxiella popilliae*, n.sp., a rickettsia causing blue disease of Japanese beetle larvae. *J. Bacteriol. 63*, 743–750.

Ehrman, L., and Kernaghan, R. P. 1971. Microorganismal basis of infectious hybrid male sterility in *Drosophilia paulistorum*. *J. Hered. 62*, 67–71.

Entwistle, P. F., and Robertson, J. S. 1968a. Rickettsiae pathogenic to two saturniid moths. *J. Invertebr. Pathol. 10*, 345–354.

Entwistle, P. F., and Robertson, J. S. 1968b. The ultrastructure of a rickettsia pathogenic to a saturnid moth. *J. Gen. Microbiol. 54*, 97–104.

Eskafi, F. M., McCoy, R. E., and Norris, R. C. 1987. Pathology of *Spiroplasma floricola* in *Galleria mellonella* larvae. *J. Invertebr. Pathol. 49*, 1–13.

Federici, B. A. 1980. Reproduction and morphogenesis of *Rickettsiella chironomi*, an unusual intracellular procaryotic parasite of midge larvae. *J. Bacteriol. 143*, 995–1002.

Federici, B. A. 1982. A new type of insect pathogen in larvae of the clover cutworm, *Scotogramma trifolii*. *J. Invertebr. Pathol. 40*, 41–54.

Giannotti, J., Vago, C., Louis, C., and Giannotti, D. 1978. Culture *in vitro* du mollicute agent de la "léthargie" des Coléoptères. *C.R. Acad. Sci. Paris Ser. D287*, 379–381.

Giannotti, J., Giannotti, D., and Vago, C. 1980. Multiplication intracellulaire et action pathogène de mollicutes isolés de plantes chez des Insectes non vecteurs. *C.R. Acad. Sci. Paris Ser. D 290*, 417–419.

Giannotti, J., Vago, C., Giannotti, D., and Legoff, C. 1981. Étude comparée *in vivo* et *in vitro* des diverses formes de l'agent mollicute de la léthargie de Coléoptères. *C.R. Acad. Sci. Paris Ser.* III *292*, 1043–1049.

Gottlieb, F. J., Goitein, R., Ehrman, L., and Inocencio, B. 1977. Interorder transfer of mycoplasma-like microorganisms between *Drosophila paulistorum* and *Ephestia kuehniella*: Tissues, dosages, and effects. *J. Invertebr. Pathol. 30*, 140–150.

Gottlieb, F. J., Simmons, G. M., Ehrman, L., Inocencio, B., Kocka, J., and Somerson, N. 1981. Characteristics of the *Drosophila paulistorum* male sterility agent in a secondary host, *Ephestia kuehniella*. *Appl. Environ. Microbiol. 42*, 838–842.

Götz, P. 1972. "*Rickettsiella chironomi*": An unusual bacterial pathogen which reproduces by multiple cell division. *J. Invertebr. Pathol. 20*, 22–30.

Granados, R. R., and Meehan, D. J. 1975. Pathogenicity of the corn stunt agent to an insect vector, *Dalbulus elimatus*. *J. Invertebr. Pathol. 26*, 313–320.

Grimont, P. A. D., and Grimont, F. 1978. The genus *Serratia*. *Annu. Rev. Microbiol. 32*, 221–248.

Grimont, P. A. D., Grimont, F., and Lysenko, O. 1979. Species and biotype identification of *Serratia* strains associated with insects. *Curr. Microbiol. 2*, 139–142.

Grimont, P. A. D., Jackson, T. A., Ageron, E., and Noonan, M. J. 1988. *Serratia entomophila* sp. nov. associated with amber disease in the New Zealand grass grub *Costelytra zealandica*. *Int. J. Syst. Bacteriol. 38*, 1–6.

Hackett, K. J., and Lynn, D. E. 1985. Cell-assisted growth of fastidious spiroplasma. *Science 230*, 825–827.

Hackett, K. J., Lynn, D. E., Williamson, D. L., Ginsberg, A. S., and Whitcomb, R. F. 1986. Cultivation of the *Drosophila* sex-ratio spiroplasma. *Science 232*, 1253–1255.

Hall, I. M., and Badgley, M. E. 1957. A rickettsial disease of larvae of species of *Stethorus* caused by *Rickettsiella stethorae*, n.sp. *J. Bacteriol. 74*, 452–455.

Henry, J. E., Street, D. A., Oma, E. A., and Goodwin, R. H. 1986. Ultrastructure of an isolate of *Rickettsiella* from the African grasshopper *Zonocerus variegatus*. *J. Invertebr. Pathol. 47*, 203–213.

Huger, A. 1964. Eine Rickettsiose der Orientalischen Schabe, *Blatta orientalis* L., verursacht durch *Rickettsiella blattae* nov. spec. *Naturwissenschaften 51*, 22.

Huger, A. M., and Krieg, A. 1967. New aspects of the mode of reproduction of *Rickettsiella* organisms in insects. *J. Invertebr. Pathol. 9*, 442–445.

Huger, A. M., Skinner, S. W., and Werren, J. H. 1985. Bacterial infections associated with the son-killer trait in the parasitoid wasp *Nasonia* (= *Mormoniella*) *vitripennis* (Hymenoptera: Pteromalidae). *J. Invertebr. Pathol. 46*, 272–280.

Hurpin, B., and Vago, C. 1958. Les maladies du hanneton commun (*Melolontha melolontha* L.) (Col. Scarabaeidae). *Entomophaga 3*, 285–330.

Iizuka, T. 1972. The pathogenic mechanism of the disease caused by *Streptococcus faecalis* AD-4 in silkworm larvae reared on the artificial diet. *J. Seric. Sci. Jpn. 41*, 333–337.

Jackson, T. A., Pearson, J. F., and Stucki, G. 1986. Control of the grass grub, *Costelytra zealandica* (White) (Coleoptera: Scarabaeidae), by application of the bacteria *Serratia* spp. causing honey disease. *Bull. Entomol. Res. 76*, 69–76.

Jarrell, K. F., and Kropinski, A. M. 1982. The virulence of protease and cell surface mutants of *Pseudomonas aeruginosa* for the larvae of *Galleria mellonella*. *J. Invertebr. Pathol. 39*, 395–400.

Jensen, D. D. 1959. A plant virus lethal to its insect vector. *Virology 8*, 164–175.

Junca, P., Saillard, C., Tully, J., Garcia-Jurado, O., Degorce-Dumas, J. -R., Mouches, C., Vignault, J. -C., Vogel, R., McCoy, R., Whitcomb, R., Williamson, D., Latrille, J., and Bove, J. M. 1980. Caractérisation de spiroplasmes isolés d'Insectes et de fleurs de France continentale, de Corse et du Maroc. Proposition pour une classification des spiroplasmes. *C.R. Acad. Sci. Paris Ser. D 290*, 1209–1212.

Kalucy, E. C., and Daniel, A. 1972. The reaction of *Anopheles annulipes* larvae to infection by *Aeromonas punctata*. *J. Invertebr. Pathol. 19*, 189–197.

Kaška, M. 1976. The toxicity of extracellular proteases of the bacterium *Serratia marcescens* for larvae of greater wax moth, *Galleria mellonella*. *J. Invertebr. Pathol. 27*, 271.

Kaška, M., Lysenko, O., and Chaloupka, J. 1976. Exocellular proteases of *Serratia marcescens* and their toxicity to larvae of *Galleria mellonella*. *Folia Microbiol. 21*, 465–473.

Kellen, W. R., Lindegren, J. E., and Hoffmann, D. F. 1972. Developmental stages and structure of a *Rickettsiella* in the navel orangeworm, *Paramyelois transitella* (Lepidoptera: Phycitidae). *J. Invertebr. Pathol. 20*, 193–199.

Kernaghan, R. P., and Ehrman, L. 1970a. Antimycoplasmal antibiotics and hybrid sterility in Drosophila paulistorum. *Science 169*, 63–64.

Kernaghan, R. P., and Ehrman, L. 1970b. An electron microscopic study of the etiology of hybrid sterility in *Drosophila paulistorum*. I. Mycoplasma-like inclusions in the testes of sterile males. *Chromosoma* (Berl.) *29*, 291–304.

Klein, M., and Purcell, A. H. 1987. Response of *Galleria mellonella* (Lepidoptera: Pyralidae) and *Tenebrio molitor* (Coleoptera: Tenebrionidae) to *Spiroplasma citri* inoculation. *J. Invertebr. Pathol. 50*, 9–15.

Koana, T., and Miyake, T. 1983. Effects of the sex ratio organism on *in vitro* differentiation of Drosophila embryonic cells. *Genetics 104*, 113–122.

Kodama, R., and Nakasuji, Y. 1971. Further studies on the pathogenic mechanism of bacterial diseases in gnotobiotic silkworm larvae. *Res. Commun. Inst. Fermentation Osaka 5*, 1–9.

Krieg, A. 1955. Licht- und elektronenmikroskopische Untersuchungen zur Pathologie der "Lorscher Erkankung" von Engerlingen und zur Zytologie der *Rickettsia melolonthae* nov. spec. *Z. Naturforsch. 10b*, 34–37.

Krieg, A. 1987. Diseases caused by bacteria and other prokaryotes. *In* "Epizootiology of Insect Diseases." (J. R. Fuxa and Y. Tanada, eds.), pp. 323–355. John Wiley and Sons, New York.

Kučera, M., and Lysenko, O. 1971. The mechanism of pathogenicity of *Pseudomonas aeruginosa*. VIII. Isolation of hemolymph proteins from *Galleria mellonella* larvae and their digestibility by the toxic protease. *J. Invertebr. Pathol. 17*, 203–210.

Kučera, M., and Lysenko, O. 1977. The mechanism of pathogenicity of *Pseudomonas aeruginosa*: Milk-clotting activity of the proteolytic enzymes toxic for *Galleria mellonella* larvae. *J. Invertebr. Pathol. 29*, 388–389.

Louis, C. 1979. Caractérisation par radio-autographie d'un stade morphologique à forte replication d'ADN chez l'agent mycoplasmique de la "léthargie des Coléoptères." *C.R. Acad. Sci. Paris Ser. D289*, 1179–1181.

Louis, C., and Plus, N. 1979. *Drosophila melanogaster* as an experimental host for study of multiplication and biology of the mycoplasma inducing the "lethargy of Coleoptera." *Ann. Microbiol.* (Inst. Pasteur) *130B*, 415–431.

Louis, C., Quiot, J. M., Giannotti, J., and Vago, C. 1978. Infection expérimentale d'une lignée cellulaire d'invertébré par le procaryote intravacuolaire de type mollicute, agent de la "léthargie des coléoptères." *Ann. Microbiol.* (Inst. Pasteur) *129B*, 621–633.

Lüthy, P., and Soper, R. S. 1969. Chalky disease, a bacterial septicemia of the cicada *Okanagana rimosa* Say. *J. Invertebr. Pathol. 14*, 158–164.

Lysenko, O. 1958. Contribution to the taxonomy of *Coccobacillus acridiorum* d'Herelle. *Folia Biol. 4*, 342–347.

Lysenko, O. 1974. Bacterial exoenzymes toxic for insects; proteinase and lecithinase. *J. Hyg. Epidemiol. Microbiol. Immunol. 18*, 347–352.

Lysenko, O. 1976. Chitinase of *Serratia marcescens* and its toxicity to insects. *J. Invertebr. Pathol. 27*, 385–386.

Lysenko, O., and Kučera, M. 1968. The mechanism of pathogenicity of *Pseudomonas aeruginosa*. VI. The toxicity of proteinases for larvae of the greater wax moth, *Galleria mellonella* L. *Folia Microbiol. Prague 13*, 295–299.

Lysenko, O., and Kučera, M. 1971. Micro-organisms as sources of new insecticidal chemicals: Toxins. In "Microbial Control of Insects and Mites." (H. D. Burges and N. W. Hussey, eds.), pp. 205–227. Academic Press, London.

Madziara-Borusiewicz, K., and Lysenko, O. 1971. The mechanism of pathogenicity of *Pseudomonas aeruginosa*. VII. The influence of toxic proteinase on hemocytes of *Galleria mellonella*. *J. Invertebr. Pathol. 17*, 138–140.

Malogolowkin, C. 1958. Maternally inherited "sex-ratio" conditions in *Drosophila willistoni* and *Drosophila paulistorum*. Genetics *43*, 274–286.

Malogolowkin, C., and Poulson, D. F. 1957. Infective transfer of maternally inherited abnormal sex-ratio in Drosophila willistoni. *Science 126*, 32.

Miyamoto, C., and Oishi, K. 1975. Effects of SR-spirochete infection on *Drosophila melanogaster* carrying intersex genes. *Genetics 79*, 55–61.

Mouches, C., Bove, J. M., Tully, J. G., Rose, D. L., McCoy, R. E., Carle-Junca, P., Garnier, M., and Saillard, C. 1983. *Spiroplasma apis*, a new species from the honey-bee *Apis mellifera*. *Ann. Microbiol.* (Inst. Pasteur) *134A*, 383–397.

Mouches, C., Bove, J. M., and Albisetti, J. 1984. Pathogenicity of *Spiroplasma apis* and other spiroplasmas for honey-bees in southwestern France. *Ann. Microbiol.* (Inst. Pasteur) *135A*, 151–155.

Nasu, S., Jensen, D. D., and Richardson, J. 1970. Electron microscopy of mycoplasma-like bodies associated with insect and plant hosts of peach Western X-disease. *Virology 41*, 583–595.

Niklas, O. F. 1957. Zur Temperaturabhängigkeit der Vertikalbewegungen Rickettsiosekranker Maikäfer-Engerlinge. *Anz. Schadlingskd. 30*, 113–116.

Nur, I., Bove, J. M., Saillard, C., Rottem, S., Whitcomb, R. M., and Razin, S. 1986. DNA probes in detection of spiroplasmas and mycoplasma-like organisms in plants and insects. *FEMS Microbiol. Lett. 35*, 157–162.

Ono, M., and Kato, S. 1968. Studies on the dissolution of peritrophic membrane in the silkworm, *Bombyx mori* L. II. On the bacterial enzyme, which decomposes peritrophic membrane, obtained from the culture filtrates of *Aeromonas*. *Bull. Seric. Exp. Stn.* (Tokyo) *23*, 9–34.

Øgaard, A. R., Bjøro, K., Bukholm, G., and Berdal, B. P. 1985. Correlation between adhesion of *Pseudomonas aeruginosa* bacteria to cell surfaces and the presence of some factors related to virulence. *Acta Pathol. Microbiol. Immunol. Scand. Sect. B. 93*, 211–216.

Podgwaite, J. D., and Cosenza, B. J. 1976. A strain of *Serratia marcescens* pathogenic for larvae of *Lymantria dispar*: Infectivity and mechanisms of pathogenicity. *J. Invertebr. Pathol. 27*, 199–208.

Poinar, G. O., Jr., Wassink, H.J.M., Leegwater-van der Linden, M. E., and van der Geest, L. P. S. 1979. *Serratia marcescens* as a pathogen of tsetse flies. *Acta Trop. 36*, 223–227.

Prinsloo, H. E. 1967. The phospholipase and gelatinase activity of gram-negative bacteria isolated from diseased insects, and typing of strains of *Serratia marcescens*. *J. Invertebr. Pathol. 9*, 420–427.

Raju, B. C., Nyland, G., Meikle, T., and Purcell, A. H. 1981. Helical, motile mycoplasmas associated with flowers and honey bees in California. *Can. J. Microbiol. 27*, 249–253.

Razin, S. 1985. Molecular biology and genetics of mycoplasmas (*Mollicutes*). *Microbiol. Rev. 49*, 419–455.

Razin, S., and Freundt, E. A. 1984. Division Tenericutes Div. Nov. (g.v. p.36). Class I. Mollicutes Edward and Freundt 1967, 267. *In* "Bergey's Manual of Systematic Bacteriology." (N. R. Krieg and J. G. Holt, eds.), Ninth Ed., Vol. 1, pp. 740–770. Williams and Wilkins, Baltimore.

Rietschel, E. Th., Brade, H., Brade, L., Kaca, W., Kawahara, K., Lindner, B., Lüderitz, T., Tomita, T., Schade, U., Seydel, U., and Zähringer, U. 1985. Newer aspects of the chemical structure and biological activity of bacterial endotoxins. *In* "Bacterial Endotoxins: Structure, Biomedical Significance, and Detection with the Limulus Amebocyte Lysate Test." (J. W. Tencate, H. R. Buller, A. Sturk, and J. Levin, eds.). *Prog. Clin. Biol. Res. 189*, 31–50.

Sakaguchi, B., and Poulson, D. F. 1961a. Some properties of the "sex-ratio" agent in Drosophilia. *Annu. Rep. Natl. Inst. Genet. 11*, 22–23.

Sakaguchi, B., and Poulson, D. F. 1961b. Distribution of "sex-ratio" agent in tissues of Drosophila willistoni. *Genetics 46*, 1665–1676.

Sakaguchi, B., and Poulson, D. F. 1963. Interspecific transfer of the "sex-ratio" condition from Drosophila willistoni to D. melanogaster. *Genetics 48*, 841–861.

Slatten, B. H., and Larson, A. D. 1967. Mechanism of pathogenicity of *Serratia marcescens*. I. Virulence for the adult boll weevil. *J. Invertebr. Pathol. 9*, 78–81.

Steinhaus, E. A. 1949. "Principles of Insect Pathology." McGraw-Hill, New York.

Steinhaus, E. A. 1959. *Serratia marcescens* Bizio as an insect pathogen. *Hilgardia 28*, 351–380.

Stucki, G., Jackson, T. A., and Noonan, M. J. 1984. Isolation and characterization of *Serratia* strains pathogenic for larvae of the New Zealand grass grub *Costelytra zealandica*. *N.Z. J. Sci. 27*, 255–260.

Thomas, G. M., and Poinar, G. O., Jr. 1983. Amended description of the genus *Xenorhabdus* Thomas and Poinar. *Int. J. Syst. Bacteriol. 33*, 878–879.

Townsend, R., Burgess, J., and Plaskitt, K. A. 1980. Morphology and ultrastructure of helical and nonhelical strains of *Spiroplasma citri*. *J. Bacteriol. 142*, 973–981.

Trought, T. E. T., Jackson, T. A., and French, R. A. 1982. Incidence and transmission of a disease of grass grub (*Costelytra zealandica*) in Canterbury. *N.Z. J. Exp. Agric. 10*, 79–82.

Tsuchiyama, S., Sakaguchi, B., and Oishi, K. 1978. Analysis of gynandromorph survivals in *Drosophila melanogaster* infected with the male-killing SR organisms. *Genetics 89*, 711–721.

Tully, J. G. 1984. Family II. Acholeplasmataceae. Edward and Freundt 1970, 1. *In* "Bergey's Manual of Systematic Bacteriology." (N. R. Krieg and J. G. Holt, eds.), Ninth Ed., Vol. 1, pp. 775–781. Williams and Wilkins, Baltimore.

Tully, J. G., Rose, D. L., Garcia-Jurado, O., Vignault, J.-C., Saillard, C., Bové, J. M., McCoy, R. E ., and Williamson, D. L. 1980. Serological analysis of a new group of spiroplasmas. *Curr. Microbiol. 3*, 369–372.

Ueda, R., Koana, T., and Miyake, T. 1989. Transient proliferation of sex ratio organisms of *Drosophila* in a primary cell culture from infected embryos. *In* "Invertebrate Cell System Applications." (J. Mitushashi, ed.), Vol. 2, pp. 77–84. CRC Press, Boca Raton, Florida.

Vasantharajan, V. N., and Munirathnamma, N. 1978. Studies on silkworm diseases—III epizootiology of a septicemic disease of silkworms caused by *Serratia marcescens*. *J. Indian Inst. Sci. 60*, 33–42.

Weiser, J. 1949. Deux nouvelles infections a virus des insectes. *Ann. Parasitol. Hum. Comp. 24*, 259–264.

Weiser, J. 1966. "Nemoci Hmyzu." Academia, Prague.

Weiser, J., and Žižka, Z. 1968. Electron-microscope studies of *Rickettsiella chironomi* in the midge *Camptochironomus tentans*. *J. Invertebr. Pathol. 12*, 222–230.

Weiss, E. 1974. Genus X. Rickettsiella Philip 1956, 267. *In* "Bergey's Manual of Determinative Bacteriology." (R. E. Buchanan and N.E. Gibbons, eds.), Eighth Ed., pp. 901–903. Williams and Wilkins, Baltimore.

Weiss, E., and Moulder, J. W. 1984. Order I. Rickettsiales Gieszczkiewicz 1939, 25. *In* "Bergey's Manual of Systematic Bacteriology." (N. R. Krieg and J. G. Holt, eds.) Vol. 1, pp. 687–74. Williams and Wilkins, Baltimore.

Weiss, E., Dasch, G. A., and Chang, K.-P. 1984. Tribe III. Wolbachieae Philip 1956, 266. *In* "Bergey's Manual of Systematic Bacteriology." (N. R. Krieg and J. G. Holt, eds.), Nineth Ed., Vol. 1, pp. 711–717. Williams and Wilkins, Baltimore.

Werren, J. H., Skinner, S. W., and Huger, A. M. 1986. Male-killing bacteria in a parasitic wasp. *Science 231*, 990–992.

Whitcomb, R. F. 1980. The genus *Spiroplasma*. *Annu. Rev. Microbiol. 34*, 677–709.

Whitcomb, R. F. 1981. The biology of spiroplasmas. *Ann. Rev. Entomol. 26*, 397–425.

Whitcomb, R. F., and Davis, R. E. 1970. *Mycoplasma* and phytarboviruses as plant pathogens persistently transmitted by insects. *Annu. Rev. Entomol. 15*, 405–464.

Whitcomb, R. F., and Tully, J. G. 1984. Family III. *Spiroplasmataceae* Skripal 1983, 408. *In* "Bergey's Manual of Systematic Bacteriology." (N. R. Krieg and J. G. Holt, eds.), Nineth Ed., Vol. 1, pp. 781–787. Williams and Wilkins Baltimore.

Whitcomb, R. F., and Williamson, D. L. 1979. Pathogenicity of mycoplasmas for arthropods. *Zentralbl. Bakteriol. Parasitenkd., Infektionskr. Hyg. Abt. 1, Orig. Reihe A 245*, 200–221.

Whitcomb, R. F., Jensen, D. D., and Richardson, J. 1968. The infection of leafhoppers by Western X-disease virus. IV. Pathology in the alimentary tract. *Virology 34*, 69–78.

Whitcomb, R. F., Tully, J. G., Bové, J. M., and Saglio, P. 1973. Spiroplasmas and acholeplasmas: Multiplication in insects. *Science 182*, 1251–1253.

Whitcomb, R. F., Williamson, D. L., Rosen, J., and Coan, M. 1974. Relationship of infection and pathogenicity in the infection of insects by wall-free prokaryotes. *Inserm 33*, 275–282.

Whitcomb, R. F., Bové, J. M., Chen, T. A., Tully, J. G., and Williamson, D. L. 1987. Proposed criteria for an interim serogroup classification for members of the genus *Spiroplasma* (Class *Mollicutes*). *Int. J. Syst. Bacteriol. 37*, 82–84.

White, G. F. 1912. The cause of European foul brood. *U.S. Dep. Agric. Bur. Entomol. Circ. 157*.

White, G. F. 1920. European foulbrood. *U.S. Dep. Agric. Bull. 810*.

Wille, H., and Martignoni, M. E. 1952. Vorläufige Mitteilung über einen neuen Krankheitstypus beim Engerling von Melolontha vulgaris F. *Schweiz. Z. All. Pathol. Bakteriol. 15*, 470–473.

Williamson, D. L. 1965. Kinetic studies of "sex ratio" spirochetes in *Drosophila melanogaster* Meigen females. *J. Invertebr. Pathol. 7*, 493–501.

Williamson, D. L., and Poulson, D. F. 1979. Sex ration organisms (spiroplasmas) of *Drosophila*. *In* "The Mycoplasmas. Plant and Insect Mycoplasmas." (R. F. Whitcomb and J. G. Tully, eds.), Vol. 3, pp. 175–208. Academic Press, New York.

Wysoki, M., and Raccah, B. 1980. A synergistic effect of two pathogenic bacteria from the

Enterobacteriaceae on the geometrid *Boarmia selenaria*. *J. Invertebr. Pathol. 35*, 209–210.

Yamada, M.-A., Nawa, S., and Watanabe, T. K. 1982. A mutant of SR organism (SRO) in *Drosophila* that does not kill the host males. *Jpn. J. Genet. 57*, 301–305.

CHAPTER 6

DNA-VIRAL INFECTIONS: BACULOVIRIDAE

I. VIRAL STRUCTURE AND REPLICATION
II. INSECT VIRAL CLASSIFICATION
III. HOSTS AND TISSUE CULTURE
IV. TYPES OF INSECT VIRUSES
V. FAMILY BACULOVIRIDAE
   A. Viral Particle
   B. Nucleocapsid
   C. Enveloped Virion
   D. Infectious Elements
VI. NUCLEAR POLYHEDROSIS VIRUSES
   A. Description
   B. Virion
   C. Viral Occlusion—Polyhedra
   D. Hosts
   E. Gross Pathology
   F. Tissue Specificity
   G. Cytopathology
   H. Pathophysiology
   I. Viral Replication
   J. Transfection of Isolated DNA
   K. Persistent Viral Infections
   L. Restriction Enzyme Analysis
   M. Genotypic Variants
   N. DNA Replication and Gene Expression
   O. Baculoviruses as Expression Vectors
VII. GRANULOSIS VIRUSES
   A. Viral Occlusion—Capsule
   B. Pathology and Pathogenesis
   C. Cytopathology
   D. Pathophysiology
   E. Viral Replication
   F. Genome and Hybrids
   G. Gene Expression
VIII. NONOCCLUDED VIRUSES
   A. *Oryctes* Nonoccluded Virus
   B. *Heliothis* Nonoccluded Virus
   C. Other Nonoccluded Viruses

The word *virus* is derived from Latin and means a slimy liquid, poison, or stench. The early definition of a virus was based on submicroscopic size and obligate pathogenicity. More recently, the definitions attempted to convey two qualities of the virus: (1) possession of its own genetic material, which inside the host cell behaved as part of the cell, and (2) presence of a submicroscopic infective stage, the virion, which served as the vehicle for introducing the viral genome into a cell

(Lwoff and Tournier 1971). These definitions, however, did not adequately separate viruses from other minute parasitic procaryotes, such as rickettsiae, mycoplasma, and chlamydia. Matthews (1991) has thoroughly discussed the characteristics of these procaryotes and their differences from the virus. He defined a virus as follows:

> A virus is a set of one or more nucleic acid template molecules, normally encased in a protective coat or coats of protein or lipoprotein, that is able to organize its own replication only within suitable host cells. Within such cells, virus replication is (i) dependent on the host's protein-synthesizing machinery, (ii) organized from pools of the required materials rather than by binary fission, (iii) located at sites that are not separated from the host cell contents by a lipoprotein bilayer membrane, and (iv) continually giving rise to variants through various kinds of change in the viral nucleic acid.

The virus must normally be transmissible and cause disease in a host.

Viral diseases are one of the most widely investigated infections in insects. These studies have resulted because of the extensive basic and applied interests in viruses and from the development of elaborate and complex equipment, including the sophisticated techniques in biochemistry, serology, pathology, tissue culture, and recombinant DNA technology. With these advances, applied insect virology has extended beyond pest control into the field of genetic engineering, where the virus serves as a vector for the expression of foreign genes to produce biochemically and pharmaceutically important products. The accomplishments in genetic engineering are due primarily to the availability of invertebrate cell lines. Up to 1989, cell lines have been formed from 55 species of invertebrates in seven orders within Arthropoda and with most of them from Lepidoptera and Diptera (Hink and Hall 1989).

I. VIRAL STRUCTURE AND REPLICATION

The viral particle (virus, virion, or vibrion) is composed of a protein shell (capsid) that surrounds the nucleic acid. The capsid provides the virus with morphological and functional properties, the nucleic acid with the genetical constituent. Each virus has only one type of nucleic acid, either deoxyribonucleic acid (DNA) or ribonucleic acid (RNA). The nucleic acids may be single- or double-stranded. The nucleic acid together with the capsid form the nucleocapsid. The simplest virus consists of a nucleic acid and a capsid. Viroids have only nucleic acids and no capsids (Diener 1983). Prions are small infectious proteins and apparently are not associated with nucleic acids (Prusiner 1984). Thus far, no viroid or prion has been reported from insects.

The design of the capsid is of two major types: (1) helical assemblage (rod-shaped) and (2) closed shell (isometric, cubic, or quasi-spherical). In some helical and cubic viruses, the nucleocapsids are surrounded by envelopes that are lipid-bilayer and may be related to components of the cell membrane. The envelope is acquired during viral replication or when the virus leaves or enters the cell. The envelope plays a role in the penetration of the virion into the cell. Some insect viruses are occluded in proteinaceous bodies that are referred to as viral occlusions, occlusions, or inclusion bodies. We agree with Goodwin (1968) that occlusion body is more descriptive and appropriate for the body containing virions and inclusion body should be a general term referring to a body with or without virions.

Viruses differ in their mode of replications and the basic types have been established for the viral families infecting humans, other vertebrates, and insects

FIGURE 6-1

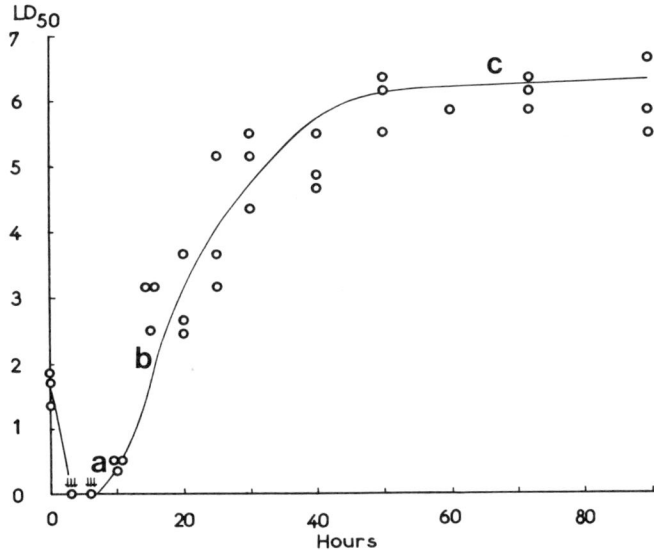

Replication of a nuclear polyhedrosis virus in the silkworm larva (LD_{50} time curve) showing (a) the latent, (b) the exponential, and (c) the stationary periods. [Courtesy of Aizawa (1959) and Academic Press.]

(Roizman 1986). The replication of all viruses involves the adsorption, uptake, and uncoating, followed by the expression and replication of the viral genome and production of viral progeny. Enzymes that are present in the viral particle or in the host cell are required for replication.

Viral replication involves three periods of development: (1) a latent or eclipse, (2) an exponential, and (3) a stationary period (Fig. 6–1). During the eclipse period, the virus is undergoing uptake, uncoating, and early stage of replication, and the virus is not infective. The exponential period is when the number of infectious virions increases exponentially until the number reaches a plateau at the stationary period.

II. INSECT VIRAL CLASSIFICATION

The first broad attempt to classify insect viruses was proposed by Holmes (1948) who listed two genera of insect viruses, *Borrelina* and *Morator*, in the family Borrelinaceae, suborder Zoophagineae, order Virales. The genus *Borrelina* contained viruses that cause polyhedrosis, wilt, and other diseases in Lepidoptera; the genus *Morator* had only one virus, the cause of sacbrood in the honey bee. Steinhaus (1949) revised Holmes' classification and based his classification on the presence, absence, or types of occlusion bodies produced by viruses. His classification had four genera: *Borrelina* with polyhedral bodies; *Bergoldia* with granular inclusions; *Paillotella* with refringent, polymorphic bodies of very irregular shapes and sizes; and *Morator* with no visible pathological inclusions. Subsequently, other workers proposed more complex systems with added genera based on differences in the location and types of viral particles and occlusions, tissue tropism, etc. (Bergold 1958; Krieg 1961a, Benz 1963a).

Several decades ago, there was extensive debate on naming viruses and the basis for their classification. The debate arose primarily because virologists desired a classification that would avoid the nomenclatorial problems associated with animal, plant and bacterial taxonomies. The International Committee on Nomenclature of Viruses (presently named the International Committee on Taxonomy of Viruses) was organized at the Eighteenth International Congress of Microbiology held in Moscow in 1966 (Wildy 1971). In the initial proposal of the classification, each virus was also defined by a cryptogram, which was subsequently eliminated because of its inflexibility to accommodate increased information on viruses and their properties. We have adopted the classification proposed in the fifth report of the International Committee on Taxonomy of Viruses (Francki et al. 1991). This classification has no hierarchical levels higher than families and does not imply phylogenetic relationships except those proven by molecular genetics.

III. HOSTS AND TISSUE CULTURE

Martignoni and Iwai (1986) list 12 insect orders as hosts of viruses; more recently, the Siphonaptera (fleas) has been reported as a host (Beard et al. 1989). There are over 1100 hosts, including a few mites and ticks (Martignoni and Iwai 1986).

As in the case of vertebrate virology, the advance in insect virology greatly accelerated with the development of insect tissue culture techniques. Even though Trager had infected cultured cells from the gonad of the silkworm with a nuclear polyhedrosis virus (NPV) as early as 1935, others were not as successful for two decades. In 1958, Grace obtained infection of ovarian tissues of emperor gum moth, *Antheraea euclaypti*, with a NPV; later, he reported limited infection of a cell line of *A. eucalypti* with a NPV of the silkworm (Grace 1967). Aizawa and Vago (1959) confirmed Trager's work with a primary ovarian culture of the silkworm. Further advances were made by Gaw et al. (1959) who succeeded in infecting a gonadal, monolayer cell line of the silkworm. However, progress was impeded even though insect cell lines became readily available. Workers could not easily infect cell lines with NPVs until Goodwin et al. (1970, 1973) described an efficient system based on the tissue culture medium that produced high infections of susceptible cells.

In 1973, Hink and Vail developed a plaque assay for the NPV of the alfalfa looper, *Autographa californica*, in a cell line of the cabbage looper, *Trichoplusia ni*. At present, the most widely used cell line for the replication of NPVs is derived from *Spodoptera frugiperda*. Up to 1983, several of the major insect pathogenic viruses had been cultured in insect cell lines except for the granulosis viruses (GVs); however, a GV was cultured in 1984 by Naser et al. in cell lines of the codling moth, *Cydia pomonella*.

IV. TYPES OF INSECT VIRUSES

More than 20 groups of viruses are known to be insect pathogens (Martigoni and Iwai 1986). They have been placed in 12 viral families, but many remain unclassified. Some are assigned to families of mainly vertebrate viruses, but others are in families with viruses specific for insects and related invertebrates (e.g., Baculoviridae, Polydnaviridae, and Ascoviridae). Viruses in the three families,

Baculoviridae, Entomopoxviridae, and Reoviridae, are unique because of the presence of occlusion bodies in which virions, at a certain stage in their development, are occluded at random. The occlusion bodies contribute to the stability and persistence of the viruses in the environment. The families of DNA viruses infecting insects are given in Fig. 6–2.

Viral occlusions are called polyhedra (singular, polyhedron) for the nuclear and cytoplasmic polyhedrosis viruses, capsule or granule for the granulosis virus, and spheroid for the entomopoxvirus. The matrix protein in the polyhedron is known as polyhedrin, that in the capsule as granulin (Summers and Egawa 1973), and in the spheroid as spheroidin (Bilimoria and Arif 1979). The polyhedrin of the NPV and the granulin of the granulosis virus (GV) belong to one group of related proteins (Rohrmann et al. 1981; Smith and Summers 1981). The occurrence of viral occlusions in three distinct families suggests that these bodies evolved independently and indicates their powerful selective advantage in the environment (Rohrmann 1986b) (Fig. 6-3).

General treatment of all types of insect and invertebrate viruses from the classification and biochemical aspects is presented in Francki et al. (1991) and from the morphological and pathological aspects in Adams and Bonami (1991).

FIGURE 6–2

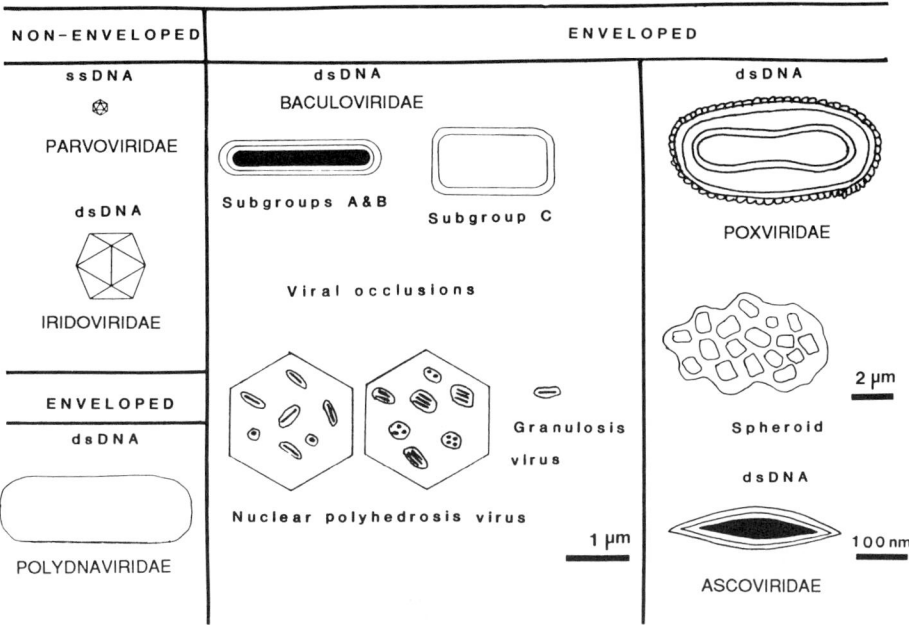

Families of DNA viruses infecting insects. The viral structures are drawn to the following scale: (1) virions, bar = 100 nm (shown by the Ascoviridae); (2) the viral occlusions—nuclear polyhedrosis virus (NPV) and granulosis virus (GV)—and the spheroid are shown at the scales indicated. Viral occlusions (polyhedron, capsule, spheroid) may vary greatly in size. The spheroid shown is of comparable dimensions to the entomopox virus reported from *Choristoneura conflicta*. Subgroup C shown is of comparable dimensions to the virus reported from *Oryctes rhinoceros,* the tail-like appendage (not shown) measures 10 × 270 nm. The polydnavirus is of comparable dimensions to those found in the Ichneumonidae. [Courtesy of Adams and Bonami (1991) and CRC Press.]

176 6 DNA-VIRAL INFECTIONS: BACULOVIRIDAE

FIGURE 6-3

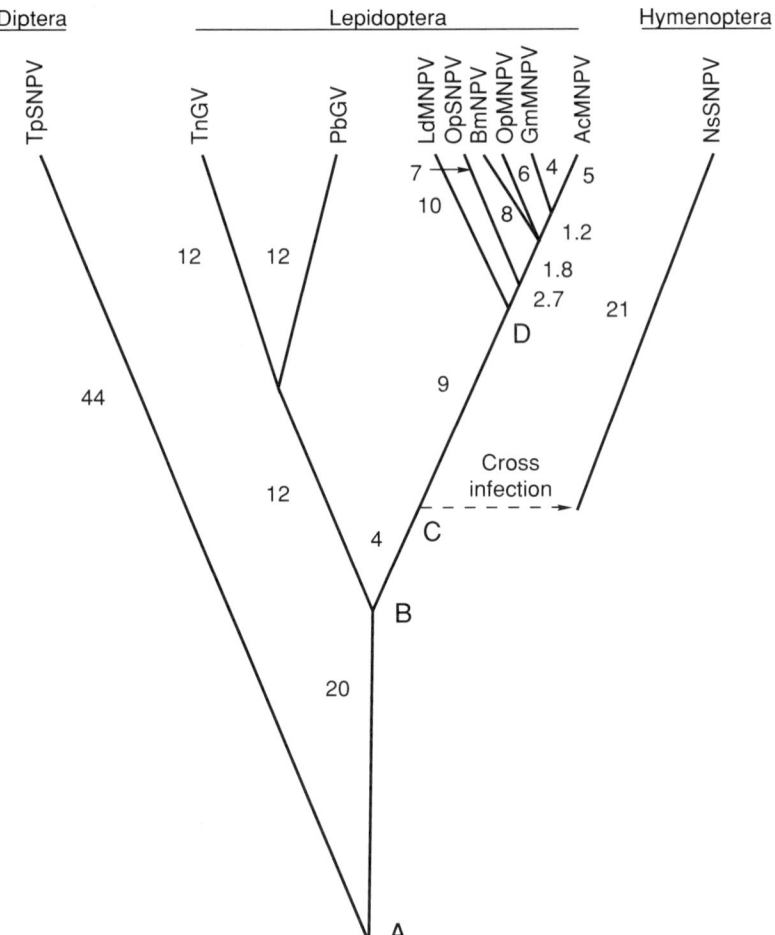

Molecular phylogeny of occluded proteins of baculoviruses. The numbers indicate the percentage of amino acid differences between various points on the tree. The letters signify major branch points in the phylogeny and include the following: (*A*) divergence of lepidopteran and dipteran lines; (*B*) divergence of granulosis viruses (GVs) from lepidopteran singly enveloped, nuclear polyhedrosis viruses (SNPVs); (*C*) development of nuclear polyhedrosis viruses (NPVs); (*D*) development of multiply enveloped, nuclear polyhedrosis viruses (MNPVs). [Courtesy of Rohrmann (1986a) and Society for General Microbiology.]

V. FAMILY BACULOVIRIDAE

Baculoviruses are the most common and most widely studied group of viruses pathogenic for insects. They are known exclusively from arthropods (Martignoni 1984). The characteristics of baculoviruses and the pathogenesis of their infections have been treated in a large number of comprehensive reviews and we shall refer to some that have been published since 1980. There are reviews on the genome and the general and molecular biology of baculoviruses (Consigli et al. 1983; Granados and Federici 1986; Doerfler and Bohm 1986; Blissard and Rohrmann 1990; Adams and McClintock 1991), the cytopathology (Tanada and Hess 1984, 1991), the techniques and methods for expression vectors and insect tissue culture (Summers and Smith

1987), the application as expression vectors (Miller 1988, 1989; Kang 1988; Luckow and Summers 1988; Maeda 1989a), and the potential of genetically engineered baculoviruses as insecticides (Wood and Granados 1991).

The family Baculoviridae previously had three subgroups: A, nuclear polyhedrosis (nucleopolyhedrosis) viruses (NPVs); B, granulosis viruses (GVs); and C, nonoccluded, rod-shaped nuclear viruses (NOVs). Francki et al. (1991) placed the first two subgroups in the subfamily Eubaculovirinae and the third subgroup in the subfamily Nudibaculovirinae. The genus NPV is characterized by the presence of polyhedral-shaped viral occlusions (polyhedra) containing randomly occluded viral particles. There are two subgenera: (1) the single-nucleocapsid NPVs (SNPV), in which only one nucleocapsid is found per envelope (type species *Bombyx mori* SNPV) and (2) multinucleocapsid NPVs (MNPV), in which several nucleocapsids (as many as 39) are enclosed in a common envelope (type species *Autographa californica* MNPV). The genus GV has one nucleocapsid per envelope and has granule-shaped viral occlusions (capsules) containing one and rarely two or more virions (type species *Plodia interpunctella* GV). The nonoccluded baculovirus has one nucleocapsid per envelope and has no occlusion bodies [type species *Heliothis zea* nonoccluded virus (NOB)].

The sequence homologies of the DNAs indicate that the members of Baculoviridae are genetically related (Smith and Summers 1982). The NPVs and GVs appear to have an ancient association with their host insects and may have evolved with them during 40 to 60 million years of existence (Rohrmann et al. 1981). The occurrence of only the SNPVs in the older insect orders and the existence of the MNPVs and the GVs only in the more recently evolved Lepidoptera suggest that the progenitor of the family Baculoviridae was a SNPV (Rohrmann 1986b). Through host-dependent evolution, the SNPVs gave rise to the MNPVs and GVs infecting Lepidoptera.

A. VIRAL PARTICLE

All baculoviruses have virions of the same basic structure: an enveloped, rod-shaped nucleocapsid in which an amorphous but definite layer exists between the nucleocapsid and the envelope (Federici 1986)(Fig. 6–4). Much confusion occurs in the older literature regarding the viral structure because different terms have been used for the same structure. The terms presently used to describe the virion are nucleoprotein core or DNA core, capsid, nucleocapsid, and envelope (Federici 1986).

The baculovirus has a single molecule of circular, supercoiled, double-stranded (ds) DNA of 80 to 220 kilobase (Miller 1988). The rod-shaped baculovirus measures from 40 to 60 × 200 to 400 nanometers (nm) (Matthews 1982). The virions are structurally complex and contain at least 10–25 polypeptides with molecular weights (MWs) ranging from 10 to 160 × 10^3. The virions are ether and heat labile.

B. NUCLEOCAPSID

The nucleocapsid is a cylindrical core of DNA and protein. The capsid is tubular, capped at both ends, and filled with DNA filaments. Very long precursors of capsids with DNA break up to form the typical nucleocapsids (Summers 1971; Hughes

FIGURE 6–4

BERGOLD 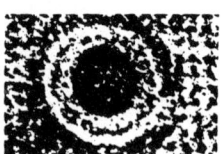 PRESENT MODEL

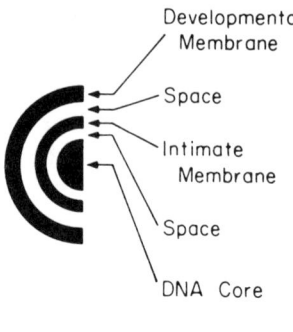 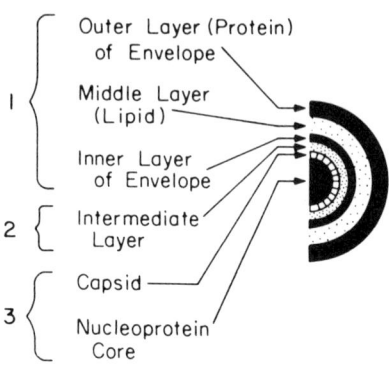

Diagrammatic representations of the baculovirus in cross section. [Courtesy of Federici (1986) and CRC Press.]

1972). Microtubules are involved in the assembly of nucleocapsids (Volkman et al. 1987; Volkman and Zaal 1990).

The capsid structure has been considered to be in helical assemblage, cross-banded, or a cylinder of stacked rings; the present concept is a structure of rings of subunits in a stacked series and the symmetry is not helical (Federici 1986). Burley et al. (1982) report that the cylindrical portion of the capsid surface is composed of 12 subunits in each ring of capsid. The ring-to-ring spacing is about 4.5 nm (Beaton and Filshie 1976). A tapering short series of subunit rings form the caps of the nucleocapsid. The protein of the cap structure is distinct from the protein that makes up the subunits in the primary rings of the capsid (Burley et al. 1982). The packaging of the DNA into the capsid appears to be associated with the cap structure (Fraser 1986a) (Fig. 6–5).

The caps at each end of the nucleocapsid differ morphologically, with a "claw and nipple" structure at one end and a "claw" at the other end (Bergold 1963a; Teakle 1969; Kobayashi 1976). The presence of different ends suggests the existence of polarity in orienting the nucleocapsid for viral envelopment, for the attachment and penetration through nuclear pores, for the emergence from the cell (exocytosis), and for the invasion through the cell membrane (Summers 1971; Kobayashi 1976; Kawamoto et al. 1977a,b).

Within the nucleocapsid, the double-stranded loop of DNA associates heterogeneously with a highly basic protein to form a cylindrical core. The basic protein is a major nucleocapsid protein and is a low molecular weight, DNA-binding protein (Monsarrat et al. 1975; Tweeten et al. 1980; Kelley et al. 1983). The gene encoding this basic protein has been located on the viral genome (Wilson et al. 1987; Maeda et al. 1991).

The capsid is made up of a major protein, 39 kilodaltons (kDa) in size, and its gene has been located on the viral genome (Pearson et al. 1988). This capsid protein occurs between the envelope and the DNA-basic DNA-binding protein complex. It has been suggested that the DNA within the capsid is in the form of a helix, coil, or supercoil (Krieg 1961b; Kozlov and Alexeenko 1967; Himeno et al. 1968), or folded as in the case of the DNA in phage T4 (Scharnhorst et al. 1977; Bud and Kelly 1980a). However, the emergence of the DNA from the capsid indicates that the DNA has been introduced in a sequence of spiralization or coils (Monsarrat et al. 1975; Revet and Guelpa 1979; Fraser 1986a) (Figs. 6–5, 6–6B).

The widths and lengths of nucleocapsids are fairly constant for each type of baculovirus. Not infrequently, however, very long forms, many times the average length, occur during viral replication, indicating the extendable nature of the nucleocapsid.

C. ENVELOPED VIRION

At certain stages in viral replication, a nucleocapsid is enclosed within an envelope and is called the mature virion. There are three types of envelopes that differ in their origin: (1) an envelope produced in the nucleus (*de novo* morphogenesis) (Figs. 6–6A,C), (2) an envelope acquired from the nuclear membrane as the nucleocapsid exits the nucleus to enter the cytoplasm (nuclear budding) (Injac et al. 1971; MacKinnon et al. 1974; Kawamoto et al. 1976), and (3) an envelope (transport membrane) that is acquired as the nucleocapsid passes out through the plasma membrane (cytoplasmic budding). The envelope formed in the nucleus has strong affinity for

FIGURE 6–5

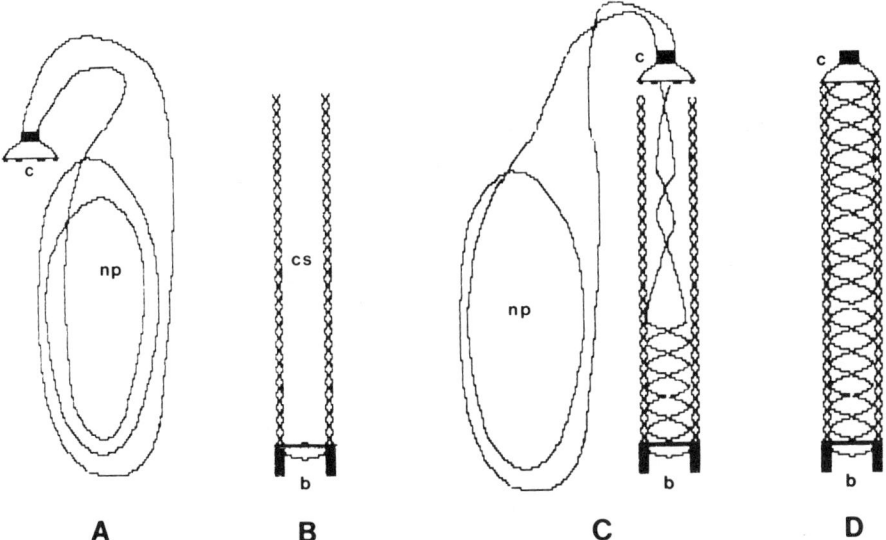

Model of nucleocapsid morphogenesis. (A) Cap (*c*) nucleoprotein (*np*) core complex associating in the virogenic stroma. (B) Base (*b*) and capsid sheath (*cs*) assembled separately in pockets of the virogenic stroma. (C) Nucleoprotein core entering preassembled capsid sheath through the cap structure, winding the circular nucleoprotein strand as it enters. (D) Fully assembled nucleocapsid. [Courtesy of Fraser (1986a) and Academic Press.]

FIGURE 6–6

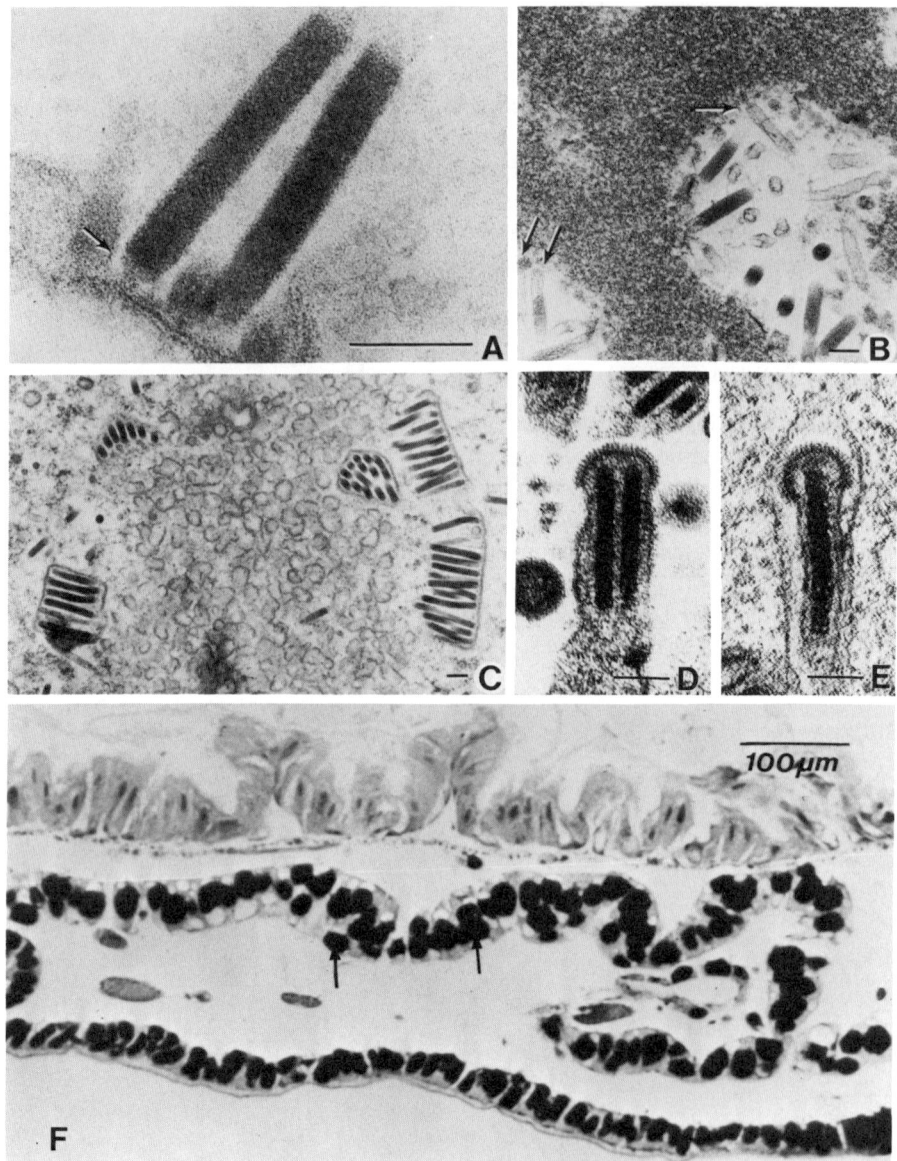

Stages in the development of nuclear polyhedrosis viruses (NPVs). (A, C) Association of nucleocapsids of *Autographa californica* NPV (AcNPV) with *de novo* envelopes, apparently through the cap structure in (A). (B) Apparent packaging of viral genomes in assembled capsids of AcNPV. [Bars in (A), (B), and (C) = 0.1 μm.] (D, E) Multiply and singly enveloped nucleocapsids of *Heliothis zea* NPVs, showing peplomers. (Bars = 0.1 μm.) (F) Longitudinal section of a NPV-infected armyworm, *Pseudaletia unipuncta*, showing a systemic infection with hypertrophied nuclei packed with polyhedra (*arrows*). [(A), (B), and (C) courtesy of Fraser (1986a) and Academic Press; (D) and (E) provided by J. R. Adams; (F) from Tanada and Hess (1984) and courtesy of Plenum Press.]

polyhedrin and granulin and is involved in the occlusion of the virions in the polyhedron and capsule. This envelope is considered to be formed *de novo* (Hughes 1972; Stoltz et al. 1973; Kawamoto et al. 1976). The second type of envelope acquired from the nuclear membrane is lost in the cytoplasm and the naked nucleocapsid obtains the third type of envelope at cytoplasmic budding. This third type of envelope, derived from the plasma membrane, has a distinctive hemispherical cap of spikes (peplomers) on one end of the virion (Kawamoto et al. 1977b; Adams et al. 1977) (Figs. 6–6D,E). The envelope fits loosely around the virion except at the anterior end with the attached peplomers, whereas the *de novo*, nuclear envelope is tightly fitted. The peplomer has a 64-kDa protein (phosphoglycoprotein or more often called glycoprotein) that functions in the attachment of virions to cells in the hemocoel prior to viropexis (adsorptive endocytosis) (Volkman and Goldsmith 1985; Keddie and Volkman 1985; Volkman 1986). The origin of the envelope is the host plasma membrane and a glycoprotein (gp64) that is encoded by the virus (Whitford et al. 1989; Blissard and Rohrmann 1989). The glycoprotein is found in the cytoplasm at early infection and moves to the plasma membrane to form the peplomer envelope (Volkman et al. 1984; Blissard and Rohrmann 1989).

The mature virion, found in an occlusion body, has a true unit membrane, from 6 to 18 nm thick, consisting of a trilaminar structure with a central lipid-containing layer bounded on each side by a layer of protein (Federici 1986). Electron micrographs have revealed an intermediate layer (6 to 12 nm thick), often overlooked or misinterpreted, between the nucleocapsid and the envelope of the mature virion (Federici 1986) (Fig. 6–4). This layer also occurs between nucleocapsids in multinucleocapsid virions. Investigators have suggested that the layer may be condensed host nucleoplasm (Harrap 1972b; Hughes 1972). Kawamoto et al. (1977a) have reported densely staining discs (22 nm in diameter) formed in the intermediate layer during the envelopment of the nucleocapsid. These discs degenerate once the virion is enveloped to become a mature virion.

The compression (or contraction) of the intermediate layer by the envelope is generally uniform around the nucleocapsid and the diameter of the mature virion is constant throughout its length. In the mosquito baculovirus, however, the degree of compression is greater at one end of the virion and the other end has a much thicker intermediate layer to form a bulbous end (Federici 1980).

D. INFECTIOUS ELEMENTS

Baculoviral infections may be initiated by isolated DNA, unenveloped virion (nucleocapsid), nuclear-enveloped virion, occluded nuclear-enveloped virion, or plasma-enveloped virion. The four virions are phenotypes that differ in virion morphology, protein composition, origin of the envelopes, mode of penetration into the cells, and relative infectivities to cultured insect cells and to different tissues in the insect.

Most recent reviews consider only two phenotypes as infectious agents, the plasma-enveloped virion (budded virus, nonoccluded virus, or extracellular virus) and the occluded nuclear-enveloped virion (occluded or polyhedra-derived virus). The plasma-enveloped or extracellular virion with peplomers is produced before the appearance of the occluded virus. This virion is the major type causing infections in tissues of the hemocoel and in cultured insect cells. It enters the cell by viropexis (endocytic pathway) with the initial attachment occurring at the peplomer end.

Most of the nuclear-enveloped virions are occluded in occlusion bodies (occluded virus). The occluded viruses are not infectious in the hemocoel, but they are the main infectious element for the horizontal transmission through the midgut of a susceptible host. The nuclear-enveloped virions that are not occluded enter the hemocoel after cell lysis and invade the cells by attachment and fusion of the envelope onto the cell plasma membrane (Tanada and Hess 1984). The nonenveloped virion (nucleocapsid) very likely enters the hemocoel after cell lysis. This nonenveloped phenotype has been reported as infectious when inoculated into the insect's hemocoel (Khosaka and Himeno 1972; Stairs 1980; Kawarabata 1974; Strokovskaya et al. 1977) and it appears to invade cells by the endocytic pathway (Kislev et al. 1969).

Investigators agree that differences exist in the infectivities of plasma-enveloped virions (extracellular virion) and of nuclear-enveloped virions (virions liberated from viral occlusions) when fed to larvae (i.e., the former is less infectious than the latter in the midgut). However, they disagree when the two phenotypes are inoculated into the insect's hemocoel. Some claim that the plasma-enveloped virions are highly infectious as compared to the nuclear-enveloped virions (Dougherty et al. 1975; Volkman and Summers 1977; Stairs 1980; Keddie and Volkman 1985). On the other hand, Ignoffo and McIntosh (1986) report that both phenotypes are equally infectious in the hemocoel when the results are based on plaque-forming units (i.e., the number of virions that are infectious rather than the total number of virions).

VI. NUCLEAR POLYHEDROSIS VIRUSES

The first virus to be detected in insects belonged to the nuclear polyhedrosis viruses (NPVs), the infection of which is called nuclear polyhedrosis (plural, nuclear polyhedroses). These viruses are easily recognized from other viruses because of the presence of unique polyhedral bodies in the cell nuclei. Maestri (1856) and Cornalia (1856, cited by Bergold 1943 and Steinhaus 1949) observed these bodies in diseased silkworm larvae (Fig. 6–7). The bodies were called "polyedrischen Körperchen" (polyhedral granules) (Fischer 1906) and the associated disease, polyederkrankheit (polyhedral disease) (Wahl 1909) or polyhedrosis (Prell 1926). The nature of the polyhedra, present in the nuclei, remained unresolved for many years. The polyhedra were considered to be protozoa by Bolle (1894) who named them *Microsporidium polyedricum*. von Prowazek (1907, 1913) demonstrated the filterability of the infectious agent of jaundice of the silkworm, but he also observed a *Coccus* that he believed to be the etiologic agent and named it *Chlamydozoon bombycis*. He and other workers considered the polyhedra to be by-products of the disease (Sasaki 1910; Glaser and Chapman 1916).

Various names have been used for the polyhedroses in lepidopteran insects (e.g., Schlaffsucht or Wipfelkrankheit of the nun moth, *Lymantria monacha*, and the wilt of the gypsy moth, *Lymantria dispar*). In the silkworm, the disease is called "jaundice" in America, "giallume" in Italy, "grasserie" in France, "nobyo" in Japan, and "Gelbsucht" or "Fettsucht" in Austria and Germany.

The concept of the viral nature of polyhedrosis was based on the filterability through porcelain filters of the infectious agents of Wipfelkrankheit of the nun moth (Escherich and Miyajima 1911; Komárek and Breindl 1924), the wilt of the gypsy

FIGURE 6-7

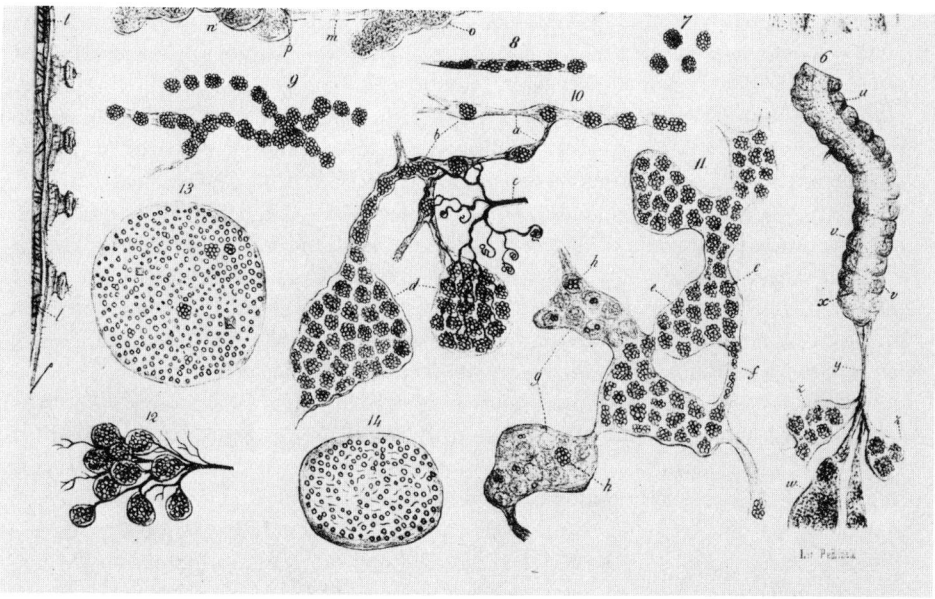

The first pictorial representation of polyhedra in the tissues of a silkworm larva infected with the virus of jaundice. [From Steinhaus (1956) and courtesy of the University of California Agricultural Experiment Station.]

moth (Glaser and Chapman 1913; Glaser 1915, 1918), and the jaundice of the silkworm (Acqua 1918). Glaser and Chapman (1916) biochemically analyzed the polyhedra and concluded that they contained nucleoprotein. Komárek and Breindl (1924) demonstrated, with special staining methods, particles in the polyhedral bodies obtained from the nun moth. This was confirmed by Paillot (1924, 1926a) who also demonstrated, with filteration, centrifugation, and dark-field microscopy, the presence of minute granules associated with the polyhedral bodies. He gave the generic name *Borrelina* to the filterable virus and *Borrelina bombycis* to the causal agent of silkworm jaundice. With the use of the electron microscope, Bergold (1947) identified rod-shaped particles from polyhedral bodies obtained from virus-infected larvae of the silkworm, gypsy moth, and nun moth. The viral particles were found to contain nucleic acids and proteins (Bergold 1947). This was confirmed by Wyatt (1952a,b) who established the presence of DNA in the NPVs and GVs. Readers interested in the historical account of baculoviruses should refer to Bergold (1943) and Benz (1986).

A. DESCRIPTION

The structure and pathology of the NPVs and GVs are very similar. First we shall describe the NPVs, and in the following section the GVs will be compared to them. The type species is the MNPV of the larva of *Autographa californica* (Matthews 1982). This virus is the most widely studied of the entomopathogenic viruses, particularly its biochemical and genetical properties.

B. VIRION

The two morphotypes of viral particles found in NPVs differ by the number of nucleocapsids within an envelope. One morphotype has a single nucleocapsid in an envelope (SNPV) (also called singly enveloped, singly embedded, single nucleocapsid, or unicapsid) (Fig. 6–6E) and the other has more than one nucleocapsid per envelope (MNPV) (also called multiply enveloped, multiply embedded, multiple nucleocapsid, or multinucleocapsid) (Fig. 6–6C). Both SNPVs and MNPVs generally occur in virus-infected insects collected in the field. The number of nucleocapsids per envelope may differ with the host and virus. For example, in a virus-infected silkworm larva, the number of nucleocapsids ranges from 1 to as many as 19 nucleocapsids per envelope (Watanabe 1975), from 1 to 24 in a larva of the almond moth, *Cadra cautella* (Adams and Wilcox 1968), and from 1 to 39 in a larva of the brown-tail moth, *Euproctis similis* (Kawamoto and Asayama 1975). The numbers may also vary with the tissues in which the viruses have replicated.

The distinctness of the SNPVs from the MNPVs has been demonstrated by cross transmission (Harper 1976; Hughes 1979); serial passages in homologous cell lines (Sohi et al. 1984), biological properties (Martignoni and Iwai 1977), biochemical, serological and restriction endonuclease analyses (Cibulsky et al. 1977a,b; Bilimoria 1983; Knell et al. 1983), and genetic recombination (Miller 1981b). However, the difference between SNPVs and MNPVs does not appear to be phylogenetically significant (Leisy et al. 1986a; Blissard and Rohrmann 1990). The pathogenicity of the SNPV and MNPV genotypes is not identical since this is dependent on the number of nucleocapsids (van Beek et al. 1988). The MNPVs are as many times more pathogenic, as the number of enveloped nucleocapsids, than the SNPVs.

C. VIRAL OCCLUSION—POLYHEDRA

Most of the enveloped nucleocapsids in the nucleus are occluded in polyhedra (Figs. 6–8, 6–9). The number of enveloped virions in a polyhedra may be as high as 200 virions per polyhedron depending on the virus, the insect host, and its tissues (Ackerman and Smirnoff 1983). The occlusion bodies of the NPV, GV, cytoplasmic polyhedrosis virus, and entomopoxvirus are resistant to adverse conditions and protect the occluded virions.

The polyhedra, in addition to enveloped nucleocapsids, may occlude naked nucleocapsids and empty or partially filled capsids. They may contain RNA (Faulkner 1962; Aizawa and Iida 1963; Kozlov et al. 1969), extraneous nuclear and cellular particles, and foreign viruses (e.g., vesicles, vacuoles, membranes, and viruslike particles that appear to be iridovirus, small icosahedral virus, and parvovirus) (Watanabe et al. 1975; Knudson and Harrap 1976; Hess et al. 1984). The capsules of GVs, the polyhedra of CPVs, and the spheroids of entomopoxviruses also occlude extraneous material and foreign viruses (Stoltz and Summers 1972; Federici et al. 1974; Hess et al. 1984).

1. Size and Structure

The size of the polyhedra may vary inversely with the number formed in a nucleus, fewer numbers are associated with large polyhedra. The diameters of the polyhedra

FIGURE 6–8

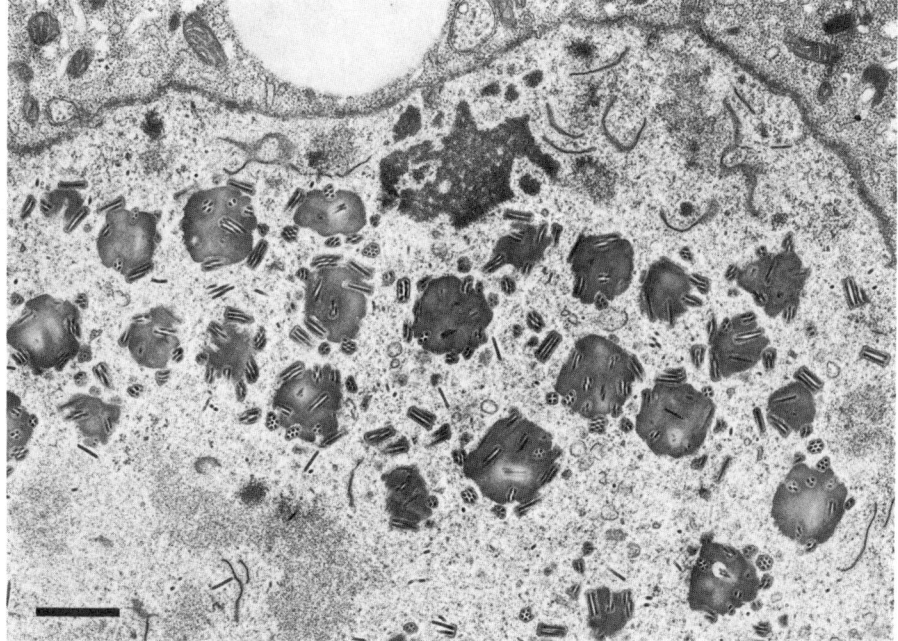

Section of a fat body cell nucleus of the beet armyworm, *Spodoptera exigua*, infected with a multiply enveloped nuclear polyhedrosis virus of *Autographa californica* at 6 days after viral inoculation. Note developing polyhedra occluding enveloped virions. (Bar = 1 μm.) (Provided by J. R. Adams.)

range from 0.5 to 15 μm. The polyhedra are heavier than water and settle to the bottom in aqueous solutions. They are not double refringent and light up strongly in a dark-field microscope.

The polyhedra matrix protein (polyhedrin) forms a distinct crystalline lattice that is not distorted by the presence of occluded virions (Morgan et al. 1955; Bergold 1963a,b) (Fig. 6–9C). The polyhedrin makes up about 95% and the virions about 5% of the protein content of the NPV polyhedra (Bergold 1963c).

Electron micrographs (Bergold 1963b; Hughes 1978; Harrap 1972a) and X-ray diffraction patterns (Engström and Kilkson 1968; Fujiwara et al. 1984) show that the protein molecules (from 40 to 102 Å in diameter) are arranged in a cubic lattice. The protein macromolecules are considered to be spherical (Bergold 1963b; Engström and Charpentier 1976), a six-armed nodal unit (Harrap 1972a), a hexagonal unit (Zhang et al. 1979), or irregular masses separated by spaces that may be greater than 3.8 nm (Hughes 1978). Federici (1986) concludes from current evidence that the matrix of the occlusion bodies is composed of a single protein (polyhedrin), the molecules of which are approximately a six-armed or nodal unit and are oriented in face-centered cubic lattice.

How the proteins crystallize to form the polyhedra has not been established. The crystallization may result from salt bridges formed by amino acid groups (lysine and arginine) with carboxyl groups (glutamine and asparagine) (Vlak and Rohrmann 1985) or divalent cations may be involved in the polyhedrin–polyhedrin interactions during the formation of polyhedra (Whitt and Manning 1988a). The

FIGURE 6–9

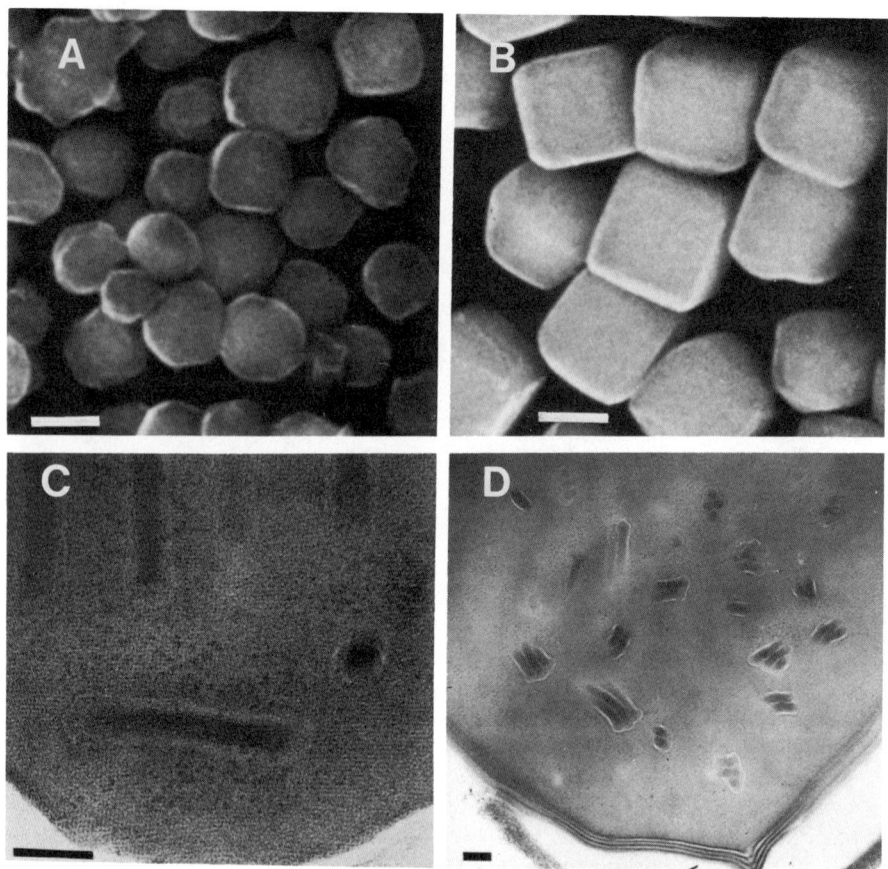

Polyhedra of nuclear polyhedrosis viruses (NPVs). (A) Polyhedra from *Heliothis zea*. (B). Polyhedra from *Autographa californica*. (C) Section of a polyhedron of a singly enveloped NPV from *Trichoplusia ni*. Note the lattice pattern of the matrix protein. (D) Section of a polyhedron of a multiply enveloped NPV from *Agrotis segetum*. [Bars in (A), (B), and (C) = 1 μm; in (D) = 0.5 μm.] (Provided by J. R. Adams.)

stability of the polyhedra may involve both ionic and hydrophobic interactions (Egawa and Summers 1972; Whitt and Manning 1987).

The envelope around the nucleocapsid has strong affinity for polyhedrin and granulin (Yamamoto et al. 1981; Blissard and Rohrmann 1990). In the few polyhedra variant, there is a deficiency of envelopes as compared with the many-polyhedra variant of the NPV of *Galleria mellonella* (Fraser and Hink 1982).

The shapes of polyhedra vary from cuboidal, dodecahedral, tetrahedral, or irregular, depending primarily on the specific virus (Fig. 6–9). Viral occlusions produced in most insects and crustacea are polyhedral, but those found in members of Diptera vary in shape from polyhedral in *Rhynchosciara angelae*, crescent-shaped or shaped like an orange segment in *Tipula paludosa*, and pleomorphic to spindle-shaped in mosquitoes (Federici 1986).

2. Production

The number of polyhedra formed in a cell is not constant, but the infected nuclei are often completely filled with polyhedra (Fig. 6–6F). Very large quantities of polyhedra are produced in some lepidopteran larvae and may comprise up to 10% of the insect's dry weight (i.e., as many as 1×10^9 polyhedra per larva). They are produced also *in vitro* (tissue culture) and are equally pathogenic to insect larvae as those produced *in vivo* (Faulkner and Henderson 1972; Vail et al. 1973; Ignoffo et al. 1974).

3. Morphotypes

Certain viral variants produce two distinct plaque morphologies, the MP (many polyhedra per cell) and FP (few polyhedra per cell) (Hink and Vail 1973; Hink and Strauss 1976). Cells infected with the FP plaques contain fewer (less than 10 per cell) but larger polyhedra, while those infected with the MP variant have smaller but many more (30 or more) polyhedra per cell.

The phenomenon of few polyhedra formation appears to be universal (Fraser and McCarthy 1984). Several MP variants produce FP variants after serial passages in cell cultures or in larvae (Potter et al. 1976; Wood 1980a; Fraser and Hink 1982). The FP variants are relatively stable genotypes and appear to arise from the spontaneous mutation of MP variants. A transposable element (*copia*-like) from the host cell is inserted in the genome of the FP variant (Miller and Miller 1982; Fraser et al. 1983; Fraser 1986b). Miller (1986) has thoroughly discussed the genetics of the FP and MP variants.

4. Dissolution

Dilute alkali and acid dissolve the polyhedra matrix proteins, but most workers use alkaline solutions (pH 9–11), particularly Na_2CO_3 + NaCl (Bergold 1947). Kawanishi et al. (1972a) have tested various solvents and have found that Na_2CO_3, originally proposed by Bergold (1947), was the only one producing intact free virions that closely resemble those processed in the insect's midgut lumen. Vaughn et al. (1989) have developed a reproducible method for releasing the virions from polyhedra by using a glycine–NaOH buffer at pH 11. Above pH 11, the nucleocapsids dissociate and lose their infectivity (Krieg 1957).

An alkaline protease present in the polyhedra assists in the dissolution of the occlusion body. The presence of the protease causes a distinct alteration in the viral proteins (Stiles et al. 1983). Some investigators have suggested that the protease is endogenous and probably virus-coded (Yamafuji and Yoshihara 1961a,b; Eppstein et al. 1975; Kozlov et al. 1978). The protease is absent in polyhedra collected from tissue-culture cells (Maruniak et al. 1979; Zummer and Faulkner 1979; Wood 1980b). This indicates that the protease is a contaminant because the polyhedra are collected usually by macerating dead or live-infected larvae, and are exposed to the larval digestive enzymes. The tissue-culture-produced polyhedra, lacking the protease, are as infectious for larvae as larva-produced polyhedra (Faulkner and Henderson 1972), and both types of polyhedra are equally effective in the field (Ignoffo et al. 1974). Thus, the protease is not essential, but it may assist in the dissolution of the polyhedra in the insect's midgut (Granados and Williams 1986).

5. Polyhedrin

Polyhedra dissociate under alkaline conditions into molecules with molecular weights estimated at 195,000 to 378,000 (Bergold 1963b; Rohrmann 1977). The basic subunit (polyhedrin) is a phosphorylated molecule of 25 to 31 kDa and most commonly reported as 29 kDa (Maruniak 1986; Blissard and Rohrmann 1990). Kozlov et al. (1986) have constructed the complete amino acid sequence of the polypeptide chain of *B. mori* NPV polyhedrin.

At one time, there was a question of the origin of polyhedrin, whether the host produced it as a defense mechanism or the virus directed its production (Bergold 1947, 1958). This was resolved when the messenger RNA for the polyhedrin was isolated, shown to be of viral origin, and the gene identified (van der Beek et al. 1980; Summers et al. 1980; Vlak et al. 1981; Adang and Miller 1982). Polyhedrin genes have been cloned and sequenced from about seven NPVs, including those of *Autographa* and *Orgyia* (van Iddekinge et al. 1983; Rohrmann 1986a).

The polyhedrin and its gene are the most extensively studied expression product and gene of baculoviruses. Polyhedrin is produced in larger amounts than perhaps any other viral product of infected eukaryotic cells (Rohrmann 1986a) [e.g., the NPV polyhedrin of *Orgyia pseudotsugata* forms up to 17.4% of the total alkali soluble protein of a virus-infected insect (Quant et al. 1984)].

The polyhedrins of the NPVs and of the cytoplasmic polyhedrosis viruses (CPVs) and the granulin of the GVs have similar molecular weights. The NPV polyhedrin and the GV granulin show similarities in peptide maps, but each protein is different and distinct (Summers and Smith 1976). Moreover, their amino acid compositions differ quantitatively and qualitatively (Maruniak 1986). The polyhedrins of NPVs and CPVs differ in amino acid compositions, tryptic peptide elution profiles from a cation-exchange resin, and N-terminal (amino-terminal) acid sequences (Rohrmann et al. 1980). A broad antigenic relationship exists, however, among the polyhedrins, the granulins, and to a lesser extent between the polyhedrins and granulins (McCarthy and Gettig 1986; Kozlov et al. 1986). The data on direct nucleotide sequences reveal a high amino acid homology at the C-terminal (carboxy-terminal) and mid-area, but less homology at the N-terminal (Akiyoshi et al. 1985; Chakerian et al. 1985; Rohrmann 1986b). Rohrmann (1986a), in his comprehensive review of polyhedrin, concludes that this protein has evolved two specialized functions: (1) formation of a protective crystal around the virus and (2) resistance to solubilization except under strong alkaline conditions similar to those found in the insect's midgut.

6. Surface Structure

A surface structure that delimits the polyhedra is evident in ultrathin sections. This envelope-like structure remains after the polyhedrin is partially dissolved with alkali (Bergold 1953; Hughes 1953) or with enzyme (Gipson and Scott 1975). The nature and origin of the structure were unresolved for some time. It was considered to be a membrane, or an artifact resulting from the denaturation of surface proteins, or a condensation of fibrous materials, or a product produced by the cell (Federici 1986).

The presence of carbohydrates in the structure (Minion et al 1979) and the isolation of a gene (p32) (Gombart et al. 1989) encoding for the protein indicate that the structure is not a true membrane but an envelope initiated by the virus. The viral protein is thiol-linked to the carbohydrate layer surrounding the occlusion body (Whitt and Manning 1988b). Moreover, Zuidema et al. (1989) have constructed a mutant virus that forms polyhedra lacking the envelope.

D. HOSTS

The NPVs infect over 400 insect species. Susceptible hosts are found mainly in Lepidoptera (34 families); to a lesser extent in Hymenoptera (Argidae, Diprionidae, Pamphiliidae, and Tenthredinidae); Diptera (Calliphoridae, Chironomidae, Culicidae, Sciaridae, Tachinidae, and Tipulidae); Coleoptera (Cerambycidae, Curculionidae, and Dermestidae); and Neuroptera (Chrysopidae and Hemerobiidae); one species each in Trichoptera (Limnephilidae), Thysanura (Praemachilidae) (Martignoni and Iwai 1986), and Siphonaptera (Pulicidae) (Beard et al. 1989). The SNPVs have been reported from all of the above insect orders, whereas the MNPVs and GVs have been found only in Lepidoptera (Rohrmann 1986b).

The MNPVs have a wide host range, although some are restricted to one or a few insect species (Ignoffo 1968). The MNPV of *Autographa californica* (AcMNPV) infects at least 33 species in 10 families (Gröner 1986); whereas the MNPV of *Colias* infects only a few species (Steinhaus 1952; Tanada 1954). Viral particles with similar general structure to baculoviruses have been reported from mites, crustacea (European crab and blue crab), and a fungus (Francki et al. 1991).

Multiple or mixed baculoviral infections in the same larva are common in the field, especially with different variants (strains). The NPVs may occur also with other viruses [e.g., cytoplasmic polyhedrosis virus, iridovirus, parvovirus, or picornavirus (Hess et al. 1984)]. The different viruses may occur in different cells or in the same cell and even in the same cellular structure (cytoplasm or nucleus). In multiple infections, there may be independent coexistence, interference (antagonism), complementation (synergism), or phenotypical mixing (Kurstak and Garzon 1975; Tanada 1976; Sherman 1985).

The infection of baculoviruses in insects generally occurs by ingestion of occluded and possibly free virions, but transmission may occur transovarially, or through the spiracle, or through cannibalism, or the ovipositor of a parasitoid. When the occluded virus is ingested, the period from infection to death in nuclear polyhedroses varies depending on several factors, mainly larval age, temperature, virulence of the viral isolate, viral dosage, and host nutrition. The ingested virion infects the exposed midgut cells and not the cells of the fore- and hindguts, which are lined with cuticle. The columnar epithelial cells of the midgut are susceptible and the goblet cells are not infected. In the cabbage looper larva (*Trichoplusia ni*), the AcMNPV simultaneously infects the columnar epithelial and regenerative midgut cells (Keddie et al. 1989).

E. GROSS PATHOLOGY

Most lepidopteran larvae infected with NPVs show no external signs or symptoms for 2 to 5 days after viral ingestion (Aizawa 1963). The initial signs are the gradual changes in color and luster of the integument with an increase in opaqueness, milkiness, and glossiness. The hemolymph turns cloudy and milky. The larva becomes less active and loses its appetite but may continue to feed up to a few days before death. The larva generally dies in 5 to 12 days, but virulent viral strains may kill very young larvae in 2 to 4 days (Ignoffo 1966). In some baculoviral infections, the larval period is prolonged, even beyond the period of a normal larval stage. The prolonged life may be caused by the EGT gene (ecdysteroid UDP-glucosyl transferase gene) located in the viral genome (O'Reilly and Miller 1989). The product expressed by this gene decomposes ecdysone, the molting hormone, and thereby increases larval life. Such prolongation of larval life would benefit viral reproduction. An unusually long period of infection occurs in the tortricid *Pandemis heparana*. Larvae infected immediately after hatching die in a few days. Death in older larvae is delayed for over 30 days, and between 60 to 100 days for larvae that entered diapause (Amargier et al. 1981).

Shortly before dying, the larvae may move away from the food, disperse, or climb an elevated location to hang from a branch or tree top by their abdominal and caudal prolegs, as in the case of "Wipfelkrankheit" of the nun moth, *Lymantria monacha* (Wahl 1909) (Figs. 6–10, 16–3).

Prior to death or shortly thereafter, the integument, if the hypodermal cells are infected, is fragile and easily torn when handled. Such a larva is in a wilted condition, typical of most nuclear polyhedroses (Figs. 6–10, 16–3). The larval body contents are a fluid mass. Although death usually occurs in the larval stage, some larvae may survive to the pupal or adult stages. The fecundity of the surviving, normal-appearing adult is unaffected, but the hatchability of the eggs may be reduced significantly (Santiago-Alvarez and Vargas-Osuna 1988).

In sawflies, the initial sign is a faint yellow discoloration particularly on the third to fifth abdominal segments resulting from the infection of midgut cells

FIGURE 6–10

Baculovirus epizootic in a population of the armyworm, *Pseudaletia unipuncta*, showing infected larvae dying at the tops of grasses.

(Bergold 1958; Aizawa 1963). The larva becomes increasingly inactive, loses its appetite, and often secretes a dark-brown fluid from the anus or vomits a milky-white fluid. As in a lepidopteran larva, the dead sawfly larva is flaccid with fragile integument, which when broken, expels the liquid body contents.

The NPVs infect several dipteran insects (e.g., mosquito, crane fly, and sciarid fly). The infection in the mosquito larva is similar to that in the sawfly larva because the infection is confined to the digestive tract (Clark et al. 1969; Hall and Fish 1974). The infected larva is sluggish and gradually stops feeding. The milky-white area of the infected hypertrophied midgut cells is visible through the integument indicating a distinct sign of infection. Rupture of the infected cells results in larval death. Larvae of three mosquito species, *Aedes epactius*, *A. atropalpus*, and *A. scutellaris*, increase in resistance to a NPV as they mature (Stiles and Paschke 1980).

In an early study in *Culex tarsalis*, Kellen et al. (1963) detected a systemic infection with tetrahedral occlusion bodies present in the hypodermal cell nuclei of developing adult legs, wings, and antennal buds. However, subsequent workers studying other *Culex* species report that viral infections occur only in cells of the digestive tract of the mosquito larva (Federici and Lowe 1972; Hall and Fish 1974). The virions are occluded in the nuclei of midgut cells. Unlike the polyhedra of other NPVs, the small proteinic occlusions gradually coalesce and grow to large fusiform or ellipsoidal occlusions (Federici and Anthony 1972). The viral morphogenesis differs from that of other baculoviruses because the nucleocapsids, approximately 180 nm, are formed by cleavage of nucleocapsids as long as 1 μm (Federici 1980). In addition, in one type of nucleocapsid envelopment, the envelopes are formed by vesicles, which subsequently dissociate. The envelopes contract, prior to and during virion occlusion, along two-thirds of the nucleocapsids resulting in the virion being bulbous at one end.

Infection of the NPV in the larva of the crane fly, *Tipula paludosa*, causes the larva to become increasingly pale until it appears chalky white, as contrasted to the earthern color of an uninfected larva (Rennie 1923). The blood and fat cells are infected. The occlusion bodies are of unusual crescent shape and are very different from those of other NPVs. They are difficult to dissolve except with an alkaline thioglycolate solution (Guelpa et al. 1977). The polyhedrin has no antigenic relationship to those of other baculoviruses.

In the sciarid fly *Rhynchosciara angelae*, the NPV causes a delay in larval development, and the larvae are smaller and paler than uninfected larvae (Diaz and Pavan 1965; Pavan et al. 1971). Infection occurs in cells of the caeca attached to the midgut and also in those of the posterior part of the midgut. Infected cells hypertrophy and form small tumors.

The primitive thysanuran is susceptible to a NPV (Larsson 1984a). The infection is restricted to the midgut epithelium and viral polyhedra are shed into the gut lumen. The singly enveloped virus, with fragile, spindle-shaped occlusion bodies, causes nuclear hypertrophy. The virus infects the bristle tail *Dilta hibernica* but not another thysanuran, *Lepisma saccharina*.

F. TISSUE SPECIFICITY

In most Lepidoptera, the NPVs cause systemic infections, multiplying in major tissues and organs (polyorganotropic), particularly the fat body, hypodermis, trachea, and blood cells (Fig. 6–6F). Other organs and tissues, such as the Malpighian tubules, reproductive organs, salivary glands, midgut, muscle, pericardium, and nervous tissues, may also be infected by certain NPVs.

Some NPVs are tissue specific or infect only a few tissues of Lepidoptera, such as the virus of *Wiseana cervinata*, which infects only tissues of mesodermal origin, the fat and muscle cells (Entwistle and Robertson 1968). The NPVs of the sawfly (Bird 1952), mosquito (Clark et al. 1969, Federici and Lowe 1972), and thysanuran (Larsson 1984a) are monorganotropic infecting only the midgut. In lepidopteran larvae, the midgut cells are often infected, but the infection is rarely severe throughout the midgut and few, if any, polyhedra are formed (Heimpel and Adams 1966).

G. CYTOPATHOLOGY

Detailed studies on the cytopathology of cells infected with NPVs have been directed primarily at viral replication. It is difficult to differentiate certain structures, (e.g., filaments, membranes, cisternae, etc.) associated with viral replication from those caused by cellular injury. In this regard, the readers should refer to Sections VI.I and VII.E on viral replication.

The cytopathology of the different baculoviruses is generally similar in various insect hosts (Tanada and Hess 1984). The following is based largely on the cytopathology of the infection in lepidopteran larvae. In a midgut cell, the first indication of cytopathology is the fusion of the enveloped virion to the microvillus cell membrane, resulting in an opening through which the naked nucleocapsid enters the microvillus and into the cytoplasm. Within the hemocoel, cell penetration is by viropexis or by fusion (Tanada et al. 1984). There is no apparent cytopathology until the viral DNA is released in the nucleus. The chromatin granules then disperse to the peripheral areas. This is the beginning of the eclipse period of viral replication. The nucleus increases in size; the nucleolus disappears about this time or shortly thereafter; a dense network [called virogenic stroma (Xeros 1956)] and dense bodies of various sizes appear. In the sawfly larva, in which the infection is restricted to the midgut epithelium, the nucleolus increases in size at an early stage of infection and then disappears (Benz 1960). In a viral infection of the armyworm, *Pseudaletia unipuncta*, marked clearing of the nucleoplasm occurs at the eclipse period (Ritter et al. 1982; Tanada et al. 1982). The virogenic stroma (viroplasm) and dense bodies form a ring (ring zone). Structures considered to be associated with viral replication (e.g., tubular profiles, envelopes, etc.) will be considered in Sections VI.I and VII.E on viral replication.

In some cases, the inner layer of a nuclear envelope shows blebbing, infolding, and cisternae (Summers and Arnott 1969; Tanada and Hess 1976) but not in others (Stoltz et al. 1973; MacKinnon et al. 1974). If the nuclear envelope layers separate, the inner layer extending into the nuclear matrix may become trilaminar (Tanada and Hess 1976). Fibrillar masses appear in the nuclei and cytoplasm of infected cells (Croizier et al 1980b). Mitochondria are enlarged, clear, often clustered together, and appear vacuolated. The endoplasmic reticulum is swollen.

The junction of the budding nucleocapsid and the plasma membrane becomes modified and forms a halo of fine filaments or spikes called peplomers (Tanada and Hess 1976; Adams et al. 1977; Kawamoto et al. 1977b). At the terminal stage of infection, even with the nucleus filled with polyhedra, the greatly hypertrophied cell exhibits only slight destruction, such as tearing of mestracheon folds (Tanada et al.

1969) or fissures between infected tracheal cells and cuticle (Injac et al. 1973). Subsequently, the nuclear membrane breaks and the nucleoplasm, which includes various stages of the virus, is released into the cytoplasm. Mitochondria are no longer visible and vacuoles are extremely numerous. The vacuoles may result from degenerated mitochondria (Tanada et al. 1969). The cell lyses shortly thereafter. In midgut columnar cells containing polyhedra, the microvilli disappear from the cell surface (Witt et al. 1977).

There is no cell multiplication or hyperplasia in some tissues, but in others—midgut cells, hypodermis, and fat cells—cellular multiplication may be extensive depending on the virus, such as the baculovirus of *Oryctes rhinoceros*.

H. PATHOPHYSIOLOGY

Baculoviral infections cause alterations in the physiology and metabolism of insects. In general, the oxygen uptake increases markedly, indicating an acceleration of host cell metabolism (Kobayashi and Kawase 1981); hemolymph proteins fluctuate with the intensity of infection and generally decrease as infection progresses (Martignoni and Milstead 1964; van der Geest and Craig 1967; Watanabe et al. 1968); but sublethal doses elicite an increase in certain hemolymph proteins (Takei and Tamashiro 1975). The rapid changes are to be expected because the fat body, a major target of baculoviruses, is the source of hemolymph proteins (Shigematsu and Noguchi 1969b).

The synthesis of RNA increases to a maximum shortly after infection and is closely followed by DNA synthesis (Benz 1963b; Himeno et al. 1976; Kobayashi and Kawase 1981). The RNA synthesis subsides to normal or subnormal levels, but DNA synthesis forms two peaks, one at an early stage and a second at a later stage of infection (Shigematsu and Noguchi 1969a). The cellular DNA polymerase activity increases markedly, followed by the activity of the viral DNA polymerase (Mikhailov et al. 1986a,b). A sequential synthesis of viral DNAs, messenger RNAs, and proteins occurs about 50 h after viral inoculation, followed by polyhedral protein synthesis (Shigematsu and Noguchi 1969c). The host DNA synthesis is reduced and when it breaks down, the viral DNA synthesis stops.

The total amino acids in the cotton leafworm, *Spodoptera littoralis*, decline from a high to a low level in a virus-infected larva, although certain amino acids (e.g., proline, lysine, aspartic acid, and histidine) increase in concentration in the hemolymph (Boctor 1980). In a silkworm larva, most of the amino acids increase up to the fourth day after inoculation, and then decline in concentrations (Kuroda and Watanabe 1983). Fats and lipids are reported to decrease in the rice moth, *Corcyra cephalonica* (Rajamohan and Jayaraj 1976), and in the spruce budworm, *Choristoneura fumiferana* (Smirnoff 1976), but increase in *S. littoralis* (Boctor 1981) and in the silkworm (Komano et al. 1966). Glycogen and glycerol levels generally decrease (Rajamohan and Jayaraj 1976; Kobayashi and Kawase 1981); the α-ketoglutaric acid increases markedly (Kuroda and Watanabe 1983); and enzymes, such as glutamic oxaloacetic transaminase, alkaline phosphatase, glutamic dehydrogenase, and isocitrate dehydrogenase, increase (Shylaja and Ramaiah 1984).

The hormonal titre in a *Spodoptera litura* larva is affected by viral infection (Subrahmanyam and Ramakrishnan 1981). The juvenile hormone titre is maintained at a high level during the last instar of the virus-infected larva, which fails to pupate. In *Heliothis virescens*, treatment with β-ecdysone significantly lowers the mortality of a virus-infected larva as compared to a larva treated with Ringer's (control) solution (Keeley and Vinson 1975). Thus, the larval hormonal balance may modify the course of viral pathogenesis. The EGT gene conjugates with ecdysone, the molting hormone, and thereby prolongs larval life (O'Reilly and Miller 1989).

After viral infection, many newly synthesized proteins and nucleic acids are closely associated with viral replication. In addition, thymidine and protein kinase, associated with the viral particle, are expressed in the infected cell (Miller et al. 1983; Wilson and Consigli 1985a). A glycoprotein, which appears to be virus-coded, may play roles in envelope acquisition, viral attachment, penetration, and uncoating (Goldstein and McIntosh 1980; Stiles and Wood 1983; Kelly and Lescott 1983). The viral particle also contains polyamines, which bind DNA and may have a role in viral multiplication and DNA assembly (Elliott and Kelly 1977, 1979).

A novel proteinaceous substance appears in the hemolymph of an armyworm larva infected with the hypertrophy strain but not in a larva infected with the typical strain of a NPV (Kaya and Tanada 1973; Hotchkin and Kaya 1983a). This protein adversely affects the internal braconid parasitoid *Glyptapanteles* (syn. *Apanteles*) *militaris* and causes the cessation of parasitoid growth, general tissue disintegration, followed by melanization of parasitoid tissues, and encapsulation by hemocytes.

I. VIRAL REPLICATION

A diagrammatic representation of the replication of a NPV is given in Figure 6–11. When the occlusion bodies are ingested, they begin to dissolve within 0.5 min and are dissolved by 3 min liberating the enveloped virions (Pritchett et al. 1982). Prolonged exposure to the midgut juice for several minutes may inactivate the virions (Watanabe 1974; Vail et al. 1979; Pritchett et al. 1984), but Ignoffo et al. (1985) found no antiviral activity in midgut juices. The liberated virions pass through the peritrophic membrane and invade the columnar epithelial cells. An enhancing factor present in the matrix of the polyhedra disrupts the peritrophic membrane by removing a 68-kDa glycoprotein and this facilitates the passage of the virion through this membrane (Derkson and Granados 1988).

Harrap and Robertson (1968) were the first to report viral invasion in midgut cells. Invasion begins by the attachment and fusion of the viral envelope to the membrane of a columnar cell microvillus (Kawanishi et al. 1972b; Tanada et al. 1975). The phospholipid and ionic charges of the viral envelope are involved in the attachment (Yamamoto and Tanada 1978a).

The nucleocapsid enters the microvillus through an opening formed at the site of fusion of the viral envelope and cell membrane (Summers 1969, 1971; Harrap 1970; Tanada et al. 1975). Attachment, fusion, and viral entry occur rapidly, within 0.25 to 4 h after the feeding of occlusion bodies (Kawanishi et al. 1972b; Granados 1978). Within the cell, the nucleocapsids are frequently adjacent to microtubules that may be involved in the vectorial movement of the virion to the nucleus (Granados 1978, Granados and Lawler 1981).

The invasion of the nucleus may occur when nucleocapsids become attached to nucleopores where uncoating (discharge of DNA) takes place (Summers 1971; Kawanishi et al. 1972b; Raghow and Grace 1974), or the nucleocapsids enter through nucleopores and uncoat in the nucleoplasm (Hirumi et al. 1975; Bassemir et al. 1983; Tanada et al. 1984). The above conclusions are based on electron micro-

FIGURE 6-11

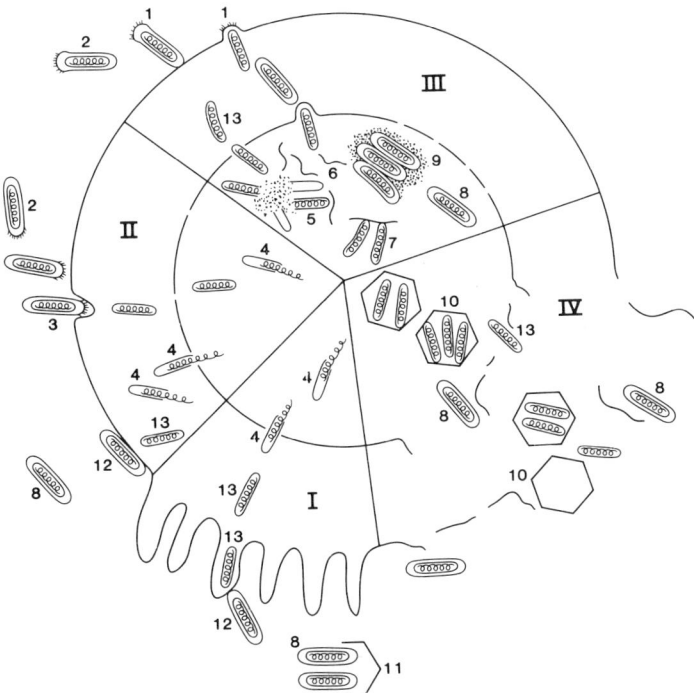

A composite diagram of four insect cells showing the penetration and replication of a baculovirus. (I) A midgut cell and (II) a mesodermal cell showing viral penetration and uncoating of DNA; (III) a mesodermal cell showing viral replication and exocytosis; (IV) a cell undergoing lysis. (1) Exocytosis; (2) plasma-enveloped extracellular virions with peplomers; (3) adsorptive endocytosis; (4) uncoating of DNA; (5) virogenic stroma; (6) *de novo* nuclear envelope; (7) envelopment of nucleocapsids; (8) nuclear-enveloped virion; (9) occlusion of enveloped nucleocapsids; (10) viral occlusions (polyhedra); (11) dissolution of polyhedron in midgut lumen; (12) fusion of nuclear-enveloped nucleocapsid; (13) nucleocapsid in cytoplasm. (Drawn by Christina L. Jordan.)

graphs, but with the use of radiolabeled virions, Wang and Kelly (1985) have reported that uncoating occurs predominantly in the cytoplasm.

The introduction of the nucleic acid into the nucleoplasm initiates the eclipse period of viral replication, and this period is terminated with the appearance of the ring zone formed by the virogenic stroma and viral capsids (Xeros 1956; Huger and Krieg 1961). Elongated tubular profiles that are considered to be precursors of viral capsids appear near the virogenic stroma (Summers 1971; Hughes 1972). Viral DNA synthesis and viral assembly are believed to occur in the virogenic stroma because of the intimate association of capsids and newly formed nucleocapsids with the stroma. Empty capsids appear, 10 h postinfection (h p.i.), shortly after the virogenic stroma (7 h p.i.), and is followed by partially and completely filled nucleocapsids (11 to 12 h p.i.) (Bassemir et al. 1983). This sequence suggests that the virus is assembled in successive stages rather than simultaneously. The viral progeny is observed at about 8 h p.i. and the developmental cycle of the virus is essentially completed in 24 h (Granados and Lawler 1981; Granados et al. 1981). During this period, the host DNA remains in a conventional chromatin structure indicating that the virus has a specific means of compartmentalizing its own DNA for RNA and DNA syntheses (Wilson and Miller 1986).

The nucleocapsids in the nucleus become closely associated with granules, 17.5 nm in diameter, and are frequently attached to fibrous filaments forming granule-filament complexes (Kobayashi 1976; Tanada and Hess 1976). The complexes appear to coil around the nucleocapsids and form the envelope, possibly *de novo*, through the action of the ribosome-like particles. Viral envelopes appear as globular structures, which are present sometimes in very large numbers. The envelopes surround single or groups of two or more nucleocapsids to form the mature enveloped virion. Enveloped virions are often not occluded in occlusion bodies in midgut cells (Tanada and Hess 1984; Granados and Williams 1986). This is in sharp contrast to the occlusion of most enveloped nucleocapsids in infected cells in the hemocoel (e.g., cells of fat body, trachea, and hypodermis).

Shortly after their formation in a midgut cell nucleus, most nucleocapsids exit through the nucleopores, openings in the nuclear envelope, or bud through the nuclear envelope (Injac et al. 1971; MacKinnon et al. 1974). The nucleocapsids that pass through the nuclear envelope acquire two loose-fitting membranes (transport vacuoles) and enter the cytoplasm (Hirumi et al. 1975; Kawamoto et al. 1976; Knudson and Harrap 1976). The transport vacuoles are lost during the transport of nucleocapsids to the basal cell membrane (Hirumi et al 1975; Granados and Lawler 1981). As the nucleocapsids pass through a basal cell membrane, they acquire a loose-fitting envelope with peplomers (Tanada and Hess 1976; Adams et al. 1977; Kawamoto et al. 1977a).

The replication of AcMNPV in midgut cells of the cabbage looper, *Trichoplusia ni*, is as follows: from 0.25 to 4 h for viral fusion and entry into a cell; from 0.5 to 6 h for the appearance of nucleocapsids in the cytoplasm; from 1 to 6 h for nucleocapsids uncoating in the cell nucleus; from 8 to 48 h for the formation of the virogenic stroma and progeny virus; from 24 to 48 h for the formation of polyhedra; from 0.5 to 48 h for the budding of nucleocapsids through the basement membrane into the hemocoel (Granados and Williams 1986).

Virions, liberated from ingested occlusion bodies, sequentially infect the susceptible insect tissues. Midgut epithelial cells are first infected followed by the midgut connective tissue sheath before the entrance of the virions into the hemocoel (Keddie et al. 1989). The succession of infections in hemocoelic tissues is initially the trachea attached to the midgut and followed by the fat body, hypodermis, and other tissues (Adams et al. 1977). Virions and infected hemocytes spread the infection throughout the hemocoel (Granados and Lawler 1981; Keddie et al. 1989).

The enveloped virion invades by means of viropexis into a susceptible cell in the hemocoel with the peplomer end facing the host cell membrane (Adams et al. 1977; Kawamoto et al. 1977b; Tanada et al. 1984). In viropexis, a vesicle (clathrin-coated pits), comparable to a phagocytic vesicle, is formed and some nucleocapsids may be destroyed by enzymatic activity (Adams et al. 1977; Tanada et al. 1984).

Enveloped virions that are produced in virus-infected cells in the hemocoel (e.g., fat body and tracheal cells) also have peplomers (Summers and Volkman 1976; Adams et al. 1977). Cup-shaped particles lacking nucleocapsids may also acquire envelopes with peplomers when they leave the cell (Kawamoto et al. 1977b). Observations indicate that the infection in insect tissue cultures is associated with peplomer-enveloped virions (Henderson et al. 1974; Dougherty et al. 1975; Raghow and Grace 1974).

Granados and Lawler (1981) reported another method of viral penetration into the insect's hemocoel. They observed the penetration of AcMNPV through the microvillus, but some nucleocapsids, instead of replicating in the nucleus, passed out through the basement plasma membrane into the hemocoel as early as 0.5 h after infection. However, Keddie et al. (1989) found no evidence of direct passage of nucleocapsids into the hemocoel.

The virions in the hemolymph of a virus-infected larva are highly infectious when inoculated into the hemocoel but are of low infectivity when introduced into the midgut lumen. In contrast, the virions obtained from occlusion bodies are highly infectious in the midgut and are less infectious when inoculated into the hemocoel (Vaughn and Faulkner 1963; Kawarabata 1974; Volkman and Summers 1977). The two types of enveloped virions can be separated serologically (Volkman 1983; Roberts 1983). The envelopes and the means of invasion may account for the differences in the infectivity of the two phenotypic forms of virions.

We shall return to the infected nuclei of cells in the hemocoel in which the nucleocapsids are being enveloped. At this stage, condensation foci of polyhedrin form small granules (from 0.2 to 0.4 μm in diameter), termed "prepolyhedra" (Federici 1986). The enveloped nucleocapsids are attracted to these foci (Fig. 6–8). As the polyhedrin accumulates, the enveloped nucleocapsids are occluded by the protein to form the polyhedron, which increases up to a certain size depending on the viral species. Polyhedra are formed about 24 to 96 h after viral ingestion. They may completely fill the nucleus without the nuclear membrane breaking down. They are released only after the lysis of the infected cell and the virions that remain occluded do not infect tissues of the hemocoel.

The viral replication described above undergoes a latent or eclipse (from 0 to 12 h), an exponential (from 12 to 72 h), and a stationary period (after 48 to 72 h) (Fig. 6–1). Yamafuji et al. (1954) made the initial observation that a latent or eclipse period occurred in baculoviral replication as in the case of other animal and plant viruses. Viral multiplication kinetics have been observed *in vivo* (Krieg 1958; Aizawa 1959) and *in vitro* (Vaughn and Faulkner 1963; Tsuda et al. 1984).

In summary, the viral replication in a cell is a biphasic process. The initial phase is the production of nucleocapsids which acquire envelopes with peplomers to form extracellular virions that infect cells in the insect's hemocoel. In the second phase, the nucleocapsids acquire *de novo* envelopes in the nucleus and are occluded in polyhedra. The occluded virions cause infection when ingested by another susceptible larva.

Exceptions to the general NPV replication in the nucleus are the replications in the cytoplasm of fat cells of the brown-tail moth, *Euproctis similis* (Asayama and Kawamoto 1975), and in the cytoplasm of cells of the midgut epithelium and epidermis of *Spodoptera exigua* (Falcon and Hess 1977). The replication in the cytoplasm is similar to that in the nucleus and includes a virogenic stroma. These unusual observations need confirmation and explanation.

J. TRANSFECTION OF ISOLATED DNA

Early workers had reported that DNAs isolated from NPVs were infectious (Onodera et al. 1965; Dobrovolskaya et al. 1965; Kok et al. 1968), but for over a decade these reports could not be confirmed. The failures were apparently associated with the technique used to obtain transfection. Oreshkin (1982) has described the procedures and conditions necessary to obtain infection of cells with viral DNA. Transfections with the DNAs of several NPVs have been obtained in larvae (Skuratovskaya et al. 1977; Strokovskaya et al. 1979), and in cell cultures (Bud and Kelly 1980b; Burand et al. 1980; Potter and Miller 1980). Analysis with restriction endonucleases has confirmed the transfections with DNA (Carstens et al. 1980).

Transfections are used in the genetic engineering of the viral genome. The procedure for recombinant viruses involves a mixture of purified NPV DNA and a recombinant transfer vector containing the foreign DNA (Kang 1988; Luckow and Summers 1988; Miller 1988; Maeda 1989a,c).

K. PERSISTENT VIRAL INFECTIONS

Persistent viral infections in insects are long-term infections in which the viruses, some of which are cytocidal, continue to multiply while the host insects or cells continue to grow or develop (Burand et al. 1986). These infections are generally classified as latent, chronic, or slow (Wood and Burand 1986). In a latent infection, the virus is usually not detectable and acute infections may occur intermittently. The virus is always detectable in a chronic infection, but the symptoms are usually inapparent. A slow infection has a long incubation period followed by a slowly progressive disease that is usually fatal. These three classes, sometimes, are difficult to separate.

A virus can persist in 3 ways: (1) without shedding infectious virions, (2) by shedding infectious virions only periodically, and (3) by replicating and releasing infectious virions continually (Burand et al. 1986). The infection of sigma virus (a RNA virus) in *Drosophila melanogaster* is a classical and thoroughly studied example of a persistent insect virus.

Most studies of persistent infections with the use of live insects lack sufficient depth because of problems in detecting the virus in insects, contamination from other pathogens, and the host condition. These conditions are minimal with the use of tissue cultures. Comprehensive studies have been conducted with two baculoviruses *in vitro*: (1) the NPV of *Spodoptera frugiperda* (SfNPV) in its homologous cell line IPLB-SF-21 and (2) the nonoccluded Hz-1 virus in IMC-Hz-1 cells (Kelly et al. 1981; Burand et al. 1986). Since the Hz-1 virus is classified as a nonoccluded virus (NOB), its persistent properties will be mainly considered later. The persistent virus (SfP) in *Spodoptera* is cytocidal. In its infection, only about 2% of the cells contain polyhedra as compared to a nonpersistent infection (McIntosh and Ignoffo 1981).

A cell culture (IPLB-SF-21) infected with SfP is governed by a mechanism different from that in a cell culture of TN-368 infected with the Hz-1 virus (Burand et al. 1986). In both types, the virions are released at low levels, but the SfP cultures grow more slowly to a higher density, with smaller cells that become detached from the culture vessel. In contrast, the TN-368 cells infected with the Hz-1 virus are nearly normal in appearance. In the persistence of the Hz-1 virus, a defective interfering viral particle functions in modulating the infection, whereas the infection in SfP culture appears to be controlled by a more general antibaculoviral factor(s), perhaps of cellular origin.

In an epizootic in an insect population, the persistent baculovirus may play an important role as a factor in the induction, transmission, and spread of the virus (Burand et al. 1986). During the peak of viral transmission and infection, defective virions may be generated and they may produce persistent infections in the insect population, thus ensuring the survival of the virus.

L. RESTRICTION ENZYME ANALYSIS

The genomes of baculoviruses are highly complex and occur as circular, supercoiled, dsDNAs of 88 to 160 kbp (kilobase pairs) (Mathews 1982; Blissard and Rohrmann 1990). Studies on baculoviral genomes accelerated when many restriction endonucleases became available. The restriction endonucleases are bacterial enzymes that act on specific sequences in the DNA and cleave the DNA at or near these sites to yield distinct fragments that are analyzed by gel electrophoresis and other methods, such as column chromatography, DNA–DNA hybridization, Southern or Northern blot techniques, etc. (Fig. 6–12). Each enzyme produces a unique profile of a DNA fragment pattern for a specific virus and a different enzyme acting on the same viral DNA would produce another profile. When the DNA fragments have common mobilities in gel electrophoresis, this suggests high genetic relatedness of the viruses; completely different mobilities would suggest that the viruses are not closely related. Double digestion with two different enzymes are used to ensure that the positions of the restriction sites as well as the sizes of the fragments are identical.

Major advantages of the use of restriction endonucleases in baculoviral taxonomy are (1) the entire genome is examined, (2) related viral strains are unequivocally identified, (3) a mixture of viruses can be resolved, and (4) genetic changes resulting from insertions and deletions can be monitored (Bilimoria 1986). These

FIGURE 6–12

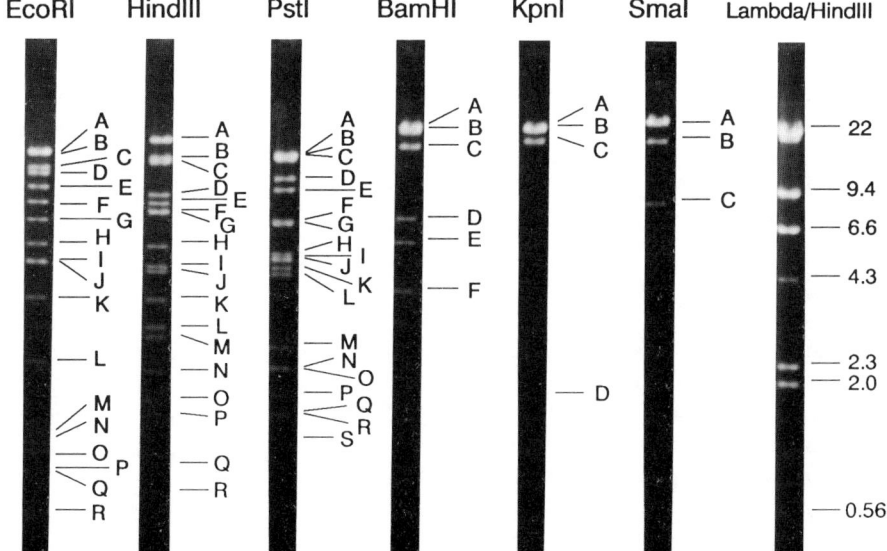

Column chromatography of *Bombyx mori* nuclear polyhedrosis virus showing cleavage patterns of DNA with six restriction endonucleases. The Lambda–HindIII column shows the molecular size patterns in kilobase pairs. [Courtesy of Maeda and Majima (1990) and Society for General Microbiology.]

enzymes have been used with increasing frequencies to identify baculoviruses and to distinguish and characterize closely related genotypic variants and species (Lee and Miller 1978; Miller and Dawes 1978a,b). For example, the viruses of *Autographa californica* and *Trichoplusia ni* were considered to be the same virus because of transmission to both hosts, but the restriction-endonuclease fragment patterns of their DNAs showed that the two viruses were closely related variants with similar fragment patterns (Miller and Dawes 1978a).

1. Physical Map

The exact location of the restriction sites in the viral DNA is determined by physical mapping. Scientists working with baculoviruses have accepted the orientation of the circular genome of *Autographa californica* MNPV (AcMNPV) as the standard (Vlak and Smith 1982) (Fig. 6–13). The zero point on the genome is the location of one of the restriction sites closest to the 5' end of the polyhedrin gene. The enzyme *Eco*RI is used to locate this site for AcNPV. The zero reference point facilitates direct comparison of data from different laboratories on the physical, genetical, transcriptional, and functional maps of AcMNPV variants, mutants, and related NPVs. The maps are arranged in a linear or a circular form.

The first physical map was constructed for the DNA of AcMNPV (Miller and Dawes 1979; Smith and Summers 1979). Other physical maps are those of the NPVs of *Spodoptera frugiperda* (Loh et al. 1981; Maruniak et al. 1984), *Heliothis zea* (Knell and Summers 1984), *Orgyia pseudotsugata* (Chen et al. 1988), *Anticarsia gemmatalis* (Johnson and Maruniak 1989), *Panolis flammea* (Possee and Kelly 1988), *Mamestra brassicae* (Wiegers and Vlak 1984, Possee and Kelly 1988), *Spodoptera littoralis* (Croizier et al. 1989), and *Bombyx mori* (Maeda and Majima 1990).

Physical maps have been used to locate changes in the DNA genomic variants (Miller and Dawes 1979; Maruniak et al. 1984) and to identify the recombination events between closely related baculoviruses (Miller 1986). A direct benefit of physical mapping studies has been the establishment of DNA restriction fragment libraries (Cochran et al. 1986). Such collections can provide DNA fragments for the purpose of transcription and translation mapping and for the development of functional maps by techniques, such as marker rescue, for the characterization of mutants.

2. AcMNPV Genome

The most thoroughly studied genome of baculoviruses (even for all insect viruses) is that of the multinucleocapsid NPV of *Autographa californica* (AcMNPV) (Fig. 6–13). The DNA genome of AcMNPV is 128 kbp or from 82 to 88 $\times$ 10^6 daltons (Smith and Summers 1978; Vlak and Odink 1979; Miller 1981b). It is believed to contain over 100 different genes (Miller 1988; Blissard and Rohrmann 1990). About 30% of the genome has been sequenced and characterized, but the functions of many genes or the sequences that apply in controlling their expression during infection are not known (Miller 1988; Blissard and Rohrmann 1990). There are repeated nucleotide sequences in the genome that function as enhancers for the expression of other genes (Guarino et al. 1986).

The gene for polyhedrin (p29) in AcMNPV is one of the first to be identified

FIGURE 6-13

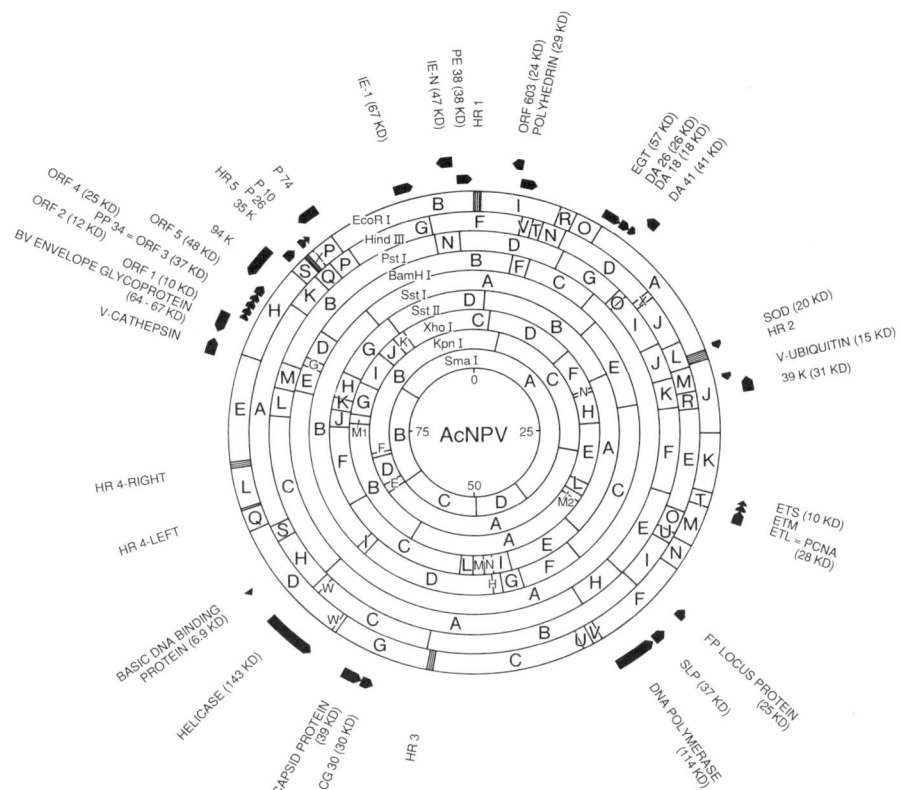

Physical and genetic map of *Autographa californica* nuclear polyhedrosis virus (prototype E2). The physical map of restriction sites was made for the enzymes *Eco*RI, *Hind*III, *Pst*I, *Bam*HI, *Sst*I, *Sst*II, *Xho*I, *Kpn*I, and *Sma*I. The first *Eco*RI site of the HR1 region is taken as the zero point of the map. Genes are indicated by their common functional name, as open reading frame (ORF) or as a code. The map runs clockwise starting with ORF603 and polyhedrin. The arrows indicate the direction of transcription. The highly repetitive regions are indicated as HR. Only those genes are incorporated of which the gene product (as protein) is known or the ORF has been studied. The molecular weights (in parentheses) of these products are designated in kilodaltons (kDa). (Provided by M. Kool and J. M. Vlak, Department of Virology, Agricultural University, Wageningen, The Netherlands.)

and located on the physical map. It is an uninterrupted, single-copy gene (about 1200 base pairs) within the 7.3-kb fragment of *Eco*RI (Smith et al. 1983a). The nucleotide sequence of the entire coding region of the polyhedrin gene has been determined (van Iddekinge et al. 1983). This gene is well conserved in many NPVs. The transcription of AcMNPV genes is very complex. The genes are usually transcribed as sets of unspliced, nested, overlapping transcripts (Crawford and Miller 1988). A single gene may be transcribed into several RNAs ,which differ in the 3' or 5' termini during infection.

M. GENOTYPIC VARIANTS

Like other organisms, each baculovirus has a number of genotypic variants. The genotypic variants of AcMNPV are considered as the model system for investigations on the molecular biology of baculoviruses (Cochran et al. 1986). Variants of AcMNPV have been isolated from successive plaque purification of each type of clones, such as the few polyhedra variant (Lee and Miller 1978; Fraser and McCarthy 1984) and from wild isolates of this virus (Smith and Summers 1978). Genotypic variants have been isolated also from many other wild isolates of NPVs collected from different geographical regions [e.g., the viruses of *Mamestra brassicae* (Vlak and Gröner 1980; Brown et al. 1981), *Heliothis* spp. (Gettig and McCarthy 1982; Hughes et al. 1983), *Spodoptera frugiperda* (Knell and Summers 1981; Loh et al. 1982), *Spodoptera littoralis* (Kislev and Edelman 1982), *S. litura* (Maeda et al. 1990), *Agrotis segetum* (Allaway and Payne 1983), and the sawfly *Neodiprion sertifer* (Brown 1982)]. In these genotypic variants, some show high DNA homology, whereas others do not. In addition to genotypic variants in wild stocks of NPVs, completely different species of NPVs may occur in the same host insect (Maeda et al. 1990).

The gene arrangement of some baculovirus genomes appears to be similar or to be colinear based on DNA homology, such as AcMNPV and *O. pseudotsugata* MNPV (Leisy et al. 1984), AcMNPV and *C. fumiferana* MNPV (Arif et al. 1985), and AcMNPV and *H. zea* SNPV (Knell and Summers 1984). Baculoviruses possess regions of repeated sequences in their genomes (Cochran et al. 1986). These reiterated sequences have been found by the presence of many *Eco*RI sites and by hybridization. Intragenic sequence homologies are found in genomes of the AcMNPV variants (Smith and Summers 1979; Cochran and Faulkner 1983; Brown et al. 1984) and with the genome of *Choristoneura fumiferana* MNPV (Arif and Doerfler 1984). Intergenic sequence homology occurs in the MNPVs of different species of *Spodoptera* (Kelly 1977; Maeda et al. 1990), in AcMNPV and *Heliothis zea* SNPV (Knell and Summers 1984), in AcMNPV and MNPV of *Orgyia pseudotsugata* (Rohrmann et al. 1982; Leisy et al. 1984), and some DNAs of 18 baculoviruses in the subgroups (Smith and Summers 1982).

Baculoviruses can recombine independently. For example, recombination occurs between AcMNPV and the virus of *Galleria mellonella* when both viruses are fed together to *Galleria* larvae to produce a mixed viral infection (Croizier et al. 1980a; Croizier and Quiot 1981). In the mixed infection, new recombinants are formed, which may replace the original parent viruses. Similar recombinations are not uncommon with viruses in tissue cultures [e.g., with temperature sensitive mutants (Brown and Faulkner 1980; Miller 1981a; Erlandson et al. 1984) and with closely related, but genetically distinguishable, variants (Summers et al. 1980; Croizier et al. 1980a)]. Kondo and Maeda (1991) have coinfected insect cell lines with variants of *Bombyx* NPV and *Autogrpha* NPV and obtained a recombinant virus with a wider host range than the two parent viruses.

In addition to recombination by viral intragenic homologous sequences, two other types of variants may occur from (1) point mutations and (2) the acquisition of host DNA inserts (Brown et al. 1985). A point mutation in the polyhedrin gene occurs when phenylalanine is substituted for leucine at amino acid 84 and this prevents the proper folding and assembly of the polyhedrin into occlusion bodies (Carstens et al. 1987).

The acquisition of host DNA sequences occurs when AcMNPV is passed serially *in vitro* (Fraser et al. 1983; Miller and Miller 1982). The host sequences are transposable elements, which may explain the origin of certain baculoviral variants and have a major role in the variations observed in baculoviruses (Fraser 1986b; Blissard and Rohrmann 1990). The transposable elements may cause a variety of genetic changes through their capacity to cause insertions, deletions, and other DNA rearrangements. Some of these elements are copia-like transposable elements named for their "copious" transcription of copia mRNAs produced in the cell. The copia-like elements are involved in the few polyhedra (FP) phenotypic mutants of AcMNPV (Miller and Miller 1982; Fraser 1986b). Moreover, the transposable elements may introduce insect promoters and enhancers to the baculoviral genome (Blissard and Rohrmann 1990). The genomic changes may cause mutations that are advantageous to the virus in virulence and host range. Such changes may account for variations in the virulence of NPVs when propagated in alternate hosts (Tompkins et al. 1988).

Recombinant DNA techniques are used to obtain genetically altered viruses, which are used as expression vectors of foreign genes (see Section VI.O, Baculoviruses as Expression Vectors).

N. DNA REPLICATION AND GENE EXPRESSION

Viral DNA replication requires the presence of DNA polymerases. Early detections of viral DNA polymerases were made in NPV infections in *G. mellonella* (Rindich et al. 1975) and in the silkworm (Himeno et al. 1976). A viral DNA polymerase has been isolated and it is a single polypeptide of MW 126,000 (Miller et al. 1981; Wang and Kelly 1983). The gene encoding a DNA polymerase has been located and sequenced (Tomalski et al. 1988). The DNA polymerase is transcribed primarily as an early gene or immediate early gene. It transcribes a 3-kb mRNA. Of the four types of DNA polymerases (α, β, γ, and δ) found in higher eukaryotes, the baculoviruses appear to utilize the δ-related replication system (Blissard and Rohrmann 1990).

The virus-specific RNA polymerase was detected first in *H. zea* larva and in *S. frugiperda* cell line infected with a NPV (Grula et al. 1981). Investigators have prepared translation maps for the viral DNA-specific, messenger RNAs (mRNAs) involved in transcribing the viral polypeptides of AcMNPV (Smith et al. 1982; Vlak and van der Krol 1982; Esche et al. 1982). At least 50 viral RNA transcripts have been mapped on the genome. The expression of the baculoviral genome (AcMNPV) in an infected cell may produce as many as 100 proteins.

The expressions of viral mRNAs occur in a sequential or ordered cascade of events in which each successive phase is dependent on the previous phase (Blissard and Rohrmann 1990). The cascade of viral gene expression (i.e., protein production) is regulated at the transcriptional level. Gene products of one temporal class of baculoviral gene can transactivate (directly or indirectly) the transcription of genes of the next temporal class. Workers have separated the above events into two to four phases (Kelly and Lescott 1981; Cochran et al. 1986; Miller 1988; Blissard and Rohrmann 1990). The temporal classes of genes, expressing the products of the various phases, are interspersed in the genome with no apparent pattern of gene arrangement (Miller 1988).

Prior to DNA synthesis, the immediate early genes, which require no viral gene products for their expression, are transcribed by the insect cell (from 0 to approximately 6 h p.i.). The host RNA polymerase(s) is responsible for the transcription of viral genes, at least at the early stage of infection (Cochran et al. 1986; Blissard and Rohrmann 1990). The immediate early gene, IE-1, can act as a transactivator of other early genes (Guarino et al. 1986; Chisholm and Henner 1988). Delayed early genes, which require other viral gene products for their transcription, are then transcribed.

During the late phase, extending approximately 6 through 18 h p.i., the late genes are first transcribed after or concurrently with the onset of viral DNA synthesis. This phase is characterized by the extensive replication of viral DNA and the formation of nucleocapsids (progeny enveloped virions), which includes the expression of the capsid and the 64-kDa glycoprotein (Miller 1988; Blissard and Rohrmann 1990). The bulk of the enveloped virions is produced within the first 18 h p.i.

The very late phase (approximately 20 through 72 h p.i.) is involved with the occlusion process. The polyadenylated transcripts from the polyhedrin gene begin to accumulate at about 18 h p.i. Shortly thereafter, the translation products appear and accumulate through 70 h p.i. All late genes have transcriptional start sites ATAAG or GTAAG, representing invariant nucleotides (Thiem and Miller 1989; Blissard and Rohrmann 1990). During this phase there is abundant polyhedrin synthesis and the formation of occlusion bodies in the nucleus. By 70 h p.i., polyhedrin is the most predominant protein in the cell (from 25 to 50% of stainable proteins) (Miller 1988). Another abundant protein, expressed at this time by the p10 gene, is also involved in the maturation of the occlusion body and possibly in the cytoskeletal structure (van der Wilk et al. 1987; Quant-Russell et al. 1987). Both the late and very late genes are apparently transcribed by a virus-induced RNA polymerase (Grula et al. 1981; Miller 1989).

Over 30 proteins and their genes have been identified (Blissard and Rohrmann 1990) (Fig. 6–13). The occlusion body proteins are the polyhedrin (29 kDa) (Rohrmann 1986a), the 10-kDa protein (also called 7.2-, 7.5-, and 8-kDa proteins) expressed at the late stage of infection and plays a role in the transport and/or attachment of the envelope to the polyhedron (Kuzio et al. 1984; Liu et al. 1986; Leisy et al. 1986b), and the 32-kDa polyhedron envelope protein (Gombart et al. 1989). By replacing the polyhedron envelope gene, a recombinant virus was formed, which produced polyhedra lacking the envelope (Zuidema et al. 1989).

Virion structural proteins are the 6.9-kDa protein, a very basic histonelike DNA-binding protein probably involved in the condensation of viral DNA prior to or during packaging (Wilson et al. 1987) and the 39-kDa protein, a major capsid protein (Thiem and Miller 1989; Blissard et al. 1989). Another protein is the virus-specific protein of the budded virus, the 64-kDa envelope glycoprotein (Whitford et al. 1989; Blissard and Rohrmann 1989).

Early genes involved in viral replication are the immediate early gene, IE-1, whose expressed proteins transactivate other genes expressed at later stages (Chisholm and Henner 1988; Guarino and Summers 1988); two early genes of ETL (acronym for *Hind* III-E-*Eco*RI-T-large open reading frame) whose products accelerate late gene expressions (Crawford and Miller 1988); a 39k delayed early gene that is activated by the gene product of IE-1 (Carson et al. 1988).

Some genes concerned with viral pathogenicity are the EGT gene whose product can conjugate ecdysone (O'Reilly and Miller 1989) and the p74 gene, located adjacent to the p10 gene, is essential for the per os infection of occluded virions (Kuzio et al. 1989).

O. BACULOVIRUSES AS EXPRESSION VECTORS

Recombinant DNA technology (genetic engineering) has been applied on baculoviruses for (1) high-efficiency eukaryotic expression vectors, (2) more efficient pest-control agents, (3) the study of baculoviral gene functions and organizations, and (4) the study of the regulation of gene expression in baculoviruses (Cochran et al. 1986). The most attractive application is the use of baculoviruses as helper-independent viral vectors for the expression of foreign genes in eukaryotes (Fig. 6–14). The expressed products of the foreign genes are biologically active (e.g., proteins having enzymatic, hormonal, or immunogenic properties) and they are not easily and satisfactorily produced in lower eukaryotic or prokaryotic expression vectors.

Miller (1981b) was the first to recognize the potential of baculoviruses as vectors. She pointed out that the following salient features of baculoviruses make them highly advantageous as a recombinant DNA vector system: (1) an extendable rod-shaped capsid (80 to 200 kb) that can accumulate additional DNA sequences, (2) a group of genes that are involved in the occlusion process but are not essential for the synthesis of extracellular nonoccluded virions and that are replaceable, (3) the availability of a strong promoter of the polyhedrin gene that is expressed following the synthesis of extracellular nonoccluded virions, (4) the nonsusceptibility of mammalian cells to baculoviruses, indicating no significant health hazard to mammals, (5) the limited host range of baculoviruses to only certain invertebrates, (6)

FIGURE 6–14

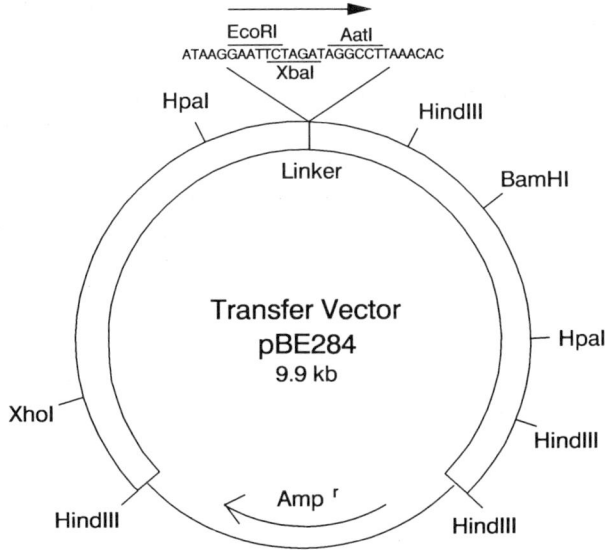

Physical map of a transfer vector used in the genetic engineering of the genome of *Bombyx mori* nuclear polyhedrosis virus. (Provided by S. Maeda.)

the mapped baculoviral genome that has been characterized with respect to its transcription activity, and (7) the expressed products of foreign genes that undergo posttranslational modifications (e.g., glycosylation, phosphorylation, and C-terminus amidation) and that are biologically active.

The following steps are taken to prepare the baculoviruses as expression vectors: (1) locate the polyhedrin gene on the baculoviral genome map, (2) clone the region of the polyhedrin gene into plasmids of the bacterium *Escherichia coli*, (3) delete the polyhedrin-coding region and insert a linker, (4) insert the passenger gene of interest into the linker, (5) use transfection procedures to introduce the transplacement plasmid and viral DNA into cells where homologous recombination (double crossover of genes) leads to the insertion of the foreign (chimera) gene into the nonessential polyhedrin region of the genome, and (6) isolate and purify the plaques of the recombinant virus (Miller et al. 1983; Cochran et al. 1986; Maeda 1989a).

A number of transplacement plasmids are available (Luckow and Summers 1988; Miller 1989; Maeda 1989c). These plasmids retain the entire polyhedrin promoter region. Baculoviral vectors have become remarkably popular in the research and industrial communities because of the high levels of gene expression, the rapidity with which such levels of achievement are obtained, and the simplicity of the viral technology employed (Miller 1989).

The polyhedrin gene is routinely substituted with foreign genes in recombinant studies because the strong polyhedrin promoter provides abundant transcription and because polyhedrin is not essential for nonoccluded-viral replication. Since the recombinant viruses lack polyhedrin, they are easily identified by the absence of polyhedra in a plaque produced in a cell monolayer. Another gene, p10, whose product is produced abundantly late in viral replication together with polyhedrin, also offers promise in foreign gene expression (Vlak et al. 1988).

Thus far, most expression studies have been conducted in insect cell cultures, but the silkworm has been used for the production of foreign gene products with the *Bombyx mori* NPV (BmNPV) (Maeda et al. 1985; Miyajima et al. 1987; Maeda 1989a). The propagation of the silkworm, a domesticated insect, is efficient, economical, and can be conducted under computerized mass-rearing systems (Maeda 1989a). Moreover, the silkworm appears to be a safe factory for the production of pharmaceuticals because it has been used as food by animals, including humans, and is less prone to induce allergy as some other insects. The expressed products of foreign genes are secreted into the silkworm's hemolymph and are as easily recovered and purified as in the case of cell culture products.

The proteins expressed in the silkworm may exceed those in cell cultures. For the mouse interleukin-3 (IL-3), the synthesis in the silkworm is more than 500-fold, which is achieved in an established cell line *B. mori* (1 mg/ml of hemolymph is synthesized per silkworm larva) (Miyajima et al. 1987). Human α interferon is produced 20 times more (about 50 μg) in a silkworm larva than in a cell culture (Maeda et al. 1985). In the production of a polyhedrin–insulin-like growth factor II (IGF-II) fusion protein, 3.6 mg have been recovered per larva compared with 0.3 mg protein per milliliter of culture medium (Marumoto et al. 1987). However, not all proteins are secreted in the larva. As a commercial venture, the cost of production *in vivo* (silkworm) and *in vitro* (cell culture) may depend on the type of product, the amount of product required, the ease of purification, the purity of the product, the relative convenience of the two methods, and the comparative costs (Miller 1988).

Baculoviral vectors have been applied for the expression of a wide variety of genes to produce (1) immune regulatory proteins, (2) DNA-binding proteins, (3) viral structural proteins, (4) insecticidal products, and (5) other proteins for biological studies. Until 1990, over 100 biologically active proteins have been expressed in cell cultures and about 30 identical or different proteins in the silkworm (Maeda 1989a).

The pharmaceutical products produced, thus far, in insect cells and in the silkworm are almost identical with those produced in humans, and their biological activities are often the same as those of the mature products formed in mammalian cells. Some proteins of medical importance to vertebrates are the interleukins and interferons that are involved in immunity (Smith et al. 1983b, 1985; Maeda et al. 1985; Miyajima et al. 1987); influenza HA protein, which elicits neutralizing protective antibodies in mammals (Kuroda et al. 1986); dengue virus proteins for immunity against dengue encephalitis (Zhang et al. 1988); and the proteins of simian rota virus VP6, parainfluenza virus HN, bluetongue virus VP2, and the human immunodeficiency virus (HIV) for use in diagnosing these diseases (Miller 1988). In 1987, the envelope proteins of HIV, produced by the AcMNPV vector in *S. frugiperda* cells, have received approval for testing as a vaccine against AIDS (acquired immunodeficiency syndrome) (Barnes 1987).

Recombinant viruses may be applied for insect control (Wood and Granados 1991). Foreign genes, which express toxins, enzymes, inhibitors, and insect hormones, may be inserted into the baculoviral genome and enhance the lethal effect of the virus. A recombinant AcNPV carrying the juvenile hormone esterase inhibits the growth of *T. ni* larvae (Hammock et al. 1990). A diuretic hormone has been inserted into the BmMNPV and effectively disturbed the larval fluid metabolism and caused earlier death than with the original virus (Maeda 1989b). The δ-endotoxin of *Bacillus thuringiensis* has been incorporated into the genome of AcMNPV (Merryweather et al. 1990).

VII. GRANULOSIS VIRUSES

Paillot (1926b) was the first to describe a granulosis viral infection (granulosis, singular; granuloses, plural) in the larva of the European cabbageworm, *Pieris brassicae*; he later described three similar diseases in the larvae of *Euxoa segetum* (Paillot 1934, 1935, 1937). The rod-shaped viral particle was detected with the electron microscope by Bergold (1948) in capsules obtained from infected larvae of the pine shoot roller, *Choristoneura* (syn. *Cacoecia*) *murinana*, and by Steinhaus et al. (1949) in capsules from the variegated cutworm, *Peridroma saucia* (syn. *P. margaritosa*). The infection is called granulosis because of the presence of minute granules (Steinhaus 1949). The virus is the granulosis virus (GV) and the occlusion body is the capsule or granule (Fig. 6–15). Wyatt (1952a,b) analyzed the amino acids and nucleotides of the GVs from *C. murinana* and *C. fumiferana* and demonstrated that the viruses contained DNA. Others showed the virus to be ds, supercoiled, covalently closed, circular DNA (Shvedchikova et al. 1969; Summers and Anderson 1972; Tweeten et al. 1977).

Advances in the GVs lagged considerably as compared with those of the NPVs

208 6 DNA-VIRAL INFECTIONS: BACULOVIRIDAE

FIGURE 6–15

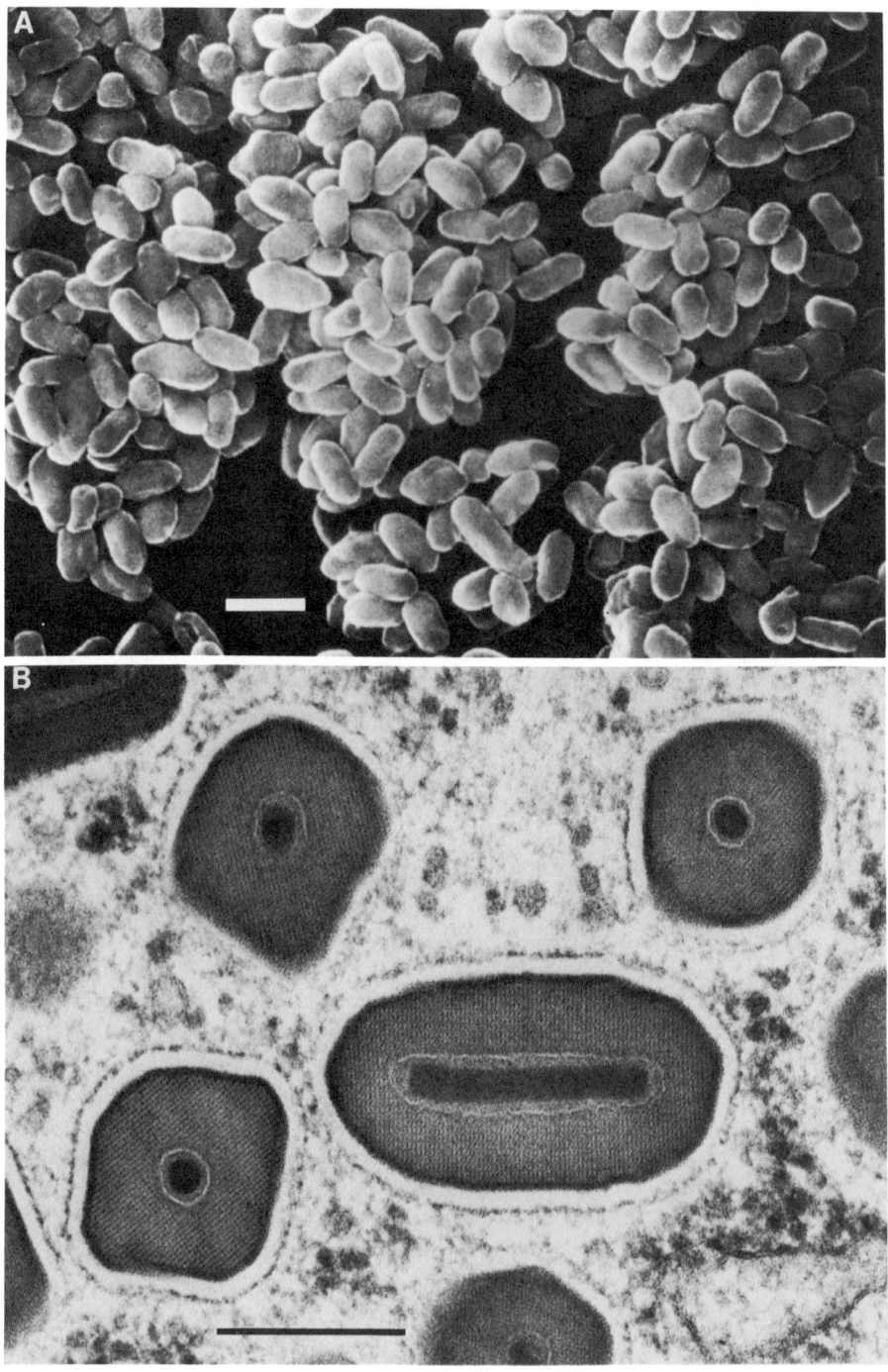

Occlusion bodies (capsules) of a granulosis virus (GV) of fall armyworm, *Spodoptera frugiperda*. (A) Scanning electron micrograph of capsules (bar = 0.5 μm). (B) Thin sections of capsules. Central capsule sectioned longitudinally, others are transversely. (Bar = 0.25 μm.) Note enveloped nucleocapsid in proteinaceous matrix. [(A) provided by J. R. Adams and (B) by C. Y. Kawanishi.]

primarily because of the lack of tissue culture systems. The early success by Vago and Bergoin (1963) with the GV of *Pieris brassicae* in *Lymantria dispar* ovarian cell lines could not be duplicated by others (Rubinstein et al. 1982; Consigli et al. 1986). Available insect cell lines that support NPV replication have been tested for GV replication without success or with the synthesis of only some of the viral-specific proteins (Consigli et al. 1986). However, the recent success by Naser et al. (1984) in propagating the GV of the codling moth in a homologous cell line should stimulate studies on the molecular biology of these viruses.

The type species is the GV of *Plodia interpunctella* (PiGV) (Francki et al. 1991). The GVs are reported to infect only members of Lepidoptera, 113 species, but there is one unconfirmed case of a Hymenoptera, Family Pamphiliidae (Martignoni and Iwai 1986). They are morphologically, biophysically, and biochemically similar to the NPVs. They differ from the NPVs mainly in (1) the number of nucleocapsids in an envelope, (2) the size and shape of the occlusion body, (3) the process of occlusion of the enveloped nucleocapsids by the capsule matrix protein (granulin), (4) the number of enveloped nucleocapsids in an occlusion body, (5) the cytopathology, and (6) to a lesser extent in tissue specificity.

The sizes and shapes of the nucleocapsids and enveloped virions of the GVs are within the range of those of the NPVs. The nucleocapsids acquire envelopes in manners similar to those of the NPVs. There are three ways of envelopment: (1) by morphogenesis or possibly *de novo* synthesis in the nucleus, though usually after the disintegration of the nuclear membrane, (2) by morphogenesis or *de novo* synthesis in the cytoplasm, or (3) by budding through the cell membrane (synhymenosis) (Robertson et al. 1974). Envelopes obtained within the cell may originate from nuclear or intracytoplasmic membranes, which proliferate during viral replication (Tanada and Hess 1984), or from smooth endoplasmic reticulum (Arnott and Smith 1968a; Hunter and Hoffmann 1970). In the NPVs, the envelope gene initiates envelope formation and a similar gene may occur in the GVs. Envelopes obtained by synhymenosis have peplomers and they differ from the envelopes of nucleocapsids that are occluded in capsules. In contrast to NPVs, there is usually one nucleocapsid per envelope and rarely two or more. Pinnock and Hess (1978) report a variant with two nucleocapsids per envelope in the GV of the fruittree leafroller, *Archips argyrospila*.

A. VIRAL OCCLUSION—CAPSULE

The capsules are much smaller (from 120 to 300 × 300 to 500 nm) than the NPV polyhedra and generally differ in shape from them (Fig. 6–15A). Each capsule has one, rarely two or more, enveloped nucleocapsids as contrasted to many enveloped nucleocapsids in the polyhedra (Fig. 6–15B). The capsules are usually ovoid or ovocylindrical, but the shapes may vary greatly, especially in an unusual host in which bizarre forms with many enveloped virions are found (Arnott and Smith 1968b; Hunter and Hoffmann 1972). In some viruses, the capsules are cuboidal, not unlike polyhedra, and may be formed by a specific virus (Stairs 1964; Bird 1976), but they still contain a single virion (Stairs et al. 1966). The capsule matrix protein (granulin), as in the case with polyhedrin, forms a crystalline lattice (Fig. 6–15B). The molecular weights of granulin range from 25,000 to 30,000, comparable to those of polyhedrin (Tweeten et al. 1981). The granulin gene has been isolated and characterized (Akiyoshi et al. 1985; Chakerian et al. 1985).

The occlusion of an enveloped nucleocapsid by granulin begins mostly at one end of the nucleocapsid (Hughes 1952; Stairs et al. 1966; Watanabe and Kobayashi 1970) and rarely at both ends or from the middle (Arnott and Smith 1968a). Like the nucleocapsid of NPV, the opposite ends of the GV nucleocapsid are morphologically different. According to Asayama (1975), the "anterior" end has no structural characteristics, but the "posterior" end has a spicular structure. It is at the anterior end that the initial condensation of granulin takes place. The granulin appears to attach to receptor sites on the viral envelope (Yamamoto et al. 1981).

When the capsule is partially dissolved in alkali, no distinct envelope is present as in the case of the polyhedra. However, an epicapsular layer has been observed in sections of capsules (Arnott and Smith 1969; Hunter and Hoffmann 1970). Moreover, Longworth et al. (1972) have detected a layer with repeating structures and with a hexagonal arrangement of holes 6 nm in diameter, typical of a membranous surface structure. The question of a membrane around the capsule may not be resolved until a capsule envelope gene is located on the GV genome.

In addition to enveloped nucleocapsids, the capsules may occlude extraneous materials [e.g., membranes, vacuoles (Pinnock and Hess 1978; Hess et al. 1984)]. A protease, similar to that present in polyhedra, is found in the capsules and it has been considered to be endogenous (Summers and Smith 1975; Tweeten et al. 1978; Langridge and Balter 1981). The protease, however, appears to be a contaminant from the midgut juice because capsules collected from a larva, whose digestive tract has been removed prior to larval maceration, do not contain protease, whereas capsules collected from the maceration of the undissected larva do contain protease (Nagata and Tanada 1983).

Within the capsules of the GVs of the armyworm, *Pseudaletia unipuncta* (Tanada and Hukuhara 1968, 1971; Tanada et al. 1975; Yamamoto and Tanada 1978c), the clover cutworm, *Scotogramma trifolii* (Stoddard 1980), and the spotted cutworm, *Xestia c-nigrum* (Goto 1990), there is an enhancing factor, a lipoprotein (from MW 93,000 to 126,000) (Yamamoto and Tanada 1978b; Hotchkin 1981). This factor enhances the infection of baculoviruses in lepidopteran larvae (Tanada 1956, 1959; Tanada and Hukuhara 1971). The phospholipid is essential for the enhancement (Yamamoto and Tanada 1978b). The synergistic factor forms about 2.5% of the capsule (Hara et al. 1976) or 5% of the total dissolved capsule components (Yamamoto and Tanada 1978c; Hotchkin 1981). This factor is not found on the surface but deep in the capsule (Yamamoto and Tanada 1980) and is associated with the viral envelope (Zhu et al. 1989).

The protein component of the synergistic factor appears to be virus-coded since it occurs in only one of two GVs that infect the armyworm (Tanada and Hukuhara 1968). The synergistic factor also enhances a few other baculoviral infections *in vitro* (Ohba and Tanada 1984a; Hukuhara and Zhu 1989). It serves as an attachment molecule of the enveloped virion to receptor sites on the cell plasma membrane as indicated by its agglutination of only certain insect cells (Ohba and Tanada 1984b) and by immunofluorescent techniques showing specific binding sites on the plasma membrane of certain cells and on the viral envelope of certain NPVs (Hukuhara and Zhu 1989).

Another enhancing protein, that is similar to the one in the armyworm GV, has been detected in the capsule of *T. ni* (Derksen and Granados 1988). The protein (101 kDa) facilitates the passage of enveloped virions through the peritrophic membrane by dissociating this membrane. The gene encoding this protein has been cloned and sequenced (Wood and Granados 1991).

B. PATHOLOGY AND PATHOGENESIS

The gross pathologies of granuloses are similar to those of nuclear polyhedroses, but differences occur depending on the types of tissues infected. The first indication of infection in the larva is a loss of appetite and a progressive color change from the normal color to a pale whitish or milky-yellow appearance, especially in the ventral side (Huger 1963). The whiteness is due to the abundance of capsules in the hypertrophied fat bodies. When the infection is limited mainly to the fat body, the larva often increases in size, becomes white, opaque, and mottled at an advanced stage of infection, and later has a brownish discoloration (Hamm and Paschke 1963). Such a larva may live longer and become larger than an uninfected one. With the change in color, the larva usually becomes progressively weaker, sluggish, and flaccid.

At an advanced stage of infection, the hemolymph is often milky-white and turbid, owing to the presence of large numbers of viral capsules discharged from disintegrated tissues. Virus-infected cells, when observed with the light microscope, appear opaque and have a yellowish to light-brown coloration (Hughes and Thompson 1951). Certain areas along the edges of the cell excrete spherical vesicles containing capsules that are agitated by Brownian movement. The spherical vesicles are characteristic for granuloses and are called "boules hyalines" (Paillot 1937; Bergold 1949). These vesicles eventually burst, liberating the capsules.

Most nuclear polyhedroses in lepidopteran larvae are systemic with infection occurring in most or all major tissues, whereas most granuloses are confined to one or a few tissues, with the fat body being the major target (Huger 1963). In some granuloses, mitotic proliferation of uninfected cells occurs in the fat body (Huger 1963; Hamm and Paschke 1963). This proliferation and hypertrophy of infected fat cells result in a bloated enlarged larva during the late stage of infection.

In the case of a systemic granulosis, the larva usually dies in a brief period, much shorter than an infection involving mainly the fat body. At death, such a larva is smaller than an uninfected larva. The integument, just prior to or at larval death, is fragile and the dead larva is wilted. This syndrome is similar to that observed in a typical nuclear polyhedrosis of Lepidoptera. Some examples of systemic GV infection are those in *Artogeia* (syn. *Pieris*) *rapae* (Tanada 1953) and in *Cydia pomonella* (Tanada 1964; Tanada and Leutenegger 1968).

In the western grape-leaf skeletonizer, *Harrisinia brillians*, the principal (probably only) site of GV infection is the midgut epithelium (monorganotropic) (Smith et al. 1956; Federici and Stern 1990). This is unlike other GVs, which, like the typical NPV, infect the midgut mainly during the initial stage of infection. Both the larval and adult midguts support complete viral replication, including the formation of capsules (Federici and Stern 1990). The infected midgut cells are discharged into the gut lumen, and virogenesis continues until the cell lysis. The larva develops diarrhea for 2 to 4 days after infection. Larval feces and adult meconium are highly contagious, resulting in horizontal transmission among larvae and vertical transmission from an adult to a larva.

C. CYTOPATHOLOGY

The description of the cytopathology of GV-infected tissues will be confined to the fat body, the major target tissue of most GVs. (Watanabe and Kobayashi 1970) (Fig. 6–16). Viral invasion into a fat cell nucleus is first recognized by the formation of

FIGURE 6-16

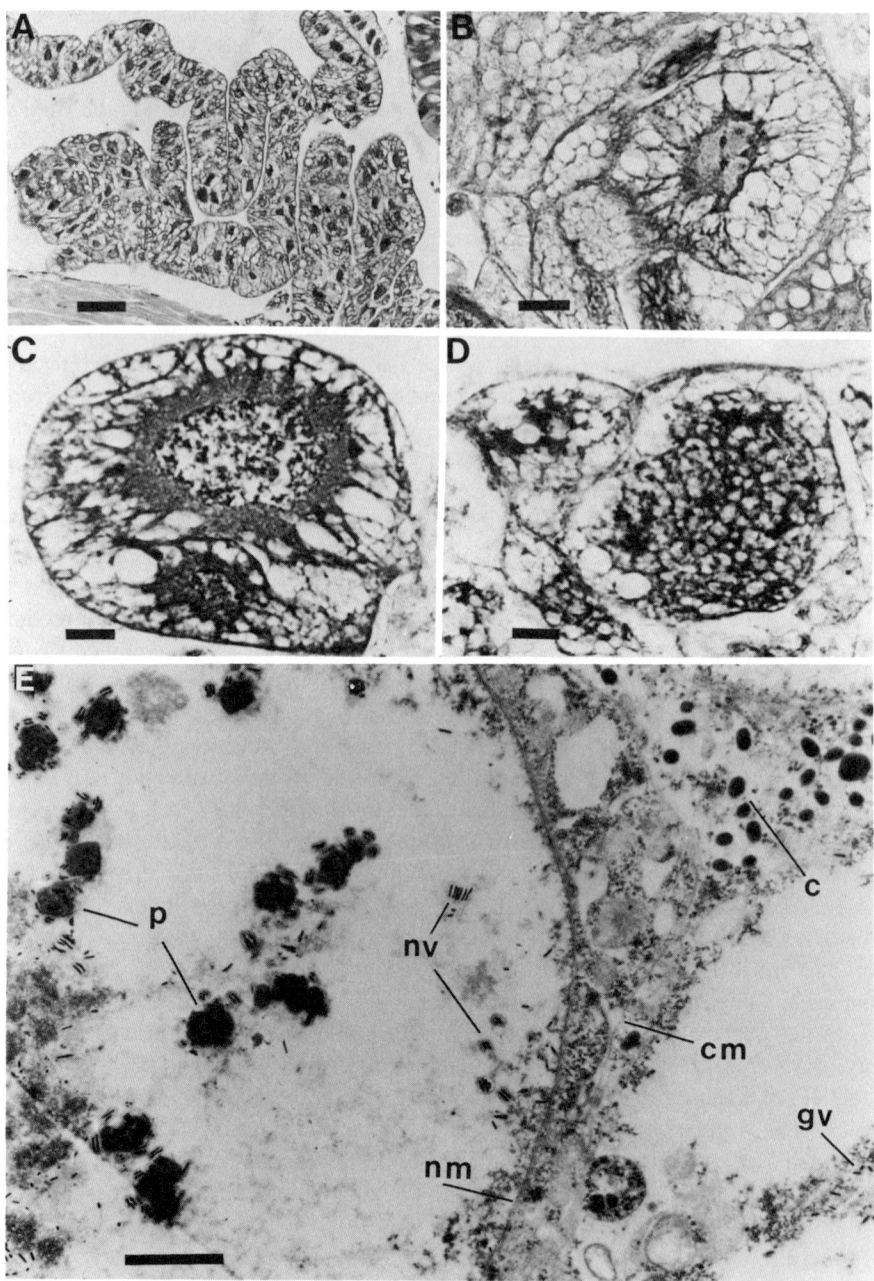

Cytopathological changes in fat body cells of the fall webworm, *Hyphantria cunea*, during the course of granulosis. (A) Two days after infection, showing cellular proliferation. (B) Three days after infection, showing increase in cell size and fat globules. (C) Four days after infection, showing degeneration of the nucleus with granulated chromatin remnants. (D) Eight days after infection, showing the development of the network in the virogenic stroma. Nuclear envelope is indistinct. (E) Section through a cell, showing the simultaneous infection with a granulosis virus in the cytoplasm and a nuclear polyhedrosis virus in the nucleus. (*cm*) cell membrane; (*nm*) nuclear membrane; (*gv*) granulosis virus; (*c*) capsules; (*nv*) nuclear polyhedrosis virus; (*p*) polyhedra. [Bar in (A) = 70 μm, (B) = 15 μm, (C) and (D) = 2 μm, and (E) = 1 μm.] [Courtesy of Watanabe and Kobayashi (1970) and Academic Press.]

small intranuclear blebs of the inner layer (membrane) of the nuclear envelope (Walker et al. 1982). The inner layer subsequently forms intranuclear extensions. The nucleolus increases in size, followed by the swelling of the nucleus (Wäger and Benz 1971). The nucleoplasm shows a less-consolidated heterochromatin, most of which are coalesced and displaced into an eccentric intranuclear position. The remainder of the nucleus has a "cleared" appearance with only small remnants of heterochromatin. Small electron-dense clumps, which are occasionally associated with nuclear pores, appear on the nuclear envelope (Pinnock and Hess 1978; Walker et al. 1982). A virogenic stroma appears and the nuclear envelope begins to break down rapidly (Fig. 6–16D). Paired, stacked cisternae that have ribosomes on both free surfaces of the cisternae are formed (Tanada and Leutenegger 1968; Hunter et al. 1975; Asayama and Inagaki 1975).

While the nuclear alterations are developing, drastic cytopathology occurs in the cytoplasm where the rough endoplasmic reticulum and the mitochondria may assume a ballon shape with fragmented cristae (Tanada and Leutenegger 1968; Asayama and Inagaki 1975). The amount of myelin whorls increases. Lipid and glycogen contents decrease. Cellular junctions, intercellular gap junctions, desmosomes, and hemidesmosomes disappear and the cell may detach from surrounding cells and basal lamina (Walker et al. 1982). The areas vacated by infected cells are apparently taken over by uninfected cells that may have been formed from cell replication.

Annulate lamellae, a characteristic of granulosis, are formed in the cytoplasm (Tanada and Hess 1984). Viral capsules appear about 3 days postinfection (d p.i.) and a high production of capsule constituents by 7 d p.i. Subsequent to the disassembly of the nuclear envelope, the only recognizable portion of the nucleoplasm is the nucleolus-like material. Masses of coiled filaments are interspersed among the capsules (Tanada and Leutenegger 1968; Arnott and Smith 1969; Walker et al. 1982). Membranous profiles and hexagonally packed, tubular membrane arrays appear late in infection. Cell replication increases at an early stage of infection or in apparently uninfected fat cells at the edges of fat lobes (Martignoni 1957; Wittig 1959).

D. PATHOPHYSIOLOGY

Limited studies on the pathophysiology of larvae infected with GVs have been conducted with radioactive isotopes to understand the relationship of DNA, RNA, and protein metabolism. Benz and Wäger (1971) reported that, prior to cytopathological changes, the first reaction of fat body cells of *C. pomonella* is a sharp increase in RNA synthesis (mostly ribosomal RNA) associated with protein synthesis and localized to the swelling nucleolus. Then the nucleolus and chromatin degenerate, followed by the concomitant decrease in RNA and DNA syntheses to normal and subnormal levels. With the appearance of the virogenic stroma, there is a second resurgence of RNA synthesis of relatively long duration (three times normal level) and a tremendous increase of DNA synthesis with a pronounced maximum (30 times normal level) at 60 to 70 h after infection. Similar RNA, DNA, and protein syntheses occur in larval fat body cells of the fall webworm, *Hyphantria cunea*, infected with a GV (Watanabe and Kobayashi 1970). Active syntheses are restricted to the nucleus at an early stage of infection. At a later stage when the nuclear envelope had disrupted, the syntheses are restricted to strands of nuclear envelope.

The electrophoretic patterns of larval hemolymph proteins in granuloses are similar to those in nuclear polyhedroses (Tanada and Watanabe 1971). The similarity between the two types of viruses is mainly because the major target of infection of both viruses is the fat body, the center of hemolymph protein synthesis.

When the Hawaiian strain of a GV infects an armyworm larva, a proteinaceous substance is formed that adversely affects the internal parasitoid *Glyptapanteles militaris* (Kaya 1970; Kaya and Tanada 1972). The parasitoid is not affected when the larva is infected by the Oregonian strain of a GV. The toxin is similar to the one present in an armyworm larva infected with the hypertrophy strain of the NPV. Other braconid and ichneumonid parasitoids are not affected as *G. militaris* (Hotchkin and Kaya 1983b). This toxic protein has a molecular weight of about 64,000 (Hotchkin and Kaya 1985).

E. VIRAL REPLICATION

In general, the mode of GV infection and replication is the same as those of NPV. The ingested GV capsules are dissolved by the highly alkaline midgut juice, and the liberated enveloped virions attach and fuse to the plasma membrane of the microvillus of the columnar midgut cell. The nucleocapsids enter a microvillus, migrate to the nucleus, and attach to the nuclear pores within 2 to 6 h p.i. Summers (1969) was the first to observe nucleocapsids attached to nucleopores and suggested that the uncoating of the nucleic acid occurred at this site. However, nucleocapsids have been observed in the nucleoplasm prior to uncoating and viral replication (Walker et al. 1982). Thus, the uncoating may occur at a nucleopore or within a nucleus. A virion-associated kinase may initiate the uncoating (Wilson and Consigli 1985b).

Virogenesis begins in a nucleus with the formation of the virogenic stroma. The eclipse period is brief with only a partial clearing of the nucleus, followed by the appearance of the virogenic stroma. Capsids appear in 6 to 12 h p.i. and are incorporated with the viral nucleoprotein core (Consigli et al. 1986). Progeny nucleocapsids are formed in 12 to 18 h p.i. in the nucleus in regions of dense, aggregated material distinct from host chromatin (Summers 1971; Hunter et al. 1975). Shortly after the appearance of the nucleocapsids, the nuclear envelope breaks down and virogenesis continues in the nucleus and cytoplasm. The envelopment of the nucleocapsids and their occlusion in capsules also occur in the nucleus and the cytoplasm. In some infected midgut cells, enveloped nucleocapsids singly or in aggregates occur in vesicles and may move toward the basement membrane. About 24 h p.i., enveloped and unenveloped nucleocapsids may occur in continuous rows in intercellular spaces between midgut cells and near the basement membrane. At a late stage, many nucleocapsids are embedded in the basement membrane or budding through the membrane and acquire envelopes with peplomers.

The movement of virions from the midgut epithelium into the hemocoel resembles that of the NPV (Granados and Lawler 1981). The infection of the fat body is by viropexis of nucleocapsids with peplomer envelopes. There is no report of attachment and fusion by nuclear-enveloped nucleocapsids, but they may occur. The occlusion of the virions occurs in the nuclear and cytoplasmic regions of the cell.

The occlusion of a nuclear-enveloped nucleocapsid occurs infrequently in midgut cells but is common in other cells just like the NPV. In capsule formation, the enveloped nucleocapsids, rather than moving to the foci of granulin condensation, are occluded by granulin condensing on the viral envelope.

Early workers separated the GVs into two types based on the site of viral replication (i.e., in the nucleus or in the cytoplasm) (Weiser 1958; Huger 1963). Subsequent studies showed that the two types were essentially the same since virogenesis was initiated in the nucleus and was completed in the nucleoplasm or in the cytoplasm depending on the time of nuclear envelope dissolution. Light autoradiography of fat body cells of *Hyphantria cunea* infected with a GV confirmed that active synthesis of DNA, RNA, and proteins was restricted to the nucleus, even though the nuclear envelope may break at an early stage in virogenesis (Watanabe and Kobayashi 1970).

There are reports of viral replication in the cytoplasm (Huger 1963; Pinnock and Hess 1978). In such an infected cell, the nucleus appears to be unaffected at an early stage of virogenesis. Schmid et al. (1983) report two types of GV infection in the fruit tortrix, *Adoxophyes orana*. Replication occurs 60% in the nucleus and 40% in the cytoplasm of fat body cells. Nuclear replication corresponds to that of the classical GV. In the cytoplasmic type, the virus replicates in cytoplasmic enclaves surrounded by a membrane; the nucleus remains uninfected, but the cytoplasm is severely disrupted. They do not report the presence of a virogenic stroma in the cytoplasm.

F. GENOME AND HYBRIDS

The genomes of GVs are similar to those of NPVs and their sizes fall within the range of the NPVs (90 to 160 kb). A number of structural proteins comparable to those of NPVs have been isolated from GVs (Consigli et al. 1986). There are only a few physical maps for GV genomes [e.g., the GVs of *Cydia pomonella* (Crook et al. 1985) and *Artogeia rapae* (Dwyer and Granados 1987; Smith and Crook 1988)]. The basic protein (VP12) responsible for DNA condensation (Tweeten et al. 1980; Wilson and Consigli 1985a) and the occlusion body matrix protein, granulin, are similar to those of NPVs. The basic VP12 protein may be acted upon by a kinase enzyme during the DNA condensation in the capsid via phosphorylation of the basic protein or during the release of DNA from a nucleocapsid (uncoating) (Wilson and Consigli 1985b).

Granulin genes of *T. ni* GV (Akiyoshi et al. 1985) and *Pieris brassicae* GV (Chakerian et al. 1985) have been cloned and sequenced. Granulin is closely related in structure and function to polyhedrin (Rohrmann 1986a,b). The two occlusion body proteins differ with the granulins having two or three additional amino acids and a conserved cysteine at the N-terminus and the polyhedrins lacking these amino acids and having a conserved proline.

There are genotypic variants of GVs. Isolates collected from different areas have revealed differences in their genomes [e.g., between the GVs of *Pieris brassicae* and *Artogeia rapae* (Crook 1981) and among variants of the GVs of the codling moth, *Cydia pomonella* (Harvey and Volkmann 1983), and *A. rapae* (Smith and Crook 1988; Belloncik et al. 1988)].

G. GENE EXPRESSION

Few studies on the expression of GV genomes have been conducted because of the lack of a suitable tissue-culture system. Cultures of GV-infected fat bodies (Maeda and Tanada 1983) and tissues from GV-infected larva (Dwyer and Granados 1988) indicate a sequential expression of the GV genome as in the case of the NPV genome. During the course of infection in *A. rapae* larva, Dwyer and Granados (1988) reported over 100 size classes of overlapping RNAs and determined the sizes, relative abundance, and map locations of over 100 transcripts. Except for granulin, they were not able to isolate the genes and establish the functions of other expressed proteins since no cross-homology occurred with those of AcMNPV.

VIII. NONOCCLUDED VIRUSES

The nonoccluded baculoviruses (NOB), subfamily Nudibaculovirinae, do not produce occlusion bodies at any stage in their reproductive cycles. The type species of NOB is the *Heliothis* NOB. The NOBs or similar viruses have been described from insects in Homoptera, Coleoptera, Diptera, Lepidoptera, and Hymenoptera and from members of other arthropods (Federici 1986). Only the NOBs of *Oryctes* and those of susceptible cell lines of *Heliothis zea* and *Estigmene acrea* have been characterized in detail. A nonoccluded filamentous virus of the parasitoid *Cotesia marginiventris* is shown in Fig. 6–17.

A. *ORYCTES* NONOCCLUDED VIRUS

The baculovirus that infects the rhinoceros beetle, *Oryctes rhinoceros*, a major pest of coconut, was first described by Huger (1965, 1966) in specimens of the rhinoceros beetle collected in Indonesia. The infection has been referred to as "Malayan disease." The virus does not produce occlusion bodies, but such bodies have been observed in the midgut epithelial cells of the beetle (Monsarrat et al. 1973a).

The DNA of *Oryctes* nonoccluded virus is circular, double stranded with MW 87×10^6 (approximately 130 kbp) and 11 structural polypeptides (Payne 1974). Viral particles are shorter and wider (220×120 nm) than those of other baculoviruses, and the nucleocapsids possess a "tail" structure at one end of the particle (Payne 1974; Payne et al. 1977). The physical map of *Oryctes rhinoceros* nonoccluded virus (OrNOB) has been prepared (Crawford et al. 1985). Isolates of OrNOB from different regions vary in pathogenicity (Zelazny 1979).

The mode of infection of OrNOB is through the mouth and possibly transovarially (Monsarrat et al. 1973a). From the midgut epithelium, the infection spreads to fat body cells. The OrNOB has been detected in the ovarian sheath and testis of the adults (Monsarrat et al. 1973a).

Larvae, pupae, and adults of the rhinoceros beetle are susceptible to OrNOB. An infected larva stops feeding, its hemolymph gradually increases in turbidity, its abdomen turns turbid, glassy, and sometimes pearly in appearance, and the larva develops diarrhea (Huger 1965, 1966). Necrotic melanized areas form in the midgut and fat body lobes (Monsarrat et al. 1973a). The successive disintegration of fat body tissues and an increase in the hemolymph volume produce a dropsy condition and the larva turns translucent. Prior to death, the larva is shiny, beige in color, and appears waxen, especially in the broadened abdominal region. An increase in turgor in the hemocoel usually causes the midgut to herniate. A terminally infected larva often develops a whitish-mottled pattern from the chalky-white bodies of clusters of disintegrating fat cells beneath the integument. Dead larva undergoes putrefaction to become either flaccid and brownish, or successively bluish or bluish black. Less frequently, some larvae may shrink and mummify.

FIGURE 6-17

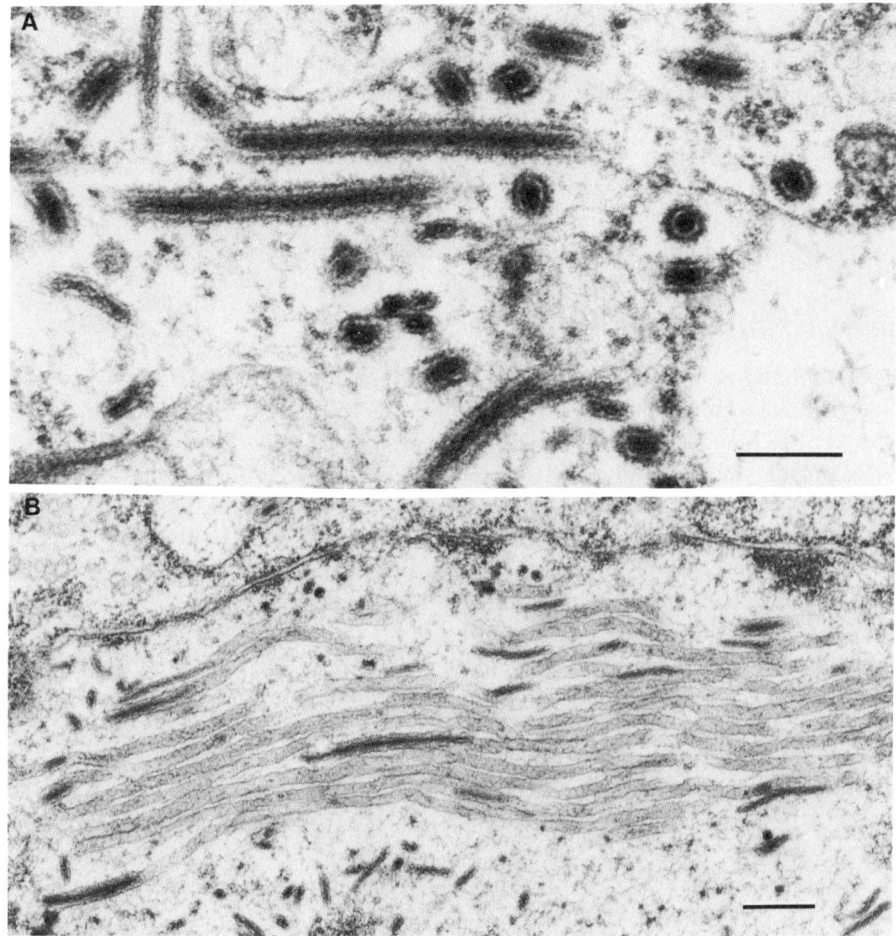

Nonoccluded filamentous virus of the parasitoid *Cotesia marginiventris*. (A) Singly enveloped viral particles in the cytoplasm of an epithelial oviduct cell. (B) Nucleocapsids and tubules in the nucleus of an epithelial calyx cell. Some tubules with nucleocapsids. (Bars = 200 nm.) [Courtesy of Hamm et al. (1990) and Academic Press.]

The OrNOB causes acute and chronic infections depending on the dosage and stage (larva or adult) of the beetle (Huger 1966). The pathologies in the larvae and adults may differ. In acute cases the infection in the larva begins in the midgut epithelium and becomes systemic, and the larva dies in about 6 days. In an acutely infected adult, the initially infected midgut develops a severe hyperplasia, and when the infection spreads to other tissues, the adult becomes immobilized and dies within a week. The infection in a chronically infected larva has been described above. Initially, a chronically infected adult develops only an infection in the midgut, and other tissues are not infected until at the very last stage. An infected adult stops feeding, is diarrheic, and excretes mainly virus. It may live as long as 30 days. The adult is a very productive viral factory that facilitates the dissemination and transmission of the virus (Zelazny 1976). Transmission among adults occurs most frequently during mating and feeding (Zelazny 1976). An infected adult female lays few or no eggs (Zelazny 1977a). These characteristics are the basis for the effective control of the rhinoceros beetle wherever the virus has been introduced (Zelazny 1977b).

The cytopathology in a fat cell of *Oryctes* is similar to that of a GV infection since the nuclear membrane disrupts when the virions appear (Monsarrat et al. 1973a,b). Virogenic stroma forms in the hypertrophied nucleus and may occur also in the cytoplasm (Huger 1966). The midguts of chronically infected adults display the most dramatic result of infection by the extensive proliferation of the replicative nidi which form cells that are infected by the virus (Huger 1972–1973). The severe hyperplasia of midgut cells suggests a malignant carcinoma.

B. *HELIOTHIS* NONOCCLUDED VIRUS

The *Heliothis* nonoccluded virus (Hz-1NOB) isolated from the *Heliothis zea* cell line (IMC-Hz-1) has circular dsDNA with MW 152 to 159 $\times$ 10^6 (approximately 230 kbp) (Granados et al. 1978; Huang et al. 1982). It has 28 structural polypeptides, 14 of which are glycosylated (Burand et al. 1983b). There are three basic genotypes (Huang et al. 1982). The genomes are distinct from those of NPVs and GVs (Langridge 1981; Ralston et al. 1981). The DNA of Hz-1NOB shares 3% homology with *Heliothis armigera* GV and from 0.1 to 1% homology with several NPVs and *Plodia interpunctella* GV (Smith and Summers 1982).

The Hz-1NOB is a persistent virus in tissue culture and is apparently not infectious to larvae of *Estigmene acrea* and *H. zea* by feeding or intrahemocoelic inoculation (Granados et al. 1978). It can be transmitted to a number of lepidopteran cell lines. One percent of the IMC-Hz-1 cells develop cytopathic effects from Hz-1NOB in the form of granulated and shriveled cells. On the other hand, Hz-1NOB causes acute infection in *Trichoplusia ni* cell line (TN-368) with 90 to 100% of the cells developing cytopathic effects (Granados et al. 1978; Ralston et al. 1981).

The replication of Hz-1NOB is generally similar to other baculoviruses, except that viral assembly in the nucleus begins and ends in membrane vesicles (Burand et al. 1986). Virions of Hz-1NOB (enveloped nucleocapsids) enter the cytoplasm after the breakdown of the nuclear envelope but do not exit (exocytosis) through the plasma membrane. This is a major difference from other baculoviruses. The virions are released only after cell lysis, which occurs in a few cells.

In the persistent infection of Hz-1NOB in IMC-Hz-1 cells, there are two types of viral particles, a standard (80 $\times$ 414 nm) and a defective particle (80 $\times$ 370 nm) (Burand et al. 1983a; Wood and Burand 1986). A population of defective viruses always contains a low percentage of standard particles that may be essential for the replication of defective particles (Burand et al. 1983a). On the other hand, the defective particle appears to interfere with the productive replication of the standard particle.

Persistently infected IMC-Hz-1 cells are resistant to superinfection by all baculoviruses tested, while the persistently infected *Trichoplusia ni* cells (TN-368) are susceptible (Wood and Burand 1986). Superinfection with its own virus or other NPVs induces the productive replication of the Hz-1 virus. On the other hand, inoculations of persistently infected TN-368 cells with AcNPV result in AcNPV replication and not Hz-1 virus activation (Ralston et al. 1981). These data indicate that the resistant nature of the IMC-Hz-1 cells is a property of cellular determination (Wood and Burand 1986).

C. OTHER NONOCCLUDED VIRUSES

The orthopteran stick insect, *Bacillus rossius*, is infected by a rod-shaped DNA virus that is not occluded (Scali et al. 1980). The infection occurs in the epithelial cell nuclei at the posterior part of the midgut. The infected nucleus is eventually extruded into the gut lumen. This virus resembles the rod-shaped virus of gyrinid beetle, *Gyrinus natator* (Gouranton 1972), but differs from the NPVs of Lepidoptera. A NOB also infects the midgut cells of the adult *G. natator* (Gouranton 1972). With infection limited to only a portion of midgut cells, the adult survives the infection without apparent symptoms.

Several species of crickets are susceptible to a NOB that resembles the *Oryctes* NOB (Huger 1985). The enveloped virion is 87 × 177 nm. It is infectious by mouth (cannibalism) and intrahemocoelic inoculation. The infected cricket molts repeatedly, is progressively uncoordinated and lethargic, and dies in the last instar. At an advanced stage of infection, the cricket is often swollen with an enormous amount of viscous and milky-opalescent hemolymph.

Nonoccluded baculoviruses have been observed in the aphids *Aphid* sp. and *Pentalonia nigronervosa* (Kitajima et al. 1978). The virions occur in fat body cells and less frequently in adjacent muscle cells. They are found mainly in the nuclei and in a few cases in the cytoplasm.

In the adult flea *Pulex simulans*, the nuclei of midgut columnar epithelial cells are infected by a NOB measuring 30 × 90 nm for the nucleocapsid and 50 × 120 nm for the enveloped virion (Beard et al. 1989). A NOB and a filamentous virus have been observed in the spotted cucumber beetle, *Diabrotica undecimpunctata* (Kim and Kitajima 1984). Each type of virus is associated with a different type of virogenic stroma. The nucleocapsids of NOB, 52 × 230 nm, are found only in hemocytes. The filamentous virus occurs in midgut nuclei and measures from 25 nm up to 2 μm. Some extremely long particles are folded like a hairpin.

The NOB of the fire ant, *Solenopsis* sp., is similar to those of *Oryctes* and *Gyrinus* (Avery et al. 1977). Infection occurs in cells of the fat tissue, mostly in the nuclei but occasionally in the cytoplasm. The honey bee is susceptible to a nonoccluded DNA virus that is filamentous and folded or coiled in the viral envelope (Clark 1978; Bailey et al. 1981). The unfolded nucleocapsid is from 40 to 60 × 3000 to 3060 nm and when folded in the envelope appears ovoid, measuring 0.1 × 0.4 μm. The disease may be endemic or may flare up to destroy a colony (Clark 1978).

Most rod-shaped DNA viruses of braconids and ichneumonids wasps are polydnaviruses, but filamentous nonoccluded viruses (some are short) have been found infecting these parasitoids, either alone or together with polydnavirus (Krell 1987; Stoltz et al. 1988; Hamm et al. 1990) (Fig. 6–17). A virus of the braconid *Cotesia* (syn. *Apanteles*) *melanoscela* replicates in both the parasitoid and its insect host larva (Stoltz and Faulkner 1978; Stoltz et al. 1988). In the braconid *Microplitis croceipes*, infection in adults causes reduced vigor, wing deformation, inability to fly, and early death (Hamm et al. 1988). The virus replicates in the nuclei of fat body cells.

In the syrphid fly *Merodon equestris*, a nonoccluded virus causes severe cellular proliferation of the salivary gland (Amargier et al. 1979). The nucleocapsid is formed in the nucleus and is enveloped in the cytoplasm. Within the intercellular space, the virus seems to have a supplementary membrane. The phantom midge, *Chaoborus crystallinus*, infected with a nonoccluded virus, has yellow discoloration and the infected midgut epithelium shows severe nuclear hypertrophy (Larsson 1984b). Individual rods are connected in chains and are arranged as paracrystalline structures that are most prominent at the nuclear periphery.

REFERENCES

Ackermann, H.-W., and Smirnoff, W. A. 1983. A morphological investigation of 23 baculoviruses. *J. Invertebr. Pathol. 41*, 269–280.

Acqua, C. 1918. Ricerche sulla malattia del giallume nel baco da seta. *Rend. Ist. Bacologico Sc. Super. Agric. Portici 3*, 245–256.

Adams, J. R., and Bonami, J. R. 1991. "Atlas of Invertebrate Viruses." CRC Press, Boca Raton, Florida.

Adams, J. R., and McClintock, J. T. 1991. Baculoviridae. Nuclear Polyhedrosis Viruses. Part 1. Nuclear polyhedrosis viruses of insects. *In* "Atlas of Invertebrate Viruses." (J. R. Adams and J.R. Bonami, eds.), pp. 87–204. CRC Press, Boca Raton, Florida.

Adams, J. R., and Wilcox, T. A. 1968. Histopathology of the almond moth, *Cadra cautella*, infected with a nuclear-polyhedrosis virus. *J. Invertebr. Pathol. 12*, 269–274.

Adams, J. R., Goodwin, R. H., and Wilcox, T. A. 1977. Electron microscopic investigations on invasion and replication of insect baculoviruses *in vivo* and *in vitro*. *Rev. Biol. Cell. 28*, 261–268.

Adang, M. J., and Miller, L. K. 1982. Molecular cloning of DNA complementary to mRNA of the baculovirus *Autographa californica* nuclear polyhedrosis virus: Location and gene products of RNA transcripts found late in infection. *J. Virol. 44*, 782–793.

Aizawa, K. 1959. Mode of multiplication of silkworm nuclear polyhedrosis virus. II. Multiplication of the virus in the early period of the LD_{50}-time curve. *J. Insect Pathol. 1*, 67–74.

Aizawa, K. 1963. The nature of infections caused by nuclear-polyhedrosis viruses. *In* "Insect Pathology: An Advanced Treatise." (E. A. Steinhaus, ed.), Vol. 1, pp. 381–412. Academic Press, New York.

Aizawa, K., and Iida, S. 1963. Nucleic acids extracted from the virus polyhedra of the silkworm, *Bombyx mori* (Linnaeus). *J. Insect Pathol. 5*, 344–348.

Aizawa, K., and Vago, C. 1959. Sur l'infection à Borrelinavirus en culture de tissus d'insectes. *Ann. Inst. Pasteur 96*, 455–460.

Akiyoshi, D., Chakerian, R., Rohrmann, G. F., Nesson, M. H., and Beaudreau, G. S. 1985. Clonng and sequencing of the granulin gene from the *Trichoplusia ni* granulosis virus. *Virology 141*, 328–332.

Allaway, G. P., and Payne, C. C. 1983. A biochemical and biological comparison of three European isolates of nuclear polyhedrosis viruses from *Agrotis segetum*. *Arch. Virol. 75*, 43–54.

Amargier, A., Lyon, J.-P., Vago, C., Meynadier, G., and Veyrunes, J.-C. 1979. Mise en évidence et purification d'une virus dans la prolifération monstrueuse glandulaire d'Insectes. Étude sur *Merodon equestris* F. (Diptère, Syrphidae). *C.R. Acad. Sci. Paris Ser. D289*, 481–484.

Amargier, A., Abol-Ela, S., Vergara, S., Meynadier, G., Martouret, D., and Croizier, G. 1981. Études histologiques et ultrastructurales des larves de *Pandemis heparana* [*Lep. Tortricidae*] au cours des stades avancés d'une baculovirose due à un nouveau virus inducteur de diapause. *Entomophaga 26*, 319–332.

Arif, B. M., and Doerfler, W. 1984. Identification and localization of reiterated sequences in the *Choristoneura fumiferana* MNPV genome. *Eur. Mol. Biol. Organ. J. 3*, 525–529.

Arif, B. M., Tjia, S. T., and Doerfler, W. 1985. DNA homologies between the genomes of *Choristoneura fumiferana* and *Autographa californica* nuclear polyhedrosis viruses. *Virus Res. 2*, 85–94.

Arnott, H. J., and Smith, K. M. 1968a. An ultrastructural study of the development of a granulosis virus in the cells of the moth *Plodia interpunctella* (Hbn.). *J. Ultrastruct. Res. 21*, 251–268.

Arnott, H. J., and Smith, K. M. 1968b. Ultrastructure and formation of abnormal capsules in a granulosis virus of the moth *Plodia interpunctella* (Hbn.). *J. Ultrastruct. Res. 22*, 136–158.

Arnott, H. J., and Smith, K. M. 1969. Ultrastructural observations on the branched rods associated with some insect granuloses. *J. Invertebr. Pathol. 13*, 345–350.

Asayama, T. 1975. Maturation process of the granulosis virus of the diamondback moth, *Plutella xylostella*. *Jpn. J. Appl. Entomol. Zool. 19*, 149–156.

Asayama, T., and Inagaki, I. 1975. Cell alternations caused by the infection with the granulosis virus in the diamondback moth, *Plutella xylostella*, and the site of appearance of nucleocapsid. *Jpn. J. Appl. Entomol. Zool. 19*, 79–84.

Asayama, T., and Kawamoto, F. 1975. An electron microscope observation on the fat body cell of the brown tail moth, *Euproctis similis* Fuessly, infected with a nucleopolyhedrosis virus. *Jpn. J. Appl. Entomol. Zool. 19*, 1–9.

Avery, S. W., Jouvenaz, D. P., Banks, W. A., and Anthony, D. W. 1977. Virus-like particles in a fire ant, *Solenopsis* sp., (Hymenoptera: Formicidae) from Brazil. *Fla. Entomol. 60*, 17–20.

Bailey, L., Carpenter, J. M., and Woods, R. D. 1981. Properties of a filamentous virus of the honey bee (*Apis mellifera*). *Virology 114*, 1–7.

Barnes, D. M. 1987. AIDS vaccine trial OKed. *Science 237*, 973.

Bassemir, U., Miltenburger, H. G., and David, P. 1983. Morphogenesis of nuclear polyhedrosis virus from *Autographa californica* in a cell line from *Mamestra brassicae* (cabbage moth). Further aspects on baculovirus assembly. *Cell Tis. Res. 228*, 587–595.

Beard, C. B., Butler, J. F., and Maruniak, J. E. 1989. A baculovirus in the flea, *Pulex simulans*. *J. Invertebr. Pathol. 54*, 128–131.

Beaton, C. D., and Filshie, B. K. 1976. Comparative ultrastructural studies of insect granulosis and nuclear polyhedrosis viruses. *J. Gen. Virol. 31*, 151–161.

Belloncik, S., Lavallée, C., Mailloux, G., and Arella, M. 1988. Characterization of granulosis virus infection in imported cabbageworm (*Artogeia rapae*) in the province of Québec. *Phytoprotection 69*, 93–98.

Benz, G. 1960. Histopathological changes and histochemical studies on the nucleic acid metabolism in the polyhedrosis-infected gut of *Diprion hercyniae* (Hartig). *J. Insect Pathol. 2*, 259–273.

Benz, G. 1963a. Arthropoden–Viren. *Vierteljahresschr. Naturforsch. Ges. Zurich 108*, 1–35.

Benz, G. 1963b. Physiopathology and histochemistry. *In* "Insect Pathology: An Advanced Treatise." (E. A. Steinhaus, ed.), Vol. 1, pp. 299–338. Academic Press, New York.

Benz, G. 1986. Introduction: Historical perspectives. *In* "The Biology of Baculoviruses." (R. R. Granados and B.A. Federici, eds.), Vol. 1, pp. 1–35. CRC Press, Boca Raton, Florida.

Benz, G., and Wäger, R. 1971. Autoradiographic studies on nucleic acid metabolism in granulosis-infected fat body of larvae of *Carpocapsa*. *J. Invertebr. Pathol. 18*, 70–80.

Bergold, G. H. 1943. Über Polyederkrankheiten bei Insekten. *Biol. Zentralbl. 63*, 1–55.

Bergold, G. H. 1947. Die Isolierung des Polyeder-Virus und die Natur der Polyeder. *Z. Naturforsch. 2b*, 122–143.

Bergold, G. H. 1948. Über die Kapselvirus-Krankheit. *Z. Naturforsch. 3b*, 338–342.

Bergold, G. H. 1953. On the nomenclature and classification of insect viruses. *Ann. N.Y. Acad. Sci. 56*, 495–516.
Bergold, G. H. 1958. Viruses of insects. *In* "Handbuch der Virusforschung." (C. Hallauer and K. F. Meyer, eds.), Vol. 4, pp. 60–142. Springer, Vienna.
Bergold, G. H. 1963a. Fine structure of some insect viruses. *J. Insect. Pathol. 5*, 111–128.
Bergold, G. H. 1963b. The molecular structure of some insect virus inclusion bodies. *J. Ultrastruct. Res. 8*, 360–378.
Bergold, G. H. 1963c. The nature of nuclear-polyhedrosis viruses. *In* "Insect Pathology: An Advanced Treatise." (E. A. Steinhaus, ed.), Vol. 1, pp. 413–456. Academic Press, New York.
Bilimoria, S. L. 1983. Genomic divergence among single-nucleocapsid nuclear polyhedrosis viruses of plusiine hosts. *Virology 127*, 15–23.
Bilimoria, S. L. 1986. Taxonomy and identification of baculoviruses. *In* "The Biology of Baculoviruses. (R.R. Granados and B.A. Federici, eds.), Vol. 1, pp. 37–59. CRC Press, Boca Raton, Florida.
Bilimoria, S. L., and Arif, B. M. 1979. Subunit protein and alkaline protease of entomopoxvirus spheroids. *Virology 96*, 596–603.
Bird, F. T. 1952. On the multiplication of an insect virus. *Biochem. Biophys. Acta. 8*, 360–368.
Bird, F. T. 1976. Effects of mixed infections of two strains of granulosis virus of the spruce budworm, *Choristoneura fumiferana* (Lepidoptera:Tortricidae), on the formation of viral inclusion bodies. *Can. Entomol. 108*, 865–871.
Blissard, G. W., and Rohrmann, G. F. 1989. Location, sequence, transcriptional mapping, and temporal expression of the gp64 envelope glycoprotein gene of the *Orgyia pseudotsugata* multicapsid nuclear polyhedrosis virus. *Virology 170*, 537–555.
Blissard, G. W., and Rohrmann, G. F. 1990. Baculovirus diversity and molecular biology. *Annu. Rev. Entomol. 35*, 127–155.
Blissard, G. W., Quant-Russell, R. L., Rohrmann, G. F., and Beaudreau, G. S. 1989. Nucleotide sequence, transcriptional mapping, and temporal expression of the gene encoding p39, a major structural protein of the multicapsid nuclear polyhedrosis virus of *Orgyia pseudotsugata*. *Virology 168*, 354–362.
Boctor, I. Z. 1980. Free amino acids of the haemolymph of the cotton leaf-worm, *Spodoptera littoralis* Boisduval full-grown larvae, infected with nuclear-polyhedrosis virus. *Experientia 36*, 638–639.
Boctor, I. Z. 1981. Studies on lipids in the hemolymph of the cotton leaf worm, *Spodoptera littoralis*, late final instar larvae, infected with nuclear polyhedrosis virus. *J. Invertebr. Pathol. 38*, 434–436.
Bolle, G. 1894. Il giallume od il mal del grasso del baco da seta. Comunicazione preliminare. *Atti Mem. I. R. Soc. Agrar. Gorizia 34*, 133–136.
Brown, D. A. 1982. Two naturally occurring nuclear polyhedrosis virus variants of *Neodiprion sertifer* Geoffr. (Hymenoptera; Diprionidae). *Appl. Environ. Microbiol. 43*, 65–69.
Brown, M., and Faulkner, P. 1980. A partial genetic map of the baculovirus, *Autographa californica* nuclear polyhedrosis virus, based on recombination studies with *ts* mutants. *J. Gen. Virol. 48*, 247–251.
Brown, D. A., Evans, H. F., Allen, C. J., and Kelly, D. C. 1981. Biological and biochemical investigations on five European isolates of *Mamestra brassicae* nuclear polyhedrosis virus. *Arch. Virol. 69*, 209–217.
Brown, S. E., Maruniak, J. E., and Knudson, D. L. 1984. Physical map of SeMNPV baculovirus DNA: An AcMNPV genomic variant. *Virology 136*, 235–240.
Brown, S. E., Maruniak, J. E., and Knudson, D. L. 1985. Baculovirus (MNPV) genomic variants: Characterization of *Spodoptera exempta* MNPV DNAs and comparison with other *Autographa californica* MNPV DNAs. *J. Gen. Virol. 66*, 2431–2441.

Bud, H. M., and Kelly, D. C. 1980a. An electron microscope study of partially lysed baculovirus nucleocapsids: The intranucleocapsid packaging of viral DNA. *J. Ultrastruct. Res. 73*, 361–368.

Bud, H. M., and Kelly, D. C. 1980b. Nuclear polyhedrosis virus DNA is infectious. *Microbiologica 3, 103–108.*

Burand, J. P., Summers, M. D., and Smith, G. E. 1980. Transfection with baculovirus DNA. *Virology 101*, 286–290.

Burand, J. P., Wood, H. A., and Summers, M. D. 1983a. Defective particles from a persistent baculovirus infection in *Trichoplusia ni* tissue culture cells. *J. Gen. Virol. 64*, 391–398.

Burand, J. P., Stiles, B., and Wood, H. A. 1983b. Structural and intracellular proteins of the nonoccluded baculovirus HZ-1. *J. Virol. 46*, 137–142.

Burand, J. P., Kawanishi, C. Y., and Huang, Y.-S. 1986. Persistent baculovirus infections. *In* "The Biology of Baculoviruses." (R. R. Granados and B. A. Federici, eds.), Vol. 1, pp. 159–175. CRC Press, Boca Raton, Florida.

Burley, S. K., Miller, A., Harrap, K. A., and Kelly, D. C. 1982. Structure of the *Baculovirus* nucleocapsid. *Virology 120*, 433–440.

Carson, D. D., Guarino, L. A., and Summers, M. D. 1988. Functional mapping of an AcNPV immediately early gene which augments expression of the IE-1 *trans*-activated 39K gene. *Virology 162*, 444–451.

Carstens, E. B., Tjia, S. T., and Doerfler, W. 1980. Infectious DNA from *Autographa californica* nuclear polyhedrosis virus. *Virology 101*, 311–314.

Carstens, E. B., Lin-Bai, Y., and Faulkner, P. 1987. A point mutation in the polyhedrin gene of a baculovirus, *Autographa californica* MNPV, prevents crystallization of occlusion bodies. *J. Gen. Virol., 68*, 901–905.

Chakerian, R., Rohrmann, G. F., Nesson, M. H., Leisy, D. J., and Beaudreau, G. S. 1985. The nucleotide sequence of the *Pieris brassicae* granulosis virus granulin gene. *J. Gen. Virol. 66*, 1263–1269.

Chen, D. D., Nesson, M. H., Rohrmann, G. F., and Beaudreau, G. S. 1988. The genome of the multicapsid baculovirus of *Orgyia pseudotsugata*: Restriction map and analysis of two sets of GC-rich repeated sequences. *J. Gen. Virol. 69*, 1375–1381.

Chisholm, G. E., and Henner, D. J. 1988. Multiple early transcripts and splicing of the *Autographa californica* nuclear polyhedrosis virus IE-1 gene. *J. Virol. 62*, 3193–3200.

Cibulsky, R. J., Harper, J. D., and Gudauskas, R. T. 1977a. Biochemical comparison of virion proteins from five nuclear polyhedrosis viruses infecting plusiine larvae (Lepidoptera: Noctuidae). *J. Invertebr. Pathol. 30*, 303–313.

Cibulsky, R. J., Harper, J. D., and Gudauskas, R. T. 1977b. Analysis of deoxyribonucleic acid from five nuclear polyhedrosis viruses infecting plusiinae larvae (Lepidoptera: Noctuidae). *J. Invertebr. Pathol. 30*, 314–317.

Clark, T. B. 1978. A filamentous virus of the honey bee. *J. Invertebr. Pathol. 32*, 332–340.

Clark, T. B., Chapman, H. C., and Fukuda, T. 1969. Nuclear-polyhedrosis and cytoplasmic-polyhedrosis virus infections in Louisiana mosquitoes. *J. Invertebr. Pathol. 14*, 284–286.

Cochran, M. A., and Faulkner, P. 1983. Location of homologous DNA sequences interspersed at five regions in the baculovirus AcMNPV genome. *J. Virol. 45*, 961–970.

Cochran, M. A., Brown, S. E., and Knudson, D. L. 1986. Organization and expression of the baculovirus genome. *In* "The Biology of Baculoviruses." (R. R. Granados and B. A. Federici, eds.), Vol. 1, pp. 239–258. CRC Press, Boca Raton, Florida.

Consigli, R. A., Tweeten, K. A., Anderson, D. K., and Bulla, L. A., Jr. 1983. Granulosis viruses, with emphasis on the GV of the Indian meal moth, *Plodia interpunctella*. *In* "Advances in Virus Research." (M. A. Lauffer and K. Maramorosch, eds.), Vol. 28, pp. 141–173. Academic Press, New York.

Consigli, R. A., Russell, D. L., and Wilson, M. E. 1986. The biochemistry and molecular biology of the granulosis virus that infects *Plodia interpunctella. Curr. Top. Microbiol. Immunol. 131*, 69–101.

Crawford, A. M., and Miller, L. K. 1988. Characterization of an early gene accelerating expression of late genes of the baculovirus *Autographa californica* nuclear polyhedrosis virus. *J. Virol. 62*, 2773–2781.

Crawford, A. M., Ashbridge, K., Sheehan, C., and Faulkner, P. 1985. A physical map of the *Oryctes* baculovirus genome. *J. Gen. Virol. 66*, 2649–2658.

Croizier, G., and Quiot, J. M. 1981. Obtention and analysis of two genetic recombinants of baculoviruses of Lepidoptera, *Autographa californica* Speyer and *Galleria mellonella* L. *Ann. Virol. 132E*, 3–18.

Croizier, G., Godse, D., and Vlak, J. 1980a. Sélection de types viraux dans les infections doubles à *Baculovirus* chez les larves de Lépidoptère. *C.R. Acad. Sci. Paris Ser. D290*, 579–582.

Croizier, G., Amargier, A., Godse, D.-B., Jacquemard, P., and Duthoit, J.-L. 1980b. Un virus de polyedrose nucléaire découvert chez le lépidoptère Noctuidae *Diparopsis watersi* (Roth.) nouveau variant du *Baculovirus* d'*Autographa californica* (Speyer). *Cot. Fib. Trop. 35*, 415–423.

Croizier, G., Boukhoudmi-Amiri, K., and Croizier, L. 1989. A physical map of *Spodoptera littoralis* B-type nuclear polyhedrosis virus genome. *Arch. Virol. 104*, 145–151.

Crook, N. E. 1981. A comparison of the granulosis viruses from *Pieris brassicae* and *Pieris rapae*. *Virology 115*, 173–181.

Crook, N. E., Spencer, R. A., Payne, C. C., and Leisy, D. J. 1985. Variation in *Cydia pomonella* granulosis virus isolates and physical maps of the DNA from three variants. *J. Gen. Virol. 66*, 2423–2430.

Derkson, A.C.G., and Granados, R. R. 1988. Alteration of a lepidopteran peritrophic membrane by baculoviruses and enhancement of viral infectivity. *Virology 167*, 242–250.

Diaz, M., and Pavan, C. 1965. Changes in chromosomes induced by microorganism infection. *Proc. Natl. Acad. Sci. U.S.A. 54*, 1321–1327.

Diener, T. D. 1983. Viroids. *In* "Advances in Virus Research." (M. A. Lauffer and K. Maramorosch, eds.), Vol. 28, pp. 241–283. Academic Press, New York.

Dobrovolskaya, G. N., Kok, I. P., Smirnova, I. A., and Chistyakova, A. V. 1965. Biological activity of DNA preparations isolated from silkworm tissues infected with nuclear polyhedrosis virus. *Mikrobiol. Zh.* (Kiev) 27, 73–77.

Doerfler, W., and Böhm, P. 1986. The molecular biology of baculoviruses. *Curr. Top. Microbiol. Immunol. 131*. Springer-Verlag, Berlin.

Dougherty, E. M., Vaughn, J. L., and Reichelderfer, C. F. 1975. Characteristics of the nonoccluded form of a nuclear polyhedrosis virus. *Intervirology 5*, 109–121.

Dwyer, K. G., and Granados, R. R. 1987. A physical map of the *Pieris rapae* granulosis virus genome. *J. Gen. Virol. 68*, 1471–1476.

Dwyer, K. G., and Granados, R. R. 1988. Mapping *Pieris rapae* granulosis virus transcripts and their in vitro translation products. *J. Virol. 62*, 1535–1542.

Egawa, K., and Summers, M. D. 1972. Solubilization of *Trichoplusia ni* granulosis virus proteinic crystal. *J. Invertebr. Pathol. 19*, 395–404.

Elliott, R. M., and Kelly, D. C. 1977. The polyamine content of a nuclear polyhedrosis virus from *Spodoptera littoralis*. *Virology 76*, 472–474.

Elliott, R. M., and Kelly, D. C. 1979. Compartmentalization of the polyamines contained by a nuclear polyhedrosis virus from *Heliothis zea*. *Microbiologica 2*, 409–413.

Engström, A., and Charpentier, R. 1976. Further X-ray studies of nuclear-polyhedrosis inclusions. *Naturwissenschaften 63*, 91.

Engström, A., and Kilkson, R. 1968. Molecular organization in the polyhedra of *Porthetria dispar* nuclear-polyhedrosis. *Exp. Cell Res. 53*, 305–310.

Entwistle, P. F., and Robertson, J. S. 1968. An unusual nuclear-polyhedrosis virus from larvae of a hepialid moth. *J. Invertebr. Pathol.* 11, 487–495.

Eppstein, D. A., Thoma, J. A., Scott, H. A., and Young, S. Y., III. 1975. Degradation of matrix protein from a nuclear-polyhedrosis virus of *Trichoplusia ni* by an endogenous protease. *Virology* 67, 591–594.

Erlandson, M., Skepasts, P., Kuzio, J., and Carstens, E. B. 1984. Genomic variants of a temperature-sensitive mutant of *Autographa californica* nuclear polyhedrosis virus containing specific reiterations of viral DNA. *Virus Res.* 1, 565–584.

Esche, H., Lübbert, H., Siegmann, B., and Doerfler, W. 1982. The translational map of the *Autographa californica* nuclear polyhedrosis virus (AcNPV) genome. *Embo J.* 1, 1629–1633.

Escherich, K., and Miyajima, M. 1911. Studies uber die Wipfelkrankheit der Nonne. *Naturwiss. Z. Forst-Landwirt.* 9, 381–402.

Falcon, L. A., and Hess, R. T. 1977. Electron microscope study on the replication of *Autographa* nuclear polyhedrosis virus and *Spodoptera* nuclear polyhedrosis virus in *Spodoptera exigua*. *J. Invertebr. Pathol.* 29, 36–43.

Faulkner, P. 1962. Isolation and analysis of ribonucleic acid from inclusion bodies of the nuclear polyhedrosis of the silkworm. *Virology* 16, 479–484.

Faulkner, P., and Henderson, J. F. 1972. Serial passage of a nuclear polyhedrosis disease virus of the cabbage looper (*Trichoplusia ni*) in a continuous tissue culture cell line. *Virology* 50, 920–924.

Federici, B. A. 1980. Mosquito baculovirus: Sequence of morphogenesis and ultrastructure of the virion. *Virology* 100, 1–9.

Federici, B. A. 1986. Ultrastructure of baculoviruses. *In* "The Biology of Baculoviruses." (R. R. Granados and B. A. Federici, eds.), Vol. 1, pp. 61–88. CRC Press, Boca Raton, Florida.

Federici, B. A., and Anthony, D.W. 1972. Formation of virion-occluding proteinic spindles in a baculovirus disease of *Aedes triseriatus*. *J. Invertebr. Pathol.* 20, 129–138.

Federici, B. A., and Lowe, R. E. 1972. Studies on the pathology of a baculovirus in *Aedes triseriatus*. *J. Invertebr. Pathol.* 20, 14–21.

Federici, B. A., and Stern, V. M. 1990. Replication and occlusion of a granulosis virus in larval and adult midgut epithelium of the western grapeleaf skeletonizer, *Harrisina brillians*. *J. Invertebr. Pathol.* 56, 401–414.

Federici, B. A., Granados, R. R., Anthony, D. W., and Hazard, E. I. 1974. An entomopoxvirus and nonoccluded virus-like particles in larvae of the chironomid *Goeldichironomus holoprasinus*. *J. Invertebr. Pathol.* 23, 117–120.

Fischer, E. 1906. Über die Ursachen der Disposition und uber Frühsymptome der Raupenkrankheiten. *Biol. Zentralbl.* 26, 534–544.

Francki, R.I.B., Fauquet, C. M., Knudson, D. L., and Brown, F. 1991. "Classification and Nomenclature of Viruses. Fifth Report of the International Committee on Taxonomy of Viruses." Arch. Virol. Suppl. 2. Springer-Verlag, Wien.

Fraser, M. J. 1986a. Ultrastructural observations of virion maturation in *Autographa californica* nuclear polyhedrosis virus infected *Spodoptera frugiperda* cell cultures. *J. Ultrastruct. Mol. Struct. Res.* 95, 189–195.

Fraser, M. J. 1986b. Transposon-mediated mutagenesis of baculoviruses: Transposon-shuttling and implications for speciation. *Ann. Entomol. Soc. Am.* 79, 773–783.

Fraser, M. J., and Hink, W. F. 1982. The isolation and characterization of the MP and FP plaque variants of *Galleria mellonella* nuclear polyhedrosis virus. *Virology* 117, 366–378.

Fraser, M. J., and McCarthy, W. J. 1984. The detection of FP plaque variants of *Heliothis zea* nuclear polyhedrosis virus grown in the IPLB-HZ 1075 insect cell line. *J. Invertebr. Pathol.* 43, 427–429.

Fraser, M. J., Smith, G. E., and Summers, M. D. 1983. Acquisition of host cell DNA sequences by baculoviruses: Relationship between host DNA insertions and FP mutants of *Autographa californica* and *Galleria mellonella* nuclear polyhedrosis viruses. *J. Virol. 47*, 287–300.

Fujiwara, T., Yukibuchi, E., Tanaka, Y., Yamamoto, Y., Tomita, K., and Hukuhara, T. 1984. X-ray diffraction studies of polyhedral inclusion bodies of nuclear and cytoplasmic polyhedrosis viruses. *Appl. Entomol. Zool. 19*, 402–403.

Gaw, Z.-Y., Liu, N. T., and Zia, T. U. 1959. Tissue culture methods for cultivation of virus Grasserie. *Acta Virol. 3*, 55–60.

Gettig, R. R., and McCarthy, W. J. 1982. Genotypic variation among wild isolates of *Heliothis* spp. nuclear polyhedrosis viruses from different geographical regions. *Virology 117*, 245–252.

Gipson, I., and Scott, H. A. 1975. An electron microscope study of effects of various fixatives and thin-section enzyme treatments on a nuclear polyhedrosis virus. *J. Invertebr. Pathol. 26*, 171–179.

Glaser, R. W. 1915. Wilt of gipsy-moth caterpillars. *J. Agric. Res. 4*, 101–128.

Glaser, R. W. 1918. The polyhedral virus of insects with a theoretical consideration of filterable viruses generally. *Science 48*, 301–302.

Glaser, R. W., and Chapman, J. W. 1913. The wilt disease of gipsy moth caterpillars. *J. Econ. Entomol. 6*, 479–488.

Glaser, R. W., and Chapman, J. W. 1916. The nature of the polyhedral bodies found in insects. *Biol. Bull. 30*, 367–391.

Goldstein, N. I., and McIntosh, A. H. 1980. Glycoproteins of nuclear polyhedrosis viruses. *Arch. Virol. 64*, 119–126.

Gombart, A. F., Pearson, M. N., Rohrmann, G. F., and Beaudreau, G. S. 1989. A baculovirus polyhedral envelope-associated protein: Genetic location, nucleotide sequence, and immunocytochemical characterization. *Virology 169*, 182–193.

Goodwin, R. H. 1968. Use of the term "virus inclusion body" and vernacular virus names. *J. Invertebr. Pathol. 12*, 479–480.

Goodwin, R. H., Vaughn, J. L., Adams, J. R., and Louloudes, S. J. 1970. Replication of a nuclear polyhedrosis virus in an established insect cell line. *J. Invertebr. Pathol. 16*, 284–287.

Goodwin, R. H., Vaughn, J. L., Adams, J. R., and Louloudes, S. J. 1973. The influence of insect cell lines and tissue-culture media on *Baculovirus* polyhedra production. *Misc. Publ. Entomol. Soc. Am. 9*, 66–72.

Goto, C. 1990. Enhancement of a nuclear polyhedrosis virus (NPV) infection by a granulosis virus (GV) isolated from the spotted cutworm, *Xestia c-nigrum* L. (Lepidoptera: Noctuidae). *Appl. Entomol. Zool. 25*, 135–137.

Gouranton, J. 1972. Development of an intranuclear nonoccluded rod-shaped virus in some midgut cells of an adult insect, *Gyrinus natator* L. (Coleoptera). *J. Ultrastruct. Res. 39*, 281–294.

Grace, T. D. C. 1958. Induction of polyhedral bodies in ovarian tissues of the tussock moth in vitro. *Science 128*, 249–250.

Grace, T. D. C. 1967. Insect cell culture and virus research. *In Vitro 3*, 104–117.

Granados, R. R. 1978. Early events in the infection of *Heliothis zea* midgut cells by a baculovirus. *Virology 90*, 170–174.

Granados, R. R., and Federici, B. A. 1986. "The Biology of Baculoviruses." Vol. 1 and 2. CRC Press, Boca Raton, Florida.

Granados, R. R., and Lawler, K. A. 1981. *In vivo* pathway of *Autographa californica* baculovirus invasion and infection. *Virology 108*, 297–308.

Granados, R. R., and Williams, K. A. 1986. In vivo infection and replication of baculoviruses. *In* "The Biology of Baculoviruses." (R. R. Granados and B. A. Federici, eds.), Vol. 1, pp. 89–108. CRC Press, Boca Raton, Florida.

Granados, R. R., Nguyen, T., and Cato, B. 1978. An insect cell line persistently infected with a baculovirus-like particle. *Intervirology 10*, 309–317.
Granados, R. R., Lawler, K. A., and Burand, J. P. 1981. Replication of *Heliothis zea* baculovirus in an insect cell line. *Intervirology 16*, 71–79.
Gröner, A. 1986. Specificity and safety of baculoviruses. *In* "The Biology of Baculoviruses." (R. R. Granados and B. A. Federici, eds.), Vol. 1, pp. 177–202. CRC Press, Boca Raton, Florida.
Grula, M. A., Buller, P. L., and Weaver, R. F. 1981. α-amanitin-resistant viral RNA synthesis in nuclei isolated from nuclear polyhedrosis virus-infected *Heliothis zea* larvae and *Spodoptera frugiperda* cells. *J. Virol. 38*, 916–921.
Guarino, L. A., and Summers, M. D. 1988. Functional mapping of *Autographa californica* nuclear polyhedrosis virus genes required for late gene expression. *J. Virol. 62*, 463–471.
Guarino, L. A., Gonzalez, M. A., and Summers, M. D. 1986. Complete sequence and enhancer function of the homologous DNA regions of *Autographa californica* nuclear polyhedrosis virus. *J. Virol. 60*, 224–229.
Guelpa, B., Bergoin, M., and Croizier, G. 1977. La protéine d'inclusion et les protéines du virion du Baculovirus du diptère *Tipula paludosa* (Meigen). *C.R. Acad. Sci. Paris Ser. D. 284*, 779–782.
Hall, D. W., and Fish, D. D. 1974. A *Baculovirus* from the mosquito *Wyeomyia smithii*. *J. Invertebr. Pathol. 23*, 383–388.
Hamm, J. J., and Paschke, J. D. 1963. On the pathology of a granulosis virus of the cabbage looper, *Trichoplusia ni* (Hübner). *J. Insect Pathol. 5*, 187–197.
Hamm, J. J., Styer, E. L., and Lewis, W. J. 1988. A baculovirus pathogenic to the parasitoid *Microplitis croceipes* (Hymenoptera: Braconidae). *J. Invertebr. Pathol. 52*, 189–191.
Hamm, J. J., Styer, E. L., and Lewis, W. J. 1990. Comparative virogenesis of filamentous virus and polydnavirus in the female reproductive tract of *Cotesia marginiventris* (Hymenoptera: Braconidae). *J. Invertebr. Pathol. 55*, 357–374.
Hammock, B. D., Bonning, B. C., Possee, R. D., Hanzlik, T. N., and Maeda, S. 1990. Expression and effects of the juvenile hormone esterase in a baculovirus vector. *Nature 344*, 458–461.
Hara, S., Tanada, Y., and Omi, E. M. 1976. Isolation and characterization of a synergistic enzyme from the capsule of a granulosis virus of the armyworm, *Pseudaletia unipuncta*. *J. Invertebr. Pathol. 27*, 115–124.
Harper, J. D. 1976. Cross-infectivity of six plusiine nuclear polyhedrosis virus isolates to plusiine hosts. *J. Invertebr. Pathol. 27*, 275–277.
Harrap, K. A. 1970. Cell infection by a nuclear polyhedrosis virus. *Virology 42*, 311–318.
Harrap, K. A. 1972a. The structure of nuclear polyhedrosis viruses. I. The inclusion body. *Virology 50*, 114–123.
Harrap, K. A. 1972b. The structure of nuclear polyhedrosis viruses. II. The virus particle. *Virology 50*, 124–132.
Harrap, K. A., and Robertson, J. S. 1968. A possible infection pathway in the development of a nuclear polyhedrosis virus. *J. Gen. Virol. 3*, 221–225.
Harvey, J. P., and Volkman, L. E. 1983. Biochemical and biological variation of *Cydia pomonella* (codling moth) granulosis virus. *Virology 124*, 21–34.
Heimpel, A. M., and Adams, J. R. 1966. A new nuclear polyhedrosis of the cabbage looper, *Trichoplusia ni*. *J. Invertebr. Pathol. 8*, 340–346.
Henderson, J. F., Faulkner, P., and MacKinnon, E. A. 1974. Some biophysical properties of virus present in tissue cultures infected with the nuclear polyhedrosis virus of *Trichoplusia ni*. *J. Gen. Virol. 22*, 143–146.
Hess, R. T., Falcon, L. A., and Morris, T. J. 1984. Electron microscope observations on the specificity of inclusion bodies of insect polyhedrosis viruses. *J. Invertebr. Pathol. 43*, 1–10.

Himeno, M., Yasuda, S., Kohsaka, T., and Onodera, K. 1968. The fine structure of a nuclear-polyhedrosis virus of the silkworm. *J. Invertebr. Pathol. 11*, 516–519.

Himeno, M., Kimura, Y., and Hayashiya, K. 1976. Nuclei acids synthesis of nuclear polyhedrosis virus in cultured embryonic cells of silkworm. *Agric. Biol. Chem. 40*, 1457–1462.

Hink, W. F., and Hall, R. L. 1989. Recently established invertebrate cell lines. *In* "Invertebrate Cell System Applications." (J. Mitsuhashi, ed.), Vol. 2, pp. 269–293. CRC Press, Boca Raton, Florida.

Hink, W. F., and Strauss, E. 1976. Replication and passage of alfalfa looper nuclear polyhedrosis virus plaque variants in cloned cell cultures and larval stages of four host species. *J. Invertebr. Pathol. 27*, 49–55.

Hink, W. F., and Vail, P. V. 1973. A plaque assay for titration of alfalfa looper nuclear polyhedrosis virus in a cabbage looper (TN-368) cell line. *J. Invertebr. Pathol. 22*, 168–174.

Hirumi, H., Hirumi, K., and McIntosh, A. H. 1975. Morphogenesis of a nuclear polyhedrosis virus of the alfalfa looper in a continuous cabbage looper cell line. *Ann. N.Y. Acad. Sci. 266*, 302–326.

Holmes, F. D. 1948. Order Virales the filterable viruses. *In* "Bergey's Manual of Determinative Bacteriology." (R. S. Breed, E.G.D. Murray and A. P. Hitchens, eds.), pp. 1125–1286. Sixth Ed., Williams & Wilkins, Baltimore.

Hotchkin, P. G. 1981. Comparison of virion proteins and granulin from a granulosis virus produced in two host species. *J. Invertebr. Pathol. 38*, 303–304.

Hotchkin, P. G., and Kaya, H. K. 1983a. Pathological response of the parasitoid, *Glyptapanteles militaris*, to nuclear polyhedrosis virus-infected armyworm hosts. *J. Invertebr. Pathol. 42*, 51–61.

Hotchkin, P. G., and Kaya, H. K. 1983b. Interactions between two baculoviruses and several insect parasites. *Can. Entomol. 115*, 841–846.

Hotchkin, P. G., and Kaya, H. K. 1985. Isolation of an agent affecting the development of an internal parasitoid. *Arch. Insect Biochem. Physiol. 2*, 375–384.

Huang, Y.-S., Hedberg, M., and Kawanishi, C. Y. 1982. Characterization of the DNA of a nonoccluded baculovirus, Hz-1V. *J. Virol. 43*, 174–181.

Huger, A. 1963. Granuloses of insects. *In* "Insect Pathology: An Advanced Treatise." (E. A. Steinhaus, ed.), Vol. 1, pp. 531–575. Academic Press, New York.

Huger, A. 1965. Ein neuer Typ von Insektenviren aus malaiischen Populationen von *Oryctes rhinoceros* (L.) (*Col., Scarabaeidae*). *Naturwissenschaften 52*, 542–543.

Huger, A. M. 1966. A virus disease of the Indian rhinoceros beetle, *Oryctes rhinoceros* (Linnaeus), caused by a new type of insect virus, *Rhabdionvirus oryctes* gen. n., sp. n. *J. Invertebr. Pathol. 8*, 38–51.

Huger, A. M. 1972–1973. Grundlagen zur biologischen Bekämpfung des Indischen Nashornkäfers, *Oryctes rhinoceros* (L.), mit *Rhabdionvirus oryctes*: Histopathologie der Virose bei Käfern. *Z. Angew. Entomol. 72*, 309–319.

Huger, A. M. 1985. A new virus disease of crickets (Orthoptera: Gryllidae) causing macronucleosis of fatbody. *J. Invertebr. Pathol. 45*, 108–111.

Huger, A., and Krieg, A. 1961. Electron microscope investigations on the virogenesis of the granulosis of *Choristoneura murinana* (Hübner). *J. Insect Pathol. 3*, 183–196.

Hughes, K. M. 1952. Development of the inclusion bodies of a granulosis virus. *J. Bacteriol. 64*, 375–380.

Hughes, K. M. 1953. The development of an insect virus within cells of its host. *Hilgardia 22*, 391–406.

Hughes, K. M. 1972. Fine structure and development of two polyhedrosis viruses. *J. Invertebr. Pathol. 19*, 198–207.

Hughes, K. M. 1978. The macromolecular lattices of polyhedra. *J. Invertebr. Pathol. 31*, 217–224.

Hughes, K. M. 1979. Some interactions of two baculoviruses of the Douglas-fir tussock moth (Lepidoptera: Lymantriidae). *Can. Entomol. 111*, 521–523.

Hughes, K. M., and Thompson, C. G. 1951. A granulosis of the omnivorous looper, *Sabulodes caberata* Guenée. *J. Infect. Dis. 89*, 173–179.

Hughes, P. R., Gettig, R. R., and McCarthy, W. J. 1983. Comparison of the time-mortality response of *Heliothis zea* to 14 isolates of *Heliothis* nuclear polyhedrosis virus. *J. Invertebr. Pathol. 41*, 256–261.

Hukuhara, T., and Zhu, Y. 1989. Enhancement of the in vitro infectivity of a nuclear polyhedrosis virus by a factor in the capsule of a granulosis virus. *J. Invertebr. Pathol. 54*, 71–78.

Hunter, D. K., and Hoffmann, D. F. 1970. A granulosis virus of the almond moth, *Cadra cautella*. *J. Invertebr. Pathol. 16*, 400–407.

Hunter, D. K., and Hoffmann, D. F. 1972. Cross infection of a granulosis virus of *Cadra cautella*, with observations on its ultrastructure in infected cells of *Plodia interpunctella*. *J. Invertebr. Pathol. 20*, 4–10.

Hunter, D. K., Hoffmann, D. F., and Collier, S. J. 1975. Observations on a granulosis virus of the potato tuberworm, *Phthorimaea operculella*. *J. Invertebr. Pathol. 26*, 397–400.

Ignoffo, C. M. 1966. Susceptibility of the first-instar of the bollworm, *Heliothis zea*, and the tobacco budworm, *Heliothis virescens*, to *Heliothis* nuclear-polyhedrosis virus. *J. Invertebr. Pathol. 8*, 531–536.

Ignoffo, C. M. 1968. Specificity of insect viruses. *Bull. Entomol. Soc. Am. 14*, 265–276.

Ignoffo, C. M., and McIntosh, A. H. 1986. Comparison of in vivo infectivity of tissue-cultured nonoccluded virus and alkali-liberated occluded virus of *Baculovirus heliothis*. *J. Invertebr. Pathol. 48*, 289–295.

Ignoffo, C. M., Hostetter, D. L., and Shapiro M. 1974. Efficacy of insect viruses propagated *in vivo* and *in vitro*. *J. Invertebr. Pathol. 24*, 184–187.

Ignoffo, C. M., McIntosh, A. H., Huettel, M. D., and Garcia, C. 1985. Relative susceptibility of *Heliothis subflexa* (Guenée) (Lepidoptera: Noctuidae) and other species of *Heliothis* to nonoccluded *Baculovirus heliothis*. *Ann. Entomol. Soc. Am. 78*, 740–743.

Injac, M., Vago, C., Duthoit, J.-L., and Veyrunes, J.-C. 1971. Libération (Release) des virions dans les polyédroses nucléaires. *C.R. Acad. Sci. Paris Ser. D273*, 439–441.

Injac, M., Duthoit, J.-L., and Amargier, A. 1973. Étude histo- et cytopathologique d'une polyédrose nucléaire de l'écaille fileuse (*Hyphantria cunea* Drury) *Lépidoptera*, *Arctiidae*. *Ann. Zool. Écol. Anim. 5*, 99–109.

Johnson, D. W., and Maruniak, J. E. 1989. Physical map of *Anticarsia gemmatalis* nuclear polyhedrosis virus (AgMNPV-2) DNA. *J. Gen. Virol. 70*, 1877–1883.

Kang, C. Y. 1988. Baculovirus vectors for expression of foreign genes. *Adv. Virus Res. 35*, 177–192.

Kawamoto, F., and Asayama, T. 1975. Studies on the arrangement patterns of nucleocapsids within the envelopes of nuclear-polyhedrosis virus in the fat-body cells of the brown tail moth, *Euproctis similis*. *J. Invertebr. Pathol. 26*, 47–55.

Kawamoto, F., Asayama, T., and Kobayashi, M. 1976. Acquisition of the envelope of nuclear polyhedrosis viruses in the Chinese oak silkworm, *Antheraea pernyi* Guer-Min, and the Japanese giant silkworm, *Dictyoploca japonica* Butler. *Appl. Entomol. Zool. 11*, 59–69.

Kawamoto, F., Kumada, N., and Kobayashi, M. 1977a. Envelopment of the nuclear polyhedrosis virus of the oriental tussock moth, *Euproctis subflava*. *Virology 77*, 867–871.

Kawamoto, F., Suto, C., Kumada, N., and Kobayashi, M. 1977b. Cytoplasmic budding of a nuclear polyhedrosis virus and comparative ultrastructural studies of envelopes. *Microbiol. Immunol. 21*, 255–265.

Kawanishi, C. Y., Egawa, K., and Summers, M. D. 1972a. Solubilization of *Trichoplusia ni* granulosis virus proteinic crystal. II. Ultrastructure. *J. Invertebr. Pathol. 20*, 95–100.

Kawanishi, C. Y., Summers, M. D., Stoltz, D. B., and Arnott, H. J. 1972b. Entry of an insect virus *in vivo* by fusion of viral envelope and microvillus membrane. *J. Invertebr. Pathol. 20*, 104–108.

Kawarabata, T. 1974. Highly infectious free virions in the hemolymph of the silkworm (*Bombyx mori*) infected with a nuclear polyhedrosis virus. *J. Invertebr. Pathol. 24*, 196–200.

Kaya, H. K. 1970. Toxic factor produced by a granulosis virus in armyworm larva: Effect on *Apanteles militaris. Science 168*, 251–253.

Kaya, H. K., and Tanada, Y. 1972. Response of *Apanteles militaris* to a toxin produced in a granulosis-virus-infected host. *J. Invertebr. Pathol. 19*, 1–17.

Kaya, H. K., and Tanada. 1973. Hemolymph factor in armyworm larvae infected with a nuclear-polyhedrosis virus toxic to *Apanteles militaris. J. Invertebr. Pathol. 21*, 211–214.

Keddie, B. A., and Volkman, L. E. 1985. Infectivity difference between the two phenotypes of *Autographa californica* nuclear polyhedrosis virus: Importance of the 64K envelope glycoprotein. *J. Gen. Virol. 66*, 1195–1200.

Keddie, B. A., Aponte, G. W., and Volkman, L. E. 1989. The pathway of infection of *Autographa californica* nuclear polyhedrosis virus in an insect host. *Science 243*, 1728–1730.

Keeley, L. L., and Vinson, S. B. 1975. β-Ecdysone effects on the development of nucleopolyhedrosis in *Heliothis* spp. *J. Invertebr. Pathol. 26*, 121–123.

Kellen, W. R., Clark, T. B., and Lindegren, J. E. 1963. A possible polyhedrosis in *Culex tarsalis* Coquillett (Diptera: Culicidae). *J. Insect Pathol. 5*, 98–103.

Kelly, D. C. 1977. The DNA contained by nuclear polyhedrosis viruses isolated from four *Spodoptera* sp. (Lepidoptera, Noctuidae): Genome size and homology assessed by DNA reassociation kinetics. *Virology 76*, 468–471.

Kelly, D. C., and Lescott, T. 1981. Baculovirus replication: Protein synthesis in *Spodoptera frugiperda* cells infected with *Trichoplusia ni* nuclear polyhedrosis virus. *Microbiologica 4*, 35–57.

Kelly, D. C., and Lescott, T. 1983. Baculovirus replication: Glycosylation of polypeptides synthesized in *Trichoplusia ni* nuclear polyhedrosis virus-infected cells and the effect of tunicamycin. *J. Gen. Virol. 64*, 1915–1926.

Kelly, D. C., Lescott, T., Ayres, M. D., Carey, D., Coutts, A., and Harrap, K. A. 1981. Induction of a nonoccluded baculovirus persistently infecting *Heliothis zea* cells by *Heliothis armigera* and *Trichoplusia ni* nuclear polyhedrosis viruses. *Virology 112*, 174–189.

Kelly, D. C., Brown, D. A., Ayres, M. D., Allen, C. J., and Walker, I. O. 1983. Properties of the major nucleocapsid protein of *Heliothis zea* singly enveloped nuclear polyhedrosis virus. *J. Gen. Virol. 64*, 399–408.

Khosaka, T., and Himeno, M. 1972. Infectivity of the components of a nuclear polyhedrosis virus of the silkworm, *Bombyx mori. J. Invertebr. Pathol. 19*, 62–65.

Kim, K. S., and Kitajima, E. W. 1984. Nonoccluded baculovirus- and filamentous virus-like particles in the spotted cucumber beetle, *Diabrotica undecimpunctata* (Coleoptera: Chrysomelid). *J. Invertebr. Pathol. 43*, 234–241.

Kislev, N., and Edelman, M. 1982. DNA restriction-pattern differences from geographic isolates of *Spodoptera littoralis* nuclear polyhedrosis virus. *Virology 119*, 219–222.

Kislev, N., Harpaz, I., and Zelcer, A. 1969. Electron-microscopic studies on hemocytes of the Egyptian cottonworm, *Spodoptera littoralis* (Boisduval) infected with a nuclear-polyhedrosis virus, as compared to noninfected hemocytes. II. Virus-infected hemocytes. *J. Invertebr. Pathol. 14*, 245–257.

Kitajima, E. W., Costa, C. L., and Sá, C. M. 1978. Baculovirus-like particles in two aphid species. *J. Invertebr. Pathol. 31*, 123–125.

Knell, J. D., and Summers, M. D. 1981. Investigation of genetic heterogeneity in wild isolates of *Spodoptera frugiperda* nuclear polyhedrosis virus by restriction endonuclease analysis of plaque-purified variants. *Virology 112*, 190–197.

Knell, J. D., and Summers, M. D. 1984. A physical map for the *Heliothis zea* SNPV genome. *J. Gen. Virol. 65*, 445–450.

Knell, J. D., Summers, M. D., and Smith, G. E. 1983. Serological analysis of 17 baculoviruses from subgroups A and B using protein blot immunoassay. *Virology 125*, 381–392.

Knudson, D. L., and Harrap, K. A. 1976. Replication of a nuclear polyhedrosis virus in a continuous cell culture of *Spodoptera frugiperda*: Microscopy study of the sequence of events of the virus infection. *J. Virol. 17*, 254–268.

Kobayashi, M. 1976. Insect viruses—replication cycle of nuclear polyhedrosis virus. *Saibo 8*, 181–192.

Kobayashi, M., and Kawase, S. 1981. Quantitative analysis of the metabolism in the isolated pupal abdomens of the silkworm, *Bombyx mori*, infected with nuclear polyhedrosis virus. *J. Invertebr. Pathol. 38*, 4–11.

Kok, I. P., Chistiakova, A. V., Gudz-Gorban, A. P., and Solomko, A. P. 1968. Infectivity and structure of DNA of the virus of nuclear polyhedrosis of the silkworm. *Proc. Thirteenth Int. Cong. Entomol., Moscow. 2*, 75–76.

Komano, T., Himeno, M., Ohno, Y., and Onodera, K. 1966. Studies on lipids in the hemolymph of the silkworm, *Bombyx mori*, during the course of nucleopolyhedrosis. *J. Invertebr. Pathol. 8*, 67–74.

Komárek, J., and Breindl, V. 1924. Die Wipfelkrankheit der Nonne und der Erreger derselben. *Z. Angew. Entomol. 10*, 99–162.

Kondo, A., and Maeda, S. 1991. Host range expansion by recombination of the baculoviruses *Bombyx mori* nuclear polyhedrosis virus and *Autographa californica* nuclear polyhedrosis virus. *J. Virol. 65*, 3625–3632.

Kozlov, É. A., and Alexeenko, I. P. 1967. Electron-microscope investigation of the structure of the nuclear-polyhedrosis virus of the silkworm, *Bombyx mori*. *J. Invertebr. Pathol. 9*, 413–419.

Kozlov, É. A., Sogulyaeva, V. M., Levitina, T. L., Sereshak, V., and Serebryanyi, S. V. 1969. Purification of polyhedron protein of the silkworm nuclear polyhedrosis virus and investigation of its association-dissociation processes in solutions. *Biokhimiya 34*, 543–548.

Kozlov, É. A., Levitina, T. L., Gusak, N. M., Larionov, G. V., Veremeichenko, S. N., and Serebryanyi, S. B. 1978. Comparative biochemical investigation of polyhedral proteins of nuclear polyhedrosis viruses. *Biochemistry* (U.S.S.R.) *43*, 1729–1734.

Kozlov, É. A., Levitina, T. L., and Gusak, N. M. 1986. The primary structure of baculovirus inclusion body proteins. Evolution and structure–function aspects. *Curr. Top. Microbiol. Immunol. 131*, 135–168.

Krell, P. J. 1987. Replication of long virus-like particles in the reproductive tract of the ichneumonid wasp *Diadegma terebrans*. *J. Gen. Virol. 68*, 1477–1483.

Krieg, A. 1957. Über Aufban und Vermehrungsmöglichkeiten von stäbchenförmigen Insekten-Viren. *Z. Naturforsch. 12b*, 120–121.

Krieg, A. 1958. Verlauf des Infektionstiters bei stäbchenförmigen Insekten-Viren. *Z. Naturforsch. 13b*, 27–29.

Krieg, A. 1961a. "Grundlagen der Insektenpathologie, Viren-, Rickettsien- und Bakterien-Infecktionen." Steinkopff Verlag, Darmstadt.

Krieg, A. 1961b. Über den Aufbau und Vermehrungmöglichkeiten von stäbchenförmigen Insekten-Viren II. *Z. Naturforsch. 16b*, 115–117.

Kuroda, K., Hauser, C., Rott, R., Klenk, H.-D., and Doerfler, W. 1986. Expression of the influenza virus haemagglutinin in insect cells by a baculovirus vector. *Eur. Mol. Biol. Organ. J. 5*, 1359–1365.

Kuroda, S., and Watanabe, H. 1983. Changes in the concentration of α-ketoglutaric acid and free amino acids in haemolymph of the silkworm *Bombyx mori*, infected with a nuclear-polyhedrosis virus. *J. Seric. Sci. Jpn. 52*, 172–176.

Kurstak, E., and Garzon, S. 1975. Multiple infections of invertebrate cells by viruses. *Ann. N.Y. Acad. Sci. 266*, 232–240.

Kuzio, J., Rohel, D. Z., Curry, C. J., Krebs, A., Carstens, E. B., and Faulkner, P. 1984. Nucleotide sequence of the p10 polypeptide gene of *Autographa californica* nuclear polyhedrosis virus. *Virology 139*, 414–418.

Kuzio, J., Jacques, R., and Faulkner, P. 1989. Identification of p74, a gene essential for virulence of baculovirus occlusion bodies. *Virology 173*, 759–763.

Langridge, W. H. R. 1981. Biochemical properties of a persistent nonoccluded baculovirus isolated from *Heliothis zea* cells. *Virology 112*, 770–774.

Langridge, W. H. R. and Balter, K. 1981. Protease activity associated with the capsule protein of *Estigmene acrea* granulosis virus. *Virology 114*, 595–600.

Larsson, R. 1984a. Insect pathological investigations on Swedish Thysanura: A nuclear polyhedrosis virus of the bristletail *Dilta hibernica*. *J. Invertebr. Pathol. 44*, 172–177.

Larsson, R. 1984b. Baculovirus-like particles in the midgut epithelium of the phantom midge, *Chaoborus crystallinus* (Diptera, Chaoboridae). *J. Invertebr. Pathol. 44*, 178–186.

Lee, H. H., and Miller, L. K. 1978. Isolation of genotypic variants of *Autographa californica* nuclear polyhedrosis virus. *J. Virol. 27*, 754–767.

Leisy, D. J., Rohrmann, G. F., and Beaudreau, G. S. 1984. Conservation of genome organization in two multicapsid nuclear polyhedrosis viruses. *J. Virol. 52*, 699–702.

Leisy, D., Nesson, M., Pearson, M., Rohrmann, G., and Beaudreau, G. 1986a. Location and nucleotide sequence of the *Orgyia pseudotsugata* single nucleocapsid nuclear polyhedrosis virus polyhedrin gene. *J. Gen. Virol. 67*, 1073–1079.

Leisy, D. J., Rohrmann, G. F., Nesson, M., and Beaudreau, G. S. 1986b. Nucleotide sequencing and transcriptional mapping of the *Orgyia pseudotsugata* multicapsid nuclear polyhedrosis virus p10 gene. *Virology 153*, 157–167.

Liu, A., Qin, J., Rankin, C., Hardin, S. E., and Weaver, R. F. 1986. Nucleotide sequence of a portion of the *Autographa californica* nuclear polyhedrosis virus genome containing the *Eco*RI site-rich region (hr_5) and an open reading frame just 5' of the p10 gene. *J. Gen. Virol. 67*, 2565–2570.

Loh, L. C., Hamm, J. J., and Huang, E.-S. 1981. *Spodoptera frugiperda* nuclear polyhedrosis virus genome: Physical maps for restriction endonucleases *Bam*HI and *Hind*III. *J. Virol. 38*, 922–931.

Loh, L. C., Hamm, J. J., Kawanishi, C., and Huang, E.-S. 1982. Analysis of the *Spodoptera frugiperda* nuclear polyhedrosis virus genome by restriction endonucleases and electron microscopy. *J. Virol. 44*, 747–751.

Longworth, J. F., Robertson, J. S., and Payne, C. C. 1972. The purification and properties of inclusion body protein of the granulosis virus of *Pieris brassicae*. *J. Invertebr. Pathol. 19*, 42–50.

Luckow, V. A., and Summers, M. D. 1988. Trends in the development of baculovirus expression vectors. *Biotechnology 6*, 47–55.

Lwoff, A., and Tournier, P. 1971. Remarks on the classification of viruses. *In* "Comparative Virology." (K. Maramorosch and E. Kurstak, eds.), pp. 1–42. Academic Press, New York.

MacKinnon, E. A., Henderson, J. F., Stoltz, D. B., and Faulkner, P. 1974. Morphogenesis of nuclear polyhedrosis virus under conditions of prolonged passage *in vitro*. *J. Ultrastruct. Res. 49*, 419–435.

Maeda, S. 1989a. Expression of foreign genes in insects using baculovirus vectors. *Annu. Rev. Entomol. 34*, 351–372.

Maeda, S. 1989b. Increased insecticidal effect by a recombinant baculovirus carrying a synthetic diuretic hormone gene. *Biochem. Biophys. Res. Comm.* 165, 1177–1183.

Maeda, S. 1989c. Gene transfer vectors of a baculovirus, *Bombyx mori* nuclear polyhedrosis virus, and their use for expression of foreign genes in insect cells. *In* "Invertebrate Cell System and Applications." (J. Mitsuhashi, ed.), Vol. 1, pp. 167–181. CRC Press, Boca Raton, Florida.

Maeda, S., and Majima, K. 1990. Molecular cloning and physical mapping of the genome of *Bombyx mori* nuclear polyhedrosis virus. *J. Gen. Virol.* 71, 1851–1855.

Maeda, S., and Tanada, Y. 1983. Protein synthesis of a granulosis virus in the larval fat body of the armyworm, *Pseudaletia unipuncta* (Noctuidae, Lepidoptera). *J. Invertebr. Pathol.* 41, 265–268.

Maeda, S., Kawai, T., Obinata, M., Fujiwara, H., Horiuchi, T., Saeki, Y., Sato, Y., and Furusawa, M. 1985. Production of human α-interferon in silkworm using a baculovirus vector. *Nature 315*, 592–594.

Maeda, S., Mukohara, Y., and Kondo, A. 1990. Characteristically distinct isolates of the nuclear polyhedrosis virus from *Spodoptera litura*. *J. Gen. Virol.* 71, 2631–2639.

Maeda, S., Kamita, S. G., and Kataoka. H. 1991. The basic DNA-binding protein of *Bombyx mori* nuclear polyhedrosis virus: The existence of an additional arginine repeat. *Virology 180*, 807–810.

Maestri, A. 1856. "Frammenti anatomici, fisiologici e patologici sul baco da seta (Bombyx mori Linn.)." Fratelli Fusi, Pavia.

Martignoni, M. E. 1957. Contributo alla conoscenza di una granulosi di *Eucosma griseana* (Hübner) (*Tortricidae, Lepidoptera*) quale fattore limitante il pullulamento dell'insetto nella Engadina alta. *Mitt. Schweitz. Anst. Forstl. Versuchsw.* 32, 371–418.

Martignoni, M. E. 1984. *Baculovirus*: An attractive biological alternative. *In* "Chemical and Biological Controls in Forestry." (W. Y. Garner and J. Harvey, Jr. eds.), ACS Symp. Ser. No. 238, pp. 55–67. Seattle, Washington.

Martignoni, M. E., and Iwai, P. J. 1977. Thermal inactivation characteristics of two strains of nucleopolyhedrosis virus (*Baculovirus* subgroup A) pathogenic for *Orgyia pseudotsugata*. *J. Invertebr. Pathol.* 30, 255–262.

Martignoni, M. E., and Iwai, P. J. 1986. A catalogue of viral diseases of insects, mites, and ticks. *U.S. Dep. Agric. For. Serv. Pac. Northwest Res. Stn. Gen. Tech. Rep. PNW-195*.

Martignoni, M. E., and Milstead, J. E. 1964. Hypoproteinemia in a noctuid larva during the course of nucleopolyhedrosis. *J. Insect Pathol.* 6, 517–531.

Marumoto, Y., Sato, Y., Fujiwara, H., Sakano, K., Saeki, Y., Agata, M., Furusawa, M., and Maeda, S. 1987. Hyperproduction of polyhedrin-IGF II fusion protein in silkworm larvae infected with recombinant *Bombyx mori* nuclear polyhedrosis virus. *J. Gen. Virol.* 68, 2599–2606.

Maruniak, J. E. 1986. Baculovirus structural proteins and protein synthesis. *In* "The Biology of Baculoviruses." (R. R. Granados and B. A. Federici, eds.), Vol. 1, pp. 129–146. CRC Press, Boca Raton, Florida.

Maruniak, J. E., Summers, M. D., Falcon, L. A., and Smith, G. E. 1979. *Autographa californica* nuclear polyhedrosis virus structural proteins compared from *in vivo* and *in vitro* sources. *Intervirology 11*, 82–88.

Maruniak, J. E., Brown, S. E., and Knudson, D. L. 1984. Physical maps of SfMNPV baculovirus DNA and its genomic variants. *Virology 136*, 221–234.

Matthews, R. E. F. 1982. Classification and nomenclature of viruses. *Intervirology 17*, 1–199.

Matthews, R. E. F. 1991. "Plant Virology." Third Ed., Academic Press, San Diego.

McCarthy, W. J., and Gettig, R. R. 1986. Current developments in baculovirus serology. *In* "The Biology of Baculoviruses." (R. R. Granados and B. A. Federici, eds.), Vol. 1, pp. 147–158. CRC Press, Boca Raton, Florida.

McIntosh, A. H., and Ignoffo, C. M. 1981. Establishment of a persistent baculovirus infection in a lepidopteran cell line. *J. Invertebr. Pathol. 38*, 395–403.

Merryweather, A. T., Weyer, U., Harris, M. P. G., Hirst, M., Booth, T., and Possee, R. D. 1990. Construction of genetically engineered baculovirus insecticides containing the *Bacillus thuringiensis* subsp. *kurstaki* HD-73 delta endotoxin. *J. Gen. Virol. 71*, 1535–1544.

Mikhailov, V. S., Ataeva, J. O., Marlyev, K. A., and Kullyev, P. K. 1986a. Changes in DNA polymerase activities in pupae of the silkworm *Bombyx mori* after infection with nuclear polyhedrosis virus. *J. Gen. Virol. 67*, 175–179.

Mikhailov, V. S., Marlyev, K. A., Ataeva, J. O., Kullyev, P. K., and Atrazhev, A. M. 1986b. Characterization of $3' \rightarrow 5'$ exonuclease associated with DNA polymerase of silkworm nuclear polyhedrosis virus. *Nucleic Acid Res. 14*, 3841–3857.

Miller, L. K. 1981a. Construction of a genetic map of the baculovirus *Autographa californica* nuclear polyhedrosis virus by marker rescue of temperature-sensitive mutants. *J. Virol. 39*, 973–976.

Miller, L. K. 1981b. A virus vector for genetic engineering in invertebrates. *In* "Genetic-Engineering in the Plant Sciences." (N. J. Panopoulos, ed.), pp. 203–224. Praeger Scientific, New York.

Miller, L. K. 1986. The genetics of baculoviruses. *In* "The Biology of Baculoviruses." (R. R. Granados and B. A. Federici, eds.), Vol. 1, pp. 217–238. CRC Press, Boca Raton, Florida.

Miller, L. K. 1988. Baculoviruses as gene expression vectors. *Annu. Rev. Microbiol. 42*, 177–199.

Miller, L. K. 1989. Insect baculoviruses: Powerful gene expression vectors. *BioEssays 11*, 91–95.

Miller, L. K., and Dawes, K. P. 1978a. Restriction endonuclease analysis to distinguish two closely related nuclear polyhedrosis viruses: *Autographa californica* MNPV and *Trichoplusia ni* MNPV. *Appl. Environ. Microbiol. 35*, 1206–1210.

Miller, L. K., and Dawes, K. P. 1978b. Restriction endonuclease analysis for the identification of Baculovirus pesticides. *Appl. Environ. Microbiol. 35*, 411–421.

Miller, L. K., and Dawes, K. P. 1979. Physical map of the DNA genome of *Autographa californica* nuclear polyhedrosis virus. *J. Virol. 29*, 1044–1055.

Miller, D. W., and Miller, L. K. 1982. A virus mutant with an insertion of a *copia*-like transposable element. *Nature 299*, 562–564.

Miller, L. K., Jewell, J. E., and Browne, D. 1981. Baculovirus induction of a DNA polymerase. *J. Virol. 40*, 305–308.

Miller, L. K., Adang, M. J., and Browne, D. 1983. Protein kinase activity associated with the extracellular and occluded forms of the baculovirus *Autographa californica* nuclear polyhedrosis virus. *J. Virol. 46*, 275–278.

Minion, F. C., Coons, L. B., and Boome, J. R. 1979. Characterization of the polyhedral envelope of the nuclear polyhedrosis virus of *Heliothis virescens*. *J. Invertebr. Pathol. 34*, 303–307.

Miyajima, A., Schreurs, J., Otsu, K., Kondo, A., Arai, K.-I., and Maeda, S. 1987. Use of the silkworm, *Bombyx mori*, and an insect baculovirus vector for high-level expression and secretion of biologically active mouse interleukin-3. *Gene 58*, 273–281.

Monsarrat, P., Meynadier, G., Croizier, G., and Vago, C. 1973a. Recherches cytopathologiques sur une maladie virale du Coléoptère *Oryctes rhinoceros* L. *C.R. Acad. Sci. Paris Ser. D276*, 2077–2080.

Monsarrat, P., Veyrunes, J.-C., Meynadier, G., Croizier, G., and Vago, C. 1973b. Purification et étude structurale du virus du Coléoptère *Oryctes rhinoceros* L. *C.R. Acad. Sci. Paris Ser. D277*, 1413–1415.

Monsarrat, P., Revet, B., and Gourevitch, I. 1975. Mise en évidence, stabilisation et purification d'une structure nucléoprotéique intracapsidaire chez le Baculovirus d'*Oryctes rhinoceros* L. *C. R. Acad. Sci. Paris Ser. D281*, 1439–1442.

Morgan, C., Bergold, G. H., Moore, D. H., and Rose, H. M. 1955. The macromolecular paracrystalline lattice of insect viral polyhedral bodies demonstrated in ultrathin sections examined in the electron microscope. *J. Biophys. Biochem. Cytol. 1*, 187–190.

Nagata, M., and Tanada, Y. 1983. Origin of an alkaline protease associated with the capsule of a granulosis virus of the armyworm, *Pseudaletia unipuncta* (Haworth). *Arch. Virol. 76*, 245–256.

Naser, W. L., Miltenburger, H. G., Harvey, J. F., Huber, J., and Huger, A. M. 1984. In vitro replication of the *Cydia pomonella* (codling moth) granulosis virus. *FEMS Microbiol. Lett. 24*, 117–121.

Ohba, M., and Tanada, Y. 1984a. *In vitro* enhancement of nuclear polyhedrosis virus infection by the synergistic factor of a granulosis virus of the armyworm, *Pseudaletia unipuncta* (*Lepidoptera: Noctuidae*). *Ann. Virol.* (Inst. Pasteur) *135E*, 167–176.

Ohba, M., and Tanada, Y. 1984b. A synergistic factor of an insect granulosis virus agglutinates insect cells. *Experientia 40*, 742–744.

Onodera, K., Komano, T., Himeno, M., and Sakai, F. 1965. The nucleic acid of nuclear-polyhedrosis virus of the silkworm. *J. Mol. Biol. 13*, 532–539.

O'Reilly, D. R., and Miller, L. K. 1989. A baculovirus blocks insect molting by producing ecdysteroid UDP-glucosyl transferase. *Science 245*, 1110–1112.

Oreshkin, E. N. 1982. Investigations of conditions for baculovirus DNA transfection in an insect cell line. *Vopr. Virusol. 27*, 235–238.

Paillot, A. 1924. Sur l'étiologie et l'épidémiologie de la "grasserie" du Ver à Soie. *C.R. Acad. Sci. Paris Ser. D179*, 229–231.

Paillot, A. 1926a. Contribution à l'étude des maladies à virus filtrant chez les insectes. Un nouveau groupe de parasites ultramicrobiens: *les Borrellina. Ann. Inst. Pasteur 40*, 314–352.

Paillot, A. 1926b. Sur une nouvelle maladie du noyau ou grasserie des chenilles de *Pieris brassicae* et un nouveau groupe de micro-organismes parasites. *C.R. Acad. Sci. Paris Ser. D182*, 180–182.

Paillot, A. 1934. Un nouveau type de maladie à ultravirus chez les insectes. *C.R. Acad. Sci. Paris Ser. D198*, 204–205.

Paillot, A. 1935. Nouvel ultravirus parasite d'*Agrotis segetum* provoquant une prolifération des tissus infectés. *C.R. Acad. Sci. Paris Ser. D201*, 1062–1064.

Paillot, A. 1937. Nouveau type de pseudo-grasserie observé chez les chenilles d'*Euxoa segetum*. *C.R. Acad. Sci. Paris Ser. D205*, 1264–1266.

Pavan, C., DaCunha, A. B., and Morsoletto, C. 1971. Virus–chromosome relationships in cells of *Rhynchosciara* (Diptera, Sciaridae). *Caryologia 24*, 371–389.

Payne, C. C. 1974. The isolation and characterization of a virus from *Oryctes rhinoceros*. *J. Gen. Virol. 25*, 105–116.

Payne, C. C., Compson, D., and de Looze, S. M. 1977. Properties of the nucleocapsids of a virus isolated from *Oryctes rhinoceros*. *Virology 77*, 269–280.

Pearson, M. N., Russell, R. L. Q., Rohrmann, G. F., and Beaudreau, G. S. 1988. p39, a major baculovirus structural protein: Immunocytochemical characterization and genetic location. *Virology 167*, 407–413.

Pinnock, D. E., and Hess, R. T. 1978. Morphological variations in the cytopathology associated with granulosis virus in the fruit-tree leaf roller, *Archips argyrospila*. *J. Ultrastruct. Res. 63*, 252–260.

Possee, R. D., and Kelly, D. C. 1988. Physical maps and comparative DNA hybridization of *Mamestra brassicae* and *Panolis flammea* nuclear polyhedrosis virus genomes. *J. Gen. Virol. 69*, 1285–1298.

Potter, K. N., and Miller, L. K. 1980. Transfection of two invertebrate cell lines with DNA of *Autographa californica* nuclear polyhedrosis virus. *J. Intertebr. Pathol. 36*, 431–432.

Potter, K. N., Faulkner, P., and MacKinnon, E. A. 1976. Strain selection during serial passage of *Trichoplusia ni* nuclear polyhedrosis virus. *J. Virol. 18*, 1040–1050.

Prell, H. 1926. Die Polyederkrankheiten der Insekten. *In* "Verl. III. Int. Entomol. Kongr. Zürich 1925." (K. Jordan and W. Horn, eds.), Vol. 2, pp. 145–168. Weimar, Austria.

Pritchett, D. W., Young, S. Y., and Yearin, W. C. 1982. Dissolution of *Autographa californica* nuclear polyhedrosis virus polyhedra by the digestive fluid of *Trichoplusia ni* (Lepidoptera: Noctuidae) larvae. *J. Invertebr. Pathol. 39*, 354–361.

Pritchett, D. W., Young, S. Y., and Yearin, W. C. 1984. Some factors involved in the dissolution of *Autographa californica* nuclear polyhedrosis virus polyhedra by digestive fluids of *Trichoplusia ni* larvae. *J. Invertebr. Pathol. 43*, 160–168.

Prusiner, S. B. 1984. Prions: Novel infectious pathogens. *In* "Advances in Virus Research." (M. A. Lauffer and K. Maramorosch, eds.), Vol 29, pp. 1–56. Academic Press, New York.

Quant, R. L., Pearson, M. N., Rohrmann, G. F., and Beaudreau, G. S. 1984. Production of polyhedrin monoclonal antibodies for distinguishing two *Orgyia pseudotsugata* baculoviruses. *Appl. Environ. Microbiol. 48*, 732–736.

Quant-Russell, R. L., Pearson, M. N., Rohrmann, G. F., and Beaudreau, G. S. 1987. Characterization of baculovirus p10 synthesis using monoclonal antibodies. *Virology 100*, 9–19.

Raghow, R., and Grace, T. D. C. 1974. Studies on a nuclear polyhedrosis virus in *Bombyx mori* cells *in vitro*. 1. Multiplication kinetics and ultrastructural studies. *J. Ultrastruct. Res. 47*, 384–399.

Rajamohan, N., and Jayaraj, S. 1976. Changes in protein, fat and glycogen in the larvae of the rice moth *Corcyra cephalonica* St. during nuclear polyhedrosis virus infection. *Indian J. Exp. Biol. 14*, 500–501.

Ralston, A. L., Huang, Y.-S., and Kawanishi, C. Y. 1981. Cell culture studies with the IMC-Hz-1 nonoccluded virus. *Virology 115*, 33–44.

Rennie, J. 1923. Polyhedral disease in *Tipula paludosa* (Meigen). *Proc. R. Soc. Edinburgh A20*, 265–267.

Revet, B. M. J., and Guelpa, B. 1979. The genome of a baculovirus infecting *Tipula paludosa* (Meig)(diptera): A high molecular weight closed circular DNA of zero superhelix density. *Virology 96*, 633–639.

Rindich, A. V., Sutugina, L. P., Kavsan, V. M., Telichuk, S. P., and Kok, I. P. 1975. DNA-polymerase activity in the nuclear polyhedrosis virus of *Galleria mellonella*. *Dagov. Akad. Nauk Ukr. RSR, Ser. B. 4*, 347–349.

Ritter, K. S., Tanada, Y., Hess, R. T., and Omi, E. M. 1982. Eclipse period of baculovirus infection in larvae of the armyworm, *Pseudaletia unipuncta*. *J. Invertebr. Pathol. 39*, 203–209.

Roberts, P. L. 1983. Neutralization studies on the *Autographa californica* nuclear polyhedrosis virus. *Arch. Virol. 75*, 147–150.

Robertson, J. S., Harrap, K. A., and Longworth, J. F. 1974. Baculovirus morphogenesis: The acquisition of the virus envelope. *J. Invertebr. Pathol. 23*, 248–251.

Rohrmann, G. F. 1977. Characterization of N-polyhedrin of two Baculovirus strains pathogenic for *Orgyia pseudotsugata*. *Biochemistry 16*, 1631–1634.

Rohrmann, G. F. 1986a. Polyhedrin structure. *J. Gen. Virol. 67*, 1499–1513.

Rohrmann, G. F. 1986b. Evolution of occluded baculoviruses. *In* "The Biology of Baculoviruses." (R. R. Granados and B. A. Federici, eds.), Vol. 1, pp. 203–215. CRC Press, Boca Raton, Florida.

Rohrmann, G. F., Bailey, T. J., Becker, R. R., and Beaudreau, G. S. 1980. Comparison of the structure of C- and N-polyhedrins from two occluded viruses pathogenic for *Orgyia pseudotsugata*. *J. Virol. 34*, 360–365.

Rohrmann, G. F., Pearson, M. N., Bailey, T. J., Becker, R. R., and Beaudreau, G. S. 1981. N-terminal polyhedrin sequences and occluded *Baculovirus* evolution. *J. Mol. Evol.* 17, 329–333.

Rohrmann, G. F., Martignoni, M. E., and Beaudreau, G. S. 1982. DNA sequence homology between *Autographa californica* and *Orgyia pseudotsugata* nuclear polyhedrosis viruses. *J. Gen. Virol.* 62, 137–143.

Roizman, B. 1986. Multiplication of viruses: An overview. *In* "Fundamental Virology." (B. N. Fields, D. M. Knipe, chief eds.), pp. 69–75. Raven Press, New York.

Rubinstein, R., Lawler, K. A., and Granados, R. R. 1982. Use of primary fat body cultures for the study of baculovirus replication. *J. Invertebr. Pathol.* 40, 266–273.

Santiago-Alvarez, C., and Vargas-Osuna, E. 1988. Reduction of reproductive capacity of *Spodoptera littoralis* males by a nuclear polyhedrosis virus (NPV). *J. Invertebr. Pathol.* 52, 142–146.

Sasaki, C. 1910. On the pathology of the jaundice (Gelbsucht) of the silkworm. *J. Tokyo Coll. Agric.* 2, 105–161.

Scali, V., Montanelli, E., Lanfranchi, A., and Bedini, C. 1980. Nuclear alterations in a baculovirus-like infection of midgut epithelial cells in the stick insect, *Bacillus rossius*. *J. Invertebr. Pathol.* 35, 109–118.

Scharnhorst, D. W., Saving, K. L., Vuturo, S. B., Cooke, P. H., and Weaver, R. F. 1977. Structural studies on the polyhedral inclusion bodies, virions, and DNA of the nuclear polyhedrosis virus of the cotton bollworm *Heliothis zea*. *J. Virol.* 21, 292–300.

Schmid, A., Cazelles, O., and Benz, G. 1983. A granulosis virus of the fruit tortrix, Adoxophyes orana F.v.R. (Lep., Tortricidae). *Mitt. Schweiz. Entomol. Ges.* 56, 225–235.

Sherman, K. E. 1985. Multiple virus interactions. *In* "Viral Insecticides for Biological Control." (K. Maramorosch and K.E. Sherman, eds.), pp. 735–753. Academic Press, New York.

Shigematsu, H., and Noguchi, A. 1969a. Biochemical studies on the multiplication of a nuclear-polyhedrosis virus in the silkworm, *Bombyx mori*. I. Nucleic-acid synthesis in larval tissues after infection. *J. Invertebr. Pathol.* 14, 143–149.

Shigematsu, H., and Noguchi, A. 1969b. Biochemical studies on the multiplication of a nuclear-polyhedrosis virus in the silkworm, *Bombyx mori*. II. Protein synthesis in larval tissues after infection. *J. Invertebr. Pathol.* 14, 301–307.

Shigematsu, H., and Noguchi, A. 1969c. Biochemical studies on the multiplication of a nuclear-polyhedrosis virus in the silkworm, *Bombyx mori*. III. Functional changes in infected cells, with reference to the synthesis of nucleic acids and proteins. *J. Invertebr. Pathol.* 14, 308–315.

Shvedchikova, N. G., Ulanov, V. P., and Tarasevich, L. M. 1969. Structure of the granulosis virus of Siberian silkworm *Dendrolimus sibiricus* Tschetw. *Mol. Biol.* (Moscow) 3, 283–287.

Shylaja, M. S., and Ramaiah, T. R. 1984. Isoenzymes of glutamate-oxalacetate transaminase in the larvae of silkworms *Bombyx mori* infected with nuclear polyhedrosis virus. *Experientia* 40, 717–718.

Skuratovskaya, I. N., Strokovskaya, L. I., Zherebtsova, E. N., and Gudz-Gorban, A. P. 1977. Supercoiled DNA of nuclear polyhedrosis virus of *Galleria mellonella* L. *Arch. Virol.* 53, 79–86.

Smirnoff, W. A. 1976. Metabolic disturbances in insect during viral and other infections. *Proc. First Int. Coll. Invertebr. Pathol. Ninth Annu. Meet. Soc. Invertebr. Pathol.* pp. 339–340. Kingston, Canada.

Smith, G. E., and Summers, M. D. 1978. Analysis of baculovirus genomes with restriction endonucleases. *Virology* 89, 517–527.

Smith, G. E., and Summers, M. D. 1979. Restriction maps of five *Autographa californica* MNPV variants, *Trichoplusia ni* MNPV, and *Galleria mellonella* MNPV DNAs with endonucleases, *Sma*I, *Kpn*I, *Bam*HI, *Sac*I, *Xho*I, and *Eco*RI. *J. Virol. 30*, 828–838.

Smith, G. E., and Summers, M. D. 1981. Application of a novel radioimmunoassay to identify baculovirus structural proteins that share interspecies antigenic determinants. *J. Virol. 39*, 125–137.

Smith, G. E., and Summers, M. D. 1982. DNA homology among subgroup A, B, and C baculoviruses. *Virology 123*, 393–406.

Smith, G. E., Vlak, J. M., and Summers, M. D. 1982. In vitro translation of *Autographa californica* nuclear polyhedrosis virus early and late mRNAs. *J. Virol. 44*, 199–208.

Smith, G. E., Vlak, J. M., and Summers, M. D. 1983a. Physical analysis of *Autographa californica* nuclear polyhedrosis virus transcripts for polyhedrin and 10,000-molecular-weight protein. *J. Virol. 45*, 215–225.

Smith, G. E., Summers, M. D., and Fraser, M. J. 1983b. Production of human beta interferon in insect cells infected with a baculovirus expression vector. *Mol. Cell. Biol. 3*, 2156–2165.

Smith, G. E., Ju, G., Ericson, B. L., Moschera, J., Lahm, H.-W., Chizzonite, R., and Summers, M. D. 1985. Modification and secretion of human interleukin 2 produced in insect cells by a baculovirus expression vector. *Proc. Natl. Acad. Sci. U.S.A, 82*, 8404–8408.

Smith, I. R. L., and Crook, N. E. 1988. Physical maps of the genomes of four variants of *Artogeia rapae* granulosis virus. *J. Gen. Virol. 69*, 1741–1747.

Smith, O. J., Hughes, K. M., Dunn, P. H., and Hall, I. M. 1956. A granulosis virus disease of the western grape leaf skeletonizer and its transmission. *Can. Entomol. 88*, 507–515.

Sohi, S. S., Percy, J., Arif, B. M., and Cunningham, J. C. 1984. Replication and serial passage of a singly enveloped baculovirus of *Orgyia leucostigma* in homologous cell lines. *Intervirology 21*, 50–60.

Stairs, G. R. 1964. Selection of a strain of insect granulosis virus producing only cubic inclusion bodies. *Virology 24*, 520–521.

Stairs, G. R. 1980. Comparative infectivity of nonoccluded virions, polyhedra, and virions released from polyhedra for larvae of *Galleria mellonella*. *J. Invertebr. Pathol. 36*, 281–282.

Stairs, G. R., Parrish, W. B., Briggs, J. D., and Allietta, M. 1966. Fine structure of a granulosis virus of the codling moth. *Virology 30*, 583–584.

Steinhaus, E. A. 1949. "Principles of Insect Pathology." McGraw-Hill, New York.

Steinhaus, E. A. 1952. The susceptibility of two species of *Colias* to the same virus. *J. Econ. Entomol. 45*, 897–899.

Steinhaus, E. A. 1956. Microbial control—the emergence of an idea. A brief history of insect pathology through the nineteenth century. *Hilgardia 26*, 107–160.

Steinhaus, E. A., Hughes, K. M., and Wasser, H. B. 1949. Demonstration of the granulosis virus of the variegated cutworm. *J. Bacteriol. 57*, 219–224.

Stiles, B., and Paschke, J. D. 1980. Midgut *p*H in different instars of three *Aedes* mosquito species and the relation between *p*H and susceptibility of larvae to a nuclear polyhedrosis virus. *J. Invertebr. Pathol. 35*, 58–64.

Stiles, B., and Wood, H. A. 1983. A study of the glycoproteins of *Autographa californica* nuclear polyhedrosis virus (AcNPV). *Virology 131*, 230–241.

Stiles, B., Dunn, P. E., and Paschke, J. D. 1983. Histopathology of a nuclear polyhedrosis infection in *Aedes epactius* with observations in four additional mosquito species. *J. Invertebr. Pathol. 41*, 191–202.

Stoddard, P. J. 1980. Persistence and Transmission of Baculoviruses in Insect Populations in Alfalfa. Ph.D. Thesis, University of California, Berkeley.

Stoltz, D. B., and Faulkner, G. 1978. Apparent replication of an unusual virus-like particle in both a parasitoid wasp and its host. *Can. J. Microbiol. 24*, 1509–1514.

Stoltz, D. B., and Summers, M. D. 1972. Observations on the morphogenesis and structure

of a hemocytic poxvirus in the midge *Chironomus attenuatus*. *J. Ultrastruct. Res. 40*, 581–598.
Stoltz, D. B., Pavan, C., and da Cunha, A. B. 1973. Nuclear polyhedrosis virus: A possible example of *de novo* intranuclear membrane morphogenesis. *J. Gen. Virol. 19*, 145–150.
Stoltz, D. B., Krell, P., Cook, D., MacKinnon, E. A., and Lucarotti, C. J. 1988. An unusual virus from the parasitic wasp *Cotesia melanoscela*. *Virology 162*, 311–320.
Strokovskaya, L. I., Skuratovskaya, I. N., Zherebtsova, E. N., and Sutugina, L. P. 1977. Purification of free polyhedrosis virus infecting *Galleria mellonella* L. *Acta Virol. 21*, 157–160.
Strokovskaya, L. I., Skuratovskaya, I. N., Gudz-Gorban, A. P., Zherebtsova, E. N., Prima, V. I., and Kok, I. P. 1979. Macromolecular structure of nuclear polyhedrosis virus genome. *Arch. Virol. 59*, 331–343.
Subrahmanyam, B., and Ramakrishnan, N. 1981. Influence of a baculovirus infection on molting and food consumption by *Spodoptera litura*. *J. Invertebr. Pathol. 38*, 161–168.
Summers, M. D. 1969. Apparent in vivo pathway of granulosis virus invasion and infection. *J. Virol. 4*, 188–190.
Summers, M. D. 1971. Electron microscopic observations on granulosis virus entry, uncoating and replication processes during infection of the midgut cells of *Trichoplusia ni*. *J. Ultrastruct. Res. 35*, 606–625.
Summers, M. D., and Anderson, D. L. 1972. Granulosis virus deoxyribonucleic acid: A closed, double-stranded molecule. *J. Virol. 9*, 710–713.
Summers, M. D., and Arnott, H. J. 1969. Ultrastructural studies on inclusion formation and virus occlusion in nuclear polyhedrosis and granulosis virus-infected cells of *Trichoplusia ni* (Hübner). *J. Ultrastruct. Res. 28*, 462–480.
Summers, M. D., and Egawa, K. 1973. Physical and chemical properties of *Trichoplusia ni* granulosis virus granulin. *J. Virol. 12*, 1092–1103.
Summers, M. D., and Smith, G. E. 1975. *Trichoplusia ni* granulosis virus granulin: A phenol-soluble, phosphorylated protein. *J. Virol. 16*, 1108–1116.
Summers, M. D., and Smith, G. E. 1976. Comparative studies of baculovirus granulins and polyhedrins. *Intervirology 6*, 168–180.
Summers, M. D., and Smith, G. E. 1987. A manual of methods for baculovirus vectors and insect cell culture procedures. *Texas Agric. Exp. Stn.*
Summers, M. D., and Volkman, L. E. 1976. Comparison of biophysical and morphological properties of occluded and extracellular nonoccluded baculovirus from in vivo and in vitro host systems. *J. Virol. 17*, 962–972.
Summers, M. D., Smith, G. E., Knell, J. D., and Burand, J. P. 1980. Physical maps of *Autographa californica* and *Rachiplusia ou* nuclear polyhedrosis virus recombinants. *J. Virol. 34*, 693–703.
Takei, G. H., and Tamashiro, M. 1975. Changes observed in hemolymph proteins of the lawn armyworm, *Spodoptera mauritia acronyctoides*, during growth, development, and exposures to a nuclear polyhedrosis virus. *J. Invertebr. Pathol. 26*, 147–158.
Tanada, Y. 1953. Description and characteristics of a granulosis virus of the imported cabbageworm. *Proc. Hawaii. Entomol. Soc. 15*, 235–260.
Tanada, Y. 1954. A polyhedrosis virus of the imported cabbageworm and its relation to a polyhedrosis virus of the alfalfa caterpillar. *Ann. Entomol. Soc. Am. 47*, 553–574.
Tanada, Y. 1956. Some factors affecting the susceptibility of the armyworm to virus infections. *J. Econ. Entomol. 49*, 52–57.
Tanada, Y. 1959. Synergism between two viruses of the armyworm *Pseudaletia unipuncta* (Haworth) (Lepidoptera, Noctuidae). *J. Insect Pathol. 1*, 215–231.
Tanada, Y. 1964. A granulosis virus of the codling moth, *Carpocapsa pomonella* (Linnaeus) (Olethreutidae, Lepidoptera). *J. Insect Pathol. 6*, 378–380.
Tanada, Y. 1976. Ecology of insect viruses. *In* "Perspectives in Forest Entomology." (J. F. Anderson and H. K. Kaya, eds.), pp. 265–283. Academic Press, New York.

Tanada, Y., and Hess, R. T. 1976. Development of a nuclear polyhedrosis virus in midgut cells and penetration of the virus into the hemocoel of the armyworm, *Pseudaletia unipuncta*. *J. Invertebr. Pathol.* 28, 67–76.

Tanada, Y., and Hess, R. T. 1984. The cytopathology of baculovirus infections in insects. In "Insect Ultrastructure." (R. C. King and H. Akai, eds.), Vol. 2, pp. 517–556. Plenum, New York.

Tanada, Y., and Hess, R. T. 1991. Baculoviridae. Granulosis Viruses. In "Atlas of Invertebrate Viruses." (J. R. Adams and J. R. Bonami, eds.), pp. 227–257. CRC Press, Boca Raton, Florida.

Tanada, Y., and Hukuhara, T. 1968. A nonsynergistic strain of a granulosis virus of the armyworm, *Pseudaletia unipuncta*. *J. Invertebr. Pathol.* 12, 263–268.

Tanada, Y., and Hukuhara, T. 1971. Enhanced infection of a nuclear-polyhedrosis virus in larvae of the armyworm, *Pseudaletia unipuncta*, by a factor in the capsule of a granulosis virus. *J. Invertebr. Pathol.* 17, 116–126.

Tanada, Y., and Leutenegger, R. 1968. Histopathology of a granulosis-virus disease of the codling moth, *Carpocapsa pomonella*. *J. Invertebr. Pathol.* 10, 39–47.

Tanada, Y., and Watanabe, H. 1971. Disc electrophoretic and serological studies of the capsule proteins obtained from two strains of a granulosis virus of the armyworm, *Pseudaletia unipuncta*. *J. Invertebr. Pathol.* 18, 307–312.

Tanada, Y., Hukuhara, T., and Chang, G. Y. 1969. A strain of nuclear-polyhedrosis virus causing extensive cellular hypertrophy. *J. Invertebr. Pathol.* 13, 394–409.

Tanada, Y., Hess, R. T., and Omi, E. M. 1975. Invasion of a nuclear polyhedrosis virus in midgut of the armyworm, *Pseudaletia unipuncta*, and the enhancement of a synergistic enzyme. *J. Invertebr. Pathol.* 26, 99–104.

Tanada, Y., Hess, R. T., and Omi, E. M. 1982. Unique virus morphogenesis and cytopathology of a baculovirus (hypertrophy strain) in larva of the armyworm, *Pseudaletia unipuncta*. *J. Invertebr. Pathol.* 40, 197–204.

Tanada, Y., Hess, R. T., and Omi, E. M. 1984. The movement and invasion of an insect baculovirus in tracheal cells of the armyworm, *Pseudaletia unipuncta*. *J. Invertebr. Pathol.* 44, 198–208.

Teakle, R. E. 1969. A nuclear-polyhedrosis virus of *Anthela varia* (Lepidoptera: Anthelidae). *J. Invertebr. Pathol.* 14, 18–27.

Thiem, S. M., and Miller, L. K. 1989. Identification, sequence, and transcriptional mapping of the major capsid protein gene of the baculovirus *Autographa californica* nuclear polyhedrosis virus. *J. Virol.* 63, 2008–2018.

Tomalski, M. D., Wu, J., and Miller, L. K. 1988. The location, sequence, transcription, and regulation of a baculovirus DNA polymerase. *Virology 167*, 591–600.

Tompkins, G. J., Dougherty, E. M., Adams, J. R., and Diggs, D. 1988. Changes in the virulence of nuclear polyhedrosis viruses when propagated in alternate noctuid (Lepidoptera: Noctuidae) cell lines and hosts. *J. Econ. Entomol.* 81, 1027–1032.

Trager, W. 1935. Cultivation of the virus of grasserie in silkworm tissue cultures. *J. Exp. Med.* 61, 501–513.

Tsuda, K., Mizuki, E., Kawarabata, T., and Aizawa, K. 1984. Replication of *Xestia c-nigrum* (Lepidoptera: Noctuidae) nuclear polyhedrosis virus in continuous cell cultures. *Appl. Entomol. Zool.* 19, 293–298.

Tweeten, K. A., Bulla, L. A., Jr., and Consigli, R. A. 1977. Supercoiled circular DNA of an insect granulosis virus. *Proc. Natl. Acad. Sci. U.S.A.* 74, 3574–3578.

Tweeten, K. A., Bulla, L. A., Jr., and Consigli, R. A. 1978. Characterization of an alkaline protease associated with a granulosis virus of *Plodia interpunctella*. *J. Virol.* 26, 702–711.

Tweeten, K. A., Bulla, L. A., Jr., and Consigli, R. A. 1980. Characterization of an extremely basic protein derived from granulosis virus nucleocapsids. *J. Virol.* 33, 866–876.

Tweeten, K .A., Bulla, L. A., Jr., and Consigli, R. A. 1981. Applied and molecular aspects of insect granulosis viruses. *Microbiol. Rev. 45*, 379–408.

Vago, C., and Bergoin, M. 1963. Développement des virus à corps d'inclusion du Lépidoptère *Lymantria dispar* en cultures cellulaires. *Entomophaga 8*, 253–261.

Vail, P. V., Jay, D. L., and Hink, W. F. 1973. Replication and infectivity of the nuclear polyhedrosis virus of the alfalfa looper, *Autographa californica*, produced in cells grown in vitro. *J. Invertebr. Pathol. 22*, 231–237.

Vail, P. V., Romine, C. L., and Vaughn, J. L. 1979. Infectivity of nuclear polyhedrosis virus extracted with digestive juices. *J. Invertebr. Pathol. 33*, 328–330.

van Beek, N.A.M., Wood, H. A., and Hughes, P. R. 1988. The number of nucleocapsids of enveloped *Autographa californica* nuclear polyhedrosis virus particles affects the survival time of neonate *Trichoplusia ni* larvae. *J. Invertebr. Pathol. 52*, 185–186.

van der Beek, C. P., Saaijer-Riep, J. D., and Vlak, J. M. 1980. On the origin of the polyhedral protein of *Autographa californica* nuclear polyhedrosis virus. Isolation, characterization, and translation of viral messenger RNA. *Virology 100*, 326–333.

van der Geest, L. P. S., and Craig, R. 1967. Biochemical changes in the larvae of the variegated cutworm, *Peridroma saucia*, after infection with a nuclear-polyhedrosis virus. *J. Invertebr. Pathol. 9*, 43–54.

van der Wilk, F., van Lent, J. W. M., and Vlak, J. M. 1987. Immunogold detection of polyhedrin, p10 and virion antigens in *Autographa californica* nuclear polyhedrosis virus-infected *Spodoptera frugiperda* cells. *J. Gen. Virol. 68*, 2615–2623.

van Iddekinge, B. J. L. H., Smith, G. E., and Summers, M. D. 1983. Nucleotide sequence of the polyhedrin gene of *Autographa californica* nuclear polyhedrosis virus. *Virology 131*, 561–565.

Vaughn, J. L., and Faulkner, P. 1963. Susceptibility of an insect tissue culture to infection by virus preparations of the nuclear polyhedrosis of the silkworm (*Bombyx mori* L.). *Virology 20*, 484–489.

Vaughn, J. L., Stone, R. D., and Zhu, G.-K. 1989. The effect of dissolution procedures on the infectivity of the nuclear polyhedrosis polyhedra derived virions for cell cultures. *In* "Invertebrate Cell System Applications." (J. Mitsuhashi, ed.), Vol. 2, pp. 23–29. CRC Press, Boca Raton, Florida.

Vlak, J. M., and Gröner, A. 1980. Identification of two nuclear polyhedrosis viruses from the cabbage moth, *Mamestra brassicae* (Lepidoptera: Noctuidae). *J. Invertebr. Pathol. 35*, 269–278.

Vlak, J. M., and Odink, K. G. 1979. Characterization of *Autographa californica* nuclear polyhedrosis virus deoxyribonucleic acid. *J. Gen. Virol. 44*, 333–347.

Vlak, J. M., and Rohrmann, G. F. 1985. The nature of polyhedrin. *In* "Viral Insecticides for Biological Control." (K. Maramorosh and K. E. Sherman, eds.), pp. 489–542. Academic Press, Orlando, Florida.

Vlak, J. M., and Smith, G. E. 1982. Orientation of the genome of *Autographa californica* nuclear polyhedrosis virus: A proposal. *J. Virol. 41*, 1118–1121.

Vlak, J. M., and van der Krol, S. 1982. Transcription of the *Autographa californica* nuclear polyhedrosis virus genome: Location of late cytoplasmic mRNA. *Virology 123*, 222–228.

Vlak, J. M., Smith, G. E., and Summers, M. D. 1981. Hybridization selection and in vitro translation of *Autographa californica* nuclear polyhedrosis virus mRNA. *J. Virol. 40*, 762–771.

Vlak, J. M., Klinkenberg, F. A., Zaal, K. J. M., Usmany, M., Klinge-Roode, E. C., Geervliet, J. B. F., Roosien, J., and Van Lent, J. W. M. 1988. Functional studies on the p10 gene of *Autographa californica* nuclear polyhedrosis virus using a recombinant expressing a p10-β-galactosidease fusion gene. *J. Gen. Virol. 69*, 765–776.

Volkman, L. E. 1983. Occluded and budded *Autographa californica* nuclear polyhedrosis virus: Immunological relatedness of structural proteins. *J. Virol. 46*, 221–229.

Volkman, L. E. 1986. The 64K envelope protein of budded *Autographa californica* nuclear polyhedrosis virus. *Curr. Top. Microbiol. Immunol. 131*, 103–118.

Volkman, L. E., and Goldsmith, P. A. 1985. Mechanism of neutralization of budded *Autographa californica* nuclear polyhedrosis virus by a monoclonal antibody: Inhibition of entry by adsorptive endocytosis. *Virology 143*, 185–195.

Volkman, L. E., and Summers, M. D. 1977. *Autographa californica* nuclear polyhedrosis virus: Comparative infectivity of the occluded, alkali-liberated, and nonoccluded forms. *J. Invertebr. Pathol. 30*, 102–103.

Volkman, L. E., and Zaal, K. J. M. 1990. *Autographa californica* M nuclear polyhedrosis virus: Microtubules and replication. *Virology 175*, 292–302.

Volkman, L. E., Goldsmith, P. A., Hess, R. T., and Faulkner, P. 1984. Neutralization of budded *Autographa californica* NPV by a monoclonal antibody: Identification of the target antigen. *Virology 133*, 354–362.

Volkman, L. E., Goldsmith, P. A., and Hess, R. T. 1987. Evidence for microfilament involvement in budded *Autographa californica* nuclear polyhedrosis virus production. *Virology 156*, 32–39.

von Prowazek, S. 1907. Chlamydozoa. II. Gelbsucht der Seidenraupen. *Arch. Protistenkd. 10*, 358–364.

von Prowazek, S. 1913. Untersuchungen über die Gelbsucht der Seidenraupen. *Centralbl. Bakteriol. Parasitenkd. Infek. 67*, 268–284.

Wäger, R., and Benz, G. 1971. Histochemical studies on nucleic acid metabolism in granulosis-infected *Carpocapsa pomonella*. *J. Invertebr. Pathol. 17*, 107–115.

Wahl, B. 1909. Über die Polyederkrankheit der Nonne (Lymantria monacha L.). *Centralbl. Gesamte Forstwes. 35*, 164–172.

Walker, S., Kawanishi, C. Y., and Hamm, J. J. 1982. Cellular pathology of a granulosis virus infection. *J. Ultrastruct. Res. 80*, 163–177.

Wang, X., and Kelly, D. C. 1983. Baculovirus replication: Purification and identification of the *Trichoplusia ni* nuclear polyhedrosis virus-induced DNA polymerase. *J. Gen. Virol. 64*, 2229–2236.

Wang, X., and Kelly, D. C. 1985. Baculovirus replication: Uptake of *Trichoplusia ni* nuclear polyhedrosis virus particles by insect cells. *J. Gen. Virol. 66*, 541–550.

Watanabe, H. 1974. Electron-microscope investigation on dissolution of polyhedra in the gut juice of the silkworm, *Bombyx mori* L. *J. Seric. Sci. Jpn. 43*, 29–34.

Watanabe, H. 1975. Variation in the number of nucleocapsid within the envelope of nuclear-polyhedrosis virus multiplied in different tissues of the silkworm, *Bombyx mori*. *J. Seric. Sci. Jpn. 44*, 497–498.

Watanabe, H., and Kobayashi, M. 1970. Histopathology of a granulosis in the larva of the fall webworm, *Hyphantria cunea*. *J. Invertebr. Pathol. 16*, 71–79.

Watanabe, H., Kobara, R., and Hosaka, M. 1968. Electrophoretic separation of the hemolymph proteins in the silkworm, *Bombyx mori* L., infected with the nuclear-polyhedrosis virus. *J. Seric. Sci. Jpn. 37*, 319–322.

Watanabe, H., Aratake, Y., and Kayamura, T. 1975. Serial passage of a nuclear polyhedrosis virus of the silkworm, *Bombyx mori*, in larvae of rice stem borer, *Chilo suppressalis*. *J. Invertebr. Pathol. 25*, 11–17.

Weiser, J. 1958. Zur Taxonomie der Insektenviren. *Česk. Parasitol. 5*, 203–211.

Whitford, M., Stewart, S., Kuzio, J., and Faulkner, P. 1989. Identification and sequence analysis of a gene encoding gp67 an abundant envelope glycoprotein of the baculovirus *Autographa californica* nuclear polyhedrosis virus. *J. Virol. 63*, 1393–1399.

Whitt, M. A., and Manning, J. S. 1987. Role of chelating agents, monovalent anion and cation in the dissociation of *Autographa californica* nuclear polyhedrosis virus occlusion body. *J. Invertebr. Pathol. 49*, 61–69.

Whitt, M. A., and Manning, J. S. 1988a. Stabilization of the *Autographa californica* nuclear polyhedrosis virus occlusion body matrix by zinc chloride. *J. Invertebr. Pathol. 51*, 278–280.

Whitt, M. A., and Manning, J. S. 1988b. A phosphorylated 34-kDa protein and a subpopulation of polyhedrin are thiol linked to the carbohydrate layer surrounding a baculovirus occlusion body. *Virology 163*, 33–42.

Wiegers, F. P., and Vlak, J. M. 1984. Physical map of the DNA of a *Mamestra brassicae* nuclear polyhedrosis virus variant isolated from *Spodoptera exigua*. *J. Gen. Virol. 65*, 2011–2019.

Wildy, P. 1971. Classification and nomenclature of viruses. *Monogr. Virol. 5*, 1–81.

Wilson, M. E., and Consigli, R. A. 1985a. Characterization of a protein kinase activity associated with purified capsids of the granulosis virus infecting *Plodia interpunctella*. *Virology 143*, 516–525.

Wilson, M. E., and Consigli, R. A. 1985b. Functions of a protein kinase activity associated with purified capsids of the granulosis virus infecting *Plodia interpunctella*. *Virology 143*, 526–535.

Wilson, M. E., and Miller, L. K. 1986. Changes in the nucleoprotein complexes of a baculovirus DNA during infection. *Virology 151*, 315–328.

Wilson, M. E., Mainprize, T. H., Friesen, P. D., and Miller, L. K. 1987. Location, transcription, and sequencing of a baculovirus gene encoding a small arginine-rich polypeptide. *J. Virol. 61*, 661–666.

Witt, D. J., Janus, C. A., and Milligan, S. E. 1977. Scanning electron microscopy of cultured *Trichoplusia ni* cells infected with *Autographa californica* nuclear polyhedrosis virus. *J. Invertebr. Pathol. 30*, 429–433.

Wittig, G. 1959. Untersuchungen über den Verlauf der Granulose bei Raupen von Choristoneura murinana (Hb.) (Lepidopt., Tortricidae). *Arch. Gesamte Virusforsch. 9*, 365–395.

Wood, H. A. 1980a. Isolation and replication of an occlusion body-deficient mutant of the *Autographa californica* nuclear polyhedrosis virus. *Virology 105*, 338–344.

Wood, H. A. 1980b. Protease degradation of *Autographa californica* nuclear polyhedrosis virus proteins. *Virology 103*, 392–399.

Wood, H. A., and Burand, J. P. 1986. Persistent and productive infections with the Hz-1 baculovirus. *Curr. Top. Microbiol. Immunol. 131*, 119–133.

Wood, H. A., and Granados, R. R. 1991. Genetically engineered baculoviruses as agents for pest control. *Annu. Rev. Microbiol. 45*, 69–87.

Wyatt, G. R. 1952a. The nucleic acids of some insect viruses. *J. Gen. Physiol. 36*, 201–205.

Wyatt, G. R. 1952b. Specificity in the composition of nucleic acids. *Exp. Cell Res. Suppl. 2*, 201–217.

Xeros, N. 1956. The virogenic stroma in nuclear and cytoplasmic polyhedroses. *Nature 178*, 412–413.

Yamafuji, K., and Yoshihara, F. 1961a. Location of polyhedral pre-viral genome containing protease-synthesizing gene. *Enzymologia 23*, 327–336.

Yamafuji, K., and Yoshihara, F. 1961b. A gene synthesizing protease and a pre-viral genome in silkworms. *Nature 192*, 782.

Yamafuji, K., Yoshihara, F., and Sato, M. 1954. Eclipse period of polyhedral disease. *Enzymologia 17*, 152–154.

Yamamoto, T., and Tanada, Y. 1978a. Biochemical properties of viral envelopes of insect baculoviruses and their role in infectivity. *J. Invertebr. Pathol. 32*, 202–211.

Yamamoto, T., and Tanada, Y. 1978b. Phospholipid, an enhancing component in the synergistic factor of a granulosis virus of the armyworm, *Pseudaletia unipuncta*. *J. Invertebr. Pathol. 31*, 48–56.

Yamamoto, T., and Tanada, Y. 1978c. Protein components of two strains of granulosis virus of the armyworm, *Pseudaletia unipuncta* (Lepidoptera, Noctuidae). *J. Invertebr. Pathol. 32*, 158–170.

Yamamoto, T., and Tanada, Y. 1980. Physicochemical properties and location of capsule components, in particular the synergistic factor, in the occlusion body of a granulosis virus of the armyworm, *Pseudaletia unipuncta*. *Virology 107*, 434–440.

Yamamoto, T., Hess, R. T., and Tanada, Y. 1981. Assembly of occlusion-body proteins around the enveloped virion of an insect baculovirus. *J. Ultrastruct. Res. 75*, 127–130.

Zelazny, B. 1976. Transmission of a baculovirus in populations of *Oryctes rhinoceros*. *J. Invertebr. Pathol. 27*, 221–227.

Zelazny, B. 1977a. *Oryctes rhinoceros* populations and behavior influenced by a baculovirus. *J. Invertebr. Pathol. 29*, 210–215.

Zelazny, B. 1977b. Occurrence of the baculovirus disease of the coconut palm rhinoceros beetle in the Philippines and in Indonesia. *FAO Plant Protect. Bull. 25*, 73–77.

Zelazny, B. 1979. Virulence of the baculovirus of *Oryctes rhinoceros* from ten locations in the Philippines and in Western Samoa. *J. Invertebr. Pathol. 33*, 106–107.

Zhang, L., Wang, X., Deng, H., and Zhang, S. 1979. Morphology and structure of the nuclear polyhedral virus of *Prodenia litura*. *Sci. Sin. 22*, 675–679.

Zhang, Y.-M., Hayes, E. P., McCarty, T. C., Dubois, D. R., Summers, P. L., Eckels, K. H., Chanock, R. M., and Lai, C.-J. 1988. Immunization of mice with dengue structural proteins and nonstructural protein NS1 expressed by baculovirus recombinant induces resistance to dengue virus encephalitis. *J. Virol. 62*, 3027–3031.

Zhu, Y., Hukuhara, T., and Tamura, K. 1989. Location of a synergistic factor in the capsule of a granulosis virus of the armyworm, *Pseudaletia unipuncta*. *J. Invertebr. Pathol. 54*, 49–56.

Zuidema, D., Klinge-Roode, E. C., van Lent, J. W. M., and Vlak, J. M. 1989. Construction and analysis of an *Autographa californica* nuclear polyhedrosis virus mutant lacking the polyhedral envelope. *Virology 173*, 98–108.

Zummer, M., and Faulkner, P. 1979. Absence of protease in baculovirus polyhedral bodies propagated in vitro. *J. Invertebr. Pathol. 33*, 383–384.

CHAPTER 7

OTHER DNA-VIRAL INFECTIONS

I. POLYDNAVIRIDAE—
   dsDNA, ENVELOPED
   A. *Ichnovirus*
   B. *Bracovirus*
   C. Genome
   D. Virus–Parasitoid–
      Lepidoptera
      Relationships
II. POXVIRIDAE—dsDNA,
    ENVELOPED
    A. Structure and Properties
    B. Inclusion Bodies
    C. Replication
    D. Pathogenesis
III. ASCOVIRIDAE—dsDNA,
     ENVELOPED
     A. Structure
   B. Replication
   C. Cytopathology
   D. Pathogenesis
IV. IRIDOVIRIDAE—dsDNA,
    NONENVELOPED
    A. Structure and Properties
    B. Replication
    C. Hosts
    D. Pathogenesis
V. PARVOVIRIDAE—ssDNA,
   NONENVELOPED
   A. Structure and Properties
   B. Hosts
   C. Pathogenesis and Viral
      Replication

The DNA viruses infecting insects occur in five other families besides Baculoviridae. The families are Polydnaviridae, Poxviridae, and Ascoviridae, all of which with double-stranded DNA (dsDNA) enveloped viruses; Iridoviridae with dsDNA nonenveloped viruses; and Parvoviridae with single-stranded DNA (ssDNA) nonenveloped viruses. Polydnaviridae has been included in the revised classification by Francki et al. (1991), and Ascoviridae has been proposed as a new family (Federici 1983). Members of these two families may be unique for insects.

I. POLYDNAVIRIDAE—dsDNA, ENVELOPED

The members of Polydnaviridae (PVs) are characterized by their polydisperse (multipartite) closed, double-stranded, superhelical DNA genomes (which gave rise to the name poly-DNA-virus) (Krell and Stoltz 1980; Krell et al. 1982; Krell 1991).

The segmented genomes are heterogeneous in size and genetic composition. They occur in parasitoids in the families Ichneumonidae and Braconidae. Two genera have been proposed: (1) *Ichnovirus* with type species *Campoletis sonorensis* CsV and (2) *Bracovirus* with type species *Cotesia melanoscela* CmV (Francki et al. 1991).

Both ichneumonid and braconid PVs replicate only in the ovarial calyx epithelium situated between the ovarioles and oviduct of a wasp (Stoltz and Vinson 1979a; Krell 1991). They cause no noticeable pathology to the wasp. The initiation of viral replication in the ichneumonid *Campoletis sonorensis* occurs in the stage II pupa (stage with only eyes, head, and thorax pigmented) (Norton and Vinson 1983). Virogenic stromata are formed in the nuclei of calyx cells. The viral particles bud from the calyx cells into the calyx lumen, where they accumulate in a dense fluid and are injected by the wasps, during oviposition, into lepidopteran hosts. However, viral-specific DNAs have been detected in other tissues of a female wasp besides the ovary and in male wasp tissues (Fleming and Summers 1986; Stoltz et al. 1986). The viruses do not replicate in lepidopteran hosts (Stoltz and Vinson 1979a,b).

A. ICHNOVIRUS

Members of *Ichnovirus* occur typically but not exclusively in the family Ichneumonidae, especially in the subfamily Campoleginae (Stoltz et al. 1981; Stoltz and Vinson 1979a). Most studies have been conducted with the type species *Campoletis sonorensis* CsV. Rotheram (1967) identifed the first virus in this group in *Nemeritis canescens*. Other ichneumonid genera with PVs are *Bathyplectes*, *Campoplex*, *Casinaria*, *Diadegma* (syn. *Horogenes*), *Eriborus*, *Glypta*, *Mesoleius*, *Olesicampe*, *Tranosema*, and *Venturia* (syn. *Devorgilla*, *Nemeritis*) (Stoltz et al. 1981). Comparative serology (immunoblot technique) of viruses from 14 species representing six genera of ichneumonid wasps has indicated that isolates from unrelated genera of wasps represent different viruses (Cook and Stoltz 1983).

Several different morphologies have been described for the viral particles. Most typical particles are ovoid in shape (130 × 350 nm) with each containing a quasicylindrical (biconvex or lenticular) nucleocapsid (85 × 330 nm) (Stoltz and Vinson 1979a; Krell 1991). The conical ends of the nucleocapsid are not identical and a distinct substructure is visible on the nucleocapsid surface. A hooklike structure is attached to one end of *C. sonorensis* PV. The nucleocapsid is surrounded by two unit membrane envelopes, the inner of which is formed *de novo* in the nucleus and the outer envelope is acquired during budding into the calyx lumen (Norton et al. 1975; Stoltz and Vinson 1979a,b). Another type of particle found in the PV from *Glypta* species has several nucleocapsids per envelope and resembles those of the braconid PV (Stoltz et al. 1981). Other viral particles from the ichneumonids *Bathyplectes curculionis* and *B. anurus* are ovoid (250 × 300 nm) and greatly differ morphologically from those described above (Hess et al. 1980).

Each envelope of a typical ichneumonid PV has a protrusion, which functions in the penetration of the basement membrane and may be involved in a subsequent membrane fusion for the entrance of the nucleocapsid into the cytoplasm of the calyx cell (Stoltz and Vinson 1979a). In an ichneumonid cell, the nucleocapsid enters presumably through a nuclear pore and uncoats in the nucleoplasm. The penetration of a PV in a lepidopteran cell is similar to that in an ichneumonid cell with the uncoating of the DNA in the nucleus, but only limited transcription takes place and without viral replication. The viral DNA persists in a lepidopteran host for up to 10 days postoviposition (Theilmann and Summers 1986).

B. *BRACOVIRUS*

The first virus of this group was reported by Vinson and Scott (1975) in the calyx region of the braconid *Cardiochiles nigriceps* (Stoltz and Vinson 1979a). Other braconid genera with PVs are *Apanteles*, *Ascogaster*, *Chelonus*, *Cotesia*, *Glyptapanteles*, *Hypomicrogaster*, *Microgaster*, *Microplitis*, *Phanerotoma*, *Pholetesor*, *Protapanteles*, and *Protomicroplitis* (Stoltz and Vinson 1979a; Krell 1991). Like the ichnoviruses, the bracoviruses are coinjected with the egg by the parasitoid during oviposition.

The genomes of bracoviruses differ from those of ichnoviruses and are morphologically similar to those of baculoviruses. The genome is packaged into cylindrical nucleocapsids, 40 nm wide and of variable lengths (from 35 to 105 nm) (Stoltz and Vinson 1979a; Krell 1991). Single or multiple nucleocapsids are surrounded by a single unit membrane, except in *Chelonus* near *curvimaculatus*, which has multiple envelopes (Jones et al. 1986). The nucleocapsid structure is asymmetrical and differs from that of baculovirus. In several braconid PVs, a long appendage (8.5 × 2 to 300 nm) is attached at one end of the nucleocapsid (Stoltz and Vinson 1977; Jones et al. 1986) (Fig. 7–1A). The appendage (tail) is morphologically similar to that of the nuclear polyhedrosis virus of *Oryctes* (Payne et al. 1977). The appendage may serve in the penetration of the PV through the basement membrane of a parasitized host larva (Krell 1991) or in penetrating through a nuclear pore to inject the viral DNA into the host nucleus (Stoltz et al. 1988). The uncoating at a nuclear pore and the persistence of the DNA require the presence of the parasitoid venom. Thus, the bracovirus uncoats at a nuclear pore, whereas the ichnovirus PV uncoats within the nucleoplasm.

As in the case of the ichneumonid PVs, the braconid PVs do not replicate in the lepidopteran hosts of the parasitoids, but the virus can penetrate into lepidopteran cells. Viral particles have been observed in the cell cytoplasm of developing muscle tissue, fat body, and nerve tissue of the lepidopteran host (Poinar et al. 1976).

C. GENOME

The genome of Polydnaviridae is a multicomponent structure consisting of a number of different sizes of circular and superhelical DNAs (Krell and Stoltz 1979, 1980; Krell et al. 1982). The number of DNAs and their sizes vary with different PV species. For example, the *Campoletis sonorensis* PV has at least 28 circular DNA segments with sizes from 6 to 20 kilobase pairs (kbp) (Theilmann and Summers 1988). The different circular DNAs occur in nonequimolar ratios and are not derived one from the other and exhibit only limited DNA homology among them (Krell and Stoltz 1980; Krell et al. 1982). However, some regions of the DNAs have homologous sequences and there is a complex pattern of cross hybridization (Theilmann and Summers 1987; Blissard et al. 1986a, 1989).

Physical and transcription maps of the genome of *C. sonorensis* PV have been constructed for some of the superhelical DNA molecules (Blissard et al. 1986b, 1989; Theilmann and Summers 1988). Viral messenger RNAs (mRNA) have been isolated (Fleming et al. 1983; Blissard et al. 1986b) and they may be expressed from

7 OTHER DNA-VIRAL INFECTIONS

FIGURE 7-1

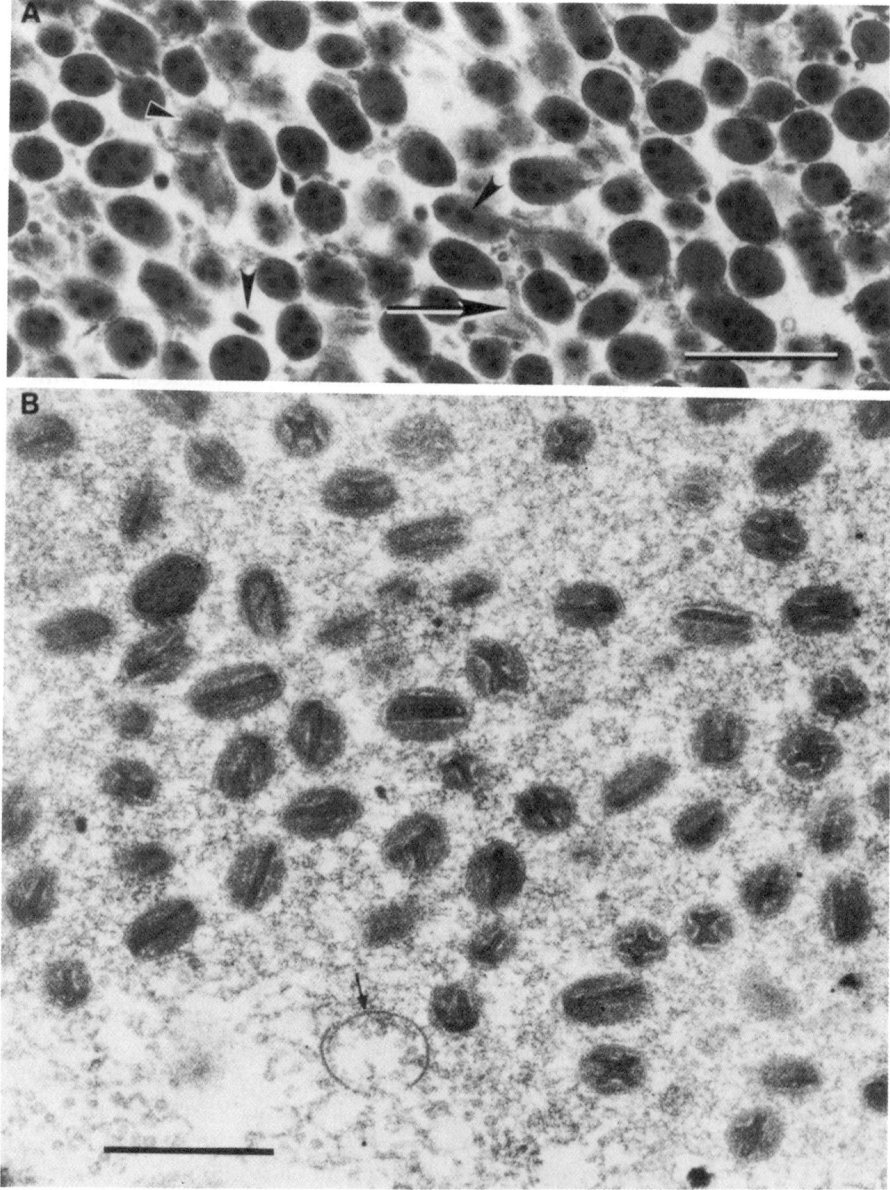

Polydnavirus and entomopoxvirus. (A) Polydnavirus in the calyx fluid in the lumen of the common oviduct of *Cotesia marginiventris*. *Triangle*, unit-length nucleocapsid; *arrowheads*, double-length nucleocapsids; *arrow*, tail of viral particle extending from the envelope. (Bar = 500 nm.) [Courtesy of Hamm et al. (1990) and Academic Press.] (B) Entomopox virus in a section of the salivary gland of the bumble bee, *Bombus impatiens*. Note, in cross section, the four lobes (vanes) extending from the nucleoid. *Arrow*, partially developed shell with small spheres. (Bar = 0.5 μm.) [From Clark (1982) and courtesy of Academic Press.]

a number of DNA molecules or from one "master" molecule (Blissard et al. 1986b). Genes encoding two closely related mRNAs have been located on a single DNA segment (Blissard et al. 1987). Some genes are expressed in a species-specific manner and others are expressed both in the wasp and lepidopteran larva (Blissard et al. 1989).

The DNAs of PVs are found in integrated chromosomal and extrachromosomal forms in the tissues of the ichneumonid and braconid wasps (Fleming and Summers 1986; Stoltz et al. 1986). The chromosomal form (i.e., integrated into the wasp genome) probably determines the maternal transmission of the viral genome to the parasitoid progeny in a Mendelian fashion (Fleming and Summers 1986, Stoltz 1990).

D. VIRUS–PARASITOID–LEPIDOPTERA RELATIONSHIPS

Major studies on the relationships of the virus, parasitoid, and lepidopteron have been conducted with the ichneumonid PVs, but most of the following apply also to the braconid PVs. The PVs do not cause pathologies to the wasps in which they replicate (Stoltz and Vinson 1979a). Although they do not replicate in the lepidopteran hosts of the parasitoids, limited transcription may occur from certain viral circular DNAs (Blissard et al. 1986a,b, 1987; Fleming and Summers 1986).

There is a mutualistic relationship between the PVs and parasitoids (Stoltz and Vinson 1979a; Edson et al. 1981; Faulkner 1982). The mutualism involves immunosuppression (i.e., the failure of the lepidopteran host to encapsulate the eggs and larvae of the parasitoid and the detrimental effects of the PV on the host lepidopteron). The immune and developmental systems of the lepidopteron are altered by the venom secretions alone (Kitano 1982, 1986), by both the venom and a PV (Guillot and Vinson 1972; Ables and Vinson 1981; Tanaka 1987), or by a PV alone (Vinson 1972; Bedwin 1979; Edson et al. 1981).

The ichneumonid PVs have debilitating effects on the development and physiology of the lepidopteran hosts of the parasitoid, such as reduced growth and retarded development (Vinson et al. 1979), reduced ecdysteroid production (Dover et al. 1987), reduced number, transformation, and altered behavior of hemocytes (Stoltz and Guzo 1986; Davies et al. 1987), and an elevated level of trehalose that is utilized by the parasitoid (Edson and Vinson 1977). The detrimental effects of the virus on the host lepidopteran larvae are beneficial to the parasitoid. The mutualistic endosymbiosis between the virus and parasitoid may explain the evolution of the polydnavirus (Whitfield 1990; Webb and Summers 1990).

II. POXVIRIDAE—dsDNA, ENVELOPED

Poxviridae is a large family containing members infecting vertebrates and invertebrates. Those infecting vertebrates cause the familiar diseases, smallpox, cowpox, fowlpox, etc. These viruses are distinguished by their size (largest of animal viruses), complex morphology (lack icosahedral symmetry), and intricate replication in the cytoplasm.

Poxviruses infecting mammals and birds were known for many years before they were found in the coleopteran *Melolontha melolontha* in 1963 by Vago. Insect poxviruses occur in widespread geographical locations and in approximately 35 insect species in the orders: Coleoptera, Lepidoptera, Diptera, Orthoptera, and

Hymenoptera (Clark 1982; Arif 1984). They are called entomopoxviruses (EPVs) because they are restricted to insects and are included in the subfamily Entomopoxvirinae with three probable genera (A, B, C) distinguished by the morphology of the virions, host range, and molecular weights of the genome (Matthews 1982; Goodwin et al. 1991). A fourth genus (D) has been proposed by Goodwin et al. (1991) for an EPV discovered in Hymenoptera by Clark (1982) (Fig. 7–1B). The genus A contains isolates from Coleoptera with the type species from *Melolontha melolontha* and an isolate from the lepidopteran species *Oncopera alboguttata*; genus B includes isolates mostly from Lepidoptera with the type species from *Amsacta moorei* and several orthopteran species; and genus C contains those from Diptera with the type species from *Chironomus luridus*. There are no serological relationships among the genera A, B, and C or with the vertebrate poxviruses. The viral particles of genus C most closely resemble vertebrate poxviruses (Goodwin et al. 1991).

A. STRUCTURE AND PROPERTIES

The entomopoxviruses have many similarities to the orthopoxviruses (vaccina subgroup) (Arif 1984; Goodwin et al. 1991). The virions are brick-shaped or oval, from 150 to 470 nm long and from 165 to 300 nm wide, with a folded outer membrane giving the appearance of a mulberry-like surface (Kurstak and Garzon 1977; Goodwin et al. 1991). The largest virions occur in Coleoptera and the smallest in Diptera and Hymenoptera. The electron-dense core (nucleoid) is surrounded by a multilayer membrane. The EPVs from Orthoptera and Lepidoptera generally contain a cylindrical core and two lateral bodies; those from Diptera contain a biconcave core and two well-developed lateral bodies; those from Coleoptera contain a unilateral concave core and one lateral body located in the cavity of the core; and those from Hymenoptera have a vaned (four lobes) core (Fig. 7–1B). Within the core, a coiled rope- or rod-like, electron lucent structure, folded into 4 to 6 segments, has been reported in some EPVs but not in others (Kurstak and Garzon 1977). Bergoin et al. (1971) report that the rope-like structure is not visible in virions free in the cytoplasm or in the process of being occluded in the spheroid; however, Hukuhara et al. (1990) have observed a similar structure in nonoccluded and occluded virions.

Viral structural proteins vary with the EPV isolates (e.g., *Euxoa auxiliaris* EPV has 24, *A. moorei* EPV has 36, and *Melanoplus sanguinipes* EPV has 39 polypeptides) (Langridge and Roberts 1982). In the *Choristoneura biennis* EPV, there are at least 40 polypeptides with molecular weights (MWs) from 12,000 to 250,000, with a major component of 59,000 (Bilimoria and Arif 1980).

The EPV genome is a linear dsDNA that constitutes approximately 5% of the viral particle (McCarthy et al. 1975a). Genome sizes of EPVs from Lepidoptera are smaller than those of EPVs from Diptera and Coleoptera. The EPV genome is similar to that of the orthopoxvirus but differs fundamentally in the low guanine + cytosine content (Arif 1984). There are four enzymes in the EPV as in the case with the vertebrate poxviruses: nucleotide phosphohydrolase, acidic and neutral deoxyribonucleases, and a DNA-dependent RNA polymerase, but others may occur (Pogo et al. 1971; McCarthy et al. 1974, 1975b).

B. INCLUSION BODIES

Two types of inclusion bodies may occur in EPVs. The first, called the spheroid, occludes viral particles, and the second, called the spindle because of its spindle shape, does not (Goodwin and Filshie 1969). Both types are found in the EPVs of Lepidoptera and Coleoptera, only spheroids are found in those of Orthoptera and Diptera, and none are found in those of Hymenoptera (Granados 1973a; Arif 1984; Goodwin et al. 1991). Both the spheroid and spindle are formed in the cytoplasm.

The spheroid has a paracrystalline proteinaceous structure and measures from 20 to 24 μm in diameter (Kurstak and Garzon 1977). The matrix protein is called spheroidin and has MW 102,000 for the *Choristoneura biennis* EPV (Bilimoria and Arif 1979) and 110,000 for the *Amsacta moorei* EPV (Langridge and Roberts 1982). Spheroidin is similar to the polyhedrin of a nuclear polyhedrosis virus and the granulin of a granulosis virus except that it is from 2 to 3 times larger than the others. The spheroidin has a high content of cysteine and methionine and requires a disulfide reducing agent, such as sodium thioglycollate or 2-mercaptoethanol, during the alkaline dissolution of the spheroids (Bergoin et al. 1970). The number of virions occluded in a spheroid varies greatly in different hosts and even in different cell types of the same host (Kurstak and Garzon 1977).

An alkaline protease has been detected in the spheroid (Bilimoria and Arif 1979; Langridge and Roberts 1982). However, it may be a contaminant from the insect's digestive tract as in the case of the alkaline protease in polyhedra and capsules of the baculoviruses because the enzyme is absent in spheroids collected from infected tissue culture cells (Langridge and Roberts 1982).

Spindles do not occlude virions. They appear in the cytoplasmic perinuclear area and develop in the interior of ergastoplasmic vesicles (Vago and Bergoin 1968; Bergoin et al. 1976). They are surrounded by a tripartite membrane. The nucleus may be involved in the synthesis of a spindle. Spindles may also be occluded, with or without the virions, by spheroids (Bird 1974; Kurstak and Garzon 1977). The proteinaceous molecular lattice of the spindle is approximately 5.8–6.0 nm (Bird 1974; Bergoin et al. 1976). Spindles range greatly in size from 0.5 to 12 μm, even within the same larva (Vago and Bergoin 1968) and some are as large as 25 μm (Kurstak and Garzon 1977).

A spindle contains the same amino acids as a spheroid, but their proportions differ, particularly in the 2.3 times higher leucine residue in a spheroid as in a spindle (Bergoin et al. 1970). The spindle and spheroid are antigenically distinct (Croizier and Veyrunes 1971). The function and significance of the spindles are unknown.

C. REPLICATION

The entomopoxvirus resembles the vertebrate poxvirus in structure and developmental cycle (Bergoin et al. 1969). When ingested by an insect, the spheroids disintegrate in the alkaline environment of the digestive tract. The liberated virions attach and fuse to the plasma membrane of a microvillus; the viral core and portion of the lateral body enter the cell (Granados 1973b). Viropexis does not appear to play a significant role in the viral infection through the midgut but is the usual mode of cell entry within the hemocoel (Devauchelle et al. 1971). After viral uncoating in an infected cell and a certain period of latency, cytoplasmic foci appear consisting of either electron-dense amorphous material (type 1 virogenic stroma or viroplasm) or aggregates of granular material interspersed with spherical vesicles (type 2). The type 1 virogenic stroma is found in most EPV infections, with both types occurring in some infections, and only type 2 virogenic stroma in others, such as in *Chironomus attenuatus* (Granados 1973a).

The first recognizable viral structures are incomplete crescentlike shells or membranes appearing at the periphery of a virogenic stroma. Membranes develop and enclose a mass of dense material to initiate viral development. Granados and Roberts (1970) interpreted the membrane or shell around an immature virion as composed of two concentric unit membranes, but others consider the shell as a unit membrane complex with a layer of spicules (Stoltz and Summers 1972; Kurstak and Garzon 1977; Hukuhara et al. 1990).

In the type 2 virogenic stroma, there are crescent or archlike envelopes associated with fibrillar material of low to moderate density containing a large number of vesicles that may play an important role in the formation of immature viral envelopes (Bergoin et al. 1969; Stoltz and Summers 1972). The incomplete viral envelope progressively encloses granulated materials that eventually condense to form immature particles found in the type 1 virogenic stroma. The matter within these particles begins to differentiate and forms the highly condensed mass of the viral nucleoid. The nucleoid at this stage has no distinct shape or particular structure. It differentiates into a mature core surrounded by a three-layered membrane and assumes a rectangular shape with a concomitant loss of the outer layer of spicules. The lateral body assumes a recognizable structural form. The outer membrane subsequently folds to give the appearance of a beaded mulberry-like structure. The replication of EPVs occurs in the cytoplasm, except for the report by Quiot et al. (1975) who observed immature virions in the nuclei of *Lymantria dispar* cells.

When an EPV is inoculated into the insect's hemocoel, the virion penetrates the cell by viropexis (endocytosis) (Devauchelle et al. 1971). The "denudation" of the virion takes place in phagosomes derived from the viropexis process, followed by the release of the nucleoid in the cytoplasm where the uncoating and liberation of the viral DNA genome take place (Kurstak and Garzon 1977).

A crystallogenic matrix in the cytoplasm is involved in the formation of spheroids (Hukuhara et al. 1990). The occluded mature virions appear to be arranged at random or radially in cross section in a spheroid (Bergoin et al. 1971; Goodwin and Filshie 1975). Occluded virions are more compact than nonoccluded ones.

Nonoccluded virions emerge from a cell by budding or exocytosis (Granados 1973a; Hukuhara et al. 1990). They are also released when a cell ruptures and the released virions infect other host cells (Hukuhara et al. 1990).

D. PATHOGENESIS

The host range of EPVs is relatively restricted with no cross-infections between different insect families (Kurstak and Garzon 1977). Some EPVs show tissue specificity being restricted to the fat body and others infect a few tissues or produce systemic infections. The hymenopteran EPV differs from other EPVs and infects only the salivary gland and hypodermis of a bumble bee (Clark 1982; Goodwin et al. 1991).

In a cell infected with EPV, the cytoplasm and nucleus hypertrophy and fat granules decrease in numbers. With the formation of spheroids in the cytoplasm, the nucleus shows changes: margination and fragmentation of the chromatin; formation of dense, amorphous masses; bundles of fibrils; and/or paracrystalline inclusions. The tissue disintegrates at an advanced stage of infection. In *Camptochironomus tentans* and *Choristoneura fumiferana*, the EPV infection results in an increased multiplication of fat body cells (Weiser 1969; Bird 1976).

Lepidoptera infected with EPVs show external symptoms of whitening and softening. Symptoms appear in 6 to 20 days and mortality occurs in 12 to 72 days after infection. Pupation is often delayed with the larva growing to more than double their normal size and weight (Bird et al. 1971). The large larval size is due to an increase in the size of fat bodies caused by cell proliferation and hypertrophy (Bird 1976).

An EPV infection in a coleopteran larva is very slow to develop (Goodwin and Filshie 1969). The first sign of infection is a progressive whiteness caused by the proliferation of infection in the fat body and epidermis. Most characteristic is the distinct white-spotted or mottled appearance of the dorsoposterior area. The dissected fat body has a brightly reflective, foamy appearance (Weiser 1965a).

In an infected fourth instar larva of the midge *Chironomus attenuatus* irregularly distributed white masses appear just beneath the integument (Stoltz and Summers 1972). The infected larva appears normal in size and is as active as an uninfected larva, but it generally dies in the late fourth instar. The virus appears to replicate only in hemocytes, which form large aggregates, either attached or in close proximity to other tissues and organs, such as the trachea, epidermis, muscle, nerve, and fat body.

III. ASCOVIRIDAE—dsDNA, ENVELOPED

The family Ascoviridae has been proposed by Federici (1983) for an unusual type of virus that infects noctuids (Fig. 7–2). The disease is characterized by a milky-white discoloration of the hemolymph, which is due to the accumulation of high concentrations of virion-containing vesicles not observed in other diseases of insects. The unique virion-containing vesicles are formed by the cleavage of host cells.

A. STRUCTURE

The ascovirus has linear dsDNA genome of about 170 kilobase pair (kbp), (Federici 1983; Federici et al. 1990, 1991). Enveloped virions are large, approximately 130 × 400 nm. The envelope exhibits a distinctive hexagonal pattern in negatively stained preparations. The isolates from *Heliothis* and *Trichoplusia* are allantoid in shape (Fig. 7–2E). Those from *Spodoptera* are bacilliform, and after formation, they are often occluded in vesiculate occlusion bodies consisting of microvesicles intermixed with protein and virions rather than containing principally a single protein, such as the polyhedrins of baculoviruses and cytoplasmic polyhedrosis viruses. There are 12 polypeptides of 10,000 to 200,000 daltons with a major protein of 50,000 daltons in the virion. The genomes of the *Trichoplusia* and *Heliothis* isolates are about 180 kbp and that from *Spodoptera* isolate is 140 kbp. Comparisons by viral peptides, DNA fragment profiles, and DNA–DNA hybridization demonstrate that the isolates from *Trichoplusia* and *Heliothis* are closely related but not identical, and they do not hybridize with the DNA of the *Spodoptera* isolate. Thus, the isolates from *Trichoplusia* and *Heliothis* are variants of the same virus, whereas the one from *Spodoptera* is a separate member of the ascovirus group indicating at least two distinct viral biotypes or species within the ascovirus group. This conclusion is supported also by the host and tissue specificities of the isolates.

7 OTHER DNA-VIRAL INFECTIONS

FIGURE 7–2

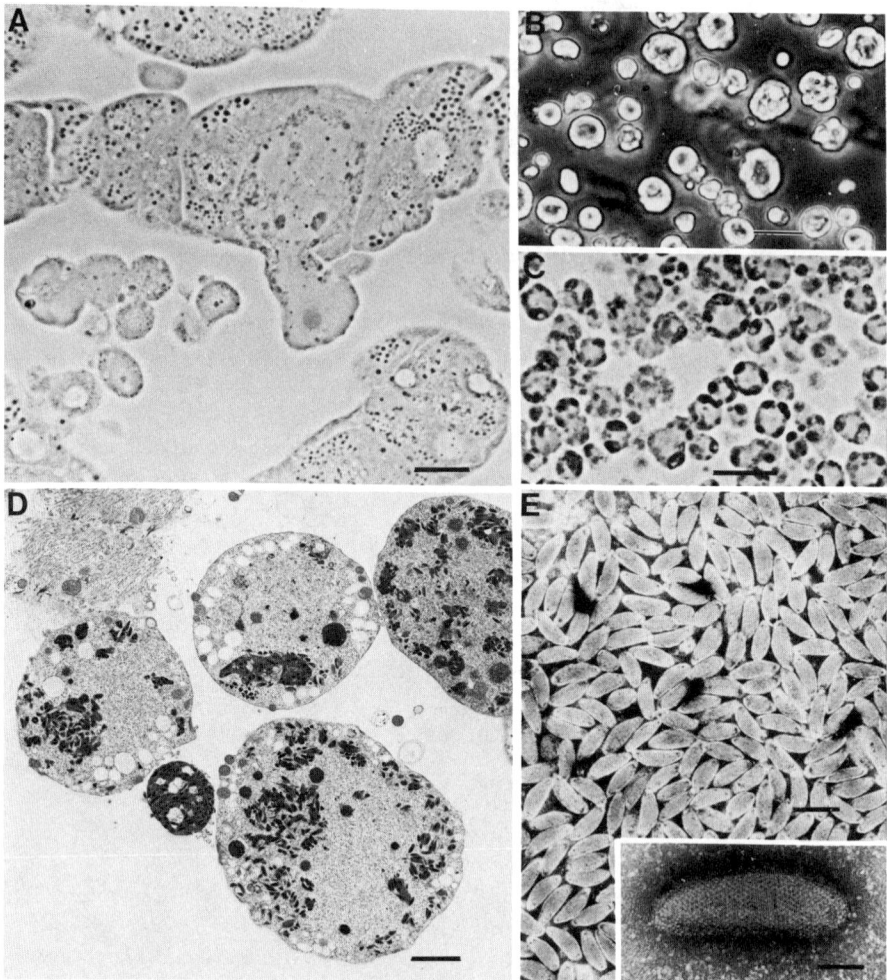

Ascovirus from noctuid larvae. (A) Thick plastic section (Epon-Araldite) under phase contrast microscopy showing viral-infected fat body cell erupting prior to vesicle formation. (B) Wet mount of hemolymph taken under phase contrast illustrating the refractile vesicles characteristic of ascoviral infection. At this stage the hemolymph becomes milky white. (C) Thick plastic section using phase contrast demonstrating the vesicles in the hemolymph. (D) Electron micrograph of vesicles illustrating internal structure of vesicles. Note viral particles. (E) Negatively stained virions illustrating allantoid shape of the particle. *Insert*, negatively stained mature virion illustrating the reticulate surface. (A), (B), (C), and (E) from *Trichoplusia ni*; (D) from *Scotogramma trifolii*. [Bars in (A), (B), (C) = 10 μm; (D) = 1 μm; (E) = 350 nm; insert = 100 nm.] (Provided by B. A. Federici.)

B. REPLICATION

Viral replication begins in the nucleus with the development of an apparent virogenic stroma (Federici 1983). Typically, the assembly of virions is not directly associated with viroplasmic centers. Viral morphogenesis is initiated after the nuclear membrane ruptures, and occurs prior to and during the cleavage of the cell into vesicles (Federici 1983; Federici et al. 1991). A multilaminar layer (unit membrane) of the inner particle is the first recognizable component of the virion. A dense nucleoprotein core forms internally to the unit membrane. As the process continues, the inner particle gradually assumes an allantoid or bacilliform shape and then is enveloped. The fully formed ascovirus consists of an internal DNA–protein core enclosed in a unit membrane, which in turn is surrounded by an envelope.

C. CYTOPATHOLOGY

In an infected cell, the nuclear envelope invaginates and the nucleus and cell are greatly hypertrophied (Federici 1983; Federici et al. 1991). Eventually, the nuclear envelope ruptures and fragments. Sheets of cytoplasmic membrane assemble throughout the cell and coalesce, partitioning the cell into a cluster of vesicles. After the vesicles are formed, they dissociate from each other and accumulate within the tissue in which they are formed. As the disease progresses, the basement membrane of the infected cell breaks down, releasing the highly refractile vesicles (from 2 to 20 μm in diameter) into the hemolymph where they accumulate in large numbers ($>10^8$ vesicles per ml) (Federici 1982; 1983) (Fig. 7–2A,B,C,D). Their presence causes the hemolymph to turn opaque white.

D. PATHOGENESIS

Ascovirus isolates from *T. ni* and *Heliothis* spp. have been trasmitted *per os* or by injection parenterally into larvae of several other species of noctuids (Federici 1983; Hamm et al. 1986). The isolate from *S. frugiperda* has been transmitted only to other species of *Spodoptera* and may be biologically distinct from other isolates (Hamm et al. 1986).

The ascovirus is highly infectious when inoculated parenterally, even with contaminated minuten pins, but not when fed to the larvae (Govindarajan and Federici 1990). The braconid parasitoid *Cotesia marginiventris* readily transmits an ascovirus, with its ovipositor, to *Spodoptera* larvae (Hamm et al. 1985). The parasitoid is adversely affected by the viral infection in the host larva, a situation similar to that reported in baculoviral infections.

After hemocoelic inoculation, the virion-containing vesicles appear within 72 h in the hemolymph, and most larvae die in 10 to 21 days (Federici 1983). Viral isolates from *Trichoplusia* and *Heliothis* infect a broad range of tissues, such as the tracheal matrix, epidermis, and fat body, whereas the isolate from *Spodoptera* infects only the fat body, which is destroyed completely (Federici and Govindarajan 1990).

The ascovirus causes a chronic infection. Infected *Trichoplusia* and *Heliothis* larvae turn opaque yellowish-white and have difficulty in molting (Browning et al. 1982; Federici 1982; Carner and Hudson 1983; Govindarajan and Federici 1990). They show reduced feeding activity and retarded growth rate with concomitantly increased larval longevity as compared with uninfected larvae.

IV. IRIDOVIRIDAE—dsDNA, NONENVELOPED

Viruses in the Iridoviridae have been called icosahedral cytoplasmic deoxyriboviruses (Kelly and Robertson 1973; Anthony and Comps 1991). The name iridovirus is derived from the characteristic iridescence observed in infected insect tissues and in centrifuged-virus pellets. The iridescence results from the Bragg reflection of visible light through microcrystals of viral particles that are arranged in a face-centered cubic lattice with appropriate periodic spacing (Williams and Smith 1957; Klug et al. 1959). The centrifuged pellet of the *Tipula* iridescent virus has an iridescent turquoise (bluish) appearance in reflected light and is orange or amber in transmitted light (Xeros 1954). The iridescence is from orange or green to blue for the mosquito iridescent virus (Clark et al. 1965; Chapman et al. 1966). However, not all insect viruses in this family produce iridescence. It does not occur in a virus isolated from *Chironomus plumosus* possibly because the long fibrils, attached to the viral capsids, prevent the formation of microcrystals (Stoltz et al. 1968). This and other chironomid larvae infected with iridoviruses do not display iridescence and they only indicate infection by markedly reduced activity (Anthony and Comps 1991).

There are five genera in this family with two of them infecting insects: the small iridescent insect virus group with viral particles of 120 to 130 nm in diameter (*Iridovirus*) and the large iridescent insect virus group with particles of approximately 180 nm in diameter (*Chloriridovirus*) (Matthews 1982). The remaining three genera occur in vertebrates. Insect iridoviruses do not replicate in vertebrate cells, except for the *Chilo* iridescent virus (CIV), which replicates in vertebrate (viper spleen) cells (McIntosh and Kimura 1974). The CIV also causes lethal toxicity when inoculated peritoneally into frog and mouse, but it does not replicate in these vertebrates (Ohba and Aizawa 1981, 1982).

The iridoviruses form the largest group of nonoccluded viruses infecting insects (Kelly 1985). The type species of *Iridovirus* is the *Chilo* iridovirus (CIV) isolated by Fukaya and Nasu (1966) from the rice stem borer, *Chilo suppressalis*. The type species of *Chloriridovirus* is the type 3, regular (R) strain of mosquito iridovirus (RMIV) isolated from *Aedes taeniorhynchus* (Clark et al. 1965; Woodard and Chapman 1968). The name *Chloriridovirus* is a misnomer when applied to RMIV, which produces yellow to orange iridescence, and it is the turquoise type (TMIV) that forms greenish to bluish iridescence.

The nomenclature of iridoviruses based on insect hosts created a problem when increasing numbers were isolated from insects, when many viruses were found to replicate in the same host, and when different viruses were found infecting the same host in nature. These isolates are serologically, physically, and biochemically distinct. Tinsley and Kelly (1970) proposed provisionally to name the iridoviruses chronologically as types. Until 1985, there were 32 isolates designated as types (Kelly 1985) and a number of others that have not been classified (Hall 1985). Types 1, 2, 6, 9, 10, and 16–29 are members of *Iridovirus* and types 3–5, 7, 8, 11–15 are members of *Choriridovirus* (Matthews 1982).

A. STRUCTURE AND PROPERTIES

The iridoviruses are icosahedral in shape. This was first demonstrated through the double-shadow casting of frozen-dried viral particles observed with the electron microscope (Williams and Smith 1958). Insect iridoviruses possess lipid, DNA, protein, and no RNA or polyamines (Kelly 1985). Viral particles have diameters of 125 to 300 nm, with one molecule of dsDNA that has MWs ranging from 100 to 250 $\times 10^6$ and forms about 12 to 30% by weight of the viral particle (Matthews 1982). The DNA is organized into a nucleosomal structure and is coiled in such a way that its main part is closely associated with the coat protein (Klump et al. 1983). The linear viral DNA is both permuted and terminally redundant (Delius et al. 1984). This is a unique genomic feature among viruses of eucaryotic cells and occurs in both vertebrate and insect viruses of this family.

Physical maps of the DNA genomes of iridovirus type 9 (from *Wiseana* spp.) and type 6 (from *Chilo suppressalis*) have been constructed (Ward and Kalmakoff 1987; Fischer et al. 1988a). Terminally repetitive DNA elements and inverted repeat DNA nucleotide sequences occur in the genome (Fischer et al. 1988a,b). The inverted repeat DNA sequence confirms the observation of stem-loop structures with the electron microscope. The repetitive DNA elements in the virus type 6 genome have an extremely complex arrangement that has no counterpart in other known DNA or RNA virus systems (Fischer et al. 1988b). The repeat sequences are important because they are associated with the regulatory functions during viral replication.

There are 13–25 structural polypeptides with MW from 10 to 250×10^3. The major structural protein of several types of iridoviruses is similar and may have a common function of forming a complete surface on the outside of the virion (Moore and Kelly 1980).

Enzymes, such as RNA polymerase, nucleotide phosphohydrolase, protein kinase, and deoxyribonuclease, may be present in the virion (Kelly and Tinsley 1973; Monnier and Devauchelle 1976; Kelley et al. 1980c; Devauchelle et al. 1985b). The unenveloped particle contains 5–9% lipids, mainly phospholipid, as an integral part of the icosahedral shell (Matthews 1982). The MWs of the virions range from 500 to 2000×10^6. Both *Iridovirus* and *Chloriridovirus* are resistant to ether whereas the other three genera are susceptible to ether and nonionic detergents.

The virion has a spherical core (nucleoid) made up of DNA and associated proteins (Devauchelle et al. 1985a). The core is surrounded by an inner membrane and an external shell (capsid shell). The external coat or shell determines the icosahedral symmetry. The viral structure is complex, consisting of two outer icosahedral shells and two inner spherical shells (Kelly 1985). The innermost shell is a highly hydrated nucleoprotein core that is enclosed by a lipid bilayer (4 nm thick), which in turn is enclosed by a dense protein shell containing almost 50% of the viral mass. Another highly hydrated protein shell is found beyond this dense protein shell. Virions emerging from a cell may acquire an envelope from the cell plasma membrane, but it is not essential for infection. The isolated core (nucleoid) of the mosquito iridovirus does not infect the mosquito larva or cell line, and the capsid is needed for infectivity (Wagner et al. 1975).

B. REPLICATION

Most studies on the replication of iridoviruses have been conducted with insect cell cultures and with viruses inoculated into the insect's hemocoel. Little is known about the invasion through the digestive tract. Within the hemocoel and in cell cultures, the virion is adsorbed to the cell plasma membrane, enters the cell through pinocytosis, and occurs in lysosomes within 1 to 2 h (Leutenegger 1967; Younghusband and Lee 1969). Phosphatase activity may occur in the lysosome (Younghusband and Lee 1970). Within the cytoplasm, several investigators have reported that some virions are enclosed in a membrane and others are free and undergoing uncoating (Kelly and Tinsley 1974a; Webb et al. 1976). Mathiesen and Lee (1981), however, have found no free virions in the cytoplasm and only virions in pinocytotic vesicles, which coalesce into lysosome-like structures packed with virions. The uncoating of a virion takes place in the pinocytotic vesicles.

Virogenic centers form in the cytoplasm at 4.5 to 6 h postinfection (h p.i.) (Leutenegger 1967; Mathiesen and Lee 1981). The DNA accumulates in the dense matrix and the newly assembled virions occur in the electron-lucent matrix of the virogenic centers. Host DNA synthesis is inhibited within 4 h p.i. and the rates of cellular RNA and protein syntheses are also affected (Mathiesen and Lee 1981). A structural protein(s) of the virion inhibits the host DNA synthesis (Goorha and Granoff 1979; Lee and Brownrigg 1982). This protein occurs in infectious virions and in noninfectious, partially filled virions and empty capsids.

During the initial 5 h p.i., the cellular RNA and protein syntheses decline rapidly and by 7 h p.i., the viral RNA and protein syntheses increase in the viroplasmic centers. The viral RNA may be synthesized initially in the nucleus and then transported to the cytoplasm as in the case of the frog iridescent virus (Goorha et al. 1978; Bladon et al. 1986).

In the electron-dense areas of the cytoplasm, virions with intact capsid shells are found in various stages of core density. The following hypotheses have been proposed for the formation of the mature virion: (1) the capsid shell is first assembled prior to the entry of the DNA (Smith 1958; Yule and Lee 1973), (2) the open membrane (capsid shell) and the nucleoprotein core are assembled concurrently and the membrane closes when filled with viral DNA (Xeros 1964), and (3) the viral DNA (nucleoid) is first formed and then is encapsidated by the protein (capsid shell) (Bird 1962; Kelly and Tinsley 1974a,b).

In most insect iridoviruses, the viral particles aggregate in the cytoplasm and then bud out through the plasma membrane, but in the mosquito iridovirus, the particles, without aggregating, bud from the cell shortly after assembly (Webb et al. 1976). The virions are also released by cell lysis and by exocytosis (i.e., extrusion of vacuoles containing clusters of viral particles) (Kelly and Tinsley 1974a).

A huge amount of iridoviral particles is produced in a single *Galleria* host [e.g., the *Tipula* iridovirus makes up no less than 25% of the dry weight of a larva (Williams and Smith 1957); the *Sericesthis* iridovirus is produced in the order of 2-mg viral protein per pupa (Day and Mercer 1964)].

Temperature is an important factor in the replication of iridoviruses, and above 32°C, no virus is produced in the inoculated larvae (Tanada and Tanabe 1965; Day and Dudzinski 1966), and at 30°C, 160,000 times more viral particles are required to cause mortality than at 20°C (Witt and Stairs 1976). At lower temperatures in *Galleria* larvae inoculated with SIV, the viral growth rate is at a maximum with 0.56 mg of virus produced per larva per day, but at 16°C, 0.030 mg of virus is produced (Day and Dudzinski 1966).

C. HOSTS

Insect iridoviruses have been isolated from Diptera, Lepidoptera, Coleoptera, Hymenoptera, and Hemiptera (Hall 1985) and recently in Orthoptera (Boucias et al. 1987). Some iridoviruses have a broad host range infecting insects in several orders, such as the *Tipula* iridovirus (TIV, type 1) (Smith et al. 1961), the *Sericesthis* iridovirus (SIV, type 2) (Steinhaus and Leutenegger 1963), and the *Chilo* iridovirus

(CIV, type 6) (Mitsuhashi 1967; Devauchelle et al. 1985b). The CIV also infects other classes of arthropods. When inoculated parenterally into terrestrial Crustacea (pill bug, *Armadillidium vulgare*, and slater, *Porcello scaber*) and into Chilopoda (house centipede, *Thereuonema higendorfi*), the CIV has infected them (Ohba and Aizawa 1979a). Moreover, the iridovirus of the pill bug (*A. vulgare*) replicates in *Galleria* larvae (Ohba et al. 1982). Other iridoviruses have a limited host range, especially those of aquatic insects, such as mosquitoes and chaoborids (Woodard and Chapman 1968; Chapman et al. 1966, 1971). Mosquito iridoviruses (MIVs) do not infect insects in other orders, but the CIV from a lepidopteran orally infects 13 mosquito species in five genera (Fukuda 1971).

D. PATHOGENESIS

Many iridoviruses are not highly infectious when ingested by larvae even in their early instars, although infection may occur by cannibalism through wounds received in fighting, by general trauma, or through parasitoids (Kelly 1985). In *Heliothis zea*, the infection rate of an iridovirus in neonate larvae is from 18 to 40%, in 5-day-old larvae from 11 to 47%, and in 9-day-old larvae from 10 to 30% (Sikorowski and Tyson 1984). On the other hand, iridoviruses can infect readily when inoculated into the hemocoel [e.g., the *Sericesthis* iridescent virus can infect at about 1 to 7 viral particles per inoculated *Galleria* larva (Day and Gilbert 1967)]. Thus, most studies with lepidopteran larvae have been conducted with parenteral inoculations in older larvae.

Iridoviruses cause systemic infections involving most insect tissues, particularly the fat body whose cells may become almost devoid of cytoplasmic organelles (Anthony and Comps 1991). The TIV infection in insect cell cultures causes cell fusion, nuclear alteration, and a rapid inhibition of host-cell macromolecular synthesis (Mathiesen and Lee 1981). *Sericesthis* iridovirus (SIV) and *Chilo* iridovirus (CIV) infections produce an initial cell contraction at 72 h p.i. to about half the normal size, extensive microtubule and paracrystal formations, and large amorphous masses (Kelly and Tinsley 1974a). In the cytoplasm, the mosquito iridovirus (MIV) infection produces tennous filaments, granulations and vacuolization; and in the nucleus, the MIV infection produces the margination of chromatin, convolution, and elongation (Webb et al. 1976).

Hemocytes of *Galleria* larva are susceptible to infection from iridoviruses and only the oenocytoids are not infected (Lea 1985). Infected plasmatocytes and adipohemocytes undergo nuclear and cytoplasmic hypertrophy, accompanied by nuclear DNA synthesis. High dosages of infective and UV-irradiated (noninfective) CIV cause an early drastic cytopathy of hemocytes, which are completely destroyed within 8 h p.i. (Ohba and Aizawa 1979b). Larvae, inoculated with a large dose of UV-irradiated CIV, do not regain the hemocyte numbers, progressively lose body weight, and are stunted.

Insect iridoviruses generally cause lethal infections, but low dosages and exposure in older larvae and pupae may result in the infected larvae and pupae developing to adults that also may be infected but capable of producing offspring (Day 1965). Infected pupae and adults are often deranged. Infected larvae do not behave abnormally until shortly before their deaths. The first sign of infection is the appearance of iridescence usually in the extremeties of structures, such as mouthparts, prolegs, and thorax and the iridescence spreads in intensity as the infection progresses. The iridescence is obvious in insects with lightly or nonpigmented integuments or in the region of the intersegmental membranes. At a late stage of infection, the larvae are lethargic and stop feeding. A moribund mosquito larva may show little or no iridescence and appears milky-white at the time of death (Clark et al. 1965).

Older mosquito larvae are much more resistant to infection when fed the mosquito iridovirus (MIV) than newly hatched and early instar larvae (Woodard and Chapman 1968; Matta and Lowe 1970). The resistance in older larvae may be due to the peritrophic membrane, which, by preventing the penetration of the viral particles, causes them to be degraded in the midgut (Stoltz and Summers 1971). How the virus enters the mosquito's hemocoel has not been established. There is one report of infection in the midgut cell of the mosquito (Hembree and Anthony 1980). The infected cell was located laterally to the invaginated foregut.

There is no evidence of vertical (transovarial) transmission of iridoviruses in insects except in the mosquitoes *Aedes taeniorhynchus* and *Psorophora ferox* (Woodard and Chapman 1968; Linley and Nielson 1968a). The regular MIV infects the ovaries of mosquitoes and may be transmitted within the egg to the progeny (Anderson 1970; Hall and Anthony 1971). Adult transmission to the progeny may occur when the virus is applied topically or inoculated parenterally to adults of *Aedes taeniorhynchus* (Fukuda and Clark 1975). The transovarial transmission in *A. taeniorhynchus* may explain the low levels of viral infection in field populations by a virus of low oral infectivity and limited persistence in the field (Linley and Nielsen 1968b).

Infected larvae of mosquito react normally until shortly before death, when they become quiescent, stop feeding, and can no longer remain on the water surface (Weiser 1965b; Anderson 1970; Matta and Lowe 1970). Iridescence first appears in the mosquito thorax or in the lateral portion of the first four abdominal segments and spreads throughout the entire larva (Chapman et al. 1966; Linley and Nielsen 1968a; Matta and Lowe 1970). Death usually occurs in the fourth instar.

Larvae of the hepialid moth, *Wiseana cervinata*, when infected with the iridovirus, are found mainly in the top 10 cm of the soil and sometimes on the surface, whereas the apparently uninfected larvae occur at depths of at least 30 cm (Fowler and Robertson 1972). In the honey bee, *Apis cerana*, an iridoviral infection causes the adult bees to cluster and crawl abnormally ("clustering disease") (Bailey and Ball 1978).

In the silkworm, the TIV produces tumors that appear as white spots on the abnormally translucent integument (Hukuhara 1964). The epidermal cells beneath the white spots are smaller with more basophilic cytoplasm than normal cells and they form multilayered folds or rounded nodules. The CIV isolated from the rice stem borer, *Chilo suppressalis* (Fukaya and Nasu 1966), forms in a fraction (15%) of the tested silkworm larvae, small areas of epidermal cells that retain their larval characteristic during metamorphosis to pupa and adult (Ono and Fukaya 1969). The larval region undergoes supernumerary larval molts from the fifth instar larva to the adult. The epidermal cells beneath the "larval" integument of the adult have viral particles, which are absent from other epidermal cells. This retention of larval characters suggests a juvenile hormonelike effect restricted to the infected epidermal cells. However, this may not be a hormonal effect because the application of juvenile hormone and ecdysone (molting hormone) to a culture of *Estigmene acrea* cells infected with TIV causes complete inhibition of viral DNA synthesis by juvenile hormone and the partial inhibition and prevention of viral capsid formation by ecdysone (Kloc et al. 1984). The viral replication continues when both hormones are removed from the cell culture.

V. PARVOVIRIDAE—ssDNA, NONENVELOPED

The family Parvoviridae has three genera: *Parvovirus*, *Dependovirus*, and *Densovirus* (Matthews 1982). This classification is no longer tenable and is up for revision by the International Committee for Taxonomy of Viruses (Tijssen and Arella 1991). These viruses rank among the smallest DNA viruses of animals. The insect pathogenic members occur in the genus *Densovirus* and are called the densonucleosis virus or densovirus (DNV). The *Densovirus* is a heterogeneous group and can be separated into at least two groups based on genome length and tissue tropism (Tijssen and Arella 1991).

The term "densonucleosis virus" is derived from the greatly hypertrophied nucleus (eosinophilic), which contains dense, voluminous, Feulgen-positive masses of virions (Amargier et al. 1965; Vago et al. 1966). The first member in this family was described from *Galleria mellonella* and is referred to as the *Galleria* DNV, the type species (Vago et al. 1964; Meynadier et al. 1964; Francki et al. 1991). In addition to Lepidoptera, similar DNVs have been described from four other insect orders, Diptera, Orthoptera, Blattodea, and Odonata (Kawase 1985).

Detailed studies have been conducted only with the DNVs from *G. mellonella* and the silkworm, *Bombyx mori*. The silkworm DNVs have been isolated from various parts of Japan and they fall into two groups: the Ina strain from Ina-shi (Ina city), Nagano prefecture (Shimizu 1975), and the Yamanashi strain (Seki and Iwashita 1983; Hu et al. 1986). The Ina strain is designated *Bombyx* DNV type 1 (DNV-1) and the Yamanashi strain is *Bombyx* DNV type 2 (DNV-2) (Watanabe et al. 1986). A third isolate from the silkworm has been reported in China (Iwashita and Chun 1982).

A. STRUCTURE AND PROPERTIES

The DNVs contain small, linear ssDNA, MW from 1.5 to 2.2×10^6, with four or six structural polypeptides (Kelly et al. 1980a; Kawase 1985). In an infected silkworm larva, 3 of the 6 polypeptides are involved in the structure of the virus (Kobayashi et al. 1988). Viral particles are nonenveloped, isometric (icosahedral) in shape and from 19 to 24 nm in diameter. In some DNVs, small quantities of polyamines have been detected (Kelly and Elliott 1977; Bando et al. 1983).

The single strands of DNA are either positive or negative, complementary, and separately encapsidated in a different viral particle (Barwise and Walker 1970; Kurstak et al. 1971, 1973). The strands come together to form a dsDNA when they are isolated from the virions *in vitro*. This has resulted in erroneously characterizing these viruses as dsDNA (Kurstak et al. 1977). Members of *Densovirus* do not require a helper virus as in the case of adeno-associated viruses (*Dependovirus*).

Bombyx DNV-1 is 23 nm in diameter with four structural polypeptides (Nakagaki and Kawase 1980b) and a single ssDNA of MW 1.7×10^6 (Nakagaki and Kawase 1980a); *Bombyx* DNV-2 is 24 nm in diameter with four or six structural polypeptides and a ssDNA of about 1.99×10^6 (Kurihara et al. 1984; Kawase et al.

1984). The four structural proteins appear to have originated from a common DNA sequence (Bando et al. 1984).

Bando et al. (1990) have sequenced completely the DNV-1 genome, including the nucleotide sequences of both 5' and 3' terminal regions. The genome map of DNV-1 is distinct but similar to those of mammalian parvoviruses with a conserved region (300 nucleotides) that is homologous indicating that an insect densovirus, a rodent parvovirus, and a human dependovirus share a common ancestor (Bando et al. 1987, 1990).

B. HOSTS

Some insect DNVs infect a number of insect species in different genera [e.g., a DNV of *Junonia coenia* infects other nymphalids and some noctuid larvae but not *Galleria mellonella* (Longworth 1978), and *Aedes* DNV infects species of *Aedes*, *Culex*, and *Culiseta* (Lebedinets et al. 1978). The *Aedes* DNV, unlike the *Galleria* DNV, produces pathology in vertebrates such as chick embryos, laboratory white mice, and rats (Lebedinets et al. 1976).

Certain DNVs have a limited host range [e.g., the *Galleria* DNV is restricted to *Galleria* (Vago et al. 1964); the *Bombyx* DNV-1 to *Bombyx* and *Glyphodes pyloalis* (Watanabe 1981); and the *Periplaneta* DNV replicates in four species of *Periplaneta* (Suto et al. 1979).] The *Bombyx* DNV-2, however, has a wide host range (Watanabe et al. 1986). The *Galleria* DNV replicates in the mouse L cell line and the extracted virions are infectious for *Galleria* larvae (Kurstak et al. 1969b), but extensive attempts to infect mice have failed (Kurstak et al. 1977). Some of the L mouse cells in the culture exposed to DNV are transformed and acquire the ability to synthesize antigens in the absence of virions (Kurstak et al. 1969a). The transformed cells do not produce virions.

C. PATHOGENESIS AND VIRAL REPLICATION

Tissues susceptible to infection vary with different DNVs and their insect hosts. The infection in *Galleria* larva is systemic with nearly all tissues being susceptible except the midgut (Kurstak and Stanislawski-Birencwajg 1968; Garzon and Kurstak 1976). Characteristic lesions occur in the nuclei, which are greatly hypertrophied and stain red with eosin or bluish-purple with Feulgen stain. The tissues breakdown in 4 to 6 days after viral inoculation and result in larval death (Vago et al. 1964; Amargier et al. 1965). There is also a dysfunction of the silk gland that contributes to larval death (Amargier 1966).

The first ultrastructural changes in DNV-infected cells of *Galleria* are an increase of free ribosomes and the formation of "microbody-like" structures arising from the accumulation of small, round bodies (from 17 to 20 nm) inside of vesicles (Garzon and Kurstak 1976). The nucleolus hypertrophies and is accompanied by a segregation of its fibrillar and granular components. The development of the granular portion coincides with the synthesis of the replicative dsDNA within the virogenic stroma. Intranuclear bodies appear and the virions are assembled in the virogenic stroma.

The complete cycle of *Densovirus* replication takes about 20 to 24 h (Tijssen and Arella 1991). The cycle can be divided into three stages, a latent period of 6 h, the appearance of viral proteins in the cytoplasm in the next 3 h, and then the appearance of proteins in the nucleus. Enough viral particles are produced in 10 to 15 h later to cause cytolysis.

In the lepidopteron *Sibine fusca*, the DNV may cause a collapse of the population (Meynadier et al. 1977a). Infected larvae lose their gregarious habit and stop feeding. Unlike *Galleria*, a marked change develops in the digestive tract of *Sibine*. The wall of the digestive tract becomes opaque, whitish, and thickened in contrast to the thin, clear, digestive wall of an uninfected larva (Amargier et al. 1979). The nucleoplasm of a midgut cell is divided into opaque virogenic masses that continue to grow with small cavities around which viral particles are formed. The mitochondria hypertrophy and burst releasing their contents and leaving vacuoles in their place. Tumorlike masses are formed in the midgut and death often results after their formation.

In the noctuid *Pseudoplusia includens*, the virus is found in the cell nuclei of hemocyte, muscle, hypodermis, and fat body (Chao et al. 1985). The larvae of the silver-spotted flambeau, *Agraulis vanillae*, infected with a DNV, are flaccid and discolored after death (Kelly et al. 1980b). Surviving prepupa often fails to shed its larval cuticle or the adult cannot eclose. Aside from the hypodermis, the infection appears to be systemic.

In the mosquito *Aedes aegypti*, the infected cells occur in the hypodermis, imaginal disc, Malpighian tubules, and in particular, the fat body (Lebedeva et al. 1973). The first ultrastructural changes in DNV-infected cells occur in the cytoplasm and nuclei (Buchatsky and Raikova 1979). In the cytoplasm, the endoplasmic reticulum and various organelles are destroyed and paracrystalline inclusions (from 18 to 20 nm in diameter) are formed. The number of inclusions increases as the infection progresses. The nuclei contain virogenic stromata and paracrystalline arrays of virions.

The DNV infection in the cockroach, *Periplaneta fuliginosa*, and the cricket, *Acheta domestica*, causes paralysis in the hind legs and the movements are uncoordinated (Suto et al. 1979; Meynadier et al. 1977b). The abdomen of a cockroach is swollen with enlarged milky-white fat body that differs from the brownish-white fat body of an uninfected cockroach. A number of individuals develop ulcers in the midgut. The DNV infection in the dragonfly, *Leucorrhinia dubia*, results in the nymphs becoming sluggish and flaccid without other external signs (Charpentier 1979).

The DNV-1 and DNV-2 infections in silkworm larvae differ from those of other insects (Kawase 1985). The major sign is the flaccid body. The flaccidity and the pale-yellow color of the alimentary tract are similar to those of infectious flacherie caused by a picornavirus. In contrast to the DNV in *Galleria*, where most tissues, except for the midgut, are infected, only the silkworm midgut is infected. The infection is in columnar cells with the goblet cells remaining relatively intact. Viral multiplication occurs in columnar cell nuclei and no histopathological changes occur in other tissues (Watanabe et al. 1976; Maeda and Watanabe 1978). Since both DNV-1 and DNV-2 infect a columnar cell, interference between the two isolates may occur (Watanabe et al. 1976; Abe and Watanabe 1987). Both DNV-1 and DNV-2 do not infect the goblet cell, but the isolate of the *Bombyx* DNV from China infects the goblet cell nuclei after an initial infection of the columnar cell nuclei (Iwashita and Chun 1982).

The DNV-1 and DNV-2 infections in silkworm larvae differ significantly in histopathology and pathogenesis (Watanabe and Kurihara 1988). DNV-1 produces an acute infection with larval death at about 7 days after inoculation, whereas the DNV-2 causes a chronic infection with time of death between 10 and 20 days and a few larvae may pupate. In DNV-1 infection, the nuclear hypertrophy of columnar cells is obvious, the cells degrade and are discharged into the gut lumen, resulting in minimally affected goblet cells forming most of the midgut. In the DNV-2 infection, the infected-midgut-columnar cells are not discharged into the gut lumen but increase in numbers and cause the epithelium to fold. The structures of the virogenic stromata also differ between the two DNVs.

There are silkworm strains that are resistant to DNV-1 or DNV-2 (Abe et al. 1987). Little is known about the resistance of silkworms to DNV-2 and most studies are with DNV-1. Two genes are involved in the nonsusceptibility of the silkworm to DNV-1: one is a recessive gene (*nsd-1*) (Watanabe and Maeda 1981; Equchi et al. 1991) and the other is a dominant gene (Eguchi et al. 1986). The recessive and dominant genes are located on different chromosomes and are not sex linked. The *nsd-1* gene is located at position 8.3 on the twenty-first chromosome of the silkworm (Eguchi et al. 1991). The recommendation to sericultural farms for the prevention of epizootics of densonucleosis is to use the silkworm strain with the nonsusceptible, homozygous recessive gene (Watanabe and Maeda 1978).

A mutant silkworm strain (*mo*), with genetic mosaicism, has the dominant *Nid-1* gene in homozygosity (Abe et al. 1990). A larva of the mutant strain has a midgut epithelium consisting of a mosaic of genetically nonsusceptible and susceptible areas. In the area infected by DNV-1, the degraded columnar cells accumulate and are attached to increased numbers of intact goblet cells to form a multiple-layered epithelium. In the uninfected areas, the single-layered epithelium has a large number of tightly arranged columnar cells with a few goblet cells among them. The increased numbers of columnar cells in the uninfected area may compensate for the deficient function of the degraded cells in the infected area. The overall effect is a prolongation of the DNV-1 infection in the larva with midgut mosaicism.

REFERENCES

Abe, H., and Watanabe, H. 1987. Double infection of two densonucleosis viruses in the silkworm, *Bombyx mori*. *Jpn. J. Appl. Entomol. Zool.*, *31*, 381–384.

Abe, H., Watanabe, H., and Eguchi, R. 1987. Genetical relationship between nonsusceptibilities of the silkworm, *Bombyx mori*, to two densonucleosis viruses. *J. Seric. Sci. Jpn.* 56, 443–444.

Abe, H., Kobayashi, M., and Watanabe, H. 1990. Mosaic infection with a densonucleasis virus in the midgut epithelium of the silkworm, *Bombyx mori. J. Invertebr. Pathol.* 55, 112–117.

Ables, J. R., and Vinson, S. B. 1981. Regulation of host larval development by the egg-larval endoparasitoid *Chelonus insularis* [Hym.: Braconidae]. *Entomophaga* 26, 453–458.

Amargier, A. 1966. Action de la densonucléose du lépidoptère Galleria mellonella L. sur la sécrétion et l'émission de substance soyeuse. *Arch. Ges. Virusforsch.* 19, 13–22.

Amargier, A., Vago, C., and Meynadier, G. 1965. Etude histopathologique d'un nouveau type de virose mise en évidense chez le lépidoptère Galleria mellonella. *Arch. Ges. Virusforsch.* 15, 659–667.

Amargier, A., Vago, C., Duthoit, J. L., and Meynadier, G. 1979. Formation tumorale d'origine parvovirale chez *Sibine fusca* [Lep.: Limacodidae]. *Entomophaga* 24, 259–271.

Anderson, J. F. 1970. An iridescent virus infecting the mosquito *Aedes stimulans*. *J. Invertebr. Pathol. 15*, 219–224.

Anthony, D. W., and Comps, M. 1991. Iridoviridae. *In* "Atlas of Invertebrate Viruses." (J. R. Adams and J. R. Bonami, eds.), pp. 55–86. CRC Press, Boca Raton, Florida.

Arif, B. M. 1984. The entomopoxviruses. *In* "Advances in Virus Research." (M. A. Lauffer and K. Maramorosch, eds.), pp. 195–213. Academic Press, Orlando, Florida.

Bailey, L., and Ball, B. V. 1978. *Apis* iridescent virus and "clustering disease" of *Apis cerana*. *J. Invertebr. Pathol. 31*, 368–371.

Bando, H., Nakagaki, M., and Kawase, S. 1983. Polyamines in densonucleosis virus from the silkworm, *Bombyx mori*. *J. Invertebr. Pathol. 42*, 264–266.

Bando, H., Kondo, N., and Kawase, S. 1984. Molecular homology among the structural proteins of densonucleosis virus from silkworm, *Bombyx mori*. *Arch. Virol. 80*, 209–218.

Bando, H., Kusuda, J., Gojobori, T., Maruyama, T., and Kawase, S. 1987. Organization and nucleotide sequence of a densovirus genome imply a host-dependent evolution of the parvoviruses. *J. Virol. 61*, 553–560.

Bando, H., Choi, H., Ito, Y., and Kawase, S. 1990. Terminal structure of a densovirus implies a hairpin transfer replication which is similar to the model for AAV. *Virology 179*, 57–63.

Barwise, A. H., and Walker, I. O. 1970. Studies on the DNA of a virus from *Galleria mellonella*. *FEBS Lett. 6*, 13–16.

Bedwin, O. 1979. The particulate basis of the resistance of a parasitoid to the defence reactions of its insect host. *Proc. R. Soc. London Ser. B205*, 267–270.

Bergoin, M., Devauchelle, G., and Vago, C. 1969. Electron microscopy study of the poxlike virus of Melolontha Melolontha L (Coleoptera, Scarabaeidae). *Arch. Ges. Virusforsch. 28*, 285–302.

Bergoin, M., Veyrunes, J. C., and Scalla, R. 1970. Isolation and amino acid composition of the inclusions of *Melolontha melolontha* poxvirus. *Virology 40*, 760–763.

Bergoin, M., Devauchelle, G., and Vago, C. 1971. Electron microscopy study of *Melolontha* poxvirus: The fine structure of occluded virus. *Virology 43*, 453–467.

Bergoin, M., Devauchelle, G., and Vago, C. 1976. Les inclusions fusiformes associées à l'Entomopoxvirus du Coléoptère *Melolontha melolontha*. *J. Ultrastruct. Res. 55*, 17–30.

Bilimoria, S. L., and Arif, B. M. 1979. Subunit protein and alkaline protease of entomopoxvirus spheroids. *Virology 96*, 596–603.

Bilimoria, S. L., and Arif, B. M. 1980. Structural polypeptides of *Choristoneura biennis* entomopoxvirus. *Virology 104*, 253–257.

Bird, F. T. 1962. On the development of the Tipula iridescent virus particle. *Can. J. Microbiol. 8*, 533–534.

Bird, F. T. 1974. The development of spindle inclusions of *Choristoneura fumiferana* (Lepidoptera: Tortricidae) infected with entomopox virus. *J. Invertebr. Pathol. 23*, 325–332.

Bird, F. T. 1976. Cell proliferation in the spruce budworm, *Choristoneura fumiferana* (Lepidoptera: Tortricidae), infected with entomopoxvirus. *Can. Entomol. 108*, 859–864.

Bird, F. T., Sanders, C. J., and Burke, J. M. 1971. A newly discovered virus disease of the spruce budworm, *Choristoneura biennis*, (Lepidoptera: Tortricidae). *J. Invertebr. Pathol. 18*, 159–161.

Bladon, T., Frosch, M., Sabour, P. M., and Lee, P. E. 1986. Association of nuclear matrix proteins with cytoplasmic assembly sites of *Tipula* iridescent virus. *Virology 155*, 524–533.

Blissard, G. W., Fleming, J. G. W., Vinson, S. B., and Summers, M. D. 1986a. *Campoletis sonorensis* virus: Expression in *Heliothis virescens* and identification of expressed sequences. *J. Insect Physiol. 32*, 351–359.

Blissard, G. W., Vinson, S. B., and Summers, M. D. 1986b. Identification, mapping, and in vitro translation of *Campoletis sonorensis* virus mRNAs from parasitized *Heliothis virescens* larvae. *J. Virol. 57*, 318–327.

Blissard, G. W., Smith, O. P., and Summers, M. D. 1987. Two related viral genes are located on a single superhelical DNA segment of the multipartite *Campoletis sonorensis* virus genome. *Virology 160*, 120–134.

Blissard, G. W., Theilmann, D. A., and Summers, M. D. 1989. Segment W of *Campoletis sonorensis* virus: Expression, gene products, and organization. *Virology 169*, 78–89.

Boucias, D. G., Maruniak, J. E., and Pendland, J. C. 1987. Characterization of an iridovirus isolated from the southern mole cricket, *Scapteriscus vicinus*. *J. Invertebr. Pathol. 50*, 238–245.

Brown, F. 1986. The classification and nomenclature of viruses: Summary of results of meetings of the International Committee on Taxonomy of Viruses in Sendai, September 1984. *Invervirology 25, 141–143*.

Browning, H. W., Federici, B. A., and Oatman, E. R. 1982. Occurrence of a disease caused by a rickettsia-like organism in a larval population of the cabbage looper, *Trichoplusia ni*, in southern California. *Environ. Entomol. 11*, 550–554.

Buchatsky, L. P., and Raikova, A. P. 1979. Electron microscope study of mosquito densonucleosis virus maturation. *Acta Virol. 23*, 170–172.

Carner, G. R., and Hudson, J. S. 1983. Histopathology of virus-like particles in *Heliothis* spp. *J. Invertebr. Pathol. 41*, 238–249.

Chao, Y.-C., Young, S. Y., III, Kim, K. S., and Scott, H. A. 1985. A newly isolated densonucleosis virus from *Pseudoplusia includens* (Lepidoptera: Noctuidae). *J. Invertebr. Pathol. 46*, 70–82.

Chapman, H. C., Clark, T. B., Woodard, D. B., and Kellen, W. R. 1966. Additional mosquito hosts of the mosquito iridescent virus. *J. Invertebr. Pathol. 8*, 545–546.

Chapman, H. C., Clark, T. B., Anthony, D. W., and Glenn, F. E., Jr. 1971. An iridescent virus from larvae of *Corethrella brakeleyi* (Diptera: Chaoboridae) in Louisiana. *J. Invertebr. Pathol. 18*, 284–286.

Charpentier, R. 1979. A nonoccluded virus in nymphs of the dragonfly *Leucorrhinia dubia* (Odonata, Anisoptera). *J. Invertebr. Pathol. 34*, 95–98.

Clark, T. B. 1982. Entomopoxvirus-like particles in three species of bumblebees. *J. Invertebr. Pathol. 39*, 119–122.

Clark, T. B., Kellen, W. R., and Lum. P. T. M. 1965. A mosquito iridescent virus (MIV) from *Aedes taeniorhynchus* (Wiedemann). *J. Invertebr. Pathol. 7*, 519–521.

Cook, D., and Stoltz, D. B. 1983. Comparative serology of viruses isolated from ichneumonid parasitoids. *Virology 130*, 215–220.

Croizier, G., and Veyrunes, J.-C. 1971. Analyse immunochimique des inclusions de la "maladie a fuseaux" du Coléoptère *Melolontha melolontha* L. *Ann. Inst. Pasteur 120*, 709–715.

Davies, D. H., Strand, M. R., and Vinson, S. B. 1987. Changes in differential haemocyte count and *in vitro* behaviour of plasmotocytes from host *Heliothis virescens* caused by *Campoletis sonorensis* polydnavirus. *J. Insect Physiol. 33*, 143–153.

Day, M. F. 1965. *Sericesthis* iridescent virus infection of the adult *Galleria*. *J. Invertebr. Pathol. 7*, 102–105.

Day, M. F., and Dudzinski, M. L. 1966. The effect of temperature on the development of *Sericesthis* iridescent virus. *Aust. J. Biol. Sci. 19*, 481–493.

Day, M. F., and Gilbert, N. 1967. The number of particles of *Sericesthis* iridescent virus required to produce infections of *Galleria* larvae. *Aust. J. Biol. Sci. 20*, 691–693.

Day, M. F., and Mercer, E. H. 1964. Properties of an iridescent virus from the beetle, *Sericesthis pruinosa*. *Aust. J. Biol. Sci. 17*, 892–902.

Delius, H., Darai, G., and Flügel, R. M. 1984. DNA analysis of insect iridescent virus 6: Evidence for circular permutation and terminal redundancy. *J. Virol. 49*, 609–614.

Devauchelle, G., Bergoin, M., and Vago, C. 1971. Etude ultrastructurale du cycle de

replication d'un entomopoxvirus dans les hémocytes de son hôte. *J. Ultrastruct. Res.* 37, 301–321.
Devauchelle, G., Stoltz, D. B., and Darcy-Tripier, F. 1985a. Comparative ultrastructure of Iridoviridae. *Curr. Top. Microbiol. Immunol.* 116, 1–21.
Devauchelle, G., Attias, J., Monnier, C., Barray, S., Cerutti, M., Guerillon, J., and Orange-Balange, N. 1985b. *Chilo* iridescent virus. *Curr. Top. Microbiol. Immunol.* 116, 37–48.
Dover, B. A., Davies, D. H., Strand, M. R., Gray, R. S., Keeley, L. L., and Vinson, S. B. 1987. Ecdysteroid-titre reduction and developmental arrest of last-instar *Heliothis virescens* larvae by calyx fluid from the parasitoid *Campoletis sonorensis*. *J. Insect Physiol.* 33, 333–338.
Edson, K. M., and Vinson, S. B. 1977. Nutrient absorption by the anal vesicle of the braconid wasp, *Microplitis croceipes*. *J. Insect Physiol.* 23, 5–8.
Edson, K. M., Vinson, S. B., Stoltz, D. B., and Summers, M. D. 1981. Virus in a parasitoid wasp: Suppression of the cellular immune response in the parasitoid's host. *Science 211*, 582–583.
Eguchi, R., Furuta, Y., and Ninaki, O. 1986. Dominant nonsusceptibility to densonucleosis virus in the silkworm, *Bombyx mori*. *J. Seric. Sci. Jpn.* 55, 177–178.
Equchi, R., Ninaki, O., and Hara, W. 1991. Genetical analysis on the nonsusceptibility to densonucleosis virus in the silkworm, *Bombyx mori*. *J. Seric. Sci. Jpn.* 60, 384–389.
Faulkner, P. 1982. A novel class of wasp viruses and insect immunity. *Nature 299*, 489–490.
Federici, B. A. 1982. A new type of insect pathogen in larvae of the clover cutworm, *Scotogramma trifolii*. *J. Invertebr. Pathol.* 40, 41–54.
Federici, B. A. 1983. Enveloped double-stranded DNA insect virus with novel structure and cytopathology. *Proc. Natl. Acad. Sci. U.S.A.* 80, 7664–7668.
Federici, B. A., and Govindarajan, R. 1990. Comparative histopathology of three ascovirus isolates in larval noctuids. *J. Invertebr. Pathol.* 56, 300–311.
Federici, B. A., Vlak, J. M., and Hamm, J. J. 1990. Comparative study of virion structure, protein composition and genomic DNA of three ascovirus isolates. *J. Gen. Virol.* 71, 1661–1668.
Federici, B. A., Hamm, J. J., and Styer, E. L. 1991. Ascoviridae. *In* "Atlas of Invertebrate Viruses." (J. R. Adams and J. R. Bonami, eds.), pp. 339–349. CRC Press, Boca Raton, Florida.
Fischer, M., Schnitzler, P., Delius, H., and Darai, G. 1988a. Identification and characterization of the repetitive DNA element in the genome of insect iridescent virus type 6. *Virology 167*, 485–496.
Fischer, M., Schnitzler, P., Scholz, J., Rösen-Wolff, A., Delius, H., and Darai, G. 1988b. DNA nucleotide sequence analysis of the *Pvu*11 DNA fragment L of the genome of insect iridescent virus type 6 reveals a complex cluster of multiple tandem, overlapping, and interdigitated repetitive DNA elements. *Virology 167*, 497–506.
Fleming, J. G. W., Blissard, G. W., Summers, M. D., and Vinson, S. B. 1983. Expression of *Campoletis sonorensis* virus in the parasitized host, *Heliothis virescens*. *J. Virol.* 48, 74–78.
Fleming, J. G. W., and Summers, M. D. 1986. *Campoletis sonorensis* endoparasitic wasps contain forms of *C. sonorensis* virus DNA suggestive of integrated and extrachromosomal polydnavirus DNAs. *J. Virol.* 57, 552–562.
Fowler, M., and Robertson, J. S. 1972. Iridescent virus infection in field populations of *Wiseana cervinata* (Lepidoptera: Hepialidae) and *Witlesia* sp. (Lepidoptera: Pyralidae) in New Zealand. *J. Invertebr. Pathol.* 19, 154–155.
Francki, R. I. B., Fauquet, C. M., Knudson, D. L., and Brown, F. 1991. "Classification and Nomenclature of Viruses. Fifth Report of the International Committee on Taxonomy of Viruses." Arch. Virol. Suppl. 2. Springer-Verlag, Wien.

Fukaya, M., and Nasu, S. 1966. A *Chilo* iridescent virus (CIV) from the rice stem borer, *Chilo suppressalis* Walker (Lepidoptera: Pyralidae). *Appl. Entomol. Zool.* 1, 69–72.

Fukuda, T. 1971. Per os transmission of *Chilo* iridescent virus to mosquitoes. *J. Invertebr. Pathol.* 18, 152–153.

Fukuda, T., and Clark, T. B. 1975. Transmission of the mosquito iridescent virus (RMIV) by adult mosquitoes of *Aedes taeniorhynchus* to their progeny. *J. Invertebr. Pathol.* 25, 275–276.

Garzon, S., and Kurstak, E. 1976. Ultrastructural studies on the morphogenesis of the Densonucleosis virus (Parvovirus). *Virology* 70, 517–531.

Goodwin, R. H., and Filshie, B. K. 1969. Morphology and development of an occluded virus from the black-soil scarab, *Orthnonius batesi*. *J. Invertebr. Pathol.* 13, 317–329.

Goodwin, R. H., and Filshie, B. K. 1975. Morphology and development of entomopoxviruses from two Australian scarab beetle larvae (Coleoptera: Scarabaeidae). *J. Invertebr. Pathol.* 25, 35–46.

Goodwin, R. H., Milner, R. J., and Beaton, C. D. 1991. Entomopoxvirinae. *In* "Atlas of Invertebrate Viruses." (J. R. Adams and J. R. Bonami, eds.), pp. 259–285. CRC Press, Boca Raton, Florida.

Goorha, R., and Granoff, A. 1979. Icosahedral cytoplasmic deoxyriboviruses. *In* "Comprehensive Virology." (H. Fraenkel-Conrat and R. R. Wagner, eds.), pp. 347–399. Plenum Press, New York.

Goorha, R., Murti, G., Granoff, A., and Tirey, R. 1978. Macromolecular synthesis in cells infected by frog virus 3. VIII. The nucleus is a site of frog virus 3 DNA and RNA synthesis. *Virology* 84, 32–50.

Govindarajan, R., and Federici, B. A. 1990. Ascovirus infectivity and effects of infection on the growth and development of noctuid larvae. *J. Invertebr. Pathol.* 56, 291–299.

Granados, R. R. 1973a. Insect poxviruses: Pathology, morphology, and development. *Misc. Publ. Entomol. Soc. Am.* 9, 73–94.

Granados, R. R. 1973b. Entry of an insect poxvirus by fusion of the virus envelope with the host cell membrane. *Virology* 52, 305–309.

Granados, R. R., and Roberts, D. W. 1970. Electron microscopy of a poxlike virus infecting an invertebrate host. *Virology* 40, 230–243.

Guillot, F. S., and Vinson, S. B. 1972. The rôle of the calyx and poison gland of *Cardiochiles nigriceps* in the host–parasitoid relationship. *J. Insect Physiol.* 18, 1315–1321.

Hall, D. W. 1985. Pathobiology of invertebrate icosahedral cytoplasmic deoxyriboviruses (Iridoviridae). *In* "Viral Insecticides for Biological Control." K. Maramorosch and K. E. Sherman, eds.), pp. 163–196. Academic Press, New York.

Hall, D. W., and Anthony, D. W. 1971. Pathology of a mosquito iridescent virus (MIV) infecting *Aedes taeniorhynchus*. *J. Invertebr. Pathol.* 18, 61–69.

Hamm, J. J., Nordlund, D. A., and Marti, O .G. 1985. Effects of a nonoccluded virus of *Spodoptera frugiperda* (Lepidoptera: Noctuidae) on the development of a parasitoid, *Cotesia marginiventris* (Hymenoptera: Braconidae). *Environ. Entomol.* 14, 258–261.

Hamm, J. J., Pair, S. D., and Marti, O. G., Jr. 1986. Incidence and host range of a new ascovirus isolated from fall armyworm, *Spodoptera frugiperda* (Lepidoptera: Noctuidae). *Fla. Entomol.* 69, 524–531.

Hamm, J. J., Styer, E. L., and Lewis, W. J. 1990. Comparative virogenesis of filamentous virus and polydnavirus in the female reproductive tract of *Cotesia marginiventris* (Hymenoptera: Braconidae). *J. Invertebr. Pathol.* 55, 357–374.

Hembree, S. C., and Anthony, D. W. 1980. Possible site of entry of the regular mosquito iridescent virus (RMIV) in *Aedes taeniorhynchus* larvae. *Mosq. News* 40, 449–451.

Hess, R. T., Poinar, G. O., Jr., Etzel, L., and Merritt, C. C. 1980. Calyx particle morphology of *Bathyplectes anurus* and *B. curculionis* (Hymenoptera: Ichneumonidae). *Acta Zool.* (Stockh.) 61, 111–116.

Hu, Y.-Y., Seki, H., and Kawase, S. 1986. Electron microscopic studies on DNA from a densonucleosis virus (Yamanashi isolate) of the silkworm, *Bombyx mori*. *Appl. Entomol. Zool.* 21, 613–619.

Hukuhara, T. 1964. Induction of epidermal tumor in *Bombyx mori* (Linnaeus) with *Tipula* iridescent virus. *J. Insect Pathol.* 6, 246–248.

Hukuhara, T., Xu, J., and Yano, K. 1990. Replication of an entomopoxvirus in two lepidopteran cell lines. *J. Invertebr. Pathol.* 56, 222–232.

Iwashita, Y., and Chun, C. Y. 1982. The development of a densonucleosis virus isolated from silkworm larvae, *Bombyx mori*, of China. *In* "The Ultrastructure and Functioning of Insect Cells." (H. Akai, R.C. King, and S. Morohoshi, eds.), pp. 161–164. Society Insect Cells Japan, Tokyo.

Jones, D., Sreekrishna, S., Iwaya, M., Yang, J.-N., and Eberely, M. 1986. Comparison of viral ultrastructure and DNA banding patterns from the reproductive tracts of eastern and western hemisphere *Chelonus* spp. (Braconidae: Hymenoptera). *J. Invertebr. Pathol.* 47, 105–115.

Kawase, S. 1985. Pathology associated with densoviruses. *In*: "Viral Insecticides for Biological Control." (K. Maramorosch and K. E. Sherman, eds.), pp. 197–231. Academic Press, New York.

Kawase, S., Cai, Y.-M., Bando, H., and Seki, H. 1984. Chemical properties of the Yamanashi isolate of the *Bombyx* densonucleosis virus. *J. Seric. Sci. Jpn.* 53, 341–347.

Kelly, D. C. 1985. Insect iridescent viruses. *Curr. Top. Microbiol. Immunol.* 116, 23–35.

Kelly, D. C., and Elliott, R. M. 1977. Polyamines contained by two densonucleosis viruses. *J. Virol.* 21, 408–410.

Kelly, D. C., and Robertson, J. S. 1973. Icosahedral cytoplasmic deoxyriboviruses. *J. Gen. Virol.* (Suppl.) 20, 17–41.

Kelly, D. C., and Tinsley, T. W. 1973. Ribonucleic acid polymerase activity associated with particles of iridescent virus types 2 and 6. *J. Invertebr. Pathol.* 22, 199–202.

Kelly, D. C., and Tinsley, T. W. 1974a. Iridescent virus replication: A microscope study of *Aedes aegypti* and *Antherea eucalypti* cells in culture infected with iridescent virus types 2 and 6. *Microbios* 9, 75–93.

Kelly, D. C., and Tinsley, T. W. 1974b. Iridescent virus replication: Patterns of nucleic acid synthesis in insect cells infected with iridescent virus types 2 and 6. *J. Invertebr. Pathol.* 24, 169–178.

Kelly, D. C., Moore, N. F., Spilling, C. R., Barwise, A. H., and Walker, I. O. 1980a. Densonucleosis virus structural proteins. *J. Virol.* 36, 224–235.

Kelly, D. C., Ayres, M. D., Spencer, L. K., and Rivers, C. F. 1980b. Densonucleosis virus 3: A recent insect parvovirus isolate from *Agraulis vanillae* (Lepidoptera: Nymphalidae). *Microbiologica* 3, 455–460.

Kelly, D. C., Elliott, R. M., and Blair, G. E. 1980c. Phosphorylation of iridescent virus polypeptides *in vitro*. *J. Gen. Virol.* 48, 205–211.

Kitano, H. 1982. Effect of the venom of the gregarious parasitoid *Apanteles glomeratus* on its hemocytic encapsulation by the host, *Pieris*. *J. Invertebr. Pathol.* 40, 61–67.

Kitano, H. 1986. The role of *Apanteles glomeratus* venom in the defensive response of its host, *Pieris rapae crucivora*. *J. Insect Physiol.* 32, 369–375.

Kloc, M., Lee, P. E., and Tajbakhsh, S. 1984. The effect of two insect hormones on the replication of *Tipula* iridescent virus in *Estigmene acrea* cells in suspension culture. *J. Invertebr. Pathol.* 43, 114–123.

Klug, A., Franklin, R. E., and Humphreys-Owen, S. P. F. 1959. The crystal structure of Tipula iridescent virus as determined by Bragg reflection of visible light. *Biochim. Biophys. Acta* 32, 203–219.

Klump, H., Beaumais, J., and Devauchelle, G. 1983. Structural and thermodynamic investigation of the Chilo iridescent virus (iridovirus type 6). *Arch. Virol.* 75, 269–276.

Kobayashi, M., Hashimoto, Y., Seki, H., and Watanabe, H. 1988. In vitro translation of

RNA from the midgut of the silkworm, *Bombyx mori*, infected with a densonucleosis virus. *J. Invertebr. Pathol. 52*, 259–267.

Krell, P. J. 1991. Polydnaviridae. *In* "Atlas of Invertebrate Viruses." (J. R. Adams and J. R. Bonami, eds.), pp. 321–338. CRC Press, Boca Raton, Florida.

Krell, P. J., and Stoltz, D. B. 1979. Unusual baculovirus of the parasitoid wasp *Apanteles melanoscelus*: Isolation and preliminary characterization. *J. Virol. 29*, 1118–1130.

Krell, P. J., and Stoltz, D. B. 1980. Virus-like particles in the ovary of an ichneumonid wasp: Purification and preliminary characterization. *Virology 101*, 408–418.

Krell, P. J., Summers, M. D., and Vinson, S. B. 1982. Virus with a multipartite superhelical DNA genome from the ichneumonid parasitoid *Campoletis sonorensis*. *J. Virol. 43*, 859–870.

Kurihara, Y., Watanabe, H., Maeda, S., and Shimizu, T. 1984. Chemical characteristics of a previously undescribed densonucleosis virus isolated from the silkworm, *Bombyx mori*. *J. Seric. Sci. Jpn. 53*, 33–40.

Kurstak, E., and Garzon, S. 1977. Entomopoxviruses (poxviruses of invertebrates). *In* "The Atlas of Insect and Plant Viruses." (K. Maramorosch, ed.), pp. 29–66. Academic Press, New York.

Kurstak, E., and Stanislawski-Birencwajg, M. 1968. Localisation du matériel antigénique du virus de la densonucléose au cours de l'infection chez *Galleria mellonella* L. *Can. J. Microbiol. 14*, 1350–1352.

Kurstak, E., Belloncik, S., and Brailovsky, C. 1969a. Transformation de cellules L de souris par un virus d'invertébrés: Le virus de la densonucléose (VDN). *C.R. Acad. Sci. Paris Ser. D269*, 1716–1719.

Kurstak, E., Côté, J.-R., Belloncik, S., Garzon, S., Trudel, M., and Chagnon, B. 1969b. Infection des cellules L de la souris par le virus de la densonucléose. *Rev. Can. Biol. 28*, 139–141.

Kurstak, E., Vernoux, J.-P., Niveleau, A., and Onji, P. A. 1971. Visualisation du DNA du virus de la densonucléose (VDN) à chaînes monocaténaires complémentaires de polarités inverse plus et moins. *C.R. Acad. Sci. Paris Ser. D272*, 762–765.

Kurstak, E., Vernoux, J. P., and Brakier-Gingras, L. 1973. Étude biophysique de l'acide désoxyribonucléique du virus de la densonucléose (VDN). II. Extraction du DNA Viral et mise en évidence de la présence de chaînes polynucléotidiques complémentaires encapsidées séparément dans les virions VDN. *Arch. Ges. Virusforsch. 40*, 274–284.

Kurstak, E., Tijssen, P., and Garzon, S. 1977. Densonucleosis viruses (Parvoviridae). *In* "The Atlas of Insect and Plant Viruses." (K. Maramorosch, ed.), pp. 67–77. Academic Press, New York.

Langridge, W. H. R., and Roberts, D. W. 1982. Structural proteins of *Amsacta moorei*, *Euxoa auxiliaris*, and *Melanoplus sanguinipes* entomopoxviruses. *J. Invertebr. Pathol. 39*, 346–353.

Lea, M. S. 1985. A *Sericesthis* iridescent virus infection of the hemocytes of the wax moth, *Galleria mellonella* (Lepidoptera). *J. Invertebr. Pathol. 46*, 219–230.

Lebedeva, O. P., Kuznetsova, M. A., Zelenko, A. P., and Gudz-Gorban, A. P. 1973. Investigation of a virus disease of the densonucleosis type in a laboratory culture of *Aedes aegypti*. *Acta Virol. 17*, 253–256.

Lebedinets, N. N., Vasilieva, V. L., and Buchatsky, L. P. 1976. The effect of *Aedes aegypti* L. mosquito densonucleosis virus on vertebrate animals. *Med. Parazitol. Bolezn. 45*, 95–97.

Lebedinets, N. N., Tsarichkova, D. B., Karpenko, L. V., Kononko, A. G., and Buchatskij, L. P. 1978. Study of the *Aedes aegypti* L. densonucleosis virus effect on preimaginal stages of different species of blood-sucking mosquitoes. *Mikrobiol. Zh. 40*, 352–356.

Lee, P. E., and Brownrigg, S. P. 1982. Effect of virus inactivation on *Tipula* iridescent virus–cell relationships. *J. Ultrastruct. Res. 79*, 189–197.

Leutenegger, R. 1967. Early events of *Seriscesthis* iridescent virus infection in hemocytes of *Gallerio mellonella* (L.). *Virology 32*, 109–116.

Linley, J. R., and Nielsen, H. T. 1968a. Transmission of a mosquito iridescent virus in *Aedes taeniorhynchus*. I. Laboratory experiments. *J. Invertebr. Pathol. 12*, 7–16.

Linley, J. R., and Nielsen, H. T. 1968b. Transmission of a mosquito iridescent virus in *Aedes taeniorhynchus*. II. Experiments related to transmission in nature. *J. Invertebr. Pathol. 12*, 17–24.

Longworth, J. F. 1978. Small isometric viruses of invertebrates. *Adv. Virus Res. 23*, 103–157.

Maeda, S., and Watanabe, H. 1978. Immunofluorescence observation of the infection of densonucleosis virus in the silkworm, *Bombyx mori*. *Jpn. J. Appl. Entomol. Zool. 22*, 98–101.

Mathiesen, W. B., and Lee, P. E. 1981. Cytology and autoradiography of *Tipula* iridescent virus infection of insect suspension cell cultures. *J. Ultrastruct. Res. 74*, 59–68.

Matta, J. F., and Lowe, R. E. 1970. The characterization of a mosquito iridescent virus (MIV). I. Biological characteristics, infectivity, and pathology. *J. Invertebr. Pathol. 16*, 38–41.

Matthews, R. E. F. 1982. Classification and nomenclature of viruses. *Intervirology 17*, 1–199.

McCarthy, W. J., Granados, R. R., and Roberts, D. W. 1974. Isolation and characterization of entomopox virions from virus-containing inclusions of *Amsacta moorei* (*Lepidoptera: Arctiidae*). *Virology 59*, 59–69.

McCarthy, W. J., Granados, R. R., Sutter, G. R., and Roberts, D. W. 1975a. Characterization of entomopox virions of the army cutworm, *Euxoa auxiliaris* (Lepidoptera: Noctuidae). *J. Invertebr. Pathol. 25*, 215–220.

McCarthy, W. J., Neser, C. F., and Roberts, D. W. 1975b. RNA polymerase activity of *Amsacta moorei* entomopox virions. *Intervirology 5*, 69–75.

McIntosh, A. H., and Kimura, M. 1974. Replication of the insect *Chilo* iridescent virus (CIV) in a poikilothermic vertebrate cell line. *Intervirology 4*, 257–267.

Meynadier, G., Vago, C., Plantevin, G., and Atger, P. 1964. Virose d'un type inhabituel chez le Lépidoptère *Galleria mellonella* L. *Rev. Zool. Agric. Appl. 63*, 207–208.

Meynadier, G., Amargier, A., and Genty, Ph. 1977a. Une virose de type densonucléose chez le lépidoptère *Sibine fusca* Stoll. *Oléagineux 32*, 357–361.

Meynadier, G., Matz, G., Veyrunes, J.-C., and Bres, N. 1977b. Virose de type densonucléose chez les orthoptères. *Ann. Soc. Entomol. Fr.* (N.S.) *13*, 487–493.

Mitsuhashi, J. 1967. Infection of leafhopper and its tissues cultivated in vitro with *Chilo* iridescent virus. *J. Invertebr. Pathol. 9*, 432–434.

Monnier, C., and Devauchelle, G. 1976. Enzyme activities associated with an invertebrate iridovirus: Nucleotide phosphohydrolase activity associated with iridescent virus type 6 (CIV). *J. Virol. 19*, 180–186.

Moore, N. F., and Kélly, D. C. 1980. A comparative study of the polypeptides of three iridescent viruses by *N*-terminal analysis, amino acid analysis, and surface labeling. *J. Invertebr. Pathol. 36*, 415–422.

Nakagaki, M., and Kawase, S. 1980a. Structural proteins of densonucleosis virus isolated from the silkworm, *Bombyx mori*, infected with the flacherie virus. *J. Invertebr. Pathol. 36*, 166–171.

Nakagaki, M., and Kawase, S. 1980b. DNA of a new parvo-like virus isolated from the silkworm, *Bombyx mori*. *J. Invertebr. Pathol. 35*, 124–133.

Norton, W. N., and Vinson, S. B. 1983. Correlating the initiation of virus replication with a specific pupal developmental phase of an ichneumonid parasitoid. *Cell Tiss. Res. 231*, 387–398.

Norton, W. N., Vinson, S. B., and Stoltz, D. B. 1975. Nuclear secretory particles associated with the calyx cells of the ichneumonid parasitoid *Campoletis sonorensis* (Cameron). *Cell Tiss. Res. 162*, 195–208.

Ohba, M., and Aizawa, K. 1979a. Multiplication of *Chilo* iridescent virus in noninsect arthropods. *J. Invertebr. Pathol. 33*, 278–283.

Ohba, M., and Aizawa, K. 1979b. In vivo insect hemocyte destruction by UV-irradiated *Chilo* iridescent virus. *J. Invertebr. Pathol. 34*, 32–40.

Ohba, M., and Aizawa, K. 1981. Lethal toxicity of arthropod iridoviruses to an amphibian, *Rana limnocharis*. *Arch. Virol. 68*, 153–156.

Ohba, M., and Aizawa, K. 1982. Mammalian toxicity of an insect iridovirus. *Acta Virol. 26*, 165–168.

Ohba, M., Mike, A., and Aizawa, K. 1982. Multiplication of a crustacean iridovirus in lepidopterous insects. *J. Invertebr. Pathol. 39*, 241–243.

Ono, M., and Fukaya, M. 1969. The juvenile-hormone-like effect of *Chilo* iridescent virus (CIV) on the metamorphosis of the silkworm, *Bombyx mori* L. (Lepidopteria: Bombycidae). *Appl. Entomol. Zool. 4*, 211–212.

Payne, C. C., Compson, D., and de Looze, S. M. 1977. Properties of the nucleocapsids of a virus isolated from *Oryctes rhinoceros*. *Virology 77*, 269–280.

Pogo, B. G. T., Dales, S., Bergoin, M., and Roberts, D. W. 1971. Enzymes associated with an insect poxvirus. *Virology 43*, 306–309.

Poinar, G. O., Jr., Hess, R., and Caltagirone, L. E. 1976. Virus- like particles in the calyz of *Phanerotoma flavitestacea* (Hymenoptera: Braconidae) and their transfer into host tissues. *Acta Zool.* (Stockh.) *57*, 161–165.

Quiot, J.-M., Bergoin, M., and Vago, C. 1975. Développement et pathogenèse d'un Entomopoxvirus en culture cellulaire de Lépidoptère. *C.R. Acad. Sci. Paris Ser. D280*, 2273–2275.

Rotheram, S. 1967. Immune surface of eggs of a parasitic insect. *Nature 214*, 700.

Seki, H., and Iwashita, Y. 1983. Histopathological features and pathogenicity of a densonucleosis virus of the silkworm, *Bombyx mori*, isolated from sericultural farms in Yamanashi prefecture. *J. Seric. Sci. Jpn. 52*, 400–405.

Shimizu, T. 1975. Pathogenicity of an infectious flacherie virus of the silkworm, *Bombyx mori*, obtained from sericultural farms in the suburbs of Ina city. *J. Seric. Sci. Jpn. 44*, 45–48.

Sikorowski, P. P., and Tyson, G. E. 1984. Per os transmission of iridescent virus of *Heliothis zea* (Lepidoptera: Noctuidae). *J. Invertebr. Pathol. 44*, 97–102.

Smith, K. M. 1958. A study of the early stages of infection with the Tipula Iridescent Virus. *Parasitology 48*, 459–462.

Smith, K. M., Hills, G. J., and Rivers, C. F. 1961. Studies on the cross-inoculation of the *Tipula* iridescent virus. *Virology 13*, 233–241.

Steinhaus, E. A., and Leutenegger, R. 1963. Icosahedral virus from a scarab (*Sericesthis*). *J. Insect Pathol. 5*, 266–270.

Stoltz, D. B. 1990. Evidence for chromosomal transmission of polydnavirus DNA. *J. Gen. Virol. 71*, 1051–1056.

Stoltz, D. B., and Guzo, D. 1986. Apparent haemocytic transformations associated with parasitoid-induced inhibition of immunity in *Malacosoma disstria* larvae. *J. Insect Physiol. 32*, 377–388.

Stoltz, D. B., and Summers, M. D. 1971. Pathway of infection of mosquito iridescent virus. I. Preliminary observations on the fate of ingested virus. *J. Virol. 8*, 900–909.

Stoltz, D. B., and Summers, M. D. 1972. Observations on the morphogenesis and structure of a hemocytic poxvirus in the midge *Chironomus attenuatus*. *J. Ultrastruct. Res. 40*, 581–598.

Stoltz, D. B., and Vinson, S. B. 1977. Baculovirus-like particles in the reproductive tracts of female parasitoid wasps II: The genus *Apanteles*. *Can. J. Microbiol. 23*, 28–37.

Stoltz, D. B., and Vinson, S. B. 1979a. Viruses and parasitism in insects. *Adv. Virus Res.* 24, 125–171.
Stoltz, D. B., and Vinson, S. B. 1979b. Penetration into caterpillar cells of virus-like particles injected during oviposition by parasitoid ichneumonid wasps. *Can. J. Microbiol.* 25, 207–216.
Stoltz, D. B., Hilsenhoff, W. L., and Stich, H. F. 1968. A virus disease in *Chironomus plumosus*. *J. Invertebr. Pathol.* 12, 118–128.
Stoltz, D. B., Krell, P. J., and Vinson, S. B. 1981. Polydisperse viral DNA's in ichneumonid ovaries: A survey. *Can. J. Microbiol.* 27, 123–130.
Stoltz, D. B., Guzo, D., and Cook, D. 1986. Studies on polydnavirus transmission. *Virology* 155, 120–131.
Stoltz, D. B., Guzo, D., Belland, E. R., Lucarotti, C. J., and MacKinnon, E. A. 1988. Venom promotes uncoating *in vitro* and persistence *in vivo* of DNA from a braconid polydnavirus. *J. Gen. Virol.* 69, 903–907.
Suto, C., Kawamoto, F., and Kumada, N. 1979. A new virus isolated from the cockroach, *Periplaneta fuliginosa* (Serville). *Microbiol. Immunol.* 23, 207–211.
Tanada, Y., and Tanabe, A. M. 1965. Resistance of *Galleria mellonella* (Linnaeus) to the *Tipula* iridescent virus at high temperatures. *J. Invertebr. Pathol.* 7, 184–188.
Tanaka, T. 1987. Effect of the venom of the endoparasitoid, *Apanteles kariyai* Watanabe, on the cellular defence reaction of the host, *Pseudaletia separata* Walker. *J. Insect Physiol.* 33, 413–420.
Theilmann, D. A., and Summers, M. D. 1986. Molecular analysis of *Campoletis sonorensis* virus DNA in the lepidopteran host *Heliothis virescens*. *J. Gen. Virol.* 67, 1961–1969.
Theilmann, D. A., and Summers, M. D. 1987. Physical analysis of the *Campoletis sonorensis* virus multipartite genome and identification of a family of tandemly repeated elements. *J. Virol.* 61, 2589–2598.
Theilmann, D. A., and Summers, M. D. 1988. Identification and comparison of *Campoletis sonorensis* virus transcripts expressed from four genomic segments in the insect hosts *Campoletis sonorensis* and *Heliothis virescens*. *Virology* 167, 329–341.
Tijssen, P., and Arella, M. 1991. Parvoviridae, structure and reproduction of densonucleosis viruses. *In* "Atlas of Invertebrate Viruses." (J. R. Adams and J. R. Bonami, eds.), pp. 41–53. CRC Press, Boca Raton, Florida.
Tinsley, T. W., and Kelly, D. C. 1970. An interim nomenclature system for the iridescent group of insect viruses. *J. Invertebr. Pathol.* 16, 470–472.
Vago, C. 1963. A new type of insect virus. *J. Insect Pathol.* 5, 275–276.
Vago, C., and Bergoin, M. 1968. Viruses of invertebrates. *In* "Advances in Virus Research." (K. M. Smith and M. A. Lauffer, eds.), Vol. 13, pp. 247–303. Academic Press, New York.
Vago, C., Meynadier, G., and Duthoit, J.-L. 1964. Étude d'un nouveau type de maladie à virus chez les Lépidoptères. *Ann. Epiphyt.* 15, 473–477.
Vago, C., Duthoit, J. L., and Delahaye, F. 1966. Les lésions nucléaires de la "Virose à noyaux denses" du Lépidoptère Galleria mellonella. *Arch. Ges. Virusforsch.* 18, 344–349.
Vinson, S. B. 1972. Factors involved in successful attack on *Heliothis virescens* by the parasitoid *Cardiochiles nigriceps*. *J. Invertebr. Pathol.* 20, 118–123.
Vinson, S. B., and Scott, J. R. 1975. Particles containing DNA associated with the oocyte of an insect parasitoid. *J. Invertebr. Pathol.* 25, 375–378.
Vinson, S. B., Edson, K. M., and Stoltz, D. B. 1979. Effect of a virus associated with the reproductive system of the parasitoid wasp, *Campoletis sonorensis*, on host weight gain. *J. Invertebr. Pathol.* 34, 133–137.
Wagner, G. W., Webb, S. R., Paschke, J. D., and Campbell, W. R. 1975. Production and

characterization of the cores of the "R" strain of mosquito iridescent virus. *Virology 64*, 430–437.

Ward, V. K., and Kalmakoff, J. 1987. Physical mapping of the DNA genome of insect iridescent virus type 9 from *Wiseana* spp. larvae. *Virology 160*, 507–510.

Watanabe, H. 1981. Characteristics of densonucleosis in the silkworm, *Bombyx mori*. *Jpn. Aric. Res. Q. 15*, 133–136.

Watanabe, H., and Kurihara, Y. 1988. Comparative histopathology of two densonucleoses in the silkworm, *Bombyx mori*. *J. Invertebr. Pathol. 51*, 287–290.

Watanabe, H., and Maeda, S. 1978. Genetic resistance to peroral infection with a densonucleosis virus in the silkworm, *Bombyx mori*. *J. Seric. Sci. Jpn. 47*, 209–214.

Watanabe, H., and Maeda, S. 1981. Genetically determined nonsusceptibility of the silkworm, *Bombyx mori*, to infection with a densonucleosis virus (*Densovirus*). *J. Invertebr. Pathol. 38*, 370–373.

Watanabe, H., Maeda, S., Matsui, M., and Shimizu, T. 1976. Histopathology of the midgut epithelium of the silkworm, *Bombyx mori*, infected with a newly-isolated virus from the flacherie-diseased larvae. *J. Seric. Sci. Jpn. 45*, 29–34.

Watanabe, H., Kawase, S., Shimizu, T., and Seki, H. 1986. Difference in serological characteristics of densonucleosis viruses in the silkworm, *Bombyx mori*. *J. Seric. Sci. Jpn. 55*, 75–76.

Webb, B. A., and Summers, M. D. 1990. Venom and viral expression products of the endoparasitic wasp *Campoletis sonorensis* share epitopes and related sequences. *Proc. Natl. Acad. Sci. U.S.A. 87*, 4961–4965.

Webb, S. R., Paschke, J. D., Wagner, G. W., and Campbell, W. R. 1976. Pathology of mosquito iridescent virus of *Aedes taeniorhynchus* in cell cultures of *Aedes aegypti*. *J. Invertebr. Pathol. 27*, 27–40.

Weiser, J. 1965a. *Vagoiavirus* gen.n., a virus causing disease in insects. *J. Invertebr. Pathol. 7*, 82–85.

Weiser, J. 1965b. A new virus infection of mosquito larvae. *Bull. Wld. Hlth. Org. 33*, 586–588.

Weiser, J. 1969. A pox-like virus in the midge *Camptochironomus tentans*. *Acta Virol. 13*, 549–553.

Whitfield, J. B. 1990. Parasitoids, polydnaviruses and endosymbiosis. *Parasitol. Today 6*, 381–384.

Williams, R. C., and Smith, K. M. 1957. A crystallizable insect virus. *Nature 179*, 119–120.

Williams, R. C., and Smith, K. M. 1958. The polyhedral form of the Tipula iridescent virus. *Biochim. Biophys. Acta 28*, 464–469.

Witt, D. J., and Stairs, G. R. 1976. Effects of different temperatures on *Tipula* iridescent virus infection in *Galleria mellonella* larvae. *J. Invertebr. Pathol. 28*, 151–152.

Woodard, D. B., and Chapman, H. C. 1968. Laboratory studies with the mosquito iridescent virus (MIV). *J. Invertebr. Pathol. 11*, 296–301.

Xeros, N. 1954. A second virus disease of the leatherjacket, *Tipula paludosa*. *Nature 174*, 562–563.

Xeros, N. 1964. Development of the *Tipula* iridescent virus (TIV). *J. Invertebr. Pathol. 6*, 261–283.

Younghusband, H. B., and Lee, P. E. 1969. Virus-cell studies of *Tipula* iridescent virus in *Galleria mellonella* (L.). I. Electron microscopy of infection and synthesis of *Tipula* iridescent virus in hemocytes. *Virology 38*, 247–254.

Younghusband, H. B., and Lee, P. E. 1970. Cytochemistry and autoradiography of *Tipula* iridescent virus in *Galleria mellonella*. *Virology 40*, 757–760.

Yule, B. G., and Lee, P. E. 1973. A cytological and immunological study of *Tipula* iridescent virus-infected *Galleria mellonella* larval hemocytes. *Virology 51*, 409–423.

CHAPTER 8

RNA-VIRAL INFECTIONS: REOVIRIDAE

I. CYTOPLASMIC
   POLYHEDROSIS VIRUSES
   A. Viral Particle
   B. Inclusion Body
   C. Replication
   D. Syndrome
   E. Transmission
   F. Histopathology
   G. Host Resistance
II. *MUSCAREOVIRUS*—HOUSE
    FLY VIRUS

Insects are susceptible to a large number of double-stranded (ds) and single-stranded (ss) RNA viruses. The families of RNA viruses infecting insects are shown in Fig. 8–1. Some of these viruses infect only insects and a few other arthropods. Although early studies were conducted with the sacbrood virus of the honey bee and the cytoplasmic polyhedrosis virus (CPV) of the Lepidoptera, the RNA viruses received less attention during the past half century as compared with the insect DNA viruses because of their low pathogenicity and the resulting limitation in insect pest control. The silkworm CPV is an exception and has been widely investigated because of its importance in sericulture and its relationship to other reoviruses that are important vertebrate pathogens.

An increasing number of single-stranded (ss) RNA viruses have been detected in insects and many are very small and ubiquitous in nature. All have been included, at one time, in the family Picornaviridae, but recently some have been placed in other families or in new families. The arthropod-borne (arboviruses) RNA viruses will not be considered in this text.

The entomopathogenic dsRNA viruses occur in the families Reoviridae and Birnaviridae. The Birnaviridae will be considered in Chapter 9. Nearly all insect viruses in the family Reoviridae are CPVs (genus *Cypovirus*, type species *Bombyx mori* CPV) (Francki et al. 1991). The CPVs share structural and biochemical similarities with other members of Reoviridae, such as an icosahedral shape with capsid from 60 to 80 nm in diameter, dsRNA from 10 to 12 discrete segments, an enzyme complex associated with the viral particles, and cytopathic changes with the formation of a viroplasm or virogenic stroma in the cytoplasm (Belloncik 1989; Hukuhara and Bonami 1991). Viruses with reolike morphology have been detected in a number of insects, and some of them fulfill the primary requisites for inclusion in

FIGURE 8–1

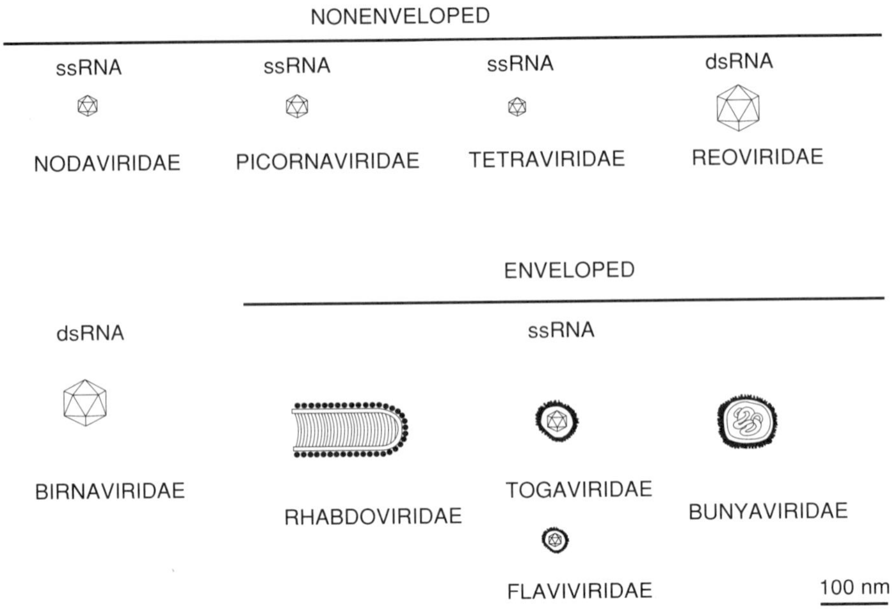

	NONENVELOPED		
ssRNA	ssRNA	ssRNA	dsRNA
NODAVIRIDAE	PICORNAVIRIDAE	TETRAVIRIDAE	REOVIRIDAE

	ENVELOPED		
dsRNA	ssRNA		
BIRNAVIRIDAE	RHABDOVIRIDAE	TOGAVIRIDAE / FLAVIVIRIDAE	BUNYAVIRIDAE

100 nm

The families of RNA viruses infecting insects. In the Reoviridae, the cytoplasmic polyhedrosis virus has viral occlusion (polyhedron) that is not shown due to space limitations. The polyhedra may range from 1 to 15 μm in diameters. The virions are drawn to scale. Recently, a member of Caliciviridae was shown to infect insects. [Courtesy of Adams and Bonami (1991) and CRC Press.]

Reoviridae. Readers interested in these viruses should refer to Hukuhara and Bonami (1991).

Comprehensive reviews on the CPV are presented in Aruga and Tanada (1971), Payne and Martens (1983), Hukuhara (1985), Belloncik (1989), and Hukuhara and Bonami (1991).

I. CYTOPLASMIC POLYHEDROSIS VIRUSES

Insect CPVs are basically similar to other reoviruses but can be distinguished by host range, morphology of the viral particle (lack of a double capsid), the sequence homologies of the genomes, and a major viral-coded protein (polyhedrin) that forms the cytoplasmic inclusion bodies or polyhedra (Martinson and Lewandowski 1974; Payne and Mertens 1983). In Reoviridae, the polyhedra are formed only in the CPVs of insects, with one exception in a freshwater daphnid (Federici and Hazard 1975). Reoviruses of vertebrates also form cytoplasmic inclusions, called "viral factories," which contain dsRNA, virus-specific polypeptides, and incomplete and complete viral particles, which may occur in crystalline arrays (Tyler and Fields 1986). These inclusion bodies, however, differ from those of CPVs in lacking a crystalline proteinaceous matrix occluding viral particles at random.

The first report of a CPV was in 1934 when Ishimori observed a polyhedrosis different from the nuclear polyhedrosis, in which the polyhedral bodies occurred in the cytoplasm rather than in the nucleus of a midgut epithelial cell of the silkworm. The importance of this CPV was first recognized in Japan where it caused major losses in sericulture (Aruga 1971). Xeros (1952) used the term "cytoplasmic polyhedrosis" to differentiate this disease from the nuclear polyhedrosis. Subsequently, the spherical (icosahedral) morphology and the RNA character of the CPV were established (Smith and Wykoff 1950; Bird and Whalen 1954; Krieg 1956). The CPV had dsRNA (Miura et al. 1968; Kalmakoff et al. 1969; Yamakawa et al. 1981) and it was placed in the Reoviridae (Fujii-Kawata et al. 1970).

The CPVs have been reported from a wide range of insects (about 250 species), with 80% of them in Lepidoptera, 16% in Diptera, 3% in Hymenoptera, and less than 1% in Coleoptera and Neuroptera (Martignoni and Iwai 1986; Hukuhara and Bonami 1991). Some CPVs have a broad host range and infect insects of different species, genera, and families (Tanaka 1971). Thus far, no CPV has been found infecting vertebrates and plants (Belloncik 1989).

The limited specificity makes identification by host difficult, and a provisional classification system is used to designate the various CPVs as types, with 12 of them proposed at present (Payne and Rivers 1976; Payne et al. 1977). The types are defined by the distinctive electrophoretic profiles (electropherotypes) of the RNA genome segments. There is a close correlation between the electropherotypes and the antigenic properties of the polyhedron proteins (polyhedrin) or the viral particle structural proteins (Mertens et al. 1989). Moreover, Galinski et al. (1983) have detected diversity in sequence homologies among different CPV types. The CPV types are Type 1 from *Bombyx mori*, Type 2 from *Inachis io*, Type 3 from *Spodoptera exempta*, Type 4 from *Actias selene*, Type 5 from *Trichoplusia ni*, Type 6 from *Biston betularia*, Type 7 from *Triphena pronuba*, Type 8 from *Abraxas grossulariata*, Type 9 from *Agrotis segetum*, Type 10 from *Aporophylla lutulenta*, Type 11 from *Spodoptera exigua*, and Type 12 from *Spodoptera exempta* (Francki et al. 1991).

The prototype CPV is that of the silkworm. There are nine strains of this virus, which are distinguished by their shapes and/or the intracellular locations of their inclusion bodies (Hukuhara and Midorikawa 1983; Hukuhara 1985). The viral genome of each strain governs the shape and cellular location of the inclusion bodies (Hukuhara and Midorikawa 1983). Hukuhara (1985) postulates that the CPV strains have arisen from the genetic reassortment of RNA segments as indicated by the differences in the strains due to not one but several traits, such as the growth parameter and site of morphogenesis of the inclusion bodies.

A. VIRAL PARTICLE

The CPV particle, icosahedral in shape, is formed in the cytoplasm. It is from 50 to 80 nm in diameter (Matthews 1982; Hukuhara and Bonami 1991) with a core of 30 to 40 nm and a single-shell capsid (Aizawa 1971; Matthews 1982). Hosaka and Aizawa (1964) have proposed two capsid shells enclosing the central core in their model, but others have found no evidence in electron micrographs for a distinct double capsid structure (Harrap and Payne 1979; Hatta and Francki 1982). The CPV particles have properties similar to those of subviral particles or "cores" of reovirus (Lewandowsky and Trayner 1972; Payne and Tinsley 1974; Payne and Harrap 1977).

The CPV particle is composed of at least four morphological components: two types of spikes, base plates, and the rest of the capsid (Hukuhara and Bonami 1991). The viral particle has 12 spikes or projections at each of the 12 vertices of the

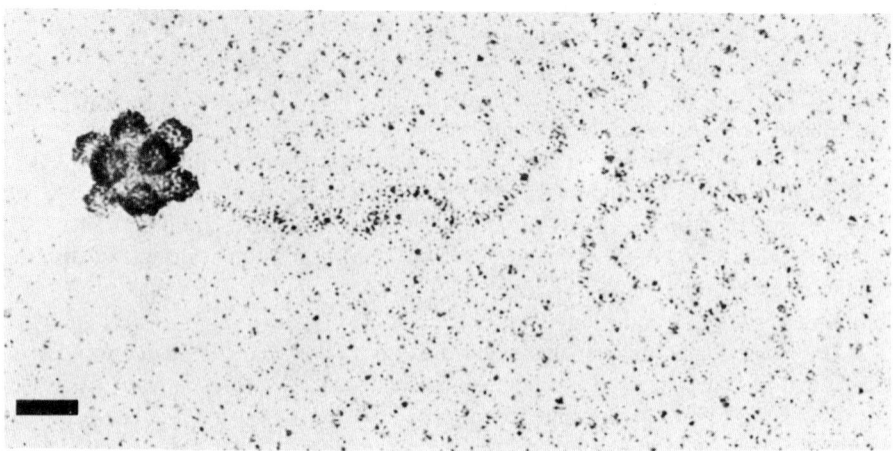

Viral particle of the cytoplasmic polyhedrosis virus of the silkworm, *Bombyx mori*, showing the release of the dsRNA from one of the vertices. (Bar = 50 nm.) [Courtesy of Yazaki and Miura (1980) and Academic Press.]

icosahedron (Hosaka and Aizawa 1964; Arnott et al. 1968; Asai et al. 1972). Each spike appears to be hollow and is attached to a base plate. The release of core material occurs through the spikes (Yazaki and Miura 1980) (Fig. 8–2). Since the spikes have been observed in all closely examined CPVs, they may be a common feature of the group.

1. Double-Stranded RNA

The dsRNA has 10 segments, with plus and minus strands, that code for the viral structural and nonstructural proteins (Harley and Rubinstein 1978; Grancher-Barry et al. 1981; Galinski et al. 1984; Pullin and Moore 1985). The molecular weights (MWs) of the 10 segments range from 0.2 to 4×10^6, with total MWs of 13 to 22×10^6 (Harley et al. 1977; Grancher-Barry et al. 1981; Galinski et al. 1984). Each of the 10 segments is transcribed independently (Smith and Furuichi 1982). In most cases, each segment of dsRNA encodes one protein (McCrae and Mertens 1983). The largest RNAs encode the viral structural proteins (Pullin and Moore 1985). There are from 3 to 10 structural proteins of MWs from 14 to 160×10^3 (Grancher-Barry et al. 1981; Landridge 1983; Galinski et al. 1984). The extracted genome segments have circular and supercoiled structures and they form genome–protein(s) complexes (Yazaki et al. 1986). Some CPVs have a high genomic stability after infection in a larva or a cell culture, while others present a high degree of variability and instability in the RNA segments, as in the case of other reoviruses (Belloncik 1989).

2. Transcription

The dsRNA is not infectious until acted upon by a complex of enzymes carried by the CPV for the synthesis of messenger RNAs (mRNA) and the 10 ssRNAs (Miura

1981). Each virion contains RNA polymerases (Lewandowski et al. 1969; Shimotohno and Miura 1973a; Kuchino et al. 1982) and dsRNA-dependent transcriptases (Furuichi 1974) for the synthesis of ssRNAs. The RNA polymerase makes a copy of one strand of each RNA segment as in the case of other reoviruses, and transcribes the mRNA (Shimotohno and Miura 1973a,b; Payne and Mertens 1983). Wu and Sun (1986) have isolated a RNA replicase with enzyme activities of both RNA polymerase and methyltransferase. The RNA replicase, which is tightly bound to each of the ds genome segments, plays a role in the transcription of dsRNA to form mRNA.

In the eucaryotic transcription system, the actively translated mRNA requires a modificaion of the 5′-terminus of mRNA and this modification is known as capping or the cap of mRNA (Furuichi 1978; Miura 1981). A complement of enzymes acts in the capping process, namely, nucleotide phosphohydrolase (Storer et al. 1974; Shimotohno and Miura 1977), RNA guanyltransferase (Furuichi 1978), and methyltransferase (Furuichi 1974; Dai et al. 1986). There appears to be one enzyme complex for each of the 10 segments of the viral genome (Smith and Furuichi 1982; Dai et al. 1982; Wu et al. 1983). The synthesis of ssRNA is spontaneous, as in other reoviruses, and the initiation of the first transcriptional cycle is synchronous for all 10 viral genome segments (Shimotohno and Miura 1973b; Smith and Furuichi 1982).

The enzyme complex occurs at the base of each spike of a virion (Yazaki and Miura 1980). The supercoiled dsRNA segments, attached to the spike protein, can be extracted with a mild disruption of the virion (Yazaki and Miura 1980, Yazaki et al. 1986) (Fig. 8–2). The segmented genomes are transcribed into ssRNAs by a passage of the dsRNA segments through the base of a spike, where the enzyme complex for mRNA synthesis is located.

B. INCLUSION BODY

In the case of CPVs, we shall use the term inclusion body (polyhedron) rather than occlusion body because some of these bodies do not occlude virions. The typical polyhedron occludes, at random, between one and several thousands of mature viral particles, but not all of them are occluded. In the mosquitoes and chironomids, each polyhedron contains a single viral particle (Hukuhara and Bonami 1991). The polyhedra, as in the case of the NPV polyhedra, are very stable, protecting the virions from adverse environmental factors and serving as vehicles for the transmission of the virus from one host to another. The size of the polyhedra varies from 0.1 to 10 μm.

The formation of inclusion bodies is a distinct characteristic of the CPVs and separates them from other reoviruses (Fig. 8–3A,B), but there are silkworm CPV strains that do not form inclusion bodies (Hukuhara 1985). In most CPVs, the inclusion bodies are formed in the cytoplasms of midgut epithelial cells, but in one mutant of the CPV of the silkworm, such bodies are also formed in the nuclei of these cells (Hukuhara and Midorikawa 1983; Hukuhara 1985) (Fig. 8–3D). Nuclear inclusions are amorphous, acicular, or hexahedral in shape, and do not occlude viral particles. In polyhedra produced in the cytoplasm, the large number of viral particles occurs randomly throughout the proteinaceous matrix. Inclusion bodies of CPVs are not bounded by an envelope as in those of the NPVs. Thus, the CPV polyhedra are more easily stained with Giemsa (Smith 1976) and other stains

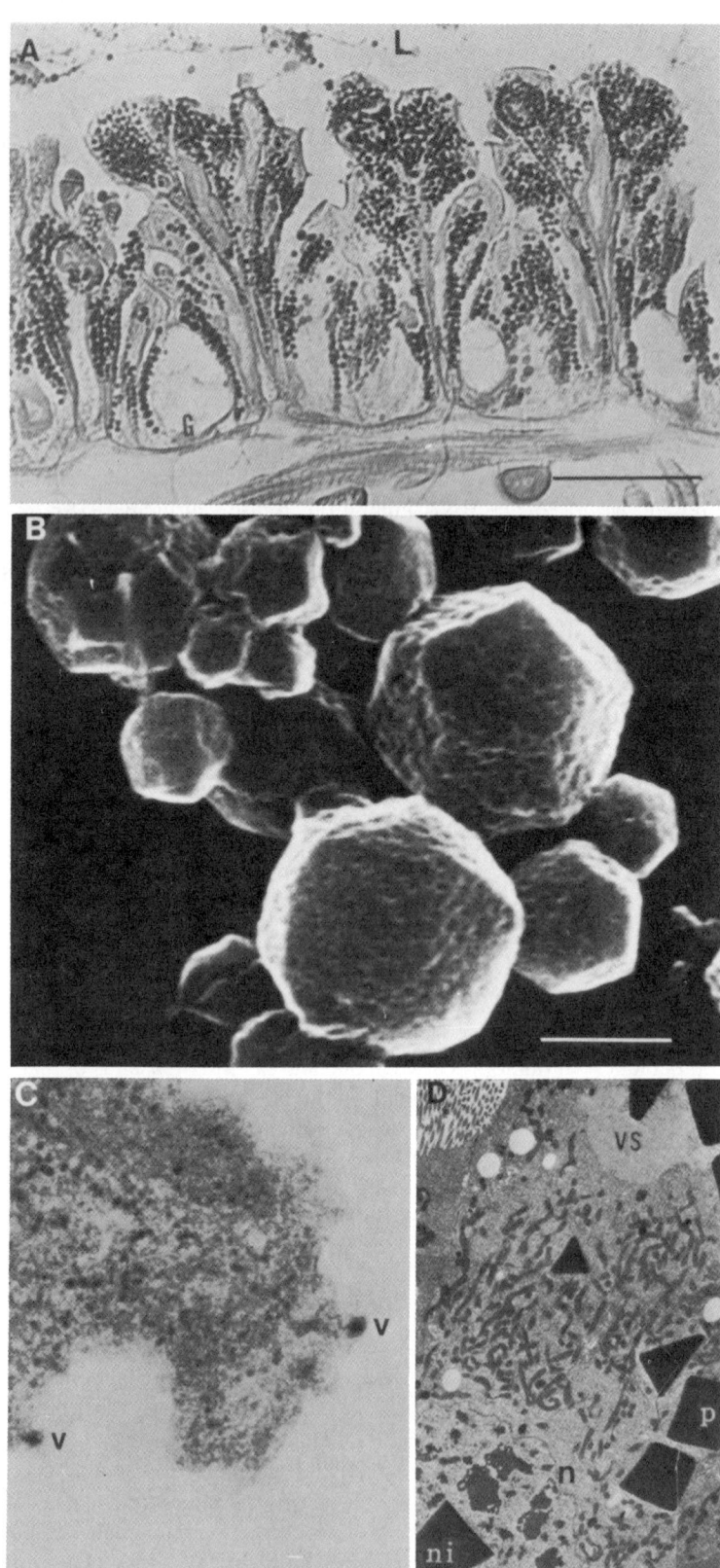

(Iwashita 1971) than the NPV polyhedra and this is due to the absence of the envelope. This difference in staining property is one way to separate the two types of polyhedrosis (Iwashita 1971; Smith 1976).

In addition to the difference in cellular location, the cytoplasmic and nuclear inclusion bodies of CPVs have proteins that differ in electrophoretic mobility and in amino acid sequence (Mori et al. 1987). The protein of the nuclear inclusion body (strain A) has tyrosine in place of the histidine residue present in the protein of the cytoplasmic inclusion body (strain H). Mori et al. (1987) have speculated that this difference may be responsible for the transportation of the inclusion body protein into a nucleus by a specific signal of the amino acid sequence.

In mosquitoes, Andreadis (1981) divided the inclusion bodies into two types: (1) anopheline mosquitoes with small cuboidal polyhedra (156 nm) with the viral particle occluded singly in the center of a polyhedron and (2) culicine mosquitoes with inclusion bodies varying widely in sizes (from 0.1 to 10 μm) and shapes (irregular and spherical) and each body containing a number of viral particles.

1. Formation and Structure

As the CPV infection progresses, the synthesis of the polyhedron matrix proteins intensifies resulting in an increase in the size of the inclusion body. The inclusion body may also increase in size by the uniting of one or more bodies to form a larger body. Thus, the sizes of inclusion bodies depend on the duration of infection in a cell (i.e., cells with advanced infection have fewer numbers but larger inclusion bodies than cells at an early stage of infection) (Aruga and Israngkul 1961). In the case of the NPV, the polyhedra do not increase in size after they are enclosed by the envelope, and they are more uniform in size than the CPV polyhedra.

The CPV cytoplasmic polyhedron is built of similar size granules, which are arranged as body-centered crystals as in the case of the NPV polyhedron (Bergold and Suter 1959; Fujiwara et al. 1984). Usually, there are two lattice patterns, a dot and a line pattern (Arnott et al. 1968; Hukuhara and Midorikawa 1983). The center-to-center spacing between rows of dots and lines is from 50 to 65 Å. Occluded viral particles do not disturb the molecular lattice of the polyhedra. On the other hand, the amorphous nuclear inclusion bodies consist of tightly packed granules of 30 to 40 Å in diameter that are not arranged in a crystalline pattern (Hukuhara and Midorikawa 1983). The amorphous inclusion bodies, less than 1 μm in diameter, first appear at multiple foci within the hypertrophied nucleus, and they later unite to form a large amorphous mass up to 14 μm in diameter, without viral particles, and may occupy most of the nucleus (Yamaguchi and Hukuhara 1973a,b). Similar amorphous spherical bodies, but occluding viral particles, occur in midgut cells of adult Queensland fruit fly, *Dacus tryoni* (Moussa 1978a).

FIGURE 8–3

Cytoplasmic polyhedrosis viruses (CPVs). (A) A CPV infection in the midgut columnar epithelium of the silkworm, *Bombyx mori*, showing polyhedra in the cytoplasm. (*L*) Lumen; (*G*) goblet cell. (Bar = 100 μm.) (Provided by H. Watanabe.) (B) Scanning electron micrograph of polyhedra of a CPV infecting the gypsy moth, *Lymantria dispar*. (Bar = 1 μm.) (Provided by J. R. Adams.) (C) A section of a portion of a silkworm midgut cell showing attached viral particles (*v*) apparently discharging the RNA. (Bar = 800 nm.) [Courtesy of Kobayashi (1971) and the University of Tokyo Press.] (D) A section of a silkworm midgut cell showing the presence of polyhedra in the cytoplasm (*P*) and in the nucleus (*ni*). (*n*) Nuclear membrane; (*vs*) virogenic stroma. (Bar = 3 μm.) (Provided by T. Hukuhara.)

Inclusion body matrices of CPVs are composed of one major polypeptide (polyhedrin) with molecular weights from 25,000 to 37,000 (Payne and Mertens 1983). The polyhedrin is coded at the direction of the smallest genomic RNA segment 10 (McCrae and Mertens 1983; Pullin and Moore 1985). The complete nucleotide sequences of the polyhedrin genes have been determined for the CPVs of *B. mori* (Arella et al. 1988; Mori et al. 1989) and of *Euxoa scandens* (Fossiez et al. 1989). Although the nucleotide sequences of polyhedrin genes of several baculoviruses have a strong homology among them (Rohrmann 1986), those of the polyhedrin genes of CPVs of several insect species indicate little or no homology (Galinski et al. 1983, 1984; Payne et al. 1986; Fossiez et al. 1989). Moreover, the nucleotide sequences of the CPV polyhedrin genes have no homology with those of several polyhedrin genes of NPVs (baculoviruses) (Rohrmann et al. 1980; Arella et al. 1988; Fossiez et al. 1989), but the amino acid compositions of the cytoplasmic (CPV) and nuclear (NPV) polyhedra are similar (Kawase 1971; Hukuhara and Midorikawa 1983).

2. Dissolution

The CPV polyhedra are highly resistant to nonionic and ionic detergents but can be dissolved at a high pH, as in the case of the NPV polyhedra. Under controlled pH and ionic strength (sodium carbonate or carbonate–bicarbonate buffers at above pH 10.5), the intact viral particles that are liberated constitute approximately 2–5% of the total weight of a polyhedron (Hukuhara and Hashimoto 1966; Payne and Mertens 1983).

Proteolytic enzymes may function in the dissolution of polyhedra in an insect's midgut. An alkaline protease has been detected in the CPV polyhedra (Mori and Kawase 1983). Since the protease is absent in polyhedra obtained from CPV-infected cultured cells, it may be a contaminant of an insect-derived enzyme (Mori et al. 1985).

3. Genetics

Characteristics of the inclusion bodies are inherited and governed by the viral genome (Hukuhara and Midorikawa 1983). The polyhedrin genes of strains H and A of the CPV of *Bombyx mori* differ slightly in the amino acid residues and in the amino acid substitution of the histidine residue by tyrosine (Mori et al. 1989). These differences may account for the strain H to form virus-occluded inclusion bodies in the cytoplasm and the strain A to produce inclusion bodies lacking virions in the nucleus. The shape of each CPV inclusion body is generally constant and remains unchanged when the virus is passed through a specific host or cell line. The shapes, however, may change in alternate hosts (Hukuhara 1985), and at high-rearing temperatures (Yamaguchi et al. 1969; Okino and Ishikawa 1971), or when the virus is transferred from cell cultures (*in vitro*) to host larvae (*in vivo*) (Granados et al. 1974; Quiot and Belloncik 1977; Longworth 1981).

4. Interference

When a mixed infection from two CPV strains occurs in a silkworm larva, there is interference between the two strains, and two types of inclusion bodies are formed in separate midgut cells and rarely in a single cell (Aruga et al. 1961; Tanaka and

Aruga 1967; Yamaguchi 1968, 1975b). Rather than interference, it is possible that the two CPV strains multiply in the same cell but with the inclusion body formation being directed by only one CPV strain (Hukuhara 1985). Most workers believe that a cell, shortly after infection with one CPV strain, is not susceptible to a second strain. Such interference does not occur between CPV and other viruses (e.g., the NPV, densonucleosis virus, or iridescent virus often infects, together with a CPV, the same midgut columnar cell) (Sato and Inoue 1978; Inoue 1981; Arella et al. 1983).

C. REPLICATION

The following description of CPV replication is based mainly on the replication in the silkworm. Ingested polyhedra are dissolved by the larval midgut juice and the liberated virions enter the midgut lumen and become attached to the midgut columnar cell by the spikes on the viral surface (Kobayashi 1971; Asai et al. 1972) (Fig. 8–3C). The viral core substance (RNA) is released, through the spike, as a filament into the cell cytoplasm (Yazaki and Miura 1980), leaving an empty viral shell on the cell surface. The inoculative force appears to result from the contraction of the elastic inner shell or membrane. A virogenic stroma (viroplasm), with micronet appearance, is formed in the cytoplasm just below the brush border, but its location may vary with larval age (Iwashita and Seki 1973). Some viral RNA synthesis may occur in the nucleus (Watanabe et al. 1967; Payne and Mertens 1983). Dense materials on the surface of the virogenic stroma condense into small particles that become enclosed within thin layers of membranes to form the corelike particles (Kobayashi 1971). Subsequently, capsidal proteins arising from the virogenic stroma occlude the corelike particles, and the spikes are attached to the capsid. As the infection progresses, several virogenic stromata grow out toward the base of the cell and fuse to form large matrices. Numerous free viral particles and small polyhedra are distributed irregularly over the surfaces of the virogenic stromata.

The above description of viral assembly proposes that a viral core is first formed and then is enclosed by the capsid (Bird 1965; Kobayashi 1971). Another proposal is that the capsid is first produced and later becomes filled with the core substance (Arnott et al. 1968).

The first polyhedra appear about 15 h postinfection (h p.i.), although polyhedrin synthesis may be detected as early as 9 h p.i. at sites near the brush border of a midgut cell (Kawase and Miyajima 1969). The formation of polyhedra is initiated by a diffusion of protein threads into a mass of viral particles located on the surface of a virogenic stroma. The loosely assembled proteins rearrange and crystallize to form a polyhedron, which grows by accretion and by the uniting of two or more polyhedra. The area surrounding the formation of polyhedra has been called the crystallogenic matrix (Arnott et al. 1968). In the case of the CPV nuclear inclusions, the inclusion body proteins are formed in the cytoplasm and then transferred into a cell nucleus to form the inclusion body (Hukuhara 1985; Mori et al. 1987).

1. Tissue Specificity

In Lepidoptera, the replication of CPV occurs mainly in the midgut epithelium and is restricted to this tissue in many insect species (Fig. 8–3A). Even after the intrahemocoelic inoculation of CPV, the target is the midgut (Yamamasu and Kawakita 1962; Aruga et al. 1963; Moussa 1978a). Cytoplasmic polyhedra, however, have been reported in the fore- and hindguts of the salt-marsh caterpillar, *Estigmene acrea* (Smith 1963); in the columnar, goblet and immature regenerative cells of the midgut epithelium; and in the muscular and tracheal cells attached to the midgut of the fall webworm, *Hyphantria cunea* (Noguchi 1984a). In insects other than Lepidoptera, the cytoplasmic polyhedra have been observed in fat body cells of coleopteran insects (Sidor 1970); in the hypodermal cells and in the developing leg, wing, and antennal buds of the mosquito *Culex tarsalis* (Kellen et al. 1966); and in the gastric caeca and midgut of *Culex salinarius* (Clark et al. 1969).

In the midgut, the CPV infection occurs mainly in columnar cells and rarely in goblet cells (Xeros 1966; Iwashita 1971; Iwashita and Seki 1973). The site of infection in a midgut varies with the host. It is the mid and hind parts of the midgut of a hymenopteron, *Anoplonyx destructor* (Longworth and Spilling 1970); initially in the hind part and spreading to the mid and anterior parts of the midgut in the silkworm and most other lepidopterons (Iwashita 1971); in the anterior and posterior thirds of the midgut with rarely the entire midgut of the black fly (Bailey et al. 1975); and in the anterior (cardiac and gastric caeca) and posterior parts of the midgut of the mosquito (Clark et al. 1969; Andreadis 1981, 1986). When the virus is inoculated into the silkworm's hemocoel, it initiates infection anywhere along the midgut (Aruga et al. 1963). The silkworm is more susceptible to the virus inoculated into the hemocoel than to the virus ingested by a larva due to the antiviral activity of the midgut (Watanabe 1966).

2. Spread of Virus

Little is known of the cell-to-cell spread of CPV along the midgut epithelium of an infected insect. Inasmuch as the cells are exposed to the gut lumen, infection would be expected to occur from a number of foci along the midgut, but this may not be the case. The early ligation experiment by Aruga et al. (1963) suggests that the movement of the virus occurs from cell to cell because the injury caused by the ligation prevents the virus to spread from the initial infection site. This also suggests that the receptor sites of infection do not occur throughout the midgut epithelium. Moreover, the young silkworm larvae in the first and second instars appear to have few receptor sites because viral interference occurs in these larvae and not in older larvae with more receptor sites available to the viruses (Aruga et al. 1961; Yamaguchi 1974, 1975a,b).

D. SYNDROME

Most cytoplasmic polyhedroses are chronic diseases and may be nonlethal to larvae that develop into infected adults. The CPV infections in dipterons have little or no adverse effect (Hukuhara and Bonami 1991). Lepidopteran larvae with lethal infection may lag in their development, and are smaller, lighter in weight, and may have an increased number of larval molts as compared with uninfected larvae. These are signs of a nutritionally deprived or starved larva (Hukuhara 1985). Young larvae (first and second instars) are more susceptible to CPV (Vail et al. 1969), and in

FIGURE 8-4

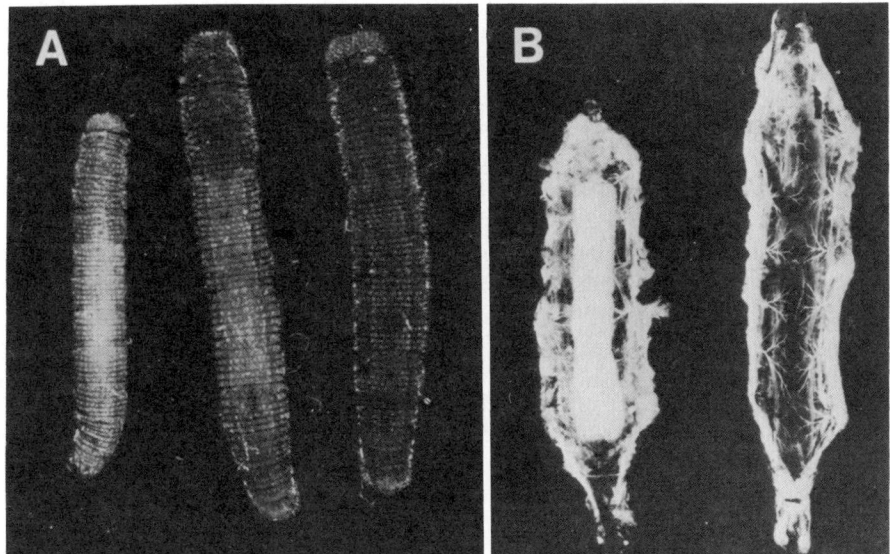

Larvae of the alfalfa caterpillar, *Colias eurytheme*, infected with a cytoplasmic polyhedrosis virus. (A) Middle and left infected larvae with color change in the central region of their bodies. (B) Larvae dissected open to show the opaque whitish-yellow appearance of the infected digestive tract (*left*) and compared with that (*right*) of an uninfected larva. [From Steinhaus and Dineen (1959) and courtesy of Academic Press.]

Hyphantria cunea, the larvae are infected only in the early instars (Yamaguchi 1976a,b).

In general, lepidopteran larvae fed the CPV polyhedra begin to show symptoms in about 4 days. They become sluggish, cease to feed, may develop diarrhea, and often vomit. Heavily infected larvae exhibit color changes, especially in lightly pigmented species, in central regions of their bodies where the whitish midguts show through the larval integuments (Fig. 8-4A).

A dissected and exposed midgut of a heavily infected larva is from opaque yellow to white in appearance as compared with the translucent, clear midgut of an uninfected larva (Fig. 8-4B). The infected midgut wall is more fragile than that of uninfected midgut and exudes a whitish fluid when broken. Infected midgut cells filled with polyhedra may break and discharge their contents or may be desquamated and enter the midgut lumen. Thus, the fecal pellets from a heavily infected larva are often whitish in color from the large number of polyhedra. Infected larvae die in about 7 to 15 days depending on the type and dosage of CPV, insect species, and larval instar at the time of infection. In larvae of the pink bollworm, *Pectinophora gossypiella*, the infection with CPV reduces the period of diapause but with an increase in larval mortality (Bell 1977).

Infected lepidopteran larvae produce pupae that are generally smaller than uninfected pupae (Neilson 1965; Vail et al. 1969). In some cases, infected pupae are malformed and do not give rise to adults (Simmons and Sikorowski 1973; Magnoler 1974; Bell and Kanavel 1976). Adults, emerging from infected pupae, may appear normal and produce viable eggs, but they are often small and malformed, especially with abnormal wings and are short lived (Sidor 1965; Vail and Gough 1970; Bellemare and Belloncik 1981). Infected adults may exhibit reduced fecundity (Vail and Gough 1970; Bullock et al. 1970; Sikorowski and Thompson 1979).

Infection in the adult Queensland fruit fly, *Dacus tryoni*, differs somewhat from that in the lepidopteran adults (Moussa 1978a). The fruit fly adult becomes sluggish and is partially paralyzed before dying. Its abdomen increases in size even though the adult has fed less than an uninfected adult. An infected midgut is swollen, fragile, dark orange to brownish, and breaks down at an advanced stage of infection. The hindgut shows a large tumorlike distension and is pale brown. The crop is filled with a clear or an orange to brownish turbid fluid. The fat body disintegrates. An adult, surviving the early postinfection period, initially may lay a few eggs but regains normal reproductive capacity after 3 to 4 days and can live for 1.5 to 2 months, but it is shorter than the normal life span of 4 to 6 months. This virus may be similar to the *Muscareovirus* described later in this chapter.

E. TRANSMISSION

The CPV infects a larva through the mouth and transmission takes place horizontally. Vertical transmission occurs commonly in Lepidoptera when a newly hatched larva ingests the virus contaminating an egg shell (transovum transmission) (Sikorowski et al. 1973; Mery and Dulmage 1975). In mosquitoes and black flies, vertical transmission occurs at a low rate and whether the transmission is transovarial or transovum has not been established (Bailey 1977; Andreadis 1986). The CPV has been detected in the hemolymph (Yamamasu and Kawakita 1962; Sikorowski et al. 1971; Miyajima 1975), but there is no evidence for transovarial transmission in these insects.

F. HISTOPATHOLOGY

In Lepidoptera, the infected epithelial cells of the midgut are enlarged at the apical ends adjacent to the lumen and the microvilli on these cells are partially or completely absent (Boucias and Nordin 1978) (Fig. 8–3A). The endoplasmic reticulum is fragmented and severely reduced in numbers (Brattsten 1987). The metabolic activity of membrane-bound cytochrome P-450 is significantly depressed. The basal ends of the cells are packed with polyhedra and the cells eventually rupture from the increasing internal pressure and release the polyhedra into the gut lumen (Iwashita 1971; Kobayashi 1971). The infected midgut cells are discharged into the gut lumen and are replaced with cells from the regenerative nidi (Iwashita and Aruga 1957; Yamaguchi and Ayuzawa 1970). Replacement of infected, but not uninfected, cells occurs generally at the larval molt (Yamaguchi 1976a, 1977, 1979; Inoue and Miyagawa 1978).

The nuclei of cells infected with the typical CPV are relatively unaffected. However, the silkworm strains that form inclusions in the nuclei cause enormous hypertrophy of the nucleoli and the aggregation of dense reticulum at multiple foci within the nuclei (Hukuhara and Bonami 1991). The nuclear inclusions appear within the areas where many masses of dense reticulum conglomerate.

G. HOST RESISTANCE

When the larva of the fall webworm, *Hyphantria cunea*, is severely infected with its virus, it is unable to molt, but when it is infected with the *Bombyx* CPV, it recovers and develops into an adult (Yamaguchi 1976a). The epithelial cells infected with *Bombyx* CPV are discharged and replaced with regenerated cells that are refractory to infection and no virus is detected in the new midgut. Moreover, the recovered fall webworm larva resists a challenge dose of *Bombyx* CPV (Yamaguchi 1979). Similar recoveries occur using the *B. mori* CPV with other insect species (e.g., *Orgyia thyellina* and *Mamestra brassicae*) (Noguchi and Yamaguchi 1982, 1984).

The resistance to CPV infection in the midgut is due to the newly regenerated midgut cells, since regenerated cells, formed after chemical injury to the midgut epithelium, are also resistant to the *B. mori* CPV (Noguchi and Yamaguchi 1985). Regenerated cells, however, are not immune to viral invasion because some cells contain small virogenic stromata with only empty capsids, suggesting incomplete viral replication (Noguchi 1984b). In the silkworm, the CPV does not infect regenerative cells in the nidi, but these cells are infected when they replace the discharged columnar cells (Iwashita and Aruga 1957).

Tumorlike bodies, which imply an immune response, may occur in certain lepidopterons that had been exposed to CPV. These bodies (from 1 to 3 mm^3) are irregularly shaped and occur in the adult abdomen usually within the midgut, but occasionally they are unattached and in the hemocoel (Neilson 1965; Magnoler 1974). These bodies are formed from larval midguts undergoing incomplete histolysis or disintegrated larval midgut epithelium enveloped within the peritrophic membrane (Hukuhara 1985).

II. *MUSCAREOVIRUS*—HOUSE FLY VIRUS

Moussa (1978b) reported the first case of a frank virus infection in the house fly, *Musca domestica*. The virus has been detected in adults and not in larvae. It infects the adult fly by ingestion and can be transmitted by inoculation into the thorax.

At the beginning of infection, adult activities increase for 3 to 4 days, with an increase in both the mating activity of males and the egg-laying capacity of females (Moussa 1978b). Then the adults stop feeding and become sluggish and shaky with feeble, trembling movements of wings and legs. They develop complete paralysis and die. Symptoms of the disease appear in 5 to 7 days and some flies survive up to 16 days.

The abdomen of an infected fly is swollen with a greatly enlarged midgut that is fragile and pale brown, unlike the translucent or white midgut of an uninfected fly. Apical portions of the hypertrophied midgut cells extend into the gut lumen and may block it in some areas. The cells are packed with dark, undulating threads in the cytoplasm. Parts of the midguts of some flies, usually at the posterior ends, are deflated and pale pink. The midguts of heavily infected flies breakdown completely.

The house fly virus is icosahedral in shape with two capsids and replicates in the hemocytes. It has been found in the cytoplasm of hemocytes together with inclusion bodies that do not occlude virions. The virus is characterized by the presence of capsomeres, 29 in number, on the periphery of the outer shell (capsid), an inner shell, a middle layer appearing as threads, a nucleoprotein core, and a viral nucleoid (Moussa 1980c). It contains dsRNA (Moussa 1980d) and belongs to the family Reoviridae but differs from other genera of this family by the possession of a middle layer, a nucleoprotein core, and a tendency to fragment during certain treatments (Moussa et al. 1982). The house fly virus has been proposed as a representative of a new genus in Reoviridae with the name *Muscareovirus* (Moussa et al. 1982).

REFERENCES

Adams, J. R., and Bonami, J. R. 1991. "Atlas of Invertebrate Viruses." CRC Press, Boca Raton, Florida.

Aizawa, K. 1971. Structure of polyhedra and virus particles of the cytoplasmic polyhedrosis. *In* "The Cytoplasmic-Polyhedrosis Virus of the Silkworm." (H. Aruga and Y. Tanada, eds.), pp. 23–36. University Tokyo Press, Tokyo.

Andreadis, T. G. 1981. A new cytoplasmic polyhedrosis virus from the salt-marsh mosquito, *Aedes cantator* (Diptera: Culicidae). *J. Invertebr. Pathol. 37*, 160–167.

Andreadis, T. G. 1986. Characterization of a cytoplasmic polyhedrosis virus affecting the mosquito *Culex restuans*. *J. Invertebr. Pathol. 47*, 194–202.

Arella, M., Devauchele, G., and Belloncik, S. 1983. Dual infection of a lepidopteran cell line with the cytoplasmic polyhedrosis virus (CPV) and the *Chilo* iridescent virus (CIV). *Ann. Virol.* (Inst. Pasteur) *134E*, 455–463.

Arella, M., Lavallée, C., Belloncik, S., and Furuichi, Y. 1988. Molecular cloning and characterization of cytoplasmic-polyhedrosis virus polyhedrin and a viable deletion mutant gene. *J. Virol. 62*, 211–217.

Arnott, H. J., Smith, K. M., and Fullilove, S. L. 1968. Ultrastructure of a cytoplasmic polyhedrosis virus affecting the monarch butterfly, *Danaus plexippus*. *J. Ultrastruct. Res. 24*, 479–507.

Aruga, H. 1971. Cytoplasmic polyhedrosis of the silkworm—historical, economical and epizootiological aspects. *In* "The Cytoplasmic-Polyhedrosis Virus of the Silkworm." (H. Aruga and Y. Tanada, eds.), pp. 3–21. University Tokyo Press, Tokyo.

Aruga, H., and Israngkul, A. 1961. Studies on the size of cytoplasmic polyhedra of the silkworm, *Bombyx mori* L. *J. Seric. Sci. Jpn. 30*, 119–125.

Aruga, H., and Tanada, Y. 1971. "The Cytoplasmic-Polyhedrosis Virus of the Silkworm." University of Tokyo Press, Tokyo.

Aruga, H., Hukuhara, T., Yoshitake, N., and Israngkul Na Ayudhya, A. 1961. Interference and latent infection in the cytoplasmic polyhedrosis of the silkworm, *Bombyx mori* (Linnaeus). *J. Insect Pathol. 3*, 81–92.

Aruga, H., Yoshitake, N., Hukuhara, T., and Owada, M. 1963. Invasion route of the cytoplasmic polyhedrosis virus in the silkworm, *Bombyx mori* L. *J. Seric. Sci. Jpn. 32*, 58–62.

Asai, J., Kawamoto, F., and Kawase, S. 1972. On the structure of the cytoplasmic-polyhedrosis virus of the silkworm, *Bombyx mori*. *J. Invertebr. Pathol. 19*, 279–280.

Bailey, C. H. 1977. Field and laboratory observations on a cytoplasmic polyhedrosis virus of blackflies (Diptera: Simuliidae). *J. Invertebr. Pathol. 29*, 69–73.

Bailey, C. H., Shapiro, M., and Granados, R. R. 1975. A cytoplasmic polyhedrosis virus from the larval blackflies *Cnephia mutata* and *Prosimulium mixtum* (Diptera: Simuliidae). *J. Invertebr. Pathol. 25*, 273–274.

Bell, M. R. 1977. Pink bollworm: Effect of infection by a cytoplasmic polyhedrosis virus on diapausing larvae. *Ann. Entomol. Soc. Am. 70*, 675–677.

Bell, M. R., and Kanavel, R. F. 1976. Effect of dose of cytoplasmic polyhedrosis virus on infection, mortality, development rate, and larval and pupal weights of the pink bollworm. *J. Invertebr. Pathol. 28*, 121–126.

Bellemare, N., and Belloncik, S. 1981. Études au laboratoire des effets d'une polyédrose cytoplasmique sur le ver gris blanc *Euxoa scandens* (Lepidoptère: Noctuidae Agrotinae). *Ann. Soc. Entomol. Quebec 26*, 28–40.

Belloncik, S. 1989. Cytoplasmic polyhedrosis viruses—Reoviridae. *Adv. Virus Res. 37*, 173–209.

Bergold, G. H., and Suter, J. 1959. On the structure of cytoplasmic polyhedra of some Lepidoptera. *J. Insect Pathol. 1*, 1–14.

Bird, F. T. 1965. On the morphology and development of insect cytoplasmic-polyhedrosis virus particles. *Can. J. Microbiol. 11*, 497–501.

Bird, F. T., and Whalen, M. M. 1954. A nuclear and a cytoplasmic polyhedral virus disease of the spruce budworm. *Can. J. Zool. 32*, 82–86.

Boucias, D. G., and Nordin, G. L. 1978. A scanning electron microscope study of *Hyphantria cunea* CPV-infected midgut tissue. *J. Invertebr. Pathol. 32*, 229–233.

Brattsten, L. B. 1987. Sublethal virus infection depresses cytochrome P-450 in an insect. *Experientia 43*, 451–454.

Bullock, H. R., Martinez, E., and Stuermer, C. W., Jr. 1970. Cytoplasmic-polyhedrosis virus and the development and fecundity of the pink bollworm. *J. Invertebr. Pathol. 15*, 109–112.

Clark, T. B., Chapman, H. C., and Fukuda, T. 1969. Nuclear-polyhedrosis and cytoplasmic-polyhedrosis virus infections in Louisiana mosquitoes. *J. Invertebr. Pathol. 14*, 284–286.

Dai, R., Wu, A., and Sun, Y. 1986. The protein subunits of the double-stranded RNA dependent RNA polymerase and methyltransferase of the cytoplasmic polyhedrosis virus of silkworm, *Bombyx mori. Sci. Sin. 29B*, 1267–1272.

Dai, R., Wu, A., Shen, X., Qian, L., and Sun, Y. 1982. Isolation of genome-enzyme complex from cytoplasmic polyhedrosis virus of silkworm *Bombyx mori. Sci. Sin. 25B*, 29–35.

Federici, B. A., and Hazard, E. I. 1975. Iridovirus and cytoplasmic polyhedrosis virus in the freshwater daphnid *Simocephalus expinosus. Nature 254*, 327–328.

Fossiez, F., Belloncik, S., and Arella, M. 1989. Nucleotide sequence of the polyhedrin gene of *Euxoa scandens* cytoplasmic polyhedrosis virus (EsCPV). *Virology 169*, 462–465.

Francki, R. I. B., Fauquet, C. M., Knudson, D. L., and Brown, F. 1991. "Classification and Nomenclature of Viruses. Fifth Report of the International Committee on Taxonomy of Viruses." Arch. Virol. Suppl. 2. Springer-Verlag, Wien.

Fujii-Kawata, I., Miura, K., and Fuke, M. 1970. Segments of genome of viruses containing double-stranded ribonucleic acid. *J. Mol. Biol. 51*, 247–253.

Fujiwara, T., Yukibuchi, E., Tanaka, Y., Yamamoto, Y., Tomita, K., and Hukuhara, T. 1984. X-ray diffraction studies of polyhedral inclusion bodies of nuclear and cytoplasmic polyhedrosis viruses. *Appl. Entomol. Zool. 19*, 402–403.

Furuichi, Y. 1974. "Methylation-coupled" transcription by virus-associated transcriptase of cytoplasmic polyhedrosis virus containing double-stranded RNA. *Nucleic Acid Res. 1*, 809–822.

Furuichi, Y. 1978. "Pretranscriptional capping" in the biosynthesis of cytoplasmic-polyhedrosis virus mRNA. *Proc. Natl. Acad. Sci. U.S.A. 75*, 1086–1090.

Galinski, M. S., Stanik, V. H., Rohrmann, G. F., and Beaudreau, G. S. 1983. Comparison of sequence diversity in several cytoplasmic polyhedrosis viruses. *Virology 130*, 372–380.

Galinski, M. S., Kingan, T., Rohrmann, G. F., Martignoni, M. E., and Beaudreau, G. S. 1984. Characterization of a cytoplasmic polyhedrosis virus infecting *Manduca sexta. Intervirology 21*, 167–173.

Granados, R. R., McCarthy, W. J., and Naughton, M. 1974. Replication of a cytoplasmic polyhedrosis virus in an established cell line of *Trichoplusia ni* cells. *Virology 59*, 584–586.

Grancher-Barray, S., Boisvert, J., and Belloncik, S. 1981. Electrophoretic characterization of proteins and RNA of cytoplasmic polyhedrosis virus (CPV) from *Euxoa scandens*. *Arch. Virol.* 70, 55–61.

Harley, E. H., and Rubinstein, R. 1978. The multicomponent genome of a different cytoplasmic polyhedrosis virus isolated from *Heliothis armigera*. *Intervirology* 10, 351–355.

Harley, E. H., Rubinstein, R., Losman, M., and Lutton, D. 1977. Molecular weights of the RNA genome segments of a cytoplasmic polyhedrosis virus determined by a new comparative approach. *Virology* 76, 210–216.

Harrap, K. A., and Payne, C. C. 1979. The structural properties and identification of insect viruses. *Adv. Virus Res.* 25, 273–355.

Hatta, T., and Francki, R. I. B. 1982. Similarity in the structure of cytoplasmic polyhedrosis virus, leafhopper A virus and Fiji disease virus particles. *Intervirology* 18, 203–208.

Hosaka, Y., and Aizawa, K. 1964. The fine structure of the cytoplasmic-polyhedrosis virus of the silkworm, *Bombyx mori* (Linnaeus). *J. Insect Pathol.* 6, 53–77.

Hukuhara, T. 1985. Pathology associated with cytoplasmic polyhedrosis viruses. *In* "Viral Insecticides for Biological Control." (K. Maramorosch and K. E. Sherman, eds.), pp. 121–162. Academic Press, New York.

Hukuhara, T., and Bonami, J. R. 1991. Reoviridae. *In* "Atlas of Invertebrate Viruses." (J. R. Adams and J. R. Bonami, eds.), pp. 393–434. CRC Press, Boca Raton, Florida.

Hukuhara, T., and Hashimoto, Y. 1966. Serological studies of the cytoplasmic- and nuclear-polyhedrosis viruses of the silkworm, *Bombyx mori*. *J. Invertebr. Pathol.* 8, 234–239.

Hukuhara, T., and Midorikawa, M. 1983. Pathogenesis of cytoplasmic polyhedrosis in the silkworm. *In* "Double-stranded RNA Viruses." (R. W. Compans and D. H. L. Bishop, eds.), pp. 405–414. Elsevior Science, Amsterdam.

Inoue, H. 1981. Double infection of midgut epithelial cells with nuclear and cytoplasmic polyhedrosis virus. *J. Seric. Sci. Jpn.* 50, 311–319.

Inoue, H., and Miyagawa, M. 1978. Regeneration of midgut epithelial cell in the silkworm, *Bombyx mori*, infected with viruses. *J. Invertebr. Pathol.* 32, 373–380.

Ishimori, N. 1934. Contribution a l'étude de la grasserie du ver a soie (*Bombyx mori*). *C.R. Soc. Biol.* 205, 1169–1170.

Iwashita, Y. 1971. Histopathology of cytoplasmic polyhedrosis. *In* "The Cytoplasmic-Polyhedrosis Virus of the Silkworm." (H. Aruga and Y. Tanada, eds.), pp. 79–101. University Tokyo Press, Tokyo.

Iwashita, Y., and Aruga, H. 1957. Mechanism of resistance to virus diseases in the silkworm, *Bombyx mori*. (III) Histological studies on the polyhedroses in the silkworm. *J. Seric. Sci. Jpn.* 26, 323–328.

Iwashita, Y., and Seki, H. 1973. The development of the cytoplasmic-polyhedrosis virus in the midgut of the embryo and the young silkworm, *Bombyx mori* L. *Bull. Coll. Agric. Univ. Utsunomiya* 8, 27–42.

Kalmakoff, J., Lewandowski, L. J., and Black, D. R. 1969. Comparison of the ribonucleic acid subunits of reovirus, cytoplasmic polyhedrosis virus, and wound tumor virus. *J. Virol.* 4, 851–856.

Kawase, S. 1971. Chemical nature of the cytoplasmic-polyhedrosis virus. *In* "The Cytoplasmic-Polyhedrosis Virus of the Silkworm." (H. Aruga and Y. Tanada, eds.), pp. 37–59. University Tokyo Press, Tokyo.

Kawase, S., and Miyajima, S. 1969. Immunoflurescence studies on the multiplication of cytoplasmic-polyhedrosis virus of the silkworm, *Bombyx mori*. *J. Invertebr. Pathol.* 13, 330–336.

Kellen, W. R., Clark, T. B., Lindegren, J. E., and Sanders, R. D. 1966. A cytoplasmic-polyhedrosis virus of *Culex tarsalis* (Diptera: Culicidae). *J. Invertebr. Pathol.* 8, 390–394.

Kobayashi, M. 1971. Replication cycle of cytoplasmic-polyhedrosis virus as observed with the electron microscope. *In* "The Cytoplasmic-Polyhedrosis Virus of the Silkworm." (H. Aruga and Y. Tanada, eds.), pp. 103–128. University Tokyo Press, Tokyo.

Krieg, A. 1956. Über die Nucleinsäuren der Polyeder-Viren. *Naturwissenschaften 43*, 537.

Kuchino, Y., Nishimura, S., Smith, R. E., and Furuichi, Y. 1982. Homologous terminal sequences in the double-stranded RNA genome segments of cytoplasmic polyhedrosis virus of the silkworm *Bombyx mori*. *J. Virol. 44*, 538–543.

Langridge, W. H. R. 1983. Characterization of a cytoplasmic polyhedrosis virus from *Estigmene acrea* (Lepidoptera). *J. Invertebr. Pathol. 42*, 259–263.

Lewandowski, L. J., and Traynor, B. L. 1972. Comparison of the structure and polypeptide composition of three double-stranded ribonucleic acid-containing viruses (diplornaviruses): Cytoplasmic polyhedrosis virus, wound tumor virus, and reovirus. *J. Virol. 10*, 1053–1070.

Lewandowski, L. J., Kalmakoff, J., and Tanada, Y. 1969. Characterization of a ribonucleic acid polymerase activity associated with purified cytoplasmic polyhedrosis virus of the silkworm *Bombyx mori*. *J. Virol. 4*, 857–865.

Longworth, J. F. 1981. The replication of a cytoplasmic polyhedrosis virus from *Chrysodeixis eriosoma* (Lepidoptera: Noctuidae) in *Spodoptera frugiperda* cells. *J. Invertebr. Pathol. 37*, 54–61.

Longworth, J. F., and Spilling, C. R. 1970. A cytoplasmic polyhedrosis of the larch sawfly, *Anoplonyx destructor*. *J. Invertebr. Pathol. 15*, 276–280.

Magnoler, A. 1974. Effects of a cytoplasmic polyhedrosis on larval and postlarval stages of the gypsy moth, *Porthetria dispar*. *J. Invertebr. Pathol. 23*, 263–274.

Martignoni, M. E., and Iwai, P. J. 1986. A catalogue of viral diseases of insects, mites, and ticks. *U.S.D.A. For. Ser. Pac. Northwest Res. Stn.*, Fourth ed. Gen. Tech. Rep. PNW-195.

Martinson, H. G., and Lewandowski, L. J. 1974. Sequence homology studies between the double-stranded RNA genomes of cytoplasmic polyhedrosis virus, wound tumor virus, and reovirus strains 1, 2 and 3. *Intervirology 4*, 91–98.

Matthews, R. E. F. 1982. Classification and nomenclature of viruses. *Intervirology 17*, 1–199.

McCrae, M. A., and Mertens, P. P. C. 1983. In vitro translation studies on and RNA coding assignments for cytoplasmic polyhedrosis viruses. *In* "Double-stranded RNA Viruses." (R. W. Compans and D. H. L. Bishop, eds.), pp. 35–41. Elsevier Science, Amsterdam.

Mertens, P. P. C., Crook, N. E., Rubinstein, R., Pedley, S., and Payne, C. C. 1989. Cytoplasmic polyhedrosis virus classification by electropherotype; validation by serological analyses and agarose gel electrophoresis. *J. Gen. Virol. 70*, 173–185.

Mery, C., and Dulmage, H. T. 1975. Transmission, diagnosis, and control of cytoplasmic polyhedrosis virus in colonies of *Heliothis virescens*. *J. Invertebr. Pathol. 26*, 75–79.

Miura, K. 1981. The cap structure in eukaryotic messenger RNA as a mark of a strand carrying protein information. *Adv. Biophys. 14*, 205–238.

Miura, K., Fujii, I., Sakaki, T., Fuke, M., and Kawase, S. 1968. Double-stranded ribonucleic acid from cytoplasmic polyhedrosis virus of the silkworm. *J. Virol. 2*, 1211–1222.

Miyajima, S. 1975. Changes in virus-infectivity titer in the hemolymph and midgut by per oral infection during the course of cytoplasmic-polyhedrosis virus of the silkworm, *Bombyx mori* L. *Res. Bull. Aichi Agric. Res. Ctr. D6*, 19–26.

Mori, H., and Kawase, S. 1983. Alkaline protease in cytoplasmic polyhedra of the silkworm, *Bombyx mori* (Lepidoptera: Bombycidae). *Appl. Entomol. Zool. 18*, 342–350.

Mori, H., Kawase, S., and Belloncik, S. 1985. Absence of an alkaline protease in cytoplasmic polyhedra obtained from the cultured cell of *Euxoa scandens* (Lepidoptera: Noctuidae). *Appl. Entomol. Zool. 20*, 498–499.

Mori, H., Sasaki, T., Minobe, Y., Miyajima, S., and Kawase, S. 1987. Difference of proteins from inclusion bodies formed in the nucleus and cytoplasm of the cytoplasmic polyhedrosis virus-infected midgut in the silkworm, *Bombyx moris. J. Invertebr. Pathol. 50*, 26–32.

Mori, H., Minobe, Y., Sasaki, T., and Kawase, S. 1989. Nucleotide sequence of the polyhedrin gene of *Bombyx mori* cytoplasmic polyhedrosis virus A strain with nuclear localization of polyhedra. *J. Gen. Virol. 70*, 1885–1888.

Moussa, A. Y. 1978a. A new cytoplasmic inclusion virus from Diptera in the Queensland fruitfly, *Dacus tryoni* (Frogg) (Diptera: Tephritidae). *J. Invertebr. Pathol. 32*, 77–87.

Moussa, A. Y. 1978b. A new virus disease in the housefly, *Musca domestica* (Diptera). *J. Invertebr. Pathol. 31*, 204–216.

Moussa, A. Y. 1980c. The housefly virus, isolation, purification and structure. *Micron 11*, 431–432.

Moussa, A. Y. 1980d. The housefly virus contains double-stranded RNA. *Virology 106*, 173–176.

Moussa, A. Y., Hawkes, R. A., Dickson, M. R., Shipp, E., and Woods, A. 1982. Serological relationships of the housefly virus and some members of the family Reoviridae. *Aust. J. Biol. Sci. 35*, 669–678.

Neilson, M. M. 1965. Effects of a cytoplasmic polyhedrosis on adult Lepidoptera. *J. Invertebr. Pathol. 7*, 306–314.

Noguchi, Y. 1984a. Comparative histopathology of the midgut epithelium of lepidopterous insects infected with a cytoplasmic-polyhedrosis virus of the fall webworm, *Hyphantria cunea. Jpn. J. Appl. Entomol. 28*, 57–62.

Noguchi, Y. 1984b. Histopathological observations on the midgut epithelia of lepidopterous insects naturally recovered from infection by a cytoplasmic-polyhedrosis virus of the silkworm, *Bombyx mori. J. Seric. Sci. Jpn. 53*, 141–145.

Noguchi, Y., and Yamaguchi, K. 1982. Development of disease in several species of lepidopterous insects subjected to cross-infection with cytoplasmic-polyhedrosis viruses. *Jpn. J. Appl. Entomol. Zool. 26*, 281–287.

Noguchi, Y., and Yamaguchi, K. 1984. An acquired resistance to virus infection in the lepidopterous insects naturally cured from *Bombyx* cytoplasmic polyhedrosis. *J. Seric. Sci. Jpn. 53*, 325–330.

Noguchi, Y., and Yamaguchi, K. 1985. Virus susceptibility of regenerated midgut cells by administration of agricultural chemicals in the fall webworm, *Hyphantria cunea. J. Seric. Sci. Jpn. 54*, 310–314.

Okino, H., and Ishikawa, Y. 1971. Change in the shape of silkworm cytoplasmic polyhedra under high temperature. I. Effect of high temperature on change in the shape of polyhedra. *Res. Bull. Aichi Agric. Res. Ctr. D2*, 65–69.

Payne, C. C., and Harrap, K. A. 1977. Cytoplasmic polyhedrosis viruses. *In* "Atlas of Insect and Plant Viruses." (K. Maramorosch, ed.), pp. 105–129. Academic Press, New York.

Payne, C. C., and Mertens, P. P. C. 1983. Cytoplasmic polyhedrosis viruses. *In* "The Reoviridae." (W. K. Joklik, ed.), pp. 425–504. Plenum Press, New York.

Payne, C. C., and Rivers, C. F. 1976. A provisional classification of cytoplasmic polyhedrosis viruses based on the sizes of the RNA genome segments. *J. Gen. Virol. 33*, 71–85.

Payne, C. C., and Tinsley, T. W. 1974. The structural proteins and RNA components of a cytoplasmic polyhedrosis virus from *Nymphalis io* (Lepidoptera: Nymphalidae). *J. Gen. Virol. 25*, 291–302.

Payne, C. C., Piasecka-Serafin, M., and Pilley, B. 1977. The properties of two recent isolates of cytoplasmic polyhedrosis viruses. *Intervirology 8*, 155–163.

Payne, C. C., Mertens, P. P. C., Pedley, S., Crook, N. E., and Rubinstein, R. 1986. Classification of cytoplasmic polyhedrosis viruses by RNA–RNA homology, electropherotype and serological analyses. *In* "Fundamental and Applied Aspects of Invertebrate Pathology." (R. A. Samson, J. M. Vlak, and D. Peters, eds.), pp. 65–68. Found. Fourth Int. Colloq. Invertebr. Pathol. Wageningen.

Pullin, J. S. K., and Moore, N. F. 1985. Gene assignments of a cytoplasmic polyhedrosis virus (type 2) from *Nymphalis io*. *Microbiologica 8*, 131–140.
Quiot, J. M., and Belloncik, S. 1977. Caratérisation d'une polyédrose cytoplasmique chez le lépidoptère *Euxoa scandens*, Riley (Noctuidae, Agrotinae). Etudes *in vivo* et *in vitro*. *Arch. Virol. 55*, 145–153.
Rohrmann, G. F. 1986. Polyhedrin structure. *J. Gen. Virol. 67*, 1499–1513.
Rohrmann, G. F., Bailey, T. J., Becker, R. R., and Beaudreau, G. S. 1980. Comparison of the structure of C- and N-polyhedrins from two occluded viruses pathogenic for *Orgyia pseudotsugata*. *J. Virol. 34*, 360–365.
Rubinstein, R., and Polsen, A. 1983. Midgut and viral associated proteases of *Heliothis armigera*. *Intervirology 19*, 16–25.
Sato, F., and Inoue, H. 1978. Double infection of viruses in the midgut of the silkworm, *Bombyx mori*. *Bull. Seric. Exp. Stn. 27*, 427–444.
Shimotohno, K., and Miura, K. 1973a. Single-stranded RNA synthesis *in vitro* by the RNA polymerase associated with cytoplasmic polyhedrosis virus containing double-stranded RNA. *J. Biochem. 74*, 117–125.
Shimotohno, K., and Miura, K. 1973b. Transcription of double-stranded RNA in cytoplasmic polyhedrosis virus in vitro. *Virology 53*, 283–286.
Shimotohno, K., and Miura, K. 1977. Nucleoside triphosphate phosphohydrolase associated with cytoplasmic polyhedrosis virus. *J. Biochem. 81*, 371–379.
Sidor, Ć. 1965. Uporedna ispitivanja osetljivosti gusenica Thaumatopoea pityocampa Schiff na specifično virozno oboljenje I viroze nekih insekatskih vrsta. *Šumarskog Lista Broj 9–10*, 381–390.
Sidor, Ć. 1970. Poliedarno virozno oboljenje topolinog krasnika (Melanophila picta Pall, coleoptera, buprestidae). *Topola Bull. Comm. Natl. Yougoslave, Beograd. God. XIV*, 77–78.
Sikorowski, P. P., and Thompson, A. C. 1979. Effects of cytoplasmic polyhedrosis virus on diapausing *Heliothis virescens*. *J. Invertebr. Pathol. 33*, 66–70.
Sikorowski, P. P., Andrews, G. L., and Broome, J. R. 1971. Presence of cytoplasmic polyhedrosis virus in the hemolymph of *Heliothis virescens* larvae and adults. *J. Invertebr. Pathol. 18*, 167–168.
Sikorowski, P. P., Andrews, G. L., and Broome, J. R. 1973. Trans-ovum transmission of a cytoplasmic polyhedrosis virus of *Heliothis virescens* (Lepidoptera: Noctuidae). *J. Invertebr. Pathol. 21*, 41–45.
Simmons, C. L., and Sikorowski, P. P. 1973. A laboratory study of the effects of cytoplasmic polyhedrosis virus on *Heliothis virescens* (Lepidoptera: Noctuidae). *J. Invertebr. Pathol. 22*, 369–371.
Smith, K. M. 1963. The cytoplasmic virus diseases. *In* "Insect Pathology: An Advanced Treatise." (E. A. Steinhaus, ed.), Vol. 1, pp. 457–497. Academic Press, New York.
Smith, K. M. 1976. "Virus–Insect Relationships." Longman Group Limited, London.
Smith, K. M., and Wyckoff, R. W. G. 1950. Structure within polyhedra associated with insect virus diseases. *Nature 166*, 861–862.
Smith, R. E., and Furuichi, Y. 1982. The double-stranded RNA genome segments of cytoplasmic polyhedrosis virus are independently transcribed. *J. Virol. 41*, 326–329.
Steinhaus, E. A., and Dineen, J. P. 1959. A cytoplasmic polyhedrosis of the alfalfa caterpillar. *J. Insect Pathol. 1*, 171–183.
Storer, G. B., Shepherd, M. G., and Kalmakoff, J. 1974. Enzyme activities associated with cytoplasmic polyhedrosis virus from *Bombyx mori*. I. Nucleotide phosphohydrolase and nuclease activities. *Intervirology 2*, 87–94.
Tanaka, S. 1971. Cross transmission of cytoplasmic polyhedrosis viruses. *In* "The Cytoplasmic-Polyhedrosis Virus of the Silkworm." (H. Aruga and Y. Tanada, eds.), pp. 201–207. University Tokyo Press, Tokyo.
Tanaka, S., and Aruga, H. 1967. Interference between the midgut nuclear-polyhedrosis virus

and the cytoplasmic-polyhedrosis virus in the silkworm, *Bombyx mori* L. *J. Seric. Sci. Jpn. 36*, 169–176.

Tyler, K. L., and Fields, B. N. 1986. Reovirus and its replication. *In* "Fundamental Virology." (B. N. Fields and D. M. Knipe, chief eds.), pp. 397–436. Raven Press, New York.

Vail, P. V., and Gough, D. 1970. Effects of cytoplasmic-polyhedrosis virus on adult cabbage loopers and their progeny. *J. Invertebr. Pathol. 15*, 397–400.

Vail, P. V., Hall, I. M., and Gough, D. 1969. Influence of a cytoplasmic polyhedrosis on various developmental stages of the cabbage looper. *J. Invertebr. Pathol. 14*, 237–244.

Watanabe, H. 1966. Some aspects on the mechanism of resistance to peroral infection by cytoplasmic-polyhedrosis virus in the silkworm, *Bombyx mori* L. *J. Seric. Sci. Jpn. 35*, 411–417.

Watanabe, H., Aruga, H., and Tanaka, S. 1967. Autoradiographic studies on the nucleic-acid synthesis in the midgut of the silkworm, *Bombyx mori* L., infected with midgut-nuclear-polyhedrosis virus. *J. Seric. Sci. Jpn. 36*, 381–387.

Wu, A.-Z., and Sun, Y.-K. 1986. Isolation and reconstitution of the RNA replicase of the cytoplasmic polyhedrosis virus of silkworm, *Bombyx mori*. *Theor. Appl. Genet. 72*, 662–664.

Wu, A., Dai, R., Shen, X., and Sun, Y. 1983. Coding assignments of each mRNA synthesized *in vitro* and protein components of CPV of silkworm, *Bombyx mori*. *Sci. Sin. 26B*, 162–166.

Xeros, N. 1952. Cytoplasmic polyhedral virus diseases. *Nature 170*, 1073.

Xeros, N. 1966. Light microscopy of the virogenic stromata of cytopolyhedroses. *J. Invertebr. Pathol. 8*, 79–87.

Yamaguchi, K. 1968. Studies on the midgut nuclear polyhedrosis in the silkworm, *Bombyx mori* L. (I) The formation site and some nature of the polyhedra. *J. Seric. Sci. Jpn. 37*, 34–42.

Yamaguchi, K. 1974. Studies on the interference between viruses in the silkworm, *Bombyx mori* L. I. Interference between typical and new strains of the cytoplasmic-polyhedrosis virus. *Bull. Saitama Seric. Exp. Stn. 46*, 78–83.

Yamaguchi, K. 1975a. Studies on the interference between viruses in the silkworm, *Bombyx mori* L. II. Interference between B, C_1 and C_2 strains of the cytoplasmic-polyhedrosis virus. *Bull. Saitama Seric. Exp. Stn. 47*, 59–61.

Yamaguchi, K. 1975b. Studies on the interference between viruses in the silkworm, *Bombyx mori* L. III Interference between A and C_1 strains of the cytoplasmic-polyhedrosis virus. *J. Seric. Sci. Jpn. 44*, 468–471.

Yamaguchi, K. 1976a. Natural cure of the fall webworm, *Hyphantria cunea*, infected with the cytoplasmic polyhedrosis virus of the silkworm, *Bombyx mori*. *J. Seric. Sci. Jpn. 45*, 60–65.

Yamaguchi, K. 1976b. Resistance of the fall webworm, *Hyphantria cunea*, to the cytoplasmic polyhedrosis virus of the silkworm, *Bombyx mori*. *J. Seric. Sci. Jpn. 45*, 377–378.

Yamaguchi, K. 1977. Regeneration of the midgut epitherial cells in the silkworm, *Bombyx mori*, infected with the cytoplasmic polyhedrosis virus. *J. Seric. Sci. Jpn. 46*, 179–180.

Yamaguchi, K. 1979. Natural recovery of the fall webworm, *Hyphantria cunea*, to infection by a cytoplasmic-polyhedrosis virus of the silkworm, *Bombyx mori*. *J. Invertebr. Pathol. 33*, 126–128.

Yamaguchi, K., and Ayuzawa, C. 1970. Studies on the midgut-nuclear polyhedrosis of the silkworm, *Bombyx mori* L. (IV) Two previously undescribed virus strains, B and C_1. *J. Seric. Sci. Jpn. 39*, 342–350.

Yamaguchi, K., and Hukuhara, T. 1973a. Studies on the midgut-nuclear polyhedrosis of the silkworm, *Bombyx mori* L. (IX) New virus strain, B_2. *J. Seric. Sci. Jpn. 42*, 239–243.

Yamaguchi, K., and Hukuhara, T. 1973b. Studies on the midgut-nuclear polyhedrosis of the silkworm, *Bombyx mori* L. (X) Electron microscopic observation on the inclusion formation of C_1 strain. *Bull. Saitama Seric. Exp. Stn. 45*, 61–65.

Yamaguchi, K., Iwashita, Y., and Inoue, K. 1969. On the midgut-nuclear polyhedrosis in the silkworm, *Bombyx mori* L. III. Effects of high temperature treatment on the shape of polyhedron of the infected larvae. *J. Seric. Sci. Jpn. 38*, 157–162.

Yamakawa, M., Shatkin, A. J., and Furuichi, Y. 1981. Chemical methylation of RNA and DNA viral genomes as a probe of in situ structure. *J. Virol. 40*, 482–490.

Yamamasu, Y., and Kawakita, T. 1962. Studies on the grasserie of silk producing insects. (IV) On the cytoplasmic polyhedrosis of pine-moth, *Dendrolimus spectabilis* Butler. *Bull. Fac. Text. Fibers Kyoto Univ. 3*, 421–443.

Yazaki, K., and Miura, K.-I. 1980. Relation of the structure of cytoplasmic polyhedrosis virus and the synthesis of its messenger RNA. *Virology 105*, 467–479.

Yazaki, K., Mizuno, A., Sano, T., Fujii, H., and Miura, K. 1986. A new method for extracting circular and supercoiled genome segments from cytoplasmic polyhedrosis virus. *J. Virol. Methods 14*, 275–283.

CHAPTER 9

OTHER RNA-VIRAL INFECTIONS

I. BIRNAVIRIDAE—BISEGMENTED, dsRNA
II. RHABDOVIRIDAE—ssRNA, ENVELOPED
   A. General Description
   B. Sigma Virus
III. PICORNAVIRIDAE—ssRNA, NONENVELOPED
   A. Cricket-Paralysis Virus
   B. *Drosophila* C Virus
   C. *Gonometa* Virus
   D. Other Picornaviruses
IV. CALICIVIRIDAE—ssRNA, NONENVELOPED
V. TETRAVIRIDAE (*Nudaurelia* β Virus Group)—ssRNA, NONENVELOPED
VI. NODAVIRIDAE—ssRNA, NONENVELOPED

Insects are susceptible to members of five other families of RNA viruses in addition to Reoviridae. One family, Birnaviridae, has double-stranded (ds) RNA. The remaining four families have single-stranded (ss) RNAs and are separated into enveloped (Rhabdoviridae) and nonenveloped viruses (Picornaviridae, Tetraviridae, and Nodaviridae). The nonenveloped viruses form the majority of ssRNA viruses infecting insects, with the largest number assigned to the family Picornaviridae. Many small, RNA insect viruses, with similarities to vertebrate picornaviruses, are ubiquitous in nature and have a wide host range (Scotti et al. 1981). Many of them cause chronic or inapparent infections, but some are virulent pathogens. More studies have been conducted with those isolated from the honey bee, *Drosophila*, and mosquito than from other insects. The RNA viruses include some arboviruses (arthropod-borne viruses), which may cause pathology in their insect vectors, but they will not be considered.

I. BIRNAVIRIDAE—BISEGMENTED, dsRNA

Members of the family Birnaviridae possess icosahedral, nonenveloped, bisegmented dsRNA genome and have been described mostly from vertebrates (fishes and birds). A virus in this family is found in laboratory populations of adult *Drosophila melanogaster*, which may die from infection. The fly develops CO_2 (anoxia) sensitivity as in the case of the infection with the sigma virus (Teninges et al. 1979).

This virus is called *Drosophila* X virus (DXV), but it has been detected, thus far, in natural populations of *Culicoides* sp. and not in those of *Drosophila* (Bonami and Adams 1991). It has two RNA segments of molecular weights (MWs) 2.3×10^6 and 2.2×10^6, five major polypeptides in the capsid, and the viral particle is 60 nm in diameter (Dobos et al. 1979; Nagy and Dobos 1984).

The DXV replicates in *Drosophila* in the cytoplasm of cells of the digestive tract, tracheal matrix, muscle, and ovarian follicles (Bonami and Adams 1991). The infected cells contain diffuse masses of material, probably a viroplasm. Viral particles have been observed in the brain.

II. RHABDOVIRIDAE—ssRNA, ENVELOPED

The viruses in this family infect vertebrates, invertebrates, and plants. Many of them are arboviruses, which infect and cause pathologies in their invertebrate carriers, mainly insects. At present one virus, the sigma virus, is restricted to insects (*Drosophila* spp.). This rhabdovirus differs from the others in (1) failure to infect and multiply in vertebrates and (2) its only known mode of transmission in nature is by the vertical route through the parents (Teninges and Bras-Herreng 1987).

A. GENERAL DESCRIPTION

Rhabdoviruses infecting animals are usually bullet-shaped (e.g., vesicular stomatitis virus, rabies virus, and sigma virus) and those infecting plants are bacilliform and/or bullet-shaped with a distinct prevalence for the bacilliform (e.g., potato yellow dwarf, lettuce necrotic yellows, and broccoli necrotic yellows viruses) (Matthews 1982). A helical nucleocapsid is enclosed by a membranous envelope whose outer surface is studded with numerous glycoprotein spikes. The nucleic acid consists of one molecule of noninfectious linear (negative sense) ssRNA with MW 3.5 to 4.6×10^6. There are four to five major polypeptides and other minor polypeptides, and the nucleocapsid contains a transcriptase and other enzymes.

B. SIGMA VIRUS

Adults of *Drosophila* spp. are commonly immobilized by exposure to CO_2 to facilitate the counting of genetic crosses. They recover from the anoxia when returned to normal air condition. L'Héritier and Teissier (1937) observed that the adults of certain strains of *Drosophila melanogaster* suffered fatal paralysis after exposure to CO_2. The larvae also reacted to CO_2 but to a lesser degree than adults (L'Héritier 1948). This trait of sensitivity to CO_2 was found to be inheritable as a cytoplasmic factor independent of the chromosome (L'Héritier and Teissier 1938). It is caused by the sigma virus in *D. melanogaster* and 10 other *Drosophila* species (Williamson 1957, 1961).

The anoxia results from an infection of the thoracic ganglia, the site of sigma virus maturation (L'Héritier 1948; Brun 1991). The sigma virus also infects other cells of *Drosophila*. The detrimental effect of CO_2 on sensitive flies depends on temperature conditions and the concentration and duration of exposure to CO_2. The CO_2 sensitivity occurs in *Drosophila* populations throughout the world. This sensitivity is the basis for the diagnosis of the sigma virus in *Drosophila* spp. However, anoxia is also caused by other viruses that multiply in *Drosophila* [e.g., all serotypes of vesicular stomatitis virus (VSV), viruses related to VSV (Piry and Chandipura viruses), two rhabdoviruses of fish, *Drosophila* X virus, and *Drosophila* iota virus of *Drosophila immigrans* (but only in males) (Brun and Plus 1980)]. In addition, a similar but delayed recovery syndrome (Texas delayed recovery strain) occurs in a CO_2-sensitive *Drosophila* strain from a chromosomally inherited semi-dominant gene (McCrady and Sulerud 1964).

1. Viral Particle

In 1965, the sigma virus was found to be bullet-shaped and was similar in size and shape to the VSV (Berkaloff et al. 1965, Fig. 9–1A). It is 70 nm in diameter and approximately 180 nm in length (Teninges 1968) with a spiral nucleocapsid of about 30 spirals and enclosed in a spiculate coat or membrane (Brun and Plus 1980). It emerges from the infected cells by budding and acquires an envelope from the cell membrane (Brun 1991). There are five virus-specific proteins with MWs 25,000, 44,000, 57,000, 68,000, and 210,000 (Richard-Molard et al. 1984). The spikes on the viral membrane are formed by a glycosylated polypeptide. The viral membrane proteins may play a major role in CO_2 sensitivity (Teninges and Bras-Herring 1987).

The nucleocapsids of the sigma virus contain nonphosphorylated proteins: a single major, one or two minor proteins, and another loosely associated protein. Teninges and Bras-Herreng (1987) have cloned the complementary DNA (cDNA) copy of the entire coding region of the glycoprotein messenger RNA (mRNA). The nucleotide sequences of the sigma virus indicate a slight but significant relationship with VSV (Indiana and New Jersey serotypes), while the relationship with the rabies virus seems much more distant. However, Francki et al. (1991) have placed the sigma virus as a probable member of the Rabies Virus Group (genus *Lyssavirus*).

2. Transmission

In nature, the sigma virus is transmitted vertically through both female and male parents (hereditary) and not by normal biological contacts between flies. It can be transmitted, however, by organ implantation and intrahemocoelic inoculation from infected (sensitive) flies (L'Héritier and Hugon de Scoeux 1947; Bernard 1968). For transmission, the infection must occur in the oocytes of the female and in the spermatocytes of the male. Viral variants differ in their ability to infect the oocytes. This is governed by the g gene of the sigma virus, with g^+ strain infecting and g^- strain not infecting the oocyte (Brun and Plus 1980). There are also temperature-sensitive viral mutants, defective strains, and strains that differ in their invasion speed and viral multiplication.

The sigma virus–*Drosophila* relationship exists in natural *Drosophila* populations as stablized and unstablized conditions (Brun and Sigot 1955; L'Héritier and Plus 1963). Under the stabilized condition, which occurs most commonly in nature, the sigma virus is transmitted by the female to all progeny that are sensitive and by the male to only a part of the progeny (Fleuriet 1982). In the unstabilized condition, which may develop from an artificial inoculation, the female transmits the virus to only part of the progeny and the male does not transmit it. Both stabilized and unstabilized flies are CO_2 sensitive.

About 10% of the natural populations of *Drosophila* are infected with the sigma

FIGURE 9-1

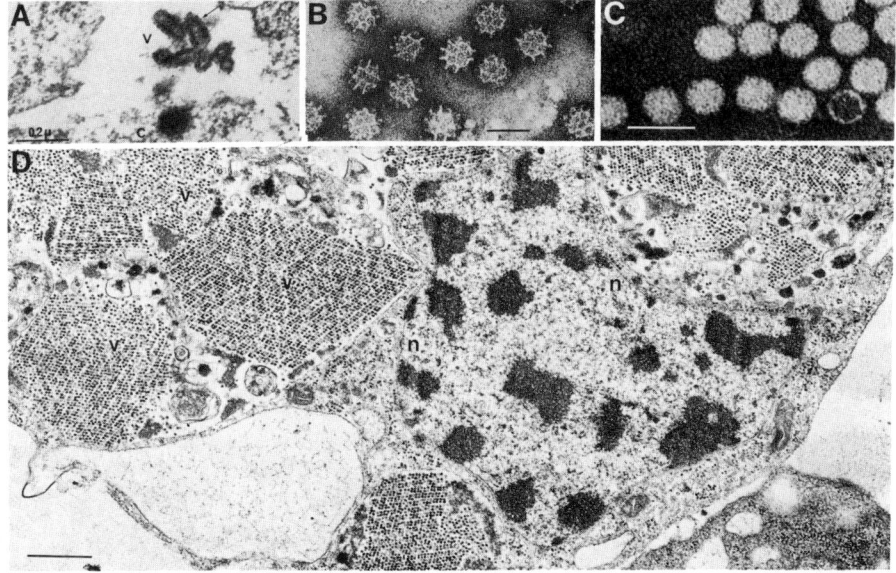

Nonoccluded RNA viruses. (A) Sigma virus (*v*) of *Drosophila* (bar = 0.2 μm). [Courtesy of Berkaloff et al. (1965) and Gauther-Villars.] (B) Calicivirus of the navel orangeworm, *Amyelois transitella* (bar = 50 nm). (Provided by D. F. Hoffmann.) (C) Flacherie virus of the silkworm, *Bombyx mori* (bar = 50 nm). (Provided by M. Matsui.) (D) Section through the nuclear region (*n*, nuclear membrane) of an infected granulocyte of the navel orangeworm showing the crystalline arrays of caliciviral particles (*v*) in the cytoplasm (bar = 0.5 μm). (Provided by D. F. Hoffmann.)

virus (Fleuriet 1986). The populations are polymorphic for two alleles, O and P, with the P allele interfering with the multiplication and transmission of the virus. The populations also have two viral types that differ on their sensitivity to the P alleles. The wild viral clones are likely to differ in their characteristics from those of laboratory strains and also from one population to another.

3. Pathogenicity

The sigma virus is generally considered to be nonpathogenic since it causes persistent infection in *Drosophila* and in cell lines where no cell lysis occurs. However, there are reports of some deleterious effects in addition to CO_2 sensitivity. Seecoff (1964) has reported premature deaths in virus-infected progeny and a longer developmental period from egg to adult. However, in a comprehensive study, Fleuriet (1981a) has found that the only significant difference between infected and uninfected females is the lower viability of the eggs laid by the former. In subsequent studies, Fleuriet (1981b) reports that the infected flies are more sensitive to over wintering than uninfected flies.

4. Host Resistance

Drosophila may vary in susceptibility to the sigma virus. Resistance to the sigma virus was first reported by Guillemain (1953) who discovered and localized a gene responsible for this resistance. At least five *ref* (refractory) genes of *Drosophila* can affect viral replication, and the gene loci and chromosomes are known (Gay 1978; Coulon and Contamine 1982).

III. PICORNAVIRIDAE—ssRNA, NONENVELOPED

The family Picornaviridae has four genera with familiar mammalian viruses: *Rhinovirus* (common cold virus), *Enterovirus* (poliovirus), *Aphthovirus* (foot-and-mouth disease virus), and *Cardiovirus* (encephalomyocarditis virus) (Francki et al. 1991). The members are among the smallest (pico = very small), positive-strand RNA viruses (Matthews 1982; Moore and Eley 1991). The virions are from 22 to 30 nm in diameter and isometric in shape with no envelope or core and no surface projections. Each viral particle contains 60 molecules of each of the four major polypeptides, three with MWs from 24 to 41×10^3 and one with MW from 5.5 to 13.5×10^3, which form a protein capsid with icosahedral symmetry. Within the viral capsid lies the ssRNA, MW from 2.5 to 3×10^6. Viral replication occurs in complexes associated with cytoplasmic membranes. Moore et al. (1985) have discussed the replication of insect picornaviruses from the biochemical and biophysical characteristics.

Many small ssRNA insect viruses have been considered as picornaviruses, but the following are assigned to the family Picornaviridae without generic assignments: cricket paralysis, *Drosophila* C and *Gonometa* viruses (Matthews 1982; Moore et al. 1985). There are about 30 small RNA viruses of unknown affinities, some of which may be assigned to new groups. Among the more widely studied are the bee acute paralysis, bee slow paralysis, bee virus X, *Drosophila* P and A, silkworm flacherie, and sacbrood viruses.

A. CRICKET-PARALYSIS VIRUS

Meynadier (1966) reported the first case of a paralysis virus in the cricket *Gryllus bimaculatus*. In 1970, the cricket-paralysis virus (CrPV) was described from other crickets, *Teleogryllus oceanicus* and *T. commodus* (Reinganum et al. 1970). The virus occurs commonly in field crickets in Australia (Reinganum et al. 1981). It has a wide host spectrum and replicates in many insect species in the orders Orthoptera, Mantodea, Heteroptera, and Lepidoptera (Reinganum 1975; Scotti et al. 1981). Strains of CrPV, which are related but not antigenically identical, have been isolated from different regions of the world.

Since CrPV infects *Drosophila*, it was initially considered to be identical to *Drosophila* C virus (DCV) to which it is similar in physicochemical and serological properties (Reinganum and Scotti 1976; Cooper et al. 1978; Reavy et al. 1983). Further detailed studies have shown that the two viruses are distinct and differ in host range, gel diffusion tests (Plus et al. 1978; Plus and Scotti 1984; Plus 1989), structural proteins expressed in infected cells, oligonucleotide maps (Pullin et al. 1982), and in RNA sequence homology (King et al. 1984). Nonetheless, the two viruses are very similar since the protease activity specified by either DCV or CrPV can perform some of the cleavages resulting in the production of the capsid protein of the other virus (Reavy and Moore 1983).

Another virus with partial serological relationship to CrPV has been isolated from an infected larva of *Pseudoplusia includens* (Chao and Young 1986). It is an icosahedral ssRNA virus of 25 nm in diameter. It infects the midgut epithelial cells.

The CrPV has been placed in *Cardiovirus* (Tinsley et al. 1984) and in *Enterovirus* (Cooper et al. 1978; Moore et al. 1980; Moore and Tinsley 1982); both genera are mammalian viruses of Picornaviridae. However, King et al. (1987) have reported that the sequence of the 3'-terminal of the 1600 nucleotide shows no homology with mammalian picornaviruses.

1. Viral Particle

The CrPV is 27 nm in diameter and has a ssRNA and four capsid proteins of MWs 8,000, 30,000, 31,000, and 33,000 (Eaton and Steacie 1980; Moore et al. 1980; Scotti et al. 1981). The DNA complementary to the ssRNA genome of CrPv has been cloned into the plasmid pBR322 by a hybrid RNA–cDNA cloning strategy (King et al. 1987).

2. Pathogenesis

The CrPV infects insects naturally through the mouth and experimentally by intrahemocoelic inoculations. In crickets, the early to midinstar nymphs are most susceptible, whereas the adults show some degree of resistance (Reinganum et al. 1970). Infected crickets develop uncoordinated movements because of a progressive paralysis of their hind legs and die rapidly. The virus replicates in the cytoplasm of cells of the epidermis, alimentary canal, and nerve ganglia. The virions tend to form crystalline aggregates. Inapparent infections, which show no signs of disease, may occur in adults of crickets and lepidopterons (Plus et al. 1978; Reinganum and Scotti 1976).

The CrPV infection in the olive fly, *Dacus oleae*, differs from that in a cricket in the presence of large numbers of particles and paracrystalline arrays of the virus in the cell cytoplasm (Manousis et al. 1988). The organs of infected cells become severely injured. The feces contain a large amount of virus through the rupturing of midgut cells.

B. *DROSOPHILA* C VIRUS

Drosophila melanogaster is susceptible to a number of small RNA viruses, such as *Drosophila* A, C, and P viruses, which are considered to be picornaviruses (Plus et al. 1976; Brun and Plus 1980). These viruses, unlike the sigma virus, are highly contagious by contact or ingestion, but they are not very pathogenic except for the *Drosophila* C virus (DCV). The DCV has been characterized adequately to be assigned to the family Picornaviridae. It is very similar in physicochemical properties to the enterovirus group (Jousset et al. 1977; Scotti et al. 1981; Moore and Tinsley 1982). It is identical in size and physical characteristics to the cricket-paralysis virus except that it has a fourth structural protein of approximate MW 10 $\times$ 10^3 (Moore et al. 1980; Plus 1989). We have already discussed the differences between DCV and CrPV in the previous section.

The DCV was first detected and isolated from laboratory and natural populations of *Drosophila* in 1972 (Jousset et al. 1972), and subsequently, from field and laboratory flies collected in widely dispersed regions of the world (Plus et al. 1975). It replicates, not only in numerous strains of *D. melanogaster*, but also in the Mediterranean fruit fly, *Ceratitis capitata*, the greater wax moth, *Galleria mellonella* (Jousset 1976), and the honey bee (Scotti et al. 1981). Different strains of DCV occur in nature (Moore et al. 1982; Clewley et al. 1983) and they show marked differences in pathogenicity resulting in early deaths to *Drosophila* and other insect species (Plus et al. 1975, 1978; Jousset 1976). Stocks of *D. melanogaster* differ in their resistance to DCV and the resistance seems to act through some cytoplasmic factors (Plus and Golubovsky 1980).

The transmission of DCV is through the mouths of larvae and adults (Jousset and Plus 1975). The virus multiplies to high titers in the cytoplasm of tracheal cells, especially those surrounding the cerebral ganglion; in contrast, the *Drosophila* A and P viruses infect the cells of the digestive tract and Malpighian tubules (Jousset et al. 1972; Plus et al. 1975). Larvae emerging from surface-contaminated eggs become infected with the virus (Jousset and Plus 1975). No transovarial transmission occurs unlike the *Drosophila* A and P viruses. *Drosophila* stocks infected with DCV have a higher tendency for lethal mutation than uninfected stocks (Golubovsky and Plus 1982).

C. *GONOMETA* VIRUS

Gonometa virus was isolated from larvae of *Gonometa podocarpi*, which were causing serious defoliation of *Pinus patula* in Uganda (Harrap et al. 1966). Infected larvae hung in a limp state by the last pair or two pairs of prolegs. The viral epizootic controlled the pest outbreak. The viral particle, 32 nm in diameter, replicates in the cytoplasm of midgut columnar and goblet cells and of fat body cells (Longworth et al. 1973). The cytoplasm is filled with virions, and the mitochondria and cell membranes are degenerated. The virus has many of the physical characteristics of a picornavirus, and Longworth et al. (1973) have proposed that it be included in the enterovirus group. This virus or a related strain infects a sympatric lasiocampid, *Pachymetana* sp.

D. OTHER PICORNAVIRUSES

Unclassified small RNA viruses of insects that have been placed within Picornaviridae include the acute bee paralysis, bee slow paralysis, bee virus X, *Drosophila* P and A, and sacbrood viruses (Matthews 1982). In addition, Moore et al. (1985) have included viruses infecting the silkworm, aphid (*Rhopalosiphum padi*), mosquito (*Mansonia uniformis*), and honey bee. We shall be concerned only with some of the viruses of the honey bee and the flacherie virus of the silkworm.

1. Honey Bee Viruses

The honey bee, *Apis mellifera*, has the largest number of diseases caused by RNA viruses with 19 of them identified, including some serologically related strains (Ball and Bailey 1991). Among them are sacbrood virus, acute bee-paralysis virus, chronic bee-paralysis virus, chronic bee-paralysis virus associate, Arkansas bee virus, bee virus X, bee virus Y, slow bee-paralysis virus, black queen-cell virus, Kashmir bee virus, cloudy-wing virus, and Egypt bee virus (Bailey and Woods

1974, 1977; Ball and Bailey 1991). These viruses form a heterogeneous group and have not been classified adequately. We shall consider only the sacbrood, acute bee paralysis, chronic bee paralysis, and chronic bee-paralysis-associate viruses.

a. Sacbrood

Sacbrood may have been known as a disease of the honey bee as early as 1857 and possibly even before beekeeping by humans (White 1917). In 1913, White named the disease "sacbrood" because the dead larva, when removed from a cell, has an appearance of a small, closed sac. He diagnosed the causal agent of sacbrood as a virus largely because of its passage through Berkfeld and Pasteur-Chamberlain filters.

The sacbrood virus is an unenveloped, isometric particle of 28 to 30 nm in diameter (Brčák et al. 1963; Bailey et al. 1964) and contains ssRNA (Lee and Furgala 1965a). In a larva, the virus develops in the cell cytoplasm of infected tissues where it is either randomly distributed or arranged in crystalline arrays (Lee and Furgala 1965b). Serious cytopathological effects occur in nearly all tissues of a severely infected larva, but in an adult, the effect is minimal (Mussen and Furgala 1977). The virus has been observed in the nuclei of infected cells, but the nucleus may represent the accumulation site of viral proteins rather than synthesis (Dall 1987). The virus multiplies in the cytoplasmic vesicular region where viral synthesis and assembly take place. In adult bees, the virus occurs in the cytoplasm of fat body cells (Lee and Furgala 1967a,b), in the hypopharyngeal gland of the worker bee, and in the brain of the drone bee (Bailey and Fernando 1972).

As in the case of foulbrood diseases caused by bacteria, the effect of sacbrood is noticed by the beekeeper on the colony as a unit (White 1917). Among the first signs are the presence of dead brood and the irregularity of the brood. A severe outbreak can noticeably weaken a colony, but frequently the strength of the colony is not diminished. Dead brood occurs mostly in capped cells and death occurs rarely in the pupal stage. Adults may remove dead larvae or puncture one or two holes in the caps covering the cells containing dead larvae. The brood combs with infected larvae have practically no odor.

An infected larva consists largely of fat tissues whose cells are comparatively large, irregular in outline, and have irregularly shaped nuclei (White 1917). Fat cells contain spherical black bodies of various sizes and are the chief cause for the granular appearance of the body fluid of a dead larva. The moribund larva turns from a bluish-white to a yellowish color and from a turgid to a flaccid condition. Shortly after death, the larva is slightly yellow and in a few days changes to brown and then to black as putrefaction continues. The body wall toughens to form a sac for the easy removal of the dead larva from a cell of the honeycomb. The dead larva dries to a mummylike remain or "scale," which does not adhere to the cell wall of the comb. The apex of the anterior third of the dead larva becomes conelike and is raised towards the roof of the cell. The larval scale is like a gondola-shaped boat. The scale is not contagious and the virus within is not infectious. Body contents of yellowish, dead larvae are highly infectious, those from brownish dead larvae are less, and those from brown to black dead larvae are not infectious.

Sacbrood virus loses its infectivity quickly within a few days outside of the living honey bee (White 1917). Even though young larvae are easily infected orally, the transmission and spread of the virus have been difficult to explain because the virus usually disappears spontaneously in a colony in late summer and the infected larvae are absent during the winter (Bailey 1975; Bailey and Woods 1977). The virus is spread generally by adult bees, in which it multiplies and causes behavioral changes but no obvious pathological signs (Bailey 1969; Bailey and Fernando 1972). Young adult, worker bees are most susceptible and acquire the virus while removing the infected larvae. The virus accumulates in the hypopharyngeal gland of the worker bee and is transmitted to uninfected larvae during feeding.

b. Acute Bee Paralysis

Paralytic signs in adult bees are an obvious manifestation of disease. Beekeepers have used the term "paralysis" to designate disorders of adult honey bees showing signs of trembling, sprawled legs and wings. Occasionally they exhibit partial hairlessness and in some cases a black shiny appearance (Burnside 1945). The death rate may be high, but the disease may be mild or transient at times.

Bailey et al. (1963) demonstrated that the adult bee paralysis was caused by several different viruses. They identified and named the first picornavirus to cause adult paralysis as acute bee-paralysis virus (ABPV) to differentiate it from the chronic bee-paralysis virus (CBPV). The ABPV is a naked, featureless, isometric particle about 30 nm in diameter (Bailey and Woods 1977). It causes a common inapparent infection in honey bees in Britain, Australia, and France (Bailey 1965; Bailey and Woods 1977; Ball and Bailey 1991). Infected adult bees usually show the first symptom of "acute" paralysis in 2 to 4 days and then die within a day, in contrast to bees with chronic bee paralysis, which die several days later following chronic paralysis. The ABPV may infect both adult bees and brood when the colony is infested by a parasitic mite that may serve as a vector in transmitting the virus from severely infected to healthy bees (Ball and Bailey 1991).

c. Chronic Bee Paralysis

The virus causing chronic bee paralysis is not a picornavirus, but it is considered in this section as a comparison with acute bee paralysis. The first experimental evidence that adult honey-bee paralysis is caused by an infectious agent was presented by Burnside in 1933. In 1945, he demonstrated that a filterable virus was involved. Bailey et al. (1963) isolated the virus and called it the chronic bee-paralysis virus (CBPV). The virus is found in honey bees throughout the world (Burnside 1933, 1945; Bailey 1965).

Adult bees infected with CBPV may exhibit one of two distinct sets of symptoms or syndromes (Ball and Bailey 1991). Adults with the first set have a bloated abdomen, which is caused by the distension of the honey sac with liquid. They do not fly; instead they crawl on the ground, exhibiting an abnormal trembling motion of the wings and bodies. Bees with the second syndrome are initially able to fly. They become almost hairless, appearing dark or almost black and shiny. Uninfected bees nibble at them and prevent their return to the colony. In a few days, they start trembling, remain flightless, and soon die.

The infected bees hold their limbs and wings slightly spread giving their bodies a flattened aspect compared to those infected with acute bee-paralysis virus or with normal bees (Bailey 1965). They seem thermophilic and remain within the cage when the entrance is opened in contrast to normal bees or those with acute bee paralysis that quickly make their escape. Bees with chronic paralysis appear to feed normally, but those with acute paralysis cannot since their mouthparts become permanently extruded and disjointed, particularly in the final stage of the disease. Chronically paralyzed bees, at 30°C, live for several days after showing symptoms, but acutely paralyzed bees die within 1 or 2 days, possibly because they are unable to feed. Bees infected with CBPV die sooner at 35°C than at 30°C, whereas those with ABPV live longer at the higher temperature. This difference in the periods of lethal infection at different temperatures can be used to diagnose the two diseases.

Under crowded condition in a hive, the CBPV may spread via the cytoplasm of broken cuticular hairs (Ball and Bailey 1991). The infection is very common and most colonies in Britain are permanently infected. The infection rate is positively correlated with the number of colonies per unit area.

The CBPV does not infect bumble bees (Bailey 1965), but the ABPV infects these bees and causes inapparent infection in some of them (Bailey and Gibbs 1964).

The CBPV particles are found in the cytoplasm of cells of the thoracic and abdominal ganglia, digestive tract, and mandibular and hypopharyngeal glands (Giauffret et al. 1966, 1970). They do not occur in the cytoplasm of fat and muscle cells (Lee and Furgala 1965c). Dense basophilic inclusions, known as Morison's cell inclusions, occur in the cytoplasm of hindgut cells of bees with chronic bee paralysis but not in those with acute bee paralysis (Bailey 1965).

Viral particles of CBPV are anisometric, mostly ellipsoidal in outline, and often with a small irregular protuberance at one end (Bailey et al. 1968; Bailey and Woods 1977), and have not been classified. It is a RNA virus and is serologically unrelated to other known viruses of the honey bee. The ellipsoidal viral particles are about 22 nm in diameter and vary considerably in lengths of about 30 to 65 nm. The capsid is composed of a single protein of 23.5 kDa.

In addition to the ellipsoidal shape, bizarre forms including rings, figures-of-eight, stalked rings, and branching rods are found with lengths up to 640 nm (Bailey et al. 1968; Bailey and Woods 1977). Some of these may be distortions resulting during specimen preparations. The viral capsids are very resistant, forming shells about 25×35 nm, with walls about 3 to 4 nm thick. The particles are insensitive to ether and contain no nucleic acid when kept for many hours in cold 1N HCl or cold KOH (Bailey et al. 1968).

A hairless black syndrome occurs in honey bees (Rothenbuhler et al. 1969). A virus has been isolated from affected bees (Rinderer 1974) and has been demonstrated as the causative agent (Rinderer and Rothenbuhler 1976). It is morphologically similar and serologically identical to the chronic bee-paralysis virus and may be a variant because the histopathology and behavior of the bees are different (Rinderer and Green 1976). Bailey (1965) has concluded that many of the so-called bee paralyses appear to be identical to chronic bee paralysis, such as Morison's bee paralysis, "Waldtrachtkrankheit," "Schwarzsucht," and "Mal Noir" of bees.

Differences between chronic and acute bee paralysis are summarized as follows (Bailey et al. 1963; Bailey 1965):

1. Viral particles differ morphologically, ABPV being isometric and CBPV being asymmetric.

2. Many more particles of CBPV than of ABPV occur in naturally paralyzed bees.
3. Adult bees die much sooner with ABPV than with CBPV after showing signs of the disease.
4. Dense basophilic inclusions (Morison's cell inclusion) occur in the hindgut cells of bees with chronic bee paralysis and are lacking in those with acute bee paralysis.
5. Bees with chronic paralysis die sooner at 35°C than at 30°C, whereas those with acute paralysis live longer at 35°C than at 30°C.
6. Both viruses differ serologically.

d. Chronic Bee-Paralysis Virus Associate

Closely associated with the chronic bee-paralysis virus (CBPV) is a small RNA virus, 17 nm in diameter, with a single polypeptide molecule of MW 15×10^3 (Bailey et al. 1980). This virus is called the chronic bee-paralysis-virus associate (CBPVA) and has been found only in bees infected with CBPV to which CBPVA is serologically unrelated.

The RNA components of CPBV and CBPVA show a relationship (Overton et al. 1982). The CBPV has five ssRNA components, consisting of two larger RNAs designated 1 (MW 1.35×10^6, 4200 nucleotides) and 2 (MW 0.9×10^6, 2800 nucleotides), and three smaller RNAs designated 3a, 3b, and 3c, each with MW 0.35×10^6 (1100 nucleotides). The CBPVA has three ssRNA components designated A, B, and C, each with MW 0.35×10^6 (1100 nucleotides).

Overton et al. (1982) have concluded that, although CBPV and CBPVA are serologically unrelated, the RNAs of CBPVA may have evolved from RNA2 of CBPV. At the present time, the CBPVA has become absolutely dependent on CBPV, which serves as a helper virus, probably in providing proteins required for the RNA replication of CBPVA (Ball et al. 1985).

2. Flacherie Virus

Flacherie is a general descriptive term used to describe a syndrome where the diseased larva is flaccid from dysenteric conditions and appears flabby, feeble, weak, withered, or loosely hanging. This syndrome has been observed in the silkworm for many centuries and has been called by numerous names, such as la maladies des morts-blancs or morts-flats, maladies de tripes, gattine, tête claire, flacherie de Pasteur, etc. The cause of the flacherie was unknown for many years and was attributed to amicrobial and microbial causes. It was Pasteur (1870) who definitely separated the dysenteries (flacherie) from fungal (muscardine), protozoan (pébrine), and baculovirus (jaundice) infections. Pasteur recognized both amicrobial and microbial causes for flacherie. The microbial flacherie was caused by the presence of a large number of certain bacteria (*Streptococcus bombycis* and *Bacillus bombycis*) in the larval digestive tract, resulting in a dysfunction of the tract and the development of the typical syndrome.

Paillot (1941) described in detail the problems associated with the etiology of flacherie and characterized the amicrobial and microbial agents. He designated the "infectious flacherie," caused by microbial agents, as "flacherie de Pasteur" or true flacherie. He concluded from his experimental studies that true flacherie was caused by a virus that replicated in the nuclei of midgut epithelial cells and a secondary invasion of *Bacillus bombycis*. Gattine was a separate disease caused by a virus and the associated *Streptococcus bombycis*. The nature of the virus described by Paillot is unclear.

In Japan, two viruses have been isolated from the silkworm with flacherie, a picornavirus replicating in the cytoplasm and a parvovirus (densonucleosis virus) in the nucleus of a midgut cell (Matsui et al. 1977). These two viruses can be separated by sucrose gradient centrifugation (Nakagaki et al. 1987). Since Paillot (1941) had reported that the virus of flacherie de Pasteur multiplied in the nucleus, this implies that the parvovirus is the etiologic agent of Pasteur's flacherie. However, we shall consider the two viruses as causing the flacherie syndrome.

The virus as the etiologic agent of the silkworm flacherie was established in Japan in 1960. During this period, a serious outbreak of flacherie occurred in silkworms cultured in the Nagano prefecture. The infection in the silkworm might have been initiated by an identical disease of the mulberry pyralid, *Glyphodes pyloalis*, a common pest of mulberry farms (Watanabe 1991). The disease was diagnosed as caused by a virus through transmission, filtration, ultracentrifugation, and electron microscopical studies (Yamazaki et al. 1960; Aizawa et al. 1964). The virus is called the infectious flacherie virus in Japan (Watanabe 1991), but we prefer the term flacherie virus (FV). The viral particle is from 26 to 27 nm in diameter (Ayuzawa 1972) (Fig. 9-1C) and contains ssRNA (Kawase et al. 1974). The ssRNA virus has properties of a picornavirus (Hashimoto and Kawase 1983), and its biochemical process of replication has been studied (Hashimoto et al. 1984). In a molting silkworm larva, the virus-specific polypeptides and mRNA change significantly in the virus-infected midgut (Choi et al. 1989).

There are reports that the FV replicates in most larval tissues, but the studies with cytological and fluorescent microscopies indicate that only midgut epithelial cells are infected (Iwashita 1965; Inoue and Ayuzawa 1972). The anterior part of the midgut is infected first. The main target is the goblet cell, which degenerates after infection and is discharged into the gut lumen. Columnar cells have been reported to contain the virus, but these cells may have ingested the discharged degenerated goblet cells. Replication of the virus takes place in the cytoplasm of a goblet cell. The FV may be transmitted from pupa to adult (Kurisu et al. 1977), but there is no evidence for transovarial transmission. The virus also infects goblet cells of *Theophila mandarina*, in which viral-specific vesicles, accompanied with a virogenic stroma, are formed in the cytoplasm (Matsui and Watanabe 1974).

IV. CALICIVIRIDAE—ssRNA, NONENVELOPED

The members of the family Caliciviridae are mainly vertebrate pathogens. The icosahedral viral particle has a ssRNA of MW from 2.6 to 2.8×10^6 (Matthews 1982). It is roughly spherical, from 35 to 39 nm in diameter, with characteristic cup-shaped surface depressions.

A calicivirus has been isolated from the navel orangeworm, *Amyelois transitella*, and has been called the chronic stunt virus (CSV) (Kellen and Hoffmann 1981). It has two morphological forms, a 38 nm cupped particle with a single major polypeptide of 70,000 MW and a smooth particle with a 29,000-MW polypeptide (Hillman et al. 1982) (Fig. 9–1B). The smooth particle is a degraded form of the cupped particle. The degradation occurs in the larval digestive tract where the cupped virions are digested by proteolytic enzymes. The two forms are found in larval frass, but the smooth form is predominant. Both forms are infectious and display similar pathogenicities (Hoffmann and Hillman 1984).

The CSV is readily transmitted through the mouth to larvae of the navel orangeworm (Kellen and Hoffmann 1981). It is highly pathogenic to neonate larvae, and older third and fourth instar larvae usually acquire a chronic infection and exhibit retarded growth and delayed mortality. Adults developing from infected larvae are stunted and the females have very small ovaries. Healthy adults live 1.5 times longer and females lay about 5- to 6-fold more eggs than diseased adults (Kellen and Hoffmann 1983). However, there is no significant difference in the percent of hatched eggs laid by healthy and diseased females.

The CSV infects granular hemocytes and the larval midgut epithelium (Kellen and Hoffmann 1981, 1987). In a hemocyte, the viral multiplication takes place in the cytoplasm, which may be filled with large paracrystalline arrays of virions (Fig. 9–1D). In the midgut, the virus causes severe damage to the epithelium resulting in the sloughing of goblet and columnar cells.

V. TETRAVIRIDAE (*Nudaurelia* β Virus Group)—ssRNA, NONENVELOPED

Members of the family Tetraviridae are confined exclusively to insects (Reinganum 1991). The first insect pathogenic virus of this family was isolated from larvae of the emperor gum moth, *Antheraea eucalypti* (Grace and Mercer 1965) and transmitted to the larvae of *A. helena* (Brzostowski and Grace 1970). A similar virus was detected in the pine tree emperor moth, *Nudaurelia cytherea cytherea* (syn. *N. cytherea capensis*), a species closely related to *A. eucalypti* (Hendry et al. 1968; Tripconey 1970). These viruses have been called the *Nudaurelia* β virus group [family Tetraviridae, type species *Nudaurelia* β virus from *N. cytherea capensis* (Francki et al. 1991)]. At present, the group consists of seven members and nine others that have not been characterized (Reinganum 1991). They infect about 15 lepidopteran insects (Greenwood and Moore 1982).

All larval stages of *A. eucalypti* are susceptible to the virus (Grace and Mercer 1965). The infected larvae become discolored and flaccid and have a constriction towards the posterior end. Moribund larvae may drop from the leaves or hang by the last pair of prolegs from branches. The onset of disease is rapid, invariably fatal, and death occurs in 6 to 8 days. Greenwood and Moore (1982), however, have reported death occurring in 10 to 12 days and the infection as chronic or inapparent. The larval midgut is obviously affected and its cells lyse completely, leaving only the peritrophic membrane that is packed with food (Grace and Mercer 1965). The hindgut contains rotting fecal pellets that the larva apparently cannot expel.

There is some question as to the site of viral replication. After observing the arrays of viral particles in the nuclei of midgut cells, Grace and Mercer (1965) believed that viral replication took place in the nucleus, but Matthews (1982) reported it to occur primarily in the cytoplasm of midgut cells.

Larvae of *Trichoplusia ni*, infected in the late instar, may show little or no effect and the virus can persist as an inapparent infection in succeeding generations (Reinganum 1991). Whether the vertical transmission is transovum or transovarial has not been established. If the adult midgut cells are infected, as in the case of the larva, the transmission is very likely transovum.

Some workers have isolated two to four different types of viral particles, from infected larvae (Brzostowski and Grace 1970; Juckes et al. 1973; Juckes 1979). Viral particles are isometric with diameters from 35 to 38 nm (Reinganum 1991). The capsid is made up of 240 copies of a polypeptide, arranged in $T = 4$ surface symmetry. This is unique among icosahedral viruses and separates the members of Tetraviridae from others (e.g., the picornaviruses with 180 polypeptide chains arranged in $T = 3$ symmetry). *Nudaurelia* β virus has one positive-sense molecule of ssRNA with MW 1.8×10^6 and one major polypeptide of MWs from 60 to 70×10^3 (Reinganum et al. 1978). The viral particle has MW 16.3×10^6.

In addition to *Nudaurelia* β virus, there are three other isolates that are serologically related to but distinguishable from each other. Juckes (1979) has compared some biophysical properties of two members, *Nudaurelia* β and ε viruses, and has detected differences between them. He has found that both viruses do exist as separate disease-causing entities.

Another member of this group has been found in mixed infection with a nuclear polyhedrosis virus in larvae of the alfalfa looper, *Autographa californica* (Morris et al. 1979). It is very similar in size and properties to *Nudaurelia* β virus. It alone causes severe stunting in an *A. californica* larva, which may weigh as little as 13 times less than an uninfected larva of the same age (Vail et al. 1983).

VI. NODAVIRIDAE—ssRNA, NONENVELOPED

The main characteristic of the family Nodaviridae is the divided (bipartite) genome that consists of two ssRNA molecules in a virion (Matthews 1982; Francki et al. 1991). The two RNA molecules have MWs of 1.0 and 0.5×10^6, respectively. Both RNA molecules are required for infection. There is one major (MW 40,000) polypeptide and one minor (MW 4.5×10^3) polypeptide. The icosahedral viral particle is about 30 nm in diameter.

The type member, Nodamura virus, was isolated from the mosquito *Culex tritaeniorhynchus* and collected at Nodamura (Noda Village) near Tokyo, Japan (Moore 1985). The virus is pathogenic for mosquitoes and mammals. Since it appears to be an arbovirus, we shall not consider it further, but there are nodaviruses that appear to be strict insect pathogens. Two members have been isolated from scarabaeid larvae. The one from the black beetle, *Heteronychus arator*, is called the black-beetle virus (BBV) (Longworth and Archibald 1975; Longworth and Carey 1976), and another from the grass grub *Costelytra zealandica* is designated as the Flock House virus (Dearing et al. 1980; Scotti et al. 1983). The Boolarra virus, which infects the lepidopteron *Oncopera intricoides*, is serologically related to the black-beetle virus and the Nodamura virus (Reinganum et al. 1985). The Arkansas bee virus has been isolated from the honey bee (Bailey and Woods 1974) and placed in this family (Matthews 1982), but further study by Lommel et al. (1985) has shown that the size of the genomic components and the presence of only a single ssRNA place it outside of this family.

The host range of nodaviruses varies from the Flock House virus infecting Coleoptera, Lepidoptera, and plants; the black-beetle virus infecting Coleoptera and Lepidoptera; and the Boolarra virus on Lepidoptera (Garzon and Charpentier 1991). These viruses may cause inapparent or persistent infection in nature.

Scarabaeids are serious pests of pasture in New Zealand. The black-beetle virus (BBV) causes substantial mortality in high black-beetle populations in New Zealand

(Longworth and Archibald 1975). Viral replication takes place in the cytoplasm of midgut and fat body cells of a larva, and the virus has been detected in pupa and adult (Longworth and Carey 1976). It has been transmitted to *Galleria mellonella* and other lepidopterons and causes a fatal paralysis after 10 to 30 days. It is not infective to mice and is serologically unrelated to *Nodamura* β virus.

The black-beetle virus, the smallest and simplest of multipartite viruses, has been extensively studied (Garzon and Charpentier 1991). Each virion contains copies of RNA1 (3105 bases), RNA2 (1399 bases), RNA3 (389 bases), and multiple copies of coat-protein alpha (47 kDa) along with its cleaved forms, proteins beta (43 kDa) and gamma (4 kDa) (Friesen and Rueckert 1981; Guarino et al 1981; Dasgupta et al. 1984; Dasmahapatra et al. 1985). Both RNA1 and RNA2 are required for virion production, with RNA1 carrying the gene(s) required for RNA polymerase functions and RNA2 as the messenger for viral coat protein (Crump and Moore 1981; Friesen and Reuckert 1982; Gallagher et al. 1983; Guarino et al. 1984; Frieson and Rueckert 1984). A scheme for the replication and particle assembly of the black-beetle virus has been proposed by Garzon and Charpentier (1991). The RNA1 is involved in the early functions (replicative) and the RNA2 in the late functions (viral assembly).

REFERENCES

Aizawa, K., Furuta, Y., Kurata, K., and Sato, F. 1964. On the etiologic agent of the infectious flacherie of the silkworm, *Bombyx mori* (Linnaeus). *Bull. Seric. Exp. Stn. 19*, 223–240.

Ayuzawa, C. 1972. Studies on the infectious flacherie of the silkworm, *Bombyx mori* L. I. Purification of the virus and its some properties. *J. Seric. Sci. Jpn. 41*, 338–344.

Bailey, L. 1965. The occurrence of chronic and acute bee paralysis viruses in bees outside Britain. *J. Invertebr. Pathol. 7*, 167–169.

Bailey, L. 1969. The multiplication and spread of sacbrood virus of bees. *Ann. Appl. Biol. 63*, 483–491.

Bailey, L. 1975. Recent research on honeybee viruses. *Bee World 56*, 55–64.

Bailey, L., and Fernando, E. F. W. 1972. Effects of sacbrood virus on adult honey-bees. *Ann. Appl. Biol. 72*, 27–35.

Bailey, L., and Gibbs, A. J. 1964. Acute infection of bees with paralysis virus. *J. Invertebr. Pathol. 6*, 395–407.

Bailey, L., and Woods, R. D. 1974. Three previously undescribed viruses from the honey bee. *J. Gen. Virol. 25*, 175–186.

Bailey, L., and Woods, R. D. 1977. Bee viruses. *In* "The Atlas of Insect and Plant Viruses, Including Mycoplasmaviruses and Viroids." (K. Maramorosch, ed.), pp. 141–156. Academic Press, New York.

Bailey, L., Gibbs, A. J., and Woods, R. D. 1963. Two viruses from adult honey bees (*Apis mellifera* Linnaeus). *Virology 21*, 390–395.

Bailey, L., Gibbs, A. J., and Woods, R. D. 1964. Sacbrood virus of the larval honey bee (*Apis mellifera* Linnaeus). *Virology 23*, 425–529.

Bailey, L., Gibbs, A. J., and Woods, R. D. 1968. The purification and properties of chronic bee-paralysis virus. *J. Gen. Virol. 2*, 251–260.

Bailey, L., Ball, B. V., Carpenter, J. M., and Woods, R. D. 1980. Small virus-like particles in honey bees associated with chronic paralysis virus and with a previously undescribed disease. *J. Gen. Virol. 46*, 149–155.

Ball, B. V., and Bailey, L. 1991. Viruses of honey bees. *In* "Atlas of Invertebrate Viruses." (J. R. Adams and J. R. Bonami, eds.), pp. 525–551. CRC Press, Boca Raton, Florida.

Ball, B. V., Overton, H. A., Buck, K. W., Bailey, L., and Perry, J. N. 1985. Relationships between the multiplication of chronic bee-paralysis virus and its associate particle. *J. Gen. Virol.* 66, 1423–1429.

Berkaloff, A., Bregliano, J.-C., and Ohanessian, A. 1965. Mise en évidence de virions dans des drosophiles infectées par le virus héréditaire σ. *C.R. Acad. Sci. Paris Ser. D260*, 5956–5959.

Bernard, J. 1968. Contribution à l'étude du virus héréditaire de la Drosophile, sigma. *Exp. Cell Res.* 50, 117–126.

Bonami, J.-R., and Adams, J. R. 1991. Birnaviridae. *In* "Atlas of Invertebrate Viruses." (J. R. Adams and J. R. Bonami, eds.), pp. 435–442. CRC Press, Boca Raton, Florida.

Brčák, J., Svoboda, J., and Králík, O. 1963. Electron microscopic investigation of sacbrood of the honey bee. *J. Invertebr. Pathol.* 5, 385–386.

Brun, G. 1991. Rhabdoviridae. *In* "Atlas of Invertebrate Viruses." (J. R. Adams and J. R. Bonami, eds.), pp. 443–460. CRC Press, Boca Raton, Florida.

Brun, G., and Plus, N. 1980. The viruses of *Drosophila*. *In* "The Genetics and Biology of Drosophila." (M. Ashburner and T. R. F. Wright, eds.), Vol. 2d, pp. 625–702. Academic Press, New York.

Brun, G., and Sigot, A. 1955. Étude de la sensibilité héréditaire au gaz carbonique chez la drosophile. II. Installation du virus σ dans la lignée germinale a la suite d'une inoculation. *Ann. Inst. Pasteur* 88, 488–513.

Brzostowski, H. W., and Grace, T. D. C. 1970. Observations of the infectivity of RNA from *Antheraea* virus (AV). *J. Invertebr. Pathol.* 16, 277–279.

Burnside, C. E. 1933. Preliminary observations of "paralysis" of honeybees. *J. Econ. Entomol.* 26, 162–168.

Burnside, C. E. 1945. The cause of paralysis of honeybees. *Am. Bee J.* 85, 354–355, 363.

Chao, Y.-C., and Young, S. Y., III. 1986. Characterization of a picornavirus isolated from *Pseudoplusia includens* (Lepidoptera: Noctuidae). *J. Invertebr. Pathol.* 47, 247–257.

Choi, H., Kobayashi, M., and Kawase, S. 1989. Changes in infectious flacherie virus-specific polypeptides and translatable mRNA in the midgut of the silkworm, *Bombyx mori*, during larval molt. *J. Invertebr. Pathol.* 53, 128–131.

Clewley, J. P., Pullin, J. S. K., Avery, R. J., and Moore, N. F. 1983. Oligonucleotide fingerprinting of the RNA species obtained from six *Drosophila* C virus isolates. *J. Gen. Virol.* 64, 503–506.

Cooper, P. D., Agol, V. I., Bachbrach, H. L., Brown, F., Ghendon, Y., Gibbs, A. J., Gillespie, J. H., Lonberg-Holm, K., Mandel, B., Melnick, J. L., Mohanty, S. B., Povery, R. C., Rueckert, R. R., Schaffer, F. L., and Tyrrell, D. A. J. 1978. Picornaviridae: Second report. *Intervirology* 10, 165–180.

Coulon, P., and Contamine, D. 1982. Role of the *Drosophila* genome in sigma virus multiplication. *Virology* 123, 381–392.

Crump, W. A. L., and Moore, N. F. 1981. *In vivo* and *in vitro* synthesis of the proteins expressed by the RNA of black beetle virus. *Arch. Virol.* 69, 131–139.

Dall, D. J. 1987. Intracellular structures associated with the multiplication of sacbrood virus. *J. Invertebr. Pathol.* 50, 261–268.

Dasgupta, R., Ghosh, A., Dasmahapatra, B., Guarino, L. A., and Kaesberg, P. 1984. Primary and secondary structure of black beetle virus RNA 2, the genomic messenger for BBV coat protein precursor. *Nucleic Acids Res.* 12, 7215–7223.

Dasmahapatra, B., Dasgupta, R., Ghosh, A., and Kaesberg, P. 1985. Structure of the black beetle virus genome and its functional implication. *J. Mol. Biol.* 182, 183–189.

Dearing, S. C., Scotti, P. D., Wigley, P. J., and Dhana, S. D. 1980. A small RNA virus isolated from the grass grub, *Costelytra zealandica* (Coleoptera: Scarabaeidae). *N.Z. J. Zool.* 7, 267–269.

Dobos, P., Hill, B. J., Hallett, R., Kells, D. T. C., Becht, H., and Teninges, D. 1979. Biophysical and biochemical characterization of five animal viruses with bisegmented double-stranded RNA genomes. *J. Virol.* 32, 593–605.

Eaton, B. T., and Steacie, A. D. 1980. Cricket paralysis virus RNA has a 3' terminal poly (A). *J. Gen. Virol. 50*, 167–171.

Fleuriet, A. 1981a. Comparison of various physiological traits in flies (*Drosophila melanogaster*) of wild origin, infected or uninfected by the hereditary rhabdovirus sigma. *Arch. Virol. 69*, 261–272.

Fleuriet, A. 1981b. Effect of overwintering on the frequency of flies infected by the rhabdovirus sigma in experimental populations of *Drosophila melanogaster. Arch. Virol. 69*, 253–260.

Fleuriet, A. 1982. Transmission efficiency of the sigma virus in natural populations of its host, *Drosophila melanogaster. Arch. Virol. 71*, 155–167.

Fleuriet, A. 1986. Perpetuation of the hereditary sigma virus in populations of its host, *Drosophila melanogaster*. Geographical analysis of correlated polymorphisms. *Genetica 70*, 167–177.

Francki, R. I. B., Fauquet, C. M., Knudson, D. L., and Brown, F. 1991. "Classification and Nomenclature of Viruses. Fifth Report of the International Committee on Taxonomy of Viruses." Arch. Virol. Suppl. 2. Springer-Verlag, Wien.

Friesen, P. D., and Rueckert, R. R. 1981. Synthesis of black beetle virus proteins in cultured *Drosophila* cells: Differential expression of RNAs 1 and 2. *J. Virol. 37*, 876–886.

Friesen, P. D., and Rueckert, R. R. 1982. Black beetle virus: Messenger for protein B is a subgenomic viral RNA. *J. Virol. 42*, 986–995.

Friesen, P. D., and Rueckert, R. R. 1984. Early and late functions in a bipartite RNA virus: Evidence for translational control by competition between viral mRNAs. *J. Virol. 49*, 116–124.

Gallagher, T. M., Friesen, P. D., and Rueckert, R. R. 1983. Autonomous replication and expression of RNA 1 from black beetle virus. *J. Virol. 46*, 481–489.

Garzon, S., and Charpentier, G. 1991. Nodaviridae. *In* "Atlas of Invertebrate Viruses." (J. R. Adams and J. R. Bonami, eds.), pp. 351–370. CRC Press, Boca Raton, Florida.

Gay, P. 1978. Les gènes de la *Drosophile* qui interviennent dans la multiplication du virus sigma. *Mol. Gen. Genet. 159*, 269–283.

Giauffret, A., Duthoit, J. L., and Caucat, M. J. 1966. Etude histologique du tissu nerveux de l'abeille atteinte de maladie noire. *Bull. Apic. Doc. Sci. Tech. Inf. 9*, 221–228.

Giauffret, A., Duthoit, J. L., and Tostain-Caucat, M. J. 1970. Ultrastructure des cellules d'abeilles infectées par le virus de la paralysie-maladie noire. Etude des inclusions cellulaires. *Bull. Apic. Doc. Sci. Tech. Inf. 13*, 115–126.

Golubovsky, M. D., and Plus, N. 1982. Mutability studies in two *Drosophila melanogaster* isogenic stocks, endemic for C picornavirus and virus-free. *Mutat. Res. 103*, 29–32.

Grace, T. D. C., and Mercer, E. H. 1965. A new virus of the saturniid *Antheraea eucalypti* Scott. *J. Invertebr. Pathol. 7*, 241–244.

Greenwood, L. K., and Moore, N. F. 1982. The *Nudaurelia* β group of small RNA-containing viruses of insects: Serological identification of several new isolates. *J. Invertebr. Pathol. 39*, 407–409.

Guarino, L. A., Hruby, D. E., Ball, L. A., and Kaesberg, P. 1981. Translation of black beetle virus RNA and heterologous viral RNAs in cell-free lysates derived from *Drosophila melanogaster. J. Virol. 37*, 500–505.

Guarino, L. A., Ghosh, A., Dasmahapatra, B., Dasgupta, R., and Kaesberg, P. 1984. Sequence of the black beetle virus subgenomic RNA and its location in the viral genome. *Virology 139*, 199–203.

Guillemain, A. 1953. Découverte et localisation d'un gène empêchant la multiplication du virus de la sensibilité héréditaire au CO_2 chez *Drosophila melanogaster. C.R. Acad. Sci. Paris Ser. D236*, 1085–1086.

Harrap, K. A., Longworth, J. F., Tinsley, T. W., and Brown, K. W. 1966. A noninclusion virus of *Gonometa podocarpi* (Lepidoptera: Lasiocampidae). *J. Invertebr. Pathol. 8*, 270–272.

Hashimoto, Y., and Kawase, S. 1983. Characteristics of structural proteins of infectious flacherie virus from the silkworm, *Bombyx mori. J. Invertebr. Pathol. 41*, 68–76.

Hashimoto, Y., Watanabe, A., and Kawase, S. 1984. Preliminary studies of terminal structures of infectious flacherie virus RNA. *Microbiologica 7*, 91–96.

Hendry, D. A., Bekker, M. F., and van Regenmortel, M. H. V. 1968. A non-inclusion virus of the pine emperor moth, *Nudaurelia cytherea capensis* Stoll. *S. Afr. Med. J. 42*, 117.

Hillman, B., Morris, T. J., Kellen, W. R., Hoffman, D., and Schlegel, D. E. 1982. An invertebrate calici-like virus: Evidence for partial virion disintegration in host excreta. *J. Gen. Virol. 60*, 115–123.

Hoffmann, D. F., and Hillman, B. 1984. Observations on the comparative pathogenicity of intact and degraded forms of a calicivirus of *Amyelois transitella. J. Invertebr. Pathol. 43*, 422–423.

Inoue, H., and Ayuzawa, C. 1972. Fluorescent antibody study of an infectious flacherie of the silkworm, *Bombyx mori. J. Invertebr. Pathol. 19*, 227–230.

Iwashita, Y. 1965. Histo- and cyto-pathological studies on the midgut epithelium of the silkworm larvae infected with infectious flacherie. *J. Seric. Sci. Jpn. 34*, 263–273.

Jousset, F.-X. 1976. Étude expérimentale du spectre d'hôtes du virus C de *Drosophila melanogaster* chez quelques diptères et lépidoptères. *Ann. Microbiol.* (Inst. Pasteur) *127A*, 529–544.

Jousset, F.-X. and Plus, N. 1975. Étude de la transmission horizontale et de la transmission verticale des picornavirus de *Drosophila melanogaster* et *Drosophila immigrans. Ann. Microbiol.* (Inst. Pasteur) *126B*, 231–249.

Jousset, F.-X., Plus, N., Croizier, M. G., and Thomas, M. 1972. Existence chez *Drosophila* de deux groupes de Picornavirus de propriétés sérologiques et biologiques différentes. *C.R. Acad. Sci. Paris Ser. D275*, 3043–3046.

Jousset, F.-X, Bergoin, M., and Revet, B. 1977. Characterization of the *Drosophila* C virus. *J. Gen. Virol. 34*, 269–283.

Juckes, I. R. M. 1979. Comparison of some biophysical properties of the *Nudaurelia* β and ϵ viruses. *J. Gen. Virol. 42*, 89–94.

Juckes, I. R. M., Longworth, J. F., and Reinganum, C. 1973. A serological comparison of some nonoccluded insect viruses. *J. Invertebr. Pathol. 21*, 119–120.

Kawase, S., Suto, C., Ayuzawa, C., and Inoue, H. 1974. Chemical properties of infectious flacherie virus of the silkworm, *Bombyx mori* L. (Lepidoptera: Bombycidae). *Appl. Entomol. Zool. 9*, 100–101.

Kellen, W. R., and Hoffmann, D. F. 1981. A pathogenic nonoccluded virus in hemocytes of the navel orangeworm, *Amyelois transitella* (Pyralidae: Lepidoptera). *J. Invertebr. Pathol. 38*, 52–66.

Kellen, W. R., and Hoffmann, D. F. 1983. Longevity and fecundity of adult *Amyelois transitella* (Lepidoptera: Pyralidae) infected by two small RNA viruses. *Environ. Entomol. 12*, 1542–1546.

Kellen, W. R., and Hoffmann, D. F. 1987. Effect of two nonoccluded viruses of *Amyelois transitella* on larvae of *Cadra figulilella* and *Ephestia elutella* (Lepidoptera: Pyralidae) in laboratory tests. *Environ. Entomol. 16*, 630–632.

King, L. A., Massalski, P. R., Cooper, J. I., and Moore, N. F. 1984. Comparison of the genome RNA sequence homology between cricket paralysis virus and strains of *Drosophila* C virus by complementary DNA hybridization analysis. *J. Gen. Virol. 65*, 1193–1196.

King, L. A., Pullin, J. S. K., Stanway, G., Almond, J. W., and Moore, N. F. 1987. Cloning of the genome of cricket paralysis virus: Sequence of the 3'-end. *Virus Res. 6*, 331–344.

Kurisu, K., Odan, T., Kurosawa, K., Manabe, Y., and Fujioka, H. 1977. Multiplication of the flacherie viruses (FVS I-FVS II) in the pupae of the silkworm, *Bombyx mori* L. *Bull. Fac. Text. Sci. Kyoto Univ. Ind. Arts Text. Fibers 8*, 11–17.

Lee, P. E., and Furgala, B. 1965a. Sacbrood virus: Some morphological features and nucleic acid type. *J. Invertebr. Pathol. 7*, 502–505.

Lee, P. E., and Furgala, B. 1965b. Electron microscopy of sacbrood virus in situ. *Virology 25*, 387–392.

Lee, P. E., and Furgala, B. 1965c. Chronic bee paralysis virus in the nerve ganglia of the adult honey bee. *J. Invertebr. Pathol. 7*, 170–174.

Lee, P. E., and Furgala, B. 1967a. Electron microscopic observations on the localization and development of sacbrood virus. *J. Invertebr. Pathol. 9*, 178–187.

Lee, P. E., and Furgala, B. 1967b. Viruslike particles in adult honey bees (*Apis mellifera* Linnaeus) following injection with sacbrood virus. *Virology 32*, 11–17.

L'Héritier, Ph. 1948. Sensitivity to CO_2 in *Drosophila*—a review. *Heredity 2*, 325–348.

L'Héritier, Ph., and Hugon de Scoeux, F. 1947. Transmission par greffe et injection de la sensibilité héréditaire au gaz carbonique chez la drosophile. *Bull. Biol. Fr. Belg. 81*, 70–91.

L'Héritier, Ph., and Plus, N. 1963. The relationship of the hereditary virus of Drosophila to its host. *In* "Biological Organization at the Cellular and Supercellular Level." (R. J. C. Harris, ed.), pp. 59–71. Academic Press, New York.

L'Héritier, Ph., and Teissier, G. 1937. Une anomalie physiologique héréditaire chez la Drosophile. *C.R. Acad. Sci. Paris Ser. D205*, 1099–1101.

L'Héritier, Ph., and Teissier, G. 1938. Un mécanisme héréditaire aberrant chez la Drosophile. *C.R. Acad. Sci. Paris Ser. D206*, 1193–1195.

Lommel, S. A., Morris, T. J., and Pinnock, D. E. 1985. Characterization of nucleic acids associated with Arkansas bee virus. *Intervirology 23*, 199–207.

Longworth, J. F., and Archibald, R. D. 1975. A virus of black beetle, *Heteronychus arator* (F.) (Coleoptera: Scarabaeidae). *N.Z. J. Zool. 2*, 233–236.

Longworth, J. F., and Carey, G. P. 1976. A small RNA virus with a divided genome from *Heteronychus arator* (F.) [Coleoptera: Scarabaeidae]. *J. Gen. Virol. 33*, 31–40.

Longworth, J. F., Payne, C. C., and Macleod, R. 1973. Studies on a virus isolated from *Gonometa podocarpi* (Lepidoptera: Lasiocampidae). *J. Gen. Virol. 18*, 119–125.

Manousis, T., Arnold, M. K., and Moore, N. F. 1988. Electron microscopical examination of tissues and organs of *Dacus oleae* flies infected with cricket paralysis virus. *J. Invertebr. Pathol. 51*, 119–125.

Matsui, M., and Watanabe, H. 1974. Electron microscope observations on the midgut of *Theophila mandarina* infected with a flacherie virus of the silkworm, *Bombyx mori*. *J. Seric. Sci. Jpn. 43*, 467–470.

Matsui, M., Maeda, S., and Watanabe, H. 1977. Some properties of a small flacherie virus of the silkworm, *Bombyx mori* L. *Jpn. J. Appl. Entomol. Zool. 21*, 79–84.

Matthews, R. E. F. 1982. Classification and nomenclature of viruses. *Intervirology 17*, 1–199.

McCrady, W. B., and Sulerud, R. L. 1964. Delayed recovery from carbon dioxide in Drosophila melanogaster. *Genetics 50*, 509–526.

Meynadier, G. 1966. Étude d'une maladie à virus chez l'Orthoptère *Gryllus bimaculatus* Geer. *C.R. Acad. Sci. Paris Ser. D263*, 742–744.

Moore, N. F. 1985. The replication schemes of insect viruses in host cells. *In* Viral Insecticides for Biological Control." (K. Maramorosch and K. E. Sherman, eds.), pp. 635–674, Academic Press, New York.

Moore, N. F., and Eley, S. M. 1991. Picornaviridae: Picornaviruses of invertebrates. *In* "Atlas of Invertebrate Viruses." (J. R. Adams and J. R. Bonami, eds.), pp. 371–386. CRC Press, Boca Raton, Florida.

Moore, N. F., and Tinsley, T. W. 1982. The small RNA-viruses of insects. Brief review. *Arch. Virol. 72*, 229–245.

Moore, N. F., Kearns, A., and Pullin, J. S. K. 1980. Characterization of cricket paralysis virus-induced polypeptides in *Drosophila* cells. *J. Virol. 33*, 1–9.

Moore, N. F., Pullin, J. S. K., Crump, W. A. L., and Plus, N. 1982. The proteins expressed by different isolates of *Drosophila* C virus. *Arch. Virol. 74*, 21–30.

Moore, N. F., Reavy, B., and King, L. A. 1985. General characteristics, gene organization and expression of small RNA viruses of insects. *J. Gen. Virol. 66*, 647–659.

Morris, T. J., Hess, R. T., and Pinnock, D. E. 1979. Physicochemical characterization of a small RNA virus associated with baculovirus infection in *Trichoplusia ni*. *Intervirology 11*, 238–247.

Mussen, E. C., and Furgala, B. 1977. Replication of sacbrood virus in larval and adult honeybees, *Apis mellifera*. *J. Invertebr. Pathol. 30*, 20–34.

Nagy, E., and Dobos, P. 1984. Synthesis of Drosophila X virus proteins in cultured *Drosophila* cells. *Virology 134*, 358–367.

Nakagaki, M., Takei, R., and Nagashima, E. 1987. Improved method of purification of the infectious flacherie virus and the *Bombyx* densonucleosis virus. *J. Seric. Sci. Jpn. 56*, 338–342.

Overton, H. A., Buck, K. W., Bailey, L., and Ball, B. V. 1982. Relationships between the RNA components of chronic bee-paralysis virus and those of chronic bee-paralysis virus associate. *J. Gen. Virol. 63*, 171–179.

Paillot, A. 1941. Les travaux de Pasteur sur la flacherie et les théories modernes sur la pathologie du tube intestinal du Bombyx du mûrier. *Ann. Epiphyt. 7*, 99–117.

Pasteur, L. 1870. "Études sur la Maladies des Vers à Soie." Tome I. Gauthier-Villars, Paris.

Plus, N. 1989. Comparative study of picornaviruses in *Drosophila* cells *in vitro*. *In* "Invertebrate Cell System Applications." (J. Mitsuhashi, ed.), Vol. 2, pp. 63–68. CRC Press, Boca Raton, Florida.

Plus, N., and Golubovsky, M. D. 1980. Resistance to *Drosophila* C virus of fifteen l(2)GL/CY stocks carrying l(2)GL lethals from different geographical origins. *Genetika 12*, 227–231.

Plus, N., and Scotti, P. D. 1984. The biological properties of eight different isolates of cricket paralysis virus. *Ann. Virol.* (Inst. Pasteur) *135E*, 257–268.

Plus, N., Croizier, G., Jousset, F.-X., and David, J. 1975. Picornaviruses of laboratory and wild *Drosophila melanogaster*: Geographical distribution and serotypic composition. *Ann. Microbiol.* (Inst. Pasteur) *126A*, 107–117.

Plus, N., Croizier, G., Veyrunes, J.-C., and David, J. 1976. A comparison of buoyant density and polypeptides of Drosophila P, C and A viruses. *Intervirology 7*, 346–350.

Plus, N., Croizier, G., Reinganum, C., and Scotti, P. D. 1978. Cricket paralysis virus and *Drosophila* C virus: Serological analysis and comparison of capsid polypeptides and host range. *J. Invertebr. Pathol. 31*, 296–302.

Pullin, J. S. K., Moore, N. F., Clewley, J. P., and Avery, R. J. 1982. Comparison of the genomes of two insect picornaviruses, cricket paralysis virus and *Drosophila* C virus, by ribonuclease T_1 oligonucleotide fingerprinting. *FEMS Microbiol. Lett. 15*, 215–218.

Reavy, B., and Moore, N. F. 1983. Cell-free translation of *Drosophila* C virus RNA: Identification of a virus protease activity involved in capsid protein synthesis and further studies on *in vitro* processing of cricket paralysis virus specified proteins. *Arch. Virol. 76*, 101–115.

Reavy, B., Crump, W. A. L., and Moore, N. F. 1983. Characterization of cricket paralysis virus- and *Drosophila* C virus-induced RNA species synthesized in infected *Drosophila melanogaster* cells. *J. Invertebr. Pathol. 41*, 397–400.

Reinganum, C. 1975. The isolation of cricket parasysis virus from the emperor gum moth, *Antheraea eucalypti* Scott, and its infectivity towards a range of insect species. *Intervirology 5*, 97–102.

Reinganum, C. 1991. Tetraviridae. *In* "Atlas of Invertebrate Viruses." (J. R. Adams and J. R. Bonami, eds.), pp. 387–392. CRC Press, Boca Raton, Florida.

Reinganum, C., and Scotti, P. D. 1976. Serological relations between twelve small RNA viruses of insects. *J. Gen. Virol. 31*, 131–134.

Reinganum, C., O'Loughlin, G. T., and Hogan, T. W. 1970. A nonoccluded virus of the field crickets *Teleogryllus oceanicus* and *T. commodus* (Orthoptera: Gryllidae). *J. Invertebr. Pathol. 16*, 214–220.

Reinganum, C., Robertson, J. S., and Tinsley, T. W. 1978. A new group of RNA viruses from insects. *J. Gen. Virol. 40*, 195–202.

Reinganum, C., Gagen, S. J., Sexton, S. B., and Vellacott, H. P. 1981. A survery for pathogens of the black field cricket, *Teleogryllus commodus*, in the Western District of Victoria, Australia. *J. Invertebr. Pathol. 38*, 153–160.

Reinganum, C., Bashiruddin, J. B., and Cross, G. F. 1985. Boolarra virus: A member of the *Nodaviridae* isolated from *Oncopera intricoides* (Lepidoptera: Hepialidae). *Intervirology 24*, 10–17.

Richard-Molard, C., Blondel, D., Wyers, F., and Dezelee, S. 1984. Sigma virus: Growth in *Drosophila melanogaster* cell culture; purification; protein composition and localization. *J. Gen. Virol. 65*, 91–99.

Rinderer, T. E. 1974. Infectivity degradation by ribonuclease and table sugar of a nonoccluded virus inoculum prepared from the honey bee. *J. Invertebr. Pathol. 24*, 120–121.

Rinderer, T. E., and Green, T. J. 1976. Serological relationship between chronic bee paralysis virus and the virus causing hairless–black syndrome in the honeybee. *J. Invertebr. Pathol. 27*, 403–405.

Rinderer, T. E., and Rothenbuhler, W. C. 1976. Characteristic field symptoms comprising honeybee hairless–black syndrome induced in the laboratory by a virus. *J. Invertebr. Pathol. 27*, 215–219.

Rothenbuhler, W. C., Kulinčević, J. M., and Stairs, G. R. 1969. An adult honeybee disease usually unrecognized. *Glean. Bee Cult. 97*, 329–331.

Scotti, P. D., Longworth, J. F., Plus, N., Croizier, G., and Reinganum, C. 1981. The biology and ecology of strains of an insect small RNA virus complex. *Adv. Virus Res. 26*, 117–142.

Scotti, P. D., Dearing, S., and Mossop, D. W. 1983. Flock House virus: A nodavirus isolated from *Costelytra zealandica* (White) (Coleoptera: Scarabaeidae). *Arch. Virol. 75*, 181-189.

Seecoff, R. L. 1964. Deleterious effects on *Drosophila* development associated with the sigma virus infection. *Virology 22*, 142–148.

Teninges, D. 1968. Mise en evidence de virions sigma dans les cellules de la lignee germinale mâle de drosophiles stabilisées. *Arch. Ges. Virusforsch. 23*, 378–387.

Teninges, D., and Bras-Herreng, F. 1987. Rhabdovirus Sigma, the hereditary CO_2 sensitivity agent of *Drosophila*: Nucleotide sequence of a cDNA clone encoding the glycoprotein. *J. Gen. Virol. 68*, 2625–2638.

Teninges, D., Ohanessian, A., Richard-Molard, C., and Contamine, D. 1979. Isolation and biological properties of *Drosophila* X virus. *J. Gen. Virol. 42*, 241–254.

Tinsley, T. W., MacCallum, F. O., Robertson, J. S., and Brown, F. 1984. Relationship of encephalomyocarditis virus to cricket paralysis virus of insects. *Intervirology 21*, 181–186.

Tripconey, D. 1970. Studies on a nonoccluded virus of the pine tree emperor moth. *J. Invertebr. Pathol. 15*, 268–275.

Vail, P. V., Morris, T. J., and Collier, S. S. 1983. An RNA virus in *Autographa californica* nuclear polyhedrosis virus preparations: Gross pathology and infectivity. *J. Invertebr. Pathol. 41*, 179–183.

Watanabe, H. 1991. Infectious flacherie virus. *In* "Atlas of Invertebrate Viruses." (J. R. Adams and J. R. Bonami, eds.), pp. 515–523. CRC Press, Boca Raton, Florida.

White, G. F. 1913. Sacbrood, a disease of bees. *U.S. Dep. Agric. Bur. Entomol. Circ. 169*.

White, G. F. 1917. Sacbrood. *U.S. Dep. Agric. Bull. 431*.

Williamson, D. L. 1957. Incidence of CO_2 sensitivity in several Drosophila species. *Drosophila Inf. Serv. 31*, 169–170.

Williamson, D. L. 1961. Carbon dioxide sensitivity in *Drosophila affinis* and *Drosophila athabasca*. *Genetics 46*, 1053–1060.

Yamazaki, H., Sakai, E., Shimodaira, M., and Yamada, T. 1960. Studies on the infectious flachery (F-type) of silkworms, *Bombyx mori* L. *Bull. Nagano Ken Seric. Exp. Stn. 61*, 1–28.

CHAPTER 10
FUNGAL INFECTIONS

I. CLASSIFICATION
II. STRUCTURE AND REPRODUCTION
III. HOSTS
    A. Spore Germination
    B. Penetration of Integument
    C. Penetration through Body Openings
    D. Replication in Hemocoel
    E. Spore Dispersal
IV. PATHOGENICITY
V. SIGNS AND SYMPTOMS
VI. EFFECT OF ENVIRONMENTAL CONDITIONS
VII. SUBDIVISION MASTIGOMYCOTINA
    A. Class Chytridiomycetes
    B. Class Oomycetes
VIII. SUBDIVISION ZYGOMYCOTINA, CLASS ZYGOMYCETES, ORDER MUCORALES
IX. SUBDIVISION ZYGOMYCOTINA, CLASS ZYGOMYCETES, ORDER ENTOMOPHTHORALES
    A. Classification
    B. Reproduction
    C. Hosts, Signs, and Symptoms
    D. Life Cycle
    E. *In Vitro* Cultivation
    F. *Massospora*
    G. *Strongwellsea*
    H. *Conidiobolus*
    I. *Basidiobolus*
X. SUBDIVISION ASCOMYCOTINA
    A. Class Hemiascomycetes, Order Endomycetales
    B. Class Plectomycetes
    C. Class Laboulbeniomycetes
    D. Class Pyrenomycetes, Order Sphaeriales, Family Clavicipitaceae
    E. Fungi of Scale Insects
    F. Order Sphaeriales, Family Hypocreaceae
    G. Class Loculoascomycetes, Orders Myriangiales and Pleosporales
XI. SUBDIVISION BASIDIOMYCOTINA
XII. SUBDIVISION DEUTEROMYCOTINA
    A. Class Hyphomycetes

The first microorganisms found to cause diseases in insects were fungi because of their conspicuous macroscopic growth on the surfaces of their hosts. Some entomopathogenic fungi, however, form no superficial growth or produce sparse, inconspicuous or minute external structures that are difficult to detect by the inexperienced investigator. Most entomogenous fungi are obligate or facultative pathogens and some are symbiotic. Their growth and development are limited mainly by the external environmental conditions, in particular, high humidity or moisture and adequate temperatures for sporulation and spore germination. The diseases caused by fungi are termed "mycoses."

The following books and reviews that we have selected on entomopathogenic fungi have appeared since 1975. This is an arbitrary cut-off date and there are many outstanding earlier publications. Most of the following references have a general treatment of fungi and also emphasize pathogenicity (Charnley 1984), taxonomic keys and illustrations (Weiser 1977; Samson et al. 1988), the use of fungi in insect pest control (Burges 1981; Hall and Papierok 1982; Ferron 1985; McCoy et al. 1988), and a broad treatment of fungus and insect associations (Roberts and Humber 1981; Wilding et al. 1989).

I. CLASSIFICATION

The higher taxa of fungi have undergone drastic revision in recent years. Up to the past decade, the higher taxa were generally considered as Phycomycetes, Ascomycetes, Basidiomycetes, and Deuteromycetes (Fungi Imperfecti). Among the several proposed revisions, we have adopted the classification by Ainsworth (1973), which separates fungi into two divisions, Myxomycota for plasmodial forms and Eumycota for nonplasmodial forms that are frequently mycelial. Entomopathogenic fungi are found in the division Eumycota and in the following subdivisions: Mastigomycotina, Zygomycotina, Ascomycotina, Basidiomycotina, and Deuteromycotina (Table 10–1). Most entomopathogenic fungi are in Zygomycotina, class Zygomycetes, order Entomophthorales; in Ascomycotina, class Pyrenomycetes, order Sphaeriales, class Laboulbeniomycetes, order Laboulbeniales; and in the Deuteromycotina, class Hyphomycetes, order Moniliales. There are no known entomopathogenic forms in the division Myxomycota (slime molds).

II. STRUCTURE AND REPRODUCTION

The fungi may consist of a single cell (e.g., yeasts and hyphal bodies) or, more commonly, branched filaments or hyphae that form a mycelium. The hyphal wall contains chitin or cellulose and other glucans or a mixture of these substances. The hyphae are uninucleate or multinucleate segments, or they are coenocytic with numerous nuclei not separated by transverse walls.

We shall briefly describe the asexual and sexual methods of reproduction in fungi, particularly those of the subdivisions Zygomycotina, Ascomycotina, and Deuteromycotina. The readers should refer to mycology texts for further detailed descriptions. Asexual (anamorphic) reproduction is by various kinds of asexual propagules. Motile asexual reproductive cells (zoospores) are present in the subdivision Mastigomycotina but not in other subdivisions. In the subdivision Zygomycotina, reproduction is by nonmotile aplanospores or conidia, which are passively carried by wind or other agents or sometimes within sporangia that are

TABLE 10–1 Major Fungal Taxa and Partial List of Genera with Entomopathogenic Species

Subdivision	Class	Order	Genera
Mastigomycotina	Chytridiomycetes	Chytridiales	*Coelomycidium* *Myiophagus*
	Chytridiomycetes	Blastocladiales	*Coelomomyces*
	Oomycetes	Lagenidiales	*Lagenidium*
	Oomycetes	Saprolegniales	*Leptolegnia* *Couchia*
Zygomycotina	Zygomycetes	Mucorales	*Sporodiniella*
	Zygomycetes	Entomophthorales	*Conidiobolus* *Entomophaga* *Entomophthora* *Erynia* *Massospora* *Meristacrum* *Neozygites*
Ascomycotina	Hemiascomycetes	Endomycetales	*Blastodendrion* *Metschnikowia* *Mycoderma* *Saccharomyces*
	Plectomycetes	Ascosphaerales	*Ascosphaera*
	Pyrenomycetes	Sphaeriales	*Cordyceps* *Torrubiella* *Nectria* *Hypocrella* *Calonectria*
	Laboulbeniomycetes	Laboulbeniales	*Filariomyces* *Hesperomyces* *Trenomyces*
	Loculoascomycetes	Myriangiales	*Myriangium*
	Loculoascomycetes	Pleosporales	*Podonectria*
Deuteromycotina			*Akanthomyces* *Aschersonia* *Aspergillus* *Beauveria* *Culicinomyces* *Engyodontium* *Fusarium* *Gibellula* *Hirsutella* *Hymenostilbe* *Metarhizium* *Nomuraea* *Paecilomyces* *Paraisaria* *Pleurodesmospora*

(continued)

TABLE 10-1 (*Continued*)

Subdivision	Class	Order	Genera
			Polycephalomyces
			Pseudogibellula
			Sorosporella
			Sporothrix
			Stilbella
			Tetranacrium
			Tilachlidium
			Tolypocladium
			Verticillium
Mycelia sterilia			*Aegerita*
Basidiomycotina	Phragmobasidiomycetes	Septobasidiales	*Filobasidiella*
			Septobasidium
			Uredinella

[a]Adapted from McCoy et al. 1988 and Samson et al. 1988.

violently projected. Asexual reproduction by means of conidia is common in Ascomycotina and Deuteromycotina, and less common in Basidiomycotina.

Various methods of sexual (teleomorphic) reproduction occur in fungi. In Mastigomycotina, there is a union of nuclei after the fusion of motile gametes (planogamy) or after a fusion of a motile male gamete with a larger stationary female gamete (oogamy). Sexual reproduction in many fungi occurs by the fusion of sexually differentiated hyphal branches (gametangia). The sizes of the male gametangia (antheridia) and the female gametangia (oogonia) may differ considerably. The oogonia contain one to several eggs (oospheres) that become thick-walled oospores after fertilization. A thick-walled zygospore is produced in Zygomycotina after the fusion between morphologically indistinguishable gametangia. Sexual fusion in some Ascomycotina and Basidiomycotina is by the transfer of a nonmotile microconidium or spermatium to a receptive hypha, but in many others, sexual fusion is between nondifferentiated hyphae. In Ascomycotina, nuclear fusion is followed by meiosis and usually mitosis to form eight haploid nuclei that develop into spores inside a sac or ascus. The nuclear fusion in Basidiomycotina occurs in a specialized cell, the basidium, in which each of the haploid nuclei enters an outgrowth to become a basidiospore. Many fungi are homothallic, and a single spore gives rise to a mycelium or thallus capable of sexual reproduction; others are heterothallic and require two different thalli to interact in sexual reproduction. When the two thalli are morphologically similar, they are said to differ in mating types.

III. HOSTS

Fungi may be associated with insects as ectoparasites and endoparasites. The ectoparasitic forms are mostly Laboulbeniales and a few genera within the Deuteromycotina. The trichomycetes (subdivision Zygomycotina) are considered as ectoparasites by some authorities, but they do not cause internal lesions in insects, and we have treated them in Chapter 2.

Insects are usually infected by spores or conidia (Zygomycotina), conidia (Deuteromycotina), zoospores (Mastigomycotina), and planonts or ascospores (Ascomycotina). Other types of infective propagules are also found, such as sclerotia or sporodochia (Fawcett 1910; Prior and Perry 1980). The fungi gain entrance into the insect mainly through the integument and in some cases through natural body openings. Transovarial transmission has not been demonstrated with fungi except in the chytrid *Coelomycidium simulii* on black flies (Tarrant and Soper 1986) and in mutualistic yeasts.

Fungi infect individuals in all orders of insects; most common are Hemiptera, Diptera, Coleoptera, Lepidoptera, Orthoptera, and Hymenoptera (David 1967; Ferron 1975). In some insect orders, the immature (nymphal or larval) stages are more often infected than the mature or adult stage; in others, the reverse may be the case. The pupal stage is infrequently attacked and the egg stage is rarely infected by fungi.

Host specificity varies considerably; some fungi infect a broad range of hosts and others are restricted to a few or a single insect species. Those with a broad host range may consist of a variety of pathotypes (McCoy et al. 1988). *Beauveria bassiana* and *Metarhizium anisopliae* infect over 100 different insect species in several insect orders, but isolates of these two fungi have a high degree of specificity (Ferron et al. 1972; Fargues 1976). Host specificity may be associated with the physiological state of the host system [i.e., insect maturation and host plant (McCoy et al. 1988)], the properties of the insect's integument and/or with the nutritional requirements of the fungus (David 1967; Kerwin and Washino 1986a,c), and the cellular defense of the host (Fargues et al. 1976, Ferron 1978). Three types of defense reaction have been observed in the insect's hemocoel: phagocytosis, cellular encapsulation (capsule, module, "giant" cell), and humoral encapsulation (Charnley 1984).

A. SPORE GERMINATION

The development of mycosis can be separated into three phases: (1) adhesion and germination of the spore on the insect's cuticle, (2) penetration into the hemocoel, and (3) development of the fungus, which generally results in the death of an insect. The adhesion of the spores onto the insect's cuticle may be a passive mechanism and involve mucilagenous material and spore surface structures (McCoy et al. 1988; Samson et al. 1988). The conidia of Entomophthorales (e.g., *Entomophthora muscae*) are covered with amorphous mucus, and those of certain deuteromycetous fungi (e.g., *Verticillium lecanii* and *Hirsutella thompsonii*) are slimy and these conidia attach to the substrate or cuticle. The dry conidia of deuteromycetous fungi have well-organized fascicles of rodlets (Boucias et al. 1988). In these conidia, the attachment is nonspecific and is due to the hydrophobicity of the rodlets and insect cuticle. The rodlets also serve to protect the conidia from desiccation and as a means of dispersal in air currents. The conidia of *Conidiobolus obscurus* have both gelatinous mucus and rodlets on the spore surface (Latgé et al. 1986b).

In some cases, spore adhesion is correlated with the aggressiveness or host specificity of a fungal species, such as with *Coelomomyces psorophorae* on mosquitoes (Zebold et al. 1979), with *Metarhizium anisopliae* on scarabaeids (Fargues et al. 1976; Fargues and Robert 1983), and with *Conidiobolus obscurus* on aphids (Latgé et al. 1987).

In some fungi, adhesion is a nonspecific phenomenon, while in others it is a specific process. Not much is known regarding the basis for host specificity and adhesion. Electrostatic forces and molecular interactions may be involved in adhesion (Samson et al. 1988). Hemagglutins, glucose, and N-acetylglucosamine are substances found on the spore surfaces. These substances occur in the mucus and surface structures (rodlets) associated with the spores. Host recognition is regulated, in part, by the physicochemical properties of primarily the insect's epicuticle. Glycoproteins may serve as specific receptors for the spores. The sugars or complex carbohydrates serve in host recognition by the motile mastigomycetous spores of *Lagenidium giganteum* and *Coelomomyces psorophorae* for mosquito larvae (Kerwin and Washino 1986b). The surface of the corn earworm larva has nutrients, primarily amino acids for the germination and growth of *Beauveria bassiana* (Woods and Grula 1984), whereas sterols and polar lipids increase the rate and level of spore germination of *Nomuraea rileyi* (Boucias and Pendland 1984).

Spore germination largely depends on the environmental humidity and temperature, and to a lesser extent light conditions and the nutritional environment. The effect of environmental conditions will be discussed later. The characteristics of the spores and their structures also affect spore germination. The germinating spore, attached to the host, forms a germ tube that serves as a penetrant hypha (Fig. 10–1). In addition to the germ tube, an appressorium may also be produced. Mucilagenous materials may coat the germ tube and appressorium (Zaccharuck 1970a; Pendland and Boucias 1984; Brey et al. 1986). These germination structures also strengthen the adhesion of the fungus. Successful germination and penetration depend, not necessarily, on the total percentage of germination but also on the duration of the ger-

FIGURE 10–1

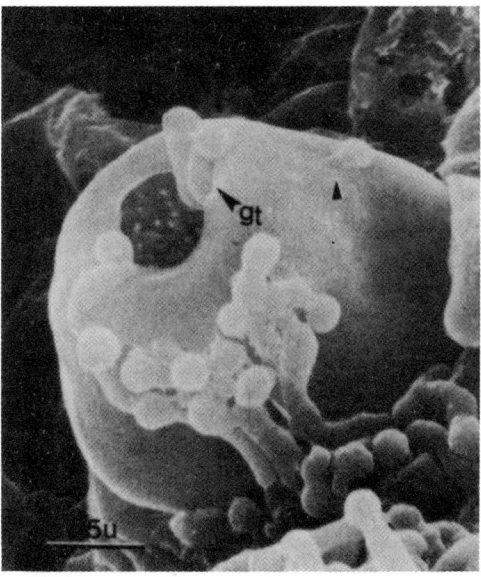

Scanning electron micrograph of the germination of conidia of the white-muscardine fungus, *Beauveria bassiana*, on the epicuticular surface of the corn earworm, *Heliothis zea*. Note germ tubes entering a natural opening in the integument. (Bar = 5 μm.) [Courtesy of McCoy et al. (1988) and CRC Press.]

mination time, mode of germination, aggressiveness of the fungus, the type of fungal spore, and host susceptibility (Samson et al. 1988).

B. PENETRATION OF INTEGUMENT

The mode of penetration mainly depends on the property of the cuticle, its thickness, sclerotization, and the presence of antifungal and nutritional substances (Charnley 1984). The newly molted larva and the newly formed pupa are more susceptible to infection than those in which the cuticle has fully hardened (Rockwood 1950; Tanada 1955; Fox 1961; Kurisu and Manabe 1978). Koidsumi (1957) reports that the medium-chain fatty acids in the epicuticle prevent the invasion of fungus through the integument. This observation has been confirmed (Champlin et al. 1981; Kerwin 1982; Smith and Grula 1982; Saito and Aoki 1983). Kerwin (1982, 1984) has found that palmitoleic (C_{16}), linoleic (C_{18}), and linolenic (C_{18}) acids are toxic to the conidia of *Erynia variabilis* (syn. *Entomophthora culicis*) over a wide range of concentrations. On the other hand, oleic acid (C_{18}), in combination with chitin and chitosan, induces vegetative and conidial germination. Thus, the response of *E. variabilis* conidia to germination is specific for certain fatty acids, their concentrations, chain length, and degree of unsaturation. Since these short-chain fatty acids, however, are absent in the orders Coleoptera and Homoptera (Champlin et al. 1981; Brey et al. 1986), the antifungal role of these lipids may be questionable (McCoy et al. 1988).

Aggressive strains of *Conidiobolus obscurus* form germ tubes in the presence of aphid cuticular extracts, whereas the nonaggressive strains do not (Latgé et al. 1987). Moreover, the cuticular extracts from nonhost insects may stimulate the spore germination (e.g., the spore of *Nomuraea rileyi*, which is nonpathogenic to aphids and is stimulated to germinate by aphid cuticular lipids), whilst extracts from *Anticarsia gemmatalis* will induce the germination of primary spores of *C. obscurus*, which cannot infect the *A. gemmatalis* larva (Boucias and Latgé 1988). Localized melanization of the integument, which commonly develops around and in front of the penetrating fungus, may prevent some fungi from infecting an insect (Brobyn and Wilding 1977; Travland 1979b).

The process of penetration through the insect's integument by a hypha germinating from a spore involves chemical (enzymatic) and physical forces. The mechanical force is noticeable at the tip of an invading hypha where the cuticular layers are distorted from pressure. The enzymes detected on germ tubes are proteases, aminopeptidases, lipase, esterase, and N-acetyl-glucosamidase (chitinase) (Ratault and Vey 1977). *In vitro* studies indicate that the digestion of the integument follows a sequential lipase–protease–chitinase process of digestion (Samšiňáková et al. 1971; Smith et al. 1981; St. Leger et al. 1986a). The proteases are the major cuticle-degrading enzymes and their activities appear to precede those of chitinases (Samšiňáková et al. 1977; Smith et al. 1981; St. Leger et al. 1987a,b, 1989b). The chitinase activity occurs mainly at the time of fungal growth, conidia formation, and sporulation of conidiophores (Coudron et al. 1984). Aggressive or virulent strains may produce large amounts of extracellular enzymes (Paris and Segretain 1978; Paris and Ferron 1979) (e.g., lipase, elastase, proteases, and α-glucanase), but in many cases no clear relationship has been established between such enzymes and aggressiveness of a fungal strain (Samson et al. 1988).

The penetration of a fungus may be restricted to certain areas of the host [e.g., thorax and abdomen of the pea aphid, *Acyrthosiphon pisum*, by the entomophthorus fungus *Conidiobolus obscurus* (Brey et al. 1986)]. In thin integuments, as found in many homopteran and lepidopteran insects, the germinating hypha (tube) penetrates directly without greatly altering its form, but in *Erynia blunckii*, it produces dual bulbs or becomes screwlike (Tomiyama and Aoki 1982). Generally, when the integument is hard, thick, or smooth, the germ tube produces an appressorium that adheres to the cuticle with mucoid substances (McCauley et al. 1968; Zacharuk 1970a,b; Schabel 1978). The appressorium may form penetration plate, penetrant tube or hypha, and hyphal bodies for invasion. Appressoria vary in sizes and shapes from unicellular clavate to multicellular ovoid, spherical, or ramifying forms. Cuticle-degrading proteases are secreted by the appressorium on the cuticle surface and by the penetrant hypha within the cuticle (Goettel et al. 1989; St. Leger et al. 1991a). These enzymes liberate monomers that can be metabolized by the germ tube in order to continue to grow into the integument (Samson et al. 1988).

When the conidia of *M. anisopliae* are applied to an elaterid larva shortly prior to molting, the fungus may or may not infect the larva. The failure to infect is due to the newly developed integument beneath the exuvium. In some cases, however, the fungus forms secondary appressoria that invade the developing new cuticle beneath the exuvial cuticle of a molting larva (Zacharuk 1973) or form blastospores in the molting fluid (Vey and Fargues 1977).

Based on their comprehensive studies, St. Leger et al. (1991b) have proposed the following events surrounding appressorial formation in *M. anisopliae*: (1) activation of a receptor on the germ-tube cell surface (St. Leger et al. 1989a, 1990a), (2) signal mediation during which a local domain is established over the site of thigmotropic induction, and (3) subsequent changes of the cell, which include the synthesis of differentiation specific mRNA (St. Leger et al. 1989a,b). An induction signal disrupts the Ca^{2+} gradient required for the maintenance of polar growth and the actin cytoskeleton, and initiates germ-tube swelling and appressorium formation (St. Leger et al. 1991a).

C. PENETRATION THROUGH BODY OPENINGS

Fungi can infect insects through the buccal cavity, spiracles, and other external openings of an insect (Fig. 10–1). Since moisture is not a problem in the alimentary tract, the spore may germinate readily in this environment; on the other hand, the digestive fluids may destroy the spore or germinating hypha. In some cases, the digestion of fungal structures may cause death by toxicosis rather than by mycosis (Charnley 1984). Preoral transmission has been reported, particularly at the buccal and esophageal cavities (Yendol and Paschke 1965; Veen 1966; Schabel 1976). Penetration by fungi through the alimentary tract has been observed in the fire ant, *Solenopsis richteri* (Broome et al. 1976), worker termite, *Reticulitermes* sp. (Kramm and West 1982), and in several species of mosquitoes (Roberts 1970; Hasan and Vago 1972; Sweeney 1975, 1977; Knight 1980).

In mosquito larvae, the fungus *Culicinomyces* sp. invades through the fore- and hindguts (Sweeney 1975). The walls of the conidia and germ tubes are coated with an irregular, electron dense, mucilagenous layer, which seems to enable the fungus to adhere to the host (Sweeney et al. 1980). The fungus causes lethal infection at 15 to 27.5°C but not when the mosquito larvae are reared at 30°C for more than 16 h per day (Sweeney 1978). At 30°C, the fungal spores germinate and the hyphae penetrate the foregut cuticle but do not invade the hemocoel; the hyphae are discarded in the next molt.

Fungi can infect insects through the spiracles and other body openings. *Metarhizium anisopliae* occasionally infects elaterid larvae through the spiracles and pores of sense organs (McCauley et al. 1968). *Beauveria bassiana* infects several mosquito species through the posterior siphon (Clark et al. 1968); in *Heliothis zea* through the spiracles (Pekrul and Grula 1979); and in the alfalfa weevil, *Hypera postica*, by way of the tracheae and not through the thick cuticle of the integument (Hedlund and Pass 1968). The anal region of the silkworm larva is most frequently infected by the fungus *Aspergillus flavus oryzae* (Aoki 1961).

D. REPLICATION IN HEMOCOEL

After the germinating hypha has penetrated the insect's integument and entered the hemocoel, it produces yeastlike hyphal bodies, essentially blastospores, that multiply by budding. In addition to hyphal bodies, hyphal strands and wall-less protoplasts may develop in the hemocoel. The dispersal throughout the hemocoel and tissue invasion vary with the fungal species. Some fungal species form both hyphal bodies and protoplasts depending on the nutritional environment.

The fungus may avoid the immune defenses of an insect by (1) developing protoplasts that are not recognized by the insect's hemocyte population (Dunphy and Nolan 1982b; Latgé et al. 1986a), (2) forming yeastlike hyphal bodies that multiply and disperse rapidly (Samson et al. 1988), and (3) producing mycotoxins.

The death of the insect ends the parasitic development of the fungus. The fungus then grows saprophytically in the hemocoel to form a mycelial mass that turns into a hard or firm sclerotium. Reproductive spores are produced within the sclerotium or on sporophores (sporangiophore and conidiophore). The sporophores and sterile hyphae emerge from the cadavers under favorable conditions, especially conditions of humidity and temperature, to form the characteristic mycelial growth on the insect's integument. The growth is initiated and occurs most predominantly at the intersegmental regions of the host. Sporulation generally occurs on cadavers but may also occur on live insects, such as in the cicada with *Massospora* spp. (Soper 1974), in the adult seed corn maggot with *Strongwellsea castrans* (Strong et al. 1960), and in mirid bugs with *Entomophthora erupta* (Ben-Ze'ev et al. 1985). Under unfavorable conditions, the fungus produces resistant or resting propagules (chlamydospores, azygospores, zygospores, or oospores). The readers should realize that fungal mycelial growth on the surface of dead insects does not necessarily mean that the fungus has killed the host, because many saprophytic fungi grow readily on dead or dying insects.

E. SPORE DISPERSAL

Spore dispersal can be an active or passive process and depends on the characteristics of the spore and sporophores. Certain fungi form conidia that are enclosed in a mucilagenous slime. Such conidia may attach to a passing insect or other invertebrates for dispersal (Fletcher 1977; Ingold 1978). Spores of many entomopathogenic fungi are forcibly discharged from sporophores and carried by the wind. Sporophores are generally positively phototrophic and negatively geotrophic so that the spores are produced in free air. This is especially the case of ascospores of *Cordyceps*, which have columnlike stromata emerging from an insect that has died in the soil. Ascospores are forcibly discharged from the perithecia occurring usually at the top of the stromata. In certain fungi (e.g., Laboulbeniales, *Massospora*, *Strongwellsea*, and a few others), the spores develop on living insects that serve as dispersal agents of the fungi.

IV. PATHOGENICITY

Fungi usually cause insect mortality by one or more of the following: nutritional deficiency, invasion and destruction of tissues, and release of toxins. Some fungi are obligate pathogens (e.g., *Coelomomyces*, Laboulbeniales, certain *Entomophthora* spp.) and their complete life cycles have not been cultured outside of a living insect. Most entomopathogenic fungi, however, are facultative pathogens capable of growing without an insect host. Some are virulent pathogens and kill the insect within a few days (e.g., in Entomophthorales); others produce chronic and prolonged infections (e.g., *Massospora* spp.).

Fungal species have numerous strains that differ in their virulence and pathogenicity. In general, the strains of a species isolated from a specific host are more virulent for that host than those isolated from other hosts. Successive transmission within a host may also result in the enhancement of virulence or the isolation of a more virulent strain. On the other hand, the *in vitro* culture often results in diminished virulence, but fungi reared on certain media retain their virulence almost as high as host passage (Fargues and Robert 1983; McCoy et al. 1988). Samšiňáková and Kálalová (1983), through monospore isolations, have obtained six spontaneous mutants that surpassed the mother strain in pathogenicity. The mutants retain their high pathogenicity both in repeated subculturing and in material subcultured after a 12-month storage.

The pathogenicity of a fungus may be associated with the production of enzymes and mycotoxins during the course of infection in an insect (i.e., at ingestion, cuticular contact, or within the hemocoel) (Ferron 1978; McCoy et al. 1988). There are numerous mycotoxins (Roberts 1981) and we shall consider some of those produced by the more common entomopathogenic fungi.

V. SIGNS AND SYMPTOMS

At an early stage of fungal infection, the insect shows little or no signs and symptoms except for a few necrotic spots which may develop at the invasion sites. At a late stage of infection, the insects generally become restless, less active, their appetites are reduced, and they lose coordination. Infected insects often move to high places or if subterranean, rise to the soil surface (McCoy et al. 1988)

The time of tissue disintegration may differ with the fungal species, mode of invasion, and host species. The cells and tissues of an infected insect may begin to disintegrate prior to the insect's death (Lefebvre 1934; Zacharuk 1971; Mohamed et al. 1978) or they may break down after death (McCauley et al. 1968). The fungal hyphae continue to grow usually resulting in mummification, and the dead insects retain their form and shape. However, in lepidopteran larvae infected with certain Entomophthorales, they become flaccid with watery contents and fragile integuments (Fig. 16-4). Shortly prior to or at death, the insect may have a characteristic color.

In several lepidopteran insects infected with *Beauveria bassiana*, the hemolymph proteins increase up to the last stages of infection and then decrease when the insects cease feeding prior to death (Gardner et al. 1979). In the silkworm larva infected with *B. bassiana*, the pH of the hemolymph drops within 1 day after infection and rises slightly in 2 to 4 days but never higher than that of uninfected hemolymph (Kusunoki and Watanabe 1982).

VI. EFFECT OF ENVIRONMENTAL CONDITIONS

Environmental conditions, particularly humidity and temperature and to a lesser extent light and air movement, are very important in infection and sporulation of entomopathogenic fungi (Müller-Kögler 1965; MacLeod et al. 1966; Roberts and Campbell 1977). Optimum temperatures for development, pathogenicity, and survival of the fungi generally fall between 20 to 30°C (McCoy et al. 1988). Resting spores may tolerate high temperatures from 80 to 100°C for 5 to 60 min and low temperatures of 7 to 10°C.

Temperature requirements vary with the fungal species and ecological niches (Samson et al. 1988). Deuteromycotina, found in tropical and subtropical areas, tend to germinate at optimal temperatures above 25°C, whereas the Entomophthorales, with a predominantly temperate distribution, have lower optimum temperatures.

Very high humidity (above 90% RH) is required for spore germination and sporulation outside of the host. A film of water may be necessary for the conidial germination of some Deuteromycotina (Hall 1981; McCoy 1981) and of *Entomophthora muscae* (Carruthers and Haynes 1986) but is unfavorable for those of other Entomophthorales (Latgé and Papierok 1988). High ambient humidity is required for the sporulation and development of the mycelium on the surface of the cadaver (Ferron 1977). The humidity of the microenvironment surrounding the spore, however, is more influential in germination than the ambient humidity. Moreover, the microclimatic moisture greatly affects conidial production (Millstein et al. 1983a,b; Nordin et al. 1983). Although high humidity is essential for sporulation, the release of the conidia of *Beauveria bassiana* and *Metarhizium anisopliae* from conidophores is stimulated by low humidity (less than 50%), darkness, and vibration (Gottwald and Tedders 1982).

The effect of light in fungal infection is not completely known. Since light often interacts with temperature and humidity, its action is difficult to evaluate separately. The short ultraviolet rays will kill spores exposed on a substratum or while being dispersed in the air. The entomophthorous genus *Conidiobolus* forms germ tubes and conidiophores that are positively phototropic, but this is not the case for *Erynia radicans* (syn. *Zoophthora radicans*, syn. *Entomophthora sphaerosperma*) (Uziel et al. 1981). More spores of many *Entomophthora* are discharged in the light, but spore germination appears less affected by light (McCoy et al. 1988). Many entomopathogenic fungi sporulate during the early morning hours before daylight, but the significance of darkness has not been evaluated. Long day length stimulates the germination of resting spores of certain *Entomophthora* and this aspect will be discussed later.

Extended exposure to far ultraviolet light generally destroys microorganisms (Krieg et al. 1981), but *Hirsutella thompsonii* survives such exposure by photoreactivation (Kenneth et al. 1979; Tuveson and McCoy 1982).

The germination of fungal spores is also dependent on the nutritional environment (Samson et al. 1988). The absence, quality, and quantity of nutrients affect germination. Cuticular lipids and other compounds in the insect's integument affect the spore germination differently (Woods and Grula 1984; Boucias and Pendland 1984; Boucias and Latgé 1988; Samson et al. 1988; St. Leger et al. 1989a). We have already discussed the favorable and unfavorable aspects of these substances. Moreover, the survival of fungal propagules varies with the surface and type of substrates (i.e., air, soil, water, foliage, chemical, host, etc.) (McCoy et al. 1988).

When an infected insect is placed under unfavorable environmental conditions (e.g., low humidity and temperature), the fungal hyphae may produce resistant structures for survival under such conditions. Generally, chlamydospores and resting spores are formed within the insect. *Nomuraea rileyi* produces three types of resistant structures: (1) thick-walled intrahyphal hyphae, (2) aerial chlamydospores, and (3) intralarval resting bodies (Pendland 1982). The thick-walled intrahyphal hyphae are also formed in the disintegrating larval tissues.

VII. SUBDIVISION MASTIGOMYCOTINA

Mastigomycotina (zygosporic fungi) include members that are arbitrarily considered together on the sole basis that the nonsexual propagative spore, which possesses one or two flagella, is adapted for locomotion in a liquid medium (Sparrow 1973a). They form perfect-state spores, typically oospores. They are almost an entirely aquatic, primitive group known collectively as water molds or aquatic phycomycetes. The simplest forms (e.g. chytrids) have no mycelial vegetative structure and produce zoospores. The entomopathogenic forms are in two classes, Chytridiomycetes and Oomycetes.

A. CLASS CHYTRIDIOMYCETES

Chytridiomycetes are primarily microscopic fungi that occur as saprobes on a wide variety of dead plants and animals and as pathogens of aquatic and terrestrial plants and animals (e.g., scale insects, mosquito larvae, and rotifers) (Sparrow 1973b). The chytrids reproduce by zoospores that have a single posteriorly directed flagellum of the whiplash type, a conspicuous nuclear cap, and sometimes a single basal mitochondrion. Sexual reproduction varies, and the zygote forms an encysted structure or a diploid planont.

Class Chytridiomycetes includes two orders with entomopathogenic forms, Chytridiales and Blastocladiales. The zoospores frequently have a conspicuous refractile droplet. The resting spore is thick-walled and generally fills its container (sporangium or a special asexually or sexually formed body), which functions as a sporangium or prosporangium.

1. Order Chytridiales

The chytrid genus *Myiophagus* has a single species *M. ucrainicus*. It infects scale insects, mealybugs, beetle larvae (*Cleonus, Anisoplia*), and dipteran pupae (Karling 1948) and causes a disease called chytridiosis. At one time, this fungus was considered to be an important pathogen of several species of armored scale insects on citrus (Fisher et al. 1949; Fisher 1950). However, changing horticultural practices and successful control with parasitoids on armored scales have greatly reduced the presence of *M. ucrainicus* in citrus orchards in Florida (McCoy et al. 1988).

Myiophagus ucrainicus transforms the body contents of insects into an orange-colored mass of globular resting spores or sporangia that may be found throughout the insect's body (Karling 1948). When mature sporangia are placed in water, zoospores with a posterior-directed flagellum emerge from one to three exit papillae formed on the sporangium. After penetrating a host, the zoospore forms a broad, unbranched germ tube, which enlarges and becomes clavate. A multinucleate thallus is formed. It becomes elongated, lobed and branched, with swellings and constrictions at regular intervals. The swellings become sporangia that mature and separate to form a mass of loose and free, orange to golden-colored spheres in the insect's hemocoel. When placed in water, the mature sporangia produce zoospores within 2 to 6 h.

2. Order Blastocladiales

The order Blastocladiales has zoospores with a centrally located "subtriangular" body of low refractivity, the so-called nuclear cap (Sparrow 1973a). A conspicuous "side body," a large mitochondrion, is found in some zoospores. The entomopathogenic forms are in the families Coelomomycetaceae and Catenariaceae. These families are holocarpic and the thallus is unwalled, branched, or lobed. The thallus is converted into a mass of thick-walled (less commonly thin-walled) resting spores or zoosporangia. The zoosporangia split or crack open and liberate the zoospores. The family Catenariaceae has several species in the genus *Catenaria* (Martin 1975a, 1978) that play major roles in the epizootics of chironomids (Martin 1984).

Coelomomycetaceae has two entomopathogenic genera, *Coelomomyces* (Keilin 1921) and *Coelomycidium* (Debaisieux 1919). *Coelomycidium simulii* infects black flies. The type species *Coelomomyces stegomyiae* was obtained from the larva of the mosquito *Aedes albopictus* (syn. *Stegomyia scutellaris*) (Keilin 1921). There are about 40 identified species widely distributed over the world (Federici 1981). Species of *Coelomomyces* are obligate pathogens of mosquitoes and a few other insects. Aquatic arthropods, other than insects, serve as alternate hosts.

a. Heteroecism

Until recently, the entire life cycle of *Coelomomyces* was unknown, and workers experienced enormous difficulty in propagating this fungus. Soil, detritus, algae, and sources of water were considered as essential factors in obtaining infection in mosquito larvae (Couch 1972). Whisler et al. (1974, 1975) made an outstanding discovery that this fungus required an alternate host, another aquatic arthropod, such as copepods and ostracods. The complex life cycle involves an obligate alternation of sexual and asexual generations between an intermediate crustacean host (gametophytic generation) and a definitive dipteran host, such as a mosquito larva (sporophytic generation) (Whisler 1979) (Fig. 10–2). This is a case of heteroecism, which was once considered among fungi to be unique for the plant-parasitic rust fungi.

The zoospores from mosquito larvae do not infect mosquito larvae but infect other aquatic arthropods (e.g., copepods) and those from copepods infect only

FIGURE 10-2

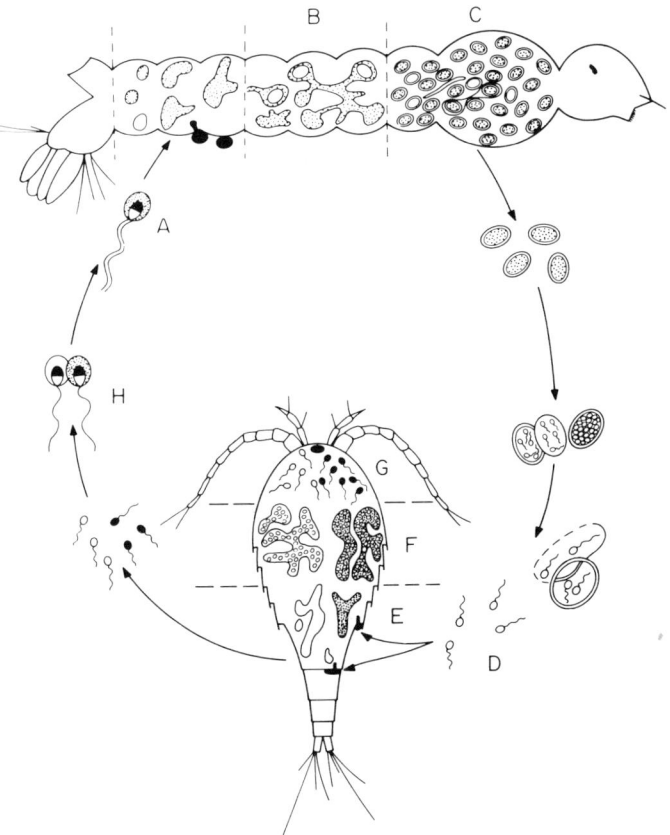

Generalized life cycle for species of *Coelomomyces*. (A) Biflagellate zygote infects hemocoel of mosquito larva producing hyphagens, which later form hyphae. (B,C) Hyphae ramify in the hemocoel and form thick-walled resistant sporangia. After liberation from larva, sporangia release meiospores (D), which infect a copepod and produce a gametophyte (E). Each gametophyte forms a gametangium (F) that releases gametes of a single-mating type (G). Opposite-mating types may be different colors, as illustrated, or unpigmented. A copepod may contain either one or both types of gametophytes, depending on the nature of the infecting meiospore. Gametes of opposite-mating type fuse inside or outside (H) copepod, forming a biflagellate zygote that completes the cycle by infecting another larva. [Courtesy of Federici (1981) and Academic Press.]

mosquito larvae (Whisler 1979; Federici 1981; Whisler et al. 1983) (Fig. 10–2). Infection of a mosquito larva is initiated by the diploid zygote, a posteriorly biflagellate zoospore. The fungus develops a diploid thallus, called a sporophyte, that produces hyphal bodies and mycelium. The swollen tips of the hyphae form resistant sporangia, which are released when the mosquito dies. Meiosis occurs in the resistant sporangium. Under appropriate conditions, the sporangia release uniflagellate zoospores (meiospores) of opposite mating types infecting the alternate host (copepod). Each zoospore develops into a thallus (gametophyte) forming gametangia that produce gametes (haploid zoospores). The gametes of opposite mating-type fuse either in or outside of the copepod to form the mosquito-infecting biflagellate zygote.

In both the mosquito and copepod, the fungus *Coelomomyces psorophorae* encysts on the cuticle and penetrates by means of an appressorium into the body cavity of the host (Zebold et al. 1979). Substances emitted from the copepod are recognized by the uniflagellate meiospore and initiate changes leading to encystment (Federici and Lucarotti 1986). *Coelomomyces psorophorae* encysts in definite patterns in both hosts primarily on the intersegmental membrane. Carbohydrates in the cuticle may be involved in the encystment of zygotes on mosquito larvae (Kerwin 1983a). Encystment occurs at temperatures of 12 to 30°C with the optimum at 25°C, at pH 5.0 to 9.5 with the optimum at pH 6.5 to 8.5 (Kerwin 1983b). The zygotes are active in water with total salinity of 5.4 parts per thousand down to double glass-distilled water.

b. Alternate Hosts

The obligate alternate host for *Coelomomyces psorophorae*, which infects the mosquito *Culiseta inornata* (Whisler et al. 1975) and *C. dodgei*, which infects *Anopheles quadrimaculatus* (Federici and Roberts 1976; Federici and Chapman 1977), is the copepod *Cyclops vernalis*; *Coelomomyces chironomi*, which is pathogenic for chironomids, utilizes the ostracod *Heterocypris incongruens* (Weiser 1976). Copepods infected with *C. dodgei* contain light amber or bright orange or a mixture of both types of gametophytes that develop into gametangia (Federici and Chapman 1977). The light-amber gametangia produce female gametes and the bright-orange ones form male gametes. The orange pigment in the gametophytes is β-carotene (Federici and Thompson 1979).

Coelomomycetes are obligate pathogens and have not been cultured *in vitro* throughout their life cycles. Attempts to culture species of *Coelomomyces* from mosquitoes on media used for vertebrate and invertebrate tissue culture and for mycoplasma culture have resulted in growth and development up to the young sporangial stage, which does not mature (Shapiro and Roberts 1976; Castillo and Roberts 1980). Cultures obtained from the alternate copepod host had a different nutritional requirement from that of cultures from the mosquito host (Castillo and Roberts 1980).

c. Hosts

Species of *Coelomomyces* have been reported from insects in two orders, Diptera and Hemiptera. The hemipteran species is the back swimmer, *Notonecta* sp., which is infected by *C. notonectae* (Bogoyavlensky 1922). In Diptera, the hosts are in three families, Chironomidae, Culicidae, and Simuliidae, with the mosquitoes being the most common host, with about 54 susceptible species (Chapman 1974). A predatory mosquito, *Toxorhynchites rutilis septentrionalis*, is infected by *Coelomomyces macleayae* (Nolan et al. 1973).

Some species of *Coelomomyces* infect only a single species of mosquito, while others have a broader host range, such as *C. indiana* with 17 hosts in three different genera of mosquitoes (Couch and Umphlett 1963) and *C. psorophorae* with seven mosquito species in three genera (Zebold et al. 1979). As a rule, Couch and Umphlett (1963) believe that each species or variety of *Coelomomyces* grows on only one species of mosquito. Since this observation was made when the life cycle of the coelomomycetes, particularly the requirement of an alternate host, was still unknown, the host specificity of these fungi should be reevaluated with further tests.

d. Hybridization

Federici (1979) hybridized *C. dodgei* and *C. punctatus*. Reciprocal crosses of gametes derived from copepods infected with each species produced hybrid zygotes that were infectious for mosquito larvae. Most of the sporangia produced by the hybrid sporophyte were not typical to either parental species. Some sporangia resembled *C. dodgei* more than *C. punctatus*, but others exhibited surface structures similar to those of *C. lativittatus*. Federici (1979) suggested that this latter species might be a naturally occurring hybrid of *C. dodgei* and *C. punctatus*. However, a subsequent study revealed that, although all stages of the sporophytic phase of the hybrid were functional, the F_1 gametophyte was inviable (Federici 1982). The meispores produced by the hybrid sporangia appeared normal, but their encystment did not result in the production of a functional gametophyte. Thus, both species are distinct. These two species are reproductively isolated because their gametangial dehiscence (gamete release) follows different diel periodicity that is dependent on photoperiod (Federici 1983).

e. Host Invasion and Replication

Coelomomyces spp. infect mosquito larvae in all four instars (Couch 1972), and the younger larvae are more susceptible than older larvae (Nnakumusane 1987a). The biflagellate zygote of *C. psorophorae*, which has emerged from a copepod, attaches apparently to specific parts of the mosquito larval body, correlated with the texture of the cuticle (head, intersegmental areas, base of anal gill, and around the anus) (Travland 1979a,b). After attachment, the zygote encysts. The cyst germinates and produces a bulb-shaped appressorium that forms a narrow tube to penetrate the larval cuticle. The appressorium, unlike those produced by other entomopathogenic fungi, injects the fungal protoplasm into a host epidermal cell. This is the only intracellular stage of the fungus. The process from the cuticular attachment to hemocoelic invasion takes about 8 h. At high temperatures (over 28°C), the rate of infection declines (Nnakumusane 1987a).

In the larval hemocoel, the earliest stages of the fungus appear as small spheres or hyphagens (down to 5.4 μm), which give rise to hyphae to form a mycelium (Martin 1969). The hyphae are not naked as believed by previous workers but bear an extracellular, granular hyphal coat of protuberances or microvilli (Powell 1976). Tips of branching hyphae form the resting sporangia. Mature resting sporangium has three distinct wall layers: an outermost pigmented layer; a second-formed, middle layer that appears relatively homogeneous; and a third-formed, thin innermost layer that may serve as the vesicle covering the spore mass just prior to zoospore release at germination (Martin 1969; Whisler et al. 1972).

The quantity of sporangia produced within a mosquito larva depends on the severity of infection. When a larva contains an abundance of hyphagens and mycelia, it is killed in a short period, and few if any sporangia are produced (Couch 1972). At a lower level of infection, abundant sporangia develop in the infected larva, which may be unusually conspicuous and striking because of the presence of large numbers of yellow to red-brown sporangia. A lightly infected larva may develop into an adult that may serve to disperse the fungus (Manalang 1930; Walker 1938; Couch 1972). The maturation of the fungus in an adult mosquito depends on host nutrition [i.e., the sporangia develop only in a female mosquito that has been fed a blood meal (Taylor et al. 1980)]. The infected female mosquito does not reproduce.

Resistant sporangia readily germinate after reaching a fully cleaved stage of development, but in some species the sporangia must undergo a period of drying and wetting (Couch and Umphlett 1963; Whisler et al. 1972; Couch 1972). The sporangium has a dehiscence line or crack through which the zoospores are discharged and become planonts.

B. CLASS OOMYCETES

The oomycetes have biflagellate zoospores with the posterior flagellum being whiplash-type and the anterior being tinsel-type with hairs along its side. Its cell wall is cellulosic. The two orders with entomopathogenic members are Lagenidiales and Saprolegniales.

1. Order Lagenidiales

The order Lagenidiales includes a heterogeneous group of microscopic simple fungi that are primarily endoparasites of algae, other aquatic fungi, higher plants, animals, and their eggs, but a few are saprobic (Sparrow 1973c). Spores are formed within or outside the sporangium and often in a vesicle.

The entomopathogenic species occur mainly in *Lagenidium*. The genus *Atkinsiella* has been included in Lagenidiales by Martin (1977), although it has been considered in the order Saprolegniales (Dick 1973). *Lagenidium* spp. are found worldwide and infect a number of species of mosquito larvae (Umphlett and Huang 1972; McCray et al. 1973; Goettel et al. 1983). Once introduced into an aquatic habitat, *Lagenidium giganteum* successfully overwinters and persists for several years (Fetter-Lasko and Washino 1983). It is also capable of surviving alternate drying and flooding of its habitat.

a. Replication

The laterally biflagellate zoospores of *Lagenidium* infect the larva through the mouth or integument (McCray et al. 1973; Lord and Roberts 1987). The zoospores form cysts producing delicate germ tubes that penetrate through the integument by proteolytic and lipolytic activities and mechanical pressure (Domnas et al. 1974). In the larval hemocoel, the fungus develops a branched, nonseptate mycelium that does not invade solid structures, such as muscles and nerves. The mycelium produces oval, spherical, and irregular segments forming sporangia, which may fill the larval hemocoel. The sporangia form zoospores, whose production is affected by environmental factors (Lord and Roberts 1985). The fungus depletes the endogenous reserves of the host, which dies from starvation (Domnas et al. 1974).

Lagenidium spp. are homothallic and sexual fusion occurs between antheridia and oogonia that are formed in adjacent cells of the same filament or in contiguous filaments (Couch and Romney 1973). Conjugation results in the formation of thick-walled oospores that are resistant to adverse environmental conditions. Under favorable conditions the oospore germinates to grow saprophytically or to produce infectious zoospores (McCoy et al. 1988). Figure 10–3 provides a diagram of the life cycle of *Lagenidium giganteum*.

FIGURE 10–3

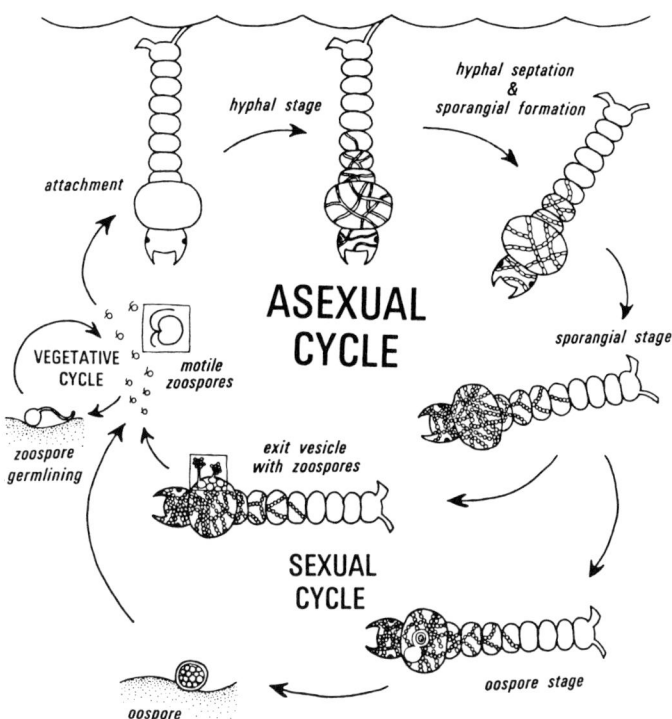

Generalized life cycle for *Lagenidium giganteum*. [Courtesy of Fetter-Lasko and Washino (1983) and Entomological Society of America.]

b. Culture and Nutrition

Lagenidium spp. can grow saprophytically, but in some species, pathogenicity is lost after several generations on agar. They have been cultured on a wide variety of nutrient media. The induction and maturation of the sexual oospores depend on the presence of sterols, fatty acids, and calcium (Willoughby 1969; Kerwin and Washino 1983, 1986a,c; Lord and Roberts 1986). There are procedures for the mass production of zoospores *in vitro* (Domnas et al. 1982; Jaronski et al. 1983).

Lagenidium giganteum appears to utilize the trehalose in the infected mosquito larva as a carbon source (Giebel and Domnas 1976). This results in the reduction of the energy reserves of the larva and leads to general physiological starvation of the host. The fungus produces an extracellular collagenolytic enzyme, which may function in invasion or development (Dean and Domnas 1983a,b).

c. Hosts

Lagenidium giganteum can grow either as a saprophyte in the aquatic environment or as a mosquito pathogen (Willoughby 1969). It infects a number of mosquito species in the genera *Aedes*, *Culex*, *Anopheles*, and *Culiseta* (Couch and Romney 1973). *Lagenidium culicidum* infects species of *Culex* and *Aedes* (Umphlett and Huang 1972; McCray et al. 1973). Generally, young larvae are more susceptible to these fungi than old larvae and pupae (Federici 1981; Lord and Roberts 1987). *Atkinsiella entomophaga* parasitizes the eggs of chironomids and, together with the chytrids, is involved in the epizootics of the chironomids (Martin 1984).

2. Order Saprolegniales

The order Saprolegniales has one large family, Saprolegniaceae, and a number of smaller ones (Dick 1973). The members have a vegetative, nonreproductive thallus that is unsegmented and only rarely septate. Entomopathogenic members are in *Leptolegnia*, *Aphanomycopsis*, and *Couchia*. *Aphanomycopsis sexualis* and *Couchia circumplexa* are pathogens of the eggs of midges (Chironomidae) (Martin 1975b, 1981). They play a role together with chytrids in the epizootics of midges (Martin 1984).

Leptolegnia chapmanii infects mosquitoes (McInnis and Zattau 1982; Seymour 1984; Zattau and McInnis 1985). The first and second instar larvae are most susceptible. The zoospores encyst either on the larval external cuticle or in the digestive tract, where they germinate and penetrate into the hemocoel. On the integument, the zoospores attach primarily to the cervical collar, intersegmental folds, anal gills, and anus. The germination tube has a bulbous collar or swelling similar to an appressorium. An established infection is fatal.

VIII. SUBDIVISION ZYGOMYCOTINA, CLASS ZYGOMYCETES, ORDER MUCORALES

The subdivision Zygomycotina has entomopathogens in the classes Zygomycetes and Trichomycetes. The Zygomycetes have two orders, Mucorales and especially Entomophthorales with insect pathogens. In the order Mucorales, the reproduction is by one to many sporangiospores (single-celled aplanospores) formed in terminal sporangia, or less often by conidia, which are usually produced singly (Hesseltine and Ellis 1973). These fungi are typically saprobic, and are less often or rarely parasites.

Sporodiniella umbellata was initially isolated from remnants of insects, but it infects insects in the family Membracidae, which are associated with cocoa plants in Ecuador (Evans and Samson 1977). *Mucor hiemalis* infects young adult bees when they are fed spores at 20°C, but the fungus generally is not a problem since the normal hive temperature is above the tolerance of the fungus (Burnside 1935). In older bees, the spores are destroyed in the digestive tracts. *Mucor hiemalis* produces mycotoxins that cause severe lesions in tissue culture cells of *Leucophaea maderae* (Vey and Quiot 1976).

IX. SUBDIVISION ZYGOMYCOTINA, CLASS ZYGOMYCETES, ORDER ENTOMOPHTHORALES

The Entomophthorales is characterized by a sporangium that has been reduced to function as a single conidium and is usually discharged forcibly at maturity. These fungi are either pathogens of animals (rarely plants) or saprophytes living in soil or dung. The mycelia of these fungi consist of coenocytic hyphae in which transverse walls occur infrequently, whereas, the hyphae commonly possess transverse walls in the subdivisions Ascomycotina, Basidiomycotina, and Deuteromycotina. The hyphal walls are thick and mainly composed of chitin that is colorless or tinted brown. The number of families in this order varies with different authorities of entomophthorous fungi, and this is discussed by Ben-Ze'ev and Kenneth (1982a). At most, three families have been proposed: Entomophthoraceae, Basidiobolaceae, and Ancylistaceae.

A. CLASSIFICATION

The classification of the entomopathogenic genera of Entomophthorales is still in a state of flux (Waterhouse 1973; Remaudière and Keller 1980; Ben-Ze'ev and Kenneth 1981; Humber 1981; King and Humber 1981; McCoy et al. 1988). Drastic revision has taken place in the genus *Entomophthora*, which has been split into four or more genera. McCoy et al. (1988) have reviewed the attempts at the reclassification of the genera of the family Entomophthoraceae (Table 10–2). Initial major revision was made by Batko in a series of papers and summarized by Batko and Weiser (1965). This was followed by the proposals of Remaudière and Keller (1980), Humber (1981), and Ben Ze'ev and Kenneth (1982a,b). The genera in the family Entomophthoraceae include *Entomophthora, Massospora, Triplosporium, Completoria, Strongwellsea, Erynia, Entomophaga,* and *Macrobiotophthora*; in the family Ancylistaceae are *Ancylistes, Conidiobolus,* and *Meristacrum*, and probably, but not yet established, *Ballocephala* and *Zygnemomyces*; in the family Basidiobolaceae is a single genus *Basidiobolus* (Ben Ze'ev and Kenneth 1982a). Some of the doubtful or rejected genera are *Culicicola, Delacroixia, Empusa, Lamia, Myiophyton, Tabanomyces, Zoophthora,* and *Zygaenobia*.

The genus *Entomophthora* was originally described and named by Cohn (1855) as *Empusa*, based on the type species *Empusa muscae*, a pathogen of the house fly. The historical account of the replacement of *Empusa* by *Entomophthora* is discussed by MacLeod (1963) and Waterhouse (1973). *Entomophthora* was at one time the largest genus of Entomophthorales with about 150 entomogenous forms, but it

TABLE 10–2 Recent Reclassifications of the Genera of the Entomophthoraceae[a]

Batko	Remaudiere and Keller	Ben Ze'ev and Kenneth	Humber
Conidiobolus	*Conidiobolus*	*Conidiobolus*	*Conidiobolus*
Entomophaga	—	*Entomophaga*	*Entomophaga*
Entomophthora	*Entomophthora*	*Entomophthora*	*Entomophthora*
Culicicola	—	—	—
Triplosporium	*Neozygites*	*Triplosporium*	*Triplosporium*
Zoophthora subg. *Zoophthora* subg. *Pandora* subg. *Erynia* subg. *Furia*	*Zoophthora* *Erynia*	*Erynia* subg. *Zoophthora* subg. *Neopandora* subg. *Erynia*	*Erynia*
Strongwellsea	—	*Strongwellsea*	*Strongwellsea*
—	—	*Meristacrum*	*Meristacrum*

[a]Courtesy of McCoy et al. (1988) and CRC Press.

has been reduced in size and many of its members have been placed in *Erynia*, others are invalid or of doubtful status (Humber and Ben Ze'ev 1981; Ben-Ze'ev and Kenneth 1982a,b). Important nomenclatorial changes are the naming of *Erynia neoaphidis* (syn. *Entomophthora aphidis sensu* Thaxter) (Remaudière and Hennebert 1980) and the synonymy of *Entomophthora virulenta* and *E. thaxteriana* with *Conidiobolus obscurus* (Latgé et al. 1980; Humber 1981).

B. REPRODUCTION

Sexual reproduction in Entomophthoraceae occurs by the fusion of mycelial fragments or hyphal bodies to form thick-walled zygospores or resting spores. Meiosis or reduction division has not been established in the life cycle (Latgé 1976). Asexual reproduction takes place by various means: (1) hypha breaking up into hyphal bodies or the budding of these bodies, (2) primary uni- or multinucleate conidia (primary spores) formed singly at the apex of simple or branched conidiophores, (3) secondary conidia (secondary spores) that are produced from the germination of primary conidia, and (4) tertiary conidia from secondary conidia. Four or more types of secondary spores are found in different species. During asexual reproduction, azygospores (resting spores) that are morphologically similar to zygospores may be formed from thick-walled hyphal bodies or single-walled chlamydospores. Latgé et al. (1982) have described the ultrastructural development during the formation and germination of azygospores. A number of genera produce protoplasts (Humber 1984). In the green peach aphid, *Myzus persicae*, *Erynia neoaphidis* produces protoplasts that do not invade tissues and acquire cell walls when the host becomes immotile (Kobayashi et al. 1984).

C. HOSTS, SIGNS, AND SYMPTOMS

Many members of Entomophthorales cause extensive and decimating epizootics in insect populations. The insect hosts of these fungi occur in over 32 families in the orders Hemiptera, Homoptera, Diptera, Lepidoptera, Coleoptera, Orthoptera, and Hymenoptera (MacLeod and Müller-Kögler 1973). Some species have a wide host range, others are restricted to a single species or a closely related group of species. The larva, pupa, and adult are infected, but in many insects, the adult is the most common stage infected. Entomophthorous species are regarded as the most important pathogens of aphids with over 10 species infecting these insects. Species of five genera are common aphid pathogens: *Conidiobolus*, *Entomophthora*, *Erynia*, *Neozygites*, and *Zoophthora* (Latgé and Papierok 1988).

An insect infected by an entomophthorous fungus generally does not display any obvious signs and symptoms at the early stage of infection. Only after the infection has spread within the body does the insect become sluggish or display a nervous restlessness. At a late stage of infection, the insect loses its power of locomotion and settles down on the underside of a branch or leaf. It may climb to the tops of plants where it dies clasping the stem with its head upwards as in grasshoppers, or the forepart of its body drops and remains suspended by the prolegs as in lepidopteran larvae (Fig. 16–4). This tendency to climb elevated places has been termed the "summit disease" syndrome (Evans 1989). Shortly before death or at death, the body may turn yellowish, but after death, it darkens as in the case of lepidopteran larvae. Subsequently, the cadavers may be covered with fungal mycelium.

The symptomatology may differ in hosts infected with the conidial form as compared with those infected with the resting-spore form. The insect infected with the resting-spore form does not have sporulating hyphae on its body surface, but its body frequently becomes brown or black, then shrivels, and the internal contents may be liquefied (Newman and Carner 1974; Kenneth 1977) (Fig. 16-4). When the infection is limited mainly to the abdomen, the insect survives for some time even in the absence of abdominal structures as in cicadas infected with *Massospora* spp. Such insects are usually sterile.

D. LIFE CYCLE

The generalized life cycle of an entomophthorous fungus is given in Figure 10–4. The conidiophore, in many species, forcibly discharges the conidium, which is covered with a mucilagenous sticky substance (Eilenberg et al. 1986). When the conidium lands on a moist substrate other than a host, it may produce secondary conidiophore and secondary conidium, which may produce tertiary conidiophore and conidium, and this process may continue until the protoplasm is depleted. Secondary and tertiary conidia may become resting spores. A special type of secondary conidium is the capilliconidium (anadhesive or capillispore), which is pro-

FIGURE 10–4

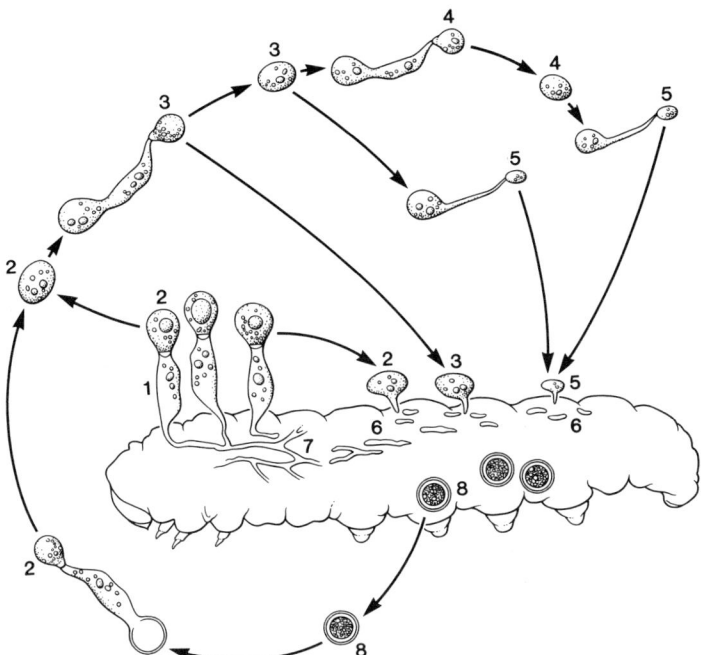

Generalized life cycle of an entomophthorous fungus. Under favorable conditions, the endostroma in the larva produces conidiophore (*1*), which forms primary conidium (*2*), which in turn may form secondary conidium (*3*), and which in turn form tertiary conidium (*4*) in the absence of a suitable host. Capilliconidium (*5*) is formed from other conidia and may be the major infective unit. Hyphal bodies (*6*) are formed in the insect's hemocoel and develop into mycelium and stroma (*7*), which produces resistant or resting spores (*8*). (Drawn by Christina L. Jordan.)

duced on a slender capillary tube (conidiophore) developing on a primary or secondary conidium (Carner 1976, 1980; Nemoto et al. 1979). The surface of the capilliconidium has sticky substances, and the spore adheres to objects with which they come in contact.

In some entomophthorous species, the secondary conidia and capilliconidia are the principal infective forms (Bitton et al. 1979; Kramer 1980a; Glare and Chilvers 1985; Mullens et al. 1987). Some secondary conidia produce long germ tubes (Kramer 1980b). Certain entomophthorous fungi on aquatic insects possess great diversity in conidial morphology and mode of germination. They may produce up to four types of conidia with two major types, two aerial and two aquatic forms (Descals et al. 1981; Descals and Webster 1981, 1984). The aquatic forms are adapted for underwater dispersal and infection. The primary conidium of the aerial type is cornute or horn-shaped; the secondary aerial type is globose, and both are discharged violently. The primary aquatic conidium is coronate, and the secondary is stellate or tetraradiate in form. Both types of aquatic conidia are not discharged violently. The stellate secondary conidium may form a cornute type of conidium possibly for aerial infection. If a conidium lands in water, it germinates and produces a hypha that grows until the protoplasm is used up. When the conidium lands on a susceptible host under favorable conditions, it produces a germ tube that penetrates the insect's integument and enters the hemocoel.

1. Stages within Insect's Hemocoel

Within the insect's hemocoel, the entomophthorous fungi form hyphal bodies, protoplasts, and hyphae. The hyphal bodies are most common and are formed by the germ tube. They multiply by budding. They are large and do not circulate freely in the hemocoel but penetrate into compact host tissues, such as muscle and fat body (Prasertphon and Tanada 1968).

The protoplasts appear early in infection and have amoeboid movement (Nolan 1985). Not all entomophthorous strains, however, produce protoplasts (Latgé and Beauvais 1987). The protoplasts multiply by budding and may occur from the time of infection to shortly before the host's death when they form hyphae (Butt et al. 1981). Under *in vitro* conditions, the formation of the hypha is by the coalition of protoplasts to form spheres or spherical cells that enlarge to become prohyphal spheres, which germinate to form hyphae (Nolan 1985). Some spheres develop thick walls to form resting spores. The mycelial wall of protoplasts is composed of linear glucan associated with low concentrations of chitin and does not contain galactose, uronic acid, and chitosan (Latgé and Beauvais 1987).

The protoplasts do not attract phagocytic hemocytes, as in the case of hyphae and hyphal bodies, and may provide a natural form of resistance against the immune response of the insect (Dunphy and Nolan 1980). The fungus multiplies and the mycelium spreads within the insect. Nearly all of the internal organs are utilized by the fungus and the insect is killed at this stage.

Certain fungi limit their attack mainly to organs in the abdomen. This is the case with *Entomophthora erupta* on the green bug (Dustan 1924); *Entomophaga kansana* on calliphorid, muscid, sarcophagid, and tachinid flies (Hutchison 1962); *Strongwellsea castrans* on the seed corn maggot, *Hylemya cilicrura* (Batko and Weiser 1965); and *Massospora* spp. on cicadas (Speare 1921). In these insects, the infected adults continue to fly actively and disseminate the fungus.

2. Period of Infection

The period from infection to the death of an insect may be as short as 3 days to as long as 12 days, with most deaths occurring at 5 to 8 days (MacLeod 1963). The period may vary with the size of the insect host. Most insects die in the late afternoon between 1500 and 1900 hours. The virulence and pathogenicity of entomophthorous fungi may be associated with the production of specific collagenolytic enzymes (Hurion et al. 1977, 1979) and mycotoxins (Prasertphon 1967; Yendol et al. 1968; Prasertphon and Tanada 1969; Dunphy and Nolan 1982a). The toxin produced by *Conidiobolus thromboides* in the culture medium is 4,4'-hydroxymethyl azoxybenzoic carboxylic acid (Claydon 1978).

3. Sporulation and Germination of Conidia

After death, the insect's body becomes hard and mummified and contains the fungal stroma. Within the stroma, resting spores (zygospores and azygospores) may be produced. Dead insects are generally attached to plants by their appendages or proboscis, or by rhizoids or pseudocystidia produced by the fungus.

When the humidity is very high and at adequate temperatures, the fungus sporulates by sending out reproductive (conidiophores) and vegetative hyphae through the insect's integument, and the cadaver is generally covered by the characteristic mass of fungal mycelium. The formation of conidiophores appears to be associated with the water-loss threshold of the infected cadaver (Millstein et al. 1983a). When large amounts of water are lost from the cadaver through diffusion into the atmosphere, only rhizoids and pseudocystidia are formed but not conidiophores. Certain conidia are discharged forcefully. In an insect attached to a broad leaf or window pane, the expelled conidia may form an aureole around the cadaver. The conidia are short-lived and may not survive over 2 weeks.

Conidiophores generally develop from hyphae within the stroma, but they may be formed directly from hyphal bodies, "prespores," or immature resting spores under conditions of low oxygen and high carbon dioxide atmosphere (Humber and Ramoska 1986). The spores are called cryptoconidia. For example, they are formed in grasshoppers infected with *Entomophaga grylli* when the abdomen is ruptured a few days prior to the insect's death.

Time, temperature, and light regimes affect the sporulation and germination of conidia and their effects may vary with the fungal species (Milner 1981). Generally, there is a diurnal periodicity with most sporulation occurring during the early morning hours when humidity is high and temperatures and light conditions are favorable (Roffey 1968; Wilding 1970; Newman and Carner 1974; Milner et al. 1984). An *Entomophthora* sp. ("grylli-" type) is capable of adjusting its lag period before conidial discharge so that most discharges from cadavers occur during the night, with the longest lag period at 0200 hours (Aoki 1981). In an *Erynia* sp. that infects the alfalfa weevil, shorter periods of showering conidia occur early in the morning (between 0100 and 0300 hours), whereas longer periods of showering occur between 0300 and 0600 hours (Millstein et al. 1982).

Newman and Carner (1975a) made an interesting biological observation with *Tarichium* (syn. *Entomophthora*) *gammae*. This fungus sporulates at relative humidity (RH) 80 to 100% and is inactivated at RH less than 50%. The conidium germinates at RH 98 to 100% and cannot survive below 75% RH. Suitable temperatures for sporulation and germination are from 10 to 26.7°C. From 32.2 to 37.8°C, the fungus cannot survive prolonged exposure. The fungus produces external conidia and internal resting spores in larvae of the soybean looper, *Pseudoplusia includens*. Larvae with the conidial stage die within precise time periods (between 1800 and 2200 hours), while those with internal resting spores die at various times of the day (Newman and Carner 1974). Since unfavorable RH (50%) and temperatures (above 32°C) occur during the summer days in South Carolina, the fungus must complete the cycle of sporulation, conidial germination, and infection of new hosts in the same night. On the other hand, the environmental conditions have no relationship with the types of spores produced by an *Entomophthora* sp. infecting aphids because a proportion of the newly infected aphids that crawl into bark crevices produce mainly resting spores, whereas those that die on leaves in the tree canopy form only conidia (Byford and Ward 1968). In other entomophthorous species, aphids that die on aerial locations form mainly conidia and those that fall onto soil form resting spores (Samson et al. 1988).

4. Resistant or Resting Structures

Entomophthorous fungi produce various types of resistant or resting structures: chlamydospore, zygospore, thick-walled hypha, and resting spore. An exceptional case is *Entomophthora floridana*, which overwinters (5°C) as conidia and does not form resting spores (Brandenburg and Kennedy 1981). The resistant structures are not infective; but under favorable conditions, they germinate and produce infective conidia (McCoy et al. 1988). The chlamydospores develop from small portions of hyphae or hyphal bodies that develop thick walls (MacLeod 1963). These spores germinate readily under favorable conditions. Zygospores are produced when two specific hyphal bodies fuse (Latgé 1976). In fermented cultures, the sexual sporulation occurs at the deceleration phase of the fungal growth, probably caused by a lack of nutritive balance (Latgé 1977). Azygospores are formed from single hyphal bodies, from buds of chlamydospores, interstitially between fungal cells, or at tips of hyphae arising from these structures.

The resistant structures persist during periods of unfavorable environmental conditions and germinate and produce infective conidia under favorable conditions. The overwintering resting spores of *Entomophaga* (syn. *Entomophthora*) *maimaiga* germinate in the spring to initiate epizootics among young gypsy moth larvae (Shimazu et al. 1987) (Fig. 16–4).

a. Formation

The formation of resting spores and the conditions under which they are produced are not completely understood. Low temperatures, short photoperiod, darkness, and unsuitable growth conditions stimulate their production. Such conditions generally occur in fall and winter months when resting spores are produced (MacLeod 1963; Müller-Kögler 1965). Transferring the mycelium, either from an insect or from an *in vitro* culture, to the soil may stimulate resting-spore formation (Latgé et al. 1979).

In semidefined liquid medium with carbon–nitrogen sources held constantly, the greatest number of resting spores is produced under darkness at 25°C with the pH near 6.5 (Latgé et al. 1978b). Resting-spore production is affected by the total concentration of nutrients and the ratio of carbon and nitrogen sources. Starvation in nitrogen and/or carbon compounds also induces resting-spore formation (Latgé 1980). The azygospores of *Conidiobolus obscurus* are produced when the infected dead aphids are held in a saturated atmosphere at 2 to 6°C (Papierok 1978). Under the same conditions but with temperatures equal to or greater than 12°C, the fungus forms conidiophores and conidia. In the case of *Entomophaga maimaiga*, the moisture content of the larval host body determines whether conidia or resting bodies are produced (Shimazu 1987). Under high larval moisture, conidia are formed and resting spores are produced at low-moisture conditions. In other cases, the spore types may vary with the physiological age of an insect (Wilding and Lauckner 1974).

The type of spore forms of *Erynia radicans* produced in the brown planthopper, *Nilaparvata lugens*, is affected by temperature and host stage (Shimazu 1979). At the temperature range from 15 to 25°C, more resting spores are produced than conidia with a decrease in temperature for each stage of the host. In the various host stages, more resting spores than conidia are produced in older than in younger nymphs; in adults than in nymphs; in macropterous than in brachypterous adults; and in the young macropterous female than in the aged female. The mortality of the planthopper is correlated with the production of resting spores by the fungus. In fungal epizootics, the prevalence of conidial forms and of resting-spore forms may vary with the progress of the epizootic and with the season.

b. Germination

In the germination of resting spores, the initial stage is the breakdown of the single large oil droplet into numerous oil droplets (Ohkawa and Aoki 1980). A new wall is formed beneath the endospore layer that gradually disintegrates possibly by enzymatic action. The germ tube perforates the epispore layer mainly by mechanical pressure. In *Erynia* (syn. *Zoophthora*) *radicans*, the epispore at the late stage of germ-tube emergence becomes uniquely roughened and dehisced (Perry et al. 1982). The germ tube forms a tiny spore ("germ conidium") at the tip (Dustan 1924; Soper et al. 1975; Tyrrell and MacLeod 1975). Multiple "germ conidia" are produced in some cases. The germ conidia are ejected forcefully from the germinating hyphae and cause infection when they contact an insect. They may also produce secondary conidia.

Workers have experienced difficulty in germinating resting spores, and this is a critical problem in the manipulation of fungi for insect pest control. The factors affecting germination differ with the fungal species. Numerous attempts were made under conditions of freezing and thawing, drying, heating, addition of nutrients, and by treatment with acids and with chitin-splitting bacteria. The results, however, have been inconsistent suggesting that a quiescent period is necessary before the resting spores are in the proper condition to initiate growth. Generally a small proportion of spores germinate readily (Hall and Halfhill 1959; Wallace et al. 1976). Under long day conditions (14:10 light:dark period), not more than 50% of the resting spores of *Erynia aphidis* germinate, and the branching of germ tubes, the production of germ conidia, and the germination of conidia occur rapidly under such conditions (Wallace et al. 1976; Matanmi and Libby 1976). The optimum temperature for the germination of *E. aphidis* resting spores is about 16.5°C (Payandeh et al. 1978). Wallace et al. (1976) suggest that a photoperiodic control may be involved.

The germination of resting spores is greatly increased with the treatment of a snail-gut enzyme, glusulase (Matanmi and Libby 1976) and with chemicals, such as c-cinnamaldehyde, d-limonene, and n-nonanol with ethanol (Soper et al. 1975). High germination of resting spores occurs when they are shaken in a liquid medium at pH 8.5 and 9.5 for 96 h (Nolan et al. 1976).

5. Conidial and Resting-Spore Generations

The formation of conidia and resting spores may not occur at the same time in an insect infected with an entomophthorus fungus. This may be determined by the moisture content of the insect, with the conidia being formed at higher moisture content and the resting spores at lower moisture content (Shimazu 1987). It may also vary with the wing forms of adults (Shimazu 1979) and with the age of the adult (Wilding and Lauckner 1974; Kramer 1979). In an unusual case, one of the two pathotypes of *Entomophaga grylli* lacks the conidial cycle (Soper et al. 1983). This pathotype also differs from the one exhibiting both the resting and conidial stages by isozyme analysis and may be a different species.

In some fungal species, there is an alternation of conidial forms and resting-spore forms in insect populations. In *Erynia bullata* that infects adult *Sarcophaga aldrichi*, one generation of conidia alternates with one generation of overwintering resting spores (MacLeod et al. 1973). In *Erynia phytonomi*, the conidia are produced in second instar larvae and resting spores in third instar larvae of the alfalfa weevil (Watson et al. 1981). This may explain the quick occurrence and disappearance of fungal epizootics in the alfalfa weevil during the spring months.

Resting spores generally appear toward the end of the active season of the insect (i.e., in late summer or in the fall). The resting spores (zygospores) of *Triplosporium fresenii* germinate in the spring in synchronization with the build up of *Aphis spiraecola* on citrus trees in Israel (Bitton et al. 1979). In populations of *Pseudoplusia includens* infected with *Tarichium gammae*, the number of individuals with conidial forms decreases and that of individuals with resting-spore forms increases as the epizootic progresses (Newman and Carner 1975b). This is not caused by short photoperiod or low temperatures but is correlated with larval size since conidial forms are predominant in small larvae (less than 1.5 cm) and the resting-spore forms in large larvae (greater than 2.5 cm). In locusts infected with *Entomophaga grylli* during the rainy season, 1% of them shows no external growth after dying and contains resting spores in their shriveled, hard cadavers (Skaife 1925). With the cessation of rains, the infected locusts die with no external conidia and their bodies contain numerous resting spores.

E. *IN VITRO* CULTIVATION

The economic potential of entomophthorous species for insect pest control has stimulated the culture of these fungi on artificial media. Unfortunately, some species have not been grown in culture (Gustafsson 1965; Latgé 1981). Early workers had little success and considered these fungi as obligate pathogens. Although later workers succeeded in culturing a number of them on natural media, such as swordfish, pork, egg yolk, or other media rich in protein (Sawyer 1929; Latgé 1981), they are probably obligate pathogens in nature.

Sawyer (1929) attempted to culture entomophthorous species on chemically defined Czapeck's medium but obtained only very feeble growth, if any. Wolf (1951) was probably the first to successfully culture entomophthorous species on a defined chemical medium. He propagated *Conidiobolus apiculata* and *C. coronata* on a dextrose–asparagine–salt synthetic medium. Both species are autotrophic and do not require vitamins and other growth factors. Perry and Latgé (1980) cultured *Conidiobolus thromboides* on a chemically defined medium that contained L-arginine, L-leucine, glycerine, and mineral salts. They obtained high growth and sporulation.

The most common medium for the culture of entomophthorous fungi is egg yolk (Müller-Kögler 1959; Gustafsson 1965). Egg yolk can be replaced by a sunflower oil–yeast extract medium (Latgé et al. 1978a). Prior to culture on egg yolk, the fungus may be more easily isolated in insect-tissue culture medium (Welton and Tyrrell 1975). In such a culture medium, some entomophthorous fungi produce protoplasts (Dunphy and Nolan 1977, 1979; MacLeod et al. 1980). At present, some entomophthorous fungi are grown on commercially available media, Sabouraud maltose or dextrose agars.

F. *MASSOSPORA*

Massospora species differ from most other entomophthorus fungi in that their growths are confined to the insects' abdomens and the conidia are not released until the hosts' bodies are ruptured (Soper 1974). The resting spores are globose, the epispore reticulated, and they are formed from hyphal bodies within the host's abdomen but not in the chambers in which the conidiophores are formed.

There are 13 species of *Massospora* collected from North and South America, southeast Asia, and Australia (Soper 1974, 1981). The type species *M. cicadina* was first isolated from adult 17-yr cicada *Magicicada septendecim* and has been found to infect all species of *Magicicada* in the eastern and midwestern United States (Soper 1974). *Massospora* species appear to infect only gregarious cicadas and none has been found from solitary cicadas. The appearance of disease in the cicada *Okanagana rimosa* is about 12 days (Soper et al. 1976b). Conidia become visible in about 6 days and are mostly gone at 9 days. They remain infectious for about 3 days.

A most intriguing question is how *Massospora* persists in an insect that remains underground for such long periods of up to 17 yr. The fungus does not infect the underground insect stages (Soper et al. 1976a,b). The fungus develops only in the adult population found above ground and persists in the soil as resting spores during the period that the insect lives underground. The cicadas are infected just prior to emergence from the soil when they come in contact with the resting spores. The early stages of infection are not clear. The coenocytic mycelium is confined to the soft posterior portion of the abdomen and destroys the insect's tissues, primarily the reproductive organs.

The mycelium of *Massospora* has cells with two or, rarely, three nuclei. The cells separate into hyphal bodies forming short binucleate conidiophores that usually produce binucleate, though sometimes uninucleate or multinucleate, conidia. As the mycelium develops, the conidiophores are crushed and the conidia lie free in groups in global pockets. The pressure of the fungus causes the terminal abdominal segment of the cicada to break away exposing the creamy-white mass of conidia, which may drop off in a single mass or crumble and fall off in aggregates. As the fungus continues to develop, successive abdominal segments are sloughed off until only the thorax and head remain of the still actively flying cicada, which disperses the conidia in the environment. Healthy cicadas become infected when they come in contact with the conidia or when they attempt to copulate with those bearing sporulating infections (Soper et al. 1976a).

Resting spores of *Massospora* are formed from hyphal bodies which are tetranucleate and larger than those that form conidiophores (Goldstein 1929). The spores (chlamydospores) are not formed within fungal cavities but are distributed throughout the insect's body cavity. They remain viable for at least 9 and up to 17 yr.

The life cycle of *M. cicadina*, in its association with the 17-yr cicada, must be the longest of any fungus. Conidial and resting-spore stages do not occur in the same cicada (Speare 1921; Soper et al. 1976a). In adult cicadas, which become infected with the resting spores upon emerging from the ground, the fungus produces only conidia. The resting spores develop in adults infected with the conidia. Resting spores are found mainly at the end of the season. This suggests that an alternation of conidial and resting-spore generations occurs in *Massospora* as in some other entomophthorous genera. Conidia formed in adult cicadas differ in appearance. The cut-off abdomens of adults with conidial infections display a white-chalky mass of conidia, whereas those infected by resting spores show a sulfur-yellow mass of spores.

Resting spores are tinged with green when young and turn dark brown when mature. They are less coherent than conidia and are scattered much more freely. A cicada infected with resting spores may have an empty cavity extending from its head to abdomen with only two to three segments missing.

In adults, the prevalence of secondary infections producing resting spores is much higher in males than in females, but this is not the case for nymphs (White and Lloyd 1983). The infected female cicadas do not oviposit, although they mate readily and disperse to new habitats (Lloyd et al. 1982). In *Magicicada cassini*, the infected and uninfected males may also disperse en masse, especially when crowded. This serves to spread the fungus, but it also enables certain individuals to escape infection.

G. *STRONGWELLSEA*

The genus *Strongwellsea* was considered doubtful by Waterhouse (1973), but Humber (1976) has presented evidence that it is a valid genus. The fungus was discovered by Strong et al. (1960) from the seed corn maggot, *Hylemya cilicrura*, and described by Batko and Weiser (1965) as *Strongwellsea castrans*. The development of this fungus continues in the live host even after sporulation as in the cases of *Massospora* spp. and *Entomophthora erupta*. Its development, however, differs from that of the others in that it is not restricted to the abdominal hemocoel but extends into the thoracic hemocoel and throughout the nervous system. The fungus neither digests the host's abdominal organs nor causes the disarticulation of the abdominal exoskeletal sclerites. The spores are formed in a deeply concave, internally enclosed hymenium that is unique in this fungus. They are forcibly discharged toward a light source and emerge from an opening on the ventral surface of the abdomen. The life span and behavior of an infected fly are not affected, but sterility may result from infection (McCoy et al. 1988). In addition to the type species, another species, *Strongwellsea magna*, has been described (Humber 1976). *Strongwellsea castrans* has been recorded from several other species in the family Anthomyidae and *S. magna* from *Fannia canicularis* in the family Muscidae.

H. *CONIDIOBOLUS*

Most of the 27 species in *Conidiobolus*, family Ancylistaceae, are saprobic, but about five species have been reported from insects (King 1976a). *Conidiobolus coronata* is a commonly isolated fungus, particularly from the soil (Tyrrell and MacLeod 1972). It infects insects, mites, nematodes, and vertebrates, including humans (King 1976b). *Conidiobolus osmodes* is pathogenic for aphids (Remaudière et al. 1976) and weevils *Hypera* spp. (Ben-Ze'ev and Kenneth 1980).

I. *BASIDIOBOLUS*

Members of the genus *Basidiobolus*, family Basidiobolaceae, are mainly saprobic (i.e., living in the soil, dung, and gut contents of amphibia and reptiles) and a few are pathogens of vertebrates. *Basidiobolus ranarum*, a common isolate from the gut of amphibia, infects the mosquito *Culex pipiens* and has been transmitted to termites and the greater wax moth (*Galleria mellonella*) (Krejzova 1978).

X. SUBDIVISION ASCOMYCOTINA

The subdivisions Ascomycotina, Basidiomycotina, and Deuteromycotina are known as the higher fungi. The ascomycetes are characterized by the sexual stage producing ascospores, usually eight in number, in a specialized cell, the ascus. A few are unicellular (yeasts), but most form complex mycelia (thalli) with fruiting bodies called ascocarps. There are no motile cells and zygospores. The entomopathogenic forms are found in the classes Hemiascomycetes, Plectomycetes, Loculoascomycetes but mostly in the Laboulbeniomycetes and Pyrenomycetes.

A. CLASS HEMIASCOMYCETES, ORDER ENDOMYCETALES

Yeasts or yeastlike forms occur mainly in the ascomycetes, but some are also found in the Basidiomycotina and Deuteromycotina. At one stage in their life cycle, the yeasts are characterized by the existence of single cells that reproduce by budding or fission. In the Endomycetales, the zygotes or single cells are directly transformed into asci. The few entomopathogenic yeasts are in Ascomycotina. Most entomogenous yeasts have been described from *Drosophila* and bark beetles in which they occur as commensals.

Certain yeasts have become dominant flora in the digestive tracts of insects fed artificial diets. In the honey bee, stress conditions caused by treatments with antibiotics, herbicides, infection, or deficient diets result in the proliferation of yeasts in the digestive tracts (Gilliam 1978). Antibiotics fed to larvae of the greater wax moth, *Galleria mellonella*, suppress the typical gut bacterium, *Streptococcus faecalis*, and simultaneously enhance the development of the yeastlike fungi *Candida guilliermondi*, *C. krusei*, and *Geotrichum candidum* (Jarosz 1979).

One of the first yeasts found to be pathogenic for insects is *Metschnikowia* (syn.

Monosporella) *unicuspidata*, which infects larvae of the biting midge, *Dasyhelea obscura* (Keilin 1920). The larval midge commonly lives in the sap that fills the infected wounds of elm or horse-chestnut trees. The elongated fungal hyphae, 30 × 2.5 μm, form asci, in each of which develops a long needle-shaped ascospore with one end sharply pointed. The yeast invades the larval body cavity, multiplies, and turns the midge milky white, particularly the posterior segments. The fat body seems to be the only organ that is completely destroyed. Even with a heavy infection, the larva continues to move but eventually dies.

The yeasts *Saccharomyces cerevisiae* and *S. ellipsoideus*, when fed to adult bees, cause dysentery (Burnside 1930). They are also highly pathogenic when inoculated into the bodies of insects. Evlakhova (1939) isolated a yeastlike fungus that was named *Blastodendrion pseudococci*. This fungus, when smeared on potato stems, infects the mealy bugs feeding on the stems. The fungus spreads from the digestive tract into the fat body and muscular systems. Several species of *Mycoderma* cause infections in insects (Metalnikov et al. 1928; Steinhaus 1949). An unidentified yeast emulsifies the fat bodies of the German cockroach, *Blattella germanica*, and causes the loss of antennal flagellomeres and the formation of uneven wings (Archbold et al. 1986). The syndrome develops over 20 days, and death occurs within 30 days postinfection. The dead insect is dark, flaccid, and has a putrid odor.

B. CLASS PLECTOMYCETES

Members of Plectomycetes have no single character that is unique to the class (Fennell 1973). Their characteristics are the production of closed ascocarps in which globose evanescent asci, borne at all levels, are formed from ascogenous hyphae. Entomopathogenic members occur in family Ascosphaeraceae, order Ascosphaerales (Skou 1972).

An entomopathogenic species is the "chalkbrood" (Kalkbrut) fungus, *Pericystis apis*, which has been renamed *Ascosphaera apis* by Spiltoir and Olive (1955). Of the two morphological forms of *A. apis*, the minor form with small cysts is usually involved in outbreaks of the chalkbrood in the bee colony. The major form with large, stalked cysts is found in secondary cases where the combs have been kept outside the hive (Gilliam 1978). Both forms do not cross with one another. Skou (1972) separated the two forms as distinct species: *A. apis* was retained for the form with small cysts, and *A. major* for the form with large cysts. *Ascosphaera major* may be primarily a facultative pathogen because it thrives on the excrements of bees.

Ascosphaera is dimorphic or heterothallic and requires two forms of mycelia for sexual reproduction in which a trichogyne fuses with a neighboring hypha, a nutriocyte (Spiltoir and Olive 1955). The nutriocyte proliferates to produce asci. Following sporogenesis, the newly developed ascospores are retained within membrane-bound chambers where they form tightly packed, spore-containing spheres, the spore balls (Spiltoir and Olive 1955; McManus and Youssef 1984). The membrane-bounded spore balls form the spore cyst.

Ascosphaera apis infects mainly drone brood and less often the worker brood of the honey bee. The adults are not affected (Bailey 1963). Outbreaks of the fungus are usually not serious (Bailey 1963), but severe losses have occurred in apiaries in California and other parts of the United States (Thomas and Luce 1972; Gilliam 1978).

Infection of *Ascosphaera* occurs when the bee larva ingests fungal spores or may occur through the external body surface (Matus and Sarbak 1974; Vandenberg and Stephen 1982). According to Bailey (1968), the larvae are most susceptible when they are 3–4 days old and then are chilled briefly. The ingested spores germinate particularly in the posterior end of the digestive tract of a chilled larva, and the hyphae penetrate into the tissues. Unless chilled, the mycelium does not develop extensively and is voided in the feces when the larva prepares to pupate (Bailey 1968). Chilling occurs mainly in the peripheral brood where the drones are normally reared. This is why the earlier literature reported that only the drone brood was susceptible. The mummified larvae are white if infected by only one strain (+ or −), but when both strains occur they become dark gray or black through the formation of fruiting bodies that appear as tiny black specks on the larvae.

Some colonies show resistance to chalkbrood. This resistance is partly correlated with the hygienic behavior of removing dead bees (as in the case with the American foulbrood) and with other unknown mechanisms (Milne 1983). The chalkbrood fungus is spread through contaminated food, tools and combs, and through the bees from infected colonies (Gilliam 1978; Hale and Menapace 1980).

A chalkbrood of honey bee in Venezuela is caused by *Arrhenosphaera cranei*, a new genus and new species (Stejskal 1974). *Ascosphaera proliperda* (Skou 1972; Youssef et al. 1984) and *A. aggregata* (Vandenberg and Stephen 1982) cause chalkbrood in solitary bees, *Megachile* spp. Isozyme characterization shows that these two species are distinct from *Ascosphaera apis* (Maghrabi and Kish 1985). The chalkbrood fungi have caused serious losses to solitary bees, which are important for the pollination of alfalfa in western United States (Stephen and Undurraga 1978). Infection in the leaf-cutting bee is similar to that of the honey bee (Youssef et al. 1984). Another fungus probably pathogenic on the leaf-cutting bee is *Microascus exsertus* (Skou 1973).

C. CLASS LABOULBENIOMYCETES

Among the entomogenous orders of fungi, the Laboulbeniales contain the largest number of species that are intimately associated with insects. These fungi, sometimes considered to be commensals, are attached to the insect's cuticle and generally do not seem to cause pathologies. Recent careful studies, however, have shown that a number cause pronounced pathologies in insects (Tavares 1985). Since such pathogenic examples will likely increase in future studies, we shall consider these fungi as pathogens. Most of the pathogenic forms penetrate the insect's cells or hemocoels to obtain their nutrients.

1. Structure

Laboulbeniales are minute, inconspicuous fungi that are named after the French entomologist Alex. Laboulbène, who was among the earliest to recognize these fungi (Benjamin 1973). Laboulbeniales have been considered as abnormal hairs or worms attached to the insect's integument. Many of them occur in restricted or specific sites on the insect or only on one sex. The position specificity may be correlated with the mating or other behavioral patterns of the host (Whisler 1968). Roland Thaxter prepared five monumental monographs between 1896 and 1931 in

FIGURE 10–5

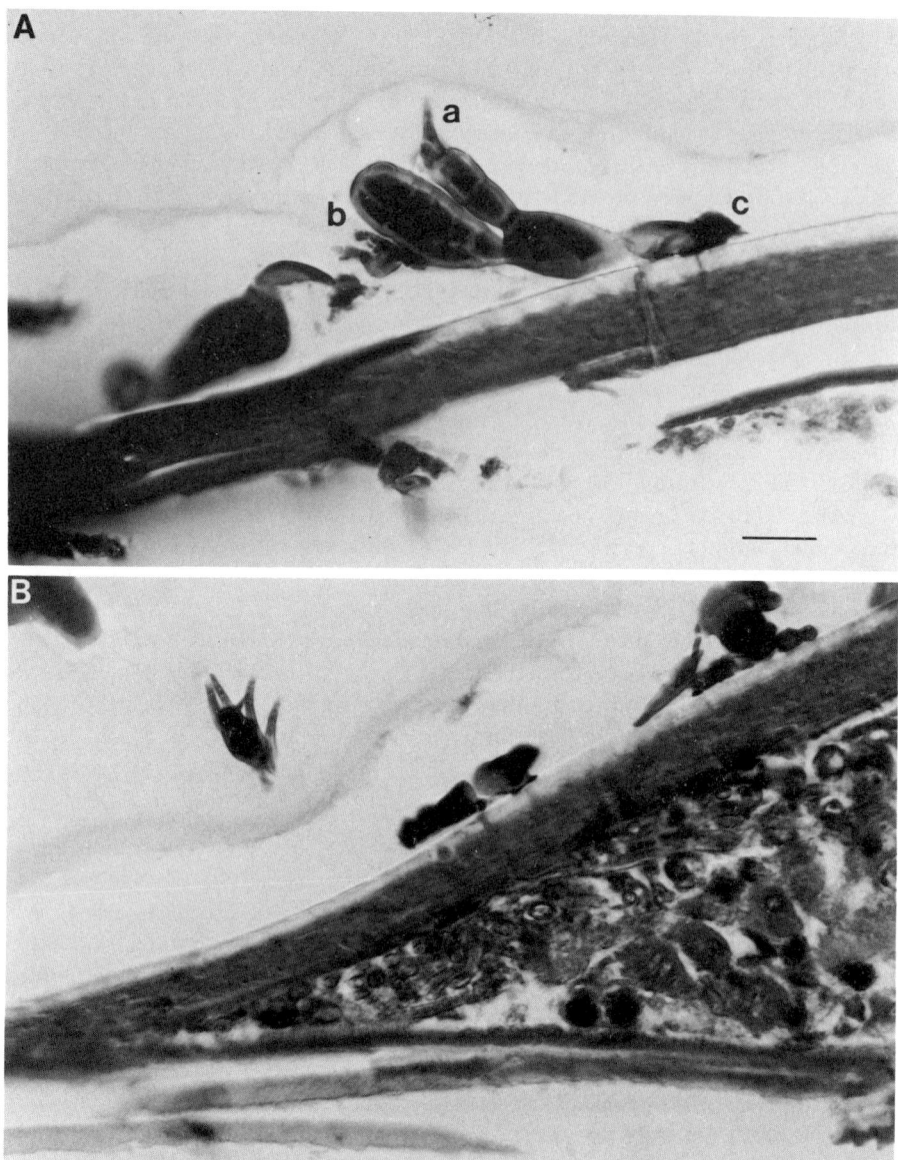

Section of prothorax of a coccinellid beetle, *Cycloneda sanguinea*, showing the infection by the laboulbeniales *Hesperomyces virescens*. (A) Thallus of fungus attached to the cuticle by the foot (*c*); note the haustorium penetrating the cuticle. (*a*) Appendage. (*b*) Perithecium. (B) Haustoria emerging from the foot and extending into the insect's hemocoel. (Bar = 20 μm.) (Provided by I. Tavares.)

which about 1340 species were considered. Extensive reviews on Laboulbeniales are by Benjamin (1971) and Tavares (1985).

Some Laboulbeniales are simple in structure with a perithecium containing ascospores, a thallus consisting of the foot and receptacle, and appendages with simple antheridia that produce spermatia (Fig. 10–5A). Others are highly complex with many primary and secondary appendages arising from a complex receptacle. The thalli develop only from ascospores and no anamorphs (conidial stages) are known (Tavares 1985). The foot is a suckerlike organ attached to the cuticle of the insect's integument. From the lower surface of the foot of some species, a haustorium arises and penetrates into the living cells of the hypodermis (Fig. 10–5A,B). The haustorium may consist of a simple or a branched rhizomycelium penetrating deeply into the host tissues.

2. Hosts

The Laboulbeniales are associated with insects, mites, and millipedes (Benjamin 1973; Tavares 1985). They are found mainly on adults, but a few occur on immature stages. The most common hosts are Coleoptera; they also occur in 11 other insect orders: Phthiraptera (suborder Anoplura), Orthoptera, Blattodea, Dermaptera, Diptera, Hemiptera, Hymenoptera, Isoptera, Mallophaga, Orthoptera, and Thysanoptera. The hosts occur in almost every known habitat (e.g., water, soil, decomposing plant and animal remains, foliage and flowers, and even bodies of living animals, including mammals and birds).

3. Replication and Pathogenicity

Laboulbeniales are obligate pathogens, and none have been cultured through a complete life cycle. Whisler (1968) has managed to grow *Fanniomyces* (syn. *Stigmatomyces*) *ceratophorus* up to the 20-cell stage, but he has not obtained mature perithecia. On an insect, the maturation of the fungus and ascospore production presumably take place only after the establishment of haustorial or other contact with the living cells of the hypodermis or fluid or tissue within the hemocoel. The developmental time in the laboratory at 24°C takes 10–12 days from ascospore to the formation of mature perithecium (Strandberg and Tucker 1974).

The fungi are transmitted by direct contact between infected and susceptible individuals, although some transmissions in the soil with burrowing insects occur (Arwidsson 1946; Lindroth 1948). Habitats where the hosts congregate in confined areas favor the transmission of Laboulbeniales.

Laboulbeniales with extensive rhizomycelium may cause severe injuries. *Trenomyces histophthorus* (originally spelled *histophtorus*) invades the body cavity and destroys the fat body and skeletal muscles of the chicken lice (Mallophaga) (Meola and DeVaney 1976). Most host cells in contact with the haustorial tubes become degenerated (Meola and Tavares 1982). In some cases, the host tissues deteriorate in advance of the haustorial tips, suggesting the production of enzymes by the fungus in this area.

Hesperomyces virescens infects the coccinellid beetle, *Chilocorus bipustulatus*, up to epizootic levels of 95% infected beetles and causes premature mortality (Kamburov et al. 1967). In the predaceous earwig, *Labidura riparia*, the infection by the fungus *Filariomyces forficulae* reduces the life span and causes differences in egg-production patterns of the insect (Strandberg and Tucker 1974). A heavy infection of Laboulbeniales on the mouthparts of the digging beetles, *Bledius* spp., inhibits feeding and results in high mortality of the beetles (Bro Larsen 1952).

D. CLASS PYRENOMYCETES, ORDER SPHAERIALES, FAMILY CLAVICIPITACEAE

The class Pyrenomyctes have unitunicate, club-shaped, or cylindrical asci in a closed ascocarp with an ostiole. The shape of the ascocarp is generally a globose or flask-shaped perithecium. Müller and von Arx (1973) grouped the Hypocreales, Diaporthales, Xylariales, and Clavicipitales together into the order Sphaeriales because there is no general agreement on the definition and delimitation of these orders. The family Clavicipitaceae contains four entomopathogenic genera, *Cordyceps*, *Podonectria*, *Hypocrella* (syn. *Hypocreella*), and *Torrubiella*. They represent the perfect stage of a number of imperfect fungi (Deuteromycotina).

The genus *Cordyceps* is the most widely known and has the largest number of entomopathogenic species in the Sphaeriales. It has been studied principally from a taxonomic standpoint on over 250 species, nearly all of them being insect pathogens (Kobayashi 1941; Mains 1958; McCoy et al. 1988). These fungi are widely distributed and infect insects in the orders Orthoptera, Isoptera, Hemiptera, Lepidoptera, Coleoptera, Hymenoptera, and Diptera, and also spiders and other fungi. Not much is known regarding their pathogenicity, pathology, and transmission (Samson et al. 1988). Some *Codyceps* spp. are obligately parasitic on ants and are important ant pathogens in the tropical forest ecosystems (de Andrade 1980; Evans and Samson 1982b, 1984). In the ant *Palthothyreus tarsatus*, *Cordyceps myrmecophila* invades only the tissues of the head and thorax (Nnakumusana 1987b). In the ants of the tropical rain forests, infection with *Cordyceps* spp. results in abnormal behavior, such as erratic uncoordinated movements, aggregation around certain trees away from any ant trail or nest, hiding beneath bark and epiphytic plants, and arboreal ants tend to descend and hide, whereas nonarboreal species tend to climb (Samson et al. 1988). Such abnormal behaviors suggest that the pathology of the fungi involves the central nervous system.

1. Historical Aspects of *Cordyceps*

Cordyceps produce unique and striking fruiting bodies that attracted the attention of early workers, and the published records of *Cordyceps* are the earliest of entomogenous fungi (Steinhaus 1956) (Fig. 10–6). The early workers referred to these fungi as "vegetable wasps" and "plant worms" (Cooke 1892; Steinhaus 1956). In 1749 a Franciscan friar, Torrubia, described dead wasps, and out of each wasp, a plant germinated from the belly and grew about five spans high (Steinhaus 1956). This plant is probably a species of *Cordyceps*. Christian Paulinus wrote at the beginning of the eighteenth century, that "certain trees in the island of Sombrero in the East Indies have large worms, attached to them under ground, in place of roots" (Steinhaus 1949).

Cordyceps have been used for many centuries as tonic food, medicine, or aphrodisiac, and in religious ceremony in China and Indonesia (Gee 1918; Hoffmann 1947; Kobayashi 1941, 1977). The ancient Chinese, about 2000 years ago, placed stone effigies of insects with cordyceps in the mouths of their dead hoping to revive them or to prevent decomposition as in the case of the fungal mummified insects (Kobayashi 1977).

Szechwan Province in China is famous as a source of cordyceps. The fungi are expensive and only the middle class or well-to-do can afford them. The Chinese

FIGURE 10-6

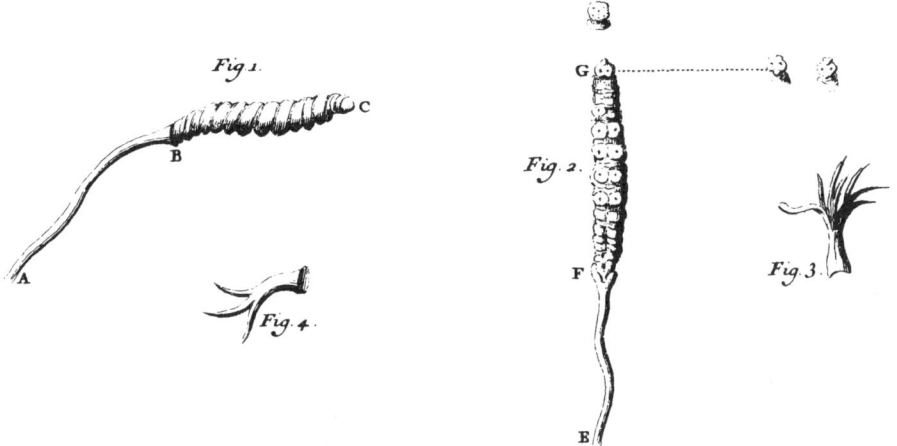

Reproduction of what is apparently the first illustration of diseased insects, *Cordyceps*-infected larvae. [From Steinhaus (1956) and courtesy of University of California Agricultural Experiment Station.]

consider that the larvae infected with *Cordyceps* have a wide-spectrum medicinal value, especially when boiled with pork, and are believed to cure opium habit and poisoning, jaundice, and even tuberculosis (Gee 1918). Even to this day, cordyceps are sold in Chinese markets. Cordycepic acid with the formula $C_7H_{12}O_6$, isolated from cordyceps, is an isomer of quinic acid (Chatterjee et al. 1957) found in cinchona bark from which alkaloids (quinine, etc.) are obtained. Cordycepin, an antibiotic and insect toxin, is used in molecular biology to block RNA synthesis (Samson et al. 1988).

2. Structure of *Cordyceps*

The stromata and synnemata of *Cordyceps* arise from the endosclerotium and emerge usually from the mouth or anus of the insect and grow toward the light source (Fig. 10–7B). They consist of compact bundles of longitudinal parallel or interwoven hyphae, and the perithecia develop on the upper part, which is often clavate or capitate and bearing the ascospores (Mains 1958). The fruiting structures may extend to 30 cm, and are simple or branched and colored yellow, orange, red, brown, ochraceous, grey, green, or black. Perithecia contain long, narrow asci with multiseptate ascospores that break at maturity into one-celled segments. Other reproductive structures are slime spores dispersed by rain, dry conidia scattered aerially, and mucoid spore balls for long-term survival (Evans 1989). Old insect cadavers are capable of regenerating fruiting bodies.

The conidial stage has not been established for most species, and only a few have been grown in culture (Mains 1958). The conidial or imperfect stages of some *Cordyceps* are species of *Spicaria*, *Sporotrichum*, *Cephalosporium*, *Hirsutella*, and others in the Deuteromycotina.

FIGURE 10–7

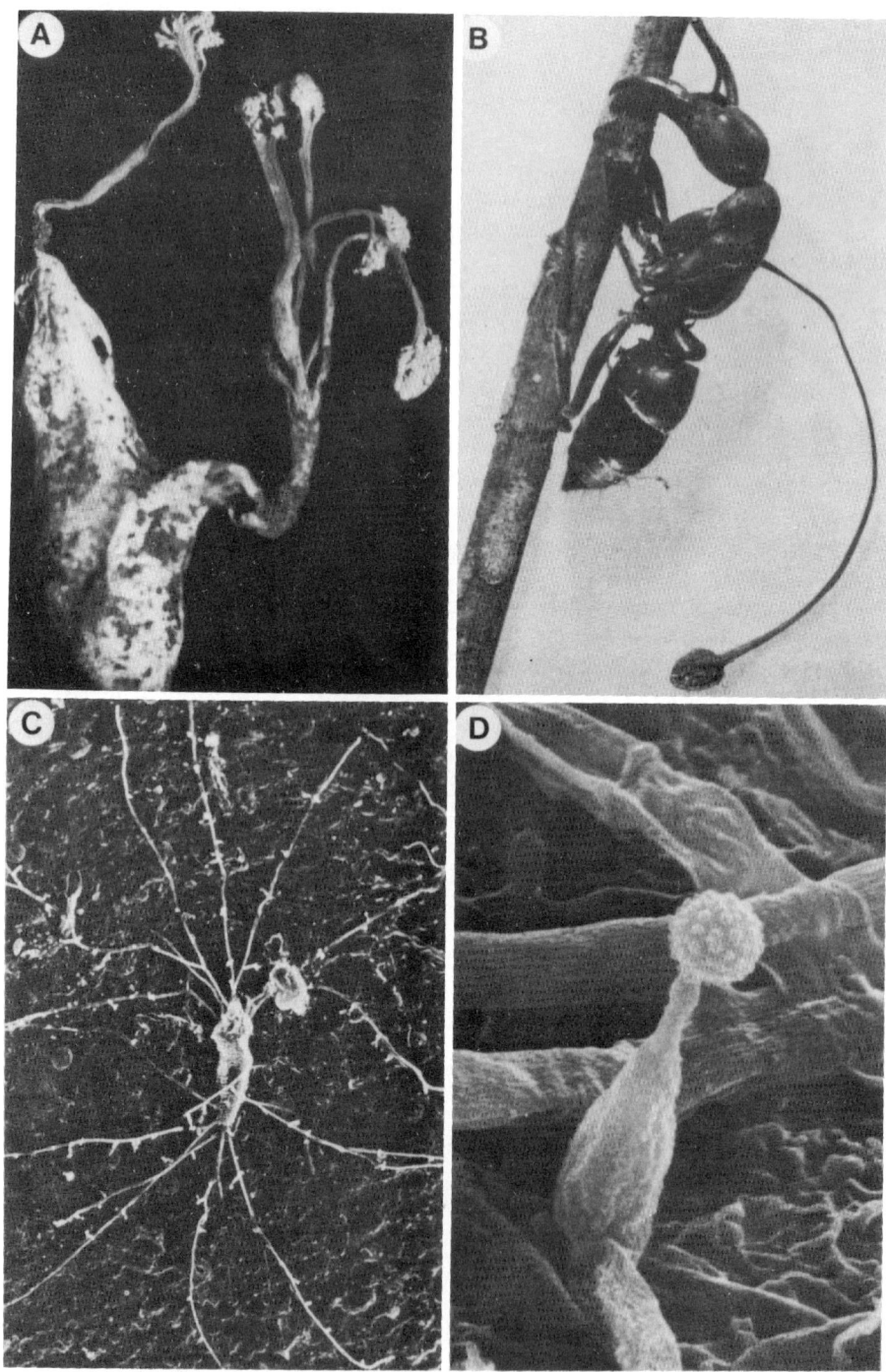

Insects infected with fungi. (A) *Paecilomyces tenuipes* on a lepidopteran larva showing synnematal growth and spore bloom at apex (magnification × 2.45). (B) *Cordyceps australis* on the ant *Paltothyreus tarsatus* (magnification × 5). (C) *Hirsutella thompsonii* on a citrus rust mite, *Phyllocoptruta oleivora*, showing sparse mycelial growth (magnification × 147). (D) Typical phialide and verrucose, globose conidium of *H. thompsonii* (magnification × 4900). [Courtesy of McCoy et al. (1988) and CRC Press.]

3. *Cordyceps militaris*

Cordyceps militaris is the most common species on insects, particularly the pupae of Lepidoptera (Mains 1958). It is widely distributed throughout the world (i.e., in North and South America, Europe, and Asia). Its form and size vary with the size and type of host. The imperfect stage has been considered to be *Paecilomyces farinosus* (syn. *Isaria farinosa*), but it is probably a species of *Cephalosporium* (Petch 1936; Kobayashi 1941; Brown and Smith 1957).

E. FUNGI OF SCALE INSECTS

In the families Clavicipitaceae and Hypocreaceae of order Sphaeriales and the family Myriangiaceae of class Loculoascomycetes, there are members that are considered to be important pathogens of scale insects. Those who are interested in fungi on scale insects should refer to Petch (1921). In the early twentieth century, some of these fungi were called the "friendly fungi" and were recommended for the control of scale insects in Florida (Watson and Berger 1937; Steinhaus 1956). These were the redheaded scale fungus (*Nectria aurantiicola*), pink scale fungus (*Nectria diploa*), whiteheaded scale fungus (*Podonectria coccicola*), and the black scale fungus (*Myriangium duriaei*). *Aschersonia aleyrodis*, *A. goldiana*, and the crimson fungi (*Verticillium cinnamomeum*) in the Deuteromycotina were also included. The successful control of citrus pests by these fungi has been questioned (Fisher et al. 1949; Fisher 1950; Samson et al. 1988).

Species of *Hypocrella*, *Podonectria*, and *Torrubiella* infect scale insects and white flies (Aleyrodidae). Some members of *Hypocrella* have their conidial forms in *Aschersonia* (Petch 1921). Some species of *Aschersonia*, which are considered to be pathogenic to soft scale insects, are the Cuban *Aschersonia* (*A. cubensis*) and the turbinate fungus (*A. turbinata*) (Watson and Berger 1937). The white-halo fungus, *Cephalosporium lecanii*, infects the green scale (*Coccus viridis*), brown soft scale (*Coccus hesperidum*), the hemispherical scale (*Saissetia hemisphaerica*), and many others.

F. ORDER SPHAERIALES, FAMILY HYPOCREACEAE

Two hypocreaceous fungi, *Cordycepioideus bisporus* and *C. octosporus*, infect termites (Blackwell and Gilbertson 1981). The genera *Nectria* and *Calonectria* occur on scale insects and are included with the friendly fungi. The imperfect stages of *Nectria* are in the genus *Fusarium* (Booth 1971; Rossman 1983).

G. CLASS LOCULOASCOMYCETES, ORDERS MYRIANGIALES AND PLEOSPORALES

The Loculoascomycetes is primarily characterized by the bitunicate asci borne in specialized stromatic locules (Luttrell 1973). Among the several orders of Loculoascomycetes, the orders Myriangiales and Pleosporales have entomopathogens. The order Myriangiales has asci scattered usually at several levels in a relatively unaltered stroma. The globose thick-walled asci are exposed individually through the crumbling of stromal tissue, and discharge their spores at the surface. This is a relatively small order containing epiphytes, parasites or hyperparasites on superficial fungi, or pathogens of scale insects. It has two families, Myriangiaceae and Saccardiaceae, with entomopathogens.

The genus *Myriangium*, family Myriangiaceae, has the most complex stroma in this order. Some members are found on stems or leaves, but others are considered to be important pathogens of scale insects throughout the tropics (McCoy et al. 1988). The mycelium of *Myriangium* forms a black cushion-shaped mat covering one or more scales; no imperfect form has been reported. Evlakhova (1974) has prepared a systematic key to the Myriangiales pathogenic on scales.

The family Saccardiaceae has asci that are restricted to a single, irregular layer in a discoid to crustose stroma (Luttrell 1973). The genus *Angatia* resembles *Myriangium* and has been reported from scale insects (Von Arx 1963; Luttrell 1973).

The order Pleosporales has asci interspersed with pseudoparaphyses or with paraphysoidal hyphae, which form an epithecium in usually middle-sized to large locules, and with a perithecium. The genus *Podonectria* has species infecting scales, with *P. coccicola* being the most common (Rossman 1978). The associated anamorphs are species of *Tetracrium* and *Tetranacrium*.

XI. SUBDIVISION BASIODIOMYCOTINA

The Basidiomycotina is characterized by the formation of basidiospores (usually four in number) from a specialized cell, the basidium. The fruiting bodies are called basidiocarps.

Only a small number of Basidiomycotina are true insect pathogens. The genera *Septobasidium* and *Uredinella* are found in Septobasidiales and both are associated with scale insects (Evans 1989). The anamorph (conidial state) for *Septobasidium clelandii*, found on female coccids, is *Harpographium corynelioides* (Coles and Talbot 1977) and for *S. pilosum*, infecting diaspine scales, is *Aegerita webberi* (Evans and Samson 1982a). The genus *Septobasidium* shows morphological similarity with the rust fungi (Dykstra 1974). We have discussed in Chapter 2 the relationship of *Septobasidium* and scale insects. At the level of a scale colony, the relationship is mutualistic, but the fungus is a pathogen at the level of the individual scale. *Septobasidium* and *Uredinella* differ in the structure of the epibasidium and their relationships with their hosts. *Septobasidium* parasitizes some members of the whole colony whereas *Uredinella* parasitizes individual insects. Both fungi form fungal mats over the insects. The infected insects generally are not killed but are sterile.

XII. SUBDIVISION DEUTEROMYCOTINA

The Deuteromycotina or Fungi Imperfecti are imperfect in the sense that most of them are recognized by their conidial (anamorph) forms since the sexual forms occur rarely or are unknown. In a hierarchical classification, these fungi are not equal to the Ascomycotina and Basidiomycotina but are subsidiary both taxonomically and by nomenclature (Ainsworth 1973). The imperfect fungi whose sexual forms have been recognized are the anamorphs of mainly Ascomycotina and a few Basidiomycotina. The imperfect state may also be associated with the Mastigomycotina and Zygomycotina (Hughes 1971). Mycologists believe that many of these fungi have lost the ability to reproduce sexually. They have developed, however, nonconventional means, such as parasexual recombination, in which nuclear fusion occurs but no proper meiosis.

When the sexual or perfect forms are known, the imperfect fungus has two names, the valid name based on the perfect stage and the imperfect name based on the conidial stage. On this basis, the imperfect fungi, even within the same genus, may be in widely separated perfect genera. The imperfect name is preceded by the word "form," thus there are form species, form genus, form family, form order, and form class. Generally, the prefix form is understood and omitted. The species included in a form genus are grouped together by the character of their conidia and conidiogenous apparatus but not necessarily by phylogeny.

The entomopathogenic Deuteromycotina are found in two classes, Hyphomycetes and Coelomycetes. Many are highly virulent pathogens and have been applied for the control of insect pests. We shall be concerned only with those that show promise in microbial control. Unlike many Entomophthorales, the imperfect fungi are easily cultured on most media used to cultivate fungi. Although the entomopathogenic Deuteromycotina usually do not have a specialized resting or overwintering structure, their thin-walled conidia are capable of surviving in the soil (Samson et al. 1988).

A. CLASS HYPHOMYCETES

A number of fungi in the class Hyphomycetes cause muscardine diseases in insects. The term "muscardine" apparently originated in the Italian language with the word "moscardino" meaning a musk comfit, grape, pear, and the like (Steinhaus 1949). Since insects infected with certain fungi become completely covered with mycelia and resemble in appearance comfits or bonbons, the French referred to such fungus-diseased insects as "muscardin." The term was first applied to the white muscardine of the silkworm caused by *Beauveria bassiana*. There are also other muscardines: the green muscardine caused by *Metarhizium anisopliae*, the yellow muscardine by *Aspergillus flavus* and *Paecilomyces farinosus*, and the red muscardine by *Sorosporella uvella* and *P. fumosa-roseus* (syn. *Spicaria fumoso-rosea*).

The genus *Isaria* previously contained a diverse assembly of species that have been transferred to other genera, such as *Beauveria*, *Metarhizium*, *Nomuraea*, *Paecilomyces*, and even to new genera.

1. *Beauveria* Species

The white muscardine was the first disease in animals shown to be caused by a fungus or any microorganism (see Chapter 1). Bassi de Lodi demonstrated the contagious and pathogenic nature of the fungus infecting silkworms and also developed measures for controlling the disease (Ainsworth 1956). In his honor, Balsamo described and named the fungus *Botrytis bassiana*. Vuillemin (1912) created the genus *Beauveria* and selected *bassiana* as the type species. The readers are encouraged to read the early historical accounts of the white muscardine narrated by Steinhaus (1949).

a. Structure and Strains

In the genus *Beauveria*, the conidiophore is characterized by bearing hyaline conidia singly on zig-zag rachis or sterigmata (Fig. 10–8D). Until 1954, a large number of species were described in the genus *Beauveria*, and MacLeod (1954) reduced 14 of these species to two: *B. bassiana* in which globose and oval spores occur in

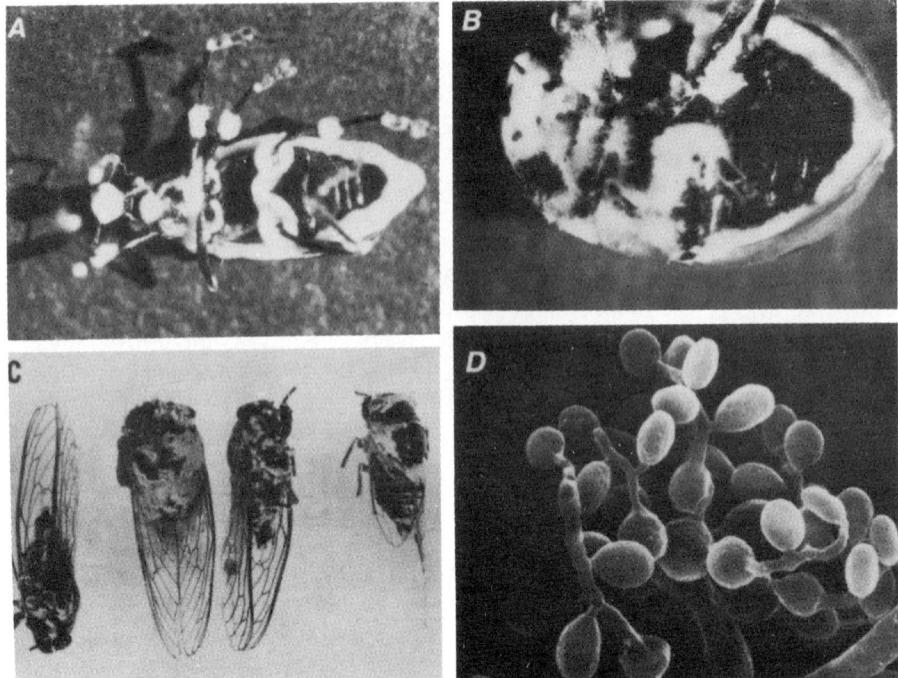

Examples of infections of the white-muscardine fungus, *Beauveria bassiana*. (A) Adult beetle, *Diaprepes abbreviatus*. (B) An unidentified ladybird beetle. (C) Adult cicadids. (D) Conidiogenous structures of *B. bassiana* depicting the typical denticulate conidia on an elongated zig-zag-shaped rhachis (magnification × 6164). [Courtesy of McCoy et al. (1988) and CRC Press.]

almost equal proportions and *B. tenella* with almost entirely oval spores. The two species are serologically distinct (Fargues et al. 1974). In 1972, De Hoog included *B. tenella* in *B. brongniartii*. Samson and Evans (1982) have described two other species, *Beauveria* (syn. *Isaria*) *amorpha* on Coleoptera in Brazil and *B. velata* on lepidopteran larvae in Ecuador.

The strains of *B. bassiana* exhibit considerable variations in virulence and pathogenicity. Such variations have been observed by Fargues (1972) with seven biotypes of this fungus. The strains of *B. bassiana*, *B. brongniartii*, and *M. anisopliae* can be distinguished by zymograms of intracellular and exocellular enzyme activities (Duriez-Vaucelle et al. 1981).

b. Hosts and Replication

Beauveria bassiana occurs worldwide. It has one of the largest host list among the imperfect fungi and also occurs in soil as a ubiquitous saprophyte. The hosts are mainly Lepidoptera, Coleoptera, and Hemiptera, but others occur in Diptera and Hymenoptera (Fig. 10–8A,B,C). This fungus has been found infecting the lungs of wild rodents, nasal passages of humans, horses, and giant tortoises (McCoy et al. 1988). Some of the major economic insect pests that are susceptible to this fungus are the European corn borer, *Ostrinia nubilalis*; the codling moth, *Laspeyresia* (syn. *Cydia*) *pomonella*; the Japanese beetle, *Popillia japonica*; the Colorado potato beetle, *Leptinotarsa decemlineata*; the chinch bug, *Blissus leucopterus*; and the European cabbageworm, *Pieris brassicae*.

The germination of the conidia of *B. bassiana* on the integument and in culture medium is dependent on certain nutritional requirements (Smith and Grula 1981; Hunt et al. 1984). The germinating conidia require a utilizable source of carbon, such as glucose, glucosamine, chitin, and starch, and also a nitrogen source is needed for hyphal growth. In submerged cultures, *B. bassiana* produces conidia or blastospores or both spore types depending on the nutritional content of the media (Thomas et al. 1987). Treatment of the blastospores with zymolase forms protoplasts (Kawamoto and Aizawa 1986).

c. Pathogenicity

Infection generally takes place through the integument. In the fire ant, *Solenopsis richteri*, however, 37% of the conidia fed to larvae germinated in the digestive tract within 72 h, and the hyphae penetrated the gut wall between 60 and 72 h (Broome et al. 1976). The injury to the digestive tract enabled the digestive juices to enter the hemocoel and change the hemolymph pH. The white-muscardine fungus also invades preferentially the larval digestive tract of the corn earworm, *Heliothis zea*, to cause starvation and nutrient depletion that may be involved in larval death (Cheung and Grula 1982).

Beauveria bassiana produces detectable mycotoxins in culture media. When inoculated into silkworm larvae, the toxins cause swelling and stiffening (Kodaira 1961b). Some toxins are proteases of high molecular weights, and they either directly damage the principal functions of the hemolymph or cause damage indirectly by producing a toxic by-product in the insect (Kučera and Samšiňáková 1968). One of the mycotoxins is beauvericin, an antibiotic that is highly toxic to the brine shrimp (*Artemia salina*) and moderately inhibiting to mosquito larvae (Hamill et al. 1969). Beauvericin causes alteration in the nuclei and affects cell migration in cell lines of *Leucophaea maderae* (Vey et al. 1973). It is not produced, however, in sufficient quantities to be involved in the pathogenesis of *B. bassiana* on *Heliothis zea* larva (Champlin and Grula 1979). Moreover, when the toxin is inoculated at 6 μg, it did not induce any deleterious effect. The cyclodepsipeptide, bassianolide, produced by *B. bassiana* and *Verticillium lecanii* is toxic to the silkworm when inoculated into the hemocoel or when fed to the larvae (Isogai et al. 1978; Kanaoka et al. 1978; Murakoshi et al. 1978). The secondary metabolites, which are antibiotics, may prevent bacterial putrefaction and thus permit fungal mummification of the insect (McCoy et al. 1988).

The composition of nutrient media influences the production of mycotoxins by *B. bassiana*. The best media for the production of the toxic proteolytic complex contain maize meal, yeast extract, and leaf extract (Kučera 1971). The presence of fructose in the medium enhances the virulence of *B. bassiana* (Samšiňáková and Hrabétová 1969).

2. *Metarhizium* Species

The green-muscardine fungus, *Metarhizium anisopliae*, is as commonly and as widely distributed as *B. bassiana* and with a wide host range. In 1879, Metchnikoff isolated the fungus from the beetle *Anisoplia austriaca* and suggested its use as a microbial agent against insect pests (Steinhaus 1949). Tulloch (1976) studied a number of species in this genus and concluded that only two species, *M. anisopliae* and *M. flavoviride*, were acceptable species. He placed the two entomopathogenic species, *M. album* and *M. brunneum*, as synonyms of *M. anisopliae*. Strains of *M. anisopliae* have been characterized by their production of esterases (de Conti et al. 1980). *Metarhizium flavoviride* also offers promise in microbial control (Ferron 1975).

a. Characteristics

The colony of *M. anisopliae* appears white when young, but as the conidia mature, the color turns to dark green. The conidiophores are branched, and the initial conidium is produced by simple abstriction at the distal end of the conidiophore. A chain of conidia is formed on each conidiophore with the youngest conidium being adjacent to the conidiophore. The mass of spore chains becomes so dense and coheres with each other to produce prismatic masses of columns of spore chains. Other strains of *Metarhizium* form different colored colonies: *album* produces white colonies and *brunneum* produces yellow to brown colonies.

Metarhizium anisopliae has two types, the short-spored form, *M. anisopliae* var. *anisopliae* (conidia 3.5–9.0 μm), and the long-spored, *M. anisopliae* var. *major* (conidia 9.0–18.0 μm) (Tulloch 1976). Strains of the variety *major* are relatively homogeneous, but those of variety *anisopliae* are very heterogeneous when tested with isoenzymes (Riba et al. 1986). These species differ in virulence and may also infect different hosts. *Oryctes* species are susceptible mainly to the *major* form (Diomandé 1969; Ferron et al. 1975; Latch 1976). On the other hand, the *major* form isolated from other beetles is not pathogenic to *Oryctes* except by intrahemocoelic infection (Diomandé 1969; Ferron et al. 1972; Fargues and Robert 1978). After the initial infection in *Oryctes*, such strains may become pathogenic when applied to the integument. Isolates of the *major* form generally are avirulent to mosquitoes (Daoust and Roberts 1982). A highly virulent mutant strain whose LD_{50} was one half that of the wild type and whose LT_{50} was one day faster was isolated by Al-Aidroos and Roberts (1978). This strain has promise in microbial control because it has early, dense sporulation, rapid *in vitro* spore germination, and enhanced production of mycotoxins. This strain degrades starch much faster than the less virulent strains (Al-Aidroos and Seifert 1980). Mutants of *M. anisopliae* selected for enhanced production of amylase are hypervirulent for larvae of the mosquito *Culex pipiens* (Robert and Messing-Al-Aidroos 1985).

b. Mycotoxins

A number of secondary metabolites act as mycotoxins and are produced by entomopathogenic fungi (Roberts 1981). Cultures of *Metarhizium anisopliae* contain the cyclodepsipeptides, destruxins A, B, C, D, and E, and desmethyldestruxin B (Kodaira 1961a, 1962; Tamura et al. 1964; Suzuki et al. 1966, 1970). Destruxins have been considered as new generation insecticides (Vey et al. 1987). They cause tetanic paralysis when inoculated into larvae of *Galleria mellonella* (Roberts 1966a,b, 1969). They are also produced in fungus-infected larvae and are important in the development of symptoms (Roberts 1966b; Suzuki et al. 1971; Vey et al.

1986). The rapid production of destruxins in the larvae causes death. The destruxins are toxic to insects only by ingestion and not through the integument (Farques et al. 1986).

These toxins affect not only lepidopteran but also dipteran larvae, including mosquito larvae (Roberts 1970; Vey et al. 1987). The toxins are released by the digestion of spores in the larval mosquito midgut (Crisan 1971). Cytopathology occurs in the midgut cells with changes in the mitochondria and endoplasmic reticulum and a strongly pycnotic nuclei (Vey et al. 1987). No tissue lesions occur at the neuromuscular level. Symptoms of intoxication and the onset of death of the mosquito larva are correlated with the total number of spores ingested by the larva. The secondary metabolites (e.g., destruxin E) may act as immunosuppressants, inhibiting the cellular and/or humoral-host defense response (Vey et al. 1985, 1986). The green-muscardine fungus also produces toxic proteolytic enzymes (Kučera 1980). Inhibitors against the proteases occur in the insect hemolymph (Kučera 1982, 1984).

Quiot et al. (1980) have made an unusual observation that destruxin E has antiviral activity against the nuclear polyhedrosis virus when inoculated at subtoxic level into virus-infected larvae of *Galleria mellonella*. The toxin appears to interfere at the site of viral synthesis and not directly on the virus itself.

3. *Paecilomyces* Species

Paecilomyces farinosus was placed at one time in the genera *Isaria* and *Spicaria* (Brown and Smith 1957; Samson 1974). It is commonly found in nature and has a wide range of hosts, mainly lepidopteran larvae. Its infection is sometimes called the yellow muscardine. Samson (1974) revised the genus *Paecilomyces* and recognized 31 species. Some other entomopathogenic fungi, besides *P. farinosus*, in this genus are *P. fumosa-roseus*, *P. amoeneroseus*, *P. javanicus*, *P. ramosus*, *P. coleopterorum*, *P. tenuipes*, *P. cicadae*, *P. lilacinus*, and *P. cinnamomeus*. Perfect stages of *Paecilomyces* spp. are in the ascomycetous genera *Byssochlamys*, *Talaromyces*, *Thermoascus*, and possibly *Torrubiella*.

4. *Nomuraea* Species

Samson (1974) described the genus *Nomuraea* in detail and retained the species *N. rileyi* and *N. atypicola*. These fungi form, on an insect, a dense white mat that, upon conidiogenesis, turns pale green or purple (McCoy et al. 1988). *Nomuraea atypicola* occurs in subtropics and tropics infecting mainly spiders (Samson and Evans 1977).

Nomuraea rileyi was at one time assigned to the genus *Spicaria* (Kish et al. 1974). *Spicaria prasina* is a synonym of this species. The fungus infects a number of lepidopteran larvae, in particular, important insect pests, such as *Heliothis zea*, *H. virescens*, and *Trichoplusia ni*. Exposure of the infected larvae to light is essential for maximal conidial production (Glare 1987). *Nomuraea rileyi* produces a mycotoxin that is toxic to the larva of the gypsy moth, *Lymantria dispar*, when applied topically and intrahemocoelically (Wasti and Hartmann 1978). Epizootics by this fungus often develop late in the growing season of crops. On soybean, the source of infection is the spores that persist on the leaves (Ignoffo et al. 1975).

5. *Sorosporella uvella*

The infection by *Sorosporella uvella* is called red muscardine, but unlike other muscardines, the fruiting bodies do not cover the exterior of the dead insect, and instead the cadaver becomes filled with reddish-colored spores (Speare 1917, 1920). The early taxonomic revision of this genus was reviewed by Steinhaus (1949). The fungus infects mainly noctuid larvae, in particular cutworms and armyworms, but other lepidopteran larvae and beetles are also susceptible. The vegetative stage is yeastlike cells or "blastocysts."

An infected larva, shortly after death, turns from creamy white to pink and increasingly red as the fungus completes its development. The cadaver remains soft and pliable, not hard and sclerotium-like as in most fungal infections. When opened, the insect's body does not exude any fluid, but the internal mycelium is a coherent mass of a shiny creamy or pink color, and of a gelatinous consistency. Later, the body shrinks, the red color increases in intensity, and the body wall is brittle and easily ruptured. Fungal spores within the cadaver are brick red, dry, and dustlike. Thus, the cadaver appears as a small sac filled with spores.

6. *Verticillium*

The genus *Cephalosporium* has been placed in synonymy with *Verticillium*, but Bałazy (1973), who studied and revised the species in *Cephalosporium*, concluded that the two genera were different. He selected to use *Cephalosporium lecanii*, but recent authors apply the name *Verticillium lecanii* (Hall 1981; McCoy et al. 1988; Samson et al. 1988). *Veticillium lecanii* is a common pathogen of scale insects in tropical and semitropical environment. It is known as the "white-halo" fungus because of the appearance of the white mycelium around the edges of the infected scale. It also infects insects other than scales.

The infection of *Verticillium lecanii* on the aphid *Macrosiphoniella sanborni* is unusual in that the fungus sporulates on the live infected aphid, which spreads the spores during the 4–6-day period of infection (Hall 1976, 1979). The sporulation occurs on the legs, antennae, or sometimes the cornicles but rarely on the thorax or abdomen until after death. The infected aphid continues to produce offspring. When the aphid is infected with high dosages of conidia or blastospores, it dies in 48 h. Abundant hyphal growth and sporulation develop over the whole surface of the aphid, but the internal tissues are not invaded as in the case of aphids infected with low dosages of conidia.

The mycelium of *Verticillium lecanii* produces the cyclodepsipeptide toxin, bassianolide, which is also produced by *B. bassiana* (Kanaoka et al. 1978). In addition, this fungus forms other insecticidal toxins, such as dipicolinic acid and C_{25} compounds (Claydon and Grove 1982).

Strains of *Verticillium lecanii* are reported to be pathogenic for the rust fungi, which cause plant diseases (Spencer and Atkey 1981). The fungi invade but do not lyse the urediniospores of the rust fungi. Some of these strains are also pathogenic for the aphid *Macrosiphoniella sanborni* (Hall 1980; Allen 1982). Such strains offer the interesting possibility of using a single fungal species to control both the pest insects and pest fungi of plants. *Verticillium fusisporum* infects several species of Homoptera but is a less dominant fungus than *V. lecanii* because of its slower rate of sporulation and fewer spores (Ekbom and Åhman 1980).

7. *Hirsutella* Species

The genus *Hirsutella* includes over 30 species that infect nearly all systematic groups of insects and certain mites. Members of this genus appear to be the imperfect stages of *Cordyceps*, *Ophiocordyceps*, or *Torrubiella* (Kendrick and Carmichael 1973). The genus *Synnematium* is a synonym of *Hirsutella* (Evans and Samson 1982b). *Hirsutella lecaniicola* is the anamorph for *Cordyceps clavulata* (Bałazy 1971), *H. stilbelliformis* for *C. kniphofioides*, and *H. ovalispora* for *C. cucumispora* (Evans and Samson 1982b). In the infection of some *Hirsutella* species, the amount of mycelium that covers the cadaver is very sparse and can be easily overlooked by the inexperienced worker (Fig. 10–7C). Some of these fungi are highly virulent (e.g., *H. thompsonii* on the citrus rust mite, *Phyllocoptruta oleivora*). They are promising candidates for microbial control, but their growth and sporulation on solid media are slow as compared with other hyphomycetes. To overcome this problem, McCoy et al. (1971, 1972) propagated *H. thompsonii* in submerged culture and successfully applied the fragmented mycelium, which sporulated on the plants under favorable conditions and controlled the mites. Conidia produced under submerged culture have verrucose walls, whereas those produced in aerial cultures have smooth walls (Van Winkelhoff and McCoy 1984) (Fig. 10–7D). The submerged conidia are virulent for adult citrus mites when applied in the field.

Samson et al. (1980) have detected three morphologically distinct groups of *Hirsutella thompsonii* and have described them as varieties *thompsonii*, *vinacea*, and *synnematosa*. Others have differentiated isolates of *Hirsutella* by electrophoretic analysis of isozyme content (Boucias et al. 1982). In general, the isozyme patterns closely follow the morphological scheme used to separate the varieties.

8. *Aspergillus* Species

There are a number of entomopathogenic species of *Aspergillus*, such as *A. flavus*, *A. parasiticus*, *A. tamarii*, *A. ochraceus*, *A. fumigatus*, *A. repens*, and *A. versicolor*. These fungi are mainly saprophytic but may infect a wide range of insect species. *Aspergillus flavus* has been studied most widely. In the honey bee, *A. flavus* and less frequently *A. fumigatus* and other species of *Aspergillus* infect the brood and adult bees. The infection in the brood is called "stonebrood" (Steinbrut) since the cadavers become hard, stonelike mummies. In general, stonebrood of honey bee is rare and is of minor importance to bee keepers (Gilliam 1978).

Several toxins have been isolated from cultures of *A. flavus*, such as kojic acid, which is low in acute toxicity but is insecticidal at high concentrations to the house fly maggot (Beard and Walton 1969). The virulence of strains of *A. flavus* to the silkworm is related to the production of kojic acid and the resistance of the strains to formalin (Kawakami and Mikuni 1969). Toscano and Reeves (1973) have isolated two toxic fractions from *A. flavus* that kill the larvae of the mosquitoes *Culex peus* and *C. tarsalis*.

The most widely known and investigated toxins of *Aspergillus flavus* and *A. parasiticus* are the aflatoxins, a group of at least 12 highly substituted polycyclic compounds. The aflatoxins are very potent carcinogens and produce tumors primarily in the liver of vertebrates, including humans (Wyllie and Morehouse 1977).

Murakoshi et al. (1971) obtained 98 isolates of *Aspergillus* from dust samples collected from sericultural farms in Japan. Most of the isolates were *A. flavus*, and from 37 of them, they detected the four aflatoxins, B_1, B_2, G_1, and G_2. Aflatoxin B_1 was highly toxic and G_1 was moderately toxic when fed to silkworm larvae. They noticed among the strains a correlation among the presence of aflatoxins and the pathogenicity to the silkworm, the tolerance to formalin, and the ability to produce pigments. Aflatoxin B_1 is produced in silkworm larva infected with *A. flavus* (Ohtomo et al. 1975).

The aflatoxins B_1, B_2, G_1, and G_2 are toxic when fed to the larvae of several insect species (e.g., *Heliothis virescens*, *H. zea*, *Spodoptera frugiperda*, *S. littoralis*, *Ostrinia nubilalis*, *Sitophilus zeamais*, and *Drosophila melanogaster*) (Gudauskas et al. 1967; McMillian et al. 1980, 1981; Llewellyn and Chinnici 1978). They act as chemosterilants when applied to the larvae of *Aedes aegypti*, the adults of *Musca domestica* and *Drosophila melanogaster*, whose egg production and egg hatch are reduced (Matsumura and Knight 1967). When applied to the last instar larvae of *Spodoptera littoralis*, aflatoxin B_1 and G_1 induce mutagenic effects on spermatogenesis, resulting in infertility (Abdou et al. 1984). The abnormalities are inherited by the progeny. When *D. melanogaster* is reared on medium containing 10 ppm of aflatoxin B_1, the larval and pupal stages are prolonged even at 24 h of exposure to the toxin (Kirk et al. 1971). The toxin appears to suppress the activity of the cerebral neurosecretory cells. Strains of *Drosophila melanogaster* vary in their susceptibility to the different aflatoxins (Chinnici and Llewellyn 1979; Gunst et al. 1982). The highly effective pathogen *Bacillus thuringiensis* is affected by the mycotoxins patulin and aflatoxin B_1 and can be used in the bioassay for the presence of these mycotoxins in food (Boutibonnes 1977; Boutibonnes et al. 1983).

A strain of *Aspergillus flavus* produces, in addition to aflatoxins, two other toxins, cyclopiazonic acid (indole-tetramic acid) and aflatrem (indole-mevalonate metabolite) (Richard and Gallagher 1979). Aflatrem is similar in structure to other tremorgenic toxins produced by *Claviceps paspali* and *Penicillium paxilli*.

Aspergillus ochraceus, which infects the silkworm among other insects, produces destruxins (Kodaira 1961a), aspochracin (Myokei et al. 1969), and ochratoxin A (Paster et al. 1983). Aspocracin, a cyclotripeptide compound, when inoculated intrahemocoelically into larvae of the silkworm and the fall webworm, *Hyphantria cunea*, is less toxic than destruxins, but when fed to the larvae, it causes, unlike destruxins, immediate knockdown followed occasionally by death. It also has some contact toxicity to eggs and first instar larvae. Its toxicity to mice is low. *Aspergillus fumigatus* produces clavine alkaloids and several tremorgens belonging to the fumitremorgen groups that are toxic to vertebrates.

9. *Culicinomyces clavisporus*

Culicinomyces clavisporus is a unique entomopathogenic hyphomycetes in that it produces conidia underwater and infects through the foregut or hindgut rather than through the exoskeleton although infection also occurs through the anal papillae (Sweeney 1975, 1979; Knight 1980). The conidial wall is covered by a mucopolysaccharide substance that is believed to be responsible for conidial adhesion to the cuticles of the fore- and hindguts (Sweeney et al. 1983). Mucoid coating also surrounds the germ tube in case the conidial wall ruptures before adhesion to the cuticle.

When mosquito larvae are exposed to low dosages of the conidia of *Culicinomyces clavisporus*, they die later than 2 days after exposure, and the sporulation of the fungus occurs on the external surface (Sweeney 1983). However, at high dosages of conidia, most larvae die within 2 days and there is no sporulation. The fungus apparently kills the larva shortly after penetrating through the digestive tract. The cause of death is unknown, and it may result from toxic substances produced by the massive invasion of the hyphae. In addition to mosquitoes, *C. clavisporus* infects larval members of Simuliidae, Ceratopogonidae, Chironomidae, Ephydridae, and Syrphidae (Knight 1980).

10. *Tolypocladium*

The genus *Tolypocladium* was described by Gams (1971) from soil fungi resembling *Beauveria*. A fungus highly pathogenic for mosquito larvae, and which was at one time called *Beauveria tenella*, has been redescribed as *Tolypocladium cylindrosporum* (Soares et al. 1979). This fungus infects mosquitoes (*Aedes* spp.) (Samson and Soares 1984; Gardner and Pillai 1986) and bibionid flies (Samson and Soares 1984). The conidium is ingested by the larval mosquito and the germinating hypha penetrates the midgut into the hemocoel (Weiser and Pillai 1981). The blastospores are more infectious to mosquito larvae than the conidia (Soares and Pinnock 1984).

11. Other Imperfect Fungi

Many other hyphomycetous fungi have been reported as entomopathogens. We shall describe only a few of them. *Aureobasidium pullulans*, a common epiphyte, infects scale insects and causes melanosis in honey bees (Poltev and Neshatayeva 1969). *Fusarium* and *Alternaria* have members that are well-known plant pathogens, but some also infect insects, particularly *Fusarium* spp. (Teetor-Barsch and Roberts 1983). Some *Fusarium* spp. are anamorphs of Ascomycetes. *Fusarium moniliforme* infects not only the corn plant but also the European corn borer, *Ostrinia nubilalis* (Vago 1958); *F. solani*, which causes root rot and stem cankers, is a weak pathogen of larval and adult bark beetles (Moore 1973; Barson 1976); and *Alternaria porri* (syn. *A. solani*), which causes early blight of potato and tomato, infects the eggs of European corn borer (Lynch and Lewis 1978). *Fusarium solani* produces fusaric acid and naphthazarin pigments that are insecticidal (Claydon et al. 1977). These mycotoxins are known to inhibit a variety of enzymic reactions by virtue of their metal-chelating properties. *Fusarium lateritium*, a pathogen of the scale insect, *Hemiberlesia rapax*, forms enniatins that are analogues of beauvericin (Grove and Pople 1980). The enniatins are less active against the mosquito *Aedes aegypti* than beauvericin but are more insecticidal against the adult blow fly, *Calliphora erythrocephala*.

Penicillium isolates produce metabolites, which are toxic to insects (Paterson et al. 1987). Some of the toxic metabolites are ochratoxin A, brevianamide A, penicillic acid, and citrinin. They cause larval weight loss or death to *Drosophila melanogaster* and *Spodoptera littoralis*.

Class Coelomycetes has two genera, *Tetranacrium* and *Aschersonia*, whose members are entomopathogens (McCoy et al. 1988). Both are included in the friendly fungi of scale insects. *Tetranacrium* is the anamorph of *Podonectria*, and *Aschersonia* is that of *Hypocrella*. *Aschersonia aleyrodis* infects the greenhouse whitefly, *Trialeurodes vaporariorum* (Fransen et al. 1987). The eggs of the whitefly are not infected, but the newly emerged larvae become infected by spores persisting on the egg surface. Older larvae are less susceptible than the young larvae, and the adults are seldom infected.

REFERENCES

Abdou, R. F., Megalla, S. E., and Azab, S. G. 1984. Mutagenic effects of aflatoxin B-1 and G-1 on the Egyptian cotton leaf-worm, *Spodoptera littoralis* (Boisd.). *Mycopathologia* 88, 23–26.

Ainsworth, G. C. 1956. Agostino Bassi, 1773-1856. *Nature 177*, 255–257.

Ainsworth, G. C. 1973. Introduction and keys to higher taxa. In "The Fungi: An Advanced Treatise." G. C. Ainsworth, F. K. Sparrow, and A. S. Sussman, eds.), Vol. IVA, pp. 1–7. Academic Press, New York.

Al-Aidroos, K., and Roberts, D. W. 1978. Mutants of *Metarhizium anisopliae* with increase virulence toward mosquito larvae. *Can. J. Gen. Cytol. 20*, 211–219.

Al-Aidroos, K., and Seifert, A. M. 1980. Polysaccharide and protein degradation, germination, and virulence against mosquitoes in the entomopathogenic fungus *Metarhizium anisopliae*. *J. Invertebr. Pathol. 36*, 29–34.

Allen, D. J. 1982. *Verticillium lecanii* on the bean rust fungus, *Uromyces appendiculatus*. *Trans. Br. Mycol. Soc. 79*, 362–364.

Aoki, J. 1961. Studies on the infection mechanism of *Aspergillus* disease in silkworm larvae, *Bombyx mori*. (II) The invading loci of causal fungus. *J. Seric. Sci. Jpn. 30*, 43–48.

Aoki, J. 1981. Pattern of conidial discharge of an *Entomophthora* species ("grylli" type) (Entomophthorales: Entomophthoraceae) from infected cadavers of *Mamestra brassicae* L. (Lepidoptera: Noctuidae). *Appl. Entomol. Zool. 16*, 216–224.

Archbold, E. F., Rust, M. K., Reierson, D. A., and Atkinson, K. D. 1986. Characterization of a yeast infection in the German cockroach (Dictyoptera: Blattellidae). *Environ. Entomol. 15*, 221–226.

Arwidsson, Th. 1946. Om svenska laboulbeniacéfynd. *Sven. Bot. Tidskr. 40*, 307–309.

Bailey, L. 1963. "Infectious Diseases of the Honey-Bee." Land Books, London.

Bailey, L. 1968. Honey bee pathology. *Annu. Rev. Entomol. 13*, 191–212.

Bałazy, S. 1971. A supplement to the morphology and nomenclature of the conidial stage of *Cordyceps clavulata* (Schw.) Ell. et Everh. *Acta Mycol. 7*, 7–12.

Bałazy, S. 1973. A review of entomopathogenic species of the genus *Cephalosporium* Corda (*Mycota*, *Hyphomycetales*). *Bull. Soc. Amis Sci. Lett. Poznań. Ser. D 14*, 101–137.

Barson, G. 1976. *Fusarium solani*, a weak pathogen of the larval stages of the large elm bark beetle *Scolytus scolytus* (Coleoptera: Scolytidae). *J. Invertebr. Pathol. 27*, 307–309.

Batko, A., and Weiser, J. 1965. On the taxonomic position of the fungus discovered by Strong, Wells, and Apple: *Strongwellsea castrans* gen. et sp. nov. (Phycomycetes; Entomophthoraceae). *J. Invertebr. Pathol. 7*, 455–463.

Beard, R. L., and Walton, G. S. 1969. Kojic acid as an insecticidal mycotoxin. *J. Invertebr. Pathol. 14*, 53–59.

Benjamin, R. K. 1971. Introduction and supplement to Roland Thaxter's contribution towards a monograph of the Laboulbeniaceae. *Bibl. Mycol. 30*, 1–155.

Benjamin, R. K. 1973. Laboulbeniomycetes. In "The Fungi: An Advanced Treatise." (G. C. Ainsworth, F. K. Sparrow, and A. S. Sussman, eds.), Vol. 4A, pp. 223–246. Academic Press, New York.

Ben-Ze'ev, I., and Kenneth, R. G. 1980. *Zoophthora phytonomi* and *Conidiobolus osmodes* [Zygomycetes: *Entomophthoraceae*], two pathogens of *Hypera* species [*Col.*: Curculionidae] conincidental in time and place. *Entomophaga 25*, 171–186.

Ben-Ze'ev, I., and Kenneth, R. G. 1981. *Zoophthora orientalis* sp. nov., a fungal pathogen of *Aphis citricola* (Homoptera: Aphididae), and two new combinations of other species of Entomophthoraceae. *Phytoparasitica 9*, 33–42.

Ben-Ze'ev, I., and Kenneth, R. G. 1982a. Features-criteria of taxonomic value in the Entomophthorales: I. A revision of the Batkoan classification. *Mycotaxon 14*, 393–455.

Ben-Ze'ev, I., and Kenneth, R. G. 1982b. Features-criteria of taxonomic value in the Entomophthorales: II. A revision of the genus *Erynia* Nowakowski 1881 (= *Zoophthora* Batko 1964). *Mycotaxon 14*, 456–475.

Ben-Ze'ev, I. S., Keller, S., and Ewen, A. B. 1985. *Entomophthora erupta* and *Entomophthora helvetica* sp. nov. (Zygomycetes: Entomophthorales), two pathogens of Miridae (Heteroptera) distinguished by pathobiological and nuclear features. *Can. J. Bot. 63*, 1469–1475.

Bitton, S., Kenneth, R. G., and Ben-Ze'ev, I. 1979. Zygospore overwintering and sporulative germination in *Triplosporium fresenii* (Entomophthoraceae) attacking *Aphis spiraecola* on citrus in Israel. *J. Invertebr. Pathol. 34*, 295–302.

Blackwell, M., and Gilbertson, R. L. 1981. *Cordycepioideus octosporus*, a termite suspected pathogen from Jalisco, Mexico. *Mycologia 73*, 358–362.

Bogoyavlensky, N. 1922. *Zografia notonectae* n.g. n.sp. *Arch. Russ. Protist. Obshch. 1*, 113–119.

Booth, C. 1971. "The genus Fusarium." Commonwealth Mycological Institute, Kew, Surrey, England.

Boucias, D. G., and Latgé, J. P. 1988. Nonspecific induction of germination of *Conidiobolus obscurus* and *Nomuraea rileyi* with host and non-host cuticle extracts. *J. Invertebr. Pathol. 51*, 168–171.

Boucias, D. G., and Pendland, J. C. 1984. Nutritional requirements for conidial germination of several host range pathotypes of the entomopathogenic fungus *Nomuraea rileyi*. *J. Invertebr. Pathol. 43*, 288–292.

Boucias, D. G., McCoy, C. W., and Joslyn, D. J. 1982. Isozyme differentiation among 17 geographical isolates of *Hirsutella thompsonii*. *J. Invertebr. Pathol. 39*, 329–337.

Boucias, D. G., Pendland, J. C., and Latgé, J. P. 1988. Nonspecific factors involved in attachment of entomopathogenic Deuteromycetes to host insect cuticle. *Appl. Environ. Microbial. 54*, 1795–1805.

Boutibonnes, P. 1977. A propos du pouvoir antibactérien de deux mycotoxines: La patuline et l'aflatoxine B_1. *Microbia 3*, 25–43.

Boutibonnes, P., Malherbe, C., Auffray, Y., Kogbo, W., and Marais, C. 1983. Mycotoxin sensitivity of *Bacillus thuringiensis*. *IRCS Med. Sci. Biochem. 11*, 430–431.

Brandenburg, R. L., and Kennedy, G. G. 1981. Overwintering of the pathogen *Entomophthora floridana* and its host, the twospotted spider mite. *J. Econ. Entomol. 74*, 428–431.

Brey, P. T. 1985. Observations of in vitro gametangial copulation and oosporogenesis in *Lagenidium giganteum*. *J. Invertebr. Pathol. 45*, 276–281.

Brey, P. T., Latge, J. P., and Prevost, M. C. 1986. Integumental penetration of the pea aphid, *Acyrthosiphon pisum* by *Conidiobolus obscurus* (Entomophthoraceae). *J. Invertebr. Pathol. 48*, 34–41.

Brobyn, P. J., and Wilding, N. 1977. Invasive and developmental processes of *Entomophthora* species infecting aphids. *Trans. Br. Mycol. Soc. 69*, 349–366.

Bro Larsen, E. 1952. On subsocial beetles from the salt-marsh, their care of progeny and adaptation to salt and tide. *Trans. Ninth Int. Congr. Entomol. Amsterdam. 1*, 502–506.

Broome, J. R., Sikorowski, P. P., and Norment, B. R. 1976. A mechanism of pathogenicity of *Beauveria bassiana* on larvae of the imported fire ant, *Solenopsis richteri*. *J. Invertebr. Pathol. 28*, 87–91.

Brown, A. H. S., and Smith, G. 1957. The genus *Paecilomyces* Bainier and its perfect stage *Byssochlamys* Westling. *Trans. Br. Mycol. Soc. 40*, 17–89.

Burges, H. D. 1981. "Microbial Control of Pests and Plant Diseases 1970–1980." Academic Press, New York.

Burnside, C. E. 1930. Fungous diseases of the honeybee. *U.S.D.A. Tech. Bull. 149*,

Burnside, C. E. 1935. A disease of young bees caused by a Mucor. *Am. Bee. J. 75*, 75–76.
Butt, T. M., Beckett, A., and Wilding, N. 1981. Protoplasts in the *in vivo* life cycle of *Erynia neoaphidis*. *J. Gen. Microbiol. 127*, 417–421.
Byford, W. J., and Ward, L. K. 1968. Effect of the situation of the aphid host at death on the type of the spore produced by Entomophthora sp. *Trans. Br. Mycol. Soc. 51*, 598–600.
Carner, G. R. 1976. A description of the life cycle of *Entomophthora* sp. in the two-spotted spider mite. *J. Invertebr. Pathol. 28*, 245–254.
Carner, G. R. 1980. *Entomophthora lampyridarum*, a fungal pathogen of the soldier beetle, *Chauliognathus pennsylvanicus*. *J. Invertebr. Pathol. 36*, 394–398.
Carruthers, R. I., and Haynes, D. L. 1986. Temperature, moisture, and habitat effects on *Entomophthora muscae* (Entomophthorales: Entomophthoraceae) conidial germination and survival in the onion agroecosystem. *Environ. Entomol. 15*, 1154–1160.
Castillo, J. M., and Roberts, D. W. 1980. In vitro studies of *Coelomomyces punctatus* from *Anopheles quadrimaculatus* and *Cyclops vernalis*. *J. Invertebr. Pathol. 35*, 144–157.
Champlin, F. R., and Grula, E. A. 1979. Noninvolvement of beauvericin in the entomopathogenicity of *Beauveria bassiana*. *Appl. Environ. Microbiol. 37*, 1122–1126.
Champlin, F. R., Cheung, P. Y. K., Pekrul, S., Smith, R. J., Burton, R. L., and Grula, E. A. 1981. Virulence of *Beauveria bassiana* mutants for the pecan weevil. *J. Econ. Entomol. 74*, 617–621.
Chapman, H. C. 1974. Biological control of mosquito larvae. *Annu. Rev. Entomol. 19*, 33–59.
Charnley, A. K. 1984. Physiological aspects of destructive pathogenesis in insects by fungi: A speculative review. In: "Invertebrate–microbial Interactions." (J. M. Anderson, A. D. M. Rayner, and D. W. H. Walton, eds.), pp. 229–270. Cambridge University Press, Cambridge.
Chatterjee, R., Srinivasan, K. S., and Maiti, P. C. 1957. *Cordyceps sinensis* (Berkeley) Saccardo: Structure of cordycepic acid. *J. Am. Pharm. Assoc. 46*, 114–118.
Cheung, P. Y. K., and Grula, E. A. 1982. In vivo events associated with entomopathology of *Beauveria bassiana* for the corn earworm (*Heliothis zea*). *J. Invertebr. Pathol. 39*, 303–313.
Chinnici, J. P., and Llewellyn, G. C. 1979. Reduced aflatoxin toxicity in hybrid crosses of aflatoxin B_1 sensitive and resistant strains of *Drosophila melanogaster* (Diptera). *J. Invertebr. Pathol. 33*, 81–85.
Clark, T. B., Kellen, W. R., Fukuda, T., and Lindegren, J. E. 1968. Field and laboratory studies on the pathogenicity of the fungus *Beauveria bassiana* to three genera of mosquitoes. *J. Invertebr. Pathol. 11*, 1–7.
Claydon, N. 1978. Insecticidal secondary metabolites from entomogenous fungi: *Entomophthora virulenta*. *J. Invertebr. Pathol. 32*, 319–324.
Claydon, N., and Grove, J. F. 1982. Insecticidal secondary metabolic products from the entomogenous fungus *Verticillium lecanii*. *J. Invertebr. Pathol. 40*, 413–418.
Claydon, N., Grove, J. F., and Pople, M. 1977. Insecticidal secondary metabolic products from the entomogenous fungus *Fusarium solani*. *J. Invertebr. Pathol. 30*, 216–223.
Cohn, F. 1855. Empusa muscae und die Krankheit der Stubenfliegen. *Dtsch. Akad. Naturforsch. Nova Acta Leopoldina 25*, 301–360.
Coles, R. B., and Talbot, P. H. B. 1977. *Septobasidium clelandii* and its conidial state *Harpographium corynelioides*. *Kew Bull. 31*, 481–488.
Cooke, M. C. 1892. "Vegetable Wasps and Plant Worms." Society Promoting Christian Knowledge, London.
Couch, J. N. 1972. Mass production of *Coelomomyces*, a fungus that kills mosquitoes. *Proc. Natl. Acad. Sci. U.S.A. 69*, 2043–2047.
Couch, J. N., and Romney, S. V. 1973. Sexual reproduction in Lagenidium giganteum. *Mycologia 65*, 250–252.
Couch, J. N., and Umphlett, C. J. 1963. Coelomomyces infections. *In* "Insect Pathology: An

Advanced Treatise." (E. A. Steinhaus, ed.), Vol. 2, pp. 149–188. Academic Press, New York.

Coudron, T. A., Kroha, M. J., and Ignoffo, C. M. 1984. Levels of chitinolytic activity during development of three entomopathogenic fungi. *Comp. Biochem. Physiol. 79B*, 339–348.

Crisan, E. V. 1971. Mechanism responsible for release of toxin by *Metarrhizium* spores in mosquito larvae. *J. Invertebr. Pathol. 17*, 260–264.

Daoust, R. A., and Roberts, D. W. 1982. Virulence of natural and insect-passaged strains of *Metarhizium anisopliae* to mosquito larvae. *J. Invertebr. Pathol. 40*, 107–117.

David, W. A. L. 1967. The physiology of the insect integument in relation to the invasion of pathogens. *In* "Insects and Physiology." (J. W. L. Beament and J. E. Treherne, eds.), pp. 17–35. Oliver and Boyd, Edinburgh.

Dean, D. D., and Domnas, A. J. 1983a. The extracellular proteolytic enzymes of the mosquito-parasitizing fungus *Lagenidium giganteum*. *Exp. Mycol. 7*, 31–39.

Dean, D. D., and Domnas, A. J. 1983b. Isolation and partial characterization of collagenolytic enzyme from the mosquito-parasitizing fungus, *Lagenidium giganteum*. *Arch. Microbiol. 136*, 212–218.

de Andrade, C. F. S. 1980. Epizootia natural causada por *Cordyceps unilateralis* (Hypocreales, Euascomycetes) em adultos de *Camponotus* sp. (Hymenoptera, Formicidae) na região de Manaus, Amazonas, Brasil. *Acta Amazonia 10*, 671–677.

Debaisieux, P. 1919. Une chytridinée nouvelle: *Coelomycidium simulli* nov. gen. nov. spec. *C.R. Soc. Biol. 82*, 899–900.

de Conti, E., Messias, C. L., de Souza, H. M. L., and Azevedo, J. L. 1980. Electrophoretic variation in esterases and phosphatases in eleven wild-type strains of *Metarrhizium anisopliae*. *Experientia 36*, 293–294.

De Hoog, G. S. 1972. The genera Beauveria, Isaria, Tritirachium, and Acrodontium gen. nov. *Stud. Mycol., 1*. 1–41.

Descals, E., and Webster, J. 1984. Branched aquatic conidia in *Erynia* and *Entomophthora* sensu lato. *Trans. Br. Mycol. Soc. 83*, 669–682.

Descals, E., Webster, J., Ladle, M., and Bass, J. A. B. 1981. Variations in asexual reproduction in species of *Entomophthora* on aquatic insects. *Trans. Br. Mycol. Soc. 77*, 85–102.

Dick, M. W. 1973. Saprolegniales. *In* "The Fungi: An Advanced Treatise." (G. C. Ainsworth, F. K. Sparrow, and A. S. Sussman, eds.), Vol. IVB, pp. 113–144. Academic Press, New York.

Diomandé, T. 1969. Contribution a l'étude du développement de la muscardine verte a *Metarrhizium anisopliae* (Metsch.) Sorokin [Fungi Imperfecti] des larves d'*Oryctes monoceros* Ol. [Coléoptère Scarabaeidae]. *Bull. I. F. A. N. Ser. A. 31*, 1381–1405.

Domnas, A., Giebel, P. E., and McInnis, T. M., Jr. 1974. Biochemistry of mosquito infection: Preliminary studies of biochemical change in *Culex pipiens quinquefasciatus* following infection with *Lagenidium giganteum*. *J. Invertebr. Pathol. 24*, 293–304.

Domnas, A. J., Fayan, S. M., and Jaronski, S. 1982. Factors influencing zoospore production in liquid cultures of *Lagenidium giganteum* (Oomycetes, Lagenidiales). *Mycologia 74*, 820–825.

Dunphy, G. B., and Nolan, R. A. 1977. Regeneration of protoplasts of *Entomophthora egressa*, a fungal pathogen of the eastern hemlock looper. *Can. J. Bot. 55*, 107–113.

Dunphy, G. B., and Nolan, R. A. 1979. Effects of physical factors on protoplasts of *Entomophthora egressa*. *Mycologia 71*, 589–602.

Dunphy, G. B., and Nolan, R. A. 1980. Response of eastern hemlock looper hemocytes to selected stages of *Entomophthora egressa* and other foreign particles. *J. Invertebr. Pathol. 36*, 71–84.

Dunphy, G. B., and Nolan, R. A. 1982a. Mycotoxin production by the protoplast stage of *Entomophthora egressa*. *J. Invertebr. Pathol. 39*, 261–263.

Dunphy, G. B., and Nolan, R. A. 1982b. Cellular immune responses of spruce budworm larvae to *Entomophthora egressa* protoplasts and other test particles. *J. Invertebr. Pathol. 39*, 81–92.

Duriez-Vaucelle, T., Fargues, J., Robert, P. H., and Popeye, R. 1981. Étude enzymatique comparée de champignons entomopathogènes des genres Beauveria et Metarhizium. *Mycopathologia 75*, 109–126.

Dustan, A. G. 1924. Studies on a new species of Empusa parasitic on the green apple bug (Lygus communis var. novascotiensis Knight) in the Annapolis Valley. *Proc. Acadian Entomol. Soc. 9*, 14–36.

Dykstra, M. J. 1974. Some ultrastructural features in the genus *Septobasidium*. *Can. J. Bot. 52*, 971–972.

Eilenberg, J., Bresciani, J., and Latgé, J.-P. 1986. Ultrastructural studies of primary spore formation and discharge in the genus *Entomophthora*. *J. Invertebr. Pathol. 48*, 318–324.

Ekbom, B. S., and Åhman, I. 1980. The fungus *Verticillium fusisporum* as an insect pathogen. *J. Invertebr. Pathol. 36*, 136–138.

Evans, H. C. 1989. Mycopathogens of insects of epigeal and aerial habitats. *In* "Insect-Fungus Interactions." (N. Wilding, N. M. Collins, P. M. Hammond and J. F. Webber, Jr., eds.), pp. 205–238. Academic Press, London.

Evans, H. C., and Samson, R. A. 1977. *Sporodiniella umbellata*, an entomogenous fungus of the Mucorales from cocoa farms in Ecuador. *Can. J. Bot. 55*, 2981–2984.

Evans, H. C., and Samson, R. A. 1982a. Entomogenous fungi from the Galápagos Islands. *Can. J. Bot. 60*, 2325–2333.

Evans, H. C., and Samson, R. A. 1982b. *Cordyceps* species and their anamorphs pathogenic on ants (Formicidae) in tropical forest ecosystems. I. The *Cephalotes* (Myrmicinae) complex. *Trans. Br. Mycol. Soc. 79*, 431–453.

Evans, H. C., and Samson, R. A. 1984. *Cordyceps* species and their anamorphs pathogenic on ants (*Formicidae*) in tropical forest ecosystems. II. The *Camponotus* (Formicinae) complex. *Trans. Br. Mycol. Soc. 82*, 127–150.

Evlakhova, A. A. 1939. A new yeast-like fungus *Blastodendrion pseudococci* nov. sp. pathogenic for mealy bugs. *Bull. Plant Prot. No.* 1 (20), pp 79–84 (*Rev. Appl. Entomol.* 1940. *A28*, 602–603).

Evlakhova, A. A. 1974. "Entomogenous Fungi. Classification, Biology, Practical Significance." (M. K. Khokhriakov, ed.), Nauka, Leningrad.

Fargues, J. 1972. Étude des conditions d'infection des larves de Doryphore, *Leptinotarsa decemlineata* Say, par *Beauveria bassiana* (Bals.) Vuill. [*Fungi Imperfecti*]. *Entomophaga 17*, 319–337.

Fargues, J. 1976. Spécificité des champignons pathogènes imparfaits [Hyphomycètes] pour les larves de Coléoptères [*Scarabaeidae* et *Chrysomelidae*]. *Entomophaga 21*, 313–323.

Fargues, J., and Robert, P.-H. 1978. Adaptabilité de deux pathotypes de *Metarrhizium anisopliae* (*Metsh.*) sor. (*Fungi imperfecti* Hyphomycètes) par culture sur milieu artificiel et par passage sur Insecte-hôte d'origine. *C. R. Acad. Sci. Paris Ser. D 287*, 165–167

Fargues, J., and Robert, P. H. 1983. Influence de l'antécédent nutritionnel sur la virulence de deux souches de l'hyphomycète entomopathogène *Metarhizium anisopliae*. *Mycopathologia 81*, 145–154.

Fargues, J., Duriez, T., Andrieu, S., and Popeye, R. 1974. Analyse sérologique comparée de deux champignons entomopathogènes, *Beauveria bassiana* (Bals.) Vuill. et *Beauveria tenella* (Delacr.) Siem. *C. R. Acad. Sci. Paris Ser. D 278*, 2245–2247.

Fargues, J., Robert, P.-H., and Vey, A. 1976. Rôle du tégument et de la défense cellulaire des Coléoptères hôtes dans la spécificité des souches entomopathogènes de *Metarrhizium anisopliae* (*Fungi Imperfecti*). *C.R. Acad. Sci. Paris Ser. D 282*, 2223–2226.

Fargues, J., Robert, P.-H., Vey, A., and Païs, M. 1986. Toxicité relative de la destruxine E pour Lépidoptère Galleria mellonella L. *C.R. Acad. Sci. Paris Ser. III 303*, 83–86.
Fawcett, H. S. 1910. An important entomogenous fungus. *Mycologia 2*, 164–168.
Federici, B. A. 1979. Experimental hybridization of *Coelomomyces dodgei* and *Coelomomyces punctatus*. *Proc. Natl. Acad. Sci. U.S.A. 76*, 4425–4428.
Federici, B. A. 1981. Mosquito control by the fungi *Culicinomyces*, *Lagenidium* and *Coelomomyces*. In "Microbial Control of Pests and Plant Diseases 1970–1980." (H. D. Burges, ed.), pp. 555–572. Academic Press, London.
Federici, B. A. 1982. Inviability of interspecific hybrids in the *Coelomomyces dodgei* complex. *Mycologia 74*, 555–562.
Federici, B. A. 1983. Species-specific gating of gametangial dehiscene as a temporal reproductive isolating mechanism in *Coelomomyces*. *Proc. Natl. Acad. Sci. U.S.A. 80*, 604–607.
Federici, B. A., and Chapman, H. C. 1977. *Coelomomyces dodgei*: Establishment of an in vivo laboratory culture. *J. Invertebr. Pathol. 30*, 288–297.
Federici, B. A., and Lucarotti, C. J. 1986. Structure and behavior of the meiospore of *Coelomomyces dodgei* during encystment on the copepod host, *Acanthocyclops vernalis*. *J. Invertebr. Pathol. 48*, 259–268.
Federici, B. A., and Roberts, D. W. 1976. Experimental laboratory infection of mosquito larvae with fungi of the genus *Coelomomyces*. II. Experiments with *Coelomomyces punctatus* in *Anopheles quadrimaculatus*. *J. Invertebr. Pathol. 27*, 333–341.
Federici, B. A., and Thompson, S. N. 1979. β-carotene in the gametophytic phase of *Coelomomyces dodgei*. *Exp. Mycol. 3*, 281–284.
Fennell, D. I. 1973. Plectomycetes; Eurotiales. In "The Fungi: An Advanced Treatise." (G. C. Ainsworth, F. K. Sparrow, and A. S. Sussman, eds.), Vol. IVA, pp. 45–68. Academic Press, New York.
Ferron, P. 1975. Les champignons entomopathogenes: Evolution des recherches au cours des dix dernieres annees. *Bul. Sci. Biol. Organ. Int. Lutte Biol.; WPRS Bul.* 1975/3.
Ferron, P. 1977. Influence of relative humidity on the development of fungal infection caused by *Beauveria bassiana* [*Fungi Imperfecti, Moniliales*] in imagines of *Acanthoscelides obtectus* [*Col.:Bruchidae*]. *Entomophaga 22*, 393–396.
Ferron, P. 1978. Biological control of insect pests by entomogenous fungi. *Annu. Rev. Entomol. 23*, 409–442.
Ferron, P. 1985. Fungal control. In "Comprehensive Insect Physiology Biochemistry and Pharmacology." (G. A. Kerkut and L. I. Gilbert, eds.), Vol. 12, pp. 313–346. Pergamon Press, Oxford.
Ferron, P., Hurpin, B., and Robert, P. H. 1972. Sur la spécificité de *Metarrhizium anisopliae* (Metsch.) Sorokin. *Entomophaga 17*, 165-178.
Ferron, P., Robert, P. H., and Deotte, A. 1975. Susceptibility of *Oryctes rhinoceros* adults to *Metarrhizium anisopliae*. *J. Invertebr. Pathol. 25*, 313–319.
Fetter-Lasko, J. L., and Washino, R. K. 1983. In situ studies on seasonality and recycling pattern in California of *Lagenidium giganteum* Couch, an aquatic fungal pathogen of mosquitoes. *Environ. Entomol. 12*, 635–640.
Fisher, F. E. 1950. Entomogenous fungi attacking scale insects and rust mites on citrus in Florida. *J. Econ. Entomol. 43*, 305–309.
Fisher, F. E., Thompson, W. L., and Griffiths, J. T., Jr. 1949. Progress report on the fungus diseases of scale insects attacking citrus in Florida. *Fla. Entomol. 32*, 2–11.
Fletcher, H. 1977. Parallel evolution in insect-dispersed fungi and insectivorous plants? *Bull. Br. Mycol. Soc. 11*, 50–51.
Fox, C. J. S. 1961. The incidence of green muscardine in the European wireworm, *Agriotes obscurus* (Linnaeus), in Nova Scotia. *J. Insect Pathol. 3*, 94–95.
Fransen, J. J., Winkelman, K., and van Lenteren, J. C. 1987. The differential mortality at various life stages of the greenhouse whitefly, *Trialeurodes vaporariorum* (Homoptera:

Aleyrodidae), by infection with the fungus *Aschersonia aleyrodis* (Deuteromycotina: Coelomycetes). *J. Invertebr. Pathol.* **50**, 158–165.

Gams, W. 1971. Tolypocladium, eine Hyphomycetengattung mit geschwollenen Phialiden. *Persoonia* **6**, 185–191.

Gardner, J. M., and Pillai, J. S. 1986. *Tolypocladium cylindrosporum* (Deuteromycotina: Moniliales), a fungal pathogen of the mosquito *Aedes australis*. *Mycopathologia* **96**, 87–90.

Gardner, W. A., Sutton, R. M., and Noblet, R. 1979. Effects of infection by *Beauveria bassiana* on hemolymph proteins of noctuid larvae. *Ann. Entomol. Soc. Am.* **72**, 224–228.

Gee, N. G. 1918. Notes on Cordyceps sinensis. *Mycol. Notes* **54**, 767–768.

Giebel, P. E., and Domnas, A. J. 1976. Soluble trehalases from larvae of the mosquito, *Culex pipiens*, and the fungal parasite, *Lagenidium giganteum*. *Insect Biochem.* **6**, 303–311.

Gilliam, M. 1978. Fungi. *In* "Honey Bee Pests, Predators, and Diseases." (R. A. Morse, ed.), pp. 78–101. Cornell University Press, Ithaca.

Glare, T. R. 1987. Effect of host species and light conditions on production of conidia by an isolate of *Nomuraea rileyi*. *J. Invertebr. Pathol.* **50**, 67–69.

Glare, T. R., Chilvers, G. A., and Milner, R. S. 1985. Capilliconidia as infective spores in *Zoophthora phalloides* (Entomophthorales). *Trans. Br. Mycol. Soc.* **85**, 463–470.

Goettel, M. S., Toohey, M. K., and Pillai, J. S. 1983. Preliminary laboratory infection trials with a Fiji isolate of the mosquito pathogenic fungus Lagenidium. *J. Invertebr. Pathol.* **41**, 1–7.

Goettel, M. S., St. Leger, R. J., Rizzo, N. W., Staples, R. C., and Roberts, D. W. 1989. Ultrastructural localization of a cuticle-degrading protease produced by the entomopathogenic fungus *Metarhizium anisopliae* during penetration of host (*Manduca sexta*) cuticle. *J. Gen. Microbiol.* **135**, 2233–2239.

Goldstein, B. 1929. A cytological study of the fungus *Massospora cicadina*, parasitic on the 17-year cicada, *Magicicada septendecim*. *Am. J. Bot.* **16**, 394–401.

Gottwald, T. R., and Tedders, W. L. 1982. Studies on the conidia release by the entomogenous fungi *Beauveria bassiana* and *Metarhizium anisopliae* (Deuteromycotina: Hyphomycetes) from adult pecan weevil (Coleoptera: Curculionidae) cadavers. *Environ. Entomol.* **11**, 1274–1279.

Grove, J. F., and Pople, M. 1980. The insecticidal activity of beauvericin and the enniatin complex. *Mycopathologia* **70**, 103–105.

Gudauskas, R. T., Davis, N. D., and Diener, U. L. 1967. Sensitivity of *Heliothis virescens* larvae to aflatoxin in *Ad libitum* feeding. *J. Invertebr. Pathol.* **9**, 132–133.

Gunst, K., Chinnici, J. P., and Llewellyn, G. C. 1982. Effects of aflatoxin B_1, aflatoxin B_2, aflatoxin G_1, and sterigmatocystin on viability, rates of development, and body length in two strains of *Drosophila melanogaster* (Diptera). *J. Invertebr. Pathol.* **39**, 388–394.

Gustafsson, M. 1965. On species of the genus *Entomophthora* Fres. in Sweden. II. Cultivation and physiology. *Lantbrukk. Ann.* **31**, 405–457.

Hale, P. J., and Menapace, D. M. 1980. Effect of time and temperature on the viability of *Ascosphaera apis*. *J. Invertebr. Pathol.* **36**, 429–430.

Hall, I. M., and Halfhill, J. C. 1959. The germination of resting spores of *Entomophthora virulenta* Hall and Dunn. *J. Econ. Entomol.* **52**, 30–35.

Hall, R. A. 1976. *Verticillium lecanii* on the aphid, *Macrosiphoniella sanborni*. *J. Invertebr. Pathol.* **28**, 389–391.

Hall, R. A. 1979. Pathogenicity of *Verticillium lecanii* conidia and blastospores against the aphid, *Macrosiphoniella sanborni*. *Entomophaga* **24**, 191–198.

Hall, R. A. 1980. Laboratory infection of insects by *Verticillium lecanii* strains isolated from phytopathogenic fungi. *Trans. Br. Mycol. Soc.* **74**, 445–446.

Hall, R. A. 1981. The fungus *Verticillium lecanii* as a microbial insecticide against aphids and scales. *In* "Microbial Control of Pests and Plant Diseases 1970–1980." (H. D. Burges, ed.), pp. 483–498. Academic Press, London.

Hall, R. A., and Papierok, B. 1982. Fungi as biological control agents of arthropods of agricultural and medical importance. *Parasitology 84*, 205–240.

Hamill, R. L., Higgens, C. E., Boaz, H. E., and Gorman, M. 1969. The structure of beauvericin, a new depsipeptide antibiotic toxic to *Artemia salina*. *Tetrahedron Lett. 49*, 4255–4258.

Hasan, S., and Vago, C. 1972. The pathogenicity of *Fusarium oxysporum* to mosquito larvae. *J. Invertebr. Pathol. 20*, 268–271.

Hedlund, R. C., and Pass, B. C. 1968. Infection of the alfalfa weevil, *Hypera postica*, by the fungus *Beauveria bassiana*. *J. Invertebr. Pathol. 11*, 25–34.

Hesseltine, C. W., and Ellis, J. J. 1973. Mucorales. *In* "The Fungi: An Advanced Treatise." (G. C. Ainsworth, F. K. Sparrow, and A. S. Sussman, eds.), Vol. IVB, pp. 187–217. Academic Press, New York.

Hoffmann, W. E. 1947. Insects as human food. *Proc. Entomol. Soc. Wash. 49*, 233–237.

Hughes, S. J. 1971. Phycomycetes, Basidiomycetes, and Ascomycetes as Fungi Imperfecti. *In* "Taxonomy of Fungi Imperfecti." (B. Kendrick, ed.), pp. 7–36. University of Toronto Press, Toronto.

Humber, R. A. 1976. The systematics of the genus *Strongwellsea* (Zygomycetes: Entomophthorales). *Mycologia 68*, 1042–1060.

Humber, R. A. 1981. An alternative view of certain taxonomic criteria used in the Entomophthorales (Zygomycetes). *Mycotaxon 13*, 191–240.

Humber, R. A. 1984. The identity of Entomophaga species (Entomophthorales: Entomophthoraceae) attacking Lepidoptera. *Mycotaxon 21*, 265–272.

Humber, R. A., and Ben-Ze'ev, I. 1981. *Erynia* (Zygomycetes: Entomophthorales): Emendation, synonymy, and transfers. *Mycotaxon 13*, 506–516.

Humber, R. A., and Ramoska, W. A. 1986. Variations in entomophthoralean life cycles: Practical implications. *In* "Fundamental and Applied Aspects of Invertebrate Pathology." (R. A. Samson, J. M. Vlak, and D. Peters, eds.), pp. 190–193. Found. Fourth Int. Colloq. Invertebr. Pathol., Wageningen, Netherlands.

Hunt, D. W. A., Borden, J. H., Rahe, J. E., and Whitney, H. S. 1984. Nutrient-mediated germination of *Beauveria bassiana* conidia on the integument of the bark beetle *Dendroctonus ponderosae* (Coleoptera: Scolytidae). *J. Invertebr. Pathol. 44*, 304–314.

Hurion, N., Fromentin, H., and Keil, B. 1977. Proteolytic enzymes of *Entomophthora coronata*. Characterization of a collagenase. *Comp. Biochem. Physiol. 56B*, 259–264.

Hurion, N., Fromentin, H., and Keil, B. 1979. Specificity of the collagenolytic enzyme from the fungus *Entomophthora coronata*: Comparison with the bacterial collagenase from *Achromobacter iophagus*. *Arch. Biochem. Biophys. 192*, 438–445.

Hutchison, J. A. 1962. Studies on a new Entomophthora attacking calyptrate flies. *Mycologia 54*, 258–271.

Ignoffo, C. M., Puttler, B., Marston, N. L., Hostetter, D. L., and Dickerson, W. A. 1975. Seasonal incidence of the entomopathogenic fungus *Spicaria rileyi* associated with noctuid pests of soybeans. *J. Invertebr. Pathol. 25*, 135–137.

Ingold, C. T. 1978. Role of mucilage in dispersal of certain fungi. *Trans. Br. Mycol. Soc. 70*, 137–140.

Isogai, A., Kanaoka, M., Matsuda, H., Hori, Y., and Suzuki, A. 1978. Structure of a new cyclodepsipeptide, Beauverilide A from *Beauveria bassiana*. *Agric. Biol. Chem. 42*, 1797–1798.

Jaronski, S., Axtell, R. C., Fagan, S. M., and Domnas, A. J. 1983. In vitro production of zoospores by the mosquito pathogen *Lagenidium giganteum* (Oomycetes: Lagenidiales) on solid media. *J. Invertebr. Pathol. 41*, 305–309.

Jarosz, J. 1979. Yeastlike fungi from greater wax moth larvae (*Galleria mellonella*) fed antibiotics. *J. Invertebr. Pathol. 34*, 257–262.

Kamburov, S. S., Nadel, D. J., and Kenneth, R. 1967. Observations on *Hesperomyces virescens* Thaxter (Laboulbeniales), a fungus associated with premature mortality of *Chilocorus bipustulatus* L. in Israel. *Isr. J. Agric. Res.* 17, 131–134.

Kanaoka, M., Isogai, A., Murakoshi, S., Ichinoe, M., Suzuki, A., and Tamura, S. 1978. Bassianolide, a new insecticidal cyclodepsipeptide from *Beauveria bassiana* and *Verticillium lecanii*. *Agric. Biol. Chem.* 42, 629–635.

Karling, J. S. 1948. Chytridiosis of scale insects. *Am. J. Bot.* 35, 246–254.

Kawakami, K., and Mikuni, T. 1969. Studies on the causative fungi of *Aspergillus* disease of the silkworm larvae. (I) Pathogenicity to the silkworm larvae and tolerance to formalin of *Aspergillus* isolates collected from co-operative rearing houses of young silkworm larvae. *Bull. Seric. Exp. Stn.* (Tokyo) 23, 327–370.

Kawamoto, H., and Aizawa, K. 1986. Formation and regeneration of protoplasts from blastospores of an entomogenous fungus, *Beauveria bassiana*. *Appl. Entomol. Zool.* 21, 531–538.

Keilin, D. 1920. On a new Saccharomycete *Monosporella unicuspidata* gen. n. nom., n. sp., parasitic in the body cavity of a dipterous larva (*Dasyhelea obscura* Winnertz). *Parasitology* 12, 83–91.

Keilin, D. 1921. On a new type of fungus: *Coelomomyces stegomyiae*, n. g., n. sp., parasitic in the body-cavity of the larva of *Stegomyia scutellaris* Walker (Diptera, Nematocera, Culicidae). *Parasitology* 13, 225–234.

Kendrick, W. B., and Carmichael, J. W. 1973. Hyphomycetes. *In* "The Fungi: An Advanced Treatise." (G. C. Ainsworth, F. K. Sparrow and A. S. Sussman, eds.), Vol. IVA, pp. 323–509. Academic Press, New York.

Kenneth, R. G. 1977. *Entomophthora turbinata* sp. n., a fungal parasite of the peach trunk aphid, Pterochloroides persicae (Lachnidae). *Mycotaxon* 6, 381–390.

Kenneth, R., Muttath, T. I., and Gerson, U. 1979. *Hirsutella thompsonii*, a fungal pathogen of mites. I. Biology of the fungus *in vitro*. *Ann. Appl. Biol.* 91, 21–28.

Kerwin, J. L. 1982. Chemical control of the germination of asexual spores of *Entomophthora culicis*, a fungus parasitic on dipterans. *J. Gen. Microbiol.* 128, 2179–2186.

Kerwin, J. L. 1983a. Biological aspects of the interaction between *Coelomomyces psorophorae* zygotes and the larvae of *Culiseta inornata*: Host-mediated factors. *J. Invertebr. Pathol.* 41, 224–232.

Kerwin, J. L. 1983b. Biological aspects of the interaction between *Coelomomyces psorophorae* zygotes and the larvae of *Culiseta inornata*: Environmental factors. *J. Invertebr. Pathol.* 41, 233–237.

Kerwin, J. L. 1984. Fatty acid regulation of the germination of *Erynia variabilis* conidia on adults and puparia of the lesser housefly, *Fannia canicularis*. *Can. J. Microbiol.* 30, 158–161.

Kerwin, J. L., and Washino, R. K. 1983. Sterol induction of sexual reproduction in *Lagenidium giganteum*. *Exp. Mycol.* 7, 109–115.

Kerwin, J. L., and Washino, R. K. 1986a. Oosporogenesis by *Lagenidium giganteum*: Induction and maturation are regulated by calcium and calmodulin. *Can. J. Microbiol.* 32, (8), 663–672.

Kerwin, J. L., and Washino, R. K. 1986b. Cuticular regulation of host recognition and spore germination by entomopathogenic fungi. *In* "Fundamental and Applied Aspects of Invertebrate Pathology." (R. A. Samson, J. M. Vlak, and D. Peters, eds.), pp. 423–425. Found. Fourth Int. Colloq. Invertebr. Pathol. Wageningen, Netherlands.

Kerwin, J. L., and Washino, R. K. 1986c. Regulation of oosporogenesis by *Lagenidium giganteum*; promotion of sexual reproduction by unsaturated fatty acids and sterol availability. *Can. J. Microbiol.* 32, 294–300.

King, D. S. 1976a. Systematics of *Conidiobolus* (Entomophthorales) using numerical taxonomy. I. Biology and cluster analysis. *Can. J. Bot.* 54, 45–65.

King, D. S. 1976b. Systematics of *Conidiobolus* (Entomophthorales) using numerical taxonomy. II. Taxonomic considerations. *Can. J. Bot.* 54, 1285–1296.

King, D. S., and Humber, R. A. 1981. Identification of the Entomophthorales. *In* "Microbial Control of Pests and Plant Diseases 1970–1980." (H. D. Burges, ed.), pp. 107–127. Academic Press, London.

Kirk, H. D., Ewen, A. B., Emson, H. E., and Blair, D. G. R. 1971. Effect of aflatoxin B_1 on development of *Drosophila melanogaster* (Diptera). *J. Invertebr. Pathol. 18*, 313–315.

Kish, L. P., Samson, R. A., and Allen, G. E. 1974. The genus *Nomuraea* Maublanc. *J. Invertebr. Pathol. 24*, 154–158.

Knight, A. L. 1980. Host range and temperature requirements of *Culicinomyces clavosporus*. *J. Invertebr. Pathol. 36*, 423–425.

Kobayashi, Y. 1941. The genus *Cordyceps* and its allies. *Sci. Rep. Tokyo Bunrika Daigaku Sect. B 5*, 53–260.

Kobayashi, Y. 1977. Miscellaneous notes on the genus *Cordyceps* and its allies. *J. Jpn. Bot. 52*, 269–272.

Kobayashi, Y., Mogami, K., and Aoki, J. 1984. Ultrastructural studies on the hyphal growth of *Erynia neoaphidis* in the green peach aphid, *Myzus persicae*. *Trans. Mycol. Soc. Jpn. 25*, 425–434.

Kodaira, Y. 1961a. Toxic substances to insects, produced by *Aspergillus ochraceus* and *Oospra destructor*. *Agric. Biol. Chem.* (Tokyo), *25*, 261–262.

Kodaira, Y. 1961b. Biochemical studies on the muscardine fungi in the silkworms, *Bombyx mori*. *J. Fac. Text. Sci. Technol. Shinshu Univ. Ser. E 29*, 1–68.

Kodaira, Y. 1962. Studies on the new toxic substances to insects, destruxin A and B, produced by *Oospora destructor*. Part I. Isolation and purification of destruxin A and B. *Agric. Biol. Chem.* (Tokyo), *26*, 36–42.

Koidsumi, K. 1957. Antifungal action of cuticular lipids in insects. *J. Insect Physiol. 1*, 40–51.

Kramer, J. P. 1979. Interactions between blow flies (Calliphoridae) and *Entomophthora bullata* (Phycomycetes: Entomophthorales). *J. N.Y. Entomol. Soc. 87*, 135–140.

Kramer, J. P. 1980a. Entomophthora sepulchralis Thaxter (Entomophthoraceae): Its rediscovery and experimentally induced infections in muscoid flies. *Mycopathologia 70*, 163–167.

Kramer, J. P. 1980b. The house-fly mycosis caused by *Entomophthora muscae*: Influence of relative humidity on infectivity and conidial germination. *N.Y. Entomol. Soc. 88*, 236–240.

Kramm, K. R., and West, D. F. 1982. Termite pathogens: Effects of ingested *Metarhizium*, *Beauveria*, and *Gliocladium* conidia on worker termites (*Reticulitermes* sp.). *J. Invertebr. Pathol. 40*, 7–11.

Krejzová, R. 1978. Taxonomy, morphology, and surface structure of *Basidiobolus* sp. isolate. *J. Invertebr. Pathol. 31*, 157–163.

Krieg, A., Gröner, A., Huber, J., and Zimmermann, G. 1981. Inaktivierung von verschiedenen Insektenpathogenen durch ultraviolette Strahlen. *Z. Pflanzenkr. Pflanzenschutz 88*, 38–48.

Kučera, M. 1971. Toxins of the entomophagous fungus *Beauveria bassiana*. II. Effect of nitrogen sources on formation of the toxic protease in submerged culture. *J. Invertebr. Pathol. 17*, 211–215.

Kučera, M. 1980. Proteases from the fungus *Metarhizium anisopliae* toxic for *Galleria mellonella* larvae. *J. Invertebr. Pathol. 35*, 304–310.

Kučera, M. 1982. Inhibition of the toxic proteases of *Metarhizium anisopliae* by extracts of *Galleria mellonella* larvae. *J. Invertebr. Pathol. 40*, 299–300.

Kučera, M. 1984. Partial purification and properties of *Galleria mellonella* larvae proteolytic inhibitors acting on *Metarhizium anisopliae* toxic protease. *J. Invertebr. Pathol. 43*, 190–196.

Kučera, M., and Samšiňáková, A. 1968. Toxins of the entomophagous fungus *Beauveria bassiana*. *J. Invertebr. Pathol. 12*, 316–320.

Kurisu, K., and Manabe, Y. 1978. Histopathological study on the relationship between the fungus invasion of *Aspergillus flavus* and the renewal of the cuticle in the mature silkworm larvae (*Bombyx mori*). *Bull. Fac. Text. Sci. Kyoto Univ. Ind. Arts Text. Fiber.* *8*, 34–40.

Kusunoki, J., and Watanabe, H. 1982. Changes in the hemolymph pH and specific gravity of the silkworm larvae, *Bombyx mori*, infected with *Beauveria bassiana*. *J. Seric. Sci. Jpn.* *51*, 447–448.

Latch, G. C. M. 1976. Studies on the susceptibility of *Oryctes rhinoceros* to some entomogenous fungi. *Entomophaga 21*, 31–38.

Latgé, J.-P. 1976. Étude morphologique et cytologique de la formation de "spores durables" chez une espèce d'Entomophthorale. *C.R. Acad. Sci. Paris Ser. D 282*, 605–608.

Latgé, J.-P. 1977. Croissance et sporulation d'*Entomophthora virulenta* Hall et Dunn en culture liquide discontinue. *C.R. Acad. Sci. Paris Ser. D 284*, 1879–1882.

Latgé, J. P. 1980. Sporulation de *Entomophthora obscura* Hall & Dunn en culture liquide. *Can. J. Microbiol. 26*, 1038–1048.

Latgé, J. P. 1981. Comparaison des exigences nutritionnelles des Entomophthorales. *Ann. Microbiol.* (Inst. Pasteur) *132*B, 299–306.

Latgé, J. P., and Beauvais, A. 1987. Wall composition of the protoplastic Entomophthorales. *J. Invertebr. Pathol. 50*, 53–57.

Latgé, J. P., and Papierok, B. 1988. Aphid pathogens. *In* "Aphids Their Biology, Natural Enemies and Control" (A. K. Minks and P. Harrewijn, eds.), Vol. 2B, pp. 323–335. Elsevier, Amsterdam.

Latgé, J. P., Remaudière, G., and Diaquin, M. 1978a. Un nouveau milieu pour la croissance et la sporulation d'Entomophthorales pathogenes d'aphides. *Ann. Microbiol.* (Inst. Pasteur) *129B*, 463–476.

Latgé, J. P., Remaudière, G., Soper, R. S., Madore, C. D., and Diaquin, M. 1978b. Growth and sporulation of *Entomophthora virulenta* on semidefined media in liquid culture. *J. Invertebr. Pathol. 31*, 225–233.

Latgé, J. P., Perry, D., Reisinger, O., Papierok, B., and Remaudière, G. 1979. Induction de la formation des spores de résistance d'*Entomophthora obscura* Hall et Dunn. *C.R. Acad. Sci. Paris Ser. D 288*, 599–601.

Latgé, J. P., King, D. S., and Papierok, B. 1980. Synonymie de Entomophthora virulenta Hall et Dunn et de Conidiobolus thromboides Drechsler. *Mycotaxon 11*, 255–268.

Latgé, J. P., Prévost, M. C., Perry, D. F., and Reisinger, O. 1982. Étude en microscopie électronique de *Conidiobolus obscurus*. I. Formation et germination des azygospores. *Can. J. Bot. 60*, 413–431.

Latgé, J. P., Beauvais, A., and Vey, A. 1986a. Wall synthesis in the Entomophthorales and its role in the immune reaction of infected insects. *Dev. Comp. Immunol. 10*, 639.

Latgé, J. P., Cole, G. T., Horisberger, M., and Prévost, M.-C. 1986b. Ultrastructure and chemical composition of the ballistospore wall of *Conidiobolus obscurus*. *Exp. Mycol. 10*, 99–113.

Latgé, J. P., Sampedro, L., Brey, P., and Diaquin, M. 1987. Aggressiveness of *Conidiobolus obscurus* against the pea aphid: Influence of cuticular extracts on ballistospore germination of aggressive and non-aggressive strains. *J. Gen. Microbiol. 133*, 1987–1997.

Lefebvre, C. L. 1934. Penetration and development of the fungus, Beauveria Bassiana, in the tissues of the corn borer. *Ann. Bot. 48*, 441–452.

Lindroth, C. H. 1948. Notes on the ecology of Laboulbeniaceae infesting carabid beetles. *Sven. Bot. Tidskr. 42*, 34–41.

Llewellyn, G. C., and Chinnici, J. P. 1978. Variation in sensitivity to aflatoxin B_1 among several strains of *Drosophila melanogaster* (Diptera). *J. Invertebr. Pathol. 31*, 37–40.

Lloyd, M., White, J., and Stanton, N. 1982. Dispersal of fungus-infected periodical cicadas to new habitat. *Environ. Entomol. 11*, 852–858.

Lord, J. C., and Roberts, D. W. 1985. Effects of salinity, *p*H, organic solutes, anaerobic conditions, and the presence of other microbes on production and survival of

Lagenidium giganteum (Oomycetes: Lagenidiales) zoospores. *J. Invertebr. Pathol. 45*, 331–338.

Lord, J. C., and Roberts, D. W. 1986. The effects of culture medium quality and host passage on zoosporogenesis, oosporogenesis, and infectivity of *Lagenidium giganteum* (Oomycetes: Lagenidiales). *J. Invertebr. Pathol. 48*, 355–361.

Lord, J. C., and Roberts, D. W. 1987. Host age as a determinant of infection rates with the mosquito pathogen *Lagenidium giganteum* (Oomycetes: Lagenidiales). *J. Invertebr. Pathol. 50*, 70–71.

Luttrell, E. S. 1973. Loculoascomycetes. *In* "The Fungi: An Advanced Treatise." (G. C. Ainsworth, F. K. Sparrow, and A. S. Sussman, eds.), Vol. IVA, pp. 135–219. Academic Press, New York.

Lynch, R. E., and Lewis, L. C. 1978. Fungi associated with eggs and first-instar larvae of the European corn borer. *J. Invertebr. Pathol. 32*, 6–11.

MacLeod, D. M. 1954. Investigations on the genera *Beauveria* Vuill. and *Tritirachium* Limber. *Can. J. Bot. 32*, 818–890.

MacLeod, D. M. 1963. Entomophthorales infections. *In* "Insect Pathology: An Advanced Treatise." (E. A. Steinhaus, ed.), Vol. 2, pp. 189–231. Academic Press, New York.

MacLeod, D. M., and Müller-Kögler, E. 1973. Entomogenous fungi: Entomophthora species with pear-shaped to almost spherical conidia (Entomophthorales: Entomophthoraceae). *Mycologia 65*, (4) 823–893.

MacLeod, D. M., Cameron, J. W. M., and Soper, R. S. 1966. The influence of environmental conditions on epizootics caused by entomogenous fungi. *Rev. Roum. Biol. 11*, 125–134.

MacLeod, D. M., Tyrrell, D., Soper, R. S., and De Lyzer, A. J. 1973. *Entomophthora bullata* as a pathogen of *Sarcophaga aldrichi*. *J. Invertebr. Pathol. 22*, 75–79.

MacLeod, D. M., Tyrrell, D., and Welton, M. A. 1980. Isolation and growth of the grasshopper pathogen, *Entomophthora grylli*. *J. Invertebr. Pathol. 36*, 85–89.

Maghrabi, H. A., and Kish, L. P. 1985. Isozyme characterization of Ascosphaerales associated with bees. I. *Ascosphaera apis*, *Ascosphaera proliperda*, and *Ascosphaera aggregata*. *Mycologia 77*, 358–365.

Mains, E. B. 1958. North American entomogenous species of Cordyceps. *Mycologia 50*, 169–222.

Manalang, C. 1930. Coccidiosis in Anopheles mosquitoes. *Philipp. J. Sci. 42*, 279–281.

Martin, W. W., III. 1969. A morphological and cytological study of development in *Coelomomyces punctatus* parasitic in *Anopheles quadrimaculatus*. *J. Elisha Mitchell Sci. Soc. 85*, 59–72.

Martin, W. W. 1975a. A new species of Catenaria parasitic in midge eggs. *Mycologia 67*, 264–272.

Martin, W. W. 1975b. *Aphanomycopsis sexualis*, a new parasite of midge eggs. *Mycologia 67*, 923–933.

Martin, W. W. 1977. The development and possible relationships of a new Atkinsiella parasitic in insect eggs. *Am. J. Bot. 64*, 760–769.

Martin, W. W. 1978. Two additional species of *Catenaria* (Chytridiomycetes, Blastocladiales) parasitic in midge eggs. *Mycologia 70*, 461–467.

Martin, W. W. 1981. *Couchia circumplexa*, a water mold parasitic in midge eggs. *Mycologia 73*, 1143–1157.

Martin, W. W. 1984. The dynamics of aquatic fungi parasitic in a stream population of the midge, *Chironomus attenuatus*. *J. Invertebr. Pathol. 44*, 36–45.

Matanmi, B. A., and Libby, J. L. 1976. The production and germination of resting spores of *Entomophthora virulenta* (Entomophthorales: Entomophthoraceae). *J. Invertebr. Pathol. 27*, 279–285.

Matsumura, F., and Knight, S. G. 1967. Toxicity and chemosterilizing activity of aflatoxin against insects. *J. Econ. Entomol. 60*, 871-872.

Matus, F., and Sarbak, I. 1974. A méhek költésmeszesedésének (acosphaeriosis) elöfordulása hazánkban. *Mag. Allator. Lapja 29*, 250–255.

McCauley, V. J. E., Zacharuk, R. Y., and Tinline, R. D. 1968. Histopathology of green muscardine in larvae of four species of Elateridae (Coleoptera). *J. Invertebr. Pathol. 12*, 444–459.

McCoy, C. W. 1981. Pest control by the fungus *Hirsutella thompsonii*. In "Microbial Control of Pests and Plant Diseases 1970–1980." (H. D. Burges, ed.), pp. 499–512. Academic Press, London.

McCoy, C. W., Selhime, A. G., Kanavel, R. F., and Hill, A. J. 1971. Suppression of citrus rust mite populations with application of fragmented mycelia of *Hirsutella thompsonii*. *J. Invertebr. Pathol. 17*, 270–276.

McCoy, C. W., Hill, A. J., and Kanavel, R. F. 1972. A liquid medium for the large-scale production of *Hirsutella thompsonii* in submerged culture. *J. Invertebr. Pathol. 19*, 370–374.

McCoy, C. W., Samson, R. A., and Boucias, D. G. 1988. Entomogenous fungi. In "CRC Handbook of Natural Pesticides. Microbial Insecticides, Part A. Entogenous Protozoa and Fungi." (C. M. Ignoffo, ed.), Vol. 5, pp. 151–236.

McCray, E. M., Jr., Umphlett, C. J., and Fay, R. W. 1973. Laboratory studies on a new fungal pathogen of mosquitoes. *Mosq. News 33*, 54–60.

McInnis, T., Jr., and Zattau, W. C. 1982. Experimental infection of mosquito larvae by a species of the aquatic fungus *Leptolegnia*. *J. Invertebr. Pathol. 39*, 98–104.

McManus, W. R., and Youssef, N. N. 1984. Life cycle of the chalk brood fungus, *Ascosphaera aggregata*, in the alfalfa leafcutting bee, *Megachile rotundata*, and its associated symptomatology. *Mycologia 76*, 830–842.

McMillian, W. W., Wilson, D. M., Widstrom, N. W., and Perkins, W. D. 1980. Effects of aflatoxin B_1 and G_1 on three insect pests of maize. *J. Econ. Entomol. 73*, 26–28.

McMillian, W. W., Widstrom, N. W., and Wilson, D. M. 1981. Rearing the maize weevil on maize genotypes when aflatoxin-producing *Aspergillus flavus* and *A. parasiticus* isolates were present. *Environ. Entomol. 10*, 760–762.

Meola, S., and DeVaney, J. 1976. Parasitism of Mallophaga by *Trenomyces histophtorus*. *J. Invertebr. Pathol. 28*, 195–201.

Meola, S., and Tavares, I. I. 1982. Ultrastructure of the haustorium of *Trenomyces histophthorus* and adjacent host cells. *J. Invertebr. Pathol. 40*, 205–215.

Metalnikov, S., Ellinger, T., and Chorine, V. 1928. A new yeast species, isolated from diseased larvae of *Pyrausta nubilalis* Hb. *Int. Corn Borer Invest., Sci. Rpts. 1927–1928. 1*, 70–71.

Millstein, J. A., Brown, G. C., and Nordin, G. L. 1982. Microclimatic humidity influence on conidial discharge in *Erynia* sp. (Entomophthorales: Entomophthoraceae), an entomopathogenic fungus of the alfalfa weevil (Coleoptera: Curculionidae). *Environ. Entomol. 11*, 1166–1169.

Millstein, J. A., Brown, G. C., and Nordin, G. L. 1983a. Microclimatic moisture and conidial production in *Erynia* sp. (Entomophthorales: Entomophthoraceae): In vivo moisture balance and conidiation phenology. *Environ. Entomol. 12*, 1339–1343.

Millstein, J. A., Brown, G. C., and Nordin, G. L. 1983b. Microclimatic moisture and conidial production in *Erynia* sp. (Entomophthorales: Entomophthoraceae): In vivo production rate and duration under constant and fluctuating moisture regimes. *Environ. Entomol. 12*, 1344–1349.

Milne, C. P., Jr. 1983. Honey bee (Hymenoptera: Apidae) hygienic behavior and resistance to chalkbrood. *Ann. Entomol. Soc. Am. 76*, 384–387.

Milner, R. J. 1981. Patterns of primary spore discharge of *Entomophthora* spp. from the blue green aphid, *Acyrthosiphon kondoi*. *J. Invertebr. Pathol. 38*, 419–425.

Milner, R. J., Holdom, D. G., and Glare, T. R. 1984. Diurnal patterns of mortality in aphids infected by entomophthoran fungi. *Entomol. Exp. Appl. 36*, 37–42.

Mohamed, A. K. A., Sikorowski, P. P., and Bell, J. V. 1978. Histopathology of *Nomuraea rileyi* in larvae of *Heliothis zea* and in vitro enzymatic activity. *J. Invertebr. Pathol. 31*, 345–352.

Moore, G. E. 1973. Pathogenicity of three entomogenous fungi to the southern pine beetle at various temperatures and humidities. *Environ. Entomol. 2*, 54–57.

Mullens, B. A., Rodriguez, J. L., and Meyer, J. A. 1987. An epizootiological study of *Entomophthora muscae* in muscoid fly populations on southern California poultry facilities, with emphasis on *Musca domestica*. *Hilgardia 55*, 1–41.

Müller, E,. and von Arx, J. A. 1973. Pyrenomycetes: Meliolales, Coronophorales, Sphaeriales. *In* "The Fungi: An Advanced Treatise." (G. C. Ainsworth, F. K. Sparrow, and A. S. Sussman, eds.), Vol. IVA, pp. 87–132. Academic Press, New York.

Müller-Kögler, E. 1959. Zur Isolierung und Kultur insektenpathogener Entomophthoraceen. *Entomophaga 4*, 261–274.

Müller-Kögler, E. 1965. "Pilzkrankheiten bei Insekten. Anwendung zur biologischen Schädlingsbekämpfung und Grundlagen der Insektenmykologie." Paul Parey, Berlin.

Murakoshi, S., Sugiyama, J., and Ohtomo, T. 1971. Studies on the *Aspergilli* from silkworm, *Bombyx mori* L. (I). The occurrence of aflatoxin-producing strains in *Aspergillus* strains isolated from sericultural farms and their toxic effects on silkworm larvae. *J. Seric. Sci. Jpn. 40*, 167–175.

Murakoshi, S., Ichinoe, M., Suzuki, A., Kanaoka, M., Isogai, A., and Tamura, S. 1978. Presence of toxin substance in fungus bodies of the entomopathogenic fungi, *Beauveria bassiana* and *Verticillium lecanii*. *Appl. Entomol. Zool. 13*, 97–102.

Myokei, R., Sakurai, A., Chang, C.-F., Kodaira, Y., Takahashi, N., and Tamura, S. 1969. Aspochracin, a new insecticidal metabolite of *Aspergillus ochraceus*. Part I. Isolation, structure and biological activities. *Agric. Biol. Chem. 33*, 1491–1500.

Nemoto, H., Kobayashi, M., and Takizawa, Y. 1979. Scanning electron microscopy of *Entomophthora* (*Triplosporium*) *floridana* (Zygomycetes: Entomophthorales) attacking the sugi spider mite, *Oligonychus hondoensis* (Acarina: Tetranychidae). *Appl. Entomol. Zool. 14*, 376–382.

Newman, G. G., and Carner, G. R. 1974. Diel periodicity of *Entomophthora gammae* in the soybean looper. *Environ. Entomol. 3*, 888–890.

Newman, G. G., and Carner, G. R. 1975a. Environmental factors affecting conidial sporulation and germination of *Entomophthora gammae Environ. Entomol. 4*, 615–618.

Newman, G. G., and Carner, G. R. 1975b. Factors affecting the spore form of *Entomophthora gammae*. *J. Invertebr. Pathol. 26*, 29–34.

Nnakumusane, E. S. 1987a. Effects of temperature on the susceptibility of *Aedes aegypti* (L.) (Diptera: Culicidae) larvae to a mosquito pathogen *Coelomomyces stegomyiae* in Uganda. *Appl. Entomol. Zool. 22*, 7–12.

Nnakumusane, E. S. 1987b. Histological studies of *Cordyceps myrmecophila* (CES) infection in the ant *Palthothyreus tarsatus* Fab. (Formicidae: Ponerinae). *Appl. Entomol. Zool. 22*, 1–6.

Nolan, R. A. 1985. Protoplasts from Entomophthorales. *In* "Fungal Protoplasts: Applications in Biochemistry and Genetics." (J. F. Peberdy and L. Ferenczy, eds.), pp. 87–112. Marcel Dekker, New York.

Nolan, R. A., Laird, M., Chapman, H. C., and Glenn, F. E., Jr. 1973. A mosquito parasite from a mosquito predator. *J. Invertebr. Pathol. 21*, 172–175.

Nolan, R. A., Dunphy, G. B., and MacLeod, D. M. 1976. In vitro germination of *Entomophthora egressa* resting spores. *Can. J. Bot. 54*, 1131–1134.

Nordin, G. L., Brown, G. C., and Millstein, J. A. 1983. Epizootic phenology of *Erynia* disease of the alfalfa weevil, *Hypera postica* (Gyllenhal) (Coleoptera: Curculionidae), in central Kentucky. *Environ. Entomol. 12*, 1350–1355.

Ohkawa, A., and Aoki, J. 1980. Fine structure of resting spore formation and germination in *Entomophthora virulenta*. *J. Invertebr. Pathol. 35*, 279–289.

Ohtomo, T., Murakoshi, S., Sugiyama, J., and Kurata, H. 1975. Detection of aflatoxin B_1 in silkworm larvae attacked by an *Aspergillus flavus* isolate from a sericultural farm. *Appl. Microbiol. 30*, 1034–1035.

Papierok, B. 1978. Obtention in vivo des azygospores d'*Entomophthora thaxteriana* Petch, champignon pathogène de Pucerons (Homoptères Aphididae). *C.R. Acad. Sci. Paris Ser. D 286*, 1503–1506.

Paris, S., and Ferron, P. 1979. Study of the virulence of some mutants of *Beauveria brongniartii* (Syn *Beauveria tenella*). *J. Invertebr. Pathol. 34*, 71–77.

Paris, S., and Segretain, G. 1978. Étude de l'activité lipasique-estérasique intracellulaire de *Beauveria tenella*. *Ann. Microbiol.* (Inst. Pasteur) *129B*, 133–145.

Paster, N., Lisker, N., and Chet, I. 1983. Ochratoxin A production by *Aspergillus ochraceus* Wilhelm grown under controlled atmospheres. *Appl. Environ. Microbiol. 46*, 1136–1139.

Paterson, R. R. M., Simmonds, M. S. J., and Blaney, W. M. 1987. Mycopesticidal effects of characterized extracts of *Penicillium* isolates and purified secondary metabolites (including mycotoxins) on *Drosophila melanogaster* and *Spodoptera littoralis*. *J. Invertebr. Pathol. 50*, 124–133.

Payandeh, B., MacLeod, D. M., and Wallace, D. R. 1978. Germination of *Entomophthora aphidis* resting spores under constant temperatures. *Can. J. Bot. 56*, 2328–2333.

Pekrul, S., and Grula, E. A. 1979. Mode of infection of the corn earworm (*Heliothis zea*) by *Beauveria bassiana* as revealed by scanning electron microscopy. *J. Invertebr. Pathol. 34*, 238–247.

Pendland, J. C. 1982. Resistant structures in the entomogenous hyphomycete, *Nomuraea rileyi*: An ultrastructural study. *Can. J. Bot. 60*, 1569–1576.

Pendland, J. C., and Boucias, D. G. 1984. Ultrastructure of conidial germination in the entomopathogenic hyphomycete *Nomuraea rileyi*. *J. Invertebr. Pathol. 43*, 432–434.

Perry, D. F., and Latgé, J.-P. 1980. Chemically defined media for growth and sporulation of *Entomophthora virulenta*. *J. Invertebr. Pathol. 35*, 43–48.

Perry, D. F., Tyrrell, D., and DeLyzer, A. J. 1982. The mode of germination of *Zoophthora radicans* zygospores. *Mycologia 74*, 549–554.

Petch, T. 1921. Fungi parasitic on scale insects. Presidential Address. *Trans. Br. Mycol. Soc. 7*, 18–40.

Petch, T. 1936. *Cordyceps militaris* and *Isaria farinosa*. *Trans. Br. Mycol. Soc. 20*, 216–224.

Poltev, V. I., and Neshatayeva, E. V. 1969. La melanose experimentale des abeilles causée par la champignon *Aureobasidium pullulans* (DeBary) Arnaud. *Bull. Apic. 12*, 189–198.

Powell, M. J. 1976. Ultrastructural changes in the cell surface of *Coelomomyces punctatus* infecting mosquito larvae. *Can. J. Bot. 54*, 1419–1437.

Prasertphon, S. 1967. Mycotoxin production by species of *Entomophthora*. *J. Invertebr. Pathol. 9*, 281–282.

Prasertphon, S., and Tanada, Y. 1968. The formation and circulation, in *Galleria*, of hyphal bodies of entomophthoraceous fungi. *J. Invertebr. Pathol. 11*, 260–280.

Prasertphon, S., and Tanada, Y. 1969. Mycotoxins of entomophthoraceous fungi. *Hilgardia 39*, 581–600.

Prior, C., and Perry, C. H. 1980. Infection of *Promecotheca papuana* with *Synnematium jonesii*. *J. Invertebr. Pathol. 35*, 14–19.

Quiot, J.-M., Vey, A., Vago, C., and Païs, M. 1980. Action antivirale d'une mycotoxine. Étude d'une toxine de l'hyphomycete *Metarhizium anisopliae* (Metsch) Sorok en culture cellulaire. *C.R. Acad. Sci. Paris Ser. D 291*, 763–766.

Ratault, C., and Vey, A. 1977. Production d'estérase et de N-acétyl-β-D glucosaminidase dans le tégument du coléoptère *Oryctes rhinoceros* par le champignon entomophthogène *Metarrhizium anisopliae*. *Entomophaga 22*, 289–294.

Remaudière, G., and Hennebert, G. L. 1980. Revision systematique de Entomophthora aphidis Hoffm. in Fres. Description de deux nouveaux pathogenes d'aphides. *Mycotaxon 11*, 269–321.

Remaudière, G., and Keller, S. 1980. Revision systematique des genres d'Entomophthoraceae a potentialite entomopathogene. *Mycotaxon 11*, 323–338.

Remaudière, G., Latgé, J.-P., Papierok, B., and Coremans-Pelseneer, J. 1976. Sur le pouvoir pathogène de quatre espèces d'Entomophthorales occasionnellement isolées d'Aphides en France. *C.R. Acad. Sci. Paris Ser. D 283*, 1065–1068.

Remaudière, G., Latgé, J.-P., and Papierok, B. 1979. Reconsidération taxonomique de *Entomophthora obscura* Hall et Dunn. *Ann. Microbiol.* (Inst. Pasteur) *130A*, 151–162.

Riba, G., Bouvier-Fourcade, I., and Caudal, A. 1986. Isoenzymes polymorphism in *Metarhizium anisopliae* (Deuteromycotina: Hyphomycetes) entomogenous fungi. *Mycopathologia 96*, 161–169.

Richard, J. L., and Gallagher, R. T. 1979. Multiple toxin production by an isolate of Aspergillus flavus. *Mycopathologia 67*, 161–163.

Robert, A., and Messing-Al-Aidroos, K. 1985. Acid production by *Metarhizium anisopliae*: Effects on virulence against mosquitoes and on detection of in vitro amylase, protease, and lipase activity. *J. Invertebr. Pathol. 45*, 9–15.

Roberts, D. W. 1966a. Toxins from the entomogenous fungus *Metarrhizium anisopliae*. I. Production in submerged and surface cultures, and in inorganic and organic nitrogen media. *J. Invertebr. Pathol. 8*, 212–221.

Roberts, D. W. 1966b. Toxins from the entomogenous fungus *Metarrhizium anisopliae*. II. Symptoms and detection in moribund hosts. *J. Invertebr. Pathol. 8*, 222–227.

Roberts, D. W. 1969. Toxins from the entomogenous fungus *Metarrhizium anisopliae*: Isolation of destruxins from submerged cultures. *J. Invertebr. Pathol. 14*, 82–88.

Roberts, D. W. 1970. *Coelomomyces, Entomophthora, Beauveria,* and *Metarrhizium* as parasites of mosquitoes. *Misc. Publ. Entomol. Soc. Am. 7*, 140–155.

Roberts, D. W. 1981. Toxins of entomopathogenic fungi. In "Microbial Control of Pests and Plant Diseases 1970–1980." (H. D. Burges, ed.), pp. 441–464. Academic Press, London.

Roberts, D. W., and Campbell, A. S. 1977. Stability of entomopathogenic fungi. *Misc. Publ. Entomol. Soc. Am. 10*, (3), 19–76.

Roberts, D. W., and Humber, R. A. 1981. Entomogenous fungi. In "Biology of Conidial Fungi." (G. T. Cole and B. Kendrick, eds.), Vol. 2, pp. 201–236. Academic Press, New York.

Rockwood, L. P. 1950. Entomogenous fungi of the genus Metarrhizium on wireworms in the Pacific northwest. *Ann. Entomol. Soc. Am. 43*, 495–498.

Roffey, J. 1968. The occurrence of the fungus *Entomophthora grylli* Fresenius on locusts and grasshoppers in Thailand. *J. Invertebr. Pathol. 11*, 237–241.

Rossman, A. Y. 1978. Podonectria, a genus in the Pleosporales on scale insects. *Mycotaxon 7*, 163–182.

Rossman, A. Y. 1983. The phragmosporous species of *Nectria* and related genera (*Calonectria, Ophionectria, Paranectria, Scoleconectria* and *Trichonectria*). *Mycol. Pap. No. 150*.

Saito, T., and Aoki, J. 1983. Toxicity of free fatty acids on the larval surfaces of two lepidopterous insects towards *Beauveria bassiana* (Bals.) Vuill. and *Paecilomyces fumoso-roseus* (Wize) Brown et Smith. (Deuteromycetes: Moniliales). *Appl. Entomol. Zool. 18*, 225–233.

Samšiňáková, A., and Hrabětová, E. 1969. Respiration of blastospores of the fungus *Beauveria bassiana* during submersed cultivation in the presence of certain sugars. *J. Invertebr. Pathol. 13*, 382–385.

Samšiňáková, A., and Kálalová, S. 1983. The influence of a single-spore isolate and repeated subculturing on the pathogenicity of conidia of the entomophagous fungus *Beauveria bassiana*. *J. Invertebr. Pathol. 42*, 156–161.

Samšiňáková, A., Mišíková, S., and Leopolod, J. 1971. Action of enzymatic systems of

Beauveria bassiana on the cuticle of the greater wax moth larvae (*Galleria mellonella*). *J. Invertebr. Pathol. 18*, 322–330.

Samšiňáková, A., Bajan, C., Kálalová, S., Kmitowa, K., and Wojciechowska, M. 1977. The effect of some entomophagous fungi on the Colorado beetle and their enzyme activity. *Bull. Acad. Pol. Sci. Cl. II. 25*, 521–526.

Samson, R. A. 1974. Paecilomyces and some allied hyphomycetes. *Stud. Mycol. 6*, 1–119.

Samson, R. A., and Evans, H. C. 1977. Notes on entomogenous fungi from Ghana. IV. The genera Paecilomyces and Nomuraea. *Proc. K. Ned. Akad. Wet. 80*, 128–134.

Samson, R. A., and Evans, H. C. 1982. Two new *Beauveria* spp. from South America. *J. Invertebr. Pathol. 39*, 93–97.

Samson, R. A., and Soares, G. G., Jr. 1984. Entomopathogenic species of the hyphomycete genus *Tolypocladium*. *J. Invertebr. Pathol. 43*, 133–139.

Samson, R. A., McCoy, C. W., and O'Donnell, K. L. 1980. Taxonomy of the acarine parasite *Hirsutella thompsonii*. *Mycologia 72*, 359–377.

Samson, R. A., Evans, H. C., and Latgé, J. P. 1988. "Atlas of Entomopathogenic Fungi." Springer-Verlag, Berlin.

Sawyer, W. H., Jr. 1929. Observations on some entomogenous members of the Entomophthoraceae in artificial culture. *Am. J. Bot. 16*, 87–121.

Schabel, H. G. 1976. Oral infection of *Hylobius pales* by *Metarrhizium anisopliae*. *J. Invertebr. Pathol. 27*, 377–383.

Schabel, H. G. 1978. Percutaneous infection of *Hylobius pales* by *Metarrhizium anisopliae*. *J. Invertebr. Pathol. 31*, 180–187.

Seymour, R. L. 1984. *Laptolegnia chapmanii*, an oomycete pathogen of mosquito larvae. *Mycologia 76*, 670–674.

Shapiro, M., and Roberts, D. W. 1976. Growth of *Coelomomyces psorophorae* mycelium *in vitro*. *J. Invertebr. Pathol. 27*, 399–402.

Shimazu, M. 1979. Resting spore formation of *Entomophthora sphaerosperma* Fresenius (Entomophthorales: Entomophthoraceae) in the brown planthopper, *Nilaparvata lugens* (Stål) (Hemiptera: Delphacidae). *Appl. Entomol. Zool. 14*, 383–388.

Shimazu, M. 1987. Effect of rearing humidity of host insects on the spore types of *Entomophaga maimaiga* Humber, Shimazu et Soper (Entomophthorales: Entomophthoraceae). *Appl. Entomol. Zool. 22*, 394–397.

Shimazu, M., Koizumi, C., Kushida, T., and Mitsuhashi, J. 1987. Infectivity of hibernated resting spores of *Entomophaga maimaiga* Humber, Shimazu et Soper (Entomophthorales: Entomophthoraceae). *Appl. Entomol. Zool. 22*, 216–221.

Skaife, S. H. 1925. The locust fungus, *Empusa grylli,* and its effects on its host. *S. Afr. J. Sci. 22*, 298–308.

Skou, J. P. 1972. Ascosphaerales. *Friesia 10*, 1–24.

Skou, J. P. 1973. *Microascus exsertus* sp. nov. associated with a leaf-cutting bee, with considerations on relationships of species in the genus *Microascus* Zukal. *Antonie van Leeuwenhoek 39*, 529–538.

Smith, R. J., and Grula, E. A. 1981. Nutritional requirements for conidial germination and hyphal growth of *Beauveria bassiana*. *J. Invertebr. Pathol. 37*, 222–230.

Smith, R. J., and Grula, E. A. 1982. Toxic components on the larval surface of the corn earworm (*Heliothis zea*) and their effects on germination and growth of *Beauveria bassiana*. *J. Invertebr. Pathol. 39*, 15–22.

Smith, R. J., Pekrul, S., and Grula, E. A. 1981. Requirement for sequential enzymatic activities for penetration of the integument of the corn earworm (*Heliothis zea*). *J. Invertebr. Pathol. 38*, 335–344.

Soares, G. O., Jr., and Pinnock, D. E. 1984. Effect of temperature on germination, growth and infectivity of the mosquito pathogen *Tolypocladium cylindrosporum* (Deuteromycotina; Hyphomycetes). *J. Invertebr. Pathol. 43*, 242–247.

Soares, G. G., Jr., Pinnock, D. E., and Samson, R. A. 1979. *Tolypocladium*, a new fungal pathogen of mosquito larvae with promise for use in microbial control. *Proc. Pap. 47 Annu. Conf. Calif. Mosq. Vector Cont. Assoc. Jan. 28–31, 1979.* pp 51–54.

Soper, R. S. 1974. The genus Massospora, entomopathogenic for cicadas, Part I, Taxonomy of the genus. *Mycotaxon 1*, 13–40.

Soper, R. S. 1981. New cicada pathogens: *Massospora cicadettae* from Australia and *Massospora pahariae* from Afghanistan. *Mycotaxon 13*, 50–58.

Soper, R. S., Holbrook, F. R., Majchrowicz, I., and Gordon, C. C. 1975. Production of Entomophthora resting spores for biological control of aphids. *Univ. Maine Agric. Exp. Stn. Tech. Bull. 76.*

Soper, R. S., Delyzer, A. J., and Smith, L. F. R. 1976a. The genus *Massospora* entomopathogenic for cicadas. Part. II. Biology of *Massospora levispora* and its host *Okanagana rimosa,* with notes on *Massospora cicadina* on the periodical cicadas. *Ann. Entomol. Soc. Am. 69*, 89–95.

Soper, R. S., Smith, L. F. R., and Delyzer, A. J. 1976b. Epizootiology of *Massospora levispora* in an isolated population of *Okanagana rimosa. Ann. Entomol. Soc. Am. 69*, 275–283.

Soper, R. S., May, B., and Martinell, B. 1983. *Entomophaga grylli* enzyme polymorphism as a technique for pathotype identification. *Environ. Entomol. 12*, 720–723.

Sparrow, F. K. 1973a. Mastigomycotina (zoosporic fungi). *In* "The Fungi: An Advanced Treatise." (G. C. Ainsworth, F. K. Sparrow, and A. S. Sussman, eds.), Vol. IVB, pp. 61–73. Academic Press, New York.

Sparrow, F. K. 1973b. Chytridiomycetes Hyphochytridiomycetes. *In* "The Fungi: An Advanced Treatise" (G. C. Ainsworth, F. K. Sparrow, and A. S. Sussman, eds.), Vol. IVB, pp. 85–110. Academic Press, New York.

Sparrow, F. K. 1973c. Lagenidiales. *In* "The Fungi: An Advanced Treatise." (G. C. Ainsworth, F. K. Sparrow, and A. S. Sussman, eds.), Vol. IVB, pp. 159–163. Academic Press, New York.

Speare, A. T. 1917. Sorosporella uvella and its occurrence in cutworms in America. *J. Agric. Res. 8*, 189–194.

Speare, A. T. 1920. Further studies of Sorosporella uvella, a fungous parasite of noctuid larvae. *J. Agric. Res. 18*, 399–440.

Speare, A. T. 1921. Massospora cicadina Peck a fungous parasite of the periodical cicada. *Mycologia 13*, 72–82.

Spencer, D. M., and Atkey, P. T. 1981. Parasitic effects of *Verticillium lecanii* on two rust fungi. *Trans. Br. Mycol. Soc. 77*, 535–542.

Spiltoir, C. F., and Olive, L. S. 1955. A reclassification of the genus Pericystis Betts. *Mycologia 47*, 238–244.

Steinhaus, E. A. 1949. "Principles of Insect Pathology." McGraw-Hill Book, New York.

Steinhaus, E. A. 1956. Microbial control—the emergence of an idea. A brief history of insect pathology through the nineteenth century. *Hilgardia 26*, 107–160.

Stejskal, M. 1974. *Arrhenosphaera cranei*, gen. et sp. nov., a beehive fungus found in Venezuela. *J. Apic. Res. 13*, 39–45.

St. Leger, R. J., Cooper, R. M., and Charnley, A. K. 1986a. Cuticle-degrading enzymes of entomopathogenic fungi: Cuticle degradation in vitro by enzymes from entomopathogens. *J. Invertebr. Pathol. 47*, 167–177.

St. Leger, R. J., Charnley, A. K., and Cooper, R. M. 1986b. Proteases as pathogenicity determinants of entomopathogenic fungi. *In* "Fundamental and Applied Aspects of Invertebrate Pathology." (R. A. Samson, J. M. Vlak and D. Peters, eds.) pp. 428–431. Found. Fourth Int. Colloq. Invertebr. Pathol. Wageningen, Netherlands.

St. Leger, R. J., Charnley, A. K., and Cooper, R. M. 1987a. Characterization of cuticle-degrading proteases produced by the entomopathogen *Metarhizium anisopliae. Arch. Biochem. Biophys. 253*, 221–232.

St. Leger, R. J., Cooper, R. M., and Charnley, A. K. 1987b. Production of cuticle-degrading enzymes by the entomopathogen *Metarhizium anisopliae* during infection of cuticles from *Calliphora vomitoria* and *Manduca sexta*. *J. Gen. Mycol.* **133**, 1371–1382.

St. Leger, R. J., Butt, T. M., Goettel, M. S., Staples, R. C., and Roberts, D. W. 1989a. Production *in vitro* of appressoria by the entomopathogenic fungus *Metarhizium anisopliae*. *Exp. Mycol.* **13**, 274–288.

St. Leger, R. J., Butt, T. M., Staples, R. C., and Roberts, D. W. 1989b. Synthesis of proteins including a cuticle-degrading protease during differentiation of the entomopathogenic fungus *Metarhizium anisopliae*. *Exp. Mycol.* **13**, 253–262.

St. Leger, R. J., Butt, T. M., Staples, R. C., and Roberts, D. W. 1990a. Second messenger involvement in differentiation of the entomopathogenic fungus *Metarhizium anisopliae*. *J. Gen. Microbiol.* **136**, 1779–1789.

St. Leger, R. J., Laccetti, L. B., Staples, R. C., and Roberts, D. W. 1990b. Protein kinases in the entomopathogenic fungus, *Metarhizium anisopliae*. *J. Gen. Microbiol.* **136**, 1401–1411.

St. Leger, R. J., Goettel, M., Roberts, D. W., and Staples, R. C. 1991a. Penetration events during infection of host cuticle by *Metarhizium anisopliae*. *J. Invertebr. Pathol.* **58**, 168–179.

St. Leger, R. J., Roberts, D. W., and Staples, R. C. 1991b. A model to explain differentiation of appressoria by germlings of *Metarhizium anisopliae*. *J. Invertebr. Pathol.* **57**, 299–310.

Stephen, W. P., and Undurraga, J. M. 1978. Chalk brood disease in the leafcutting bee. *Oreg. State Univ. Agric. Exp. Stn. Bull.* 630.

Strandberg, J. O., and Tucker, L. C. 1974. *Filariomyces forficulae*: Occurrence and effects on the predatory earwig, *Labidura riparia*. *J. Invertebr. Pathol.* **24**, 357–364.

Strong, F. E., Wells, K., and Apple, J. W. 1960. An unidentified fungus parasitic on the seed-corn maggot. *J. Econ. Entomol.* **53**, 478–479.

Suzuki, A., Kuyama, S., Kodaira, Y., and Tamura, S. 1966. Structural elucidation of destruxin A. *Agric. Biol. Chem.* (Tokyo), **30**, 517–518.

Suzuki, A., Taguchi, H., and Tamura, S. 1970. Isolation and structure elucidation of three new insecticidal cyclodepsipeptides, destruxins C and D and desmethyldestruxin B, produced by *Metarrhizium anisopliae*. *Agric. Biol. Chem.* (Tokyo), **34**, 813–816.

Suzuki, A., Kawakami, K., and Tamura, S. 1971. Detection of destruxins in silkworm larvae infected with *Metarrhizium anisopliae*. *Agric. Biol. Chem.* (Tokyo), **35**, 1641–1643.

Sweeney, A. W. 1975. The mode of infection of the insect pathogenic fungus *Culicinomyces* in larvae of the mosquito *Culex fatigans*. *Aust. J. Zool.* **23**, 49–57.

Sweeney, A. W. 1977. Infection of aseptically reared mosquito larvae with *Culicinomyces* sp. *J. Invertebr. Pathol.* **30**, 273.

Sweeney, A. W. 1978. The effects of temperature on the mosquito pathogenic fungus *Culicinomyces*. *Aust. J. Zool.* **26**, 47–53.

Sweeney, A. W. 1979. Infection of mosquito larvae by *Culicinomyces* sp. through anal papillae. *J. Invertebr. Pathol.* **33**, 249–251.

Sweeney, A. W. 1983. The time-mortality response of mosquito larvae infected with the fungus *Culicinomyces*. *J. Invertebr. Pathol.* **42**, 162–166.

Sweeney, A. W., Wright, R. G., and Van der Lubbe, L. 1980. Ultrastructural observations on the invasion of mosquito larvae by the fungus *Culicinomyces*. *Micron* **11**, 487–488.

Sweeney, A. W., Inman, A. O., Bland, C. E., and Wright, R. G. 1983. The fine structure of *Culicinomyces clavisporus* invading mosquito larvae. *J. Invertebr. Pathol.* **42**, 224–243.

Tamura, S., Kuyama, S., Kodaira, Y., and Higashikawa, S. 1964. The structure of destruxin B, a toxic metabolite of *Oospora destructor*. *Agric. Biol. Chem.* (Tokyo), **28**, 137–138.

Tanada, Y. 1955. Susceptibility of the imported cabbageworm to fungi: *Beauveria* spp. *Proc. Hawaii. Entomol. Soc.* **15**, 617–622.

Tarrant, C. A., and Soper, R. 1986. Evidence for the vertical transmission of *Coelomycidium simulii* (Myceteae (Fungi): Chytridiomycetes). *In* "Fundamental and Applied Aspects of Invertebrate Pathology." (R. A. Samson, J. M. Vlak, and D. Peters, eds.), p. 212. Found. Fourth Int. Colloq. Invertebr. Pathol., Wageningen, Netherlands.

Tavares, I. I. 1985. "Laboulbeniales (Fungi, Ascomycetes)." J. Cramer Publisher, Braunschweig, Germany.

Taylor, B. W., Harlos, J. A., and Brust, R. A. 1980. *Coelomomyces* infection of the adult female mosquito *Aedes trivittatus* (Coquillett) in Manitoba. *Can. J. Zool.* 58, 1215–1219.

Teetor-Barsch, G. H., and Roberts, D. W. 1983. Entomogenous *Fusarium* species. *Mycopathologia* 84, 3–16.

Thaxter, R. 1896. Contribution towards a monograph of the Laboulbeniaceae, Part I. *Mem. Am. Acad. Arts Sci.* 12, 187–249.

Thaxter, R. 1908. Contribution towards a monograph of the Laboulbeniaceae, Part II. *Mem. Am. Acad. Arts Sci.* 13, 217–469.

Thaxter, R. 1924. Contribution towards a monograph of the Laboulbeniaceae, Part III. *Mem. Am. Acad. Arts Sci.* 14, 309–426.

Thaxter, R. 1926. Contribution towards a monograph of the Laboulbeniaceae, Part IV. *Mem. Am. Acad. Arts Sci.* 15, 309–426.

Thaxter, R. 1931. Contribution towards a monograph of the Laboulbeniaceae, Part V. *Mem. Am. Acad. Arts Sci.* 16, 309–426.

Thomas, G. M., and Luce, A. 1972. An epizootic of chalk brood, *Ascosphaera apis* (Massen ex Claussen) Olive and Spiltoir in the honey bee, *Apis mellifera* L. in California. *Am. Bee J.* 112, 88–90.

Thomas, K. C., Khachatourians, G. G., and Ingledew, W. M. 1987. Production and properties of *Beauveria bassiana* conidia cultivated in submerged culture. *Can. J. Microbiol.* 33, 12–20.

Tomiyama, H., and Aoki, J. 1982. Infection of *Erynia blunckii* (Lak. ex Zimm.) Rem. et Henn. (Entomophthorales: Entomophthoraceae) in the diamond-back moth, *Plutella xylostella* L. (Lepidoptera: Yponomeutidae). *Appl. Entomol. Zool.* 17, 375–384.

Toscano, N. C., and Reeves, E. L. 1973. Effect of *Aspergillus flavus* mycotoxin in *Culex* mosquito larvae. *J. Invertebr. Pathol.* 22, 55–59.

Travland, L. B. 1979a. Structures of the motile cells of *Coelomomyces psorophorae* and function of the zygote in encystment on a host. *Can. J. Bot.* 57, 1021–1035.

Travland, L. B. 1979b. Initiation of infection of mosquito larvae (*Culiseta inornata*) by *Coelomomyces psorophorae*. *J. Invertebr. Pathol.* 33, 95–105.

Tulloch, M. 1976. The genus *Metarhizium*. *Trans. Br. Mycol. Soc.* 66, 407–411.

Tuveson, R. W., and McCoy, C. W. 1982. Far-ultraviolet sensitivity and photoreactivation of *Hirsutella thompsonii*. *Ann. Appl. Biol.* 101, 13–18.

Tyrrell, D., and MacLeod, D. M. 1972. A taxonomic proposal regarding *Delacroixia coronata* (Entomophthoraceae). *J. Invertebr. Pathol.* 20, 11–13.

Tyrrell, D., and MacLeod, D. M. 1975. In vitro germination of *Entomophthora aphidis* resting spores. *Can. J. Bot.* 53, 1188–1191.

Umphlett, C. J., and Huang, C. S. 1972. Experimental infection of mosquito larvae by a species of the aquatic fungus *Lagenidium*. *J. Invertebr. Pathol.* 20, 326–331.

Uziel, A., Schwartz, A., and Kenneth, R. G. 1981. Positive phototropism and hydromorphogenesis in germination structures of *Conidiobolus* species. *Isr. J. Bot.* 30, 75–80.

Vago, C. 1958. Virulence cryptogamique simultanée vis-à-vis d'un végétal et d'un insecte. *C.R. Acad. Sci. Paris Ser. D* 247, 1651–1653.

Vandenberg, J. D., and Stephen, W. P. 1982. Etiology and symptomatology of chalkbrood in the alfalfa leafcutting bee, *Megachile rotundata*. *J. Invertebr. Pathol.* 39, 133–137.

van Winkelhoff, A. J., and McCoy, C. W. 1984. Conidiation of *Hirsutella thompsonii* var. *synnematosa* in submerged culture. *J. Invertebr. Pathol.* 43, 59–68.

Veen, K. H. 1966. Oral infection of second-instar nymphs of *Schistocerca gregaria* in *Metarrhizium anisopliae*. *J. Invertebr. Pathol. 8*, 254–256.

Vey, A., and Fargues, J. 1977. Histological and ultrastructural studies of *Beauveria bassiana* infection in *Leptinotarsa decemlineata* larvae during ecdysis. *J. Invertebr. Pathol. 30*, 207–215.

Vey, A., and Quiot, J.-M. 1976. Action toxique du champignon *Mucor hiemalis* sur les cellules d'insectes en culture *in vitro*. *Entomophaga 21*, 275–279.

Vey, A., Quiot, J.-M., and Vago, C. 1973. Mise en évidence et étude de l'action d'une mycotoxine, la beauvericine, sur des cellules d'insectes cultivées *in vitro*. *C.R. Acad. Sci. Paris Ser. D 276*, 2489–2492.

Vey, A., Quiot, J.-M., Vago, C., and Fargues, J. 1985. Effect immunodépresseur de toxines fongiques: Inhibition de la réaction d'encapsulement multicellulaire par les destruxines. *C.R. Acad. Sci. Paris Ser. III 300*, 647–651.

Vey, A., Quiot, J.-M., and Pais, M. 1986. Toxémie d'origine fongique chez les Invertébrés et ses conséquences cytotoxiques: Étude sur l'infection à *Metarhizium anisopliae* (Hyphomycète, Moniliales) chez les Lépidoptères et les Coléoptères. *C.R. Soc. Biol. 180*, 105–112.

Vey, A., Quiot, J.-M., and Vago, C. 1987. Mode d'action insecticide d'une mycotoxine, la Destruxine E, sur les diptères vecteurs et disséminateurs de germes. *C.R. Acad. Sci. Paris Ser. III, 304*, 229–234.

von Arx, J. A. 1963. Die Gattungen der Myriangiales. *Persoonia 2*, 421–475.

Vuillemin, P. 1912. *Beauveria*, nouveau genre de Verticilliacées. *Paris Soc. Bot. Fr. Bull. 59*, pp. 34–40.

Walker, A. J. 1938. Fungal infections of mosquitoes, especially of *Anopheles costalis*. *Ann. Trop. Med. Parasitol. 32*, 231–241.

Wallace, D. R., MacLeod, D. M., Sullivan, C. R., Tyrrell, D., and DeLyzer, A. J. 1976. Induction of resting spore germination in *Entomophthora aphidis* by long-day light conditions. *Can. J. Bot. 54*, 1410–1418.

Wasti, S. S., and Hartmann, G. C. 1978. Host–parasite interactions between larvae of the gypsy moth, *Lymantria dispar* (L.) (Lepidoptera: Lymantridae) and the entomogenous fungus, *Nomuraea rileyi* (Farlow) Samson (Moniliales: Moniliaceae). *Appl. Entomol. Zool. 13*, 23–28.

Waterhouse, G. M. 1973. Entomophthorales. *In* "The Fungi: An Advanced Treatise." (G. C. Ainsworth, F. K. Sparrow and A. S. Sussman, eds.), Vol. IVB, pp. 219–229. Academic Press, New York.

Watson, J. R., and Berger, E. W. 1937. Citrus insects and their control. *Bull. Fla. Agric. Ext. Ser. 88*.

Watson, P. L., Barney, R. J., Maddox, J. V., and Armbrust, E. J. 1981. Sporulation and mode of infection of *Entomophthora phytonomi*, a pathogen of the alfalfa weevil. *Environ. Entomol. 10*, 305–306.

Weiser, J. 1976. The intermediary host for the fungus *Coelomomyces chironomi*. *J. Invertebr. Pathol. 28*, 273–274.

Weiser, J. 1977. "An Atlas of Insect Diseases." Second Ed., Dr. W. Junk, B. V., Hague.

Weiser, J., and Pillai, J. S. 1981. *Tolypocladium cylindrosporum* [Deuteromycetes, Moniliaceae] a new pathogen of mosquito larvae. *Entomophaga 26*, 357–361.

Welton, M. A., and Tyrrell, D. 1975. A note on the isolation of *Entomophthora* species on artificial media. *J. Invertebr. Pathol. 26*, 405.

Whisler, H. C. 1968. Experimental studies with a new species of *Stigmatomyces* (Laboulbeniales). *Mycologia 60*, 65–75.

Whisler, H. C. 1979. The fungi versus the arthropods. *In* "Insect–Fungus Symbiosis Nutrition, Mutualism, Commensalism." (Batra, L. R., ed.), pp. 1–32. Allanheld, Osmun; Montclair, New Jersey.

Whisler, H. C., Shemanchuk, J. A., and Travland, L. B. 1972. Germination of the resistant sporangia of *Coelomomyces psorophorae*. *J. Invertebr. Pathol.* 19, 139–147.

Whisler, H. C., Zebold, S. L., and Shemanchuk, J. A. 1974. Alternate host for mosquito parasite *Coelomomyces*. *Nature* 251, 715–716.

Whisler, H. C., Zebold, S. L., and Shemanchuk, J. A. 1975. Life history of *Coelomomyces psorophorae*. *Proc. Natl. Acad. Sci. U.S.A.* 72, 693–696.

Whisler, H. C., Wilson, C. M., Travland, L. B., Olson, L. W., Borkhardt, B., Aldrich, J., Therrien, C. D., and Zebold, S. L. 1983. Meiosis in *Coelomomyces*. *Exp. Mycol.* 7, 319–327.

White, J., and Lloyd, M. 1983. A pathogenic fungus, *Massospora cicadina* Peck (Entomophthorales), in emerging nymphs of periodical cicadas (Homoptera: Cicadidae). *Environ. Entomol.* 12, 1245–1252.

Wilding, N. 1970. Entomophthora conidia in the air-spora. *J. Gen. Microbiol.* 62, 149–157.

Wilding, N., and Lauckner, F. B. 1974. Entomophthora infecting wheat bulb fly at Rothamsted, Hertfordshire, 1967–76. *Ann. Appl. Biol.* 76, 161–170.

Wilding, N., Collins, N. M., Hammond, P. M., and Webber, J. F. 1989. "Insect–Fungus Interactions." Academic Press, New York.

Willoughby, L. G. 1969. Pure culture studies on the aquatic phycomycete, *Lagenidium giganteum*. *Trans. Br. Mycol. Soc.* 52, 393–410.

Wolf, F. T. 1951. The cultivation of two species of Entomophthora on synthetic media. *Bull. Torrey Bot. Club* 78, 211–220.

Woods, S. P., and Grula, E. A. 1984. Utilizable surface nutrients on *Heliothis zea* available for growth of *Beauveria bassiana*. *J. Invertebr. Pathol.* 43, 259–269.

Wyllie, T. D., and Morehouse, L. G. 1977. "Mycotoxic Fungi, Mycotoxins, Mycotoxicoses: An Encyclopedic Handbook." Vol. 1. Marcel Dekker, New York.

Yendol, W. G., and Paschke, J. D. 1965. Pathology of an *Entomophthora* infection in the eastern subterranean termite, *Reticulitermes flavipes* (Kollar). *J. Invertebr. Pathol.* 7, 414–422.

Yendol, W. G., Miller, E. M., and Behnke, C. N. 1968. Toxic substances from entomophthoraceous fungi. *J. Invertebr. Pathol.* 10, 313–319.

Youssef, N. N., Roush, C. F., and McManus, W. R. 1984. In vivo development and pathogenicity of *Ascosphaera proliperda* (Ascosphaeraceae) to the alfalfa leafcutting bee, *Megachile rotunda*. *J. Invertebr. Pathol.* 43, 11–20.

Zacharuk, R. Y. 1970a. Fine structure of the fungus *Metarrhizium anisopliae* infecting three species of larval Elateridae (Coleoptera). II. Conidial germ tubes and appressoria. *J. Invertebr. Pathol.* 15, 81–91.

Zacharuk, R. Y. 1970b. Fine structure of the fungus *Metarrhizium anisopliae* infecting three species of larval Elateridae (Coleoptera). III. Penetration of the host integument. *J. Invertebr. Pathol.* 15, 372–396.

Zacharuk, R. Y. 1971. Ultrastructural changes in tissues of larval Elateridae (Coleoptera) infected with the fungus *Metarrhizium anisopliae*. *Can. J. Microbiol.* 17, 281–289.

Zacharuk, R. Y. 1973. Penetration of the cuticular layers of elaterid larvae (Coleoptera) by the fungus *Metarrhizium anisopliae*, and notes on a bacterial invasion. *J. Invertebr. Pathol.* 21, 101–106.

Zattau, W. C., and McInnis, T., Jr. 1985. Life cycle and mode of infection of *Leptolegnia chapmanii* (Oomycetes) parasitizing *Aedes aegypti*. *J. Invertebr. Pathol.* 50, 134–145.

Zebold, S. L., Whisler, H. C., Shemanchuk, J. A., and Travland, L. B. 1979. Host specificity and penetration in the mosquito pathogen, *Coelomomyces psorophorae*. *Can. J. Bot.* 57, 2766–2770.

Zoebelein, G., and Kniehase, U. 1985. Laboratory, greenhouse and field trials on the effect of nikkomycins on insects and mites. *Pflanzenschutz-Nachr.* 38, 203–304.

CHAPTER 11

PROTOZOAN INFECTIONS: ZOOMASTIGINA, RHIZOPODA, AND CILIOPHORA

I. HISTORICAL ASPECTS
II. RELATION OF PROTOZOA TO INSECTS
III. PORTALS OF ENTRY
IV. PATHOGENESIS, SIGNS, AND SYMPTOMS
V. ZOOMASTIGINA (FLAGELLATE) INFECTIONS
   A. Life Cycle
   B. Transmission
   C. Pathology
   D. Entomopathogenic Flagellates
   E. Insects as Vectors of Trypanosomatids
VI. RHIZOPODA (AMOEBIC) INFECTIONS
   A. Amoebiasis of Honey Bee
   B. Amoebiasis of Grasshopper
   C. Amoebiasis in Other Insects
VII. CILIOPHORA (CILIATE) INFECTIONS
   A. Penetration into Hemocoel
   B. Ciliatoses by *Tetrahymena* spp.
   C. Ciliatoses by *Lambornella* spp.
   D. Other Ciliatoses

The protozoa ("first animals") are a heterogeneous group of microorganisms of very diverse characters, behavior, and life cycles (Hall 1953; Kudo 1966). They were grouped together at one time as a "catch all" of minute, animallike organisms that had no place to go. Included in this unnatural assemblage of microorganisms are some that have plantlike characteristics, such as possession of photosynthetic pigment (chlorophyll) or "cellular" cell wall resembling those of higher plants or chitinous secretions similar to those of fungi. For a long time protozoologists have recognized this anomalous situation in the previous four classes established by Bütschli (1880–1889). The four classes were separated primarily on the basis of the locomotor organelles: class Mastigophora (Flagellata) move by means of flagella, Sarcodina by means of pseudopodia, Infusoria (Ciliophora) by means of cilia or are sessile in the adult stage, and Sporozoa, which, if motile, move without the aid of special locomotory organelles.

With increased knowledge of the phylogenetics and systematics of these microorganisms, major changes in the higher classification were adopted by the Committee on Systematics and Evolution of the Society of Protozoologists in 1980 (Levine et al. 1980). The Committee has elevated the Protozoa to a subkingdom under kingdom Protista and established seven phyla: Sarcomastigophora, Labyrinthomorpha, Apicomplexa, Microspora, Ascetospora, Myxospora, and Ciliophora. Recently, the protists (small eukaryotes with a single or only a few cells) have been reorganized into a broader category, the protoctists (include small fungi and algae) to form a fifth kingdom Protoctista alongside the other four kingdoms, Monera (Prokaryotae), Fungi, Animalia, and Plantae (Margulis 1990). The 36 phyla in Protoctista are separated into four groups, with the former subkingdom Protozoa distributed among these groups. We have adopted the classification proposed in the *Handbook of Protoctista* (Margulis et al. 1990).

The protozoa are minute unicellular organisms, sometimes designated as being acellular or noncellular. They are not comparable to a simple metazoan cell but are more complex and appear to have evolved along separate evolutionary lines. They possess organelles whose functions are similar to the more complex, multicellular organs of higher animals. Members of protozoa are found in nearly all habitats, both aquatic and terrestrial, except possibly the aerial habitat. They vary in shapes, sizes, and colors, and in most morphological characteristics. They also differ in their habits, behaviors, and life cycles. Reproduction takes place by sexual and asexual means, but some forms appear to lack sexual and others lack asexual reproduction. In some cases, autogamy or self-fertilization takes place when two nuclei fuse within a cell. Their association with insects range from commensalistic to pathogenic. A list of entomogenous protozoa is given in Table 11–1. Readers who are interested in the general characteristics and classification of protozoa should refer to the excellent treatise, *An Illustrated Guide to the Protozoa* (Lee et al. 1985) and the broad-based treatment, *Handbook of Protoctista* (Margulis et al. 1990). We also recommend to those who are involved in the applied aspects of protozoa in insect control the review by Henry (1981) and the comprehensive treatment by Brooks (1988).

I. HISTORICAL ASPECTS

The protozoa, because of their minute size, remained unobserved until the development of the microscope. Antony van Leeuwenhoek (1632–1723), who produced lenses and built microscopes, discovered free-living, fresh-water protozoa (Dobell 1932). From the descriptions of the animalcules provided by van Leeuwenhoek, Dobell (1932) believes that he also observed coccidians in cats and flagellates in the digestive tract of the horse fly (tabanid). van Leeuwenhoek is generally recognized as the father of protozoology for his observations on numerous protozoa.

We shall touch briefly on some historical aspects of protozoa that are pathogens of insects. The flagellates seen by van Leeuwenhoek in the gut of the horse fly were probably *Crithidia*, *Blastocrithidia*, or *Trypanosoma* (Wallace 1966). In 1828, Dufour reported the presence of gregarines in the digestive tracts of several beetles and an earwig. He proposed the generic name *Gregarina* and the species *conica* for the gregarine found in the beetles and *ovata* for the one in the earwig. Burnett (1851) observed flagellates of the genus *Bodo* (presumably *Herpetomonas muscarum*) in the digestive tract of the house fly, *Musca domestica*. In 1856, Leidy described *Herpetomonas* (syn. *Bodo*) *muscarum* from the digestive tract of the house fly; *H*.

TABLE 11-1 Classification Scheme of Entomogenous Protozoa[a]

Taxa[b]	Representative genera
Phylum Zoomastigina	
Class Kinetoplastida	*Herpetomonas, Crithidia, Leptomonas*
Class Retortamonadida	*Retortamonas, Chilomastix*
Class Diplomonadida	*Octomitus*
Class Pyrsonymphida	*Oxymonas, Pyrsonympha*
Class Parabasalia	*Trichomonas, Devoscovinia, Trichonympha*
Phylum Rhizopoda	
Class Lobosea	
Subclass Gymnamoebia	
Order Amoebida	*Malpighiella, Malameba, Malpighamoeba*
Phylum Apicomplexa	
Class Gregarinia	
Order Eugregarinida	*Gregarina, Ascogregarina*
Order Neogregarinida	*Mattesia, Farinocystis, Ophryocystis*
Class Coccidia	
Order Adeleida	*Adelina, Legerella, Barrouxia*
Phylum Microspora	
Class Microsporea	
Order Minisporida	*Chytridiopsis, Hessea*
Order Microsporida	*Nosema, Pleistophora, Amblyospora*
Phylum Haplosporidia	
Class Haplosporea	
Order Haplosporida	*Haplosporidium*
Phylum Ciliophora	
Subphylum Postciliodesmatophora	
Class Spirotrichea	
Order Heterotrichida	*Nyctotherus*
Subphylum Rhabdophora	
Class Litostomatea	*Balantidium*
Subphylum Cyrtophora	
Class Phyllopharyngea	
Subclass Suctoria	
Order Evaginogenida	*Discophrya, Rhynchophrya*
Class Oligohymenophorea	
Subclass Hymenostomia	
Order Hymenostomatida	*Tetrahymena, Lambornella*
Subclass Peritrichia	
Order Mobilida	*Epistylis, Opercularia*

[a]Prepared by W. M. Brooks.
[b]Compiled primarily from Margulis et al. 1990.

melolonthae from the scarab beetles, *Melolontha quercina* and *M. brunnea*; and the ciliate *Nyctotherus ovalis* from the intestine of the cockroach *Blatta orientalis*.

During the early part of the nineteenth century, the silkworm industry in Europe was devastated by the disease pébrine in the silkworm, and a number of investigators, particularly Pasteur, became involved in the study of this disease. We have already discussed (Chapter 1) the exploits of Pasteur in controlling pébrine. The organism causing pébrine was described by Naegeli (1857) and named *Nosema bombycis*. Both Pasteur and Naegeli were not aware that the organism was a protozoan.

According to Hall (1953), the group Ciliata was set up by Perty in 1852 and the Flagellata by Cohn in 1853. In 1879, Leuckart included the gregarines and coccidians in the newly created group Sporozoa. Balbiani (1882) recognized the uniqueness of the microsporidia, which at that time included only a single-named species *Nosema bombycis*, and proposed the new category Microsporidies. The orders Myxosporidia (Bütschli 1881), Microsporidia, and Sarcosporidia (Balbiani 1882) were included with Coccidia and Gregarinida in the class Sporozoa. Balbiani (1882) succeeded in transmitting *Nosema bombycis* only to a few insect species and not to many others. He also observed in the orthopteran, *Platycleis grisea*, another microsporidium that developed in the epithelium of the insect's digestive tract.

In 1897, Schaudinn and Siedlecki proposed the terms micro- and macrogametes for gametes in protozoa and described their formation. They showed that the zygote production in coccidians resulted from the fusion of these gametes (anisogamy). Siedlecki (1899) reported further on the life cycle of coccidians. From 1880 to the early twentieth century, the protozoologists paid increasing attention to the protozoan infections of humans and other vertebrates that were transmitted by insects and other arthropods (e.g., malaria, babesiosis, African sleeping sickness, etc.). The use of microsporidia as biological control agents was recommended as early as 1905 by Pérez against the green crab, *Carcinus maenas*, a serious predator of oysters, but no steps were taken in this direction. Taylor and King (1937) might have been the first to use protozoa for insect control when they applied the amoeba *Malameba locustae* for the control of grasshoppers. Subsequently, an increasing number of attempts to control insect pests were made with protozoa, in particular the microsporidia.

II. RELATION OF PROTOZOA TO INSECTS

There are about 1200 species of protozoa, out of about 15,000 described species, that are associated with insects (Lipa 1963). The entomogenous protozoa are commonly found in the digestive tracts of insects as commensals or they are in a mutualistic association with insects. Some insects serve as vectors of protozoan diseases of vertebrates and plants. In many of these cases the protozoa multiply in the insect vectors and may even cause harm to some vectors. A great number of protozoa are pathogenic to insects. They are members of the protozoan phyla: Rhizopoda, Haplosporidia, Microspora, Zoomastigina, Ciliophora, Apicomplexa, Chytridiomycota, and Oomycota. The Chytridiomycota and Oomycota are regarded as fungi in the present text. The majority of the highly pathogenic forms occur in Apicomplexa and Microspora, particularly those that invade the hemocoel and develop intracellularly.

III. PORTALS OF ENTRY

The majority of protozoa enter the insects by way of the mouth and digestive tract. Penetration through the integument occurs in the ciliates *Lambornella* spp. through invasion cysts attached to the cuticle (Clark and Brandl 1976). The infective stage is generally a spore or cyst, but it may be the vegetative or noncysted reproductive form (e.g., in flagellates and ciliates). Those protozoa that remain in the lumen of the digestive tract are attached to the epithelium, or enter appendages associated with the digestive tract and generally cause no obvious pathology. These forms are mainly ciliates, flagellates, and gregarines. Others penetrate into the hemocoel and exist extracellularly in the hemolymph or intracellularly within the cells of various tissues and organs, and cause pathologies. They are mainly the apicomplexans and microsporidia.

Vertical transmission from parent to offspring occurs in many protozoa, especially the microsporidia. The transmission is transovum by way of the ovary (transovarial) or by surface-contaminated eggs. The egg surfaces are contaminated from spore-containing feces of females with protozoan-infected digestive tracts. These types of transovum transmissions are in addition to the common *per os* route, and their significance, as a means of vertical transmission, varies greatly with the protozoa and their hosts (Brooks 1988). Transmission through surface-contaminated eggs is probably not important in regulating insect populations, but transovarial transmission is often highly significant in the transmission of microsporidia. There are few cases of the transmission by insect vectors, which themselves may be infected by protozoa (e.g., microsporidia).

IV. PATHOGENESIS, SIGNS, AND SYMPTOMS

Most entomopathogenic protozoa have low virulence and cause a chronic infection that often does not kill an insect. Such a chronically infected insect frequently does not exhibit marked external signs and symptoms (e.g., color changes and abnormal movement or behavior). Some protozoa, however, are highly virulent, and depending on the type of tissues attacked, the infection may be acute and fatal. In some cases, the infected insects become chlorotic or whitish, are reduced in size, and remain in the immature stages much longer than the uninfected individuals. The enormous numbers of protozoan spores in the fat, midgut, or hemolymph may cause these structures to turn milky white. The integument of dead insects (mainly larvae) generally remains firm and does not readily disintegrate.

The intracellular forms usually occur in the cytoplasm. No toxins have been detected in protozoan infections in insects, but Weiser (1961) has suggested that toxins may be produced by microsporidia that cause tumorlike growths and inflammatory responses in insects. Some protozoa exhibit tissue tropism and infect only certain tissues or organs (e.g., certain microsporidia and neogregarines infect only midgut epithelium or fat tissues). Others invade nearly all major tissues and organs to cause a systemic infection.

V. ZOOMASTIGINA (FLAGELLATE) INFECTIONS

The phylum Zoomastigina includes protozoa that generally possess flagella and/or pseudopodia (some forms with both organelles), a single type of nucleus, and typically do not form spores. The flagellates (mastigotes) have one or more flagella or undulipodia[1] and usually reproduce asexually by longitudinal binary fission.

---

[1] The term *flagellum* is ambiguous because it has been used for both prokaryotic and eukaryotic organelles even though they differ structurally. Margulis (1990) recommends using flagellum for prokaryotes (bacteria) and undulipodium (pl. undulipodia) for the motility of eukaryotes.

Some are autotrophic forms with photosynthetic capabilities and others are heterotrophic forms dependent on large complex molecules for growth and reproduction.

The entomogenous flagellates are in the classes Retortamonadida, Diplomonadida, Pyrsonymphida, Parabasalia, and Kinetoplastida. The members lack chromatophores and are predominantly parasitic. Those found in the digestive tracts of termites, roaches, and a few other insects are largely mutualistic or commensalistic. Class Kinetoplastida has members, generally referred to as trypanosomatids, that are elongate and slender in shape, with a single flagellum either free or attached to the body by an undulating membrane. The flagellum that arises within a reservoir (pocket) is associated with a contractile vacuole and a kinetoplast (also called kinetonucleus, micronucleus, blepharoplast, or parabasal body). The kinetoplast is a DNA-rich organelle associated with the mitochondria. The size and shape of the kinetoplast and its position in the flagellate are of taxonomic and genetic significance. In the Kinetoplastida, the family Trypanosomatidae contains most of the entomogenous flagellates and the family Bodonidae has only a few entomogenous species that are commensals.

The genera of family Trypanosomatidae are characterized on the basis of the adult morphology and on the forms that occur at different periods of their development (Vickerman 1990). The most characteristic form is used to name the genus. The generic names of the trypanosomatids are often used as adjectives to describe the various body forms. The developmental forms may vary by the lengths of the flagellum and by the position of the kinetoplast–kinetosome–flagellar pocket complex. For example, *Leptomonas* and *Leishmania* have both leptomonad and leishmanial forms. *Blastocrithidia raabei* has five body forms, leishmanial, leptomonad, crithidial, blastocrithidial, and encysted forms (Lipa 1966).

Since the generic adjectives have created, at times, confusion in the description and nomenclature of flagellates, Hoare and Wallace (1966) proposed a terminology for the body forms to replace the use of generic names. The terminology is based on the arrangement of the flagellum, as determined by its starting point in the vicinity of the kinetoplast, by its course or path through the organism, and by its point of emergence from the body. The proposed terms are (1) *amastigote*, round to oval body, flagellum short, not emerging from pocket; (2) *promastigote*, body elongate with kinetoplast at anterior end and flagellum emerging anteriorly and unattached; (3) *opisthomastigote*, similar, but kinetoplast is postnuclear and flagellar pocket forms long canal to the anterior end; (4) *epimastigote*, with prenuclear kinetoplast, the flagellum emerges from a pocket part way along the body and is attached to the body along its anterior part; (5) *trypomastigote*, similar, but kinetoplast and flagellar pocket are postnuclear; (6) *choanomastigote*, with peculiar "barleycorn" form, the kinetoplast just in front of nucleus, and flagellum emerges anteriorly. The term *endomastigote* has been applied to the form in which a large flagellum is doubled and coiled within the cell but not protruding to the outside (Yoshida et al. 1978). In addition, other proposed terms are haptomonad for the attached form, nectomonad for the feeding active stage, and spheromastigote for a rounded body with an emerging flagellum attached to it.

Wallace et al. (1983) have presented important guidelines for the description of species primarily of the monoxenous "lower" trypanosomatids. They also point out criteria that may be applied to the heteroxenous trypanosomatids with alternate hosts, usually a vertebrate and an invertebrate. Electrophoretic analyses have been applied to distinguish species of trypanosomatids (Goncalves de Lima et al. 1979;

Camargo et al. 1982; Soares et al. 1986). The analyses were conducted on fragments of kinetoplast DNA, with esterase isoenzymes, and with surface proteins.

About 400 species of Trypanosomatidae are associated with insects, but only a minority of them are pathogens; the majority are commensals or live only periodically in their insect vectors (Lipa 1963). There are 348 species of insects that serve as hosts of the monoxenous lower trypanosomatids (Wallace et al. 1983). These lower trypanosomatids offer rich material for the analysis of parasitism as an evolutionary process. Wallace (1966, 1979) has compiled a list of trypanosomatids from insects, except for the genera *Trypanosoma*, *Leishmania*, and *Phytomonas*.

The trypanosomatids vary greatly in external shape, but they are usually elongate, slender, and either pointed or rounded at the axis. Frequently, one or more spiral twists occur in the body. The body ranges from 4 to 385 μm in length. The nucleus is normally found in the middle third of the body length. The cytoplasm contains basophilic granules that vary in size and number. There is a single flagellum. However, it may be short or absent in certain attached forms, and in cysts, and in certain species, such as *Herpetomonas roubaudi*, *Leptomonas arctocorixae*, and *L. craggi* (Wallace 1966). Associated with the flagellum are centriole, kinetoplast, reservoir, and contractile vacuole. The external openings of the reservoirs are at the anterior tip (*Leptomonas*, *Herpetomonas*, *Crithidia*) or along the body side (*Blastocrithidia*, *Trypanosoma*).

A. LIFE CYCLE

The life cycle of an individual flagellate is simply longitudinal fission followed by the growth of newly formed individuals (Wallace 1979). In a few cases, the division is unequal. The division of the kinetoplast and the formation of a new flagellum generally precede nuclear division.

The flagellates may occur in various forms during the life cycle. These forms develop as population cycles that may be complex in heteroxenous flagellates or much less marked or absent in the monoxenous lower trypanosomatids (Wallace 1979). A major problem is to distinguish between stage differences that are due to an obligate cycle and differences in the form or behavior that are impressed on the organism by external factors. For example, in *Crithidia* of mosquitoes, the flagellates are short, stubby, and motionless while attached to wall of the host's intestine but become promptly elongated when the gut is teased apart in water (Ross 1906). On the other hand, the promastigote–opisthomastigote transformation appears to be a natural part of the life cycle.

A cyst stage may occur in a flagellate life cycle. Wallace (1979) has described three kinds of cysts: flagella cysts ("straphangers"), cysts formed by internal buds, and cysts formed in an enclosure.

Thus far, sexual reproduction has been observed rarely in trypanosomatids. In a species of *Leptomonas* that infects the silkworm, the promastigotes aggregate in a rosette during encystment and become stumpy in form (Abe 1980). These promastigotes have two nuclei and two kinetoplasts, suggesting that fusion of two organisms has taken place in the rosettes. Subsequently, each individual flagellate isolates itself from the cluster and acquires a thick coat to form a cyst.

B. TRANSMISSION

In most cases, insects become infected by trypanosomatids through the ingestion of organisms, not necessarily cysts, which have come directly or indirectly from the feces, or from animal blood and plant sap (Wallace 1979). The cannibalistic water scorpion, *Nepa cinerea*, may acquire *Leptomonas* (syn. *Herpetomonas*) *jaculum* by feeding on infected individuals (Porter 1909b). There are reports of flagellates invading the ovaries of their insect hosts, such as *L. jaculum* in *N. cinerea* (Porter 1909b); *Crithidia gerridis* in the water bug, *Gerris paludum* (Porter 1909a); and *Herpetomonas rhinoestri* in the oestrid fly, *Rhinoestrus nivarletti* (Rodhain 1926). The presence of flagellates in the ovary suggests that vertical (transovarial) transmission may occur in insects, but in the above examples, no flagellate was found in the embryo or larva from an infected female insect. The flagellates may be carried from a larva to a pupa and to an adult insect, as for example, *Crithidia fasciculata* in mosquitoes (Clark et al. 1964) and *Herpetomonas muscarum* in the eye gnat, *Hippelates pusio* (Bailey and Brooks 1972b). In the transmission of *C. fasciculata*, the infections in larval mosquitoes are lost at molting and are reacquired early in subsequent instars. The infection in a pupa begins when the flagellates enter the midgut while the peritrophic membrane is being shed prior to pupation. By the time pupation is completed, the flagellates are immotile but continue to multiply in the pupal midgut; they become motile again in the newly emerged adult mosquito.

Insect trypanosomatids exhibit little host specificity. They are easily transmitted under laboratory conditions through oral, rectal, and intrahemocoelic inoculations to a large number of different insect species. Wallace (1966) assembled a list of oral transmissions of flagellates to insects. Only a few cases were refractory. In some insects, such as the wax moth larva, certain flagellates inoculated into the hemocoel produced a fatal infection.

C. PATHOLOGY

The flagellate infection in animals and plants is known as "flagellatosis" or "flagellosis." The majority of insect flagellates occurs in the digestive tracts and others in the Malpighian tubules (Wallace 1979). None appears to be strictly limited to a particular region of the digestive tract, but some are more common in certain parts. Species of *Crithidia* occur more frequently in the hindgut and rectum; some *Blastocrithidia* and *Herpetomonas* are found mainly in the midgut. Many species are attached to the intestinal wall, often in a continuous carpetlike layer, while others occur free in the gut lumen. They may occur external or internal to the peritrophic membrane. Flagellates attach to midgut walls by weaving their flagella between the microvilli or attach to Malpighian tubules by engulfing several microvilli with an external flagellar membrane (Rowton et al. 1981; Tieszen et al. 1986). The flagellates in the digestive tracts cause mild infections.

Some flagellates occur in the salivary gland and within the hemocoel. Those that invade tissues and organs in the hemocoel cause more virulent and, at times, fatal infections. The flagellates invade a host cell with the flagellum foremost (Ellis et al. 1980; Evans and Ellis 1983).

D. ENTOMOPATHOGENIC FLAGELLATES

The flagellates occur especially in two orders of insects, Hemiptera with 35% and Diptera with 55% of the recorded flagellates (Wallace 1966). The flagellates display little host specificity and are cross transmissible by *per os* or by intrahemocoelic inoculations to various species of the host family as well as those of different families and even orders of insects. *Herpetomonas muscarum* has been found in 26 species of various families of dipterons and *Crithidia fasciculata* from 21 species of mosquitoes.

The strictly entomogenous trypanosomatids belong to the four genera: *Herpetomonas*, *Crithidia*, *Blastocrithidia*, and *Rhynchoidomonas* (Brooks 1974; Wallace 1979). Inasmuch as only a few species of *Rhynchoidomonas* are found in insects, we shall not discuss them in detail. They are parasites of flies and occur in the Malpighian tubules and intestines (Wallace 1966). Other genera associated with insects are *Leptomonas*, which are found in insects, other protozoa, nematodes, and molluscs; *Leishmania* and *Trypanosoma*, which contain pathogens of vertebrates and are transmitted by insects and other blood-sucking invertebrates; and *Phytomonas*, which live in plants and may be transmitted by insects. The vertebrate pathogens *Leishmania* and *Trypanosoma* cause pathology in their insect vectors.

1. *Herpetomonas* Flagellatoses

In 1963, Lipa reported that there were more than 40 species of *Herpetomonas* from insects, the majority were from Diptera. *Herpetomonas muscarum*, the most widely known species, is found in more than 200 species of flies. It inhabits mainly the digestive tract where it exists as opisthomastigote and promastigote forms. When it invades and develops in large numbers in the hemocoel of a house fly, it kills the fly (Kramer 1961). In the eye gnat, *Hippelates pusio*, heavy infection by *H. muscarum* causes death, especially in larvae in which 48% mortality occurs as compared to 6% in pupae (Bailey and Brooks 1972a,b). Larvae become infected when they ingest contaminated adult feces or feed on adults killed by the flagellate. Adult to adult transmission occurs through fecal-contaminated food and water. The infection in a larva may persist through the pupa and into the adult. The flagellates may penetrate the gut wall, probably through the gastric caeca or anterior midgut, and develop in the hemolymph without invading the organs and tissues. As the rearing temperature of the infected gnats is increased from 18.5 to 35.0°C, the mortality of the flies decreases.

Herpetomonas swainei, which infects the sawfly, *Neodiprion swainei*, was described by Smirnoff and Lipa (1970). The flagellate infects larvae, eonymphs, and adults. It inhabits the digestive tract, Malpighian tubules, and hemolymph and causes about 15% mortality in sawfly larvae and a higher prevalence in pupae. Infected individuals display no obvious external signs and symptoms, but when they are infected at an early larval stage, about 20% of them perish. A generation of flagellates is produced in 8 to 12 days. The flagellate has been transmitted to 10 other sawfly species in the families Diprionidae and Tenthredinidae (Smirnoff 1971). It also infects some of the hymenopteran parasitoids of sawflies.

2. *Crithidia* and *Blastocrithidia* Flagellatoses

Crithidia fasciculata was isolated and described from the mosquito *Anopheles maculipennis* (Léger 1902). Larval and adult mosquitoes are easily infected, but an infected larva does not carry the flagellates to the adult stage (Wallace 1943). Adult to adult transmission of the flagellate occurs readily and probably results from the ingestion of contaminated feces and food (Wallace 1943; Garnham 1959). Even in heavily infected adult mosquitoes, the flagellate causes little, if any, pathology. On the other hand, in the tsetse fly (*Glossina* spp.), *C. fasciculata* produces an intense hemocoelic infection resulting in the death of flies in 4 to 9 days (Ibrahim and Molyneux 1987).

Members of the genus *Blastocrithidia*, like those in *Leptomonas*, have a cyst stage ("straphangers") that is attached to the flagellum of an epimastigote by a "cytoplasmic bridge" (Tieszen et al. 1985; Schaub and Böker 1986). After the straphangers detach, the bridges remain as remnants on the flagellum. Species of *Blastocrithidia* mainly inhabit the insect's hindgut and its glands. It is an impressive sight to see the digestive tract covered by a carpet of flagellates (Schaub and Böker 1986). *Blastocrithidia raabei* is found not only in the digestive tract but also in the hemocoel of the coreid bug, *Mesocerus marginatus* (Lipa 1966).

Blastocrithidia triatomae is a promising candidate for the microbial control of triatomines, the vectors of Chagas' disease, because of its pathogenicity and capacity to form highly resistant cysts *in vitro* (Schaub 1990a,b, 1991; Schaub et al. 1990). It colonizes the intestinal tract like *Trypanosoma cruzi*, the etiologic agent of Chagas' disease, but unlike most trypanosomatids, it is highly pathogenic. *Blastocrithidia triatomae* causes, to its triatomine hosts, disturbances in excretion, digestion, and sclerotization (Schaub 1991). It reduces the starvation resistance, cellular immune response, adult life span, and reproduction rate of its hosts. It increases larval developmental times and larval mortality rates.

Blastocrithidia caliroae appears, thus far, to be the only flagellate to cause significant epizootics and collapse of an insect population. This species was described by Lipa et al. (1977) from the pear slug, *Caliroa cerasi* (Hymenoptera, Tenthredinidae), a pest of fruit and other trees and shrubs in Europe, New Zealand, and the United States. In the field, *B. caliroae* infects the eonymphs and all larval stages, except for the first instar. At a late, heavily infected stage, the larvae are easily recognized by the dry appearance of their mucous coat. Their color is a dull dark brown to blackish brown, sometimes with yellow spots where the dried-up yellow slime has peeled off. Larvae generally stop feeding, die, and dry up before forming cocoons. Larvae, infected shortly before maturity, may carry over the infections to eonymphs that die before changing to the pronymphal stage. Lipa et al. (1977) postulated that the flagellate was responsible for the collapse of field populations in possibly three outbreaks of the pear slug.

3. *Leptomonas* Flagellatoses

There are about 40 species of *Leptomonas* reported from insects, mainly Hemiptera, Diptera, and Siphonaptera (Lipa 1963). At least 23 species of *Leptomonas* have been reported from 66 species of Hemiptera (Wallace 1977). A cyst stage has been reported for the genera *Leptomonas* and *Blastocrithidia* (Tieszen et al. 1985).

Leptomonas pyrrhocoris is a common inhabitant of the digestive tracts of larvae and adults of the hemipteran *Pyrrhocoris apterus* (Zotta 1921). Occasionally, the flagellate invades the salivary gland and hemocoel (Lipa 1963). The flagellate occurs in the gut as promastigote and encysted amastigote forms, but in the hemolymph, the promastigotes are most common. It causes rather weak pathologies. When it inhabits the digestive tract, the insect only develops diarrhea, but when it invades the hemocoel and salivary gland, the activity of the insect is reduced. An abundance of flagellates in the hemocoel causes the hemolymph to thicken and turn whitish in color.

Leptomonas pyrrhocoris easily infects insects of various orders when inoculated into the hemocoel, such as *Galleria mellonella* (Lepidoptera), *Calliphora* sp. (Diptera), *Tenebrio molitor* (Coleoptera), and *Naucoris cimicoides* and *Notonecta glauca* (Hemiptera) (Zotta 1921). It develops well in the hemolymphs of those insects, but in the hemolymph of *Carausius morosus* (Orthoptera), it degenerates and dies after several hours or a few days. The flagellate causes significant changes in the hemogram of infected insects (Zotta and Teodoresco 1933). The numbers of prohemocytes (proleucocytes) increase three times and those of the plasmatocytes and granulocytes (macro- and micronucleocytes) decrease markedly. *Leptomonas pyrrhocoris* develops also as a saprophyte in the flowers of *Colchicum autumnale* (Galli-Valerio 1920). This may be a source of contamination from which the insect acquires the flagellates.

Leptomonas serpens inhabits the gut and salivary gland of the pentatomid bug *Nezara viridula*, which feeds on the sap of tomato plants (Gibbs 1957). The infected bug transmits the metacyclic forms of the flagellate into the plant sap. Transmission to uninfected bugs occurs when they suck the flagellate-containing sap of such plants or by consuming the flagellates contaminating the plant surface. The association of flagellates, such as *L. pyrrhocoris* and *L. serpens*, with insects and plants creates a problem in distinguishing *Leptomonas* and *Phytomonas*, which are separated mainly on the basis of their hosts.

A number of leptomonads have been described from fleas, but Lipa (1963) believes that they are probably identical with *Leptomonas ctenocephali*, which was described from the dog flea, *Ctenocephalides canis* (Fantham 1912). *Leptomonas ctenocephali* inhabits the hindgut and Malpighian tubules and rarely the midgut of a flea. The majority of the forms in the hindgut are promastigotes, which are attached to the epithelium. The amastigote and encysted forms are prevalent in the rectum. Healthy flea larvae and adults become infected when they feed on contaminated feces.

The number of *Leptomonas* species in Lepidoptera is limited. *Leptomonas pyraustae* was described by Paillot (1928) from a few individuals of the European corn borer, *Ostrinia* (syn. *Pyrausta*) *nubilalis*. It inhabits the gut and Malpighian tubules, and the tubules become slightly hypertrophied and lose their transparency. A *Leptomonas* species infects the silkworm, *Bombyx mori*, and its ultrastructure has been examined (Abe 1978; Abe and Iwashita 1989). In the pupa, it multiplies intracellularly as a leishmanial form (amastigote) and extracellularly as a leptomonad form (promastigote). In the hemolymph, the flagellate forms rosettes and encysts. The cysts are also found in the lumen of the larval digestive tract and are expelled with the feces.

E. INSECTS AS VECTORS OF TRYPANOSOMATIDS

Insects serve as vectors of trypanosomatid flagellates, which infect vertebrates and plants. These flagellates are usually temporary residents of the insect's digestive tract. They are members of the genera *Phytomonas*, *Leishmania*, and *Trypanosoma* and are referred to as heteroxenous flagellates. *Phytomonas* is similar to *Leptomonas* except that the former is mainly in plants and the latter in insects. Members of *Leishmania* have two forms, the promastigote, which develops in the insect's digestive tract, and the amastigote, which develops intracellularly in the vertebrate host. Species of *Trypanosoma* have four body forms: epimastigote, promastigote, and amastigote are found in an insect vector, and trypomastigote in a vertebrate host.

The insect trypanosomatids that we have described previously do not infect vertebrates, whereas a few species of vertebrate *Trypanosoma* and *Leishmania* cause pathologies in their insect vectors (Kramer 1963; Lipa 1963; Brooks 1974). Some vertebrate flagellates multiply to very large numbers in their insect vectors, especially in the digestive tract, without causing severe injury. This is the case with *Trypanosoma rhodesiense*, which is transmitted by the tsetse flies (*Glossina* spp.) to cause the African sleeping sickness. The flagellates exist extracellularly in the digestive tract and salivary gland. They cause no apparent pathology aside from the salivary gland, which turns opaque from its normal translucency and may be hypertrophied (Buxton 1955). This is also generally the case with *Trypanosoma cruzi* and its relationship with reduviid bugs.

Most species of vertebrate *Trypanosoma* live extracellularly in the insect's digestive tract, but some develop intracellularly and cause harm to the insect vector. *Trypanosoma lewisi* develops in the gut epithelium of its flea vector, *Xenopsylla cheopis*, and causes mortality when the infection occurs in newly emerged adult fleas but not in older fleas (Garnham 1955). *Trypanosoma rangeli*, which causes transient infections in humans, produces severe pathologies in the reduviid bug *Rhodnius prolixus*, which becomes sluggish, light in color, and translucent (Grewal 1957; Watkins 1971a,b). The severely infected, first instar nymph is not able to molt (Grewal 1957). The flagellate infects most tissues of the reduviid bug, particularly the gut muscles, fat body, epidermis, and salivary gland (Watkins 1971a). It inhibits the actinomycete symbiote *Nocardia rhodnii* and causes molting deformities and death of the reduviid. Another prime cause of death is an autointoxication from toxins produced by the damaged muscles, nerves, and tracheal ruptures, and by inhibited excretion.

Species of *Leishmania* live in the digestive tract of the sand fly vector *Phlebotomus* spp. and, in general, cause no harm to their insect vectors. However, when found in large numbers, the flagellate *Leishmania donovani*, the cause of Indian Kala-azar, may completely block the lumen of the pharynx of a heavily infected fly (*Phlebotomus argentipes*) (Smith et al. 1941). The fly cannot ingest a second blood meal, and dies in a few days. This blocking phenomenon, which resembles the blockage in fleas infected with the plague bacillus, apparently does not occur in other species of sand flies (Kramer 1963).

VI. RHIZOPODA (AMOEBIC) INFECTIONS

The amoebic protozoa normally possess one of four types of pseudopodia: (1) lobopodium is broad and blunt, (2) filopodium is slender and needle-sharp and may be branched but does not form a reticulum, (3) axopodium is slender, normally unbranched, and is supported by some central "skeletal" structure, and (4) reticulopodium is slender, repeatedly branching and anastomosing. The entomopathogenic amoebae have lobopodia for locomotion and ingestion of food. The cytoplasm is generally divided into a granular endoplasm containing inclusions and a hyaline ectoplasm. The life cycle of the amoeba is simple, without special stages, except for cyst formation. Reproduction is by binary and multiple fission.

Nearly all entomogenous forms occur in the families Amoebidae and Endamoebidae of the order Amoebida. In the family Amoebidae, there are three strictly entomogenous genera: *Malameba*, *Malpighamoeba*, and *Malpighiella* (Brooks 1988). There are also several other genera, *Amoeba* and *Hartmanella*, in which a few entomogenous species are found. The genera of the family Endamoebidae, which are associated with insects, are *Entamoeba*, *Endamoeba*, *Endolimax*, and *Dobellina*. Most amoebae are commensals in the digestive tracts of insects, but some are pathogens and the diseases they cause are referred to as amoeboses or amoebiasis. The most widely studied pathogenic amoebae are those of the honey bee and grasshoppers.

The life cycle of an entomopathogenic amoeba generally is simple with primary and secondary trophozoites and cysts. The development of *Malamoeba scolyti* in the bark beetle, *Dryocoetes autographus*, however, has two phases in the life cycle: (1) large multinucleate amoebae (trophozoites) developing in the digestive tract and (2) a parasitic stage with cysts in the Malphigian tubules (Purrini and Žižka 1983). The vegetative stages are multinucleate, variable in shape but generally spherical.

A. AMOEBIASIS OF HONEY BEE

The amoeba infecting the honey bee, *Apis mellifera*, was named *Vahlkampfia mellificae* by Prell (1926) but is now placed in the genus *Malpighamoeba*. The adult honey bee appears to be the only known host of this amoeba (Bailey 1968a).

The life cycle of *Malpighamoeba mellificae* consists of primary and secondary trophozoites and cysts. The primary trophozoites are long and slender with an average diameter of 3.68 μm (Liu 1985a). Their surfaces have some wrinkles and the cytoplasm contains some vesicles. The secondary trophozoites vary in size and their shapes are highly irregular. Their smooth surfaces have numerous protrusions and small pits. Some of them have pseudopodia. Mature cysts are small, oval-shaped, and usually with smooth surfaces, but some have wrinkles.

The ingested cysts give rise to primary trophozoites that penetrate into the midgut epithelial cells and multiply and grow (Liu 1985b). Subsequently, they emerge from the cells, enter the lumen of the digestive tract, and migrate into the lumen of the Malpighian tubules to feed on the brush border but do not penetrate into the hemocoel. The cysts are found in the intestine and feces. The life cycle from the cyst to cyst stage takes about 21 days (Bailey 1968a).

The pathology caused by *Malpighamoeba mellificae* is generally chronic. However, when the Malpighian tubules contain large numbers of amoebae, the tubules are transparent and distended, their epithelia are destroyed, their lumens are obstructed, and their functions are inhibited. Severe infections are fatal to adult bees. The disease is evident mainly in spring and decreases in summer (Bailey 1968a).

Amoebiasis may occur together with other diseases, such as nosematosis, acarine disease, May sickness, and viral paralysis. Such mixed infections are more serious than only the amoebiasis. In mixed infections of the amoeba and *Nosema apis*, there are more bee colonies infected with the two protozoa than colonies infected with *Nosema* alone (Bailey 1968b).

Control of amoebiasis includes sanitary measures and proper beekeeping practices, the removal of dead bees, a supply of fresh food and water, the transfer of bees to clean hives, and the disinfection of contaminated hives with vapors of acetic acid or with 1 or 2% carbolic acid solution (Morison 1931; Bailey 1955).

B. AMOEBIASIS OF GRASSHOPPER

Malpighamoeba locustae was first described by King and Taylor (1936) from grasshoppers of the genus *Melanoplus*, but it was subsequently placed in a new genus and renamed *Malameba locustae* because it differed considerably from the honey bee amoeba, *Malpighamoeba mellificae* (Taylor and King 1937). Some workers have spelled the genus as *Malamoeba*, but we shall retain the original spelling, *Malameba*. This amoeba infects about 50 species of grasshoppers and locusts (Brooks 1988). Aside from grasshoppers, the only other recorded host is a thysanuran, *Lepisma saccharina* (Larsson 1976).

In grasshoppers, the amoeba occurs primarily in the lumens of the Malpighian tubules and may completely pack the tubules, which swell enormously (King and Taylor 1936; Taylor and King 1937). The trophozoites and cysts may also occur in the hemocoel (King and Taylor 1936; Taylor and King 1937; Prinsloo 1961; Henry 1968). Most infections of *M. locustae* have been from laboratory-cultured grasshoppers and the natural infection in the field is low (Brooks 1988).

The life cycle of *M. locustae* has been studied by Evans and Elias (1970) and by Harry and Finlayson (1975, 1976), but their accounts differ somewhat. According to Evans and Elias (1970), the ingested cysts give rise to primary trophozoites that penetrate the midgut epithelium and multiply intracellularly. By binary and multiple fissions, the secondary trophozoites are formed and they enter the hemocoel to infect the Malpighian tubules. However, Harry and Finlayson (1975, 1976) report that the primary trophozoites, which enter the epithelial cells of the midgut and caeca, do not penetrate into the hemocoel. They are extruded into the gut lumen of the digestive tract when their host cells degenerate. They enter the Malpighian tubules as secondary trophozoites and feed, as extracellular pathogens, on the brush border of tubule cells. On the other hand, Hanrahan (1975) found no evidence of the amoebae attacking the microvilli or the epithelial cells in her ultrastructural study.

The secondary trophozoites divide very rapidly and double their numbers every 24 h so that by day 20–22, the lumens of the tubules are packed with cysts and trophozoites. The number of cysts discharged with the feces ranges from 2 to 4 million. Transmission takes place by the ingestion of cysts and by cannibalism among some grasshoppers. The mechanical injury caused by the amoebae results in reduced adult longevity and fecundity.

The sizes of *Malameba* vary with different hosts and with methods of preparation of the amoeba (Henry 1968). The trophozoites have been reported as 4–50 μm in diameter. The thick-walled uninucleate cysts are oval or slightly elongate, from 8.5 to 19 μm in length and from 4.6 to 6.2 μm in width.

Grasshoppers, when lightly infected, display little or no external signs or symptoms; however, as the infection progresses, an infected individual becomes less active and loses its appetite (King and Taylor 1936; Taylor and King 1937). Severe infections in the Malpighian tubules affect their function and may result in an accumulation of toxins (Henry 1968). Cellular deterioration is severe in the Malpighian tubules and in the adjacent intestine of a moribund grasshopper. Heavily infected individuals are very sluggish, lethargic, and just prior to their death, undergo tetanic convulsions of their jumping legs and mouthparts. The amoebiasis causes reduced fecundity and egg viability in laboratory cultures of grasshoppers. This reduction is associated with an inhibitory effect of the amoeba on the normal bacterial flora of the reproductive organs of the female grasshopper (Prinsloo 1961).

Since amoebiasis is not uncommon in laboratory cultures of grasshoppers, attempts have been made to control the amoeba. Sulfa drugs and the antibiotic Thipyrameth fed to grasshoppers in their diets reduce the number of cysts (Henry 1968).

C. AMOEBIASIS IN OTHER INSECTS

In cockroaches, *Endamoeba blattae* is found in *Blatta orientalis*, and *Entamoeba thomsoni* and *Endolimax blattae* in *B. orientalis* and *Periplaneta americana* (Lucas 1927). In chironomid flies, the vegetative and cyst stages of *Amoeba chironomi* occur in the alimentary tracts and feces (Porter 1909c). *Dobellina mesnili* often occurs in large masses in the space between the peritrophic membrane and gut epithelium of various species of *Trichocera* (Keilin 1917; Bishop and Tate 1939). This species was first named *Entamoeba mesnili* by Keilin (1917) from specimens in the midgut of *Trichocera hiemalis*. *Malpighiella refringens* inhabits the Malpighian tubules of the rat flea, *Ceratophyllus* (syn. *Nosopsyllus*) *fasciatus* (Minchin 1910). This is a large amoeba, from 8 to 12 μm, resembling *Malpighamoeba mellificae*, but it differs in the cyst, which is tetranucleate.

Amoebae pathogenic to vertebrates also develop in insects. For example, *Endamoeba histolytica*, the well-known cause of dysentery in humans, has been observed in the digestive tracts of many insects, such as *Musca domestica*, *Periplaneta americana*, *Calliphora erythrocephala*, *Fannia canicularis*, etc. (Lipa 1967).

VII. CILIOPHORA (CILIATE) INFECTIONS

The ciliates possess (1) simple cilia or compound ciliary organelles (used in locomotion and feeding), (2) generally two types of nuclei, (3) sexual process of conjugation replacing syngamy, (4) presence of cortical infraciliature, (5) homothetogenic (transverse) (as opposed to symmetrogenic longitudinal mirrow image) fission, (6) commonly with a cytostome (cell mouth) and cytopharynx, and (7) possession of pellicular alveoli and specialized ribbons of subpellicular microtubules associated with the somatic kinetosomes (ciliary basal bodies) (Corliss 1982; Small and Lynn 1985).

Most ciliates associated with insects are commensals and only a few are pathogens (Steinhaus 1946; Corliss 1960; Corliss and Coats 1976). The commensals are found in the digestive tracts of insects, particularly those of termites and cockroaches. In aquatic insects, a number of ciliates are epibionts (ectocommensals or ectoparasites) in habit. The pathogenic forms penetrate and multiply in the insect's hemocoel, killing the host in the process. The hosts are mosquitoes and black flies (Brooks 1988). Most ciliates appear to be restricted to a single or few host species. The disease caused by ciliates is called ciliatosis.

The majority of ciliates pathogenic for insects are found in two genera, *Tetrahymena* and *Lambornella*, family Tetrahymenidae, order Hymenostomatida, class Oligohymenophorea (formerly order Holotricha) (Corliss and Coats 1976). These ciliates are characterized by uniform body ciliature and a ventral and well-defined buccal cavity with the ciliature composed of a single undulating membrane and an adoral zone of three membranelles. Over 3000 papers have been published on the biology of one genus *Tetrahymena* (Corliss 1973a). This genus contains most of the insect pathogenic forms. Corliss (1960) has synonymized many species described in the genera *Balantidium*, *Glaucoma*, *Lambornella*, *Leptolegnia*, *Paraglaucoma*, *Protobalantidium*, and *Turchiniella* as a few species of *Tetrahymena*. The pathogenic ciliates cause disease mainly in aquatic insects (e.g., chironomids and mosquitoes). They enter the insect's hemocoel and often kill their hosts. It is an impressive sight to see hundreds of ciliates swimming in the hemolymphs of infected mosquito and chironomid larvae (Fig. 11–1C).

FIGURE 11–1

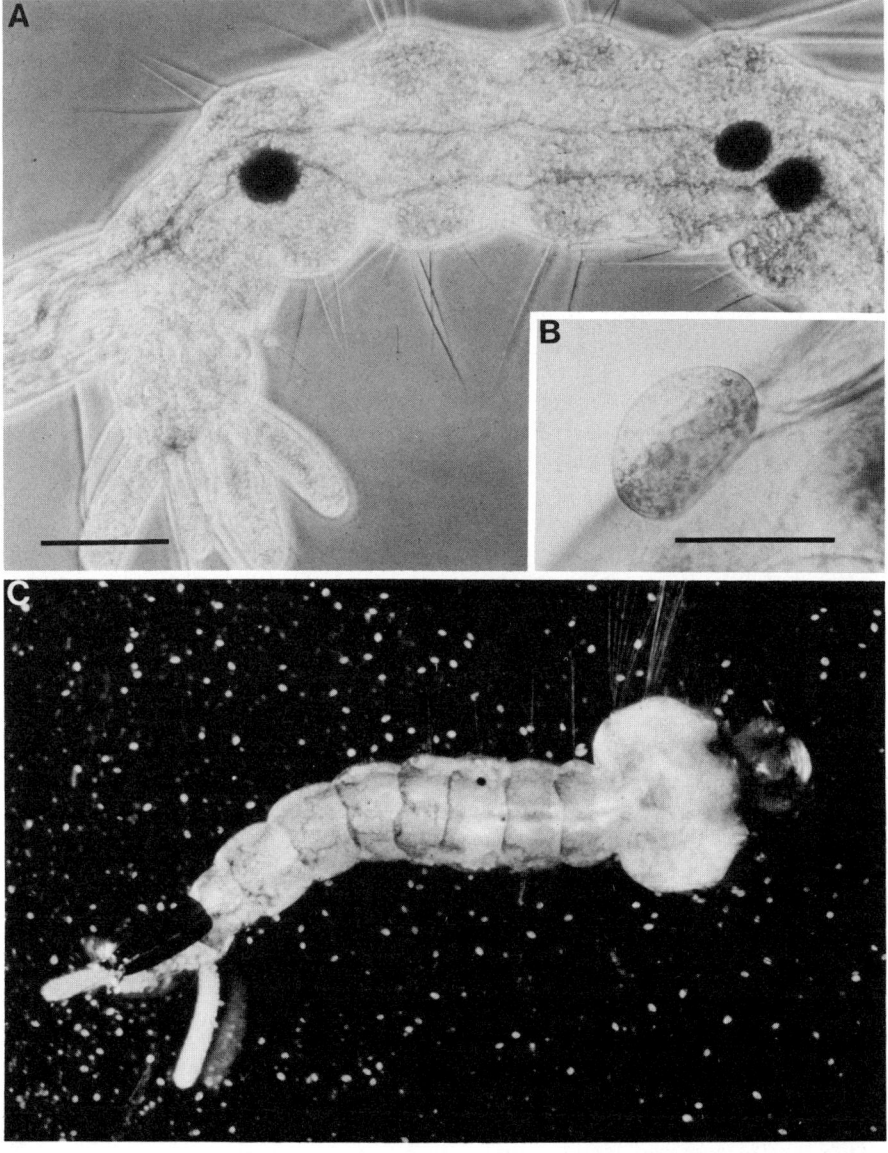

Infection of mosquitoes, *Aedes* spp., by the ciliate *Lambornella clarki*. (A) Newly hatched first instar larva of *A. sierrensis* with three invasion, cuticular cysts (stained with amide black dye). (Bar = 0.1 mm.) (B) Invasion cyst on the cuticle of third instar larva of *A. albopictus*. (Bar = 50 μm.) (C) Moribund fourth instar larva of *A. sierrensis* packed with ciliates. Note the free-swimming trophonts that have escaped through breaks in the cuticle of the anal papillae. (Provided by J. O. Washburn.)

A. PENETRATION INTO HEMOCOEL

The mode of entrance by ciliates into the insect's hemocoel is unknown for most species. There are four possible routes of entry: mouth, cuticle, artificial breaks or wounds in the body wall, and "natural breaks or weaknesses" during molting (Corliss 1960; Corliss and Coats 1976). There is no case of entrance by way of the mouth and digestive tract into the hemocoel. Invasion of *Tetrahymena* spp. occurs through abrasions of the integuments of midge larva (Corliss 1960), of fish, and of various vertebrate larvae (Hoffman et al. 1975). *Tetrahymena rostrata* invades through the integument of a dorsal pouch of the gray garden slug, *Deroceras reticulatum* (Brooks 1968).

The invasion of *Lambornella* by means of cuticular cysts was suggested by Muspratt (1947) and subsequently proven by Clark and Brandl (1976) (Fig. 11–1A,B). The cysts were observed on the integument of mosquito larva as early as 1921 by Keilin. When mosquito larvae are attacked by *Lambornella clarki*, melanized spots develop in the cuticle where unsuccessful and successful penetrations have taken place (Clark and Brandl 1976) (Fig. 11–1A). Successful penetration and infection depend on host activity, host age, and cuticle thickness. The first and second instar larvae are most susceptible to infection. The penetration of a single cyst can result in the infection of a host larva. The holes made by cuticular cysts may enable an opportunistic fungus, *Pythium flevoense*, to invade through the opening and kill the host larva (Washburn et al. 1988a).

B. CILIATOSES BY *TETRAHYMENA* SPP.

Tetrahymena pyriformis (syn. *Glaucoma pyriformis*) is the type species and best known member of the genus *Tetrahymena* (Corliss 1973b). It is normally pyriform, but the plasticity of the cuticle produces considerable distortion. It averages 30 × 50 μm, ranging from 10 to 11 μm minimum to over 90 μm in length. The number of kineties (ciliary meridians or rows of cilia) ranges from 15 to 25, usually from 17 to 21. It has no caudal cilium.

Tetrahymena pyriformis is widely distributed throughout the world and has many ecological habitats. Presumably, it is a microphagous organism (bacteria feeder) in nature. It is implicated fairly often in infections of vertebrates and invertebrates (Corliss 1960). Some cases of insect infections, however, may be improper identification for such forms as *Lambornella stegomyiae* and *L. clarki*. It thrives when inoculated into the hemocoels of many different insects and causes their deaths (Lwoff 1924; Corliss 1954; McLaughlin 1959; Thompson 1958). Its route of infection is unknown (very likely not through the mouth) and may be accidental through wounds or breaks or other kinds of artificial openings (Corliss 1960).

The infectivity of *T. pyriformis* varies with the species of mosquitoes. When present in high concentrations, it killed 76% of *Culex tarsalis* and less than 3% of *Aedes aegypti* (Grassmick and Rowley 1973). When inoculated into the larvae of the wax moth, *Galleria mellonella*, at a dose of 200 ciliates per larva, 75% died in 5 days and 100% died in 12 days (McLaughlin 1959). Dosages of 100,000 to 200,000 ciliates per larva caused 100% mortality in 2 days. Pupae inoculated with 60 to 80 ciliates died in 3 to 5 days, and dead pupae contained from 65,000 to 1,500,000 ciliates.

Tetrahymena chironomi was described from *Chironomus plumosus* (Corliss

1960). It can be easily reared on nonliving media. Infection in the chironomid larva is lethal. Corliss (1960) has reported a prevalence of about 9% infection in 2000 larvae. This truncate ciliate is one of the smallest species of *Tetrahymena* and measures 23 by 40 μm, with upper limits of length only slightly over 50 μm. It has from 23 to 28 kineties and does not possess a caudal cilium. The spherical micronucleus is relatively large. The ciliate does not exhibit dimorphism or cystic stages in its life cycle. Conjugation occurs in any size host, but only when the numbers of ciliates are very high, in the thousands, and usually takes place about 2 days before the host dies. It multiplies rapidly in the body cavity of a chironomid larva and produces dense populations of as many as 250,000 ciliates per larva. Heavy infection causes death in about 8 days, a light one in about 41 days. Generally, an infected larva does not pupate. The basis for the pathology is not known, but the ciliates probably utilize the hemoglobulin and fat body of the host. The route of invasion is unknown and attempts to infect through the mouth have failed. Little is known about the free-living stage of this ciliate.

Corliss (1960) made an interesting observation on the infection of *T. chironomi* in the chironomid. After the ciliates had consumed the various nutrient substances and brought the host's life nearly to the end, they entered into an act of sexuality that spells their doom! They undergo a fatal type of self-conjugation shortly before or after the death of the host. Such a suicidal association should be studied for its effect on the evolution of the ciliate as an insect pathogen.

The three ciliates *Tetrahymena dimorpha* (Batson 1983), *T. sialidos* (Batson 1985), and *Lambornella clarki* (Washburn et al. 1988b) are polymorphic with distinct forms associated with the parasitic (intrahemocoelic) and free-living habits. *Tetrahymena dimorpha* infects black flies (Simuliidae) and occurs in the hemocoels of larvae, pupae, and adults (Batson 1983). In a larva, the ciliate population (Hyde form) never exceeds above 240 and the infection is benign. When an infected larva matures to pupa and to adult, there is a dramatic increase in numbers up to 19,000 ciliates per fly; the fly soon dies from the heavy burden of ciliates. Associated with the increase in ciliate population during host metamorphosis is a marked morphological transformation of the ciliates in an adult host. These ciliates (Jekyll form) are small and pyriform in shape and the cortex is greatly modified; the total number of kineties is considerably reduced and has a limited range of 19 to 22. Most significantly, the kineties are arranged in a typical tetrahymenine precision. In contrast, the Hyde forms, in a larval host, are large, broadly oval, and possess an unusually wide range of kineties, from 30 to 66, that are characteristically disorganized, being incomplete, meandering, or branched.

Tetrahymena sialidos infects the larva of the alder-fly, *Sialis lutaria* (Megaloptera) (Batson 1985). In nature, from 40 to 70% of larvae are infected. The infection leads to inevitable death, about 11 to 12 months after infection when the number of ciliates has reached about 55,000 per infected host. The ciliates emerging from moribund or dead larvae undergo a distinct morphological transformation followed by a period of synchronous conjugation. The death of the host and the emergence of ciliates coincide with the appearance of a new generation of host larvae.

C. CILIATOSES BY *LAMBORNELLA* SPP.

Lambornella stegomyiae was discovered in larvae of the mosquito *Aedes scutellaris*, (syn. *Stegomyia scutellaris*, probably *A. albopictus*) by Lamborn (1921) and was named by Keilin (1921). Corliss (1960) placed the species in the genus *Tetrahymena* but later (Corliss and Coats 1976) resurrected the genus *Lambornella* when Clark and Brandl (1976) proved that *L. clarki* produced "invasion" cysts to infect a mosquito larva. Corliss and Coats (1976) thoroughly studied *Lambornella* and found that the major characters differentiating this genus from *Tetrahymena* are the cuticular invasion cysts and the larger number of kineties than in *Tetrahymena*. They redescribed the species *L. stegomyiae*. It averages 78 by 22 μm in size and is somewhat spindle-shaped with the anterior more strongly tapered than the posterior end. The modal number of kineties is about 30. The hemispherical cuticular cysts are from 30 to 40 μm in diameter and about 20 μm in height when found on mosquito larva. The hosts are tree-hole-breeding species of several subgenera (including *Stegomyiae*) belonging to *Aedes* and *Culex* from Malaysia, Rhodesia, and South Africa. The ciliate reported by Muspratt (1947) as *Glaucoma pyriformis* is probably *L. stegomyiae* (Corliss and Coats 1976).

When *Lambornella stegomyiae* gains entrance into the hemocoel, it multiplies rapidly in the mosquito larva and affects the fat body and destroys some fat lobes. An infected larva is pale with pearly-white opalescence of the tracheal gills and its cuticle is less transparent than that of an uninfected larva. The ciliates emerge from an infected larva through ruptures in the body wall, such as the breakage of an anal gill (Fig. 11–1C).

Lambornella clarki averages 71 by 45 μm in size and is generally broadly pyriform in shape though it is occasionally tailed (Corliss and Coats 1976). The modal number of kineties is 46, with a range of 44 to 52. The hemispherical cuticular invasion cyst is from 40 to 60 μm in diameter and from 25 to 40 μm in height when found in mosquito larvae. Its host is *Aedes sierrensis* in central California.

The prevalence of *L. clarki* in field populations of *Aedes sierrensis* may be as high as 100% (Egerter and Anderson 1985; Washburn and Anderson 1986). The mortality rates range from 21 to 76% and the infection does not always result in larval death, and 77% of the emerging adults may be infected. In an adult mosquito, the ciliates are restricted to the hemocoel, except in older females, in which they invade the ovaries, resulting in parasitic castration. The castrated, infected females deposit ciliates rather than eggs, and they serve in the dispersal of the ciliates (Egerter and Anderson 1985; Egerter et al. 1986; Washburn et al. 1989). The ovipositional behavior of these infected females is prolonged and mimics that of a hormonal gravid female in its gonotropic cycle. This ciliate has significant biological control potential because of the parasite-induced dispersal by hosts, the desiccation-resistant cysts, an active host-searching infective stage, and high infection rate and mortality (Washburn and Anderson 1991).

The life cycle of *L. clarki* is tightly linked to *Aedes sierrensis* larvae (Washburn et al. 1991a). The mosquito serves as both a predator and a host, while the ciliate serves as a pathogen and as a prey (Washburn et al. 1988b; Washburn and Anderson 1991; Washburn et al. 1991a,b). The ciliate is dimorphic with a free-living trophont stage (pear-shaped) and a theront stage (spherical in shape), which is obligately pathogenic for mosquito larvae. The trophont stage produces theronts under two conditions: (1) some trophonts released from moribund larvae divide and produce theronts or (2) chemical cues released from feeding larvae induce trophonts to undergo cellular division and the daughter cells differentiate into theronts. The mosquito larvae feed on the trophonts, and in the absence of theronts, they can eliminate the ciliates. However, the larvae release chemical cues that induce trophonts to form pathogenic theronts. The theronts, likewise, can eliminate the mosquito population in the tree hole but cannot survive in the absence of larvae.

Food for the *A. sierrensis* larvae plays a significant role in the ciliate–mosquito interactions (Washburn et al. 1991a,b). The regulatory role of *L. clarki* is modified by the availablity of food for the mosquito larvae, competition among the larvae, and changes in larval feeding behavior. Under certain conditions (e.g., resource limitations), the natural enemies, such as *L. clarki*, may actually increase the fitness of the adult mosquito by allowing for more or larger adults, or both, to complete development.

D. OTHER CILIATOSES

Lichtenstein (1921) described *Ophryoglena collini* from the ephemerid *Baetis* sp. It is an oval-shaped ciliate, from 200 to 300 μm in length and from 120 to 230 μm in diameter. It obtains nourishment from the blood, muscular and adipose tissues, and especially the gonads of the ephemerid.

Mercier and Poisson (1923) observed an infection of a *Colpoda* sp. in a nymph of the water bug, *Nepa cinerea*. A tumor the size of a pinhead that developed on the metathorax consisted of external and internal enlargements, with one section filled with numerous ciliates, hemocytes, and tracheae and the other with degenerating muscles. The presence of ciliates suggested that these microorganisms invaded through the integument possibly through a wound and caused the tumor.

Entomogenous ciliates of the order Peritrichida are epibionts, but some may cause pathologies to their hosts. Some peritrichs are host specific on species of aquatic beetles and bugs (hemipterans) (Neuninger 1948; Lust 1950). *Operculariella parasitica* occurs in colonial forms in the esophagus of the water beetle, *Dytiscus*, and other genera (Stammer 1948).

Only a small number of Suctoria have been reported from insects. Most of them are epibionts, but some may cause pathologies, especially when they are attached to an insect and affect its life processes, such as respiration (Lipa 1967).

REFERENCES

Abe, Y. 1978. The life cycle of *Leptomonas* sp., a protozoan parasite of the silkworm, *Bombyx mori* L. *J. Seric. Sci. Jpn.* 47, 421–426.

Abe, Y. 1980. On the encystment of *Leptomonas* sp. (Kinetoplastida: Trypanosomatidae), a parasite of the silkworm, *Bombyx mori* Linnaeus. *J. Protozool.* 27, 372–374.

Abe, Y., and Iwashita, Y. 1989. Ultrastructures of *Leptomonas* sp. LA-SES-AA7601 strain pathogenic to the silkworm, *Bombyx mori*. *J. Seric. Sci. Jpn.* 58, 443–447.

Bailey, C. H., and Brooks, W. M. 1972a. Histological observations on larvae of the eye gnat, *Hippelates pusio* (Diptera: Chloropidae), infected with the flagellate *Herpetomonas muscarum*. *J. Invertebr. Pathol.* 19, 342–353.

Bailey, C. H., and Brooks, W. M. 1972b. Effects of *Herpetomonas muscarum* on development and longevity of the eye gnat, *Hippelates pusio* (Diptera: Chloropidae). *J. Invertebr. Pathol.* 20, 31–36.

Bailey, L. 1955. Control of amoeba disease by the fumigation of combs and by fumagillin. *Bee World* 36, 162–163.

Bailey, L. 1968a. Honey bee pathology. *Annu. Rev. Entomol.* 13, 191–212.

Bailey, L. 1968b. The measurement and interrelationships of infections with *Nosema apis* and *Malpighamoeba mellificae* of honey-bee populations. *J. Invertebr. Pathol.* 12, 175–179.

Balbiani, E. G. 1882. Sur les microsporidies ou psorospermies des Articulés. *C.R. Acad. Sci. Paris Ser. D 95*, 1168–1171.

Batson, B. S. 1983. *Tetrahymena dimorpha* sp nov. (Hymenostomatida: Tetrahymenidae), a new ciliate parasite of Simuliidae (Diptera) with potential as a model for the study of ciliate morphogenesis. *Philo. Trans. R. Soc. London B301*, 345–363.

Batson, B. S. 1985. A paradigm for the study of insect–ciliate relationships: *Tetrahymena sialidos* sp. nov. (Hymenostomatida: Tetrahymenidae), parasite of larval *Sialis lutaria* (Linn.) (Megaloptera: Sialidae). *Philo. Tran. R. Soc. London B310*, 123–144.

Bishop, A., and Tate, P. 1939. The morphology and systematic position of *Dobellina mesnili* nov. gen. (*Entamoeba mesnili* Keilin, 1917). *Parasitology 31*, 501–510.

Brooks, W. M. 1968. Tetrahymenid ciliates as parasites of the gray garden slug. *Hilgardia 39*, 205–276.

Brooks, W. M. 1974. Protozoan infections. *In* "Insect Diseases." (G. E. Cantwell, ed.), Vol. 1, pp. 237–300. Marcel Dekker, New York.

Brooks, W. M. 1988. Entomogenous protozoa. *In* "CRC Handbook of Natural Pesticides. Microbial Insecticides, Part A. Entomogenous Protozoa and Fungi." (C. M. Ignoffo, ed.), Vol. 5, pp. 1–149. CRC Press, Boca Raton, Florida.

Burnett, W. J. 1851. The organic relations of some Infusoria, including investigations concerning the structure and nature of the genus *Bodo* (Ehr.). *Proc. Boston Soc. Nat. Hist. 4*, 124–125.

Bütschli, O. 1881. X. Einzelne Thiergruppen. *A. Protozoa. Zool. Jahrb.* (1880). *1*, 122–173.

Bütschli, O. 1880–1889. Protozoa. Sarkodina und Sporozoa. (Abt. I, 1880–1882); Mastigophora (Abt. II, 1883–1887); Infusoria und System der Radiolaria (Abt. III, 1887–1889). *In* "Klassen und Ordnungen des Thier-Reichs." (H. G. Bronn, ed.), *I*, 1–616; *II*, 617–1097; *III*, 1098–2035. C. F. Winter, Leipzig.

Buxton, P. A. 1955. "The Natural History of Tsetse Flies. An Account of the Biology of the Genus *Glossina* (Diptera)." H. K. Lewis, London.

Camargo, E. P., Mattei, D. M., Barbieri, C. L., and Morel, C. M. 1982. Electrophoretic analysis of endonuclease-generated fragments of k-DNA, of esterase isoenzymes, and of surface proteins as aids for species identification of insect trypanosomatids. *J. Protozool. 29*, 251–258.

Clark, T. B., and Brandl, D. G. 1976. Observations on the infection of *Aedes sierrensis* by a tetrahymenine ciliate. *J. Invertebr. Pathol. 28*, 341–349.

Clark, T. B., Kellen, W. R., Lindegren, J. E., and Smith, T. A. 1964. The transmission of *Crithidia fasciculata* Leger 1902 in *Culiseta incidens* (Thomson). *J. Protozool. 11*, 400–402.

Corliss, J. O. 1954. The literature on *Tetrahymena*: Its history, growth, and recent trends. *J. Protozool. 1*, 156–169.

Corliss, J. O. 1960. *Tetrahymena chironomi* sp. nov., a ciliate from midge larvae, and the current status of facultative parasitism in the genus *Tetrahymena*. *Parasitology 50*, 111–153.

Corliss, J. O. 1973a. Guide to the literature on *Tetrahymena*: A companion piece to Elliott's "General Bibliography." *Trans. Am. Microsc. Soc. 92*, 468–491.

Corliss, J. O. 1973b. History, taxonomy, ecology, and evolution of species of *Tetrahymena*. *In* "Biology of *Tetrahymena*." (A. M. Elliott, ed.), pp. 1–55. Dowden, Hutchinson & Ross, Stroudsburg, Pennsylvania.

Corliss, J. O. 1982. Ciliophora. *In* "Synopsis and Classification of Living Organisms." (S. P. Parker, ed.), Vol. I, pp. 603–637. McGraw-Hill, New York.

Corliss, J. O., and Coats, D. W. 1976. A new cuticular cyst-producing tetrahymenid ciliate, *Lambornella clarki* n. sp., and the current status of ciliatosis in culicine mosquitoes. *Trans. Am. Microsc. Soc. 95*, 725–739.

Dobell, C. 1932. "Antony van Leeuwenhoek and his "little animals."" Harcourt, Brace, New York.

Dufour, L. 1828. Sur la Grégarine, nouveau genre de ver qui vit en troupeau dans les intestins de divers insectes. *Ann. Sci. Nat. Ser. 1. 13*, 366–368.

Egerter, D., and Anderson, J. R. 1985. Infection of the western treehole mosquito, *Aedes sierrensis* (Diptera: Culicidae), with *Lambornella clarki* (Ciliophora: Tetrahymenidae). *J. Invertebr. Pathol. 46*, 296–304.

Egerter, D. E., Anderson, J. R., and Washburn, J. O. 1986. Dispersal of the parasitic ciliate *Lambornella clarki*: Implications for ciliates in the biological control of mosquitoes. *Proc. Natl. Acad. Sci. U.S.A. 83*, 7335–7339.

Ellis, D. S., Evans, D. A., and Stamford, S. 1980. The penetration of the salivary glands of *Rhodnius prolixus* by *Trypanosoma rangeli*. *Z. Parasitenkd. 62*, 63–74.

Evans, D. A., and Ellis, D. S. 1983. Recent observations on the behaviour of certain trypanosomes within their insect hosts. *In* "Advances in Parasitology." (J. R. Baker and R. Muller, eds.), Vol. 22, pp. 1–42. Academic Press, New York.

Evans, W. A., and Elias, R. G. 1970. The life cycle of *Malamoeba locustae* (King et Taylor) in *Locusta migratoria migratoides* (R. et F.). *Acta Protozool. 7*, 229–241.

Fantham, H. B. 1912. Some insect flagellates and the problem of the transmission of *Leishmania*. *Brit. Med. J. 2*, 1196–1197.

Galli-Valerio, B. 1920. Le cycle évolutif probable de l'Herpetomonas pyrrhocoris Zotta et Galli-Valerio. *Schweiz. Med. Wochenschr. 21*, 401–402.

Garnham, P. C. C. 1955. The comparative pathogenicity of protozoa in their vertebrate and invertebrate hosts. *Symp. Soc. Gen. Microbiol. 5*, 191–206.

Garnham, P. C. C. 1959. Some natural protozoal parasites of mosquitoes with special reference to Crithidia. *Trans. First Int. Conf. Insect Pathol. Biol. Control* pp. 287–294. Praha 1958.

Gibbs, A. J. 1957. *Leptomonas serpens* n.sp., parasitic in the digestive tract and salivary glands of *Nezara viridula* (Pentatomidae) and in the sap of *Solanum lycopersicum* (tomato) and other plants. *Parasitology 47*, 297–303.

Goncalves de Lima, V. M. Q., Roitman, I., and Kilgour, V. 1979. Five trypanosomatid species of insects distinguished by isoenzymes. *J. Protozool. 26*, 648–652.

Grassmick, R. A., and Rowley, W. A. 1973. Larval mortality of *Culex tarsalis* and *Aedes aegypti* when reared with different concentrations of *Tetrahymena pyriformis*. *J. Invertebr. Pathol. 22*, 86–93.

Grewal, M. S. 1957. Pathogenicity of *Trypanosoma rangeli* Tejera, 1920 in the invertebrate host. *Exp. Parasitol. 6*, 123–130.

Hall, R. P. 1953. "Protozoology." Prentice-Hall, New York.

Hanrahan, S. A. 1975. Ultrastructure of *Malameba locustae* (K. & T.), a protozoan parasite of locusts. *Acrida 4*, 237–249.

Harry, O. G., and Finlayson, L. H. 1975. Histopathology of secondary infections of *Malpighamoeba locustae* (Protozoa, Amoebidae) in the desert locust, *Schistocerca gregaria* (Orthoptera, Acrididae). *J. Invertebr. Pathol. 25*, 25–33.

Harry, O. G., and Finlayson, L. H. 1976. The life-cycle, ultrastructure and mode of feeding of the locust amoeba *Malpighamoeba locustae*. *Parasitology 72*, 127–135.

Henry, J. E. 1968. *Malameba locustae* and its antibiotic control in grasshopper cultures. *J. Invertebr. Pathol. 11*, 224–233.

Henry, J. E. 1981. Natural and applied control of insects by protozoa. *Annu. Rev. Entomol. 26*, 49–73.

Hoare, C. A., and Wallace, F. G. 1966. Developmental stages of trypanosomatid flagellates: A new terminology. *Nature 212*, 1385–1386.

Hoffman, G. L., Landolt, M., Camper, J. E., Coats, D. W., Stookey, J. L., and Burek, J. D. 1975. A disease of freshwater fishes caused by *Tetrahymena corlissi* Thompson, 1955, and a key for identification of holotrich ciliates of freshwater fishes. *J. Parasitol. 61*, 217–223.

Ibrahim, E. A., and Molyneux, D. H. 1987. Pathogenicity of *Crithidia fasciculata* in the haemocoele of *Glossina*. *Acta Tropica* 44, 13–22.

Keilin, D. 1917. Une nouvelle entamibe, *Entamoeba mesnili* n.sp., parasite intestinal d'une larve d'un Diptère. *C.R. Soc. Biol.* 80, 133–136.

Keilin, D. 1921. On a new ciliate: *Lambornella stegomyiae* n. g., n. sp., parasitic in the body-cavity of the larvae of *Stegomyia scutellaris* Walker (Diptera, Nematocera, Culicidae). *Parasitology 13*, 216–224.

King, R. L., and Taylor, A. B. 1936. *Malpighamoeba locustae*, n. sp. (Amoebidae), a protozoan parasitic in the Malpighian tubes of grasshoppers. *Trans. Am. Microsc. Soc.* 55, 6–10.

Kramer, J. P. 1961. *Herpetomonas muscarum* (Leidy) in the haemocoel of larval *Musca domestica* L. *Entomol. News 72*, 165–166.

Kramer, J. P. 1963. Pathogens of vertebrates and plants as pathogens of their acarine and insect vectors. *In* "Insect Pathology: An Advanced Treatise." (E. A. Steinhaus, ed.), Vol. 1, pp. 251-272. Academic Press, New York.

Kudo, R. R. 1966. "Protozoology." Fifth Ed. Charles C. Thomas, Springfield, Illinois.

Lamborn, W. A. 1921. A protozoan pathogenic to mosquito larvae. *Parasitology 13*, 213–215.

Larsson, R. 1976. Insect pathological investigations on Swedish Thysanura 1. Observations on *Malamoeba locustae* (Protozoa, Amoebidae) from *Lepisma saccharina* (Thysanura, Lepismatidae). *J. Invertebr. Pathol.* 28, 43–46.

Lee, J. J., Hunter, S. H., and Bovee, E. C. 1985. "An Illustrated Guide to the Protozoa." Society of Protozoologists, Lawrence, Kansas.

Léger, L. 1902. Sur un flagellé parasite de l'*Anopheles maculipennis*. *C.R. Soc. Biol. 54*, 354–356.

Leidy, J. 1856. A synopsis of Entozoa and some of their ecto-congeners observed by the author. *Proc. Acad. Nat. Sci. Philadelphia 8*, 42–58.

Leuckart, R. 1879-1886. "Die Parasiten des Menschen und die von ihnen herrührenden Krankheiten." Second Ed. C. F. Winter'sche, Leipzig.

Levine, N. D., Corliss, J. O., Cox, F. E. G., Deroux, G., Grain, J., Honigberg, B. M., Leedale, G. F., Loeblich, A. R., III, Lom, J., Lynn, D., Merinfeld, E. G., Page, F. C., Poljansky, G., Sprague, V., Vávra, J., and Wallace, F. G. 1980. A newly revised classification of the Protozoa. *J. Protozool.* 27, 37–58.

Lichtenstein, J.-L. 1921. *Ophryoglena collini* n. sp. parasite coelomique des larves d'Ephémères. *C.R. Soc. Biol. 85*, 794–796.

Lipa, J. J. 1963. Infections caused by Protozoa other than Sporozoa. *In* "Insect Pathology: An Advanced Treatise." (E. A. Steinhaus, ed.), Vol. 2, pp. 335–361. Academic Press, New York.

Lipa, J. J. 1966. *Blastocrithidia raabei* sp. n., a flagellate parasite of *Mesocerus marginatus* L. (Hemiptera: Coreidae). *Acta Protozool.* 4, 19–23.

Lipa, J. J. 1967. "An Outline of Insect Pathology." PW Ril, Warszawa (English translation by H. Markiewicz 1975).

Lipa, J. J., Carl, K. P., and Valentine, E. W. 1977. *Blastocrithidia caliroae* sp. n., a flagellate parasite of *Caliroa cerasi* (L.) (*Hymenoptera*: *Tenthredinidae*) and notes on its epizootics in host field populations. *Acta. Protozool.* 16, 121–129.

Liu, T. P. 1985a. Scanning electron microscope observations on the pathological changes of Malpighian tubules in the worker honeybee, *Apis mellifera*, infected by *Malpighamoeba mellificae*. *J. Invertebr. Pathol.* 46, 125–132.

Liu, T. P. 1985b. Scanning electron microscopy of developmental stages of *Malpighamoeba mellificae* Prell in the honey bee. *J. Protozool.* 32, 139–144.

Lucas, C. L. T. 1927. Two new species of amoeba found in cockroaches: With notes on the cysts of *Nyctotherus ovalis* Leidy. *Parasitology 19*, 223–235.

Lust, S. 1950. Symphorionte Peritrichen auf Käfern und Wanzen. *Zool. Jahrb. Abt. Systematik, Ökol. Geograph. Tiere 79*, 353–436.

Lwoff, A. 1924. Infection expérimentale à *Glaucoma piriformis* (infusoire) chez *Galleria mellonella* (lépidoptère). *C.R. Acad. Sci. Paris Ser. D 178*, 1106–1108.

Margulis, L. 1990. Introduction. *In* "Handbook of Protoctista." (L. Margulis, J. O. Corliss, M. Melkonian, and D. J. Chapman, eds.), pp. xi-xxiii. Jones and Bartlett, Boston.

Margulis, L., Corliss, J. O., Melkonian, M., and Chapman, D. J. 1990. "Handbook of Protoctista." Jones and Bartlett, Boston.

McLaughlin, R. E. 1959. Experimental infection of larval and pupal stages of *Galleria mellonella* (L.) (Pyralidae, Lepidoptera) by the ciliate *Tetrahymena pyriformis* (Ehrbg.). *J. Protozool. 6*, (Suppl.), 27.

Mercier, L., and Poisson, R. 1923. Un cas de parasitisme accidentel d'une Nèpe par un Infusorie. *C.R. Acad. Sci., Paris Ser. D 176*, 1838–1841.

Minchin, E. A. 1910. On some parasites observed in the rat-flea (*Ceratophyllus fasciatus*). *Festschr. 60, Geburstag R. Hertwigs 1*, 289–302.

Morison, G. D. 1931. Amoebic disease of bees in Great Britain. *Bee World 12*, 56.

Muspratt, J. 1947. Note on a ciliate protozoon, probably *Glaucoma pyriformis*, parasitic in culicine mosquito larvae. *Parasitology 38*, 107–110.

Naegeli, C. 1857. Über die neue Krankheit der Seidenraupe und verwandte Organismen. *Bot. Zeitung. 15*, 760–761.

Neuninger, U. 1948. Die Peritrichen der Umgebung von Erlangen mit besonderer Berücksichtigung ihrer Wirtsspezifität. *Zool. Jahrb. Abt. Systematik. Okol. Geograph. Tiere 77*, 169–266.

Paillot, A. 1928. On the natural equilibrium of Pyrausta nubilalis Hb. *Int. Corn Borer Invest. Sci. Rep. 1*, 77–106.

Pérez, Ch. 1905. Microsporidies parasites des crabes d'Arcachon. *Soc. Sci. Arcachon Trav. Lab. 8*, 15–36.

Porter, A. 1909a. The morphology and life-history of *Crithidia gerridis* as found in the British water-bug, *Gerris paludum*. *Parasitology 2*, 348–366.

Porter, A. 1909b. The life-cycle of *Herpetomonas jaculum* (Léger), parasitic in the alimentary tract of *Nepa cinerea*. *Parasitology 2*, 367–391.

Porter, A. 1909c. *Amoeba chironomi*, nov. sp., parasitic in the alimentary tract of the larva of a *Chironomus*. *Parasitology 2*, 32–41.

Prell, H. 1926. Beiträge zur Kenntnis einer Amöbenseuche der Honigbiene. *Z. Angew. Entomol. 12*, 163–168.

Prinsloo, H. E. 1961. Die beperkende invloed van'n amebiese parasiet op die bakteriese flora van die voortplantingsorgane van die bruinsprinkaanwyfie. *Suid-Afrik. Tydskr. Landbouwetensk. 4*, 553–556.

Purrini, K., and Žižka, Z. 1983. More on the life cycle of *Malamoeba scolyti* (Amoebidae: Sarcomastigophora) parasitizing the bark beetle *Dryocoetes autographus* (Scolytidae, Coeloptera). *J. Invertebr. Pathol. 42*, 96–105.

Rodhain, J. 1926. Le mode de transmission de *Herpetomonas rhinoestri* Rod. *C.R. Soc. Biol. 95*, 1128–1130.

Ross, R. 1906. Note on the parasites of mosquitoes found in India between 1895 and 1899. *J. Hyg. 6*, 101–109.

Rowton, E. D., Lushbaugh, W. B., and McGhee, R. B. 1981. Ultrastructure of the flagellar apparatus and attachment of *Herpetomonas ampelophilae* in the gut and malpighian tubules of *Drosophila melanogaster*. *J. Protozool. 28*, 297–301.

Schaub, G. A. 1990a. The effect of *Blastocrithidia triatomae* (Trypanosomatidae) on the reduviid bug *Triatoma infestans*: Influence of group size. *J. Invertebr. Pathol. 56*, 249–257.

Schaub, G. A. 1990b. Influence of starvation of the reduviid bug *Triatoma infestans* on the pathological effects of *Blastocrithidia triatomae* (Trypanosomatidae) and the coprophagic infection rate of the bug. *Z. Angew. Zool. 77*, 381–394.

Schaub, G. A. 1991. Flagellatoses of insects with emphasis on Blastocrithidia triatomae. *Proc. Fifth Int. Colloq. Invertebr. Pathol.*, pp. 502–506. Adelaide, Australia, 1990.

Schaub, G. A., and Böker, C. A. 1986. Scanning electron microscopic studies of *Blastocrithidia triatomae* (Trypanosomatidae) in the rectum of *Triatoma infestans* (Reduviidae). *J. Protozool. 33*, 266–270.

Schaub, G. A., Reduth, D., and Pudney, M. 1990. The peculiarities of *Blastocrithidia triatomae*. *Parasitol. Today 6*, 361–363.

Schaudinn, F., and Siedlecki, M. 1897. Beiträge zur Kenntnis der Coccidien. *Verh. Dtsch. Zool. Ges. Seventh year*, pp 192–210.

Siedlecki, M. 1899. Étude cytologique et cycle évolutif de *Adelea ovata* Schneider. *Ann. Inst. Pasteur 13*, 169–192.

Small, E. B., and Lynn, D. H. 1985. Phylum Ciliophora Doflein, 1901. In "An Illustrated Guide to the Protozoa." (J. J. Lee, S. H. Hunter, and E. C. Bovee, eds.), pp. 393–575. Society of Protozoologists, Lawrence, Kansas.

Smirnoff, W. A. 1971. Adaptation of the flagellate *Herpetomonas swainei* sp. n. on ten species of Tenthredinidae and Diprionidae. *Proc. Fourth Int. Colloq. Insect Pathol.* pp 58–59. College Park, Md., 25–28 August 1970.

Smirnoff, W. A., and Lipa, J. J. 1970. *Herpetomonas swainei* sp. n., a new flagellate parasite of *Neodiprion swainei* (Hymenoptera: Tenthredinidae). *J. Invertebr. Pathol. 16*, 187–195.

Smith, R. O. A., Halder, K. C., and Ahmed, I. 1941. Further investigations on the transmission of kala-azar. Part IV. The duration of life and other observations on 'blocked' files. *Indian J. Med. Res. 29*, 783–787.

Soares, M. J., Brazil, R. P., Tanuri, A., and de Souza, W. 1986. Some ultrastructural aspects of *Crithidia guilhermei* n. sp. isolated from *Phaenicia cuprina* (Diptera: Calliphoridae). *Can. J. Zool. 64*, 2837–2842.

Stammer, H. J. 1948. Eine neue eigenartige entoparasitische Peritriche *Operculariella parasitica* n.g., n.sp. *Zool. Jahrb. Abt. Systematik Ökol. Geograph. Tiere. 77*, 163–168.

Steinhaus, E. A. 1946. "Insect Microbiology." Comstock, Ithaca, New York.

Taylor, A. B., and King, R. L. 1937. Further studies on the parasitic amoebae found in grasshoppers. *Trans. Am. Microsc. Soc. 56*, 172–176.

Thompson, J. C., Jr. 1958. Experimental infections of various animals with strains of the genus *Tetrahymena*. *J. Protozool. 5*, 203–205.

Tieszen, K., Molyneux, D. H., and Abdel-Hafez, S. K. 1985. Ultrastructure of cyst formation in *Blastocrithidia familiaris* in *Lygaeus pandurus* (Hemiptera: Lygaeidae). *Z. Parasitenkd. 71*, 179–188.

Tieszen, K. L., Molyneux, D. H., and Abdel-Hafez, S. K. 1986. Host–parasite relationships of *Blastocrithidia familiaris* in *Lygaeus pandurus* Scop. (Hemiptera: Lygaeidae). *Parasitology 92*, 1–12.

Tuzet, O., and Laporte, M. 1965. *Blastocrithidia vagoi*, n. sp. parasite de l'Hemiptère Hétéroptère *Eurydema ventralis* Kol. *Arch. Zool. Exp. Gén. Paris 105*, 77–81.

Vickerman, K. 1990. Phylum Zoomastigina Class Kinetoplastida. In "Handbook of Protoctista." (L. Margulis, J. O. Corliss, M. Melkonian, and D. J. Chapman, eds.), pp. 215–238. Jones and Bartlett, Boston.

Wallace, F. G. 1943. Flagellate parasites of mosquitoes with special reference to *Crithidia fasciculata* Léger, 1902. *J. Parasitol. 29*, 196–205.

Wallace, F. G. 1966. The trypanomatid parasites of insects and arachnids. *Exp. Parasitol. 18*, 124–193.

Wallace, F. G. 1977. *Leptomonas seymouri* sp. n. from the cotton stainer *Dysdercus suturellus*. *J. Protozool. 24*, 483–484.

Wallace, F. G. 1979. Biology of the Kinetoplastida of Arthropods. In "Biology of the Kinetoplastida." (W. H. R. Lumsden and D. A. Evans, eds.), Vol. 2, pp. 213–240. Academic Press, New York.

Wallace, F. G., Camargo, E. P., McGhee, R. B., and Roitman, I. 1983. Guidelines for the description of new species of lower trypanosomatids. *J. Protozool. 30*, 308–313.

Washburn, J. O., and Anderson, J. R. 1986. Distribution of *Lambornella clarki* (Ciliophora: Tetrahymenidae) and other mosquito parasites in California treeholes. *J. Invertebr. Pathol. 48*, 296–309.

Washburn, J. O., and Anderson, R. J. 1991. Insect ciliates: Potential for container-breeding mosquitoes. *Proc. Fifth Int. Colloq. Invertebr. Pathol.* pp 507–511. Adelaide, Australia, 1990.

Washburn, J. O., Egerter, D. E., Anderson, J. R., and Saunders, G. A. 1988a. Density reduction in larval mosquito (Diptera: Culicidae) populations by interactions between a parasitic ciliate (Ciliophora: Tetrahymenidae) and an opportunistic fungal (Oömycetes: Pythiaceae) parasite. *J. Med. Entomol. 25*, 307–314.

Washburn, J. O., Gross, M. E., Mercer, D. R., and Anderson, J. R. 1988b. Predator-induced trophic shift of a free-living ciliate: Parasitism of mosquito larvae by their prey. *Science 240*, 1193–1195.

Washburn, J. O., Anderson, J. R., and Mercer, D. R. 1989. Emergence characteristics of *Aedes sierrensis* (Diptera: Culicidae) from California treeholes with particular reference to parasite loads. *J. Med. Entomol. 26*, 173–182.

Washburn, J. O., Anderson, J. R., and Mercer, D. R. 1991a. Parasitism of newly-hatched *Aedes sierrensis* (Diptera: Culicidae) larvae by *Lambornella clarki* (Ciliophora; Tetrahymenidae) following habitat flooding. *J. Invertebr. Pathol. 58*, 67–74.

Washburn, J. O., Mercer, D. R., and Anderson, J. R. 1991b. Regulatory role of parasites: Impact on host population shifts with resource availability. *Science 253*, 185–188.

Watkins, R. 1971a. Histology of *Rhodnius prolixus* infected with *Trypanosoma rangeli*. *J. Invertebr. Pathol. 17*, 59–66.

Watkins, R. 1971b. *Trypanosoma rangeli*: Effect on excretion in *Rhodnius prolixus*. *J. Invertebr. Pathol. 17*, 67–71.

Weiser, J. 1961. Die Mikrosporidien als Parasiten der Insekten. *Monograph Angew. Entomol. Beih. Z. Angew. Entomol.* No. 17.

Yoshida, N., Freymüller, E., and Wallace, F. G. 1978. *Herpetomonas mariadeanei* sp. n. (Protozoa, Trypanosomatidae) from *Muscina stabulans* (Fallén, 1816) (Diptera, Muscidae). *J. Protozool. 25*, 421–425.

Zotta, G. 1921. Sur la transmission expérimentale du *Leptomonas pyrrhocoris* Z. chez des insectes divers. *C.R. Soc. Biol. 85*, 135–137.

Zotta, G., and Teodoresco, A.-M. 1933. Formule leucocytaire de la chenille de *Galleria mellonella* infectée par le *Leptomonas pyrrhocoris*. *C.R. Soc. Biol. 114*, 314–316.

CHAPTER 12

PROTOZOAN INFECTIONS: APICOMPLEXA, MICROSPORA

I. APICOMPLEXA, GREGARINIA
    A. Eugregarinida
    B. Neogregarinida
II. APICOMPLEXA, COCCIDIA
    A. Life Cycle of *Adelina*
    B. *Adelina* Species
III. MICROSPORA, MICROSPOREA
    A. Microsporidian Taxonomy
    B. Hosts
    C. Transmission
    D. Gross Pathology
    E. Cellular Pathology
    F. Signs and Symptoms
    G. Spore
    H. Polar-Filament Function
    I. Polar-Filament Extrusion
    J. Structure of Presporal Stages and Spore Formation
    K. General Life Cycle
    L. Factors Affecting the Life Cycle
    M. Life Cycle of Family Amblyosporidae
    N. Proliferative Sequences
    O. Sexuality in Microsporidia

We shall consider in this chapter the members of the old "Sporozoa," particularly the gregarines, coccidia (singular coccidian), and microsporidia (singular microsporidium). This group comprises the majority of the entomopathogenic species. The Haplosporidia and Helicosporidia have also been placed in Sporozoa, but their members, which are insect pathogens, have been separated in "protista insertae sedis" meaning that their status has not been finalized (Sprague 1979). The so-called insect haplosporidia (singular haplosporidian) infect the digestive tract, fat body, oenocytoids, and Malpighian tubules of insects (Weiser 1963; Woolever 1966). Among the haplosporidia, *Nephridiophaga blattellae* is the best known (Woolever 1966). The helicosporidia (singular helicosporidian) have been considered as both protozoa and fungi. There are several species in insects, and *Helicosporidium parasiticum* has received the most attention (Kellen and Lindegren 1974; Lindegren and Hoffmann 1976). Because their pathogenic status remains obscure, we shall not discuss any further the haplosporidia and helicosporidia.

The former class Sporozoa is a heterogeneous assembly of organisms. Workers had recognized this anomalous situation for many years, but it was the increased information from ultrastructural studies with the electron microscope that paved the way to their drastic reorganization. The Committee of the Society of Protozoologists assigned the members of the Sporozoa to four phyla: Apicomplexa, Microspora, Myxozoa, and Ascetospora (Levine et al. 1980). *The Handbook of Protoctista* has retained these phyla, except Haplosporidia is used rather than Ascetophora (Margulis et al. 1990). However, we shall continue to use sporozoa as a general term denoting these spore-forming protoctists.

Various types of associations occur between protozoa and insects. Some of these protozoa are pathogens of both vertebrates and insects (e.g., haemosporidia causing malaria); others are commensals or weakly virulent in the insect's digestive tract (e.g., gregarines); still, others are highly virulent forms (e.g., microsporidia).

Most of the entomopathogenic protozoa occur in the phyla Apicomplexa and Microspora. The life cycles of the members of the two phyla show certain similarities. The major groups differ in their life cycles by the absence of certain stages and the specialized development of other stages. The diverse terminology describing the stages and forms in the life cycles of various groups has led to much confusion. Levine (1971) has made a major attempt to establish a uniform terminology. Excellent glossaries are found in *An Illustrated Guide to the Protozoa* (Lee et al. 1985) and the *Handbook of Protoctista* (Margulis et al. 1990).

One of the most seriously misused terms is schizogony which is a type of multiple fission that occurs in the formation of offspring cells in apicomplexa, microsporidia, and myxozoa. Schizogony has often been limited to mean merogony, but three types of schizogony may occur in a life cycle: (1) sporogony forming sporozoites, (2) merogony forming merozoites, and (3) gametogony producing gametes. In this text, we shall use schizogony in the broadest sense meaning multiple budding.

Members of the phylum Apicomplexa have typical fine structures, primarily the "apical complex," that occur in at least one stage in the life cycle (Levine 1971, 1982; Scholtyseck 1979). These fine structures can be observed only with the electron microscope. They generally include the polar ring(s), micronemes, rhoptries, conoid, and subpellicular microtubules. Micropores, used for feeding, are generally present at some stage. The members lack cilia and reproduce sexually by syngamy. They form spores, reproduce by sexual and asexual means, and move by body flexion or gliding. When pseudopodia are present, they are used for feeding and not for locomotion. Some groups have flagellated microgametes. The conoid, if present, forms a complete cone. The oocysts contain infective sporozoites that result from sporogony.

The Apicomplexa are separated into three classes: Gregarinia, Coccidia, and Hematozoa (Vivier and Desportes 1990). The difference between gregarines and coccidia is in their gamogony; the female gamont of the gregarine gives rise to a number of gametes and that of coccidian to a single gamete. The Hematozoa have gamogony similar to that of the Coccidia but are distinct because of their sporogony and cytological features. We shall not discuss the Hematozoa because they are vertebrate pathogens even though they do cause pathology in their insect vectors.

I. APICOMPLEXA, GREGARINIA

The class Gregarinia, commonly referred to as gregarines, has mature gamonts (trophozoites) that are large and extracellular and are found in the digestive tracts and body cavities of invertebrates. They may possess attachment organelles (the mucron and epimerite), have generally similiar gametes (isogametes), and the gamonts undergo syzygy. The zygotes form oocysts within gametocysts, and the life cycle characteristically consists of gametogony and sporogony. This class has four orders: Blastogregarinida, Archigregarinida, Eugregarinida, and Neogregarinida (Vivier and Desportes 1990). Since Blastogregarinida and Archigregarinida are parasites of annelids, sipunculids, enteropneustids, and ascidians, we shall not consider these orders. The species of Eugregarinida are largely of low virulence to insects, whereas those of Neogregarinida are much more virulent and may cause lethal infections.

Gregarines form a very large group with about 1450 named species in three orders (Levine 1982). During their initial reproductive stage, the organisms line themselves end to end, two or more in numbers, to form an association known as syzygy. This characteristic association was used by Dufour (1828) to coin the term "gregarines." The hosts are marine invertebrates, earthworms, and arthropods that often show extremely heavy infections (Watson 1916; Foerster 1938; Levine 1977a). Some that are found in insects are closely associated with and adapted to the life cycles of insects.

A. EUGREGARINIDA

Most eugregarines are pathogens of annelids and arthropods, but some are considered to be commensals or mutualists. Some are restricted to a single or a few insect hosts (Brooks 1988). Their life cycle lacks merogony, and reproduction is sexual only (Fig. 12–1). Some are relatively large, about 10 mm in length and visible to the naked eye; others are among the smallest protozoa, for example, *Lipotropha*, a parasite of blow fly larvae (Manwell 1977).

1. Structure

The order Eugregarinida contains mainly aseptate gregarines with over 400 species and septate gregarines with about 900 species (Levine 1977a,b). In the aseptate gregarines (acephaline), the trophozoite has one compartment; whereas in the septate forms (cephaline) there are several compartments. The first compartment, called the protomerite, may possess an epimerite, which may have hooks or processes used to attach the gregarine to the host cell. The protomerite is connected to the deutomerite, which contains the nucleus. The deutomerite may be separated further in some species. In the aseptate gregarine, the gamont attaches to the host cell by means of a mucron. The epimerite differs from the mucron in that it appears to be separated by a septum from the rest of the body, when observed with the light microscope, while the mucron is not. When examined with the electron microscope, no septum appears to separate the epimerite (Levine 1977a), which, however, is devoid of all contractile elements, while the mucron contains fibers that are probably contractile (Vávra 1969). Both the epimerite and the mucron appear to be modified conoids (Levine 1977a).

The structure of gregarines shows considerable differentiation. The outer or ectoplasm layer is often divided into the epicyte (cortical), sarcocyte (intermediate), and mycocyte (inner) layers. The epicyte is trilaminar and has two cytomembranes that lie just beneath the plasma membrane (Vivier 1968; Vávra 1969; Philippe and

FIGURE 12-1

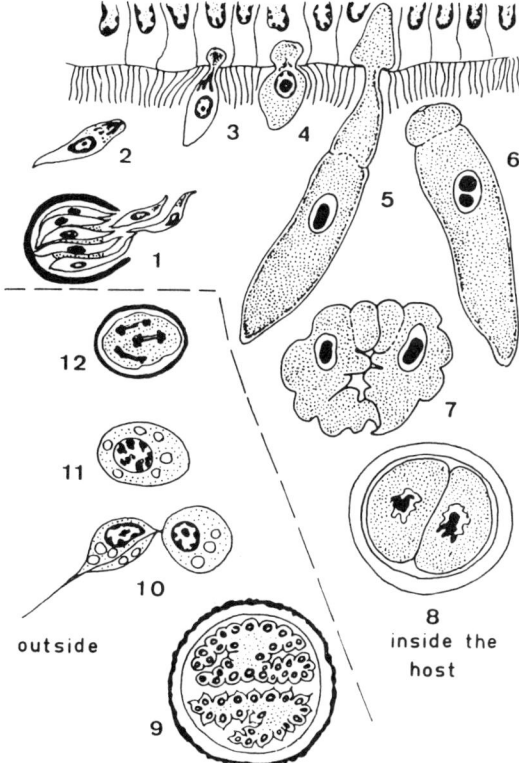

Life cycle of gregarine *Stylocephalus* sp. (*1–8*) Developmental stages occurring in the intestine of the host (a tenebrionid beetle): (*1*) sporozoite escaping from the spore ingested by the host; (*2*) free sporozoite; (*3*) differentiation of the epimerite; (*4–5*) growing gamonts (syn. trophozoites); (*6*) mature gamont detached from the intestinal epithelium; (*7*) pairing of gamonts; (*8*) encystment. (*9–12*) Developmental stages occurring outside the host: (*9*) differentiation of gametes (female gametes in the upper gamont, male ones in the lower); (*10*) fertilization; (*11*) zygote; (*12*) meiosis (*10*, *11*, and *12* occur under the cyst wall). [Courtesy of Vivier and Desportes (1990) and by permission of Jones and Bartlett Publishers. Drawing by S. Manion-Artz.]

Schrével 1982; Schrével et al. 1983). The mycocyte contains contractile fibrils or myonemes, and in certain species, fibrils that may have skeletal function. The endoplasm contains a nucleus, which may have a nucleolus and special granular forms of glycogen, "paraglycogen," a reserve food characteristic of gregarines. The paraglycogen gives the gregarine a greyish, opaque appearance and stains with iodine to an ochre brown.

2. Life Cycle

The life cycle of gregarines varies greatly and is often complex (Fig. 12–1). The insect ingests a mature oocyst, which is acted upon by the digestive juices in the insect's alimentary tract. Several sporozoites, eight or rarely four, emerge from the oocyst and penetrate into the midgut epithelial cells or into the hemocoel and become trophozoites. After reaching a certain stage in development, the trophozoites emerge from the cells and destroy them in the process. In the lumen of the digestive tract, they develop subsequently into gamonts. In some gregarines the sporozoites do not penetrate the cell but remain attached as trophozoites to the cell with epimerites or mucrons. A septate gregarine loses its epimerite when it becomes a mature gamont. The gamonts undergo syzygy in pairs or in longer groups [Fig. 12–1(7)]. A membrane develops around the associated gamonts to produce a cyst (gametocyst). One gamont produces microgametes and the other member produces macrogametes, or the gamonts form isogametes. The gametes fuse to form zygotes, which are the only diploid stage in the life cycle of gregarines (Grassé 1953). The zygotes secrete a thick membrane to form the oocysts. The zygotes first undergo meiotic divisions, followed by mitotic divisions, to produce haploid sporozoites. The oocysts of certain families of gregarines emerge by bursting the gametocyst wall. In other families, the gametocysts have special tubes called sporoducts; these tubes evert, and the oocysts are expelled through the sporoducts.

In the life cycle of the eugregarine, there is no merogony, and the number that develops in a host is no more than the number of sporozoites that emerge from the ingested oocysts. Accordingly, the number of eugregarines in a host is generally low and the pathogenesis is usually weak or chronic.

3. Interaction with Host Development

In some eugregarines, the rate of development is more related to the host developmental stage than to the duration of infection. With *Diplocystis tipulae*, a parasite of the crane fly, *Tipula paludosa*, sporogony occurs in the pupal and adult stages (Sherlock 1979).

The life cycle of the gregarine, *Schneideria schneiderae*, and the development of its sciarid host, *Trichosia pubescens*, are synchronized (Malavasi et al. 1976; da Cunha and Jurand 1978). The molting hormone ecdysone, produced by the sciarid, acts as a signal for the gregarines to emerge from the cells of the intestinal caeca, pair together, and start the sexual stage. When an adult fly emerges, the gregarines are in the sporozoite stage in the insect's digestive tract. The fly disseminates the sporozoites when it lays its eggs or defecates. The fly larva becomes infected by ingesting the sporozoites. The infected host cell hypertrophies and becomes a nurse cell to feed its guest. The polytene chromosomes of the infected cell attain enormous size, dissociate, and form a huge polyploid nucleus (da Cunha et al. 1968; da Cunha and Jurand 1978). The adaptation of *S. schneiderae* to the nonhabitual host *Plastosciara* sp. is not as intimate and does not cause cellular and chromosomal changes in the host (da Cunha et al. 1978).

4. Pathology

Some gregarines inhabiting the digestive tracts of invertebrates are regarded as commensals. Others should be considered as pathogens because they destroy individual cells; however, the damage to the midgut is easily repaired, and the pathology is generally weak. Gregarines cause their most severe pathology during their intracellular and intercellular developments, for example, *Leidyana ephestiae* (Lipa 1967). The affected cells generally develop cavities, their nuclei degenerate, and the cells are strongly basophilic upon staining. Other examples of pathogenic midgut gregarines are *Gregarina rostrata* in *Lagria hirta*, *Stictospora provincialis* in *Amphimallon solstitialis*, and *Gregarina cuneata* in *Tenebrio molitor*.

The overall effect of gregarines in the digestive tracts has been questioned by investigators who reached different conclusions with the same gregarine and host insect. The eugregarine *Gregarina polymorpha*, a common inhabitant in the larval midgut of the mealworm, *Tenebrio molitor*, may occur in large numbers of up to 6,000 gregarines and completely block the midgut (Harry 1967). Sumner (1936) considers the gregarines to be in a mutualistic association because their presence increases the growth rate and reduces the death rate of mealworm larvae fed swine thyroid gland. Harry (1967), however, believes the gregarine to be parasitic because it has a detrimental effect on the development of the mealworm grown on suboptimal diet.

The eugregarine *Pyxinia frenzeli* commonly attaches to larval midgut cells of the black carpet beetle, *Attagenus megatoma*. Dunkel and Boush (1969) observed that infected last instar larvae deprived of food for 42 days lost weight significantly faster than uninfected larvae; the infected adult females weighed less than the uninfected individuals. On the other hand, Schwalbe and Baker (1976) placed the same gregarine and beetle species under nutritional stress conditions. They observed that the fresh weight loss and the utilization of the stored protein, glycogen, or lipid were the same between the infected and uninfected larvae, but the infected larvae had a higher moisture content. They concluded that *P. frenzeli* was a commensal or a mutualist, rather than a pathogen. The difference between the results obtained by Dunkel and Boush (1969) and those of Schwalbe and Baker (1976) is possibly caused by the number of gregarines harbored by the mealworm larvae tested by these two groups of workers. The intensity of infection may affect the host response to the gregarine, especially in the case of chronic infections.

Gregarines of the genus *Ancyrophora* are found only in the digestive tracts of predaceous insects, such as Plecoptera, Trichoptera, and Odonata (Baudoin and Mouthon 1976). This association suggests that *Ancyrophora* have become adapted to the diet of insect predators.

Gregarines that inhabit the gastric caeca generally produce more severe pathology than those in the midgut. They may cause the cells of the gastric caeca to hypertrophy and, in many cases, rupture the walls of the caeca, leading to secondary infection and bacterial septicemia. Some gregarines inhabiting the gastric caeca are *Gregarina chrysomelae* in *Chrysomela polita*, *G. macrocephalia* in *Aphodius depressus*, and *Didymophyes ontophagi* in *Ontophagus fracticornis*.

Coelomic, septate gregarines, which penetrate into the insect's hemocoel, cause injuries and often kill the insect. The sporozoites penetrate into the epithelial cells or enter the tunica of the connective tissue and the muscles around the digestive tract (Weiser 1963). With the growth of the gregarine, the cell ruptures and the gregarine enters the hemocoel and produces cysts. *Diplocystis major* infects the cricket *Gryllus campestris*, and *D. schneideri* infects the American cockroach, *Periplaneta americana*. The genus *Monocystis* produces cysts in the body cavities of Carabidae and Scarabaeidae. According to Huger and Lenz (1976), an unidentified coelomic gregarine forms whitish cysts in the hemocoel of the American termite, *Coptotermes acinaciformis*.

5. Aseptate Gregarines

The type species *Ascogregarina* (syn. *Ascocystis*) *culicis* is a well-known aseptate gregarine found in the mosquito genera *Aedes*, *Anopheles*, *Armigeres*, and *Culex* (Levine 1977a; Ward et al. 1982). The infection of *A. culicis* in larvae of *Aedes aegypti* may reduce the development of the dog heartworm, *Dirofilaria immitis*, in the mosquito (Sneller 1979). Other species of *Ascogregarina* infecting mosquitoes are *A. taiwanensis* and *A. armigerei* (Lien and Levine 1980). These species of *Ascogregarina* are specific to a certain degree for various species of mosquitoes. The pathology caused by most *Ascogregarina* is minimal but may result in death in some cases (Walsh and Olson 1976).

Other aseptate gregarines of insects are found in the genera *Kofoidina*, *Enterocystis*, *Gamocystis*, *Monocystis*, and *Diplocystis*. Species of *Enterocystis* are found in the digestive tracts of Ephemeroptera (Codreanu 1940) (e.g., *E. racovitzai* and *E. fungoides* in *Baetis vernus*). A *Monocystis* species is found in the body cavity of *Hoplia* sp. (Weiser and Wille 1960). It produces white cysts, 0.5 mm in diameter, in which the gametes produce zygotes that form navicular spores with 8 sporozoites each. An infected larva has from 6 to 15 cysts and may die before pupation.

6. Septate Gregarines

Most of the entomopathogenic septate gregarines are in the families Gregarinidae, Leidyanidae, Didymorphidae, Stylocephalidae, and Actinocephalidae. One of the best known species is *Gregarina blattarum*, a parasite of cockroaches (e.g., *Blattella germanica*, *Blatta orientalis*, and *Periplaneta americana*). *Gregarina blattarum* lives for a short period in the midgut epithelium but matures as gamonts in the gut lumen (Sprague 1941). Mature gamonts, in syzygy, are from 500 to 1100 μm in length and from 160 to 400 μm in width. Gametocysts are spherical or oval. They discharge through the sporoducts long chains of oocysts, as many as 10,000 per chain. The oocysts are ellipsoidal to cylindrical with rounded ends, 8.8 μm long and 4.3 μm wide. They are covered with a mucoid sheath giving them a truncate appearance. *Retractocephalus raphidopalpi* occurs in the midgut of the beetle, *Raphidopalpa* (syn. *Aulacophora*) *foveicollis* (Haldar and Chakraborty 1976). The initial development of this septate gregarine is intracellular in midgut cells, followed by extracellular existence as gamonts. The epimerite, a simple symmetrical globular structure, is retractile into the protomerite. The dehiscence of the cyst is by simple rupture, and the barrel-shaped oocysts are extruded in chains.

B. NEOGREGARINIDA

The gregarines of the order Neogregarinida are called neogregarines or schizogregarines. The neogregarines are considered to be more primitive than the eugregarines (Manwell 1977). Unlike the eugregarines, which multiply by gametogony and sporogony, the neogregarines have an additional schizogony, the merogony, which takes place once or twice in the insect (Table 12–1). Whereas the number of eugregarines in the host is governed by the number of ingested oocysts and their sporozoites, in the neogregarines, their numbers are not restricted this way but increase through merogony. Multiplication by merogony compensates for the small number of oocysts ingested by the insect. Merogony occurs intracellularly or extracellularly. The high numbers result in a more virulent infection as compared to that produced by eugregarines.

A number of entomopathogenic neogregarines produce lethal infections in important insect pests in the orders Diptera, Coleoptera, and Hemiptera. Most neogregarines infect a single or a few insect species, but some, such as *Farinocystis*

TABLE 12-1 Pathogenesis of Eugregarinida and Neogregarinida

Character	Eugregarines	Neogregarines
Anatomical location in host	Gamonts usually in midgut, with heavy infections in pyloric region, esophagus, crop and rectum; rarely in hemocoel.	Gamonts mainly in hemocoel in various organs, primarily fat body.
Numbers produced per host	Numbers depend on the number of ingested oocysts. No merogony and host; generally not heavily infected.	Merogony occurs, and the numbers may be considerable.
Pathology	Generally cause inapparent infection; only a few individual cells destroyed, mainly nonlethal.	Invade several organs, e.g., fat body, Malpighian tubule, and midgut epithelium. Infection often chronic but at times lethal.

tribolii and *Mattesia grandis*, attack a large number of insect species (Purrini 1976, 1977).

The neogregarines infect various organs and tissues, such as the intestinal epithelium (e.g., *Caulleryella*, *Schizocystis*), Malpighian tubules (e.g., *Ophryocystis*, *Machadoella*, *Mattesia*), and, in particular, the fat body (e.g., *Mattesia*, *Menzbieria*, *Lipocystis*, *Farinocystis*, *Syncystis*, *Lipotropha*) (Weiser 1963). The pathogenesis in some species involves another complicating factor, autoinfection, in which the oocysts germinate and produce sporozoites causing reinfection of the same host (Žižka 1972, 1977).

1. Structure

Oocysts of neogregarines are mostly navicular, oval or spherical in shape (Weiser 1955). The oocyst wall is two or three layered (Žižka 1977). Some oocysts have one or two pores (poles) covered by a cap or plug, whose dissolution enables the wormlike sporozoites to emerge (Weiser 1963). In others, the wall is thin and bursts in the insect's digestive tract to liberate the sporozoites. In the navicular oocyst, the sporozoites are alternately attached to both poles by means of a thin stalk or disc. In other types, the sporozoites are attached to only one pole and resemble a cluster of bananas.

2. Life Cycle

The life cycle of neogregarines is much more complex than that of eugregarines (Weiser 1955). We shall consider the life cycle of *Mattesia trogodermae*, which infects the khapra beetle, *Trogoderma granarium*, as described by Canning (1964). The oocyst, after ingestion by the larva, is acted upon by digestive juices dissolving the polar plug. The sporozoites emerge through the poles and enter the hemocoel possibly by way of the Malpighian tubules. A sporozoite invades a fat cell and develops in the cytoplasm into a multinucleate meront that is plasmodium-like. The meront undergoes merogony to form actively motile, micronuclear merozoites, from 5 to 8 by 2.5 to 3 μm in size, which are responsible for the spread of infection throughout the insect. They invade fat cells to become meronts (maximum size 20 by 15 μm) and undergo macronuclear merogony to form macronuclear merozoites (from 4 to 5 μm in diameter). These are motile and remain extracellular, free in the hemocoel. The macronuclear merozoites become gamonts (gametocytes) and associate in pairs. A cyst wall is produced around the associated gamonts to produce a gametocyst (from 6.0 to 8.5 μm in diameter). Each gamont produces four gametes, which fuse in pairs to form two zygotes. Around each zygote a spore wall develops to form a lemon-shaped oocyst. The zygote nuclei divide to produce eight nuclei in each oocyst, and the cytoplasm splits longitudinally to form eight sporozoites. Each pole of the oocyst is sealed with a plug that is soluble in the digestive juices of the host. Oocysts in fresh preparations measure 13 by 7.7 μm. Since meiosis does not occur during gamete production, a zygotic meiosis with a haploid life cycle is assumed by Canning (1964).

The number of merogonies differs with the genera (e.g., *Mattesia, Farinocystis, Menzbieria, Lipocystis*, and *Ophryocystis* have two merogonies whereas other genera, e.g., *Caulleryella, Tipulocystis, Merogregarina, Lipotropha, Selenidium, Meroselenidium, Syncystis, Spirocystis, Schizocystis*, and *Siedlecka* have one merogony) (Weiser 1955, 1963). The number of oocysts per gametocyst varies from 1 to 8200, depending on the genus.

3. Hosts

Neogregarines are transmitted to their hosts through food contaminated with oocysts. In some species (e.g., *Syncystis mirabilis*, which infects the water scorpion, *Nepa cinerea*) transmission may occur through cannibalism on infected individuals (Weiser 1963). When neogregarines occur in the digestive tract or Malpighian tubules, their oocysts are defecated and contaminate the eggs or food of an insect. Oocysts are also dispersed upon death and decomposition of the insect hosts. Many neogregarines are pathogens of stored-product insects, and such products as flour, grain, etc., are often contaminated with their oocysts. In areas where food is stored under conditions encouraging insect build up (e.g., in old colonies of bark beetles or in small ponds) the neogregarines may cause massive infection.

Infection by neogregarines may reduce the resistance of the insects to physical and chemical agents. The infected insects are less tolerant than uninfected individuals to changes in oxygen content in the water and to exposure to X-rays and chemical insecticides (Weiser 1963).

In the flour beetle, *Tribolium castaneum*, infection by *Farinocystis tribolii* causes a juvenile hormonal effect and produces larval–pupal and pupal–adult intermediate forms (Rabindra et al. 1981). A similar effect is produced when ether extracts of spores (oocysts) are applied to pupae of *T. castaneum* and to final instar nymphs of the bug *Dysdercus cingulatus*.

The genus *Ophryocystis* has over 10 species that attack the Malpighian tubules of beetles, mostly Tenebrionidae. *Ophryocystis mesnili* is a fairly well-known species infecting the mealworm, *Tenebrio molitor*. The genus *Mattesia* has about eight species with *M. dispora* as the type species. All species infect the fat body, except for *M. povolnyi*, which attacks the Malpighian tubules (Weiser 1952) and *M. geminata* the hypodermis (Jouvenaz and Anthony 1979). *Mattesia dispora* (syn.

Coelogregarina ephestiae) infects several insect species (e.g., *Ephestia kuehniella*, *E. elutella*, and *Plodia interpunctella*). It invades primarily the fat body and causes the cells to lyse. An infected larva dies before pupation.

Although *Coelogregarina ephestiae* has been placed in synonymy with *M. dispora* (Weiser 1954; Purrini 1977), Ghélélovitch (1948) differentiates *Coelogregarina* from *Mattesia* by the presence of one oocyst per gametocyst in the former and two oocysts in the latter. Canning (1964) concurs with Ghélélovitch that the monosporous *Coelogregarina* should be separated from the disporous *Mattesia*.

Mattesia grandis infects perorally the larva and adult of the cotton boll weevil, *Anthonomus grandis* (McLaughlin 1965a). The nutritional status of the host greatly influences the development of *M. grandis*. When little or no lipid material is deposited in the fat tissue of an adult weevil because of a diet deficiency, the development of the neogregarine is greatly retarded (McLaughlin 1965b). The dietary deficiency appears to be polyunsaturated fatty acids, for which the neogregarine competes with the host cell (Thompson and McLaughlin 1977).

Mattesia povolnyi develops epicellularly with two merogonies and one gametogony in the Malpighian tubules of the European sunflower moth, *Homeosoma nebulellum* (Weiser 1952). Heavy infection blocks the tubules and may cause the insect's death. The infection of *Mattesia geminata* in the tropical fire ant, *Solenopsis geminata*, is unusual and occurs in the hypodermis (Jouvenaz and Anthony 1979). This neogregarine causes the disruption of developing eyes, the melanization of the cuticle, and pupal death. *Farinocystis tribolii* infects the fat body of *Tribolium confusum* and *T. castaneum* (Weiser 1953). The fine structures of the various stages of this neogregarine have been studied by Žižka (1977, 1978a,b,c).

Species of *Caulleryella* are pathogens of Diptera, particularly of mosquitoes (Weiser 1963). *Caulleryella pipientes* attacks the midgut of *Culex pipiens* larva. *Caulleryella pipientis* and *C. anophelis* infect other mosquitoes besides *C. pipiens* and *Anopheles* spp.

Syncystis mirabilis is a pathogen of the water bug, *Nepa cinerea* (Lipa 1967). It develops in fat body cells. *Menzbieria chalcographi* infects the bark beetle, *Pityogenes chalcographus*, a pest of fruit trees (Weiser 1963). Members of the genus *Machadoella* are important pathogens of the neotropical bloodsucking bugs in the family Reduviidae.

II. APICOMPLEXA, COCCIDIA

Less than 1% of all known species of the subclass Coccidia infect only insects. The entomogenous genera are the following: *Adelina*, *Chagasella*, *Legerella*, *Ithania*, and *Barrouxia* (Levine 1983; Brooks 1988). A new genus, *Rasajeyna*, has been proposed for a coccidian infecting craneflies, *Tipula paludosa* and *T. vittata* (Beesley 1977, 1978). The entomogenous coccidians appear to be host specific, but a few have a fairly wide host range (Brooks 1988). Most coccidians are pathogens of vertebrates, with some of them being transmitted by insect vectors.

The coccidians differ from gregarines in many characteristics, but the essential difference is their small mature gamonts, which are intracellular and without mucron or epimerite, whereas in gregarines the gamonts are large and extracellular and possess a mucron. The life cycle of coccidians involves merogony, gamogony, and sporogony. The gametes are characteristically anisogamous. Syzygy, if present, involves markedly anisogamous gametes. The life cycle is mainly a haploid cycle.

The diploid stage occurs only in the zygote, which undergoes meiosis within a short period. There is, therefore, no chromosomal reduction during gamete formation.

A. LIFE CYCLE OF *ADELINA*

We shall describe the life cycle of the important entomopathogenic genus *Adelina*. Members of this genus occur in Coleoptera, Lepidoptera, Orthoptera, Embioptera, Diptera, Collembola, and other invertebrates. *Adelina* develops mainly in the fat body, but some species infect most tissues in the hemocoel. Three different types of schizogony occur in the life cycle, and through these forms of multiplication, the coccidian can increase in numbers until the host dies.

The mature cyst (oocyst), which contains the spores, is ingested by the insect, and the spores are liberated in the digestive tract. The vermiform sporozoites emerge from the spores and penetrate through the midgut wall into the hemocoel to become young meronts that resemble sporozoites. The invasion of the sporozoites into cells takes place in three phases: (1) attachment and orientation of the coccidian, (2) induction of a parasitophorous vacuole in the host cell, and (3) translocation of the coccidian into the vacuole (Russell 1983). A close membrane to membrane association between the host cell and coccidian is maintained throughout the invasion. The sporozoite appears to enter the parasitophorous vacuole by "capping" the host–coccidian junction down to its body and thus entering the host cell.

The meronts enter the hemocoel and infect fat bodies, which lie, in particular, near the digestive tract. The intracellular meront grows into a multinucleate spherical or oval body and undergoes the first merogony by fragmenting into a mass of bacilliform merozoites that invade other cells. The merozoite again produces oval multinucleate bodies resulting in two types of schizogony: a merogony as in the previous generation to form merozoites, which may undergo several merogonies, and a gametogony to produce gametocytes (gameteblasts). Some gametocytes develop into large, globular or oval female gametocytes, which become macrogametes; others become slender male gametocytes. The male gametocyte adheres to the female gametocyte as a crescent-shaped body. A common membrane is formed around the pair to form a gametocyst. The nucleus of the male gametocyte undergoes two divisions and forms four microgametes, one of which fertilizes the macrogamete and the remaining three lie unused between the gametocyst and the oocyst, which is formed shortly after fertilization. The zygote nucleus divides repeatedly to produce a multinucleated sporont with 3 to 30 nuclei that enter fingerlike protrusions on the surface of the plasmodial structure. This is the beginning of sporogony, the third schizogony in the life cycle. The nuclei are budded off to produce the sporoblasts. The cyst is now called an oocyst. Each sporoblast divides to form two sporozoites in a small spherical spore (sporocyst). The number of spores in an oocyst varies from 3 to 30 spores even within the same and different insect species.

B. *ADELINA* SPECIES

The development of *Adelina cryptocerci* in the wood cockroach, *Cryptocercus punctulatus*, occurs during the early stages in the fat body around the gut, and in advanced heavy infections, the coccidian develops in nearly all tissues throughout the insect (Yarwood 1937). *Adelina melolonthae* infects the European cockchafer,

Melolontha melolontha. The first schizogony is a merogony and occurs in blood cells to produce merozoites (schizozoites) (Tuzet et al., 1965). A macromerogony and a micromerogony take place in fat cells to produce the merozoites. The micromerogony gives rise to the future gamonts. *Adelina mesnili* is a pathogen of the webbing clothes moth, *Tineola bisselliella* (Pérez 1899, 1903) and the beetle *Tribolium ferrugineum* (Bhatia 1937). *Adelina sericesthis* infects the Australian scarabaeid, *Sericesthis pruinosa*, and several other species (Weiser and Beard 1959).

III. MICROSPORA, MICROSPOREA

The members of class Microsporea in the phylum Microspora are commonly called microsporidia. The disease they cause is called microsporidiosis. Microsporidia rank among the smallest of eukaryotes and have an obligate intracellular habit. They possess unicellular spores, containing a uninucleate or binucleate sporoplasm, and an extrusion apparatus always with a polar filament and polar cap (Fig. 12–2). They do not have mitochondria. Their life-cycle stages are also ultrastructurally unique and distinct from other spore-forming protozoa and play a critical role in taxonomic determinations (Larsson 1986). Even though they are eukaryotic microorganisms, their ribosomal RNAs have prokaryotic properties (Ishihara and Hayashi 1968; Curgy et al. 1980). Moreover, the sequence of ribosomal RNA suggests that the microsporidia are extremely ancient eukaryotes and had separated very early from other eukaryotes (Vossbrinck et al. 1987).

Class Microsporea contains two orders, Minisporida and Microsporida. We shall not discuss the order Minisporida because only a few members are insect pathogens causing very little pathology and little is known about their biology. There are about 800 described species of Microsporida. Pertinent references on microsporidia that have appeared after 1975 are Vávra 1976a,b; Sprague 1977a,b, 1982; Weiser 1985; Lacey and Undeen 1986; Issi 1986; Larsson 1986, 1988; Brooks 1988; and Canning 1990.

As a group, the microsporidia are the most important protozoan pathogens of insects. They form the majority of the protozoa pathogenic to insects and cause economically serious diseases in beneficial and pest insects. They are the most promising protozoa for use in microbial control. All microsporidia are obligate pathogens and multiply only in living cells. Many cause severe acute infections in insects, but some produce only inapparent or chronic infections, which nonetheless may play an important role in regulating insect populations. Most of the massive protozoan epizootics in insects result from microsporidioses.

A. MICROSPORIDIAN TAXONOMY

The taxonomy of microsporidia has undergone drastic revision during the past two decades. Recognizing the uniqueness of microsporidia, Weiser (1977, 1985), Sprague (1977a, 1982), and later Issi (1986) have elevated these organisms to the level of a distinct phylum, the Microsporidia or Microspora (Table 12–2). These classifications radically depart from the closed and rigid system that was formerly used with this group of organisms. Most microsporidiologists follow the system developed by Sprague (Brooks 1988; Canning, 1990) and we have adopted this classification. Comprehensive lists of microsporidian genera and species have been

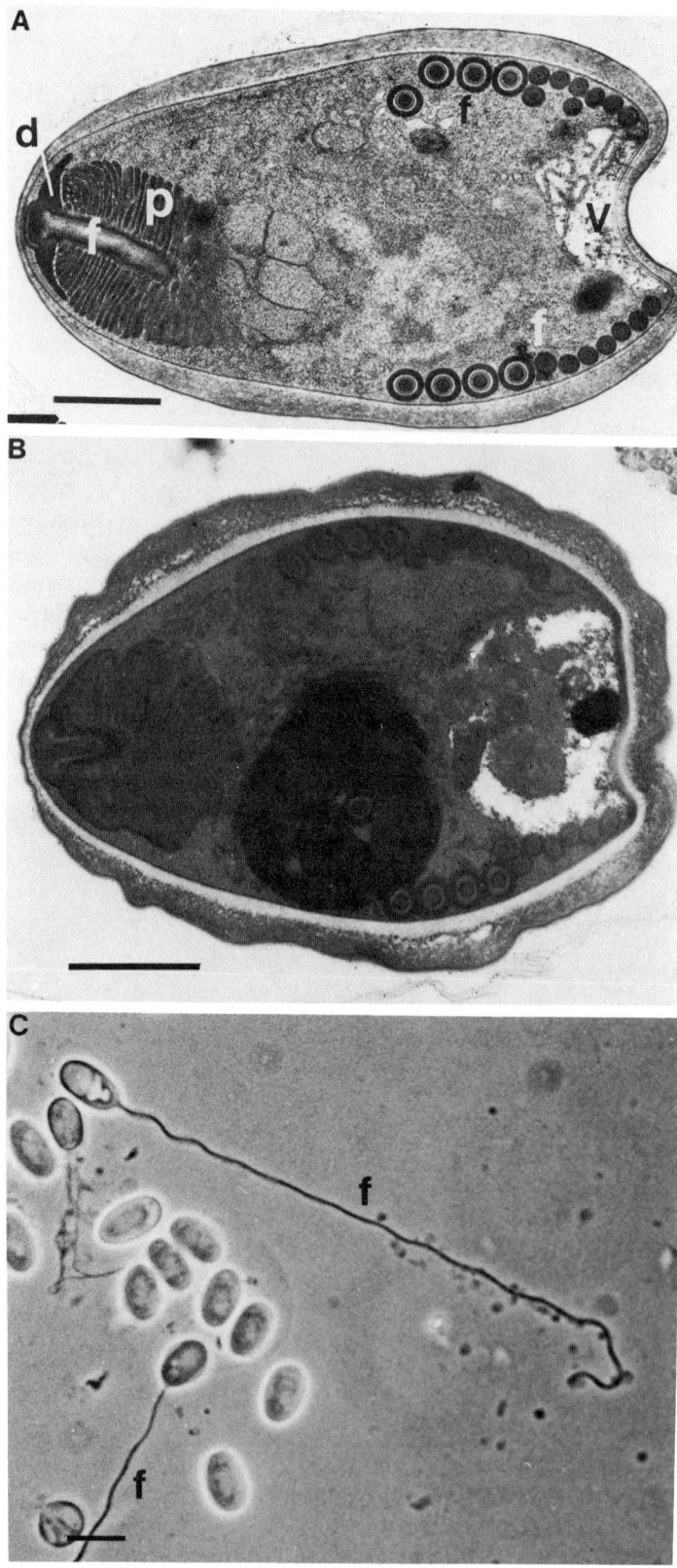

prepared by Sprague (1977b), Issi (1986), Brooks (1988), and Canning (1990). Those who are interested in the evolution of microsporidian classification should refer to the excellent historical review by Sprague (1977a) and Issi (1986).

The classification of certain genera (e.g., *Nosema*, *Perezia*, *Glugea*, and *Encephalitozoon*) has been changed considerably. In 1905, Pérez redefined the genus *Nosema* and introduced the term "monosporous," which was interpreted by subsequent workers to mean one spore from one sporont. This has led to confusion in *Nosema* for a half-century (Sprague 1978). In *Nosema bombycis*, the type species, the sporont gives rise to two spores (Ishihara 1969; Cali 1971), as in the case of other disporous genera *Glugea*, *Perezia*, and *Encephalitozoon*. The student should refer to the discussions on the clarification of this problem and of various genera by Sprague and Vernick (1971), Sprague (1977a, 1978), and Vávra et al. (1981). However, the genus *Nosema* contains the largest number of species and still represents a heterogeneous group from which many species will have to be removed to other genera.

Another group of microsporidia that have been reorganized are those closely associated with the genus *Thelohania* (Hazard and Oldacre 1976). Generally, the sporonts of these species give rise, after three sporogonic divisions, to eight uninucleate spores enclosed by a membrane, the sporophorous vesicle. Occasionally, a few sporonts fail to complete the normal number of divisions and produce macrospores varying in numbers from one to four within the sporophorous vesicle. Some species, in addition to octospores, have another kind of sporogony producing a variable number (6–40) of binucleate spores that are not enclosed by a sporophorus vesicle. Such species have been placed in several new genera (e.g., *Amblyospora* and *Parathelohania*). *Pleistophora* produces a variable and a large number of sporoblasts from a sporont and also appears to be multigenic.

The taxonomy of microsporidian genera and species is based on differences in morphology, life cycle, and parasite–host relations. These may be inadequate in some cases, and there is a need for other criteria (e.g., biochemical analyses and serology) to differentiate genera and species (Knell and Zam 1978; Sato et al. 1981; Vávra et al. 1981).

B. HOSTS

The hosts of microsporidia are found in all five classes of vertebrates: fishes, amphibians, reptiles, birds, and mammals (Canning 1990). They also include almost all invertebrate phyla and even some proctoctists, such as ciliates, myxozoans, and gregarines. The most common hosts are arthropods and then fish. About 700 species have been recorded from these hosts (Canning 1977; Sprague 1977b). Insects in nearly all taxonomic orders are susceptible to microsporidia (Sprague 1977c) and over half of the hosts occur in two orders, Lepidoptera and Diptera. The

FIGURE 12–2

Electron micrographs of sporoblast and spores of the microsporidium *Amblyospora* sp., a pathogen of mosquitoes, *Aedes* spp. (A) Mature sporoblast in which all organelles except the spore wall are fully differentiated. (*d*) Disc; (*p*) polaroplast; (*f*) polar filament; (*v*) posterior vacuole. (Bar = 1 μm.) (B) Spore with exospore and endospore layers of the spore wall. (Bar = 1 μm.) (C) Spores with evaginated polar filaments. (Bar = 5 μm.) [(A) and (B) courtesy of Andreadis (1983) and Society of Protozoologists; (C) provided by T. G. Andreadis.

TABLE 12–2 Classification Systems for the Microsporidia[a]

System I[b]	System II[c]	System III[d]
Phylum Microspora	Phylum Microspora	Phylum Microsporidia
Class Rudimicrosporea	Class Metchnikovellidea	Class Microsporidea
Order Metchnikovellida	Order Metchnikovellida	Subclass Metchnikovellidea
Family Metchnikovellidae	Family Metchnikovellidae	Order Metchnikovellida
Class Microsporea	Order Chytridiopsida	Family Metchnikovellidae
Order Minisporida	Family Chytridiopsidae	Subclass Chytridiopsidea
Family Chytridiopsidae		Order Chytridiopsida
Family Hesseidae	Order Hesseida	Family Chytridiopsidae
Family Burkeidae	Family Hesseidae	Family Buxtehudeidae
Family Buxtehudeidae		
	Class Microsporididea	Subclass Cylindrosporidea
Order Microsporida	Order Pleistophoridida	Order Cylindrosporida
Suborder Pansporoblastina	Family Pleistophoridae	Family Striatosporidae
Family Pleistophoridae	Family Thelohaniidae	Family Cylindrosporidae
Family Pseudopleistophoridae	Family Amblyosporidae	
Family Duboscqiidae	Family Culicosporidae	Subclass Nosematidea
Family Thelohaniidae		Order Culicosporida
	Order Nosematidida	

Family Nosematidae
Family Mrazekidae

Family Culicosporidae
Family Golbergiidae

Order Glugeida
    Family Pereziidae
    Family Glugeidae
    Family Thelohaniidae

Order Nosematida
    Family Amblyosporidae
    Family Burenellidae
    Family Spraguidae
    Family Nosematidae
        Subfamily Nosematinae
        Subfamily Pseudopleistophorinae

Family Burenellidae
Family Amblyosporidae
Family Culicosporidae
Family Gurleyidae
Family Telomyxidae
Family Tuzetiidae

Suborder Apanosporoblastina
    Family Glugeidae
    Family Spraguidae
    Family Unikaryonidae
    Family Pereziidae
    Family Cougourdellidae
    Family Caudosporidae
    Family Nosematidae
    Family Mrazekiidae

---

[a] Courtesy of Brooks (1988) and CRC Press
[b] According to Sprague (1977a, 1982)
[c] According to Weiser (1977, 1985)
[d] According to Issi (1986)

insect hosts usually have chewing mouthparts in their susceptible stages, but there are a few with sucking mouth parts, such as those in Anoplura, Hemiptera, and Homoptera.

At one time, microsporidia were considered to be host specific and were identified on the basis of their hosts. Such generalization was discarded when many microsporidia were found to infect a wide range of hosts. Most of them, however, are restricted within a class or order and some to genera of animals. No entomogenous microsporidium is known to infect warm-blooded vertebrates, and no vertebrate microsporidium infects insects. However, transient infections have developed when spores of *Nosema algerae*, a mosquito pathogen, are inoculated subcutaneously into the cooler body sites of the tail, ears, and feet of a white mouse (Undeen and Alger 1976), and opportunistic microsporidian infections have occurred in humans, especially with impaired immune system (Cali and Owen 1988). Body temperatures appear to be one of the limiting factors because the microsporidium can grow in a vertebrate cell line at temperatures below those of warm-blooded animals (from 35 to 37°C) (Ishihara 1968a; Undeen 1975).

In some cases, an insect species that is resistant through oral inoculation may become susceptible when the microsporidium is inoculated into the hemocoel (Fisher and Sanborn 1962; Undeen and Maddox 1973). In this case, the midgut is an effective barrier to the microsporidium. Different orders of insects may be susceptible to the same microsporidium. Thus, hymenopteran parasitoids may be susceptible to the microsporidia of their lepidopteran hosts (Brooks 1973). An unusual case is the infection of the entomogenous microsporidium *Nosema algerae* in the larval stages of schistosomes that are in a different phylum from that of insects (Lai and Canning 1980). In addition, the microsporidia in the genus *Amblyospora* require intermediate hosts (class Crustacea) in order to complete their life cycle (Sweeney et al. 1985; Andreadis 1985a).

Under *in vitro* conditions, Tsang et al. (1982) have observed that the host-specificity barrier may be crossed by adjusting the osmotic pressure and K^+ and Na^+ concentrations of the tissue culture medium. Under low osmotic pressure and high $K^+:Na^+$ ratio, they have succeeded in infecting the embryonic fibroblast-like cells of the German cockroach with *Nosema disstria* obtained from the cell line of the forest tent caterpillar, *Malacosoma disstria*.

C. TRANSMISSION

Microsporidia invade insects through three natural portals of entry: oral, cuticular, and ovarial pathways (Kramer 1976). Venereal transmission between sexes is rare but has been reported in *Plodia interpunctella* (Kellen and Lindegren 1971) and coccinellid beetles (Toguebaye and Marchand 1984). Entrance by the oral and cuticular portals results in horizontal transmission, and by the ovarial portal in vertical transmission. These types of transmissions are believed to occur commonly in the same host generation, but in the European corn borer, *Ostrinia nubilalis*, the microsporidium *Nosema pyrausta* is transmitted primarily by transovarial infection (vertical) in the first generation and by vertical and horizontal transmissions in the second generation (Siegel et al. 1988).

The oral portal is the most common route through which the microsporidium gains access to host tissues. *Nosema apis*, which infects the digestive tracts of adult bees, appears to infect only through the oral route (Bailey 1963). Transmission through feces contaminated with microsporidian spores commonly occurs in susceptible hosts (Andreadis 1987). The larval silk glands are often infected but have not been considered as a route of horizontal transmission except in the case of a *Nosema* sp. of the gypsy moth, *Lymantria dispar* (Jeffords et al. 1987). The heavily infected silk glands of the gypsy moth larvae excrete abundant spores together with the silk. Gypsy moth larvae follow the silk trails and are likely to become infected.

The ovarial portal is commonly used with lepidopteran and mosquito hosts. In this case, the microsporidium is incorporated into the egg or embryo within the females' reproductive tracts, and the progeny from such infected females is also infected. In some cases, even though the microsporidium is within an egg, transovarial transmission does not occur. In the winter moth, *Operophtera brumata*, the embryo within an egg is not infected nor is the larva one day after emergence, but the larva subsequently becomes infected when the microsporidian spores are ingested together with the remains of the yolk as the larva eats its way through the eggshell (Canning et al. 1985). Thus, the infection is similar to that of a larva feeding on spores contaminating the surface of the eggshell. Some microsporidia, such as *Amblyospora* species (syn. *Thelohania* spp.) infecting mosquitoes (Kellen and Wills 1962; Kellen et al. 1965), *Nosema kingi* in drosophilids (Armstrong 1976), and others in Lepidoptera, utilize both the oral and ovarial routes.

Entry via the cuticular portal involves the inoculation of the microsporidium by the ovipositor of a parasitoid (Brooks 1973). Certain lepidopteran hosts, which have hymenopteran parasitoids, are invaded through all three portals of entry.

In the tropical fire ant, *Solenopsis geminata*, the adult ants serve as vectors by acquiring the spores of the microsporidium *Burenella dimorpha* through cannibalism of ruptured infected pupae (Jouvenaz et al. 1981). The adult is not infected since it does not pass the ingested spores into its midgut but retains them together with the particulate food in an infrabuccal pouch. It feeds a pellet of food and spores to the larva in the fourth instar, which is the only stage in the insect's life cycle that is vulnerable to infection.

D. GROSS PATHOLOGY

Microsporidia are intracellular pathogens, usually in the cytoplasm and infrequently in the nucleus. They produce infections that range from subacute and chronic to acute and virulent. Some microsporidia infect only a single tissue or organ, such as the midgut epithelium, muscle, or fat body. Tissue specificity, however, may vary with the host infected by the microsporidium. *Nosema algerae* infects the brain and nerve ganglia of *Aedes aegypti*, but in *Anopheles quadrimaculatus*, it infects nearly all tissues (Hazard and Lofgren 1971). Moreover, the infection in mosquitoes may differ with the sexes. The concentration of spore dosage may also determine the types of tissue infected. A low dosage of *Vairimorpha necatrix* results in a chronic infection of mainly the fat body and some muscular tissues of *Trichoplusia ni*, whereas a high dosage produces an acute infection of mainly the midgut tissues (Chu and Jaques 1979). In some cases, sporogony is confined to a specific tissue, and the type of sporogony may differ with the infected tissues (Issi 1986).

The fat-body tissue is the most frequently infected by microsporidia. Infection in the fat body alone generally causes a chronic infection, and an infected larva may survive to adulthood. When a systemic infection occurs, the pathology is severe and may result in the death of an insect after a short period of infection. Fatal infections result when vital organs or tissues are severely attacked.

The period of lethal infection varies with the virulence of the microsporidium and the extent of infection in the host. The infection, in some cases, is very prolonged and extends over several months to a year. It remains dormant in diapausing insects and continues only when the insect regains its activity (Weiser 1961; Nordin 1975). On the other hand, the development of microsporidia in diapausing insects is shorter than in nondiapausing insects (Issi 1986). The infection in a diapausing or hibernating pupa increases the rate of respiration to nearly that of a nondiapausing pupa and also affects cold hardiness such that it cannot withstand supercooling to the extent of an uninfected pupa (Metspalu 1976; Metspalu and Hiiesaar 1980).

Sublethal infections by microsporidia may cause severe pathologies in some insects. In the corn earworm, *Heliothis zea*, with a sublethal infection by *Nosema heliothidis*, the following pathologies result: in the adult, a reduction in fecundity, longevity, and mating success; in the pupa, a reduction in diapause and an increase in deformities; in the larva, a retardation in growth (Gaugler and Brooks 1975). In the cabbage looper, *Trichoplusia ni*, the reproductive capacity of an adult moth varies with the intensity of the larval infection by the microsporidium (Tanabe and Tamashiro 1967). Light infection in a larva had no affect on fecundity.

According to Fisher and Sanborn (1964), microsporidian infection in larvae of *Tribolium castaneum* caused a juvenile hormonelike effect. They reported that the larvae grew faster, had supernumerary molts, and rarely, if ever, pupated. The surgical transplants of the *Nosema* to allatectomized cockroaches also prevented their development to adults. Milner (1972), however, could not repeat and confirm the above results. On the other hand, there are cases of irregular larval molts, failure to pupate, and disturbed diapause in microsporidian infections (Weiser 1977), but there is no definitive study that the microsporidia are capable of regulating the hormonal metabolism of the host.

The effect of environmental factors on microsporidian infections has been described in detail by Issi (1986). He has considered such factors as host density, temperature, insect nutrition depending on the host plant, and increased susceptibility of infected insects to stressors, such as chemical and bacterial insecticides.

E. CELLULAR PATHOLOGY

Within a cell, the microsporidium usually occurs in the cytoplasm. There are a few cases of their occasional presence in the cell nuclei [e.g., *Nosema apis* in the honey bee (Steche 1965a), *Nosema* (syn. *Perezia*) *fumiferanae* in the spruce budworm, *Choristoneura fumiferana* (Percy 1973), and *Nosema bombycis* in the silkworm (Takizawa et al. 1973)].

Weissenberg (1976) classifies the interaction of a microsporidium within a host cell into three biological types: (1) the microsporidium develops under favorable conditions and spreads without limitation throughout the entire cytoplasm and eventually displaces the cytoplasm and destroys the cell. This is the most common type in insects. (2) The second type is rare and has been observed only in the infection of ganglion cells of the angler fish, *Lophius*, by *Spraguea* (syn. *Nosema*) *lophii*. The microsporidium develops only in the region from which the axon extends from the nerve cell. The infected cell is greatly hypertrophied and loses its functions. (3) In the third type, the cell invasion is localized, but the host cells and developing microsporidia form a syncitial xenoma that resembles a biological unit. Xenomas are common in fishes, but certain microsporidia produce xenomas in insects [e.g., sciarid flies (Pavan et al. 1971) and black flies (Weiser 1976)].

In insects, the initial phase of microsporidian infection is a stimulation of the cell to form a concentration of endoplasmic reticula around the pathogen, an increase in the number of mitochondria, and stimulation of nucleic acid synthesis (Issi 1986). The destruction of the cell begins when the microsporidium goes into sporogony. At early stages of infection, there is a reduction in the number of endoplasmic reticula and an increased vacuolation in the cell (Tanabe 1971; Liu and Davies 1972; Issi 1986). At the terminal stage, only the nucleus and mitochondria remain. In the sciarid flies, the microsporidium causes marked ultrastructural changes in the cells of the salivary gland (Jurand et al. 1967). In these cells, the ribosomes are depleted, the endoplasmic reticula are disarranged, the mitochondria decrease in size but increase in numbers, and vacuoles are formed in the cytoplasm. Some microsporidia cause hypertrophy of the nucleus and an increase in the size of the chromosomes (Diaz and Pavan 1965; Jurand et al. 1967; Roberts et al. 1967; Martins and Perondini 1977).

In some cases, there is little, if any, hypertrophy of the infected cell, but in others, there is great hypertrophy, as in the case of a microsporidium infecting the fat cells of the crane fly (Weissenberg 1976), or *Nosema acridophagus* producing tumorlike growths at the infection foci in grasshoppers (Henry 1967). Inflammatory responses develop in some insects (Brooks 1971). Such responses, in severe cases, produce nodules that are infiltrated with hemocytes and melanized, and the infected organ becomes deranged. In the honey bee, *Nosema apis* does not infect the hypopharyngeal glands but causes severe cytopathology, such as the disintegration and vacuolation of the cytoplasm of the glands (Wang and Moeller 1971).

The infection by *Vairimorpha* (syn. *Nosema*) *plodiae* in *Barathra brassicae* and *Galleria mellonella* causes enzymatic disturbances (Kučera and Weiser 1975). The activity of lactate dehydrogenase isoenzyme increases substantially by the fifth day of infection, but other isoenzymes are not affected. In the honey bee, infection by *Nosema apis* depresses RNA synthesis in midgut cells (Hartwig and Przełęcka 1971). The drug fumagillin, which has been used to control microsporidian infections, arrests DNA replication in the pathogen and inhibits its development and this, in turn, restores RNA synthesis in the bee's intestine.

F. SIGNS AND SYMPTOMS

An insect infected with a microsporidium generally is altered in color, size, form, and activity, depending on the tissues and organs infected. A translucent larva turns increasingly opaque or dull, milky white as the infection progresses. The integument usually remains intact throughout infection. In pébrine of the silkworm, the dark mottled areas or dark-brown spots on the integument are a pathognomic sign of this disease. In some infections, the growth of a larva is retarded or the larva is reduced in size prior to death. In others, the larva is swollen or distended in form. The activity of an infected larva is often reduced at an advanced stage of infection. Such reduced activity is most pronounced when the infection occurs in the muscle and nervous tissues.

Microsporidian infections may alter the behavior of the honey bee. The infection by *Nosema apis* substantially reduces the hoarding behavior of adult bees and may affect the overall production of a colony (Rinderer and Elliott 1977a). Abnormal behavior also results through the indirect effect of the microsporidium on the hypopharyngeal glands, making them unable to secrete royal jelly (Wang and Moeller 1970). *Nosema*-infected bees are apparently physiologically older than healthy bees of the same age. They start guarding, foraging, dancing, and orientation flying earlier than healthy bees. Since they often do not attend or feed the queen because of the atrophy of their hypopharyngeal glands, this may contribute to the queen's frequent escape from *Nosema* infection.

G. SPORE

The most distinct, unique, and complicated structure of the microsporidium is the spore, which is probably the smallest eucaryotic cell. The unicellular spore is the only stage in the microsporidian life cycle capable of surviving outside of the host. It has a characteristic sporoplasm-extrusion apparatus that includes the polar filament. This apparatus is uniquely adapted for the inoculation of the sporoplasm into a host. The spore appears to lack reserve substances (Maurand and Loubès 1973). Microsporidian spores have three basic components: sporoplasm, extrusion apparatus, and spore envelope (Fig. 12–2).

1. Size, Shape, and Surface of Spore

Microsporidian species generally possess spores of uniform sizes and shapes. Most spores are in the size range of 3 to 6 by 2 to 4 μm, but some are over 20 μm in length (Sprague 1982). The sizes, at times, range greatly within a single species (*Stempellia*, *Pleistophora*) or there may be two distinct sizes (dimorphic) (*Vairimorpha*, *Burenella*, *Parathelohania*, *Amblyospora*) (Issi 1986). The sizes of spores of *Vavraia* (syn. *Pleistophora*) *culicis* vary when produced in different mosquito species (Weiser and Coluzzi 1972).

The shapes of microsporidian spores are usually oval, ovoidal, ovocylindrical, or pyriform, but occasionally they are spherical, reniform, tubular, bacilliform, spiral, crescent-shaped, or comma-shaped. Some genera (e.g., *Amblyospora*, *Parathelohania*, *Toxoglugea*, *Weiseria*, etc.) have unique spore shapes, which are useful to characterize the genera. The surface sculpturings of certain spores, when observed with the scanning electron microscope, are distinct and characteristic and may also serve in classifying the microsporidia (Fowler and Reeves 1975). Some microsporidia (e.g., *Caudospora* and *Weiseria*), have filamentous, sabre- or rodlike appendages usually situated at one or both poles of the spore (Issi 1986). Certain microsporidia from aquatic hosts have spores covered with a layer of mucus ("mucocalyx") that consists of a thick coat of long, very fine fibers oriented perpendicular to the spore surface (Lom and Vávra 1963c; Vávra and Barker 1980). Such appendages and mucus layer, which have been reported from only species infecting aquatic hosts, serve to connect the spores in the sporophorous vesicle, to assist in flotation (Vávra 1963), or to attach to the substratum as tendrils (Overstreet and Weidner 1974). Appendages occur on the membrane that surrounds the spores of *Trichodubosqia epeori*, and serve to attach the sporophorous vesicle to a substrate, such as an alga, fed upon by the aquatic nymphs of the host mayfly, *Rhithrogena iridina* (Batson 1982).

2. Spore Wall

The spore wall is imperforate and consists of three layers: an outer proteinaceous exospore; a middle chitinous endospore; and an inner plasma membrane, the plasmolemma (Vávra 1976a; Sprague 1982; Larsson 1986). The exospore layer has no apparent substructure but may appear granular or fibrillar. Its surface is sometimes smooth but is usually finely corrugated, or may be ornamented with ridges, ribs, or filamentous appendages. It is generally uniformly thick all over the spore but is usually thinner than the endospore layer. The exospore is soluble in strong alkali (Vávra 1968; Maurand and Loubès 1973). In some microsporidia, the exospore layer is subdivided [e.g., three layers in *Pleistophora* (syn. *Plistophora*) *hyphessobryconis* (Lom and Corliss 1967) and four or five layers in *Nosema locustae* (Huger 1960)].

The endospore layer varies among microsporidian species (Vávra 1976a). It is generally thick but is thin in some microsporidia and absent in others. Chitin, which occurs as a protein–chitin complex, is the main chemical component of this layer. The endospore layer provides the spore wall with a resistance sufficient to withstand the internal pressure generated during polar filament extrusion.

The cytoplasmic membrane, the plasmolemma, is the third and innermost layer of the spore wall and is a unit membrane similar to other plasma membranes. The inner face of the spore membrane of *Amblyospora* (syn. *Thelohania*) *bracteata* is covered by studlike projections and granular subunits that are complementary to the depressions on the concave face of the cytoplasmic membrane (Liu and Davies 1973). These projections presumably prevent the outer spore layer from spinning around during exsporulation.

3. Sporoplasm

The sporoplasm has one or two nuclei and endoplasmic reticula with ribosomes. It is not separated by a bounding membrane from the polar filament, extrusion apparatus. The nucleus is generally located in the widest part of the spore and often in the posterior part. The limiting membrane that surrounds the extruded sporoplasm is acquired during the passage of the sporoplasm through the everted filament (Vávra 1976a). The sporoplasm is formed by most, if not all, of the cytoplasm in the spore.

4. Extrusion Apparatus

The extrusion apparatus of the spore is unique for the microsporidia. It is not highly developed in some microsporidia or may be completely lacking in others (Canning 1990), but when present, it is a complex structure. It occupies most of the spore and consists of the polar filament, with its anchoring apparatus, and the polaroplast (Fig. 12–2). The posterior vacuole may also be involved in the extrusion process. The Golgi apparatus in the spore is greatly hypertrophied and takes part in the formation of an assembly of organelles that belong to the extrusion apparatus (Vávra 1976a; Götz 1981; Larsson 1986).

At the anterior part of the spore, the anchoring disc ("polar cap") [Fig. 12–

2(d)], to which the basal end of the polar filament is attached, is enclosed in the polar sac or polaroplast (Huger 1960; Lom 1972; Larsson 1986). The disc functions as a hinge when the filament is everted. The polaroplast, a system of saclike or lamellar components (Huger 1960; Weidner 1970; Larsson 1986) [Fig. 12–2(p)], occupies from 25 to 35% of the mature spore volume (Vávra 1976a). It is closely associated with the manubrial portion of the polar filament and both are considered as parts of the same membrane system ("polar filament–polaroplast complex") (Lom and Corliss 1967). The polaroplast is regarded as a specialized derivative of the endoplasmic reticulum or more likely a reoriented package of remnants of the Golgi membrane complex (Vávra 1976a; Sprague and Vernick 1969; Jensen and Wellings 1972). It is capable of changing in volume, apparently by swelling, which increases the intrasporal pressure.

A membrane-lined area with a clear or spongy content is located close to the posterior pole of most microsporidian spores (Vávra 1976a; Larsson 1986) [Fig. 12–2(v)]. This area is usually called the posterior vacuole, but it appears to be the remains of the Golgi apparatus (Sprague and Vernick 1969). The posterior vacuole differs in shapes and sizes among various species or it is absent (Vávra 1976a). It plays a role during the ejection of the polar filament (Lom and Vávra 1963a,b).

5. Polar Filament

The polar filament or polar tube, which serves as an "inoculating needle," is a conspicuous and characteristic organelle of the microsporidian spore [Fig. 12–2(f)]. Its presence in spores places an organism in the phylum Microspora (Issi 1986). The polar filament is typically threadlike and is anchored at the anterior portion (polaroplast complex) of the spore wall by a mushroom-shaped, multilayered anchoring disc (Vávra 1976a; Larsson 1986). The attached filament extends posteriorly and either terminates in a funnel or forms coils, which wind around the inner surface of the spore wall, often covering most of the posterior one-third or two-thirds of the spore. The straight basal (manubrial) portion of the filament near the attachment site is thicker than its other parts.

The number of coils of the filament within the spores may vary in a single species, but variations in numbers are greater among different species. Two types of spores are found in *Nosema* spp., one with a few coils and a second type with many coils (Iwano and Ishihara 1991). Most microsporidia have about 11 coils, but the number may range from 3 to 5 in spores of *Encephalitozoon* (syn. *Nosema*) *cuniculi* (Petri and Shiodt 1966) to as many as 44 in those of *Nosema apis* (Scholtyseck and Danneel 1962). In addition to the number of coils, the angle of tilt of the coils is of taxonomic value (Burges et al. 1974).

The polar filament is very elastic, stretching to three times the length of the coiled state when extended from the spore. It is also elastic transversally so that its diameter increases twofold at the site of passage of the sporoplasm (Lom and Vávra 1963a). The extruded polar filament is about 0.1 μm in diameter when seen with the light phase microscope, and in some cases up to 400 μm long. The length measured with the light microscope has been used to characterize a species, but such measurements have little value in classification because there is no way to check if complete extrusion has occurred (Larsson 1988).

Studies with the electron microscope show that (1) the filament consists of several concentric layers, (2) the proportion of the individual layers varies along the longitudinal axis, and (3) the layering is very similar in all microsporidian polar filaments (Vávra 1976a). In cross section, there are as many as 11 different layers (Schubert 1969). The polar filament is considered to be a composite structure of two tubes, one within the other, which appear to be membranous in cross section (Vávra 1976a), or to have a single membrane enveloping a proteinaceous matrix (Weidner 1972, 1976).

The lumen of the polar filament contains electron dense material leading to the initial interpretation of a solid structure. The dense core of the polar filament consists of a single, low molecular weight polypeptide (Weidner 1976). The function of the core substance is believed to enhance the structural rigidity of the filament, serve as the vehicle for the movement of the surrounding membrane, lubricate the filament interior during extrusion, or form the filament.

The origin of the polar filament has been ascribed to the following cell structures: (1) the Golgi zone, (2) the endoplasmic reticulum, (3) the nucleus, (4) the ribosomes, or (5) the posterosome or posterior body. The best data suggest its origin from the Golgi zone (Jensen and Wellings 1972; Walker and Hinsch 1972; Götz 1981) or the posterosome (Weiser and Žižka 1975).

H. POLAR-FILAMENT FUNCTION

There have been questions on the function of the polar filament and how the sporoplasm emerged from the spore. The controversy was concerned mainly with the structure of the polar filament, whether it was solid or tubular. Ohshima (1937) was the first to conclude that the polar filament was a tube for the passage of sporoplasm to the exterior of the spore. These questions have been answered by electron microscopical observations showing the tubular nature of the polar filament and the passage of the sporoplasm through the filament (Ishihara 1968b; Weidner 1972; Fowler and Reeves 1975). Accordingly, the polar filament serves as an inoculating organelle to transport the sporoplasm from a spore into a host cell (Fig. 12–2C).

I. POLAR-FILAMENT EXTRUSION

The mode of polar-filament extrusion is not completely established (Larsson 1986). It is generally accepted to occur as follows (Lom and Vávra 1963a; Lom 1972; Undeen 1978). The ingested spore is acted upon by the digestive juices of an insect's midgut. The layers of the anchoring disc mediate the stimuli. The polaroplast located at the anterior pole of the spore absorbs water and swells. The posterior vacuole also swells. The swelling exerts an internal pressure as great as 30–60 atmospheres. The polar filament breaks through the apical part of the anchoring disc and the spore wall at its thinnest part. It everts like the fingers of a glove. The rapid eversion and uncoiling of the filament cause the spore to vibrate violently. For a brief moment, the everted polar filament becomes rigid because of the great pressure exerted within the spore, and the sporoplasm is shot through and out of the filament canal.

Another version of polar-filament ejection is proposed by Weidner (1976, 1982). He suggests that, rather than filament eversion, the filament grows during extrusion by incorporating proteinaceous material emerging from the growing tip to form a wall of the filament, which has a cylinder-within-a-cylinder profile.

Several environmental parameters affect spore germination, such as the pH of the spore environment, type of ions present, osmotic pressure around spores, temperature conditions, inhibiting substances, sequence of germination events, and biochemical properties of the spores (Undeen 1990). Undeen (1990) has proposed a hypothesis on the mechanism of microsporidian germination based upon (1) the concentrations of trehalose, a disaccharide, in germinated and ungerminated spores, (2) stimulation by anions and cations, and (3) the high osmostic pressure. He proposes that (1) monovalent ions diffuse passively into the spores; (2) this stimulates changes that cause trehalase to act on trehalose, possibly through an intermediary calcium exchange or release; (3) the trehalose is degraded into higher concentrations of smaller molecules; (4) this results in increased osmotic pressure and the rapid movement of water into the spore; and (5) the extrusion of the polar filament occurs.

The polar filaments can be extruded from the spores, outside of their hosts, by numerous methods (e.g., application of mechanical pressure, freezing and thawing, drying and wetting, or treatment with various salts and acids or bases in suitable concentrations) (Larsson 1986). The method may vary for different microsporidian species. In certain species (e.g., *Nosema algerae* and *Vavraia culicis*), the spores taken directly from a mosquito larva are inactive and cannot be forced to germinate and extrude the polar filaments before pretreatment (priming) and incubation in water (Undeen 1978, 1983). The pretreatment step requires a high pH in a dilute salt solution and the germination step requires electrolytes at an optimal pH of about 9.5 to 10.0. The K^+ is most effective for most microsporidian species. Inasmuch as the spores of certain microsporidia germinate in both susceptible and nonsusceptible hosts, the stimulus for germination appears to be primarily a protective function against inadvertant germination by environmental changes and only secondarily an adaptation for the infection of a specific host (Undeen 1990).

J. STRUCTURE OF PRESPORAL STAGES AND SPORE FORMATION

The developmental stages, sporoplasms, meronts, and sporonts, are structurally nonspecialized simple cells (Vávra 1976a) (Fig. 12–3). The sporoblast, the stage between the simple cells and the spore, contains some organelles destined to be in the future spore. The simple cell of the developmental stages has a one or two nuclei (diplokaryon), or it is a plasmodial cell with several nuclei or chain of nuclei. There are no centrioles, but a specialized area on the nuclear membrane is a primitive centrosome ("nuclear plaque" or a "centriolar plaque") and serves as an attachment for the spindle fibers (Vávra and Undeen 1970; Walker and Hinsch 1972). The diplokaryon consists of two closely adjacent nuclei that are structurally identical (Vávra 1976a). Both nuclei behave synchronously during the cell cycle with the nuclear membrane remaining intact at nuclear division.

The cytoplasm is homogeneously granular with free ribosomes and weakly developed endoplasmic reticulum (Larsson 1986). It lacks reserve substances and structures comparable to kinetosomes, peroxisomes, typical lysosomes, and mitochondria (Vávra 1976a).

The cell membrane changes with such regularity during the microsporidian life cycle that the different stages can be identified according to its structure (Vávra 1976a). The sporoplasm within the spore is not limited by a membrane, but after emergence through the polar filament, it is enclosed within a double membrane that

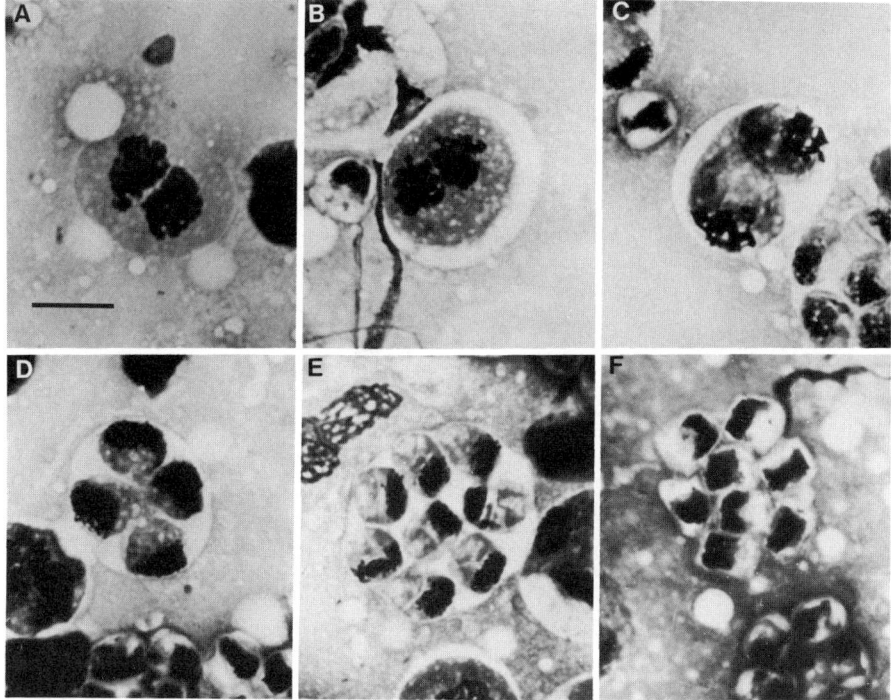

Light micrographs of a portion of the replicative stages of the microsporidium *Amblyospora californica* in the larva of the mosquito *Culex tarsalis*. (A) Binucleate meront; (B) early sporont; (C) binucleate sporont; (D) quadrinuclear sporont; (E) eight sporoblasts; (F) eight spores. The sporonts and sporoblasts are enclosed in a sporophorous vesicle, which is not apparent around the spores. (Bar = 10 μm.) (Provided by J. J. Becnel.)

is derived from the polaroplast membrane (Weidner et al. 1984). When inoculated into a host cell, one of the envelopes disappears and the sporoplasm is surrounded by a single membrane (Ishihara 1968b; Cali 1971; Weidner 1972).

The meront is generally surrounded by a single-unit membrane that actively interacts in endocytosis with the host cell (Vávra 1976a, 1976b). The membrane surface can be substantially increased by vesicular or tubular projections that interdigitate with the host cell cytoplasm (Vávra 1976a).

Two significant changes occur during the transformation from meront to sporont: (1) the rough endoplasmic reticulum increases and the cisternae are arranged in concentric layers around the nucleus and (2) the cell envelope grows in thickness and complexity by the addition of electron-dense material external to the plasma membrane (Cali 1971; Vávra 1976a; Larsson 1986). In many species, the sporont transforms into a multinucleate plasmodium that cleaves off individual fingerlike sporoblasts (Fig. 12–3). In the suborder Apansporoblastina, the morphogenesis of the sporont membrane ends with the formation of a dark extramembranous coat, whereas in the suborder Pansporoblastina, an additional membrane forms the pansporoblast. The use of the term pansporoblast has been questioned since it differs from the pansporoblast of the Myxozoa, which has a different origin and is composed of living cells. Larsson (1986) prefers the use of "sporophorous vesicle," as proposed by Canning and Hazard (1982), for the pansporoblast, and we have adopted this terminology. During merogony, the envelope of the sporophorous vesicle is generally produced at the beginning of sporogony or in a few genera (Larsson 1988) (Fig. 12–3). It appears as a sachet enshrouding the sporoblasts or spores originating from a single sporont, but it is often fragile and difficult to see in a microscope smear.

The products of the sporont replication are the sporoblasts whose cell walls consist of two principal layers, the outer proteinaceous dense exospore and the chitinous endospore. As the sporoblast matures to a spore, the polar filament and polaroplast are formed. A Golgi apparatus is found in the center of a sporogonial plasmodium and it is divided among the sporoblasts (Larsson 1985). The Golgi apparatus forms the polar filament, but in *Buxtehudea scaniae* and some other microsporidia with spherical spores, the polar filament is formed from diffuse cytoplasmic vesicles (Larsson 1980, 1986). The polar sac develops around the anterior tip of the polar filament. Within the polar sac, the complex anchoring apparatus is formed.

K. GENERAL LIFE CYCLE

The microsporidian life cycle has two distinct sequences: (1) merogony, the vegetative phase (previously called schizogony) and (2) sporogony, the production of spores (Fig. 12–4). The mother cell for merogony is called the meront and for

FIGURE 12–4

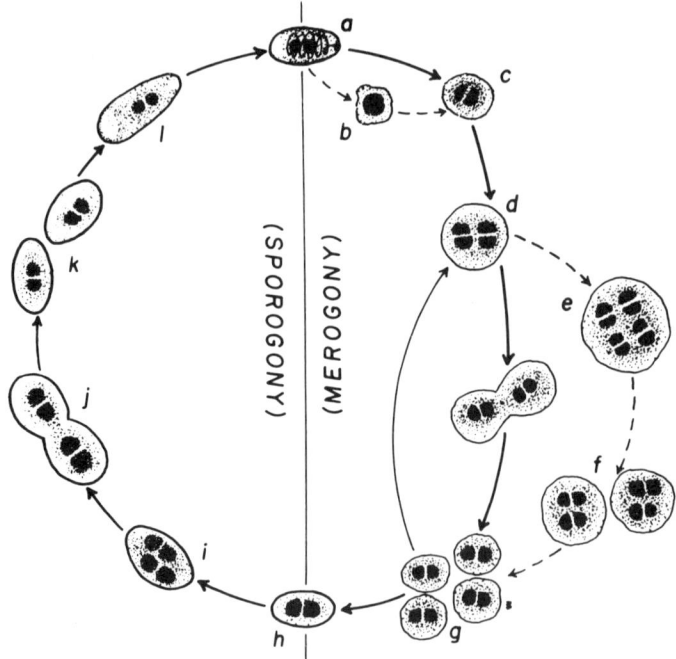

Schematic drawing of the life cycle of *Nosema algerae* showing the merogonic and sporogonic sequences. (*a*) Spore; (*b*) uninucleate meront; (*c*,*g*) binucleate meronts; (*d*,*f*) tetranucleate meronts; (*e*) octonucleate meront; (*h*) binucleate sporont; (*i*) tetranucleate sporont; (*j*) sporont undergoing cytokinesis; (*k*,*l*) binucleate sporoblasts. [Courtesy of Brooks (1988) and CRC Press.]

sporogony, the sporont. In most cases, the spore-to-spore cycle takes place inside one cell (Issi 1986). Microsporidia have basically similar life cycles, but specific variations (e.g., the mode of division and the number of daughter cells) are of taxonomic value in classifying these organisms (Vávra 1976b). Some genera have three different sporogonic sequences leading to marked spore polymorphism. Canning (1990) has described the complex and diverse types of sporogony found in different microsporidian genera.

During merogony, the microsporidium multiplies rapidly by binary fission, plasmotomy (fragmentation of a multinucleate cell or multiple fission), or multiple budding. Additional multiplication occurs during the sporulation phase and terminates in the formation of spores (sporogony). The stages in the merogonial sequence are (1) the sporoplasm, the invasive stage, and (2) the meront, the main multiplicative stage. The daughter cells may remain united as chains. The sporogonial stages are (1) the sporont, the initial stage for additional multiplication by binary or multiple fission and (2) the sporoblast, a nondividing morphogenesis stage to form the spore. During these stages, the cells may have one (unikaryon) or two coupled nuclei (diplokaryon) that are closely adjacent but separated by their membranes, or several nuclei in a plasmodium.

Microsporidia differ in the number of generations formed during merogony and sporogony. In some, merogony is relatively short and sporogonial multiplication is extensive (*Pleistophora*). In others, the merogony is more extensive and prominent, whereas sporogony is short and produces a limited number of spores from single sporonts (*Nosema, Encephalitozoon*). Both merogony and sporogony may occur in the same host tissue (Gray et al. 1968); in different tissues of the same insect host (Hazard and Fukuda 1974; Hazard et al. 1984), or in different generations of the same insect species [e.g., *Amblyospora* spp. (Andreadis 1983)]. Merogony has not been described in many species and is even reported to be absent in some species (Canning 1990).

We shall discuss the general life cycle beginning with the ingestion of a spore by an insect. The spore is acted upon by digestive juices and the polar filament is extruded. The sporoplasm is inoculated into a host cell by the polar filament. This ensures that the microsporidium initially lies directly within the cell cytoplasm and not in a parasitophorus vacuole derived from the host plasma membrane (Canning 1990). This situation provides protection against the lytic action that may occur in a vacuole. Within a host cell, the sporoplasm grows and develops cytoplasmic organelles and becomes a meront. In the mosquito *Anopheles albimanus*, the sporoplasms of *Nosema algerae* may be found in the thorax and first three abdominal segments at 1 h after the initial ingestion of the spores, and the first merogonial division occurs between 30 and 36 h (Avery and Anthony 1983).

The meront is initially a round cell, several microns in size, with a compact round nucleus (unikaryon) or with two nuclei (diplokaryon). As the meront grows in size, it divides by binary fission into single cells or only its nucleus divides to form a plasmodial cell, which later separates into single cells. The meronts continue to divide to start a new merogonial cycle. The number of divisions may depend on the size and type of the infected host cell [i.e., in small muscle and epidermal cells, the meront may have two divisions, whereas in fat body and gastric caeca cells, it may have at least three divisions (Avery and Anthony 1983)]. In some species, there are two structurally different generations of meronts, with the first generation represented by small cells with compact nuclei and rounded plasmodia, and the second by long ribbonlike plasmodial stages with several large nuclei (Weiser 1961).

How microsporidia spread from cell to cell, or from organ to organ, has not been firmly established. Most investigators believe that the meront is incapable of active movement (Vávra 1976b). Cali and Briggs (1967) have suggested that meronts are transported by phagocytic cells, which incorporate them into the host tissues. Kurtti et al. (1983) have proposed that degenerating cells extrude vegetative forms, enclosed in vesicles, which spread the microsporidium. In the silkworm, Abe (1978) has reported the presence of two forms of *Nosema bombycis*, an acid and an amoeboid form. The acid form resembles a sporoplasm and is responsible for the initial spread and the amoeboid form for the subsequent distribution of the microsporidium throughout the insect. The most plausible hypothesis is the movement of small binucleate forms as in *Nosema bombycis* (Ishihara 1969). These forms are sporoplasms, which are discharged by a specific type of spore in an infected cell (Iwano and Ishihara 1991). The polar filaments of these specific spores are able to penetrate into adjacent cells and inoculate the sporoplasms into them. In *Amblyospora*, the diplokaryotic microsporidian cells spread from the oenocytoids to adipose tissues in the male mosquito (Andreadis 1983). In the female mosquito, the spores in the oenocytoids extrude their polar filaments and sporoplasms to infect the oocytes.

At a certain stage in the life cycle, the meront gives rise to the sporont. Both cell types resemble each other very closely and differ mainly in the structure of their cell membranes. What triggers the transition from merogonial to sporogonial stages is unknown, but Ishihara (1969) suggests that certain conditions of the host cell may initiate this transition.

Sporogony varies in different microsporidia. The sporont nucleus may divide once or several times. The first division is meiotic in some species. Further divisions are mitotic. In a simple life cycle, the first nuclear division is followed by cytokinesis and the two cells, which become sporoblasts, are thus formed from a single sporont (e.g., *Encephalitozoon* and *Nosema*). When the nucleus divides one or more times without cytokinesis, a sporogonial plasmodium is produced. This plasmodium later produces unicellular sporoblasts by budding, by plasmotomy, by breaking up directly into such cells, and by other means of division. In this case, a number of sporoblasts (and of spores) are formed from each sporont. This number is very regular in some families (e.g., one sporont forming two spores in Telomyxidae; four spores in Gurleyidae, eight spores in Thelohaniidae; and 16 spores in Duboscqiidae). There are some enigmatic species in families in which a variable number of spores are formed from a single sporont (Hazard and Oldacre 1976).

Sporoblasts are single cells arising from the division of the sporont or of the sporogonial plasmodium. They form a clearly defined stage preceding the spore. At this stage, the spore organelles start to form and the microsporidian cell, for the first time, acquires definite polarity (Vávra 1976b). When seen with the light microscope, they are recognized by their dense cytoplasm limited by a thick wall.

Spore morphogenesis, resulting from the transformation of the sporoblast, begins first with the differentiation of the polar filament and polar sac and followed by the polaroplast and posterior vacuole (Fig. 12–2A). The final stage is the formation of the thick endospore layer of the spore wall.

Generally, the formation of spores terminates the life cycle of a microsporidium within the host, and the spores begin a new cycle only after ingestion by another host. There are reports, however, of spore germination within the host cell, such as in cultured cells infected with *Nosema* spp. (Iwano and Ishihara 1989, 1991) and in mosquito larvae infected with *Nosema algerae* (Vávra and Undeen 1970; Avery and Anthony 1983). In a dimorphic *Nosema* sp., the first type of spores, possessing a few coils of polar filament, discharges a short polar filament with germination occurring within a host cell; the second type of spores, possessing many coils of filament, discharges a long polar filament with germination occuring in the midgut lumen and not in the cell (Iwano and Ishihara 1991). The first type of spores may be important in the dissemination of the microsporidium within an insect.

L. FACTORS AFFECTING THE LIFE CYCLE

Temperature and the physiological condition of the honey bee affect the life cycle of *Nosema apis* (Steche 1965b). During the active period of spring and summer months, the life cycle is relatively simple, whereas in autumn and winter, it is more complex. The sporonts formed in summer differ in ultrastructure from those produced in winter (Youssef and Hammond 1971). High-protein food fed to infected bees increases their longevity and also increases the *Nosema* spore development (Rinderer and Elliott 1977b).

The host cell condition is important in regulating the microsporidian life cycle. Ishihara (1969) believes that sporulation occurs only under certain cell conditions. When a microsporidium occurs in an insect egg or in a hibernating insect, its development is arrested, and development continues only when the embryo hatches or when the larva emerges from hibernation (Weiser 1961).

The length of the microsporidian life cycle varies and can be as short as 24 h to produce a generation (McLaughlin 1969). In *Nosema bombycis*, sporulation begins as early as 48 h after infection (Ishihara 1970). The generation time varies with the host and temperature.

M. LIFE CYCLE OF FAMILY AMBLYOSPORIDAE

An extremely unique life cycle occurs in dimorphic species of *Amblyospora* isolated from the mosquito larvae (Fig. 12–5). The elucidation of this life cycle is one of the major accomplishments in insect protozoology. The life cycle of *Amblyospora* remained incomplete and unsolved for some time because the haploid spores produced in male mosquito larvae are not infectious to other mosquito larvae. Workers suspected that these haploid spores infected an alternate host (Hazard et al. 1979; Undeen and Avery 1984). Sweeney et al. (1984, 1985) were the first to present clear evidence of a two-host life cycle. The haploid spores from mosquito larvae infect an intermediate copepod host, in which spores infectious for mosquito larvae are produced. *Parathelohania* spp., which are dimorphic, also require a copepod intermediate host (Avery and Undeen 1990).

Andreadis (1985a,b) has described in detail the life cycle of *Amblyospora*. Larvae of the mosquito *Aedes cantator* produce haploid meiospores that directly infect an alternate copepod host, *Acanthocyclops vernalis*. The microsporidium develops in the fat body and kills the copepod. The developmental sequences are unikaryotic, and there is no evidence of a sexual cycle or the formation of a diploid condition. Spores from the copepod differ from those produced in mosquitoes and are infectious only for the mosquito larvae.

In the mosquito, *Amblyospora* has two different proliferative sequences depending on the sex of the mosquito (Hazard and Weiser 1968; Hazard and Oldacre

FIGURE 12-5

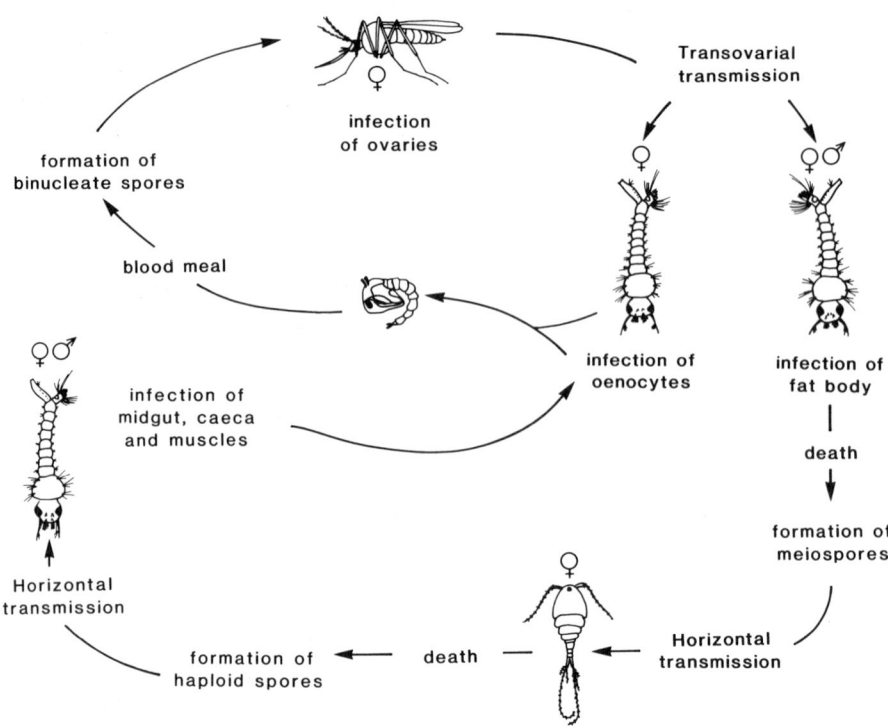

Schematic drawing of the life cycle of *Amblyospora connecticus* in the mosquito *Aedes cantator* and the copepod *Acanthocyclops vernalis*. [Courtesy of Andreadis (1990) and the Society of Protozoologists.]

1976; Andreadis and Hall 1979; Andreadis 1983). In a male mosquito larva, the diplokaryotic microsporidian cells spread from oenocytes to adipose tissues, where vegetative reproduction is succeeded by a sporogonic sequence that produces haploid octospores confined in a sporophorous vesicle membrane (Andreadis and Hall 1979; Andreadis 1983). The patent infection in the male larva is usually lethal because of the rapid and massive multiplication of the microsporidium. In a female larva, the infection is restricted to oenocytes. Microsporidian multiplication is limited and does not kill the larva that develops into an adult. The diplokaryotic vegetative stages give rise to binucleate spores within hypertrophied oenocytes. The spores extrude their sporoplasms infecting the developing host oocytes, and the microsporidium is transmitted to the next mosquito generation (vertical transmission). The efficient vertical transmission of the microsporidium through the female mosquito and the horizontal transmission through the copepod intermediate host provide the means for the persistence of the microsporidium in the natural environment (Avery and Undeen 1990).

The transmission of an *Amblyospora* sp. from an infected female *Culex salinarius* to her progeny is dependent on the synchronization of the microsporidian sporulation and the ovarian development of the host (Lord and Hall 1983). Both sporulation development and ovarian development are initiated when the mosquito takes a blood meal. However, Lord and Hall (1983) have succeeded in inducing sporulation in the absence of a blood meal by the injection of 20-hydroxyecdysone, a hormone whose precursor, ecdysone, is produced by the ovary of blood-fed mosquito. They have prevented sporulation in response to a blood meal by ovariectomy of the adult prior to blood feeding and by decapitation immediately after feeding.

N. PROLIFERATIVE SEQUENCES

The microsporidian life cycle may be altered by irregularities in the sporogonial and sporulation sequences. These sequences may be complex. For example, in *Culicosporella lunata*, the sporonts enter three different sporulation sequences, two in the larva and one in the adult mosquito (Hazard et al. 1984; Becnel and Fukuda 1991). One sequence in the larva involves abortive meiosis and rarely ends with meiospores; the other predominant sequence involves diplokaryotic sporogonial plasmodia and produces lanceolate, binucleate spores; the third in adult mosquitoes results in a second type of binucleate spore. Temperature may also affect the sequences in *Vairimorpha* spp. (Tanabe 1971; Fowler and Reeves 1974; Pilley 1976; Maddox and Sprenkel 1978) and in *Burenella dimorpha* (Jouvenaz and Lofgren 1984). At high temperatures, these microsporidia produce binucleate single spores (*Nosema*-like) and at low temperatures, form uninucleate octospores in an envelope (*Thelohania*-like).

During merogony and sporogony, at least five developmental cycles can be distinguished by the nuclear types of the microsporidia. Larsson (1986) has described these cycles.

Dimorphic spores with diplokaryotic or unikaryotic nuclei are found in members of families Amblyosporidae, Burenellidae, Culicosporidae (genus *Hazardia*), Culicosporellidae, and Spraguidae (Sprague 1982; Becnel and Fukuda 1991). The newly created family Burenellidae is characterized by two sequences of sporogony that have morphologically different sporonts and spores (Jouvenaz and Hazard 1978). One sequence of sporogony has sporonts that sporulate in the hypodermis to produce diplokaryotic spores not enclosed in a membrane-bounded vesicle; the second sequence has multinucleated sporonts in fat cells that produce eight unikaryotic spores enclosed in a membrane-bounded sporophorous vesicle. The sporophorous vesicle membrane enclosing the spores is not persistent, and octets of spores are not seen in wet mounts.

As described previously, the sex of adult mosquitoes affects the development of species of *Amblyospora*, in which the sporogony is suppressed in the female mosquito (Kellen et al. 1965). In other dimophic species, the sporogony causing a lethal infection occurs in both sexes (Undeen et al. 1984). There are other irregularities in the life cycle of species of *Amblyospora* infecting mosquitoes. When one of the parents of a hybrid mosquito is an unsuitable host, the microsporidium develops normally during merogony and early sporulation, but spore morphogenesis proceeds abnormally to produce aberrant spores (Kellen et al. 1966).

In *Stempellia simulii*, there are two types of sporogonial plasmodia (sporoblasts) (Maurand and Manier 1967). One type produces invariably 12 spores, the other forms 8 or 16 spores. In *S. mutabilis*, variations in the number and size of spores depend on different sporogonial processes (Desportes 1976). Eight microspores are produced after three divisions from small sporonts, whereas four macrospores are formed after two divisions.

O. SEXUALITY IN MICROSPORIDIA

The sexuality in microsporidia has been in question for many years. Early workers believed that they had observed gametic conjugation followed by nuclear fusion in the microsporidian life cycle. Since these observations were made on stained smears of very minute organisms, there was considerable doubt whether fusion could be differentiated from the division of cells and nuclei. Among the early workers, Mercier (1909) and Debaisieux (1928) suggested that similar gametes (meronts) conjugated to form binucleate cells, whose nuclei subsequently fused to form the sporont. The nucleus of the sporont was considered as a zygote nucleus. Debaisieux (1928) also proposed that in the *Nosema*-type of life cycle, after the conjugation of meronts, the nuclear fusion took place in the spore or perhaps in the sporoplasm after its liberation. Later workers believed that, rather than the conjugation of two cells, autogamy occurred. According to Weiser (1961) and Maurand (1966), the diplokaryotic meronts, which appear at the end of merogony in some species, underwent autogamy and changed into sporonts. Kudo (1924) believed that the binucleate cells derived from tetranucleate meronts contained two "cousin" nuclei, rather than "daughter" nuclei, which fuse to form the "sporont mother cell." On the

FIGURE 12–6

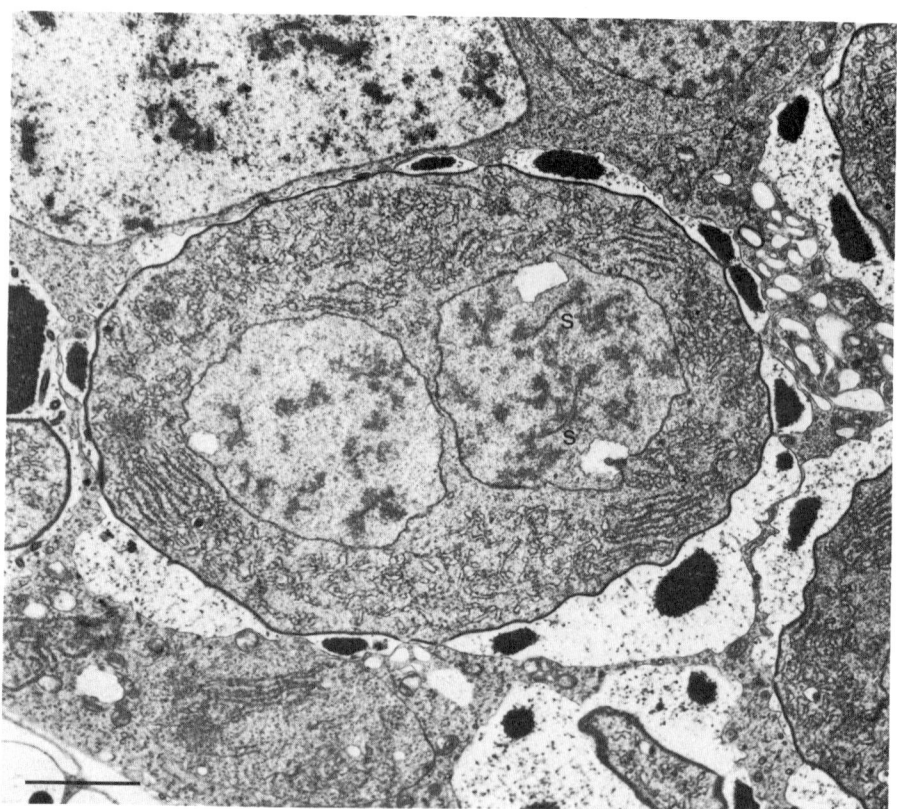

Electron micrograph of a diplokaryotic sporont of the microsporidium *Amblyospora californica* in the mosquito *Culex tarsalis*. The sporont is undergoing meiosis as indicated by the synaptonemal complex (s). (Bar = 10 μm.) (Provided by J. J. Becnel.)

other hand, Ohshima (1973) claimed that, about 20 h after the binucleate sporoplasm had emerged from the spore and entered the host cell, the nuclei fused in autogamy. Consequently, studies with the light microscope did not produce satisfactory evidence for sexuality in the microsporidia.

Observations with the electron microscope clearly established sexuality based on the occurrence of meiosis in microsporidia (Fig. 12–6). Initial reports of meiosis were made by Loubès et al. (1976) and Joyon and Vávra (in Vávra 1976b) and subsequently by Loubès (1979) in five other microsporidian genera. Others followed with meiosis being observed in *Amblyospora*, *Parathelohania* (Hazard et al. 1979), *Ameson* (Vivares and Sprague 1979), *Vairimorpha* (Malone and Canning 1982), *Napamichum* (syn. *Chapmanium*) (Larsson 1984, 1990), and *Culicosporella* (Becnel and Fukuda 1991). These workers based their conclusions on structures associated with meiosis, such as synaptonemal complexes, axial cores, and the configuration of chromosomes (Fig. 12–6).

Meiosis occurs in early sporogony, and the sporogonic cycle is the haploid phase. Thus, karyogamy must occur during merogony. The first clear evidence of gametogenesis was presented by Hazard and co-workers in *Hazardia milleri*, *Culicospora magna*, and *Amblyospora* sp. infecting mosquitoes (Hazard and Brookbank 1984; Hazard et al. 1985; Becnel et al. 1987). The nuclear fusion took place in diplokaryotic meronts followed by an unusual meiosis-like process in young sporonts. They confirmed karyogamy by ultrastructural observations and by measuring the DNA content in the nuclei of the meronts. The haploid spores, produced by the haploid sporonts, initiated a developmental sequence of gametogenesis followed by plasmogamy, which resulted in diplokaryotic stages.

REFERENCES

Abe, Y. 1978. Propagative reproduction of *Nosema bombycis* Naegeli in larvae of the silkworm. *J. Seric. Sci. Jpn. 47*, 279–284.

Andreadis, T. G. 1983. Life cycle and epizootiology of *Amblyospora* sp. (Microspora: Amblyosporidae) in the mosquito, *Aedes cantator*. *J. Protozool. 30*, 509–518.

Andreadis, T. G. 1985a. Experimental transmission of a microsporidian pathogen from mosquitoes to an alternate copepod host. *Proc. Natl. Acad. Sci. U.S.A. 82*, 5574–5577.

Andreadis, T. G. 1985b. Life cycle, epizootiology, and horizontal transmission of *Amblyospora* (Microspora: Amblyosporidae) in a univoltine mosquito, *Aedes stimulans*. *J. Invertebr. Pathol. 46*, 31–46.

Andreadis, T. G. 1987. Horizontal transmission of *Nosema pyrausta* (Microsporida: Nosematidae) in the European corn borer, *Ostrinia nubilalis* (Lepidoptera: Pyralidae). *Environ. Entomol. 16*, 1124–1129.

Andreadis, T. G. 1990. Epizootiology of *Amblyospora connecticus* (Microsporida) in field populations of the saltmarsh mosquito, *Aedes cantator*, and the cyclopoid copepod, *Acanthocyclops vernalis*. *J. Protozool. 37*, 174–182.

Andreadis, T. G., and Hall, D. W. 1979. Development, ultrastructure, and mode of transmission of *Amblyospora* sp. (Microspora) in the mosquito. *J. Protozool. 26*, 444–452.

Armstrong, E. 1976. Transmission and infectivity studies on *Nosema kingi* in *Drosophila willistoni* and other drosophilids. *Z. Parasitenkd. 50*, 161–165.

Avery, S. W., and Anthony, D. W. 1983. Ultrastructural study of early development of *Nosema algerae* in *Anopheles albimanus*. *J. Invertebr. Pathol. 42*, 87–95.

Avery, S. W., and Undeen, A. H. 1990. Horizontal transmission of *Parathelohania anophelis* to the copepod, *Microcyclops varicans*, and the mosquito, *Anopheles quadrimaculatus*. *J. Invertebr. Pathol. 56*, 98–105.

Bailey, L. 1963. "Infectious Diseases of the Honey-bee." Land Books, London.

Batson, B. S. 1982. A light and electron microscopical study of *Trichoduboscqia epeori* Léger (Microspora: Duboscqiidae). *J. Protozool. 29*, 202–212.

Baudoin, J., and Mouthon, J. 1976. Nouvel exemple de corrélation entre la présence d'une Grégarine du genre *Ancyrophora* Léger, 1892 et le régime alimentaire de l'hôte. *C.R. Acad. Sci. Paris Ser. D 282*, 1285–1287.

Becnel, J. J., and Fukuda, T. 1991. Ultrastructure of *Culicosporella lunata* (Microsporida: Culicosporellidae fam. n.) in the mosquito *Culex pilosus* (Diptera: Culicidae) with new information on the developmental cycle. *Eur. J. Protistol. 26*, 319–329.

Becnel, J. J., Hazard, E. I., Fukuda, T., and Sprague, V. 1987. Life cycle of *Culicospora magna* (Kudo, 1920) (Microsporida: Culicosporidae) in *Culex restuans* Theobald with special reference to sexuality. *J. Protozool. 34*, 313–322.

Beesley, J. E. 1977. The life-cycle of *Rasajeyna nannyla* n. gen., n. sp., a coccidian pathogen of *Tipula paludosa* Meigen. *Parasitology 74*, 273–283.

Beesley, J. E. 1978. Seasonal abundance of three life cycle stages of *Rasajeyna nannyla* (Coccidia) in *Tipula paludosa* and *T. vittata*. *J. Invertebr. Pathol. 31*, 255–259.

Bhatia, M. L. 1937. On *Adelina tribolii*, a coccidian parasite of *Tribolium ferrugineum* F. *Parasitology 29*, 239–246.

Brooks, W. M. 1971. The inflammatory response of the tobacco hornworm, *Manduca sexta*, to infection by the microsporidian, *Nosema sphingidis*. *J. Invertebr. Pathol. 17*, 87–93.

Brooks, W. M. 1973. Protozoa: Host–parasite–pathogen interrelationships. *Misc. Publ. Entomol. Soc. Am. 9*, 105–111.

Brooks, W. M. 1988. Entomogenous protozoa. In "CRC Handbook of Natural Pesticides. Microbial Insecticides, Part A, Entomogenous Protozoa and Fungi." (C. M. Ignoffo, ed.), Vol. 5, pp. 1–149. CRC Press, Boca Raton, Florida.

Burges, H. D., Canning, E. U., and Hulls, I. K. 1974. Ultrastructure of *Nosema oryzaephili* and the taxonomic value of the polar filament. *J. Invertebr. Pathol. 23*, 135–139.

Cali, A. 1971. Morphogenesis in the genus *Nosema*. *Proc. Fourth Int. Colloq. Insect Pathol.* pp. 431–438. College Park, Maryland, Aug. 25–28, 1970.

Cali, A., and Briggs, J. D. 1967. The biology and life history of *Nosema tracheophila* sp. n. (Protozoa:Cnidospora:Microsporidea) found in *Coccinella septempunctata* Linnaeus (Coleoptera: Coccinellidae). *J. Invertebr. Pathol. 9*, 515–522.

Cali, A., and Owen, R. L. 1988. Microsporidiosis. In "Laboratory Diagnosis of Infectious Diseases Principles and Practice." (A. Balows, W. J. Hausler, Jr., M. Ohashi, and A. Turano, eds.), Vol. 1, pp. 929–950. Springer-Verlag, New York.

Canning, E. U. 1964. Observations on the life history of *Mattesia trogodermae* sp. n., a schizogregarine parasite of the fat body of the khapra beetle, *Trogoderma granarium* Everts. *J. Insect Pathol. 6*, 305–317.

Canning, E. U. 1977. Microsporida. In "Parasitic Protozoa." (J. P. Kreier, ed.), Vol. 4, pp. 155–196. Academic Press, New York.

Canning, E. U. 1990. Phylum Microspora. In "Handbook of Protoctista." (L. Margulis, J. O. Corliss, M. Melkonian, and D. J. Chapman, eds.), pp. 53–72. Jones and Bartlett, Boston.

Canning, E. U., and Hazard, E. I. 1982. Genus *Pleistophora* Gurley, 1893: An assemblage of at least three genera. *J. Protozool. 29*, 39–49.

Canning, E. U., Barker, R. J., Page, A. M., and Nicholas, J. P. 1985. Transmission of microsporida, especially *Othosoma operophterae* (Canning, 1960) between generations of winter moth *Operophtera brumata* (L) (Lepidoptera; Geometridae). *Parasitology 90*, 11–19.

Chu, W. H., and Jaques, R. P. 1979. Pathologie d'une microsporidiose de l'arpenteuse du chou, *Trichoplusia ni* [*Lep.*: *Noctuidae*], par *Vairimorpha necatrix*. *Entomophaga 24*, 229–235.

Codreanu, M. 1940. Sur quatre grégarines nouvelles du genre *Enterocystis* parasites des éphémères torrenticoles. *Arch. Zool. Exp. Gen. 81*, 113–122.

Curgy, J.-J., Vávra, J., and Vivares, C. 1980. Presence of ribosomal RNAs with prokaryotic properties in *Microsporidia*, eukaryotic organisms. *Biol. Cell. 38*, 49–51.

da Cunha, A. B., and Jurand, A. 1978. Ultrastructural study of the relationships between the gregarine *Schneideria schneiderae* Da Cunha et al. 1975 and the cells of the host *Trichosia pubescens* Morgante 1969 (Sciaridae, Diptera). *Arch. Protistenkd. 120*, 233–254.

da Cunha, A. B., Morgante, J. S., Pavan, C., and Garrido, M. C. 1968. Studies on cytology and differentiation in Sciaridae. I. Chromosome changes induced by a gregarine in *Trichosia sp.* (Diptera, Sciaridae). *Caryologia 21*, 271–282.

da Cunha, A. B., Morgante, J. S., Marques, J., and Garrido, M. C. 1978. O desenvolvimento da gregarina *Schneideria schneiderae* (da Cunha et al., 1975) no hospedeiro heterólogo *Plastosciara* sp. e o problema da adaptação. *Rev. Bras. Biol. 38*, 885–893.

Debaisieux, P. 1928. Études cytologiques sur quelques Microsporidies. Le cycle des Microsporidies. *La Cellule 38*, 439–441.

Desportes, I. 1976. Ultrastructure of *Stempellia mutabilis* Leger et Hesse, microsporidie parasite de l'éphémère *Ephemera vulgata* L. *Protistologica 12*, 121–150.

Diaz, M., and Pavan, C. 1965. Changes in chromosomes induced by microorganism infection. *Proc. Natl. Acad. Sci. U.S.A. 54*, 1321–1327.

Dufour, L. 1828. Note sur la Grégarine, nouveau genre de ver qui vit en troupeau dans les intestins de divers insectes. *Ann. Sci. Nat. Ser. 1. 13*, 366–368.

Dunkel, F. V., and Boush, G. M. 1969. Effect of starvation on the black carpet beetle, *Attagenus megatoma*, infected with the eugregarine *Pyxinia frenzeli*. *J. Invertebr. Pathol. 14*, 49–52.

Fisher, F. M., Jr., and Sanborn, R. C. 1962. Observations on the susceptibility of some insects to Nosema (Microsporidia: Sporozoa). *J. Parasitol. 48*, 926–932.

Fisher, F. M., Jr., and Sanborn, R. C. 1964. *Nosema* as a source of juvenile hormone in parasitized insects. *Biol. Bull. 126*, 235–252.

Foerster, H. 1938. Beobachtungen über das Auftreten von Gregarinen in Insekten. *Z. Parasitenkd. 10*, 644–673.

Fowler, J. L., and Reeves, E. L. 1974. Spore dimorphism in a microsporidan isolate. *J. Protozool. 21*, 538–542.

Fowler, J. L., and Reeves, E. 1975. Microsporidan spore structure as revealed by scanning electron microscopy. *J. Invertebr. Pathol. 26*, 1–6.

Gaugler, R. R., and Brooks, W. M. 1975. Sublethal effects of infection by *Nosema heliothidis* in the corn earworm, *Heliothis zea*. *J. Invertebr. Pathol. 26*, 57–63.

Ghélélovitch, S. 1948. *Coelogregarina ephestiae*, schizogrégarine parasite d'*Ephestia kühniella* Z. (Lépidoptère). *Arch. Zool. Exp. Gen. 85*, No. 3, 155–168.

Götz, P. 1981. Homology of the manubrium of *Mrazekia brevicauda* and the polar filament of other Microsporidia. *Z. Parasitenkd. 64*, 321–333.

Grassé, P.-P. 1953. Sous-embranchement des Sporozoaires (Sporozoa Leuckart, 1879; Rhabdogeniae Delarge et Hérouard, 1896; Telosporidia Schaudinn, 1900). *Traite Zool. 1*, Fasc. 2, 545–690.

Gray, F. H., Cali, A., and Briggs, J. D. 1968. Intracellular stages in the life cycle of the microsporidian *Nosema apis*. *J. Invertebr. Pathol. 14*, 391–394.

Haldar, D. P., and Chakraborty, N. 1976. *Retractocephalus*—A new genus of cephaline gregarines (Protozoa: Sporozoa) from insects. *Curr. Sci. 45*, 668–669.

Harry, O. G. 1967. The effect of a eugregarine *Gregarina polymorpha* (Hammerschmidt) on the mealworm larva of *Tenebrio molitor* (L.). *J. Protozool. 14*, 539–547.

Hartwig, A., and Przełęcka, A., 1971. Nucleic acids in intestine of *Apis mellifica* infected with *Nosema apis* and treated with Fumagillin DCH: Cytochemical and autoradiographic studies. *J. Invertebr. Pathol. 18*, 331–336.

Hazard, E. I., and Brookbank, J. W. 1984. Karyogamy and meiosis in an *Amblyospora* sp. (Microspora) in the mosquito, *Culex salinarius*. *J. Invertebr. Pathol.* 44, 3–11.

Hazard, E. I., and Fukuda, T. 1974. *Stempellia milleri* sp. n. (Microsporida: Nosematidae) in the mosquito *Culex pipiens quinquefasciatus* Say. *J. Protozool.* 21, 497–504.

Hazard, E. I., and Lofgren, C. S. 1971. Tissue specificity and systematics of a *Nosema* in some species of *Aedes*, *Anopheles*, and *Culex*. *J. Invertebr. Pathol.* 18, 16–24.

Hazard, E. I., and Oldacre, S. W. 1976. Revision of Microsporida (Protozoa) close to *Thelohania*, with descriptions of one new family, eight new genera, and thirteen new species. *U.S. Dep. Agric. Tech. Bull. 1530*.

Hazard, E. I., and Weiser, J. 1968. Spores of *Thelohania* in adult female *Anopheles*: Development and transovarial transmission, and redescriptions of *T. legeri* Hesse and *T. obesa* Kudo. *J. Protozool.* 15, 817–823.

Hazard, E. I., Andreadis, T. G., Joslyn, D. J., and Ellis, E. A. 1979. Meiosis and its implications in the life cycles of *Amblyospora* and *Parathelohania* (Microspora). *J. Parasitol.* 65, 117–122.

Hazard, E. I., Fukuda, T., and Becnel, J. J. 1984. Life cycle of *Culicosporella lunata* (Hazard & Savage, 1970) Weiser, 1977 (Microspora) as revealed in the light microscope with a redescription of the genus and species. *J. Protozool.* 31, 385–391.

Hazard, E. I., Fukuda, T, and Becnel, J. J. 1985. Gametogenesis and plasmogamy in certain species of Microspora. *J. Invertebr. Pathol.* 46, 63–69.

Henry, J. E. 1967. *Nosema acridophagus* sp. n., a microsporidian isolated from grasshoppers. *J. Invertebr. Pathol.* 9, 331–341.

Huger, A. 1960. Electron microscope study on the cytology of a microsporidian spore by means of ultrathin sectioning. *J. Insect Pathol.* 2, 84–105.

Huger, A. M., and Lenz, M. 1976. Nachweis einer Cölomgregarine (Protozoa: Sporozoa) bei der australischen Termite Coptotermes acinaciformis (Isoptera: Rhinotermitidae). *Z. Angew. Entomol.* 81, 252–258.

Ishihara, R. 1968a. Growth of *Nosema bombycis* in primary cell cultures of mammalian and chicken embryos. *J. Invertebr. Pathol.* 11, 328–329.

Ishihara, R. 1968b. Some observations on the fine structure of sporoplasm discharged from spores of a microsporidian, *Nosema bombycis*. *J. Invertebr. Pathol.* 12, 245–258.

Ishihara, R. 1969. The life cycle of *Nosema bombycis* as revealed in tissue culture cells of *Bombyx mori*. *J. Invertebr. Pathol.* 14, 316–320.

Ishihara, R. 1970. Fine structure of *Nosema bombycis* (Microsporidia, Nosematidae) developing in the silkworm (*Bombyx mori*) - I. *Bull. Coll. Agric. Vet. Med. Nihon Univ.* 27, 84–91.

Ishihara, R., and Hayashi, Y. 1968. Some properties of ribosomes from the sporoplasm of *Nosema bombycis*. *J. Invertebr. Pathol.* 11, 377–385.

Issi, I. V. 1986. Microsporidia as a phylum of parasitic protozoa. *Protozoology* (Leningrad) 10, 6–136.

Iwano, H., and Ishihara, R. 1989. Intracellular germination of spores of a *Nosema* sp. immediately after their formation in cultured cell. *J. Invertebr. Pathol.* 54, 125–127.

Iwano, H., and Ishihara, R. 1991. Dimorphism of spores of *Nosema* spp. in cultured cell. *J. Invertebr. Pathol.* 57, 211–219.

Jeffords, M. R., Maddox, J. V., and O'Hayer, K. W. 1987. Microsporidian spores in gypsy moth larval silk: A possible route of horizontal transmission. *J. Invertebr. Pathol.* 49, 332–333.

Jensen, H. M., and Wellings, S. R. 1972. Development of the polar filament–polaroplast complex in a microsporidian parasite. *J. Protozool.* 19, 297–305.

Jouvenaz, D. P., and Anthony, D. W. 1979. *Mattesia geminata* sp. n. (Neogregarinida: Ophrocystidae) a parasite of the tropical fire ant, *Solenopsis geminata* (Fabricius). *J. Protozool.* 26, 354–356.

Jouvenaz, D. P., and Hazard, E. I. 1978. New family, genus, and species of microsporida

(Protozoa: Microsporida) from the tropical fire ant, *Solenopsis geminata* (Fabricius) (Insecta: Formicidae). *J. Protozool.* 25, 24–29.

Jouvenaz, D. P., and Lofgren, C. S. 1984. Temperature-dependent spore dimorphism in *Burenella dimorpha* (Microspora: Microsporida). *J. Protozool.* 31, 175–177.

Jouvenaz, D. P., Lofgren, C. S., and Allen, G. E. 1981. Transmission and infectivity of spores of *Burenella dimorpha* (Microsporida: Burenellidae). *J. Invertebr. Pathol.* 37, 265–268.

Jurand, A., Simões, L. C. G., and Pavan, C. 1967. Changes in the ultrastructure of salivary gland cytoplasm in *Sciara ocellaris* (Comstock, 1882) due to microsporidian infection. *J. Insect Physiol.* 13, 795–803.

Kellen, W. R., and Lindegren, J. E. 1971. Modes of transmission of *Nosema plodiae* Kellen and Lindegren, a pathogen of *Plodia interpunctella* (Hübner). *J. Stored Prod. Res.* 7, 31–34.

Kellen, W. R., and Lindegren, J. E. 1974. Life cycle of *Helicosporidium Parasiticum* in the navel orangeworm, *Paramyelois transitella*. *J. Invertebr. Pathol.* 23, 202–208.

Kellen, W. R., and Wills, W. 1962. The transovarian transmission of *Thelohania californica* Kellen and Lipa in *Culex tarsalis* Coquillet. *J. Insect Pathol.* 4, 321–326.

Kellen, W. R., Chapman, H. C., Clark, T. B., and Lindegren, J. E. 1965. Host–parasite relationships of some *Thelohania* from mosquitoes (Nosematidae: Microsporidia). *J. Invertebr. Pathol.* 7, 161–166.

Kellen, W. R., Clark, T. B., Lindegren, J. E., and Sanders, R. D. 1966. Development of *Thelohania californica* in two hybrid mosquitoes. *Exp. Parasitol.* 18, 251–254.

Knell, J. D., and Zam, S. G. 1978. A serological comparison of some species of Microsporida. *J. Invertebr. Pathol.* 31, 280–288.

Kramer, J. P. 1976. The extra-corporeal ecology of microsporidia. In "Comparative Pathobiology." (L. A. Bulla, Jr. and T. C. Cheng, eds.), Vol. 1, pp. 127–135. Plenum Press, New York.

Kučera, M., and Weiser, J. 1975. Lactate dehydrogenase isoenzymes in the larvae of *Barathra brassicae* and *Galleria mellonella* during microsporidan infection. *J. Invertebr. Pathol.* 25, 109–114.

Kudo, R. 1924. Studies on microsporidia parasitic in mosquitoes. III. On *Thelohania legeri* Hesse (= *Th. illinoisensis* Kudo). *Arch. Protistenkd.* 49, 147–162.

Kurtti, T. J., Tsang, K. R., and Brooks, M. A. 1983. The spread of infection by the microsporidan, *Nosema disstriae*, in insect cell lines. *J. Protozool.* 30, 652–657.

Lacey, L. A., and Undeen, A. H. 1986. Microbial control of black flies and mosquitoes. *Annu. Rev. Entomol.* 31, 265–296.

Lai, P. F., and Canning, E. U. 1980. Infectivity of a microsporidium of mosquitoes (*Nosema algerae*) to larval stages of *Schistosoma mansoni* in *Biomphalaria glabrata*. *Int. J. Parasitol.* 10, 293–301.

Larsson, R. 1980. Insect pathological investigations on Swedish Thysanura, II. A new microsporidian parasite of *Petrobius brevistylis* (Microcoryphia, Machilidae); description of the species and creation of two new genera and a new family. *Protistologica 16*, 85–101.

Larsson, R. 1984. Ultrastructural study and description of *Chapmanium dispersus* n. sp. (Microspora, Thelohaniidae), a microsporidian parasite of *Endochironomus* larvae (Diptera, Chironomidae). *Protistologica 20*, 547–563.

Larsson, R. 1985. On the cytology, development and systematic position of *Thelohania asterias* Weiser, 1963, with creation of the new genus *Bohuslavia* (Microspora, Thelohaniidae). *Protistologica 21*, 235–248.

Larsson, R. 1986. Ultrastructure, function, and classification of microsporidia. *Progr. Protistol.* 1, 325–390.

Larsson, R. 1988. Identification of microsporidian genera (Protozoa, Microspora)—a guide with comments on the taxonomy. *Arch. Protistenkd.* 136, 1–37.

Larsson, J. I. R. 1990. Description of a new microsporidium of the water mite *Limnochares aquatica* and establishment of the new genus *Napamichum* (Microspora, Thelohaniidae). *J. Invertebr. Pathol.* 55, 152–161.

Lee, J. J., Hunter, S. H., and Bovee, E. C. 1985. "An Illustrated Guide to the Protozoa." Society of Protozoologists, Allen Press, Lawrence, Kansas.

Levine, N. D. 1971. Uniform terminology for the protozoan subphylum Apicomplexa. *J. Protozool.* 18, 352–355.

Levine, N. D. 1977a. Revision and checklist of the species (other than *Lecudina*) of the aseptate gregarine family Lecudinidae. *J. Protozool.* 24, 41–52.

Levine, N. D. 1977b. Checklist of the species of the aseptate gregarine families Aikinetocystidae, Diplocystidae, Allantocystidae, Schaudinnellidae, Ganymedidae, and Enterocystidae. *J. Invertebr. Pathol.* 29, 175–181.

Levine, N. D. 1982. Apicomplexa. *In* "McGraw Hill Synopsis and Classification of Living Organisms." (S. P. Parker, ed.), Vol. 1, pp. 571–587. McGraw-Hill, New York.

Levine, N. D. 1983. The genera *Barrouxia*, *Defretinella*, and *Goussia* and of the coccidian family Barrouxiidae (Protozoa, Apicomplexa). *J. Protozool.* 30, 542–547.

Levine, N. D., Corliss, J. O., Cox, F. E. G., Deroux, G., Grain, J., Honigberg, B. M., Leedale, G. F., Loeblich, A. R., III, Lom, J., Lynn, D., Merinfeld, E. G., Page, F. C., Poljansky, G., Sprague, V., Vávra, J., and Wallace, F. G. 1980. A newly revised classification of the Protozoa. *J. Protozool.* 27, 37–58.

Lien, S.-M., and Levine, N. D. 1980. Three new species of *Ascocystis* (Apicomplexa, Lecudinidae) from mosquitoes. *J. Protozool.* 27, 147–151.

Lindegren, J. E., and Hoffmann, D. F. 1976. Ultrastructure of some developmental stages of *Helicosporidium* sp. in the navel orangeworm, *Paramyelois transitella*. *J. Invertebr. Pathol.* 27, 105–113.

Lipa, J. J. 1967. Studies on gregarines (*Gregarinomorpha*) of arthropods in Poland. *Acta Protozool.* 5, 97–179.

Liu, T. P., and Davies, D. M. 1972. Ultrastructure of the cytoplasm in fat-body cells of the blackfly, *Simulium vittatum*, with microsporidian infection; a freeze-etching study. *J. Invertebr. Pathol.* 19, 208–218.

Liu, T. P., and Davies, D. M. 1973. Ultrastructural architecture and organization of the spore envelope during development in *Thelohania bracteata* (Strickland, 1913) after freeze-etching. *J. Protozool.* 20, 622–630.

Lom, J. 1972. On the structure of the extruded microsporidian polar filament. *Z. Parasitenkd.* 38, 200–213.

Lom, J., and Corliss, J. O. 1967. Ultrastructural observations on the development of the microsporidian protozoon *Plistophora hyphessobryconis* Schaperclaus. *J. Protozool.* 14, 141–152.

Lom, J., and Vávra, J. 1963a. The mode of sporoplasm extrusion in microsporidian spores. *Acta Protozool.* 1, 81–89.

Lom, J., and Vávra, J. 1963b. Fine morphology of the spore in Microsporidia. *Acta Protozool.* 1, 279–283.

Lom, J., and Vávra, J. 1963c. Mucous envelopes of spores of the subphylum Cnidospora (Doflein, 1901). *Věst. Česk. Spol. Zool.* 27, 4–6.

Lord, J. C., and Hall, D. W. 1983. Sporulation of *Amblyospora* (Microspora) in female *Culex salinarius*: Induction by 20-hydroxyecdysone. *Parasitology* 87, 377–383.

Loubés, C. 1979. Recherches sur la Méiose chez les Microsporidies: Conséquences sur les cycles biologiques. *J. Protozool.* 26, 200–208.

Loubés, C., Maurand, J., and Rousset-Galangau, V. 1976. Présence de complexes synaptonématiques dans le cycle biologique de *Gurleya chironomi* Loubès et Maurand, 1975: Un argument en faveur d'une sexualité chez les Microsporidies? *C.R. Acad. Sci. Paris Ser. D* 282, 1025–1027.

Maddox, J. V., and Sprenkel, R. K. 1978. Some enigmatic microsporidia of the genus *Nosema. Misc. Publ. Entomol. Soc. Am. 11*, 65–84.

Malavasi, A., da Cunha, A. B., Morgante, J. S., and Marques, J. 1976. Relationships between the gregarine *Schneideria schneiderae* and its host *Trichosia pubescens* (Diptera, Sciaridae). *J. Invertebr. Pathol. 28*, 363–371.

Malone, L. A., and Canning, E. U. 1982. Fine structure of *Vairimorpha plodiae* (Microspora, Burenellidae), a pathogen of *Plodia interpunctella* (Lepidoptera, Phycitidae) and infectivity of the dimorphic spores. *Protistologica 18*, 503–516.

Manwell, R. D. 1977. Gregarines and haemogregarines. *In* "Parasitic Protozoa, Gregarines, Haemogregarines, Coccidia, Plasmodia, and Haemoproteids." (J. P. Kreier, ed.), Vol. 3, pp. 1–32. Academic Press, New York.

Margulis, L., Corliss, J. O., Melkonian, M., and Chapman, D. J. 1990. "Handbook of Protoctista." Jones and Bartlett, Boston.

Martins, R. R., and Perondini, A. L. P. 1977. Effects of microsporidia on the striated parietal muscle of *Rhynchosciara angelae* (Diptera: Sciaridae). *J. Invertebr. Pathol. 30*, 422–428.

Maurand, J. 1966. *Plistophora simulii* (Lutz et Splendore 1904), Microsporidie parasite des larves de *Simulium*; cycle, ultrastructure, ses rapports avec *Thelohania bracteata* (Strickland 1913). *Bull. Soc. Zool. Fr. 91*, 621–630.

Maurand, J., and Loubes, C. 1973. Recherches cytochimiques sur quelques Microsporidies. *Bull. Soc. Zool. Fr. 98*, 373–383.

Maurand, J., and Manier, J.-F. 1967. Une Microsporidie nouvelle pour les larves de simulies. *Protistologica 3*, 445–449.

McLaughlin, R. E. 1965a. *Mattesia grandis* n. sp., a sporozoan pathogen of the boll weevil, *Anthonomus grandis* Boheman. *J. Protozool. 12*, 405–413.

McLaughlin, R. E. 1965b. Some relationships between the boll weevil, *Anthonomus grandis* Boheman, and *Mattesia grandis* McLaughlin (Protozoa: Neogregarinida). *J. Invertebr. Pathol. 7*, 464–473.

McLaughlin, R. E. 1969. *Glugea gasti* sp. n., a microsporidan pathogen of the boll weevil *Anthonomus grandis*. *J. Protozool. 16*, 84–92.

Mercier, L. 1909. Contribution à l'étude de la sexualité chez les myxosporidies et chez les microsporidies. *Mem. Acad. R. Belg. Cl. Sci. 2*, 1–52.

Metspalu, L. 1976. On the effect of microsporidiosis on hibernating pupae of noctuids. *Bioloogia 25*, 13–19.

Metspalu, L., and Hiiesaar, K. 1980. Effect of microsporidiosis on the hibernating pupae of *Pieris brassicae* L. and *Pieris rapae* L. *Bioloogia 29*, 328–335.

Milner, R. J. 1972. *Nosema whitei*, a microsporidan pathogen of some species of *Tribolium*. III. Effect on *T. castaneum*. *J. Invertebr. Pathol. 19*, 248–255.

Nordin, G. L. 1975. Transovarial transmission of a *Nosema* sp. infecting *Malacosoma americanum*. *J. Invertebr. Pathol. 25*, 221–228.

Ohshima, K. 1937. On the function of the polar filament of *Nosema bombycis*. *Parasitology 29*, 220–224.

Ohshima, K. 1973. On the autogamy of nuclei and the spore formation of *Nosema bombycis* Nägeli. *Annot. Zool. Jpn. 46*, 30–44.

Overstreet, R. M., and Weidner, E. 1974. Differentiation of microsporidian spore-tails in *Inodosporus spraguei* Gen. et Sp. N. *Z. Parasitenkd. 44*, 169–186.

Pavan, C., Biesele, J., Reiss, R. W., and Wertz, A. V. 1971. Changes in the ultrastructure of *Rhynchosciara* cells infected by Microsporidia. *In* "Studies in Genetics, VI." (M. R. Wheeler, ed.), pp. 241–271. Univ. Texas Publication 7103.

Percy, J. 1973. The intranuclear ocurrence and fine structural details of schizonts of *Perezia fumiferanae* (Microsporida: Nosematidae) in cells of *Choristoneura fumiferana* (Clem.) (Lepidoptera: Tortricidae). *Can. J. Zool. 51*, 553–554.

Pérez, C. 1899. Sur une coccidie nouvelle *Adelea Mesnili* (n. sp.), parasite coelomique d'un lépidoptère. *C.R. Mem. Soc. Biol. 51*, 694–696.

Pérez, C. 1903. Le cycle évolutif de l'*Adelea Mesnili*, Coccidie coelomique parasite d'un Lépidoptère. *Arch. Protistenkd. 2*, 1–12.

Pérez, C. 1905. Microsporidies parasites des Crabes d'Arcachon. *Soc. Sci. Arcachon Trav. Lab. 8*, 15–36.

Petri, M., and Shiodt, T. 1966. On the ultrastructure of *Nosema cuniculi* in the cells of the Yoshida rat ascites sarcoma. *Acta Pathol. Microbiol. Scand. 66*, 437–446.

Philippe, M., and Schrével, J. 1982. The three cortical membranes of the gregarines (parasitic protozoa). Characterization of the membrane proteins of *Gregarina blaberae*. *Biochem. J. 201*, 455–464.

Pilley, B. M. 1976. A new genus, *Vairimorpha* (Protozoa: Microsporida), for *Nosema necatrix* Kramer 1965: Pathogenicity and life cycle in *Spodoptera exempta* (Lepidoptera: Noctuidae). *J. Invertebr. Pathol. 28*, 177–183.

Purrini, K. 1976. Zwei Schizogregarinen-Arten (Protozoa, Sporozoa) bei vorratsschädlichen Insekten in jugoslawischen Mühlen. *Anz. Schadlingskd. Pflanz. Umweltschutz. 49*, 83–85.

Purrini, K. 1977. Über eine neue Schizogregarinen-Krankheit der Gattung Mattesia Naville (Sporoz., Dischizae) des Zottigen Fichtenborkenkäfers, Dryocoetes autographus Ratz. (Coleopt., Scolytidae). *Anz. Schadlingskd. Pflanz. Umweltschutz 50*, 132–135.

Rabindra, R. J., Balasubramanian, M., and Jayaraj, S. 1981. The effects of *Farinocystis tribolii* on the growth and development of the flour beetle *Tribolium castaneum*. *J. Invertebr. Pathol. 38*, 345–351.

Rinderer, T. E., and Elliott, K. D. 1977a. Influence of nosematosis on the hoarding behavior of the honeybee. *J. Invertebr. Pathol. 30*, 110–111.

Rinderer, T. E., and Elliott, K. D. 1977b. Worker honey bee response to infection with *Nosema apis*: Influence of diet. *J. Econ. Entomol. 70*, 431–433.

Roberts, P. A., Kimball, R. F., and Pavan, C. 1967. Response of Rhynchosciara chromosomes to microsporidian infection. *Exp. Cell Res. 47*, 408–422.

Russell, D. G. 1983. Host cell invasion by Apicomplexa: An expression of the parasite's contractile system? *Parasitology 87*, 199–209.

Sato, R., Kobayashi, M., Watanabe, H., and Fujiwara, T. 1981. Serological discrimination of several kinds of microsporidian spores isolated from the silkworm, *Bombyx mori*, by an indirect fluorescent antibody technique. *J. Seric. Sci. Jpn. 50*, 180–184.

Scholtyseck, E. 1979. "Fine Structure of Parasitic Protozoa: An Atlas of Micrographs, Drawings and Diagrams." Springer-Verlag, Berlin.

Scholtyseck, E., and Danneel, R. 1962. Über die Feinstruktur der Spore von *Nosema apis*. *Dtsch. Entomol. Z. 9*, 471–476.

Schrével, J., Caigneaux, E., Gros, D., and Philippe, M. 1983. The three cortical membranes of the gregarines. I. Ultrastructural organization of *Gregarina blaberae*. *J. Cell Sci. 61*, 151–174.

Schubert, G. 1969. Ultracytologische Untersuchungen an der Spore der Mikrosporidienart, *Heterosporis finki*, gen. n., sp. n. *Z. Parasitenkd. 32*, 59–79.

Schwalbe, C. P., and Baker, J. E. 1976. Nutrient reserves in starving black carpet beetle larvae infected with the eugregarine *Pyxinia frenzeli*. *J. Invertebr. Pathol. 28*, 11–15.

Sherlock, P. L. 1979. *Diplocystis tipulae* sp. nov. (Sporozoa: Eugregarinorida), a parasite of *Tipula paludosa* Meigen (Diptera: Tipulidae). *Parasitology 78*, 207–220.

Siegel, J. P., Maddox, J. V., and Ruesink, W. G. 1988. Seasonal progress of *Nosema pyrausta* in the European corn borer, *Ostrinia nubilalis*. *J. Invertebr. Pathol. 52*, 130–136.

Sneller, V.-P. 1979. Inhibition of *Dirofilaria immitis* in gregarine-infected *Aedes aegypti*: Preliminary observations. *J. Invertebr. Pathol. 34*, 62–70.

Sprague, V. 1941. Studies on *Gregarina blattarum* with particular reference to the chromosome cycle. *Ill. Biol. Monogr. 18*, 57 pp.
Sprague, V. 1977a. Classification and phylogeny of the microsporidia. *In* "Comparative Pathobiology. Systematics of the Microsporidia." (L. A. Bulla, Jr., and T. C. Cheng, eds.), Vol. 2, pp. 1–30. Plenum Press, New York.
Sprague, V. 1977b. Annotated list of species of microsporidia. *In* "Comparataive Pathobiology. Systematics of the Microsporidia." (L. A. Bulla, Jr. and T. C. Cheng, eds.), (Vol. 2, pp. 31–334, 447–461. Plenum Press, New York.
Sprague, V. 1977c. The zoological distribution of the microsporidia. *In*: Comparative Pathobiology. Systematics of the Microsporidia." (L. A. Bulla, Jr. and T. C. Cheng, eds.), Vol. 2, pp. 335–446. Plenum Press, New York.
Sprague, V. 1978. Characterization and composition on the genus *Nosema*. *Misc. Publ. Entomol. Soc. Am. 11*, 5–16.
Sprague, V. 1979. Classification of the Haplosporidia. *Mar. Fish. Rev. 41 (1/2)*, 40–44.
Sprague, V. 1982. Microspora. *In* "Synopsis and Classification of Living Organisms." (S. P. Parker, ed.) Vol. 1, pp. 589–594. McGraw-Hill, New York.
Sprague, V., and Vernick, S.H. 1969. Light and electron microscope observations on *Nosema nelsoni* Sprague, 1950 (Microsporida, Nosematidae) with particular reference to its Golgi complex. *J. Protozool. 16*, 264–271.
Sprague, V., and Vernick, S.H. 1971. The ultrastructure of *Encephalitozoon cuniculi* (Microsporida, Nosematidae) and its taxonomic significance. *J. Protozool. 18*, 560–569.
Steche, W. 1965a. Zur Ontogonie von *Nosema apis* Zander im Mitteldarm der Arbeitsbiene. *Bull. Apic. 8*, 181–209.
Steche, W. 1965b. Observations sur l'ontogenese de *Nosema apis* Zander dans l'intestin moyen de l'abeille ouvriere. *Bull. Apic. 8*, 210–212.
Steinhaus, E. A. 1949. "Principles of Insect Pathology." McGraw-Hill, New York.
Sumner, R. 1936. Relation of gregarines to growth and longevity in the mealworm, Tenebrio molitor L. *Ann. Entomol. Soc. Am. 29*, 645–648.
Sweeney, A. W., Hazard, E. I., and Graham, M. F. 1984. Life cycle of a microsporidium (*Amblyospora* sp.) infecting the mosquito *Culex annulirostris*. Proc. Fourth Aust. Appl. Entomol. Res. Conf. (P. Bailey and D. Swincer, eds.) pp 382–386. Adelaide 24–28 Sept. 1984.
Sweeney, A. W., Hazard, E. I., and Graham, M. F. 1985. Intermediate host for an *Amblyospora* sp. (Microspora) infecting the mosquito, *Culex annulirostris*. *J. Invertebr. Pathol. 46*, 98–102.
Takizawa, H., Vivier, E., and Petitprez, A. 1973. Développement intranucléaire de la microsporidie *Nosema bombycis* dans les cellules de Vers à soie après infestation expérimentale. *C. R. Acad. Sci. Paris Ser. D 277*, 1769–1772.
Tanabe, A.M. 1971. The pathology of two microsporida in the armyworm, *Pseudaletia unipuncta* (Haworth) (Lepidoptera, Noctuidae). Ph.D. Thesis, University of California, Berkeley.
Tanabe, A. M., and Tamashiro, M. 1967. The biology and pathogenicity of a microsporidian (*Nosema trichoplusiae* sp. n.) of the cabbage looper, *Trichoplusia ni* (Hubner) (Lepidoptera: Noctuidae). *J. Invertebr. Pathol. 9*, 188–195.
Thompson, A. C., and McLaughlin, R. E. 1977. Comparison of the lipids and fatty acids of *Mattesia grandis* and the fat body of the host, *Anthonomus grandis*. *J. Invertebr. Pathol. 30*, 108–109.
Toguebaye, B. S., and Marchand, B. 1984. Etude histopathologique et cytopathologique d'une microsporidiose naturelle chez la Coccinelle des Cucurbitacées d'Afrique, *Henosepilachna elaterii* [*Col.*: *Coccinellidae*]. *Entomophaga 29*, 421–429.
Tsang, K. R., Brooks, M. A., and Kurtti, T. J. 1982. Culture conditions regulating the infection of cells by an intracellular microorganism. *In* "Invertebrate Cell Culture Appli-

cations." (K. Maramorosch and J. Mitsuhashi, eds.), pp. 125–157. Academic Press, New York.

Tuzet, O., Vago, C., Ormières, R., and Robert, P. 1965. *Adelina melolonthae* n. sp., coccidie parasite des larves de *Melolontha melolontha*. *Arch. Zool. Exp. Gen. 106*, 513–521.

Undeen, A. H. 1975. Growth of *Nosema algerae* in pig kidney cell cultures. *J. Protozool. 22*, 107–110.

Undeen, A. H. 1978. Spore-hatching processes in some *Nosema* species with particular reference to *N. algerae* Vávra and Undeen. *Misc. Publ. Entomol. Soc. Am. 11*, 29–49.

Undeen, A. H. 1983. The germination of *Vavraia culicis* spores. *J. Protozool. 30*, 274–277.

Undeen, A. H. 1990. A proposed mechanism for the germination of microsporidian (Protozoa: Microspora) spores. *J. Theor. Biol. 142*, 223–235.

Undeen, A. H., and Alger, N. E. 1976. *Nosema algerae*: Infection of the white mouse by a mosquito parasite. *Exp. Parasitol. 40*, 86–88.

Undeen, A. H., and Avery, S. W. 1984. Germination of experimentally nontransmissible microsporidia. *J. Invertebr. Pathol. 43*, 299–301.

Undeen, A. H., and Maddox, J. V. 1973. The infection of nonmosquito hosts by injection with spores of the microsporidan *Nosema algerae*. *J. Invertebr. Pathol. 22*, 258–265.

Undeen, A. H., Vávra, J., and Rothfels, K. H. 1984. The sex of larval simuliids infected with microsporidia. *J. Invertebr. Pathol. 43*, 126–127.

Vávra, J. 1963. Spore projections in *Microsporidia*. *Acta Protozool. 1*, 153–155.

Vávra, J. 1968. Ultrastructural features of *Caudospora simulii* Weiser (Protozoa, Microsporidia). *Folia Parasitol. (Praha) 15*, 1–9.

Vávra, J. 1969. *Lankesteria barretti* n. sp. (Eugregarinida, Diplocystidae), a parasite of the mosquito *Aedes triseriatus* (Say) and a review of the genus *Lankesteria* Mingazzini. *J. Protozool. 16*, 546–570.

Vávra, J. 1976a. Structure of the microsporidia. In "Comparative Pathobiology. Biology of the Microsporidia." (L. A. Bulla, Jr., and T. C. Cheng, eds.), Vol. 1, pp. 1–85. Plenum Press, New York.

Vávra, J. 1976b. Development of the microsporidia. In "Comparative Pathobiology. Biology of the Microsporidia." (L. A. Bulla, Jr. and T. C. Cheng, eds.), Vol. 1, pp. 87–109. Plenum Press, New York.

Vávra, J., and Barker, R. J. 1980. The microsporidian mucocalyx as seen in the scanning electron microscope. *Fol. Parasitol. (Praha) 27*, 19–21.

Vávra, J., and Undeen, A. H. 1970. *Nosema algerae* n. sp. (Cnidospora, Microsporida) a pathogen in a laboratory colony of *Anopheles stephensi* Liston (Diptera, Culicidae). *J. Protozool. 17*, 240–249.

Vávra, J., Canning, E. U., Barker, R. J., and Desportes, I. 1981. Characters of microsporidian genera. *Parasitology 82*, Pt. 4, 131–142.

Vivares, C. P., and Sprague, V. 1979. The fine structure of *Ameson pulvis* (Microspora, Microsporida) and its implications regarding classification and chromosome cycle. *J. Invertebr. Pathol. 33*, 40–52.

Vivier, E. 1968. L'Organisation ultrastructurale corticale de la gregarine *Lecudina pellucida*; ses rapports avec l'alimentation et la locomotion. *J. Protozool. 15*, 230–246.

Vivier, E., and Desportes, I. 1990. Phylum Apicomplexa. In "Handbook of Protoctista." (L. Margulis, J. O. Corliss, M. Melkonian, and D. J. Chapman, eds.), pp. 549–573. Jones and Bartlett, Boston.

Vossbrinck, C. R., Maddox, J. V., Friedman, S., Debrunner-Vossbrinck, B. A., and Woese, C. R. 1987. Ribosomal RNA sequence suggests microsporidia are extremely ancient eukaryotes. *Nature 326*, 411–414.

Walker, M. H., and Hinsch, G. W. 1972. Ultrastructural observations of a microsporidian protozoan parasite in *Libinia dubia* (Decapoda). I. Early spore development. *Z. Parasitenkd. 39*, 17–26.

Walsh, R. D., and Olson, J. K. 1976. Observations on susceptibility of certain culicine mosquito species to infection by *Lankesteria culicis* (Ross). *Mosq. News 36*, 154–160.

Wang, Der-I., and Moeller, F. E. 1970. The division of labor and queen attendance behavior of nosema-infected worker honey bees. *J. Econ. Entomol. 63*, 1539–1541.

Wang, Der-I., and Moeller, F. E. 1971. Ultrastructural changes in the hypopharyngeal glands of worker honey bees infected by *Nosema apis*. *J. Invertebr. Pathol. 17*, 308–320.

Ward, R. A., Levine, N. D., and Craig, G. B., Jr. 1982. *Ascogregarina* nom. nov. for *Ascocystis* Grassé, 1953 (Apicomplexa, Eugregarinorida). *J. Parasitol. 68*, 331.

Watson, M. E. 1916. Studies on gregarines including descriptions of twenty-one new species and a synopsis of the eugregarine records from the Myriapoda, Coleoptera and Orthoptera of the world. *Ill. Biol. Monogr. 2*, No. 3, 1–258.

Weidner, E. 1970. Ultrastructural study of microsporidian development. I. *Nosema* sp. Sprague, 1965 in *Callinectes sapidus* Rathbun. *Z. Zellforsch. 105*, 33–54.

Weidner, E. 1972. Ultrastructural study of microsporidian invasion into cells. *Z. Parasitenkd. 40*, 227–242.

Weidner, E. 1976. The microsporidian spore invasion tube. The ultrastructure, isolation, and characterization of the protein comprising the tube. *J. Cell. Biol. 71*, 23–34.

Weidner, E. 1982. The microsporidian spore invasion tube. III. Tube extrusion and assembly. *J. Cell. Biol. 93*, 976–979.

Weidner, E., Byrd, W., Scarborough, A., Pleshinger, J., and Sibley, D. 1984. Microsporidian spore discharge and the transfer of polaroplast organelle membrane into plasma membrane. *J. Protozool. 31*, 195–198.

Weiser, J. 1952. Cizopasníci housenek zavíječe slunečnicového Homeosoma nebulellum Hbn. se zvláštním zřetelem na druh Mattesia povolnyi sp. n. *Zool. Entomol. Listy. 15*, 252–264.

Weiser, J. 1953. Schizogregariny z hmyzu škodícího zásobám mouky. *Acta Soc. Zool. Bohemoslov. Vest. Cesk. Spol. Zool. 17*, 199–211.

Weiser, J. 1954. Zur systematischen Stellung der Schizogregarinen der Mehlmotte, *Ephestia kühniella* Z. *Arch. Protistenkd. 100*, 127–142.

Weiser, J. 1955. A new classification of the Schizogregarina. *J. Protozool. 2*, 6–12.

Weiser, J. 1961. Die Mikrosporidien als Parasiten der Insekten. *Monogr. Angew. Entomol. 17*, 1–149.

Weiser, J. 1963. Sporozoan infections. *In* "Insect Pathology: An Advanced Treatise." E. A. Steinhaus, ed.), Vol. 2, pp. 291–334. Academic Press, New York.

Weiser, J. 1976. The *Pleistophora debaisieuxi* xenoma. *Z. Parasitenkd. 48*, 263–270.

Weiser, J. 1977. Contribution to the classification of Microsporidia. *Vest. Cesk. Spol. Zool. 41*, 308–321.

Weiser, J. 1985. Phylum Microspora Sprague, 1969. *In* "An Illustrated Guide to the Protozoa." (J. J. Lee, S. H. Hutner, and E. C. Bovee, eds.), pp. 375–383. Society of Protozoologists. Allen Press, Lawrence, Kansas.

Weiser, J., and Beard, R. L. 1959. *Adelina sericesthis* n.sp., a new coccidian parasite of scarabaeid larvae. *J. Insect Pathol. 1*, 99–106.

Weiser, J., and Coluzzi, M. 1972. The microsporidian Plistophora culicis Weiser, 1946 in different mosquito hosts. *Folia Parasitol. 19*, 197–202.

Weiser, J., and Wille, H. 1960. Über eine Gregarine aus der Leibshöhle der Engerlinge von *Hoplia* sp. aus der Schweiz. *Cesk. Parasitol. 7*, 351–354.

Weiser, J., and Žižka, Z. 1975. Stages in sporogony of *Plistophora debaisieuxi* Jírovec (*Microsporidia*). *Acta Protozool. 14*, 185–194.

Weissenberg, R. 1976. Microsporidian interactions with host cells. *In* "Comparative Pathobiology." (L. A. Bulla, Jr. and T. C. Cheng, eds.), Vol. 1, pp. 203–237. Plenum Press, New York.

Woolever, P. 1966. Life history and electron microscopy of a haplosporidian, *Nephridio-*

phaga blattellae (Crawley) n. comb., in the Malpighian tubules of the German cockroach, *Blattella germanica* (L.). *J. Protozool. 13*, 622–642.

Yarwood, E. A. 1937. The life cycle of *Adelina cryptocerci* sp. nov., a coccidian parasite of the roach *Cryptocercus punctulatus*. *Parasitology 29*, 370–390.

Youssef, N. N., and Hammond, D. M. 1971. The fine structure of the developmental stages of the microsporidian *Nosema apis* Zander. *Tissue Cell. 3*, 283–294.

Žižka, Z. 1972. An electron microscope study of autoinfection in neogregarines (Sporozoa, Neogregarinida). *J. Protozool. 19*, 275–280.

Žižka, Z. 1977. Fine structure of the neogregarine *Farinocystis tribolii* Weiser, 1953. Developmental stages in sporogony and parasite-host relations. *Z. Parasitenkd. 54*, 217–228.

Žižka, Z. 1978a. Fine structure of the neogregarine *Farinocystis tribolii* Weiser, 1953. Developmental stages in merogony. *J. Protozool. 25*, 50–56.

Žižka, Z. 1978b. Fine structure of the neogregarine *Farinocystis tribolii* Weiser, 1953. Free gametocytes. *Acta Protozool. 17*, 255–259.

Žižka, Z. 1978c. Fine structure of the neogregarine *Farinocystis tribolii* Weiser, 1953. Syzygy and gametes formation. *Protistologica 14*, 209–215.

CHAPTER 13

NEMATODES, NEMATOMORPHS, AND PLATYHELMINTHES

I. HISTORICAL REVIEW
II. TAXONOMY
III. GENERAL LIFE CYCLE OF NEMATODES
IV. ASSOCIATIONS WITH INSECT HOSTS
V. TYPES OF INSECT–NEMATODE ASSOCIATIONS
VI. HOST SPECIFICITY
VII. MODE OF INFECTION
VIII. HOST RESISTANCE
IX. PATHOLOGY
    A. External Effects
    B. Internal Effects
    C. Behavioral Effects
X. BIOLOGY AND LIFE CYCLE OF SELECTED NEMATODES
    A. Tetradonematidae
    B. Mermithidae
    C. Rhabditidae
    D. Steinernematidae
    E. Heterorhabditidae
    F. Diplogasteridae
    G. Allantonematidae
    H. Sphaerulariidae
    I. Aphelenchoididae
    J. Entaphelenchidae
    K. Oxyuridae
XI. NEMATOMORPHA
XII. PLATYHELMINTHES

Nematodes, commonly referred to as roundworms, eelworms, or threadworms, are translucent, usually elongate, and more or less cylindrical throughout their body length. The body is covered by a noncellular elastic cuticle that differs chemically from the chitinous cuticle of arthropods. Although transverse striae or external annulations often occur on the cuticle, these are superficial and not true segmentations. Nematodes have excretory, nervous, digestive, reproductive, and muscular systems but lack circulatory and respiratory systems. The alimentary canal consists of a mouth situated terminally, followed by the stoma or buccal cavity, an esophagus, intestine, and rectum with the anus opening ventrally.

Sexes are usually separate. The male's reproductive system opens ventrally into the rectum forming a cloaca. The adult male is distinguished by the presence of one or two testes and by spicules in association with the cloaca. The female's reproductive system involves one or two ovaries with the vulva located ventrally, usually near the middle of the body. Inasmuch as the taxonomy of the nematodes is largely based on the sexual structures, the immature nematodes are difficult to sex because there are no apparent genital structures. For more detailed information on the morphology and embryology of nematodes, the reader should refer to Thorne (1961), Goodey (1963), Chitwood and Chitwood (1974), Maggenti (1981), and Poinar (1983).

Nematodes are diverse organisms and, except for arthropods, have habitats more varied than any other group of animals. They are found from the arid desert to the frozen tundra, from fresh to salt water, and from hot springs to thawed-out Arctic ice. Nematodes occur as free-living organisms and as facultative or obligate parasites (pathogens) of plants and animals. Inasmuch as insects are the largest group of animals, it is not surprising that nematodes are widely associated with them. The insect–nematode associations range from accidental to obligatory and from commensal to parasitic.

Insects serve as vectors or intermediate hosts for a number of nematode parasites of vertebrates (Levine 1968; Schmidt and Roberts 1985). For example, mosquitoes transmit *Dirofilaria immitis*, causal agent of the dog heartworm, and *Wuchereria bancrofti*, causal agent of filariasis in humans, and coprophagus beetles serve as intermediate hosts for *Spirocerca lupi*, a tumor-causing nematode in dogs. Close association also exists with insects and plant nematodes. *Rhadinaphelenchus cocophilus*, causative agent of red-ring disease of coconut palm, is transmitted in part by palm weevils (Fenwick 1969), and *Bursaphelenchus xylophilus*, causative agent of wilting disease of pine trees, is transmitted by cerambycid beetles (Mamiya and Enda 1972; Mamiya 1984). In some instances, the insect vector is adversely affected by the presence of the nematode that induces high insect mortality. We shall not discuss this aspect of insects serving as intermediate hosts or vectors.

I. HISTORICAL REVIEW

One of the earliest reports of an insect-parasitic nematode was made by Reamur in 1742 when he described a nematode that was later named *Sphaerularia bombi* (Poinar 1975a; Nickle and Welch 1984). Shortly thereafter, in 1747, Gould described the detrimental effects of mermithids on ants (Nickle 1974). In 1826, Kirby, who wrote the first comprehensive work on insect diseases (see Chapter 1), ended his chapter with an interesting account of the infection of insects with worms. Since these findings, published reports on insect–nematode associations became more frequent, and several workers have published lists of insects harboring nematodes. The most recent compilation is by Poinar (1975a) who has listed nematode parasites and associates of insects. In addition, Shephard (1974) has prepared an extensive literature on arthropods as final hosts for nematodes and nematomorphs from 1900 to 1972, and Gaugler and Kaya (1990) have edited a book on steinernematid and heterorhabditid nematodes.

In 1929, R. W. Glaser found the nematode *Steinernema* (syn. *Neoaplectana*) *glaseri* infecting the Japanese beetle, *Popillia japonica*, and was the first to culture this parasitic nematode on artificial media and use it in field tests against the beetle (Glaser 1932; Glaser and Farrell 1935). Between 1932 and 1942, significant reduction in beetle populations was obtained in several field tests. Later, Dutky and Hough (1955) found another steinernematid known as the DD-136 strain of *Steinernema carpocapsae* and tested it on the codling moth. Others also applied this nematode against a number of insect pests in laboratory and field trials with encouraging results (Poinar 1979; Kaya 1985; Klein 1990). The use of *S. glaseri* and *S.*

carpocapsae in biological control was accelerated because of the imagination of their discoverers and the ease in producing great numbers of nematodes on an artificial medium or a suitable insect host. In addition, *Heterorhabditis* spp., similar in action to steinernematids, have been isolated and described (Poinar 1975b; Khan et al. 1976). These nematodes are currently being applied against agricultural and turf pests (Kaya 1985; Gaugler 1987; Klein 1990).

In 1976, *Romanomermis culicivorax* (syn. *Reesimermis nielseni*) became commercially available as a biological control agent against mosquitoes. Unfortunately, this commercial venture failed in part because of the difficulty in the production, storage, and transport of the nematode and the more effective control with *Bacillus thuringiensis* subspecies *israelensis*. This mermithid is still used on a small scale for mosquito larval control in many parts of the world.

II. TAXONOMY

The classification of nematodes is continually undergoing revision, including several groups of insect-parasitic (entomogenous) nematodes. Maggenti (1981, 1991) has summarized the higher classification of nematodes. The major groups of entomogenous nematodes are given in Table 13–1. There are two classes of nematodes, Adenophorea and Secernentea, both of which contain important insect-parasitic nematodes. Major groups of entomogenous nematodes are in the orders Stichosomida, Rhabditida, Diplogasterida, and Tylenchida. Major reviews on taxonomy and general descriptions of entomogenous nematodes that may be of interest to the readers are by Gaugler (1987), Gaugler and Kaya (1990), Kaya (1987), Nickle (1984), and Poinar (1975a, 1979). In addition, Massey (1974) has published a handbook of nematodes associated with bark beetles. Specific treatises on insect-parasitic Tylenchida have been presented by Wachek (1955), Rühm (1956), Nickle (1967a), and Remillet and Laumond (1991) and on Mermithidae by Petersen (1985) and Kaiser (1991).

Identification of entomogenous nematodes to species is difficult or not possible with immature stages; the adult stages are often required for proper identification. The specimens, in properly preserved condition, should be sent to nematode taxonomists. Special methods are required for killing and fixing insect nematodes. For most nematodes, hot (80°C) Ringer's solution or 1% sodium chloride solution is excellent for killing them. They are preserved in either TAF [7 ml formaldehyde (38%), 2 ml triethanolamine, and 91 ml distilled water] or an aqueous solution of 3% formalin [3 ml formaldehyde (38%) and 97 ml distilled water]. Ethyl alcohol, a general insect preservative, causes distortion of the internal anatomy of nematodes and is not recommended.

III. GENERAL LIFE CYCLE OF NEMATODES

Most nematodes have simple life cycles and undergo three main stages of development: egg, juvenile (immature stages of nematodes are called juveniles to avoid confusion with larval stages of insects), and adult. In a simple life cycle, the mated female deposits her eggs in the environment and the juvenile usually undergoes one molt in the egg and emerges as a second-stage juvenile. The majority of nematode species molt four times before becoming adults. These molts may occur in the egg, free in the environment, or in the insect host. Some insect nematodes have a

TABLE 13–1 Classification of the Major Insect-Parasitic Nematode Groups[a]

Classification of insect-parasitic nematodes

Phylum: Nemata (syn. Nematoda)
  Class: Adenophorea (syn. Aphasmida)
    Order: Stichosomida (Mermithida in part)
      Family: Tetradonematidae
      Family: Mermithidae

  Class: Secernentea (syn. Phasmidia)
    Order: Rhabditida
      Family: Rhabditidae
      Family: Steinernematidae
      Family: Heterorhabditidae
      Family: Oxyuridae

    Order: Diplogasterida
      Family: Diplogasteridae

    Order: Tylenchida
      Family: Allantonematidae[b]
      Family: Sphaerulariidae

    Order: Aphelenchida
      Family: Aphelenchoididae
      Family: Entaphelenchidae

[a]Modified after Maggenti (1981, 1991).
[b]Includes species from the families Neotylenchidae (Fortuner and Raski 1987) and Iotonchiidae (Remillet and Laumond 1991). Because of instability among various families within the superfamily Sphaerularioidea, a conservative approach is used and species in these families are placed with the Allantonematidae.

resistant stage called the "dauer juvenile" or "dauer." The dauer juvenile is the third-stage nematode, which is usually ensheathed in the second-stage cuticle (Fig. 13–1) and occurs commonly in the rhabditids. The word *dauer*, first used by Fuchs (1915), means durability or permanence in German and is not synonymous with the term infective juvenile. Many noninfectious (free-living) nematodes produce dauer juveniles. Immature nematodes are like the adults in appearance and structure, and therefore their development is analogous to ametabolous insects. Most nematodes are amphigonus (male and female are separate individuals) and mating is required to produce offspring.

Some entomogenous nematodes have complex life cycles, which include an alternation of gametogenetic and parthenogenetic generations. These complex life cycles will be discussed later with specific examples.

IV. ASSOCIATIONS WITH INSECT HOSTS

Van Zwaluwenburg (1928) listed 16 orders of insects in which 749 species had associations with nematodes, gordian worms (nematomorphs), and spiny-headed

FIGURE 13-1

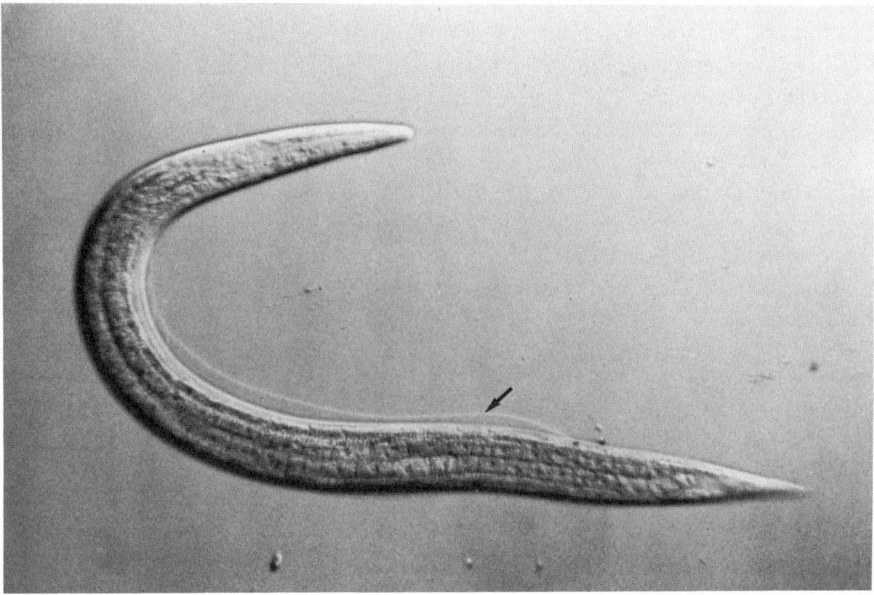

Dauer juvenile of *Steinernema carpocapsae* ensheathed in the second-stage cuticle (*arrow*). Juvenile measures 600 × 25 μm. (Photographed by P. Timper.)

worms (acanthocephalans). Of these, about 420 were nematode associations. Since then, our knowledge of nematode–insect associations has increased considerably. Poinar (1975a) listed 19 insect orders with 3142 insect–nematode associations, which included phoretic, facultative, and obligate parasitic relationships as well as those that utilized insects as intermediate hosts. The number of insect–nematode associations will undoubtedly increase with more intensive and extensive studies. For example, Giblin et al. (1981), Giblin and Kaya (1984), and Giblin-Davis et al. (1990) isolated aphelenchoidids and diplogasterids from the reproductive tracts of soil-nesting bees and reported their phoretic association with these bees.

V. TYPES OF INSECT–NEMATODE ASSOCIATIONS

Relationships between nematodes and insects vary from fortuitous association to obligatory parasitism. Entomogenous nematodes have been classified into various groups. Van Zwaluwenburg (1928) grouped nematodes into five classifications: primary parasitism, secondary parasitism, mechanical association (internal), mechanical association (external), and commensalism. Filipjev and Schuurmans-Stekhoven (1941) placed insect nematodes into those living within the alimentary tracts of insects and those living in the body cavity of insects, whereas Welch (1963) classified nematodes as to whether they were external or internal parasites of insects. Steinhaus (1949) divided nematodes associated with invertebrates into three groups: nematode parasites of the insect gut, nematodes semiparasitic in insects, and nematode parasites of the body cavity and tissues of insects. Poinar's (1975a) classification, which will be used in this text, placed insect–nematode associations into three groups: phoretic relationship, facultative parasitism, and obligate parasitism.

Phoretic relationships, a form of commensalism in which one organism associates with another species in order to obtain transportation, are common between insects and nematodes. These occur in many rhabditids, which are carried from one habitat to another by insects frequenting decomposing matter. Many bark beetles serve as transport hosts for rhabditids, diplogasterids, and aphelenchoidids, which feed on microorganisms occurring in the beetle galleries (Massey 1974). Usually the dauer juvenile is transported from habitat to habitat. The nematodes are carried on various parts of the insect's body, such as the wings or elytra, intersegmental folds, genitalia, or gut. These nematodes have little or no pathological effect on the host and benefit by being carried to a new environment. Care must be taken when examining an insect to establish the true nature of the relationship because the phoretic nematode will often invade and feed on a dead transport host and may be mistaken as a facultative or obligate parasite.

Some nematodes are able to infect healthy insects as well as having the ability to complete their life cycle as free-living organisms. These nematodes are facultative parasites of insects. Some rhabditids, diplogasterids, and aphelenchoidids are facultative parasites (Poinar 1975a).

Obligate parasites cannot complete their life cycle in nature without a living insect. Except for the oxyurids and a few other groups that are found in the insect's gut, the obligate nematodes, such as the mermithids, tetradonematids, allantonematids, sphaerulariids, and entaphelenchids, usually occur in the hemocoel. Most of these nematodes castrate, debilitate, or kill their hosts. Although they can be cultured on artificial media (Bedding 1981, 1984b; Friedman 1990), nematodes in the families Steinernematidae and Heterorhabditidae are classified as obligate parasites of insects.

VI. HOST SPECIFICITY

Host specificity of entomogenous nematodes ranges from those associated with a single insect species to those with a wide host range. Specificity of nematodes can be categorized into three groups: monoxenous (highly specific or species specific), oligoxenous (moderately specific or limited host range), and polyxenous (little specificity or wide host range). Little attention has been given to specificity of entomogenous nematodes (Stoffolano 1973), but some generalizations can be drawn.

Nematodes that are obligate parasites of insects tend to be monoxenous or oligoxenous. The mermithid *Perutilimermis* (syn. *Agamomermis*) *culicis* appears to be specific to the mosquito *Aedes sollicitans* (Petersen et al. 1967), and the tetradonematid *Tetradonema plicans* has only been recorded from sciarid flies (Hudson 1974a). Examples of oligoxenous species include the mermithid *Mermis nigrescens*, which infects several species of grasshoppers (Christie 1937), and the allantonematid *Heterotylenchus* (syn. *Paraiotonchium*) *autumnalis*, which appears to infect only a few species of muscid flies (Stoffolano 1973). The mermithid *Romanomermis culicivorax* infects 17 species of mosquitoes in nature and 40 species under laboratory conditions (Petersen and Chapman 1979). The facultative parasites *Deladenus* (syn. *Beddingia*) spp. are oligoxenic and infect siricid wood wasps, a beetle associated with the wood wasps, and several species of *Rhyssa*, which are hymenopteran parasitoids of the wood wasps (Bedding 1984a). In contrast, some steinernematids and heterorhabditids are polyxenic. For example, *Steinernema carpocapsae* infects more than 250 species of insects from several different orders under laboratory conditions (Poinar 1979), and studies with *Heterorhabditis bacteriophora* (syn. *heliothidis*) indicate that it has a broad host range (Khan et al. 1976; Poinar 1979). (This wide host range is unnatural. In nature, the nematodes are restricted to soil insects.) Others, such as *S. kushidai*, tend to have a restricted host range (Mamiya 1989).

VII. MODE OF INFECTION

Insect-parasitic nematodes parasitize their hosts by directly penetrating through the cuticle into the hemocoel or by entering through natural openings (spiracles, mouth, and anus). Some insect-parasitic nematodes possess a spear or stylet that is used to pierce the cuticle. Species in the order Tylenchida possess a stylet, and those in the order Mermithida possess a spear, which is technically called an odontostyle. Rhabditids, diplogasterids, and oxyurids do not have spears or stylets. Some rhabditids can penetrate into the hemocoel, and *Heterorhabditis* has an anterior tooth that is used to scrape and rupture the host's cuticle (Bedding and Molyneux 1982). Accordingly, the absence of a stylet does not preclude penetration of the nematode into the hemocoel.

Nematodes infect their insect hosts passively or actively. Passive infection occurs when a mermithid deposits its eggs on the host's food. The eggs are ingested by an insect, and the nematodes hatch, bore through the midgut, and enter the hemocoel. Christie (1937) has described hatching and penetration by *Mermis nigrescens* through the midgut of a grasshopper. About 2 h after ingestion, the egg hatches at the posterior end of the midgut near the region where the Malpighian tubules are attached. The infective juvenile uses its spear to penetrate through the midgut into the hemocoel within 20 to 30 min.

Active infection occurs when the nematodes seek their hosts and penetrate directly through the integument into the hemocoel. Poinar and Doncaster (1965) observed the penetration process of the sphaerulariid *Tripius sciarae* into the dipteran host *Bradysia paupera*. The infective adult female, ensheathed in the fourth-stage cuticle, produces an adhesive mass about its head. This secretion digests the anterior portion of the ensheathed cuticle and adheres the nematode to the host. The attached nematode uses its stylet and possibly some enzymes to penetrate into the host. The penetration process may take from 10 min to 2 h, and the wound is sealed by the adhesive substance after the nematode has entered the insect. In *Deladenus siricidicola*, no secretions are produced during the penetration process (Bedding 1984a). The infective adult nematode adopts a position for entry with its head at a right angle to the larval wood wasp and then begins to thrust its stylet in and out at the rate of 60 to 100 thrusts per min. Initial penetration of the integument takes from 30 sec to 6 min, and the nematode completely enters its host after 40 to 60 min.

Host finding by infective juveniles of steinernematid and heterorhabditid nematodes can be an active process in response to physical and chemical cues. For example, *Steinernema carpocapsae* forms aggregations in response to chemical and bacterial gradients (Pye and Burman 1981), host fecal components (Schmidt and All 1978, 1979), plant roots (Bird and Bird 1986), and carbon dioxide (Gaugler et al.

1980). Host finding can be enhanced more than 20-fold after 13 rounds of genetic selection (Gaugler et al. 1989a) without a loss of overall fitness (Gaugler et al. 1990). Host-finding ability is positively correlated with carbon dioxide production by the host (Gaugler et al. 1991). In addition, infective juveniles can actively search for their hosts (Choo et al. 1989) or wait for a suitable host to enter their habitat (Gaugler et al. 1989b; Kaya 1990). Once a suitable host is found, *S. carpocapsae* invades by entering the mouth or the anus and then mechanically penetrating through the midgut into the hemocoel (Poinar and Himsworth 1967). It also infects its host by entering through the spiracles in response to carbon dioxide (Gaugler et al. 1991) and penetrating into the hemocoel (Triggiani and Poinar 1976).

VIII. HOST RESISTANCE

An insect resists nematode infection through behavioral, physical, or physiological means. Behavioral resistance occurs when the insect actively avoids or repels the nematode. Petersen (1975a) reported that extremely active mosquito species had a lower prevalence of infection by the mermithid *Romanomermis culicivorax* than less active ones. Scarab larvae may avoid infection by wiping nematodes away from the mouth (Akhurst 1986b). Physical resistance occurs when the nematode cannot penetrate the integument or the cocoon of a host insect. *R. culicivorax* has difficulty in penetrating the integument of older mosquito larvae (Petersen and Willis 1970). Dauer juveniles of *Steinernema carpocapsae* cannot penetrate the silken cocoons of hymenopteran parasitoids (Kaya and Hotchkin 1981), but if a hole is made in the cocoon, infection occurs (Kaya 1978a,b). Spiracular openings are portals of entry for nematodes (Triggiani and Poinar 1976), but sieve plates over the spiracles, especially with scarab larvae, may deny nematodes access through this entry point (Akhurst 1986b). Finally, younger instars of black fly larvae are resistant to infection by *Steinernema carpocapsae* because the comparatively large nematode is excluded from the insect's mouth (Gaugler and Molloy 1981).

Physiological resistance to infection involves the destruction of the nematode by digestive enzymes in the insect's alimentary tract and the melanization and encapsulation of the nematode within the hemocoel. Although there is no example of nematode destruction by digestive enzymes, this probably occurs when a nematode is ingested by a less than suitable host. There are numerous examples of melanization and encapsulation of nematodes within the insect's hemocoel (Poinar 1969; Nappi 1975; Stoffolano 1986; Nappi and Christensen 1987; Jackson and Brooks 1989). Encapsulation of filarial nematodes in mosquitoes that serve as intermediate hosts has been known since the early 1900s (Brug 1932), but the first report of humoral encapsulation of an entomogenous nematode was made by Welch (1960) who observed encapsulated mermithid juveniles in mosquito larvae. Subsequently, Welch and Bronskill (1962) reported that melanotic encapsulation of *Steinernema carpocapsae* occurred in larvae of several mosquito species. Complete encapsulation of this nematode usually occurs within 5 h after its penetration into the hemocoel of *Aedes aegypti* (Bronskill 1962). Although the nematode is encapsulated, the majority of the larvae die of septicemia caused by the bacterium *Xenorhabdus nematophilus*, which is mutualistically associated with this nematode. Surviving mosquito larvae usually die as pupae or shortly after adult emergence. In some cases, the encapsulated nematodes are partially or completely expelled from the host's body cavity at the time of molting.

IX. PATHOLOGY

Pathology in insect hosts caused by nematode infection may be manifested externally, internally, or behaviorally. External pathological effects are expressed by morphological changes, whereas internal effects involve alterations in morphology and physiology. Insects infected with nematodes often show aberrant behavior. In some instances, such as a mermithid or a steinernematid infection, the host insect is killed; in others, such as an allantonematid or a sphaerulariid infection, the host insect becomes sterile or has reduced fecundity.

A. EXTERNAL EFFECTS

Insects generally show little external evidence of a nematode infection. Even when external anomalies are observed, a dissection of the host or the emergence of the nematode is necessary to confirm a nematode infection. The number of nematodes infecting an insect or the age of the host at the time of infection often determines whether external anomalies occur. In mermithids, the infections usually manifest greater external morphological changes than in the other nematode groups. The alfalfa weevil, *Hypera postica*, infected with the mermithid *Hexamermis arvalis* may have malformed elytra (Poinar and Gyrisco 1962). The infected weevil larvae are yellow-green compared with the dark green of uninfected larvae. In black fly larvae, the mermithid infection may cause distorted abdomens (Phelps and De-Foliart 1964). When fourth instar nymphs of the desert locust, *Schistocerca gregaria*, are infected with 40 individuals of *Mermis nigrescens*, the adults developing from the infected nymphs have deformed wings (Craig and Webster 1974).

Perhaps the most interesting manifestations caused by mermithids are on primary and secondary sexual characteristics in social insects, resulting in intercastes (Wheeler 1928) or intersex in chironomids (Rempel 1940; Wülker 1961, 1964). Intersex is distinguished from intercaste; the former shows characteristics of both sexes in an individual, whereas the latter shows characteristics of various castes that are usually of the same sex.

Wheeler (1910, 1928) has described intercastes caused by an undescribed mermithid in a number of ant species. The mermithized ants differ only indistinctly from normal individuals in having a slightly distended gaster or a slight change in color. On the other hand, there may be marked changes in the external morphology so that the mermithized ant is not identical to any normal caste but shows female, worker or soldier characteristics in varying degrees. Several types of intercastes occur, and these have been classified as mermithigates, mermithogynes, and mermithostratioles. Memithigates occur in the *Pheidole* ants and have female, worker and soldier characters. Vandel (1927) restricts this term to intercastes among the worker ants and uses mermithostratioles to characterize modified soldiers. Mermithogynes are found in *Lasius* ants, where infected females resemble normal females but have smaller heads, shorter wings, and partly distended gaster. Although these anomalies occur, they are not common because in 30 years of collecting *Pheidole* ants, Wheeler (1928) found only a few mermithized colonies.

Rempel (1940) was the first to characterize the morphological intersex alterations in chironomids parasitized by mermithids. Parasitized female chironomids often have male genital appendages, but parasitized males usually retain the male genital appendages with female modifications occurring in the tarsi, antennae, and wings.

Another example of external effects is wounds caused by the penetration of the nematode through the cuticle. A wound is produced when *Deladenus siricidicola* penetrates the integument of a siricid wood wasp larva (Bedding 1984a). The wound bleeds for a few seconds and then the hemolymph melanizes. A scar, from 0.05 to 0.15 mm in diameter and light orange-brown to dark brown, is formed.

B. INTERNAL EFFECTS

Nematodes in the hemocoels of insects cause many physiological changes, and these changes have been recorded from a number of different insect groups. The presence of mermithids results in sterility and ultimately death of the adult insects. Adult grasshoppers infected with *Agamermis decaudata* or *Mermis nigrescens* have markedly reduced ovaries and are usually sterile (Christie 1937). The physiological basis for parasitic castration has been examined by Gordon et al. (1973b) who showed that vitellogenesis proceeds normally up through the beginning of yolk deposition in female desert locust, *Schistocerca gregaria*, infected with *M. nigrescens*. After this initial period, vitellogenesis proceeds rapidly in uninfected female locusts but is greatly depressed in infected individuals. Consequently, *M. nigrescens* sterilized the locusts by preventing yolk deposition. Moreover, certain vitellogenic proteins are significantly reduced in the hemolymph of infected locusts. Although the fat body is not degenerated, the mermithid apparently stimulates catabolism of the fat-body protein and then utilizes the amino acids for its growth. Accordingly, the total fat-body protein level becomes depleted by *M. nigrescens* infection (Gordon and Webster 1971).

Molting of the desert locust can be inhibited by *M. nigrescens* infection (Craig and Webster 1974). Third instar locusts containing 20 or more mermithids and fourth instars containing 30 or more mermithids cannot molt to the next instar. Inasmuch as the levels of ecdysone in infected locusts are not significantly different from uninfected ones, *M. nigrescens* probably inhibits molting by decreasing the rate of fat-body protein synthesis, and thereby depleting the proteins required in the molting process. When 20 or more *M. nigrescens* occur in the hemocoel of adult migratory locust, *Locusta migratoria*, there is an increase in blood uric acid together with an overall decrease in fecal uric acid levels (Condon and Gordon 1977a). These results demonstrate that the Malpighian tubules are not functioning properly.

Black fly larvae infected with mermithids have reduced fat-body contents (Phelps and DeFoliart 1964), but the presence of mermithids does not significantly affect head width or body length (Condon and Gordon 1977b). Black fly adults infected with mermithids (*Isomermis* sp.) show retarded development of the reproductive organs in both sexes (Anderson and Shemanchuk 1987). Larvae of sciarid flies infected with *Tetradonema* sp. show a reduction in fat bodies (Hungerford 1919). Similarly, larvae of the mosquito *Aedes aegypti* infected with *Romanomermis* become depleted of fat bodies, leg rudiments, and other preadult structures (Petersen et al. 1968). The depletion of the fat bodies and other preadult structures does not occur until 4 to 6 days postparasitization, which is just prior to nematode emergence from the mosquito larvae (Bailey and Gordon 1973). At this time, the mermithids have their most rapid growth and therefore have the greatest effect on their hosts. Alfalfa weevil infected with *Hexamermis arvalis* has depleted fat bodies, and the silk glands (modified Malpighian tubules) are greatly reduced and nonfunctional (Poinar and Gyrisco 1962).

The infective stage of the mermithid *Filipjevimermis* (syn. *Oesophagomermis*) *leipsandra* enters the hemocoel of the *Diabrotica* beetle and penetrates one of the ganglia of the central nervous system (Poinar 1968). Observations indicate that the mermithid forces its head through the neural lamella and perineurium and enters the ganglion coming to rest in the neuropile, but the cells and fibers in the ganglion do not seem to be adversely affected. The growth of the mermithid in the ganglion causes the neural lamella to stretch. About 12 days after infection, the neural lamella ruptures freeing the nematode into the hemocoel. The beetle is not killed by this rupture but dies when the nematode emerges from its host about 22 days after infection.

The allantonematids *Deladenus wilsoni* and *D. siricidicola* infect and sterilize siricid wood wasps (Bedding 1984a). The female wood wasps are sterilized when the juvenile nematodes invade the developing eggs, but the males remain fertile even though the juveniles penetrate the testes. The fertility of the male is not affected because most spermatozoa have passed into the vesiculae seminale early in pupation before the nematodes penetrate the testes. Another allantonematid, *Heterotylenchus autumnalis*, sterilizes the female face fly, *Musca autumnalis*, by invading the ovaries (Fig. 13–2A,B). Testes are not destroyed by the nematode although the accessory glands are lacking in heavily infected males (Nappi 1973). The mating process is unaffected, but males often cannot pass viable spermatozoa to the female.

Bark beetles infected by allantonematids may also be sterilized or show a reduced reproductive capacity. When bark beetles are heavily parasitized by *Sulphuretylenchus elongatus*, the oocytes are shrunken and the females are sterile (Ashraf and Berryman 1970a,b). In light or moderate infections, the beetles have a reduced reproductive capacity. The nematodes rarely invade the ovaries or testes. Other pathological effects include greatly reduced epithelial cells and muscle layers of the fore- and hindguts, and the midgut cells show cellular disintegration and necrosis. Hemolymph composition and oocyte development are affected when the Douglas-fir beetle, *Dendroctonus pseudotsugae*, is infected with the nematode *Contortylenchus reversus* (Thong and Webster 1975a). Trehalose levels are not affected by the presence of the nematodes, but hemolymph proteins are reduced. A consequence of protein depletion is a reduction in oocyte size. When the bark beetle, *Ips typographus*, is infected with *Parasitylenchus typographi*, reduction in fat bodies occurs (Rühm 1956).

Palm (1948) investigated the pathological effect of *Sphaerularia bombi* on several hymenopteran *Bombus* species. Infected queens have degenerate ovaries and are sterile. Ovarian development is also inhibited in queens with dead nematodes.

A rhabditid, *Diploscapter lycostoma*, causes damage to the pharyngeal glands of the Argentine ant, *Iridomyrmex humilis* (Markin and McCoy 1968). Other than reduced production of glandular secretions, the nematode apparently has no other effect on the worker ant.

C. BEHAVIORAL EFFECTS

Insects infected with nematodes often have abnormal behavior compared with uninfected individuals. However, abnormal behavior *per se* may be the result of other

FIGURE 13-2

(A) Ovaries of the face fly, *Musca autumnalis*. Ovaries on the right contain mature eggs of the face fly, and those on the left are infested with the juveniles of *Heterotylenchus autumnalis*. (B) Close-up of an ovary infested with the nematodes.

anomolies and should not be used as the sole criterion for a nematode infection. Abnormal behavior in insects is usually manifested late in nematode infection. Alfalfa weevils infected with *Hexamermis* have difficulty in crawling and orienting themselves (Poinar and Gyrisco 1962). Worker ants infected with mermithids show a high degree of negative phototropism compared with uninfected ants (Wheeler 1928). Infected ants do not feed immature ants and appear to be in a chronic state of hunger. Superinfection of mermithids affects the flight of the desert locust (Weis-Fogh 1956) and the activity of mosquito larvae (Welch 1960). Bark beetles infected with allantonematids have reduced gallery size (Nickle 1971; Thong and Webster 1975b) or constructed aberrant galleries (Ashraf and Berryman 1970a). In heavily infected bark beetles, their flights are also impaired.

Queen bees belonging to the genus *Bombus* show aberrant behavior when infected with the nematode *Sphaerularia bombi* (Poinar and van der Laan 1972). Infected queens fly near the ground, often alighting and crawling under fallen leaves and digging in the soil to deposit juvenile nematodes. Infected queens do not make nests, whereas uninfected individuals forage and initiate new colonies. Females of the face fly, *Musca autumnalis*, infected with *Heterotylenchus autumnalis* initially show similar behavior as uninfected flies by visiting the faces of cattle to obtain protein from the lachrymal secretions. However, once the ovaries of the face flies are invaded by the nematodes, the flies generally do not return to cattle faces but only visit fresh cattle dung (Kaya et al. 1979). Apparently, the ovaries of these infected flies are continually invaded by the nematodes causing distention, and such flies have a continual need to "nemaposit" nematodes in dung, whereas uninfected flies without mature eggs in their ovaries need to return to cattle faces to obtain more protein for egg production.

X. BIOLOGY AND LIFE CYCLE OF SELECTED NEMATODES

A. TETRADONEMATIDAE

Tetradonema plicans, an internal obligate parasitic nematode of sciarid flies, can be an important biological control agent in the greenhouse (Hudson 1974a). The life cycle of this nematode is not completely understood. Hungerford (1919) suggests that nematode eggs ingested by a sciarid maggot hatch in the midgut, and the infective juveniles penetrate through the midgut into the hemocoel. On the other hand, Hudson (1974b) believes that the eggs hatch in the soil, and the infective juveniles, which are ingested by the maggot, penetrate the midgut and invade the hemocoel. In either case, both authors concur that nematode penetration occurs through the midgut and not through the integument. Once in the hemocoel, the nematodes grow quickly and usually kill the maggot before pupation. When an older maggot is infected, it pupates and emerges as an adult fly that is usually sterile.

Normally, maggots are killed by a nematode infection between 5 and 18 days depending on the number of nematodes and the maggots' age at the time of infection (Hudson 1974b). Both male and female nematodes must be present in the same maggot for egg production. The female nematode exits from the maggot and deposits eggs in the soil (Ferris and Ferris 1966), or it deposits eggs in maggots or retains the fertilized eggs in her body. The eggs are liberated into the soil after the decay of the maggot or female nematode. Dispersal of nematodes occurs by wind or water or by infected adult flies dying in uninfested soil (Hungerford 1919).

B. MERMITHIDAE

Unlike the tetradonematids, most mermithids do not reach adulthood in the hemocoel of the host. They emerge from the host as postparasitic juveniles, molt to the adult stage, mate, and produce progeny. The parasitic stages are short lived in comparison with the free-living stages.

In most cases, an infective mermithid juvenile will seek out and penetrate through the host's integument. In a few cases, different modes of infection occur. For example, the egg of *Mermis nigrescens* is ingested and hatches in the midgut of the grasshopper host (Webster and Thong 1984; Kaiser 1991). The infective mermithid then penetrates into the hemocoel. *Pheromermis pachysoma* infects paratenic hosts, such as mosquitoes, caddisflies, and crane flies, and encysts as a second-stage nematode in various tissues (Poinar et al. 1976a). The final host is a vespid that preys upon the paratenic host.

1. *Romanomermis culicivorax*

One of the most studied mermithids is *Romanomermis culicivorax*, a nematode parasite of mosquitoes (Figs. 13–3 and 13–4). This nematode, an effective biological control agent of mosquitoes under certain situations, will parasitize a number of different mosquito species (Petersen and Chapman 1979). The life cycle of *R. culicivorax* is similar to most aquatic mermithids (Fig. 13–5). There is one molt within the egg, and the juvenile emerges in the second stage. This juvenile, which is the infective stage (also referred to as the preparasitic stage), seeks out a mosquito larva and penetrates directly through the integument into the hemocoel. The nematode does not increase appreciably in length for the first 3 days after infection, but between 4 to 5 days, it increases greatly in size (Gordon et al. 1974) and emerges from the host about 7 days after infection (Petersen and Willis 1972). The time of emergence of *R. culicivorax* from its mosquito host is temperature depen-

FIGURE 13–3

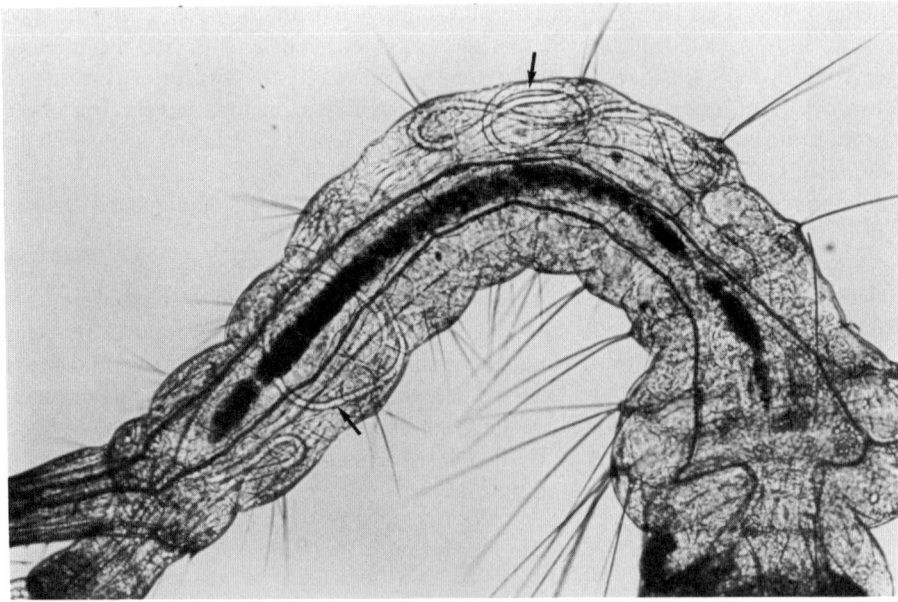

The mosquito *Culex tarsalis* infected by the nematode *Romanomermis culicivorax*. Arrows point to nematodes within the mosquito larva. (Courtesy of E. A. Platzer.)

FIGURE 13-4

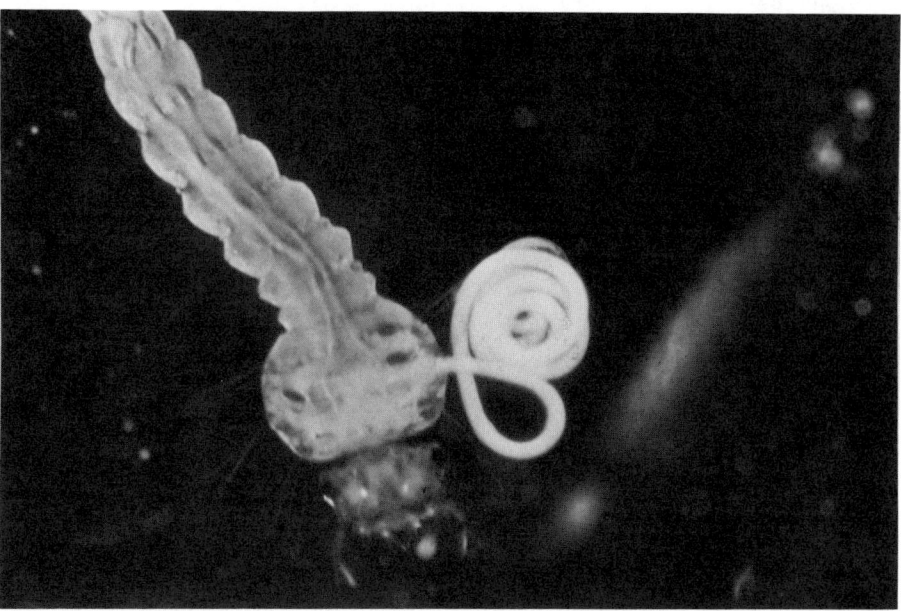

Romanomermis culicivorax emerging from a mosquito larva. (Courtesy of J. J. Petersen.)

dent (Hughes and Platzer 1977). At 20°C, the nematode takes 12 days to emerge, at 27°C, 7 days, and at 32°C, 6 days.

The postparasitic juveniles molt to adults between 10 and 50 days after emergence from their hosts, with the males maturing a few days earlier than the females (Petersen 1975b). After mating, the females, which average 2500 eggs over their

FIGURE 13-5

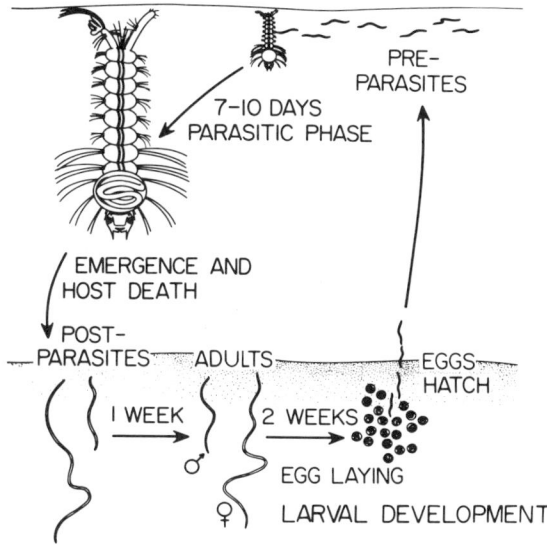

Life cycle of *Romanomermis culicivorax*. (Courtesy of E. A. Platzer and Marcel Dekker.)

lifetime, begin to deposit their eggs from 25 to 30 days after emergence from their hosts, and most females deposit eggs within 140 days.

R. culicivorax has become established and has recycled after field application (Petersen and Willis 1975; Walker et al. 1985). Establishment has occurred with postparasites in areas where the mosquito breeds in permanent pools or in pools that become dry for a short period of time (Westerdahl et al. 1982). Not all areas can be colonized by the mermithid as other ecological factors limit its establishment. An important limiting factor is sodium chloride. As sodium chloride concentration levels increase from 0 to 0.04 M, the percent infection is greatly reduced (Petersen and Willis 1970); thus, the mermithid's ability to colonize brackish waters is restricted.

2. Black Fly Mermithids

A number of black fly mermithids are potential biological control agents (Phelps and DeFoliart 1964; Gordon et al. 1973a), but their life cycles have not been studied in great detail. In part, this is because of the difficulty in working with black flies as well as the mermithids. Molloy and Jamnback (1975) demonstrated that *Neomesomermis flumenalis* penetrated through the integument of black flies. Once in the hemocoel, parasite development occurs. The postparasitic juveniles emerge by boring through the host's integument, molt to the adult stage in the stream bed, mate, and lay eggs that embryonate and eventually hatch to release infective juveniles (Gordon et al. 1973a). Bailey et al. (1977) found that the free-living stages of *N. flumenalis* can be maintained in the laboratory by placing 50 males and 50 females in wet sand. Under these conditions the majority of the females becomes gravid.

3. *Agamermis decaudata*

Males and females of *A. decaudata* are found coiled from 5 to 15 cm below the soil surface (Christie 1936). The females deposit their eggs in the soil during the summer and early fall and stop laying eggs with the onset of cold temperatures. The eggs do not hatch until the following spring. Between May and July, the infective juveniles hatch, disperse to the soil surface, climb up the vegetation, and enter the hemocoels of recently hatched grasshoppers by penetrating directly through their integument. It takes about 10 min to penetrate a host. The mermithid remains in the host from 1 to 3 months to complete its parasitic development. Usually only one nematode is found per host. The postparasitic juvenile emerges from the grasshopper, enters the soil during the summer and early fall, and remains as a postparasitic juvenile until the following summer when it molts to the adult stage and mates. Soon thereafter, the female begins to produce eggs. With the onset of cold weather, the female stops laying eggs until the following summer and then dies. Thus, her progeny hatches over two summers. The shortest time for the life cycle from egg to egg is 2 yr.

4. *Mermis nigrescens*

The life cycle of *M. nigrescens* differs from *A. decaudata* (Christie 1937). Gravid females of *M. nigrescens* migrate from the soil surface and climb up vegetation to deposit eggs containing infective second-stage juveniles. Deposition of eggs occurs only during daylight hours after a night rain or during rainy days. When the rain stops or the foliage becomes dry, the females drop to the ground and enter the soil. Maximum egg deposition occurs during June and July.

Grasshoppers become infected by consuming the nematode eggs on the foliage. The egg hatches in the midgut, and the infective juvenile bores into the hemocoel. It is not unusual to find more than one mermithid in a grasshopper (Christie 1929). The parasitic development of the mermithid varies from 1 to 3 months. The postparasitic juvenile emerges from its host by penetrating through the integument, a process that is lethal to the grasshopper. Emergence occurs throughout summer and early fall depending on the time of infection. After emerging, the postparasitic juvenile enters the soil to a depth of 15 to 45 cm and overwinters. The following spring, the mermithids molt to adults and mating occurs. However, mating is not a prerequisite for viable egg production because an unmated female can produce viable eggs (Christie 1929). Eggs develop in the body of the female throughout summer, but oviposition does not occur until the following June. The females may live for several years if they are prevented from depositing eggs.

The ratio of male to female mermithids is influenced by the number of nematodes in the same host (Christie 1929). Generally, hosts with three or less *M. nigrescens* produce females, whereas hosts with more than 24 mermithids produce only males. Between 4 and 23 mermithids, males and females may emerge from the same host. Similar observations have been made with *Romanomermis* (Petersen 1984).

C. RHABDITIDAE

A number of rhabditid species are found associated with insects, and some of these are facultative parasites, whereas others have phoretic relationships (Poinar 1975a). *Rhabditis insectivora*, a facultative parasite of a cerambycid beetle, can develop continuously in the environment without its host. When hosts are present, the dauer juveniles enter the hemocoel of their host via the midgut and mature to an adult. The adult nematodes leave the beetle and enter the beetle's environment, and then the females deposit their eggs. The beetles do not seem to be adversely affected when their hemocoels contain only a few nematodes, but if great numbers occur, the beetles are weakened and die.

D. STEINERNEMATIDAE

Representatives of this family offer much promise as biological control agents because of their high virulence and broad host range. *Steinernema carpocapsae*, for example, kills its hosts within 48 h and will infect many insect species in the laboratory and field (Poinar 1979). A number of *Steinernema* species have been described from natural infections of insects, and all have a mutualistic association with bacteria (Dutky 1959; Thomas and Poinar 1979). The bacteria, in the genus *Xenorhabdus* (syn. *Achromobacter*), have been studied in great detail (Poinar and Thomas 1965, 1966; Akhurst 1980, 1982, 1986a). As an example, the bacterial symbiont *Xenorhabdus nematophilus* produces two colony forms, designated as phase one (i.e., primary form) and phase two (i.e., secondary form) (Akhurst and Boemare 1990). Phase one is isolated from infective nematodes and tends to be unstable, converting to phase two. Phase two is isolated from old insect cadavers or *in vitro* culture of nematodes. Other differences between phases are detected in biochemical tests, the production of antibiotics in phase one, and nematode reproduction *in vitro* (Akhurst and Boemare 1990).

Differences of opinions have been expressed on the systematics of this group of nematodes. Wouts et al. (1982) consider the genus *Neoaplectana* to be a junior synonym of *Steinernema*. Poinar (1984) did not agree with this nomenclature and retained *Neoaplectana* as a valid genus, but in 1990, he suggested that *Steinernema* should be used until further systematic studies are conducted. Table 13-2 lists the species and their symbiotic bacteria.

The life cycles of steinernematids are similar, and *S. carpocapsae* will serve as a model. Dauer juveniles are ingested by a susceptible host and enter the hemocoel by penetrating directly through the midgut or enter through the spiracles and penetrate the tracheae (Fig. 13-6). Once in the hemocoel, the dauer juveniles release the associated bacterium through their anus. The bacterium multiplies rapidly in the host's body and causes host death by septicemia within 48 h. The nematodes develop rapidly to the adult stage, mate, and produce eggs. First-stage juveniles emerge from the eggs and become adults; in small hosts, however, they may become dauer juveniles (Fig. 13-1, 13-7). The progeny from these first generation adults (or later generations in larger hosts) becomes dauer juveniles. Cues for dauer juvenile formation appear to be high nematode density, suggesting the presence of a pheromone, and low-soluble nutrient status in the cadaver (Popiel et al. 1989). If adequate moisture is present, the dauer juveniles leave their host. In the laboratory, they exit from the host 8-14 days after infection. The dauer juveniles can remain alive for a long period of time and, under laboratory conditions, some will live for 5 yr.

Matricidal endotoky occurs (i.e., in older adult females). The eggs hatch within

TABLE 13-2 Species of *Steinernema* and Their Mutualistically Associated Bacterium, *Xenorhabdus*[a]

Nematodes	Bacteria
Steinernema affinis	*Xenorhabdus bovienii*
S. anomali	unidentified
S. carpocapsae[b]	*X. nematophilus*
S. feltiae (syn. *bibionis*)	*X. bovienii*
S. glaseri	*X. poinarii*
S. intermedia	*X. bovienii*
S. kushidai	unidentified
S. rara	unidentified
S. scapterisci	unidentified

[a]Modified after Akhurst and Boemare (1990).
[b]Also referred to as *Steinernema feltiae*. See Poinar (1990) for clarification.

X. BIOLOGY AND LIFE CYCLE OF SELECTED NEMATODES   477

FIGURE 13-6

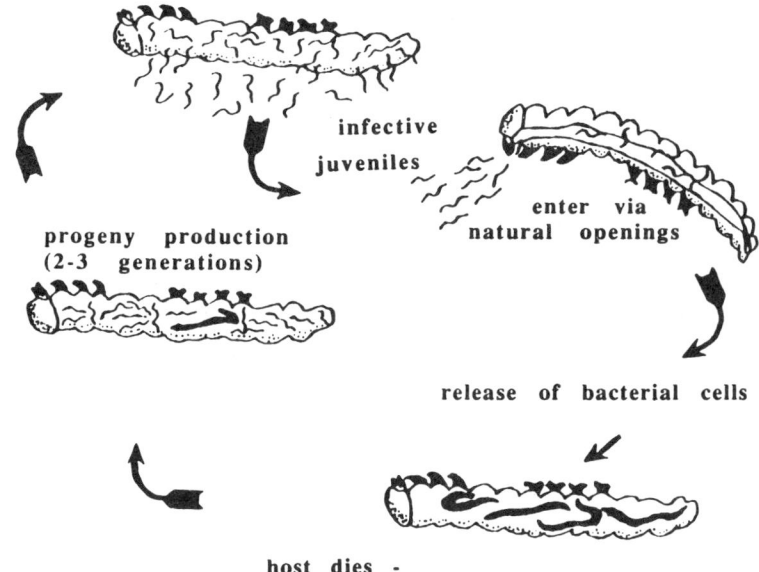

Life cycle of *Steinernema carpocapsae*.

FIGURE 13-7

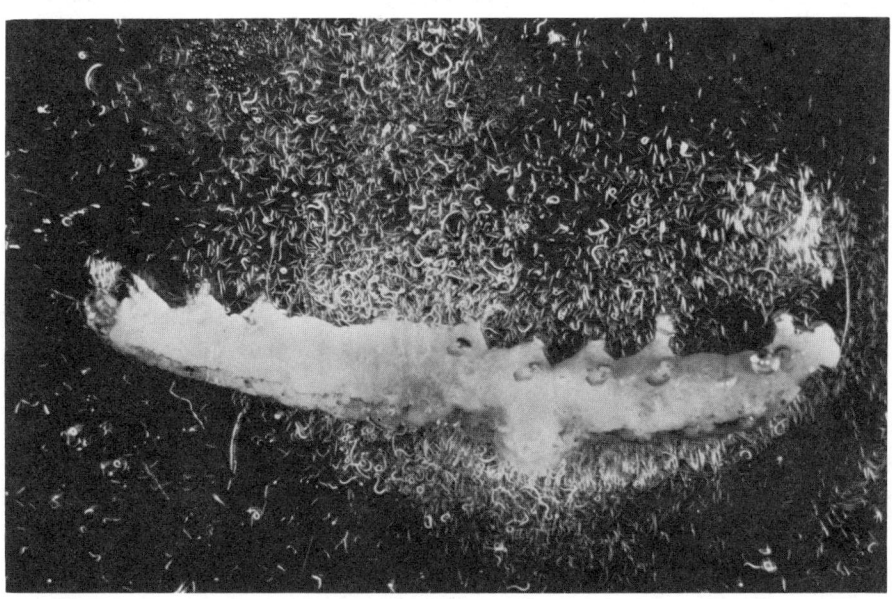

Larva of *Spodoptera exigua* infected by *Steinernema carpocapsae* and dissected to reveal nematode reproduction. (Photographed by A. H. Hara.)

the body, and the juveniles feed on the body contents of the mother. These dauer juveniles escape through the oral and anal openings of the dead female.

The nematode needs the associated bacterium for its growth and development, and the bacterium needs the nematode to invade an insect (Dutky 1959). The nematode can kill its host without its associated bacterium but is unable to reproduce, and the bacterium cannot invade the hemocoel of its host without the nematode.

S. carpocapsae along with its associated bacterium has been grown on media with pork kidney–peptone agar (Dutky et al. 1964), macerated pork kidney (Bedding 1981), dog food (House et al. 1965), chicken offal (Bedding 1984b), and semidefined diet in fermenters (Friedman 1990) without loss of infectivity. Because the bacterium is present, the nematode is grown monoxenically. Glaser (1931) was the first to culture a nematode on an artificial medium. This nematode, *Steinernema glaseri*, is now grown axenically on a defined diet and useful information has been obtained on its dietary requirements (Jackson and Platzer 1974).

Insects infected with steinernematids are slightly discolored and flaccid and remain intact. Upon dissection, the tissues have a gummy consistency and do not have a putrid odor. Secondary microorganisms are usually not present in great numbers because the associated bacterium produces an antibiotic (Akhurst 1982).

E. HETERORHABDITIDAE

Heterorhabditids have a similar life cycle to the steinernematids, but major differences also exist. Dauer juveniles, which invade the hemocoel, release the bacterium *Xenorhabdus luminescens*, killing the host within 48 h, and reach adulthood rapidly (Fig. 13–8). These adults resulting from the dauer juveniles are hermaphrodites rather than being male or female as in *S. carpocapsae*, and therefore only one dauer heterorhabditid juvenile is needed to enter the host for progeny production. Eggs laid by the hermaphrodites produce juveniles that develop into males and females. These mate and produce eggs that hatch and become dauer juveniles. If adequate moisture is present, the dauer juveniles will leave the host and infect another insect host.

The bacterium associated with heterorhabditids is similar in many ways to *X. nematophilus*, but it does have major differences (Akhurst and Boemare 1990). Infected host turns brick red and luminesces in the dark. The tissues take on a gummy characteristic as in *S. carpocapsae* infections (Poinar et al. 1977).

F. DIPLOGASTERIDAE

Most diplogasterids are free living and their associations with insects are usually phoretic. A few are facultative parasites of insects (Bovien 1937; Poinar 1975a). *Eudiplogaster aphodii* is a facultative parasite of the scarabaeid beetle, *Aphodius fimetarius* (Poinar et al. 1976b). The third-stage juveniles emerge from *A. fimetarius* probably via the alimentary and reproductive tracts into cattle dung. The nematodes establish successive generations in the dung. As the dung dries, the third-stage infective juveniles are formed. About this time, scarabaeid larvae occur in the dung, and the juveniles enter the insects' midguts and penetrate into the hemocoels. Apparently, no visible harm is done to the beetle larvae, which continue to develop

FIGURE 13-8

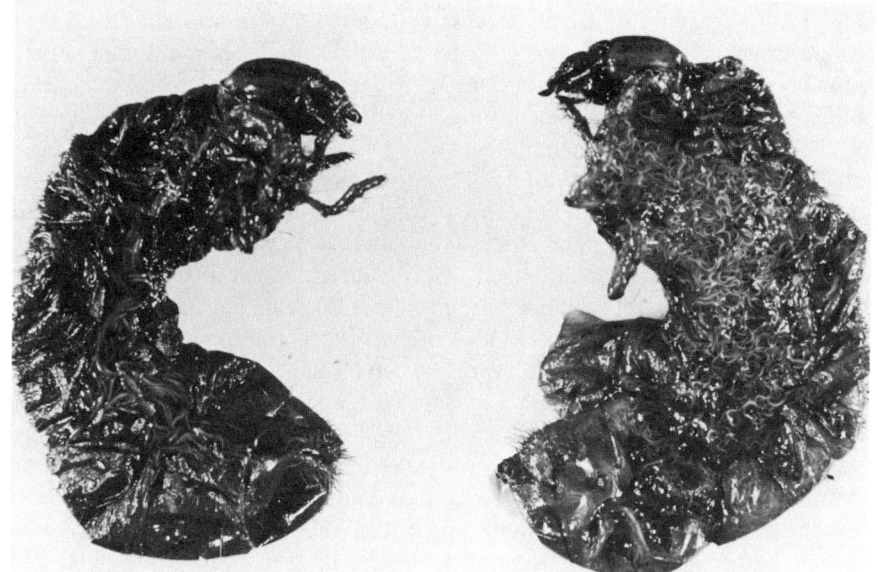

Scarabaeid larvae with the cuticle removed to expose infection by *Heterorhabditis megidis*. (Courtesy of M. G. Klein.)

and become adults. The nematodes reproduce in the hemocoels after the beetles reach adulthood.

G. ALLANTONEMATIDAE

The allantonematids represent a diverse famly of facultative and obligate nematodes infecting insects in several orders. The genus *Deladenus* was originally placed in the family Neotylenchidae, but it has been recently reclassified with the allantonematids (Fortuner and Raski 1987). Further reclassification has been proposed by Remillet and Laumond (1991) within this group; however, we have taken a more conservative approach and will keep all species in the family Allatonematidae.

1. *Deladenus siricidicola*

This nematode infects the siricid wood wasp, *Sirex noctilio*, a serious pest of Monterey pine plantations in Australia and New Zealand (Bedding 1984a). *Deladenus* (syn. *Beddingia*) *siricidicola* has two life cycles. One life cycle is free living and the nematode reproduces oviparously while feeding on the symbiotic fungus *Amylostereum areolatum* (Fig. 2–2). This fungus and *Sirex* wood wasp have a mutualistic association. The wood wasp feeds on the fungus, and the fungus, which infects and kills pine trees, utilizes the wood wasp for dispersal. The other life cycle is parasitic with the nematode reproducing ovoviviparously within the hemocoel of the *Sirex* wood wasp. Each life cycle may continue indefinitely without the intervention of the other.

The fungal or mycetophagus life cycle is simple. When juvenile nematodes are placed on young cultures of *A. areolatum*, the adult nematodes are found within 5 days at 22°C. After mating, the females deposit eggs that hatch after 4 to 5 days.

These juveniles feed on the fungus and become adults in 7 days, and providing young mycelial growth is present, successive generations can be maintained. When no young mycelial growth is present, the nematode produces a different form of adult male and female. After mating, the male dies, and the infective female enters siricid wood wasp larvae by penetrating directly through the integument into the hemocoel to begin the parasitic life cycle. Once in the hemocoel, the female nematode grows, but the reproductive system remains immature until the host begins to pupate (the life cycle of *Sirex* is long and it may be 2 yr before the host pupates). At pupation, the nematode's reproductive system initiates its growth. Clearly, the nematode is synchronized with the development of its host. The eggs are released in the insect's hemocoel and hatch, but hatching also occurs in the mother (matricidal endotoky) in which case juveniles emerge by penetrating her body. The juveniles migrate to the ovaries and enter the host eggs before the end of pupation. Consequently, the adult female wood wasp is sterile upon emergence from the pupa. Transmission occurs when the female wood wasp deposits eggs containing nematodes. Uninfected *Sirex* females oviposit in the same pine trees, and the nematodes infect the immature siricids to continue the parasitic life cycle. In males, the testes are also invaded in the pupal stage, but the spermatozoa are formed very early before nematode invasion of the testes and stored in the vesiculae seminale. The infected males mate normally and fertilize the females. The nematodes that cannot escape from the males die together with their hosts, and the males are considered "dead-end" hosts.

2. *Contortylenchus* spp.

Contortylenchus spp., nematode parasites of bark beetles, have a simple life cycle (Thong and Webster 1973; Nickle 1967a; Kaya 1984). The gravid female of *C. reversus* lays its eggs in the hemocoel of larva, pupa, or adult beetle depending on the age of the host at the time of invasion. The eggs hatch, and the newly emerged nematodes develop into fourth-stage juveniles, which penetrate from the insect's hemocoel into the hindgut and are defecated in the bark beetles' galleries (Fig. 13–9). The juveniles molt to adults, which mate within the galleries. The males die in the galleries, but the fertilized females seek and infect beetle larvae. The mode of infection is not really known but probably occurs by direct penetration through the cuticle. The life cycle of *C. elongatus* is similar except that nematode eggs are not deposited in the hemocoel until the bark beetles become adults.

3. *Heterotylenchus autumnalis*

A complex life cycle is found in *Heterotylenchus* (syn. *Paraiotonchium*) *autumnalis*, a nematode parasite of the face fly, *Musca autumnalis* (Stoffolano and Nickle 1966; Nickle 1967b). The life cycle involves an alternation of gamogenetic and parthenogenetic stages. The gravid (gamogenetic) female deposits eggs in the hemocoel of an adult female face fly, and the eggs hatch producing parthenogenetic females. The parthenogenetic females lay thousands of eggs that develop into immature males and females; these nematodes invade the ovaries of the face fly (Fig. 13–2B). The infected female face fly, in its attempts to oviposit, deposits nematodes in cattle dung. The nematodes mate, and the fertilized female nematodes seek out face fly larvae. The nematodes penetrate the face fly larvae directly through

FIGURE 13-9

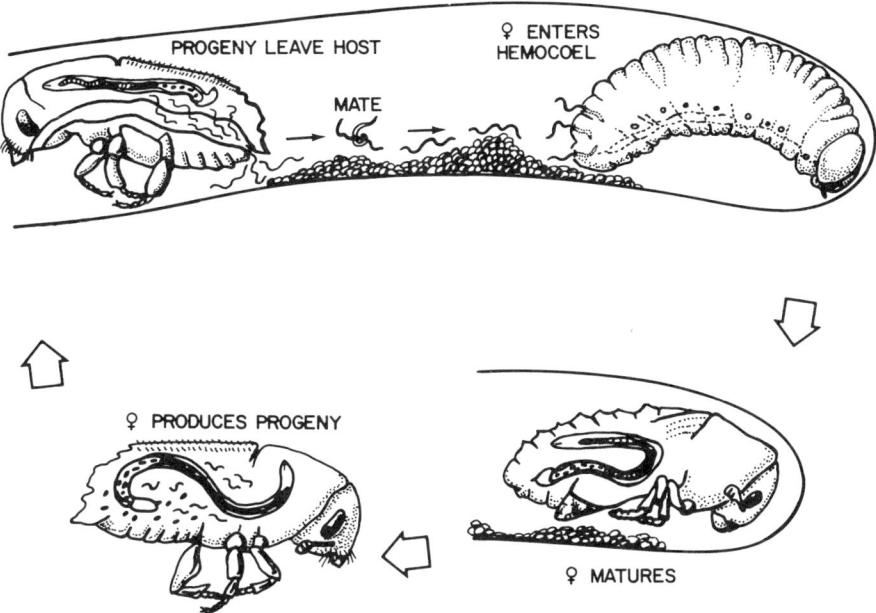

Life cycle of *Contortylenchus* species. (Drawn by R. M. Giblin-Davis; courtesy of Marcel Dekker.)

the integument. The gamogenetic female does not deposit eggs until the face fly emerges as an adult. Male face flies are infected but because the nematodes have no means of escape, they are dead-end hosts and do not participate in the natural distribution of nematodes (Nappi 1973).

H. SPHAERULARIIDAE

The life cycle of *Sphaerularia bombi* is not completely understood although this nematode was among the first to be described from insects (Poinar 1975a). Poinar and van der Laan (1972) clarified some aspects of the life cycle of this nematode. An infected *Bombus* queen deposits third-stage juveniles in the soil. Apparently, the nematodes are discharged from the hindgut, but the possibility of nematodes being deposited from the reproductive system cannot be eliminated. The nematodes become adults and mate, and the fertilized females are ready to enter new hosts. The mode of infection, whether it is by cuticular penetration or by oral ingestion, has not been established. *Bombus* queens are infected during the fall when they enter the soil for hibernation. The nematode deposits eggs in the hemocoel. The juvenile undergoes two molts within the egg and hatches as the third-stage nematode.

I. APHELENCHOIDIDAE

Most aphelenchoidids have phoretic relationships with insects, though some are parasitic. Rühm (1956) reports that *Parasitaphelenchus* develops in the hemocoel of bark beetles. The third-stage juveniles enter the bark beetles, molt to the fourth-stage juveniles, and leave the beetles. Nematodes molt to adults in the beetles' galleries and deposit eggs, and their progenies, which feed on fungal mycelia in the galleries, develop until the third stage and begin to seek a new beetle host. The effect on the host is slight.

Noctuidonema guyanense appears to be an obligate ectoparasite on adult noctuid moths (Remillet and Silvain 1988; Rogers et al. 1990). The nematode occurs on the intersegmental membranes of the posterior two abdominal segments, and less frequently more anterior on the abdomen, among the hair pencils, and on the male claspers (Rogers et al. 1990). Very little is known about the biology of this nematode.

J. ENTAPHELENCHIDAE

This small family presently comprises four genera (Poinar 1975a). All members, so far described, are found in the hemocoels of beetles belonging to three families. In general, the life cycle of *Entaphelenchus* is similar to *Contortylenchus*. The female nematode deposits her eggs in the hemocoel of a beetle, the eggs hatch, the nematodes develop into fourth-stage juveniles that emerge out of the hemocoel and into the hindgut, are defecated, and enter the soil where they molt into adults. The nematodes mate, and the fertilized female infects another beetle.

K. OXYURIDAE

Oxyurids are found in the alimentary tracts of invertebrates. These nematodes produce no known detrimental effects and have an obligate relationship with insects and other invertebrates (Poinar 1975a). Susceptible hosts acquire the nematodes by ingesting the eggs that hatch in the alimentary tracts. All nematode development occurs in the gut, and the only stage found free in the environment is the egg.

XI. NEMATOMORPHA

Members of the phylum Nematomorpha (Gordiacea) superficially resemble nematodes and are called hair snakes, horsehair worms, or gordian worms. They differ from nematodes by their asymmetrical esophagus and their gonads opening at the posterior end of the body through a cloaca in both sexes. As adults, the nematomorphs are usually larger and more uniformly cylindrical than nematodes and have a faintly to darkly colored cuticle.

The life cycle of *Chordodes japonensis* in mantids has been studied by Inoue (1958, 1960, 1962). Adults emerge from mantids in September and October and enter streams or ponds to overwinter. Mating occurs in water from October through June. In May, the fertilized females lay eggs for about 3 weeks. The juveniles emerge from the eggs about 30 days later and are ingested by aquatic immature insects (paratenic hosts), such as mayflies, chironomids, and mosquitoes. Within 48 h, the juveniles are in the hemocoel of an aquatic insect and the majority is encysted. Apparently, the juvenile extrudes the cyst material through its anus from the bilobed intestinal gland situated in the posterior portion of the intestine (Poinar and Doelman 1974). Inoue (1960) also observed that the cysts are surrounded by hemocytes. Young mantids become infected when they ingest adult mayflies, chironomids, or mosquitoes containing the encysted juveniles. The juvenile escapes from the cyst directly or the cyst is digested by the host midgut juices, and the freed juvenile penetrates into the hemocoel of the mantid. About 3 months later, the mature worm emerges from the adult mantid. Inoue (1962) found that a higher percentage of mantids became infected when they were fed mayflies than chironomids or mosquitoes. Accordingly, mayflies appear to be more efficient paratenic hosts for this gordian worm than chironomids or mosquitoes. Additionally, survival of the encysted juveniles in the paratenic hosts is maximal for 2 weeks after encystment and begins to decline thereafter.

The life cycle of *Gordius robustus* in Mormon crickets has been outlined by Thorne (1940). Overwintering *Gordius* females emerge from loose gravelly sand or from under sticks, stones, and rubbish during April and May to enter shallow water where eggs are deposited. Eggs hatch shortly after deposition, and the small juveniles are probably ingested by young crickets drinking contaminated water or eating infested aquatic plants. Inoue (1962) does not rule out the possibility that paratenic hosts may also be involved. Once ingested, the juvenile enters the hemocoel of the cricket and growth takes place rapidly. By late June, the full-grown worms emerge from the crickets by breaking through the abdominal wall. After emergence, mating occurs and the females overwinter until the following spring. *G. robustus* is considered an important biological control agent of the Mormon crickets and up to 50% in a given population may be infected. The effect of the worm on the cricket is usually sterilization.

XII. PLATYHELMINTHES

The phylum Platyhelminthes contains helminths of lowest organization. Most platyhelminths are flattened from the dorsal to the ventral, hence the common name flatworm. Flatworms have no body cavity, a simple digestive tract usually with only one opening, and a very simple nervous system. A few species of flatworms utilize insects as intermediate hosts. The trematodes (flukes) and cestodes (tapeworms) are all parasitic. On the other hand, the tubellarians (planaria) are for the most part free living. Some of the free-living forms are predaceous, feeding on a variety of small invertebrates, including many insects.

The predaceous planarian, *Dugesia dorotocephala*, shows potential as a biological control of mosquitoes and chironomids (Medved and Legner 1974; Legner and Yu 1975; Lacey and Lacey 1990). *D. dorotocephala* will feed on eggs, larvae, and pupae of mosquitoes. It secretes a sticky mucous from the epidermal glands to capture prey. Often, more than one planarian will feed on the same prey. In some flatworms in the genus *Mesostoma*, toxic substances are believed to be secreted in the mucous as ensnarled mosquito larvae show symptoms of paralysis. When mosquito larvae brush against these worms, the larvae become paralyzed and die (Case and Washino 1979). In small field trials, this planarian has significantly reduced mosquito populations.

REFERENCES

Akhurst, R. J. 1980. Morphological and functional dimorphism in *Xenorhabdus* spp., bacteria symbiotically associated with the insect pathogenic nematodes *Neoaplectana* and *Heterorhabditis*. *J. Gen. Microbiol.* 121, 303–309.

Akhurst, R. J. 1982. Antibiotic activity of *Xenorhabdus* spp., bacteria symbiotically associated with insect pathogenic nemtaodes of the families Heterorhabditidae and Steinernematidae. *J. Gen. Microbiol. 128*, 3061–3065.

Akhurst, R. J. 1986a. *Xenorhabdus nematophilus* subsp. *beddingii* (Enterobacteriaceae): A new subspecies of bacteria mutualistically associated with entomopathogenic nematodes. *Int. J. Syst. Bacteriol. 36*, 454–457.

Akhurst, R. J. 1986b. Controlling insects in soil with entomopathogenic nematodes. *In* "Fundamental and Applied Aspects of Invertebrate Pathology." (R. A. Samson, J. M. Vlak, and D. Peters, eds.), pp. 265–267. Found. Fourth Int. Colloq. Invertebr. Pathol., Wageningen, The Netherlands.

Akhurst, R. A., and Boemare, N. E. 1990. Biology and taxonomy of *Xenorhabdus*. *In* "Entomopathogenic Nematodes in Biological Control." (R. Gaugler and H.K. Kaya, eds.), pp. 75–90. CRC Press, Boca Raton, Florida.

Anderson, J. F., and Shemanchuk, J. A. 1987. The biology of *Simulium articum* Malloch in Alberta. Part II. Seasonal parity structure and mermithid parasitism of populations attacking cattle and flying over the Athabasca River. *Can. Entomol. 119*, 29–44.

Ashraf, M., and Berryman, A. A. 1970a. Biology of *Sulphuretylenchus elongatus* (Nematoda: Sphaerulariidae), and its effect on its host, *Scolytus ventralis* (Coleoptera: Scolytidae). *Can. Entomol. 102*, 197–213.

Ashraf, M., and Berryman, A. A. 1970b. Histopathology of *Scolytus ventralis* (Coleoptera: Scolytidae) infected by *Sulphuretylenchus elongatus* (Nematoda: Sphaerulariidae). *Ann. Entomol. Soc. Am. 63*, 924–930.

Bailey, C. H., and Gordon, R. 1973. Histopathology of *Aedes aegypti* (Diptera: Culicidae) larvae parasitized by *Reesimermis neilseni* (Nematoda: Mermithidae). *J. Invertebr. Pathol. 22*, 435–441.

Bailey, C. H., Gordon, R., and Mills, C. 1977. Laboratory culture of the free-living stages of *Neomesomermis flumenalis*, a mermithid nematode parasite of Newfoundland blackflies (Diptera: Simuliidae). *Can. J. Zool. 55*, 391–397.

Bedding, R. A. 1981. Low cost *in vitro* mass production of *Neoaplectana* and *Heterorhabditis* species (Nematoda) for field control of insect pests. *Nematologica 27*, 109–114.

Bedding, R. A. 1984b. Large scale production, storage and transport of the insect-parasitic nematodes *Neoaplectana* spp. and *Heterorhabditis* spp. *Ann. Appl. Biol. 104*, 117–120.

Bedding, R. A. 1984b. Large scale production, storage and transport of the insect-parasitic nematodes *Neoaplectana* spp. and *Heterorhabditis* spp. *Ann. Appl. Biol. 101*, 117–120.

Bedding, R. A., and Molyneux, A. S. 1982. Penetration of insect cuticle by infective juveniles of *Heterorhabditis* spp. (Heterorhabditidae: Nematoda). *Nematologica 28*, 354–359.

Bird, A. F., and Bird, J. 1986. Observations on the use of insect parasitic nematodes as a means of biological control of root-knot nematodes. *Int. J. Parasitol. 16*, 511–516.

Bovien, P. 1937. Some types of association between nematodes and insects. *Vidensk. Medd. Dan. Naturhis. Foren. Khobenhavn. Bd. 101*, 1–114.

Bronskill, J. F. 1962. Encapsulation of rhabditoid nematodes in mosquitoes. *Can. J. Zool. 40*, 1269–1275.

Brug, S. L. 1932. Chitinization of parasites in mosquitoes. *Bull. Entomol. Res. 23*, 229–231.

Case, T. J., and Washino, R. K. 1979. Flatworm control of mosquito larvae in rice fields. *Science 206*, 1412–1414.

Chitwood, B. G. and Chitwood, M. B. 1974. "Introduction to Nematology." University Park Press, Baltimore, Maryland.

Choo, H. Y., Kaya, H. K., Burlando, T. M., and Gaugler, R. 1989. Entomopathogenic nematodes: Host-finding ability in the presence of plant roots. *Environ. Entomol. 18*, 1136–1140.

Christie, J. R. 1929. Some observations on sex in the Mermithidae. *J. Exp. Zool. 53*, 59–76.

Christie, J. R. 1936. Life history of *Agamermis decaudata*, a nematode parasite of grasshoppers and other insects. *J. Agric. Res. 52*, 161–198.

Christie, J. R. 1937. *Mermis subnigrescens*, a nematode parasite of grasshoppers. *J. Agric. Res.* 55, 353–364.

Condon, W. J. and Gordon, R. 1977a. Effects of the mermithid nematode *Mermis nigrescens* on the levels of hemolymph and fecal uric acid in its host, the migratory locust *Locusta migratoria*. *Can. J. Zool.* 55, 690–692.

Condon, W. J., and Gordon, R. 1977b. Some effects of mermithid parasitism on the larval blackflies *Prosimulium mixtum fuscum* and *Simulium venustum*. *J. Invertebr. Pathol.* 29, 56–62.

Craig, S. M., and Webster, J. M. 1974. Inhibition of molting of the desert locust, *Schistocerca gregaria*, by the nematode parasite *Mermis nigrescens*. *Can. J. Zool.* 52, 1535–1539.

Dutky, S. R. 1959. Insect microbiology. *Adv. Appl. Microbiol.* 1, 175–200.

Dutky, S. R., and Hough, W. S. 1955. Note on a parasitic nematode from codling moth larvae *Carpocapsa pomonella* (Lepidoptera, Olethreutidae). *Proc. Entomol. Soc. Wash.* 57, 244.

Dutky, S. R., Thompson, J. V., and Cantwell, G. E. 1964. A technique for the mass propagation of the DD-136 nematode. *J. Insect Pathol.* 6, 417–422.

Fenwick, W. W. 1969. Red ring disease of the coconut palm. *In* "Nematodes of Tropical Crops." (J. E. Peachey, ed.), pp. 89–98. Techn. Comm. No. 40 Commonwealth Bureau of Helminth., St. Albans, Hert, England.

Ferris, J. M., and Ferris, V. R. 1966. Observations on *Tetradonema plicans*, an entomoparasitic nematode, with a key to the genera of the family Tetradonematidae (Nematoda: Trichosyringida). *Ann. Entomol. Soc. Am.* 58, 964–971.

Filipjev, I. N., and Schuurmans Stekhoven, J. H., Jr. 1941. "A Manual of Agricultural Helminthology." E. J. Brill, Leiden, Holland.

Fortuner, R. and Raski, D. J. 1987. A review of Neotylenchoidea Thorne, 1941 (Nemata: Tylenchida). *Rev. Nematol.* 10, 257–267.

Friedman, M. J. 1990. Commercial production and development. *In* "Entomopathogenic Nematodes in Biological Control." (R. Gaugler and H. K. Kaya, eds.), pp. 153–172. CRC Press, Boca Raton, Florida.

Fuchs, G. 1915. Die Naturgeschichte der Nematoden und einiger anderer Parasiten; I. des *Ips typographus*, L.: 2. des *Hylobius abietis* L. *Zool. Jahrb.* 38, 109–222.

Gaugler, R. 1987. Entomogenous nematodes and their prospects for genetic improvement. *In* "Biotechnology in Invertebrate Pathology and Cell Culture." (K. Maramorosch, ed.), pp. 457–484. Academic Press, New York.

Gaugler, R., and Kaya, H. K. 1990. "Entomopathogenic Nematodes in Biological Control." CRC Press, Boca Raton, Florida.

Gaugler, R., and Molloy, D. 1981. Instar susceptibility of *Simulium vittatum* (Diptera: Simuliidae) to the entomogenous nematode *Neoaplectana carpocapsae*. *J. Nematol.* 13, 1–5.

Gaugler, R., Lebeck, L., Nakagaki, B., and Boush, G. M. 1980. Orientation of the entomogenous nematode *Neoaplectana carpocapsae* to carbon dioxide. *Environ. Entomol.* 9, 649–652.

Gaugler, R., Campbell, J. F., and McGuire, T. R. 1989a. Selection for host-finding in *Steinernema feltiae*. *J. Invertebr. Pathol.* 54, 363–372.

Gaugler, R., McGuire, T., and Campbell, J. 1989b. Genetic variability among strains of the entomopathogenic nematode *Steinernema feltiae*. *J. Nematol.* 21, 247–253.

Gaugler, R., Campbell, J. F., and McGuire, T. R. 1990. Fitness of a genetically improved entomopathogenic nematode. *J. Invertebr. Pathol.* 56, 106–116.

Gaugler, R., Campbell, J. F., and Gupta, P. 1991. Characterization and basis of enhanced host-finding in a genetically improved strain of *Steinernema carpocapsae*. *J. Invertebr. Pathol.* 57, 234–241.

Giblin, R. M., and Kaya, H. K. 1984. Associations of halictid bees with the nematodes,

Aduncospiculum halicti (Diplogasterida: Diplogasteroididae) and *Bursaphelenchus kevini* (Aphelenchida: Aphelenchoididae). *J. Kansas Entomol. Soc. 57*, 92–99.

Giblin, R., Kaya, H. K., and Brooks, R. W. 1981. Occurrence of *Huntaphelenchoides* sp. (Aphelenchoididae) and *Acrostichus* sp. (Diplogasteridae) in the reproductive tracts of soil nesting bees (Hymenoptera: Apoidea). *Nematologica 27*, 20–27.

Giblin-Davis, R. M., Norden, B. B., Batra, S. W. T., and Eickwort, G. C. 1990. Commensal nematodes in the glands, genitalia, and brood cells of bees (Apoidea). *J. Nematol. 22*, 150–161.

Glaser, R. W. 1931. The cultivation of a nematode parasite of an insect. *Science 73*, 614.

Glaser, R. W. 1932. Studies on *Neoaplectana glaseri*, a nematode parasite of the Japanese beetle *(Popillia japonica). N. J. Dep. Agric. Circ. No. 211.*

Glaser, R. W., and Farrell, C. C. 1935. Field experiments with the Japanese beetle and its nematode parasite. *J. N.Y. Entomol. Soc. 43*, 345–371.

Goodey, T. 1963. "Soil and Freshwater Nematodes." Second Edition." Methuen, London.

Gordon, R., and Webster, J. M. 1971. *Mermis nigrescens*: Physiological relationship with its host, the adult desert locust *Schistocerca gregaria. Exp. Parasitol. 29*, 66–79.

Gordon, R., Ebsary, B. A., and Bennett, G. F. 1973a. Potentialities of mermithid nematodes for the biocontrol of blackflies (Diptera: Simuliidae)—A review. *Exp. Parasitol. 33*, 226–238.

Gordon, R., Webster, J. M., and Hislop, T. G. 1973b. Mermithid parasitism, protein turnover and vitellogenesis in the desert locust, *Schistocerca gregaria* Forskål. *Comp. Biochem Physiol. 46B*, 575–593.

Gordon, R., Bailey, C. H., and Barber, J. M. 1974. Parasitic development of the mermithid nematode *Reesimermis neilseni* in the larval mosquito *Aedes aegypti. Can. J. Zool. 52*, 1293–1302.

House, H. L., Welch, H. E., and Cleugh, T. R. 1965. Food medium of prepared dog biscuit for the mass-production of the nematode DD136 (Nematoda; Steinernematidae). *Nature 206*, 847.

Hudson, K. E. 1974a. Regulation of greenhouse sciarid fly populations using *Tetradonema plicans* (Nematoda: Mermithoidea). *J. Invertebr. Pathol. 23*, 85–91.

Hudson, K. E. 1974b. Biology and tetrad development of the insect parasite *Tetradonema plicans* (Nematoda: Tetradonematidae). *Nematologica 20*, 455–468.

Hughes, D. S., and Platzer, E. G. 1977. Temperature effects on the parasitic phase of *Romanomermis culicivorax* in *Culex pipiens. J. Nematol. 9*, 173–175.

Hungerford, H. B. 1919. Biological notes on *Tetradonema plicans*, Cobb, a nematode parasite of *Sciara coprophila* Lintner. *J. Parasitol. 5*, 186–192.

Inoue, I. 1958. Studies on the life history of *Chordodes japonensis*, a species of Gordiacea. I. The development and structure of the larva. *Jpn. J. Zool. 12*, 203–218.

Inoue, I. 1960. Studies on the life history of *Chordodes japonensis*, a species of Gordiacea. II. On the manner of entry into the aquatic insect-larvae of *Chordodes* larvae. *Annot. Zool. Jpn. 33*, 132–141.

Inoue, I. 1962. Studies on the life history of *Chordodes japonensis*, a species of Gordiacea. III. The mode of infection. *Annot. Zool. Jpn. 35*, 12–19.

Jackson, G. J., and Platzer, E. G. 1974. Nutritional biotin and purine requirements, and the folate metabolism of *Neoaplectana glaseri. J. Parasitol. 60*, 453–457.

Jackson, J. J., and Brooks, M. A. 1989. Susceptibility and immune response of western corn rootworm larvae (Coleoptera: Chrysomelidae) to the entomogenous nematode, *Steinernema feltiae* (Rhabditida: Steinernematidae). *J. Econ. Entomol. 82*, 1073–1077.

Kaiser, H. 1991. Terrestrial and semiterrestrial Mermithidae. *In* "Manual of Agricultural Nematology." (W. R. Nickle, ed.), pp. 899–965. Marcel Dekker, New York.

Kaya, H. K. 1978a. Interaction between *Neoaplectana carpocapsae* (Nematoda: Steinernematidae) and *Apanteles militaris* (Hymenoptera: Braconidae), a parasitoid of the armyworm, *Pseudaletia unipuncta. J. Invertebr. Pathol. 31*, 358–364.

Kaya, H. K. 1978b. Infectivity of *Neoaplectana carpocapsae* and *Heterorhabditis heliothidis* to pupae of the parasite *Apanteles militaris*. *J. Nematol. 10*, 241–244.

Kaya, H. K. 1984. Nematode parasites of bark beetles. *In* "Plant and Insect Nematodes." (W. R. Nickle, ed.), pp. 727–754. Marcel Dekker, New York.

Kaya, H. K. 1985. Entomogenous nematodes for insect control in IPM systems. *In* "Biological Control in Agricultural IPM Systems." (M. A. Hoy and D. C. Herzog, eds.), pp. 283–302. Academic Press, New York.

Kaya, H. K. 1987. Diseases caused by nematodes. *In* "Epizootiology of Insect Diseases." (J. R. Fuxa and Y. Tanada, eds.), pp. 453–470. John Wiley & Sons, New York.

Kaya, H. K. 1990. Soil ecology. *In* "Entomopathogenic Nematodes in Biological Control." (R. Gaugler and H. K. Kaya, eds.), pp. 93–115. CRC Press, Boca Raton, Florida.

Kaya, H. K., and Hotchkin, P. G. 1981. The nematode *Neoaplectana carpocapsae* Weiser and its effect on selected ichneumonid and braconid parasites. *Environ. Entomol. 10*, 474–478.

Kaya, H. K., Moon, R. D., and Witt, P. L. 1979. Influence of the nematode, *Heterotylenchus autumnalis*, on the behavior of face fly, *Musca autumnalis*. *Environ. Entomol. 8*, 537–540.

Khan, A., Brooks, W. M., and Hirschmann, H. 1976. *Chromonema heliothidis* n. gen., n. sp. (Steinernematidae, Nematoda), a parasite of *Heliothis zea* (Noctuidae, Lepidoptera), and other insects. *J. Nematol. 8*, 159–168.

Klein, M. G. 1990. Efficacy against soil-inhabiting insect pests. *In* "Entomopathogenic Nematodes in Biological Control." (R. Gaugler and H. K. Kaya, eds.), pp. 195–214. CRC Press, Boca Raton, Florida.

Lacey, L. A., and Lacey, C. M. 1990. The medical importance of riceland mosquitoes and their control using alternatives to chemical insecticides. *J. Am. Mosq. Control Assoc. Suppl. 2.* 93.

Legner, E. F., and Yu, H. S. 1975. Larvicidal effects on mosquitoes of substances secreted by the planarian, *Dugesia dorotocephala* (Woodworth). *Proc. Calif. Mosq. Control Assoc. 43*, 128–131.

Levine, N. D. 1968. "Nematode Parasites of Domestic Animals and of Man." Burgess Publ. Co., Minneapolis, Minnesota.

Maggenti, A. 1981. "General Nematology." Springer-Verlag, New York.

Maggenti, A. R. 1991. Nemata: Higher classification. *In* "Manual of Agricultural Nematology." (W. R. Nickle, ed.), pp. 147–187. Marcel Dekker, New York.

Mamiya, Y. 1984. The pine wood nematode. *In* "Plant and Insect Nematodes." (W. R. Nickle, ed.), pp. 589–626. Marcel Dekker, New York.

Mamiya, Y. 1989. Comparison of the infectivity of *Steinernema kushidai* (Nematode: Steinernematidae) and other steinernematid and heterorhabditid nematodes for three different insects. *Appl. Entomol. Zool. 24*, 302–308.

Mamiya, Y., and Enda, N. 1972. Transmission of *Bursaphelenchus lignicolus* (Nematoda: Aphelenchoididae) by *Monochamus alternatus* (Coleoptera: Cerambycidae). *Nematologica 18*, 159–162.

Markin, G. P. and McCoy, C. W. 1968. The occurrence of a nematode, *Diploscapter lycostoma*, in the pharyngeal glands of the Argentine ant, *Iridomyrmex humilis*. *Ann. Entomol. Soc. Am. 61*, 505–509.

Massey, C. L. 1974. Biology and Taxonomy of Nematode Parasites and Associates of Bark Beetles in the United States. *Agric. Handbook 446*.

Medved, R. A., and Legner, E. F. 1974. Feeding and reproduction of the planarian, *Dugesia dorotocephala* (Woodworth), in the presence of *Culex peus* Speiser. *Environ. Entomol. 3*, 637–641.

Molloy, D., and Jamnback, H. 1975. Laboratory transmission of mermithids parasitic in blackflies. *Mosq. News 35*, 337–342.

Nappi, A.J. 1973. Effects of parasitization by the nematode, *Heterotylenchus autumnalis*, on mating and oviposition in the host, *Musca autumnalis*. *J. Parasitol. 59*, 963–969.

Nappi, A. J. 1975. Parasite encapsulation in insects. *In* "Invertebrate Immunity." (K. Maramorosch and R. E. Shope, eds.), pp. 293–326. Academic Press, New York.

Nappi, A. J., and Christensen, B. M. 1987. Insect immunity and mechanisms of resistance by nematodes. *In* "Vistas on Nematology." (J. A. Veech and D. W. Dickson, eds.), pp. 285–291. E. O. Painter Printing Co., DeLeon Springs, Florida.

Nickle, W. R. 1967a. On the classification of the insect parasitic nematodes of the Sphaerulariidae Lubbock, 1861 (Tylenchoidea: Nematoda). *Proc. Helminth. Soc. Wash.* 34, 72–94.

Nickle, W. R. 1967b. *Heterotylenchus autumnalis* sp. n. (Nematoda: Sphaerulariidae), a parasite of the face fly, *Musca autumnalis* De Geer. *J. Parasitol.* 53, 398–401.

Nickle, W. R. 1971. Behavior of the shothole borer, *Scolytus rugulosus*, altered by the nematode parasite *Neoparasitylenchus rugulosi*. *Ann. Entomol. Soc. Am.* 64, 751.

Nickle, W. R. 1974. Nematode infections. *In* "Insect Diseases." (G. E. Cantwell, ed.), Vol. 2, pp. 327–376. Marcel Dekker, New York.

Nickle, W. R. 1984. "Plant and Insect Nematodes." Marcel Dekker, New York.

Nickle, W. R., and Welch, H. E. 1984. History, development, and importance of insect nematology. *In* "Plant and Insect Nematodes." (W. R. Nickle, ed.), pp. 627–653. Marcel Dekker, New York.

Palm, N. B. 1948. Normal and pathological histology of the ovaries in *Bombus* Latr. (Hymenopt.). *Opuscula Entomol. Suppl. VII*, 101.

Petersen, J. J. 1975a. Penetration and development of the mermithid nematode *Reesimermis neilseni* in eighteen species of mosquitoes. *J. Nematol.* 7, 207–210.

Petersen, J. J. 1975b. Development and fecundity of *Reesimermis neilseni*, a nematode parasite of mosquitoes. *J. Nematol.* 7, 211–214.

Petersen, J. J. 1984. Nematode parasites of mosquitoes. *In* "Plant and Insect Nematodes." (W. R. Nickle, ed.), pp. 797–820. Marcel Dekker, New York.

Petersen, J. J. 1985. Nematodes as biological control agents: Part I. Mermithidae. *Adv. Parasitol.* 24, 307–344.

Petersen, J. J., and Chapman, H.C. 1979. Checklist of mosquito species tested against the nematode parasite *Romanomermis culicivorax*. *J. Med. Entomol.* 15, 468–471.

Petersen, J. J., and Willis, O. R. 1970. Some factors affecting parasitism by mermithid nematodes in southern house mosquito larvae. *J. Econ. Entomol.* 63, 175–178.

Petersen, J. J., and Willis, O. R. 1972. Procedures for the mass rearing of a mermithid parasite of mosquitoes. *Mosq. News 32*, 226–230.

Petersen, J. J., and Willis, O. R. 1975. Establishment and recycling of a mermithid nematode for the control of mosquito larvae. *Mosq. News 35*, 526–532.

Petersen, J. J., Chapman, H. C., and Woodard, D. B. 1967. Preliminary observations on the incidence and biology of a mermithid nematode of *Aedes sollicitans* (Walker) in Louisiana. *Mosq News 27*, 493–498.

Petersen, J. J., Chapman, H. C., and Woodard, D. B. 1968. The bionomics of a mermithid nematode of larval mosquitoes in southwestern Louisiana. *Mosq. News 28*, 346–352.

Phelps, R. J., and DeFoliart, G. R. 1964. Nematode parasitism of Simuliidae. *Univ. Wisconsin Res. Bull.* 245.

Poinar, G. O., Jr. 1968. Parasitic development of *Filipjevimermis leipsandra* Poinar and Welch (Mermithidae) in *Diabrotica u. undecimpunctata* (Chrysomelidae). *Proc. Helminth. Soc. Wash.* 35, 161–169.

Poinar, G. O., Jr. 1969. Arthropod immunity to worms. *In* "Immunity to Parasitic Animals." (G. J. Jackson, R. Herman, and I. Singer, eds.), Vol. 1, pp. 173–210. Appleton-Century-Crofts, New York.

Poinar, G. O., Jr. 1975a. "Entomogenous Nematodes. A Manual and Host List of Insect–Nematode Associations." E. J. Brill, Leiden, The Netherlands.

Poinar, G. O., Jr. 1975b. Description and biology of a new insect parasitic rhabditoid, *Heterorhabditis bacteriophora* n. gen., n. sp. (Rhabditida: Heterorhabditidae n. fam.). *Nematologica 21*, 463–470.

Poinar, G. O., Jr. 1979. "Nematodes for Biological Control of Insects." CRC Press, Boca Raton, Florida.

Poinar, G. O., Jr. 1983. "The Natural History of Nematodes." Prentice-Hall, Englewood Cliffs, New Jersey.

Poinar, G. O., Jr. 1984. On the nomenclature of the genus *Neoaplectana* Steiner, 1929 (Steinernematidae: Rhabditida) and the species *N. carpocapsae* Weiser, 1955. *Rev. Nematol. 7*, 199–200.

Poinar, G. O., Jr. 1990. Biology and taxonomy of Steinernematidae and Heterorhabiditidae. *In* "Entomopathogenic Nematodes in Biological Control." (R. Gaugler and H. K. Kaya, eds.), pp. 23–61. CRC Press, Boca Raton, Florida.

Poinar, G. O., Jr., and Doelman, J. J. 1974. A reexamination of *Neochordodes occidentalis* (Montg.) comb. n. (Chrododidae: Gordioidea): Larval penetration and defense reaction in *Culex pipens* L. *J. Parasitol. 60*, 327–335.

Poinar, G. O., Jr., and Doncaster, C. C. 1965. The penetration of *Tripius sciarae* (Bovien)(Sphaerulariidae: Aphelenchoidea) into its insect host, *Bradysia paupera* Tuom. (Mycetophilidae: Diptera). *Nematologica 11*, 73–78.

Poinar, G. O., Jr., and Gyrisco, G. G. 1962. Studies on the bionomics of *Hexamermis arvalis* Poinar and Gyrisco, a mermithid parasite of alfalfa weevil, *Hypera postica* (Gyllenhal). *J. Insect Pathol. 4*, 469–483.

Poinar, G. O., Jr., and Himsworth, P. T. 1967. *Neoaplectana* parasitism of larvae of the greater wax moth, *Galleria mellonella*. *J. Invertebr. Pahtol. 9*, 241–246.

Poinar, G. O., Jr., and Thomas, G. M. 1965. A new bacterium, *Achromobacter nematophilus* sp. nov. (Achromobacteriaceae: Eubacteriales) associated with a nematode. *Int. Bull. Bacteriol. Nomen. Taxon. 15*, 249–252.

Poinar, G. O., Jr., and Thomas, G. M. 1966. Significance of *Achromobacter nematophilus* Poinar and Thomas (Achromobacteraceae: Eubacteriales) in the development of the nematode, DD-136 (*Neoaplectana* sp. Steinernematidae). *Parasitology 56*, 385–390.

Poinar, G. O., Jr., and van der Laan, P. A. 1972. Morphology and life history of *Sphaerularia bombi*. *Nematologica 18*, 239–252.

Poinar, G. O., Jr., Lane, R. S., and Thomas, G. M. 1976a. Biology and redescription of *Pheromermis pachysoma* (V. Linstow) n. gen., n. comb. (Nematoda: Mermithidae), a parasite of yellowjackets (Hymenoptera: Vespidae). *Nematologica 22*, 360–370.

Poinar, G. O., Jr., Triggiani, O., and Merritt, R. W. 1976b. Life history of *Eudiplogaster aphodii* (Rhabditida: Diplogasteridae), a facultative parasite of *Aphodius fimetarius* (Coleoptera: Scarabaeidae). *Nematologica 22*, 79–86.

Poinar, G. O., Jr., Thomas, G. M., and Hess, R. 1977. Characteristics of the specific bacterium associated with *Heterorhabditis bacteriophora* (Heterorhabditidae: Rhabditida). *Nematologica 23*, 97–102.

Popiel, I., Grove, D. L., and Friedman, M. J. 1989. Infective juvenile formation in the insect parasitic nematode *Steinernema feltiae*. *Parasitology 99*, 77–81.

Pye, A. E., and Burman, M. 1981. *Neoaplectana carpocapsae*: nematode accumulations on chemical and bacterial gradients. *Exp. Parasitol. 51*, 13–20.

Remillet, M., and Laumond, C. 1991. Sphaerularioid nematodes of importance in agriculture. *In* "Manual of Agricultural Nematology." (W. R. Nickle, ed.), pp. 967–1024. Marcel Dekker, New York.

Remillet, M., and Silvain, J.-F. 1988. *Noctuidonema guyanense* n.g., n. sp. (Nematoda: Aphelenchoididae) ectoparasite de noctuelles du genre *Spodoptera* (Lepidoptera: Noctuidae). *Rev. Nematol. 11*, 21–24.

Rempel, J. G. 1940. Intersexuality in Chironomidae induced by nematode parasitism. *J. Exp. Zool. 84*, 261–289.

Rogers, C. E., Marti, O. G., Simmons, A. M., and Silvain, J. F. 1990. Host range of *Noctuidonema guyanense* (Nematoda: Aphelenchoididae): An ectoparasite of moths in French Guiana. *Environ. Entomol. 19*, 795–798.

Rühm, W. 1956. "Die Nematoden der Ipiden." *Parasitol. Schriftenreihe, Jena, 6*.

Schmidt, G. S., and Roberts, L. S. 1985. "Foundations of Parasitology." Third Ed. Times Mirror/Mosby College Publ., St. Louis, Missouri.

Schmidt, J., and All, J. N. 1978. Chemical attraction of *Neoaplectana carpocapsae* (Nematoda: Steinernematidae) to insect larvae. *Environ. Entomol.* 7, 605–607.

Schmidt, J., and All, J. N. 1979. Attraction of *Neoaplectana carpocapsae* (Nematoda: Steinernematidae) to common excretory products of insects. *Environ. Entomol.* 8, 55–61.

Shephard, M. R. N. 1974. Anthropods as final hosts of nematodes and nematomorphs. An annotated bibliography 1900–1972. Techn. Comm. No. 45. Commonwealth Inst. Helminth., St. Albans, Commonwealth Agr. Bur., Farnham Royal, Slough, U. K.

Steinhaus, E. A. 1949. "Principles of Insect Pathology." McGraw-Hill, New York.

Stoffolano, J. G., Jr. 1973. Host specificity of entomophilic nematodes—a review. *Exp. Parasitol.* 33, 263–284.

Stoffolano, J. 1986. Nematode-induced host responses. *In* "Humoral and Cellular Immunity in Arthropods." (A. P. Gupta, ed.), pp. 117–155. John Wiley and Sons, New York.

Stoffolano, J. G., Jr., and Nickle, W. R. 1966. Nematode parasite (*Heterotylenchus* sp.) of face fly in New York State. *J. Econ. Entomol.* 59, 221–222.

Thomas, G. M., and Poinar, G. O., Jr. 1979. *Xenorhabdus* gen. nov., a genus of entomopathogenic, nematophilic bacteria of the family Enterobacteriaceae. *Int. J. Syst. Bacteriol.* 29, 352–360.

Thong, C. H. S., and Webster, J. M. 1973. Morphology and the post-embryonic development of the bark beetle nematode *Contortylenchus reversus* (Sphaerulariidae). *Nematologica 19*, 159–168.

Thong, C. H. S., and Webster, J. M. 1975a. Effects of *Contortylenchus reversus* (Nematoda: Sphaerulariidae) on hemolymph composition and oocyte development in the beetle *Dendroctonus pseudotsugae* (Coleoptera: Scolytidae). *J. Invertebr. Pathol.* 26, 91–98.

Thong, C. H. S., and Webster, J. M. 1975b. Effects of the bark beetle nematode, *Contortylenchus reversus*, on gallery construction, fecundity, and egg viability of the Douglas fir beetle, *Dendroctonus pseudotsugae* (Coleoptera: Scolytidae). *J. Invertebr. Pathol.* 26, 235–238.

Thorne, G. 1940. The hairworm, *Gordius robustus* Leidy, as a parasite of the Mormon cricket, *Anabrus simplex* Holdeman. *J. Wash. Acad. Sci.* 30, 219–231.

Thorne, G. 1961. "Principles of Nematology." McGraw-Hill, New York.

Triggiani, O., and Poinar, G. O., Jr. 1976. Infection of adult Lepidoptera by *Neoaplectana carpocapsae* (Nematoda). *J. Invertebr. Pathol.* 27, 413–414.

Vandel, A. 1927. Modifications determinées par un nématode du genre "Mermis" chez les ouvrières et les soldats de la fourmi "*Pheidole pallidula*" Nyl. *Bull. Biol. Fr. Belgique.* 61, 38–48.

Van Zwaluwenburg, R. H. 1928. The interrelationships of insects and roundworms. *Bull. Exp. Stn. Hawaii. Sugar Planters' Assoc., Entomol. Ser. Bull.* 20.

Wachek, F. 1955. Die entoparasitischen Tylenchiden. *Parasitol. Schriftenreihe 3*, 1–119.

Walker, T. W., Meek, C. L., and Wright, V. L. 1985. Establishment and recycling of *Romanomermis culicivorax* (Nematoda: Mermithidae) in Louisiana ricelands. *J. Am. Mosq. Control Assoc. 1*, 468–473.

Webster, J. M., and Thong, C. H. S. 1984. Nematode parasites of orthopterans. *In* "Plant and Insect Nematodes." (W. R. Nickle, ed.), pp. 697–726. Marcel Dekker, New York.

Weis-Fogh, T. 1956. Biology and physics of locust flight. II. Flight performance of the desert locust (*Schistocerca gregaria*). *Phil. Trans. R. Soc. Ser. B 239*, 459–510.

Welch, H. E. 1960. *Hydromermis churchillensis* n. sp. (Nematoda: Mermithidae) a parasite of *Aedes communis* (DeG.) from Churchill, Manitoba, with observations of its incidence and bionomics. *Can. J. Zool.* 38, 465–474.

Welch, H. E. 1963. Nematode infections. *In* "Insect Pathology: an Advanced Treatise." (E. A. Steinhaus, ed.), Vol. 2, pp. 363–392. Academic Press, New York.

Welch, H. E., and Bronskill, J. F. 1962. Parasitism of mosquito larvae by the nematode, DD136 (Nematoda: Neoaplectanidae). *Can. J. Zool. 40*, 1263–1275.

Westerdahl, B. B., Washino, R. K., and Platzer, E. G. 1982. Successful establishment and subsequent recycling of *Romanomermis culicivorax* (Mermithidae: Nematoda) in a California rice field following postparasite application. *J. Med. Entomol. 19*, 34–41.

Wheeler, W. R. 1910. The effects of parasitic and other kinds of castration in insects. *J. Exp. Zool. 8*, 377–438.

Wheeler, W. M. 1928. *Mermis* parasitism and intercastes among ants. *J. Exp. Zool. 50*, 165–237.

Wouts, W. M., Mráček, Z., Gerdin, S., and Bedding, R. A. 1982. *Neoaplectana* Steiner, 1929 a junior synonym of *Steinernema* Travassos, 1927 (Nematoda; Rhabditida). *Syst. Parasitol. 4*, 147–154.

Wülker, W. 1961. Untersuchungen uber die Intersexualitat der Chironomiden (Dipt.) nach *Paramermis*-Infektion. *Arch. Hydrobiol. Suppl. 25*, 127–181.

Wülker, W. 1964. Parasite-induced changes of internal and external sex characters in insects. *Exp. Parasitol. 15*, 561–597.

CHAPTER 14

HOST RESISTANCE

I. RESISTANCE AT THE SPECIES LEVEL
II. INSECT AGE AND STAGE
III. NUTRITIONAL FACTORS
   A. Dietary Factors
   B. Exogenous Antimicrobial Factors
IV. PHYSICAL FACTORS
V. MORPHOLOGICAL AND PHYSIOLOGICAL DEFENSES
   A. Integument
   B. Glandular Secretions
   C. Respiratory System
   D. Digestive Tract
VI. HEMATOLOGY
   A. Hemocytes
   B. Function of Hemocytes
VII. BLOOD CLOTTING
VIII. CELLULAR IMMUNITY
   A. Phagocytosis
   B. Cellular Encapsulation
   C. Giant Cells, Nodules, and Tumors
   D. Responses of Hemocytes to Biotic Agents
IX. INSECT PLASMA AND HUMORAL IMMUNITY
X. INNATE HUMORAL IMMUNITY
   A. Hemagglutinins
   B. Humoral Encapsulation
   C. Other Humoral Factors
XI. ACQUIRED IMMUNITY
   A. Antibacterial and Toxic Substances
   B. Lysozymes
   C. Cecropins, Attacins, and Diptericins
   D. Phenoloxidase System
   E. Complement
   F. Other Immunity-Related Factors in Hemolymph
   G. Inhibitors Acting on Immune Process
XII. PASSIVELY ACQUIRED IMMUNITY

The reader, after going through the preceding chapters, has learned that insects are susceptible to a large number of pathogenic microorganisms. Fortunately, most of these pathogens are not infectious to humans, other vertebrates, or plants. The converse is generally true that most pathogens of humans, other vertebrates, and plants do not infect insects. This type of innate or natural immunity is known as "phylogenetic resistance" (or "phylogenetic immunity"). As one ascends the phylogenetic classification from species to kingdom, the difference in resistance of the species increases (i.e., the greater the diversity between species, the more they are apt to differ in their resistance to a specific pathogen).

Early investigators could not infect insects with such well-known human bacterial pathogens as *Pasteurella pestis* of plague, *Corynebacterium diphtheriae* of diphtheria, *Mycobacterium leprae* of leprosy, *Clostridium tetani* of tetanus, and *Mycobacterium tuberculosis* of tuberculosis (Steinhaus 1949). Similarly, when the cockroach *Blabera fusca* was fed the following bacteria, *Mycobacterium tuberculosis* or *Salmonella typhimurium* (both human pathogens), *Mycobacterium phlei* (nonpathogenic hay bacterium), *Bacillus larvae* (cause of American foulbrood of honey bee), or *Bacillus popilliae* (cause of type A milky disease of the Japanese beetle), none of them caused infection (Clot and Vago 1970). The bacteria passed through the digestive tract in 3 to 5 days, and the eliminated bacteria retained their pathogenicity to their original hosts.

A decreasing number of cases of cross transmission of insect pathogens are apparent as one proceeds from the lower taxa (the genera and species) to the higher taxa. Within the insect class Hexapoda (syn. Insecta), specific diseases are found only in certain orders, such as the nuclear polyhedrosis virus (NPV) disease, which occurs mainly in Lepidoptera, less frequently in Hymenoptera, and rarely in Diptera and Orthoptera. The only other arthropod thus far found infected with NPV is a crustacean (Couch 1974). The granulosis virus (GV) is much more specific and has been reported infecting only the Lepidoptera (Ignoffo 1968) and one unconfirmed case of a Hymenoptera (Martignoni and Iwai 1986). The milky-disease bacteria, which include several species, infect only Coleoptera in the family Scarabaeidae. We refer the reader to each pathogen chapter for other examples of host specificity and resistance to disease.

I. RESISTANCE AT THE SPECIES LEVEL

At the species level, great variations in the susceptibility of insect strains and varieties to pathogens occur. Such differences are genetically determined and have been studied most extensively in the silkworm and honey bee to develop resistant strains. We shall discuss some examples from these two insects.

Many silkworm varieties or strains have almost homozygous genetic constitution and have been used to study the basis for the genetic resistance to disease. There are strains that are resistant to insect viruses [e.g., cytoplasmic polyhedrosis viruses (CPV) and NPV, densonucleosis virus, and flacherie virus] (Aruga and Watanabe 1964; Aratake 1973a,b; Watanabe, 1965, 1967; Watanabe and Maeda 1978, 1981). The resistance, in general, is controlled by polygenes. Strain Daizo, however, is a special case with a completely dominant major gene for resistance to CPV (Watanabe 1965, 1967). The resistance to densonucleosis virus is inherited as a completely recessive nonsex-linked character (Watanabe and Maeda 1981).

In the honey bee, four mechanisms of resistance are involved against pathogens: (1) hygienic behavior, (2) larval resistance, (3) efficiency in straining spores from the nectar in the honey sac (proventriculus), and (4) presence of antibiotics in the food (Rothenbuhler et al. 1968). The first three mechanisms are genetically determined. There are resistant strains for *Bacillus larvae* of the American foulbrood (Rothenbuhler and Thompson 1956; Rothenbuhler 1964), the virus of chronic bee paralysis (Bailey 1965; Kulinčević and Rothenbuhler 1975), the fungus *Ascosphaera apis* of chalkbrood (Milne 1982, 1983), and the protozoan *Nosema apis* (Sidorov et al. 1975). The hygienic behaviors of uncapping and removal are determined by two homozygous recessive genes, with both genes effective for the American foulbrood (Rothenbuhler 1964). Uncapping behavior leads to the removal of caps from cells containing dead brood and discarding them. In the case of chalkbrood, the removal behavior of dead uncapped brood is involved, but capped dead brood are not removed (Milne 1983). Resistance against *Nosema apis* is due to heterosis, probably of a polygenic system (Sidorov et al. 1975).

Certain "wild insects" have strains that are more or less resistant to baculoviruses (NPVs and GVs). Isolates of the endemic California oak moth, *Phryganidia californica* (Martignoni and Schmid 1961), the fall armyworm, *Spodoptera frugiperda* (Reichelderfer and Benton 1974; Fuxa 1987), the light-brown apple moth, *Epiphyas postvittana* (Briese et al. 1980), and the velvetbean caterpillar, *Anticarsia gemmatalis* (Boucias et al. 1980), differ in their susceptibility to NPVs. The resistance in the fall armyworm is not sex linked and is due to a single gene or genes that lack dominance (Reichelderfer and Benton 1974). A dominant autosomal gene, which segregates according to simple Mendelian ratios, is responsible for the resistance of the potato tuber moth, *Phthorimaea operculella*, to its GV (Briese 1982). The survivors of an epizootic of GV of the larch bud moth, *Eucosma griseana*, were resistant to the virus by about 38 times (Martignoni 1957). Since the dosage–mortality slope was very steep, the viral epizootic appeared to have removed most of the more susceptible individuals from the larch bud moth population.

In *Drosophila melanogaster*, certain natural populations are resistant to the sigma virus, which produces CO_2 sensitivity in adult flies (Fleuriet 1976). A gene, ref (2) p^p allele, on the second chromosome is known to interfere with the multiplication of the virus, and this gene partly explains the resistance occurring in natural populations. There are other alleles (e.g., M^m and D^p) that decrease the probability of initiating infection by the sigma virus (Coulon and Contamine 1982). Five genes are known to act specifically on the multiplication of the virus. Coulon and Contamine (1982) propose the hypothesis that a mutation affecting the viral protein–genome complex may modify viral interactions with the host genes.

Attempts have been made to select strains resistant to various pathogens by exposing continuous generations of the survivors to the pathogens. House flies have become resistant to *Bacillus thuringiensis* after the 27th and 50th generations (Harvey and Howell 1965). The resistance acts primarily against the β-exotoxin (flyfactor) produced by the bacillus. Early attempts to select for resistance to the spore–crystal protein (δ-endotoxin) complex of this bacillus have failed, but McGaughey (1985) has increased the resistance of the Indianmeal moth, *Plodia interpunctella*, by nearly 30-fold in two generations fed the bacillus spores and crystals, and after 15 generations reached a plateau 100 times higher than the control level. The resistance was inherited as a partially recessive trait (McGaughey and Beeman 1988). Moreover, McGaughey (1985) has reported the first case of an insect pest developing resistance to the application of a microbial insecticide. The *P. interpunctella* strains collected from treated grain bins were more resistant to *B. thuringiensis* than strains from untreated bins. The mechanism of resistance is correlated with a 50-fold reduction in affinity of the membrane receptor on midgut cells to the spore–crystal protein (Van Rie et al. 1990). Subsequently, Tabashnik et al. (1990) have reported resistance of field populations of the diamondback moth to *B. thuringiensis*. In mosquitoes, Goldman et al. (1986) have reported a small shift in resistance against one of two wild strains of *Aedes aegypti* exposed to 14 generations under LC_{50} selection pressure of *B. thuringiensis* subspecies *israelensis*.

Silkworm strains have been selected for resistance to viral diseases. After 19 generations, selection under cold (5°C) treatment produced a strain resistant to the induction of viral diseases, such as the nuclear and cytoplasmic polyhedroses and flacherie (Aizawa et al. 1961). After exposure to a flacherie virus, resistant silkworm strains have been obtained after five generations (Uzigawa and Aruga 1966; Watanabe 1967). A selection pressure greater than 60% mortality is effective in inducing and retaining the resistance of the selected generation. The hybrids from crosses between selected strains are more resistant than those from crosses of unselected strains. Briese and Mende (1983) have serially exposed a susceptible laboratory strain of the potato tuber moth to a GV. After exposure for over six generations of the moth, the larval resistance to the virus was increased by 140-fold.

Information on insect resistance to fungi, protozoa, and nematodes is not as well documented. We have already discussed some cases on the silkworm and honey bee. Certain races of the moths *Antheraea pernyi* and *Platysamia cecropia* are more resistant to microsporidia than others (Weiser 1969). In fungi, differences in susceptibility occur in isolates of the wireworm, *Conoderus falli*, to the greenmuscardine fungus, *Metarhizium anisopliae* (Bell and Hamalle 1971), and the resistance of the clones of the aphid *Acyrthosiphon pisum* to *Entomophthora obscura* has been reported for the first time (Papierok and Wilding 1979). Mosquitoes, susceptible to infection by nematodes, have developed resistance when exposed to them for successive generations (Woodard and Fukuda 1977; Petersen 1978).

Thus far, insects that have developed resistance after continuous application of chemical insecticides are not equally resistant to pathogens. There are numerous studies in this area. An example is that of Ignoffo and Roush (1986) who reported that permethrin- and methomyl-resistant strains of *Heliothis virescens* are as susceptible to a NPV, *Bacillus thuringiensis*, and a microsporidium (*Vairimorpha necatrix*) as nonresistant strains.

II. INSECT AGE AND STAGE

Some insects are susceptible to diseases only at certain stages in their life cycles. All pathogens of the brood diseases of the honey bee were primarily considered to infect only the immature stages, the larva and pupa. This is still the case with the American and European foulbroods, but the sacbrood virus has been found to infect young adult bees (Lee and Furgala 1967; Bailey 1969). The *Nosema* disease of the honey bee, caused by the protozoan *Nosema apis*, generally develops only in adults and not in larvae and pupae, but a strain of *Nosema* in South Africa infects the brood and not the adult bees (Buys 1972, Clark 1980).

Insect viruses, especially the baculoviruses and CPV, were considered at one time to infect only the immature and not the adult stages, but there are increasing reports that the adults are also susceptible but usually to a lesser degree. In most cases, the virus-infected adults show little or no symptoms. The adult cabbage looper, *Trichoplusia ni*, when infected by intrahemocoelic inoculation with NPV, displays no ill effects with respect to longevity, mating, and oviposition (Vail and Hall 1969). Some lepidopteran adults infected with CPV display deformities, reduced fertility, and tumors (Neilson 1965; Vail and Gough 1970). In these cases, the larvae that become infected with the CPV show few signs and symptoms, and the infection only becomes apparent in adults.

When the fifth instar larvae of the codling moth, *Cydia pomonella*, are induced for pupation or for diapause, the individuals become significantly more tolerant to a GV (Camponovo and Benz 1984). The prepupal stage of the European spruce sawfly, *Diprion hercyniae*, is immune to infection by a NPV that attacks the midgut epithelium (Bird 1953). In the prepupa, the temporary midgut epithelium consists entirely of embryonic (regenerative) cells that are resistant to the virus.

There is ample evidence that different larval stages vary in their susceptibility to pathogens. In some cases, the larva increases in resistance as it develops, and such resistance is referred to as "maturation immunity." Honey bee larvae younger than 24 h are most susceptible to American foulbrood and can be infected with 10 or fewer spores of the etiologic agent *Bacillus larvae* (Bailey 1968). The basis for the increasing resistance as the larva develops is still unknown, but it is associated with the germination of spores and the development of the bacterium in the insect's digestive tract. The early larval instars of the almond moth, *Cadra cautella*, and the Indianmeal moth are more susceptible to *Bacillus thuringiensis* than the older instars (McGaughey 1978).

Bioassay studies indicate that the younger larval instars of the silkworm are more susceptible to a CPV than the older instars (Aruga and Watanabe 1964). In the forest tent caterpillar, *Malacosoma disstria*, the larva markedly increases in resistance to a NPV from the first to the last instar (Stairs 1965a). The first instar larva is almost 1000 times more susceptible than the third, and 68,000 times more than the fourth instar larva. In *Heliothis armigera*, there is maturation immunity against the NPV (Whitlock 1977; Teakle et al. 1985) but not for the GV (Whitlock 1977). The maturation immunity in *Heliothis* may involve the midgut and factor(s) in the hemocoel (Teakle et al. 1986). The larva of the armyworm, *Pseudaletia separata*, increases in resistance to a NPV as it develops (Neelgund and Mathad 1974). The virus also infects pupa and adult, but the resistance of these stages to infection is much higher than that in the larval stage. The LD_{50} of a NPV in 8-day-old larvae of the corn earworm, *Heliothis zea*, is 250 times higher than in 3-day-old larvae (Allen and Ignoffo 1969). This difference, however, is less than twofold when the dose is calculated on the basis of larval body weight. On the other hand, a quantitative study, based on body weight and involving the fall webworm, *Hyphantria cunea*, and NPV and GV, indicates that distinct maturation immunity occurs in the larval stages (Boucias and Nordin 1977). When the mortality data are converted on the basis of a dose per mg body weight, the decrease in susceptibility is only partly due to differences in larval weights. There is a difference in the inter-instar susceptibility to the two types of viruses. Evans (1981, 1983) has reported similar results with *Mamestra brassicae* and its NPV, where there is a 34,000-fold difference in LD_{50} values from the first to fifth larval instars. Maturation immunity to the NPV, however, varies with different insect species (Burgerjon et al. 1981).

With microsporidia, maturation immunity occurs in the stored-product beetle, *Tribolium castaneum*, infected with *Nosema whitei* (Milner 1973). The LD_{50} increases consistently from the first to the fifth instar. In contrast, Gaugler and Molloy (1981) demonstrated that younger instars of black fly larvae are resistant to nematode infection whereas older instars are susceptible. The mechanism is due to the physical exclusion of the nematodes from the mouth of the black fly larvae in the younger instars.

The reader should be aware that even though larval susceptibility to a pathogen may be correlated with larval weight, this is not due to dilution as in the case of chemical insecticides, because the pathogen can multiply in the insect. It is more likely that other humoral and cellular factors are involved.

In the case of fungal infections, the reports are varied (Müller-Kögler 1965). Certain fungi infect all stages of host insects, but in some cases, the egg and pupal stages are more resistant than the larval and adult stages. The eggs of noctuids *Mamestra brassicae* and *Spodoptera littoralis* are infected by *Paecilomyces fumosaroseus* but not by *Nomuraea* (syn. *Spicaria*) *rileyi* (Fargues and Rodriguez-Rueda 1980). The larval stages may show maturation immunity. The younger larval instars of the cabbage looper, *Trichoplusia ni* (Getzin 1961; Ignoffo et al. 1975), and of the Colorado potato beetle, *Leptinotarsa decemlineata* (Schaerffenberg 1957), are more susceptible to *Nomuraea rileyi* and *B. bassiana*, respectively, than the older instars. Feng et al. (1985) have also reported that the first instar larva of the European corn borer is the most susceptible stage to infection with *B. bassiana*. On the other hand, Mohamed et al. (1977) have reported that larvae of the corn earworm, *Heliothis zea*, in the third to fifth instars are more susceptible to the infection of *N. rileyi* than those in the first and second instars. Thorvilson et al. (1985) found no significant differences among the larval instars of the noctuid *Plathypena scabra* to *N. rileyi*. Rizzo (1977) has examined the relationship between the ages of various adult Diptera and their susceptibility to infection with *B. bassiana* and *Metarhizium anisopliae*. The ages of the house fly, *Musca domestica*, the black blow fly, *Phormia regina*, and the onion fly, *Hylemya antiqua*, have no significant effect on the virulence of the two fungi.

The critical period in the development of the larva to infectious diseases is shortly after molting, when its newly regenerated gut epithelium and the newly formed integument are more susceptible to fungal penetration. The newly formed pupae of some Lepidoptera are susceptible to fungal infections but not the older pupae (Sussman 1951; Tanada 1955). This resistance may be associated with the development of the waxy epicuticular layer on the integument as well as sclerotization.

III. NUTRITIONAL FACTORS

A healthy insect is generally more resistant to infection than a stressed one. Stresses brought about by malnutrition, metabolic imbalances, physical, and other factors may result in infection by potential pathogens or by the activation of a chronic to an acute infection. Certain stressors may enhance chronic infections or activate latent (occult) viral infections. Studies along this line, however, are limited even with the availability of artificial diets for most insects.

A. DIETARY FACTORS

There are early reports that nutritional factors are responsible for the outbreaks of diseases in the honey bee and the silkworm, but such observations need to be documented with quantitative studies. Hare and Andreadis (1983) reported that the susceptibility of the Colorado potato beetle, *Leptinotarsa decemlineata*, to the fungus *Beauveria bassiana* varied with the host plants. Larvae least susceptible to the fungus developed on host plants most suitable for beetle survival in the field and on field-grown plants rather than on glasshouse-grown plants. Moreover, larvae reared on tomato plants in August were much more susceptible than those reared on the same plants in June. The susceptibility of the gypsy moth, *Lymantria dispar*, to a NPV varies with the type of host plant fed to the larvae (Keating and Yendol 1987). For further discussion, see Section III.H., Host Resistance, in Chapter 15.

The larvae of the greater wax moth, *Galleria mellonella*, are much more susceptible to a NPV when fed high-nitrogen and high-carbohydrate diets than on diets with a lower concentration of these nutrients (Shvetsova 1950). This has been confirmed by Pimentel and Shapiro (1962) with the high-nitrogen diet but not with a high-carbohydrate diet. In the spruce budworm, *Choristoneura fumiferana*, infected with *Nosema fumiferanae*, diseased larvae reared on 2.5% nitrogen had significantly higher mortality than those reared on 4.5% nitrogen diet (Bauer and Nordin 1988). The elevated levels of dietary nitrogen also improved the growth and survival of the protozoan within the insect. The larvae of *Calophasia lunula*, when fed snapdragon plants, are undernourished, and the impoverished larvae, already infected with a CPV, experience increased mortality (Bucher and Harris 1968). In the cabbageworm, *Pieris brassicae*, the omission or reduction of sucrose or casein in formulated diets causes an increase in the prevalence of infections by a GV (David et al. 1972; David and Taylor 1977). Since the sugar content in plants fluctuates diurnally and during different seasons, this nutrient may play a role in the viral epizootics in populations of the cabbageworm.

In the imported cabbageworm, *Pieris rapae*, dehydrated artificial diet (approximately 50% moisture) increases larval mortality caused by a GV from less than 10% to about 40% at 25°C and to about 62% at 4°C (Biever and Wilkinson 1978). The larvae showing external signs of viral infection recover to an apparently healthy condition when their diet is changed from a dehydrated to a fresh normal diet. The absence of a source of sterol in the diet of the oriental house fly, *Musca domestica vicina* (syn. *M. vicina*), produces flacciform larvae that are unable to resist the infection from pathogenic bacteria (Silverman and Levinson 1954). The optimum vitamin C (ascorbic acid) content in the diet of the codling moth, *Cydia pomonella*, is from 0.6 to 0.8%. Suboptimal and supraoptimal concentrations of this vitamin result in a decrease in the number of circulating hemocytes and an increase in the susceptibility of the larva to *Bacillus thuringiensis* and *Beauveria bassiana* (Pristavko and Dovzhenok 1974).

B. EXOGENOUS ANTIMICROBIAL FACTORS

Antimicrobial factors in the insect's food can suppress the development of certain unfavorable microorganisms in the digestive tract. Honey has been known for many years to have antimicrobial activity. The factor against bacteria is called "inhibine." White et al. (1963) believe that the antibacterial activity of inhibine is due to hydrogen peroxide that is produced together with gluconic acid from glucose acted upon by glucose oxidase. The royal jelly, secreted by workers and fed to the queen, also has antimicrobial properties. It is more bacteriostatic against gram-positive (*Staphylococcus aureus* and *Bacillus metiens*) than against gram-negative bacteria (*Escherichia coli* and *Eberthella typhosa*) (McCleskey and Melampy 1939). The active component can be extracted with ethyl alcohol and acetone in the form of crystals.

The brood food affects the susceptibility of honey bee lines to American foulbrood. In resistant lines, the food is not only more effective in inhibiting the germination of spores of the causal agent *Bacillus larvae*, but is also more effective in reducing the number of vegetative cells of the bacillus than brood food from susceptible lines (Rose and Briggs 1969).

Plants contain antimicrobial compounds that inhibit the growth of microorganisms. These compounds are mainly terpenoids (particularly β-caryophyllene, myrcene, and pinene), phenols, and flavonoids, including catechin and tannins. The action of these substances varies with the bacterial species (Smirnoff 1972; Hedin et al. 1978). The feces of silkworm contain two antibacterial phenolic acids, protocatechuic acid, and p-hydroxybenzoic acid (Iizuka et al. 1974). In addition, caffeic acid, which is derived from chlorogenic acid in mulberry leaves, is converted in the silkworm's digestive tract into the antibacterial caffeoquinone (Nakano et al. 1987). Substances that suppress pathogenic protozoa (microsporidia) have also been detected in plants (Smirnoff 1967). A red fluorescent protein that exhibits antiviral properties has been isolated from the digestive juices of the silkworm larva (Hayashiya et al. 1976a,b). Precursors in mulberry leaves are converted by digestive enzymes to form the protein. It has a molecular weight of 27,000 to 28,000 (Uchida et al. 1984). It precipitates the NPV *in vitro* and is presumed to inactivate the virus in the larva in a manner analogous to the serological neutralization by serum proteins.

The importance of antimicrobial substances in the natural food of an insect has been demonstrated in the boll weevil, *Anthonomus grandis*. When the adult boll weevil is fed bacterially contaminated artificial diet, its midgut deteriorates and becomes progressively less functional with continued feeding and results in starvation and desiccation (McLaughlin and Sikorowski 1978). When the adult is transferred to cotton, its natural food, the midgut recovers if the deterioration is not too advanced. This recovery is associated with the suppressants—gossypol, caryophyllene, gallic acid, and tannins—that are found in the cotton plant (Hedin et al. 1978). A diet containing a high gossypol fraction from cottonseed suppresses the growth of gut bacteria and slightly improves the egg hatch of the boll weevil.

IV. PHYSICAL FACTORS

Although there is much information on the effect of physical factors on the physiological condition of the insect, there is still a serious lack of knowledge on how these factors affect insect resistance. Abnormal physical conditions may stress the insect and increase its susceptibility to pathogens. Since the physical factors generally affect both the insect and the pathogen, there is difficulty, at times, to separate the effect on the insect from that on the pathogen.

Most infections are favored by high temperatures and high humidities. The period of infection is generally shortened under these conditions because of an increased susceptibility of the insect or an enhanced pathogenicity of the pathogen. In some cases, however, the insect, when exposed to certain viruses and protozoa, recovers from the infection when it is transferred to high temperatures (thermal therapy) (Tanada 1967). Whether this apparent recovery is caused by an increased resistance of the insect or by a detrimental effect on the pathogen has not been completely resolved. In the case of the NPVs of the armyworm, *Pseudaletia unipuncta*, and the silkworm, *Bombyx mori*, the high temperature appears to inhibit the development of the virus (Watanabe and Tanada 1972; Kobayashi et al. 1981), although the immune host responses cannot be completely ruled out. Both factors appear to be involved in the case of the flacherie virus of the silkworm (Inoue and Tanada 1977). The flacherie virus infects the goblet cells but not the regenerative cells of the midgut epithelium. When fourth instar larvae are fed the virus and held for 5 days at 27°C and then transferred to a high temperature (37°C) for 1 to 3 days, the virus does not spread in the midgut cells. The infected goblet cells are shed from the epithelium and discharged with gut contents. The larva survives because of the discharge of virus-infected cells and the escape of newly regenerated cells from infection. Thermal therapy is also effective in the silkworm against the CPV and the small flacherie virus, both of which also infect the midgut epithelium (Inoue 1977).

Thermal therapy can adversely affect protozoans infecting insects (Weiser 1961). Eggs as well as pupae infected with microsporidians have been exposed to high temperatures with marked reduction, but not complete eradication, of the protozoans. The spore production of *Nosema fumiferanae* is reduced when the infected larva of the spruce budworm, *Choristoneura fumiferana*, is reared at an elevated temperature (30°C) (Wilson 1979). Spores are able to survive, but vegetative stages succumb to thermal treatment. The thermal susceptibility of vegetative stages has been confirmed in the case of microsporidia infecting insect cell lines held at 35°C (Ishihara 1968; Undeen 1975; Wilson and Sohi 1977). This observation suggests that the insect-pathogenic microsporidia are unlikely to survive in warm-blooded mammals because of their high body temperatures.

There are a few cases of insects resisting the infection of fungus when reared at high temperatures. The cabbage looper, *Trichoplusia ni* (Getzin 1961), the velvet-bean caterpillar, *Anticarsia gemmatalis* (Boucias et al. 1984), and the corn earworm, *Heliothis zea* (Gardner 1985), when reared at 30°C, resist the infection of *Nomuraea rileyi* after the apparent invasion of the fungus. Since the fungus can be cultured at 30°C, the insect may possess immune factors that become effective at high temperatures. With entomopathogenic nematodes, temperatures above 30°C inhibit the development of *Steinernema carpocapsae* and *Heterorhabditis bacteriophora* (Kaya 1977; Milstead 1981). When the honey bee brood in a hive is exposed to *S. carpocapsae*, no infection occurs (Kaya et al. 1982). The lack of infection to the brood is attributed to the high temperatures (34–35°C) occurring in the hive.

The effect of light on insect resistance to pathogens has received little attention. Watanabe and Takamiya (1976) report that silkworm larvae reared continuously in the dark are more susceptible when fed NPV or CPV. The effect is more conspicuous on larvae reared on mulberry leaves than on artificial diet.

V. MORPHOLOGICAL AND PHYSIOLOGICAL DEFENSES

Morphological and anatomical structures of an insect form natural barriers to infection. The major structures are the integument, the respiratory system consisting of spiracles and tracheae, and the digestive tract, including the intestinal epithelium and the peritrophic membrane. They form the "first line of defense." They are effective barriers, but we still do not know completely how they prevent the invasion of pathogens. Because it is difficult at times to separate morphological and physiological defense factors, we will discuss both aspects in this section.

The physiology and structure of the cell may also prevent pathogen infection. In many diseases, pathogens confine their invasion and replication to certain tissues (i.e., the tissue specificity may be associated with cell resistance). This aspect has hardly been investigated. An unusual antiviral reaction occurs in insect tissue culture. When the cell cultures of the gypsy moth, *Lymantria dispar*, are infected by the CPV isolated from the white cutworm, *Euxoa scandens*, the cells develop an antiviral reaction and expel the viroplasms, virions, and polyhedra by cell budding (Quiot et al. 1980).

A. INTEGUMENT

In general, only nematodes and fungi are capable of penetrating the integument or exoskeleton of an insect with relative ease; viruses, bacteria, and protozoa generally cannot invade the insect's hemocoel through the integument unless the integument is broken, or unless they are transmitted by vectors. The exoskeleton of insects not only serves as an effective waterproof, physical barrier, but it also contains chemical substances that inhibit the growth and penetration of microorganisms (David 1967). When the integument is treated with fat solvents or is scratched with abrasives, it is much more readily penetrated by fungi. The waxy epicuticular layer of the silkworm integument contains free medium-chain saturated fatty acids, presumably caprylic, capric, and linoleic acids, that inhibit the invasion of fungi (Koidsumi 1957; Champlin et al. 1981). Lipids in the epicuticle inhibit the invasion of the fungi *Beauveria bassiana* in *Eurygaster integriceps* (Evlakhova and Chekhourina 1963), *Aspergillus flavus* in *Platysamia cecropia* (Sussman 1951), and *B. bassiana* and *Paecilomyces fumosa-roseus* in *Bombyx mori* and *Hyphantria cunea* (Saito and Aoki 1983). In the alkali bee, *Nomia melanderi*, the integument of the prepupa contains bacteriostatic substances that inhibit the growth of gram-positive and gram-negative bacteria (Bienvenu et al. 1968). These substances are primarily saturated 16- and 18-carbon fatty acids and a monosaturated 18-carbon chain. Since these short-chain fatty acids are not found in the cuticles of all insect orders, Samson et al. (1988) questions the role of these compounds in insect resistance. The role of lipids may be more subtle and may involve their structural configurations.

In contrast to the antimicrobial substances in the larval integument, the surfaces of the larvae of *Heliothis zea* have amino acids and glucosamine that initiate the germination and growth of *Beauveria bassiana* (Woods and Grula 1984). The amino acid composition on the larval surfaces varies among instars and with the time after molting. Amines and peptides are also present but do not inhibit the germination of fungi.

B. GLANDULAR SECRETIONS

Glandular secretions of certain insects exhibit antimicrobial activity when tested in the laboratory, but the question has not been resolved whether such secretions function in the resistance of insects to these microorganisms (Maschwitz 1967; Maschwitz et al. 1970). Gochnauer et al. (1979) have reported that geraniol and citrol present in the Nassanoff scent-gland secretions of adult worker honey bee inhibit or kill the chalkbrood fungus, *Ascosphaera apis*. The secretion of the mandibular gland of the mature worker ant, *Calomyrmex* sp., acts as both inhibitor against soil fungi and bactericide against soil bacteria (Brough 1983). In other ants, the metapleural glands secrete an active antibiotic agent, phenylacetic acid, which may protect body surfaces and nests from microorganisms (Maschwitz et al. 1970).

C. RESPIRATORY SYSTEM

The vertebrate respiratory system is a major site of entry of pathogens, but in insects, the respiratory system offers limited access for them. Although fungi are known to infect through the spiracles (Pekrul and Grula 1979; McCauley et al. 1968; Clark et al. 1968), in most cases, this mode of infection is the exception rather than the rule. However, steinernematid and heterorhabditid nematodes utilize the respiratory system as one of the major entry points to infect insects (Triggiani and Poinar 1976; Georgis and Hague 1981). Moreover, Gaugler et al. (1980, 1991) demonstrated that steinernematid nematodes are attracted to carbon dioxide, which may serve as a short-range attractant for eventual penetration through the spiracles. Insects have evolved mechanisms to limit or prevent nematode invasion through this portal of entry. The spiracles may be protected by sieve plates (Bedding and Molyneux 1982; Galbreath 1976), be small in diameter (Bedding and Molyneux 1982), or release only small quantities of carbon dioxide (Gaugler et al. 1991).

D. DIGESTIVE TRACT

Digestive tracts of insects consist of three major parts, the fore-, mid-, and hindguts. Fore- and hindguts are lined with cuticle, which serves as a mechanical barrier. Only a few pathogens gain entrance into the hemocoel through these two parts of the digestive tract. The midgut has a single layer of epithelial cells, which consist of columnar and occasional regenerative cells; in larvae of Lepidoptera and Trichoptera, the goblet cells are also components of the epithelium. Distal surfaces of cells, bordering the gut lumen, are lined by a chitinous peritrophic membrane that is attached at the anterior end of the midgut. The midgut is the portal of entry of many virulent bacteria, viruses, and protozoa, but the number that is able to invade through the midgut is relatively few compared to those that are kept out. This is evident because healthy insects often contain potential bacterial pathogens in their digestive tracts. These potential pathogens cause no harm and invade only when the midgut is injured (e.g., by glass particles) (Weiser and Lysenko 1956; Steinhaus 1958) or when the resistance of the insect is lowered by stress conditions. In the digestive tracts of grasshoppers, there are potentially pathogenic bacteria, such as Cloaca type B, *Proteus* and *Serratia* groups of Enterobacteriaceae, *Pseudomonas fluorescens*, and *Bacillus cereus*, which gain entrance into the hemocoel when there is a physical rupture of the gut caused at the time of molting or by protozoan gregarines (Bucher 1959). These bacteria often kill the grasshoppers once they gain entrance into the hemocoel. Such microorganisms can infect when inoculated into the hemocoel but not when fed to insects. This is also true for certain plant and animal viruses transmitted by insects (Tinsley 1975).

1. Peritrophic Membrane

The peritrophic membrane is porous and permits minute particles to pass through it, but it is still an effective barrier against microorganisms. The ultrastructure of the peritrophic membrane varies with insect species (Fig. 14–1). The mosquito iridescent virus, one of the largest "spherical" viruses, does not penetrate readily through the peritrophic membrane of a mosquito larva (Stoltz and Summers 1971). In the cabbage looper, *Trichoplusia ni*, the membrane has large discontinuities (Paschke

FIGURE 14-1

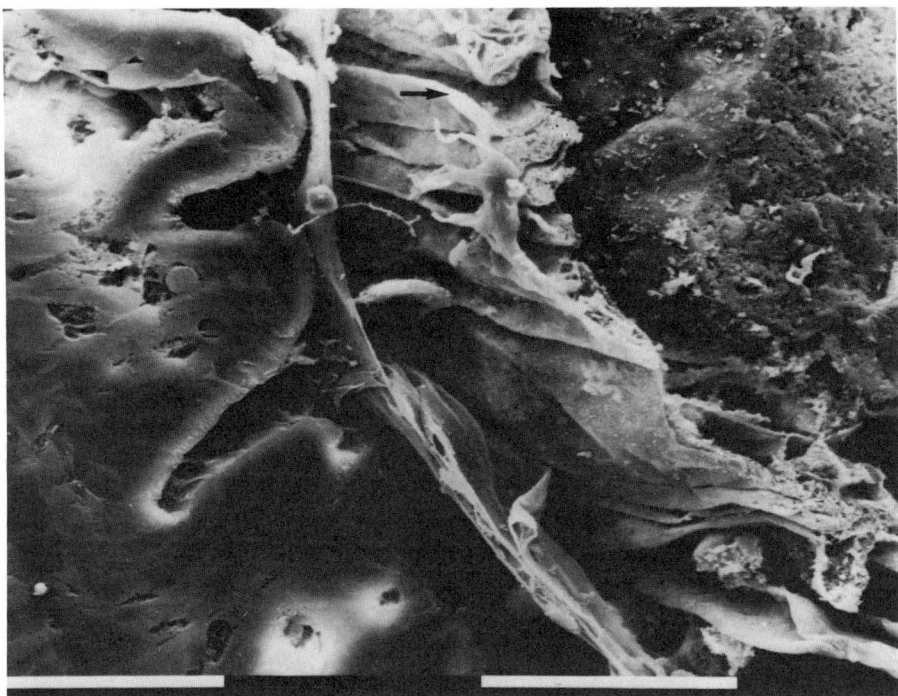

Scanning electron micrograph of the peritrophic membrane (*arrow*) from a scarab larva, *Phyllophaga hirticula* (white bar = 0.1 mm). (Courtesy of B. T. Forschler and Academic Press.)

and Summers 1975), but in the Douglas-fir tussock moth, *Orgyia pseudotsugata*, the membrane shows no pores or other discontinuities through which a bacterial or viral pathogen could penetrate (Brandt et al. 1978). The cabbage looper larvae fed NPV or GV have peritrophic membranes that are fragile compared with those of the untreated controls (Derksen and Granados 1988). The fragility is caused by a factor present in the occlusion bodies of these baculoviruses. The specific biochemical and structural changes in the peritrophic membrane enhance the infectivity of the baculoviruses. The peritrophic membrane returns to normal at 2 to 4 h postfeeding of the baculoviruses. The data suggest that this factor could be used to overcome the physical barriers of the peritrophic membrane and improve the efficacy of viral pesticides and other microbial pathogens.

The presence or absence of the peritrophic membrane does not determine the resistance of the honey bee larva to the American foulbrood caused by *Bacillus larvae*, but changes in composition and thickness of the membrane may restrict the bacillus to the lumen of the digestive tract (Davidson 1970). The honey sac (proventriculus) in the adult honey bee plays a role in its resistance to American foulbrood. Resistant strains of the honey bee filter the bacterial spores of *Bacillus larvae* more efficiently than susceptible strains (Plurad and Hartman 1965).

2. Midgut Epithelium

The midgut epithelium, in addition to its physical and chemical properties, has two other probable roles in resisting infections: it forms an electrically charged barrier to pathogen entry and regulates the solute composition of the midgut contents, the intracellular fluids of the midgut epithelium, and the hemolymph fluid (Harvey and Blankemeyer 1975). Junctional elements between cells regulate the intracellular passage of substances. The viral receptor sites and electrical charges on the cell surfaces appear to affect the attachment of enveloped virions of baculoviruses to the cell membrane. When the negative charges on the envelopes of the baculoviruses are neutralized by cationic detergents, there is an increased infection of the baculovirus fed to larvae of the armyworm, *Pseudaletia unipuncta* (Yamamoto and Tanada 1978).

The midgut epithelium has high regenerative capacity and, thereby, may resist minor infections since cells of the regenerative nidi are generally not infected by pathogens. When midgut cells are invaded by microorganisms, there is an increased mitotic activity in the regenerative nidi, the infected cells are extruded (desquamation) into the lumen of the digestive tract, and portions, if not all, of the cells are excreted. Subsequently formed midgut cells may be free of infection. This has been observed in infections with a microsporidium in larvae of the oak moth, *Phryganidia californica* (Lipa and Martignoni 1960); with the flacherie virus (Inoue 1974a; Inoue and Tanada 1977), the CPV, and a small RNA virus (Inoue and Miyagawa 1978) in the silkworm; with the silkworm CPV in the fall webworm, *Hyphantria cunea* (Yamaguchi 1979); with the NPVs in the greater wax moth, *Galleria mellonella*, and the spruce budworm, *Choristoneura fumiferana* (Stairs 1965b); with *Bacillus thuringiensis* in the rice moth, *Corcyra cephalonica* (Chiang et al. 1986), and with the milky-disease bacterium, *Bacillus popilliae*, in the European chafer, *Rhizotrogus majalis* (Splittstoesser et al. 1978).

In the European chafer, the invasion of *B. popilliae* in the midgut results in abnormal desquamation and wound repair and in the formation of a hemocyte capsule (Splittstoesser et al. 1978; Kawanishi et al. 1978) (Fig. 14–2A). The mesenteric epithelial cells of the European chafer exhibit bactericidal activity and are capable of destroying some of the vegetative rods of *Bacillus popilliae* that have entered the cytoplasm (Kawanishi et al. 1978)(Fig. 14–2B). The degree of bacterial alteration is correlated with the tightness of enclosure of the bacilli within the phagocytic vacuoles. When the bacillus is loosely enclosed, the vacuole is flaccid, the contact between vacuolar membrane and bacillus is minimal, and the ultrastructure of the bacillus is either normal or only minimally altered.

3. pH

The pH of the midgut may affect the survival and pathogenicity of microorganisms. Insects that are susceptible to *Bacillus cereus* have midgut pH less than 9, and close to the pH (6.6–7.4) that is optimum for the toxic enzyme, lecithinase (phospholipase C), produced by the bacillus (Heimpel 1955). In contrast, insects that are susceptible to *Bacillus thuringiensis* have midgut pHs between pH 9.0 and 10.5

FIGURE 14–2

(A) Masses of hemocytes (capsule) containing vegetative rods of *Bacillus popilliae* on hemocoelic surface of midgut adjacent to area of bacterial proliferation. (*M*) Midgut cells; (*C*) capsule. (Bar = 40 μm.) (B) A mesenteric cell of the European chafer containing degraded vegetative rods (*arrows*) of *B. popilliae* (bar = 1 μm). (Courtesy of C. Y. Kawanishi and Academic Press.)

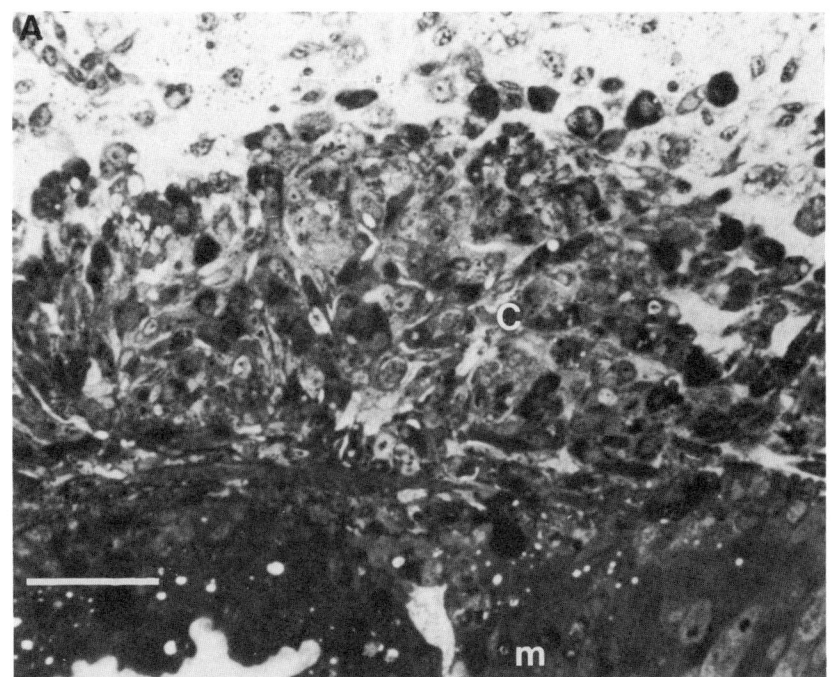

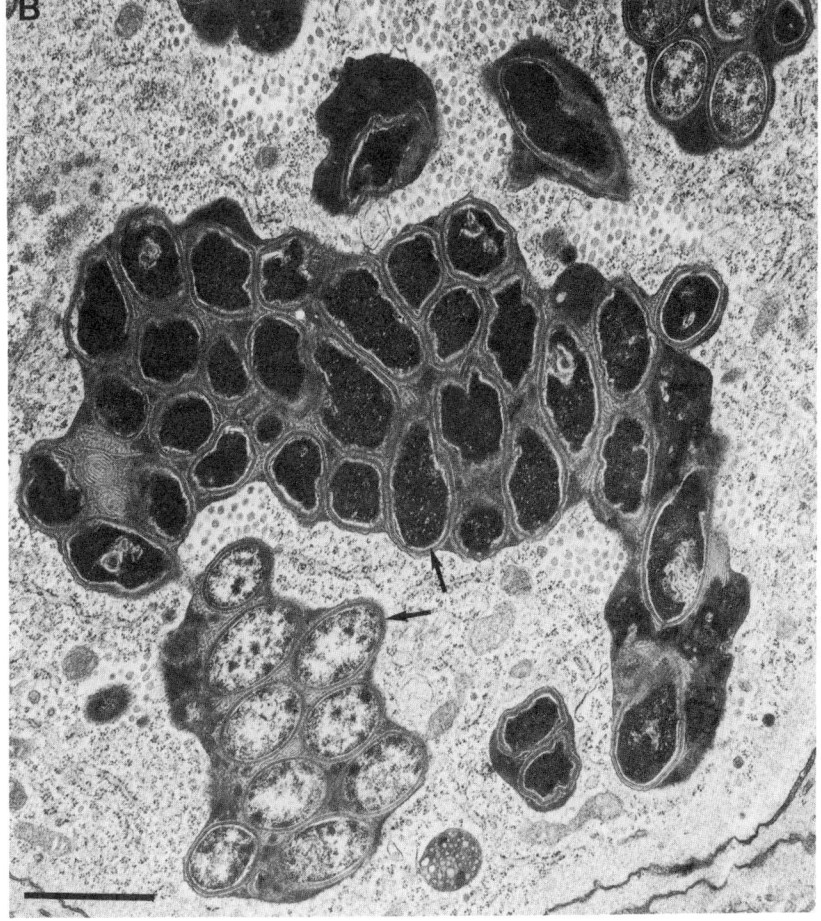

(Angus 1956). The high alkalinity is needed to dissolve the crystal endotoxin produced by this bacillus. In the case of baculoviruses, the occlusion bodies containing the virions are dissolved by the action of protease under alkaline conditions of the midgut. In mosquitoes, the resistance to a NPV is related to the high midgut pH present in older larvae (Stiles and Paschke 1980).

4. Oxygen

The oxygen tension in the midgut may affect the germination of spores. Low oxygen tension or anaerobic conditions favor the spore germination of *Bacillus larvae* (Bailey 1981) and of the chalkbrood fungus, *Ascosphaera apis* (Bailey 1967), but such conditions are unfavorable for vegetative growth. In the case of the chalkbrood, the brood is most susceptible when chilled immediately after it has been capped (Bailey 1967). In addition, the low-oxygen condition in the digestive tract lumen favors anaerobic and facultative anaerobic bacteria (e.g., *Clostridium* spp.).

5. Endogenous Antimicrobial Substances

Within the midgut, antimicrobial substances have been detected. In the silkworm larva, Masera (1954) has reported bactericidal, bacteriostatic, fungicidal, and other factors that act on pathogenic bacteria, NPV, and the microsporidium causing pébrine. Aizawa (1962) and Aruga and Watanabe (1964) have also detected in the midguts of silkworm larvae antiviral substances that are considered to be enzymes. Some of these antimicrobial compounds may have been derived from food plants. We have discussed such products in Section III.B., Exogenous Antimicrobial Factors.

In the digestive tracts of many insects (Odonata, Orthoptera, Hymenoptera, Coleoptera, and Lepidoptera), lysozyme is found in increasing concentrations from the esophagus to the anterior part of the intestine and in decreasing concentrations from the posterior portion of the midgut to the hindgut (Maschwitz 1967; Mohrig and Messner 1968). The lysozyme appears to be excreted by cells of the digestive tract and may protect an insect from the numerous bacteria eaten with food. It may also regulate the nature of the bacterial flora in the digestive tract.

6. Midgut Microflora

The composition of the microflora in the midgut may influence the susceptibility of an insect to certain pathogens. In larvae of *Galleria mellonella*, the predominant and usual bacterium in the midgut is *Streptococcus faecalis* (Jarosz 1979a). This bacterium suppresses other bacterial species ingested with food by an antibiotic (bacteriocin) and by releasing a lysozyme-like enzyme. The presence of this bacterium suppresses larval infection by other bacteria, such as *Pseudomonas aeruginosa*, *Proteus mirabilis*, *Bacillus thuringiensis*, and *Serratia marcescens*. In the desert locust, *Schistocerca gregaria*, the bacterial flora in the digestive tract inhibits the germination and viability of the conidia of *Metarhizium anisopliae*, possibly by the production of antifungal toxin (Dillon and Charnley 1986).

VI. HEMATOLOGY

In addition to the defense mechanisms described above, an insect is able to resist the development of disease even after the microorganism has entered its hemocoel. Such resistance is associated with cellular reactions and acellular (humoral) responses. The cellular and humoral immunity are interrelated (e.g., cells may secrete certain humoral factors, and in addition, humoral factors may be involved in cellular activities, such as phagocytosis and encapsulation). Since cellular immunity is mainly involved with blood cells (hemocytes), we shall briefly describe the hemocytes.

Blood or hemolymph comprises about 10 to 40% of the volume of an insect's body (Wyatt 1975; Wyatt and Pan 1978). It is mainly composed of two components, the fluid portion or plasma and blood cells or hemocytes.

The process of cell multiplication and differentiation is called hemopoiesis. Other terms that have been used for this process are hematopoiesis, hemocytopoiesis, and hemocytogenesis (Feir 1979). Usually very few circulating cells in the hemolymph are seen undergoing mitosis except under stress conditions, such as injury. The cells that divide are mainly prohemocyte and plasmatocyte, and possibly granulocyte, adipohemocyte, and spherulocyte (Arnold 1974; Jones 1977; Feir 1979). Aside from the prohemocyte, there is very little information on the multiplication of other circulating hemocytes.

A. HEMOCYTES

Hemocytes are a complex of several types of cells that circulate within the hemolymph but are sometimes attached loosely to other tissues or are enmeshed within them. They do not transport oxygen. They are nucleate cells, and their forms are sometimes amoeboid and pleomorphic (polymorphic). Even though they superficially resemble the leucocytes of vertebrates, they differ from the latter in origin, and in some aspects of development, multiplication, morphology, or function (Arnold 1974). There are no hemocytes comparable to the vertebrate erythrocytes (Jones 1962).

The classification of insect hemocytes has been a complex and controversial subject (Arnold 1979; Jones 1979). Our present knowledge is limited and based on studies of hemocytes of not more than 200 insect species in about 100 genera (Jones 1970; Arnold 1979). Hemocytes have been studied mostly in Lepidoptera, Hymenoptera, Coleoptera, and Diptera (Gupta 1985a). The confusion results partly from natural causes, such as the inherent variability of hemocytes within a species, differences between species, reactions of various cells of different species and different ages, various methods employed to observe and preserve them, etc.; partly from differences in opinions concerning the mutability of hemocyte categories and the mutability of various parameters that are used to delimit different types of hemocytes (Arnold 1974; Feir 1979).

There are three well-defined types of hemocytes in most insects (prohemocytes, plasmatocytes, and granulocytes) and one or more of four other types in many insects (coagulocytes, spherulocytes, adipohemocytes, and oenocytoids) (Fig. 14–3). Only these seven types have been identified ultrastructurally in insects (Gupta 1979, 1985a). Other highly specialized types occur infrequently or only in a few

FIGURE 14–3

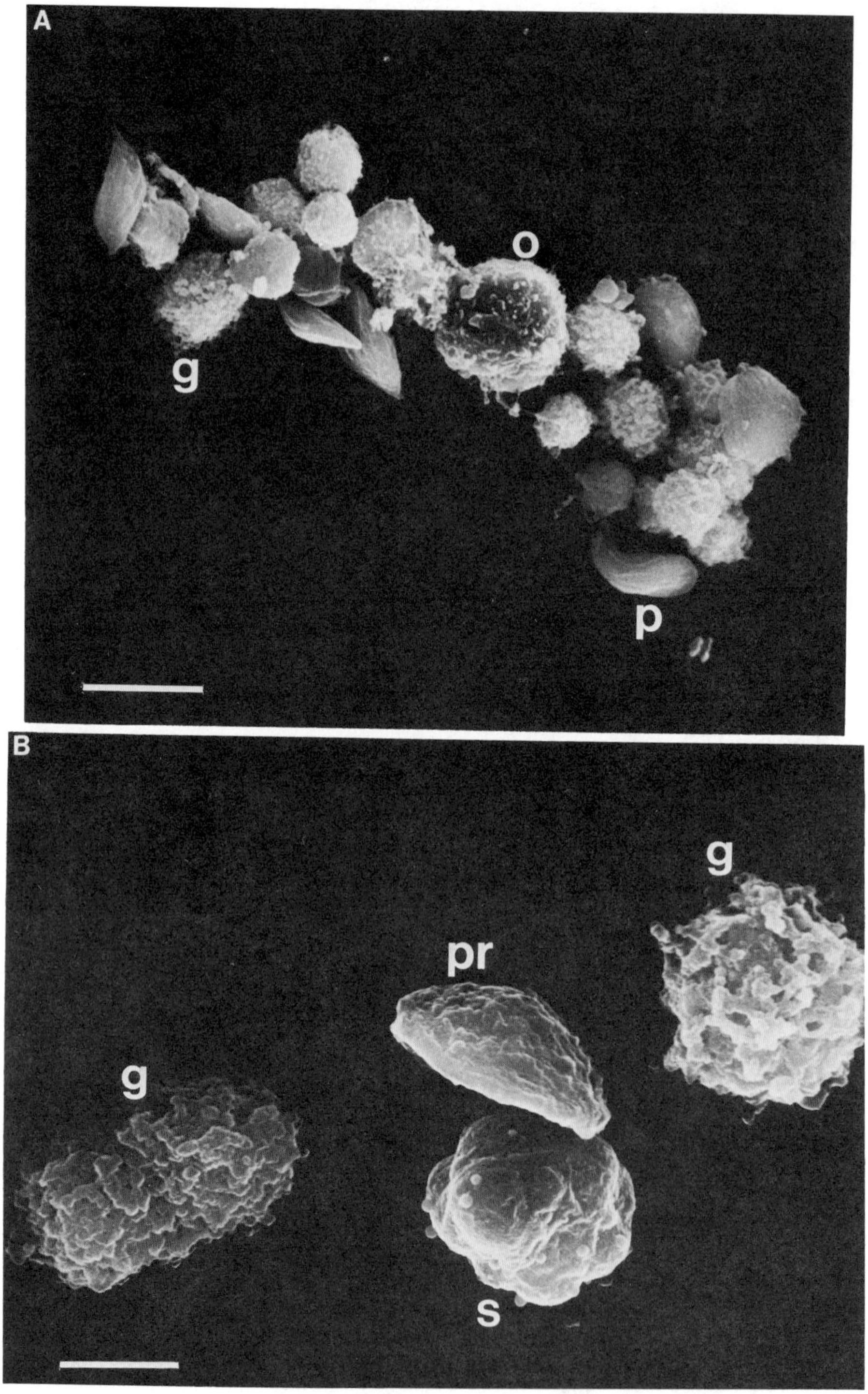

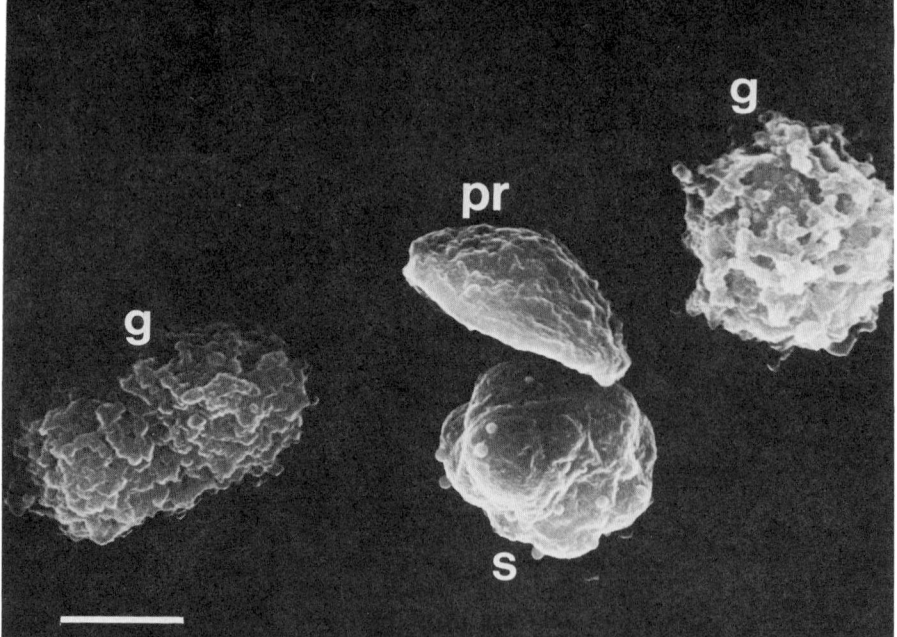

Scanning electron micrographs of the hemocytes from a fifth instar larva of the silkworm, *Bombyx mori*. (A) *g* is the granulocyte, *o*, the oenocytoid, and *p*, the plasmatocyte. (Bar = 20 μm.) (B) *g* is the granulocyte, *pr*, the prohemocyte, and *s*, the spherulocyte. (Bar = 4 μm.) [Courtesy of Akai (1976) and University of Tokyo Press.]

insect species (podocytes, crescent cells, and vermicytes) but have not been identified as distinct types with the electron microscope.

1. Prohemocytes

Prohemocytes (Arnold 1972; Gupta 1979) are small, usually round or ellipsoidal cells (6 to 10 μm wide by 6 to 14 μm long) with a large, round, centrally located nucleus (3.6 to 12 μm) surrounded by a thin rim of cytoplasm. They are germinal cells and are considered to be the main source of postembryonic hemocyte multiplication. They are the stem cells that give rise to the other main categories of hemocytes. They occur, in most stages of an insect, as only a small part of the total hemocyte population.

2. Plasmatocytes

Plasmatocytes vary greatly in size (3.3 to 5 μm by 3.3 to 40 μm) and are highly pleomorphic. The round to ovoid nucleus is generally centrally located. Scattered chromatin masses and a nucleolus may be present. Plasmatocytes are usually prominent in size and abundant in numbers and form the basic type of hemocyte in most insects (Gupta and Sutherland 1966). They generally resemble the prohemocytes that have acquired a quantity of cytoplasm, and transitional forms between the two are common. Bréhelin and Zachary (1983) recognize two types of plasmatocytes: typical and macrophagous. The typical plasmatocyte is often regular, sometimes discoid, and poorly differentiated. In the hemocoel, it forms capsules around large foreign bodies but does not phagocytize particles. It seems to occur only in Lepidoptera and Diptera. The macrophagous plasmatocyte is a large polymorphic cell with numerous digitations (pseudopodia). It is phagocytic and rapidly engulfs foreign material injected into the blood as well as tissue debris. It sometimes occurs in capsules when part of the foreign material can be phagocytized.

3. Granulocytes

Granulocytes (granular hemocytes) are compact cells of variable sizes (10 to 45 μm by 4 to 32 μm) and shapes, generally round or disc-shaped, with relatively small, usually centrally located, round or ovoid nuclei (2 to 8 μm by 2 to 7 μm). Micropapillae may occur on the plasma membrane. The large volume of cytoplasm characteristically contains many, mostly small, round granules. Granulocytes have some ultrastructural features similar to plasmatocytes but are characterized mainly by membrane-bounded, electron-dense, or granular bodies in the cytoplasm. Several types of membrane-bound granules occur in various insects (Costin 1975; Gupta 1985a). Some granules are structureless and electron dense, others are structured and contain microtubules. Granulocytes may contain lipid droplets, especially in older cells. The cytoplasm of the granulocyte has abundant free ribosomes (polysomes), Golgi bodies, both smooth and rough endoplasmic reticula, and lysosomes. Generally, mitochondria are few in number. They vary in insects from amoeboid to nonmotile forms and are easily misidentified. In some insects, they are phagocytic but not in others.

4. Adipohemocytes

Adipohemocytes are typically compact, round, or ovoid cells with usually small, rough, or ovoid nuclei situated centrally or eccentrically in a large volume of cytoplasm. The cell (from 7 to 45 μm in diameter) and the nucleus (from 4 to 10 μm in diameter) vary in size. The nucleus may be concave, biconvex, punctate, or lobate. The plasma membrane may have micropapillae, filopodia, or other irregular processes. The cytoplasm is characterized by the presence of many, small to large, sharply outlined, refringent, fat droplets. The adipohemocytes are generally nonmotile.

5. Coagulocytes

Coagulocytes (cystocytes) are polymorphous fragile cells, with generally eccentric, round nuclei with chromatin granules arranged in a cartwheel pattern. They vary from small to large (from 3 to 30 μm) with relatively small nuclei (from 5 to 11 μm). The hyaline cytoplasm is rich in polyribosomes, moderately developed endoplasmic reticulum, and few mitochondria. The coagulocytes have features of both granulocytes and oenocytoids. They are usually identified *in vitro* when they disintegrate and produce a granular precipitate. As the cells partially lyse, the nuclei, at times, are shifted to the cells' edges or may be completely ejected. They have been reported to act in phagocytosis, cellular encapsulation, and nodule formation (Rowley and Ratcliffe 1981).

6. Spherulocytes

Spherulocytes are ovoid or round cells (9 to 25 μm by 5 to 10 μm) with generally small nuclei (5 to 9 μm by 2.5 μm) rich in chromatin bodies and located centrally or eccentrically. Micropapillae, filopodia, or other processes may occur on the plasma membrane. They are characterized by the large, spherular or sometimes elliptical inclusions (spherules) that often distend the cell and conceal the centrally located nucleus. The cells release the material in the spherules by exocytosis. The cytoplasm contains rough endoplasmic reticulum, Golgi bodies, elongated mitochondria, polyribosomes, lysosomes, and numerous microtubules. They usually represent a relatively small portion of the hemocytes or are absent, but they are very striking in higher Diptera and Lepidoptera. They are nonmotile. They seem not to divide and appear to be another specialized cell within the granular hemocyte complex.

7. Oenocytoids

Oenocytoids are nonmotile cells of variable sizes and shapes, with small, round, usually eccentric nuclei, and large amounts of superficially homogeneous cytoplasm. At times, two nuclei are found. The nuclear size ranges from 3 to 15 μm in diameter. The shapes of oenocytoids vary from triangles, scimitars, or crooked-neck cell gourds. The size varies from very small to large (from 16 to over 54 μm). These cells are not present in all insects, nor in all stages of a given insect (Jones 1977). The generally thick cytoplasm may vary in complexity and may contain several kinds of plate-, rod-, or needlelike inclusions. The so-called crescent cells in the cockroach are oenocytoids (Gupta 1985b).

The oenocytoids contain polyribosomes, an abundance of large mitochondria, and other organelles, such as Golgi bodies and rough endoplasmic reticulum, and are poorly developed. Lysosomes appear to be absent. The cells are very unstable *in vitro*, breakdown, and release material into the hemolymph, but this action does not cause plasma gelation. They are not phagocytes. Their origin and relationship have not been established, but they seem to be related to the granular cell complex and may be derived from plasmatocytes. In mosquitoes, the oenocytoids are involved in melanin production (Drif and Brehélin 1983).

B. FUNCTION OF HEMOCYTES

A variety of functions has been attributed to hemocytes, but most of them have not been well documented (Jones 1964, 1970; Arnold 1974). The functions that are generally accepted are blood coagulation, phagocytosis, encapsulation, detoxification, and storage and distribution of nutritive materials. We shall be concerned with the first three functions. Phagocytosis and encapsulation comprise aspects of cellular immunity, which also includes giant cell, nodule, and tumor formation.

VII. BLOOD CLOTTING

Hemolymph coagulation in insects is affected by developmental, nutritional, seasonal, and ecological factors, and pathological conditions (Grégoire 1974). In infected or starved insects, the hemolymph may not clot or the pattern of coagulation is altered. The importance of hemolymph coagulation in insects remains to be demonstrated, but blood clotting may be considered as an immune reaction because it is involved in wound healing and may exclude the invasion of pathogens. Substances present in the plasma or excreted by blood cells during coagulation may also inhibit or destroy some microorganisms.

Insect hemolymph coagulation consists of two physiologically distinct processes that can occur independently or together: (1) cellular agglutination or coagulation and (2) plasma coagulation (Grégoire 1974; Crossley 1975). In simple cellular coagulation, plasmatocytes are mostly involved, but some other cell types may be included (Jones 1970).

Coagulation resulting from lysis of hemocytes with a granular plasma precipitate is currently attributed to a specific type of hemocyte, the coagulocyte. They are identified on the basis of their function in coagulation. These cells undergo very rapid and selective changes during coagulation. The reaction of the coagulocyte plasma membrane to foreign surfaces differs from those of other hemocytes (Goffinet and Grégoire 1979). The plasma membrane tends to develop microruptures, accompanied by the ejection of cytoplasmic material and the clotting reaction of the surrounding plasma. This difference may explain the functional contrast between active coagulocytes and other hemocytes, at least at the beginning of coagulation. Coagulocytes tend to break down when blood is removed from an insect, and only the nuclei remain intact on the slide. Other hemocytes that break down but may not function in coagulation are the spherulocytes, granulocytes, and oenocytoids.

The agent(s) responsible for blood coagulation in insects is still not known, but it seems to originate from hemocytes. Barwig and Bohn (1980) have detected in the blood of the cockroach *Leucophaea maderae* two proteinaceous coagulogens, one contained in hemocytes and the other in plasma. They suggest that clotting is initiated by the disintegration of hemocytes and the release of their coagulogen, which forms a soft gel around the cells. Immediately after the release of the hemocyte coagulogen, it reacts with the plasma coagulogen to form a rigid clot.

VIII. CELLULAR IMMUNITY

When cells participate in the protection of an insect against infectious diseases, the resistance is called cellular immunity. Some authors prefer the term constitutive cellular factor in the hemolymph rather than cellular immunity (Boman and Hultmark 1987; Dunphy and Thurston 1990). In either case, this type of immunity is observed mainly as phagocytosis and encapsulation. Pinocytosis is considered to be a form of phagocytosis. Associated with cellular reactions are capsules, giant cells, nodules, tumors, and melanotic lesions. The cellular reaction is usually a recognition of "self" and "nonself," but the hemocytes also take part in the elimination of toxins and abnormal or dead tissues.

A. PHAGOCYTOSIS

In 1884, Metschnikoff observed certain cells in the crustacean *Daphnia* engulfing the fungus *Blastomyces* (syn. *Monospora*) *bicuspidata*. He named the cells phagocytes (cells which devour) and the process as phagocytosis. Phagocytosis occurs in all animals, from the unicellular protozoa to higher animals. It is the most apparent and easily detected defense mechanism in insects. The phagocytes enter epidermis, endocrine glands, tracheal epithelium, salivary glands, parts of the alimentary canal, Malpighian tubules, pericardial cells, fat bodies, light organs, and muscles. Both circulating and fixed phagocytes are active in removing foreign particulates from the hemolymph. They engulf viruses, bacteria, fungi, protozoa, autolyzing tissues, erythrocytes, carmine ink, mycetomes, polystyrene particles, etc. (Shapiro 1979).

The term *endocytosis* has been used to combine the concepts of phagocytosis and pinocytosis (Crossley 1975). Endocytosis is the process by which the plasma membrane is infolded, leading to vesiculation and transfer of membrane and enclosed substances into the cell without actual penetration of the membrane. At one extreme, large insoluble particles are said to be phagocytized, and at the other extreme, minute or solute particles are said to be pinocytized. Besides pinocytosis, veripexis has also been used to designate the entrance (or ingestion) of viral particles into a cell (Fazekas de St. Groth 1948). Phagocytosis plays a part both in morphogenesis and in resistance to pathogens; only the latter aspect will be treated in detail in this chapter. In morphogenesis, phagocytic hemocytes are involved in the rapid breakdown of tissues and in the reutilization of cellular components during insect metamorphosis.

Phagocytic cells in insects are sometimes called arthrocytes, which are mesodermal cells capable of absorbing and selectively segregating various colloidal substances (Jones 1977). Arthrocytes include the circulating hemocytes and cells of various phagocytic organs. The specialized cells in phagocytic organs are considered by some workers as hemocytes. Arthrocytes are known by a variety of names, such as those in the head are called cephalic nephrocytes, those in pericardial sinus are called pericardial cells, and those in the esophagus are called garland cells. Specialized fat cells with an abundance of urates are also arthrocytes. Jones

(1977) has segregated the arthrocytes into three major groups based on their locations in an insect: (1) dorsal (pericardial cells, endocardial cells, cells within phagocytic organs), (2) ventral (subesophageal body cells of the embryo, garland cells of postembryonic insects, cephalic nephrocytes), and (3) circulating or temporary sessile arthrocytes (hemocytes). In addition, other cell types have been called phagocytes because they ingest specific substances, such as sperm cells by integumentary epithelial cells and follicular cells in Hemiptera, and symbionts by mycetocytes (Carayon 1977).

The competence of cellular immunity depends on the ability of the cells to recognize self and nonself components in an insect. In mammals, this competence is highly developed and is due to specific serum proteins (immunoglobulins) that act as recognition factors (opsonins or antibodies). Mammalian phagocytes more actively ingest foreign substances that are opsonized with antibody and components of complement (van Oss 1978). Although in the past the general consensus was that there is little or no evidence of the presence of opsonins in insects (Scott 1971; Crossley 1975; Ratcliffe et al. 1976; Anderson 1977), recent studies indicate that opsonins may be involved in phagocytosis (Brookman et al. 1988).

Mohrig et al. (1979a,b) have detected a phagocytosis-stimulating factor in the hemolymph of *Galleria mellonella* larva that has been inoculated with phagocytosable latex beads. This factor, which does not occur in an untreated larva, can be transferred by inoculation to other larvae. According to Ratcliffe et al. (1984), the recognition of nonself material is enhanced by the activation of the prophenoloxidase system in *G. mellonella* larva. Their conclusion is based on tests with monolayer preparations of hemocytes of *G. mellonella* treated with bacteria or bacterial endotoxin. The granulocytes discharge granules to induce hemolymph melanization and coagulation around foreign bodies, and the plasmatocytes respond to the coated (melanin) surface of these bodies. Wago (1980) has observed humoral factors (not opsonins) that promote the elongation of the filopodia (pseudopodia) of granulocytes of the silkworm. The filopodia play an essential role in the attachment of the blood cell to foreign bodies.

Hemagglutinins (lectins) serve as opsonins in mammals and may also be involved in insects as hemolymph recognition factors (Gupta 1985a). An agglutinin specific for certain erythrocytes is found in cockroaches (Scott 1971; Anderson et al. 1972). The hemagglutinins in *Leucophaea maderae* are present on the plasma membranes, in the cytoplasm, and in cytoplasmic vacuoles of granulocytes and spherulocytes (Amirante and Mazzalai 1978). In the American cockroach, hemoagglutination is due to a heat labile, nondialyzable, euglobulin-type protein, not a globulin, that has lipid, saccharide, and esterase activities (Scott 1971, 1972). The agglutinin or other opsonic humoral factor, however, does not enhance the hemocytes of *Blaberus craniifer* to phagocytize several species of bacteria (Anderson et al. 1973).

1. Process

Phagocytes ingest particles in at least three ways: (1) by formation of pinocytotic vesicles to engulf fluid that contains small particles, (2) by encircling the particles with pseudopodia, and (3) by close contact and spreading of the plasma membrane (Salt 1970). The process of phagocytosis involves the following steps (Whitcomb et al. 1974; Götz and Boman 1985). (1) Recognition by hemocytes of the presence of foreign bodies occurs in insects, but how the hemocytic activity is induced is still unknown. (2) Chemotactic attraction is involved where hemocytes are attracted to foreign bodies. The existence of this stage is still not completely resolved. (3) The attachment phase occurs when the particulate matter adheres to the cell surface. This may involve physicochemical forces or the receptor molecules of foreign bodies attaching to binding sites on the hemocyte surface (Young 1985). Opsonins may also be involved (Brookman et al. 1988; Dunphy and Chadwick 1989). (4) Ingestion occurs with pseudopodia or by membrane invagination to surround and enclose the particle. In the cytoplasm, the particle is contained in a phagocytic vacuole, the wall of which is the original cell membrane. (5) Degranulation results from the release of enzymes from granules (lysosomes) into the phagocytic vacuole. (6) The digestion of particles occurs by lysosomal enzymes, and the undigestible particles are released by exocytosis. Evidence for steps 1 and 2 is limited (see previous paragraphs). Steps 3–6 have been observed in insect cell cultures (Ratcliffe et al. 1976; Brookman et al. 1988).

According to Ratcliffe and Rowley (1974), when hemocytes of the greater wax moth larva phagocytize the bacterium *Escherichia coli in vitro*, attachment and invasion occur with the participation of pseudopodia, as in the case of vertebrate cells. Most often no vacuoles or lysozymes are apparent in the hemocytes, but sometimes ingestion occurs by the formation of phagocytic vacuoles. Kim (1981) has examined the ultrastructure of plasmatocytes and granulocytes during phagocytosis. Coated vesicles are the sites of plasma membrane invaginations to form phagosomes (phagocytic vacuoles).

Anderson (1974) has studied *in vitro* the metabolic activity of hemocytes of the cockroach *Blaberus craniifer* during phagocytosis. Characteristic alterations, such as increased respiration, hexose monophosphate shunt stimulation, and hydrogen peroxide production, which occur in phagocytizing mammalian leukocytes, are absent in phagocytizing insect hemocytes. The only metabolic alterations detectable in insect hemocytes are associated with glycolysis. Anderson (1974) concludes that the myeloperoxidase–halide–hydrogen–peroxide antibacterial system of mammalian leukocytes does not function in insect hemocytes. Despite the absence of mammalian metabolic activities, the *Blaberus* hemocytes, in *in vitro* systems, are phagocytically active and capable of killing the following bacterial species: *Staphylococcus aureus*, *S. albus*, *Streptococcus faecalis*, *Serratia marcescens*, and *Proteus mirabilis*. In the greater wax moth larva, the typical lysosomal enzymes (i.e., acid phosphatase, β-glucuronidase, esterase, and lysozyme) are found in both plasmatocytes and granulocytes (Chain and Anderson 1983). Rowley and Ratcliffe (1979) also report that, occasionally within phagocytic plasmatocytes of the greater wax moth larva, there is acid phosphatase activity associated with primary lysosomes that fuse with phagosomes that enclose phagocytized latex particles.

2. Effectiveness

All foreign particles are not ingested equally by phagocytes from any given species. The effectiveness of phagocytosis depends on such factors as the phagocytic index (number of phagocytizing cells), availability of circulating hemocytes, nature of the pathogen, and frequency of attack or virulence of the pathogen. Many pathogens are phagocytized and digested when their numbers are below the lethal infectious concentrations [e.g., the NPV in the silkworm (Inoue 1974b)]. Some virulent pathogens, however, are able to overcome the phagocytic activity and destroy the phagocytes [e.g., *Bacillus thuringiensis* (Wittig 1965), *Xenorhabdus luminescens* (Dunphy and Webster 1988), and *Beauveria bassiana* (Hou and Chang 1985)]. In such a case of cellular destruction, endocytosis may be a means by which pathogens gain entrance into cells.

B. CELLULAR ENCAPSULATION

Encapsulation generally occurs when a particle, pathogen, or parasite is too large to be engulfed by an individual cell. The hemocytes collect and surround the large particle and then adhere to its surface, usually forming a dense covering. There is generally an intra- and extracellular deposition of the black pigment, melanin. The entire structure is called a capsule.

Encapsulation takes place in two distinct phases (Ratcliffe et al. 1976; Schmit and Ratcliffe 1978; Ratner and Vinson 1983; Ennesser and Nappi 1984; Gupta 1985a). First, hemocytes (granulocytes and/or coagulocytes) aggregate around a foreign object; second, the hemocytes lyse and release substances in the hemolymph attracting other hemocytes (mainly plasmatocytes) that complete the formation of the capsule. A "surface recognition" phenomenon of self and nonself, as in the case of phagocytosis, may be involved in attachment and encapsulation (Salt 1960, 1968). Habitual parasitoids, which are not normally encapsulated in an insect, become encapsulated when they are wounded or their surfaces are altered. Intraspecific organ or tissue transplants are usually not encapsulated, but cut ends of tracheae or damaged areas of organs are quickly encapsulated. This surface phenomenon has been recognized widely with encapsulation (Nappi 1975; Vinson 1977; Ratcliffe and Rowley 1979). Adhesion and encapsulation may involve the surface charge of the substratum (Lackie 1986). Agarose beads injected into the locust *Schistocerca gregaria* are not encapsulated until the charge on the beads approaches neutrality. The nonself recognition, however, may also involve substances with opsonin-like properties.

Surface recognition is also involved in the encapsulation of the fungus *Entomophthora egressa* by insect hemocytes (Dunphy and Nolan 1980a, 1981, 1982; Boucias and Latgé 1988). The granulocytes readily encapsulate fungal hyphae but do not adhere to rod-shaped, spherical, or germinating hyphal bodies (protoplasts) that have surface proteins protecting them from the attachment of hemocytes. In the case of hyphae, chitin or its subunits in the wall may initiate the encapsulation.

In most insects, hemocytes are involved in encapsulation, but in certain Diptera (e.g., chironomids, mosquitoes, and psychodids) only humoral factors (see Section X.B., Humoral Encapsulation) produce the capsules (Götz et al. 1977). Tissues other than blood have been observed forming capsules around foreign bodies. When the nematode *Physaloptera hispida* invades and becomes associated with the epithelium of either the colon or rectum of the German cockroach, *Blattella germanica*, it becomes enveloped by a layer of connective tissues that forms an enucleate capsule membrane (Schell 1952). There is no evidence of hemocyte activity. When the nematode *Tunicamermis melolonthae* attaches itself to the integumental, proctodael, or tracheal epithelium of the European cockchafer, *Melolontha melolontha*, it becomes surrounded by actively developing epithelial cells that form a thin-walled sac (Couturier 1963).

The basis for the activation of hemocytes to encapsulate a foreign object, such as a parasitoid, is still not established. There are two theories. The first proposed by Salt (1970) states that the cell-mediated responses are initiated only after fortuitous contacts of host hemocytes with a parasitoid, rather than the cells being attracted to the parasitoid from a distance. The second theory developed by Nappi (1975) states that the cell-mediated immune reactions are activated by changes in the normal concentration of host hormones when the insect is invaded by a parasitoid. The hormonal imbalance causes changes in cell-membrane permeability and brings about premature differentiation and migration of hemocytes. The altered cell permeability causes lysis of hemocytes at the surface of the parasitoid and the release of substances that result in encapsulation. The readers should refer to the reviews by Nappi (1975) and Vinson (1977) for further discussion pertaining to these theories.

1. Cellular and Sheath Capsules

Salt (1970) has differentiated two kinds of capsules, the cellular and sheath types. Cellular capsules consist of a hemocyte layer of 50 cells or more in thickness and slight melanization may occur (Fig. 14–2A). Sheath capsules result when the object is covered by a relatively thin layer of hemocytes, and melanization is severe. The degree of melanization varies with the insect species (Götz and Boman 1985). Besides hemocytes, occasionally fat bodies, tracheoles, muscles, and Malpighian tubules are attached to or embedded in the capsule, apparently by chance (Salt 1970). Hemocytic capsules of several insects, such as those in the cockroach *Leucophaea maderae* are surrounded by a neohypodermis that secretes a cuticle (Matz et al. 1971). The neohypodermis originates from tracheal and integumentary epithelia.

The size of the hemocytic capsule depends on the number of circulating hemocytes and the number of sessile hemocytes that can be brought into circulation (Jones 1977). The hemocytes forming the capsules multiply only in a few cases. The types of hemocytes that form the capsule vary among different insects. In some cases only one type of hemocyte is involved and in others several types. The most common are plasmatocytes and granulocytes. The first *in vitro* encapsulation was reported by Vey (1969) and Vey and Vago (1971). They observed the successive aggregation of micro- and macroplasmatocytes around the introduced fungi.

The cellular capsule usually consists of one or more layers of cells. A two-layered capsule has a relatively thin layer next to the object or parasite and an outer, less transparent, layer that is composed largely of flattened hemocytes. The inner layer may be jellylike and may contain mucopolysaccharides (Bronskill 1960). The mucopolysaccharide seems to be involved in the initial adhesion and the eventual cementing of cells together. Ratcliffe (1975) has reported a sticky surface on the spherule cell, possibly from the secretion of acid mucopolysaccharide, which binds bacteria and hemocytes to the cell surface. Since spherulocytes are involved with plasmatocytes in capsule formation *in vivo*, they may function in encapsulation by providing the mucopolysaccharide found in the inner layer of the capsule (Ratcliffe and Rowley 1975). The inner layer was believed, at one time, to be syncitial in nature, but electron microscopic observation indicates that the cells do not break down to form a syncitium (Grimstone et al. 1967; Salt 1970; Whitcomb et al. 1974; Baerwald 1979). The middle and outer layers contain numerous microtubules, mitochondria, ribosomes, and have intercellular cell junctions, such as desmosomes (local "weld" spots without channels for intercellular communication) or gap junctions (E-type with cell-to-cell communicating channels) (Baerwald 1979). Some cells in the outer layer may retain their power of movement and may alter the shape of the capsule and even disperse from it after the death of the parasitoid (Salt 1956). In time-lapse cinematography of the encapsulation of fungal hyphae in a hemocyte culture, the hemocytes that first gather around the fungus appear to attract other cells and induce the adhesion of these cells to produce the capsule (Vey et al. 1975).

In a larva of *Ephestia kuehniella*, a three-layered capsule is formed around pieces of Araldite (Götz and Boman 1985). The innermost layer consists of tightly packed, but not visibly flattened, hemocytes that show pronounced signs of necrotic and autolytic alterations. The middle layer consists of extremely flattened hemocytes. The outer layer has cells that appear normal in shape. Cellular elements, such as microtubules and desmosomes, are predominant in the middle region. Electron-dense material fills the intercellular spaces between the flattened cells. Plasmatocytes are the only hemocytes in the cellular envelope.

Sheath capsules (Salt 1970) are a variation of the cellular types. They are much thinner and are composed of a tough, brown (melanized) envelope overlain by relatively few layers of hemocytes. An intermediate form of capsule occurs in larvae of the beetle *Diabrotica* parasitized by the mermithid *Filipjevivermis leipsandra* (Poinar et al. 1968). The capsule is composed of an inner melanized layer, including disintegrated cellular material, a thin layer of necrotic cells, a layer of flattened cells, and a layer of normal-appearing hemocytes. In rapid melanization, this type resembles a sheath capsule, but in the development of flattened cells and in gross appearance it is like a cellular capsule (Salt 1970).

2. Inhibition of Capsule Formation

Nonself components may not be encapsulated. In some braconid and ichneumonid parasitoids, a virus associated with the calyx fluid appears to confer immunity to the parasitoid egg and larva against encapsulation by the host larva (see Chapter 7). In the braconid parasitoid, *Apanteles glomeratus*, the calyx fluid is not involved, but the venom that is injected with the egg protects the parasitoid from encapsulation by the host *Pieris rapae crucivora* (Kitano 1982; Wago and Kitano 1985). The venom, however, has no effect on the encapsulation of other foreign objects and on melanization.

Toxins produced by fungi also inhibit encapsulation. The mycotoxin destruxin E_1 increases the susceptibility of *Galleria mellonella* to *Aspergillus niger* through the immunodepressive effect on encapsulation (Vey et al. 1985). Destruxins A and B do not have such an effect. With the fungus *Nomuraea rileyi*, hyphal bodies do not elicit a strong cellular defense response, whereas mycelial fragments do (Boucias and Latgé 1988). Apparently, the hyphal-body stage does not release high levels of the β-1,3-glucan molecules, which are exocellular polysaccharide molecules, or does not contain the receptor complex suitable for opsonin or hemocyte attachment. The β-1,3-glucan molecules elicit a cellular immune response. Other carbohydrates similar to β-1,3-glucan molecules that are important cell wall components of fungi do not activate a cellular defense response.

C. GIANT CELLS, NODULES, AND TUMORS

Giant cells have been observed in insects into which bacteria pathogenic to vertebrates had been inoculated (Metalnikov and Toumanoff 1923; Metalnikov 1927; Rooseboom 1937). The bacteria are taken up by the giant cells. These cells are multinucleate, but it is still unclear whether such structures arise from a fusion or from an incomplete division of hemocytes (Whitcomb et al. 1974). Some giant cells are syncitial in nature. When the spirurid nematode *Abbreviata caucasica* enters epithelial cells of the colon wall of the German cockroach, *Blattella germanica*, it becomes enclosed in a syncitial giant cell (Poinar, 1974). Hemocytes that phagocytize microsporidia, such as *Nosema lymantriae*, *Thelohania hyphantriae*, etc., coalesce to form giant cells or megalocytes (Weiser 1969).

Nodules are formed when high doses of microorganisms or suspensions of nonliving particles are inoculated into an insect's hemocoel (Ratcliffe and Rowley 1979; Götz and Boman 1985). They are formed very rapidly, within 24 h after the inoculation of bacteria. Sometimes nodules are difficult to separate from capsules, and both types may occur within one insect invaded by the same parasite.

There are two hypotheses on nodule formation. The first proposes that nodules result from an initial phagocytosis (Salt 1970; Götz 1973). After ingestion of the pathogens, the phagocytes become necrotic, clump together, and are encapsulated by other hemocytes. The second, which appears to have replaced the first hypothesis, maintains that the hemocytes do not phagocytize the foreign bodies, but the granule-containing hemocytes (granulocytes and/or coagulocytes) cause the coagulation of the foreign bodies. Subsequently, other hemocytes (mainly plasmatocytes) form sheaths (encapsulation) around the coagulum (Ratcliffe and Rowley 1979). In this hypothesis, Gunnarsson and Lackie (1985) propose that the amount of granules exocytosed by hemocytes depends on the number of membrane-bound receptors on the hemocyte surface that are activated by the foreign agent. Thus, individual molecules activate only a few receptors and stimulate only a minor degranulation, whereas a particle, such as a spore, activates many receptors and produces heavy degranulation, causing a large localized clot that entraps hemocytes to form the center of a nodule.

Melanotic lesions or tumors are produced in response to various stimuli. According to Matz (1965), lesion formation consists of two reactions: (1) exaggerated proliferation of epithelial cells and (2) encapsulation of abnormal tissue by hemocytes, which then undergo cancerous alteration. In cockroaches and locusts, the tumor starts with a mitotic proliferation of embryonic epithelial cells in the midgut wall as well as in the epithelium of ectodermal organs (Matz 1983). The increased number of cells forms several layers. The cells remain embryonic. Defensive mechanisms are stimulated (e.g., hemocytes appear and surround and invade the growth to form the typical tumor). In many cases, the epithelial cells or a portion of them remain alive and undergo mitosis, but in others, the cells die.

Tumors are produced in insects by various factors, such as viruslike particles, mycoplasma-like organisms, microsporidia, acellular filtrates or RNA extracts of tumors, hormonal disturbances, and nerve severance (Harshbarger and Taylor 1968; Krieg 1973; Burdette 1974). Most tumors in insects are benign, but malignancy has been observed only in *Drosophila* and in cockroaches.

Tsang and Brooks (1980a,b) have reported the first case of *in vitro* malignant transformation of insect cells. The cells of an embryonic-cockroach cell line in its 290th passage, when injected into hemocoels of homologous species *Blattella germanica*, cause paralysis, progressive loss of appendages, and premature death. The cause of the pathologies is not known, but the injected cells are metastatic and invasive, producing tumors, especially in the fat body. The tumors are transplantable and malignant in secondary hosts. The malignant cells can be recovered from the tumors and grown again in culture. No virus or other microorganisms have been detected in the cell line. Three other embryonic cell lines from *B. germanica* also cause malignant neoplasms after injection into homologous hosts (Tsang and Brooks 1981). As few as 10 cells ultimately kill the host. Three other species of cockroaches, *Periplaneta americana*, *Supella longipalpa*, and *Leucophaea maderae*, are also susceptible to the malignant cells of *B. germanica*.

Certain viruses cause "transmissible" tumors in insects. Parvovirus-like viruses (20 nm) cause tumors in the cockroaches *Leucophaea maderae* (Matz and Bergoin 1984) and *Periplaneta fuliginosa* (Suto et al. 1978), and in the lepidopteran, *Sibine fusca* (Amargier et al. 1979).

Tumors in animals (fishes, in particular) infected with microsporidia are called xenomas, in which the pathogens and the hypertrophying cells exist in a complex comparable to plant galls (Weissenberg 1968). According to Weiser (1976), there are two main types of xenomas in insects. (1) Syncitial xenoma in which membranes of infected cells dissolve and result in a multinuclear plasmodium, sometimes accompanied by nuclear hypertrophy. The number of nuclei remains the same or decreases in infected tissues. (2) Neoplastic xenoma, which is analogous to tumor formation because of increased cell numbers, is further separated into two groups. One in which the invaded cells divide so that each microsporidium (e.g., *Pleistophora debaisieuxi* in black flies) occurs within its own host cell. The number of infected cells increases from 10 to 30 times that of the original number. The other in which the nucleus of the primary infected cell fragments to form a tumor as in the xenoma produced by *Glugea* in fishes. Weiser (1976) points out that it is difficult to differentiate whether the host reaction to the microsporidium is a neoplasm or tumor from the necrotic process or repair of infected tissue by an insect. Nonetheless, the changes brought about in the tissues indicate that the microsporidium excretes substances that cause the insect to adapt to the pathogen.

In addition to xenomas, certain microsporidia may produce inflammatory responses in insects. In the tobacco hornworm, *Manduca sexta*, infected with *Nosema sphingidis*, the posterior region of the midgut epithelium produces a nodule or tumor within 24 h after being exposed to microsporidian spores (Brooks 1971). By 120 h, striking inflammatory responses appear with the production of extensive nodule formations at sites of infected cells in the midgut, labial gland, Malpighian tubules, and fat tissues. At a late stage of infection, posterior midgut cells become deranged and there is an intense hemocyte infiltration and melanization of the nodule. The grasshopper *Melanoplus sanguinipes*, when infected with *Nosema acridophagus*, develops cellular proliferation in the regenerative nidi of the midgut and gastric caeca within 24 h followed by the migration of these cells to tumors caused by the infection (Henry 1969). As the tumor grows, the epithelia of the midgut and caeca shows pronounced vacuolization and occasional degradation in the vicinity of tumors due to a functional change of the nidi cells. The cells in some large tumors are arranged in a characteristic swirling pattern around an occasional melanotic core. Most of the tumor cells originate from regenerative nidi.

D. RESPONSES OF HEMOCYTES TO BIOTIC AGENTS

There are wide variations in the effectiveness of cellular immunity in protecting insects from pathogens. When nonliving materials are injected, the mean number of cells that ingest such particles is generally high (phagocytic index), but with pathogens, the phagocytic index may vary significantly. When small numbers of bacteria are inoculated into the hemocoel, they are rapidly removed from the hemolymph by adhesive hemocytes that are attached to tissues throughout an insect (Gagen and Ratcliffe 1976; Faye 1978; Ratcliffe and Walters 1983). When large numbers of bacteria are inoculated, they are dramatically reduced in a short time by nodule formation (Dunn and Drake 1983; Horohov and Dunn 1983; Ratcliffe and Walters 1983; Walters and Ratcliffe 1983). Some bacteria are destroyed by phagocytosis of the circulating hemocytes. Cellular immunity, however, appears to be ineffective, especially with a massive invasion of pathogens. The variability may be associated with the physiological state and condition of an insect and with the dosage, virulence, and type of pathogen. Other factors, such as temperature, radiation, and anti-inflammatory or antihemocytic sera, may also influence cellular immunity (Rooseboom 1937; Whitcomb et al. 1974; Lynn and Vinson 1977).

Cheung et al. (1978) conducted a comparative study on the phagocytosis of the insect pathogenic *Bacillus thuringiensis* and the nonpathogenic *Micrococcus lysodeikticus* in the corn earworm, *Heliothis zea*. When the bacteria were inoculated into the larval hemolymph, there was rapid phagocytosis and extensive removal of the organisms within 2 h. The bacteria that survived the initial clearance initiated a new round of growth that was evident in 6 to 8 h after inoculation. In the case of *M. lysodeikticus*, a second period of clearance occurred 8–12 h after inoculation, and the larva survived the infection. With *B. thuringiensis*, 60–80% of the bacteria were phagocytized and lysed in the initial clearance, but the second wave of clearance did not occur; instead, the bacteria multiplied extensively and caused death in 12 to 16 h after inoculation.

Any condition that reduces the quantity of phagocytes or restricts their multiplication would be expected to depress cellular immunity. One of the ways that the insect parasitoids overcome the cellular defense of their hosts is through their rapid feeding, which depletes the host's blood both of cells and nutrients and thereby prevents blood formation and encapsulation (Salt 1968). Cellular defense may be overcome also by blocking the activity of phagocytes. When Bettini et al. (1951) inoculated vertebrate erythrocytes into the American cockroach, *Periplaneta americana*, they observed that phagocytes engulfed the erythrocytes and that the cockroach became more susceptible to bacteria. Stairs (1964) inoculated India ink into the greater wax moth larvae, *Galleria mellonella*, and observed that over 70% of the cells had ingested the ink particles within 2 h. Larvae pretreated with ink were more susceptible to infection by free virions of a nuclear polyhedrosis virus that had been inoculated into the hemocoel. The increased susceptibility of the larva to viral infection was a result of blockage of phagocytosis by the India ink.

When the phagocytic hemopoietic tissue of the grasshopper is blocked by the inoculation of iron saccharate, the insect becomes infected by a low dose of *B. thuringiensis*, which normally would have been removed by the phagocytic tissue (Hoffmann 1976). Moreover, the blockage of the phagocytes leads to a necrosis and encapsulation of such cells and affects the formation of hemocytes by the hemopoietic tissue. Anderson et al. (1973) succeeded in demonstrating phagocytosis of several bacterial species by *in vitro* cultures of hemocytes of the cockroach *Blaberus craniifer*. Prior exposure of the hemocytes, especially to killed bacteria and to latex particles, causes a partial blockage of the bactericidal capacity of the cells.

Some parasites and pathogens remain alive even though they are encapsulated within their insect hosts. Such is the case with the immature stages of some

Acanthocephala, Trematoda, Cestoda, and Nematoda (Poinar 1969; Götz and Boman 1985) and fungi (Vey 1969). The fungus *Mucor hiemalis* may be enveloped in a multilayered nodule in the larval hemocoel of the greater wax moth, but it often multiplies and invades the fat body (Vey 1968). Strains of *Metarhizium anisopliae* vary in their ability to escape through the wall of the capsule (granuloma) produced in hemocoels of the beetles *Oryctes rhinoceros* and *Cetonia aurata* (Fargues et al. 1976). *Beauveria bassiana* is also capable of overcoming phagocytosis and encapsulation (Hou and Chang 1985; Bidochka and Khachatourians 1987).

Virulence plays a role in the successful encapsulation of fungi. Weak pathogens (e.g., *Aspergillus niger*) are encapsulated by plasmatocytes and destroyed, whereas virulent forms (e.g., *Beauveria bassiana*) are able to resist encapsulation and produce toxins that kill the insect (Vey and Vago 1971).

The dosage and virulence of the pathogen may affect the cellular response of an insect. Certain bacteria of low virulence may cause death at high dosages even after all of them have been phagocytized (Metalnikov 1927). The cause of death in such cases has not been established and may be due to excessive loss of metabolic energy and to the production of toxins by the bacteria. Some highly pathogenic bacteria cause lethal infection even at a low dose of less than 10 bacterial cells per larva. Even in these bacteria, there is usually a minimum infective dose, below which the host defense reactions are effective. At low dosages of *Bacillus thuringiensis* in the armyworm, *Pseudaletia unipuncta* (Wittig 1965), and in the grasshopper *Locusta migratoria* (Hoffmann et al. 1974), the phagocytes are capable of protecting the insects from lethal infection. This does not happen at high dosages of the bacterium. In the case of *Xenorhabdus* spp., their pathogenicity is dependent on the insect host species (Akhurst and Boemare 1990).

There were extensive debates during the early twentieth century regarding the importance of cellular versus humoral immunities in insects. Such debates border on irrationality because both types of immunities are involved in protecting insects, and in some cases with both participating at the same time. We are still ignorant of the roles played by each type of immunity in most diseases of insects.

IX. INSECT PLASMA AND HUMORAL IMMUNITY

The plasma is the fluid, noncellular portion of the hemolymph. It is a biochemically rich solution, very different from vertebrate plasma, with widely varying proportions of inorganic ions, high levels of free amino acids, usually a rather high level of trehalose, and sometimes substantial amounts of other solutes, such as organic phosphates, citrate, glycerol, and peptides (Wyatt and Pan 1978).

The immune factors that occur in the plasma are the basis for humoral immunity, but some of these factors are secreted by certain hemocytes. In this respect, humoral immunity may be closely associated with cellular immunity. In vertebrates, humoral immunity involves antibodies, opsonins, complement, properdin, interferon, bactericides, and other factors. Reports demonstrating the presence of these factors in insects are controversial issues. There are early claims that serological techniques (in particular, agglutination), which are used with vertebrate systems, are also applicable to insects to demonstrate the presence of antibody. These claims, however, resulted from faulty interpretation and from artifacts (Briggs 1958). No globulins similar to those of vertebrates have been detected in insects. This implies that antibodies, which are globulins in nature, do not occur in insects. Nonetheless, there is abundant evidence that humoral immunity occurs in insects.

In humoral and, to a lesser extent, cellular immunities, certain terminology is used to describe various types of immunity based on methods of acquisition. An immunity is said to be "acquired" when an organism has gained the immunity during its lifetime, as opposed to "normal" or "natural" immunity. *Acquired immunity* generally is associated with the activity of antibodies or substances produced by the organism in reponse to an antigen, which may be living or nonliving. An "innate" or natural immunity does not depend on antibodies or similar immune substances. At times, it is difficult to differentiate acquired from innate immunities when the immune factors, which occur naturally in an insect, are enhanced and augmented after the invasion of a pathogen. In this case, there is an overlapping of innate and acquired immunities; but in acquired immunity, there is a persistence of the immune factors at high levels that protects insects from subsequent invasion by pathogens.

Acquired immunity may be *naturally acquired* as a result of a natural infection of a pathogen or *artificially acquired* when the pathogen or vaccine is inoculated into the organism. Each of these types of acquired immunity may be either *active* or *passive* in character. In an active immunity, the organism participates directly or actively in developing its immunity by producing antibody or other immune substances. This is usually accompanied by increased cellular activity. Passive immunity is conferred to an organism that has been inoculated with antibody or immune substance produced in another individual of the same or different species. The immunized organism does not participate in the production of protective substances.

Antigen and antibody are usually defined in terms of the other: antigens are substances that stimulate the production of antibodies, and antibodies are substances produced by an animal's body in response to the stimulus of the introduced antigen. As mentioned previously, the vertebrate antibodies are composed of globulins, and on this basis, no antibodies occur in insects. However, insects develop immune response when infected by a pathogen, and this reaction is comparable to the antigen–antibody reaction of vertebrates.

X. INNATE HUMORAL IMMUNITY

In the noncellular (plasma) portion of the insect blood, there are natural immune factors. These factors, as pointed out earlier, are not chemically comparable to antibodies of vertebrates although some may function in a similar manner. The two main types of natural immune factors in insects are hemagglutinins (lectins) and phenoloxidases (Götz and Boman 1985; Boman and Hultmark 1987). Others are bactericidal factors, lysozymes (muramidase), and factors that resemble certain humoral factors in mammals (Boman and Hultmark 1987; Dunn 1990).

A. HEMAGGLUTININS

Hemagglutinins, presently considered to be lectins, the noncatalytic carbohydrate-binding proteins, have been reported from several insects, such as the lepidopteran *Citheronia regalis* (Bernheimer et al. 1952), the cockroaches *Periplaneta americana* (Scott 1971, 1972; Lackie 1981), *Blaberus craniifer* (Anderson et al. 1972; Donlon and Wemyss 1976), and *Leucophaea maderae* (Amirante and Mazzalai 1978), the locust *Schistocerca gregaria* (Lackie 1981), and the bug *Rhodnius prolixus* (Pereira et al. 1981). The hemagglutinins of different insect species differ in properties (Lackie 1981). In the American cockroach, the hemagglutinin is a heat-labile, nondialyzable, euglobulin-type protein, which is electrophoretically a slow-moving fraction of hemolymph proteins. The hemagglutinin is not a gamma-globulin, does not fix complement, and does not function as an opsonin to facilitate phagocytosis (Lackie 1981).

In a comparative study of electrophoretic mobilities of serum and hemolymph proteins of a wide range of animals from invertebrates to vertebrates, Nelstrop et al. (1970) have detected proteins migrating in the globulin range in the hemolymph of the grasshopper *Locusta*. However, they point out that all evidence indicates the lack of ability to form circulating antibodies in invertebrates. In the hemolymph of *Blaberus*, there is a humoral factor, which functions similarly to a serum factor, C3 proactivator found in other invertebrates, in causing the lysis of erythrocytes (Anderson et al. 1972).

The function of lectins is not completely known (Yeaton 1981, 1983). If lectins play a role in insect immunity, they can act in various ways: (1) as membrane-bound receptors, (2) as humoral opsonic factors, or (3) by acting independently of hemocytes and agglutinate invading bacteria (Lackie 1981). Adult honey bees inoculated with killed *Bacillus larvae* have produced agglutinating substances in the hemolymph, which has a high titer for *B. larvae* and a low titer for the related *B. subtilis*, but none for the bacterium *Salmonella thompson* (Gilliam and Jeter 1970). The agglutinating substances are absent in the nonvaccinated normal hemolymph of an adult bee. In the locust *Schistocerca gregaria* and the American cockroach, *Periplaneta americana*, glycoproteins, or proteins that occur naturally in the hemolymph, agglutinate flagellate protozoa (Seaman and Robert 1968; Ingram et al. 1984). Prior inoculation of flagellates also increases the agglutinating titers in these insects. These agglutinins may be lectins. In *Rhodnius prolixus*, Pereira et al. (1981) have detected three lectins in the crop, midgut, and hemolymph, respectively, that agglutinate the protozoan *Trypanosoma cruzi*. The lectins show specificity for certain vertebrate erythrocytes and the epimastigote stage of the protozoan, indicating that the lectins recognize specific receptor sites.

Insect lectins have been purified from the flesh fly, *Sarcophaga peregrina* (Komano et al. 1980, 1981), from *Hyalophora cecropia* (Götz and Boman 1985), and from *Antheraea pernyi* (Qu et al. 1987). They are specific for galactose. The active lectin of the flesh fly has a molecular weight of 190,000 and consists of α (32 kDa) and β (30 kDa) subunits (Komano et al. 1980; Natori 1987). The active lectin does not occur in the hemolymph and is formed when the insect is injured. The injured *Sarcophaga peregrina* larva contains a factor that is necessary for the activation of the lectin gene in the insect's fat body. This factor, not found in uninjured larvae, is a small nonpeptide molecule (<1000 daltons) (Shiraishi and Natori 1989). Part of the α subunit, which is present normally in the hemolymph, is converted to the β subunit, probably after partial proteolysis by a protease activated in response to the injury. This results in the formation of the active lectin consisting of the α and β subunits at a molar ratio of 2:1 (Komano et al. 1981). The *Sarcophaga* lectin has been shown to participate in the elimination of sheep erythrocytes injected into *Sarcophaga* larvae (Komano and Natori 1985). Takahashi et al. (1985) have cloned and determined the nucleotide sequence of the gene for the *Sarcophaga* lectin.

B. HUMORAL ENCAPSULATION

In certain Diptera, such as mosquitoes (Culicidae), midges (Chironomidae), and Chaoboridae, the plasma melanizes and encapsulates the microorganisms (nematodes, fungi, and bacteria), and hemocytes may or may not participate in capsule formation (Bronskill 1962; Götz 1969, 1973; Poinar 1974). The noncellular capsules are called humoral capsules and are formed only against live parasites (Gupta 1985a). This unusual type of encapsulation is fast (capsules are formed within minutes), highly efficient, and occurs *in vivo* as well as *in vitro*. It occurs only in insects with small numbers of hemocytes (Götz and Boman 1985). Plasma encapsulation indicates that the tyrosine–phenoloxidase system is of primary importance in encapsulation. Phenoloxidase inhibitors, such as phenylthiourea and reduced glutathione, which prevent melanization, inhibit or retard the formation of encapsulation (Brewer and Vinson 1971; Vey and Götz 1975). The readers are referred to Section XI.D., Phenoloxidase System, for the process and significance of this system in immunity.

C. OTHER HUMORAL FACTORS

Several workers have detected bactericidal factors of unknown composition (Briggs 1958; Stephens 1963b; Kawarabata and Aizawa 1968; Hink and Briggs 1968; Bakula 1970). Stephens (1963b) in her study with seven field-collected insect species reported that the hemolymphs of six of them had bactericidal activity against *Shigella dysenteriae* and three against *Salmonella typhosa*. Both bacteria are nonpathogenic to insects. In the case with *Pseudomonas aeruginosa*, three insect species, and with *Bacillus cereus*, only one insect species had bactericidal reaction against these pathogenic bacteria. These results indicate that the hemolymph of a number of insects found in nature shows no strong bactericidal activity against entomopathogenic bacteria but may exhibit varying degrees of activity against nonpathogenic forms. Hink and Briggs (1968) characterized the bactericidal factors in normal *Galleria* larvae as heat stable with a molecular weight of about 7000, which appears not to be a protein or peptide. These factors may be protein inhibitors that are found in the hemolymphs of *Galleria* and other insects (Hanschke and Hanschke 1975).

XI. ACQUIRED IMMUNITY

Acquired immunity results when immune factors are produced or stimulated by the invasion of a foreign body or pathogen into the hemocoel of an insect. Injury may also induce the initial stage of the immune response of protein synthesis, suggesting a similarity in the responses of an insect to injury and to microorganisms (Götz and Boman 1985). Moreover, Cherbas (1973) has isolated a factor, haemokinin, with molecular weight 50,000, which is formed after an injury. This factor causes the plasmatocytes to become amoeboid and adhesive and may thereby play a role in humoral immunity.

Insects have been immunized by methods used in immunizing mammals (Stephens 1963a). They include injections of: (1) toxoids, produced from soluble toxins, whose toxic activity has been destroyed without the loss of their immunologic activity, (2) vaccines prepared from microorganisms killed by physical or chemical methods, (3) substances isolated from infectious microorganisms, and (4) living attenuated microorganisms. Immune factors that are stimulated are both cellular and, particularly, acellular (humoral) elements. The activation of cellular immunity is usually not discussed with acquired immunity, but we shall include it because it is closely associated with acquired humoral immunity.

During phagocytosis and encapsulation, there is generally an increase in the number of hemocytes in an insect (Whitcomb et al. 1974). In the grasshopper *Locusta migratoria*, the inoculation of sublethal doses of *Bacillus thuringiensis* results initially in a significant decrease of circulating hemocytes, followed 2 days later by a sharp increase in hemocyte count (Hoffmann et al. 1974). One to two days after inoculation, the hemopoietic tissues show signs of considerable hypertrophy, and both polymorphic reticular cells and maturing blood clusters increase in numbers. This is accompanied by significant changes in the differential and total hemocyte counts. The significance of these changes to acquired resistance in insects has not been clarified.

A. ANTIBACTERIAL AND TOXIC SUBSTANCES

There are numerous studies demonstrating acquired humoral immunity against bacteria and toxins in insects. Most of the studies conducted prior to 1950 produced conflicting results and were difficult to reproduce (Stephens 1963a). Readers interested in these historical investigations should refer to the excellent review by Briggs (1964). Subsequent studies have confirmed the presence of antibacterial and antitoxic factors in immunized insects, but the identity of these substances has not been completely established (Briggs 1958; Stephens 1959; Stephens and Marshall 1962; Hink and Briggs 1968; Hoffmann 1980).

Endotoxins from *P. aeruginosa* and several other bacteria produce immune responses that are similar to those produced by the whole bacterial vaccine (Chadwick 1968; Chadwick and Vilk 1969). The inoculation of the endotoxin protects the larva against infection from *P. aeruginosa* but not against some other pathogenic gram-negative species. The polysaccharide component of the endotoxin is responsible for the induced immunity (Chadwick 1971).

A bacteriostatic principle may occur in the immunized larval hemolymph of the greater wax moth because the antibacterial activity requires an induction period of 15 to 30 min (Chadwick 1977). Chadwick (1977) first treated the bacterium *Serratia marcescens* with immunized hemolymph, and when the hemolymph was washed off, she found the bacterium still alive. On the other hand, the result with *Pseudomonas aeruginosa* treated with immunized hemolymph was different and appeared to be bactericidal. Anderson et al. (1973) also noticed that phagocytes of *Blaberus craniifer* killed certain entomopathogenic species of bacteria but did not destroy other phagocytized bacteria (e.g., *Pseudomonas aeruginosa*, *Escherichia coli*, *Salmonella typhosa*, and *Diplococcus pneumoniae*). The variations in the susceptibility of bacteria to immune hemolymph suggest that diverse mechanisms are operating in killing different bacterial species (Rasmuson and Boman 1977; Boman and Steiner 1981).

Specific humoral immunity to soluble toxins or venoms occurs in the American cockroach (Karp and Rheins 1980; Rheins and Karp 1984). The honey bee toxoid, phospholipase A_2, generates in the cockroach a specific primary response that peaked in 10 days and gradually subsided by the fifth week. A second booster dose of the toxoid causes a secondary immune response. The response is specific and is not induced by injecting an unrelated toxoid. The humoral factor is a protein that is sensitive to trypsin (Rheins and Karp 1982). It behaves like a precipitating antibody-like molecule, since it forms specific precipitin bands with the homologous antigens in immunodiffusion (Ouchterlony) gels. Rheins and Karp (1982) conclude that the activity is analagous to that of vertebrate antibody molecules, and that this activity is dependent on the protein composition of the humoral factor. In the silkworm, protease inhibitors present in the hemolymph may prevent injury from toxic protease produced by invading fungi (Eguchi 1982).

In summary, the general characteristics of induced immunity in insects against bacteria are (1) it is rapidly acquired, (2) a single inoculation of live or dead bacteria is adequate for immunity to develop, and immunity may be enhanced by repeated inoculations, (3) nonbacterial substances may also stimulate immune responses, (4) the immunity generally is of brief duration but may last for 2 weeks, (5) it is relatively nonspecific and is not associated with immunoglobulins, and (6) cell-free hemolymph shows antibacterial activity. The nature of the antibacterial principles in insects has been ascribed to bacteriostatic substances and bacteriolytic or bactericidal substances, such as lysozyme, cecropin, attacin, phenoloxidase, complement, etc. These substances may form a complex, multicomponent, inducible hemolymph protein system (Wyatt and Pan 1978; Hoffmann 1980; Boman and Steiner 1981).

B. LYSOZYMES

Lysozymes (muramidases) are mucolytic enzymes located in the lysosomes, the acid phosphatase positive organelles found in hemocytes, pericardial cells, and ecdysial (prothoracic) glands (Gupta 1985a). These enzymes play an active role in insect resistance, particularly to bacterial pathogens. They have been detected in the hemolymphs of a large number of insects, such as cockroaches, grasshoppers, silkworm, house fly, greater wax moth, cecropia moth, and blow fly (Malke 1965; Powning and Davidson 1973; Jarosz 1979b; Hultmark et al. 1980).

Lysozymes obtained from different insects are structurally and antigenically different (Croizier and Croizier 1978). This may account for the variations in the bacteriolytic responses of different insect species. Schneider (1985) has purified three lysozymes from the hemolymph of the cricket *Gryllus bimaculatus* inoculated with sublethal doses of *Bacillus thuringiensis*.

Certain cell cultures of the American cockroach are resistant to bacterial infections, and the medium of such culture contains a lytic enzyme presumably produced by the insect cells (Landureau and Jolles 1970). This lytic enzyme is more like a chitinase than a true lysozyme, for it is a poor muramidase and an excellent chitinase with *M. lysodeikticus* cells and chitopentose as substrates. Moreover, Cheung et al. (1978) have reported that the lysozyme that lyses *M. lysodeikticus* appears to be different from the bacteriolysin that acts on vegetative cells of *Bacillus thuringiensis*. This bacteriolysin may be the *Chalaropsis* B enzyme that acts as *N, O*-diacetylmuramidase and *N*-acetylmuramidase (Shih and Hash 1971). Kingan and Ensign (1968) have observed that the *Chalaropsis* B enzyme rapidly and completely dissolves the cell wall of *B. thuringiensis*, whereas lysozyme and myxobacter AL-1 enzyme show little or no activity on the cell wall.

In the hemolymph of the healthy silkworm, Kinoshita and Inoue (1977) have reported that the bactericidal principle, which kills the bacterium *Escherichia coli*, consists of at least two components: a lysozyme-like enzyme and a cofactor that is anionic and of low molecular weight. Egg-white lysozyme cannot replace the enzyme in the insect plasma to produce the bactericidal activity. A lysin that occurs in the hemolymph of the spruce budworm, *Choristoneura fumiferana*, acts on protozoa *Paramecium caudatum* and to a lesser extent on *Euglena gracilis* (Dunphy and Nolan 1980b).

Some workers consider that the major antibacterial principle in acquired immunity in insects is the enhanced bacteriolysis by enzymes, such as lysozyme and other factors (Malke 1965; Kawarabata and Aizawa 1968; Mohrig and Messner 1968; Powning and Davidson 1973, 1976; Hoffmann et al. 1977). Lysozyme is present in insect hemolymph and increases with the inoculation of foreign substances, including bacteria. Its specific source varies and may be associated with the hemopoietic tissues of *Locusta* (Hoffmann et al. 1977), phagocytic hemocytes (Crossley 1975; Anderson and Cook 1979), or fat-body cells (Croizier and Croizier 1980). In *Calliphora* larvae injected with live bacteria, however, the rising lysozyme titer has been attributed to a release from pericardial cells rather than from hemocytes (Crossley 1972). With the use of metabolic inhibitors, actinomycin D and cycloheximide, Croizier and Croizier (1980) have shown that the production of lysozyme is under the control of only one mechanism, the synthesis of messenger RNA. The increase in lysozyme in the hemolymph is due to protein synthesis and does not result from the release of a preformed enzyme.

There is some question of the role of lysozyme in insect immunity (Chadwick 1970; Boman 1981). Chadwick (1970) reports that, even though lysozyme content increases when *Galleria* larvae are inoculated with *Pseudomonas aeruginosa*, the increase is not correlated with acquired immunity since the lysozyme content continues to be high after the acquired immunity has subsided. Moreover, when lysozyme is inoculated into *Galleria* larvae, it does not increase the resistance of a larva to infection from *Bacillus cereus* and *Pseudomonas aeruginosa* (Chadwick 1977). The lysozyme-like activity may be only one component of a more complicated immune response (Anderson and Cook 1979). Both cecropins and lysozymes may be needed for a complete and effective destruction of bacterial cell walls since cecropins act on the bacterial cytoplasmic membrane and lysozymes on the murein skeletons (Götz and Boman 1985; Boman and Hultmark 1987).

In primitive orthopteran insects, only lysozyme or similar compounds have been reported as antibacterial factors. Lambert and Hoffmann (1985) have reported a first case of antibacterial activity in the hemolymph of the locust *Locusta migratoria* inoculated with *Pseudomonas aeruginosa*. They have isolated a 1.5-kDa antibacterial factor that was active against gram-negative bacteria.

C. CECROPINS, ATTACINS, AND DIPTERICINS

In the hemolymphs of saturniid pupae (*Samia cynthia*, *Hyalophora cecropia*) that have been injected with *Escherichia coli* and *Enterobacter cloacae*, there are inducible antibacterial activities that lyses *E. coli* and *Micrococcus lysodeikticus* (Boman et al. 1974; Spies et al. 1986b). In sodium dodecyl sulfate (SDS) polyacrylamide gel electrophoresis of immune *Hyalophora cecropia* hemolymph, the overall pattern of newly synthesized proteins was demonstrated as nine bands (P1–P9) (Faye et al. 1975; Pye and Boman 1977; Boman and Hultmark 1987). Some of the isolated polypeptides, P5, P7, P9A, and P9B, have antibacterial activity (Hultmark et al. 1980, 1983; Boman and Steiner 1981; Hoffmann et al. 1981; Lee et al. 1983). These proteins appear to result from *de novo* synthesis of ribonucleic acid and proteins in acquired immunity and are produced in the pupal fat body (Faye and Wyatt 1980).

Hultmark et al. (1980) consider P7 to be a lysozyme because its pH-rate profile, bacteriolytic specificity, and amino acid composition are very similar to those of the lysozyme of *Galleria mellonella*. Fractions P9A and P9B are highly potent against *E. coli* and other gram-negative bacteria. They are heat stable, very basic proteins with molecular weights of 4005 for fraction A and 4036 for fraction B (Steiner et al. 1981). These two fractions may be comparable to the low molecular weight, heat-stable, bactericidal factors reported by various workers. They are also similar to the cofactor of lysozyme isolated by Kinoshita and Inoue (1977) and to the two basic proteins detected by Jarosz (1979b) in the blood of *Galleria mellonella* immunized with *Pseudomonas aeruginosa* (Hoffmann et al. 1981).

The P9A and P9B fractions are not of the lysozyme class of enzymes and represent a separate type of bacteriolytic proteins named cecropins (P9A is called cecropin A, and P9B as cecropin B) (Hultmark et al. 1980; Steiner et al. 1981). The molecular cloning, cDNA sequencing, and chemical synthesis of cecropin B have been accomplished (van Hofsten et al. 1985). Other antibacterial proteins are called cecropins C, D, E, and F, and the factor G (Hultmark et al. 1982). Cecropin D is a major cecropin, and its amino acid sequence shows homology to cecropins A and B. The cecropins are small, basic proteins with a known sequence of 37 amino acids (Steiner et al. 1981; Hultmark et al. 1982; Qu et al. 1982). Hultmark et al. (1982) conclude that the three major cecropins, A, B, and D, are products of three different genes that are derived from a common ancestor. All three are efficient against a number of gram-positive and gram-negative bacteria, including *Escherichia coli*, *Serratia marcescens*, *Pseudomonas aeruginosa*, *Xenorhabdus nematophilus*, *Bacillus megaterium*, *B. subtilis*, *B. thuringiensis*, and *Streptococcus faecalis* (Götz and Boman 1985; Boman and Hultmark 1987).

Andreu et al. (1985) have synthesized six analogues of cecropin A. They have proposed an amphipathic α-helix model as the basis for the activity of the cecropins against bacteria. They have explored the cecropin structure and its relationship to antibacterial activity (i.e., the involvement of cecropin with the bacterial membrane).

Cecropins C, E, and F occur in very low amounts and their primary structures have not been established, but they appear to be related to other cecropins (Götz and Boman 1985). The primary structure of factor G is also undetermined, but its amino acid sequence differs from those of the cecropins. In the Chinese oak silk moth, *Antheraea pernyi*, the major antibacterial factor is cecropin D (Qu et al. 1982). The cecropins B and D of *Antheraea* differ from those in *Hyalophora cecropia* by four and three consecutive amino acid replacements, respectively. The homology of the cecropins in the two insect species suggests that these compounds originated from a single ancestral gene. Since the cecropins are produced in a number of Lepidoptera, Steiner et al. (1981) consider them as the main antibacterial components so far identified in immune hemolymph that act against potential insect pathogens. In *Trichoplusia ni*, induced proteins related to attacins, cecropins, and lysozyme are isolated when last instar larvae are injected with the bacterium *Enterobacter cloacae* (Andersons et al. 1990). However, the injection of NPV does not cause an induction any greater than that caused by an injection of saline. In fact, viral infection causes some breakdown of the antibacterial proteins.

The P5 fraction from the immune blood of cecropia pupae contains six closely related antibacterial proteins, designated attacins A–F (Hultmark et al. 1983; Engström et al. 1984a,b). The six attacins form two groups according to their amino acid composition and amino-terminal sequences (i.e., attacins A–D constitute a basic, and attacins E and F constitute an acidic group). Their molecular weights are about 23,000, which is much heavier than those of cecropins. Only two different genes may be involved in their production (Kockum et al. 1984). Attacins act on the permeability of the outer membrane (cell envelope), and the bacteria become sensitive to cecropins, antibiotics, and lysozyme (Engström et al. 1984a). The complete amino acid sequence of attacin F has been determined by Engström et al. (1984b). It is derived from attacin E. These two are the only major native attacins present in immune hemolymph. The attacins can kill *E. coli* and two other gram-negative bacteria isolated from the digestive tract of a silkworm larva but not other gram-negative and gram-positive bacteria.

In addition to Lepidoptera, attacin- and cecropin-like antibacterial proteins have been detected in Diptera and Coleoptera. When the adult darkling beetle, *Eleodes* sp., *Drosophila melanogaster* (Flyg et al. 1987), and the tsetse fly, *Glossina morsitans* (Kaaya et al. 1987), are inoculated with bacteria, antibacterial proteins are produced in the hemolymphs (Spies et al. 1986a).

The flesh fly, *Sarcophaga peregrina*, produces several antibacterial proteins (diptericins) when injured with a needle (Natori 1977; Okada and Natori 1983). Three of these proteins, which are almost identical in the primary structure of 39 amino acid residues, are called sarcotoxins IA, IB, and IC (Okada and Natori 1985a). Sarcotoxin IA has been cloned and the nucleotide sequence has been determined (Matsumoto et al. 1986b). There is some homology in the amino acid residues of sarcotoxin IA and those of cecropins, but there is much less in the cleavage sites of these two proteins. Sarcotoxins I are bactericidal for gram-positive and gram-negative bacteria at a concentration of less than 0.5 µg/ml. They act on the bacterial cytoplasmic membrane and interact with liposomes (Okada and Natori 1984, 1985b; Nakajima et al. 1987). Also in *Sarcophaga peregrina* there are groups of attacin-like proteins termed sarcotoxins II (see Dunn 1990) and a novel group of proteins called sarcotoxins III (Baba et al. 1987). These proteins are synthesized in the fat body in response to injury and are eventually secreted into the hemolymph (Takahashi et al. 1985; Matsumoto et al. 1986a; Ando and Natori 1988).

D. PHENOLOXIDASE SYSTEM

The phenoloxidase (phenol metabolism) system may be a primitive mechanism in insects. It is present in nonimmune and immune insects and is involved in the phagocytic process, melanin production around wounds, and the formation of humoral and cellular capsules (Brookman et al. 1988; Nappi and Christensen 1987; Boucias and Latgé 1988). This system is capable of accelerating the oxidation of phenolic compounds that are commonly present in the plasma. The products of the phenoloxidase reactions (melanization) include quinones that may be toxic to invading microorganisms.

The phenoloxidases vary in different insects [e.g., the enzymes in the silkworm is 70 kDa and probably occurs as a dimer (Ashida Dohke 1980); in the house fly, it ranges from 311 to 340 kDa (Yamaura et al. 1980)]. The phenoloxidases are highly reactive and occur as inert proenzymes (prophenoloxidases) that may be activated by many substances, including proteolytic enzymes, carbohydrates, organic solvents, and detergents (Crossley 1975; Götz and Boman 1985). Microbial products activate prophenoloxidases in the plasma of immunized *Galleria* larva (Pye 1974) and in the hemocyte monolayer of this insect (Ratcliffe et al. 1984). The bacterial cell wall contains peptidoglycan, the active principle initiating the activation of prophenoloxidase to phenoloxidase (Yoshida and Ashida 1986).

Prophenoloxidases are formed by insect hemocytes. Iwana and Ashida (1986) have detected a prophenoloxidase with polyclonal antibody in the oenocytoids of the silkworm.

When insects are challenged with the proper foreign body or wounded, the prophenoloxidase system is activated resulting in a cascade reaction. The cascade is initiated when the granulocytes exocytose and release degranulation factors into the hemolymph (Boucias and Latgé 1988). After activation, the phenoloxidase is very sticky and binds tightly to foreign surfaces. In addition to phenoloxidase, this cascade reaction is believed to result in the release of other factors responsible for eliciting further granulocytes to exocytose and produce cell-adhesive proteins. Activated prophenoloxidase alone or the combination of prophenoloxidase and these cell-adhesive proteins act as an opsonin, enhancing phagocytosis of bacterial cells and stimulating the recruitment and adhesion of plasmatocytes to form multicellular sheets around large foreign bodies (Gunnarsson and Lackie 1985; Boucias and Latgé 1988).

An obvious reaction of the phenoloxidase system is melanization, which may be extracellular or intracellular (Nappi 1975). In the mosquito *Aedes sollicitans*, intracellular melanization occurs in the Malpighian tubule cell invaded by the nematode *Dirofilaria immitis* (Bradley and Nayar 1985), but in most other situations, extracellular melanization occurs. The phenoloxidase system of Diptera, particularly mosquitoes and midges, in which hemocytes generally do not participate in encapsulation, is an effective defense reaction against nematodes and fungi (Götz 1969; Götz and Vey 1974). When fungal spores are injected into the hemocoels of the midges *Chironomus thummi* and *C. luridus*, the encapsulation completely prevents the development of *Aspergillus niger*, and partially inhibits *Mucor hiemalis* (Götz and Vey 1974). Encapsulation proceeds rapidly, and within 5 min, most spores are enclosed in a solid capsule of melanin. Infrequently, distintegrating hemocytes may participate in capsule formation. Even though the spores of *Beauveria bassiana* are not encapsulated in the hemocoel, its penetrating hypha may be encapsulated in the cuticle of the integument. In the silkworm infected with the microsporidium *Nosema bombycis*, a pathognomic sign of the disease is the "pepperlike" spots appearing in the integument. These brown spots are groups of blackened spores, presumably killed by the phenoloxidase system, that are pushed into the newly formed integument during molting (Weiser 1969). These spots are eliminated at the next ecdysis.

E. COMPLEMENT

Complement or complement-like substance has been proposed as an antibacterial factor in insects (Faye et al. 1975; Aston et al. 1976). The identification of complement is difficult because of its multifactorial nature and its organization into different functional groups of proteins. The presence of complement or complement-like factors may be established with the use of inhibitors, such as cobra-venom factor, zymosan, and inulin. Inoculation of these inhibitors has reduced the resistance of insects immunized against such bacteria as *Pseudomonas aeruginosa*, *E. coli*, and *Serratia marcescens* (Stephens 1962; Kawarabata 1970; Aston et al. 1976). Faye et al. (1975), from their study on the chemical nature and behavior of the antibacterial factor, conclude that the antibacterial activity is complement-like in nature and similar to that found in vertebrates, but that it differs as follows: (1) in insects, it is inducible, but in humans it is normally constitutive, (2) the cycloheximide concentration turns over rapidly, but this is not known in humans, and (3) in insects, the activity is from 100 to 1000 times more effective in killing *E. coli* than in humans. Despite this evidence, Chadwick and Aston (1976) caution that the proposals for an insect complement system are at best tentative. The complement inhibitors may be exerting their efforts, directly or indirectly, on other effects or mechanisms or at other phases of the protective response.

F. OTHER IMMUNITY-RELATED FACTORS IN HEMOLYMPH

Protease inhibitors occur in the hemolymphs of a number of insects in seven insect orders (Hanschke and Hanschke 1975). The level of protease inhibitor increases significantly after the inoculation of inactivated bacteria into the hemocoel of a greater wax moth larva. Since the inoculated larva exhibits increased resistance against lethal concentrations of proteolytic enzymes (e.g., trypsin, chymotrypsin, pronase P, and extracellular protease produced by *Pseudomonas aeruginosa*), the protease inhibitor may be one of the factors acting in the antimicrobial defense system in the hemolymph. Inhibitors in the wax moth larva are also formed in response to toxic proteases produced in infections with *Metarhizium anisopliae* (Kučera 1982, 1984) and with microsporidia (Kučera and Weiser 1985). The inhibitors are peptides.

An antiviral protection develops in greater wax moth larvae when they are inoculated intrahemocoelically with inactivated insect virus, vertebrate virus, physiological saline, and even when injured with a cutaneous wound (Odier and Vago 1973; Odier 1975). The wounded larvae are much more resistant to infection, either through ingestion or intrahemocoelic inoculation, by the densonucleosis virus (*Parvovirus*) than the uninjured larvae. The nonspecific immunity increases progressively after the injury but is limited to a short period of time (about 8 to 11 days).

There are several reports of antiviral activity resulting after the vaccination of an insect with inactivated virus preparations (Raheja and Brooks 1971). Aizawa (1954) used inactivated NPV to vaccinate silkworm pupae. The pupae were inoculated with the virus vaccine from 1 to 3 times at intervals of 24 h and then challenged with the active virus 32 h after the inoculation. Aizawa reported that the vaccination was effective in preventing some of the pupae from dying of viral infection. He (1954, 1970) also detected a "virus-inactivating principle," which was present in the hemolymph of a silkworm infected with a NPV but not in the hemolymph of uninfected larvae. The antiviral principle was nearly completely destroyed by heating at 70°C for 15 min, and was resistant to trypsin and periodic acid (Kawarabata 1970). The partially purified principle had a low protein content. It did not provide protection by injection into larvae either before or after the virus inoculation, and was therefore not analogous to interferon (Aizawa 1967). As mentioned above, Odier and Vago (1973) also reported antiviral immunity against the densonucleosis virus (*Parvovirus*) in greater wax moth larvae. On the other hand, Rohel et al. (1980) were not able to protect greater wax moth larvae from the cricket-paralysis virus by vaccination with heat- or formalin-inactivated virus.

Interferon is widely recognized as an effective antiviral agent that is produced in vertebrates, especially mammals and birds. Several attempts have been made to detect interferon in insects, but most of them have been unsuccessful. Garzon and Kurstak (1969) have reported, in greater wax moth larva infected with a NPV, a substance with biological functions similar to those of interferon. The synthetic polynucleotide, polyinosinic-polycytidylic acid (Poly I C), is a potent inducer of interferon in vertebrate cells. Bergold and Ramirez (1972) conclude that this polynucleotide produces an interferon-like substance in two mosquito cell lines and possibly in adult mosquitoes in the presence and absence of arboviruses. However, Kalmakoff et al. (1977) could not detect any antiviral response, such as the production of interferon, in cultured *Aedes aegypti* cells with the use of ^{32}P-labeled double-stranded RNA from reovirus and with Poly I C.

Enzmann (1973) has detected an interferon-like antiviral substance in cell cultures of *Aedes albopictus* that have been persistently infected with the Sindbis virus. The virus-free media, harvested from persistently infected cultures, protect mosquito cells from infection. Actinomycin D inhibits the synthesis of this factor. Enzmann (1973) concludes from these observations, together with the stability of the factor in acid, that the factor is interferon-like and maintains the cell–virus equilibrium in the persistently infected culture. Riedel and Brown (1977, 1979), with the use of the same virus and mosquito-cell line, however, believe that the suppression of viral synthesis is not the result of a classical interferon-like agent, because the antiviral agent is virus specific as well as cell specific. The agent is of low molecular weight (from 12,000 to 15,000) and is produced by the cells in detectable levels 3 days after infection followed by increasing levels of concentration.

The hemolymph of the silkworm pupa inhibits the development of the flacherie virus. This inhibition is reduced by the inoculation of antibiotics (e.g., mitomycin C, chromomycin A3, and actinomycin D) and is enhanced by the inoculation of Poly I C (Kurisu and Odan 1978). Such an increase in resistance by Poly I C suggests that an interferon-like substance is produced in the pupa.

G. INHIBITORS ACTING ON IMMUNE PROCESS

There are inhibitors produced by pathogens and parasitoids that act on the immune process. Certain microorganisms [e.g., *Serratia marcescens* (Flyg et al. 1980), *Bacillus thuringiensis* (Siden et al. 1979; Dalhammar and Steiner 1984), and the nematode, *Steinernema carpocapsae* (Götz et al. 1981; Boemare et al. 1982)] produce inhibitors that affect attacins and/or cecropins. These inhibitors may be proteases (Flyg and Xanthopoulos 1983). *Bacillus thuringiensis* produces two factors, inhibitors A and B, that block the humoral defense system against *E. coli* and *B. cereus*, respectively (Dalhammar and Steiner 1984). Inhibitor A is an exoprotease that is formed at the beginning of the stationary growth phase of the bacillus. It selectively destroys cecropins and attacins. Attempts to induce humoral immunity in the form of cecropins, lysozymes, and other immune inhibitors have so far been unsuccessful in insects known to have symbionts (Boman and Steiner 1981).

The calyx fluid from the venom apparatus of the braconid *Apanteles glomeratus* prevents the encapsulation of the braconid eggs in the larva of *Pieris rapae crucivora* (Kitano 1986). The polydnavirus associated with the parasitoid may also inhibit egg encapsulation (see Chapter 7).

XII. PASSIVELY ACQUIRED IMMUNITY

Passive immunity has been demonstrated in certain insect species. In the greater wax moth, the hemolymphs of larvae immunized with attenuated vaccines of bacterial pathogens provide protection to unvaccinated larvae against the lethal infection of the bacteria (Zernoff 1927, Stephens 1959; De Verno et al. 1983). Gingrich (1964) has conducted similar studies with the milkweed bug, *Oncopeltus fasciatus*. The strength of an actively acquired or passively acquired immunity in the bug is essentially equal. In the greater wax moth larva, the transfer of whole hemolymph, cell-free hemolymph or hemocytes from insects previously immunized with the lipopolysaccaride of *Pseudomonas aeruginosa* results in the immunization of the larva against the bacterium (De Verno et al. 1983). Whole immune hemolymph, injected into the larva as early as 3 h after vaccination, confers protection for as long as 40 h, whereas hemocytes alone confer good protection when transferred as early as 30 min and up to about 4 h after vaccination. Zernoff (1927) reports that the passive immunity in the greater wax moth larva against *Salmonella enteritidis* (syn. *Bacillus danysz*) lasts as long as 5 days.

Besides passively acquired immunity against bacteria, such immunity has also been demonstrated against ciliates in the American cockroach (Seaman and Robert 1968) and against the filarial worm (*Onchocerca lienalis*) in the black flies (*Simulium* spp.) (Ham 1986).

REFERENCES

Aizawa, K. 1954. Immunological studies of the silkworm jaundice virus. (3) Experiments on the defence of infection in the silkworm jaundice. *Virus 4*, 245–248.

Aizawa, K. 1962. Antiviral substance in the gut-juice of the silkworm, *Bombyx mori* (Linnaeus). *J. Insect Pathol. 4*, 72–76.

Aizawa, K. 1967. Mode of multiplication of the nuclear-polyhedrosis virus of the silkworm. *J. Seric. Sci. Jpn. 36*, 327–331.

Aizawa, K. 1970. Defense reactions of the silkworm, *Bombyx mori* against the nuclear polyhedrosis. "Proc. Fourth Int. Colloq. Insect Pathol." pp. 352–356. College Park, Maryland.

Aizawa, K., Furuta, Y., and Nakamura, K. 1961. Selection of a resistant strain to virus induction in the silkworm, *Bombyx mori*. *J. Seric. Sci. Jpn. 30*, 405–412.

Akai, H. 1976. "Ultrastructural Morphology of Insects." University of Tokyo Press, Tokyo.

Akhurst, R. J., and Boemare, N. E. 1990. Biology and taxonomy of *Xenorhabdus*. In "Entomopathogenic Nematodes in Biological Control." (R. Gaugler and H. K. Kaya, eds.), pp. 75–90. CRC Press, Boca Raton, Florida.

Allen, G. E., and Ignoffo, C. M. 1969. The nucleopolyhedrosis virus of *Heliothis*: Quantitative in vivo estimates of virulence. *J. Invertebr. Pathol. 13*, 378–381.

Amargier, A., Vago, C., Duthoit, J. L., and Meynadier, G. 1979. Formation tumorale d'origine parvovirale chez *Sibine fusca* [Lep.: Limacodidae]. *Entomophaga 24*, 259–271.

Amirante, G. A., and Mazzalai, F. G. 1978. Synthesis and localization of hemagglutinins in hemocytes of the cockroach *Leucophaea maderae* L. *Dev. Comp. Immunol. 2*, 735–740.

Anderson, R. S. 1974. Metabolism of insect hemocytes during phagocytosis. *Contemp. Top. Immunobiol. 4*, 47–54.

Anderson, R. S. 1977. Biochemistry and physiology of invertebrate macrophages *in vitro*. In "Comparative Pathobiology, Invertebrate Immune Responses." (L. A. Bulla, Jr. and T. C. Cheng, eds.), Vol. 3, pp. 1–20. Plenum Press, New York.

Anderson, R. S., and Cook, M. L. 1979. Induction of lysozymelike activity in the hemolymph and hemocytes of an insect, *Spodoptera eridania*. *J. Invertebr. Pathol.* 33, 197–203.

Anderson, R. S., Day, N. K. B., and Good, R. A. 1972. Specific hemagglutinin and a modulator of complement in cockroach hemolymph. *Infect. Immun.* 5, 55–59.

Anderson, R. S., Holmes, B., and Good, R. A. 1973. In vitro bactericidal capacity of *Blaberus craniifer* hemocytes. *J. Invertebr. Pathol.* 22, 127–135.

Andersons, D., Gunne, H., Hellers, M., Johansson, H., and Steiner, H. 1990. Immune response in *Trichoplusia ni* challenged with bacteria or baculoviruses. *Insect Biochem.* 20, 537–543.

Ando, K., and Natori, S. 1988. Molecular cloning, sequencing, and characterization of cDNA for sarcotoxin IIA, an inducible antibacterial protein of *Sarcophaga peregrina* (flesh fly). *Biochemistry 27*, 1715–1721.

Andreu, D., Merrifield, R. B., Steiner, H., and Boman, H. G. 1985. N-terminal analogues of cecropin A: Synthesis, antibacterial activity, and conformational properties. *Biochemistry 24*, 1683–1688.

Angus, T. A. 1956. The reaction of certain lepidopterous and hymenopterous larvae to *Bacillus sotto* toxin. *Can. Entomol.* 88, 280–283.

Aratake, Y. 1973a. Difference of the resistance to infectious flacherie virus between the strains of the silkworm, *Bombyx mori* L. *Sansi-Kenkyu (Acta Serol.)* 86, 48–57.

Aratake, Y. 1973b. Strain difference of the silkworm, *Bombyx mori* L., in the resistance to a nuclear-polyhedrosis virus. *J. Seric. Sci. Jpn.* 42, 230–238.

Arnold, J. W. 1972. A comparative study of the haemocytes (blood cells) of cockroaches (Insecta: Dictyoptera: Blattaria), with a view of their significance in taxonomy. *Can. Entomol. 104*, 309–348.

Arnold, J. W. 1974. The hemocytes of insects. *In* "The Physiology of Insecta." (M. Rockstein, ed.), Second Ed. Vol. 5, pp. 201–254. Academic Press, New York.

Arnold, J. W. 1979. Controversies about hemocyte types in insects. *In* "Insect Hemocytes." (A. Gupta, ed.), pp. 231–258. Cambridge University Press, Cambridge.

Arnold, J. W., and Hinks, C. F. 1983. Haemopoiesis in Lepidoptera. III. A note on the multiplication of spherule cells and granular haemocytes. *Can. J. Zool. 61*, 275–277.

Aruga, H., and Watanabe, H. 1964. Resistance to per os infection with cytoplasmic-polyhedrosis virus in the silkworm, *Bombyx mori* (Linnaeus). *J. Insect Pathol.* 6, 387–394.

Ashida, M., and Dohke, K. 1980. Activation of pro-phenoloxidase by the activating enzyme of the silkworm, *Bombyx mori*. *Insect Biochem. 10*, 37–47.

Aston, W. P., Chadwick, J. S., and Henderson, M. J. 1976. Effect of cobra venom factor on the in vivo immune response in *Galleria mellonella* to *Pseudomonas aeruginosa*. *J. Invertebr. Pathol.* 27, 171–176.

Baba, K., Okada, M., Kawano, T., Komano, H., and Natori, S. 1987. Purification of sarcotoxin III, a new antibacterial protein of *Sarcophaga peregrina*. *J. Biochem. 102*, 69–74.

Baerwald, R. J. 1979. Fine structure of hemocyte membranes and intercellular junctions formed during hemocyte encapsulation. *In* "Insect Hemocytes." (A. P. Gupta, ed.), pp. 155–188. Cambridge University Press, Cambridge.

Bailey, L. 1965. Paralysis of the honey bee, *Apis mellifera* Linnaeus. *J. Invertebr. Pathol.* 7, 132–140.

Bailey, L. 1967. The effect of temperature on the pathogenicity of the fungus, Ascosphaera apis, for larvae of the honey bee, Apis mellifera. *In* "Proc. Int. Coll. Insect Pathol. Microb. Contr." (P. A. van der Laan, ed.), pp. 162–167. North-Holland, Amsterdam.

Bailey, L. 1968. Honey bee pathology. *Annu. Rev. Entomol. 13*, 191–212.

Bailey, L. 1969. The multiplication and spread of sacbrood virus of bees. *Ann. Appl. Biol. 63*, 483–491.

Bailey, L. 1981. "Honey Bee Pathology." Academic Press, London.
Bakula, M. 1970. Antibacterial compounds in the cell-free haemolymph of *Drosophila melanogaster*. *J. Insect Physiol.* 16, 185–197.
Barwig, B., and Bohn, H. 1980. Evidence for presence of two clotting proteins in insects. *Naturwissenschaften* 67, 47–48.
Bauer, L. S., and Nordin, G. L. 1988. Nutritional physiology of the eastern spruce budworm, *Choristoneura fumiferana*, infected with *Nosema fumiferanae*, and interactions with dietary nitrogen. *Oecologia* 77, 44–50.
Bedding, R. A., and Molyneux, A. S. 1982. Penetration of insect cuticle by infective juveniles of *Heterorhabditis* spp. (Heterorhabditidae: Nematoda). *Nematologica* 28, 354–359.
Bell, J. V., and Hamalle, R. J. 1971. Comparative mortalities between field-collected and laboratory-reared wireworm larvae. *J. Invertebr. Pathol.* 18, 150–151.
Bergold, G. H., and Ramirez, N. 1972. Replication of arboviruses in mosquitoes. *In* "Moving Frontiers in Invertebrate Virology." (T. W. Tinsley and K. A. Harrap, eds.) *Monogr. Virol.* 6, 56–59.
Bernheimer, A. W., Caspari, E., and Kaiser, A. D. 1952. Studies on antibody formation in caterpillars. *J. Exp. Zool.* 119, 23–35.
Bettini, S., Sarkaria, D. S., and Patton, R. L. 1951. Observations on the fate of vertebrate erythrocytes and hemoglobin injected into the blood of the American cockroach (*Periplaneta americana* L.). *Science* 113, 9–10.
Bidochka, M. J., and Khachatourians, G. G. 1987. Hemocytic defense response to the entomopathogenic fungus *Beauveria bassiana* in the migratory grasshopper *Melanoplus sanguinipes*. *Entomol. Exp. Appl.* 45, 151–156.
Bienvenu, R. J., Atchison, F. W., and Cross, E. A. 1968. Microbial inhibition by prepupae of the alkali bee, *Nomia melanderi*. *J. Invertebr. Pathol.* 12, 278–282.
Biever, K. D., and Wilkinson, J. D. 1978. A stress-induced granulosis virus of *Pieris rapae*. *Environ. Entomol.* 7, 572–573.
Bird, F. T. 1953. The effect of metamorphosis on the multiplication of an insect virus. *Can. J. Zool.* 31, 300–303.
Boemare, N., Laumond, C., and Luciani, J. 1982. Mise en évidence d'une toxicogenèse provoquée par le Nématode axénique entomophage *Neoaplectana carpocapsae* Weiser chez l'Insecte axénique *Galleria mellonella* L. *C.R. Acad. Sci. Paris Ser. III* 295, 543–546.
Boman, H. G. 1981. Insect responses to microbial infections. *In* "Microbial Control of Pests and Plant Diseases 1970–1980." (H. D. Burges, ed.), pp. 769–784. Academic Press, New York.
Boman, H. G., and Hultmark, D. 1987. Cell-free immunity in insects. *Annu. Rev. Microbiol.* 41, 103–126.
Boman, H. G., and Steiner, H. 1981. Humoral immunity in Cecropia pupae. *Curr. Top. Microbiol. Immunol.* 94/95, 75–91.
Boman, H. G., Nilsson-Faye, I., Paul, K., and Rasmuson, T., Jr. 1974. Insect immunity. I. Characteristics of an inducible cell-free antibacterial reaction in hemolymph of *Samia cynthia* pupae. *Infect. Immun.* 10, 136–145.
Boucías, D., and Latgé, J-P. 1988. Fungal elicitors of invertebrate cell defense system. *In* "Fungal Antigens." (E. Drouhet, G. T. Cole, L. de Repentigny, J-P. Latgé, and B. Dupont, eds.), pp. 121–137. Plenum, New York.
Boucias, D. G., and Nordin, G. L. 1977. Interinstar susceptibility of the fall webworm, *Hyphantria cunea*, to its nucleopolyhedrosis and granulosis viruses. *J. Invertebr. Pathol.* 30, 68–75.
Boucias, D. G., Johnson, D. W., and Allen, G. E. 1980. Effects of host age, viral dosage, and temperature on the infectivity of a nucleopolyhedrosis virus against velvetbean caterpillar, *Anticarsia gemmatalis*, larvae. *Environ. Entomol.* 9, 59–61.

Boucias, D. G., Bradford, D. L., and Barfield, C. S. 1984. Susceptibility of the velvetbean caterpillar and soybean looper (Lepidoptera: Noctuidae) to *Nomuraea rileyi*: Effects of pathotype, dosage, temperature, and host age. *J. Econ. Entomol.* 77, 247–253.

Bradley, T. J., and Nayar, J. K. 1985. Intracellular melanization of the larvae of *Dirofilaria immitis* in the Malpighian tubules of the mosquito, *Aedes sollicitans*. *J. Invertebr. Pathol.* 45, 339–345.

Brandt, C. R., Adang, M. J., and Spence, K. D. 1978. The peritrophic membrane: Ultrastructural analysis and function as a mechanical barrier to microbial infection in *Orgyia pseudotsugata*. *J. Invertebr. Pathol.* 32, 12–24.

Brehélin, M., and Zachary, D. 1983. About insect plasmatocytes and granular cells. *Dev. Comp. Immunol.* 7, 683–686.

Brewer, F. D., and Vinson, S. B. 1971. Chemicals affecting the encapsulation of foreign material in an insect. *J. Invertebr. Pathol.* 18, 287–289.

Briese, D. T. 1982. Genetic basis for resistance to a granulosis virus in the potato moth *Phthorimaea operculella*. *J. Invertebr. Pathol.* 39, 215–218.

Briese, D. T., and Mende, H. A. 1983. Selection for increased resistance to a granulosis virus in the potato moth, *Phthorimaea operculella* (Zeller)(Lepidoptera: Gelechiidae). *Bull. Entomol. Res.* 73, 1–9.

Briese, D. T., Mende, H. A., Grace, T. D. C., and Geier, P. W. 1980. Resistance to a nuclear polyhedrosis virus in the light-brown apple moth *Epiphyas postvittana* (Lepidoptera: Tortricidae). *J. Invertebr. Pathol.* 36, 211–215.

Briggs, J. D. 1958. Humoral immunity in lepidopterous larvae. *J. Exp. Zool.* 138, 155–188.

Briggs, J. D. 1964. Immunological responses. *In* "The Physiology of Insecta." (M. Rockstein, ed.), Vol. 3, pp. 259–283. Academic Press, New York.

Bronskill, J. F. 1960. The capsule and its relation to the embryogenesis of the ichneumonid parasitoid Mesoleius tenthredinis Morl. in the larch sawfly, Pristiphora erichsonii (Htg.) (Hymenoptera: Tenthredinidae). *Can. J. Zool.* 38, 769–775.

Bronskill, J. F. 1962. Encapsulation of rhabditoid nematodes in mosquitoes. *Can. J. Zool.* 40, 1269–1275.

Brookman, J. L., Ratcliffe, N. A., and Rowley, A. F. 1988. Optimization of a monolayer phagocytosis assay and its application for studying the role of the prophenoloxidase system in the wax moth, *Galleria mellonella*. *J. Insect Physiol.* 34, 337–345.

Brooks, W. M. 1971. The inflammatory response of the tobacco hornworm, *Manduca sexta*, to infection by the microsporidian, *Nosema sphingidis*. *J. Invertebr. Pathol.* 17, 87–93.

Brough, E. J. 1983. The antimicrobial activity of the mandibular gland secretion of a formicine ant, *Calomyrmex* sp. (Hymenoptera: Formicidae). *J. Invertebr. Pathol.* 42, 306–311.

Bucher, G. E. 1959. Bacteria of grasshoppers of western Canada: III. Frequency of occurrence, pathogenicity. *J. Insect Pathol.* 1, 391–405.

Bucher, G. E., and Harris, P. 1968. Virus diseases and their interaction with food stress in *Calophasia lunula*. *J. Invertebr. Pathol.* 10, 235–244.

Burdette, W. J. 1974. Tumors in Drosophila and antibacterial immunity. *Contemp. Top. Immunobiol.* 4, 283–288.

Burgerjon, A., Biache, G., Chaufaux, J., and Petré, Z. 1981. Sensibilité comparée, en fonction de leur âge, des chenilles de *Lymantria dispar*, *Mamestra brassicae* et *Spodoptera littoralis* aux virus de la polyédrose nucléaire. *Entomophaga* 26, 47–58.

Buys, B. 1972. Nosema in brood. *S. Afr. Bee J.* 44, No. 6, 2–4.

Camponovo, F., and Benz, G. 1984. Age-dependent tolerance to *Baculovirus* in last larval instars of the codling moth, *Cydia pomonella* L., induced either for pupation or for diapause. *Experientia* 40, 938–939.

Carayon, J. 1977. Les cellules capables de phagocytose chez les Insectes. *Ann. Parasitol.* (Paris). 52, 63–65.

Chadwick, J. S. 1968. Some aspects of immune responses in insects. *In Vitro 3*, 120–128.

Chadwick, J. S. 1970. Relation of lysozyme concentration to acquired immunity against *Pseudomonas aeruginosa* in *Galleria mellonella*. *J. Invertebr. Pathol. 15*, 455–456.

Chadwick, J. S. 1971. An assessment of the ability of individual moieties of *Pseudomonas aeruginosa* endotoxin to induce immunity in larvae of *Galleria mellonella*. *J. Invertebr. Pathol. 17*, 299–300.

Chadwick, J. S. 1977. Induction and effector mechanism in insect immunity. *In* "Comparative Pathobiology, Invertebrate Immune Responses." (L. A. Bulla, Jr. and T. C. Cheng, eds.), Vol. 3, pp. 85–102. Plenum Press, New York.

Chadwick, J. S., and Aston, W. P. 1976. Effector mechanisms involved in the protective response in *Galleria mellonella* towards bacterial pathogens. *In* "Proc. First Int. Colloq. Invertebr. Pathol. Queen's University." pp 204–209. Kingston, Canada.

Chadwick, J. S., and Vilk, E. 1969. Endotoxins from several bacterial species as immunizing agents against *Pseudomonas aeruginosa* in *Galleria mellonella*. *J. Invertebr. Pathol. 13*, 410–415.

Chain, B. M., and Anderson, R. S. 1983. Observations on the cytochemistry of the hemocytes of an insect, *Galleria mellonella*. *J. Histochem. Cytochem. 31*, 601–607.

Champlin, F. R., Cheung, P. Y. K., Pekrul, S., Smith, R. J., Burton, R. L., and Grula, E. A. 1981. Virulence of *Beauveria bassiana* mutants for the pecan weevil. *J. Econ. Entomol. 74*, 617–621.

Cherbas, L. 1973. The induction of an injury reaction in cultured haemocytes from saturniid pupae. *J. Insect Physiol. 19*, 2011–2023.

Cheung, P. Y. K., Grula, E. A., and Burton, R. L. 1978. Hemolymph responses in *Heliothis zea* to inoculation with *Bacillus thuringiensis* or *Micrococcus lysodeikticus*. *J. Invertebr. Pathol. 31*, 148–156.

Chiang, A. S., Yen, D. F., and Peng, W. K. 1986. Defense reaction of midgut epithelial cells in the rice moth larva (*Corcyra cephalonica*) infected with *Bacillus thuringiensis*. *J. Invertebr. Pathol. 47*, 333–339.

Clark, T. B. 1980. A second microsporidian in the honeybee. *J. Invertebr. Pathol. 35*, 290–294.

Clark, T. B., Kellen, W. R., Fukuda, T., and Lindegren, J. E. 1968. Field and laboratory studies on the pathogenicity of the fungus *Beauveria bassiana* to three genera of mosquitoes. *J. Invertebr. Pathol. 11*, 1–7.

Clot, J., and Vago, C. 1970. Recherches sur le passage de bactéries pathogènes pour les invertébrés et les vertébrés (homme, animal), à travers le tube digestif de dictyoptères disséminateurs (expériences sur *Blabera fusca*). *Ann. Rech. Veter. 1*, 31–40.

Costin, N. M. 1975. Histochemical observations of the haemocytes of *Locusta migratoria*. *Histochem. J. 7*, 21–43.

Couch, J. A. 1974. Free and occluded virus, similar to *Baculovirus*, in hepatopancreas of pink shrimp. *Nature 247*, 229–231.

Coulon, P., and Contamine, D. 1982. Role of the *Drosophila* genome in sigma virus multiplication. II. Host spectrum variants among the haP mutants. *Virology 123*, 381–392.

Couturier, A. 1963. Recherches sur des Mermithidae, Nématodes parasites du Hanneton commun (*Melolontha melolontha* L. Coléopt. Scarab.). *Ann. Epiphy. 14*, 203–267.

Croizier, G., and Croizier, L. 1978. Purification et comparaison immunologique de 2 lysozymes d'Insectes. *C.R. Acad. Sci., Paris Ser. D 286*, 469–472.

Croizier, G., and Croizier, L. 1980. Étude du phénomène de l'induction de facteurs bactériolytiques chez les lépidoptères: Inhibition de la production du lysozyme chez les larves de *Galleria mellonella* [*Lep.: Pyralidae*] par l'Actinomycine D et la Cycloheximide. *Entomophaga 25*, 219–228.

Crossley, A. C. S. 1972. The ultrastructure and function of pericardial cells and other nephrocytes in an insect: *Calliphora erythrocephala*. *Tissue Cell 4*, 529–560.

Crossley, A. C. 1975. The cytophysiology of insect blood. In "Advances in Insect Physiology." (J. E. Treherne, M. J. Berridge, and V. B. Wigglesworth, eds.), Vol. 11, pp. 117–221. Academic Press, New York.

Dalhammar, G., and Steiner, H. 1984. Characterization of inhibitor A, a protease from *Bacillus thuringiensis* which degrades attacins and cecropins, two classes of antibacterial proteins in insects. *Eur. J. Biochem. 139*, 247–252.

David, W. A. L. 1967. The physiology of the insect integument in relation to the invasion of pathogens. In "Insects and Physiology." (J. W. L. Beament and J. E. Treherne, eds.), pp. 17–35. Oliver and Boyd, London.

David, W. A. L., and Taylor, C. E. 1977. The effect of sucrose content of diets on susceptibility to granulosis virus disease in *Pieris brassicae*. *J. Invertebr. Pathol. 30*, 117–118.

David, W. A. L., Ellaby, S., and Taylor, G. 1972. The effect of reducing the content of certain ingredients in a semisynthetic diet on the incidence of granulosis virus disease in *Pieris brassicae*. *J. Invertebr. Pathol. 20*, 332–340.

Davidson, E. W. 1970. Ultrastructure of peritrophic membrane development in larvae of the worker honey bee (*Apis mellifera*). *J. Invertebr. Pathol. 15*, 451–454.

Derksen, A. C. G., and Granados, R. R. 1988. Alteration of a lepidopteran peritrophic membrane by baculoviruses and enhancement of viral infectivity. *Virology 167*, 242–250.

De Verno, P. J., Aston, W. P., and Chadwick, J. S. 1983. Transfer of immunity against *Pseudomonas aeruginosa* P11-1 in *Galleria mellonella* larvae. *Dev. Comp. Immunol. 7*, 423–434.

Dillon, R. J., and Charnley, A. K. 1986. Inhibition of *Metarhizium anisopliae* by the gut bacterial flora of the desert locust, *Schistocerca gregaria*: Evidence for an antifungal toxin. *J. Invertebr. Pathol. 47*, 350–360.

Donlon, W. C., and Wemyss, C. T. 1976. Analysis of the hemagglutinin and general protein element of the hemolymph of the West Indian leaf cockroach, *Blaberus craniifer*. *J. Invertebr. Pathol. 28*, 191–194.

Drif, L., and Brehélin, M. 1983. The circulating hemocytes of *Culex pipiens* and *Aedes aegypti*: Cytology, histochemistry, hemograms and functions. *Dev. Comp. Immunol. 7*, 687–690.

Dunn, P. E. 1990. Humoral immunity in insects. *BioScience 40*, 738–744.

Dunn, P. E., and Drake, D. R. 1983. Fate of bacteria injected into naive and immunized larvae of the tobacco hornworm *Manduca sexta*. *J. Invertebr. Pathol. 41*, 77–85.

Dunphy, G. B., and Chadwick, J. S. 1989. Effects of selected carbohydrates and the contribution of the prophenoloxidase cascade system to the adhesion of strains of *Pseudomonas aeruginosa* and *Proteus mirabilis* to hemocytes of nonimmune larval *Galleria mellonella*. *Can. J. Microbiol. 35*, 524–527.

Dunphy, G. B., and Nolan, R. A. 1980a. Response of eastern helmlock looper hemocytes to selected stages of *Entomophthora egressa* and other foreign particles. *J. Invertebr. Pathol. 36*, 71–84.

Dunphy, G. B., and Nolan, R. A. 1980b. Protozoan lysins in the larval eastern spruce budworm hemolymph. *J. Invertebr. Pathol. 36*, 433–437.

Dunphy, G. B., and Nolan, R. A. 1981. A study of the surface proteins of *Entomophthora egressa* protoplasts and of larval spruce budworm hemocytes. *J. Invertebr. Pathol. 38*, 352–361.

Dunphy, G .B., and Nolan, R. A. 1982. Cellular immune responses of spruce budworm larvae to *Entomophthora egressa* protoplasts and other test particles. *J. Invertebr. Pathol. 39*, 81–92.

Dunphy, G. B., and Thurston, G. S. 1990. Insect immunity. In "Entomopathogenic Nematodes in Biological Control." (R. Gaugler and H. K. Kaya, eds.), pp. 301–323. CRC Press, Boca Raton, Florida.

Dunphy, G. B., and Webster, J. M. 1988. Virulence mechanisms of *Heterorhabditis heliothidis* and its bacterial associate, *Xenorhabdus luminescens*, in non-immune larvae of the greater wax moth, *Galleria mellonella*. *Int. J. Parasitol. 18*, 729–737.

Eguchi, M. 1982. Inhibition of the fungal protease by haemolymph protease inhibitors of the silkworm, *Bombyx mori* L. (Lepidoptera: Bombycidae). *Appl. Entomol. Zool. 17*, 589–590.

Engström, P., Carlsson, A., Engström, Å., Tao, Z.-j., and Bennich, H. 1984a. The antibacterial effect of attacins from the silk moth *Hyalophora cecropia* is directed against the outer membrane of *Escherichia coli*. *EMBO J. 3*, 3347–3351.

Engström, Å., Engström, P., Tao, Z.-j., Carlsson, A., and Bennich, H. 1984b. Insect immunity. The primary structure of the antibacterial protein attacin F and its relation to two native attacins from *Hyalophora cecropia*. *EMBO J. 3*, 2065–2070.

Ennesser, C. A., and Nappi, A. J. 1984. Ultrastructural study of the encapsulation response of the American cockroach, *Periplaneta americana*. *J. Ultrastruct. Res. 87*, 31–45.

Enzmann, P.-J. 1973. Induction of an interferon-like substance in persistently infected Aedes albopictus cells. *Arch. Ges. Virusforsch. 40*, 382–389.

Evans, H. F. 1981. Quantitative assessment of the relationships between dosage and response of the nuclear polyhedrosis virus of *Mamestra brassicae*. *J. Invertebr. Pathol. 37*, 101–109.

Evans, H. F. 1983. The influence of larval maturation on responses of *Mamestra brassicae* L. (Lepidotpera: Noctuidae) to nuclear polyhedrosis virus infection. *Arch. Virol. 75*, 163–170.

Evlakhova, A. A., and Chekhourina, T. A. 1963. Antifungal action of the cuticle of *Eurygaster integriceps* Put. *Dokl. Akad. Nauk SSSR 148*, 977–978. (English translation *48*, 199–201).

Fargues, J., and Rodriguez-Rueda, D. 1980. Sensibilité des oeufs des Noctuides *Mamestra brassicae* et *Spodoptera littoralis* aux Hyphomycètes *Paecilomyces fumoso-roseus* et *Nomuraea rileyi*. *C.R. Acad. Sci. Paris Ser. D 290*, 65–68.

Fargues, J., Robert, P.-H., and Vey, A. 1976. Rôle du tégument et de la défense cellulaire des Coléoptères hôtes dans la spécificité des souches entomopathogènes de *Metarrhizium anisopliae* (Fungi Imperfecti). *C.R. Acad. Sci. Paris Ser. D 282*, 2223–2226.

Faye, I. 1978. Insect immunity: Early fate of bacteria injected in saturniid pupae. *J. Invertebr. Pathol. 31*, 19–26.

Faye, I., and Wyatt, G. R. 1980. The synthesis of antibacterial proteins in isolated fat body from Cecropia silkmoth pupae. *Experientia 36*, 1325–1326.

Faye, I., Pye, A., Rasmuson, T., Boman, H. G., and Boman, I. A. 1975. Insect immunity. II. Simultaneous induction of antibacterial activity and selective synthesis of some hemolymph proteins in diapausing pupae of *Hyalophora cecropia* and *Samia cynthia*. *Infect. Immun. 12*, 1426–1438.

Fazekas de St. Groth, S. 1948. Viropexis, the mechanism of influenza virus infection. *Nature 162*, 294–295.

Feir, D. 1979. Multiplication of hemocytes. *In* "Insect Hemocytes." (A. P. Gupta, ed.), pp. 67–82. Cambridge University Press, Cambridge.

Feng, Z., Carruthers, R. I., Roberts, D. W., and Robson, D. S. 1985. Age-specific dose-mortality effects of *Beauveria bassiana* (Deuteromycotina: Hyphomycetes) on the European corn borer, *Ostrinia nubilalis* (Lepidoptera: Pyralidae). *J. Invertebr. Pathol. 46*, 259–264.

Fleuriet, A. 1976. Presence of the hereditary rhabdovirus sigma and polymorphism for a gene for resistance to this virus in natural populations of *Drosophila melanogaster*. *Evolution 30*, 735–739.

Flyg, C., and Xanthopoulos, K. G. 1983. Insect pathogenic properties of *Serratia marcescens*. Passive and active resistance to insect immunity studied with protease-deficient and phage-resistant mutants. *J. Gen. Microbiol. 129*, 453–464.

Flyg, C., Kenne, K., and Boman, H. G. 1980. Insect pathogenic properties of *Serratia marcescens*: Phage-resistant mutants with a decreased resistance to *Cecropia* immunity and a decreased virulence to *Drosophila*. *J. Gen. Microbiol.* **120**, 173–181.

Flyg, C., Dalhammar, G., Rasmuson, B., and Boman, H. G. 1987. Insect Immunity. Inducible antibacterial activity in *Drosophila*. *Insect Biochem.* **17**, 153–160.

Fuxa, J. R. 1987. *Spodoptera frugiperda* susceptibility to nuclear polyhedrosis virus isolates with reference to insect migration. *Environ. Entomol.* **16**, 218–223.

Gagen, S. J., and Ratcliffe, N. A. 1976. Studies on the in vivo cellular reactions and fate of ingested bacteria in *Galleria mellonella* and *Pieris brassicae* larvae. *J. Invertebr. Pathol.* **28**, 17–24.

Galbreath, R. A. 1976. Spiracle structure and function in *Costelytra zealandica* larvae (Coleoptera: Scarabaeidae). *N.Z. J. Zool.* **3**, 333–337.

Gardner, W. A. 1985. Effects of temperature on the susceptibility of *Heliothis zea* larvae to *Nomuraea rileyi*. *J. Invertebr. Pathol.* **46**, 348–349.

Garzon, S., and Kurstak, E. 1969. Interférence sélective au niveau de tissus entre le virus de la polyédrie nucléaire (VPN) et le virus de la densonucléose (VDN) et présence d'une substance de type interféron chez un arthropode. *Rev. Can. Biol.* **28**, 89–94.

Gaugler, R., and Molloy, D. 1981. Instar susceptibility of *Simulium vittatum* (Diptera: Simuliidae) to the entomogenous nematode *Neoaplectana carpocapsae*. *J. Nematol.* **13**, 1–5.

Gaugler, R., Lebeck, L., Nakagaki, B., and Boush, G. M. 1980. Orientation of the entomogenous nematode *Neoaplectana carpocapsae* to carbon dioxide. *Environ. Entomol.* **9**, 649–652.

Gaugler, R., Campbell, J. F., and Gupta, P. 1991. Characterization and basis of enhanced host-finding in a genetically improved strain of *Steinernema carpocapsae*. *J. Invertebr. Pathol.* **57**, 234–241.

Georgis, R., and Hague, N. G. M. 1981. A neoaplectanid nematode in the larch sawfly *Cephalcia lariciphila* (Hymenoptera: Pamphiliidae). *Ann. Appl. Biol.* **99**, 171–177.

Getzin, L. W. 1961. *Spicaria rileyi* (Farlow) Charles, an entomogenous fungus of *Trichoplusia ni* (Hübner). *J. Insect Pathol.* **3**, 2–10.

Gilliam, M., and Jeter, W. S. 1970. Synthesis of agglutinating substances in adult honeybees against *Bacillus larvae*. *J. Invertebr. Pathol.* **16**, 69–70.

Gingrich, R. E. 1964. Acquired humoral immune response of the large milkweed bug, *Oncopeltus fasciatus* (Dallas), to injected materials. *J. Insect Physiol.* **10**, 179–194.

Gochnauer, T. A., Boch, R., and Margetts, V. J. 1979. Inhibition of *Ascosphaera apis* by citral and geraniol. *J. Invertebr. Pathol.* **34**, 57–61.

Goffinet, G., and Grégoire, C. 1979. Gross and ultrastructural changes in coagulocytes of mole-cricket, *Gryllotalpa* sp. (Orthoptera: Gryllotalpidae) during *in vitro* hemolymph coagulation. *Int. J. Insect Morphol. Embryol.* **8**, 59–66.

Goldman, I. F., Arnold, J., and Carlton, B. C. 1986. Selection for resistance to *Bacillus thuringiensis* subspecies *israelensis* in field and laboratory populations of the mosquito *Aedes aegypti*. *J. Invertebr. Pathol.* **47**, 317–324.

Götz, P. 1969. Die Einkapselung von Parasiten in der Hämolymphe von Chironomus-Larven (Diptera). *Zool. Anz. Suppl. Verh. Zool. Ges.* **33**, 610–617.

Götz, P. 1973. Immunreaktionen bei Insekten. *Naturwiss. Rundsch.* **26**, 367–375.

Götz, P., and Boman, H. G. 1985. Insect immunity. *In* "Comprehensive Insect Physiology, Biochemistry and Pharmacology. Integument, Respiration and Circulation." (G. A. Kerkut and L. I. Gilbert, eds.), Vol. 3, pp. 453–485. Pergamon Press, New York.

Götz, P., and Vey, A. 1974. Humoral encapsulation in Diptera (Insecta): Defence reactions of *Chironomus* larvae against fungi. *Parasitology* **68**, 193–205.

Götz, P., Roettgen, I., and Lingg, W. 1977. Encapsulement humoral en tant que réaction de défense chez les Diptères. *Ann. Parasitol.* **52**, 95–97.

Götz, P., Boman, A., and Boman, H. G. 1981. Interactions between insect immunity and an insect-pathogenic nematode with symbiotic bacteria. *Proc. R. Soc. London B 212*, 333–350.

Grégoire, C. 1974. Hemolymph coagulation. *In* "The Physiology of Insecta." (M. Rockstein, ed.), Second Ed. Vol. 5, pp. 309–360. Academic Press, New York.

Grimstone, A. V., Rotheram, S., and Salt, G. 1967. An electron-microscope study of capsule formation by insect blood cells. *J. Cell Sci. 2*, 281–292.

Gunnarsson, S. G. S., and Lackie, A. M. 1985. Hemocytic aggregation in *Schistocerca gregaria* and *Periplaneta americana* as a response to injected substances of microbial origin. *J. Invertebr. Pathol. 46*, 312–319.

Gupta, A. P. 1979. Hemocyte types: Their structures, synonymies, interrelationships, and taxonomic significance. *In* "Insect Hemocytes." (A. P. Gupta, ed.), pp. 85–127. Cambridge University Press, Cambridge.

Gupta, A. P. 1985a. Cellular elements in the hemolymph. *In* "Comprehensive Insect Physiology Biochemistry and Pharmacology. Integument, Respiration and Circulation." (G. A. Kerkut and L. I. Gilbert, eds.), Vol. 3, pp. 401–451. Pergamon Press, New York.

Gupta, A. P. 1985b. The identity of the so-called crescent cell in the hemolymph of the cockroach, *Gromphadorhina portentosa* (Schaum)(Dictyoptera: Blaberidae). *Cytologia 50*, 739–746.

Gupta, A. P., and Sutherland, D. J. 1966. *In vitro* transformations of the insect plasmatocyte in some insects. *J. Insect Physiol. 12*, 1369–1375.

Ham, P. J. 1986. Acquired resistance to *Onchocerca lienalis* infections in *Simulium ornatum* Meigen and *Simulium lineatum* Meigen following passive transfer of haemolymph from previously infected simuliids (Diptera, Simuliidae). *Parasitology 92*, 269–277.

Hanschke, R., and Hanschke, M. 1975. Untersuchungen zum Vorkommen und zur Funktion eines Proteinaseninhibitors in der Hämolymphe von Insekten. *Acta Biol. Med. Ger. 34*, 531–537.

Hare, J. D., and Andreadis, T. G. 1983. Variation in the susceptibility of *Leptinotarsa decemlineata* (Coleoptera: Chrysomelidae) when reared on different host plants to the fungal pathogen, *Beauveria bassiana* in the field and laboratory. *Environ. Entomol. 12*, 1891–1896.

Harshbarger, J. C., and Taylor, R. L. 1968. Neoplasms of insects. *Annu. Rev. Entomol. 13*, 159–190.

Harvey, T. L., and Howell, D. E. 1965. Resistance of the house fly to *Bacillus thuringiensis* Berliner. *J. Invertebr. Pathol. 7*, 92–100.

Harvey, W. R., and Blankemeyer, J. T. 1975. Epithelial structure and function. *In* "Invertebrate Immunity. Mechanisms of Invertebrate Vector-Parasite Relations." (K. Maramorosch and R. E. Shope, eds.), pp. 3–23. Academic Press, New York.

Hayashiya, K., Nishida, J., and Kawamoto, F. 1971. On the biosynthesis of the red fluorescent protein is [sic] the digestive juice of the silkworm larvae. *Jpn. J. Appl. Entomol. Zool. 15*, 109–114.

Hayashiya, K., Nishida, J., and Uchida, Y. 1976a. The mechanism of formation of the red fluorescent protein in the digestive juice of silkworm larvae—the formation of chlorophyllide-a. *Jpn. J. Appl. Entomol. Zool. 20*, 37–43.

Hayashiya, K., Uchida, Y., and Nishida, J. 1976b. Comparison of anti-viral activities of the silkworm larvae reared in light and in darkness in relation to the formation of red fluorescent protein (RFP). *Jpn. J. Appl. Entomol. Zool. 20*, 139–143.

Hedin, P. A., Lindig, O. H., Sikorowski, P. P., and Wyatt, M. 1978. Suppressants of gut bacteria in the boll weevil from the cotton plant. *J. Econ. Entomol. 71*, 394–396.

Heimpel, A. M. 1955. The pH in the gut and blood of the larch sawfly, *Pristiphora erichsonii* (Htg.), and other insects with reference to the pathogenicity of *Bacillus cereus* Fr. and Fr. *Can. J. Zool. 33*, 99–106.

Henry, J. E. 1969. Early morphogenesis of tumours induced by *Nosema acridophagus* in *Melanoplus sanguinipes*. *J. Insect Physiol. 15*, 391–394.

Hink, W. F., and Briggs, J. D. 1968. Bactericidal factors in haemolymph from normal and immune wax moth larvae, *Galleria mellonella*. *J. Insect Physiol.* 14, 1025–1034.

Hoffmann, D. 1976. Rôle de la phagocytose et mise en jeu de facteurs antibactériens solubles dans l'immunisation expérimentale chez *Locusta migratoria*. *C.R. Acad. Sci. Paris Ser. D 282*, 1021–1024.

Hoffmann, D. 1980. Induction of antibacterial activity in the blood of the migratory locust *Locusta migratoria* L. *J. Insect Physiol.* 26, 539–549.

Hoffmann, D., Brehelin, M., and Hoffmann, J. A. 1974. Modifications of the hemogram and of the hemocytopoietic tissue of male adults of *Locusta migratoria* (Orthoptera) after injection of *Bacillus thuringiensis*. *J. Invertebr. Pathol.* 24, 238–247.

Hoffmann, D., Brehélin, M., and Hoffmann, J. A. 1977. Premiers résultats sur les réactions de défense antibactériennes de larves et d'imagos de *Locusta migratoria*. *Ann. Parasitol.* 52, 87–88.

Hoffmann, D., Hultmark, D., and Boman, H. G. 1981. Insect immunity: *Galleria mellonella* and other Lepidoptera have Cecropia-P9-like factors active against gram negative bacteria. *Insect Biochem.* 11, 537–548.

Horohov, D. W., and Dunn, P. E. 1983. Phagocytosis and nodule formation by hemocytes of *Manduca sexta* larvae following injection of *Pseudomonas aeruginosa*. *J. Invertebr. Pathol.* 41, 203–213.

Hou, R. F., and Chang, J. 1985. Cellular defense response to *Beauveria bassiana* in the silkworm, *Bombyx mori*. *Appl. Entomol. Zool.* 20, 118–125.

Hultmark, D., Steiner, H., Rasmuson, T., and Boman, H. G. 1980. Insect immunity. Purification and properties of three inducible bactericidal proteins from hemolymph of immunized pupae of *Hyalophora cecropia*. *Eur. J. Biochem.* 106, 7–16.

Hultmark, D., Engström, Å., Bennich, H., Kapur, R., and Boman, H. G. 1982. Insect immunity: Isolation and structure of cecropin D and four minor antibacterial components from Cecropia pupae. *Eur. J. Biochem.* 127, 207–217.

Hultmark, D., Engström, Å., Andersson, K., Steiner, H., Bennich, H., and Boman, H. G. 1983. Insect immunity. Attacins, a family of antibacterial proteins from *Hyalophora cecropia*. *EMBO J.* 2, 571–576.

Ignoffo, C. M. 1968. Specificity of insect viruses. *Bull. Entomol. Soc. Am.* 14, 265–276.

Ignoffo, C. M., and Roush, R. T. 1986. Susceptibility of permethrine- and methomyl-resistant strains of *Heliothis virescens* (Lepidoptera: Noctuidae) to representative species of entomopathogens. *J. Econ Entomol.* 79, 334–337.

Ignoffo, C. M., Puttler, B., Marston, N. L., Hostetter, D. L., and Dickerson, W. A. 1975. Seasonal incidence of the entomopathogenic fungus *Spicaria rileyi* associated with noctuid pests of soybeans. *J. Invertebr. Pathol.* 25, 135–137.

Iizuka, T. 1983. Studies on the bacterial flora in the midgut and on the antibacterial activity in the digestive juice of larvae of the silkworm *Bombyx mori* L. *Sericologia 23*, 227–244.

Iizuka, T., Koike, S., and Mizutani, J. 1974. Antibacterial substances in feces of silkworm larvae reared on mulberry leaves. *Agric. Biol. Chem.* 38, 1549–1550.

Ingram, G. A., East, J., and Molyneux, D. H. 1984. Naturally occurring agglutinins against trypanosomatid flagellates in the haemolymph of insects. *Parasitology 89*, 435–451.

Inoue, H. 1974a. Multiplication of an infectious-flacherie virus in the resistant and susceptible strains of the silkworm, *Bombyx mori*. *J. Seric. Sci. Jpn.* 43, 318–324.

Inoue, H. 1977. Thermal therapy of virus diseases of the silkworm, *Bombyx mori*. *J. Seric. Sci. Jpn.* 46, 306–312.

Inoue, H., and Miyagawa, M. 1978. Regeneration of midgut epithelial cell in the silkworm, *Bombyx mori*, infected with viruses. *J. Invertebr. Pathol.* 32, 373–380.

Inoue, H., and Tanada, Y. 1977. Thermal therapy of the flacherie virus disease in the silkworm, *Bombyx mori*. *J. Invertebr. Pathol.* 29, 63–68.

Inoue, K. 1974b. Phagocytosis of nuclear polyhedra and the locarization (sic) of the acid phosphatase in the granular cell of the silkworm, *Bombyx mori* L. *J. Seric. Sci. Jpn. 43*, 394–400.

Ishihara, R. 1968. Growth of *Nosema bombycis* in primary cell cultures of mammalian and chicken embryos. *J. Invertebr. Pathol.* 11, 328–329.

Iwama, R., and Ashida, M. 1986. Biosynthesis of prophenoloxidase in hemocytes of larval hemolymph of the silkworm, *Bombyx mori*. *Insect Biochem.* 16, 547–555.

Jarosz, J. 1979a. Gut flora of *Galleria mellonella* suppressing ingested bacteria. *J. Invertebr. Pathol.* 34, 192–198.

Jarosz, J. 1979b. Simultaneous induction of protective immunity and selective synthesis of hemolymph lysozyme protein in larvae of *Galleria mellonella*. *Biol. Zentralbl.* 98, 459–471.

Jones, J. C. 1962. Current concepts concerning insect hemocytes. *Am. Zool.* 2, 209–246.

Jones, J. C. 1964. The circulatory system of insects. *In* "The Physiology of Insecta." (M. Rockstein, ed.), Vol. 3, pp. 1–107. Academic Press, New York.

Jones, J. C. 1970. Hemocytopoiesis in insects. *In* "Regulation of Hematopoiesis." (A. S. Gordon, ed.), Vol. 1, pp. 7–65. Appleton-Century-Crofts, New York.

Jones, J. C. 1977. "The Circulatory System of Insects." Thomas, Springfield, Illinois.

Jones, J. C. 1979. Pathways and pitfalls in the classification and study of insect hemocytes. *In* "Insect Hemocytes Development, Forms, Functions, and Techniques." (A. P. Gupta, ed.), pp. 279–300. Cambridge University Press, Cambridge.

Kaaya, G. P., Flyg, C., and Boman, H. G. 1987. Insect Immunity. Induction of cecropin and attacin-like antibacterial factors in the haemolymph of *Glossina morsitans morsitans*. *Insect Biochem.* 17, 309–315.

Kalmakoff, J., Williams, B. R. G., and Austin, F. J. 1977. Antiviral response in insects? *J. Invertebr. Pathol.* 29, 44–49.

Karp, R. D., and Rheins, L. A. 1980. Induction of specific humoral immunity to soluble proteins in the American cockroach (*Periplaneta americana*). II. Nature of the secondary response. *Dev. Comp. Immunol.* 4, 629–639.

Kawanishi, C. Y., Splittstoesser, C. M., and Tashiro, H. 1978. Infection of the European chafer, *Amphimallon majalis*, by *Bacillus popilliae*: Ultrastructure. *J. Invertebr. Pathol.* 31, 91–102.

Kawarabata, T. 1970. Studies of an acquired resistance on microbial infections in the silkworm (*Bombyx mori* L.). *Kyushu Univ. Fac. Agric. Sci. Bull.* 24, 231–254.

Kawarabata, T., and Aizawa, K. 1968. Immunologic principles in microbial infections in insects. *Proc. Joint U.S.-Jpn. Semin. Microb. Control Insect Pests.* pp. 143–145. Fukuoka, April 21–23, 1967.

Kaya, H. K. 1977. Development of the DD-136 strain of *Neoaplectana carpocapsae* at constant temperatures. *J. Nematol.* 9, 346–349.

Kaya, H. K., Marston, J. M., Lindegren, J. E., and Peng, Y. S. 1982. Low susceptibility of the honey bee, *Apis mellifera* L. (Hymenoptera: Apidae), to the entomogenous nematode, *Neoaplectana carpocapsae* Weiser. *Environ. Entomol.* 11, 920–924.

Keating, S. T., and Yendol, W. G. 1987. Influence of selected host plants on gypsy moth (Lepidoptera: Lymantriidae) larval mortality caused by a baculovirus. *Environ. Entomol.* 16, 459–462.

Kim, K. S. 1981. Ultrastructure of bean leaf beetle (*Ceratoma trifurcata*) hemocytes and their phagocytic activities on tobacco mosaic virus, a non-beetle-transmitted virus. *J. Ultrastruct. Res.* 75, 300–313.

Kingan, S. L., and Ensign, J. C. 1968. Chemical composition of the cell wall of *Bacillus thuringiensis* var. *thuringiensis*. *J. Bacteriol.* 95, 724–726.

Kinoshita, T., and Inoue, K. 1977. Bactericidal activity of the normal, cell-free hemolymph of silkworms (*Bombyx mori*). *Infect. Immun.* 16, 32–36.

Kitano, H. 1982. Effect of the venom of the gregarious parasitoid *Apanteles glomeratus* on its hemocytic encapsulation by the host, *Pieris*. *J. Invertebr. Pathol.* 40, 61–67.

Kitano, H. 1986. The role of *Apanteles glomeratus* venom in the defensive response of its hosts, *Pieris rapae crucivora*. *J. Insect Physiol.* 32, 369–375.

Kobayashi, M., Inagaki, S., and Kawase, S. 1981. Effect of high temperature on the development of nuclear polyhedrosis virus in the silkworm, *Bombyx mori*. *J. Invertebr. Pathol. 38*, 386–394.

Kockum, K., Faye, I., Hofsten, P. v., Lee, J.-Y., Xanthopoulos, K. G., and Boman, H. G. 1984. Insect immunity. Isolation and sequence of two cDNA clones corresponding to acidic and basic attacins from *Hyalophora cecropia*. *EMBO J. 3*, 2071–2075.

Koidsumi, K. 1957. Antifungal action of cuticular lipids in insects. *J. Insect Physiol. 1*, 40–51.

Komano, H., and Natori, S. 1985. Participation of *Sarcophaga peregrina* humoral lectin in the lysis of sheep red blood cells injected into the abdominal cavity of larvae. *Dev. Comp. Immunol. 9*, 31–40.

Komano, H., Mizuno, D., and Natori, S. 1980. Purification of lectin induced in the hemolymph of *Sarcophaga peregrina* larvae on injury. *J. Biol. Chem. 255*, 2919–2924.

Komano, H., Mizuno, D., and Natori, S. 1981. A possible mechanism of induction of insect lectin. *J. Biol. Chem. 256*, 7087–7089.

Krieg, K. 1973. "Invertebraten in der Geschwulstforschung." Theodor Steinkopff, Dresden.

Kučera, M. 1982. Inhibition of the toxic proteases from *Metarhizium anisopliae* by extracts of *Galleria mellonella* larvae. *J. Invertebr. Pathol. 40*, 299–300.

Kučera, M. 1984. Partial purification and properties of *Galleria mellonella* larvae proteolytic inhibitors acting on *Metarhizium anisopliae* toxic protease. *J. Invertebr. Pathol. 43*, 190–196.

Kučera, M., and Weiser, J. 1985. Different course of proteolytic inhibitory activity and proteolytic activity in *Galleria mellonella* larvae infected by *Nosema algerae* and *Vairimorpha heterosporum*. *J. Invertebr. Pathol. 45*, 41–46.

Kulinčević, J. M., and Rothenbuhler, W. C. 1975. Selection for resistance and susceptibility to hairless–black syndrome in the honeybee. *J. Invertebr. Pathol. 25*, 289–295.

Kurisu, K., and Odan, T. 1978. Inhibiting factor of flacherie-virus multiplication in the pupae of the silkworm (*Bombyx mori*). *Bull. Fac. Text. Sci. Kyoto Univ. Ind. Arts Text. Fibers 8* (3), 41–45.

Lackie, A. M. 1981. The specificity of the serum agglutinins of *Periplaneta americana* and *Schistocerca gregaria* and its relationship to the insects' immune response. *J. Insect Physiol. 27*, 139–143.

Lackie, A. M. 1986. The role of substratum surface-charge in adhesion and encapsulation by locust hemocytes in vivo. *J. Invertebr. Pathol. 47*, 377–378.

Lambert, J., and Hoffmann, D. 1985. Mise en évidence d'un facteur actif contre des bactéries Gram négatives dans le sang de *Locusta migratoria*. *C.R. Acad. Sci. Paris Ser. III 300*, 425–430

Landureau, J. C., and Jollès, P. 1970. Lytic enzyme produced *in vitro* by insect cells: lysozyme or chitinase? *Nature 225*, 968–969.

Lee, J.-Y., Edlund, T., Ny, T., Faye, I., and Boman, H. G. 1983. Insect immunity. Isolation of cDNA clones corresponding to attacins and immune protein P4 from *Hyalophora cecropia*. *EMBO J. 2*, 577–581.

Lee, P. E., and Furgala, B. 1967. Viruslike particles in adult honey bees (*Apis mellifera* Linnaeus) following injection with sacbrood virus. *Virology 32*, 11–17.

Lipa, J. J., and Martignoni, M. E. 1960. *Nosema phryganidiae* n. sp., a microsporidian parasite of *Phryganidia californica* Packard. *J. Insect Pathol. 2*, 396–410.

Lynn, D. C., and Vinson, S. B. 1977. Effects of temperature, host age, and hormones upon the encapsulation of *Cardiochiles nigriceps* eggs by *Heliothis* spp. *J. Invertebr. Pathol. 29*, 50–55.

Malke, H. 1965. Über das Vorkommen von Lysozym in Insekten. *Z. Allgem. Mikrobiol. 5*, 42–47.

Martignoni, M. E. 1957. Contributo alla conoscenza di una granulosi di *Eucosma griseana* (Hübner) (Tortricidae, Lepidoptera) quale fattore limitante il pullulamento dell'insetto nella Engadina alta. *Mitt. Schweiz. Anst. Forstl. Versuchswes. 32*, 371–418.

Martignoni, M. E., and Iwai, P. J. 1986. A catalogue of viral diseases of insects, mites, and ticks. U.S. Dep. Agric. For. Serv., Pac. Northwest Res. Stn. Gen. Tech. Rep. PNW–195.

Martignoni, M. E., and Schmid, P. 1961. Studies on the resistance to virus infections in natural populations of Lepidoptera. *J. Insect Pathol. 3*, 62–74.

Maschwitz, U. 1967. Eine neuartige Form der Abwehr von Mikroorganismen bei Insekten. *Naturwissenschaften 54*, 649.

Maschwitz, U., Koob, K., and Schildknecht, H. 1970. Ein Beitrag zur Function der Metathoracaldrüse der Ameisen. *J. Insect Physiol. 16*, 387–404.

Masera, E. 1954. Sul contenuto microbico intestinale del baco da seta e sull'etiologia della flaccidezza. *Agric. Venezie 8*, 714–735.

Matsumoto, N., Nakanishi, Y., and Natori, S. 1986a. Homologies of nucleotide sequences in the 5′-end regions of two developmentally regulated genes of *Sarcophaga peregrina*. *Nucleic Acid Res. 14*, 2685–2698.

Matsumoto, N., Okada, M., Takahashi, H., Ming, Q. X., Nakajima, Y., Nakanishi, Y., Komano, H., and Natori, S. 1986b. Molecular cloning of a cDNA and assignment of the C-terminal of sarcotoxin IA, a potent antibacterial protein of *Sarcophaga peregrina*. *Biochem. J. 239*, 717–722.

Matz, G. 1965. Morphogenese tumorale chez les insectes. *J. Insect Physiol. 11*, 637–639.

Matz, G. 1983. Cellular defence mechanisms in *Leucophaea maderae* Fabr. and *Locusta migratoria* L. tumors (Insecta). *Dev. Comp. Immunol. 7*, 699–701.

Matz, G., and Bergoin, M. 1984. Experimental tumors in *Leucophaea maderae* (Dictyoptera): A viral etiology. *J. Invertebr. Pathol. 43*, 424–426.

Matz, G., Monier, Y., and Vago, C. 1971. Une réaction de défense cellulaire chez les insectes: L'enkystement épithélial. *Bull. Soc. Zool. Fr. 96*, 209–215.

McCauley, V. J. E., Zacharuk, R. Y., and Tinline, R. D. 1968. Histopathology of green muscardine in larvae of four species of Elateridae (Coleoptera). *J. Invertebr. Pathol. 12*, 444–459.

McCleskey, C. S., and Melampy, R. M. 1939. Bactericidal properties of royal jelly of the honeybee. *J. Econ. Entomol. 32*, 581–587.

McGaughey, W. H. 1978. Effects of larval age on the susceptibility of almond moths and Indianmeal moths to *Bacillus thuringiensis*. *J. Econ. Entomol. 71*, 923–925.

McGaughey, W. H. 1985. Insect resistance to the biological insecticide *Bacillus thuringiensis*. *Science 229*, 193–194.

McGaughey, W. H., and Beeman, R. W. 1988. Resistance to *Bacillus thuringiensis* in colonies of Indianmeal moth and almond moth (Lepidoptera: Pyralidae). *J. Econ. Entomol. 81*, 28–33.

McLaughlin, R. E., and Sikorowski, P. P. 1978. Observations of boll weevil midgut when fed natural food or on bacterially contaminated artificial diet. *J. Invertebr. Pathol. 32*, 64–70.

Messner, B. 1974. Einige Aspekte der Abwehrmechanismen bei wirbellosen Tieren. *Biol. Rundsch. 12*, 272–276.

Métalnikov, S. 1927. "L'infection microbienne et l'immunité chez la Mite de Abeilles *Galleria mellonella*." Monogr. Inst. Pasteur, Masson, Paris.

Metalnikov, S., and Toumanoff, K. 1923. La lèpre chez les Insectes. *C.R. Soc. Biol. 89*, 935–936.

Metschnikoff, E. 1884. Ueber eine Sprosspilzkrankheit der Daphnien. Beitrag zur Lehre über den Kampf der Phagocyten gegen Krankheitserreger. *Arch. Pathol. Anat. Physiol. Klin. Med. 96*, 177–195.

Milne, C. P., Jr. 1982. Laboratory measurement of brood disease resistance in the honeybee. 1. Uncapping and removal of freeze-killed brood by newly emerged workers in laboratory test cages. *J. Apic. Res. 21*, 111–114.

Milne, C. P., Jr. 1983. Honey bee (Hymenoptera: Apidae) hygienic behavior and resistance to chalkbrood. *Ann. Entomol. Soc. Am. 76*, 384–387.

Milner, R. J. 1973. *Nosema whitei*, a microsporidan pathogen of some species of *Tribolium*.

IV. The effect of temperature, humidity and larval age on pathogenicity for *T. castaneum*. *Entomophaga 18*, 305–315.

Milstead, J. E. 1981. Influence of temperature and dosage on mortality of seventh instar larvae of *Galleria mellonella* (Insecta: Lepidoptera) caused by *Heterorhabditis bacteriophora* (Nematoda: Rhabditoidea) and its bacterial associate *Xenorhabdus luminescens*. *Nematologica 27*, 167–171.

Mohamed, A. K. A., Sikorowski, P. P., and Bell, J. V. 1977. Susceptibility of *Heliothis zea* larvae to *Nomuraea rileyi* at various temperatures. *J. Invertebr. Pathol. 30*, 414–417.

Mohrig, W., and Messner, B. 1968. Immunreaktionen bei Insekten. II. Lysozym als antimikrobielles Agens in Darmtrakt von Insekten. *Biol. Zentralbl. 87*, 705–718.

Mohrig, W., Schittek, D., and Hanschke, R. 1979a. Immunological activation of phagocytic cells in *Galleria mellonella*. *J. Invertebr. Pathol. 34*, 84–87.

Mohrig, W., Schittek, D., and Hanschke, R. 1979b. Investigations on cellular defense reactions with *Galleria mellonella* against *Bacillus thuringiensis*. *J. Invertebr. Pathol. 34*, 207–212.

Müller-Kögler, E. 1965. "Pilzkrankheiten bei Insekten." Paul Parey, Berlin.

Nakajima, Y., Qu, X. M, and Natori, S. 1987. Interaction between liposomes and sarcotoxin IA, a potent antibacterial protein of *Sarcophaga peregrina* (flesh fly). *J. Biol. Chem. 262*, 1665–1669.

Nakano, H., Tahara, S., Iizuka, T., and Mizutani, J. 1987. A defense mechanism against pathogenic bacteria in the digestive tracts of silkworm larvae—*in vitro* evidence of the formation of caffeoquinone, a true antibacterial substance, and synergism of amino compounds. *Agric. Biol. Chem. 51*, 549–555.

Nappi, A. J. 1975. Parasite encapsulation in insects. *In* "Invertebrate Immunity, Mechanisms of Invertebrate Vector–Parasite Relationships." (K. Maramorosch and R. E. Shope, eds.), pp. 293–326. Academic Press, New York.

Nappi, A. J., and Christensen, B. M. 1987. Insect immunity and mechanisms of resistance by nematodes. *In* "Vistas on Nematology." (J. A. Veech and D. W. Dickson, eds.), pp. 285–291. E. O. Painter, DeLeon Springs, Florida.

Natori, S. 1977. Bactericidal substance induced in the haemolymph of *Sarcophaga peregrina* larvae. *J. Insect Physiol. 23*, 1169–1173.

Natori, S. 1987. Hemolymph proteins participating in the defence system of *Sarcophaga peregrina*. *In* "Molecular Entomology." (J. H. Law, ed.), pp. 369–378. Alan R. Liss, New York.

Neelgund, Y. F., and Mathad, S. B. 1974. Studies on the maturation immunity of the armyworm, *Pseudaletia separata* Walker, infected with nuclear polyhedrosis virus. *J. Karnatak Univ. 19*, 14–21.

Neilson, M. M. 1965. Effects of a cytoplasmic polyhedrosis on adult Lepidoptera. *J. Invertebr. Pathol. 7*, 306–314.

Nelstrop, A. E., Taylor, G., and Collard, P. 1970. Comparative studies of serum and haemolymph proteins. *Comp. Biochem. Physiol. 35*, 191–199.

Odier, F. 1975. Influence d'une blessure cutanée sur l'évolution de la densonucléose chez les larves du Lépidoptère *Galleria mellonella* L. contaminées par voie buccale. *Entomophaga 20*, 241–244.

Odier, F., and Vago, C. 1973. Mise en évidence d'une immunité antivirale chez les insectes. *C.R. Acad. Sci. Paris Ser. D 277*, 1257–1260.

Okada, M., and Natori, S. 1983. Purification and characterization of an antibacterial protein from haemolymph of *Sarcophaga peregrina* (flesh-fly) larvae. *Biochem. J. 211*, 727–734.

Okada, M., and Natori, S. 1984. Mode of action of a bactericidal protein induced in the haemolymph of *Sarcophaga peregrina* (flesh-fly) larvae. *Biochem. J. 222*, 119–124.

Okada, M., and Natori, S. 1985a. Primary structure of sarcotoxin I, an antibacterial protein induced in the hemolymph of *Sarcophaga peregrina* (flesh fly) larvae. *J. Biol. Chem. 260*, 7174–7177.

Okada, M., and Natori, S. 1985b. Ionophore activity of sarcotoxin I, a bactericidal protein of *Sarcophaga peregrina*. *Biochem. J.* 229, 453–458.

Papierok, B., and Wilding, N. 1979. Mise en évidence d'une différence de sensibilité entre 2 clones du Puceron du Pois, *Acyrthosiphon pisum* Harr. (Homoptères *Aphididae*), exposés à 2 souches du Champignon Phycomycète: *Entomophthora obscura* Hall & Dunn. *C.R. Acad. Sci. Paris Ser. D 288*, 93–95.

Paschke, J. D., and Summers, M. D. 1975. Early events in the infection of the arthropod gut by pathogenic insect viruses. *In* "Invertebrate Immunity." (K. Maramorosch and R. E. Shope, eds.), pp. 75–112. Academic Press, New York.

Pereira, M. E. A., Andrade, A. F. B., and Ribeiro, J. M. C. 1981. Lectins of distinct specificity in *Rhodnius prolixus* interact selectively with *Trypanosoma cruzi*. *Science* 211, 597–600.

Pekrul, S., and Grula, E. A. 1979. Mode of infection of the corn earworm (*Heliothis zea*) by *Beauveria bassiana* as revealed by scanning electron microscopy. *J. Invertebr. Pathol.* 34, 238–247.

Petersen, J. J. 1978. Development of resistance by the southern house mosquito to the parasitic nematode *Romanomermis culicivorax*. *Environ. Entomol.* 7, 518–520.

Pimentel, D., and Shapiro, M. 1962. The influence of environment on a virus–host relationship. *J. Insect Pathol.* 4, 77–87.

Plurad, S. B., and Hartman, P. A. 1965. The fate of bacterial spores ingested by adult honey bees. *J. Invertebr. Pathol.* 7, 449–454.

Poinar, G. O., Jr. 1969. Arthropod immunity to worms. *In* "Immunity to Parasitic Animals." (G. J. Jackson, R. Herman, and I. Singer, eds.), Vol. 1, pp. 173–210. Appleton-Century-Crofts, New York.

Poinar, G. O., Jr. 1974. Insect immunity to parasitic nematodes. *Contemp. Top. Immunobiol.* 4, 167–178.

Poinar, G. O., Jr., Leutenegger, R., and Götz, P. 1968. Ultrastructure of the formation of a melanotic capsule in *Diabrotica* (Coleoptera) in response to a parasitic nematode (Mermithidae). *J. Ultrastruct. Res.* 25, 293–306

Powning, R. F., and Davidson, W. J. 1973. Studies on insect bacteriolytic enzymes—I. Lysozyme in haemolymph of *Galleria mellonella* and *Bombyx mori*. *Comp. Biochem. Physiol.* 45B, 669–681.

Powning, R. F., and Davidson, W. J. 1976. Studies on insect bacteriolytic enzymes—II. Some physical and enzymatic properties of lysozyme from haemolymph of *Galleria mellonella*. *Comp. Biochem. Physiol.* 55B, 221–228.

Powning, R. F., and Irzykiewicz, H. 1967. Lysozyme-like action of enzymes from the cockroach *Periplaneta americana* and from some other sources. *J. Insect Physiol.* 13, 1293–1299.

Pristavko, V. P., and Dovzhenok, N. V. 1974. Ascorbic acid influence on larval blood cell number and susceptibility to bacterial and fungal infection in the codling moth, *Laspeyresia pomonella* (Lepidoptera: Tortricidae). *J. Invertebr. Pathol.* 24, 165–168.

Pye, A. E. 1974. Microbial activation of prophenoloxidase from immune insect larvae. *Nature* 251, 610–613.

Pye, A. E., and Boman, H. G. 1977. Insect immunity. III. Purification and partial characterization of immune protein P5 from hemolymph of *Hyalophora cecropia* pupae. *Infect. Immun.* 17, 408–414.

Pye, A. E., and Yendol, W. G. 1972. Hemocytes containing polyphenoloxidase in *Galleria* larvae after injections of bacteria. *J. Invertebr. Pathol.* 19, 166–170.

Qu, X.-M., Steiner, H., Engström, Å., Bennich, H., and Boman, H. G. 1982. Insect immunity: Isolation and structure of cecropins B and D from pupae of the Chinese oak silk moth, *Antheraea pernyi*. *Eur. J. Biochem.* 127, 219–224.

Qu, X.-M., Zhang, C.-F., Komano, H., and Natori, S. 1987. Purification of a lectin from the hemolymph of Chinese oak silk moth (*Antheraea pernyi*) pupae. *J. Biochem.* 101, 545–551.

Quiot, J.-M., Vago, C., and Belloncik, S. 1980. Réaction antivirale par bourgeonnement cellulaire. Étude en culture de cellules de Lépidoptère infectée par un *Reovirus* de polyédrose cytoplasmique. *C.R. Acad. Sci. Paris Ser. D 291*, 481–483.

Raheja, A. K., and Brooks, M. A. 1971. Inability of the forest tent caterpillar, *Malacosoma disstria*, to acquire resistance to viral infection. *J. Invertebr. Pathol. 17*, 136–137.

Rasmuson, T., and Boman, H. G. 1977. The assay and the specificity problem in insect immunity. *In* "Developmental Immunobiology." (J. B. Solomon and J. D. Horton, eds.), pp. 83–90. Elsevier/North-Holland (Biomedical Press), Amsterdam.

Ratcliffe, N. A. 1975. Spherule cell-test particle interactions in monolayer cultures of *Pieris brassicae* hemocytes. *J. Invertebr. Pathol. 26*, 217–223.

Ratcliffe, N. A., and Rowley, A. F. 1974. In vitro phagocytosis of bacteria by insect blood cells. *Nature 252*, 391–392.

Ratcliffe, N. A., and Rowley, A. F. 1975. Cellular defense reactions of insect hemocytes in vitro: Phagocytosis in a new suspension culture system. *J. Invertebr. Pathol. 26*, 225–233.

Ratcliffe, N. A., and Rowley, A. F. 1979. Role of hemocytes in defense against biological agents. *In* "Insect Hemocytes. Development, Forms, Functions, and Techniques." (A. P. Gupta, ed.), pp. 331–414. Cambridge University Press, Cambridge.

Ratcliffe, N. A., and Walters, J. B. 1983. Studies on the *in vivo* cellular reactions of insects: Clearance of pathogenic and non-pathogenic bacteria in *Galleria mellonella* larvae. *J. Insect Physiol. 29*, 407–415.

Ratcliffe, N. A., Gagen, S. J., Rowley, A. F., and Schmit, A. R. 1976. Studies on insect cellular defence mechanisms and aspects of the recognition of foreignness. *Proc. First Int. Colloq. Invertebr. Pathol.* pp. 210–214. Queen's University, Kingston, Canada.

Ratcliffe, N. A., Leonard, C., and Rowley, A. F. 1984. Prophenoloxidase activation: Nonself recognition and cell cooperation in insect immunity. *Science 226*, 557–559.

Ratner, S., and Vinson, S. B. 1983. Encapsulation reactions *in vitro* by haemocytes of *Heliothis virescens*. *J. Insect Physiol. 29*, 855–863.

Reichelderfer, C. F., and Benton, C. V. 1974. Some genetic aspects of the resistance of *Spodoptera frugiperda* to a nuclear polyhedrosis virus. *J. Invertebr. Pathol. 23*, 378–382.

Rheins, L. A., and Karp, R. D. 1982. An inducible humoral factor in the American cockroach (*Periplaneta americana*): Precipitin activity that is sensitive to a proteolytic enzyme. *J. Invertebr. Pathol. 40*, 190–196.

Rheins, L. A., and Karp, R. D. 1984. The humoral immune response in the American cockroach, *Periplaneta americana*: Reactivity to a defined antigen from honeybee venom, phospholipase A_2. *Dev. Comp. Immunol. 8*, 791–801.

Riedel, B., and Brown, D. T. 1977. Role of extracellular virus in the maintenance of the persistent infection induced in *Aedes albopictus* (mosquito) cells by Sindbis virus. *J. Virol. 23*, 554–561.

Riedel, B., and Brown, D. T. 1979. Novel antiviral activity found in the media of Sindbis virus-persistently infected mosquito (*Aedes albopictus*) cell cultures. *J. Virol. 29*, 51–60.

Rizzo, D. C. 1977. Age of three dipteran hosts as a factor governing the pathogenicity of *Beauveria bassiana* and *Metarrhizium anisopliae*. *J. Invertebr. Pathol. 30*, 127–130.

Rohel, D. Z., Chadwick, J., and Faulkner, P. 1980. Tests with inactivated cricket paralysis virus as a possible immunogen against a virus infection of *Galleria mellonella* larvae. *Intervirology 14*, 61–68.

Rooseboom, M. 1937. Contribution à l'étude de la cytologie du sang de certains insectes, avec quelques considérations générales. *Arch. Neerl. Zool. 2*, 432–559.

Rose, R.I., and Briggs, J. D. 1969. Resistance to American foulbrood in honey bees. IX. Effects of honey-bee larval food on the growth and viability of *Bacillus larvae*. *J. Invertebr. Pathol. 13*, 74–80.

Rothenbuhler, W. C. 1964. Behavior genetics of nest cleaning in honey bees. IV. Responses of F_1 and backcross generations to disease-killed brood. *Am. Zool. 4*, 111–123.

Rothenbuhler, W. C., and Thompson, V. C. 1956. Resistance to American foulbrood in honey bees. I. Differential survival of larvae of different genetic lines. *J. Econ. Entomol. 49*, 470–475.

Rothenbuhler, W. C., Kulinčević, J. M., and Kerr, W. E. 1968. Bee genetics. *Annu. Rev. Genet. 2*, 413–438.

Rowley, A. F., and Ratcliffe, N. A. 1979. An ultrastructural and cytochemical study of the interaction between latex particles and the haemocytes of the wax moth *Galleria mellonella* in vitro. *Cell Tissue Res. 199*, 127–137.

Rowley, A. F., and Ratcliffe, N. A. 1981. Insects. *In* "Invertebrate Blood Cells." (N. A. Ratcliffe and A. F. Rowley, eds.), Vol. 2, pp. 421–488. Academic Press, New York.

Saito, T., and Aoki, J. 1983. Toxicity of free fatty acids on the larval surfaces of two lepidopterous insects towards *Beauveria bassiana* (Bals.) Vuill. and *Paecilomyces fumoso-roseus* (Wize) Brown et Smith (Deuteromycetes: Moniliales). *Appl. Entomol. Zool. 18*, 225–233.

Salt, G. 1956. Experimental studies in insect parasitism. IX. The reactions of a stick insect to an alien parasite. *Proc. R. Soc. London B 146*, 93–108.

Salt, G. 1960. Experimental studies in insect parasitism. XI. The haemocytic reaction of a caterpillar under varied conditions. *Proc. R. Soc. London B 151*, 446–467.

Salt, G. 1968. The resistance of insect parasitoids to the defence reactions of their hosts. *Biol. Rev. 43*, 200–232.

Salt, G. 1970. "The Cellular Defense Reactions of Insects." Cambridge Monogr. Exp. Biol. 16. University Printing House, Cambridge, England.

Samson, R. A., Evans, H. C., and Latgé, J. P. 1988. "Atlas of Entomopathogenic Fungi." Springer-Verlag, Berlin.

Schaerffenberg, B. 1957. *Beauveria bassiana* (Vuill) Link als Parasit des Kartoffelkäfers (*Leptinotarsa decemlineata* Say). *Anz. Schädlingskd. 30*, 69–74.

Schell, S. C. 1952. Tissue reactions of *Blattella germanica* L. to the developing larva of *Physaloptera hispida* Schell, 1950 (Nematoda: Spiruroidea). *Trans. Am. Microsc. Soc. 71*, 293–302.

Schmit, A. R., and Ratcliffe, N. A. 1978. The encapsulation of araldite implants and recognition of foreignness in *Clitumnus extradentatus*. *J. Insect Physiol. 24*, 511–521.

Schneider, P. M. 1985. Purification and properties of three lysozymes from hemolymph of the cricket, *Gryllus bimaculatus* (De Geer). *Insect Biochem. 15*, 463–470.

Scott, M. T. 1971. A naturally occurring hemagglutinin in the hemolymph of the American cockroach. *Arch. Zool. Exp. Gen. 112*, 73–80.

Scott, M. T. 1972. Partial characterization of the hemagglutinating activity in hemolymph of the American cockroach (*Periplaneta americana*). *J. Invertebr. Pathol. 19*, 66–71.

Seaman, G. R., and Robert, N. L. 1968. Immunological response of male cockroaches to injection of Tetrahymena pyriformis. *Science 161*, 1359–1361.

Shapiro, M. 1979. Changes in hemocyte populations. *In* "Insect Hemocytes Development, Forms, Functions, and Techniques." (A. P. Gupta, ed.), pp. 475–523. Cambridge University Press, Cambridge.

Shih, J. W.-K, and Hash, J. H. 1971. The N, O-diacetylmuramidase of *Chalaropsis* species. III. Amino acid composition and partial structural formula. *J. Biol. Chem. 246*, 994–1006.

Shiraishi, A., and Natori, S. 1989. Humoral factor activating the *Sarcophaga* lectin gene in cultured fat body. *Insect Biochem. 19*, 261–267.

Shvetsova, O. I. 1950. The polyhedral disease of the greater wax moth (*Galleria mellonella* L.) and the role of nutritional factors in virus diseases of insects. *Mikrobiologiya 19*, 532–542.

Sidén, I., Dalhammar, G., Telander, B., Boman, H. G., and Somerville, H. 1979. Virulence factors in *Bacillus thuringiensis*: Purification and properties of a protein inhibitor of immunity in insects. *J. Gen. Microbiol. 114*, 45–52.

Sidorov, N. G., Kuptev, V. S., and Mugalimov, I. A. 1975. Resistance of honey bees to *Nosema* disease and a genetic method of control. *Veterinariya* (Moscow). *7*, 63–65.

Silverman, P. H., and Levinson, Z. H. 1954. Lipid requirements of the larvae of the housefly *Musca vicina* (Macq.) reared under non-aseptic conditions. *Biochem. J. 58*, 291–294.

Smirnoff, W. A. 1967. Effects of some plant juices on the ugly-nest caterpillar, *Archips cerasivoranus*, infected with Microsporidia. *J. Invertebr. Pathol. 9*, 26–29.

Smirnoff, W. A. 1972. Effects of volatile substances released by foliage of *Abies balsamea*. *J. Invertebr. Pathol. 19*, 32–35.

Spies, A. G., Karlinsey, J. E., and Spence, K. 1986a. The immune proteins of the darkling beetle, *Eleodes* (Coleoptera: Tenebrionidae). *J. Invertebr. Pathol. 47*, 234–235.

Spies, A. G., Karlinsey, J. E., and Spence, K. D. 1986b. Antibacterial hemolymph proteins of *Manduca sexta*. *Comp. Biochem. Physiol. 83B*, 125–133.

Splittstoesser, C. M., Kawanishi, C. Y., and Tashiro, H. 1978. Infection of the European chafer, *Amphimallon majalis* by *Bacillus popilliae*: Light and electron microscope observations. *J. Invertebr. Pathol. 31*, 84–90.

Stairs, G. R. 1964. Changes in the susceptibility of *Galleria mellonella* (Linnaeus) larvae to nuclear-polyhedrosis virus following blockage of the phagocytes with India ink. *J. Insect Pathol. 6*, 373–376.

Stairs, G. R. 1965a. Quantitative differences in susceptibility to nuclear-polyhedrosis virus among larval instars of the forest tent caterpillar, *Malacosoma disstria* (Hübner). *J. Invertebr. Pathol. 7*, 427–429.

Stairs, G. R. 1965b. The effect of metamorphosis on nuclear-polyhedrosis virus infection in certain Lepidoptera. *Can. J. Microbiol. 11*, 509–512.

Steiner, H., Hultmark, D., Engström, Å., Bennich, H., and Boman, H. G. 1981. Sequence and specificity of two antibacterial proteins involved in insect immunity. *Nature 292*, 246–248.

Steinhaus, E. A. 1949. "Principles of Insect Pathology." McGraw-Hill, New York.

Steinhaus, E. A. 1958. Stress as a factor in insect disease. *Proc. Tenth Int. Congr. Entomol., 1956. 4*, 725–730.

Stephens, J. M. 1959. Immune responses of some insects to some bacterial antigens. *Can. J. Microbiol. 5*, 203–228.

Stephens, J. M. 1962. Bactericidal activity of the blood of actively immunized wax moth larvae. *Can. J. Microbiol. 8*, 491–499.

Stephens, J. M. 1963a. Immunity in insects. *In* "Insect Pathology: An Advanced Treatise." (E. A. Steinhaus, ed.), Vol. 1, pp. 273–297. Academic Press, New York.

Stephens, J. M. 1963b. Bactericidal activity of hemolymph of some normal insects. *J. Insect Pathol. 5*, 61–65.

Stephens, J. M., and Marshall, J. H. 1962. Some properties of an immune factor isolated from the blood of actively immunized wax moth larvae. *Can. J. Microbiol. 8*, 719–725.

Stiles, B., and Paschke, J. D. 1980. Midgut pH in different instars of three *Aedes* mosquito species and the relation between pH and susceptibility of larvae to a nuclear polyhedrosis virus. *J. Invertebr. Pathol. 35*, 58–64.

Stoltz, D. B., and Summers, M. D. 1971. Pathway of infection of mosquito iridescent virus. I. Preliminary observations on the fate of ingested virus. *J. Virol. 8*, 900–909.

Sussman, A. S. 1951. Studies of an insect mycosis. I. Etiology of the disease. *Mycologia 43*, 338–350.

Suto, C., Kawamoto, F., and Kumada, N. 1978. Pathological studies on the cockroach. I. Spontaneous occurrence of hindgut ulcer in the smoky-brown cockroach, *Periplaneta fuliginosa* (Serville). *Eiseidobutsu 29*, 197–204.

Tabashnik, B. E., Cushing, N. L., Finson, N., and Johnson, M. W. 1990. Field development of resistance to *Bacillus thuringiensis* in diamondback moth (Lepidoptera: Plutellidae). *J. Econ. Entomol. 83*, 1671–1676.

Takahashi, H., Komano, H., Kawaguchi, N., Kitamura, N., Nakanishi, S., and Natori, S. 1985. Cloning and sequencing of cDNA of *Sarcophaga peregrina* humoral lectin induced on injury of the body wall. *J. Biol. Chem. 260*, 12228–12233.

Tanada, Y. 1955. Susceptibility of the imported cabbageworm to fungi: *Beauveria* spp. *Proc. Hawaii. Entomol. Soc. 15*, 617–622.

Tanada, Y. 1967. Effect of high temperatures on the resistance of insects to infectious diseases. *J. Seric. Sci. Jpn. 36*, 333–339.

Teakle, R. E., Jensen, J. M., and Giles, J. E. 1985. Susceptibility of *Heliothis armiger* to a commercial nuclear polyhedrosis virus. *J. Invertebr. Pathol. 46*, 166–173.

Teakle, R. E., Jensen, J. M., and Giles, J. E. 1986. Age-related susceptibility of *Heliothis punctiger* to a commercial formulation of nuclear polyhedrosis virus. *J. Invertebr. Pathol. 47*, 82–92.

Thorvilson, H. G., Pedigo, L. P., and Lewis, L. C. 1985. Soybean leaf consumption by *Nomuraea rileyi* (Fungi: Deuteromycotina)-infected *Plathypena scabra* (Lepidoptera: Noctuidae) larvae. *J. Invertebr. Pathol. 46*, 265–271.

Tinsley, T. W. 1975. Factors affecting virus infection of insect gut tissue. *In* "Invertebrate Immunity. Mechanisms of Invertebrate Vector–Parasite Relations. (K. Maramorosch and R. E. Shope, eds.), pp. 55–63. Academic Press, New York.

Triggiani, O., and Poinar, G. O., Jr. 1976. Infection of adult Lepidoptera by *Neoaplectana carpocapsae* (Nematoda). *J. Invertebr. Pathol. 27*, 413–414.

Tsang, K. R., and Brooks, M. A. 1980a. Dose response of insects to malignant transformed cells. *In Vitro 16*, 469–474.

Tsang, K. R., and Brooks, M. A. 1980b. Neoplasias caused by cultured insect cells. *In* "Invertebrate Systems in Vitro." (E. Kurstak, K. Maramorosch, and A. Dübendorfer, eds.), pp. 535–550. Elsevier/North-Holland (Biomedical Press), Amsterdam.

Tsang, K. R., and Brooks, M. A. 1981. Malignant transformation of insect cells *in vitro*. *In* "Phyletic Approaches to Cancer." (C. J. Dawe, J. C. Harshbarger, S. Kondo, T. Sugimura, and S. Takayama, eds.), pp. 267–274. Japan Science Society Press, Tokyo.

Uchida, Y., Kawamoto, F., Himeno, M., and Hayashiya, K. 1984. A virus-inactivating protein isolated from the digestive juice of the silkworm, *Bombyx mori*. *J. Invertebr. Pathol. 43*, 182–189.

Undeen, A. H. 1975. Growth of *Nosema algerae* in pig kidney cell cultures. *J. Protozool. 22*, 107–110.

Uzigawa, K., and Aruga, H. 1966. On the selection of resistant strains to the infectious flacherie virus in the silkworm, *Bombyx mori* L. *J. Seric. Sci. Jpn. 35*, 23–26.

Vail, P. V., and Gough, D. 1970. Effects of cytoplasmic-polyhedrosis virus on adult cabbage loopers and their progeny. *J. Invertebr. Pathol. 15*, 397–400.

Vail, P. V., and Hall, I. M. 1969. The influence of infections of nuclear-polyhedrosis virus on adult cabbage loopers and their progeny. *J. Invertebr. Pathol. 13*, 358–370.

van Hofsten, P., Faye, I., Kockum, K., Lee, J.-Y., Xanthopoulos, K. G., Boman, I. A., Boman, H. G., Engström, Å., Andreu, D., and Merrifield, R. B. 1985. Molecular cloning, cDNA sequencing, and chemical synthesis of cercropin B from *Hyalophora cecropia*. *Proc. Natl. Acad. Sci. U.S.A. 82*, 2240–2243.

van Oss, C. J. 1978. Phagocytosis as a surface phenomenon. *Annu. Rev. Microbiol. 32*, 19–39.

Van Rie, J., McGaughey, W. H., Johnson, D. E., Barnett, B. D., and Van Mellaert, H. 1990. Mechanisms of insect resistance to the microbial insecticide *Bacillus thuringiensis*. *Science 247*, 72–74.

Vey, A. 1968. Réactions de défense cellulaire dans les infections de blessures à *Mucor hiemalis* Wehmer. *Ann. Epiphy. 19*, 695–702.

Vey, A. 1969. Étude *in vitro* des réactions hémocytaires anticryptogamiques des larves de Lépidoptères. *Ann. Zool. Ecol. Anim. 1*, 93–100.

Vey, A., and Götz, P. 1975. Humoral encapsulation in Diptera (Insecta): Comparative studies *in vitro*. *Parasitology 70*, 77–86.

Vey, A., and Vago, C. 1971. Réaction anticryptogamique de type granulome chez les insectes. *Ann. Inst. Pasteur 121*, 527–532.

Vey, A., Bouletreau, M., Quiot, J. M., and Vago, C. 1975. Étude *in vitro* en microcinématographie des réactions cellulaires d'invertébrés vis-a-vis d'agents bactériens et cryptogamiques. *Entomophaga 20*, 337–351.

Vey, A., Quiot, J.-M., Vago, C., and Fargues, J. 1985. Effet immunodépresseur de toxines fongiques: Inhibition de la réaction d'encapsulement multicellulaire par les destruxines. *C.R. Acad. Sci. Paris Ser. III 300*, 647–651.

Vinson, S. B. 1977. Insect host responses against parasitoids and the parasitoid's resistance: With emphasis on the Lepidoptera-Hymenoptera association. *Comp. Pathobiol. 3*, 103–125.

Wago, H. 1980. Humoral factors promoting the adhesive properties of the granular cells and plasmatocytes of the silkworm, *Bombyx mori*, and their possible role in the initial cellular reactions to foreignness. *Cell. Immunol. 54*, 155–169.

Wago, H., and Kitano, H. 1985. Effects of the venom from *Apanteles glomeratus* on the hemocytes and hemolymph of *Pieris rapae crucivora*. *Appl. Entomol. Zool. 20*, 103–110.

Walters, J. B., and Ratcliffe, N. A. 1983. Studies on the *in vivo* cellular reactions of insects: Fate of pathogenic and non-pathogenic bacteria in *Galleria mellonella* nodules. *J. Insect Physiol. 29*, 417–424.

Watanabe, H. 1965. Resistance to peroral infection by the cytoplasmic-polyhedrosis virus in the silkworm, *Bombyx mori* (Linnaeus). *J. Invertebr. Pathol. 7*, 257–258.

Watanabe, H. 1967. Development of resistance in the silkworm, *Bombyx mori*, to peroral infection of a cytoplasmic-polyhedrosis virus. *J. Invertebr. Pathol. 9*, 474–479.

Watanabe, H., and Maeda, S. 1978. Genetic resistance to peroral infection with a densonucleosis virus in the silkworm, *Bombyx mori*. *J. Seric. Sci. Jpn. 47*, 209–214.

Watanabe, H., and Maeda, S. 1981. Genetically determined nonsusceptibility of the silkworm, *Bombyx mori*, to infection with a densonucleosis virus (*Densovirus*). *J. Invertebr. Pathol. 38*, 370–373.

Watanabe, H., and Takamiya, K. 1976. Susceptibility of the silkworm larvae, *Bombyx mori*, reared under different light conditions to polyhedrosis viruses. *J. Seric. Sci. Jpn. 45*, 403–406.

Watanabe, H., and Tanada, Y. 1972. Infection of a nuclear-polyhedrosis virus in armyworm, *Pseudaletia unipuncta* Haworth (Lepidoptera: Noctuidae), reared at a high temperature. *Appl. Entomol. Zool. 7*, 43–51.

Weiser, J. 1961. Die Mikrosporidien als Parasiten der Insekten. Monogr. Angew. Entomol. No. 17. Paul Parey, Hamburg.

Weiser, J. 1969. Immunity of insects to protozoa. *In* "Immunity to Parasitic Animals." (G. J. Jackson, R. Herman, and I. Singer, eds.), Vol. 1, pp. 129–147. Appleton-Century-Crofts, New York.

Weiser, J. 1976. The *Pleistophora debaisieuxi* xenoma. *Z. Parasitenkd. 48*, 263–270.

Weiser, J., and Lysenko, O. 1956. Septikemie bource morušového. *Cesk. Mikrobiol. 1*, 216–222.

Weissenberg, R. 1968. Intracellular development of the microsporidan *Glugea anomala* Moniez in hypertrophying migratory cells of the fish *Gasterosteus aculeatus* L., an example of the formation of "xenoma" tumors. *J. Protozool. 15*, 44–57.

Whitcomb, R. F., Shapiro, M., and Granados, R. R. 1974. Insect defense mechanisms against microorganisms and parasitoids. *In* "The Physiology of Insecta." (M. Rockstein, ed.), Second Ed. Vol. 5, pp. 447–536. Academic Press, New York.

White, J. W., Jr., Subers, M. H., and Schepartz, A. I. 1963. The identification of inhibine, the antibacterial factor in honey, as hydrogen peroxide and its origin in a honey glucose-oxidase system. *Biochem. Biophys. Acta 73*, 57–70.

Whitlock, V. H. 1977. Effect of larval maturation on mortality induced by nuclear polyhedrosis and granulosis virus infections of *Heliothis armigera*. *J. Invertebr. Pathol. 30*, 80–86.

Wilson, G. G. 1979. Reduced spore production of *Nosema fumiferanae* (Microsporida) in spruce budworm (*Choristoneura fumiferana*) reared at elevated temperature. *Can. J. Zool. 57*, 1167–1168.

Wilson, G. G., and Sohi, S. S. 1977. Effect of temperature on healthy and microsporidia-infected continuous cultures of *Malacosoma disstria* hemocytes. *Can. J. Zool. 55*, 713–717.

Wittig, G. 1965. Phagocytosis by blood cells in healthy and diseased caterpillars. I. Phagocytosis of *Bacillus thuringiensis* Berliner in *Pseudaletia unipuncta* (Haworth). *J. Invertebr. Pathol. 7*, 474–488.

Woodard, D. B., and Fukuda, T. 1977. Laboratory resistance of the mosquito *Anopheles quadrimaculatus* to the mermithid nematode *Diximermis peterseni*. *Mosq. News 37*, 192–195.

Woods, S. P., and Grula, E. A. 1984. Utilizable surface nutrients on *Heliothis zea* available for growth of *Beauveria bassiana*. *J. Invertebr. Pathol. 43*, 259–269.

Wyatt, G. R. 1975. Hemolymph in insects and arachnids—some biochemical features. *In* "Invertebrate Immunity. Mechanisms of Invertebrate Vector–Parasite Relations." (K. Maramorosch and R. E. Shope, eds.), pp. 225–240. Academic Press, New York.

Wyatt, G. R., and Pan, M. L. 1978. Insect plasma proteins. *Annu. Rev. Biochem. 47*, 779–817.

Yamaguchi, K. 1979. Natural recovery of the fall webworm, *Hyphantria cunea*, to infection by a cytoplasmic-polyhedrosis virus of the silkworm, *Bombyx mori*. *J. Invertebr. Pathol. 33*, 126–128.

Yamamoto, T., and Tanada, Y. 1978. Biochemical properties of viral envelopes of insect baculoviruses and their role in infectivity. *J. Invertebr. Pathol. 32*, 202–211.

Yamaura, I., Yonekura, M., Katsura, Y., Ishiguro, M., and Funatsu, M. 1980. Purification and some physico-chemical properties of phenoloxidase from the larvae of housefly. *Agric. Biol. Chem. 44*, 55–59.

Yeaton, R. W. 1981. Invertebrate lectins: I. Occurrence. *Dev. Comp. Immunol. 5*, 391–402.

Yeaton, R. W. 1983. Wound responses in insects. *Am. Zool. 23*, 195–203.

Yoshida, H., and Ashida, M. 1986. Microbial activation of two serine enzymes and prophenoloxidase in the plasma fraction of hemolymph of the silkworm, *Bombyx mori*. *Insect Biochem. 16*, 539–545.

Young, J. D.-E. 1985. Role of ionic events in the triggering of phagocytosis. *J. Theor. Biol. 116*, 475–478.

Zernoff, V. 1927. L'immunité passive chez, *Galleria mellonella*. *C.R. Soc. Biol. 97*, 1697–1699.

Ziprin, R. L. 1978. Immune response of the greater wax moth, *Galleria mellonella*, induced by the marine pseudomonad B-16. *J. Invertebr. Pathol. 32*, 396–397.

CHAPTER 15

MICROBIAL CONTROL

I. APPROACHES TO MICROBIAL CONTROL
II. ECONOMIC THRESHOLD
III. FACTORS AFFECTING EFFICACY
    A. Dosage
    B. Timing of Application
    C. Coverage
    D. pH
    E. Antimicrobiosis
    F. Target Pest
    G. Compatibility
    H. Host Resistance
IV. SAFETY
V. MOLECULAR GENETICS
VI. MICROBIAL PESTICIDES
VII. MASS PRODUCTION
VIII. FORMULATIONS
IX. STANDARDIZATION
X. STORAGE
XI. APPLICATION TECHNOLOGY
XII. ADVANTAGES AND DISADVANTAGES OF MICROBIAL CONTROL
XIII. EXAMPLES OF LONG-TERM CONTROL
    A. Viruses
    B. Bacteria
    C. Fungi
    D. Protozoa
    E. Nematodes
XIV. EXAMPLES OF SHORT-TERM CONTROL
    A. Viruses
    B. Bacteria
    C. Fungi
    D. Protozoa
    E. Nematodes
XV. EXAMPLES OF THE IPM APPROACH
XVI. DESTRUCTION OF SYMBIONTS

The concept of microbial control originated over a hundred years ago (Steinhaus 1956). By the late 1800s and early 1900s, many attempts were made to control insects with pathogens (see Chapter 1). Some attempts were successful, others were partially successful, and some were failures. The early unpredictable results with insect pathogens in field applications and the subsequent development and successful use of broad spectrum chemical pesticides to reduce pest populations slowed the development of microbial pesticides. Broad spectrum chemical pesticides continue to play an important role in insect and mite control in modern day society for agriculture (food and fiber production) and forestry (Matsumura 1985). They are also used to suppress pests of public health concerns, of stored products, and in and around dwellings.

Although chemical control has been efficacious, the drawbacks—pesticide resistance; resurgence of the target organism or emergence of secondary pests to pest status because of the destruction of parasitoids and predators; impact on the nontarget organisms, including humans; environmental pollution through the accumulation of pesticides in soil, water, and air; and residues on the agricultural products and animals—have necessitated the development of more selective control methods compatible with the environment. Moreover, legislation now and in the future will limit or eliminate many chemical pesticides because of the aforementioned drawbacks. Insect pathogens overcome many problems of chemical pesticides, but they are not extensively used despite their many positive features (Falcon 1985). The reasons for their lack of use vary, but one key for greater acceptance of these biological-control agents is education. People have been conditioned to the "quick kill" of chemical pesticides and expect similar results with pathogens. Very few pathogens kill very quickly. In this chapter, we shall examine the role of pathogens in suppressing pest populations and the factors that must be considered in developing microbial pesticides.

The manipulation of pathogens or their by-products to bring about the reduction of pest populations has been referred to as microbial control. Microbial control is defined as that part of biological control concerned with the use of microorganisms or their by-product(s) for the control and reduction of insect pests or other pestiferous organisms. Microbial pesticides, sometimes referred to as biorational pesticides, include bacteria, fungi, viruses, protozoans, and nematodes. Biorational pesticide is defined as chemical or living microscopic pest-control agents, which are inherently different from conventional chemical pesticides.

The literature on microbial control of insects and mites is voluminous and is widely scattered in many different journals and books. As a starting point for more detailed information, the reader should consult Burges and Hussey (1971), Burges (1981a), Kurstak (1982), Fuxa (1987), Ignoffo (1988), and Baker and Dunn (1990).

I. APPROACHES TO MICROBIAL CONTROL

Terminology and approaches in microbial control have been adopted primarily from biological control. One approach is conservation, where management systems enhance and conserve existing naturally occurring pathogens. The other is augmentative, where pathogens are propagated and periodically released into the field to suppress the pest. The augmentative approach includes inundative or inoculative release. Inundative releases, also known as short-term or temporary control, are conducted to suppress a pest by releasing natural enemies and not by the progeny. This approach is similar to the use of chemical insecticides because the mortality of the pest is more or less immediate and no prolonged interaction between the natural enemy and pest occurs. The natural enemy (i.e., pathogen) may survive in the environment but either occurs in a different habitat without an effective mechanism of transmission or occurs below the effective concentration level to reduce the pest below the economic threshold.

Inoculative releases, also referred to as long-term or permanent control, are dependent upon progeny being produced for more than one generation following the colonization of the natural enemy for pest suppression. Control is dependent on the progenies and their ability to persist and spread in the pest population. Periodic reintroduction of the pathogen may be required to augment the field inoculum. In some situations, especially where the pathogen is protected from environmental extremes and in a favorable niche, it may survive, providing long-term control without significant progeny production. Thus, these two types of releases intergrade depending on the economic threshold of the insect pest and the distinction can be arbitrary. In addition, novel methods to control pests, such as the disruption of symbionts or the activation of chronic or latent infections, have been attempted but with limited success.

Fuxa (1987) states that several approaches have evolved for the use of pathogens in integrated pest management (IPM). These include permanent introduction and establishment (long-term control), inundative augmentation (short-term control), inoculative augmentation (long-term control often limited to a single season), and environmental manipulation (conservation). Environmental manipulation involves the enhancement of naturally occurring pathogens for pest control. In this chapter, the terms short- and long-term control will be used.

Pathogens may be used in conjunction with other control tactics. They may be used in combination with chemical pesticides, other biological-control agents, cultural techniques, mass trapping of adults, and resistant plant varieties.

II. ECONOMIC THRESHOLD

Pathogens can be applied against insect pests with a high or low economic threshold level. The approaches differ depending on the cost–benefit ratio. Long-term control is most feasible against insects on plants or products that can sustain a high level of damage and/or are of low economic value (i.e., high economic threshold). Long-term control has been very successful against some forest, turf, and pasture insects. It has also been successful where the pathogen is already present, and the additional application of the pathogen can initiate an epizootic earlier than would occur naturally. This approach has been particularly effective with certain fungal pathogens in citrus, soybean, and glasshouse systems. In the case of storage bins for grain, *Bacillus thuringiensis*, because of its persistence, has potential for long-term control of lepidopteran pests (Kinsinger and McGaughey 1976).

Short-term control is most feasible against insects on plants or products that are of high economic value or cannot sustain a high level of damage (i.e., low economic threshold). Pathogens are used similarly as chemical pesticides and reapplication is often necessary. For example, *B. thuringiensis* used against lepidopteran insects on cole crops and lettuce has been widely accepted because of its effectiveness on these pests and can be applied up to the harvest date (no tolerance level). Short-term control approaches have been used against a number of forest defoliators because there is little or no negative environmental impact although some recent evidence suggests otherwise (see Section IV., Safety).

At one time, doubts existed whether pathogens could be effectively applied for short-term control of insects of low economic threshold. These doubts resulted from the general belief that pathogens take a long time to kill their hosts (long period of lethal infection), and the insects cause damage to the crop before control is attained. However, highly virulent pathogens can kill quickly. *B. thuringiensis* kills some mosquito hosts within 1 or 2 h and lepidopteran hosts within 24 h; some baculoviruses kill young hosts within 48 h; high concentrations of microsporidia may kill quickly because of septicemia caused by the extruded polar filament damaging gut cells; and entomopathogenic nematodes in the families Steinernematidae and Heterorhabditidae cause mortality within 48 h. Examples of control of low economic threshold insects are the use of nuclear polyhedrosis virus (NPV) and *B. thuringiensis* against *Heliothis zea* in sweet corn (Tanada and Reiner 1962), NPV against *H. zea* on cotton (Stacey et al. 1977), granulosis virus against the codling moth on fruits (Falcon et al. 1968; Huber and Dickler 1977; Jaques et al. 1981), *Steinernema carpocapsae* against the artichoke plume moth on artichokes (Bari and Kaya 1984), and *B. thuringiensis* and baculoviruses against lepidopteran insects on cabbage (Sears et al. 1983).

Pathogens are compatible with many pesticides and their combined application may give better protection. Because pest resistance to some chemical pesticides occurs, pathogens can be integrated with other control tactics to prevent or slow the resistance mechanism and prolong the use of chemical pesticides. This approach allows more tools to suppress the pest population. In using microbial pesticides, the cost of the material is not the only consideration; the benefit to the environment must be considered as well.

III. FACTORS AFFECTING EFFICACY

A. DOSAGE

In theory, one infective unit (spore, virion, conidium, egg, infective juvenile) should be sufficient to infect a host, but in actual situations, a minimum number of units are required to initiate an infection. The concepts of lethal dose (LD), regularly used with chemical pesticides, have been discussed in Chapter 3. Briefly, dose or dosage refers to the actual number of infective units necessary to cause the death of the host, whereas concentration refers to the number of infective units exposed to the host to cause death. In the former case, the number of infective units ingested or contacted by the host is known; in the latter case, the number of infective units exposed to but not necessarily ingested or contacted by the host is known. Thus, dosage is more precise than concentration.

Infective dose (ID) or infective concentration (IC) refers to the actual number of infective units needed to initiate an infection or the number of infective units exposed to the host to cause infection, respectively. Lethal time (LT) is the time of pathogen exposure to the host until its death. These concepts can also be expressed in terms of 50 or 95% values that affect a given population.

The number of infective units applied in the field against a host will affect the degree of control. It is important to apply the desired concentration of the pathogen, properly placed and at the right time, to obtain good control of the pest.

B. TIMING OF APPLICATION

Timing is an important principle in the application of any pesticide, particularly pathogens. If the pest is not present at a susceptible stage or weather conditions are not suitable for pathogen survival or persistence, the microbial pesticide will fail to control the pest.

Pathogens are inactivated by sunlight, especially the ultraviolet (UV) component (Jaques 1985). UV protectants or applications in the late afternoon or evening will benefit the pathogen. Against foliage pests, the application of most pathogens (bacteria, protozoa, viruses) should be made in weather conditions favorable for spraying (i.e., the absence of rain and wind). If fungi or nematodes are being applied, the humidity should be high. Temperature may be a limiting factor for pathogens, particularly when it exceeds 30°C. On the other hand, high field temperatures may accelerate the infectious process and result in quicker mortality, whereas lower temperatures may have the opposite effect. As with chemical insecticides, the best times to apply pathogens are during the early morning or late evening hours because meteorological conditions, such as inversions or winds, are at a minimum.

C. COVERAGE

Thorough coverage, especially for short-term control agents, is important. Unlike many chemical insecticides that have fumigant, contact, and stomach activities, pathogens must be ingested in the case of baculoviruses, bacteria, protozoans, or some nematodes or must come into contact with the cuticle as with fungi or certain nematodes. Most pathogens cannot actively disperse on their own with the exception of fungi and nematodes, and therefore, require delivery to the site of the target insect. The pest needs to contact the pathogen, or the pathogen must rely on other biological and physical agents for dispersal. Furthermore, as plants grow, the pathogen is diluted in space or removed from the feeding site and reapplication may be necessary. In the aquatic environment, the pathogen must be present where the target organism occurs. In running water, the flow rate and volume of water determine the distribution of the pathogen and its effective range.

D. pH

When applying microbial insecticides in an aqueous suspension, the pH of the water should be near 7.0 because the δ-endotoxin (crystal protein toxin) of *B. thuringiensis* and the occlusion bodies of baculoviruses are dissolved by alkaline conditions that reduce their effectiveness. Buffers are added to overcome the pH problems.

With baculoviruses, the occlusion bodies may be partially dissolved by the alkaline pH of the exudates on cotton leaves (Falcon 1971) or the presence of dew on leaves (Andrews and Sikorowski 1973; Young et al. 1977). The dissolution of the occlusion bodies reduces the activity of the virions.

E. ANTIMICROBIOSIS

The host plant of insect herbivores can significantly affect their susceptibility to disease. Differences in susceptibility to a given pathogen may be due to dietary stress (Noguchi and Yamaguchi 1984; Felton and Dahlman 1984) or direct antimicrobial activity of the plant (Kushner and Harvey 1962; Smirnoff 1972; Morris 1972; Felton et al. 1987; Keating et al. 1988). An example of dietary stress is the addition of L-canavanine, the guanidinooxy analog of arginine synthesized by numerous leguminous plants, to insect diet to increase the susceptibility of the hornworm, *Manduca sexta*, to *B. thuringiensis*. Although canavanine has antimicrobial activity, the mode of action of *B. thuringiensis* is through the ingestion of the crystal protein toxin. Possibly, this plant allelochemical alters the midgut cell membranes, which, combined with the action of the bacterial endotoxin, enhances gut permeability and increases the susceptibility of the hornworm to disease.

Many plants produce antimicrobial compounds, which inhibit the activity of microbial-control agents. For example, extracts from conifers and deciduous trees inhibit the growth of *B. thuringiensis* (Kushner and Harvey 1962; Morris 1972). More specifically, terpenoids from balsam fir (Smirnoff 1972), terpenoids and phenolics from cotton (Hedin et al. 1978), and nicotine (Krischik et al. 1988) inhibit the growth of *B. thuringiensis*. Allelochemicals may have negative effects on microsporidian development because the infected insects take longer to die when fed certain plant extracts (Smirnoff 1967). The alkaloid, tomatine from tomato plants, inhibits colony formation and growth of the fungal pathogen *Beauveria bassiana* (Costa and Gaugler 1989a). With viruses, rutin and chlorogenic acid present in tomato plants show antiviral activity and reduce the infectivity of the NPV of *Heliothis zea* (Felton et al. 1987). These studies suggest that insects tolerating high concentrations of allelochemicals may be protected from certain pathogens.

Many allelochemicals occur in plants, and their effects may or may not be evident on a given pathogen when the insect herbivore is feeding on the entire plant. As pointed out above, the allelochemical(s) may have a negative effect on the pathogen and increase the lethal dosage or lethal time for the insect host. Thus, chlorogenic acid is oxidized to chlorogenoquinone when the foliage of tomato plants is damaged during feeding by *H. zea* (Felton and Duffey 1990). Chlorogenoquinone, an oxidizing agent, binds to the occlusion bodies of the NPV and reduces their solubility under alkaline conditions. This binding may interfere with the infection process by impairing the release of infective virions in the midgut. In the gypsy moth, *Lymantria dispar*, larvae fed NPV-contaminated oak leaves are less susceptible to viral infection when compared with those fed contaminated aspen leaves (Keating and Yendol 1987). Decreased viral pathogenicity is correlated with increased acidity and tannin contents of oak leaves (Keating et al. 1988). In other cases, there is no effect on the pathogen's ability to infect the insect host. Costa and Gaugler (1989b) have shown that the Colorado potato beetle reared on five solanaceous plant species with different amounts of glycoalkaloid content is equally susceptible to *B. bassiana* when it is applied directly to the insect's cuticle.

In addition to antimicrobiosis, other plant characteristics can affect the efficacy of a pathogen. *B. thuringiensis* is more efficacious against the spruce budworm feeding on balsam fir than on white spruce (Morris 1982). The reasons for the better efficacy on balsam fir may be due, in part, to the different needle size, the impinging characteristics of the spray, and differences in the time of foliage flush in spring.

F. TARGET PEST

The age and condition of the pest are important factors affecting the effectiveness of microbial pesticides. Generally, earlier instars of the pest are more susceptible to pathogens. Pests under stress, such as those weakened by another pathogen, food quality, or crowding, are easier to control than those not stressed.

G. COMPATIBILITY

Insect pathogens are often compatible with other control tactics. One of the advantages of insect pathogens is that they tend to be host specific and therefore, are often compatible with parasitoids, predators, and other pathogens. The combination of insect pathogens with chemical pesticides or other biological-control agents can provide a synergistic or antagonistic effect (Jaques and Morris 1981). A case-by-case evaluation of their compatibility must be made before the pathogens are combined with a chemical pesticide or another biological-control agent. It should be readily apparent that bactericides, fungicides, and nematicides are not compatible with bacterial, fungal, and nematode pathogens, respectively. If the pathogen and other control agents are not compatible, the pathogen can still be used in IPM programs by properly manipulating the time of its applications.

H. HOST RESISTANCE

Host specificity observed with many insect pathogens demonstrates that insect species are naturally resistant to these microorganisms. Can an insect species susceptible to a pathogen become resistant to it? Indeed, insects that are susceptible to a pathogen can show resistance through maturation immunity and environmental and genetic factors (see Chapters 3 and 14). In maturation immunity, the younger insects are more susceptible to a pathogen than older ones. Environmental factors affecting susceptibility include nutrition, temperature, and chemical stresses. Finally, there is significant genetic variability in the insects' ability to respond to all major groups of pathogens, and selection pressure can lead to increased changes in their response to the pathogens (Burges 1971; Briese 1986a). Specifically, selective breeding of silkworms and honey bees has resulted in resistance to viral, bacterial, and protozoan pathogens (Burges 1971).

Genetic resistance to bacterial (McGaughey 1985, 1990) and viral (Briese 1986b; Fuxa et al. 1988) pathogens has been recorded against pestiferous insects. High levels of resistance to the δ-endotoxin of *B. thuringiensis* subspecies *kurstaki* have been recorded for the Indianmeal moth, *Plodia interpunctella*, a pest of stored-grain and -cereal products (McGaughey 1990). The stored-grain habitat where *B. thuringiensis* is not exposed to environmental extremes provides an excellent situation for the selection for insect resistance. There is continual breeding of the pest that is in constant contact with the bacterial toxin. Accordingly, naturally resistant populations of the Indianmeal moth have been isolated from grain storage bins. The resistant trait is incompletely recessive, and several alleles or genes are believed to be involved, and resistance in this insect is linked to an alteration in toxin–membrane binding of the midgut cells (Van Rie et al. 1990). Resistance has also been recorded for the mosquito *Aedes aegypti* toward *B. thuringiensis* subspecies *israelensis* (Goldman et al. 1986), the diamondback moth, *Plutella xylostella*, to *B. thuringiensis* subspecies *kurstaki* (Tabashnik et al. 1990), and the tobacco budworm, *Heliothis virescens*, toward the genetically engineered *Pseudomonas fluorescens* containing the δ-endotoxin of *B. thuringiensis* subspecies *kurstaki* (Stone et al. 1989).

Baculovirus resistance has been observed in inbred insect populations in the laboratory, but whether it will occur extensively to compromise the application of these viruses against field populations remains to be seen (Briese 1986b). Very few field studies have addressed this problem. After a granulosis virus epizootic in *Eucosma griseana* populations, there was an increase in the LD_{50} and in the regression slope for this insect the year after the epizootic (Martignoni 1957). Briese (1986b) suggested that the epizootic affected the susceptible population and the residual populations did not show true resistance. Resistance to baculoviruses has been observed in field populations of *Spodoptera frugiperda* (Fuxa et al. 1988). At the beginning of the season, the larvae are susceptible to the NPV, but later in the season, there is a trend towards reduced susceptibility and increased heterogeneity after exposure to the virus.

IV. SAFETY

Safety of microbial-control agents to vertebrates, including humans, has been reviewed in detail by Burges (1981b), Rogoff (1982), and Laird et al. (1990). Pathogens are considered for pest control because of their specificity to a target organism or their specificity within the class Insecta or Acarina. Pathogens infecting a wider host range beyond Insecta and Acarina are usually eliminated from consideration as microbial-control agents or require greater scrutiny in safety testing against nontarget organisms. Conversely, if the pathogen has no detectable affect on vertebrates, does it mean that the pathogen or its by-product(s) is harmless? The harmful effect may occur years later in the form of a carcinoma or teratological condition. Pathogens, therefore, must be tested to ensure their safety.

To some extent, these safety tests have occurred naturally because vertebrates such as birds and mammals are exposed to pathogens during epizootics (Lautenschlager and Podgwaite 1979; Lautenschlager et al. 1980). Humans have been exposed to baculoviruses occurring on food (Heimpel et al. 1973), and antibodies against baculoviruses have been detected in human sera (Roder and Gröner 1980). In these instances, no records of adverse effects on birds, mammals, and humans are known. If there is an effect, however slight, then are the benefits of using the microbial-control agent outweighed by the risks? Burges (1981b) believes " . . . that a pathogen should be registered as safe when there is reasonable evidence that it is so and in the absence of concrete evidence that it is not."

In the United States, the Environmental Protection Agency (EPA) requires that safety tests on candidate viruses, bacteria, protozoa, and fungi must be conducted before registration for microbial control (Betz et al. 1990). Nematodes are exempt from registration under current EPA regulations. EPA has developed protocols for safety tests for pathogens (Table 15–1). These protocols require Tier I tests, which include thorough product identification and human health and environmental effects. If the insect pathogen fails the Tier I tests, Tier II–IV tests are required, but the chances for registration are diminished. None of the currently registered microbial-control agents have required testing beyond Tier I.

Although human health and safety are the primary concerns in these safety tests (Siegel and Shadduck 1990), the impact on nontarget organisms and the environment is also of great importance. With most registered microbial-control agents, no direct infection of parasitoids and predators occurs. Parasitoids and predators, however, can be indirectly affected by the microbial-control agent by reducing the availability of host or by killing the immature parasitoid within a susceptible host (Gröner 1990; Melin and Cozzi 1990). There are a few cases where parasitoids are susceptible to *B. thuringiensis* (Tanada 1984; Melin and Cozzi 1990). Some pathogens, such as protozoa (Brooks 1973), fungi (Goettel et al. 1990), and nematodes (Akhurst 1990), may directly infect parasitoids and predators. These pathogens must be examined on a case-by-case basis to determine their effects on nontarget organisms (Vinson 1990). No detrimental effects have been noted with the use of *B*.

TABLE 15-1 Data Needed for Registration of Microbial-Control Agents by the U.S. Environmental Protection Agency[a]

Data needed	Requirement
I. Product analysis	
A. Product identification and manufacturing process	Required
B. Discussion of formation of unintentional ingredients	Required
C. Analysis of samples	Conditionally required
D. Certification of limits	Required
E. Analytical methods	Required
F. Physical and chemical properties	Conditionally required
G. Submittal of samples	Conditionally required
II. Residue analysis	
III. Toxicology—Tier I[b]	
A. Acute oral exposure	Required
B. Acute dermal exposure	Required
C. Acute pulmonary exposure	Required
D. Acute intravenous exposure	Required
E. Primary eye irritation	Required
F. Hypersensitivity incidents	Required
G. Tissue culture (viruses only)	Required
IV. Nontarget organisms and environmental fate—Tier I	
A. Avian oral	Required
B. Avian inhalation	Conditionally required
C. Wild animal	Conditionally required
D. Freshwater fish	Required
E. Freshwater aquatic invertebrate	Required
F. Estuarine and marine animal	Required
G. Nontarget plants	Required
H. Nontarget insects	Required
I. Honeybee	Required

[a]Modified after Betz et al. (1990).
[b]Tier II–Tier III tests (not shown) are required when Tier I tests are positive. No currently registered microbial-control agents have required testing beyond Tier I.

thuringiensis to the abiotic environment, but the application of *B. thuringiensis* against the gypsy moth in an oak forest in Oregon has reduced species richness in the guild of nontarget leaf-feeding Lepidoptera (Miller 1990). The widespread use of *B. thuringiensis* may have negative impact on the populations of susceptible insects serving as biological-control agents of noxious weeds, on the reduction of these insects serving as food for birds, and on the rare or endangered species of insects. Other microbial agents have not been used as extensively, and no evidence exists to indicate that environmental pollution will occur with these pathogens.

The safety of genetically engineered insect pathogens in field applications has not been established (Fuxa 1990a,b) although engineered NPVs have been tested in the field under strict guidelines (Wood and Granados 1991). The likelihood of the transfer of genetically engineered genes being incorporated into native organisms is remote. Nonetheless, field tests to demonstrate no adverse effects of engineered organisms are currently being conducted. EPA is taking a cautious approach on this issue and examines each genetically engineered pathogen on a case-by-case basis. *Pseudomonas fluorescens* that are killed but contain the δ-endotoxin gene (Feitelson et al. 1990) have been approved by EPA as a delivery system for insect control. Because the genetically altered *P. fluorescens* are nonliving, the environmental concerns associated with this engineered microorganism are nonexistent with the exception of host resistance (Stone et al. 1989). More approvals for testing will undoubtedly occur, and larger field tests providing safety and efficacy data should eventually lead to the registration of genetically altered microorganisms.

V. MOLECULAR GENETICS

The expression of foreign genes in heterologous living organisms has opened new avenues for insect pathogens. Advances have been made with two major groups of insect pathogens, the bacteria and baculoviruses. The underlying premise for the incorporation of foreign genes of insect pathogens into other organisms or of foreign genes into insect pathogens is to improve the pathogens' efficacy against target pests.

A major focus for recombinant DNA technology has been the δ-endotoxin of *B. thuringiensis*. After sporulation, the spore and the δ-endotoxin become separated because the sporangium that surrounds them is very fragile. This results in the δ-endotoxin being exposed to the elements after application, reducing its effectiveness as an insecticide (Feitelson et al. 1990). Through genetic engineering, it is possible to repackage the δ-endotoxin into a more desirable form. Thus, the gene for the δ-endotoxin that occurs on a plasmid (Kronstad et al. 1983) has been cloned into the gram-negative bacterium *Pseudomonas fluorescens*, and indications are that it will be commercially successful (Feitelson et al. 1990). Field tests show that these genetically engineered cells are more efficacious than *B. thuringiensis* preparations or a chemical pesticide standard (Feitelson et al. 1990). *P. fluorescens* cells do not lyse, are non-spore-formers, and occur naturally on plants. Because of public concerns and government regulations on the release of genetically engineered organisms, *P. fluorescens* cells containing the δ-endotoxin are killed and stabilized before they are formulated. The use of δ-endotoxin alone without its associated bacterium, however, is actually an application of a chemical insecticide. The organism (*B. thuringiensis*) plays a role in pathogenicity (see Chapter 4).

The δ-endotoxin gene has been cloned and expressed in a number of plants, including tobacco, tomato, potato, and cotton (Leemans et al. 1990). These transgenic plants are protected from attack by lepidopteran larvae (Fig. 15–1). Because not all plant parts are attacked by insects, research efforts are ongoing to limit expression to specific plant structures (i.e., root, leaf, stem, fruit, etc.). Such efforts should retard development of resistance of insects to the δ-endotoxin and minimize its occurrence in edible plant parts.

Baculoviruses generally show low virulence, have poor field persistence, possess a limited host range, and have a long period of lethal infection. The introduction of appropriate foreign genes into the baculovirus genomes may increase pathogenicity or insecticidal activity (Wood and Granados 1991). Maeda (1989a) suggests that the insertion of genes producing toxins, enzymes, inhibitors, or hormones specific to insects that block neuronal function or hormone production may be possible. Infection with such recombinant baculoviruses may cause immediate toxicity, deranged behavior, or arrested growth in the target insect. When a recombinant gene for the diuretic hormone is introduced into the nuclear polyhedrosis virus (NPV) of *Bombyx mori*, the hormone gene is expressed in *B. mori* larvae, causing

FIGURE 15-1

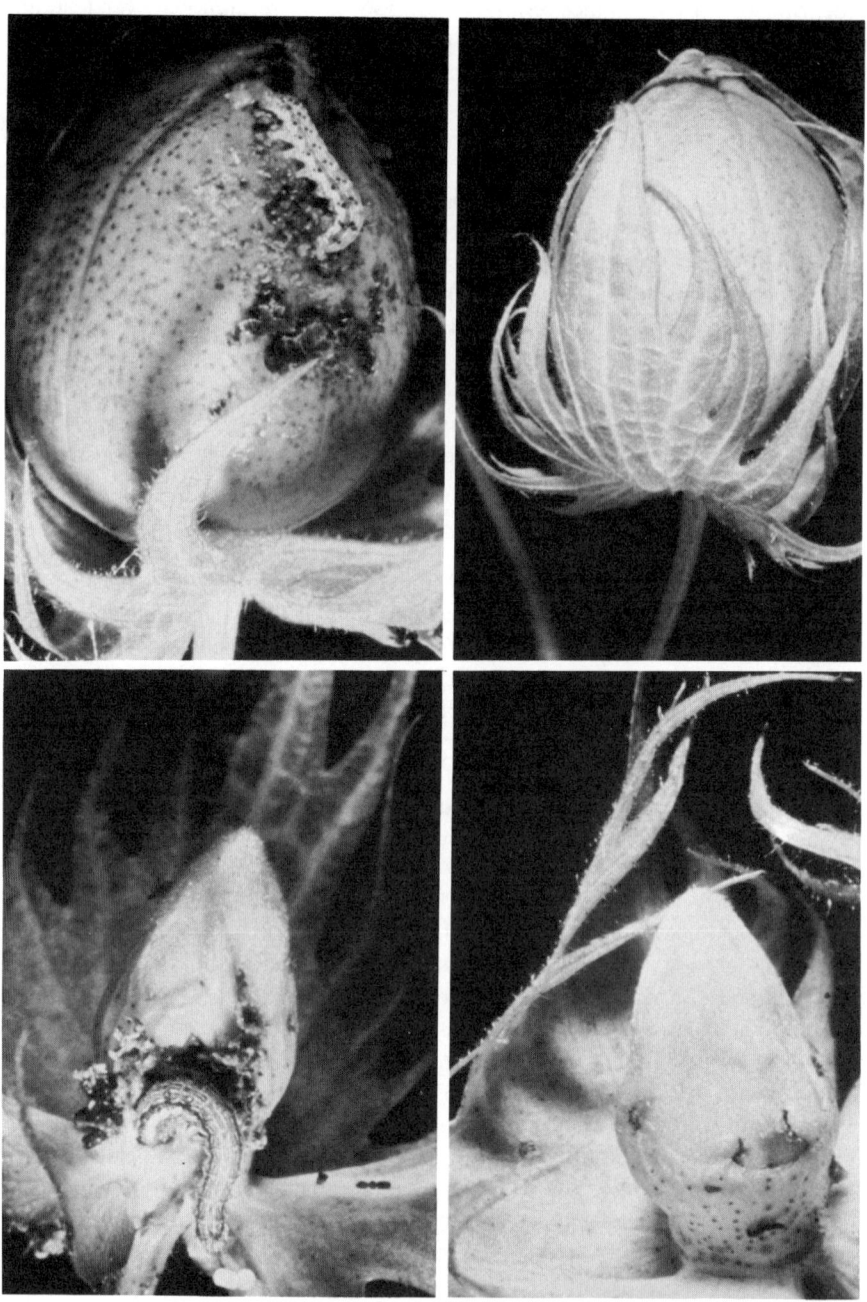

Transgenic cotton bolls containing δ-endotoxin of *Bacillus thuringiensis* and showing no feeding damage by *Heliothis zea* (*right*). Nontransgenic cotton bolls with *H. zea* actively feeding (*left*). (Provided by S. R. Sims, Monsanto Company, St. Louis, Missouri.)

an alteration in larval fluid metabolism (Maeda 1989b). The recombinant virus kills infected larvae 20% faster than the original virus. The scorpion toxin gene has been inserted in the NPVs of *Autographa californica* (Stewart et al. 1991) and *Bombyx mori* (Maeda et al. 1991), killing their respective hosts in less than 86 h postinfection. In addition, a recombinant NPV of *A. californica* containing a mite neurotoxin gene causes paralysis in *Trichoplusia ni* larvae in 48 h postinjection into the hemocoel (Tomalski and Miller 1991). In each case, mortality by the recombinant NPV was increased significantly over the wild-type NPV.

VI. MICROBIAL PESTICIDES

Several microbial pesticides have been successfully registered by commercial companies or by government agencies in the United States and other countries. Table 15–2 shows the chronology of microbial agents registered in the United States. The registered microbial pesticides include viruses, bacteria, fungi, and protozoa. Be-

TABLE 15-2 Chronology of Microbial-Control Agents Registered in the United States

Year	Pathogen group	Microbial agent	Target insect(s)
1948	Bacterium	*Bacillus popilliae*	Japanese beetle
1948	Bacterium	*Bacillus lentimorbus*	Japanese beetle
1961	Bacterium	*Bacillus thuringiensis*[a]	Lepidopteran larvae
1975	Virus	*Heliothis* NPV[b]	Cotton bollworm, *Heliothis*
1976	Virus	*Orgyia* NPV[b]	Douglas-fir tussock moth
1978	Virus	*Lymantria* NPV[b]	Gypsy moth
1980	Protozoan	*Nosema locustae*	Grasshoppers
1981	Fungus	*Hirsutella thompsonii*	Citrus rust mite
1981	Bacterium	*Bacillus thuringiensis* subsp. *aizawai*	Wax moth
1981	Bacterium	*Bacillus thuringiensis* subsp. *israelensis*	Mosquito and black fly larvae
1983	Virus	*Neodiprion* NPV[b]	European pine sawfly
1988	Bacterium	*Bacillus thuringiensis* subsp. *tenebrionis*[c]	Colorado potato beetle, elm leaf beetle
1991	Bacterium	Encapsulated Bt toxin in *Pseudomonas fluorescens*[d]	Lepidopteran larvae
1991	Fungus	*Lagenidium giganteum*	Mosquito larvae

[a]*B. thuringiensis* subsp. *thuringiensis* until 1971 and since then commercial companies use *B. thuringiensis* subsp. *kurstaki*.
[b]NPV (nuclear polyhedrosis virus).
[c]*B. thuringiensis* subsp. *tenebrionis* is identical to subsp. *san diego* (Krieg et al. 1987).
[d]Bacterium is killed and stabilized. The process leaves the endotoxin effectively encapsulated within a bacterial cell.

TABLE 15-3 Some Insect Pathogens That Show Potential as Microbial-Control Agents for Commercial Development

Pathogen group	Pathogen	Target insect(s)
Viruses	Codling moth GV[a,b]	Codling moth
	Plodia GV[a]	Indianmeal moth
	Autographa NPV[c]	Many lepidopteran species
	Spodoptera NPV[c]	Spodoptera complex
Bacteria	Bacillus sphaericus	Mosquito larvae
	Serratia entomophila[d]	White grubs
Fungi	Beauveria bassiana[e]	Many insect species
	Metarhizium anisopliae[e]	Many insect species
	Nomuraea rileyi	Many lepidopteran species
	Paecilomyces fumosa-roseus	Many insect species
	Tolypocladium cylindrosporum	Mosquito larvae
	Verticillium lecanii[e]	Aphids and whiteflies
Protozoa	Nosema pyrausta	European corn borer
	Vairimorpha necatrix	Many lepidopteran species

[a]GV (granulosis virus).
[b]Registered in some European countries.
[c]NPV (nuclear polyhedrosis virus).
[d]Registered in New Zealand (Jackson 1989).
[e]Registered or used commercially in countries other than the United States (McCoy 1990).

sides these registered pathogens, others show considerable promise as microbial-control agents and may eventually become available for use (Table 15–3).

Many principles governing the utilization of chemical pesticides, parasitoids, and predators to control insects and other arthropods are applicable to pathogens. Insect pathologists and agricultural entomologists have adopted many of the principles and methods used in chemical control for pathogens (Tanada 1959, 1967). Therefore, the use of microorganisms to control arthropod pests has a close interrelationship with biological and chemical controls.

Before a microbial-control agent is considered for field use, a number of factors need to be considered. These include safety (covered in the previous section); mass production, storage, formulation, standardization (bioassay), and application technology.

VII. MASS PRODUCTION

Microbial-control agents, many of which are facultative pathogens, can be produced on artificial media in large quantities. *B. thuringiensis* is produced in large submerged fermentation tanks (Couch and Ross 1980). Fungi such as *Beauveria bassiana*, *Metarhizium anisopliae*, and *Hirsutella thompsonii* are produced by fermentation procedures (Weiser 1982). In certain countries, *M. anisopliae* is produced on sterilized rice in plastic bags (Weiser 1982). Submerged cultures to produce fungal blastospores have also been utilized, but blastospores are more susceptible to environmental factors, and the more resistant conidia are favored. The nematode

Steinernema (syn. *Neoaplectana*) *carpocapsae* can be mass produced *in vitro* on polyurethane sponge containing a protein and sterol source and the inoculation of the primary form of the bacterium *Xenorhabdus nematophilus* (Bedding 1984a) or in liquid culture in fermentation tanks (Friedman 1990).

Obligate pathogens require a living insect host for mass production. *Heliothis* NPV is produced in larvae of *Heliothis zea* (Ignoffo and Anderson 1979; Shieh 1989). Similarly, the microsporidium *Nosema locustae* is produced in its grasshopper host (Henry and Oma 1981); the bacterium *B. popilliae* is produced in Japanese beetle grubs (Ignoffo and Hink 1971); and the nematode *Romanomermis culicivorax* is produced in its mosquito host (Petersen 1984). In some cases, pathogens can be mass produced in factitious hosts. *Nosema algerae*, a microsporidian pathogen of mosquitoes, can be produced in larvae of the corn earworm, *Heliothis zea* (Undeen and Maddox 1973). A *Heliothis* larva produces 2000 times more spores than a mosquito larva, and no loss in virulence or pathogenicity of the microsporidium occurs in the factitious host (Undeen and Alger 1975).

Tissue or cell culture techniques for the mass production of obligate pathogens have not yet materialized although viruses produced in this manner are effective under field conditions (Dougherty et al. 1982; Jaques 1977). The cost of utilizing cell culture for mass-producing insect viruses at this time is prohibitive (Hink 1982; Shieh 1989). Moreover, the repeated subculturing of viruses in cell lines has resulted in reduced production of occlusion bodies (Granados 1976). Advances are being made in this field, and perhaps someday this technology will be available for the mass production of insect pathogens (Stockdale and Priston 1981).

VIII. FORMULATIONS

Formulation refers to the resultant composition when a pesticide is mixed with an ingredient(s). Although any material added to a microbial-control agent makes a formulated product, we generally restrict discussion of formulations to microbial agents that are marketed commercially or from an experimenter's point of view, the form in which the microbial-control agent is applied. The ingredients should contribute to the viability, stability, virulence, and efficacy of the microbial-control agent and the acceptance of the product by the user (Couch and Ignoffo 1981). The user must find the product easy to mix and apply. The basic formulations of microbial agents, like most chemical pesticides, attempt to enhance the ease of application, persistence of the product, wettability or adhesiveness to the host plant and/or insect, and attractiveness to the insects. They are prepared as wettable powders, liquids (aqueous or emulsifiable suspensions), dusts, granules, and baits.

The addition of an enhancing agent can be incorporated in the formulation. The formulated products usually contain inert carriers (various oils, emulsifiers, extenders, suspending agents, buffers, or diluents). Ingredients to improve flowability, anticaking, and convenience of handling and mixing are also added. Each ingredient must be tested to ensure that there is no detrimental effect on the pathogens or on the plants and nontarget organisms.

At the time of spray application, other additives may be incorporated to increase the efficacy of the pathogen. These additives include spreaders, stickers, UV protectants, thickening agents to prevent rapid evaporation (humectants), and gustatory stimulants (phagostimulants).

IX. STANDARDIZATION

Standardization is an essential step in the production of a reliable, consistent microbial insecticide. For *B. thuringiensis* subspecies *kurstaki*, the standard method approved and adopted on a worldwide basis is the international unit. The international unit (IU) is defined as a quantity of a biological toxin or its equivalent based on a bioassay that produces a particular biological effect agreed upon internationally. The biological effect for *B. thuringiensis* is death of the bioassay insect.

The original reference standard was prepared by the Pasteur Institute in France, using *B. thuringiensis* subspecies *thuringiensis* (Burges 1967). The standard was arbitrarily given a potency of 1000 IU/mg and the test insect was *Anagasta kuehniella*. Subsequently, the subspecies *thuringiensis* was replaced by the subspecies *kurstaki* in the United States and a new reference standard was prepared. The *kurstaki* standard was 18 times more active than the *thuringiensis* standard when bioassayed against the cabbage looper, *Trichoplusia ni*, and was given a value of 18,000 IU/mg (Dulmage 1975). In the United States, *B. thuringiensis* subspecies *kurstaki* is bioassayed against a reference standard to ensure the commercial production of a reliable consistent product (Dulmage et al. 1981). Other lepidopteran assay insects include *Spodoptera exigua* (expressed in *Spodoptera* units) for the *B. thuringiensis* subspecies *kurstaki* and the diamondback moth, *Plutella xylostella*, (expressed in diamondback moth units) for the transgenic *Pseudomonas fluorescens* containing the δ-endotoxin.

The IU for *B. thuringiensis* subspecies *israelensis* uses the mosquito *Aedes aegypti* as the bioassay insect. A reference standard has been prepared by the Pasteur Institute and assigned a potency of 1000 IU/mg (Davidson 1982). Similarly, for *B. thuringiensis* subspecies *tenebrionis* (*san diego*), the Colorado potato beetle International Unit (CPB IU) with a reference standard (10% technical powder 5653) assigned a potency of 50,000 CPB IU has been proposed (Ferro and Gelernter 1989). The second instar of the Colorado potato beetle would serve as the bioassay insect.

Standardization for baculoviruses has been attempted using different methods. The most common technique for NPV standardization is to use "viral units" or "larval equivalents" (Ignoffo 1966, Pinnock 1975). One viral unit equals 10^9 polyhedral occlusion bodies (POB) and one larval equivalent equals 6×10^9 POB. (In the literature, polyhedral inclusion body or PIB is also used.) The limited stability in storage and difficulties encountered in obtaining reproducible bioassays have hampered progress (Dulmage and Burgerjon 1977). Another approach to standardization has been designed with the NPV of the Douglas-fir tussock moth (Martignoni and Iwai 1978). This standard is based on the response of a population of susceptible test insects in a standardized bioassay. This response standard is valid only if the test insects are genetically uniform and if the assays are repeated under the same conditions with the same materials each time. Currently, this procedure has only been developed for the tussock moth.

Universally acceptable standards for fungi, protozoa, and nematodes remain to be developed. For fungi, viable conidia or blastospores are based on colony-forming units (CFU) on artificial media. Spore counts are used for microsporidia, and counts of living infective stage or assay of one nematode-one insect (Georgis 1990) is used for nematodes.

X. STORAGE

Great variation in the storage capacity of pathogens exists. Commercial products should be held routinely at cool temperatures and protected from sunlight. Exposure of the packaged product to sun or extremely low or high temperatures will result in the loss of virulence or death of the microorganism. When *B. thuringiensis* is held at cool temperatures, its shelf life can be extended for several years.

Pathogens formulated as a wettable powder or dust should be kept dry. If these products become wet during storage, there will not only be a loss of virulence but also difficulty in application because of lumping and caking of the carrier. In contrast, formulations of the nematodes *R. culicivorax*, *S. carpocapsae*, or *Heterorhabditis bacteriophora* must be maintained in a moist environment. Producers of microbial pesticides are cognizant of the proper storage conditions for their products. The problem of virulence loss usually occurs because the ultimate seller or user fails to store the products under the proper conditions.

Some pathogens maintained under laboratory conditions can survive prolonged storage. *B. popilliae* has retained its infectivity for over 30 years under laboratory conditions (Dunbar and Beard 1975), whereas microsporidian spores can be stored for prolonged periods in an aqueous medium or in cadavers at low temperatures. However, microsporidian spores may lose some of their activity. Henry and Oma (1974) showed that fresh spores of *Nosema locustae* resulted in higher infection rates than spores stored in water or in cadavers at $-10°C$ for 1 yr.

Most fungal spores can be lyophilized and maintained for long periods in the cold; the nematodes *S. carpocapsae* and *S. glaseri* can be maintained at 5°C for several years. Eggs of *R. culicivorax* can be maintained in moist sand at room temperatures for several months (Petersen 1984).

XI. APPLICATION TECHNOLOGY

Application technology involves the mechanics of proper placement of a desirable concentration of active agents on the target site to obtain maximum effective control of the target insect. Thorough coverage of the target site is particularly important in short-term control because microbial pesticides such as *B. thuringiensis*, baculoviruses, protozoans, and fungi must be ingested or contact their hosts to be effective. The short residual activity of these agents also requires that the pest consumes or contacts the pathogen soon after application, and this is accomplished by thorough coverage at the site of insect feeding.

Application of microbial pesticides has utilized equipment and technology developed for chemical pesticides (Tanada 1967). Most microbial pesticides in forest and agricultural ecosystems are still applied using conventional ground and aerial equipment (Fig. 15–2). However, comparisons of ground and aerial spray equipment indicate that different nozzles affect the distribution and effectiveness of microbial pesticides (Smith and Bouse 1981). Therefore, the method of application varies with a given microbial pesticide and its target insect.

Application of pathogens through overhead sprinkler irrigation systems shows promise (Hamm and Hare 1982). *Nomuraea rileyi*, *Vairimorpha* spp., or NPVs applied with this system produce infections in the target insects. Moreover, when

FIGURE 15-2

Ground and aerial applications of insect pathogens. In some cases, protective clothing is needed to apply entomopathogens. When applying entomopathogens, comply with labeled restrictions and state regulations. (A) Aerial application of *Bacillus thuringiensis* subspecies *tenebrionis* to control Colorado potato beetle on potato plants. (Provided by W. D. Gelernter, Mycogen Corporation, San Diego, California.) (B) Application of *Bacillus thuringiensis* subspecies *tenebrionis* to control Colorado potato beetle larvae on potato plants. (Provided by W. D. Gelernter, Mycogen Corporation, San Diego, California.) (C) Application of steinernematid nematodes to control cutworms on grape vines.

continuous infestations of the target insect are present, an epizootic of the NPV has developed.

Application technology for long-term control programs can utilize conventional spray equipment but is also amenable to other approaches as well. For example, the NPV of the European pine sawfly, *Neodiprion sertifer*, has been applied with conventional spray equipment resulting in virus establishment and maintaining the sawfly population below the economic threshold level (Bird 1955). The dissemination of the pathogen can be done also by infecting or contaminating an arthropod (insect or mite) carrier and releasing it into the field (i.e., autodissemination). The pathogens, themselves, are not disseminating but are being distributed by the carrier. Gard and Falcon (1978) demonstrated that adult moths attracted to a light source can be contaminated with a formulated *Heliothis* NPV and spread this pathogen downwind from the light source. Zelazny (1977) released baculovirus-contaminated or infected adults of *Oryctes* that spread the virus, significantly reducing the rhinoceros beetle population. Pathogens, especially fungi, have been released by placing dead infected larvae into the field (Sprenkel and Brooks 1975). Thus, pathogen introduction does not need to rely solely on spray technology.

XII. ADVANTAGES AND DISADVANTAGES OF MICROBIAL CONTROL

Advantages of using pathogens for controlling pests are

1. Specificity to target organisms or to a limited number of host species.
2. Harmless to vertebrates and plants.

3. No toxic residues.
4. Little or no environmental pollution.
5. Little or no development of resistance by the target organism. (However, resistance to *Bacillus thuringiensis* in field populations of insects has been observed.)
6. No secondary pest outbreak.
7. Compatible with many chemical pesticides, parasitoids, predators, and other pathogens.
8. Possibility of long-term control.
9. Ease of application of pathogen with conventional spray equipment.
10. Mass production capability with facultative pathogens.
11. Adaptable to genetic modification through biotechnology.

Although the advantages of microbial control are many, these same advantages can be disadvantageous. The disadvantages are

1. Specificity only to target organism.
2. Pathogen or its by-product(s) is harmful to nontarget organisms.
3. Strict timing of application is necessary for maximal effect.
4. Good coverage is essential for pathogen ingestion or contact by target organism.
5. Time of infection till death is not immediate.
6. Susceptibility to inactivation by environmental conditions.
7. Loss of virulence and pathogenicity by frequent subculturing.
8. Short shelf life and/or requirements for special handling.
9. Obligate pathogen is difficult or expensive to mass produce.
10. Uneconomical except for high-value crops.
11. Fear of pathogens by the public.
12. Risks associated with genetically engineered pathogen (host range modification, gene exchange to other organisms, and genetic stability). Hochberg and Waage (1991) suggest that engineered organisms modified to decrease the lethal time may adversely affect parasitoid populations and cause pest resurgence.

Before embarking on a microbial-control program, careful considerations must be given to the type of pests occurring in the target area. If more than one type of pest occurs, consideration should be given to an IPM approach.

XIII. EXAMPLES OF LONG-TERM CONTROL

Long-term control has been accomplished primarily against pests having a high economic threshold (forest, turf, and pasture) and with pathogens persisting in either the biotic or physical environments or in both environments. In most cases, the pathogen causes chronic or prolonged infections and produces large numbers of progeny that persist in the environment. The host population and other biotic factors (i.e., alternate hosts, parasitoids, and predators) play an important role in the persistence and spread of the pathogen. Each situation, however, needs to be examined carefully because the factors allowing for long-term control vary from pest to pest and from crop to crop.

A. VIRUSES

1. Sawflies

The NPV of sawflies in the forest ecosystem seems particularly suited for long-term control (Kaya 1976; Cunningham and Entwistle 1981; Cunningham 1982). The European pine sawfly, *Neodiprion sertifer*, the red-headed pine sawfly, *N. lecontei*, Swaine's jack pine sawfly, *N. swainei*, and European spruce sawfly, *Gilpinia hercyniae*, have been successfully controlled with one or two introductions of their respective NPVs into field populations. Cunningham (1982) and Cunningham and Entwistle (1981) have summarized the field tests with the NPVs of sawflies. In addition, Bedford (1981) has summarized the studies against the rhinoceros beetle, *Oryctes rhinoceros*, with the nonoccluded baculovirus. Long-term control of lepidopteran insects with baculoviruses has been implied particularly in the forest ecosystem, but due to natural occurrences of these viruses, the role of the applied viruses has been obscure.

The European pine sawfly, a colony-feeding species, occurs throughout pine forests in Eurasia and North America. It was found in New Jersey in 1925 without its natural enemies. In 1949, Bird (1953) received a sample of NPV from Sweden and applied the virus into sawfly populations from 1951 to 1953, which resulted in short-term control. More importantly, Bird (1955) found that the introduction of small amounts of NPV can initiate epizootics in sawfly populations. Three years after the introduction of the NPV into 40 hectares of Scots pine, the sawfly population was controlled. The virus was introduced into other North American forests and was effective in reducing sawfly populations to subeconomic levels for up to 5 yr (Kaya 1976). In Eurasia, the native home of the sawfly, the augmentation of the NPV with the naturally occurring virus has been successful. In some areas of Europe, sawfly larvae were being killed 1–7 yr after application (Cunningham 1982).

The method of viral application in sawfly populations did not affect the outcome. Applications with hand-held sprayers, mist blowers, other ground equipment, or aircraft were successful. Thorough coverage, an important principle in microbial control, is not as important with the NPV because contaminated adults, parasitoids, predators, scavengers, and birds disseminate the virus. Thus, viral application made every third row of trees in a plantation resulted in an epizootic, keeping the population levels low. For maximum effectiveness, the NPV should be applied before the sawfly reaches the fourth instar.

In summary, the factors for the successful use of this virus in long-term control are (1) viral application before the fourth instar, (2) the gregarious feeding habits of the larvae, (3) viral spread from one area to another by birds and insect parasitoids, predators, and scavengers, and (4) viral persistence in the environment.

The red-headed pine sawfly, *Neodiprion lecontei*, is a native insect of North America. Outbreaks of this insect have been controlled by applying the NPV, resulting in short-term and long-term control. Application of the NPV resulted in mortality within 14 to 48 days. Examination of treated plots the following year showed little or no presence of the sawfly (Cunningham 1982). This insect has a colonial habit, and if one larva becomes infected, the virus is likely to spread throughout the colony. In addition, the virus is probably spread by a complex of insect parasitoids, predators, and scavengers. The application of NPV of the Swaine's jack pine sawfly, *Neodiprion swainei*, has given similar control patterns as seen for the red-headed pine sawfly and European pine sawfly. The virus spreads several meters in the year of application, possibly due to larval migration (Smirnoff 1961). Smirnoff (1962) has advocated spraying low concentrations of the NPV or disseminating contaminated sawfly cocoons into field populations to initiate epizootics.

Another example of long-term control in the forest ecosystem has been documented with the European spruce sawfly, *Gilpinia hercyniae*. This insect, introduced into North America during the early 1900s, occurred in outbreak proportions. Parasitoids and predators were imported from Europe and, apparently, the sawfly NPV was accidentally introduced with these importations. Once the NPV was introduced, it spread rapidly and kept the population, along with other biological-control agents, below the economic threshold level. Bird and Burk (1961) applied this NPV on spruce trees in Ontario, Canada in 1950 and observed viral epizootics for 9 consecutive years. Unlike other sawflies that have been discussed, the larvae of the European spruce sawfly are not colonial feeders. The success of this NPV as a long-term control agent is due to several factors. Transovum transmission of NPV (Bird 1961) and the contamination of foliage with NPV by adult sawflies (Neilson and Elgee 1968) have been postulated as effective mechanisms of spread and transmission. Evans and Entwistle (1982) have summarized the studies conducted in England and surmised that the overwintering NPV on the foliage is particularly important. A high carry-over of NPV inoculum occurs in dense sawfly populations, but as the populations fall, this carry-over is greatly reduced.

Birds serve as a principal factor in NPV dispersal because samples of bird feces collected throughout the year contained polyhedra with infective virions (Entwistle et al. 1977). They acquire the NPV by consuming virus-killed larvae present on trees throughout the year. Soil is a possible reservoir as viable NPV has been isolated from the surface down to 13 cm. The sawfly overwinters as an eonymph at the junction between the forest litter and soil, and when the adult emerges, it may be contaminated with the NPV (Evans and Entwistle 1982).

2. Rhinoceros Beetle

The rhinoceros beetle, *Oryctes rhinoceros*, is a major pest of coconut and other palms in the South Pacific and Southeast Asia (Bedford 1980). The adult beetles cause damage when they fly to the central crown of a palm and bore and feed through the heart into unopened fronds. Palms may be killed by repeated or heavy beetle attacks on the growing point or the damage by the beetles provides entry points for other insects or plant pathogens. If beetle populations are low, the damage is slight and the palms recover. The larvae live in dead palms and in compost and do not cause economic damage.

A nonoccluded baculovirus has provided long-term control of this insect (Bedford 1981). The simplest, most economical and direct method of viral introduction into the field is by autodissemination. Adults from the field or reared in the laboratory are infected by immersing them in a viral suspension and then allowed to crawl through sawdust mixed with the virus. The infected adult midgut cells produce large numbers of viral particles that are disseminated with the feces. The infected adults do not feed but are attracted to breeding and feeding sites. The virus is also transmitted during mating when the uninfected adult ingests virus-laden fecal material from its sexual partner (Zelazny 1976). The virus does not persist well in the environment and transmission from adults to larvae must occur in a short period of time. When an infected adult beetle visits the breeding site for oviposition, it defecates and the larvae already present in the breeding site become infected upon ingestion of the contaminated fecal matter. Subsequently, other larvae may become infected from previously infected larvae.

The virus has been released on many islands in the South Pacific between 1967 and 1975. The results of the viral application have been excellent with significant reductions in damage to palms within 12 to 18 months after the establishment of the virus in the beetle population. The success of this program has been achieved because of the excellent method of viral transmission and the reduced longevity, fecundity, and feeding of infected adult beetles (Zelazny 1973). The virus and the beetle population have established an equilibrium level that in most cases is below the economic threshold.

3. Lepidopteran Insects

Viruses of lepidopteran insects are not as easily evaluated in long-term control programs because of natural epizootics occurring in the populations and the cyclical nature of outbreaks of these insects. When a NPV was applied against the gypsy moth, *Lymantria dispar*, in Sardinia in 1966 and 1972, the postspray mortality was over 85% while no NPV mortality was found in untreated plots (Magnoler 1968). The NPV, found in areas adjacent to the treated plots, was believed to have been spread by the migration of diseased larvae. However, the documentation of long-term control of gypsy moth with the applied NPV is open to question at this time. Similarly, the application of baculoviruses for the Douglas-fir tussock moth, *Orgyia pseudotsugata*, the white-marked tussock moth, *O. leucostigma*, the eastern spruce budworm, *Choristoneura fumiferana*, the forest tent caterpillar, *Malacosoma disstria*, the wattle bagworm, *Kotochalia junodi*, and others has not resulted in long-term control, or the results have been obscured by natural viral epizootics (Cunningham 1982).

The granulosis virus (GV) of *Plodia interpunctella*, a pest of stored grain and grain products, has potential for long-term control in storage bins (Kinsinger and McGaughey 1979). The GV can persist and protect the grain for 1 yr or longer. In this case, the GV is protected from environmental extremes and persists in the stored grain.

B. BACTERIA

The Japanese beetle, *Popillia japonica*, was accidentally introduced into New Jersey in 1916. In dense populations, the grubs cause substantial damage to lawns and pastures by feeding on grass roots, and adults cause feeding damage to a wide variety of ornamental and agricultural plants. In sparse populations, the damage by the adults and grubs is tolerable.

Dutky (1940) described two species of bacteria causing milky disease in Japanese beetle grubs: the more virulent and dominant bacterial species *Bacillus popilliae* and the less virulent *B. lentimorbus*. These bacteria were mass produced *in vivo* in Japanese beetle grubs, formulated in talc, and distributed throughout 14 eastern states and the District of Columbia between 1939 and 1951 (Fleming 1968).

These bacteria, primarily *B. popilliae*, have been successful microbial-control agents because of their capacity to persist in the host larval environment, the soil. Infected adult beetles and other animals disperse the bacterial spores naturally but slowly. The best method of establishing the bacteria is by applying them directly into beetle-infested soil. Once spores are established in the soil, they will persist through recycling for many years (Klein 1981). At one colonization site, the bacteria were still present 25 yr later (Ladd and McCabe 1967). Inasmuch as spores stored in soil for 7 yr lose some of their infectivity (St. Julian et al. 1978), the presence after 25 yr indicates that the bacteria recycled in the environment. Thus, the persistence and propagation of the bacteria are dependent on the immigration and deposition of eggs by adult beetles in contaminated soil.

Highly virulent strains of milky-disease organisms are not preferred because such strains kill the host rapidly and thereby do not multiply and sporulate in sufficient numbers (Dutky 1963; St. Julian et al. 1970). Populations of the beetle declined during the late 1940s, and the Japanese beetle is usually an inconsequential pest in older areas of infestations. Occasionally, resurgence of beetles has occurred, but these are in isolated areas. In 1973 and 1974, a resurgence of the Japanese beetle occurred in Connecticut, which is an old area of infestation (Dunbar and Beard 1975), and an abnormal strain that is less virulent and infectious than the typical *B. popilliae* appears to be part of the reason for the resurgence. Both virulent and abnormal strains were isolated from diseased grubs in Connecticut. The development of this less virulent strain in the Connecticut population may be related to the passage through another scarab species.

Overall, the milky-disease organisms have been successful in long-term control of the Japanese beetle. The high economic threshold of turf and pasture, the persistence of the bacteria in the soil, and the equilibrium between the bacteria and the grub populations account for the success. As this insect spreads into new areas, the population levels will be high until the milky-disease organisms are applied and become firmly established as long-term control agents.

Bacillus thuringiensis can serve as a long-term control agent of lepidopteran pests in protected situations, such as bins of stored grain (Kinsinger and McGaughey 1979). The bacterium can persist and protect the grain in farm bins for at least a year. Unfortunately, this situation has resulted in resistant insects (McGaughey 1985).

C. FUNGI

Fungal epizootics occur naturally in many arthropod populations (Carruthers and Hural 1990; McCoy et al. 1988), but these natural epizootics usually occur after considerable pest damage has been done. The initiation of artificial epizootics has been accomplished for long-term control, especially in areas or situations where high-humidity conditions prevail. In many cases, the long-term control is usually for one growing season, and the reapplication of fungi is required the following season to increase the inoculum. By introducing the fungal pathogen into the host population early, the epizootic is initiated earlier and prevents or reduces damage by the pest.

Hall and Burges (1979) demonstrated that the fungus *Verticillium lecanii* controlled the green peach aphid, *Myzus persicae*, on chrysanthemum in glasshouses in England. One application of the fungus maintained excellent control for 3 months, which is the duration for chrysanthemum cultivation. Another aphid, *Aphis gos-*

sypii, has been effectively controlled with *V. lecanii* in spite of the natural reinfestations of this aphid in the glasshouse (Hall 1981). Infected aphids serve as an additional source of inoculum, and the conidia disperse very well in glasshouse situations (Hall and Burges 1979).

The fungus *Hirsutella thompsonii* is a specific fungal pathogen of Acarina, especially eriophyid and tetranychid mites. Much research with this fungus has been conducted against the citrus rust mite, *Phyllocoptruta oleivora*. An epizootic reducing the mite populations to low levels occurs naturally in Florida citrus groves, usually from July through December. Unfortunately, fungicides to control the greasy leafspot disease on citrus decrease the natural inoculum of *H. thompsonii*, resulting in secondary outbreaks of the citrus rust mite. Consequently, a practical method of reintroducing *H. thompsonii* in orange groves is needed.

Since 1969, McCoy and his co-workers (see McCoy 1981) have conducted laboratory and field tests with *H. thompsonii* with the objective of initiating epizootics in citrus rust mite populations. The results have varied because of climatic conditions, but at high humidity, the application of fragmented mycelia of *H. thompsonii* significantly reduces high citrus rust mite populations on fruit and foliage for 6 to 12 months. A wettable powder conidial formulation contains a carrier that functions as a substrate for fungal growth. After application, sporulation occurs on this substrate and increases the inoculum in the field. When environmental conditions are optimal for fungal growth, the mite population is reduced within 4 weeks. The future of *H. thompsonii* as a mycoacaricide is dependent on the development of formulations that protect the conidia or mycelia from environmental stresses and enable them to persist in the environment.

Another fungus offering promise in long-term control is *Nomuraea rileyi*, a pathogen of several noctuid pests of pasture, row crops, and grain. Epizootics have been initiated in the soybean ecosystem by the direct application of conidia (Ignoffo et al. 1976) or the distribution of pieces of larvae killed by the fungus (Sprenkel and Brooks 1975). These approaches induce epizootics against foliage feeding noctuid larvae earlier in the soybean growing season and reduce damage. This application is only for one growing season, and reapplication is necessary the following season.

In the aquatic ecosystem, the fungus *Lagenidium giganteum* has been effective against several species of mosquito larvae. The zoospore of this oomycete is short-lived, but the continual presence of hosts in stagnant water allows fungal recycling (Guzman and Axtell 1987). The oospore, on the other hand, can survive seasonal drought and persist in the environment (Washino 1982; Jaronski and Axtell 1983). Moreover, this ability to persist indicates that the fungus has the potential for long-term control in some mosquito breeding sites.

D. PROTOZOA

Of all protozoans infecting arthropods, one of the most common is the microsporidia. Most microsporidia have a chronic, debilitating effect on their hosts, and consequently, the period of lethal infection is long. Brooks (1988) has reviewed the use of protozoans for microbial control.

Only one microsporidium, *Nosema locustae*, is registered to control insects in the United States. *Nosema locustae* can provide long-term control against grasshoppers and crickets on rangeland and pasture (Henry and Oma 1981). When 25–50% of a pasture was sprayed, up to 40% of the grasshoppers were infected. In the following year, 40% of the progeny over a widespread area were infected indicating that good spread and transmission of the *Nosema* had occurred. The *Nosema* probably spread by transovum or transovarial transmission or by cannibalism of diseased individuals (Henry 1972). Although the effect of *N. locustae* infection is debilitating rather than rapid mortality, the infection reduces fertility of the sperm in males and fecundity in females (Henry and Oma 1981). Thus, the potential for long-term control of grasshoppers and crickets on rangeland and pasture appears feasible.

Nosema fumiferanae introduced into populations of the spruce budworm has remained at high levels (Wilson 1981). Similarly, *Nosema pyrausta*, a naturally occurring microsporidium of the European corn borer, when applied as a foliar spray, reduces the number of larvae per plant in the year of the application; it also suppresses host populations the following year. Both microsporidia are transmitted transovarially.

E. NEMATODES

Long-term control has been accomplished in New Zealand and Australia with the nematode *Deladenus* (syn. *Beddingia*) *siricidicola* against the wood wasp, *Sirex noctilio*, an introduced Eurasian pest of Monterey pine, *Pinus radiata*. The control of the wood wasp was complicated by the association of a mutualistic fungus, *Amylostereum areolatum*. *S. noctilio* females attack living pine trees to oviposit and implant toxic mucus as well as spores of *A. areolatum* (Talbot 1977). The mucus and fungus kill the trees, and wood wasp larvae feed on the wood containing the fungal mycelia. The life cycle of the wood wasps may take 1, 2, or 3 yr.

D. siricidicola was initially found in New Zealand and probably originated from Europe (Bedding 1984b). The nematode does not kill its wood wasp host but rather sterilizes the female wood wasp by invading the ovaries. These female wood wasps deposit eggs containing nematodes into trees with healthy wood wasp larvae. Because the nematode has a mycetophagous life cycle and a parasitic one, it can persist in the environment (Bedding 1984b). Bedding and Akhurst (1974) developed mass-production techniques for *Deladenus* by growing the nematodes in flasks using wheat and water as a substrate for *A. areolatum*. The nematodes are harvested and placed into a gelatin solution in oxygenated water. This solution forms a foam and allows for a high degree of nematode survival. The foam containing the nematode is injected with a syringe into holes in trees made with a special tool. This procedure resulted in a 99% infection rate of female *S. noctilio* in treated trees. Zondag (1979) stated that *D. siricidicola* is the most important biological-control agent of the wood wasp on the North Island of New Zealand, with infection levels as high as 90%. The nematode reduces the reproductive potential of *S. noctilio* and in conjunction with other introduced parasitoids and predators continues to keep the population of the wood wasp below the economic threshold.

The mosquito nematode *Romanomermis culicivorax* can recycle in the environment (Petersen 1984). However, the full potential of this nematode as a long-term control agent of mosquitoes has not yet been realized.

XIV. EXAMPLES OF SHORT-TERM CONTROL

Short-term control with pathogens has been accomplished against pests in the forest and agricultural ecosystems as well as those of public health importance. Much of the successful and practical short-term control has been with highly virulent pathogens such as *B. thuringiensis* against lepidopteran and dipteran insects and baculoviruses against sawflies and lepidopteran insects. Successful use of these pathogens is based on favorable economic and efficacy comparisons with chemical pesticides. Experimentally, short-term control has been demonstrated with fungi, protozoa, baculoviruses of Lepidoptera, and nematodes. These pathogens, some of which are registered with EPA, have not been extensively used as microbial-control agents because of the inconsistent results, the failure to obtain comparable results with chemical pesticides, the cost of producing and marketing pathogens, and the host specificity of many pathogens. Other reasons for their lack of acceptance may be the disadvantages of microbial-control agents outlined earlier in this chapter.

A. VIRUSES

Short-term control has been accomplished with the NPVs of sawflies, such as the loblolly pine sawfly, *Neodiprion taedae linearis* (Yearian et al. 1973), and the jack pine sawfly, *N. pratti banksianae* (Bird 1955, 1961). When the NPV is applied against early instars, larval mortalities are in excess of 95% for both species, especially at high application rates.

The occurrence of natural viral epizootics in populations of forest lepidopteran defoliators has masked the results of field trials for short- or long-term control. Against gypsy moth, the reduction of egg masses and some foliage protection have been recorded in NPV-treated plots (Yendol et al. 1977; Wollam et al. 1978). Generally, the treated plots have provided acceptable levels of foliage protection (Cunningham 1982). For successful use of the gypsy moth NPV, two aerial applications of 2.5×10^{11} polyhedral occlusion bodies/ha are recommended (Cunningham 1982). The first should be applied when the leaves are partially expanded and the insects are in the first and second instars. The second application should be made 5–10 days after the first but before the larvae reach the fourth instar.

Tests conducted with the multinucleocapsid strain of the Douglas-fir tussock moth NPV have provided foliage protection and population reduction (Stelzer et al. 1977). On the other hand, the NPV of the eastern spruce budworm, *Choristoneura fumiferana*, has caused high population reduction but little to no foliage protection in the year of application (Cunningham 1982). The reasons for the lack of foliage protection are unknown but may be related to finding an efficient delivery method and the timing of viral application, a more virulent strain of virus, a virus that persists better in the environment, as well as other factors.

The application of baculoviruses in agroecosystems has shown considerable promise in field tests but has not attained the desired degree of success or acceptance by the growers. Currently, only one baculovirus, the NPV of *Heliothis* spp., is registered for use in agriculture. This NPV has been extensively tested in cotton with some testings in corn, tobacco, and grain.

Results in the 1960s with *Heliothis* NPV on cotton were not consistent. In some situations, the results were comparable with standard insecticides when applied at 2- to 5-day intervals (Ignoffo et al. 1965); in other tests, the NPV treatments resulted in increased cotton yields over the untreated plots but less than the standard insecticides (McGarr 1968). Promising results were obtained in corn, especially when *Heliothis* adults laid their eggs on corn silk to which the NPV was applied (Tanada and Reiner 1962). Although the results were not always consistent, the NPV showed sufficient potential that private industry began the development of *Heliothis* NPV as a microbial insecticide in 1966. Testing of various NPV formulations produced by industry continued into the 1970s with results that varied from adequate to excellent (Andrews et al. 1975). New formulations of the NPV were eventually developed that exhibited better shelf life and greater activity than earlier formulations.

Treatments with these new viral formulations gave consistently good cotton yields over the control except when the *Heliothis* populations were high. In such situations, the NPV was usually less effective than the chemical standard (Yearian et al. 1980). Unfortunately, grower reluctance, poor economics, and the restricted host range of the NPV resulted in the eventual withdrawal of this product from the market.

Other NPVs, such as those of the velvetbean caterpillar, *Anticarsia gemmatalis*, the beet armyworm, *Spodoptera exigua*, and the soybean looper, *Pseudoplusia includens*, have shown promise for microbial control in agriculture (Yearian and Young 1982). More tests are required to determine if they can become commercially feasible.

A NPV that has a wide host range will be attractive as a microbial insecticide. Most NPVs found in nature have a restricted host range, but Vail and Jay (1973) showed that the NPV isolated from the alfalfa looper, *Autographa californica*, has a wide host range within Lepidoptera. A higher concentration is often necessary for infection with this virus against many pest species.

GV offers potential for the microbial control of the codling moth in apple orchards. Falcon et al. (1968) demonstrated that repeated applications of the GV against the first instar codling moth protected the fruit. Similar results were obtained by Huber and Dickler (1977) and Jaques et al. (1981). Unlike many GVs, the one from codling moth kills its insect host very rapidly and makes it a more effective microbial-control agent (Tanada 1964). This GV is registered for use in some European countries.

B. BACTERIA

Bacillus thuringiensis is a classical example of a microbial insecticide. It is being produced by a number of commercial companies for use against pests of medical and veterinary importance, stored products, agriculture, and forestry. *B. thuringiensis* subspecies *israelensis* (Bti) is effective against larval mosquitoes (Goldberg and Margalit 1977) and black flies (Gaugler and Finney 1982); *B. thuringiensis* subspecies *tenebrionis* against Colorado potato beetle, *Leptinotarsa decemlineata* (Zehnder and Gelernter 1989); and *B. thuringiensis* subspecies *kurstaki* (Btk) against many lepidopteran larvae. In the United States, Btk has been used against agricultural and forestry pests. In other countries, different subspecies of *B. thuringiensis* are used against lepidopteran pests (Lüthy et al. 1982). For example, in Russia the subspecies *thuringiensis*, *galleriae*, and *dendrolimus* are being used for insect control.

Although Bti is active against many mosquito species, certain genera of mosquitoes are more susceptible to this bacterium than others [e.g., *Culex*, *Aedes*, and *Uranotaenia* are more susceptible than *Anopheles* (Goldberg and Margalit 1977)]. Other factors that influence the efficacy of Bti include the instar of the target mosquito larva, the concentration of the active ingredient in the water, and the water quality. Early instars are more susceptible than older ones, higher concentrations and longer exposure times increase host susceptibility, and free chlorine and water pollution decrease bactericidal activity (Davidson 1982). Salinity, pH between 4.3 and 10, and sunlight have little effect on the pathogen.

A number of field studies with commercial preparations of Bti demonstrate that mosquitoes can be effectively controlled. Species occurring in salt marshes, pastures, ponds, and rice fields have been controlled with various Bti formulations.

Field applications against black fly larvae have been effective (Gaugler and Finney 1982). The Bti suspensions or formulated materials are placed into a flowing stream and the effectiveness of the bacterial preparations is evaluated downstream. Using this method, black fly larvae have been suppressed in streams and rivers immediately downstream. In West Africa, Lacey et al. (1982) obtained 100% reduction of *Simulium damnosum* 19 km downriver from the point of application and a partial reduction for an additional 15 km. The greatest population reduction is obtained nearest the site of application and less control is achieved as the distance from the point of application increases. The lack of control at further distances is related in part to the dilution of the bacterial suspension in the stream water and the flow rate of the stream (Gaugler and Finney 1982).

Bacillus thuringiensis subspecies *tenebrionis* is most effective against first and second instar larvae of the Colorado potato beetle (Zehnder and Gelernter 1989). When exposed to the bacterium for 24 h, the third instar larvae and adult beetles stop feeding but are not killed. These stages resume feeding within 48 to 72 h after treatment. In the field, spray application will have optimum effect when small larvae are the predominant life stage (Fig. 15-3).

Bacillus thuringiensis subspecies *kurstaki* (Btk) is widely used against lepidopteran insects on various crops. A few examples include *Colias* on alfalfa, *Pieris* and *Trichoplusia* on crucifers, *Manduca* on tomato and tobacco, *Heliothis* on cotton and tomato, and *Platynota* and *Argyrotaenia* on grape. Because Btk has no tolerance levels, it can be used shortly before crop harvest without any restrictions and is applied as a cleanup spray for lepidopteran insects. Btk is also applied early in the cropping season to reduce damage by lepidopteran pests and to prevent the occurrence of secondary pest outbreaks by protecting parasitoids and predators of these pests. During the course of the growing season, the target pest populations may increase with the influx of dispersing insects. Then, IPM strategies are used to reduce pest numbers to levels below the economic threshold.

The use of Btk in forestry has been reviewed by Morris (1982). Forests are stable ecosystems with diverse communities that are easily upset by broad-spectrum chemical pesticides. Btk reduces the impact upon these communities, but nontarget lepidopteran insects are affected (Miller 1990).

Btk is effective in the forest against a number of Lepidoptera. In Canada and the United States, it is applied against the spruce budworm over large acreage by aircraft (Morris 1982). Similarly, it is used against the gypsy moth throughout much of its range. In the early 1970s, Btk was not efficacious against these insects, but improved formulations, increasing concentrations, and a better understanding of application technology have provided foliage protection. In addition to the above insects, Btk is effective against other forest pests, including geometrids, tent caterpillars, lymantriids, arctiids, and tortricids and has been used successfully to eradicate the gypsy moth from infestations in California (Brown et al. 1984).

FIGURE 15-3

Colorado potato beetle larvae. The top larva is healthy and the bottom larva has been killed by *Bacillus thuringiensis* subspecies *tenebrionis*. (Provided by W. D. Gelernter, Mycogen Corporation, San Diego, California.)

C. FUNGI

The use of entomogenous fungi as biological insecticides has been recently reviewed (McCoy et al. 1988; McCoy 1990). In Brazil, *Metarhizium anisopliae* has been used effectively against spittlebugs on sugar cane and the costs are lower than the usual chemical treatments (Ferron 1981). *Beauveria bassiana*, *B. brongniartii*, and *M. anisopliae* also have potential use against soil-inhabiting insects, such as curculionids, scarabs, and lepidopterons (Ferron 1981).

Verticillium lecanii has controlled aphids and whiteflies on cucumbers for 2 weeks with one application (Hall 1981). However, unlike the long-term control situation on chrysanthemums, the fungus does not spread and reapplication is necessary. Application of *Entomophthora* spp. produced *in vitro* against aphids and mites on glasshouse plants resulted in 95% mortality within 24 h, suggesting the presence of fungal toxins (Wilding 1981).

The application of *Nomuraea rileyi* has not resulted in economic control of several lepidopteran pests (Ignoffo 1981). Larvae, such as *Trichoplusia ni*, *Heliothis zea*, and *H. virescens*, have been treated with this fungus in the field. Although significant reduction of larvae is attained, the fungus fails to protect the crops from economic damage.

D. PROTOZOA

Short-term control of insects with protozoans has shown, in general, that infection will occur, but population reduction is not sufficient or rapid enough to prevent damage or the results are not consistent (Brooks 1988). For example, *Vairimorpha necatrix* has been applied against the tobacco budworm and corn earworm with mixed results (Fuxa and Brooks 1979). Aqueous sprays do not adequately reduce the population density or damage to plants even though a majority of the insects are infected with the *Vairimorpha*. Conversely, a cornmeal formulation of spores is comparable with chemical pesticides.

Nosema locustae has reduced grasshopper populations by 50% within 4 weeks after application and has caused 30–50% infection in the surviving population (Henry 1971). *Nosema* (syn. *Glugea*) *gasti* and *Mattesia grandis* have been successfully introduced in a bait formulation into field populations of the boll weevil and have infected a high percentage of the weevils (McLaughlin et al. 1969). In addition, *Nosema algerae* has been field tested against mosquito larvae (Anthony et al. 1978). This microsporidium has infected 86% of the mosquitoes, indicating that it has potential as a short-term microbial insecticide.

E. NEMATODES

Kaya (1985) and Klein (1990) recently reviewed nematodes as short-term control agents. Field applications of steinernematids and heterorhabditids have shown high efficacy for controlling pest species when applied under favorable conditions. These conditions include the application of these nematodes in soil and cryptic habitats where ample moisture occurs. Although moisture is required for survival, the use of these nematodes in aquatic situations has not been efficacious because they are not adapted to this environment. Foliar applications are usually not successful even when adequate moisture is present (Begley 1990).

Steinernema spp. are effective against plant-boring insects (Bedding and Miller 1981a; Miller and Bedding 1982; Lindegren and Barnett 1982). *Steinernema carpocapsae* has controlled carpenterworms in fig trees and *S. feltiae* (syn. *bibionis*) against sesiids in currants. *Heterorhabditis* is effective against black vine weevils in potted plants (Bedding and Miller 1981b).

XV. EXAMPLES OF THE IPM APPROACH

The compatibility of microbial agents with other control tactics makes them excellent candidates for integrated pest management schemes (Jaques and Morris 1981). *B. thuringiensis* is compatible with many chemical pesticides, other pathogens, pheromones, or parasitoids and predators (Tanada 1984). In field studies, it has been mixed with the organophosphate insecticide, acephate, for the control of the spruce budworm (Morris 1977b). This combination has provided higher mortality of the spruce budworm than the bacterium alone whereas acephate has given no control at all. Smirnoff et al. (1973) have recommended the combination of *B. thuringiensis* and chitinase against the spruce budworm, but the results of such combinations have been mixed (Morris 1982). Against the Colorado potato beetle in potatoes, *B. thuringiensis* subspecies *tenebrionis* has been effectively integrated with low rates of aldicarb (Ferro and Gelernter 1989). Aldicarb, which is a systemic insecticide, controls the leaf hoppers and aphids, and the bacterium controls the Colorado potato beetle. However, the timing of application must coincide with egg hatch or when young instars are predominant because older instars and adult beetles are not effectively controlled by the bacterium.

Combinations of *B. thuringiensis* and entomopathogenic viruses have produced conflicting results. *B. thuringiensis*–NPV combination against the Douglas-fir tussock moth did not increase mortality when compared with the bacterium or NPV alone (Stelzer et al. 1975). In Japan, *B. thuringiensis*–cytoplasmic polyhedrosis virus combination produced higher larval mortality of *Dendrolimus spectabilis* than either agent alone (Katagiri and Iwata 1976). When *B. thuringiensis* was applied and the insect parasitoid *Cotesia melanoscela* (syn. *Apanteles melanoscelus*) was released into plots infested with gypsy moth, the combined treatments consistently provided greater population reduction and foliage protection than either agent alone (Wollam and Yendol 1976). Moreover, Weseloh et al. (1983) demonstrated that plots sprayed with the bacterium against the gypsy moth had a high natural prevalence of parasitism with *A. melanoscelus* than unsprayed plots. They attributed these findings to sublethal effects of *B. thuringiensis* upon gypsy moth larvae that remained at a younger instar.

When the organophosphate insecticide, fenitrothion, was applied first against the spruce budworm followed by an application of NPV on the same day, some foliage protection and population reduction were obtained with NPV alone and NPV plus the chemical (Morris 1977a). When the plots were checked the following year, foliage protection and population reduction were also noted in the NPV alone and NPV plus chemical treatments.

Baculoviruses are compatible with many chemical pesticides (Yearian and Young 1982). The efficacy of *T. ni* and *H. zea* NPVs is not adversely affected by most carbamates, organophosphates, and chlorinated hydrocarbons.

The combination of chemical pesticides and microsporidia has not received much attention. The carbamate, carbaryl, and *Nosema locustae* have been used in combination against grasshoppers (Henry and Oma 1981). Carbaryl is applied first as a bait and then followed with bait application of *Nosema*. The population of the grasshopper is reduced by the carbaryl bait, and the residual grasshoppers are infected or killed by the microsporidium.

Verticillium lecanii and *Hirsutella thompsonii* appear to be compatible with parasitic and predatory arthropods (Hall 1981; McCoy 1981). Although many chemical insecticides and fungicides adversely affect the conidia germination of *V. lecanii*, *H. thompsonii*, and *N. rileyi* under laboratory conditions, the fungi can be integrated with some pesticides. For example, by waiting 24 h after pesticide application, *Verticillium* can be effectively used against aphid pests on chrysanthemums (Hall 1981). In general, the effects of the pesticides are less in the field than in the laboratory.

Nematodes are compatible with most chemical and biological insecticides (Kaya 1985). As expected, chemical insecticides that have nematicidal or nematostatic activity can impair the nematode's ability to infect its host (Hara and Kaya 1983). Juvenile hormones or their analogs do not have an adverse effect on *Romanomermis culicivorax* (Petersen 1984). Bari and Kaya (1984) applied a combination of *B. thuringiensis* and *Steinernema carpocapsae* against the artichoke plume moth. The nematode alone treatment provided better results than the bacterium alone or bacterium plus nematode treatment, suggesting that antagonism between the two biological-control agents occurred.

XVI. DESTRUCTION OF SYMBIONTS

The possibility of insect control by the destruction of their symbionts is not yet feasible, but the removal of symbionts has detrimental effects on the host (Krieg 1971). There are a number of methods to obtain aposymbiotic hosts, such as chemotherapy, extirpation of mycetomes, and thermoinactivation. Another method is to immunize vertebrates against symbionts of blood-feeding insects. When the insects feed on these vertebrates, they become aposymbiotic (Nogge 1978).

The most practical method of using this approach for insect control is chemotherapy (Krieg 1971). Chemicals, such as sodium hypochlorite and mercuric chloride, have been used against insects with symbiotic microorganisms on the external surface of eggs. Antibiotics have been used for insects with internal symbionts although this approach must remove all symbionts to be effective. Antibiotics for insect control may not be an effective method because their use may lead to insects with resistant symbionts, get into human or livestock food chains, and result in other resistant microorganisms.

REFERENCES

Akhurst, R. J. 1990. Safety to nontarget invertebrates of nematodes of economically important pests. *In* "Safety of Microbial Insecticides." (M. Laird, L. A. Lacey, and E. W. Davidson, eds.), pp. 233–240. CRC Press, Boca Raton, Florida.

Andrews, G. L., and Sikorowski, P. P. 1973. Effects of cotton leaf surfaces on the nuclear polyhedrosis virus of *Heliothis zea* and *Heliothis virescens* (Lepidoptera: Noctuidae). *J. Invertebr. Pathol. 22*, 290–291.

Andrews, G. L., Harris, F. A., Sikorowski, P. P., and McLaughlin, R. E. 1975. Evaluation of *Heliothis* nuclear polyhedrosis virus in a cottonseed oil bait for control of *Heliothis virescens* and *H. zea* on cotton. *J. Econ. Entomol. 68*, 87–90.

Anthony, D. W., Savage, K. E., Hazard, E. I., Avery, S. W., Boston, M. D., and Oldacre, S. W. 1978. Field tests with *Nosema algerae* Vavra and Undeen (Microsporida, Nosematidae) against *Anopheles albimanus* Wiedemann in Panama. *Misc. Publ. Entomol. Soc. Am. 11*, 17–27.

Baker, R. R., and Dunn, P. E. 1990. "New Directions in Biological Control." Alan R. Liss, New York.

Bari, M. A., and Kaya, H. K. 1984. Evaluation of the entomogenous nematode *Neoaplectana carpocapsae* (syn. *Steinernema feltiae*) Weiser (Rhabditida: Steinernematidae) and the bacterium *Bacillus thuringiensis* Berliner var. *kurstaki* for suppression of the artichoke plume moth (Lepidoptera: Pterophoridae). *J. Econ. Entomol. 77*, 225–229.

Bedding, R. A. 1984a. Large scale production, storage and transport of the insect-parasitic nematodes *Neoaplectana* spp. and *Heterorhabditis*. *Ann. Appl. Biol. 104*, 117–120.

Bedding, R. A. 1984b. Nematode parasites of Hymenoptera. *In* "Plant and Insect Nematodes." (W. R. Nickle, ed.), pp. 755–795. Marcel Dekker, New York.

Bedding, R. A., and Akhurst, R. J. 1974. Use of the nematode *Deladenus siricidicola* in the biological control of *Sirex noctilio* in Australia. *J. Austr. Entomol. Soc. 13*, 129–135.

Bedding, R. A., and Miller, L. A. 1981a. Disinfesting blackcurrant cuttings of *Synanthedon tipuliformis*, using the insect parasitic nematode, *Neoaplectana bibionis*. *Environ. Entomol. 10*, 449–453.

Bedding, R. A., and Miller, L. A. 1981b. Use of a nematode, *Heterorhabditis heliothidis*, to control black vine weevil, *Otiorhynchus sulcatus*, in potted plants. *Ann. Appl. Biol. 99*, 211–216.

Bedford, G. O. 1980. Biology, ecology, and control of palm rhinoceros beetles. *Annu. Rev. Entomol. 25*, 309–339.

Bedford, G. O. 1981. Control of the rhinoceros beetle by baculovirus. In "Microbial Control of Pests and Plant Diseases 1970–1980." (H. D. Burges, ed.), pp. 409–426. Academic Press, New York.

Begley, J. W. 1990. Efficacy against insects in habitats other than soil. In "Entomopathogenic Nematodes in Biological Control." (R. Gaugler and H. K. Kaya, eds.), pp. 215–231. CRC Press, Boca Raton, Florida.

Betz, F. S., Forsyth, S. F., and Stewart, W. E. 1990. Registration requirements and safety considerations for microbial pest control agents in North America. In "Safety of Microbial Insecticides." (M. Laird, L. A. Lacey and E. W. Davidson, eds.), pp. 3–10. CRC Press, Boca Raton, Florida.

Bird, F. T. 1953. The use of a virus disease in the biological control of the European pine sawfly, *Neodiprion sertifer* (Geoffr.). *Can. Entomol. 85*, 437–446.

Bird, F. T. 1955. Virus diseases of sawflies. *Can. Entomol. 87*, 124–127.

Bird, F. T. 1961. Transmission of some insect viruses with particular reference to ovarial transmission and its importance in the development of epizootics. *J. Insect Pathol. 3*, 352–380.

Bird, F. T., and Burk, J. M. 1961. Artificially disseminated virus as a factor controlling the European spruce sawfly, *Diprion hercyniae* (Htg.) in the absence of introduced parasites. *Can. Entomol. 93*, 228–238.

Briese, D. T. 1986a. Host resistance to microbial control agents. In "Biological Plant and Health Protection." (J. M. Franz, ed.), Vol. 32, pp. 233–256. G. Fisher Verlag, New York.

Briese, D. T. 1986b. Insect resistance to baculoviruses. In "The Biology of Baculoviruses." (R. R. Granados and B. A. Federici, eds.), Vol. 2, pp. 237–263. CRC Press, Boca Raton, Florida.

Brooks, W. M. 1973. Protozoa: Host–parasite–pathogen interrelationships. *Misc. Publ. Entomol. Soc. Am. 9*, 105–111.

Brooks, W. M. 1988. Entomogenous protozoa. In "Handbook of Natural Pesticides, Microbial Pesticides Part A. Entomogenous Protozoa and Fungi." (C. M. Ignoffo, ed.), Vol. 5, pp. 1–149, CRC Press, Boca Raton, Florida.

Brown, L. R., Kaya, H. K., Reardon, R. C., and Fusco, R. A. 1984. The Santa Barbara gypsy moth eradication effort. *Calif. Agric. 38* (3/4), 4–7.

Burges, H. D. 1967. The standardization of products based on *Bacillus thuringiensis*. Proc. Int. Colloq. Insect Pathol. Microbial Control, Wageningen, The Netherlands, 1966, pp. 306–312. North-Holland Publ. Co., Amsterdam.

Burges, H. D. 1971. Possibilities of pest resistance to microbial control agents. In "Microbial Control of Insects and Mites." (H. D. Burges and N. W. Hussey, eds.), pp. 445–457. Academic Press, New York.

Burges, H. D. 1981a. "Microbial Control of Pests and Plant Diseases 1970–1980." Academic Press, New York.

Burges, H. D. 1981b. Safety, safety testing and quality control of microbial pesticides. In "Microbial Control of Pests and Plant Diseases 1970–1980." (H. D. Burges, ed.), pp. 737–767. Academic Press, New York.

Burges, H. D., and Hussey, N. W. 1971. "Microbial Control of Insects and Mites." Academic Press, New York.

Carruthers, R. I., and Hural, K. 1990. Fungi as naturally occurring entomopathogens. In "New Directions in Biological Control." (R. R. Baker and P. E. Dunn, eds.), pp. 115–138. Alan R. Liss, New York.

Costa, S. D., and Gaugler, R. R. 1989a. Sensitivity of *Beauveria bassiana* to solanine and tomatine: Plant defensive chemicals inhibit an insect pathogen. *J. Chem. Ecol. 15*, 697–706.

Costa, S. D., and Gaugler, R. 1989b. Influence of *Solanum* host plants on Colorado potato beetle (Coleoptera: Chrysomelidae) susceptibility to the entomopathogen *Beauveria bassiana*. *Environ. Entomol.* 18, 531–536.

Couch, T. L., and Ignoffo, C. M. 1981. Formulation of insect pathogens. In "Microbial Control of Pests and Plant Diseases 1970–1980." (H. D. Burges, ed.), pp. 621–634. Academic Press, New York.

Couch, T. L., and Ross, D. A. 1980. Production and utilization of *Bacillus thuringiensis*. *Biotechnol. Bioeng.* 22, 1297–1304.

Cunningham, J. C. 1982. Field trials with baculoviruses: Control of forest insect pests. In "Microbial and Viral Pesticides." (E. Kurstak, ed.), pp. 335–386. Marcel Dekker, New York.

Cunningham, J. C., and Entwistle, P. F. 1981. Control of sawflies by baculovirus. In "Microbial Control of Pests and Plant Diseases 1970–1980." (H. D. Burges, ed.), pp. 379–407. Academic Press, New York.

Davidson, E. W. 1982. Bacteria for the control of arthropod vectors of human and animal disease. In "Microbial and Viral Pesticides." (E. Kurstak, ed.), pp. 289–315. Marcel Dekker, New York.

Dougherty, E. M., Cantwell, G. E., and Kuchinski, M. 1982. Biological control of the greater wax moth (Lepidoptera: Pyralidae), utilizing in vivo- and in vitro-propagated baculovirus. *J. Econ. Entomol.* 75, 675–679.

Dulmage, H. T. 1975. The standardization of formulations of the δ-endotoxins produced by *Bacillus thuringiensis*. *J. Invertebr. Pathol.* 25, 279–281.

Dulmage, H., and Burgerjon, A. 1977. Industrial and international standardization of microbial pesticides—II. Insect viruses. *Entomophaga* 22, 131–139.

Dulmage, H. T., et al. 1981. Insecticidal activity of isolates of *Bacillus thuringiensis* and their potential for pest control. In "Microbial Control of Pests and Plant Diseases 1970–1980." (H. D. Burges, ed.), pp. 193–222. Academic Press, New York.

Dunbar, D. M., and Beard, R. L. 1975. Present status of milky disease of Japanese and oriental beetles in Connecticut. *J. Econ. Entomol.* 68, 453–457.

Dutky, S. R. 1940. Two new spore-forming bacteria causing milky diseases of Japanese beetle larvae. *J. Agric. Res.* 61, 57–68.

Dutky, S. R. 1963. The milky diseases. In "Insect Pathology: An Advanced Treatise." (E. A. Steinhaus, ed.), Vol. 2, pp. 75–115. Academic Press, New York.

Entwistle, P. F., Adams, P. H. W., and Evans, H. F. 1977. Epizootiology of a nuclear-polyhedrosis virus in European spruce sawfly, *Gilpinia hercyniae*: Birds as dispersal agents of the virus during winter. *J. Invertebr. Pathol.* 30, 15–19.

Evans, H. F., and Entwistle, P. F. 1982. Epizootiology of the nuclear polyhedrosis virus of European spruce sawfly with emphasis on persistence of virus outside the host. In "Microbial and Viral Pesticides." (E. Kurstak, ed.), pp. 449–461. Marcel Dekker, New York.

Falcon, L. A. 1971. Microbial control as a tool in integrated control programs. In "Biological Control." (C. B. Huffaker, ed.), pp. 346–364. Plenum, New York.

Falcon, L. A. 1985. Development and use of microbial insecticides. In "Biological Control in Agricultural IPM Systems." (M. A. Hoy and D. C. Herzog, eds.), pp. 229–242. Academic Press, New York.

Falcon, L. A., Kane, W. R., and Bethell, R. S. 1968. Preliminary evaluation of a granulosis virus for control of the codling moth. *J. Econ. Entomol.* 61, 1208–1213.

Feitelson, J. S., Quick, T. C., and Gaertner, F. 1990. Alternate hosts for *Bacillus thuringiensis* delta-endotoxin genes. In "New Directions in Biological control." (R. R. Baker and P. E. Dunn, eds.), pp. 561–571. Alan R. Liss, New York.

Felton, G. W., and Dahlman, D. L. 1984. Allelochemical induced stress: Effects of L-canavanine on the pathogenicity of *Bacillus thuringiensis* in *Manduca sexta*. *J. Invertebr. Pathol.* 44, 187–191.

Felton, G. W., and Duffey, S. S. 1990. Inactivation of baculovirus by quinones formed in insect-damaged plant tissues. *J. Chem. Ecol. 16*, 1221–1236.

Felton, G. W., Duffey, S. S., Vail, P. V., Kaya, H. K., and Manning, J. 1987. Interaction of nuclear polyhedrosis virus with catechols: Potential incompatibility for host-plant resistance against noctuid larvae. *J. Chem. Ecol. 13*, 947–957.

Ferro, D. N., and Gelernter, W. D. 1989. Toxicity of a new strain of *Bacillus thuringiensis* to Colorado potato beetle (Coleoptera: Chrysomelidae). *J. Econ. Entomol. 82*, 750–755.

Ferron, P. 1981. Pest control by the fungi Beauveria and Metarhizium. *In* "Microbial Control of Pests and Plant Diseases 1970–1980." (H. D. Burges, ed.), pp. 465–482. Academic Press, New York.

Fleming, W. E. 1968. Biological control of the Japanese beetle. *U.S. Dep. Agric. Tech. Bull. No. 1383.*

Friedman, M. J. 1990. Commercial production and development. *In* "Entomopathogenic Nematodes in Biological Control." (R. Gaugler and H. K. Kaya, eds.), pp. 153–172. CRC Press, Boca Raton, Florida.

Fuxa, J. R. 1987. Ecological considerations for the use of entomopathogens in IPM. *Annu. Rev. Entomol. 32*, 225–251.

Fuxa, J. R. 1990a. Environmental risks of genetically engineered entomopathogens. *In* "Safety of Microbial Insecticides." (M. Laird, L. A. Lacey, and E. W. Davidson, eds.), pp. 203–207. CRC Press, Boca Raton, Florida.

Fuxa, J. R. 1990b. New directions for insect control with baculoviruses. *In* "New Directions in Biological Control." (R. R. Baker and P. E. Dunn, eds.), pp. 97–113. Alan R. Liss, New York.

Fuxa, J. R., and Brooks, W. M. 1979. Effects of *Vairimorpha necatrix* in sprays and corn meal on *Heliothis* species in tobacco, soybeans, and sorghum. *J. Econ. Entomol. 72*, 462–467.

Fuxa, J. R., Mitchell, F. L., and Richter, A. R. 1988. Resistance of *Spodoptera frugiperda* (Lep.: Noctuidae) to a nuclear polyhedrosis virus in the field and laboratory. *Entomophaga 33*, 55–63.

Gard, I. M., and Falcon, L. A. 1978. Auto-dissemination of entomopathogens: Viruses. *In* "Microbial Control of Insect Pests: Future Strategies in Pest Management Systems." (G. E. Allen, C. M. Ignoffo, and R. P. Jaques, eds.), pp. 46–54. N.S.F.-U.S. Dep. Agric. Univ. Fl. Workshop.

Gaugler, R., and Finney, J. R. 1982. A review of *Bacillus thuringiensis* var. *israelensis* (Serotype 14) as a biological control agent of black flies (Simuliidae). *Misc. Publ. Entomol. Soc. Am. 12*(4), 1–17.

Georgis, R. 1990. Formulation and application technology. *In* "Entomopathogenic Nematodes in Biological Control." (R. Gaugler and H. K. Kaya, eds.), pp. 173–191. CRC Press, Boca Raton, Florida.

Goettel, M. S., Poprawski, T. J., Vandenberg, J. D., Li, Z., and Roberts, D. W. 1990. Safety to nontarget invertebrates of fungal biocontrol agents. *In* "Safety of Microbial Insecticides." (M. Laird, L. A. Lacey and E. W. Davidson, eds.), pp. 209–231. CRC Press, Boca Raton, Florida.

Goldberg, L. J., and Margalit, J. 1977. A bacterial spore demonstrating rapid larvicidal activity against *Anopheles sergentii*, *Uranotaenia unguiculata*, *Culex univitattus*, *Aedes aegypti* and *Culex pipiens*. *Mosq. News 37*, 355–358.

Goldman, I. F., Arnold, J., and Carlton, B. C. 1986. Selection for resistance to *Bacillus thuringiensis* subspecies *israelensis* in field and laboratory populations of the mosquito *Aedes aegypti*. *J. Invertebr. Pathol. 47*, 317–324.

Granados, R. R. 1976. Infection and replication of insect pathogenic viruses in tissue culture. *Adv. Virus Res. 20*, 189–236.

Gröner, A. 1990. Safety to nontarget invertebrates of baculoviruses. *In* "Safety of Microbial

Insecticides." (M. Laird, L. A. Lacey, and E. W. Davidson, eds.), pp. 135–147. CRC Press, Boca Raton, Florida.

Guzman, D. R., and Axtell, R. C. 1987. Population dynamics of *Culex quinquefasciatus* and the fungal pathogen *Lagenidium giganteum* (Oomycetes: Lagenidiales) in stagnant water pools. *J. Am. Mosq. Control Assoc. 3*, 442–449.

Hall, R. A. 1981. The fungus *Verticillium lecanii* as a microbial insecticide against aphids and scales. *In* "Microbial Control of Pests and Plant Diseases 1970–1980." (H. D. Burges, ed.), pp. 483–498. Academic Press, New York.

Hall, R. A., and Burges, H. D. 1979. Control of aphids in glasshouses with the fungus *Verticillium lecanii*. *Ann. Appl. Biol. 93*, 235–246.

Hamm, J. J., and Hare, W. W. 1982. Application of entomopathogens in irrigation water for control of fall armyworms and corn earworms (Lepidoptera: Noctuidae) on corn. *J. Econ. Entomol. 75*, 1074–1079.

Hara, A. H., and Kaya, H. K. 1983. Development of the entomogenous nematode, *Neoaplectana carpocapsae* (Rhabditida: Steinernematidae), in insecticide-killed beet armyworm (Lepidoptera: Noctuidae). *J. Econ. Entomol. 76*, 423–426.

Hedin, P. A., Lindig, O. H., Sikorowski, P. P., and Wyatt, M. 1978. Suppressants of gut bacteria in the boll weevil from the cotton plant. *J. Econ. Entomol. 71*, 394–396.

Heimpel, A. M., Thomas, E. D., Adams, J. R., and Smith, L. J. 1973. The presence of nuclear polyhedrosis viruses of *Trichoplusia ni* on cabbage from the market shelf. *Environ. Entomol. 2*, 72–75.

Henry, J. E. 1971. Experimental application of *Nosema locustae* for control of grasshoppers. *J. Invertebr. Pathol. 18*, 389–394.

Henry, J. E. 1972. Epizootiology of infections by *Nosema locustae* Canning (Microsporida: Nosematidae) in grasshoppers. *Acrida 1*, 111–120.

Henry, J. E., and Oma, E. A. 1974. Effect of prolonged storage of spores on field applications of *Nosema locustae* (Microsporida: Nosematidae) against grasshoppers. *J. Invertebr. Pathol. 23*, 371–377.

Henry, J. E., and Oma, E. A. 1981. Pest control by *Nosema locustae*, a pathogen of grasshoppers and crickets. *In* "Microbial Control of Pests and Plant Diseases 1970–1980." (H. D. Burges, ed.), pp. 573–586. Academic Press, New York.

Hink, W. F. 1982. Production of *Autographa californica* nuclear polyhedrosis virus in cells from large-scale suspension cultures. *In* "Microbial and Viral Pesticides." (E. Kurstak, ed.), pp. 493–506. Marcel Dekker, New York.

Hochberg, M. E., and Waage, J. K. 1991. Control engineering. *Nature 352*, 16–17.

Huber, J., and Dickler, E. 1977. Codling moth granulosis virus: Its efficiency in the field in comparison with organophosphorous insecticides. *J. Econ. Entomol. 70*, 557–561.

Ignoffo, C. M. 1966. Standardization of products containing insect viruses. *J. Invertebr. Pathol. 8*, 547–548.

Ignoffo, C. M. 1981. The fungus *Nomuraea rileyi* as a microbial insecticide. *In* "Microbial Control of Pests and Plant Diseases 1970–1980." (H. D. Burges, ed.), pp. 513–538. Academic Press, New York.

Ignoffo, C. M. 1988. "Microbial Insecticides. Part A Entomogenous Protozoa and Fungi. CRC Handbook of Natural Pesticides, Vol. 5." CRC Press, Boca Raton, Florida.

Ignoffo, C. M., and Anderson, R. F. 1979. Bioinsecticides. *In* "Microbial Technology" (H. J. Peppler and D. Perlman, eds.), Second Ed. Vol. 1, pp. 1–28. Academic Press, New York.

Ignoffo, C. M., and Hink, W. F. 1971. Propagation of arthropod pathogens in living systems. *In* "Microbial Control of Insects and Mites." (H. D. Burges and N. W. Hussey, eds.), pp. 541–580. Academic Press, New York.

Ignoffo, C. M., Chapman, A. J., and Martin, D. F. 1965. The nuclear-polyhedrosis virus of *Heliothis zea* (Boddie) and *Heliothis virescens* (Fabricius). III. Effectiveness of the virus against field populations of *Heliothis* on cotton, corn, and grain sorghum. *J. Invertebr. Pathol. 7*, 227–235.

Ignoffo, C. M., Marston, N. L., Hostetter, D. L., Puttler, B., and Bell, J. V. 1976. Natural and induced epizootics of *Nomuraea rileyi* in soybean caterpillars. *J. Invertebr. Pathol. 27*, 191–198.

Jackson, T. A. 1989. Development of *Serratia entomophila* as an inoculative biological control agent for the grass grub, *Costelytra zealandica*. Proc. Fifth Australasian Conf. Grassland Invertebr. Ecol. P.P. Stahle, ed. pp. 55–62. D & D Printing, Victoria, Australia.

Jaques, R. P. 1977. Stability of entomopathogenic viruses. *Misc. Publ. Entomol. Soc. Am. 10*, 99–116.

Jaques, R. P. 1985. Stability of insect viruses in the environment. *In* "Viral Insecticides for Biological Control." (K. Maramorosch and K. E. Sherman, eds.), pp. 285–360. Academic Press, New York.

Jaques, R. P., and Morris, O. N. 1981. Compatibility of pathogens with other methods of pest control and with different crops. *In* "Microbial Control of Pests and Plant Diseases 1970–1980." (H. D. Burges, ed.), pp. 695–715. Academic Press, New York.

Jaques, R. P., Laing, J. E., MacLellan, C. R., Proberbs, M. D., Sanford, K. H., and Trottier, R. 1981. Apple orchard tests on the efficacy of the granulosis virus of the codling moth, *Laspeyresia pomonella* (Lep.: Olethreutidae). *Entomophaga 26*, 111–117.

Jaronski, S., and Axtell, R. C. 1983. Persistence of the mosquito fungal pathogen *Lagenidium giganteum* (Oomycetes; Lagenidiales) after introduction into natural habitats. *Mosq. News 43*, 332–337.

Katagiri, K., and Iwata, Z. 1976. Control of *Dendrolimus spectabilis* with a mixture of cytoplasmic polyhedrosis virus and *Bacillus thuringiensis*. *Appl. Entomol. Zool. 11*, 363–364.

Kaya, H. K. 1976. Insect pathogens in natural and microbial control of forest defoliators. *In* "Perspectives in Forest Entomology." (J. F. Anderson and H. K. Kaya, eds.), pp. 251–263. Academic Press, New York.

Kaya, H. K. 1985. Entomogenous nematodes for insect control in IPM systems. *In* "Biological Control in Agricultural IPM Systems." (M. A. Hoy and D. C. Herzog, eds.), pp. 283–302. Academic Press, New York.

Keating, S. T., and Yendol, W. G. 1987. Influence of selected host plants on gypsy moth (Lepidoptera: Lymantriidae) larval mortality caused by a baculovirus. *Environ. Entomol. 16*, 459–462.

Keating, S. T., Yendol, W. G., and Schultz, J. C. 1988. Relationship between susceptibility of gypsy moth larvae (Lepidoptera: Lymantriidae) to a baculovirus and host plant foliage constituents. *Environ. Entomol. 17*, 952–958.

Kinsinger, R. A., and McGaughey, W. H. 1976. Stability of *Bacillus thuringiensis* and a granulosis virus of *Plodia interpunctella* on stored wheat. *J. Econ. Entomol. 69*, 149–154.

Kinsinger, R. A., and McGaughey, W. H. 1979. Susceptibility of populations of Indianmeal moth and almond moth to *Bacillus thuringiensis*. *J. Econ. Entomol. 72*, 346–349.

Klein, M. G. 1981. Advances in the use of *Bacillus popilliae* for pest control. *In* "Microbial Control of Pests and Plant Disease 1970–1980." (H. D. Burges, ed.), pp. 183–192. Academic Press, New York.

Klein, M. G. 1990. Efficacy against soil-inhabiting insect pests. *In* "Entomopathogenic Nematodes in Biological Control." (R. Gaugler and H. K. Kaya, eds.), pp. 195–214. CRC Press, Boca Raton, Florida.

Krieg, A. 1971. Aposymbiosis, a possible method for antimicrobial control of arthropods. *In* "Microbial Control of Insects and Mites." (H. D. Burges and N. W. Hussey, eds.), pp. 673–677. Academic Press, New York.

Krieg, A., Huger, A. M., and Schnetter, W. 1987. "*Bacillus thuringiensis* var. *san diego*" Stamm M-7 ist identisch mit dem zuvor in Deutschland isolierten käferwirksamen *B. thuringiensis* subsp. *tenebrionis* Stamm BI 256–82. *J. Appl. Entomol. 104*, 417–424.

Krischik, V. A., Barbosa, P., and Reichelderfer, C. F. 1988. Three trophic level interactions: Allelochemicals, *Manduca sexta* (L.), and *Bacillus thuringiensis* var. *kurstaki* Berliner. *Environ. Entomol. 17*, 476–482.

Kronstad, J. W., Schnepf, H. E., and Whiteley, H. R. 1983. Diversity of locations for *Bacillus thuringiensis* crystal protein genes. *J. Bacteriol. 154*, 419–428.

Kurstak, E. 1982. "Microbial and Viral Pesticides." Marcel Dekker, New York.

Kushner, D. J., and Harvey, G. T. 1962. Antibacterial substances in leaves: Their possible role in insect resistance to disease. *J. Insect Pathol. 4*, 155–184.

Lacey, L. A., Escaffre, H., Philippon, B., Sékétéli, A., and Guillet, P. 1982. Large river treatment with Bacillus thuringiensis (H-14) for the control of Simulium damnosum s.l. in the Onchocerciasis Control Programme. *Tropenmed. Parasitol. 33*, 97–101.

Ladd, T. L., Jr., and McCabe, P. J. 1967. Persistence of spores of *Bacillus popilliae*, the causal organism of type A milky disease of Japanese beetle larvae, in New Jersey soils. *J. Econ. Entomol. 60*, 493–495.

Laird, M., Lacey, L. A., and Davidson, E. W. 1990. "Safety of Microbial Insecticides." CRC Press, Boca Raton, Florida.

Lautenschlager, R. A., and Podgwaite, J. D. 1979. Passage of nucleopolyhedrosis virus by avian and mammalian predators of the gypsy moth, *Lymantria dispar* L. *Environ. Entomol. 8*, 210–214.

Lautenschlager, R. A., Podgwaite, J. D., and Watson, D. E. 1980. Natural occurrence of the nucleopolyhedrosis virus of the gypsy moth, *Lymantria dispar* (Lep.: Lymantriidae) in wild birds and mammals. *Entomophaga 25*, 261–267.

Leemans, J., Reynaerts, A., Hofte, H., Peferoen, M., Van Mellaert, H., and Joos, H. 1990. Insecticidal crystal proteins from *Bacillus thuringiensis* and their use in transgenic crops. *In* "New Directions in Biological Control." (R. R. Baker and P. E. Dunn, eds.), pp. 573–581. Alan R. Liss, New York.

Lindegren, J. E., and Barnett, W. W. 1982. Applying parasitic nematodes to control carpenterworms. *Calif. Agric. 36* (11/12), 7–8.

Lüthy, P., Cordier, J.-L., and Fischer, H.-M. 1982. *Bacillus thuringiensis* as a bacterial insecticide: Basic considerations and application. *In* "Microbial and Viral Pesticides." (E. Kurstak, ed.), pp. 35–74. Marcel Dekker, New York.

Maeda, S. 1989a. Expression of foreign genes in insects using baculovirus vectors. *Annu. Rev. Entomol. 34*, 351–372.

Maeda, S. 1989b. Increased insecticidal effect by a recombinant baculovirus carrying a synthetic diuretic hormone gene. *Biochem. Biophys. Res. Comm. 165*, 1177–1183.

Maeda, S., Volrath, S. L., Hanzlik, T. N., Harper, S. A., Majima, K., Maddox, D. W., Hammock, B. D., and Fowler, E. 1991. Insecticidal effects of an insect-specific neurotoxin expressed by a recombinant baculovirus. *Virology 184*, 777–780.

Magnoler, A. 1968. Laboratory and field experiments on the effectiveness of purified and non-purified nuclear polyhedral virus of *Lymantria dispar* L. *Entomophaga 13*, 335–344.

Martignoni, M. E. 1957. Contributo alla conoscenza di una granulosi di *Eucosma griseana* (Hübner) (*Tortricidae, Lepidoptera*) quale fattore limitante il pullulamento dell'insetto nella Engadina alta. *Mitt. Schweiz. Anstalt. Forstl. Versuchswesen. 32*, 371–418.

Martignoni, M. E., and Iwai, P. J. 1978. Activity standardization of technical preparations of Douglas-fir tussock moth *Baculovirus*. *J. Econ. Entomol. 71*, 473–476.

Matsumura, F. 1985. "Toxicology of Insecticides." Second Ed. Plenum, New York.

McCoy, C. W. 1981. Pest control by the fungus *Hirsutella thompsonii*. *In* "Microbial Control of Pests and Plant Diseases 1970–1980." (H. D. Burges, ed.), pp. 499–512. Academic Press, New York.

McCoy, C. W. 1990. Entomogenous fungi as microbial pesticides. *In* "New Directions in Biological Control." (R. R. Baker and P. E. Dunn, eds.), pp. 139–159. Alan R. Liss, New York.

McCoy, C. W., Samson, R. A., and Boucias, D. G. 1988. Entomogenous fungi. *In* "CRC

Handbook of Natural Pesticides, Microbial Insecticides, Part A. Entomogenous Protozoa and Fungi." (C. M. Ignoffo, ed.), Vol. 5, pp. 151–236. CRC Press, Boca Raton, Florida.

McGarr, R. L. 1968. Field tests with a nuclear polyhedral virus against the bollworm and tobacco budworm, 1964–1966. *J. Econ. Entomol. 61*, 342.

McGaughey, W. H. 1985. Insect resistance to the biological insecticide *Bacillus thuringiensis*. *Science 229*, 193–195.

McGaughey, W. H. 1990. Insect resistance to *Bacillus thuringiensis* δ-endotoxin. *In* "New Directions in Biological Control." (R. R. Baker and P. E. Dunn, eds.), pp. 583–598. Alan R. Liss, New York.

McLaughlin, R. E., Cleveland, T. C., Daum, R. J., and Bell, M. R. 1969. Development of the bait principle for boll weevil control. IV. Field tests with a bait containing a feeding stimulant and the sporozoans *Glugea gasti* and *Mattesia grandis*. *J. Invertebr. Pathol. 13*, 429–441.

Melin, B. E., and Cozzi, E. M. 1990. Safety to nontarget invertebrates of lepidopteran strains of *Bacillus thuringiensis* and their β-exotoxins. *In* "Safety of Microbial Insecticides." (M. Laird, L. A. Lacey and E. W. Davidson, eds.), pp. 149–167. CRC Press, Boca Raton, Florida.

Miller, J. C. 1990. Field assessment of the effects of a microbial pest control agent on nontarget Lepidoptera. *Am. Entomol. 36*, 135–139.

Miller, L. A., and Bedding, R. A. 1982. Field testing of the insect parasitic nematode, *Neoaplectana bibionis* (Nematoda: Steinernematidae) against the current borer moth, *Synanthedon tipuliformis* (Lep.: Sesiidae) in blackcurrants. *Entomophaga 27*, 109–114.

Morris, O. N. 1972. Inhibitory effects of foliage extracts of some forest trees on commercial *Bacillus thuringiensis*. *Can. Entomol. 104*, 1357–1361.

Morris, O. N. 1977a. Long-term effects of aerial applications of virus–fenitrothion combinations against the spruce budworm, *Choristoneura fumiferana* (Lepidoptera: *Tortricidae*). *Can. Entomol. 109*, 9–14.

Morris, O. N. 1977b. Long term study of the effectiveness of aerial application of *Bacillus thuringiensis*-acephate combinations against the spruce budworm, *Choristoneura fumiferana* (Lepidoptera: Tortricidae). *Can. Entomol. 109*, 1239–1248.

Morris, O. N. 1982. Bacteria as pesticides: Forest applications. *In* "Microbial and Viral Pesticides." (E. Kurstak, ed.), pp. 239–287. Marcel Dekker, New York.

Neilson, M. M., and Elgee, D. E. 1968. The method and role of vertical transmission of a nucleopolyhedrosis virus in the European spruce sawfly, *Diprion hercyniae*. *J. Invertebr. Pathol. 12*, 132–139.

Nogge, G. 1978. Aposymbiotic tsetse flies, *Glossiana morsitans morsitans* obtained by feeding on rabbits immunized specifically with symbionts. *J. Insect Physiol. 24*, 299–304.

Noguchi, Y., and Yamaguchi, K. 1984. Effects of food plants and stress agents on the development of disease in lepidopterous insects subjected to cross-infection with cytoplasmic-polyhedrosis viruses. *Jpn. J. Appl. Entomol. Zool. 28*, 9–13.

Petersen, J. J. 1984. Nematode parasites of mosquitoes. *In* "Plant and Insect Nematodes." (W. R. Nickle, ed.), pp. 797–820. Marcel Dekker, New York.

Pinnock, D. E. 1975. Pest populations and virus dosage in relation to crop productivity. *In* "Baculoviruses for Insect Pest Control: Safety Considerations." (M. Summers, R. Engler, L. A. Falcon, and P. V. Vail, eds.), pp. 145–154. Am. Soc. Microbiol. Publ., Washington, D.C.

Röder, A., and Gröner, A. 1980. Immunreaktion von menschlichen und tierischen Seren mit Baculoviren. *Naturwissenschaften 67*, 49–50.

Rogoff, M. H. 1982. Regulatory safety data requirements for registration of microbial pesticides. *In* "Microbial and Viral Pesticides." (E. Kurstak, ed.), pp. 645–679. Marcel Dekker, New York.

Sears, M. K., Jaques, R. P., and Laing, J. E. 1983. Utilization of action thresholds for microbial and chemical control of lepidopterous pests (Lepidoptera: Noctuidae, Pieridae) on cabbage. *J. Econ. Entomol. 76*, 368–374.

Shieh, T. R. 1989. Industrial production of viral pesticides. *Adv. Virus Res. 36*, 315–343.

Siegel, J. P., and Shadduck, J. A. 1990. Safety of microbial insecticides to vertebrates–humans. *In* "Safety of Microbial Insecticides." (M. Laird, L. A. Lacey, and E. W. Davidson, eds.), pp. 101–113. CRC Press, Boca Raton, Florida.

Smirnoff, W. A. 1961. A virus disease of *Neodiprion swainei* Middleton. *J. Insect Pathol. 3*, 29–46.

Smirnoff, W. A. 1962. Trans-ovum transmission of virus of *Neodiprion swainei* Middleton (Hymenoptera, Tenthredinidae). *J. Insect Pathol. 4*, 192–200.

Smirnoff, W. A. 1967. Effects of some plant juices on the ugly-nest caterpillar, *Archips cerasivoranus*, infected with microsporidia. *J. Invertebr. Pathol. 9*, 26–29.

Smirnoff, W. A. 1972. Effects of volatile substances released by foliage of *Abies balsamea*. *J. Invertebr. Pathol. 19*, 32–35.

Smirnoff, W. A., Fettes, J. J., and Desaulniers, R. 1973. Aerial spraying of a *Bacillus thuringiensis*-chitinase formulation for control of the spruce budworm (Lepidoptera: Tortricidae). *Can. Entomol. 105*, 1535–1544.

Smith, D. B., and Bouse, L. F. 1981. Machinery and factors that affect the application of pathogens. *In* "Microbial Control of Pests and Plant Diseases 1970–1980." (H. D. Burges, ed.), pp. 635–653. Academic Press, New York.

Sprenkel, R. K., and Brooks, W. M. 1975. Artificial dissemination and epizootic initiation of *Nomuraea rileyi*, an entomogenous fungus of lepidopterous pests of soybeans. *J. Econ. Entomol. 68*, 847–850.

Stacey, A. L., Yearian, W. C., and Young, S. Y., III. 1977. Evaluation of *Baculovirus heliothis* with feeding stimulants for control of *Heliothis* larvae on cotton. *J. Econ. Entomol. 70*, 779–784.

Steinhaus, E. A. 1956. Microbial control—the emergence of an idea. *Hilgardia 26*, 107–160.

Stelzer, M. J., Neisess, J., and Thompson, C. G. 1975. Aerial applications of a nucleopolyhedrosis virus and *Bacillus thuringiensis* against the Douglas fir tussock moth. *J. Econ. Entomol. 68*, 269–272.

Stelzer, M., Neisess, J., Cunningham, J. C., and McPhee, J. R. 1977. Field evaluation of baculovirus stocks against Douglas-fir tussock moth in British Columbia. *J. Econ. Entomol. 70*, 243–246.

Stewart, L. M. D., Hirst, M., Ferber, M. L., Merryweather, A. T., Cayley, P. J., and Possee, R. D. 1991. Construction of an improved baculovirus insecticide containing an insect-specific toxin gene. *Nature 352*, 85–88.

St. Julian, G., Sharpe, E., and Rhodes, R. A. 1970. Growth pattern of *Bacillus popilliae* in Japanese beetle larvae. *J. Invertebr. Pathol. 15*, 240–246.

St. Julian, G., Bulla, L. A., Jr., and Detroy, R. W. 1978. Stored *Bacillus popilliae* spores and their infectivity against *Popillia japonica* larvae. *J. Invertebr. Pathol. 32*, 258–263.

Stockdale, H., and Priston, R. A. J. 1981. Production of insect viruses in cell culture. *In* "Microbial Control of Pests and Plant Diseases, 1970–1980." (H. D. Burges, ed.), pp. 313–328. Academic Press, New York.

Stone, T. B., Sims, S. R., and Marrone, P. G. 1989. Selection of tobacco budworm for resistance to a genetically engineered *Pseudomonas fluorescens* containing the δ-endotoxin of *Bacillus thuringiensis* subsp. *kurstaki*. *J. Invertebr. Pathol. 53*, 228–234.

Tabashnik, B. E., Cushing, N. L., Finson, N., and Johnson, M. W. 1990. Field development of resistance to *Bacillus thuringiensis* in diamondback moth (Lepidoptera: Plutellidae). *J. Econ. Entomol. 83*, 1671–1676.

Talbot, P. H. B. 1977. The *Sirex–Amylostereum–Pinus* association. *Annu. Rev. Phytopathol. 15*, 41–54.

Tanada, Y. 1959. Microbial control of insect pests. *Annu. Rev. Entomol. 4*, 277–302.

Tanada, Y. 1964. A granulosis virus of the codling moth, *Carpocapsa pomonella* (Linnaeus) (Olethreutidae, Lepidoptera). *J. Insect Pathol. 6*, 378–380.

Tanada, Y. 1967. Microbial pesticides. *In* "Pest Control Biological, Physical, and Selected Chemical Methods." (W. W. Kilgore and R. L. Doutt, eds.), pp. 31–88. Academic Press, New York.

Tanada, Y. 1984. *Bacillus thuringiensis*: Integrated control—past, present, and future. *In* "Comparative Pathobiology." (T. C. Cheng, ed.), Vol. 7, pp. 59–90. Plenum, New York.

Tanada, Y., and Reiner, C. 1962. The use of pathogens in the control of the corn earworm, *Heliothis zea* (Boddie). *J. Invertebr. Pathol. 4*, 139–154.

Tomalski, M. D., and Miller, L. K. 1991. Insect paralysis by baculovirus-mediated expression of a mite neurotoxin gene. *Nature 352*, 82–85.

Undeen, A. H., and Alger, N. E. 1975. The effect of the microsporidan, *Nosema algerae*, on *Anopheles stephensi*. *J. Invertebr. Pathol. 25*, 19–24.

Undeen, A. H., and Maddox, J. V. 1973. The infection of nonmosquito hosts by injection with spores of the microsporidan *Nosema algerae*. *J. Invertebr. Pathol. 22*, 258–265.

Vail, P. V., and Jay, D. L. 1973. Pathology of a nuclear polyhedrosis virus of the alfalfa looper in alternate hosts. *J. Invertebr. Pathol. 21*, 198–204.

Van Rie, J., McGaughey, W. H., Johnson, D. E., Barnett, B. D., and Van Mellaert, H. 1990. Mechanism of insect resistance to the microbial insecticide *Bacillus thuringiensis*. *Science 247*, 72–74.

Vinson, S. B. 1990. Potential impact of microbial insecticides on beneficial arthropods in the terrestrial environment. *In* "Safety of Microbial Insecticides." (M. Laird, L. A. Lacey, and E. W. Davidson, eds.), pp. 43–64. CRC Press, Boca Raton, Florida.

Washino, R. K. 1982. Biocontrol of mosquitoes associated with California rice fields with special reference to the recycling of *Lagenidium giganteum* Couch and other microbial agents. *In* "Biocontrol of Medical and Veterinary Pests." (M. Laird, ed.), pp. 122–139. Praeger, New York.

Weiser, J. 1982. Persistence of fungal insecticides: Influence of environmental factors and present and future applications. *In* "Microbial and Viral Pesticides." (E. Kurstak, ed.), pp. 531–557. Marcel Dekker, New York.

Weseloh, R. M., Andreadis, T. G., Moore, R. E. B., Anderson, J. F., Dubois, N. R., and Lewis, F. B. 1983. Field confirmation of a mechanism causing synergism between *Bacillus thuringiensis* and the gypsy moth parasitoid, *Apanteles melanoscelus*. *J. Invertebr. Pathol. 41*, 99–103.

Wilding, N. 1981. Pest control by Entomophthorales. *In* "Microbial Control of Pests and Plant Disease 1970–1980." (H. D. Burges, ed.), pp. 539–554. Academic Press, New York.

Wilson, G. G. 1981. *Nosema fumiferanae*, a natural pathogen of a forest pest: Potential for pest management. *In* "Microbial Control of Pests and Plant Diseases 1970–1980." (H. D. Burges, ed.), pp. 595–601. Academic Press, New York.

Wollam, J. D., and Yendol, W. G. 1976. Evaluation of *Bacillus thuringiensis* and a parasitoid for suppression of the gypsy moth. *J. Econ. Entomol. 69*, 113–118.

Wollam, J. D., Yendol, W. G., and Lewis, F. B. 1978. Evaluation of aerially-applied nuclear polyhedrosis virus for suppression of the gypsy moth, *Lymantria dispar* L. *U.S. Dep. Agric. For. Ser. Res. Pap. NE-396*.

Wood, H. A., and Granados, R. R. 1991. Genetically engineered baculoviruses as agents for pest control. *Annu. Rev. Microbiol. 45*, 69–87.

Yearian, W. C., and Young, S. Y. 1982. Control of insect pests of agricultural importance by viral insecticides. *In* "Microbial and Viral Pesticides." (E. Kurstak, ed.), pp. 387–423. Marcel Dekker, New York.

Yearian, W. C., Young, S. Y., and Livingston, J. M. 1973. Field evaluation of a nuclear polyhedrosis virus of *Neodiprion taedae linearis*. *J. Invertebr. Pathol. 22*, 34–37.

Yearian, W. C., Luttrell, R. G., Stacey, A. L., and Young, S. Y. 1980. Efficacy of *Bacillus thuringiensis* and *Baculovirus heliothidis*-chlordimeform spray mixtures against *Heliothis* spp. on cotton. *J. Ga. Entomol. Soc. 15*, 260–271.

Yendol, W. G., Hedlund, R. C., and Lewis, F. B. 1977. Field investigation of a baculovirus of the gypsy moth. *J. Econ. Entomol. 70*, 598–602.

Young, S. Y., Yearian, W. C., and Kim, K. S. 1977. Effect of dew from cotton and soybean foliage on activity of *Heliothis* nuclear polyhedrosis virus. *J. Invertebr. Pathol. 29*, 105–111.

Zehnder, G. W., and Gelernter, W. D. 1989. Activity of the M-ONE formulation of a new strain of *Bacillus thuringiensis* against the Colorado potato beetle (Coleoptera: Chrysomelidae): Relationship between susceptibility and insect life stage. *J. Econ. Entomol.* 82: 756–761.

Zelazny, B. 1973. Studies on *Rhabdionvirus orcytes*. II. Effect on adults of *Oryctes rhinoceros*. *J. Invertebr. Pathol. 22*, 122–126.

Zelazny, B. 1976. Transmission of a baculovirus in populations of *Oryctes rhinoceros*. *J. Invertebr. Pathol. 27*, 221–227.

Zelazny, B. 1977. *Oryctes rhinoceros* populations and behavior influenced by a baculovirus. *J. Invertebr. Pathol. 29*, 210–215.

Zondag, R. 1979. Control of *Sirex noctilio* F. with *Deladenus siricidicola* Bedding. Part II. Introductions and establishments in the South Island 1968–75. *N.Z. J. For. Sci. 9*, 68–76.

CHAPTER 16
EPIZOOTIOLOGY

I. DEFINITION OF SOME EPIZOOTIOLOGICAL TERMS
II. MODELING
III. KEY FACTORS IN EPIZOOTIOLOGY
    A. Transmission
    B. Pathogen Population
    C. Host Population
IV. THE ENVIRONMENT
    A. Agricultural Ecosystem
    B. Rangeland, Pasture, and Turfgrass Ecosystems
    C. Forest Ecosystem
    D. Aquatic Ecosystem
V. PRESENT STATUS

Epizootiology is defined as the science of causes and forms of mass phenomena of diseases at all levels of intensity in an animal host population. The study of insect epizootiology, linked to the broader science of ecology, includes diseases caused by noninfectious (amicrobial) and infectious (microbial) agents.

The principles of epizootiology of insect diseases were first presented by Steinhaus (1949). Subsequently, he examined the theories and concepts of disease as they affected insect populations and addressed questions concerning the characteristics of disease in natural control (Steinhaus 1954). Accordingly, insect epizootiology has theoretical and practical aspects. The theoretical aspects explain the patterns of disease in host populations, advance the knowledge of disease and pathogen dynamics, and make insect epizootiology a predictable science. The practical aspects utilize the explanations of disease patterns to increase disease levels in insect pest populations or to decrease its levels in beneficial insect populations.

The principles of epizootiology of insect diseases are based on epidemiology, a science that deals with disease behavior in human populations (Tanada 1963, 1964). Although insect epizootiology is concerned with diseases caused by noninfectious and infectious agents, the primary emphasis is on those caused by pathogens (Gaugler 1987). Noninfectious agents probably will not be an important component of study in disease dynamics by insect pathologists in the near future. In contrast, epidemiology currently emphasizes diseases of humans of noninfectious origins, such as cardiovascular, cancer, emphysema, and genetic disorders; but some of infectious nature, such as acquired immune deficiency syndrome (AIDS), Lyme disease, and malaria, are also being studied with great intensity. Our focus in this chapter will be on the insect epizootiology of diseases caused by pathogens. For more in-depth discussion of insect epizootiology, the reader is referred to Fuxa and Tanada (1987a).

I. DEFINITION OF SOME EPIZOOTIOLOGICAL TERMS

A clear understanding of the terminology used in epizootiology is essential but sometimes difficult because many of the terms are not precisely defined. For example, epizootic is defined as an unusually large number of cases of disease in a host population. This definition does not state what is meant by "unusually large number" and a relatively few cases of disease in a population could be an epizootic if none is expected (Fuxa and Tanada 1987b).

Enzootic disease refers to one that is usually of low prevalence and is constantly present in the host population. Enzootic diseases occur over a long duration whereas epizootic diseases are sporadic, limited in time, and characterized by a sudden change in prevalence.

An epizootic normally shows a form of variation in time called the epizootic wave. The epizootic wave can be divided into the preepizootic, epizootic, and postepizootic phases (Steinhaus 1949). The epizootic wave is only part of a larger curve that includes the enzootic state (Fig. 16–1).

Prevalence is an important measure in insect epizootiology. Prevalence, defined as the number of hosts (or portion of hosts at risk) expressing a disease at any given point in time, is dependent on the proportion of hosts with the disease and the duration of the disease. On the other hand, incidence is defined as the number of new cases of disease in a defined population in a given time interval. Incidence then refers to new cases occurring in a given population at a given time, whereas prevalence refers to both old and new cases of disease at a given time. Incidence is a more precise term and is often misused for prevalence.

FIGURE 16–1

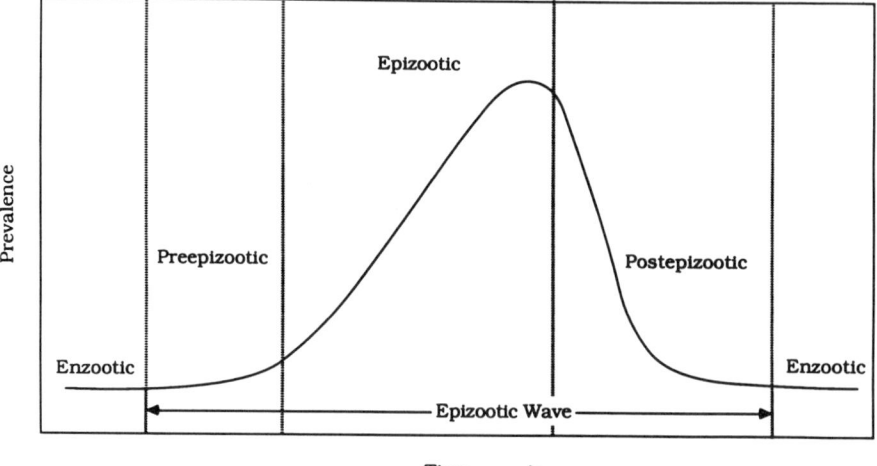

A curve showing the epizootic wave. It is divided into the enzootic, preepizootic, epizootic, and postepizootic phases. (Modified after Steinhaus 1949.)

II. MODELING

Insect epizootiologists have only recently utilized modeling to objectively compare their perception about a given system with known facts. Research models are used to organize and guide research of a particular system. Brown (1987) states that "researchers can formally specify hypotheses, inferences, and data and thus objectively compare their perceptions of a system with what is actually known." Moreover, models identify what additional information is needed to further understand a given system. For example, an alfalfa system would include the crop, its pests (insects and mites, other animals, plant pathogens, and weeds), natural enemies of the pests, and environmental factors. The system can have subsystems, such as the interaction between an insect pest and a pathogen. The result is a system approach defined as an imitation and representation of the real system. A typical model consists of three parts: a biological–conceptual framework formulated from a database accumulated through the years, a mathematical representation of that framework, and a computer that implements the mathematics. Basic mathematical theory dealing with epizootics and insect pathogens has been described by Brown (1987), and a review on epizootiological models is presented by Onstad and Carruthers (1990). More in-depth mathematical concepts are discussed by Anderson and May (1981) and Anderson (1982).

Brown (1987) defines the term "epizootiological dynamics" as the study of the change in host and pathogen numbers over time and space and the intrinsic and extrinsic factors responsible for the change. Changes dealing with time form temporal dynamics and those over space comprise spatial dynamics. Intrinsic factors are genetic effects that are difficult to measure; these factors, therefore, refer to fixed and constant effects of a specific population or system, such as a host population's intrinsic growth rate. Extrinsic factors are nongenetic effects and are variable or independent effects of the population's environment, such as weather. Accordingly, epizootiology is concerned with the host, the pathogen, and the environment over time and space.

Before a simple model can be constructed, a few assumptions must be made. Consider a hypothetical insect population in a constant environment with all stages (from eggs to adults) equally susceptible to a horizontally transmitted pathogen. The pathogen population is estimated simply by the number of infected hosts rather than by counting the pathogens. [However, in some cases a direct count of the pathogens is desirable (e.g., in persistence evaluation).] With these assumptions, the dynamics of the system can be represented by describing the susceptible and infected populations independently as follows:

$$\begin{pmatrix} \text{No. of susceptible} \\ \text{hosts tomorrow} \end{pmatrix} = \begin{pmatrix} \text{No. of susceptible} \\ \text{hosts today} \end{pmatrix} - \begin{pmatrix} \text{No. of susceptible} \\ \text{hosts infected today} \end{pmatrix}$$

and,

$$\begin{pmatrix} \text{No. of infected} \\ \text{hosts tomorrow} \end{pmatrix} = \begin{pmatrix} \text{No. of infected} \\ \text{hosts alive today} \end{pmatrix} + \begin{pmatrix} \text{No. of new infected} \\ \text{hosts today} \end{pmatrix} - \begin{pmatrix} \text{No. of infected hosts} \\ \text{that die today} \end{pmatrix}$$

The number of susceptible hosts infected on any day depends on the number of contacts between susceptible (S) and infected (I) hosts. This contact may or may not result in transmission. An assumption is made that some constant proportion, p, of these contacts results in transmission. This constant can be incorporated into the word equation and is referred to as the transmission efficiency or pathogen transmissibility. (In these descriptions, contact is between uninfected susceptible and infected hosts. Transmission of disease in insects usually involves contact between a susceptible host and an infectious, resistant stage of the pathogen).

Another term needs to be incorporated into the word equation. This is the number of infected hosts dying on any given day, which would be the mortality rate. If infected hosts only die from the pathogen, then the average mortality rate, m, is inversely proportional to the period of lethal infection. If the period of lethal infection is 2 days (time $= t$), an average of 50% of the infected hosts die each day and $m = 0.50$. If the period of lethal infection is 4 days, then an average of 25% of the infected hosts die each day and $m = 0.25$, and so on. Therefore,

$$\text{No. of infected that die today} = m \cdot I_t$$

This parameter, m, is the pathogenicity or virulence of the pathogen and is expressed algebraically:

$$S_{t+1} = S_t - pS_tI_t$$
$$I_{t+1} = I_t + pS_tI_t - mI_t$$

If $p = 0.005$ (transmission efficiency is $\frac{1}{2}$% each day) and $m = 0.25$ (period of lethal infection is 4 days) with 100 susceptible individuals and one infected host in this population, the equations can be used to calculate the daily number of susceptible and infected hosts. The equations demonstrate three important principles. First, because the rate at which individuals become infected depends on the number of individuals, the growth of the epizootic is a density-dependent process. Second, an epizootic does not cease because of the lack of susceptible hosts for there are still susceptible hosts remaining after the epizootic has diminished. Third, the infected population peaks when the rate of new infections is precisely equal to the death rate of infected individuals and this occurs when $S_t = m/p$ or stated in words, when the susceptible population equals the pathogenicity or virulence, m, divided by the transmission efficiency, p. This ratio is called the threshold density; all epizootics have a host threshold density defined by m/p or pathogenicity over transmissibility. When the susceptible host population exceeds this threshold, disease prevalence increases, and when the host population is below this threshold, prevalence decreases.

From these kinds of mathematical equations, more complex models have been developed. These include models developed by Anderson and May (1981) and Anderson (1982) who assumed that the infective stages of the pathogen would be homogeneously distributed in large numbers in the environment and would be mixing with the susceptible host population. They incorporated transovarial and transovum transmissions where both susceptible and infected individuals reproduced (Fig. 16–2). On the other hand, Hochberg (1989) assumed that the infective stage was not homogeneously distributed and that a reservoir (soil, host cadavers on trees, or live hosts) of protected, enduring infective stages existed. He showed that

FIGURE 16-2

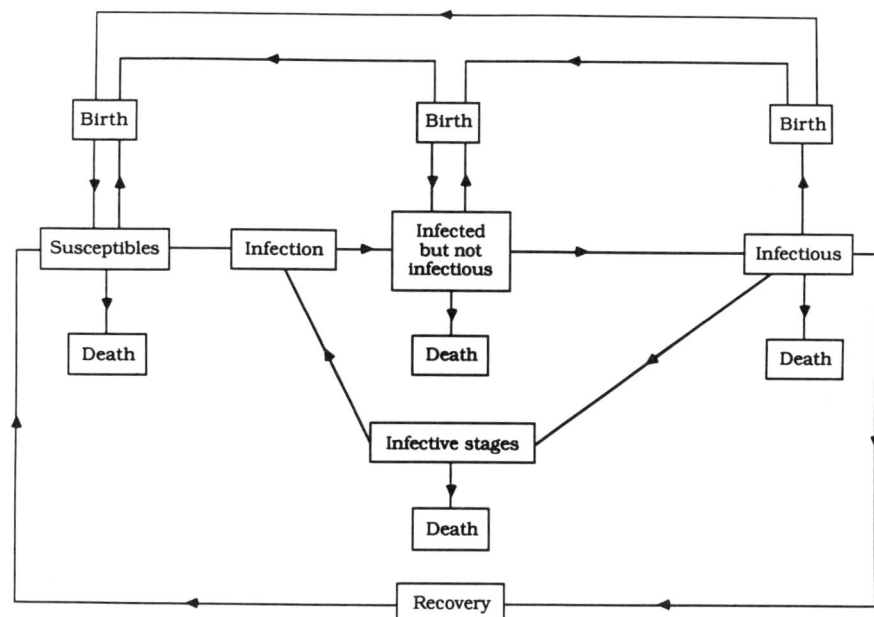

A conceptual host–pathogen model showing the interactions of the host and pathogen. (Modified after Anderson 1982.)

host populations may be regulated to low and relatively consistent densities when sufficient numbers of pathogens are dispersed from pathogen reservoirs to habitats where transmission can occur.

As more variables are added to the model, it becomes a dynamic simulation model. Simulation models represent broad classes of host–pathogen relationships and attempt to be realistic. These models can serve as a preliminary framework, which can be modified to explain or account for most biological idiosyncrasies. They can be a useful guide to obtain information necessary for a model of a particular system or used in the evaluation of the potential impact of specific pest management practices on the host–pathogen dynamics.

Models dealing with specific insect–pathogen systems are few in number. Basically, the models can be divided into two groups: to guide theoretical epizootiological research or to evaluate a particular ecological system. The latter models are designed to increase ecological understanding or to develop pest management systems (Onstad and Carruthers 1990).

Epizootic research models have been developed to predict *Nomuraea rileyi* prevalence on velvetbean caterpillar, *Anticarsia gemmatalis*, in soybean fields (Kish and Allen 1978), to describe the aerial dispersal of *Erynia* conidia (Brown and Nordin 1986), and to describe the effects of host stress and age-specific susceptibility of the western tent caterpillar, *Malacosoma californicum*, to a nuclear polyhedrosis virus (Wellington et al. 1975).

Onstad and Maddox (1989, 1990) created two of the most complex models dealing with the life histories of healthy and microsporidian-infected hosts. The European corn borer–*Nosema pyrausta* model was created to study the mechanisms of the population dynamics and to determine whether the *Nosema* can regulate the borer population when distributed homogeneously or heterogeneously in space (Onstad and Maddox 1989). When *N. pyrausta* persists, it can regulate the borer population below the carrying capacity of the corn environment. The *Tribolium confusum* and *Nosema whitei* model performs well until the flour medium becomes conditioned by chemicals secreted or excreted by the beetle (Onstad and Maddox 1990). Once the flour medium is conditioned, the predictions deviated from the observed data. However, the observations do support the predicted changes in prevalence through time.

Examples of epizootic models in pest management systems are (1) the use of *Bacillus thuringiensis* as a microbial insecticide (Brand and Pinnock 1981), (2) the onion maggot (*Delia antiqua*)–*Entomophthora muscae* model that was developed to aid in improved integrated pest management (IPM) strategies (Carruthers et al. 1985), and (3) the alfalfa weevil (*Hypera postica*)–*Erynia* model that examined various control strategies to take advantage of the fungal epizootic (Brown and Nordin 1982).

III. KEY FACTORS IN EPIZOOTIOLOGY

Three primary factors contribute to the cause and development of epizootics of infectious diseases. These are an effective means of pathogen transmission, the pathogen population, and the host population. The factors are interrelated and are closely associated with the biotic and abiotic environments that may affect the increase or decrease in the prevalence of disease in a host population. In most epizootiological studies, only a single pathogen species is considered for a host population. An insect population, however, is usually affected by several pathogen species together with a complex of insect parasitoids and predators. These biological agents may occur simultaneously within the population or may be separated by time or life stages. The pathogen(s) may not only infect individuals in the host population but also the parasitoid and predator complex and other insect species associated with the host population.

A. TRANSMISSION

Transmission is the process by which a pathogen or parasite is passed from a source of infection to a new host (Anderson and May 1982; Andreadis 1987). Direct transmission is when the pathogen is transferred from an infected to a susceptible host without the intervention of any other living agent; and indirect transmission involves one or more species of intermediate hosts or vectors. Among insect pathogens, transmission is usually direct and follows a passage from host to host or host to environment to host pathway. Transmission can also be horizontal or vertical. Horizontal transmission is the transfer of the pathogen from individual to individual but not directly from parent to offspring (Canning 1982). Vertical transmission is the transfer of the pathogen from the parent to the progeny (Fine 1975). Vertical transmission transfers the pathogen from one host generation to the next and has also been referred to as congenital, parental, or hereditary (Andreadis 1987).

Insect pathogens have a variety of mechanisms and adaptations to perpetuate themselves. A fundamental aspect of epizootiology is the transmission pathways adapted by these pathogens to ensure their survival. The continual occurrence of a pathogen in a host population demonstrates its ability to be transmitted, and the mechanisms of transmission determine the changes in the host population and the spread of a disease (Anderson 1982).

In horizontal transmission, the normal route into the host is through the mouth for bacteria, viruses, and protozoa or through the integument for fungi and nematodes. Oral transmission occurs by ingestion of food contaminated with the infective stage of the pathogen or by predation or cannibalism of infected insects. Some fungi (Schabel 1976; Sweeney 1977) and nematodes (Poinar 1979) infect through the mouth. In addition, spores of *Beauveria bassiana* and *Metarhizium anisopliae* can infect through the spiracles (Hedlund and Pass 1968; Pekrul and Grula 1979), whereas steinernematid and heterorhabditid nematodes can enter through the spiracles or anus and penetrate into the hemocoel to infect their insect host (Poinar 1979). One protozoan, an aquatic ciliate in the genus *Lambornella* pathogenic to larval mosquitoes, can penetrate through the cuticle by the formation of "invasion" cysts (Corliss and Coats 1976).

Another method of transmission through the integument is during oviposition by hymenopteran parasitoids (Kaya 1982). A few examples document such transmission experimentally in the laboratory, but its significance in the initiation of epizootics is unknown. Some bacteria (Bucher 1963a), microsporidia (Brooks 1973), and baculoviruses (Irabagon and Brooks 1974; Beegle and Oatman 1975) can be transmitted in this fashion.

In vertical transmission, direct transfer of pathogens from parents to their progeny occurs primarily with viruses, bacteria, and protozoa. In most insects, vertical transmission is through the mother and is referred to as matroclinal or maternal-mediated (Fine 1975). The pathogen can be on the surface of the egg (transovum transmission) or within the egg via the ovary (transovarial transmission). Transovum transmission is common with baculoviruses (Zelazny 1976; Evans 1986), cytoplasmic polyhedrosis viruses (Payne 1981), and bacteria (Bucher 1963b). The progeny are infected by ingestion of the virus or bacterium on the egg chorion contaminated by fecal or meconial discharges or substances used to cover the eggs. Transovarial transmission occurs with an iridescent virus of mosquitoes (Linley and Nielsen 1968) and many microsporidia (Canning 1982; Brooks 1988). The pathogen gains entry into the egg when tissues in the reproductive tract, such as ovaries and accessory glands, are infected. Infection occurs by the direct invasion of the embryo (Andreadis 1990a,b) or by ingestion of the pathogen in the yolk shortly before eclosion (Canning 1982).

Paternal-mediated vertical transmission is not common. It has been documented in a few cases; for example, direct parental infection of ova from infected spermatazoa as in the sigma virus of *Drosophila* (Seecof 1968) or by venereal transfer to the female as in some microsporidia (Thomson 1958; Kellen and Lindegren 1971) and nuclear polyhedrosis viruses (NPV) (Hamm and Young 1974).

Transstadial transmission is the transfer of a pathogen from one host stage to the next throughout all or part of the host's life cycle. This type of transmission is characteristic among pathogens that produce chronic infections (low pathogenicity) and are often transmitted vertically. Infection typically occurs in the larval stage through vertical or horizontal routes. Due to the chronic nature of the infection, the hosts may survive to adulthood and may transmit the pathogens to their progeny.

B. PATHOGEN POPULATION

An epizootic is more apt to develop at high host densities and, therefore, is host density dependent. However, it may occur at low host densities depending on the pathogen density and the spatial distribution or dispersion of the pathogen. The properties of individual pathogens affect the development of an epizootic. These properties vary among and within the different types of pathogens (i.e., viruses, bacteria, fungi, protozoa, and nematodes). The properties of significance in epizootiology are (1) infectivity and virulence, (2) capacity to survive, and (3) capacity to disperse. In addition, the spatial distribution or dispersion and the numbers of pathogens play significant roles in epizootics.

1. Infectivity and Virulence

Infectivity, the ability to cause infection, varies with different pathogens. The mode of invasion is one factor involved with infectivity. A pathogen with more than one mode of invasion may increase its chances to initiate an epizootic. As previously discussed, a number of pathogens have multiple invasion routes.

Virulence, the disease-producing power of the pathogen, is often measured by the response of the host to a known pathogen inoculum. This response can be measured using the median lethal dose (LD_{50}) or median lethal time (LT_{50}). A lesser number of pathogens required to kill a host indicate a more virulent form than one of the same species that requires a greater number. The pathogen may also infect a host and cause a chronic nonlethal infection. A quick-killing pathogen is more virulent than one that takes a considerable amount of time to kill. A pathogen that replicates rapidly is generally more virulent than one that replicates slowly in the host. Besides replication, other factors—tissue susceptibility, toxin, and enhancing factors—may play important roles in the virulence of the pathogen. Virulence is an alterable biological property, but the loss or gain of virulence during an epizootic has received very little attention.

Pathogenicity, a term nearly synonymous to virulence, refers to the disease-producing ability of the pathogen and is applied to groups or species of microorganisms. Pathogenicity, therefore, is genetically determined.

Many insect pathogens produce toxins that enhance their pathogenicity. Toxins have been identified in fungi (Roberts 1981) and bacteria (Lysenko 1981), but little is known of toxins in viral, protozoan, and nematode infections. The classic example of toxin production by an insect pathogen is the δ-endotoxin of *Bacillus thuringiensis* (see Chapter 4).

Infectivity can be increased by factors produced by pathogens or present in the environment. Although toxins play a role in enhancing infectivity and pathogenicity, there are also nontoxic-enhancing factors produced by pathogens. Tanada (1985) summarized the research with the synergistic factor associated with the granulosis virus (GV) of the armyworm, *Pseudaletia unipuncta*. This factor enhances other baculovirus infections. Derksen and Granados (1988) isolated a factor from the GV capsule of *Trichoplusia ni* and NPV polyhedra of *Autographa californica* that caused specific biochemical and structural changes in the peritrophic membrane of *T. ni* larvae, which enhanced viral infection. Chemicals, especially mucin (Stephens 1959) and boric acid (Doane and Wallis 1964; Shapiro and Bell 1982), bolster infectivity of some bacteria and viruses under laboratory conditions.

Genetic changes in insect pathogens (Al-Aidroos and Roberts 1978; Faust and Travers 1981; Wood et al. 1981; Samšiňáková and Kálalová 1983) have been observed in the laboratory suggesting that natural selection may occur in the field and may influence epizootics. Geographical isolates of entomopathogens have shown biochemical differences, but these results have not been related to any biological parameters, such as virulence (Tanada and Fuxa 1987).

Passage of pathogens through an insect host is known to increase or decrease their virulence in the laboratory (Steinhaus 1949), but little information is available on this subject under field conditions. In one study, Dunbar and Beard (1975) reported that *Bacillus popilliae* was less effective against the Japanese beetle (*Popillia japonica*) population in Connecticut because of the apparent appearance of less virulent strains. Further studies showed that the variant strain was morphologically different, did not sporulate well in Japanese beetle grubs, and had diminished virulence (Bulla et al. 1978). A "normal" *B. popilliae* strain was also found, but grubs seemed to be resistant to the variant as well as the normal strain. The factors affecting the loss of virulence have not been established although Klein (1981) speculates that strains from different grub species may account for this loss.

2. Capacity to Survive

Survival of the pathogen may occur in the abiotic or biotic environments of the host's habitat. Nonresistant and resistant forms of the pathogens may survive and persist in both environments. Usually the nonresistant form is more common in the biotic, whereas the resistant form is more common in the abiotic environment. The ability of insect pathogens, particularly the resistant forms, to survive and persist in the environment is well documented. Studies with the resistant stages, such as polyhedra and granules of baculoviruses, polyhedra of cytoplasmic polyhedrosis viruses, spores of bacteria and protozoa, conidia and spores of fungi, and dauer juveniles and eggs of nematodes, have been made in relation to the abiotic environment.

Most pathogens can survive longer only under conditions that protect them from environmental extremes. The half-lives of pathogens exposed to the environment are less than 1 day for occluded and nonoccluded viruses, less than 4 h for microsporidian spores, less than 3 days for *Bacillus thuringiensis* spores, about 7 days for the δ-endotoxin, and less than 7 days for the conidia of *Nomuraea rileyi* (Ignoffo and Hostetter 1977).

Low-moisture conditions do not ordinarily affect the persistence of occluded viruses and bacterial spores but greatly affect fungi, protozoa, and nematodes. Field temperatures do not affect the survivability of most pathogens, but ultraviolet light is detrimental to pathogens (Gaugler and Boush 1978; Dunkle and Shasha 1989; Ignoffo et al. 1977b). Pathogens can survive on the host plants provided they are protected from the detrimental effects of the physical factors. They can survive in bark crevices, insect debris on host plants, and stomata of leaves (Reed 1971; Doane 1975).

Soil is a natural reservoir for many insect pathogens; yet information on the role and effect of soil on pathogen persistence is limited. The reason for this lack of information is the complexity of the soil system. Soil varies in chemical composition and physical structure. It is a dynamic system, and combined with its physical, biological, and chemical complexity, this dynamic state makes soil a difficult medium to conduct epizootiological research. A major constraint is the inability to observe the insect *in situ* without disturbing the soil environment (Villiani and Wright 1990). The soil environment offers an opportune site for insect–pathogen interactions; more than 90% of insect pests spend part of their life cycle in the soil (Gaugler 1988).

The physicochemical properties and soil biota affect the survival and persistence of the pathogen. *Bacillus popilliae* spores persist in the soil for 2 to 7 yr and over 25 yr in colonization sites indicating that infected beetle larvae are responsible for the recycling of the bacterium (Klein 1981). Viruses overwinter in soil (Tanada and Omi 1974; Ignoffo and Hostetter 1977) and accumulate in the litter and soil beneath trees (Hukuhara and Namura 1972; Evans and Entwistle 1982) and beneath field crops (Jaques 1970), whereas certain facultative pathogens can grow in the soil if a proper substrate is present (Gottwald and Tedders 1984; Studdert and Kaya 1990).

Surface contamination of eggs by pathogens commonly occurs. In the gypsy moth, the egg surface and the scales surrounding the egg mass may be contaminated with NPV and serve as an inoculum to infect the newly hatched larvae as the larvae chew their way through the chorion (Doane 1975, 1976). Recent investigations suggest, however, that contamination of NPV in the surrounding environment plays a more significant role than previously thought (Murray and Elkinton 1989). With another forest defoliator, the Douglas-fir tussock moth, *Orgyia pseudotsugata*, the cocoons containing pupae dead from NPV serve as a source to contaminate egg masses when rains and melting snow leach the virus from the cocoons onto the egg masses (Thompson 1978). The occurrence of such virus-contaminated egg masses has been used to predict an impending epizootic and population collapse.

Pathogens can survive and persist in infected and uninfected primary, secondary, and alternate hosts. In the primary hosts, chronic but lethal infections are generally more advantageous than acute infections for pathogen survival because the pathogen has a longer period to replicate. This has been shown for the *Oryctes* baculovirus (Zelazny 1973), the fungus *Massospora* (Soper et al. 1976), *B. popilliae* (Klein 1981), and microsporidia (Weiser 1963). Pathogens may survive in secondary hosts, such as parasitoids, predators, or sarcophagous feeders of the primary host. There are many examples of microsporidia from primary hosts infecting insect parasitoids (Brooks 1973). These infected parasitoids can transmit the microsporidia back to the primary host.

Some insect pathogens require an alternate host to complete their development. Development of the pathogen in a dissimilar, alternate, obligate host is referred to as heteroecism. There are two good examples in insect pathology. Whisler et al. (1975) discovered that fungi in the genus *Coelomomyces* required a primary host and an alternative copepod host to complete their life cycle. The definitive host is the mosquito larva. A similar situation occurs with microsporidia in the genus *Amblyospora*, which infect mosquito larvae and require copepod hosts to complete their development (Sweeney et al. 1985; Andreadis 1985). Epizootics of the fungi and protozoa appear to be synchronized with the mosquito and copepod populations.

The nematode *Deladenus* (syn. *Beddingia*) *siricidicola* survives by having a free-living mycetophagous phase that feeds on the fungus *Amylostereum areolatum* (Bedding 1984). When the juvenile female nematode comes into close proximity of its wood wasp host, *Sirex noctilio*, instead of developing into the mycetophagous adult, it differentiates into an infective female adult.

Nonsusceptible invertebrates and vertebrates may serve as carriers when they occur in the same environment as the pathogens. These carriers distribute the pathogens throughout the hosts' habitat. A more detailed discussion is covered in the next section on pathogen dispersal.

3. Capacity to Disperse

Most pathogens have limited capacity to disperse by their own actions and must rely on physical and biotic factors for their dispersal. Fungi and nematodes can disperse on their own over a limited area, but most bacteria, protozoa, and viruses depend on biological and physical agents for dispersal. Some fungi in the Entomophthorales possess conidiophores that forcibly discharge conidia and if the primary conidia land on a nonhost substrate, they produce secondary conidiophores and conidia (Roberts and Humber 1981). Steinernematid and heterorhabditid nematodes can orient to their hosts. They show directed movement to carbon dioxide (Gaugler et al. 1980) or to excreta from their hosts (Schmidt and All 1979).

Wind, rain, and moving waterways (physical factors) are important in dispersing pathogens. Wind disperses fungal conidia. Dust containing NPV can be blown from the ground onto foliage (Thompson 1978; Olofsson 1988), and rain is an important factor in the distribution of baculoviruses within a forest canopy (Bird 1961; Thompson and Scott 1979). Streams carry pathogens of aquatic insects from one area to another or irrigation water can serve as a means to move pathogens from the soil to the leaf surface.

Infected primary, secondary, and alternate hosts and contaminated carriers (i.e., biotic factors) can disperse pathogens into new environments. Newly hatched larvae of the gypsy moth climb to the tops of trees and are dispersed by the wind. Acquisition of an infectious dose of NPV by these newly hatched, dispersing larvae aids in the distribution of the pathogen (Doane 1976). Epizootics have been initiated by the introduction of infected hosts into their natural populations [e.g., the NPV of the forest tent caterpillar, *Malacosoma disstria* (Stairs 1965a), the baculovirus of the rhinoceros beetle, *Oryctes rhinoceros* (Bedford 1981), and the fungus *Nomuraea rileyi* of lepidopterons (Sprenkel and Brooks 1975)].

The behavior of infected hosts assists in the dispersal of some pathogens (Benz 1963). Lepidopteran and sawfly larvae infected with NPV show negative geotropism and climb to elevated places just prior to death, and the remains from dead larvae contaminate the environment below. Viral occlusion bodies may remain viable for several years if they are protected from environmental extremes. The gypsy moth (Doane 1970), nun moth, *Lymantria monacha* (Komarek and Breindl 1924), and sawflies (Bird 1961), for example, show this type of behavioral and transmission pattern. Negative geotropism has also been observed with insects infected with bacterial and entomophthorous fungi (Benz 1963). Grasshoppers killed by *Entomophaga* often are attached to upper parts of plants, and this behavior allows the conidia to be dispersed by the wind (MacLeod 1963). Some soil-inhabiting insects infected with rickettsia or iridescent viruses tend to disperse to the soil surface (Niklas 1964; Fowler and Robertson 1972). This behavior exposes the infected insects to predators and other dispersal agents. Some insects are gregarious when healthy but disperse when infected with the NPV (Smirnoff 1960; Evans and Allaway 1983) or the cytoplasmic polyhedrosis virus (Tamashiro and Huang 1963). This dispersal behavior by infected insects allows for a wider distribution of pathogens.

Movement of chronically infected insects is important in pathogen dispersal. The fungus *Massospora cicadina* infects the 17-yr cicada when the nymph emerges from the soil to transform to the adult (Soper et al. 1976). The infected adult cicada is not killed and continues to fly and distributes the conidia from the abdomen, which is exposed when the segments fall off. These conidia infect other adults in which resting spores are formed and are dispersed in a similar manner as the conidia. Adult rhinoceros beetles infected with the *Oryctes* baculovirus effectively spread the virus in their contaminated feces (Bedford 1981). Nematodes are dispersed into new habitats by infected insects. Face flies, *Musca autumnalis*, infected with the nematode *Heterotylenchus* (syn. *Paraiotonchium*), become terminal dung seekers and deposit nematodes in fresh cow dung where healthy hosts occur (Kaya et al. 1979). Nematode parasites of bark beetles (Kaya 1984) and siricid wood wasps (Bedding 1984) are distributed into the larval breeding habitat by infected insects.

Adult insect hosts can disperse pathogens that are present on their body surfaces. Adults of the tent caterpillar disperse a NPV over wide areas (Stairs 1972). Experimentally, the NPV of *Heliothis* was dispersed by adult moths when they became contaminated at a light source (Gard and Falcon 1978) or the NPV of *Colias* was transmitted to eggs when the genitalia of adult butterflies were contaminated (Martignoni and Milstead 1962). *Oryctes rhinoceros* often transmits a baculovirus during mating (Zelazny and Alfiler 1991).

Many insects, other invertebrates, and vertebrates serve as carriers of insect pathogens. They usually become carriers when they encounter diseased insects in an epizootic (Entwistle 1982). Parasitoids and predators, through their contaminated body parts, especially the mouthparts and ovipositors, serve as vectors or carriers to susceptible hosts (Andreadis 1987). Sarcophagid flies (Hostetter 1971; Stairs 1966) and mites (Szalay-Marzso and Vago 1975) feeding on virus-killed insects can spread the virus. Predatory assassin bugs feeding on microsporidian-infected insects can pass viable spores through their alimentary tracts (Kaya 1979). Moreover, mammals and birds feeding on contaminated foliage and/or infected insects can deposit viable pathogens in their feces and serve as agents of dispersal (Crawford and Kalmakoff 1977; Entwistle et al. 1977a,b; Lautenschlager et al. 1980).

Parasitoids and predators have been implicated in the initiation of viral epizootics. The adult of the parasitoid *Sarcophaga aldrichi*, after feeding on NPV-killed forest tent caterpillar, contaminates foliage with its mouthpart and legs (Stairs 1966). The parasitoid can travel considerable distances to create new foci of infection. The pentatomid predator *Podisus maculiventris* contaminated with the NPV of the cabbage looper, *Trichoplusia ni*, has initiated an epizootic in looper populations (Biever et al. 1982). The NPV of the red-headed pine sawfly, *Neodiprion lecontei*, a native insect, spreads more rapidly than the NPV of the European pine sawfly, *N. sertifer*, an introduced species into North America, because the former has a greater number of parasitoids and predators that are believed to be important agents in viral dispersal and transmission (Bird 1964). An epizootic seems to develop most rapidly when a high parasitoid–predator population and high viral density occur in the host population (Bird 1961; Stairs 1966).

4. Pathogen Density

In theory, a single pathogen should be capable of infecting its host, but in nearly all cases, a minimum number of pathogens are required to cause an infection. The pathogen population is very important in the initiation of epizootics (Anderson and May 1980, 1981; Anderson 1982), and the characteristics intrinsic to the pathogen population need to be understood. There are three such characteristics: density, dispersion, and composition or quality of the individuals, that are measured as group characteristics (Tanada and Fuxa 1987).

The density of the pathogen population is one of the most important factors determining the initiation of an epizootic. Closely associated with the pathogen density is spatial distribution. Accordingly, high density and a widespread distribution of the pathogen population in the host habitat increase the probability of pathogen–host contact. Conversely, low pathogen density or sparse distribution decreases the probability of pathogen–host contact, which keeps the disease at an enzootic level. However, the enzootic disease may intensify to an epizootic depending on various factors, including pathogenicity and virulence, transmission, and properties of the host population. Pathogens may be concentrated in limited areas and occur at a relatively low density in relation to the entire host population. The disease may be maintained in the enzootic or epizootic state. In the rhinoceros beetle, the *Oryctes* baculovirus does not persist well in the external environment but does replicate and persist in the digestive tracts of adult rhinoceros beetles. The virus is transmitted during copulation (Zelazny and Alfiler 1991) or is deposited in the feeding and breeding sites of the beetle and is maintained in the beetle population in the enzootic or epizootic state (Bedford 1981).

High pathogen density usually occurs during and after an epizootic has decimated a large host population in which the pathogen has replicated extensively. The NPVs of the Douglas-fir tussock moth (Thompson 1978), the gypsy moth (Doane 1976), the European pine sawfly (Olofsson 1988), and the European spruce sawfly (Balch and Bird 1944) occur at high densities in the environment after a severe epizootic. With the gypsy moth, the population in the area of a severe NPV epizootic remains at low nondamaging levels for 8 to 10 yr. The NPV survives on the environmental surfaces from one generation to the next (Podgwaite et al. 1979).

The critical inoculum threshold for a pathogen population that allows for an epizootic to develop remains obscure. Such information can only be deduced from laboratory studies (e.g., LD_{50}) or from field applications of microbial-control agents. In the former case, the LD_{50} provides evidence for the importance of pathogen population density. In the latter case, the deliberate introduction of pathogens into host populations has initiated epizootics although the minimum number of pathogens to accomplish this has not been established. Successful and outstanding examples of pathogen introduction have been obtained with *Bacillus popilliae* (Fleming 1968), the NPVs of sawflies (Bird and Burk 1961; Entwistle et al. 1983), and the *Oryctes* baculovirus of the rhinoceros beetle (Bedford 1981).

An increase in pathogen population density may not necessarily result in an epizootic. Once a certain threshold of pathogen numbers is reached, further numbers will not significantly increase host contact with the pathogens. Thus, no increase in infection or mortality occurs with an increase in the concentrations of microsporidia (Wilson and Kaupp 1976), viruses (Podgwaite et al. 1984), or nematodes (Sosa and Beavers 1985).

Pathogen density can affect the manifestation of disease and, consequently, affect epizootics. High pathogen density can cause mortality more quickly (Kellen and Hoffmann 1982; Sheppard and Stairs 1977) than low pathogen density. Rapid host mortality reduces the replication of the pathogen and the subsequent pathogen density. For example, certain microsporidia cause acute disease with little or no spore production at high doses but cause chronic disease with high spore production at low doses (Brooks 1988).

Pathogen and host population densities are closely aligned in epizootiology. Generally, pathogen population is host density dependent, but density dependence must be defined within the context of space–time framework (Tanada and Fuxa 1987). Epizootics do not occur spontaneously in high host-population densities because field studies have shown that at a low host density an enzootic is already occurring and an increase in host density results in an epizootic (White and Dutky 1940; Tanada 1961; Doane 1970; Thompson 1978).

Pathogen–host relationship is independent of host density when the pathogen threshold density is high and widely distributed. This situation occurs when pathogens are applied as microbial-control agents or after an extensive epizootic. When there is a subsequent increase or invasion of hosts in the habitat, the pathogen may act as a host-density-dependent factor (Doane 1970; Thompson 1978).

5. Dispersion

The spatial distribution or dispersion of insect pathogens in relation to epizootiology has received little attention. The distribution of the pathogen population determines whether the pathogen acts as a host-dependent or -independent factor (Tanada and Fuxa 1987). Host-density dependency results from the multiplication of the pathogen and transmission (vertical or horizontal) to increasing numbers as the host density rises. When the pathogen population is widely distributed at or above the threshold density of infection, an epizootic develops rapidly regardless of the host population (Steinhaus 1954).

After a severe viral epizootic in populations of the European spruce sawfly, the NPV was effective in keeping the sawfly at very low host densities (Bird and Elgee 1957). Similarly, a viral epizootic destroyed a large outbreak of the tussock moth, and the viral inoculum in the forest habitat was so high and widely distributed that subsequent generations died mainly in the early instars (Thompson 1978). Doane (1976) believes that the high viral inoculum in the forest habitat after a viral epizootic in gypsy moth populations is an important factor in keeping this insect at low levels. After the virus is degraded by environmental factors, the potential for another gypsy moth outbreak is greatly enhanced.

In most cases the pathogen population is usually at a low density or discontinuously distributed in the host environment and must be reintroduced into the host population. The pathogen may be introduced by vectors, immigration of diseased or contaminated individuals, wind, or water. An enzootic may ensue which may remain at this level or increase to an epizootic depending on the density and properties of the host population and efficiency of transmission.

Information on the spatial distribution of pathogen units in the host habitat is essential because environmental contamination is the major route through which most insect pathogens infect their hosts (Tanada and Fuxa 1987). If a host population is aggregated, the degree of aggregation or dispersion of the pathogen in the host's habitat would partially determine the number of encounters between the two organisms. Unfortunately, the most rigorous description of spatial distribution or dispersion, namely mathematical, is rare in the study of insect pathogen populations. Dispersion of insect pathogens in the host's habitat has been described for two fungal pathogens and in both cases the distribution was clumped and aggregated (Soper and MacLeod 1981; Fuxa 1984).

Nonmathematical descriptions of entomopathogen distributions are common. In microbial control, studies on the distribution of pathogen are often done. Droplet size (i.e., degree of clumping) affects the control of forest lepidopteran pests by *B. thuringiensis* (Thompson et al. 1977), and the clumping of pathogen units in bait formulations increases the percentage of infection (Fuxa and Brooks 1979; Henry et al. 1978).

Temporal distribution of entomopathogens in the host's habitat is an important element in the development of epizootics. A susceptible host must contact an infectious agent in both space and time for infection to occur. Pathogens have been sampled or estimated over time periods; they have been related to environmental conditions, production at various densities of host population, subsequent disease prevalences, or a combination of these factors (Tanada and Fuxa 1987). Concentrations of a virus (Jaques and Harcourt 1971) and a nematode (Mracek 1982) in the soil vary seasonally in relation to infected host densities and seasonal temperatures. Certain Entomophthorales have characteristic daily (Wilding 1970; Newman and Carner 1974; Mullens et al. 1987) and seasonal patterns (Harper et al. 1984). The diel periodicity depends on environmental conditions, mainly light and humidity (Newman and Carner 1974). The conidia are produced during the early morning or evening hours when light conditions, humidity, and temperature are optimal. Because the conidia are sensitive to sunlight and temperature, they must encounter a susceptible host in a relatively short period of time.

C. HOST POPULATION

Individual insects possess many qualities that are expressed ultimately as population phenomena. The demography (size and density, growth, distribution, and migration) of the population and its dynamics are influenced by the quantity and quality of the progeny produced by the female. The genetic make-up of individuals and populations under varying conditions modifies the demography and dynamics of populations. Moreover, the quantity and quality of resources affect population distribution and abundance. These factors influence the epizootiology of disease in insect populations. The insect as an individual has been discussed in Chapter 14, which covers factors such as age and stage, physiological condition, and immunity. Here, the host population factors considered are (1) susceptibility, (2) density, and (3) behavior.

1. Susceptibility

Individuals from two or more discrete insect populations of a species may respond differently to pathogens (Watanabe 1987). For example, the NPV of the wattle bagworm, *Kotochalia junodi*, from one population located more than 320 km away was more infectious than the NPV isolated from a local bagworm population (Ossowski 1960). Similarly, the LD_{50} of the Indianmeal moth GV showed more than a sevenfold difference between a Georgia and California Indianmeal moth populations (Hunter and Hoffmann 1973). In the honey bee, *Apis mellifera*, larvae fed spores of *Bacillus larvae*, the etiologic agent of American foulbrood, caused differential mortality among different colonies (Rothenbuhler and Thompson 1956) and bee strains (Hoage and Rothenbuhler 1966). Except for the honey bee, the basis for the differences in susceptibility of the above examples remains unexplored though it is probably due to their exposure or lack of exposure to the pathogen over numerous generations. With the honey bee, the inability of the bacterium to multiply in the larval midgut of the older brood and the thickness of the peritrophic membrane may account for the lack of penetration into the hemocoel (Sutter et al. 1968). In addition, recent evidence suggests that insects continually exposed over several generations to a given pathogen can become resistant. This aspect is discussed in Chapters 14 and 15.

After a natural epizootic, the residual population may be less susceptible to a pathogen. Martignoni (1957) showed that larvae of the larch bud moth, *Eucosma griseana*, required 38 times more GV to obtain an LD_{50}. The dosage–mortality regression slope was very steep, suggesting that the epizootic removed most of the susceptible individuals from the field.

2. Density

Epizootics of infectious diseases develop or are most evident at high host densities (Watanabe 1987). Pathogens generally act in a density-dependent manner, destroying more hosts as the host density increases. High host densities enhance the contact between healthy and infected individuals and between individuals and the pathogens. Furthermore, crowded conditions stress individuals in a population, which may affect the prevalence of disease (Steinhaus 1958). More hosts become infected, which generate more pathogens and initiate an epizootic. However, epizootics also occur at low host densities, particularly when the pathogen densities are high and widely dispersed (Bird and Elgee 1957; Tanada 1961).

High densities of the gypsy moth population ultimately collapse because of NPV infections. Doane (1970) proposed that density-dependent transmission serves to initiate NPV epizootics. In dense populations, newly hatched larvae become infected when they ingest the NPV contaminating the chorion (Doane 1970) or the area surrounding the egg mass (Murray and Elkinton 1989). When these infected larvae die on the foliage, they serve as an inoculum of NPV for older larvae. Woods and Elkinton (1987) support this hypothesis and show a bimodal temporal pattern of mortality caused by the NPV. The peak mortality from NPV occurs during the late larval instars with the highest mortality at high-density populations. Some sawfly NPVs are not disseminated effectively in populations of low host densities and occur at the enzootic level; in contrast, epizootics of NPVs commonly occur in high-density populations (Bird 1953, 1961). With the armyworm, *Pseudaletia unipuncta*, epizootics of GV and NPV appeared to be host density dependent (Tanada 1961).

Microsporidia are major mortality factors in dense populations of the gypsy moth in the Soviet Union (See Elkinton and Liebhold 1990) and of the green tortrix, *Tortrix viridana*, in forests of Germany (Franz and Huger 1971). In Japan (Shimazu and Soper 1986) and the United States (Andreadis and Weseloh 1990), the fungus *Entomophaga maimaiga* has decimated dense populations of the gypsy moth under humid conditions.

3. Behavior

Insect behavior can affect epizootics by spreading pathogens throughout the habitat. The behavioral aspects of infected insects have been discussed in an earlier section (see Section III.B.3., Capacity to Disperse).

Newly hatched larvae of the gypsy moth may acquire an infectious dose of NPV from contaminated egg masses (transovum transmission) (Doane 1970). Uninfected larvae and those that acquire an infectious viral dose climb to the top of trees to feed. The larvae are concentrated, particularly in a dense population, and when a larva dies from NPV, the possibility for larval-to-larval spread of the virus is enhanced.

Larvae of the western tent caterpillar, *Malacosoma californicum pluviale*, are colonial and live in silken tents. Wellington (1962) described the larval behavior of these caterpillars and demonstrated their effect on NPV transmission. Larvae were categorized as active (four categories of active) and inactive. Active colonies are characterized by large elongate tents, whereas inactive colonies have small elongate tents. Moreover, larvae in an active colony are vigorous and construct several tents during the first few instars, whereas larvae in the inactive colony are sluggish and have low vitality, and their chances for survival even in the absence of NPV are low. Individuals of active and inactive colonies are susceptible to NPV that can be transmitted through transovum transmission. Because the larvae from an inactive colony are highly aggregated, larval-to-larval transmission of NPV is enhanced. These larvae serve to amplify the viral load in the environment. When larvae in an active colony become infected with NPV, some larvae escape infection because several other tents are available and chances for larval-to-larval transmission are diminished. However, the virus persists in the population, and, apparently, sufficient individuals in the active colony become infected to transmit the virus on the egg mass.

Ovipositional behavior may affect epizootics. In the European spruce sawfly, *Gilpinia hercyniae*, and the European pine sawfly, *Neodiprion sertifer*, ovipositional habit influences epizootics of NPV through transovum transmission (Bird 1961). The European spruce sawfly female lays her eggs singly, and the NPV-infected females will disperse widely, establishing opportunities for new foci of infection from which the disease may spread through larval-to-larval contact. The European pine sawfly female lays her eggs in a cluster, and although all individuals from a contaminated egg mass may die from NPV infection, the spread of the virus is limited and dependent on parasitoids, predators, scavengers, and rain.

Other behavioral traits, including grooming, "walling off," cannibalism, and "house cleaning," may affect the initiation of epizootics. In the termite *Reticulitermes* sp., grooming of individuals exposed to the fungus *Metarhizium anisopliae* resulted in the spread of the pathogen to healthy individuals in the population (Kramm et al. 1982). Once individuals died from the fungal disease, however, healthy termites avoided the cadavers. In the Formosan subterranean termite, *Coptotermes formosanus*, diseased individuals are walled off from the healthy colony, which prevents the spread of the pathogen (Fujii 1975). With *Tipula* iridescent virus, transmission is believed to occur through cannibalism (Carter 1973); first instars consumed diseased fourth instars that survived the previous generation. Finally, hives containing honey bee workers, which remove dead brood with the American foulbrood organism, *Bacillus larvae*, within a few days after capping of the brood, have a lower prevalence of this disease than those that remove the brood after the bacterium has formed many infective spores (Rothenbuhler 1964). This hygienic or house cleaning behavior is controlled genetically.

IV. THE ENVIRONMENT

Individuals of the same species live collectively to form a population; populations of different species inhabit a geographic area to form a community; and the community is influenced by its physical environment (Price and Waldbauer 1982). This complex system of abiotic and biotic factors forms an ecosystem. The individual components of epizootiology—transmission, pathogen population, and host population—comprise the main elements of an epizootic, but another important consideration is the environment in which the host and pathogen interactions occur. The environment can be viewed as biotic and abiotic factors that impinge directly upon the main elements of an epizootic or can constitute an ecosystem in which an epizootic occurs. Benz (1987) has reviewed the individual environmental components but warns that they do not act discretely, and the whole environment must be kept constantly in mind. Carruthers and Soper (1987) and Maddox (1987), on the other hand, evaluate epizootics by ecosystems. Ecosystems can be designated in many different ways and are diverse although common environmental factors often link them together. Here, we shall utilize the ecosystem approach—agricultural, rangeland, pasture and turfgrass, forest, and aquatic—and provide selected examples of natural epizootics from each. Although numerous epizootics of diseases have been observed in these ecosystems, only a few natural epizootics have been documented with quantitative data throughout their entire course. We remind the reader that insect pest species can occur in more than one ecosystem and that examples of induced epizootics are presented in Chapter 15 (see Section XIII., Examples of Long-Term Control).

A. AGRICULTURAL ECOSYSTEM

Agricultural ecosystems (agroecosystems) are intensively manipulated by humans and are represented by biological and physical homogeneity. This homogeneity is maintained through intense cultivation practices and by the use of fertilizers and pesticides to reduce diversity. Agroecosystems, particularly in developed countries, are unstable and characterized by uniformity in planting, plant development and density, and maturation of a single plant species. Because the diversity of plant and insect species is lacking and sudden permutations can be imposed by weather and human activities, outbreaks of one or a few insect pest species are not uncommon. In the agroecosystem, we commonly find viral, fungal, and protozoan epizootics.

1. Fall Armyworm

The larvae of the fall armyworm, *Spodoptera frugiperda*, feed on corn, sorghum, millet, and other grasses, including those that occur in pastures and on soybeans in the absence of grasses. Although the fall armyworm is susceptible to a number of pathogens, a 2-yr study showed that the NPV accounts for more mortality than any other pathogen or parasitoid (Fuxa 1982). During the growing season, the NPV infection increases more quickly in pastures than in corn or sorghum fields, but the infection rates at the time of harvest of corn and sorghum are similar to that in pasture. The higher prevalence of NPV in pasture during the early growing season is probably due to the host plant because the initial reservoir of NPV in the soil is identical for each crop. Rain and other physical agents more easily contaminate the low-growing grasses with the NPV than the corn or sorghum, which grows upright. In addition, the larval behavior on corn and sorghum may account for the early season differences. Larvae feed in a more protected location, the whorl in corn and sorghum, and do not move from plant to plant, whereas they are more mobile on grasses in the pasture and are more likely to contact infected individuals or occlusion bodies. The behavior of NPV-infected individuals to climb to high places to die on the host plant may account for the increased infection in corn and sorghum later in the season.

A number of factors affect epizootics in the fall armyworm and these factors can interact in complex ways (Fuxa 1990). In pasture, the grasses occur near the soil level, and the amount of virus in the soil, rainfall, presence of cattle, and tilling of the soil affect the prevalence of NPV (Fuxa and Geaghan 1983). On the other hand, Mitchell and Fuxa (1990) showed that the NPV prevalence varies widely between corn fields in two different nearby locations. At one location, abiotic factors, where NPV occurs at high densities in the soil, are correlated with the prevalence of infection, whereas at another, biotic factors, where little virus is present in soil, are more important. Rainfall and solar radiation (abiotic factors) have negative effects on prevalence, probably because they affect viral persistence on the insect's food surface. Host density and host plant size (biotic factors) are correlated with disease prevalence. The prevalence of the NPV begins to increase after the larval population approaches one per corn plant. Once this population level is attained, the increase in infection is continuous regardless of population fluctuations. Tall plants have a negative effect on prevalence because the NPV is not dispersed to the upper part of the plant by abiotic factors. Parasitoids do not influence the prevalence of the NPV (Mitchell and Fuxa 1990).

The host plant can influence the susceptibility of the fall armyworm to the NPV (Richter et al. 1987). Larvae feeding on a less suitable host (soybean), and therefore stressed, are more susceptible to the NPV than those feeding on corn. In fact, larvae reared on signalgrass, millet, or corn are less susceptible to the NPV than those reared on sorghum or the common bermuda grass. In the field, early-season prevalence of NPV is greater on signalgrass than on bermuda grass, which is opposite of what is to be expected from laboratory studies. Even though host plants do affect epizootics (Fuxa 1982), structural parameters of the plant, such as the height of leaves above the soil reservoir of NPV or the leaf area of the plant, also play a significant role. The role of antiviral factors in plants appears to be minimal.

The fall armyworm is widely distributed from central South America to the southeastern United States. Geographical isolates of the fall armyworm or its NPV may affect epizootics (Fuxa 1987). For example, larval populations from Louisiana and Texas are more susceptible to the NPV than those from Brazil. Although the Brazilian fall armyworm is somewhat isolated from other populations, the adults are highly mobile, and it is possible that genes transferring NPV resistance can occur in susceptible populations. The LD_{50} of the NPV increases as the distance between the host and the NPV isolate increases. These results differ from those obtained with the wattle bagworm where the NPV from a distant source was more virulent than the NPV isolated from where the insects were collected (Ossowski 1960).

2. Aphids

Because aphids have sucking mouthparts, they are not particularly susceptible to entomopathogenic microorganisms except for the fungi that penetrate directly through their cuticle. One of the most common fungal pathogens causing epizootics in aphid populations is the Entomophthorales. In spite of many observations demonstrating that these fungi cause considerable mortality in aphid populations, it is not clear what limits their effectiveness in population suppression (Carruthers and Soper 1987).

The most important environmental factor that influences the course of epizootics in aphid populations is moisture. In the pea aphid, *Entomophthora* infection is positively correlated to the average rainfall recorded 12 days before observing the disease (Wilding 1975). For an epizootic of an *Entomophthora* species, there must be a minimum of 90% relative humidity for at least 8 h per day (Missonnier et al. 1970). The spotted alfalfa aphid populations in Israel are decimated yearly by epizootics caused by *Erynia radicans* (Carruthers and Soper 1987) even though this part of Israel is hot and dry. The moisture is provided by heavy dewfall and irrigation within the canopy of the plants.

Temperature can also limit disease prevalence of the Entomophthorales (Carruthers and Soper 1987). It regulates the developmental rates of both fungi and hosts. If the temperature favors the development of the host and is above or below the optimum for the pathogen, the aphid population will increase. If the temperature favors the fungi, epizootics can occur providing the moisture requirements are met.

3. Lepidopteran Larvae in Soybeans

Epizootics of the hyphomycetes fungus *Nomuraea rileyi* occur regularly in noctuid larvae in soybeans (Kish and Allen 1978; Ignoffo 1981; Fuxa 1984). These epizootics devastate susceptible host populations but often too late in the soybean growing season to prevent economic damage. *N. rileyi* appears to overwinter in the soil in soybean fields, and the amount of overwintering inoculum is one of the key factors in the development of epizootics (Ignoffo et al. 1977b).

Larvae infected by the overwintering inoculum spread the disease. Movement of infected host larvae in the soybean canopy and the airborne dispersal of infective conidia are responsible for the secondary spread of the fungus. Infected larvae initially occur in limited loci, but the fungus can spread throughout the field rapidly (Fuxa 1984). Dry windy conditions bolster conidial dispersal but limit conidial survival, germination, and infection (Ignoffo et al. 1977a; Gardner et al. 1977). High moisture, especially elevated relative humidities and dew, is essential for infection (Kish and Allen 1978). Rainfall over an extended period of time is detrimental to an epizootic development because the infective conidia are washed from the leaf surfaces. Alternation of wet and dry environmental conditions appears to provide optimal conditions for the spread of the fungus and the development of epizootics.

4. European Corn Borer

The European corn borer, *Ostrinia nubilalis*, is a major pest of corn in the United States. The microsporidium *Nosema pyrausta* is one of the most important natural mortality factors present in corn borer populations (Maddox 1987). This protozoan is efficiently transmitted vertically and horizontally. Transovarial and transovum transmissions are common (Kramer 1959). Infected females lay infected eggs or the spores from the infected accessory gland contaminate the external surface of the eggs. These infected larvae or larvae, which become infected when they consume the spores on the chorion during hatching, tunnel into the corn plant stalk. Horizontal transmission may occur because the feces from infected larvae contain spores, and healthy larvae, which may enter the same tunnel, consume the contaminated plant parts and become infected (Zimmack and Brindley 1957; Kramer 1959). Fecal material with spores occurs at the tunnel entrances; wind can blow such material to another plant, especially into the whorl, and healthy young larvae, which prefer the whorl, can become infected. Newly hatched infected larvae may move to adjacent plants providing further means of pathogen dispersal (Lewis 1978). A parasitoid, *Macrocentrus grandii*, may spread the *Nosema* horizontally (Andreadis 1982; Siegel et al. 1986), but the importance of this transmission method in initiating epizootics is unknown (Maddox 1987). High host density and widespread spatial *Nosema* distribution are requisites for epizootic development of this microsporidium in field populations of the European corn borer (Hill and Gary 1979).

B. RANGELAND, PASTURE, AND TURFGRASS ECOSYSTEMS

Rangeland and pasture ecosystems are represented by a diversity of plant species, but in physical make-up and structure, they tend toward a homogeneous environment. Rangeland constitutes a more stable environment because it is less managed and has a greater diversity of plant species compared with a pasture. In the rangeland, grasses are the primary plant species because they thrive in areas where moisture is below levels for sustaining most tree species. The pasture ecosystem, dominated by grasses and legumes, is managed by humans more so than the rangeland and has many characteristics of the agroecosystems. However, the plants are perennial, and catastrophic alterations are rare in rangeland and pasture environments. Because of the high economic threshold of the plants, particularly in the rangeland environment, insect pest populations are usually not subjected to artificial control tactics. The population reduction in this ecosystem is often caused by disease. Insect pests of animals are also present in these habitats and are subject to epizootics.

Turfgrass ecosystem is highly managed and is more characteristic of agroecosystems. Turfgrass consists of uniform stands of grass or a mixture of grasses maintained at a relatively low height and is used for ornamental, recreational, or functional purposes (Tashiro 1987). Immature stages of many pests of turfgrass and pasture reside in the soil, and the difficulty of conducting studies in such an ecosystem has been alluded to previously (see Section III.B.2, Capacity to Survive).

1. Grasshoppers

Grasshoppers are one of the major herbivores of rangeland, and a fungus, *Entomophaga grylli*, infects many different grasshopper species throughout the world (Roffey 1968; Milner 1978; MacLeod and Müller-Kögler 1973). Epizootics of this fungal disease are not uncommon (Pickford and Riegert 1964), and high prevalence of disease is associated with warm, moist environmental conditions. A major constraint in the study of these epizootics is the high mobility of grasshoppers.

In controlled grasshopper studies, the combination of host and fungal density-dependent relationships, spatial dynamics, and varying environmental conditions regulates disease prevalence (Carruthers and Soper 1987). From 3 to 5 yr are required to change from low host and pathogen densities to grasshopper outbreak levels; these outbreaks are followed by fungal epizootics, and the grasshoppers return to low host densities.

Nosema locustae is another important pathogen of grasshoppers. It has a wide host range, infecting many different species throughout the world (Brooks 1988). Both vertical and horizontal transmissions occur. In Idaho rangelands, the prevalence of infection ranged from 2 to 8% in three species of grasshoppers (Henry 1972). The grasshopper species have overlapping life cycles with two species hatching early in the season followed by the third species hatching later. Infection of the early season grasshopper species ensures that the third species are also infected. These grasshoppers species are cannibalistic, which is the main means of horizontal transmission. The diseased third species provides the inoculum for the early season grasshoppers for the following year.

2. Face Flies

The face fly, *Musca autumnalis*, is a pest of cattle on rangeland and pasture. This insect, native to Europe, is infected by the nematode *Heterotylenchus* (syn. *Paraiotonchium*) *autumnalis*. The nematode overwinters in infected diapausing adults and is continually present in face fly populations from year to year at an enzootic level (Kaya and Moon 1979, Geden and Stoffolano 1984). Behavior of the infected female fly enhances transmission by continually visiting cattle dung (Kaya et al. 1979). The infected female fly deposits nematodes rather than eggs because the nematodes invade the ovaries, which alter the fly's behavior. Healthy female flies oviposit in the infested dung, and the maggots are invaded by nematodes. Infected maggots are not killed, and they develop to adults. Although the nematode does not appear to regulate the fly population, the prevalence of infection can be high (Kaya and Moon 1979).

3. *Wiseana*

The univoltine species of *Wiseana* (Lepidoptera: Hepialidae) are important pests of pasture in New Zealand. As young larvae, they live on the soil surface and feed on grasses; the later instars dig a vertical burrow and spend the rest of their larval lives in the same burrow, emerging at night to feed on grasses.

Larvae of *Wiseana* spp. are susceptible to a number of viral diseases, including a NPV, a GV, an entomopoxvirus, and an iridescent virus. The most prevalent virus is the NPV (Kalmakoff and Crawford 1982). Young larvae become infected when they ingest viral occlusions present on the soil surface, on the underside of grass leaves, or in pasture debris. Surviving larvae make burrows, but some may feed on contaminated foliage and become infected with the NPV. These infected larvae usually die outside their burrows, where they are consumed by birds or their deaths contribute to the soil reservoir. In addition to birds dispersing the NPV through their feces, wind blowing dry debris, mechanical movement by grazing sheep and farm equipment, and surface run-off water may spread the viral occlusions.

4. White Grubs

Larvae, often referred to as white grubs, belonging to the coleopteran family, Scarabaeidae, can cause severe damage to turfgrass and grasses in pasture. They are injurious because they feed on roots of grasses, usually below the soil–thatch interface (Tashiro 1987). Larvae are susceptible to a number of pathogens, including the milky-disease organisms (i.e., *Bacillus popilliae* and other closely related species), amber disease (syn. honey disease) organism (*Serratia entomophila*), blue disease organisms (*Rickettsiella* spp.), fungi, protozoans, and nematodes (Klein 1988). Much of the long-term control studies with *B. popilliae* have been discussed in Chapter 15.

The white grub receiving the most attention in turfgrass has been the Japanese beetle, *Popillia japonica* (Fleming 1968). Because it is difficult to follow epizootics of soil insects, most studies report the prevalence of diseases in the Japanese beetle populations (Dunbar and Beard 1975; Hanula and Andreadis 1988). In Connecticut, the prevalence of *B. popilliae* is low, but the prevalence of infection of the microsporidium *Ovavesicula popilliae* exceeds 50% (Hanula 1990). This microsporidium can reduce the fecundity of adult beetles by 50% under laboratory conditions. However, no differences are detected in the weight or development of the field-collected larvae. Hanula (1990) suggests that *O. popilliae* may reduce Japanese beetle populations through environmental stress, such as low humidity, temperature extremes, and soil texture.

C. FOREST ECOSYSTEM

The natural forest ecosystem may appear homogeneous, but they are diverse in plant species complex and in habitat. This ecosystem, less frequently affected by catastrophic events compared with the agricultural and rangeland, pasture, and turfgrass ecosystems, is relatively stable because of its heterogeneity. This heterogeneity is due to the terrain, the diversity of plant species, and the environmental conditions in a forest. Moreover, spatial interactions between insect hosts and pathogens are complex because of the forest canopy, understory growth, tree bole, forest litter, and soil. These factors also create microclimatic differences that affect insect populations and pathogens. Outbreaks of insect pests tend to be cyclical, and pathogens serve an important role in regulating a number of forest pest populations.

1. Tent Caterpillars

Tent caterpillars, *Malacosoma* spp., which are univoltine, are distributed throughout much of the world. In North America, the forest tent caterpillar, *M. disstria*, is the most widely spread and destructive, feeding on aspen, maples, oaks, and tupelo gum. Although it is referred to as a tent caterpillar, this species does not construct a tent. They form silken mats on trunks and branches on which they congregate, particularly during the early instars. In the last instar, they disperse and feed individually before pupating.

A number of pathogens have been isolated from *M. disstria*, including the fungi *B. bassiana* and *Entomophthora* sp., a microsporidium, and a NPV (Stairs 1972). *Entomophthora* has been isolated from many larval populations, but epizootics have been observed only in oak–maple stands on wet sites, suggesting that this fungus is best suited to humid areas (Stairs 1972).

The NPV is prevalent in many populations of *Malacosoma* species. Viral epizootics commonly occur after host densities have been at high levels for several generations. The NPV persists until the host population decreases to a low level. Younger larvae are highly susceptible to the NPV, whereas older larvae are resistant, requiring more viral occlusions to initiate an infection (Stairs 1965b). Most infected larvae will die unless they are infected late in the last larval instar. These infected last instar larvae can pupate and give rise to infected adults. Males may transmit viral occlusions with their spermatozoa, and the females may contaminate the eggs during oviposition (transovum transmission) (Stairs 1966).

A single NPV-infected larva can spread the virus to uninfected individuals because *M. disstria* is gregarious throughout much of its larval life. As infected individuals die, they increase the inoculum in the immediate environment and serve as foci from which the virus can spread. The dispersal agents include wind, rain, birds, parasitoids, predators, scavengers, and the host larvae themselves (Stairs 1972). For example, the adults of the dipteran parasitoid *Sarcophaga aldrichi* are efficient transmitters of the NPV, and epizootics develop more rapidly when high populations of flies and the virus occur together (Stairs 1966).

2. Gypsy Moth

The gypsy moth, *Lymantria dispar*, is a serious defoliator of deciduous trees in the northeastern United States, Europe, and Asia. This insect is susceptible to a number of pathogens, of which the NPV has been studied the most. Other pathogens include the cytoplasmic polyhedrosis virus (Magnoler 1974); the bacterium *Streptococcus faecalis* (Doane 1970); microsporidia (Jeffords et al. 1988); the fungi *Beauveria bassiana*, *Paecilomyces farinosus* (Majchrowicz and Yendol 1973), and *Entomophaga maimaiga* (Aoki 1974); and a mermithid nematode (Schaefer and Ikebe 1982).

High population densities of the gypsy moth usually collapse because of the actions of pathogens, especially the NPV (Doane 1970; Elkinton and Liebhold 1990). Viral transmission appears to be density dependent (Doane 1970; Woods and Elkinton 1987), and the epizootics are initiated when larvae hatch from egg masses contaminated with the NPV through transovum transmission (Doane 1969; Shapiro and Robertson 1987) or more likely acquiring the NPV from environmentally contaminated surfaces (Murray and Elkinton 1989). Neonate larvae may acquire lethal infection of the NPV from contaminated bark or from pupal mats after the larvae leave the egg mass (Weseloh and Andreadis 1986; Woods et al. 1989). NPV-infected larvae migrate to the top of trees and die (Fig. 16–3). These dead larvae contaminate the foliage, which serves as foci of infection for uninfected larvae (Woods and Elkinton 1987).

Vertical transmission of NPV is not as important as the horizontal transmission (Murray and Elkinton 1989). Environmental contamination appears to be the main factor in horizontal transmission (Podgwaite et al. 1979). Although parasitoids may

FIGURE 16-3

A gypsy moth larva killed by a nuclear polyhedrosis virus showing the characteristic behavior of hanging by its prolegs on the upper part of a tree. (Provided by T. G. Andreadis.)

serve as vectors, their role in spreading the NPV in gypsy moth populations remains unknown (Podgwaite et al. 1981).

The fungus *E. maimaiga* is a highly virulent pathogen and is considered to be an important natural mortality of the gypsy moth in Japan (Aoki 1974, Shimazu et al. 1987). The fungus was introduced into North America between 1910 and 1911, but surveys indicated that it had not established itself (Andreadis and Weseloh 1990). However, in 1989, the fungus was isolated from gypsy moth populations throughout the Northeast, and the fungus is believed to have persisted from the original introduction made in the early 1900s (Andreadis and Weseloh 1990; Hajek et al. 1990). Gypsy moth larvae killed by the fungus or the NPV exhibit almost identical syndromes (Fig. 16-4), and that the fungus mortality, in the past, was mistaken possibly for the NPV mortality.

The quantity and frequency of rainfall appear to be the primary factors responsible for initiating and maintaining the fungal epizootic. Epizootics are initiated by the infection of early instar gypsy moth larvae in the spring (Shimazu et al. 1987).

FIGURE 16–4

Fifth and sixth instar gypsy moth larvae killed by the fungus *Entomophaga maimaiga* on tree trunk. Dead larvae resemble nuclear polyhedrosis virus mortality. (Provided by T. G. Andreadis.)

Resting spores of the fungus occur in the leaf litter and soil. Newly hatched larvae, which are found on the ground, come into contact with germinating resting spores and become infected. These larvae die, and if moisture conditions are adequate, conidia are produced that discharge and infect other larvae.

3. Spruce Budworm

The spruce budworm, *Choristoneura fumiferana*, is one of the most destructive forest pests in North America. This univoltine insect feeds on spruce-fir trees. A number of pathogens have been isolated from this insect (Stairs 1972), but the most important is the microsporidium *Nosema fumiferanae* (Wilson 1981).

Although larval-to-larval transmission occurs, vertical transmission of *N. fumiferanae* is the most important. Infected adult females readily transmit the microsporidium through their eggs. This microsporidium causes a chronic and debilitating disease in spruce budworm larvae. Heavily infected larvae may die, but infections are usually nonlethal. In fact, the infected larvae, pupae, and adults often do not show signs or symptoms of disease (Thomson 1958). *Nosema fumiferanae* commonly occurs in epizootic proportions (Wilson 1981). In one study conducted between 1955 and 1959, Thomson (cited in Wilson 1981) reported an increase of infected larvae of 36% in 1955 to 81% in 1959. In 1959, there was a reduction in egg numbers and a subsequent low larval population, which was attributed to direct larval mortality and the reduction in fertility of females caused by the microsporidium. These data demonstrate that *N. fumiferanae* can decimate populations of the spruce budworm, but from an epizootiological point of view, the lack of information on host population density presents problems in interpreting its proper impact on the spruce budworm (Maddox 1987).

D. AQUATIC ECOSYSTEM

Aquatic ecosystems represent diverse habitats because of the physical movement of water (lentic versus lotic habitats), the temporary or permanent nature of the aquatic habitat, the chemical nature of water, and the biological components in water. The effects of temperature and sunlight are minimal in most aquatic habitats, but factors, such as pH, salinity, and inorganic and organic solutes, may directly or indirectly affect the pathogen. Insects and pathogens have coevolved in this habitat, which has resulted, in some cases, in very complex life cycles of the pathogens. The prevalence of diseases has been reported for many aquatic insects, but the environmental factors necessary for the development of epizootics are not clearly understood. Epizootics have been documented particularly against insects of medical or veterinary importance. We shall focus our attention on the microsporidium *Amblyospora connecticus* infecting mosquitoes.

A complex life cycle occurs in the genus *Ambylospora* because of the occurrence of alternating generations in different hosts (Andreadis 1985, 1990b; Sweeney et al. 1985). For example, the primary host of *Amblyospora connecticus* is the coastal salt marsh mosquito, *Aedes cantator*, and the intermediate host is the copepod *Acanthocyclops vernalis* (Andreadis 1988, 1990a). In the spring, this microsporidium is horizontally transmitted from the overwintering copepod to the mosquito. The mosquitoes become infected when they ingest spores from dead copepods. The mosquitoes develop a benign infection, and the females transmit the microsporidium to their progeny via infected eggs. The eggs are deposited in temporary pools during the summer and hatch in early fall when the pools are flooded by rain or high tides. Infected mosquito larvae die, and the life cycle is complete when the spores are ingested by the copepods. The microsporidium persists by surviving in two different hosts.

This microsporidium causes seasonal epizootics in larval mosquito populations (Andreadis 1990a). Epizootics occur each fall in the mosquito population from the synchronized hatching of transovarially infected eggs and again in the fall and winter or first few weeks in the spring in the copepod population. The copepod aestivates during the summer.

V. PRESENT STATUS

Progress in the epizootiology of insect diseases has been slow because of the lack of pertinent information on the primary factors: pathogen population, host population, and environment. Most early studies have been qualitative in nature. The complexity of the discipline has discouraged quantitative studies in-depth. Recently, however, a number of encouraging quantitative studies have been conducted by modeling the insect populations. Computers have made it possible to deal with the voluminous data that necessarily accumulate from quantitative studies on the interactions of the primary factors.

The increased emphasis in the use of microbial insecticides will result in studies that should provide valuable basic information on insect epizootics in all ecosystems. Such information will augment those obtained from natural outbreaks of disease. Pathogens are unlikely to eradicate an insect pest, and the goal of insect epizootiology is to determine how, when, and where the pathogens regulate the insect population. This goal has been partially attained in the epizootics of fungal, protozoan, and viral diseases in insect populations.

REFERENCES

Al-Aidroos, K., and Roberts, D. W. 1978. Mutants of *Metarhizium anisopliae* with increased virulence toward mosquito larvae. *Can. J. Genet. Cytol. 20*, 211–219.

Anderson, R. M. 1982. Theoretical basis for the use of pathogens as biological control agents of pest species. *Parasitology 84*, 3–33.

Anderson, R. M., and May, R. M. 1980. Infectious diseases and population cycles of forest insects. *Science 210*, 658–661.

Anderson, R. M., and May, R. M. 1981. The population dynamics of microparasites and their invertebrate hosts. *Philos. Trans. R. Soc. London Ser. B. 291*, 451–524.

Anderson, R. M., and May, R. M. 1982. "Population Biology of Infectious Diseases." Springer-Verlag, Berlin.

Andreadis, T. A. 1982. Impact of *Nosema pyrausta* on field populations of *Macrocentrus grandii*, an introduced parasite of the European corn borer, *Ostrinia nubilalis*. *J. Invertebr. Pathol. 38*, 298–302.

Andreadis, T. G. 1985. Experimental transmission of a microsporidian pathogen from mosquitoes to an alternate copepod host. *Proc. Natl. Acad. Sci. U.S.A. 82*, 5574–5577.

Andreadis, T. G. 1987. Transmission. *In* "Epizootiology of Insect Diseases." (J. R. Fuxa and Y. Tanada, eds.), pp. 159–176. John Wiley & Sons, New York.

Andreadis, T. G. 1988. *Amblyospora connecticus* sp. nov. (Microsporida: Amblyosporidae): Horizontal transmission studies in the mosquito, *Aedes cantator* and formal description. *J. Invertebr. Pathol. 52*, 90–101.

Andreadis, T. G. 1990a. Epizootiology of *Amblyospora connecticus* (Microsporida) in field populations of the saltmarsh mosquito, *Aedes cantator*, and the cyclopoid copepod, *Acanthocyclops vernalis*. *J. Protozool. 37*, 174–182.

Andreadis, T. G. 1990b. Polymorphic Microsporidia of mosquitoes: Potential for biological control. *In* "New Directions in Biological Control." (R. R. Baker and P. E. Dunn, eds.), pp. 177–188. Alan R. Liss, New York.

Andreadis, T. G., and Weseloh, R. M. 1990. Discovery of *Entomophaga maimaiga* in North American gypsy moth, *Lymantria dispar*. *Proc. Natl. Acad. Sci. U.S.A. 87*, 2461–2465.

Aoki, J. 1974. Mixed infection of the gypsy moth, *Lymantria dispar japonica* Motschulsky (Lepidoptera: Lymantriidae), in a larch forest by *Entomophthora aulicae* (Reich.) Sorok. and *Paecilomyces canadensis* (Vuill.) Brown et Smith. *Appl. Entomol. Zool. 9*, 185–190.

Balch, R. E., and Bird, F. T. 1944. A disease of the European spruce sawfly, *Gilpinia hercyniae* (Htg.), and its place in natural control. *Sci. Agric. 25*, 65–80.

Bedding, R. A. 1984. Nematode parasites of Hymenoptera. *In* "Plant and Insect Nematodes." (W. R. Nickle, ed.), pp. 755–795. Marcel Dekker, New York.

Bedford, G. O. 1981. Control of the rhinoceros beetle by baculovirus. *In* "Microbial Control of Pests and Plant Diseases 1970–1980." (H. D. Burges, ed.), pp. 409–426. Academic Press, New York.

Beegle, C. C., and Oatman, E. R. 1975. Effect of a nuclear polyhedrosis virus on the relationship between *Trichoplusia ni* (Lepidoptera: Noctuidae) and the parasite *Hyposoter exiguae* (*Hymenoptera*: Ichneumonidae). *J. Invertebr. Pathol. 25*, 59–71.

Benz, G. 1963. Physiopathology and histochemistry. *In* "Insect Pathology: An Advanced Treatise." (E. A. Steinhaus, ed.), Vol. 1, pp. 299–338. Academic Press, New York.

Benz, G. 1987. Environment. *In* "Epizootiology of Insect Diseases." (J. R. Fuxa and Y. Tanada, eds.), pp. 177–214. John Wiley & Sons, New York.

Biever, K. D., Andrews, P. L., and Andrews, P. A. 1982. Use of a predator, *Podisus maculiventris*, to distribute virus and initiate epizootics. *J. Econ. Entomol. 75*, 150–152.

Bird, F. T. 1953. The use of a virus disease in the biological control of the European pine sawfly, *Neodiprion sertifer* (Geoffr.). *Can. Entomol. 85*, 437–446.

Bird, F. T. 1961. Transmission of some insect viruses with particular reference to ovarial transmission and its importance in the development of epizootics. *J. Insect Pathol. 3*, 352–380.

Bird, F. T. 1964. The use of viruses in biological control. *Entomophaga Mem. Hors. Ser. No. 2*, 465–473.

Bird, F. T., and Burk, J. M. 1961. Artificially disseminated virus as a factor controlling the European spruce sawfly, *Diprion hercyniae* (Htg.) in the absence of introduced parasites. *Can. Entomol. 93*, 228–238.

Bird, F. T., and Elgee, D. E. 1957. A virus disease and introduced parasites as factors controlling the European spruce sawfly, *Diprion hercyniae* (Htg.), in central New Brunswick. *Can. Entomol. 89*, 371–378.

Brand, R. J., and Pinnock, D. E. 1981. Application of biostatistical modelling to forecasting the results of microbial control trials. *In* "Microbial Control of Pests and Plant Diseases 1970–1980." (H. D. Burges, ed.), pp. 667–693. Academic Press, New York.

Brooks, W. M. 1973. Protozoa: Host–parasite–pathogen interrelationships. *Misc. Publ. Entomol. Soc. Am. 9*, 105–111.

Brooks, W. M. 1988. Entomogenous Protozoa. *In* "Handbook of Natural Pesticides, Vol. 5, Microbial Insecticides, Part A, Entomogenous Protozoa and Fungi." (C. M. Ignoffo, ed.), pp. 1–149. CRC Press, Boca Raton, Florida.

Brown, G. C. 1987. Modeling. *In* "Epizootiology of Insect Diseases." (J. R. Fuxa and Y. Tanada, eds.), pp. 43–68. John Wiley & Sons, New York.

Brown, G. C., and Nordin, G. L. 1982. An epizootic model of an insect–fungal pathogen system. *Bull. Math. Biol. 44*, 731–739.

Brown, G. C., and Nordin, G. L. 1986. Evaluation of an early harvest approach for induction of *Erynia* epizootics in alfalfa weevil populations. *J. Kan. Entomol. Soc. 59*, 446–453.

Bucher, G. E. 1963a. Transmission of bacterial pathogens by the ovipositor of a hymenopterous parasite. *J. Insect Pathol. 5*, 277–283.

Bucher, G. E. 1963b. Survival of populations of *Streptococcus faecalis* Andrewes and Horder in the gut of *Galleria mellonella* (Linnaeus) during metamorphosis, and transmission of the bacteria to the filial generation of the host. *J. Insect Pathol. 5*, 336–343.

Bulla, L. A., Jr., Costilow, R. N., and Sharpe, E. S. 1978. Biology of *Bacillus popilliae*. *Adv. Appl. Microbiol. 23*, 1–18.

Canning, E. U. 1982. An evaluation of protozoal characteristics in relation to biological control of pests. *Parasitology 84*, 119–149.

Carruthers, R. I., and Soper, R. S. 1987. Fungal diseases. *In* "Epizootiology of Insect Diseases." (J. R. Fuxa and Y. Tanada, eds.), pp. 357–416. John Wiley & Sons, New York.

Carruthers, R. I., Haynes, D. L., and MacLeod, D. M. 1985. *Entomophthora muscae* (Entomophthorales: Entomophthoraceae) mycosis of the onion fly, *Delia antiqua* (Diptera: Anthomyiidae). *J. Invertebr. Pathol. 45*, 81–93.

Carter, J. B. 1973. The mode of transmission of *Tipula* iridescent virus. II. Route of infection. *J. Invertebr. Pathol. 21*, 136–143.

Corliss, J. O., and Coats, D. W. 1976. A new cuticular cyst-producing tetrahymenid ciliate, *Lambornella clarki* n. sp., and the current status of ciliatosis in culicine mosquitoes. *Trans. Am. Microsc. Soc. 95*, 725–739.

Crawford, A. M., and Kalmakoff, J. 1977. A host–virus interaction in a pasture habitat *Wiseana* spp. (Lepidoptera: Hepialidae) and its baculoviruses. *J. Invertebr. Pathol. 29*, 81–87.

Derksen, A. C. G., and Granados, R. R. 1988. Alteration of a lepidopteran peritrophic membrane by baculoviruses and enhancement of viral infectivity. *Virology 167*, 242–250.

Doane, C. C. 1969. Trans-ovum transmission of a nuclear-polyhedrosis virus in the gypsy moth and inducement of virus susceptibility. *J. Invertebr. Pathol. 14*, 199–210.

Doane, C. C. 1970. Primary pathogens and their role in the development of an epizootic in the gypsy moth. *J. Invertebr. Pathol. 15*, 21–33.

Doane, C. C. 1975. Infectious sources of nuclear polyhedrosis virus persisting in natural habitats of the gypsy moth. *Environ. Entomol. 4*, 392–394.

Doane, C. C. 1976. Ecology of pathogens of the gypsy moth. *In* "Perspectives in Forest Entomology." (J. F. Anderson and H. K. Kaya, eds.), pp. 285–293. Academic Press, New York.

Doane, C. C., and Wallis, R. C. 1964. Enhancement of the action of *Bacillus thuringiensis* var. *thuringiensis* Berliner on *Porthetria dispar* (Linnaeus) in laboratory tests. *J. Insect Pathol. 6*, 423–429.

Dunbar, D. M., and Beard, R. L. 1975. Present status of milky disease of Japanese and Oriental beetles in Connecticut. *J. Econ. Entomol. 68*, 453–457.

Dunkle, R. L., and Shasha, B. S. 1989. Response of starch-encapsulated *Bacillus thuringiensis* containing ultraviolet (UV) screens to sunlight. *Environ. Entomol. 18*, 1035–1041.

Elkinton, J. S., and Liebhold, A. M. 1990. Population dynamics of gypsy moth in North America. *Annu. Rev. Entomol. 35*, 571–596.

Entwistle, P. F. 1982. Passive carriage of baculoviruses in forests. *In* "Proc. Third Int. Colloq. Invertebr. Pathol." pp. 344–351. Univ. Sussex, Brighton, United Kingdom.

Entwistle, P. F., Adams, P. H. W., and Evans, H. F. 1977a. Epizootiology of a nuclear-polyhedrosis virus in European spruce sawfly (*Gilpinia hercyniae*): The status of birds as dispersal agents of the virus during the larval season. *J. Invertebr. Pathol. 29*, 354–360.

Entwistle, P. F., Adams, P. H. W., and Evans, H. F. 1977b. Epizootiology of a nuclear-polyhedrosis virus in European spruce sawfly, *Gilpinia hercyniae*: Birds as dispersal agents of the virus during winter. *J. Invertebr. Pathol. 30*, 15–19.

Entwistle, P. F., Adams, P. H. W., Evans, H. F., and Rivers, C. F. 1983. Epizootiology of a nuclear polyhedrosis virus (Baculoviridae) in European spruce sawfly (*Gilpinia hercyniae*): Spread of disease from small epicentres in comparison with spread of baculovirus diseases in other hosts. *J. Appl. Ecol. 20*, 473–487.

Evans, H. F. 1986. Ecology and epizootiology of baculoviruses. *In* "The Biology of Baculoviruses." (R. R. Granados and B. A. Federici, eds.), Vol. II. pp. 89–132. CRC Press, Boca Raton, Florida.

Evans, H. F., and Allaway, G. P. 1983. Dynamics of baculovirus growth and dispersal in *Mamestra brassicae* L. (Lepidoptera: *Noctuidae*) larval populations introduced into small cabbage plots. *Appl. Environ. Microbiol. 45*, 493–501.

Evans, H. F., and Entwistle, P. F. 1982. Epizootiology of the nuclear polyhedrosis virus of European spruce sawfly with emphasis on persistence of virus outside the host. *In* "Microbial and Viral Pesticides." (E. Kurstak, ed.), pp. 449–461. Marcel Dekker, New York.

Faust, R. M., and Travers, R. S. 1981. Occurrence of resistance to neomycin and kanamycin in *Bacillus popilliae* and certain serotypes of *Bacillus thuringiensis*: Mutation potential in sensitive strains. *J. Invertebr. Pathol. 37*, 113–116.

Fine, P. E. M. 1975. Vectors and vertical transmission: An epidemiologic perspective. *Ann. N.Y. Acad. Sci. 266*, 173–194.

Fleming, W. E. 1968. Biological control of the Japanese beetle. *U.S. Dep. Agric. Tech. Bull. 1383*.

Fowler, M., and Robertson, J. S. 1972. Iridescent virus infection in field populations of *Wiseana cervinata* (Lepidoptera: Hepialidae) and *Witlesia* sp. (Lepidoptera: Pyralidae) in New Zealand. *J. Invertebr. Pathol. 19*, 154–155.

Franz, J. M., and Huger, A. M. 1971. Microsporidia causing the collapse of an outbreak of the green tortrix (*Tortrix viridana* L.) in Germany. *In* "Proc. Fourth Int. Colloq. Insect Pathol." pp. 48–53. College Park, Maryland.

Fujii, J. K. 1975. Effects of an entomogenous nematode, *Neoaplectana carpocapsae* Weiser, on the Formosan subterranean termite, *Coptotermes formosanus* Shiraki, with ecological and biological studies on *C. formosanus*. Ph.D. Thesis, University of Hawaii, Honolulu.

Fuxa, J. R. 1982. Prevalence of viral infections in populations of fall armyworm, *Spodoptera frugiperda*, in southeastern Louisiana. *Environ. Entomol. 11*, 239–242.

Fuxa, J. R. 1984. Dispersion and spread of the entomopathogenic fungus *Nomuraea rileyi* (Moniliales: Moniliaceae) in a soybean field. *Environ. Entomol. 13*, 252–258.

Fuxa, J. R. 1987. *Spodoptera frugiperda* susceptibility to nuclear polyhedrosis virus isolates with reference to insect migration. *Environ. Entomol. 16*, 218–223.

Fuxa, J. R. 1990. New directions for insect control with baculoviruses. *In* "New Directions in Biological Control." (R. R. Baker and P. E. Dunn, eds.), pp. 97–113. Alan R. Liss, New York.

Fuxa, J. R., and Brooks, W. M. 1979. Effects of *Vairimorpha necatrix* in sprays and corn meal on *Heliothis* species in tobacco, soybeans, and sorghum. *J. Econ. Entomol. 72*, 462–467.

Fuxa, J. R., and Geaghan, J. P. 1983. Multiple-regression analysis of factors affecting prevalence of nuclear polyhedrosis virus in *Spodoptera frugiperda* (Lepidoptera: Noctuidae) populations. *Environ. Entomol. 12*, 311–316.

Fuxa, J. R., and Tanada, Y. 1987a. "Epizootiology of Insect Diseases." John Wiley & Sons, New York.

Fuxa, J. R., and Tanada, Y. 1987b. Epidemiological concepts applied to insect epizootiology. *In* "Epizootiology of Insect Diseases." (J. R. Fuxa and Y. Tanada, eds.), pp. 3–21. John Wiley & Sons, New York.

Gard, I. E., and Falcon, L. A. 1978. Autodissemination of entomopathogens: Virus. *In* "Microbial Control of Insect Pests: Future Strategies in Pest Management Systems." (G. E. Allen, C. M. Ignoffo, and R. P. Jaques, eds.), pp. 46–51. N.S.F.-U.S.D.A., University of Florida, Gainesville.

Gardner, W. A., Sutton, R. M., and Noblet, R. 1977. Persistence of *Beauveria bassiana*, *Nomuraea rileyi*, and *Nosema necatrix* on soybean foliage. *Environ. Entomol. 6*, 616–618.

Gaugler, R. R. 1987. Noninfectious diseases. *In* "Epizootiology of Insect Diseases." (J. R. Fuxa and Y. Tanada, eds.), pp. 245–255. John Wiley & Sons, New York.

Gaugler, R. 1988. Ecological considerations in the biological control of soil-inhabiting insects with entomopathogenic nematodes. *Agric. Ecosys. Environ.* 24, 351–360.

Gaugler, R., and Boush, G. M. 1978. Effects of ultraviolet radiation and sunlight on the entomogenous nematode, *Neoaplectana carpocapsae*. *J. Invertebr. Pathol.* 32, 291–296.

Gaugler, R., LeBeck, L., Nakagaki, B., and Boush, G. M. 1980. Orientation of the entomogenous nematode *Neoaplectana carpocapsae* to carbon dioxide. *Environ. Entomol.* 9, 649–652.

Geden, C. J., and Stoffolano, J. G., Jr. 1984. Nematode parasites of other dipterans. *In* "Plant and Insect Nematodes." (W. R. Nickle, ed.), pp. 849–898. Marcel Dekker, New York.

Gottwald, T. R., and Tedders, W. L. 1984. Colonization, transmission, and longevity of *Beauveria bassiana* and *Metarhizium anisopliae* (Deuteromycotina: Hyphomycetes) on pecan weevil larvae (Coleoptera: Curculionidae) in the soil. *Environ. Entomol.* 13, 557–560.

Hajek, A. E., Humber, R. A., Elkinton, J. S., May, B., Walsh, S. R. A., and Silver, J. C. 1990. Allozyme and restriction fragment length polymorphism analyses confirm *Entomophaga maimaiga* responsible for 1989 epizootics in North American gypsy moth populations. *Proc. Natl. Acad. Sci. U.S.A.* 87, 6979–6982.

Hamm, J. J., and Young, J. R. 1974. Mode of transmission of nuclear-polyhedrosis virus to progeny of adult *Heliothis zea*. *J. Invertebr. Pathol.* 24, 70–81.

Hanula, J. L. 1990. Epizootiological investigations of the microsporidium *Ovavesicula popilliae* and bacterium *Bacillus popilliae* in field populations of the Japanese beetle (Coleoptera: Scarabaeidae). *Environ. Entomol.* 19, 1552–1557.

Hanula, J. L., and Andreadis, T. G. 1988. Parasitic microorganisms of Japanese beetle (Coleoptera: Scarabaeidae) and associated scarab larvae in Connecticut soils. *Environ. Entomol.* 17, 709–714.

Harper, J. D., Herbert, D. A., and Moore, R. E. 1984. Trapping patterns of *Entomophthora gammae* (Weiser) (Entomophthorales: Entomophthoraceae) conidia in a soybean field infested with the soybean looper, *Pseudoplusia includens* (Walker) (Lepidoptera: Noctuidae). *Environ. Entomol.* 13, 1186–1190.

Hedlund, R. C., and Pass, B. C. 1968. Infections of the alfalfa weevil, *Hypera postica*, by the fungus *Beauveria bassiana*. *J. Invertebr. Pathol.* 11, 25–34.

Henry, J. E. 1972. Epizootiology of infections by *Nosema locustae* Canning (Microsporidia: Nosematidae) in grasshoppers. *Acrida 1*, 111–120.

Henry, J. E., Oma, E. A., and Onsager, J. A. 1978. Relative effectiveness of ULV spray applications of spores of *Nosema locustae* against grasshoppers. *J. Econ. Entomol.* 71, 629–632.

Hill, R. E., and Gary, W. J. 1979. Effects of the microsporidium, *Nosema pyrausta*, on field populations of European corn borers in Nebraska. *Environ. Entomol.* 8, 91–95.

Hoage, T. R., and Rothenbuhler, W. C. 1966. Larval honey bee response to various doses of *Bacillus larvae* spores. *J. Econ. Entomol.* 59, 42–45.

Hochberg, M. E. 1989. The potential role of pathogens in biological control. *Nature* 337, 262–265.

Hostetter, D. L. 1971. A virulent nuclear polyhedrosis virus of the cabbage looper, *Trichoplusia ni*, recovered from the abdomens of sarcophagid flies. *J. Invertebr. Pathol.* 17, 130–131.

Hostetter, D. L., and Ignoffo, C. M. 1977. Environmental stability of microbial insecticides. *Misc. Publ. Entomol. Soc. Am.* 10 (3), 1–119.

Hukuhara, T., and Namura, H. 1972. Distribution of a nuclear-polyhedrosis virus of the fall webworm, *Hyphantria cunea*, in soil. *J. Invertebr. Pathol.* 19, 308–316.

Hunter, D. K., and Hoffmann, D. F. 1973. Susceptibility of two strains of Indian meal moth to a granulosis virus. *J. Invertebr. Pathol.* 21, 114–115.

Ignoffo, C. M. 1981. The fungus *Nomuraea rileyi* as a microbial insecticide. *In* "Microbial Control of Pests and Plant Diseases 1970–1980." (H. D. Burges, ed.), pp. 513–538. Academic Press, New York.

Ignoffo, C. M., Garcia, C., Hostetter, D. L., and Pinnell, R. E. 1977a. Laboratory Studies of the entomophthorus fungus *Nomuraea rileyi*: Soil-borne contamination of soybean seedlings and dispersal of diseased larvae of *Trichoplusia ni*. *J. Invertebr. Pathol.* 29, 147–152.

Ignoffo, C. M., Hostetter, D. L., Sikorowski, P. P., Sutter, G., and Brooks, W. M. 1977b. Inactivation of representative species of entomopathogenic viruses, a bacterium, fungus, and protozoan by an ultraviolet light source. *Environ. Entomol.* 6, 411–415.

Irabagon, T. A., and Brooks, W. M. 1974. Interaction of *Campoletis sonorensis* and a nuclear polyhedrosis virus in larvae of *Heliothis virescens*. *J. Econ. Entomol.* 67, 229–231.

Jaques, R. P. 1970. Natural occurrence of viruses of the cabbage looper in field plots. *Can. Entomol.* 102, 36–41.

Jaques, R. P., and Harcourt, D. G. 1971. Viruses of *Trichoplusia ni* (Lepidoptera: Noctuidae) and *Pieris rapae* (Lepidoptera: Pieridae) in soil in fields of crucifers in southern Ontario. *Can. Entomol.* 103, 1285–1290.

Jeffords, M. R., Maddox, J. V., McManus, M. L., Webb, R. E., and Wieber, A. 1988. Egg contamination as a method for the inoculative release of exotic microsporidia of the gypsy moth. *J. Invertebr. Pathol.* 51, 190–196.

Kalmakoff, J., and Crawford, A. M. 1982. Enzootic virus control of *Wiseana* spp. in the pasture environment. *In* "Microbial and Viral Pesticides." (E. Kurstak, ed.), pp. 435–448. Marcel Dekker, New York.

Kaya, H. K. 1979. Microsporidian spores: Retention of infectivity after passage through the gut of the assassin bug, *Zelus exsanguis* (Stål). *Proc. Hawaii. Entomol. Soc.* 23, 91–94.

Kaya, H. K. 1982. Parasites and predators as vectors of insect diseases. *In* "Proc. Third. Int. Colloq. Invertebr. Pathol." pp. 39–44. University of Sussex, Brighton, United Kingdom.

Kaya, H. K. 1984. Nematode parasites of bark beetles. *In* "Plant and Insect Nematodes." (W. R. Nickle, ed.), pp. 727–754. Marcel Dekker, New York.

Kaya, H. K., and Moon, R. D. 1979. The nematode *Heterotylenchus autumnalis* and face fly *Musca autumnalis*: A field study in northern California. *J. Nematol.* 10, 333–341.

Kaya, H. K., Moon, R. D., and Witt, P. L. 1979. Influence of the nematode, *Heterotylenchus autumnalis*, on the behavior of face fly, *Musca autumnalis*. *Environ. Entomol.* 8, 537–540.

Kellen, W. R., and Hoffmann, D. F. 1982. Dose-mortality and stunted growth responses of larvae of the navel orangeworm, *Amyelois transitella*, infected by chronic stunt virus. *Environ. Entomol.* 11, 214–222.

Kellen, W. R., and Lindegren, J. E. 1971. Modes of transmission of *Nosema plodiae* Kellen and Lindegren, a pathogen of *Plodia interpunctella* (Hübner). *J. Stored Prod. Res.* 7, 31–34.

Kish, L. P., and Allen, G. E. 1978. The biology and ecology of *Nomuraea rileyi* and a program for predicting its incidence on *Anticarsia gemmatalis* in soybean. *Fl. Agric. Exp. Stn. Bull.* 795.

Klein, M. G. 1981. Advances in the use of *Bacillus popilliae* for pest control. *In* "Microbial Control of Pests and Plant Diseases 1970–1980." (H. D. Burges, ed.), pp. 183–192. Academic Press, New York.

Klein, M. G. 1988. Pest management of soil-inhabiting insects with microorganisms. *Agric. Ecosys. Environ.* 24, 337–349.

Komárek, J., and V. Breindl. 1924. Die Wipfelkrankheit der Nonne und der Erreger derselben. *Z. Angew. Entomol.* 10, 99–162.

Kramer, J. P. 1959. Some relationships between *Perezia pyraustae* Paillot (Sporozoa, Nosematidae) and *Pyrausta nubilalis* (Hübner) (Lepidoptera, Pyralidae). *J. Insect Pathol. 1*, 25–33.

Kramm, K. R., West, D. F., and Rockenbach, P. G. 1982. Termite pathogens: Transfer of the entomopathogen *Metarhizium anisopliae* between *Reticulitermes* sp. termites. *J. Invertebr. Pathol. 40*, 1–6.

Lautenschlager, R. A., Podgwaite, J. D., and Watson, D. E. 1980. Natural occurrence of the nucleopolyhedrosis virus of the gypsy moth, *Lymantria dispar* (Lep.: Lymantriidae) in wild birds and mammals. *Entomophaga 25*, 261–267.

Lewis, L. C. 1978. Migration of larvae of *Ostrinia nubilalis* (Lepidoptera: Pyralidae) infected with *Nosema pyrausta* (Microsporida: Nosematidae) and subsequent dissemination of this microsporidium. *Can. Entomol. 110*, 897–900.

Linley, J. R., and Nielsen, H. T. 1968. Transmission of a mosquito iridescent virus in *Aedes taeniorhynchus*. II. Experiments related to transmission in nature. *J. Invertebr. Pathol. 12*, 17–24.

Lysenko, O. 1981. Principles of pathogenesis of insect bacterial diseases as exemplified by the nonsporeforming bacteria. *In* "Pathogenesis of Invertebrate Microbial Diseases." (E. W. Davidson, ed.), pp. 163–188. Allanheld, Osmun Publ., Totowa, New Jersey.

MacLeod, D. M. 1963. Entomophthorales infections. *In* "Insect Pathology: An Advanced Treatise." (E. A. Steinhaus, ed.) Vol. 2, pp. 189–231. Academic Press, New York.

MacLeod, D. M., and Müller-Kögler, E. 1973. Entomogenous fungi: Entomophthora species with pear-shaped to almost spherical conidia (Entomophthorales: Entomophthoraceae). *Mycologia 65*, 823–893.

Maddox, J. V. 1987. Protozoan diseases. *In* "Epizootiology of Insect Diseases." (J. R. Fuxa and Y. Tanada, eds.), pp. 417–452. John Wiley & Sons, New York.

Magnoler, A. 1974. Effects of a cytoplasmic polyhedrosis on larval and postlarval stages of the gypsy moth, *Porthetria dispar*. *J. Invertebr. Pathol. 23*, 263–274.

Majchrowicz, I., and Yendol, W. G. 1973. Fungi isolated from the gypsy moth. *J. Econ. Entomol. 66*, 823–824.

Martignoni, M. E. 1957. Contributo alla conoscenza di una granulosi di *Eucosma griseana* (Hübner) (*Tortricidae*, *Lepidoptera*) quale fattore limitante il pullulamento dell'insetto nella Engadina alta. *Mitt. Schweiz. Anst. Versuchw. 32*, 371–418.

Martignoni, M. E., and Milstead, J. E. 1962. Trans-ovum transmission of the nuclear polyhedrosis virus of *Colias eurytheme* Boisduval through contamination of the female genitalia. *J. Insect Pathol. 4*, 113–121.

Milner, R. J. 1978. On the occurrence of *Entomophthora grylli*, a fungal pathogen of grasshoppers in Australia. *J. Aust. Entomol. Soc. 17*, 293–296.

Missonnier, J., Robert, Y., and Thoizon, G. 1970. Circonstances épidémiologiques semblant favoriser le developpement des mycoses a Entomophthorales chez trois aphides, *Aphis fabae* Scop., *Capitophorus horni* Börner et *Myzus persicae* (Sulz.). *Entomophaga 15*, 169–190.

Mitchell, F. L., and Fuxa, J. R. 1990. Multiple regression analysis of factors influencing a nuclear polyhedrosis virus in populations of fall armyworm (Lepidoptera: Noctuidae) in corn. *Environ. Entomol. 19*, 260–267.

Mráček, Z. 1982. Horizontal distribution in soil, and seasonal dynamics of the nematode *Steinernema kraussei*, a parasite of *Cephalcia abietis*. *Z. Angew. Entomol. 94*, 110–112.

Mullens, B. A., Rodriguez, J. L., and Meyer, J. A. 1987. An epizootiological study of *Entomophthora muscae* in muscoid fly populations on southern California poultry facilities, with emphasis on *Musca domestica*. *Hilgardia 55*, 1–41.

Murray, K. D., and Elkinton, J. S. 1989. Environmental contamination of egg masses as a major component of transgenerational transmission of gypsy moth nuclear polyhedrosis virus (LdMNPV). *J. Invertebr. Pathol. 53*, 324–334.

Newman, G. G., and Carner, G. R. 1974. Diel periodicity of *Entomophthora gammae* in the soybean looper. *Environ. Entomol. 3*, 888–890.

Niklas, O. F. 1964. Vertikalbewegungen Rickettsiose-kranker Larven von *Amphimallon solstitiale* (Linnaeus), *Anomala dubia aenea* (De Geer) und *Maladera brunnea* (Linnaeus) (Col., *Lamellicornia*). *Anz. Schaedlingskd. 37*, 22–24.

Olofsson, E. 1988. Environmental persistence of the nuclear polyhedrosis virus of the European pine sawfly in relation to epizootics in Swedish Scots pine forests. *J. Invertebr. Pathol. 52*, 119–129.

Onstad, D. W., and Carruthers, R. I. 1990. Epizootiological models of insect diseases. *Annu. Rev. Entomol. 35*, 399–419.

Onstad, D. W., and Maddox, J. V. 1989. Modeling the effects of the microsporidium, *Nosema pyrausta*, on the population dynamics of the insect, *Ostrinia nubilalis*. *J. Invertebr. Pathol. 53*, 410–421.

Onstad, D. W., and Maddox, J. V. 1990. Simulation model of *Tribolium confusum* and its pathogen, *Nosema whitei*. *Ecol. Model. 51*, 143–160.

Ossowski, L. L. J. 1960. Variation in virulence of a wattle bagworm virus. *J. Insect Pathol. 2*, 35–43.

Payne, C. C. 1981. Cytoplasmic polyhedrosis virus. *In* "Pathogenesis of Invertebrate Microbial Diseases." (E. W. Davidson, ed.), pp. 61–100. Allanheld, Osmun, Totowa, New Jersey.

Pekrul, S., and Grula, E. A. 1979. Mode of infection of the corn earworm (*Heliothis zea*) by *Beauveria bassiana* as revealed by scanning electron microscopy. *J. Invertebr. Pathol. 34*, 238–247.

Pickford, R., and Riegert, P. W. 1964. The fungous disease caused by *Entomophthora grylli* Fres., and its effects on grasshopper populations in Saskatchewan in 1963. *Can. Entomol. 96*, 1158–1166.

Podgwaite, J. D., Shields, K. S., Zerillo, R. T., and Bruen, R. B. 1979. Environmental persistence of the nucleopolyhedrosis virus of the gypsy moth, *Lymantria dispar*. *Environ. Entomol. 8*, 528–536.

Podgwaite, J. D., Shields, K. S., and Lautenschlager, R. A. 1981. Epidemiology. *In* "The Gypsy Moth: Research Toward Integrated Pest Management." (C. C. Doane and M. L. McManus, eds.), pp. 482–487. *U.S. Dep. Agric. Tech. Bull. 1584*.

Podgwaite, J. D., Rush, P., Hall, D., and Walton, G. S. 1984. Efficacy of the *Neodiprion sertifer* (Hymenoptera: Diprionidae) nucleopolyhedrosis virus (Baculovirus) product, Neochek-S. *J. Econ. Entomol. 77*, 525–528.

Poinar, G. O., Jr. 1979. "Nematodes for Biological Control of Insects." CRC Press, Boca Raton, Florida.

Price, P. W., and Waldbauer, G. P. 1982. Ecological aspects of pest management. *In* "Introduction to Insect Pest Management." (R. L. Metcalf and W. H. Luckmann, eds.), pp. 33–68. John Wiley & Sons, New York.

Reed, E. M. 1971. Factors affecting the status of a virus as a control agent for the potato moth (*Phthorimaea operculella* (Zell.) (Lep., Gelechiidae)). *Bull. Entomol. Res. 61*, 207–222.

Richter, A. R., Fuxa, J. R., and Abdel-Fattah, M. 1987. Effect of host plant on the susceptibility of *Spodoptera frugiperda* (Lepidoptera: Noctuidae) to a nuclear polyhedrosis virus. *Environ. Entomol. 16*, 1004–1006.

Roberts, D. W. 1981. Toxins of entomopathogenic fungi. *In* "Microbial Control of Pests and Plant Diseases 1970–1980." (H. D. Burges, ed.), pp. 442–464. Academic Press, New York.

Roberts, D. W., and Humber, R. A. 1981. Entomogenous fungi. *In* "Biology of Conidial Fungi." (G. T. Cole and B. Kendrick, eds.), Vol. 2, pp. 201–236. Academic Press, New York.

Roffey, J. 1968. The occurrence of the fungus *Entomophthora grylli* Fresenius on locusts and grasshoppers in Thailand. *J. Invertebr. Pathol. 11*, 237–241.

Rothenbuhler, W. C. 1964. Behavior genetics of nest cleaning in honey bees. IV. Responses of F1 and backcross generations to disease-killed brood. *Am. Zool. 4*, 111–123.

Rothenbuhler, W. C., and Thompson, V. C. 1956. Resistance to American foulbrood in honey bees. I. Differential survival of larvae of different genetic lines. *J. Econ. Entomol. 49*, 470–475.

Samšiňáková, A., and Kálalová, S. 1983. The influence of a single-spore isolate and repeated subculturing on the pathogenicity of conidia of the entomophagous fungus *Beauveria bassiana*. *J. Invertebr. Pathol. 42*, 156–161.

Schabel, H. G. 1976. Oral infection of *Hylobius pales* by *Metarrhizium anisopliae*. *J. Invertebr. Pathol. 27*, 377–383.

Schaefer, P. W., and Ikebe, K. 1982. Recovery of *Hexamermis* sp. (Nematoda: Mermithidae), parasitizing gypsy moth, *Lymantria dispar* (L.) in Hokkaido, Japan. *Environ. Entomol. 11*, 675–680.

Schmidt, J., and All, J. N. 1979. Attraction of *Neoaplectana carpocapsae* (Nematoda: Steinernematidae) to common excretory products of insects. *Environ. Entomol. 8*, 55–61.

Seecof, R. 1968. The sigma virus infection of Drosophila melanogaster. *In* "Current Topics in Microbiology and Immunology." (K. Maramorosch, ed.), Vol. 42, pp. 59–93. Springer-Verlag, New York.

Shapiro, M., and Bell, R. A. 1982. Enhanced effectiveness of *Lymantria dispar* (Lepidoptera: Lymantriidae) nucleopolyhedrosis virus formulated with boric acid. *Ann. Entomol. Soc. Am. 75*, 346–349.

Shapiro, M., and Robertson, J. L. 1987. Yield and activity of gypsy moth (Lepidoptera: Lymantriidae) nucleopolyhedrosis virus recovered from survivors of viral challenge. *J. Econ. Entomol. 80*, 901–905.

Sheppard, R. F., and Stairs, G. R. 1977. Dosage-mortality and time-mortality studies of a granulosis virus in a laboratory strain of the codling moth, *Laspeyresia pomonella*. *J. Invertebr. Pathol. 29*, 216–221.

Shimazu, M., and Soper, R. S. 1986. Pathogenicity and sporulation of *Entomophthora maimaiga* Humber, Shimazu, Soper and Hajek (Entomophthorales: Entomophthoraceae) on larvae of the gypsy moth, *Lymantria dispar* L. (Lepidoptera: Lymantriidae). *Appl. Entomol. Zool. 21*, 589–596.

Shimazu, M., Koizumi, C., Kushida, T., and Mitsuhashi, J. 1987. Infectivity of hibernated resting spores of *Entomophaga maimaiga* Humber, Shimazu et Soper (Entomophthorales: Entomophthoraceae). *Appl. Entomol. Zool. 22*, 216–221.

Siegel, J. P., Maddox, J. V., and Ruesink, W. G. 1986. Impact of *Nosema pyrausta* on a braconid, *Macrocentrus grandii*, in central Illinois. *J. Invertebr. Pathol. 47*, 271–276.

Smirnoff, W. A. 1960. Observations on the migration of larvae of *Neodiprion swainei* Midd. (Hymenoptera: Tenthredinidae). *Can. Entomol. 92*, 957–958.

Soper, R. S., and MacLeod, D. M. 1981. Descriptive epizootiology of an aphid mycosis. *U.S. Dep. Agric. Tech. Bull.* 1632.

Soper, R. S., Smith, L. F. R., and Delyzer, A. J. 1976. Epizootiology of *Massospora levispora* in an isolated population of *Okanagana rimosa*. *Ann. Entomol. Soc. Am. 69*, 275–283.

Sosa, O., Jr., and Beavers, J. B. 1985. Entomogenous nematodes as biological control organisms for *Ligyrus subtropicus* (Coleoptera: Scarabaeidae) in sugarcane. *Environ. Entomol. 14*, 80–82.

Sprenkel, R. K., and Brooks, W. M. 1975. Artificial dissemination and epizootic initiation of *Nomuraea rileyi*, an entomogenous fungus of lepidopterous pests of soybeans. *J. Econ. Entomol.* **68**, 847–851.

Stairs, G. R. 1965a. Artificial initiation of virus epizootics in forest tent caterpillar populations. *Can. Entomol.* **97**, 1059–1062.

Stairs, G. R. 1965b. Quantitative differences in susceptibility to nuclear-polyhedrosis virus among larval instars of the forest tent caterpillar, *Malacosoma disstria* (Hübner). *J. Invertebr. Pathol.* **7**, 427–429.

Stairs, G. R. 1966. Transmission of virus in tent caterpillar populations. *Can. Entomol.* **98**, 1100–1104.

Stairs, G. R. 1972. Pathogenic microorganisms in the regulation of forest insect populations. *Annu. Rev. Entomol.* **17**, 355–372.

Steinhaus, E. A. 1949. "Principles of Insect Pathology." McGraw-Hill, New York.

Steinhaus, E. A. 1954. The effects of disease on insect populations. *Hilgardia* **23**, 197–261.

Steinhaus, E. A. 1958. Stress as a factor in insect disease. *Proc. Int. Congr. Entomol. Tenth Congr.* (Montreal) **4**, 725–730.

Stephens, J. M. 1959. Mucin as an agent promoting infection by *Pseudomonas aeruginosa* (Schroeter) Migula in grasshoppers. *Can. J. Microbiol.* **5**, 73–77.

Studdert, J. P., and Kaya, H. K. 1990. Water potential, temperature, and soil type on the formation of *Beauveria bassiana* soil colonies. *J. Invertebr. Pathol.* **56**, 380–386.

Sutter, G. R., Rothenbuhler, W. C., and Raun, E. S. 1968. Resistance to American foulbrood in honey bees. VII. Growth of resistant and susceptible larvae. *J. Invertebr. Pathol.* **12**, 25–28.

Sweeney, A. W. 1977. Infection of aseptically reared mosquito larvae with *Culicinomyces* sp. *J. Invertebr. Pathol.* **30**, 273.

Sweeney, A. W., Hazard, E. I., and Graham, M. F. 1985. Intermediate host for an *Amblyospora* sp. (Microspora) infecting the mosquito, *Culex annulirostris*. *J. Invertebr. Pathol.* **46**, 98–102.

Szalay-Marzsó, L., and Vago, C. 1975. Transmission of baculovirus by mites. Study of granulosis virus of codling moth (*Laspeyresia pomonella* L.). *Acta Phytopathol. Acad. Sci. Hung.* **10**, 113–122.

Tamashiro, M., and Huang, S.-S. 1963. A cytoplasmic polyhedrosis of *Cactoblastis cactorum* (Berg). *J. Insect Pathol.* **5**, 397–399.

Tanada, Y. 1961. The epizootiology of virus diseases in field populations of the armyworm, *Pseudaletia unipuncta* (Haworth). *J. Insect Pathol.* **3**, 310–323.

Tanada, Y. 1963. Epizootiology of infectious diseases. *In* "Insect Pathology: An Advanced Treatise." (E. A. Steinhaus, ed.), Vol. 2, pp. 423–475. Academic Press, New York.

Tanada, Y. 1964. Epizootiology of insect diseases. *In* "Biological Control of Insect Pests and Weeds." (P. DeBach and E. I. Schlinger, eds.), pp. 548–578. Chapman and Hall, London.

Tanada, Y. 1985. A synopsis of studies on the synergistic property of an insect baculovirus: A tribute to Edward A. Steinhaus. *J. Invertebr. Pathol.* **45**, 125–138.

Tanada, Y., and Fuxa, J. R. 1987. The pathogen population. *In* "Epizootiology of Insect Diseases." (J. R. Fuxa and Y. Tanada, eds.), pp. 113–187. John Wiley & Sons, New York.

Tanada, Y., and Omi, E. M. 1974. Epizootiology of virus diseases in three lepidopterous insect species of alfalfa. *Res. Popul. Ecol.* **16**, 59–67.

Tashiro, H. 1987. "Turfgrass Insects of the United States and Canada." Cornell University Press, Ithaca, New York.

Thompson, C. G. 1978. Nuclear polyhedrosis epizootiology. *In* "The Douglas-Fir Tussock Moth: A Synthesis." (M. H. Brookes, R. W. Stark, and R. W. Campbell, eds.), pp. 136–140. U.S. Government Printing Office, Washington D. C.

Thompson, C. G., and Scott, D. W. 1979. Production and persistence of the nuclear polyhedrosis virus of the Douglas-fir tussock moth, *Orgyia pseudotsugata* (Lepidoptera: Lymantriidae), in the forest ecosystem. *J. Invertebr. Pathol. 33,* 57–65.

Thompson, C. G., Neisess, J., and Batzer, H.O. 1977. Field tests of *Bacillus thuringiensis* and aerial application strategies on western mountainous terrain. *U.S. Dep. Agric. For. Serv. Res. Pap. PNW-230.*

Thomson, H. M. 1958. Some aspects of the epidemiology of a microsporidian parasite of the spruce budworm, Choristoneura fumiferana (Clem.). *Can. J. Zool. 36,* 309–316.

Villani, M. G., and Wright, R. J. 1990. Environmental influences on soil macroarthropod behavior in agricultural systems. *Annu. Rev. Entomol., 35,* 249–269.

Watanabe, H. 1987. The host population. *In* "Epizootiology of Insect Diseases." (J. R. Fuxa and Y. Tanada, eds.), pp. 71–112. John Wiley & Sons, New York.

Weiser, J. 1963. Sporozoan infections. *In* "Insect Pathology: An Advanced Treatise." (E. A. Steinhaus, ed.), Vol. 2, pp. 291–334. Academic Press, New York.

Wellington, W. G. 1962. Population quality and maintenance of nuclear polyhedrosis between outbreaks of *Malacosoma pluviale* (Dyar). *J. Insect Pathol. 4,* 285–305.

Wellington, W. G., Cameron, P. J., Thompson, W. A., Vertinsky, I. B., and Landsberg, A. S. 1975. A stochastic model for assessing the effects of external and internal heterogeneity on an insect population. *Res. Popul. Ecol. 17,* 1–28.

Weseloh, R. M., and Andreadis, T. G. 1986. Laboratory assessment of forest microhabitat substrates as sources of the gypsy moth nuclear polyhedrosis virus. *J. Invertebr. Pathol. 48,* 27–33.

Whisler, H. C., Zebold, S. L., and Shemanchuk, J. A. 1975. Life history of *Coelomomyces psorophorae. Proc. Natl. Acad. Sci. U.S.A. 72,* 693–696.

White, R. T., and Dutky, S. R. 1940. Effect of the introduction of milky diseases on populations of Japanese beetle larvae. *J. Econ. Entomol. 33,* 306–309.

Wilding, N. 1970. *Entomophthora* conidia in the air-spora. *J. Gen. Microbiol. 62,* 149–157.

Wilding, N. 1975. Entomophthora species infecting pea aphis. *Trans. R. Entomol. Soc. 127,* 171–183.

Wilson, G. G. 1981. *Nosema fumiferanae*, a natural pathogen of a forest pest: Potential for pest management. *In* "Microbial Control of Pests and Plant Diseases 1970–1980." (H. D. Burges, ed.), pp. 595–601. Academic Press, New York.

Wilson, G. G., and Kaupp, W. J. 1976. Application of *Nosema fumiferanae* and *Pleistophora schubergi* (Microsporida) against spruce budworm in Ontario, 1976. *Can. For. Serv. Inf. Rep. No. IP-X-15.*

Wood, H. A., Hughes, P. R., Johnston, L. B., and Langridge, W. H. R. 1981. Increased virulence of *Autographa californica* nuclear polyhedrosis virus by mutagenesis. *J. Invertebr. Pathol. 38,* 236–241.

Woods, S. A., and Elkinton, J. S. 1987. Bimodal patterns of mortality from nuclear polyhedrosis virus in gypsy moth (*Lymantria dispar*) populations. *J. Invertebr. Pathol. 50,* 151–157.

Woods, S. A., Elkinton, J. S., and Podgwaite, J. D. 1989. Acquisition of nuclear polyhedrosis virus from tree stems by newly emerged gypsy moth (Lepidoptera: Lymantriidae) larvae. *Environ. Entomol. 18,* 298–301.

Zelazny, B. 1973. Studies on *Rhabdionvirus oryctes*. II. Effect on adults of *Oryctes rhinoceros. J. Invertebr. Pathol. 22,* 122–126.

Zelazny, B. 1976. Transmission of a baculovirus in populations of *Oryctes rhinoceros. J. Invertebr. Pathol. 27,* 221–227.

Zelazny, B., and Alfiler, A. R. 1991. Ecology of baculovirus-infected and healthy adults of *Oryctes rhinoceros* (Coleoptera: Scarabaeidae) on coconut palms in the Philippines. *Ecol. Entomol. 16,* 253–259.

Zimmack, H. L., and Brindley, T. A. 1957. The effect of the protozoan parasite *Perezia pyraustae* Paillot on the European corn borer. *J. Econ. Entomol. 50,* 637–640.

INDEX

Abbreviata caucasica, 518
Abraxas grossulariata, 277
Acanthocephala, 462–463, 521
Acanthocyclops vernalis, 443–444, 621
Acarapis woodi, 65
Acarina, 561, 576
Acarine disease, 400
Acheta domestica, 263
Acholeplasma, 158
Acholeplasmataceae, 158
Achromobacter, 36. See also *Xenorhabdus*
Achromobacter eurydice, 152
Achromobacter nematophilus. See *Xenorhabdus nematophilus*
Acinetobacter calcoaceticus var. *anitratus*, 111
Acquired immune deficiency syndrome (AIDS), 207, 595
Acromyrmex, 19
Actias selene, 277
Actinocephalidae, 420
Actinomycetes, 29–30, 153
Acute bee paralysis, 303–304. See also Honey bee, pathogens of
Acute bee-paralysis virus, 302, 304–306
Acyrthosiphon kondoi, 35
Acyrthosiphon pisum, 35, 325, 495. See also Pea aphid
Adeleida, 390
Adelina 390, 423
  hosts of, 424–425
  life cycle of, 424
Adelina cryptocerci, 424
Adelina melolonthae, 424
Adelina mesnili, 425
Adelina sericesthis, 425
Adenophorea, 461–462
Adipohemocytes, 507, 510
  characteristics of, 510
Adoxophyes orana, 215
Aedes, 37, 110, 114, 122, 365, 403, 406, 419, 427, 579
Aedes aegypti, 15, 55
  bacterial infection of, 122
  as bioassay insect, 568

  effect of fungal toxins, 364–365
  nematode infection of, 468
  protozoan infection of, 404, 419–420, 431
  resistance in, 466, 495, 560
  viral infection of, 262–263
Aedes albopictus, 330, 403, 405, 532
Aedes atropalpus, 191
Aedes cantator, 443
Aedes densovirus, 262
Aedes dorsalis, 115
Aedes epactius, 191
Aedes scutellaris, 191, 405
Aedes sierrensis, 403, 406, 407
Aedes sollicitans, 464, 530
Aedes taeniorhynchus, 256, 260
Aegerita, 6, 321
Aegerita webberi, 356
Aerobacter aerogenes var. *acridiorum*. See *Enterobacter aerogenes*
Aeromonas punctata, 84, 151
Aeromonas sp., 152
Aflatoxins, 363–364
African sleeping sickness, 391, 399
Agamermis culicis. See *Perutilimermis culicis*
Agamermis decaudata, 468, 474
Agaricus, 21
Agraulis vanillae, 263. See also Silver-spotted flambeau
Agria affinis, 64
Agrobacterium, 102
Agrotis ipsilon, 95. See also Black cutworm
Agrotis segetum, 22, 186, 202, 277. See also *Scotia segetum*
Ailanthus triphysa, 122
Akanthomyces, 320
Alfalfa caterpillar, 285. See also *Colias eurytheme*
Alfalfa looper, 174, 309, 579. See also *Autographa californica*
Alfalfa weevil, 38, 326, 341, 344, 467, 469–470, 600. See also *Hypera postica*
Allantonematidae, 462
  biology, life cycle, 479–481

Almond moth, 69, 184, 496. See also *Cadra cautella*
Altenaria, 365
Altenaria porri, 365
Altenaria solani. See *A. porri*
Alternate host, 443, 604. See also Heteroecism
Alternation of generations, 346, 462, 480, 621
Amastigote, 393, 397–399
Amber disease, 150, 617
Amblyospora, 390, 427, 430–431, 434, 443–445, 604, 621. See also *Thelohania*
  life cycle of, 441, 443
  meiosis, 447
Amblyospora bracteata, 435
Amblyospora californica, 439, 446
Amblyospora connecticus, 444, 621
Amblyosporidae, 428–429
  proliferative sequence, 445
  transmission of, 444–445
Ambrosia beetle, 17–19, 22, 38. See also Fungi, insect cultivated
Ambrosia fungi, 17–18. See also Fungi, insect cultivated
Ambrosiella, 18
American cockroach, 419, 513, 520, 523, 525–526, 533. See also *Periplaneta americana*
American foulbrood, 152–153, 349, 610–612. See also *Bacillus larvae*
  host resistance in, 493–496, 498–499, 503
  syndrome of, 119–120
  vernacular names of, 119
American termite. See *Coptotermes acinaciformis*
Ameson, 447
Amoeba, 401–402
  characteristics of, 399
  entomogenous forms, 399–400
Amoeba chironomi, 402
Amoebiasis
  of honey bee, 400
  of grasshopper, 401–402
Amoebic infections. See Rhizopoda infections
Amoebida, 390, 400
Amoebidae, 399–400
Amoebidium parasiticum, 15
Amoeboses. See Amoebiasis
Amphimallon solstitiales, 418
Amsacta mourei, 250–251
Amyelois transitella, 307. See also Navel orangeworm
Amylostereum, 19
Amylostereum areolatum, 479, 577, 604
Amylostereum chailletti, 19
Anacystis nidulans, 114
Anaeroplasmataceae, 158
Anagasta kuehniella, 88, 108–109, 568. See also Mediterranean flour moth
Ancylistaceae, 337, 347
Ancylistes, 337

Ancyrophora, 419
Angatia, 356
Animalia, kingdom, 389
Anisoplia, 329
Anisoplia austriaca, 359
Anopheles, 114, 335, 419, 423, 579
Anopheles albimanus, 441
Anopheles annulipes, 84, 151
Anopheles gambiae, 115
Anopheles maculipennis, 396
Anopheles quadrimaculatus, 332, 431
Anopheles stephensi, 122
Anoplonyx destructor, 284
Anoplura, 351, 430
Anoxia, 296–297
Ant
  fungus-growing, 19, 21–22
  leaf-cutting, 19, 22
Antagonism, 73, 560, 583. See also Interference
Antheraea eucalypti, 174, 308. See also Emperor gum moth
Antheraea helena, 308
Antheraea pernyi, 109, 495, 523. See also Chinese oak silkworm
Anthomyidae, 347
Anthonomus grandis, 25, 150, 423, 499. See also Boll weevil
Antibacterial substances, 22, 525
Antibiosis, 14
Antibody, 2, 513, 521–523, 525, 561
  antibiotic, 359, 478, 529–530
Anticarsia gemmatalis, 200, 324, 494, 500. See also Velvet bean caterpillar
Antifreeze compound, 55. See also Cryoprotectant
Antimicrobial compounds, 498–499, 501, 558–559
Antiviral principal, 531
Apanosporoblastina, 429, 439
Apanteles, 247
Apanteles glomeratus, 517, 532
Apanteles melanoscelus. See *Cotesia melanoscela*
Apanteles militaris. See *Glyptapanteles militaris*
Aphanomycopsis, 336
Aphanomycopsis sexualis, 336
Aphasmida. See Adenophorea
Aphelenchida, 462
Aphelenchoididae, 462
  biology and life cycle, 481–482
Aphid
  endosymbionts of, 34–35
  epizootics in, 614
  microbial control of, 566, 575–576, 581, 583
  pathogens of
    fungus, 322, 324, 338, 342–343, 347, 362
    virus, 219, 302
  resistance in, 495
Aphid nonoccluded baculovirus, 219

Aphis gossypii, 575–576
Aphis spiraecola, 344
Aphodius depressus, 419
Aphodius fimetarius, 478
Aphodius tasmaniae, 118
Aphthovirus, 300
Apicomplexa, 389–391, 423
　characteristics of, 415
　classes of, 415
Apis cerana, 260
Apis mellifera, 63, 302, 400. See also Honey bee
Aporia crataegi, 7
Aporophylla lutulenta, 277
Aposymbiotic hosts, 27, 37,
　methods of obtaining, 38–39, 584
Appressorium, 323, 325, 332–333, 336
Arbovirus, 275, 296–297, 309, 532
Archigregarinida, 416
Archips argyrospila, 209. See also Fruittree leafroller
Argentine ant, 469. See also *Iridomyrmex humilis*
Argidae, 189
Argyrotaenia, 580
Aristotle, 3
Arkansas bee virus, 302, 309
Armadillidium vulgare, 259
Armigeres, 419
Armyworm. See also *Pseudaletia unipuncta*
　epizootic in, 190, 610
　pathogen complementation in, 73, 210, 602
　pathogen interference in, 73–74
　resistance in, 499, 504, 521
　viral infection of, 180, 192
　toxin, 214
Arrhenosphaera cranei, 349
Artemia salina, 359
Arthrocytes, 512–513
Artichoke plume moth, 557, 583
Artogeia rapae, 108, 211, 215. See also Imported cabbageworm
Artogeia rapae granulosis virus, 215
Ascetospora, 389
Aschersonia, 6, 320, 355, 365
Aschersonia aleyrodis, 355, 365
Aschersonia cubensis, 355. See also Cuban *Aschersonia*
Aschersonia goldiana, 355
Aschersonia turbinata, 355. See also Turbinate fungus
Ascocystis culicis. See *Ascogregarina culicis*
Ascogaster, 247
Ascogregarina, 390, 420
Ascogregarina armigerei, 420
Ascogregarina culicis, 419
Ascogregarina taiwanensis, 420
Ascoidea, 18

Ascomycetes, 319, 347, 365
Ascomycotina, 18, 319–321, 336, 347, 356
Ascosphaera, 320, 348–349
Ascosphaera aggregata, 349
Ascosphaera apis, 348–349, 493, 501, 506. See also Chalkbrood fungus
Ascosphaera cranei, 349
Ascosphaera major, 348
Ascosphaera proliperda, 349
Ascosphaeraceae, 348
Ascosphaerales, 320, 348
Ascoviridae, 174–175, 245
　characteristics of, 253
　cytopathology of, 255
　pathogenesis of, 255
　replication of, 254–255
　structure of, 253
Ascovirus, 253–255. See also Ascoviridae
Aspergillus, 320
　species of, 363
　toxins of, 363–364
Aspergillus flavus, 357, 363–364, 501. See also Yellow muscardine fungus
Aspergillus flavus oryzae, 326
Aspergillus fumigatus, 363–364
Aspergillus niger, 517, 521, 530
Aspergillus ochraceus, 363–364
Aspergillus parasiticus, 363
Aspergillus repens, 363
Aspergillus tamarii, 363
Aspergillus versicolor, 363
Aspidiotus, 17
Atkinsiella, 334
Atkinsiella entomophaga, 336
Atta, 19, 21
Attacin, 526, 528, 532
　gene of, 529
Attagenus megatoma, 419. See also Black carpet beetle
Attamyces, 21
Attamyces bromatificus, 21
Attini, 19
Aulacophora foveicollis. See *Raphidopalpa foveicollis*
Aureobasidium pullulans, 365
Auriculariaceae, 15
Autogamy, 446–447
Autographa californica, 69, 174, 186, 309. See also Alfalfa looper
　nuclear polyhedrosis virus of
　　DNA replication and gene expression, 202–204, 216
　　genome of, 200–201
　　host range of, 189
　　as microbial control agent, 566
　　polyhedra of, 186, 188, 602
　　as recombinant virus, 207, 565
　　replication of, 180, 185, 196–197
　　type species of, 177, 183

Babesiosis, 391
Bacillaceae, 86–122
Bacillus, 86
Bacillus alvei, 121–122, 152
Bacillus anthracis, 101
Bacillus apisepticus, 149
Bacillus bombycis, 306–307
Bacillus brevis, 121–122
Bacillus cereus, 119, 524
    characteristics of, 68, 86–87, 97
    crystal genes of, 101
    host resistance to, 527, 532
    pheromone production in, 25
    as potential pathogen, 502
    toxin production in, 84, 86, 91, 94, 110, 504
    transformation of, 102
Bacillus danysz. See *Salmonella enteritidis*
Bacillus firmus, 122
Bacillus fribourgensis, 118
Bacillus larvae. See also American foulbrood
    characteristics of, 119
    epizootiology of, 610–612
    host resistance to, 121, 493, 496, 499, 503, 506, 523
    replication of, 119
Bacillus laterosporus, 122, 152
Bacillus lentimorbus, 8, 115, 565, 574
    parasporal body of, 118
    syndrome of, 118
Bacillus lentimorbus var. *melolontha*, 118
Bacillus megaterium, 102, 122, 523, 528
    crystal of, 97
    genetic engineering with, 97
Bacillus melolonthae liquefaciens, 149
Bacillus metiens, 498
Bacillus moritai, 119
Bacillus noctuarum, 150
Bacillus pluton. See *Melissococcus pluton*
Bacillus popilliae
    characteristic of, 86
    crystal of, 117
    epizootiology of, 603–604, 607, 617
    historical account, 8, 574
    host resistance to, 493, 504
    as microbial control agent, 565, 567, 574–575
    replication of, 115–118
Bacillus popilliae var. *lentimorbus*, 118
Bacillus popilliae var. *melolontha*, 118
Bacillus popilliae var. *popilliae*, 118
Bacillus pulvifaciens, 122
Bacillus rossius, 219
Bacillus sphaericus, 566
    crystal protein gene of, 114
    host and pathology of, 114–115
    parasporal crystal of, 113–114
    strains of, 112
    toxins of, 113–114
Bacillus sphingidis, 149

Bacillus subtilis, 101–102, 119, 122, 523, 528
    crystal of, 97
    genetic engineering with, 97
Bacillus thuringiensis. (See also specific subspecies)
    crystal of
        formation, 87, 96–97
        number and type, 93, 97–98, 151
        protein, 88, 99
    epizootiology of, 600, 603
    historical account, 7–8, 87–88
    hosts of, 114
    host resistance to, 111, 494–496, 498, 504–506, 520–521, 523, 525–526, 528, 532
    IPM approach, 582–583
    life cycle of, 87
    as microbial control agent, 556–562, 565–566, 575, 578–580, 609
        application, 569–571
        formulation, 91
        standardization, 91, 568
        storage, 569
    pathogenicity to vertebrates, 104, 111–112
    recombinant DNA, 102, 563–564
    subspecies of
        pathotype, 89
        serotype, 88–89
    toxins of
        δ-endotoxin, 96
        α-exotoxin, 91, 94
        β-exotoxin, 94–96
        phospholipase C (lecithinase), 91, 94
        others, 110–111
    virulence and pathogenicity, 68–70, 84–86, 364, 602
Bacillus thuringiensis aizawai, 90–92, 99–100, 102, 104, 565
    crystal protein gene of, 108
    β-exotoxin, 94
Bacillus thuringiensis alesti, 90, 92
Bacillus thuringiensis amagiensis, 91
Bacillus thuringiensis berliner, 100
Bacillus thuringiensis cameroun, 91
Bacillus thuringiensis canadensis, 90
Bacillus thuringiensis colmeri, 90
Bacillus thuringiensis coreanensis, 91
Bacillus thuringiensis dakota, 90
Bacillus thuringiensis darmstadiensis, 89–91, 104, 108
    exotoxin and endotoxin genes of, 94
    parasporal body of, 98
    production of β-exotoxin, 94
Bacillus thuringiensis dendrolimus, 101, 108, 579
Bacillus thuringiensis entomocidus, 90, 100, 102, 104
Bacillus thuringiensis finitimus, 90, 97
Bacillus thuringiensis fukuokaensis, 89–90
Bacillus thuringiensis galleriae, 90–92, 94, 103–104, 579

Bacillus thuringiensis indiana, 90
Bacillus thuringiensis israelensis, 90, 92
   crystal of, 98, 103–104
      gene, 100, 104
   historical account, 89
   host resistance to, 495, 560
   hosts of, 89, 91, 106, 114
   as microbial control agent, 461, 565, 579
      reference standard, 91, 568
   mode of action, 107
   recombinant DNA, 102
   toxin of, 103–104
   vertebrate pathogenicity of, 111
Bacillus thuringiensis japonensis, 90
Bacillus thuringiensis kenyae, 90, 94, 103
Bacillus thuringiensis konkukian, 91
Bacillus thuringiensis kumamotoensis, 90
Bacillus thuringiensis kurstaki, 90, 92
   crystal of, 97, 103
      gene, 100
   host resistance to, 560
   hosts of, 89, 91, 104, 106, 110
   as microbial control agent, 565, 579–580
      commercial formulation, 91
      reference standard, 91, 568
   recombinant DNA, 102
   vertebrate pathogenicity of, 111
Bacillus thuringiensis kyushuensis, 89–90, 104
Bacillus thuringiensis leesis, 91
Bacillus thuringiensis medellin, 91
Bacillus thuringiensis mexicanensis, 91
Bacillus thuringiensis monterrey, 91
Bacillus thuringiensis morrisoni, 89–90, 94, 100, 104
Bacillus thuringiensis neoleonensis, 90
Bacillus thuringiensis nigeriensis, 90
Bacillus thuringiensis ostriniae, 90
Bacillus thuringiensis pakistani, 90
Bacillus thuringiensis pondicheriensis, 90
Bacillus thuringiensis san diego, 89, 100. See also *Bacillus thuringiensis tenebrionis*
Bacillus thuringiensis shandongiensis, 90
Bacillus thuringiensis silo, 91
Bacillus thuringiensis sotto, 90, 92
   crystal of, 97
      gene, 100
   historical account, 87–88
   host syndrome, 71, 108
Bacillus thuringiensis subtoxicus, 91
Bacillus thuringiensis sumiyoshiensis, 90
Bacillus thuringiensis tenebrionis. See also *Bacillus thuringiensis san diego*
   crystal of, 91, 98
      gene, 100
   hosts of, 89, 91, 108
   as microbial control agent, 565, 568, 570, 579–582
   recombinant DNA, 102
   reference standard, 91, 568

Bacillus thuringiensis thompsoni, 90, 94
Bacillus thuringiensis thuringiensis, 71, 90, 92, 94, 100, 103, 565, 568, 579
Bacillus thuringiensis tochigiensis, 90
Bacillus thuringiensis toguchini, 91
Bacillus thuringiensis tohokuensis, 90
Bacillus thuringiensis tolworthi, 90, 94, 103–104
Bacillus thuringiensis toumanoffi, 90, 94
Bacillus thuringiensis wuhanensis, 91, 101
Bacillus thuringiensis yunnanensis, 90, 97
Bacteria, 29–30
   characteristics of, 83
   as endosymbiont, 35–36
   entomopathogenic families of, 85
   internal microbiota of, 25
   symbiont of flagellate, 28–29
   taxonomy of, 83
Bacterial infections, 86–122, 147–162
   entomopathogenic bacteria, types of, 85
   pathology of, 84–85
   portals of entry, 84–85
   symptoms of, 84
Bacteremia, definition of, 84
Bactericidal activity (factor), 506, 524–529, 531
Bacteriocin, 110, 506
Bacteriocytes, definition of, 14
Bacteriotome, definition of, 14
Bacterium eurydice. See *Achromobacter eurydice*
Bacteroides termitidis, 25
Baculoviridae, 174, 182–220, 245
   DNA sequence homology, 177
   infectious elements of, 181–182
   subgroups of, 177
   viral particles of
      enveloped virion, 179–181
      nucleocapsid, 177, 179
Baculovirus
   complementation in, 73, 189
   epizootiology of, 601–607
   historical aspects, 306
   host resistance to, 494–495, 560–561
   infectious element of, 181–182
   infectivity of, 69, 182, 504, 506
   interference in, 74
   genes, 201–205, 247, 281
   genetics, molecular, 563, 565
   as microbial control agent, 556–558, 568–570, 572, 578, 583
   molecular phylogeny, 176–177
   mutant (variant) 70, 201–203
   nonoccluded, 177, 216–220, 572–574
   nucleocapsid morphogenesis of, 179
   phenotype of, 181–182
   safety of, 561
   transmission of, 182
   vector of, 2, 203, 205–207
   virion of, 178, 181
Baetis, 407

Baetis vernus, 420
Balantidium, 390, 403
Ballocephala, 337
Barathra brassicae, 433
Bark beetles. See also *Dendroctonus*
  fungal culture by, 17–18
  fungal infection of, 347, 365
  internal microbiota of, 25
  nematode infection of, 464, 469, 471, 480–481, 606
  protozoan infection of, 400, 422–423
Barrouxia, 390, 423
Basidiobolaceae, 337, 347
Basidiobolus, 337, 347
Basidiobolus ranarum, 347
Basidiomycetes, 319
Basidiomycotina, 15, 18, 21, 319, 321, 336, 347, 356
Bassi de Lodi, 3–4, 357
Bassianolide, 362
Bathyplectes, 246
Bathyplectes anurus, 246
Bathyplectes curculionis, 246
Beauveria, 320, 365
  hosts of, 358–359
  pathogenicity of, 359
  replication of, 359
  strains of, 357–358
  structure of, 357
Beauveria amorpha, 358
Beauveria bassiana. See also White-muscardine fungus
  characteristics of, 357
  historical account, 3, 6
  host resistance to, 497–498, 501, 515, 521, 530, 559
  hosts of, 322, 358, 618
    hemolymph, 328
  infection route of, 323, 326
  as microbial control agent, 566
  nomenclature of, 357
  pathogenicity of, 359
  strains of, 70, 357
  toxins of, 359, 362
Beauveria brongniartii, 358
Beauveria tenella, 358, 365
Beauveria velata, 358
Beauvericin, 359, 365
Beddingia siricidicola. See *Deladenus siricidicola*
Beddingia spp. See *Deladenus*
Bee acute paralysis virus, 300
Bee slow paralysis virus, 300, 302
Bee virus X, 300, 302
Bee virus Y, 302
Beet armyworm, 95, 185, 579. See also *Spodoptera exigua*
Benign infection, 621
Benzene hexachloride (BHC), 59
Bergoldia, 173

Biorational pesticide, 555
Birnaviridae, 275–276, 296–297
Biston betularia cytoplasmic polyhedrosis virus, 277
Blabera fusca, 493
Blaberus, 514, 523
Blaberus craniifer, 513–514, 520, 522, 525
Black beetle, 309. See also *Heteronychus arator*
Black-beetle virus, 309
  characteristics of, 310
  pathology of, 310
  replication of, 310
Black blow fly, 497. See also *Phormia regina*
Black carpet beetle, 419. See also *Attagenus megatoma*
Black cutworm, 95. See also *Agrotis ipsilon*
Black fly. See also Simuliidae
  bacterial infection of, 89, 109, 112
  fungal infection of, 321, 330
  microbial control of, 565, 579–580
  nematode infection of, 467–468, 474, 533
  protozoan infection of, 402, 405
  resistance in, 466, 496, 519
  viral infection of, 284, 286
Black queen-cell virus, 302
Black scale fungus, 355. See also *Myriangium duriaei*
Blastocladiales, 320, 329–334
Blastocrithidia, 389, 394–397
Blastocrithidia caliroae, 397
Blastocrithidia raabei, 393, 397
Blastocrithidia triatomae, 397
Blastodendrion, 320
Blastodendrion pseudococci, 348
Blastogregarinida, 416
Blastomyces bicuspidata, 512
Blatta orientalis, 390, 402, 420
Blattabacterium, 36
Blattabacterium cuenoti, 33
Blattella germanica, 348, 420, 515, 518–519. See also German cockroach
Blattidae, symbiont of, 33
Blattodea, 261, 351
Bledius, 351
Blissus leucopterus, 359. See also Chinch bug
Blue disease, 154, 617
Blue tongue virus VP2, 207
Boarmia selenaria, 151
Bodo, 389
Bodo muscarum. See *Herpetomonas muscarum*
Bodonidae, 393
Boll weevil, 150, 499, 582. See also *Anthonomus grandis*
Bombus impatiens, 248
Bombyx, 105, 262
Bombyx DNV-1, 261–263
Bombyx DNV-2, 261–263
Bombyx mori. See also Silkworm
  bacterial infection of, 97, 105, 109
  hemocytes of, 508

protozoan infection of, 398
recombinant gene, 563, 565
resistance in, 501
viral infection of, 205, 261, 278, 281–282, 287
Bombyx mori cytoplasmic polyhedrosis virus, 275, 277, 287
Bombyx mori nuclear polyhedrosis virus, 188, 199, 200, 202, 206–207
Boolarra virus, 309
Boric acid, 69–70, 602
Borrelina, 173, 183
Borrelina bombycis, 183
Borrelinaceae, 173
Bostrychid beetle, 39
Botrytis bassiana, 357
Boules hyalines, 211
Bovicola bovis, 111
Bovicola crassipes, 111
Bovicola limbatus, 111
Bovicola ovis, 111
Brachypelta aterrima, 39
Brachytosis, 84, 122, 147, 153
Braconidae, 246
Bracovirus, 247. See also Polydnaviridae
Bradysia paupera, 465
Brown planthopper, 36, 343. See also *Nilaparvata lugens*
Brown soft scale, 355. See also *Coccus hesperidum*
Brown-tail moth, 184. See also *Euproctis similis*
Buchnera, 35
Bunyaviridae, 276
Burenella, 434
Burenella dimorpha, 431, 445
Burenellidae, 429, 445
Burkeidae, 428
Bursaphelenchus xylophilus, 460
Buxtehudea scaniae, 440
Buxtehudeidae, 428
Byssochlamys, 361

Cabbage looper. See also *Trichoplusia ni*
epizootics in, 606
microbial control of, 568, 581
pathogens of
bacteria, 71, 151, 157
protozoa, 432
virus, 174, 189, 196
resistance in, 495, 497, 500, 502
Cacoecia murinana. See *Choristoneara murinana*
Cadra cautella, 69, 184, 496. See also Almond moth
Caliciviridae, 276, 307
Calicivirus, 299, 307–308
California oak moth, 494. See also *Phryganidia californica*

Caliroa cerasi, 397. See also Pear slug
Calliphora, 398, 527
Calliphora erythrocephala, 15, 365, 402
Calliphoridae, 189
Calomyrmex, 501
Calonectria, 320, 355
Calophasia lunula, 498
Camnula pellucida, 148
Campoleginae, 246
Campoletis sonorensis, 246
Campoletis sonorensis polydnavirus, 246–247
Campoplex, 246
Camptochironomus tentans, 155, 252
Canavanine, 558
Candida guilliermondi, 347
Candida krusei, 347
Capilliconidium, 339–340. See also Capillispore
Capillispore, 339. See also Capilliconidium
Capsule, 181, 207–214, 251. See also Granulosis virus
dissolution of, 210
enhancing factor in, 210
formation of, 214
shapes of, 209
structure, 209
virion occlusion, 210
Capsules, blood, 322, 509, 512, 515, 524, 529–530. See also Hemocyte capsule
Carabidae, 419
Carausius morosus, 398
Carcinoma, 561. See also Tumors
Carcinus maenas, 391
Cardiochiles nigriceps, 247
Cardiovirus, 300–301
Casinaria, 246
Catenaria, 330
Catenariaceae, 330
Caudospora, 434
Caudosporidae, 429
Caulleryella, 421–423
Caulleryella anophelis, 423
Caulleryella pipientes, 423
Cecropia, 53
Cecropin, 526–529, 532
Cell line, IMC-H$_2$-1, 198, 218
Cell line, IPLB-SF-21, 198
Cephalosporium, 18, 353, 355, 362
Cephalosporium lecanii, 355, 362. See also White-halo fungus
Cerambycidae, 36, 189
Ceratitis capitata, 302
Ceratocystis, 18
Ceratophyllus fasciatus, 402
Ceratopogonidae, 89, 365
Cerecine, 86, 110
Cerococcus, 17
Cestoda, 521
Cetonia aurata, 521
Chagas' disease, 397
Chagasella, 423

Chalkbrood, 348, 493–494
Chalkbrood fungus, 501, 506. See also *Ascosphaera apis*
Chalky disease, 153
Chaoboridae, 523
Chaoborus crystallinus, 220
Chapmanium. See *Napamichum*
Chelonus, 247
Chelonus near *curvimaculatus*, 247
Chemical insecticide, 495, 497, 554–558, 560, 563, 566–567, 569, 571, 578–580, 582–583. See also Injuries, chemical agents
Chermes. See *Kermes*
Chilo iridovirus, 256–260
Chilocorus bipustulatus, 351
Chilomastix, 390
Chilopoda, 259
Chinch bug, 6, 359. See also *Blissus leucopterus*
Chinese oak silkworm, 109, 528. See also *Antheraea pernyi*
Chinese plantworm, 3
Chionaspis, 17
Chironomidae, 89, 189, 332, 365, 523
Chironomus attenuatus, 251, 253
Chironomus decorus, 157
Chironomus frommeri, 157
Chironomus luridus, 250, 530
Chironomus plumosus, 256, 404
Chironomus thummi, 157, 530
Chitinase, 84, 110, 150–151, 324, 526, 582
Chlamydia, 154, 157, 172
Chlamydozoon bombycis, 182
Chloriridovirus, 256–257
Choanomastigote, 393
Chordodes japonensis, 482
Choristoneura biennis, 251
Choristoneura conflicta, 175
Choristoneura fumiferana. See also Eastern spruce budworm
  epizootics in, 620–621
  microbial control of, 574, 578
  pathogens of
    bacteria, 108
    protozoa, 67, 432
    virus, 74, 193, 202, 207, 252
  resistance in, 498, 500, 504, 527
Choristoneura murinana, 207. See also Pine shoot roller
Chronic bee-paralysis virus 302
  characteristics of, 304–305
  differ from acute bee paralysis, 305–306
  host resistance to, 493
  pathogenesis of, 305
  syndrome of, 304–305
  transmission of, 305
Chronic bee-paralysis virus associate, 302–303, 306
Chronic stunt virus, 307–308

Chrysomela polita, 419
Chrysomphalus, 17
Chrysopa carnea, 26, 95
Chrysopidae, 189
Chytrid, 321, 336
Chytridiales, 320, 329–330
Chytridiomycetes, 320, 329
Chytridiomycota, 391
Chytridiopsida, 428
Chytridiopsidae, 428
Chytridiopsis, 390
Ciliate, 17, 390–391
  characteristics of, 402
  ciliatoses, 404–407
  as entomopathogens, 402–403
  penetration into hemocoel, 403–404
Ciliatosis by *Lambornella*, 405–407
Ciliatosis by *Tetrahymena*, 404–405
Ciliophora, 388–391
  characteristics of, 402
  entomogenous forms of, 402–403
  penetration into hemocoel, 403–404
Cimex lectularis, 31
Citrobacter, 25
Citrobacter freundii, 25
Citrus red mite, 95. See also *Panonychus citri*
Citrus rust mite, 354, 363, 576. See also *Phyllocoptruta oleivora*
Cixiinae, 31
Claviceps paspali, 364
Clavicipitaceae, 352, 355
Clavicipitales, 352
Cleonus, 329
Cleons punctiventris, 6. See also Sugar-beet weevil
Cloaca cloacae var. *acridiorum*, 149
Cloaca type A. See *Coccobacillus acridiorum*
Cloaca type B. See *Serratia liquefaciens*
Clostridium, 68, 86, 117, 122, 153, 506
Clostridium brevifaciens, 122
Clostridium malacosomae, 122
Clostridium perfringes, 94
Clostridium tetani, 493
Cloudy-wing virus, 302
Clover cutworm, 157, 210. See also *Scotogramma trifolii*
Clustering disease, 260
Coagulocytes, 507, 510–511, 515, 518
Coccidia, 390–391, 415, 423–424
Coccobacillus, 149
Coccobacillus acridiorum, 6–7, 149
Coccus (virus), 182
Coccus hesperidum, 355. See also Brown soft scale
Coccus viridis, 355. See also Green scale
Cochliomyia hominivorax, 61. See also Screwworm fly
Cockroach, wood-eating, 27–30. See also *Cryptocercus*

Codling moth, 174, 358, 460, 498, 557, 566, 579. See also *Cydia pomonella*
Codling moth granulosis virus, 209, 496, 566, 579
Coelogregarina, 423
Coelogregarina ephestiae. See *Mattesia dispora*
Coelomomyces, 320, 327, 604
    heteroecism in, 330–332
    hosts of, 330, 332
    hybridization of, 333
    replication of, 333–334
    species of, 330
Coelomomyces chironomi, 332
Coelomomyces dodgei, 332–333
Coelomomyces indiana, 332
Coelomomyces lativittatus, 333
Coelomomyces macleayae, 332
Coelomomyces notonectae, 332
Coelomomyces psorophorae, 322–323, 331–333
Coelomomyces punctatus, 333
Coelomomyces stegomyiae, 330
Coelomomycetaceae, 330
Coelomomycetes, 332
Coelomycetes, 357, 365
Coelomycidium, 320, 330
Coelomycidium simulii, 321–322, 330
Colchicum autumnale, 398
Cold-hardiness, 54–56, 432
Coleoptera, 17, 30, 86, 89, 99, 111, 149, 158–159, 189, 216, 249–250, 258, 277, 309, 322, 324, 338, 351–352, 358, 398, 420, 424, 493, 506–507, 529
Colias, 580, 606
Colias eurytheme, 285. See also Alfalfa caterpillar
Colladonus montanus, 158
Collembola, 32, 424
Colorado potato beetle, 160, 358–359, 497, 559, 565, 568, 570, 579–582, See also *Leptinotarsa decemlineata*
Colpoda, 407
Commensalism, definition of, 13
Comperia merceti, 26
Complement, 521, 523, 526, 530–531
Complementation (synergism), 73, 189
Completoria, 337
Conidiobolus, 320, 328, 337–338, 347
Conidiobolus apiculata, 345
Conidiobolus coronata, 345, 347
Conidiobolus obscurus, 322, 324–325, 338, 343
Conidiobolus osmodes, 347
Conidiobolus thromboides, 341, 345
Conoderus falli, 495
Conoid, 415–416
Contortylenchus, 480–482
Contortylenchus elongatus, 480
Contortylenchus reversus, 469, 480
Coptosoma scutellatum, 39

Coptotermes acinaciformis, 419. See also American termite
Coptotermes formosanus, 25, 29. See also Formosan subterranean termite
Coptotermes lacteus, 25
Corcyra cephalonica, 193, 504
Cordycepin, 353
Cordycepioideus bisporus, 355
Cordycepioideus octosporus, 355
Cordyceps, 3, 320, 327 363
    historical aspects, 352–353
    structure of, 353
Cordyceps australis, 354
Cordyceps clavulata, 363
Cordyceps cucumispora, 363
Cordyceps kniphofioides, 363
Cordyceps militaris, 355
Cordyceps myrmecophila, 352
Coreomycetopsis oedipus, 15
Corn earworm, 69, 323, 359, 432, 496–497, 500, 520, 567, 582. See also *Heliothis zea*
Corynebacterium, 35–36
Corynebacterium diphtheriae, 493
Corynebacterium okanaganae, 153
Costelytra zealandica, 25, 150, 309. See also New Zealand grass grub
Cotesia, 247
Cotesia marginiventris, 216–217, 248, 255
Cotesia melanoscela, 219, 246, 583
Cotton boll weevil, 423. See also *Anthonomus grandis*
Cotton bollworm, 565
Cotton leafworm, 193. See also *Spodoptera littoralis*
Couchia, 320, 336
Couchia circumplexa, 336
Cougourdellidae, 429
Crescent cells, 510
Cricket paralysis virus, 531
    host of, 300
    pathogenesis of, 301
    viral particle of, 301
Crimson fungi, 355. See also *Verticillium cinnamomeum*
Crithidia, 389–390, 394–396
Crithidia fasciculata, 395–396
Crithidia flagellatoses, 396–397
Crithidia gerridis, 395
Crustacea, 259
Cryoprotectant, 55. See also Antifreeze compound
Cryptocercus, 27, 29. See also Cockroach, wood-eating
Cryptocercus punctulatus, 27, 29–30, 424. See also Wood cockroach
Cryptolaemus montrouzieri, 64
Crystal protein toxin, 558. See also Endotoxin, δ-
Ctenocephalides canis, 398. See also Dog flea

Cuban *Aschersonia*, 355. See also *Aschersonia cubensis*
Cubitermes severus, 30
Culex, 37, 114, 122, 191, 262, 335, 406, 419, 579
Culex australicus, 36
Culex fatigans, 122
Culex globocoxitus, 36
Culex peus, 363
Culex pipiens, 347, 360, 423
   endosymbionts of, 36
   microbiota, internal, 24
   reproductive incompatibility in, 37
Culex quinquefasciatus, 115
Culex salinarius, 284, 444
Culex tarsalis, 191, 284, 363, 404, 439, 446, 472
Culex tritaeniorhynchus, 309
Culicicola, 337
Culicidae, 89, 189, 332, 523
Culicinomyces, 320, 325
Culicinomyces clavisporus, 364–365
Culicoides, 297
Culicospora magna, 447
Culicosporella, 447
Culicosporella lunata, 445
Culicosporellidae, 445
Culicosporida, 428
Culicosporidae, 428–429, 445
Culiseta, 262, 335
Culiseta incidens, 112
Culiseta inornata, 332
Curculionidae, 36, 189
Cyclocephala, 116
Cyclocephala hirta, 116
Cycloneda sanguinea, 350
Cyclops vernalis, 332
Cydia pomonella, 174, 211, 213, 215, 496, 498. See also *Laspeyresia pomonella*, codling moth
Cylindrosporidae, 428
Cylindrosporidea, 428
Cypovirus, 275
Cyrtophora, 390
Cysts, cuticular invasion, 391, 403–404, 406, 601
Cytoplasmic polyhedrosis, 284
Cytoplasmic polyhedrosis virus, 175, 188–189, 253, 281, 285, 498, 504, 601, 603, 606, 618
   characteristics of, 275–276
   histopathology of, 286–287
   host resistance to, 287, 493, 495–496, 504
   hosts of, 277
   interference, 74
   replication of, 283–284
   syndrome of, 284–286
   transcription of, 278–279
   transmission of, 286

   types of, 277
   viral particle of, 277–279
Cytoplasmic polyhedrosis virus inclusion body, 184
   characteristics of, 279–280
   dissolution of, 281
   formation and structure of, 279–281
   genetics of, 282
   interference in, 282

Dacus, 24
Dacus oleae, 20, 24, 301. See also Olive fly
Dacus tryoni, 281, 286. See also Queensland fruit fly
Dahlbominus fuscipennis, 54
Daphnia, 512
Dasyhelea obscura, 348
Dauer juveniles, 462–463, 475–476, 478, 603
DDT, 58–59, 95, 337
Deladenus, 464, 479
Deladenus siricidicola, 465, 468–469, 577, 604
   biology and life cycle of, 479–480
Deladenus wilsoni, 469
Dendroctonus, 25. See also Bark beetle
Dendroctonus frontalis, 18
Dendroctonus pseudotsugae, 469
Dendroides canadensis, 55
Dendrolimus sibiricus, 110. See also Siberian silkworm
Dendrolimus spectabilis, 110
Dengue virus, 207
Densonucleosis, 264
Densonucleosis virus, 261, 283, 307, 493, 531. See also Densovirus
Densovirus, 261–263. See also Densonucleosis virus
Dependovirus, 261–262
Dermaptera, 32, 351
Dermestidae, 189
Deroceras reticulatum, 404
Desiccation. See Injuries, from physical agents, moisture
Desmethyldestruxin, 360
Destruxins, 360–361, 364, 517
Deuteromycotina, 18, 319–321, 328, 336, 347, 352–353. See also Imperfect fungi
   characteristics of, 356–357
   entomopathogens of, 357
Deutomerite, 416
Devoscovinia, 390
Diabrotica, 469, 517
Diabrotica undecimpunctata, 219. See also Spotted cucumber beetle
Diadegma, 246
Diagnosis, 75–76
Diamondback moth, 494, 560, 568. See also *Plutella xylostella*
Diaporthales, 352

Diaprepes abbreviatus, 358
Didymophyes ontophagi, 419
Didymorphidae, 420
Diel periodicity, 333, 341, 609
Dilta hibernica, 191
Diplococcus pneumoniae, 525
Diplocystis, 420
Diplocystis major, 419
Diplocystis schneideri, 419
Diplocystis tipulae, 418
Diplogasterida 461–462
Diplogasteridae, 462, 478–479
Diplomonadida, 390, 393
Diploscapter lycostoma, 469
Diprion hercyniae, 496. See also European spruce sawfly
Diprionidae, 189, 396
Diptera, 89, 94, 98–99, 111, 149, 155, 158–159, 186, 189, 216, 249–250, 258, 261, 277, 322, 332, 338, 351–352, 358, 395–398, 420, 423–424, 427, 493, 497, 507, 509–510, 515, 523, 527, 529–530
Dirofilaria immitis, 420, 460, 530. See also Dog heartworm
Discophrya, 390
Diseases
  amicrobial, caused by
    biological agents, 60–61
    chemical agents, 57–60
    genetic factors, 61–62
    hormonal disruption, 64–65
    mechanical agents, 53–54
    nutritional factors, 62–64
    physical agents, 54–57
  classification of, 52–53
  microbial
    course of infection, 71–72
    dosage of pathogen, 70
    infectivity of pathogen, 67–68
    interactions among microorganisms, 73–74
    pathogens, potential, facultative, and obligate, 65–66
    portals of entry, 66–67
    signs, symptoms, and syndromes of, 70–71
    toxins, microbial, 66
    types of, 65
    virulence and pathogenesis of pathogen, 68–70
Diuretic hormone, 207, 563
Dixidae, 89
DNA polymerase gene, 203
DNA uncoating, 194–195. See also Uncoating of DNA
DNA-viral infections, 171–220, 245–264. See also DNA viruses
DNA viruses, 175, 245, 275. See also DNA-viral infections
Dobellina, 400
Dobellina mesnili, 402

Dog flea, 398. See also *Ctenocephalides canis*
Dog heartworm, 420. See also *Dirofilaria immitis*
Douglas fir beetle. See *Dendroctonus pseudotsugae*
Douglas-fir tussock moth, 469, 503, 583, 604. See also *Orgyia pseudotsugata*
  long-term control of, 574, 578, 607–608
  microbial control of, 565
  standardization of virus, 568
Drosophila, 26, 61, 95, 110–111, 158, 160–161, 296–302, 347, 518, 601
Drosophila A virus, 300–302
Drosophila C virus, 300–302
Drosophila iota virus, 298
Drosophila melanogaster, 63, 529
  anoxia in, 296–298
  fungal infections of, 364–365
  hybrid sterility in, 159
  inhibitor of bacillus, 111
  normal bacteria in, 22, 24
  RNA viruses of, 301–302
  sigma virus infection of, 57, 67, 198, 494
Drosophila P virus, 300–302
Drosophila paulistorum, 159
Drosophila pseudoobscura, 161
Drosophila X virus, 297–298
Dryocoetes autographus, 400
Duboscqiidae, 428, 442
Dugesia dorotocephala, 483
Dysdercus cingulatus, 422
Dytiscidae, 17
Dytiscus, 407

Eastern spruce budworm, 574, 578, 580, 582. See also *Choristoneura fumiferana*
Eberthella typhosa, 498
Ecdysone, 27, 64, 190, 194, 204, 260, 418, 445, 468
Ecdysone, β-, 193
Ecdysteroid UDP-glucosyl transferase gene (EGT gene), 190, 194, 204
Eclipse (latent) period, 192, 195, 197, 214
Economic threshold, 555–557, 570–575, 577, 580
Egypt bee virus, 302
Eleodes sp., 529
Elmidae, 17
Embioptera, 32, 424
Emperor gum moth, 174, 308. See also *Antheraea eucalypti*
Empusa, 337
Empusa muscae, 337
Encapsulation, 322, 466, 507, 510, 530, 532
  phases of, 515
  surface recognition in, 515
  theories of, 516, 518, 521
Encephalitozoon, 427, 441–442

Encephalitozoon cuniculi, 436
Endamoeba, 400
Endamoeba blattae, 402
Endamoeba histolytica, 402
Endamoebidae, 399–400
Endocytobiosis, 14, 32
Endocytosis, 195, 439. *See also* Viropexis
    definition of, 512
Endolimax, 400
Endolimax blattae, 402
Endomycetales, 320
    hosts of, 347–348
    pathogenesis of, 348
    types of, 347
Endomycopsis, 18
Endosymbiont. *See* Microbiota, intracellular
Endosymbiosis, theory of, 32–33
Endosymbiont. *See* Microbiota, intracellular
Endotoxin, δ-, 68, 87–88, 207, 494, 558, 560, 568, 602–603. *See also Bacillus thuringiensis*, crystal protein
    formation of, 66
    gene of, 562–563
    host range and syndrome of, 107–110
    mode of action, 106–107
    pathological and cytological effects of, 104–106
Engyodontium, 320
Enhancing factor, 150, 194, 565, 602
Entamoeba, 400
Entamoeba mesnili, 402
Entamoeba thomsoni, 402
Entaphelenchidae, 462, 482
Entaphelenchus, 482
Enterobacter, 85, 151
Enterobacter aerogenes, 149, 151
Enterobacter agglomerans, 25
Enterobacter cloacae, 149, 527–528
Enterobacteriaceae, 34–35, 85, 148–151, 161, 502
Enterocystis, 420
Enterocystis fungoides, 420
Enterocystis racovitzai, 420
Enterovirus, 300–301
Entomophaga, 320, 337, 605
Entomophaga grylli, 341, 344, 615
Entomophaga kansana, 340
Entomophaga maimaiga, 342–343, 610, 618–620
Entomophthora, 320, 327–328, 337, 341, 581, 614, 618
Entomophthora anisopliae. *See Metarhizium anisopliae*
Entomophthora aphidis. *See Erynia neoaphidis*
Entomophthora culicis. *See Erynia variabilis*
Entomophthora egressa, 515
Entomophthora erupta, 326, 340, 346
Entomophthora floridana, 342
Entomophthora gammae. *See Tarichium gammae*

Entomophthora grylli. *See Entomophaga grylli*
Entomophthora maimaiga. *See Entomophaga maimaiga*
Entomophthora muscae, 322, 328, 600
Entomophthora obscura, 495
Entomophthora sphaerosperma. *See Erynia radicans*
Entomophthora thaxteriana, 338
Entomophthora virulenta, 338
Entomophthoraceae, 337–338
Entomophthorales, 4, 319–320, 322, 327–328, 357, 609, 614
    characteristics of, 336
    classification of, 337–338
    hosts of, signs and symptoms, 338
    in vitro cultivation of, 344–348
    life cycle of
        conidia-sporulation and germination, 341–342
        conidial and resting spore generations, 344
        period of infection, 341
        resistant or resting structures, 342–344
        stages within insect's hemocoel, 340–341
    reproduction of, 338
Entomopoxviridae, 175
Entomopoxvirinae, 175, 248, 616
    genera of, 250
    hosts of, 250
    inclusion bodies of, 174, 184, 250–251
    pathogenesis of, 252–253
    replication of, 251–252
    structure and properties of, 250
Entomopoxvirus, 249, 616
Environmental Protection Agency (EPA), 561, 562, 578
Enzootic, 607–608, 610–616
    definition of, 596
Ephemeroptera, 420
Ephestia, 105
Ephestia cautella, 38. *See also* Almond moth
Ephestia elutella, 423
Ephestia kuehniella, 159, 423, 517
Ephydridae, 365
Epimerite, 416–418, 420, 423
Epiphyas postvittana, 494
Epistylis, 390
Epizootic, 5, 561, 612–616, 618–619, 621–622
    bacterial, 6, 110, 155
    definition of, 596
    fungal, 330, 336, 338, 343–344, 348, 361, 575–576, 600, 618–619
    initiation of, 605, 607
    phase of, 596
    protozoan, 397, 425
    viral, 190, 198, 263, 302, 494, 498, 560, 570, 573–574, 578
    wave of, 596
Epizootiology, 2, 6
    definition of terms, 596

environment in
  agricultural ecosystem, 612–615, 617
  aquatic ecosystem, 621
  forest ecosystem, 617–621
  pasture, 616–617
  rangeland, 615–616
  turf, 617
host population, 609–612, 622
modeling, 597–600, 622
pathogen population, 602–609
present status, 622
principles of, 595
transmission of pathogens, 600–602
Eriborus, 246
Erwinia carotovora, 24
Erynia, 320, 337–338, 341, 599–600
Erynia aphidis, 343
Erynia blunckii, 325
Erynia bullata, 344
Erynia neoaphidis, 338
Erynia phytonomi, 344
Erynia radicans, 328, 343, 614
Erynia variabilis, 324
Escherichia coli, 22, 26, 97, 102, 111, 206, 498, 514, 525–532
Estigmene acrea, 216, 218, 260, 284. *See also* Salt-marsh caterpillar
Eubacteriales, 34
Eubaculovirinae, 177
Eucosma griseana, 494, 560. *See also* Larch bud moth
Eudiplogaster aphodii, 478
Euglena gracilis, 527
Eugregarinida, 390
  interaction with host development, 418
  life cycle of, 417–418
  pathology of, 418–419
  species of, 416
  structure of, 416–417
Eumycota, 319
Euproctis chrysorrhoea, 73
Euproctis similis, 184, 197. *See also* Brown-tail moth
European cabbageworm, 207, 359. *See also Pieris brassicae*
European cockchafer, 118, 424–425, 504, 515. *See also Melolontha melolontha*
European corn borer, 7, 63. *See also Ostrinia nubilalis*
  insect-pathogen model, 599–600
  as microbial control target, 566
  pathogens of
    bacteria, 88, 95, 108, 122
    fungi, 358, 365, 497
    protozoa, 398, 430, 577, 614–615
European foulbrood, 68, 120–121, 495. *See also Melissococcus pluton*
  etiology of, 152
  syndrome of, 152–153

European pine sawfly, 9. *See also Neodiprion sertifer*
  epizootic in, 606–607, 611
  microbial control of, 6, 570, 572–573
  as microbial control target, 565
European spruce sawfly. *See also Diprion hercyniae*
  epizootic in, 607–608, 611
  microbial control of, 9, 572–573
  resistance to virus, 496
European sunflower moth, 423. *See also Homeosoma nebulellum*
Eurygaster integriceps, 501
Euscelus incisus, 33, 36, 38
Euscelus plebejus. See Euscelus incisus
Euxoa auxiliaris, 250
Euxoa scandens, 281, 501. *See also* White cutworm
Euxoa segetum, 207
Evaginogenida, 390
Exeristes comstockii, 150
Exosymbiont. *See* Microbiota, extracellular
Exotoxin, α-, 91–94
Exotoxin, β-. *See also* Thuringiensin
  characteristics of, 94–95
  formation of, 94
  host resistance to, 111, 494
  host spectrum of, 94–95
  names for, 94
  pathology of, 95–96
  serotypes, produced in, 94
  syndrome of, 109
Exotoxin, γ-, 91

Face fly. *See also Musca autumnalis*
  epizootic of, 616
  nematode infection in, 469–471, 480
  behavior, 606
Factor G, 528
Facultative pathogen. *See* Diseases, microbial
Fall armyworm, 95, 208, 494, 612–613. *See also Spodoptera frugiperda*
Fall webworm, 212–213, 284, 364. *See also Hyphantria cunea*
  resistance to virus, 287, 496, 504
  virus interference in, 74
Fannia, 119
Fannia canicularis, 95, 347, 402
Fanniomyces ceratophorus, 351
Farinocystis, 390, 421–422
Farinocystis tribolii, 420–423
Father of insect pathology. *See* Bassi de Lodi
Father of invertebrate pathology, 8
Filariasis, 460
Filariomyces, 320
Filariomyces forficulae, 351
Filipjevimermis leipsandra, 469, 517
Filobasidiella, 321

Fire ant, 325. See also *Solenopsis richteri*
Flacherie, 5, 87, 152
   history of, 306
   syndrome of, 306–307
Flacherie de Pasteur, 306
Flacherie virus, 263, 299–300, 302, 532
   characteristics of, 306
   host resistance to, 493, 495, 500, 504
   pathogenesis of, 306–307
   transmission of, 307
Flagellata, 388, 391
Flagellate, 27–28, 389. See also Zoomastigina
   characteristics of, 392–393
   entomogenous forms, 393–394
   entomopathogens, 395–398
   flagellatoses of, 396–398
   insect vectors of, 398–399
   life cycle of, 394
   pathology of, 395
   transmission of, 394–395
Flagellatosis, 395
Flagellosis. *See* Flagellatosis
Flavibacterium, 36
Flaviviridae, 276
Flock House virus, 309
Forest tent caterpillar, 430, 574, 605–606, 617. See also *Malacosoma disstria*
Formosan subterranean termite, 611. See also *Coptotermes formosanus*
Foulbrood, 3, 303
Friendly fungi, 6, 355, 365
Frog iridescent virus, 258
Fruittree leafroller, 209. See also *Archips argyrospila*
Fungal infections, 318–366, 497, 622
Fungi
   entomopathogenic
      classification of, 319–320
      effect of environment, 328–329
      hosts of, 321–322
         penetration of integument, 324–325
         replication in hemocoel, 326
         spore dispersal, 326–327
         spore germination, 322–324
      pathogenicity of, 327
      signs and symptoms of, 327–328
      structure and reproduction of, 319, 321
   insect cultivated
      ambrosia beetles, 17–18
      ants, 19, 21
      termites, 21
      wood wasps, 19
   kingdom, 389
Fungi Imperfecti, 356. See also Deuteromycotina
Fungi as internal microbiota, 26
Fungi in symbiosis, 13–14
Fungi as external microbiota, 14–17
Fusarium, 18, 320, 355, 365
Fusarium lateritium, 365

Fusarium moniliforme 365
Fusarium solani, 365

Galleria, 105, 150, 202, 258–259, 263, 524, 527
Galleria densonucleosis virus, 261–263
Galleria mellonella. See also Greater wax moth
   bacterial infection of, 85, 95, 109, 148, 150, 160
   fungal infection of, 347, 360–361
   protozoan infection of, 398, 404, 433
   resistance in, 498, 504, 506, 513, 517, 520, 528–529
   viral infection of, 186, 202–203, 261–263, 302, 310, 361
Gametogony (gamogony), 415–416, 420, 423
Gametophytic generation, 330–331, 333
Gamocystis, 420
Gattine, 306
Genetic engineering of organisms, 61, 97, 102, 172, 198, 205, 207, 560, 562–565, 571
Genotypic variants. (*See* specific baculoviruses)
Geocoris punctipes, 94
Geotrichum candidum, 347
German cockroach, 38–39, 63, 348, 430, 515, 518. See also *Blattella germanica*
Gerris paludum, 395
Giant cell, 322, 512, 517–518
Gibellula, 320
Gilpinia hercyniae. See *Diprion hercyniae*
Glaucoma, 403
Glaucoma pyriformis. See *Tetrahymena pyriformis*
Glossina, 150, 396, 399. See also Tsetse fly
Glossina morsitans, 24, 529. See also Tsetse fly
Glyphodes pyloalis, 262, 307
Glugea, 427, 519
Glugea gasti. See *Nosema gasti*
Glugeida, 429
Glugeidae, 429
Glypta, 246
Glyptapanteles, 247
Glyptapanteles militaris, 194, 214
Gnathotrichus, 18
Golbergiidae, 429
Gonometa podocarpi, 302
Gonometa virus, 300, 302
Gordiacea, 482
Gordius, 483
Gordius robustus, 483
Granulin, 188, 209, 214–215, 251
   amino acid composition of, 188
   gene of, 209
   nucleotide sequence of, 188
   peptide map of, 188
Granulocytes, 299, 398, 507–508, 511, 514–515, 518, 530
   characteristics of, 509
   function of, 509

Granulosis, 211, 213–214
Granulosis viruses
  complementation in, 73, 602
  cytopathology of, 211–213
  differ from nuclear polyhedrosis virus, 209
  envelope gene of, 209
  epizootiology of, 602, 609–610, 616
  gene expression in, 215–216
  genome and hybrids of, 215
  historical account, 207–209
  host resistance to, 493, 495–496
  hosts of, 174, 189
  interference in, 74
  as microbial control agent, 557, 574, 579
  occlusion of, 175, 184, 209–210, 251. See also Capsules
  pathology and pathogenesis of, 211
  pathophysiology of, 213–214
  physical map of, 215
  replication of, 214–215
Grass grub. See *Costelytra zealandica*
Grasshoppers
  epizootics in, 605, 615–616
  historical account, 6–7
  IPM approach, 583
  microbial control of, 565, 567, 576–577, 582
  pathogens of
    bacteria, 70, 112, 148–149, 151
    fungus, 338, 341
    nematode, 464–465, 468, 471, 474–475
    protozoa, 69, 391, 400–402
  resistance in, 502, 519–521, 523, 526
Greater wax moth, 22, 395, 498, 504, 514, 520–521, 526, 531, 533. See also *Galleria mellonella*
Green-muscardine fungus, 5–7, 357, 359, 361, 495. See also *Metarhizium anisopliae*
Green peach aphid, 338, 575. See also *Myzus persicae*
Green rice leafhopper, 36. See also *Nephotettix cincticeps*
Green scale, 355. See also *Coccus viridis*
Green tortrix, 610. See also *Tortrix viridana*
Greenhouse whitefly, 31, 365. See also *Trialeurodes vaporariorum*
Gregarina, 389–390
Gregarina blattarum, 420
Gregarina chrysomelae, 419
Gregarina conica, 389
Gregarina cuneata, 418
Gregarina macrocephalia, 419
Gregarina ovata, 389
Gregarina polymorpha, 419
Gregarina rostrata, 418
Gregarine aseptate (acephaline)
  pathology of, 420
  species of, 419–420
  structure of, 420
Gregarinia, 390, 415
Gregarinida, 391

Gregarinidae, 420
Gryllus bimaculatus, 300
Gryllus campestris, 419
Gurleyidae, 429, 442
Gymnamoebia, 390
Gypsy moth. See also *Lymantria dispar*
  epizootics in, 604–605, 607–608, 610, 618–620
  microbial control of, 562, 565, 574, 578, 580
  pathogens of
    bacteria, 7, 84, 150, 153
    fungus, 361
    protozoa, 431
    virus, 182–183, 281
  resistance in, 70, 501, 559
Gyrinidae, 17
Gyrinus natator, 219
Gyrinus nonoccluded baculovirus, 219

Haemokinin, 524
Haemotobia, 89
Haematobia irritans, 110. See also Horn fly
Hairless black syndrome, 305
Haliplidae, 17
Haplosporea, 390
Haplosporida, 390
Haplosporidia, 390–391, 414–415
Haplosporidium, 390
Haptomonad, 393
Harpographium cornelioides, 356
Harrisinia brillians, 211. See also Western grape-leaf skeletonizer
Hartmanella, 400
Hazardia, 445
Hazardia milleri, 447
Helicosporidia, 414
Helicosporidium parasiticum, 414
Heliothis, 255, 496, 578, 580, 606
Heliothis armigera, 218, 496
Heliothis ascovirus, 253, 255
Heliothis nonoccluded baculovirus, 177
  characteristics of, 218
  genotypes of, 218
  persistent infection of, 218–219
  replication of, 218
Heliothis nuclear polyhedrosis virus, 202, 565, 567, 570, 578–579, 606
Heliothis virescens. See also Tobacco budworm
  microbial control of, 560, 581
  resistance in, 111, 495
  pathogens of
    bacteria, 95, 108, 110
    fungi, 361, 364
    virus, 193
Heliothis zea. See also Corn earworm,
  IPM approach to, 583
  microbial control of, 557, 559, 567, 581
  pathogens of
    bacteria, 94

Heliothis zea (*cont.*)
    fungi, 323, 359, 361, 364
    protozoa, 69, 432
    virus, 186
    resistance in, 496–497, 500–501, 520
Heliothis zea cell line (IMC-H$_z$-1), 216, 218
Heliothis zea nuclear polyhedrosis virus, 180, 202
Hemagglutinins, 513, 522–523. See also Lectin
Hematozoa, 415
Hemerobiidae, 189
Hemiascomycetes, 320, 347
Hemiberlesia rapax, 365
Hemiptera, 23, 94, 111, 159, 258, 322, 332, 338, 351–352, 358, 395, 397–398, 420, 430, 513
Hemispherical scale, 355. See also *Saissetia hemisphaerica*
Hemocyte capsule, 504. See also Capsule, blood
Hemocyte count, 525
Hemocytogenesis, 507
Hemocytopoiesis, 507
Hemolysins, 110
Hemopoiesis, 507
Hemopietic tissues, 520, 525, 527
Hepialidae, 616
Herpetomonas 390, 394–396
Herpetomonas flagellatoses, 396
Herpetomonas jaculum. See *Leptomonas jaculum*
Herpetomonas melolonthae, 389–390
Herpetomonas muscarum, 389, 395–396
Herpetomonas rhinoestri, 395
Herpetomonas roubaudi, 394
Herpetomonas swainei, 396
Hesperomyces, 320
Hesperomyces virescens, 350–351
Hessea, 390
Hesseida, 428
Hesseidae, 428
Hessian fly, 62. See also *Mayetiola destructor*
Heterocypris incongruens, 332
Heteroecism, 330–332, 604. See also Alternate host
Heteronychus arator, 309. See also Black beetle
Heteroptera, 17, 300
Heterorhabditidae, 151, 462, 464, 478, 557
Heterorhabditis 151, 461, 465, 582
Heterorhabditis bacteriophora, 465, 500, 569
Heterorhabditis heliothidis. See *Heterorhabditis bacteriophora*
Heterorhabditis megidis, 479
Heterosis, 494
Heterotrichida, 390
Heterotylenchus, 606, 616
Heterotylenchus autumnalis, 464, 469–470, 480–481
Hexamermis, 470
Hexamermis arvalis, 467, 469

Hexapoda, 493
Hippelates pusio, 395–396
Hirsutella, 320, 353, 363
Hirsutella lecaniicola, 363
Hirsutella ovalispora, 363
Hirsutella stilbelliformis, 363
Hirsutella thompsonii, 322, 328, 354, 363, 565–566, 576, 583
Hirsutella thompsonii var. *synnematosa*, 363
Hirsutella thompsonii var. *thompsonii*, 363
Hirsutella thompsonii var. *vinacea*, 363
Hodotermitidae, 27
Holotricha, 402
Holst test, 121
Homeosoma nebulellum, 423. See also European sunflower moth
Homoptera 31, 216, 324, 338, 362, 430
Honey bee. See also *Apis mellifera*
    epizootics in, 610–612
    freezing injury, 55
    genetic disease of, 61
    historical account, 2–3
    microbiota, internal, 26
    mite parasite of, 65
    nutritional disease of, 63
    pathogens of
        bacteria, 68, 119–122, 149, 152, 160, 493
        fungus, 336, 347–349, 363, 365
        protozoa, 400–401, 432–434, 443
        virus, 173, 219, 260, 275, 296, 302–306, 309
    resistance in, 495, 497–499, 500–501, 503, 523, 525
Honey disease. See Amber disease
Hooke, Robert, 30
Hoplia, 420
Hormonal disruption. See Diseases, amicrobial
Hormone, diuretic, 563. See also Diuretic hormone
Horn fly, 110. See also *Haematobia irritans*
Horogenes. See *Diadegma*
Host population. See also Epizootiology
    behavior of, 611–612, 613, 616
    density of, 598, 602, 608, 610, 613, 615–616, 618, 621
    susceptibility of, 609–610
House fly. See also *Musca domestica*
    microbiota, internal, 24, 26
    pathogens of
        bacteria, 95, 119, 148
        fungus, 363
        protozoa, 389–396
        virus, 287–288
    resistance in, 111, 494, 497, 529
House fly virus. See *Muscareovirus*
Human body louse, 32
Human immunodeficiency virus (HIV), 207
Humoral encapsulation, 322, 523–524
Humoral immunity, 507, 521–524. See also Immunity, humoral

Hyalophora cecropia, 157, 523, 527–528
Hydraenidae, 17
Hydrophilidae, 17
Hylemya, 119
Hylemya antiqua, 15, 497. See also Onion fly
Hylemya cilicrura, 340, 346. See also Seedcorn maggot
Hylemya platura, 24. See also Seedcorn maggot
Hylobius pales, 14
Hymenoptera, 32, 60, 86, 94, 149, 159, 162, 189, 209, 216, 250, 258, 277, 322, 338, 351–352, 358, 397, 493, 506–507
Hymenostilbe, 320
Hymenostomatida, 402
Hymenostomia, 390
Hypera, 347
Hypera postica, 38, 326, 467. See also Alfalfa weevil
Hyphantria cunea, 74, 212–213, 215, 284, 287, 364, 496, 501, 504. See also Fall webworm
Hyphomycetes, 319, 357, 364–365
Hypocreaceae, 355
Hypocreales, 352
Hypocreella. See *Hypocrella*
Hypocrella, 320, 352, 355, 365
Hypomicrogaster, 247

Ichneumonid polydnaviruses, 246, 249. See also *Ichnovirus*
Ichneumonidae, 175, 246
Ichnovirus, 246–247. See also Polydnaviridae
Icosahedral cytoplasmic deoxyriboviruses. See Iridoviridae
Immunity
  acquired, 524
    antibacterial and toxic substances in, 525–526
    attacins, 528–529
    cecropins, 527–528
    complement, 530–531
    diptericins, 529
    inhibitors, 532
    lysozymes, 526–527
    other immunity-related factors, 531–532
    phenoloxidase system, 529–530
  cellular, 507, 521–522, 524
    encapsulation, cellular, 515–516
      cellular and sheath capsules, 516–517
      inhibition of capsule formation, 517
      giant cells, nodules, and tumors, 517–519
    phagocytosis
      effectiveness, 514–515
      process, 513–514
  humoral (innate). See also Humoral immunity
    encapsulation, 523–524
    innate, 522
      hemagglutinins, 522–523

insect plasma, 521–522
other factors, 524
induced, 525
innate, 522
passive, 522, 533
phylogenetic, 492–493
Imperfect fungi, 352, 356–357, 365. See also Deuteromycotina
Imported cabbageworm, 108, 498. See also *Artogeia rapae*, *Pieris rapae*
Inachis io, 277
Inapparent infection, 301, 308–309
Incidence, definition of, 596
Inclusion body, 276–277, 279–283, 286–287. See also Cytoplasmic polyhedrosis virus
Incubation period, definition of, 71
Indian Kala-azar, 399
Indianmeal moth, 67, 106, 494, 496, 560, 566, 609–610. See also *Plodia interpunctella*
Infectious flacherie. See Flacherie
Infectivity, definition of, 67–68, 602
Inflammatory response, 392, 433, 519
Influenza HA protein, 207
Infusoria, 388
Inhibine, 498
Inhibitor, protease, 526
Inhibitor of immune response, 110, 111, 532
Injuries
  from biological agents, 60–61
  from chemical agents. See also Chemical insecticides
    insecticides, 57–60
    poisons, 60
  from genetic factors
    engineered, 61–62
    inherited, 61
  from mechanical agents, 53–54
  from physical agents
    high temperature, 54
    low temperature, 54–56
    moisture, 56
    oxygen and carbon dioxide, 56–57
    radiation, 57
Injury factor, hemokinin, 53–54
Inquilinism, definition of, 13
Insect iridoviruses. See Iridoviridae
Insect pathology
  applications of, 1–2
  brief history, 2–9
Insect Pathology Research Institute, 9
Insect pest management (IPM), 7, 556, 560, 571, 580, 582–583, 600
Insect poxvirus. See Entomopoxvirus
Insect vectors, 391, 396, 399, 423, 460, 501
Insecta, 561. See also Hexapoda
Interference, 73–74, 189. See also Antagonism
Interferon, 2, 207, 521, 531–532
Interleukin, 2, 206
International Committee for Taxonomy of Viruses, 174, 261

International corn borer investigation, 7
Iotonchiidae, 462
Ips grandicollis, 25
Ips paraconfusus, 25
Ips typographus, 469
Iridescent virus, 283, 601, 605, 616. See also *Iridovirus*
Iridomyrmex humilis, 469. See also Argentine ant
Iridoviridae, 245. See also Iridescent virus
 hosts of, 256, 258–259
 nomenclature of, 255–256
 pathogenesis of, 259–261
 replication of, 257–258
 structure and properties of, 256–257
Iridovirus, 184, 189, 255–261. See also Iridoviridae
Isaria, 3, 357, 361
Isaria amorpha. See *Beauveria amorpha*
Isaria farinosa. See *Paecilomyces farinosus*
Isle of Wight disease, 65
Isomermis, 468
Isoptera, 94, 351–352
Ithania, 423
Itoplectis conquisitor, 85

Jack pine sawfly, 578. See also *Neodiprion pratti banksianae*
Japanese beetle. See also *Popillia japonica*
 damage of, 574
 epizootics in, 603–604, 617
 historical account, 7–8
 microbial control of, 565, 567, 575
 pathogens of
  bacteria, 115
  fungus, 115, 358, 493
 resistance in, 105
Japanese Beetle Laboratory, 7
Jaundice, 182–183, 306
Junonia coenia, 262
Juvenile hormone, 64–65, 193, 207, 260, 422, 432, 583
Juvenile nematode, 469, 471, 473, 479–480, 482

Kalotermitidae, 27
Kashmir bee virus, 302
Kermes, 17
Khapra beetle, 421. See also *Trogoderma granarium*
Kinetoplastida, 390, 393
Koch's postulates, 74
Kofoidina, 420
Kotochalia junodi. See Wattle bagworm

Labidura riparia, 351
Laboulbeniales, 4, 15, 319–321, 327, 349–350

Laboulbeniomycetes, 319–320
 hosts of, 351
 pathology of, 349
 replication and pathogenicity of, 351
 structure of, 349, 351
Laboulbeniopsis termitarius, 15
Labyrinthomorpha, 389
Lachnosterna, 85
Lagenidiales, 320, 334–336
Lagenidium, 320
 culture and nutrition of, 335
 hosts of, 335–336
 life cycle figure of, 335
 replication of, 334
Lagenidium culicidum, 335
Lagenidium giganteum, 323, 334–335, 565, 576
Lagria hirta, 418
Lambornella, 390–391, 402–404, 406, 601
Lambornella clarki, 403–405
 characteristics of, 406
 dimorphism of, 406
 life cycle of, 406–407
 regulatory role of, 407
Lambornella stegomyiae, 404–406
Lamia, 337
Laodelphax striatellus, 36, 39
Larch bud moth, 86, 494, 610. See also *Eucosma griseana*
Lasius, 467
Laspeyresia pomonella, 358. See also *Cydia pomonella*
Latent viral infections, 497, 556. See also Occult viral infections
Lecithinase, 86, 94, 110, 504. See also Phospholipase C
Lectin, 513, 522–523. See also Hemagglutinin
 gene of, 523
Legerella, 390, 423
Leidyana ephestiae, 418
Leidyanidae, 420
Leishmania, 393–394, 396, 398–399
Leishmania donovani, 399
Lepidoptera, 86, 89, 94, 98–99, 111, 149, 157–158, 173, 177, 189, 191–192, 209, 211, 216, 219, 249–250, 253, 258, 261, 275, 277, 283, 286, 300, 309, 322, 338, 352, 355, 358, 398, 424, 427, 431, 493, 497, 502, 506–507, 509–510, 528–529, 580, 616
Lepidosaphes, 17
Lepiota, 21
Lepisma saccharina, 191, 401
Leptinotarsa decemlineata, 160, 359, 497. See also Colorado potato beetle
Leptolegnia, 320, 336, 403
Leptolegnia chapmanii, 336
Leptomonas, 390, 393–394, 396–398
Leptomonas arctocorixae, 394
Leptomonas craggi, 394
Leptomonas ctenocephali, 398

Leptomonas jaculum, 395
Leptomonas pyraustae, 398
Leptomonas pyrrhocoris, 397-398
Leptomonas serpens, 398
Léthargie, 159
Leucoagaricus, 21
Leucocoprinus, 21
Leucophaea maderae, 336, 359, 511, 513, 516, 519, 522
Leucorrhinia dubia, 263
Limnephilidae, 189
Lipocystis, 421-422
Lipotropha, 416, 421-422
Litostomatea, 390
Loblolly pine sawfly, 578. See also *Neodiprion taedae linearis*
Lobosea, 390
Loculoascomycetes, 320, 347, 355-356
Locusta, 523
Locusta migratoria, 25, 468, 521, 525, 527
Locustana pardalina, 85
Lophius, 432
Lorscher Krankheit, 155
Lucilia, 119
Lucilia sericata. See *Phenicia sericata*
Lymantria, 105
Lymantria dispar, 159, 182, 252, 281, 361, 431, 498, 501, 559. See also Gypsy moth
Lymantria dispar granulosis virus, 209
Lymantria monacha, 73, 182, 190. See also Nun moth
Lymantria nuclear polyhedrosis virus, 565
Lyme disease, 595-596
Lymexylidae, 17
Lysin, 526
Lysosome, 257, 509-510, 514
Lysozyme, 506, 514, 522, 526-529, 532
Lyssavirus, 298

Machadoella, 421, 423
Macrobiotophthora, 337
Macrocentrus grandii, 615
Macrosiphoniella sanborni, 362
Macrotermitinae, 21
Magicicada, 345
Magicicada cassini, 346
Magicicada septendecim, 345
Malacosoma, 68, 150, 617-618. See also Tent caterpillar
Malacosoma americanum, 122
Malacosoma californicum pluviale. See Western tent caterpillar
Malacosoma disstria, 122, 430, 496, 618. See also Forest tent caterpillar
Malacosoma pluviale, 122
Maladies des morts-blancs, 306
Maladies de tripes, 306
Malameba, 390, 400-401
Malameba locustae, 391, 401-402

Malamoeba scolyti, 400
Malaria, 391, 596
Malayan disease, 216
Malignant carcinoma, 218. See also Tumor, malignant
Mallophaga, 351
Mal Noir disease, 305
Malpighamoeba, 390, 400
Malpighamoeba locustae. See *Malameba locustae*
Malpighamoeba mellificae, 400-402
Malpighiella, 390, 400
Malpighiella refringens, 402
Mamestra brassicae, 148, 200, 202, 287, 496-497
Manduca, 580
Manduca sexta, 102, 105, 109-110, 519, 558. See also Tobacco hornworm
Mansonia, 114
Mansonia uniformis, 302
Mantodea, 300
Massospora, 320, 326-327, 337, 340, 604
  life cycle of, 345-346
  pathogenesis of, 345-346
  species of, 345
Massospora cicadina, 345-346, 606
Mastigomycotina, 319-321, 329, 356
Mastigophora, 388
Mastotermes darwiniensis, 25
Mastotermitidae, 27
Matricidal endotoky, 476, 480
Mattesia, 390, 421-422
Mattesia dispora, 422-423
Mattesia geminata, 422-423
Mattesia grandis, 421, 423, 582
Mattesia povolnyi, 422-423
Mattesia trogodermae, 421
Maturation immunity, 322, 496-497, 559-560, 580, 583, 618
May disease, 160, 400
Mayetiola destructor, 62. See also Hessian fly
Mecoptera, 32
Mediterranean flour moth, 88, 108. See also *Anagasta kuehniella*
Mediterranean fruit fly. See *Ceratitis capitata*
Megachile, 349
Megaloptera, 405
Meiosis
  in coccidian, 424
  in gregarine, 417, 422
  in microspordium, 442, 445-447
Melanization, 56, 324, 466, 513, 516, 524, 529-530
Melanoplus bivittatus, 148
Melanoplus sanguinipes, 250, 519
Melanotic lesions, 512
Melissococcus pluton, 68, 121, 152. See also European foulbrood
Melolontha, 6
Melolontha brunnea, 390

Melolontha melolontha. See also European cockchafer
  pathogens of
    bacteria, 118, 149, 154–156, 159
    nematode, 515
    protozoa, 424–425
    virus, 249–250
  resistance in, 515
Melolontha quercina, 390
Melolontha vulgaris, 154
Membracidae, 336
Menzbieria, 421–422
Menzbieria chalcographi, 423
Meristacrum, 320, 337
Mermis nigrescens, 464–465, 467–468, 472, 474–475
Mermithida, 462, 465
Mermithidae, 461–462, 471–475
Merodon equestris, 220
Merogony, 415, 418, 420–423, 439–441, 445, 447
Merogregarina, 422
Meroselenidium, 422
Mesocerus marginatus, 397
Mesoleius, 246
Mesostoma, 483
Metarhizium, 320, 357
  characteristics of, 360
  mycotoxins of, 360–361
  species of, 359–360
Metarhizium album, 360
Metarhizium anisopliae. See also Green-muscardine fungus
  antibiosis for, 14
  characteristics of, 357, 360
  historical account, 5–6
  host resistance to, 325, 495, 497, 506, 521, 531, 611
  hosts of, 359
  as microbial control agent, 566, 581
  mycotoxins of, 360–361
  portal of entry, 326, 601
  spore release of, 328
  taxonomy of genus, 360
Metarhizium anisopliae var. *anisopliae*, 360
Metarhizium anisopliae var. *major*, 360
Metarhizium brunneum, 360
Metarhizium flavoviride, 360
Metaseiulus occidentalis, 89, 95
Metchnikovellida, 428
Metchnikovellidae, 428
Metchnikovellidea, 428
Metschnikowia, 320
Metschnikowia unicuspidata, 347–348
Microascus exsertus, 349
Microbial control, 2–3, 7, 9, 88, 112, 149, 207, 357, 360, 363, 391, 397, 406, 425, 461, 471, 483, 607–609
  advantages and disadvantages of, 570–571
  application technology in, 566, 569–570
  approaches to, 555–556
  destruction of symbionts, 584
  economic threshold, 556–557
  factors affecting efficacy
    antimicrobiosis, 558–559
    compatibility, 559–560
    coverage, 558, 569, 572
    dosage, 557
    host resistance, 560–561
    pH, 558
    target pest, 559
    timing of application, 557–558
  history of, 554–555
  inundative or inoculative method, 555–556
  IPM approach, 582–584
  long-term control, 555–556, 570–571, 578, 581
    bacterial, 574–575
    fungal, 575–576
    nematode, 577
    protozoan, 576–577
    viral, 572–574
  microbial pesticides
    formulation, 566–568
    mass production, 566–567
    standardization, 566, 568–569
    storage, 566, 569
  molecular genetics in, 563–565
  safety of, 561–563, 566
  short-term control, 555–556, 558, 569, 572, 577
    bacterial, 579–581
    fungal, 581
    nematode, 582
    protozoan, 581–582
    viral, 578–579
Microbial insecticides, 554–555, 557–559, 565, 567–569, 579, 581, 622
Microbial toxins. *See* Diseases, microbial
Microbiota, external, 13
  bacteria, 14–15
  effect of host morphology, 14–15
  fungi, 15–17
  protozoa, 17
Microbiota, internal, 13, 21–22
  bacteria, 23–26
  fungi, 26
  host effects on, 27
  protozoa, 26–27
  termite. *See* Termite, internal microbiota of
  viruses, 26
Microbiota, intrahemocoelic and intracellular
  insect hosts of, 34–36
    homopteran and hemipteran, 34–35
  loss of endosymbionts, 36–37
  mutualism, intracellular, 32–34
    location of, 13, 33
    symbionts, isolation of, 33
    symbionts, protection of, 34

symbionts, regulation of, 33–34
reproductive incompatibility, 37–38
Micrococcaceae, 85
Micrococcus, 15
Micrococcus lysodeikticus, 520, 526–527
Micrococcus muscae, 26
Micrococcus nigrofaciens, 85
Microgaster, 247
Microorganisms, nonpathogenic, 12–40
Microplitis, 247
Microplitis croceipes, 219
Microspora, 389–391, 415, 425, 428, 436
Microsporea, 390, 425, 428
Microsporidea, 428
Microsporidia, 391–392
  cellular pathology of, 432–433
  characteristics of, 425
  factors affecting life cycle, 443
  gross pathology of, 431–432
  host resistance to, 495–496, 500, 504, 518
  hosts of, 427–430
  importance of, 425
  life cycle of, 440–443
  orders of, 425
  phylum, 428
  presporal stages and spore formation of, 438–440
  proliferative sequences in, 445
  sexuality of, 446–447
  syndrome of, 433–434
  taxonomy of, 425–428
  transmission of, 430–431
Microsporidian spore
  extrusion apparatus of, 425, 434–436
  polar filament of, 425, 434, 436–438, 440–442
  extrusion, 435, 437–438
  function, 437
  size, shape, and surface of, 434–435
  spore wall of, 435
  sporoplasm of, 435
Microsporidies, 391
Microsporidiosis, 425
Microsporidium polyedricum, 182
Milkweed bug, 533. See also *Oncopeltus fasciatus*
Milky diseases of Scarabaeidae, 8, 115–118, 493, 574
Minisporida, 390, 425, 428
Mobilida, 390
Modeling. See Epizootiology
Moisture. See Injuries, from physical agents
Mollicutes, 85, 158–162
  characteristics of, 158
  entomopathogens of, 158
  taxonomy of, 158
Monera, kingdom, 389
Moniliales, 319
Monocystis, 419–420

Monorganotropic disease, 192, 211
Monosporella bicuspidata. See *Blastomyces bicuspidata*
Monosporella unicuspidata. See *Metschnikowia unicuspidata*
Monosymbiosis, 32
Morator, 173
Morison's bee paralysis, 305
Mosquito
  endosymbionts of, 33
  epizootics in, 601, 604
  external microbiota of, 15
  genetic disease of, 61–62
  pathogens of
    bacteria, 7, 84, 89, 95, 104, 109, 112–115, 122, 556, 560, 565–568, 579
    fungus, 322–323, 325, 329–336, 347, 359–360, 363–365
    nematode, 460–461, 466, 471–473, 482–483, 577
    platyhelminthes, 483
    protozoa, 394–396, 402–404, 406–407, 419, 423, 425, 430–431, 434, 438–439, 441, 444–445, 567, 576, 582, 621
    virus, 186, 191–192, 257, 260, 263, 279, 281, 284, 286, 296, 302, 309
  reproductive incompatibility in, 37–38
  resistance in, 65, 494–495, 502, 506, 511, 515, 523, 530
Mosquito iridovirus, 256–257, 259–260, 502
Mrazekiidae, 429
Mucin, 69–70, 602
Mucor hiemalis, 336, 521, 530
Mucorales, 320, 336
Mucron, 416, 418
Multiple nucleocapsid nuclear polyhedrosis virus (MNPV), 177, 183–184
Muramidase, 526
Musca autumnalis, 469–471, 480. See also Face fly
Musca domestica, 24, 95, 287, 364, 389, 402, 497. See also House fly
Musca domestica vicina, 498. See also Oriental house fly
Muscardine, 4, 306, 357, 362
Muscareovirus, 286–288
Muscidae, 347
Mutualism
  definition of, 13
  forms of, 14
  importance of, 2, 12–13
  intracellular and intrahemocoelic forms in, 30–37
  symbionts
    algae, 14
    bacteria, 15, 23–25, 29–39, 159
    fungus, 13–21, 26, 35–36, 39, 322, 356, 577
    lichens, 14

Mutualism (*cont.*)
   nematode, 14, 73, 475, 479
   protozoa, 27–30, 39, 391, 393, 416, 419
   virus, 36
Mycangia, 18, 20, 22
Mycelia sterilia, 321
Mycetocyte, 14, 31–34, 36
Mycetome, 14, 22, 31–33, 38
Mycobacterium leprae, 493
Mycobacterium phlei, 493
Mycobacterium tuberculosis, 493
Mycoderma, 320, 348
Mycoplasma, 85, 158–159, 172
   characteristics of, 159
Mycoplasmataceae, 158
Mycosis, 319, 322, 325
Mycotoxins, 326–327, 336, 341, 360–365
Myiophagus, 320, 329
Myiophagus ucrainicus, 329–330
Myiophyton, 337
Myriangiaceae, 355–356
Myriangiales, 320, 355
Myriangium, 6, 320, 356
Myriangium duriaei, 355. See also Black scale fungus
Myrmicinae, 19
Myxomycota, 319
Myxospora, 389
Myxosporidia, 391
Myxozoa, 415, 439
Myzus persicae, 338. See also Green peach aphid

Napamichum, 447
Nasonia vitripennis, 162
Nasutitermes exitiosus, 25, 30
Naucoris cimicoides, 398
Navel orangeworm, 157, 299, 307–308. See also *Amyelois transitella*
Nectomonad, 393
Nectria, 6, 320, 355
Nectria aurantiicola, 355. See also Redheaded scale fungus
Nectria diploa, 355. See also Pink scale fungus
Nemata, 462. See also Nematoda
Nematoda, 521. See also Nemata
Nematode, 17, 347, 500
   association with insects, 460, 462
      types of, 463–464
   characteristics of, 459–460
   classification of, 462
   host resistance to, 466
   host specificity in, 464–465
   mode of infection, 465–466
   pathology of
      behavioral effects, 469–471
      external effects, 467–468
      internal effects, 468–469
Nematode infection, 459–482
Nematomorpha, 482
Nematomorpha infection, 482–483
Nemeritis canescens, 246
Neoaplectana, 476
Neoaplectana carpocapsae. See *Steinernema carpocapsae*
Neoaplectana glaseri. See *Steinernema glaseri*
Neodiprion nuclear polyhedrosis virus, 565
Neodiprion pratti banksianae, 578. See also Jack pine sawfly
Neodiprion sertifer, 202. See also European pine sawfly
Neodiprion swainei, 396. See also Swaine's jack pine sawfly
Neodiprion taedae linearis, 578. See also Loblolly pine sawfly
Neogregarines, 392, 420. See also Schizogregarines
Neogregarinida, 390, 416
   hosts of, 420–423
   life cycle of, 420–422
   structure of, 421
Neomesomermis flumenalis, 474
Neotylenchidae, 462, 479
Neozygites, 320, 337–338
Nepa cinerea, 395, 407, 422–423
Nephotettix cincticeps, 35–36. See also Green rice leafhopper
Nephridiophaga blattellae, 414
Neuroptera, 32, 94, 189, 277
New Zealand grass grub, 150. See also *Costelytra zealandica*
Nezara viridula, 398
Nicotine, 559
Nilaparvata lugens, 343. See also Brown planthopper
Nitrogen metabolism in termites. See Termite, nutrition of
Nobel Prize, 6, 74
Nocardia rhodnii, 399
Noctuidonema guyanense, 482
Nodamura virus, 309–310
Nodavirus, 309
Nodaviridae, 276, 296
   characteristics of, 309
   entomopathogens in, 309–310
Nodule, 433, 510, 512
   formation hypothesis, 518
Nomia melanderi, 501
Nomuraea, 320, 357, 361
Nomuraea atypicola, 361
Nomuraea rileyi
   epizootic model of, 599, 614
   host resistance to, 500, 517
   hosts of, 361, 497

as microbial control agent, 566, 569, 576,
    581, 583, 603
  nomenclature of, 361
  spore of, 323, 329
Nonoccluded baculovirus. See Baculovirus, non-
    occluded
Nosema, 390, 400
  classification of, 427
  life cycle of, 441–443, 445
  resistance to, 495
  polar filament of, 436
  sexuality in, 446
  spore of, 434
  transmission of, 431–432, 615
Nosema acridophagus, 69, 433, 519
Nosema algerae
  life cycle of, 440–441, 443
  as microbial control agent, 567, 582
  pathology of, 431
  polar filament extrusion in, 438
  as vertebrate pathogen, 430
Nosema apis, 400
  life cycle of, 443
  pathology of, 432–434
  polar filament of, 436
  resistance to, 493–495
  transmission of, 430–431
Nosema bombycis
  historical aspects, 390–391
  pathognomic sign of, 530
  pathology of, 432–433
  sporoplasm of, 442
  type species, 427
Nosema cuneatum, 69
Nosema cuniculi. See Encephalitozoon cuniculi
Nosema disstria, 430
Nosema fumiferanae, 67, 432, 498, 500, 577,
    621
Nosema gasti, 582
Nosema heliothidis, 432
Nosema kingi, 431
Nosema locustae, 435, 565, 567, 569, 576–577,
    582–583, 616
Nosema lophii. See Spraguea lophii
Nosema lymantriae, 73, 518
Nosema plodiae, 67. See also Vairimorpha
    plodiae
Nosema pyrausta, 430, 566, 577, 599–600,
    614–615
Nosema sphingidis, 519
Nosema whitei, 496, 600
Nosematidida, 428
Nosematidae, 429
Nosematidea, 428
Nosematinae, 429
Nosematosis, 400
Nosopsyllus fasciatus. See Ceratophyllus fas-
    ciatus

Notonecta, 332
Notonecta glauca, 398
Nuclear polyhedrosis, 182, 211, 214, 277
  vernacular names of, 182
Nuclear polyhedrosis virus, 9, 209, 211–212,
    214–216, 219, 251, 282, 309
  baculoviruses as expression vectors, 205–207
  cytopathology of, 192–193
  description of, 176–177, 183–184
  DNA replication and gene expression in, 203–
    205
  epizootics of, 599, 601–602, 604–613, 616,
    618
  genotypic variants of, 201–203
  gross pathology of, 189–191
  historical account, 182–183
  host resistance to, 493–496, 498, 514, 520,
    528, 531
  hosts of, 174, 189
  as microbial control agent, 557, 559, 561–
    563, 565, 568–569, 572–574,
    578–579, 583
  occlusion (polyhedron) of, 175, 184–189
  pathophysiology of, 193–194
  as persistent viral infection, 198–199
  replication of, 180, 194–197
  restriction enzyme analysis of, 199–200
    AcMNPV genome, 200–201
    physical map, 200
  tissue specificity of, 191–192
  transfection of isolated DNA, 197–198
  virion of, 184
  virulence of, 69–70, 73–74
Nudaurelia β virus, 308–309. See also
    Tetraviridae
Nudaurelia cytherea capensis. See N. cytherea
    cytherea
Nudaurelia cytherea cytherea, 308. See also
    Pine tree emperor moth
Nudaurelia ε virus, 309
Nudibaculovirinae, 177
Nun moth, 6, 73, 182–183, 190, 605. See also
    Lymantria monacha
Nutritional diseases, 62–64
Nyctotherus, 390
Nyctotherus ovalis, 389–390
Nymphalis antiopa, 109

Obligate pathogen. See Diseases, microbial,
    definition of
Occlusion body. (See also of other insect viruses)
  definition of, 172
  of cytoplasmic polyhedrosis virus, poly-
    hedron, 279–281
  of entomopoxvirus, spheroid, 250–251
  of granulosis virus, capsule, 208–212
  of nuclear polyhedrosis, polyhedron, 184–189

Occult viral infections, 497. *See also* Latent viral infections
Octomitus, 390
Odonata, 261, 419, 506
Oenocytoids, 414, 442, 444, 507–508, 530
 characteristics of, 510–511
Oesophagomermis leipsandra. See *Filipjevimermis leipsandra*
Okanagana rimosa, 153, 345
Olesicampe, 246
Oligohymenophorea, 390, 402
Olive fly, 20, 24, 39, 301. *See also Dacus oleae*
Onchocerca lienalis, 533
Oncopeltus fasciatus, 148, 533. *See also* Milkweed bug
Oncopera alboguttata, 250
Oncopera intricoides, 309
Onion fly, 497. See also *Hylemya antiqua*
Onion maggot, 600. See also *Delia antiqua*
Ontophagus fracticornis, 419
Oomycetes, 320, 329, 334
Oomycota, 391
Opercularia, 17, 390
Operculariella parasitica, 407
Operophtera brumata, 431. *See also* Winter moth
Ophiocordyceps, 363
Ophryocystis, 390, 421–422
Ophryocystis mesnili, 422
Ophryoglena collini, 407
Ophyra leucostoma, 119
Opisthomastigote, 393–394, 396
Opsonins, 513–514, 521, 523
Orbopercularia, 17
Orgyia nuclear polyhedrosis virus, 188, 565
Orgyia leucostigma. See White-marked tussock moth
Orgyia pseudotsugata, 188, 503. *See also* Douglas-fir tussock moth
Orgyia pseudotsugata nuclear polyhedrosis virus, 188, 200, 202
Orgyia thyellina, 287
Oriental house fly, 498. See also *Musca domestica vicina*
Orthopoxvirus, 250
Orthoptera, 94, 111, 149, 249–250, 258, 261, 300, 322, 338, 351–352, 398, 424, 493, 506
Oryctes, 218, 360, 570
Oryctes baculovirus, 193, 604, 606–607. See also *Oryctes* nonoccluded virus
Oryctes nonoccluded virus, 219, 247
 characteristics of, 216
 pathology of, 216–218
 transmission of, 217–218
Oryctes rhinoceros, 175, 521, 606
Ostrinia nubilalis, 364–365, 398, 430. *See also* European corn borer
 bacterial infection of, 95, 108, 122
 epizootics in, 614–615
 fungal infection of, 358, 364–365
 microbial control of, 88, 577
 nutritional disease of, 63
 protozoan infection of, 398, 430, 614
Ovavesicula popilliae, 617
Oxymonas, 390
Oxyuridae, 482

Pachymetana, 302
Paecilomyces, 320, 357, 361. *See also* Yellow muscaridine fungus
Paecilomyces amoeneroseus, 361
Paecilomyces coleopterorum, 361
Paecilomyces cicadae, 361
Paecilomyces cinnamomeus, 361
Paecilomyces farinosus, 355, 357, 361, 618
Paecilomyces fumosa-roseus, 357, 361, 497, 501, 566
Paecilomyces javanicus, 361
Paecilomyces lilacinus, 361
Paecilomyces ramosus, 361
Paecilomyces tenuipes, 354, 361
Paillotella, 173
Paltothyreus tarsatus, 352, 354
Pamphiliidae, 189, 209
Pandemis heparana, 190
Panolis flammea, 200
Panonychus citri, 95. *See also* Citrus red mite
Pansporoblast, 439–440. *See also* Sporophorous vesicle
Pansporoblastina, 428, 439
Papilio demoleus, 104
Parabasalia, 390, 393
Paracolobactrum rhyncoli, 149–150
Paraglaucoma, 403
Paraglycogen, 417
Parainfluenza virus HN, 207
Paraiotonchium autumnalis. See *Heterotylenchus autumnalis*
Paraisaria, 320
Paramecium caudatum, 527
Paramyelois transitella, 157. *See also* Navel orangeworm
Parasitaphelenchus, 481
Parasite, definition of, 65
Parasitism, definition of, 13
Parasitoid, 2, 7, 330
 bacterial infection in, 150
 as biological control agent, 555, 560–561, 566, 571–573, 577, 580, 582
 effect of viral toxin, 194, 214
 epizootics, role in, 600–601, 604, 606, 611, 615, 618
 mutualism with virus, 249
 nematode infection in, 466

nutritional disease of, 64
protozoan infection in, 396, 430
resistance in, 515–517, 520, 532
as vectors of pathogens, 85, 189, 255, 259, 431
viral infection in, 217, 219, 246–249
Parasitylenchus typographi, 469
Parasporal body. *See also* Crystal
  characteristics of, 96
  formation of, 96–97
  number of, 97–98
  in species other than *B. thuringiensis*, 112–113, 117, 122, 151
  types of, 98
Parathelohania, 427, 434, 443, 447
Parathion, 58
Parthenogenesis, arrhenotokus, 38
Parvoviridae, 245
  genera of, 261
  hosts of, 262
  pathogenesis of, 262–264
  replication of, 262–264
  structure and properties of, 261–262
Parvovirus, 184, 189, 261–264, 307, 519, 531. *See also* Parvoviridae
Pasteur, 4–5, 87
Pasteurella pestis, 493
Pathogen, definition of, 65
Pathogen population, 602–609, 612, 614–615, 618, 622
Pathogenicity,
  definition of, 68, 602
  variations in, 71
Pea aphid, 325, 614. *See also Acyrthosiphon pisum*
Pear slug, 397. *See also Caliroa cerasi*
Pébrine, 4, 306, 433, 506
Pectinophora gossypiella, 284. *See also* Pink bollworm
Penicillium, 365
Penicillium paxilli, 364
Pentalonia nigronervosa 219
Peplomer, 180–181, 192, 195–197, 209, 214
Perezia, 427
Perezia fumiferanae. See *Nosema fumiferanae*
Pereziidae, 429
Pericystis apis. See *Ascosphaera apis*
Peridroma margaritosa, 207. See *Peridroma saucia*
Peridroma saucia, 207. *See also* Variegated cutworm
Period of lethal infection, definition of, 71
Periplaneta, 262
Periplaneta americana, 402, 419–420, 519–520, 522–523. *See also* American cockroach
Periplaneta densovirus, 262
Periplaneta fuliginosa, 263, 519

Peritrichia, 390
Peritrichida, 407
Persistent virus, 218, 298, 309. *See also* Baculovirus
Perutilimermis culicis, 464
Phagocytosis, 322, 507, 510, 523, 525
  effectiveness of, 514–515
  process of, 513–514
Phanerotoma, 247
Pheidole, 467
Phenicia sericata, 24
Phenoloxidase system, 513, 524, 526, 529–530
Pheromermis pachysoma, 472
Pheromones, 18, 25
Philosamia ricini, 109
Phlebotomus, 89, 399. *See also* Psychodidae
Phlebotomus argentipes, 399
Pholetesor, 247
Phormia regina, 63, 497. *See also* Black blow fly
Phospholipase A, 84, 151, 525
Phospholipase C, 86, 93–94, 504. *See also* Lecithinase
Photoperiod, 343–344
Photoreactivation, 328
Phryganidia californica, 494, 504. *See also* California oak moth
Phthiraptera, 351
Phthorimaea operculella, 494. *See also* Potato tuber moth
Phycomycetes, 319
Phyllocoptruta oleivora, 354, 363. *See also* Citrus rust mite
Phyllopharyngea, 390
Physaloptera hispida, 515
Phytomonas, 394, 396, 398
Picornaviridae, 275, 296
  genera of, 300
  insect viruses of, 300–307
  replication of, 300
  viral particle of, 300
Picornavirus, 189, 263, 300–302, 304, 307
Pieris, 105, 580
Pieris brassicae, 108, 207, 209, 215, 359, 498. *See also* European cabbageworm
Pieris rapae, 211, 498. See also *Artogaeia rapae*
Pieris rapae crucivora, 517, 532
Pimpla turionella, 36
Pine shoot roller, 207. *See also Choristoneura murinana*
Pine tree emperor moth, 308. *See also Nudaurelia cytherea cytherea*
Pink bollworm, 285. *See also Pectinophora gossypiella*
Pink scale fungus, 355. *See also Nectria diploa*
Pinocytosis, definition of, 512
Pinus patula, 302

Pinus radiata, 577
Pityogenes chalcographus, 423
Plant worms, 352
Plantae, 389
Plasmatocytes, 398, 507–509, 515–516, 518, 530
  characteristics of, 509
  function of, 509
Plastosciara, 418
Plathypena scabra, 497
Platycleis grisea, 391
Platyhelminthes infection, 483
Platynota, 580
Platypodidae, 17
Platysamia cecropia, 495, 501
Plecoptera, 32, 419
Plectomycetes, 318, 347–348
Pleistophora, 390, 434, 441
Pleistophora culicis. See *Vavraia culicis*
Pleistophora debaisieuxi, 519
Pleistophora hyphessobryconis, 435
Pleistophoridae, 428
Pleistophoridida, 428
Pleosporales, 320, 355
Pleurodesmospora, 320
Plistophora hyphessobryconis. See *Pleistophora hyphessobryconis*
Plodia granulosis virus, 566
Plodia interpunctella, 67, 106, 111, 423, 430, 560, 574. See also Indianmeal moth
Plodia interpunctella granulosis virus, 177, 209, 218, 609
Plurisymbiosis, 32
Plutella xylostella, 560. See also Diamondback moth
Podisus maculiventris, 606
Podonectria, 6, 320, 352, 355–356, 365
Podonectria coccicola, 355–356. See also Whiteheaded scale fungus
Point mutation, 202
Polycephalomyces, 321
Polydnaviridae, 174. See also Polydnavirus
  Bracovirus, 247
  characteristics of, 245–246
  genome of, 247–249
  Ichnovirus, 246–247
  virus–parasitoid–lepidoptera relationships, 249
Polydnavirus, 175, 219, 248, 532. See also Polydnaviridae
  as endosymbiont, 36
  mutualism of, 249
  pathology of, 249
Polyederkrankheit, 182
Polyedrischen Körperchen, 182
Polyhedrin, 185, 197, 202, 204. See also Polyhedron
  of cytoplasmic polyhedrosis virus, 253, 281–282
  of nuclear polyhedrosis virus, 197, 202, 204, 209, 251, 253
Polyhedrin gene, 202, 204–206, 281–282
  figure of, 201
  identification of, 200–201
Polyhedron envelope gene, 189, 204
Polyhedron of cytoplasmic polyhedrosis virus. See also Occlusion body; inclusion body
  description and types of, 276–277, 279–281
  dissolution of, 184, 281
  formation and structure of, 281–282, 283, 286
  genetics of, 282
  historical account, 276–277
  polyhedrin of, 276–277, 281
Polyhedron of nuclear polyhedrosis virus. See also Occlusion body; Inclusion body
  definition and type of, 175
  dissolution of, 184, 187, 195
  morphotypes of, 187
  polyhedrin of, 181, 185–186, 188, 251
  production of, 187, 195, 212
  size and structure of, 180–186
  surface structure of, 188–189, 281
Polyorganotropic disease, 191
Polypedilum vanderplanki, 56,
Polyxenous, definition of, 464
Popillia japonica, 105, 115, 154, 358, 460. See also Japanese beetle
Poplar sawfly, 69. See also *Trichiocampus viminalis*
Porcello scaber, 259
Portals of entry, types of, 66–67
Postciliodesmatophora, 390
Potato tuber moth, 494–495. See also *Phthorimaea operculella*
Potential pathogen. See Diseases, microbial
Poxviridae, 245
  characteristics of, 249–250
  pathogenesis of, 252–253
  replication of, 251–252
  types of, 248–250
Praemachilidae, 189
Predator, 2, 7
  as biological control agent, 555, 560–561, 566, 571–573, 577, 580, 582
  epizootics, role in, 600–661, 604–606, 611, 618
  nutritional disease of, 64
  in transmission of pathogens, 85
Prevalence, definition of, 596
Prions, 172
Pristiphora erichsonii, 86. See also Larch sawfly
Proctodael feeding, 27–28, 39
Proctotists, 415, 427
Procubitermes aburiensis, 30
Prohemocytes, 398, 507–508
  characteristics of, 509
Prokaryotae, 389

Promastigote, 393–394, 396–398
Properdin, 521
Protapanteles, 247
Protease, 187, 251, 282, 324–325
Protease inhibitor, 531
Proteobacteria, 35
Proteus, 85, 151, 502
Proteus mirabilis, 85, 151, 506, 514
Proteus rettgeri, 151
Proteus vulgaris, 151
Protista, 389
Protobalantidium, 403
Protoctista, 389
Protomerite, definition of, 416
Protomicroplitis, 247
Protoparce quinquemaculata, 149
Protoparce sexta, 149. See also *Manduca sexta*
Protophormia terrae-novae, 63
Protozoa
  characteristics of, 388–389
  classification of, 389–390
  historical aspects, 389, 391
  pathogenesis, signs, and symptoms of, 392
  portals of entry, 391–392
  relation to insects, 26, 29, 391, 415
Protozoan infections, 388–407, 414–447, 622
Protura, 32
Pseudaletia separata, 496
Pseudaletia unipuncta, 73–74, 180, 190, 192, 210, 499, 504, 521. See also Armyworm
Pseudogibellula, 321
Pseudomonadaceae infections, 85, 147–148
Pseudomonas, 85, 151
Pseudomonas aeruginosa
  characteristics of, 147
  host resistance to, 524–525, 527–528, 530–531, 533
  interference in, 74, 506
  pathogenicity of, 70, 147–148
  strains of, 148
  toxic exoenzymes of, 148
Pseudomonas aureofaciens, 148
Pseudomonas chlororaphis, 94, 148
Pseudomonas fluorescens, 111, 148, 502, 560, 562–563, 565, 568
Pseudomonas noctuarum, 150
Pseudomonas putida, 148
Pseudomonas savastanoi, 24
Pseudomonas septica, 148
Pseudomonas syringae pathovar *savastanoi*, 24
Pseudopleistophoridae, 428
Pseudopleistophorinae, 429
Pseudoplusia includens, 263, 301, 342, 344. See also Soybean looper
Pseudovitellus. See Mycetome
Psocoptera, 32
Psorophora ferox, 260
Psychodidae, 89. See also *Phlebotomus*
Psylla buxi, 31

Pulex simulans, 219
Pulicidae, 189
Pyrausta nubilalis. See *Ostrinia nubilalis*
Pyrenomycetes, 319–320, 347, 352
Pyrethroids, 59
Pyrrhocoris apterus, 397
Pyrsonympha, 390
Pyrsonymphida, 390, 393
Pythium flevoense, 404
Pyxinia frenzeli, 419

Queensland fruit fly, 281. See also *Dacus tryoni*

Rabies virus, 297–298
Raphidopalpa foveicollis, 420
Rasajeyna, 423
Recombinant DNA technology, 172, 198, 203, 205–207, 563
Red muscardine, 357, 362
Red-headed pine sawfly, 572–573, 606. See also *Neodiprion lecontei*
Redheaded scale fungus, 355. See also *Nectria aurantiicola*
Red-ring disease of coconut, 460
Reduviidae, 423
Reesimermis nielseni. See *Romanomermis culicivorax*
Reoviridae, 175, 275–288, 296, 532. See also Cytoplasmic polyhedrosis virus
Reproductive incompatibility, 158. See also Microbiota, intrahemocoelic and intracellular
Resistance, host. See also Encapsulation, immunity
  blood clotting, 511–512
  hematology, 507–511
    function of hemocytes, 511
    hemocytes, 507–511
    response to biotic agents, 519–521
  insect age and stage, 495–497
  morphological and physiological defenses, 500
    digestive tract, 502–507
    glandular secretions, 501–502
    integument, 501
    respiratory system, 502
  nutritional factors in,
    dietary, 497–498
    exogenous antimicrobial factors, 498–499
  physical factors in, 499–500
  at species level, 264, 349, 493–495, 560, 575
Resistance, phylogenetic, 492–493
Resistant or resting structures of Entomophthorales
  formation of, 342–343
  germination of, 343–344
Restriction enzyme analysis. See Nuclear polyhedrosis viruses

Reticulitermes, 28–29, 325, 611
Reticulitermes flavipes, 23, 25, 29
Retortamonadida, 390, 393
Retortamonas, 390
Retractocephalus raphidopalpi, 420
Rhabditida, 461–462
Rhabditidae, 462, 475
Rhabditis insectivora, 475
Rhabdophora, 390
Rhabdoviridae, 276, 296
  general description of, 297
  sigma virus, 297
    host resistance to, 299–300
    pathogenicity of, 299
    transmission of, 298–299
    viral particle of, 298
Rhabdovirus, 297–298
Rhadinaphelenchus cocophilus, 460
Rhinoceros beetle, 216, 570, 572–574, 606–607. See also *Oryctes rhinoceros*
Rhinoestrus nivarletti, 395
Rhinotermitidae, 27, 29
Rhinovirus, 300
Rhithrogena iridani, 435
Rhizopoda, 390–391, 399–402. See also Amoeba
Rhizopoda infections, 399–402
Rhizotrogus majalis, 504
Rhodnius prolixus, 399, 523
Rhopalosiphum padi, 302
Rhynchoidomonas, 396
Rhynchophrya, 390
Rhynchosciara angelae, 186, 191
Rhyncolus porcatus, 150
Rhyssa, 464
Rice stem borer, 256, 260. See also *Chilo suppressalis*
Rice weevil, 33. See also *Sitophilus oryzae*
Rickettsiae, 172, 605
  entomopathogens of, 153–154
  hosts of, 154–158
  life cycle of, 154–157
  reproductive incompatibility in insects, 37, 158
  vertebrate pathogenicity in, 158
Rickettsiales, 36, 85
Rickettsiella, 154, 158, 617
Rickettsiella armadillidii, 154
Rickettsiella blattae, 154
Rickettsiella cetonidarum, 154
Rickettsiella chironomi, 154, 157
Rickettsiella grylli, 154
Rickettsiella melolonthae, 153–155, 157–158
Rickettsiella popilliae, 153–154
Rickettsiella schistocercae, 154
Rickettsiella stethorae, 157
Rickettsiella tenebrionis, 154
Rickettsiella tipulae, 154
RNA-viral infections, 275–288, 296–310
  families of RNA viruses, 276

Romanomermis, 468, 475
Romanomermis culicivorax
  biology and life cycle of, 472–474
  figures of, 472–473
  host resistance to, 466
  hosts of, 464
  as microbial control agent, 461, 474, 567, 569, 577, 583
Rudimicrosporea, 428

Sabouraud dextrose agar, 345
Sabouraud maltose agar, 345
Sacbrood, 173, 303
Sacbrood virus, 275, 300, 302
  host resistance to, 495
  pathogenesis of, 303–304
  viral particle of, 303
Saccardiaceae, 356
Saccharomyces, 320
Saccharomyces cerevisiae, 348
Saccharomyces ellipsoideus, 348
Saissetia hemisphaerica, 355. See also Hemispherical scale
Salmonella enteritidis, 533
Salmonella paratyphi, 26
Salmonella thompson, 523
Salmonella typhimurium, 493
Salmonella typhosa, 26, 524–525
Salt-marsh caterpillar, 284. See also *Estigmene acrea*
Salt-marsh mosquito, 621. See also *Aedes cantator*
Samia cynthia, 527
Samia cynthia x *ricini*, 157
Saprolegniaceae, 336
Saprolegniales, 320, 334, 336
Sarcina flava, 96
Sarcodina, 388
Sarcomastigophora, 389
Sarcophaga, 119, 523
Sarcophaga aldrichi, 344, 606, 618
Sarcophaga peregrina, 523, 529
Sarcosporidia, 391
Sarcotoxins, 529
Scale insects
  endosymbionts of, 35–36
  external microbiota of, 15–17
  fungal infections of, 6, 355–356, 362, 365
Scarabaeidae, 419, 493, 617
Schistocerca americana, 6
Schistocerca gregaria, 467–468, 506, 515, 523
Schistocerca pallens, 149
Schizocystis, 421, 422
Schizogony, definition of, 415
Schizogregarines, 420. See also Neogregarines
Schlaffsucht, 182
Schneideria schneiderae, 418
Schwarzsucht, 305
Sciaridae, 89, 189

Scolytidae, 17
Scolytus multistriatus, 85
Scolytus scolytus, 150
Scotia segetum, 22. See also *Agrotis segetum*
Scotogramma trifolii, 157, 210, 254. See also Clover cutworm
Screwworm fly, 61. See also *Cochliomyia hominivorax*
Secernentea, 461
Seedcorn maggot, 24, 326, 340, 346. See also *Hylemya cilicrura*, *H. platura*
Selenephera lunigera, 110
Selenidium, 422
Septicemia, definition of, 84
Septobasidiales, 356
Septobasidium, 15–17, 321, 356
Septobasidium clelandii, 356
Septobasidium pilosum, 356
Sericesthis iridovirus, 258–259
Sericesthis pruinosa, 425
Serratia
 characteristics of, 149
 hosts of, 149–151
 stress on, 85, 151–152
 as microbial control agent, 150
 strains of, 149–150
 transmission of, 85, 150
Serratia entomophila, 150, 566, 617
Serratia liquefaciens, 149–150
Serratia marcescens
 complementation with, 73
 host resistance to, 514, 525, 528, 530, 532
 hosts of, 149–150
 stress on, 506
 pathogenicity of, 150
 repellency of, 151
 transmission of, 85, 150
Serratia piscatorum, 150
Serratia proteamaculans, 150. See also *Serratia liquefaciens*
Serritermitidae, 27
Sex-ratio disease of spiroplasma, 160–162
Shigella dysenteriae, 524
Sialis lutaria, 405
Siberian silkworm, 110. See also *Dendrolimus sibiricus*
Sibine, 263
Sibine fusca, 263, 519
Siedlecka, 422
Sigma virus. See also Rhabdoviridae
 characteristics of, 198, 297–298
 host of, 57, 297
 host resistance to, 299–300, 494
 pathology, anoxia of, 296–297, 299, 301
 transmission of, 67, 298–299, 601
 viral particle of, 298
Signs, definition of, 70–71
Silkworm. See also *Bombyx mori*
 bacterial infection of, 71, 87–88, 105, 108–109, 150–151, 153

 fungal infection of, 326, 328, 357, 363–364
 genetic disease of, 61
 historical account, 2–5
 protozoan infection of, 394, 398, 432–433, 442
 resistance in, 493, 495–497, 499, 501, 504, 506, 508, 514, 526, 529–532
 viral infection of, 69, 173–174, 182–184, 193, 206, 260–261, 263–264, 281, 282–283, 287, 299, 302, 306–307
Silver-spotted flambeau, 263. See also *Agraulis vanillae*
Simian rota virus VP6, 207
Simuliidae, 89, 332, 365, 405. See also Black flies
Simulium, 533
Simulium bivittatum, 122
Simulium damnosum, 580
Simulium vittatum, 109. See also Black fly
Sindbis virus, 532
Single nucleocapsid nuclear polyhedrosis virus, 177, 180
 description of, 184
Siphonaptera, 174, 189, 397
Sirex, 479–480. See also Wood wasp
Sirex cyaneus, 19–20
Sirex gigas, 20
Sirex noctilio, 479. See also Siricid wood wasp
Siricid wood wasp, 464–465, 468–469, 479, 606. See also *Sirex noctilio*
Siricidae, 19
Sitophilus granarius, 38
Sitophilus oryzae, 33
Sitophilus zeamais, 364
Sitotroga cerealella, 95
Slow bee-paralysis virus, 302
Solenopsis, 219
Solenopsis geminata, 423, 431. See also Tropical fire ant
Solenopsis richteri, 325, 359. See also Fire ant
Sorbic acid, 70
Sorosporella, 321
Sorosporella uvella, 357, 362
Sotto disease bacillus, 87
Southern pine beetle, 18. See also *Dendroctonus frontalis*
Soybean looper, 342, 579. See also *Pseudoplusia includens*
Sphaeriales, 319–320, 352, 355
Sphaerularia bombi, 460, 469, 471, 481
Sphaerulariidae, 462, 481
Sphaerularioidea, 462
Spheroid, 184, 250–251
Spheroidin, 251
Spheromastigote, 393
Spherulocytes, 507–508, 510–511, 516
 description of, 510
Spicaria, 353, 361
Spicaria fumoso-rosea. See *Paecilomyces fumosa-roseus*

Spicaria prasina, 361
Spicaria rileyi. See *Nomuraea rileyi*
Spindle, 250–251
Spirocerca lupi, 460
Spirocystis, 422
Spiroplasma, 85, 158–162
Spiroplasma apis, 160
Spiroplasma citri, 160
Spiroplasma melliferum 160
Spiroplasmataceae, 158–159
Spirotrichea, 390
Spodoptera, 198, 202, 255, 566
Spodoptera ascovirus, 253, 255
Spodoptera exempta, 277
Spodoptera exigua. See also Beet armyworm
 bacterial infection of, 95
 microbial control of, 568
 nematode infection of, 477
 viral infection of, 69, 185, 197, 277
Spodoptera frugiperda. See also Fall armyworm
 bacterial infection of, 95, 108
 cell line of, 174, 203, 207
 epizootics in, 612–613
 fungal infection of, 364
 resistance in, 494, 561
 viral infection of, 198, 200, 208, 255
Spodoptera littoralis. See also Cotton leafworm
 bacterial infection of, 108–109, 193, 200, 202
 fungal infection of, 364–365
 resistance in, 497
Spodoptera litura 70, 105, 193, 202
Spore-crystal protein. See *Bacillus thuringiensis*, *B. sphaericus*
Sporodiniella, 320
Sporodiniella umbellata, 336
Sporogony
 in Apicomplexa, 415–416, 420, 423
 in Microsporea, 427, 440–442, 445, 447
 stage of, 415
Sporophorous vesicle, 427, 434, 440, 444–445
 description of, 439
Sporophytic generation. See *Coelomomyces*, heteroecism
Sporothrix, 321
Sporotrichum, 353
Sporotrichum globuliferum. See *Beauveria bassiana*
Sporozoa, 27, 388, 391, 414
Sporozoite. See Apicomplexa life cycle
Spotted alfalfa aphid, 614
Spotted cucumber beetle, 219. See also *Diabrotica undecimpunctata*
Spotted cutworm, 210. See also *Xestia c-nigrum*
Spraguea lophii, 432
Spraguidae, 429, 445
Spruce budworm. See also *Choristoneura fumiferana*
 epizootiology in, 620–621
 outbreak of, 8
 protozoan infection of, 67, 432, 577
 resistance in, 498, 500, 504, 527, 559
 viral infection of, 74, 193
Stable fly, 95. See also *Stomoxys calcitrans*
Staphylococcus, 38, 150
Staphylococcus albus, 514
Staphylococcus aureus, 498, 514
Stegomyia scutellaris. See *Aedes albopictus, A. scutellaris*
Steinernema, 475–476, 582
Steinernema affinis, 476
Steinernema anomali, 476
Steinernema bibionis. See *Steinernema feltiae*
Steinernema carpocapsae,
 historical aspects, 460–461
 host resistance to, 466, 500, 532
 hosts of, 465, 475
 in IPM approach, 582–583
 life cycle of, 463, 476–478
 as microbial control agent, 557, 567, 569, 582
 mutualism with bacteria, 73, 151, 475–476, 478
Steinernema feltiae, 476, 582
Steinernema glaseri, 8, 460, 476, 478, 569
Steinernema intemedia, 476
Steinernema rara, 476
Steinernema kushidai, 465, 476
Steinernema scapterisci, 476
Steinernematidae, 151, 464, 557
 biology and life cycle of, 475–478
Stempellia, 434
Stempellia mutabilis, 445
Stempellia simulii, 445
Stereum. See *Amylostereum*
Sterile male technique, 57, 61, 65
Stethorus, 157
Stichosomida, 461–462
Stictospora provincialis, 418
Stigmatomyces ceratophorus. See *Fanniomyces ceratophorus*
Stilbella, 321
Stomoxys, 119
Stomoxys calcitrans, 95. See also Stable fly
Stonebrood (Steinbrut), 363
Strepsiptera, 32
Streptococcaceae, 85, 152
Streptococcus, 15, 25, 147
Streptococcus apis, 152
Streptococcus bombycis, 306–307
Streptococcus faecalis
 entomopathogenic serotypes of, 153
 host resistance to, 347, 506, 514, 528
 hosts of, 153, 618
 stress, 85
 as internal microbiota, 22
 syndrome, brachytosis of, 153

Streptococcus faecalis-S. faecium, 150
Streptococcus pluton. See Melissococcus pluton
Stressors, 85, 497, 502, 558, 610
    definition of, 72
Strongwellsea, 327, 337, 346–347
Strongwellsea castrans, 326, 340, 346–347
Strongwellsea magna, 347
Stylocephalidae, 420
Stylocephalus, 417
Suctoria, 17, 390, 407
Sugar-beet weevil, 6. See also *Cleonus punctiventris*
Sulphuretylenchus elongatus, 469
Summit disease, 338, 613, 618
Supella longipalpa, 519
Swaine's jack pine sawfly, 572. See also *Neodiprion swainei*
Symbionin, 35
Symbiont. *See also* Mutualism
    host and location of, 13–14, 32–33
    in host resistance, 532
    in insect taxonomy, 2
    in microbial control, 584
    protection of, 34
    regulation of, 33–35
    role of, 37, 63
    transmission of, 39–40
    types of, 32
Symbiontes, 36
Symbiosis, definition of, 13
Symbiote. *See* Symbiont
Symbiotic organ. *See* Mycetome
Symptoms, definition of, 70–71
Syncystis, 421–422
Syncystis mirabilis, 422–423
Syndrome, definition of, 70–71
Synergism, 73, 560. *See also* Complementation
Synnematium, 363
Syrphidae, 365
Syzygy, definition of, 416

Tabanomyces, 337
Tachinidae, 189
Talaromyces, 361
Tarichium gammae, 341, 344
Teleogryllus commodus, 300
Teleogryllus oceanicus, 300
Telomyxidae, 429, 442
Temperature. *See also* Injuries from physical agents
    in epizootiology, 603, 609, 614, 617
    fungi, effect on, 332, 341–343
    high temperature effect on microorganisms, 38, 258, 282, 499–500
    in microbial control, 558, 569
    nematode, effect on, 472–473
    protozoa, effect on, 396, 443, 445
    virus, effect on, 258, 297

Tenebrio molitor, 398, 418–419, 422
Tenebrionidae, 422
Tenericutes, 158
Tent caterpillar, 68, 606, 617. See also *Malacosoma*
Tenthredinidae, 189, 396–397
Tephritidae, 24
Termite
    association with fungi, 21
    cellulose digestion in, 28–30
    fungal infection of, 325, 347, 355
    internal microbiota of,
        bacteria, 25, 29–30
        flagellate, 27–29
        protozoa, 393, 402, 419
    nutrition of, 30, 39
Termitidae, 21, 27, 29
Termitomyces, 21, 29
Tête claire, 306
Tetracrium, 356
Tetradonema, 468
Tetradonema plicans, 464, 471
Tetradonematidae, 462, 471
Tetrahymena, 390, 402–404, 406
Tetrahymena chironomi, 404–405
Tetrahymena dimorpha, 405
Tetrahymena pyriformis, 404, 406
Tetrahymena rostrata, 404
Tetrahymena sialidos, 405
Tetrahymenidae, 402
Tetranacrium, 321, 356, 365
Tetranychus pacificus, 95
Tetraviridae, 276, 296, 308–309. See also *Nudaurelia* β virus
Thelohania, 427, 431, 445. See also *Amblyospora*
Thelohania bracteata. See *Amblyospora bracteata*
Thelohania hyphantriae, 518
Thelohania similis, 73
Thelohaniidae, 428–429, 442
Theophila mandarina, 307
Thereuonema higendorfi, 259
Thermal therapy, *See* High temperature
Thermoascus, 361
Thuricin, 110
Thuringiensin, 94–96. See also Exotoxin, β-
Thuringiolysin, 110
Thysanoptera, 32, 351
Thysanura, 32, 189
Tilachlidium, 321
Tineola bisselliella, 425. See also Webbing clothes moth
Tipula iridescent virus. See *Tipula* iridovirus
Tipula iridovirus, 256, 258, 260, 611
Tipula paludosa, 118, 156, 186, 191, 418, 423
Tipula vittata, 423
Tipulidae, 155, 189
Tipulocystis, 422

Tobacco budworm, 110–111, 560, 582. See also *Heliothis virescens*
Tobacco hornworm, 102, 109–110, 519. See also *Manduca sexta*
Togaviridae, 276
Tolypocladium, 321, 365
Tolypocladium cylindrosporum, 365, 566
Tomatine, 559
Torrubiella, 320, 352, 355, 361, 363
Tortrix viridana, 610. See also Green tortrix
Torulopsis, 26
Toxemia, definition of, 84
Toxoglugea, 434
Toxorhynchites rutilis septentrionalis, 332
Trachmyrmex, 16
Tranosema, 246
Transfection of viral DNA, 197–198, 206
Transfer vector, physical map of, 205
Transgenic organism, 563–564
Transmission, transovarial
 definition of, 67
 in bacteria, 85
 in epizootiology, 600–601, 615, 621
 in fungus, 322
 in microbial control, 577, 598
 in protozoa, 4, 392, 395, 430–431, 444
 in virus, 189, 260, 286, 297–298, 307–308
Transmission, transovum
 definition of, 67
 in epizootiology, 601, 604, 611, 615, 618
 in microbial control, 573, 577, 598
 in protozoa, 392
 in virus, 286, 302, 308
Transmission, venereal, 67, 430
Trematoda, 521
Trenomyces, 320
Trenomyces histophthorus, 351
Trenomyces histophtorus. See *Trenomyces histophthorus*
Trialeurodes vaporariorum, 365. See also Greenhouse whitefly
Tribolium castaneum, 422–423, 432, 496
Tribolium confusum, 38, 423, 600
Tribolium ferrugineum, 425
Trichiocampus irregularis, 69
Trichiocampus viminalis, 69. See also Poplar sawfly
Trichocera, 402
Trichocera hiemalis, 402
Trichodubosqia epeori, 434
Trichomonas, 390
Trichomycetes, 15, 26, 321, 336
Trichonympha, 390
Trichoplusia, 580
Trichoplusia ni. See also Cabbage looper
 cell line (TN-368), 198, 218
 recombinant toxin on, 565
 pathogens of
  bacteria, 104, 122, 151, 157
  fungus, 361, 581
  protozoa, 431–432
  virus, 69, 189, 196, 254–255, 308, 602
 resistance in, 497, 500, 502, 528
Trichoplusia ascovirus, 253–258
Trichoplusia ni cytoplasmic polyhedrosis virus, 277
Trichoplusia ni granulosis virus, 210
Trichoplusia ni nuclear polyhedrosis virus, 186, 200
Trichoptera, 32, 189, 419, 502
Trichosia pubescens, 418
Triphena pronuba, 277
Tripius sciarae, 465
Triplosporium, 337
Triplosporium fresenii, 344
Trogoderma granarium, 421. See also Khapra beetle
Tropical fire ant, 423, 431. See also *Solenopsis geminata*
Trypanosoma, 389, 394, 396, 398–399
Trypanosoma cruzi, 397, 399, 523
Trypanosoma lewisi, 399
Trypanosoma rangeli, 399
Trypanosoma rhodesiense, 399
Trypanosomatidae, 393–394
Trypomastigote, 393, 399
Tsetse fly, 24, 55, 150, 396, 399, 529. See also *Glossina*, *G. morsitans*
Tumors
 by irradiation, 57, 512, 518
 malignant, 61, 518. See also Carcinoma
 by nutritional disease, 63–64
 by protozoa, 392, 407, 433
 by virus, 287, 495, 519
 xenoma, 519
Tunicamermis melolonthae, 515
Turbinate fungus, 355. See also *Aschersonia turbinata*
Turchiniella, 403
Tuzetiidae, 429
Tylenchida, 461–462, 465
Type A milky disease organism. See *Bacillus popilliae*
Type B milky disease organism. See *Bacillus lentimorbus*
Tyrosine-phenyloxidase system, 524

Uncoating of DNA, 195, 214, 246–247, 251, 258. See also DNA uncoating
Unikaryonidae, 429
Uranotaenia, 579
Uredinella, 321, 356

Vaccine, 2, 524–525, 531, 533
Vaccinia subgroup, 250
Vahlkampfia mellificae. See *Malpighamaeba mellificae*
Vairimorpha, 434, 445, 447, 569

Vairimorpha necatrix, 431, 495, 566, 582
Vairimorpha plodiae, 433. See also *Nosema plodiae*
Van Leeuwenhoek, 389
Vanessa cardui, 110
Vanessa urticae, 7
Variegated cutworm, 207. See also *Peridroma saucia*
Vavraia culicis, 434, 438
Vegetable wasps, 352
Velvetbean caterpillar, 494, 500, 579, 599. See also *Anticarsia gemmatalis*
Venturia, 246
Verbenol, 18, 25
Verbenone, 18
Vertical transmission. See Transmission, transovarial, transovum, venereal
Verticillium, 6, 321, 362
Verticillium cinnamomeum, 355. See also Crimson fungi
Verticillium fusisporum, 362
Verticillium lecanii. See also White-halo fungus
  conidia of, 322
  hosts of, 362
  as microbial control agent, 566, 575–576, 581, 583
  nomenclature of, 362
  nutritional effect on, 69
  as pathogen of rust fungus, 362
  toxin of, 359, 362
Vesicular stomatitis virus, 297–298
Vibrionaceae infections, 151–152
Viral infections, 171–220, 245–264, 275–288, 296–310. (*See* specific viruses)
Viral paralysis, 400
Viral toxin, 214
Virales, 173
Virogenic stroma
  of ascovirus, 254
  of cytoplasmic polyhedrosis virus, 275, 281, 283, 287, 501
  of densovirus, 262–264
  of *Drosophila* X virus, 297
  of entomopoxvirus, 251
  of flacherie virus, 307
  of granulosis virus, 212–214
  of iridovirus, 258
  of nonoccluded baculovirus, 219
  of nuclear polyhedrosis virus, 179, 192, 195–197
  of polydnavirus, 246
Viroid, 172
Viropexis. *See also* Endocytosis, 181, 192, 196, 214, 251
Viroplasm. *See* Virogenic stroma
Virulence
  cause of variations in, 69–70, 327, 603
  definition of, 68, 602

effect of nutrients, 69
in epizootiology, 6, 602–603
Viruses as endosymbionts, 26, 36. *See also* Polydnavirus mutualism

Waldtrachtkrankheit, 305
Wattle bagworm, 574, 609, 613
Webbing clothes moth, 425. See also *Tineola bisselliella*
Weiseria, 434
Western grape-leaf skeletonizer, 211. See also *Harrisinia brillians*
Western tent caterpillar, 599, 611
Western X-disease, 158
Wheat cockchafer, 5–6. See also *Anisoplia austriaca*
White cutworm, 501. See also *Euxoa scandens*
White-halo fungus, 355, 362. See also *Cephalosporium lecanii*
Whiteheaded scale fungus, 355. See also *Podonectria coccicola*
White-marked tussock moth, 574
White-muscardine fungus, 3, 5, 70, 323, 357–359. See also *Beauveria bassiana*
Wilting disease of pine, 460
Winter moth, 431. See also *Operophtera brumata*
Wipfelkrankheit, 182, 190
Wiseana, 616
Wiseana cervinata, 192, 260
Wolbachia, 153
  characteristics of, 158
  as endosymbiont, 36
  replication of, 158
  reproductive incompatibility, cause of, 38
Wolbachia pipientis, 37
Wolbachia postica, 38
Wolbachieae, 153
Wood cockroach, 424. See also *Cryptocercus punctulatus*
Wood wasp, 19, 577, 605. See also *Sirex*
Wound healing, 511
Wuchereria bancrofti, 460

Xenoma, syncitial, 433, 519
Xenopsylla cheopis, 399
Xenorhabdus, 151, 475–476, 521. See also *Achromobactor*
Xenorhabdus bovienii, 476
Xenorhabdus luminescens, 151, 478, 514
Xenorhabdus nematophilus, 73, 466, 475–476, 528, 567
Xenorhabdus poinarii, 476
Xestia c-nigrum, 210. *See also* Spotted cutworm
Xiphydriidae, 19
Xylariales, 352
Xyleborus ferrugineus, 38

Yeasts, 18, 26, 32, 34, 36, 347,
Yellow muscardine fungus, 357, 361. See also *Aspergillus flavus*; *Paecilomyces farinosus*

Zoomastigina, phylum, 390–391
Zoomastigina infections, 392–399. *See also* Flagellate

Zoophagineae, 173
Zoophthora, 337–338
Zoophthora radicans. See *Erynia radicans*
Zootermopsis nevadensis, 30
Zoraptera, 32
Zygaenobia, 337
Zygnemomyces, 337
Zygomycetes, 319–320, 336
Zygomycotina, 319–321 336, 356

ISBN 0-12-683255-2